THE SIBLEY GUIDE TO BIRDS

WRITTEN AND ILLUSTRATED BY

DAVID ALLEN SIBLEY

Second Edition

Alfred A. Knopf, New York

2014

For Joan, Evan, and Joel

THIS IS A BORZOI BOOK
PUBLISHED BY ALFRED A. KNOPF

www.aaknopf.com

Knopf, Borzoi Books, and the colophon are registered trademarks of Random House LLC.

ISBN: 978-0-307-95790-0

This edition prepared and produced by
Scott & Nix, Inc.
150 West 28th Street, Suite 1900
New York, NY 10001

www.scottandnix.com

Printed and bound by
C & C Offset Printing Co. Ltd., China

A revision of *The Sibley Guide to Birds*, copyright © 2000 by Chanticleer Press, Inc. An Andrew Stewart Publishing Edition, published by Alfred A. Knopf, a division of Random House LLC, New York, in October 2000.

Second Edition, March 2014

FSC
www.fsc.org
MIX
Paper from responsible sources
FSC® C008047

The paper of this book is FSC® certified.

Key to the Group Accounts

Each group of related species is introduced by a summary of the characteristics of the qroup. These accounts show all species in a family (or subfamily or order) for comparison. Look here to see the range of variation in the group, as well as fundamental similarities and differences among genera.

Common name and scientific name of the family or group.

This paragraph highlights the general characteristics of the group and gves the nuber of species and genera covered in the guide. Overview of typical behaviors (feeding, nesting, etc.) is included.

Swallows

FAMILY: HIRUNDINIDAE

12 species in 7 genera (4 very rare species are not shown here). Highly aerial, with pointed wings, notched tails, and smooth, flowing flight, all swallows feed almost exclusively on insects captured in sustained flight. Often seen perched on twigs or wires or on the ground, sometimes in groups of hundreds or thousands. Nesting habits vary by species and between genera. Two genera, including Cliff, Cave, and Barn Swallows, build mud nests on cliffs or buildings; Tree, Violet-green, and Purple Martin nest in cavities of birdhouses; Bank and Northern Rough-winged nest in burrows in sand banks. Adult females are shown.

Genus *Stelgidopteryx*

Northern Rough-winged Swallow, page 391

Genus *Progne*

Purple Martin, page 389

Genus *Tachycineta*

Tree Swallow, page 390

Violet-green Swallow, page 390

Genus *Riparia*

Bank Swallow, page 391

Genus *Hirundo*

Barn Swallow, page 393

Genus *Petrochelidon*

Cliff Swallow, page 392

Cave Swallow, page 392

All species in the group, except the very rare, are pictured. Grouped by genus, the birds are reproduced in relative scale. In most cases, immature plumages or comparitively drab females are shown, as these present the greatest identification challenge.

Key to the Range Maps

The range maps show the complete distribution of each main species in our area. Bear in mind that within the mapped range, each species occurs in appropriate habitats and at variable density (common to scarce).

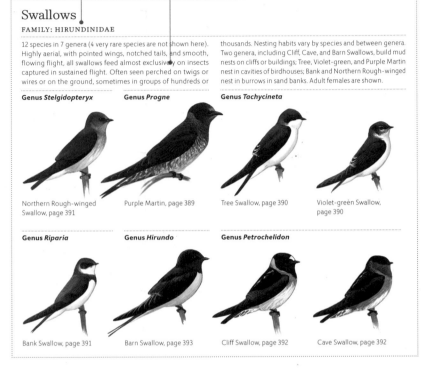

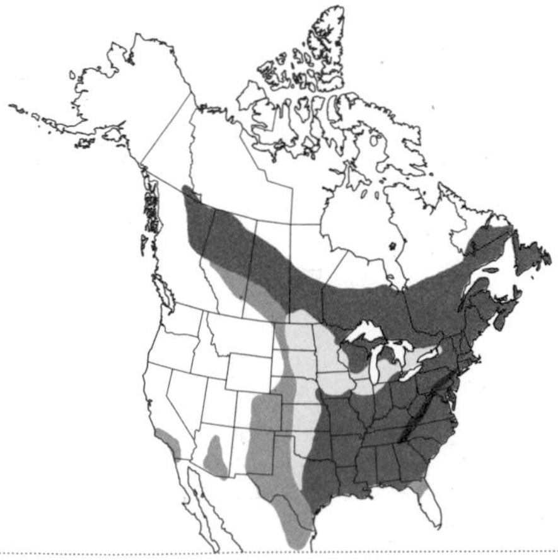

Winter Shows the normal winter distribution of the species, Many are somewhat nomadic in winter, occupying only parts of the mapped range at any given time.

Summer For virtually all species, this is the breeding range and is more consistently and uniformly occupied than the winter range.

Year-round Indicates that the species can be found all year in this area, even though winter and summer populations may involve different individual birds. Only a few species are truly resident.

Migration Main migratory routes are shown, as well as areas of regular dispersal and post-breeding wandering. Note that migration also passes through summer and winter ranges.

Rare Shows areas of rare occurrence (may be a single or up to a few records per year).

The Sibley Guide to Birds

A Scott & Nix Edition

Contents

Preface to the Second Edition

Since I began work on this new edition of *The Sibley Guide to Birds*, I've heard the question many times, "Why do a revision? Have the birds changed?" No, they haven't. Their distribution has changed a little, even in just fifteen years, and a number of the range maps required updating. A few species, such as Pink-footed Goose and Green-breasted Mango, were once essentially unrecorded in North America and have since increased in frequency, so they needed to be added to the guide. But that's about it for changes in the birds themselves.

The plan for a revised edition was driven mostly by my own ideas about things that I wanted to change. I thought the text could be improved, to make more direct comparisons between challenging species and to include information about status and habitat. And I wanted to revise the artwork.

A few revisions were needed to correct real errors in the paintings—the flying adult male Rose-throated Becard lacked a rose throat-patch in the first edition. Also, a few images needed minor revisions to make them more typical. For example, the displaying male Ruffed Grouse was originally illustrated with a broken dark tail-band, which is shown by only a small percentage of males.

Many revisions of artwork were minor—fixing shading, adjusting colors and contrasts, and making minor changes to color patterns. About twenty percent of the total images in the guide received more substantive revisions. This included major changes in colors or patterns, changing bill shape or tail shape. All was the result of new information (at least to me). In many cases, more personal experience with a species and/or better reference material made it possible to fine tune patterns and colors that I didn't know very well the first time around. In other cases, it was simply the result of the intervening years of birding and studying, seeing the artwork with new eyes and noticing small flaws that were not apparent when I originally painted them.

The need for a revision to the guide was not a surprise to me; I've been planning it since before the first edition was printed in 2000. As I said in the preface of the first edition, I still learn new things every time I go birding. It's one of the things that makes birding so enjoyable and satisfying, but it means that even in the weeks between finishing the work for this revised edition and writing this preface, I have learned things that will not make it into this guide. I'll have to start planning the next revision.

As an author this can be frustrating, but as a birder it is invigorating. Our understanding of birds and nature is always advancing, and discoveries of global significance can be made in a city park or a backyard. There is lots of information in this book, but behind every image and every bit of text are questions to be explored, and new discoveries to be made. I hope that this book will provide an entry for you to further your own exploration of the natural world, and to share in that sense of discovery.

—David Sibley
October 2013
Concord, Massachusetts

Acknowledgments

I should begin by thanking everyone who has ever taken the time to talk with me about birds, and especially everyone who has published anything about bird identification. Their ideas, in one form or another, have been incorporated into this book. I wish it were possible to name more of them, as this book represents the combined work of many hundreds of people. Below, I name only the people who were directly consulted on this project (with sincere apologies to anyone I have overlooked), but the contributions of all the others are no less significant.

In the intricately connected internet world today, especially, an idle question or a posted photo can prompt an investigation that leads to some new discovery. The science of bird identification has become a global collaboration with countless conversations happening simultaneously. I make no pretense of being able to keep up with all of it, and obviously it is impossible to thank all of the participants by name. I can merely offer a generic thanks to all who continue to add to our knowledge of bird identification.

I have been fortunate to live at different times at Point Reyes, California, and Cape May, New Jersey, and to have worked for WINGS, Inc., making friends and sharing ideas with many of the most knowledgeable and insightful birders in North America. Thanks to all.

For answering queries, reviewing drafts, and providing photos, reprints, and other material assistance, thanks to Per Alstrom, John Arvin, Richard C. Banks, Barb Beck, Bob Behrstock, Chris Benesh, Craig Benkman, Louis Bevier, Bryan Bland, Tony Bledsoe, Rick Bowers, Eric Breden, Ned Brinkley, Rick Cech, Patrick Comins, Richard Crossley, Mike Danzenbaker, Jon Dunn, Pete Dunne, Barny Dunning, Vince Elia, Chris Elphick, Ted Eubanks, Shawneen Finnegan, Frank Gallo, Kimball Garrett, Frank Gill, Paul Guris, Lucas Hale, Bob Hamlin, Keith Hansen, Paul Holt, Steve Holzman, Eric Horvath, Steve N. G. Howell, Rich Hoyer, Alvaro Jaramillo, Kevin Karlson, Kenn Kaufman, Greg Lasley, Sheila Lego, Paul Lehman, Tony Leukering, Ian Lewington, Jerry Liguori, Bruce Mactavish, Bob Maurer, Brad Meiklejohn, Shaibal Mitra, Steve Mlodinow, Joe Morlan, Killian Mullarney, Marleen Murgitroyde, Frank Nicolletti, Michael O'Brien, Dennis Paulson, Brian Patteson, Wayne Petersen, Ron Pittaway, Noble Proctor, Peter Pyle, Martin Reid, Gary Rosenberg, Margaret Rubega, Will Russell, Ray Schwartz, Debra Love Shearwater, Ray Smart, Mitchell Smith, P. William Smith, Rich Stallcup, Clay and Pat Sutton, Thede Tobish, Jeremiah Trimble, Declan Troy, Joe, Sandy, Catherine, and Michelle Usewicz, John Vanderpoel, Dave Ward, Paige Warren, Sophie Webb, Scott Weidensaul, Sheri Williamson, Tom Wood, Gail Diane Yovanovich, and John Cameron Yrizarry.

Staff at the following institutions provided logistical support and/or access to collections: Museum of Comparative Zoology, Harvard University, Cambridge, Massachusetts; Slater Museum, University of Puget Sound; California Academy of Sciences, San Francisco; VIREO and the Academy of Natural Sciences, Philadelphia; American Museum of Natural History, New York; United States National Museum, Washington, D.C.; Peabody Museum, Yale University, New Haven, Connecticut; University of California, Berkeley and Irvine; North Carolina State Museum, Raleigh; Louisiana State University, Baton Rouge; Cornell University, Ithaca, New York; Library of Natural Sounds, Cornell University Laboratory of Ornithology; University of Alaska, Fairbanks; New Jersey Audubon Society's Cape May Bird Observatory; Point Reyes Bird Observatory, Stinson Beach, California; Manomet Bird Observatory, Manomet, Massachusetts; University of Georgia, Athens; Savannah River Ecology Laboratory, Aiken, South Carolina; and University of Arizona, Tucson.

My family has supported my desire to draw birds since preschool days; the love and support of my parents, Fred and Peggy Sibley, has carried me through all these years, and the practical benefits of having an ornithologist father cannot be overstated.

I thank my agent, Russell Galen, for smoothing the way.

Special thanks are due to George Scott and Charles Nix of Scott & Nix, Inc. for their expert and persistent work on the second edition.

Thanks to Andy Hughes, Kathy Zuckerman, and Sonny Mehta, and all of the publishing team at Alfred A. Knopf.

Thanks to the editing and design team of the first edition at Chanticleer Press (Amy K. Hughes, George Scott, Lisa R. Lester, Drew Stevens, Anthony Liptak, Melissa Martin, Vincent Mejia, and Bernadette Vibar), and Thumb Print (Areta Buk, James Waller, and Jeffrey Edelstein).

I thank John Cameron Yrizarry for reviewing all the original artwork and for offering many helpful suggestions.

Thanks to Vince Elia and Shawneen Finnegan for help in preparing the range maps in the first edition, and to Paul Lehman for reviewing all the maps for the second edition.

Thanks to technical reviewers of the first edition Will Russell, Kimball Garrett, Jon Dunn, Steve N. G. Howell, Chris Elphick, and Frank Gill. Will Russell provided generous support and advice at all stages of this project, from indulging my wanderings in the early years to the careful reading of the entire manuscript.

I must single out the influence of the late Ray Schwartz, whose companionship and support through the formative years of this project were a constant source of strength.

Last, I thank my wife, Joan Walsh, and my sons, Evan and Joel, who really made me believe that I could do this—and then managed to put up with the lifestyle that developed as a result. It's done…again!

Introduction

This book covers the identification of 923 species (plus many regional forms) found in the continent of North America north of Mexico, including the United States and Canada and all adjacent islands, but excluding Hawaii, Bermuda, and Greenland. Offshore waters are included to a distance of 200 miles (320 km) or halfway to the nearest land that is not part of the North American region, whichever is closer. These boundaries conform to the American Birding Association's established definition of the North American region.

I have included all species that occur regularly within this area, including rare but regular visitors, introduced species with established populations, and many exotic species that are seen regularly in the wild. Rare but regular visitors are defined loosely as those species recorded ten or more times in the last twenty-five years, and for species recorded less frequently, decisions were made to include some and exclude others. Species that occur regularly only in western Alaska, and nowhere else in North America, are more likely to be excluded, while species that occur in areas with more observers, or that seem to be increasing in frequency, are more likely to be included.

Established introduced species such as Rock Pigeon, Mute Swan, and Chukar are included, as well as species with small breeding populations (but not yet added to the official list) such as Red-vented Bulbul and Red Junglefowl. Also included are many exotic species that are seen only occasionally in the wild, including many waterfowl, upland game birds, parrots, mynas, and finches.

Recognizable regional variations of species are included, such as the subspecies groups of Dark-eyed Junco and Brant, and identifiable subspecies that are rare visitors to North America are also included, such as European and Asian Whimbrels.

Classification of Birds

Birds, like all other living things, are classified by scientists in a system of groups and subgroups, each with a unique scientific name. Birds make up the class Aves. The class is subdivided into orders, each order into families, each family into genera, each genus into species, and species into subspecies. For example, the Red-tailed Hawk is in the class Aves, order Falconiformes, family Accipitridae, genus *Buteo*, species *jamaicensis*. The scientific name is written as "Genus species" or *Buteo jamaicensis*. Subspecies are described for many species, and designate regional variations within the species. Red-tailed Hawks in most of western North America are the subspecies *calurus* and have the full name *Buteo jamaicensis calurus* (although in this book they are given regional names, which often include multiple subspecies).

The taxonomy of families, genera, and species as well as all common and scientific names used in this book follow the seventh edition (1998) of the American Ornithologists' Union (AOU) *Check-list of North American Birds*, including changes to the list through the 54th supplement (2013).

Sequence of species

In this guide, the sequence of bird families and the grouping of species within genera, as well as genera within families, closely follows the most recent AOU *Check-list*. However, the sequence of some families, and of species within many families deviates from the list in order to place similar species in proximity for better comparison.

Experienced birders who have used earlier bird guides will notice new changes in group sequence, most notably that the falcons are now placed between woodpeckers and flycatchers, instead of immediately after the hawks. These changes are based on advances in research on the evolutionary history of birds, which emphasizes the fundamental similarities and differences between groups. Considering these relationships can enhance our understanding of the birds, and their identification.

Definition of a species

Among ornithologists there is an ongoing debate about the definition of a species, and related debate over the classification of certain species. Most species are distinctive and strongly differentiated, but there is a continuum from distinct species, such as Red-breasted and White-breasted nuthatches, to weakly-differentiated populations, some of which are classified as species (e.g. Mallard and American Black Duck, or Western and Glaucous-winged Gulls) and some as subspecies (e.g. Yellow-rumped Warbler).

In this gray area the potential for splitting one species into two or more, and for lumping two or more species into one, is a constant topic of discussion among birders. The *Check-list* committee of the AOU makes final decisions in such matters and its decisions are based on published and carefully considered scientific research, but still reflect prevailing opinion of a very ambiguous subject. The list will always be a work in progress.

Subspecies

In this book I have tried to illustrate all distinctive regional populations, emphasizing variation rather than the species unit, on the premise that any distinguishable population is noteworthy, whether it is classified as a species or not. Many of these regional populations are not safely identified outside their known range, or are identifiable with only a low level of confidence; further study will increase those levels in many cases, and some of these populations will undoubtedly be elevated to full species status.

I have intentionally avoided using scientific names for subspecies, as these can be more confusing than helpful to the nonspecialist. Instead, recognizable subspecies populations are named by the geographic region in which they breed (and sometimes by established English names). These regional populations as defined here may include one or several named subspecies. Most subspecies variation is determined by climate and follows a few simple rules; thus, the boundaries between regional populations are surprisingly consistent (see map on page *xi*). For more details on the distribution, identification, and nomenclature of subspecies, refer to sibleyguides.com, the AOU's 1957 *Check-list* (fifth edition), and Peter Pyle's *Identification Guide to North American Birds* (Slate Creek Press, Bolinas, CA, 1997, 2008).

Equipment for Birding

You can watch birds with no equipment at all, but binoculars are essential for seeing details at a distance and for watching birds without disturbing them. Seven or eight power binoculars are best for birding, and certain models are designed specifically for birding with a wide field of view, close focus distance, and other factors. Read reviews, ask other birders, and visit a specialty store to help choose the best binoculars in your price range.

Many birders also buy a telescope and tripod for studying birds, especially waterfowl and sandpipers, which are often seen at a distance. A telescope can also be very useful for seeing details of birds at close range.

A small pocket notebook and pencil is the best way to record details of your observations, and the process of taking notes and/or sketching is an excellent way to learn. Digital cameras and audio recorders also make it very easy to capture a record of your observations, and allow you to share the recorded images or sounds with other birders later to confirm an identification.

Many smartphone apps are available that can help with identification, sounds, recent sightings, lists, and note taking. Many regions with an active birding community have developed ways to quickly communicate news of bird sightings, either through text messages, Twitter, Facebook, or an email listserv group. All of these methods make the smartphone a more and more essential part of a birder's toolkit.

Learning to Identify Birds

If you are reading this book, you presumably have an interest in learning how to identify birds. A few suggestions for getting started are outlined below, but all build on the same theme: to become an expert birder, you must study birds. Sketching and taking notes in the field are exercises that will force you to look more closely, reinforce your memory, and greatly increase the rate at which you learn. The joy of small discoveries is part of the great appeal of birding, and patient study is always rewarded.

The first rule is simple: *Look at the bird.* Don't fumble with a book, because by the time you find the right picture the bird will most likely be gone. Start by looking at the bird's bill and facial markings. Watch what the bird does, watch it fly away, and only then try to find it in your book.

Browse the book at home. Spend time just flipping through the pages, watch how the characteristics of the birds change with each family, learn the difference between a thrush, a wren, an oriole, and birds in other groups.

Recognize patterns. There is order in the universe, and birds are no exception. All the minutiae of variation (appearance, behavior, occurrence) fit into predictable patterns, and as you gain experience these patterns coalesce into a framework of knowledge.

Pay attention to taxonomy. Learning the families and genera of birds provides one of the most powerful frameworks for understanding patterns of variation. Ducks, cormorants, and grebes are in three different families, and the birds in each family share a number of very distinctive traits.

Practice seeing details. Much of the skill involved in identifying birds is being able to sort out the details of the bird's appearance at a distance. To build an understanding of how plumage patterns work, practice with the Bird Topography diagrams beginning on page *xvi*. Study details at close range whenever possible.

Study shapes. Experienced birders use shape as an important clue for identifying broad groups, such as genera, as well as individual species. Paying attention to shape may also reveal differences in posture, actions, and even plumage.

Study habits. Although behavior is never absolutely reliable for identification, habits provide strong supporting clues and shortcuts to identification. Birds spend virtually all their active moments searching for food, so almost all their actions and habitat choices relate to food.

Pay attention to seasons. The occurrence of birds is closely tied to the changing seasons. Detailed information on the seasonal occurrence of birds is available for most areas online (see the excellent eBird.org). A simple reference to the calendar may be all that is needed to resolve some difficult identification problems.

Use multiple field marks when trying to identify a bird. Birds are variable, and there are exceptions to every rule. Don't be discouraged by this, but do look for several different field marks on any bird. If one doesn't match up, the bird may require more study. Experience will allow you to judge the reliability of different characteristics.

Beware of misjudging size. Size is very difficult to judge at a distance, and for the beginner, size is nearly useless for identification except when making direct comparisons. Experienced observers do find size a helpful indicator, but only in familiar surroundings or with reference points that allow an accurate judgment of distance.

Meet other birders. Your best source of information on local birds and birding will be other birders. In many areas, nature centers, sanctuaries, local Audubon Society chapters, Facebook groups, and bird clubs offer regular field trips and meetings, and these can be extremely useful sources of information and camaraderie. The American Birding Association (aba.org), a national organization serving avid birders, publishes the bimonthly magazine *Birding*, as well as birding guides for many states and regions.

Psychological Effects and Mistakes

We all make mistakes, and misidentifications of birds are common. Almost none of these mistakes can be blamed on abnormal birds; they are almost always the result of flaws in how we see them. Our perception of size, color, and shape is all relative, and easily skewed by factors we are not even aware of. Often the illusion is fleeting, we catch these mistakes quickly and move on. In other cases, the illusion is more persistent, or the sighting is brief, and we are left with an incorrect impression and no chance to change it.

Expectations are a very powerful psychological effect, and can even alter the way we perceive colors and shapes. In the broad sense, expectations include everything from wishful thinking (a desire or hope to see a certain species) to a comment from another birder ("I think this might be a...") to personal experience ("I've seen...here before"). This leads to misidentifications. Expectations make it very hard to notice the unexpected, and overlooking a rare species is probably much more frequent than mistaking a common bird for a rare one. These effects are operating all the time, and we are usually unaware of any bias.

Being aware of the potential for bias is one of the best

ways to avoid these errors: always being willing to step back, reconsider your identification from another angle, and check for other possibilities. One of the main values of experience is that it teaches us when we need to be wary of flaws in perception.

Variation in Appearance

You should not expect any bird seen in the field to match exactly the picture of its species in this field guide. Every bird is an individual, and each species is variable (and no field guide's pictures are absolutely accurate). Variation occurs in every aspect of a bird's appearance, behavior, and voice, and only experience can teach you to recognize the normal range of variation within a species.

Many characteristics are presented in this book as average differences. It is important to understand this concept. Average differences are noticeable in a series of individuals but can be contradicted (sometimes dramatically so) by single individuals. One good example involves the bill length of Semipalmated and Western Sandpipers. On average, Western Sandpipers are longer-billed than Semipalmated, but there is some overlap, so while very long-billed and very short-billed birds can be safely identified, a large number of individuals have ambiguous bill length. Likewise, among the large gulls, most species show average differences in shape and size. For example, Iceland Gulls average smaller and shorter-billed than Herring Gulls, but some Herring Gulls are quite small and short-billed, and some Icelands are large and long-billed. This extensive overlap makes it unwise to attempt an identification based solely on silhouette.

The overall impression you receive from the combination of subtle differences is often referred to as gestalt or jizz (from G.I.S.S.—"general impression of shape and size"). It is a useful concept to understand—that the sum of several vague average differences can together make for a fairly reliable identification—but it is a "soft" characteristic and should not be overemphasized. Observers should beware of using jizz as a substitute for careful study and thought.

Most variation in birds' appearance can be accounted for by the age and sex of individual birds and its subspecies group. Typical examples showing the range of these variations are illustrated in this book. Within this basic range, however, there are many other variations that may be encountered.

North America, showing regions used to define subspecies populations. Main regions are shown in color, while dotted lines outline subregions. The Pacific and Interior West regions together form the Western region. These regions offer a rough guide to subspecies distribution, but the distribution of each population differs somewhat from the regions outlined. Note that populations living in more humid climates (especially the Pacific Northwest) tend to be darker than those in arid regions (especially the Southwest).

Geographic Variation and Subspecies

Many regional variations within species are illustrated in this book. These cover the spectrum from distinctive populations that may eventually be considered full species themselves to subtle and inconsistent variations that are evident only upon careful comparison. Much of the variation in subspecies populations is gradual, or clinal, across a species' range. Such clinal variation makes it impossible to divide a species into well-defined subspecies populations, even though the birds at the ends of the clines may be strikingly different in appearance. For example, the white spotting on the wing coverts of Downy and Hairy Woodpeckers is most extensive in the East, and least extensive in the Pacific Northwest. The extremes are easily recognized, and large regions are occupied by birds of relatively uniform appearance; but between each of these regions there is a broad area of clinal change. Within the transitional areas many intermediate birds occur that cannot be easily assigned to either population.

Polymorphism

Some species, particularly some herons, geese, hawks, and jaegers, occur in a variety of plumage colors that have no relation to age or sex. Typically, color morphs are not evenly distributed throughout the range of a species but are found mainly or exclusively in certain populations.

Wear and Fading

From the moment they grow, a bird's feathers are constantly subjected to the combined effects of abrasion and bleaching. Given how much wear and tear feathers sustain, it is remarkable that most feathers are worn for a full year, and some for even longer.

With practice it is possible to recognize worn plumage even when the species is unknown. Several important effects of plumage wear on a bird's appearance can be considered and may aid in identification:

- Worn plumage looks frayed, ragged, and colorless; paler and drabber than fresh plumage.
- Light-colored feathers (or light-colored parts of feathers) wear more quickly than dark, because melanin (dark pigment) actually strengthens the feather.
- The cinnamon or buff color seen on many juvenile shorebirds, terns, and hawks fades to white within weeks.
- Juvenal feathers are particularly weak and wear faster and fade more than adult feathers under the same conditions. The most faded and worn gulls are 1st summer birds, which still retain much of their juvenal plumage.

Changes in Posture and Head Shape

Shape is a very important clue for identification. In most species, all individuals have the same general shape and proportions regardless of age, sex, or season. But while certain aspects of shape are consistent, others are extremely variable. Shapes of bills and feet exhibit little variation, and primary projection and spacing, tail shape, and tail length are all relatively constant features of plumage. Body and head shape, however, depend largely on the position of the overlying feathers and can change dramatically as a bird ruffles or smooths its feathers (see topography page *xvi*).

Light and Atmospheric Conditions

Light—its variation at different times of day and under different atmospheric conditions—can play tricks on the birder. Colors and contrasts appear to change, and even the size and proportions of a bird may seem to alter as it passes from a back-lit silhouette to a well-lit portrait. Strong light washes out colors. Birds in fog appear larger than normal; birds flying at twilight appear smaller.

Cosmetic Coloration and Staining

Certain gulls and terns show a variable pinkish blush on body feathers, related to diet. Geese and cranes (and other waterbirds in some situations) often become stained with iron, which gives their feathers a rusty or orange color. Occasional birds are seen with chemical stains—most often waterbirds with dark brown or blackish oil stains on their undersides.

In some research studies, birds are trapped and dyed (usually yellow-orange) or marked with plastic wing tags or leg flags so that large-scale movement patterns can be monitored. These markers are usually brightly colored and very conspicuous and should rarely cause any identification problems.

Aberrant Plumages

From time to time you may encounter birds that are truly aberrant. Though rare, a few plumage conditions appear with some regularity.

The most frequent abnormal plumage condition is the reduction or absence of melanin. A total lack of melanin produces a true albino, with entirely white feathers as well as pinkish eyes, bill, and legs. True albinos are rare in the wild and usually do not survive long.

Birds with reduced melanin are partial albinos or incomplete albinos. Melanin can be completely lacking from all feathers (producing an all-white bird), or from scattered feathers (producing patches of white or a pied plumage). Or melanin can be present but reduced in all feathers (producing a dilute plumage that looks faded or washed out).

Partial albinism often follows feather groups, e.g., white spectacles or an entirely white head might appear, and it can be symmetrical or nearly so. Certain species, including Red-tailed Hawk, Willet, gulls, American Robin, blackbirds, and grackles, seem more prone to albinism than others.

The term leucistic is often applied to birds with reduced melanin, but it has been used to specify both pied and dilute patterns. The term "partial albino" is more widely used to describe any bird with reduced melanin. These conditions have many different causes, and can be genetic or environmental, temporary or permanent. It is generally impossible for an observer in the field to determine the underlying cause of the abnormality.

In some cases melanin is lacking from parts of individual feathers (see American Crow, page 384, for an example of this producing a white wingstripe). Note that an absence of melanin can occur while other pigments, e.g. red or yellow, are unaffected and albinism can reveal other pigments that are normally hidden by black.The opposite condition—an excess of melanin, or melanism—is much less frequent, and results in birds that are overall sooty blackish or chestnut-brown. This can be very difficult to distinguish from staining. Other color aberrations, involving an excess or absence of particular pigments, are very rare.

Yellow to red colors in bird plumage are the result of carotenoid pigments. These chemicals come from the bird's diet, and are important for immune function, so bright red and yellow colors signal a bird that is healthy and well-fed. Unusually pale or limited red to yellow color is seen occasionally and usually indicates a bird that is less healthy or lacking a proper diet.

It is important to remember that most birds molt their feathers over a period of weeks, and then wear those feathers for up to a year. The color and condition of the feathers reflects the health of the bird at the time the feathers were grown, and not necessarily at the time of observation.

Variations of carotenoid pigments present during the molt can produce feathers with strikingly different colors. In Cedar Waxwing (page 460) the tail tip is normally yellow, but if the birds feed on certain exotic Asian honeysuckle berries while they are growing new feathers, the yellow pigment is replaced by orange. Similarly, Baltimore Oriole, Yellow-breasted Chat, and other species have been recorded with brighter, more reddish, feathers than normal.

Bill Deformities

Birds' bills are constantly growing—like human fingernails—but they are subjected to constant wear, which maintains them at their proper length. An injury or slight deformity can result in abnormal growth or an unusually long or twisted bill.

Hybrids

Hybrids are the offspring of parents of two different species. The offspring of parents of two different subspecies (of the same species) are known as intergrades rather than hybrids (e.g., Yellow-rumped Warbler).

Hybrids occur frequently between certain closely related species and are often overlooked by birders. The most regularly seen crosses are illustrated in this book, but many other crosses are nearly as frequent, and a great number of other hybrids have been recorded. The groups most prone to hybridization are geese, ducks, grouse, gulls, and hummingbirds.

It is important to point out that hybrids are extremely variable—even siblings within a single brood of hybrids can look quite different from one another—and the illustrations included in this book only begin to cover the subject. Most hybrid birds are fertile, and mate with one of their parent species or with another hybrid to produce backcrosses, which can then go on to produce more backcrosses, etc., leading to a complete spectrum of variation between the two pure parental types (e.g., Blue-winged and Golden-winged Warblers, page 472).

Hybrids can show unexpected characteristics not present on either parent, and it is often impossible to determine whether an individual bird is in fact a hybrid or to say with certainty which two species are the parents. The standard terminology applied to all such cases is "apparent hybrid."

Habitat and Habits

The typical habitat of each species is briefly mentioned in the text beside the map in the species accounts of this guide. This information is useful primarily as a hint for where to look for a species, and may be helpful but not diagnostic for identification.

When habitat is useful for identification it is generally in the form of probabilities. Greater and Lesser Scaup differ in their preferred habitat choice, and in many regions there are well-known ponds or bays where only one species is expected. The generalization is that Greater prefers deeper water, larger bodies of water, and (not surprisingly) is much more likely to occur on salt water.

But while one particular lake might be favored by Lesser Scaup and another nearby lake favored by Greater, this knowledge provides very little predictive power for which species of scaup might show up on a third lake nearby.

Birds are so mobile, particularly during migration, they often end up in unusual habitats. Western Tanager's "preferred" habitat (for nesting) is coniferous or mixed forests in mountainous terrain, but during migration they can often be found in lowland cottonwood groves, backyard shade trees, even in low willow thickets in the desert or prairie.

The bottom line is: use habitat as a guide to help find birds and to suggest the most likely species, but identify the bird by other field marks. Be wary of identifying birds to species by habitat alone.

Learning Songs and Calls

Experienced birders can identify most species by sound. In forests, where birds are often difficult to see, the majority of birds are located and identified by sound. The value of knowing bird sounds is clear, but no aspect of birding is more difficult to master than voice identification. As with so many other skills, there is no substitute for personal experience. The first step is simply to pay attention to bird voices. Learning bird songs can be compared to learning a foreign language: the keys are repetition and (for rapid learning) total immersion.

Written voice descriptions, such as those included in this book, can be helpful for pointing out differences between similar songs and calls or for bringing to mind songs and calls already heard. However, words at best provide a very feeble sound impression. A great deal of the variation in songs and calls cannot be put into words, and no one should expect to learn bird songs by reading about them. Listening to recordings is much more useful (especially in conjunction with reading descriptions). Nothing, however, can replace actual field experience: hearing a song, tracking down the singer, and watching it sing. The best way to remember these experiences is to keep notes: listen to the song, try to imitate it, and describe it in your own words.

Understanding the Context of Bird Sounds

Songs are vocalizations used to advertise either to a potential mate (in courtship) or to a potential rival (as a territorial display). These tend to be more complex sounds, given from a conspicuous perch and only during breeding season, and are usually stereotyped (i.e., an individual male has a repertoire of one or several songs and performs them the same way every time). Calls, on the other hand, tend to be shorter vocalizations that convey information about location, alarm, etc. Most calls are used year-round, and subtle variations in volume, pitch, and other aspects communicate information to other birds.

Some species substitute non-vocal sounds for song. For example, woodpeckers pound their bill very rapidly on a resonant tree, producing a "drumming" sound that functions as a song (see page 321). Snipe and woodcock both perform displays in flight, with complex sounds produced by air rushing over wing or tail feathers.

Songs are the most distinctive vocalizations of most species. Calls are generally shorter and simpler than songs, and each species has a variety of different calls used for different communication purposes. The most frequent calls, heard year-round, are referred to as the contact call and the flight call (even though so-called flight calls are often given by perching birds). These sounds can be very useful for identification, but only for a birder with sharp ears and a good deal of experience.

The written descriptions of vocalizations in this guide follow some simple rules. Vowel sounds represent pitch from lower to higher: *oo, oh, ah, eh, ee*. Consonants represent "hard" (e.g., *t, k, p*, etc.) or "soft" (e.g., *s, l, th*, etc.) sounds. The letter *z* indicates a buzzy sound.

The most important characteristics to listen for in bird vocalizations are pitch (high or low, rising or falling), quality (harsh, clear, liquid, buzzy, etc.), and rhythm (fast, slow, choppy, singsong, etc.). The length of a song, the length of time between songs, and whether subsequent songs are the same or different can also be useful in identifying species.

With experience you will begin to recognize the basic qualities of birds' voices—for example a thrush-like or flycatcher-like quality. This kind of broad determination can be a very helpful first step toward identification.

Finding Rare Birds

Most birders who find rare birds are looking for rare birds. This is not to say that one can find a rarity simply by looking for it, but the observer who is prepared will find rare species far more often than the observer who is not. An intimate knowledge of the common species is essential (an equivalent knowledge of the rare species always helps, of course). The expert will keep an open mind, so that a mere flash of color, a slightly different shape, or just an impression of "something not quite right" may focus attention and lead to the detection of a rare bird. Knowing patterns of occurrence also improves your chances, as rare species tend to occur at predictable times and places.

While rarity-hunting is one of the most exciting aspects of birding, you must be prepared to defend your identification, and to this end it is worthwhile to take extensive notes and attempt to photograph or record a bird you think may be a rarity. Use extra care when identifying rare birds, considering alternatives such as aberrant plumages of common species. Contact other birders as soon as possible so that they have an opportunity to see the bird. Don't be afraid to propose an identification or to question the identifications of others; everyone makes mistakes (even the experts). Do be cautious, be honest with yourself and others, and when mistakes are made try to learn from them.

Ethics

As more people take up birding, the behavior of every individual birder becomes more important. In all situations you must first consider the welfare of the birds. Avoid making a disturbance, especially at roosting and nesting sites. Tread lightly and encourage others to do the same. Be respectful and helpful to others, especially to nonbirders (who may not appreciate someone trampling lawns or blocking traffic just to see a bird).

Playback

The widespread and easy availability of birdsong recordings makes it easy for anyone to play recordings in the field, which often elicits a response from the birds. This artificially alters the bird's behavior and therefore has an undeniable impact on the birds. In certain cases, the impact may be significant, either distracting the bird from more important chores, using energy that it needs for other things, or even luring it into the open where it can be captured by a predator. In addition, using playback can be very disruptive to other birders.

In most cases, the effect on birds is probably not significant, and may be preferable to other tactics for seeing a bird, but playback should be used sparingly and with an appreciation of the disruption that it causes. Using playback is illegal in all US National Parks and National Wildlife Refuges, and in many local parks. In general it is good practice to avoid using playback in any popular public birding spot, or on any individual bird that is being sought by multiple birders.

Observing

Whether you are observing or photographing birds, it is important to avoid disturbing them whenever possible. Obey any posted rules and property boundaries, watch from a distance, and do not intentionally flush birds. Keeping these simple courtesies in mind will benefit the birds, and allow you to enjoy their unaltered natural behavior.

Extinct Species

Sadly, the following North American species and subspecies are all presumed extinct. The year and location of the last confirmed record in the wild is given.

Labrador Duck (*Camptorhynchus labradorius*). 1878 in New York.

Eskimo Curlew (*Numenius borealis*). 1962 in Texas; 1963 in Barbados.

Great Auk (*Pinguinus impennis*). 1830s off Newfoundland; 1844 off Iceland.

Passenger Pigeon (*Ectopistes migratorius*). 1900 in Ohio.

Carolina Parakeet (*Conuropsis carolinensis*). 1905 in Missouri.

Ivory-billed Woodpecker (*Campephilus principalis*). 1944 in Louisiana; 1987 in Cuba.

Bachman's Warbler (*Vermivora bachmanii*). 1962 in South Carolina.

Two distinctive populations usually considered subspecies are now extinct:

Heath Hen (*Tympanuchus cupido cupido*), a northeastern coastal population of Greater **Prairie-Chicken** (*Tympanuchus cupido*); 1931 in Massachusetts.

Dusky Seaside Sparrow (*Ammodramus maritimus nigrescens*). An isolated and very local population of Seaside Sparrow (*Ammodramus maritimus*); 1980 in Florida.

The greatest threat facing North American birds today, as in the past, is habitat alteration; coastal dunes, freshwater wetlands, and grasslands are in the greatest danger. To learn more—and to help—contact any of the local, national, or international conservation organizations, including the following:

National Audubon Society
225 Varick Street
New York, NY 10014

212-979-3000

audubon.org

The Nature Conservancy
4245 North Fairfax Drive
Arlington, VA 22203

703-841-5300

nature.org

American Bird Conservancy
4249 Loudoun Avenue
The Plains, VA 20198

888-247-3624

abcbirds.org

Bird Topography

Birders deal with feathers. The actual flesh and bones of a bird's body are barely visible, and the varied shapes and colors of birds are almost entirely created by the feathers. Even though the colors and shapes of individual feathers are so variable, the arrangement of feathers on a bird's body is similar across all species.

Northern Cardinal showing changes in overall body shape from movement of feathers. Raised or fluffed out (above) as seen during cold weather or when relaxed; feathers compressed (below) typically seen during hot weather or stress.

Birds' bodies are not uniformly covered with feathers. The feathers grow in discrete groups (within several tracts), leaving other parts of the body bare. Knowing the basic feather groups and how the feathers in each group are arranged may be the most important basic information for a birder trying to identify a bird by its appearance. Even on a uniformly black species, such as a crow, one can identify all the basic feather groups described below. Studying a common bird—a House Sparrow, a gull, even a pet parakeet—is one way to familiarize yourself with feather groups.

Within each feather group, feathers grow in orderly rows, overlapping like shingles on a roof. The shape and arrangement of all feathers is perfectly coordinated to cover the bird's entire body with a streamlined and waterproof insulating coat.

Plumage patterns almost invariably follow the contours of feather groups, and most markings can be described by reference to the relatively simple framework that feather groups provide. Learning the basis of common markings (e.g., wingbars, eyering, etc.) will greatly enhance your understanding of birds' appearance.

One of the most frequent patterns seen on feathers is a simple dark streak along the shaft of each feather. Such streaks on the breast and flank feathers line up to form the long rows of streaked feathers characteristic of many songbirds. Differences among species are entirely the result of differences in the color and shape of feather markings, not of any difference in the shape or arrangement of feathers.

Parts of a Passerine

This figure shows the basic parts of a passerine, or songbird.

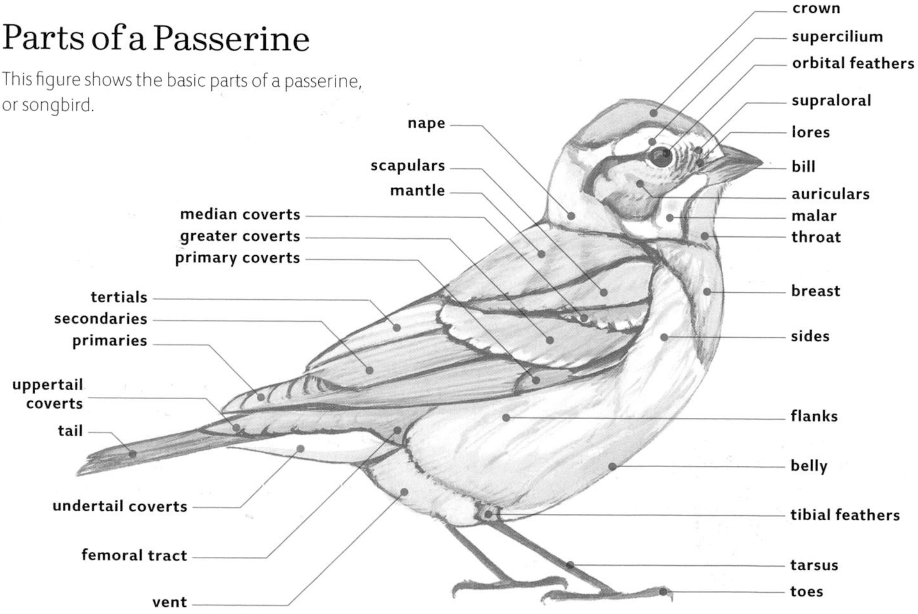

Head Feather Groups

The head feathers can be divided into five main groups radiating back from the base of the bill.

Other subdivisions and details of head feathers:

Crown Covers top of head.

Eyebrow or **Supercilium** Side of head above eye.

Throat Spans underside of lower jaw.

Malar Feathers along side of lower jaw. Sometimes called submoustachial stripe, when "malar" is used for dark lateral throat-stripe.

Auriculars or **cheeks** Complex set of feathers that channels sound into ear. Feathers at rear border of auriculars are short, sturdy, and densely colored. Feathers over ear opening are lacy and unpatterned.

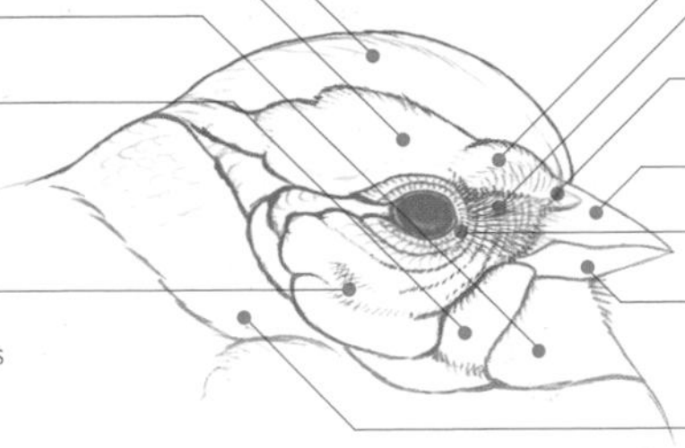

Supraloral or **fore-supercilium** Front end of supercilium, just above lores.

Lores Tiny feathers between eye and bill.

Nasal tuft or **nasal bristles** Well developed in some families.

Upper mandible Upper half of bill.

Orbital feathers Several rows of tiny feathers encircling eye.

Lower mandible Lower half of bill.

Nape Back of neck, wrapping around the back of the crown, supercilium, and cheeks. The area just behind the cheeks may be referred to as the sides of the neck.

Head Feather Patterns

Supercilium or **eyebrow stripe** Variable in width and shape, not necessarily following the outline of this feather group.

Eye-arcs or **broken eyering** Formed by pale color on some of the orbital feathers.

Eyeline or **eyestripe** Variable marking following upper border of auriculars.

Lateral throat-stripe Often referred to as malar stripe; does not include any malar feathers.

Lateral crown-stripe

Median crown-stripe

Mustache stripe or **moustachial stripe** Follows lower border of auriculars.

White-throated Sparrow

More head patterns

In the birds shown below, the divisions between feather groups have been enhanced to allow comparison with the White-throated Sparrow above. Notice the similarity of the basic feather groups, even though the color patterns are very different on these distantly-related species.

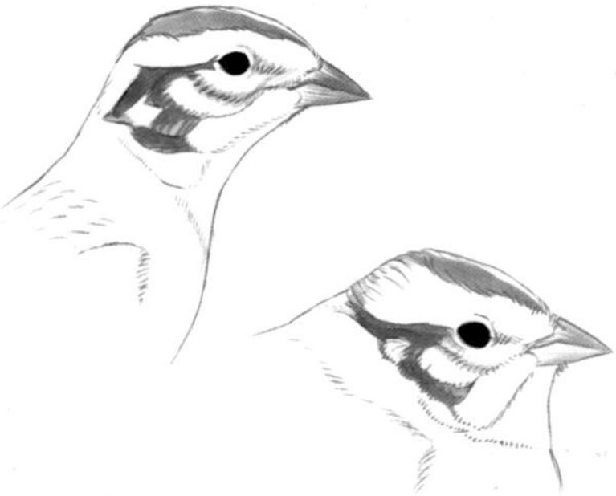

Red-eyed Vireo

Changes in posture create changes in pattern. These illustrations show the head of a sparrow in alert and relaxed postures. Note how the feather groups on the front of the head maintain their shape, while the feather groups on the neck and back of the head expand and contract dramatically to maintain coverage.

Northern Flicker

Body Feathers

Basic passerine body feathers, from front.

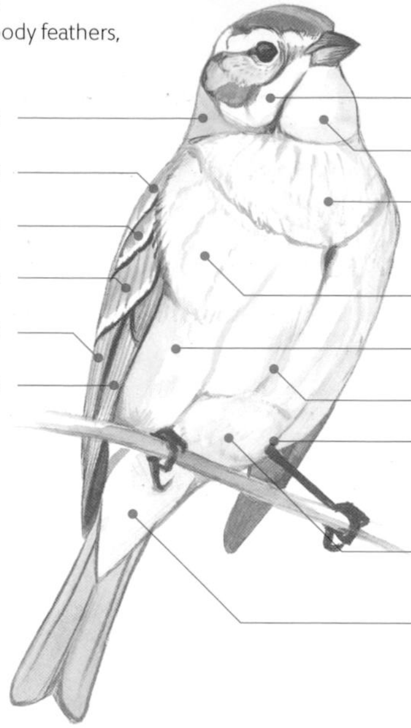

Malar Feathers along side of lower jaw.

Throat Spans underside of lower jaw.

Breast Feathers continuous across front of body. Actually attached to neck; entire group expands and contracts markedly as neck is raised and lowered.

Sides Overlapping bend of wing.

Flanks Long feathers along side of body.

Belly Actually unfeathered; covered by long feathers growing inward from flanks.

Tibial feathering Tiny feathers covering upper leg; barely visible.

Vent Several small groups of feathers between belly and undertail coverts.

Undertail coverts Overlap base of tail, below.

Basic passerine body feathers, from behind, with feathers fluffed.

Nape Back of neck.

Mantle Center of back, often patterned as streaks. A pair of contrasting pale stripes on the mantle on some species are called braces.

Scapulars Overlap base of wing, above. Mantle and scapulars are together referred to as the back.

Rump Rump feathers lie under the folded wings; arbitrarily divided into upper rump and lower rump. The contrasting rump-patch of many species involves uppertail coverts as well as rump feathers.

Femoral tract Sides of rump; mostly hidden by longest flank feathers.

Uppertail coverts Overlap base of tail, above.

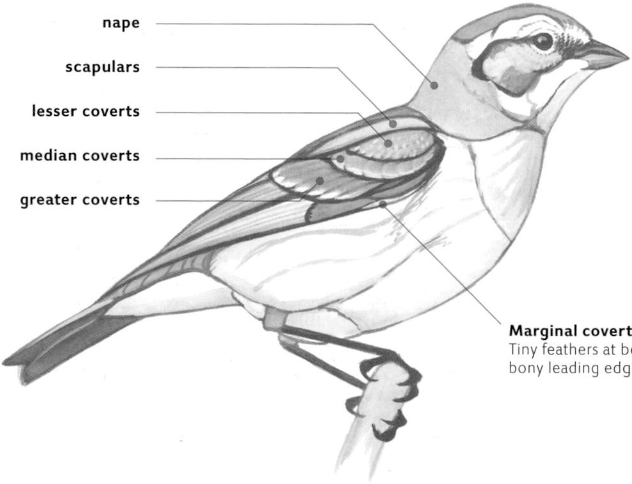

Passerine in alert, sleeked posture. In this pose, with the wings held out from the body, the usually concealed lesser coverts and marginal coverts are easily seen.

Marginal coverts
Tiny feathers at bend of wing covering bony leading edge of "hand."

Wing Feathers

Right wing of a passerine, closed but held loosely, viewed from behind. Note how the feathers stack up, with the innermost tertial on top and the outermost primary on the bottom. With the wing folded against the body, only the outer edges of the remiges are visible. Secondaries and primaries are numbered from center of wing (same order in which most species molt).

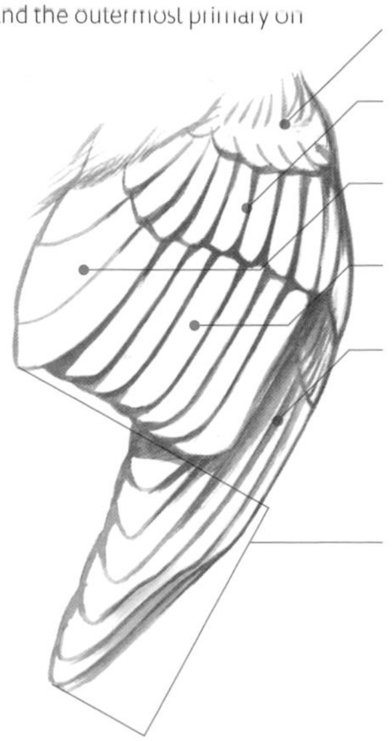

Median secondary coverts Overlap bases of greater coverts. Pale tips form *upper wingbar*.

Greater secondary coverts Overlap bases of secondaries. Pale tips form *lower wingbar*.

Tertials Three innermost secondaries. On folded wing these broad feathers rest on top of other secondaries.

Secondaries Six on most songbirds (plus three tertials); long flight feathers growing from "forearm" bones.

Primaries Nine or ten long flight feathers growing from "hand" bones and forming lower border of folded wing.

Remiges Primaries, secondaries, and tertials together. Remiges and tail feathers (*rectrices*) are collectively called *flight feathers*.

Primary projection Projection of primary tips beyond tertial tips. Length of primaries relative to tail is also a useful measure in identifying many species.

Passerine in flight, from above.
Note that outer webs of flight feathers are visible.

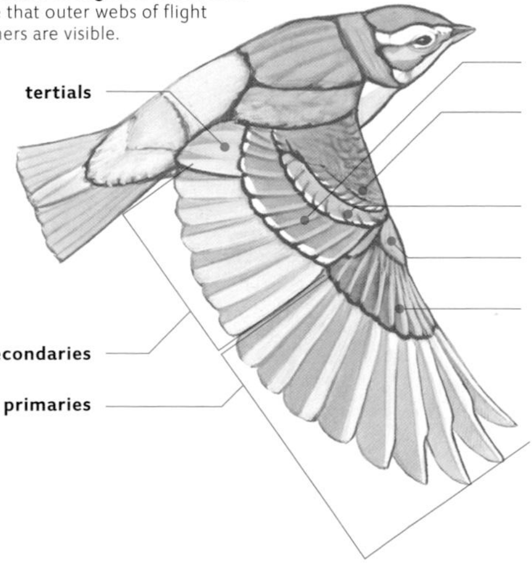

Greater secondary coverts

Lesser secondary coverts Overlap bases of median coverts. Rarely visible on passerines; usually concealed by scapulars and sides when wing is folded.

Median secondary coverts

Alula Three feathers on the "thumb."

Greater primary coverts Overlap bases of primaries.

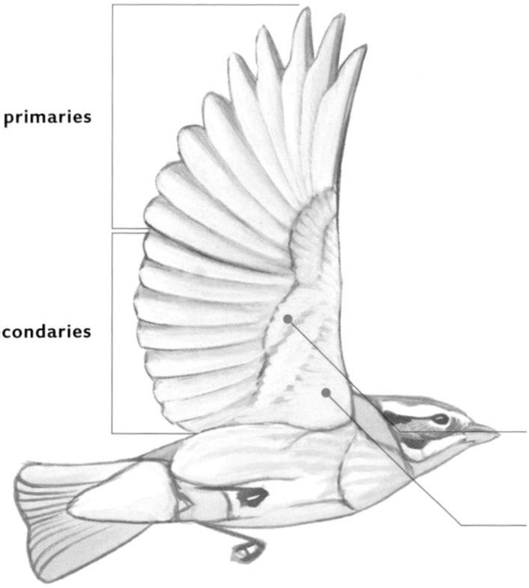

Tail

Tail feathers, also known as rectrices (singular: rectrix), form a simple fan that can be spread open or folded closed. When folded the feathers stack up with the central tail feathers on top and outermost tail feathers on the bottom.

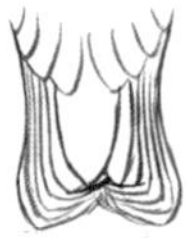

Underwing coverts The "wing linings." Rows of feathers corresponding to upperwing coverts, but less easily distinguished.

Axillaries The "armpit." Overlap base of wing, below.

Parts of a Shorebird

Top: Small shorebird, relaxed. This typical shorebird differs significantly from passerines in wing structure and in its two distinguishable groups of scapulars, which are much more prominent than the scapulars on passerines. The scapulars hang loosely when relaxed, covering most of the wing. (They are often pulled up when active, exposing the wing coverts.) The secondaries and primaries are nearly or entirely concealed when the wings are folded. Note the many rows of lesser coverts (bottom illustration). The pale V on the back of many shorebirds is formed by pale edges on the mantle and upper scapular feather groups.

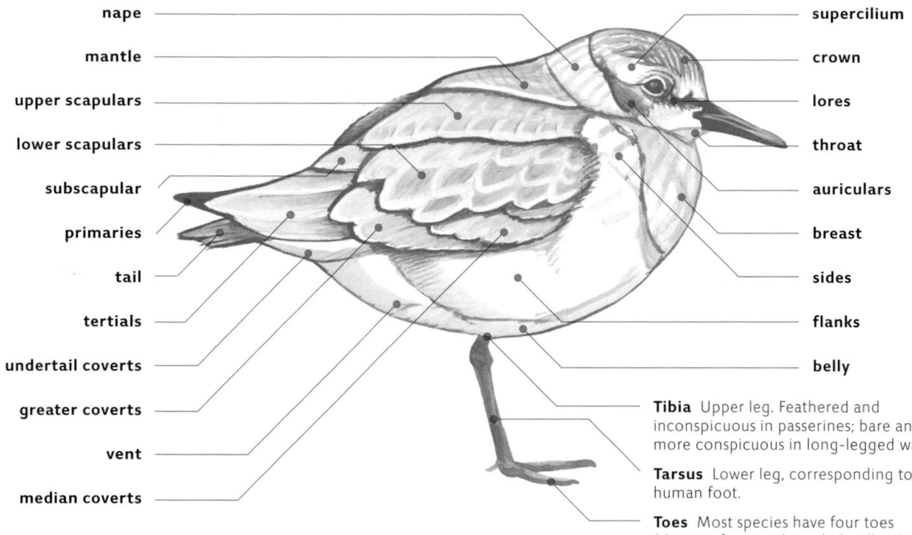

Tibia Upper leg. Feathered and inconspicuous in passerines; bare and more conspicuous in long-legged waders.

Tarsus Lower leg, corresponding to the human foot.

Toes Most species have four toes (three in front and one behind), which may be joined by webbing (*palmation*) or partial webbing (*semipalmation*).

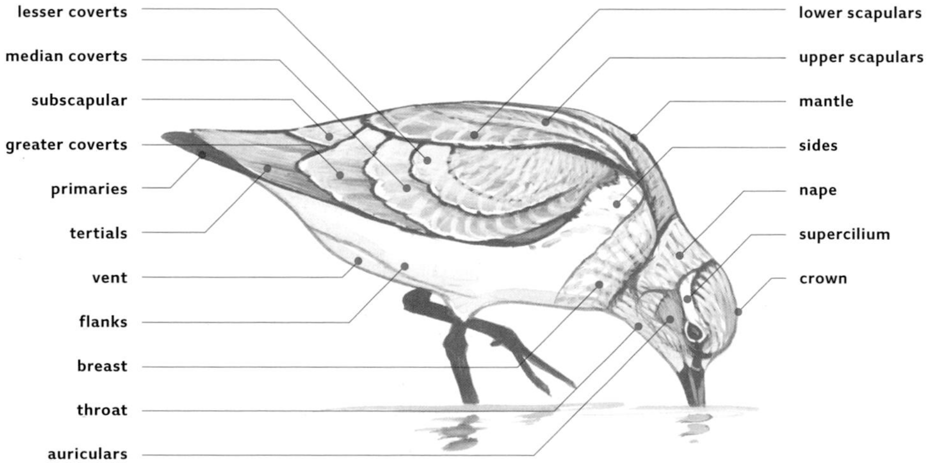

Parts of a Duck

Duck, swimming. Note that waterbirds such as ducks and shorebirds are more or less uniformly covered with feathers to create seamless waterproofing. It may be difficult to distinguish the feather groups on any part of a bird that is normally in contact with water.

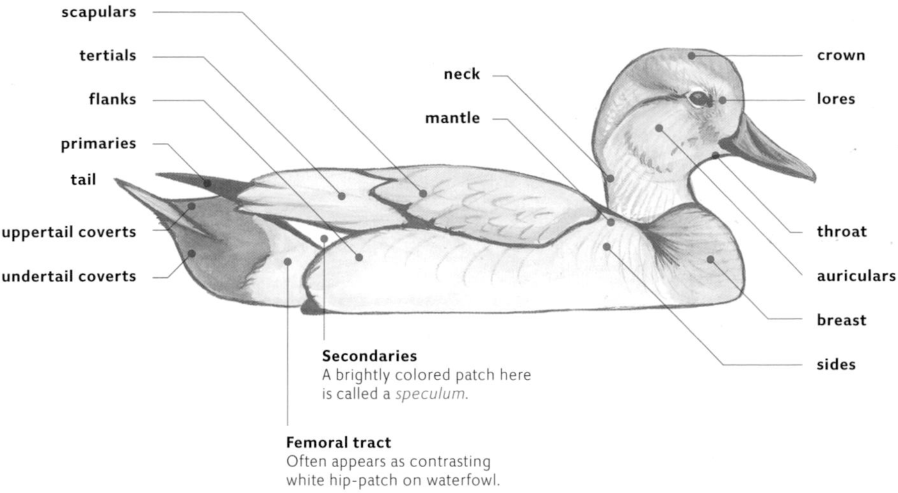

Parts of a Gull

Large Gull, standing. All feather groups are essentially the same as on a shorebird.

Gull in flight, showing feathers typical of most long-winged species. The pale *wingstripe* seen on many shorebirds is formed by pale bases of secondaries and/or primaries (often combined with white tips on greater coverts). *Windows* are translucent patches on flight feathers, visible on a flying bird, where lightly pigmented areas allow light to pass through. A *carpal-bar* is a contrastingly colored band on the upperwing, extending along a diagonal line from tertials to carpal joint, or "wrist" (not always the entire distance). A dark carpal-bar forms part of the M pattern seen on many species. Interestingly, this bar crosses all the rows of wing coverts and does not follow the contours of any single feather-tract.

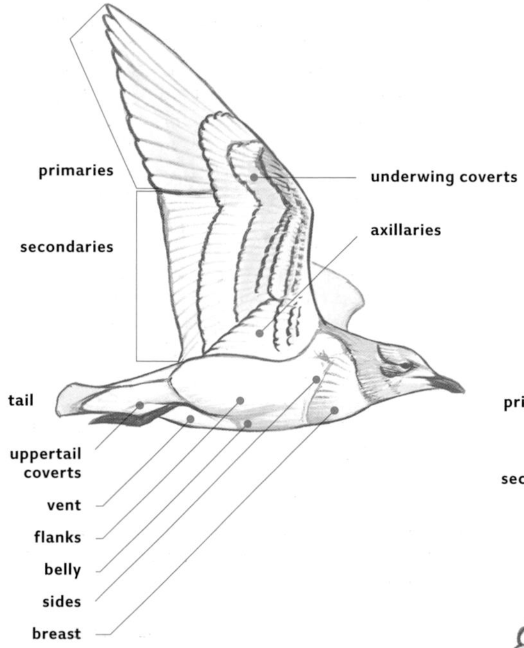

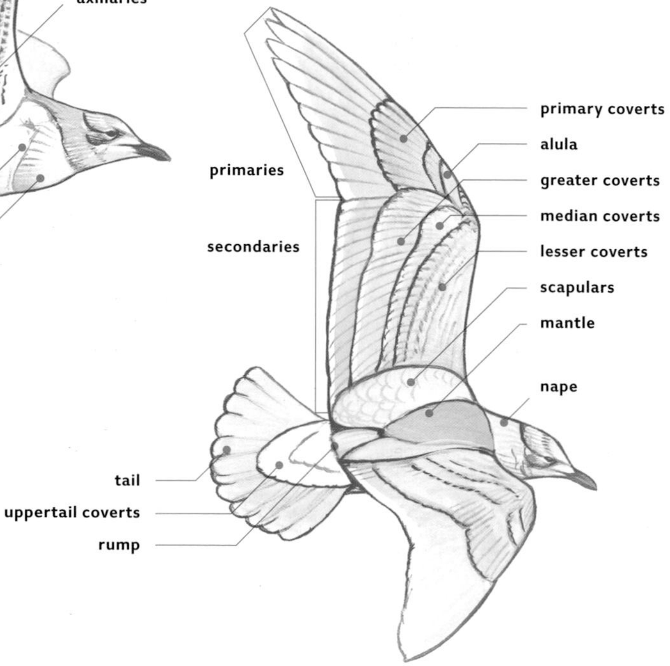

Bare Parts

Head of large gull, showing bare parts that are important in gull identification.

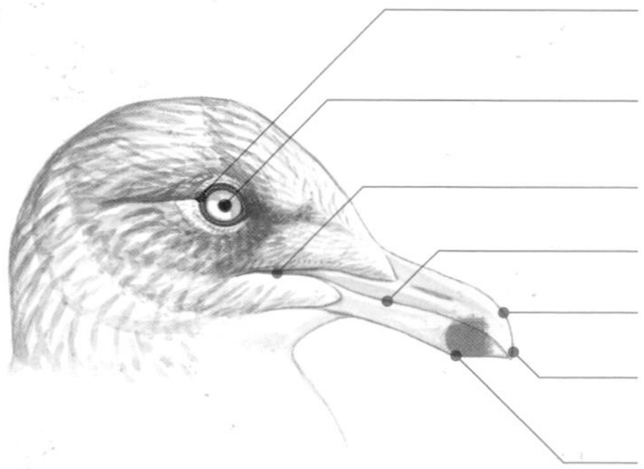

Orbital ring Or "eyering." Unfeathered ring immediately around eye. Present on all birds, brightly colored on some.

Iris Color is often useful for identification and age determination. Pupil is black in all bird species.

Gape Fleshy edges at corners of mouth.

Tomia (sing., tomium) Cutting edges of bill.

Culmen Upper edge of bill; visible in profile.

Nail Tip of upper mandible; distinctly set off on some species (but not on gulls).

Gonydeal angle Visible in profile. Most large gulls show a bright gonydeal spot.

Molt and Plumage

For birds, molting is the process of dropping old feathers and growing new ones in their place, and all birds molt at least once a year. Observers can nearly always identify a bird without reference to molt, but a basic understanding of molt can be helpful for determining the age of a bird, for assessing plumage variation, and for identification.

The process of growing an entirely new set of feathers is so demanding that it usually does not coincide with other demanding activities such as nesting or migrating. Each species (or subspecies) has evolved a schedule that allows it to molt at a time when food is readily available and other demands are not too high. In general, long-distance migrants molt after fall migration, other species molt before; this schedule differs among some closely related species (e.g., Semipalmated and Western Sandpipers, Common and Lesser Nighthawks, etc.).

Age terminology can be confusing, as several different systems are used by birders. In this book, age of immature birds is described using the *life-year system*, which calculates age in the same way that we figure our own ages. *1st year* applies to a bird during its first year of life, from fledgling through the 1st winter and 1st summer, ending around its first "birthday," when it begins the molt to 2nd year plumage. Most passerines molt into adult plumage at one year of age, but many nonpasserines have distinguishable 2nd, 3rd, and even 4th year plumages before reaching adulthood.

Seasonal plumages of adults are described in this book as *breeding* and *nonbreeding*. A few species have a third plumage either before or after breeding, called a supplemental plumage.

It is important to realize that the labels used in this book describe the bird's appearance and do not always have a connection to molt. For example, a 1st summer appearance or adult breeding appearance (e.g., Snow Bunting) may simply result from the fading and abrasion of feathers grown in a previous molt, and not from any new molt.

Alternative terminologies include the *calendar-year system*, which describes immature birds as 1st calendar year (or hatching year or HY) from hatching until December 31, 2nd calendar year (or second year or SY) from January 1 until the next December 31, etc.

A third set of terms, the *Humphrey-Parkes system*, is essential for any in-depth study of molt, as it is purposely devoid of references to calendars and breeding cycles. It is best suited for describing individual feathers rather than the overall appearance of the plumage, and is not recommended for birders. For further details consult Peter Pyle's *Identification Guide to North American Birds* (Slate Creek Press, 1997, 2008).

Molt Cycle of a Typical Passerine

In this hypothetical example, changing colors indicate newly molted feathers. Note that flight feathers are retained for a full year (in most species), while body feathers may be molted two or three times a year. The dates and feather replacements shown are examples only. There is tremendous variation among species in the timing and extent of each molt, but virtually all species have a complete molt once each year, and this usually takes place just after nesting.

Labels compare three different terminologies. Listed first is the life-year (LY) system, the terminology used in this book, second is the calendar-year (CY) system, third is the Humphrey-Parkes (HP) system.

Life-year	**Calendar-year**	**Humphrey-Parkes**
juvenile	1st calendar Jul–Aug	juvenal plumage
1st winter	1st calendar Aug to 2nd calendar Mar	1st basic plumage
1st summer	2nd calendar Apr–Jul	1st alternate plumage
adult nonbreeding	at least 2nd calendar Aug to 3rd calendar Mar	definitive basic plumage
adult breeding	at least 3rd calendar Apr–Jul	definitive alternate plumage

Key to the Species Accounts

The information and illustrations for each species are arranged in similar ways on nearly every page, with size, flight, voice, and other topics in the same position. Comparing species to another involves scanning horizontally across the pages.

Common name

Scientific name

Key structural features and field marks are highlighted along with comparative text to similar species.

Both the underside and upperside of birds in flight are included, often with more than one example of each species to show variation in wing and tail pattern or characteristic shapes.

Seasonal plumages are labeled with typical date spans (but some individuals show plumage outside of the listed dates). Other plumages are worn year-round.

All significant age, sex, and seasonal variation are shown and labeled accordingly. Significant geographic variations are illustrated and labeled with simple regional names (or sometimes with established common names).

Maps show the complete North American range, including migration routes and areas of vagrant records. In a few cases, maps of subspecies/regional populations are included.

Magnolia Warbler

Setophaga magnolia

L 5" WS 7.5" WT 0.3 oz (8.7 g)

Small and relatively long-tailed; tail often raised and fanned, showing off unique pattern. Drab individuals can be similar to Canada or Prairie Warblers; note thin wingbars, yellow rump, white undertail coverts, and bright yellow breast.

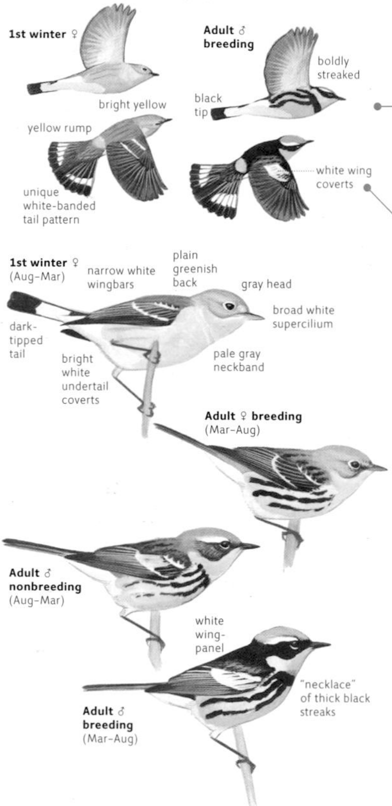

VOICE: Song short, musical but rather weak and simple *sweeter sweeter SWEETEST*; variable; most similar to Hooded but much less emphatic. Call a unique, tinny, hoarse *vint* or *chuif.* Flight call a very high, weak, lightly trilled, rather soft *zzip.*

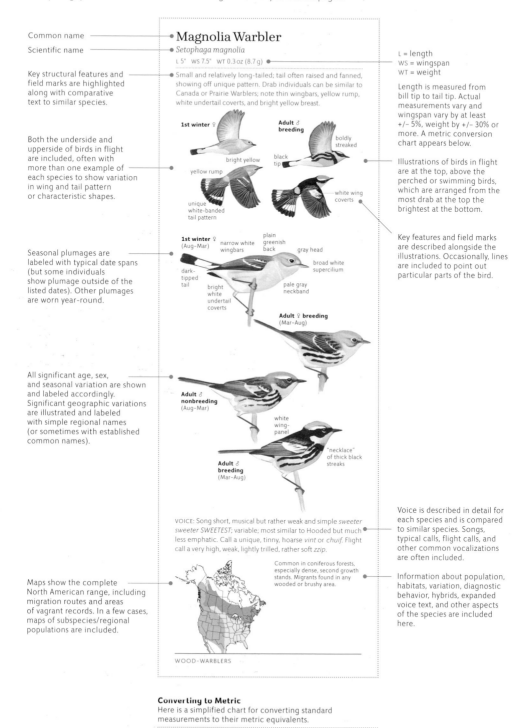

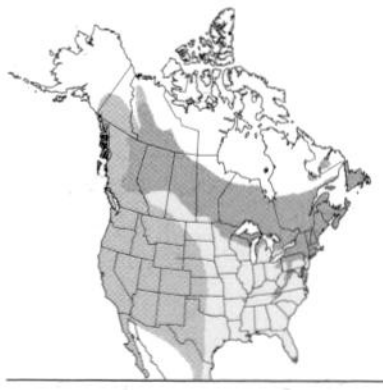

Common in coniferous forests, especially dense, second growth stands. Migrants found in any wooded or brushy area.

WOOD-WARBLERS

L = length
WS = wingspan
WT = weight

Length is measured from bill tip to tail tip. Actual measurements vary and wingspan vary by at least +/– 5%, weight by +/– 30% or more. A metric conversion chart appears below.

Illustrations of birds in flight are at the top, above the perched or swimming birds, which are arranged from the most drab at the top the brightest at the bottom.

Key features and field marks are described alongside the illustrations. Occasionally, lines are included to point out particular parts of the bird.

Voice is described in detail for each species and is compared to similar species. Songs, typical calls, flight calls, and other common vocalizations are often included.

Information about population, habitats, variation, diagnostic behavior, hybrids, expanded voice text, and other aspects of the species are included here.

Converting to Metric

Here is a simplified chart for converting standard measurements to their metric equivalents.

inches	► millimeters	multiply by	25.0
inches	► centimeters	multiply by	2.5
feet	► meters	multiply by	0.3
ounces	► grams	multiply by	28.3
pounds	► kilograms	multiply by	0.45

Key to the Group Accounts

Each group of related species is introduced by a summary of the characteristics of the group. These accounts show all species in a family (or subfamily or order) for comparison. Look here to see the range of variation in the group, as well as fundamental similarities and differences among genera.

Common name and scientific name of the family or group.

This paragraph highlights the general characteristics of the group and gves the nuber of species and genera covered in the guide. Overview of typical behaviors (feeding, nesting, etc.) is included.

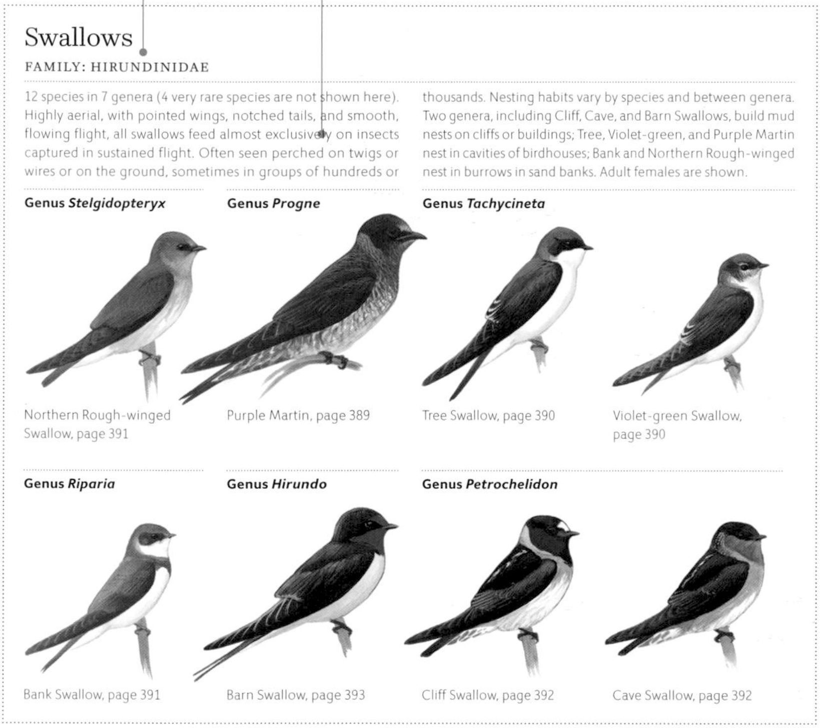

Swallows
FAMILY: HIRUNDINIDAE
12 species in 7 genera (4 very rare species are not shown here). Highly aerial, with pointed wings, notched tails, and smooth, flowing flight, all swallows feed almost exclusively on insects captured in sustained flight. Often seen perched on twigs or wires or on the ground, sometimes in groups of hundreds or thousands. Nesting habits vary by species and between genera. Two genera, including Cliff, Cave, and Barn Swallows, build mud nests on cliffs or buildings; Tree, Violet-green, and Purple Martin nest in cavities of birdhouses; Bank and Northern Rough-winged nest in burrows in sand banks. Adult females are shown.
Genus Stelgidopteryx
Genus Progne
Genus Tachycineta
Northern Rough-winged Swallow, page 391
Purple Martin, page 389
Tree Swallow, page 390
Violet-green Swallow, page 390
Genus Riparia
Genus Hirundo
Genus Petrochelidon
Bank Swallow, page 391
Barn Swallow, page 393
Cliff Swallow, page 392
Cave Swallow, page 392

All species in the group, except the very rare, are pictured. Grouped by genus, the birds are reproduced in relative scale. In most cases, immature plumages or comparitively drab females are shown, as these present the greatest identification challenge.

Key to the Range Maps

The range maps show the complete distribution of each main species in our area. Bear in mind that within the mapped range, each species occurs in appropriate habitats and at variable density (common to scarce).

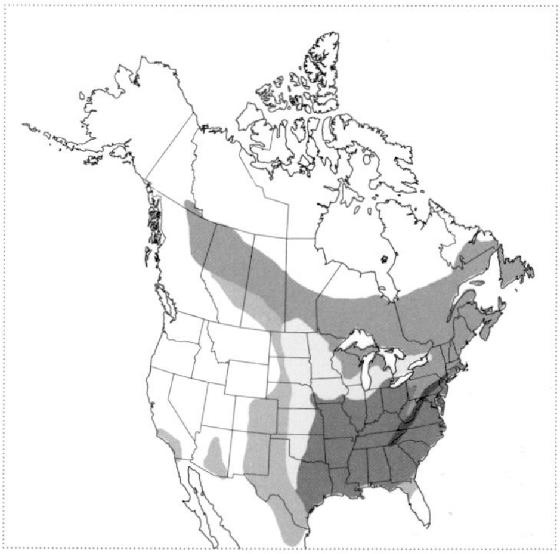

Winter Shows the normal winter distribution of the species, Many are somewhat nomadic in winter, occupying only parts of the mapped range at any given time.

Summer For virtually all species, this is the breeding range and is more consistently and uniformly occupied than the winter range.

Year-round Indicates that the species can be found all year in this area, even though winter and summer populations may involve different individual birds. Only a few species are truly resident.

Migration Main migratory routes are shown, as well as areas of regular dispersal and post-breeding wandering. Note that migration also passes through summer and winter ranges.

Rare Shows areas of rare occurrence (may be a single or up to a few records per year).

The Sibley Guide to Birds

Ducks, Geese, and Swans

FAMILY: ANATIDAE

51 species occur regularly, and many additional species occur as rare visitors or escapes.

GEESE AND SWANS: 12 species occur regularly; all are large, with muted colors and little plumage variation, often in large flocks, calling loudly in flight. Walk easily on land, pairs remain together year-round. Swans (1 genus) are largest and all-white. Geese (in 3 genera) smaller, white, gray, or brown. Adults are shown (not to scale with ducks).

Genus *Anser*

Pink-footed Goose, page 5

Greater White-fronted Goose, page 4

Genus *Chen*

Snow Goose, page 10

Ross's Goose, page 11

Emperor Goose, page 9

Genus *Branta*

Brant, page 8

Barnacle Goose, page 9

Cackling Goose, page 6

Canada Goose, page 7

Genus *Cygnus*

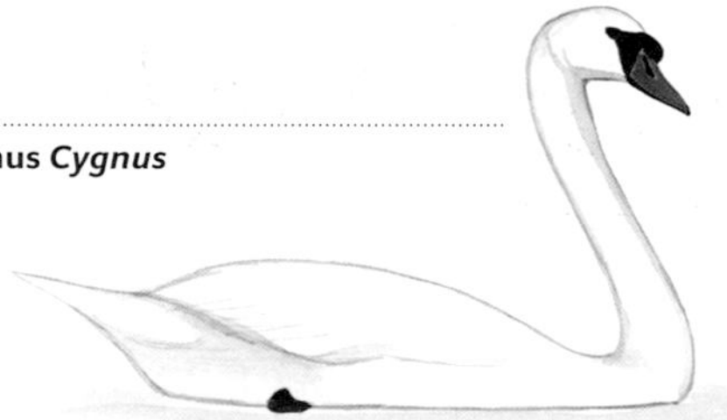

Mute Swan, page 12

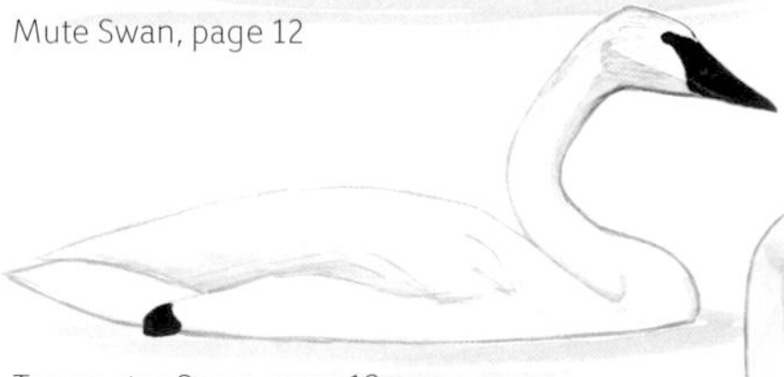

Trumpeter Swan, page 12

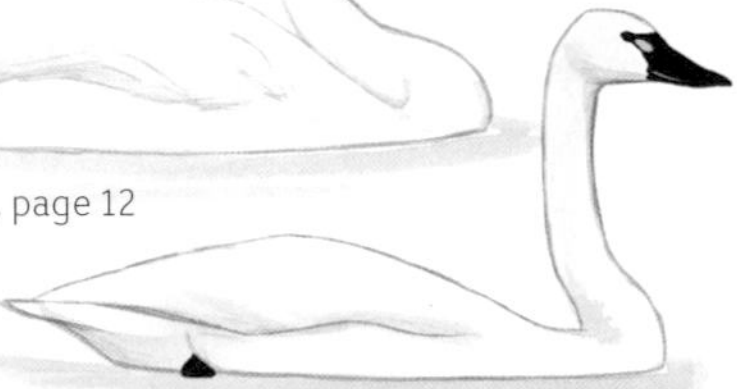

Tundra Swan, page 13

DABBLING DUCKS: 16 species (in 4 genera) occur regularly. These species rarely dive, feeding mainly by dabbling bill in water or tipping up. Most take off directly from water, without running. Whistling-Ducks are more goose-like, with long legs and neck, run to take off from water, call frequently in flight.

Genus *Dendrocygna*

Black-bellied Whistling-Duck, page 14

Fulvous Whistling-Duck, page 14

Genus *Cairina*

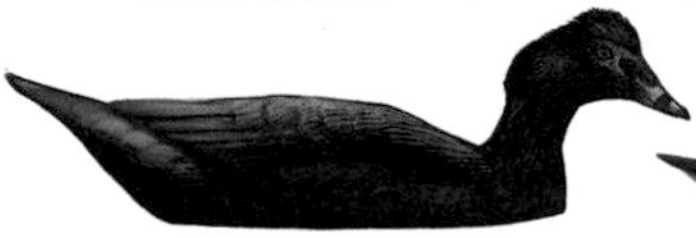

Muscovy Duck, page 15

Genus *Aix*

Wood Duck, page 15

Genus *Anas*

Eurasian Wigeon, page 21

American Wigeon, page 21

American Black Duck, page 19

Mallard, page 18

Mottled Duck, page 19

Gadwall, page 20

Northern Shoveler, page 23

Northern Pintail, page 20

Blue-winged Teal, page 22

Cinnamon Teal, page 22

Garganey, page 23

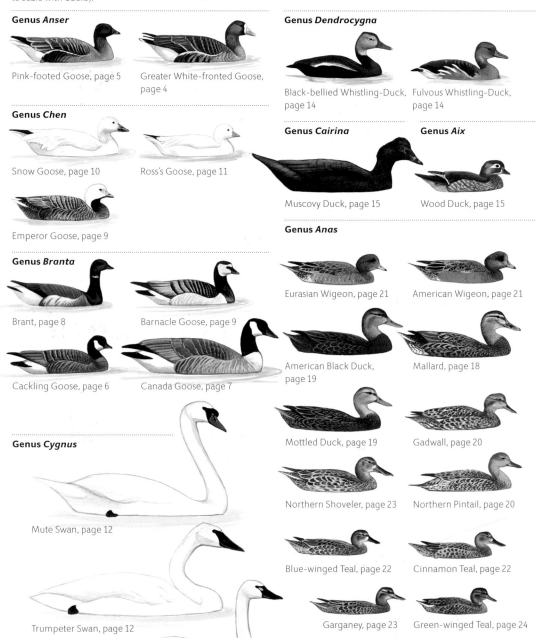

Green-winged Teal, page 24

DIVING DUCKS: 23 species occur regularly (in 11 genera). A more diverse group than dabblers, these species usually dive underwater for food (beware that all ducks are capable of diving), and frequent deeper water than dabblers. Diving ducks have relatively smaller wings and heavier bodies than dabblers, with quicker wingbeats and usually require a running start before taking off from the water. Other duck-like swimming birds include loons, grebes, cormorants, coots and gallinules, and alcids. Adult females are shown.

Genus *Aythya*

Canvasback, page 26

Redhead, page 26

Ring-necked Duck, page 27

Tufted Duck, page 27

Greater Scaup, page 28

Lesser Scaup, page 28

Genus *Polysticta*

Steller's Eider, page 32

Genus *Somateria*

Spectacled Eider, page 31

King Eider, page 31

Common Eider, page 30

Genus *Histrionicus*

Harlequin Duck, page 32

Genus *Melanitta*

Surf Scoter, page 34

White-winged Scoter, page 35

Black Scoter, page 34

Genus *Clangula*

Long-tailed Duck, page 33

Genus *Bucephala*

Common Goldeneye, page 36

Barrow's Goldeneye, page 36

Bufflehead, page 37

Genus *Lophodytes*

Hooded Merganser, page 37

Genus *Mergus*

Common Merganser, page 38

Red-breasted Merganser, page 38

Genus *Nomonyx*

Masked Duck, page 39

Genus *Oxyura*

Ruddy Duck, page 39

Greater White-fronted Goose

Anser albifrons

L 28" WS 53" WT 4.8 lb (2,200 g) ♂>♀

Brownish-gray overall with pale gray upperwing, white tip on tail, and pale pink/orange bill; adult has white face and (usually) dark bars on belly. Orange legs very conspicuous. Beware confusion with similar barnyard Graylag Goose.

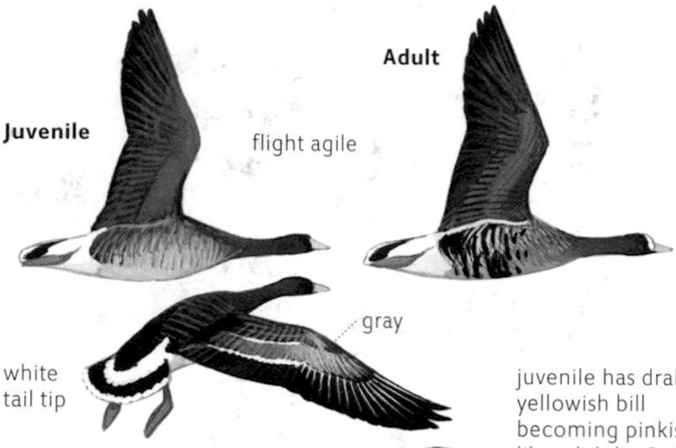

Juvenile

Adult

flight agile

gray

white tail tip

juvenile has drab yellowish bill becoming pinkish like adult by Oct

Tundra Greater White-fronted Goose

Juvenile (Aug–Dec)

white frontal feathering begins to appear by Sep; most immatures are white-fronted by Apr

orange legs

pinkish bill

Adult

speckled belly (variable)

white flank-line

Adult

VOICE: Common honk a distinctive quick, high-pitched laughing or yelping of two or three rising syllables *ho-leeleek* or *kilik*. Flock noise higher, clearer, with more rapid syllables than other geese; feeding flock gives low, buzzing chorus.

Generally uncommon and local even in its favored winter range. Forages in grain fields, meadows, and marshes, roosts at night on ponds and lakes. In summer nests on marshy tundra.

Greenland Greater White-fronted Goose

Juvenile (Aug–Dec)

Adult

white front

Rare winter visitor to Northeast, usually with flocks of Canada Geese. Definite records south to North Carolina but difficulty of identification makes true range uncertain.

Taiga Greater White-fronted Goose

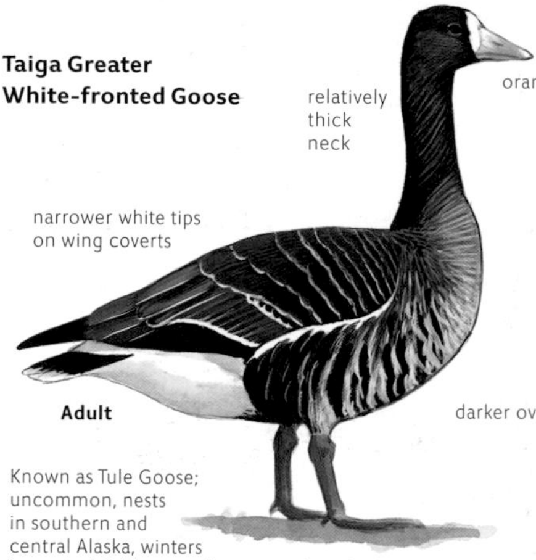

relatively thick neck

large orange bill

narrower white tips on wing coverts

Adult

darker overall

Known as Tule Goose; uncommon, nests in southern and central Alaska, winters in central California, where it tends to be found in smaller and more wooded ponds than Tundra form. Averages larger and darker (especially on neck) than Tundra form, bill long and more often orange-tinged.

Greater White-fronted × Canada Goose hybrid

these hybrids seen occasionally

compare hybrid domestic geese p. 16

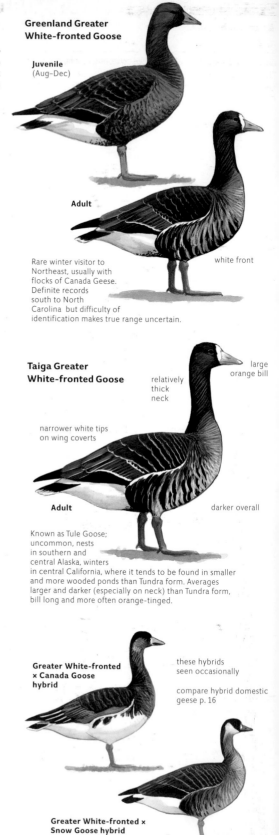

Greater White-fronted × Snow Goose hybrid

Pink-footed Goose

Anser brachyrhynchus

L 28" WS 53" WT 4.8 lb (2,200 g) ♂>♀

Related to Greater White-fronted Goose, and usually found with Canada Goose. Note all-brown head, mostly dark bill, pale gray sheen on back, and mostly white tail. Also compare dark juvenile Snow Goose and very rare bean-geese.

VOICE: A high, two-syllabled honk *HEE-lerk*, very similar to Greater White-fronted but slightly lower and harsher. Also higher short squeaking notes.

Very rare but increasing visitor from Greenland and Europe to the Northeast Oct–Mar, usually with flocks of Canada Geese.

Graylag Goose

Anser anser

L 32.5" WS 65" WT 6.9 lb (3,130 g) ♂>♀

Orange bill and pink legs similar to Greater White-fronted Goose, but note heavier body, larger bill, plain pale belly and lack of white flank line. Domestic Geese derived from Graylag Goose and Swan Goose, often partly white, are a common sight in city parks and farmyards (see page 16).

VOICE: Very loud, raucous braying or honking, the familiar barnyard goose sound.

Common and widespread as a domestic bird in city parks and farmyards, sometimes joining flocks of Canada Geese. Many individuals are hybrids with Swan Goose or Canada Goose. Two definite records of wild vagrants (Alaska, Newfoundland), and other records are possible as nesting population increases in Greenland.

Tundra Bean-Goose (*Anser serrirostris*) and Taiga Bean-Goose (*Anser fabalis*) are very similar, closely related to each other and to Pink-footed Goose. Distinguished from adult Greater White-fronted Goose by mostly dark bill, dark face, plain pale belly. Distinguishing Tundra from Taiga Bean-Goose requires careful attention to size, proportions, and bill pattern.

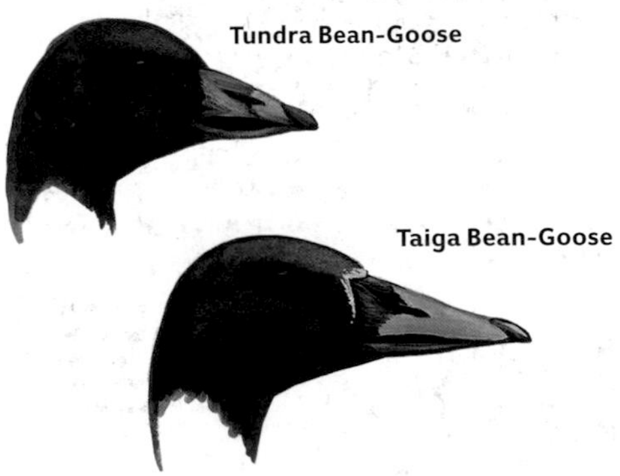

VOICE: Very similar to Greater White-fronted Goose but slightly lower-pitched.

Both are very rare visitors to western Alaska with a few records farther south and east.

Cackling Goose

Branta hutchinsii

L 25" WS 43" WT 3.5 lb (1,600 g) ♂>♀

Very similar to Canada Goose and only recently split. Distinguished by smaller size, relatively short neck and small bill, but some overlap with smallest Canada Geese. Western populations (wintering in Pacific coast states) have shortest bill and dark brown breast.

Richardson's nests in northern Alaska and Canada, winters mainly in mid-continent, with small numbers to Pacific coast and very few to Atlantic coast. Pacific forms nest in western Alaska, winter in Pacific states.

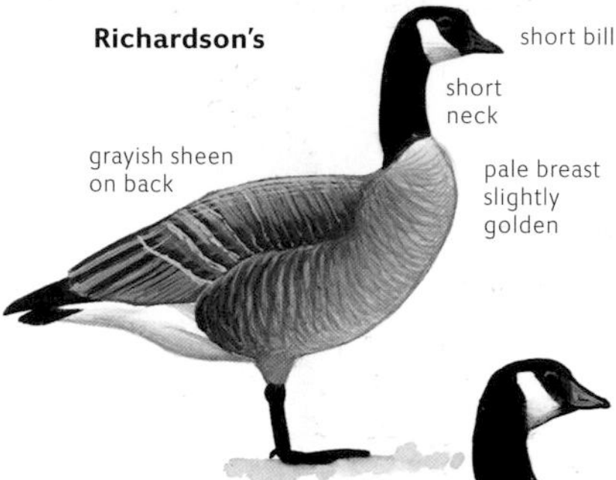

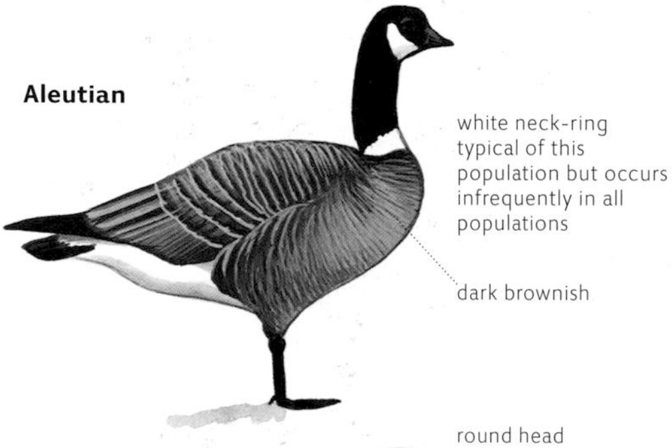

When grazing, shorter neck of Cackling Goose is held straighter, longer neck of Canada Goose is more strongly arched and curved.

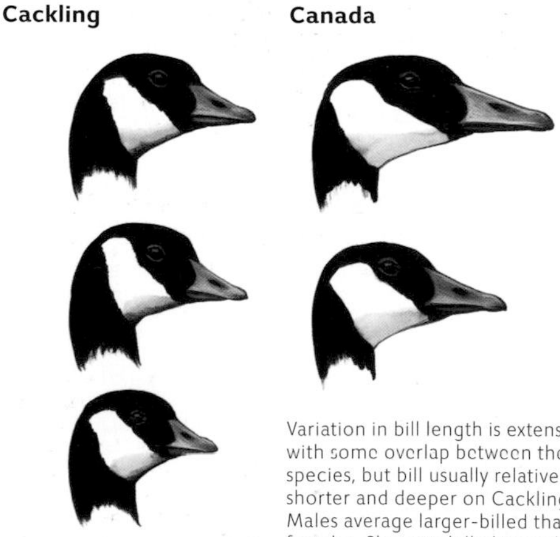

Variation in bill length is extensive, with some overlap between the species, but bill usually relatively shorter and deeper on Cackling. Males average larger-billed than females. Shortest-billed Canada Geese (Lesser) and longest-billed Cackling Geese (Richardson's) are found together mid-continent

VOICE: Calls of Pacific and Aleutian forms distinctive, high-pitched squeaking or yelping *yeek* or *uriik*, higher and sharper than Canada Goose. Richardson's gives honking call similar to Canada Goose but on average slightly higher-pitched.

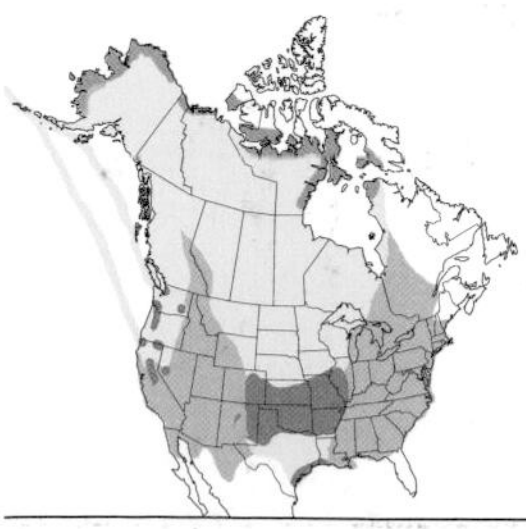

Locally common at a few key wintering and staging areas, elsewhere uncommon to rare. Forms pure flocks where common, otherwise mixes with flocks of Canada Geese. In winter, forages in farmland, wet meadows, shallow ponds and marshes, roosts on open water.

Canada Goose

Branta canadensis

L 45" WS 60" WT 9.8 lb (4,500 g) ♂>♀

Brownish body, pale breast, black neck, and white cheek shared only by the smaller Cackling Goose. Flocks fly high in well-defined V formation. Subspecies variation in Canada Goose involves mainly size. The widespread Typical form is large and pale, with pale breast.

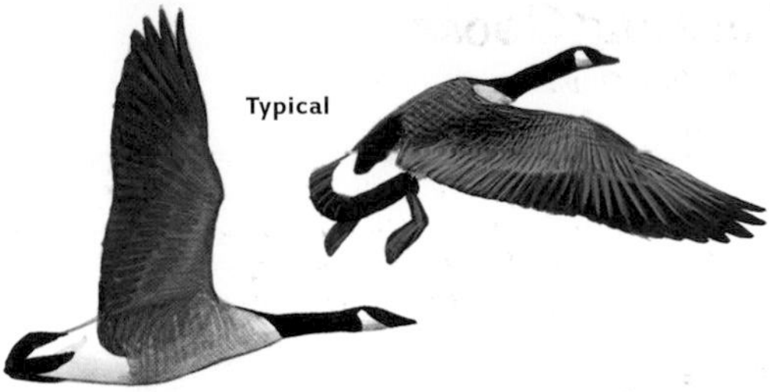

Typical Canada Goose

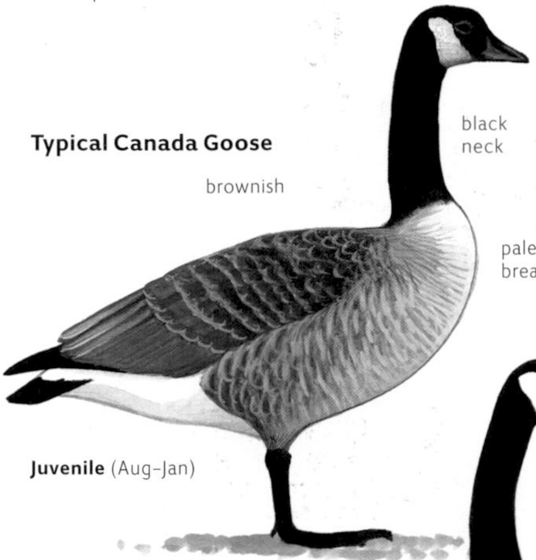

Dusky Canada Goose

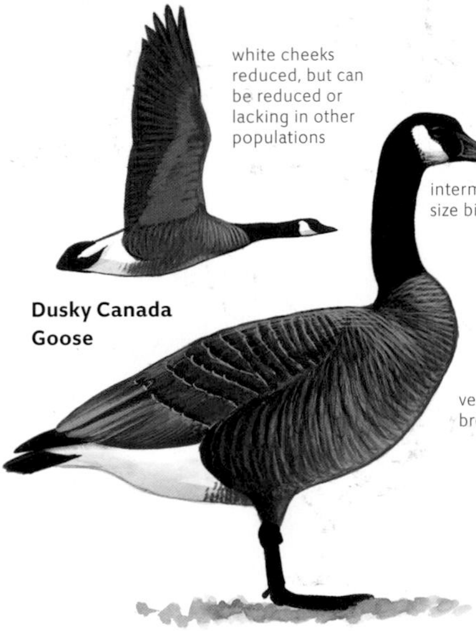

Typical form is found throughout most of range. Darker subspecies such as Dusky nest along Pacific coast from southern Alaska to British Columbia, winter south to California. Lesser nests in northwestern part of range, winters mainly mid-continent.

Lesser Canada Goose

occasional variants of all populations have dark cheeks or white forehead or white neck

VOICE: Familiar call a loud, resonant, and musical honk *h-ronk* and *h-lenk*; flock chorus gentle, slow-paced, mellow; no harsh or sharp notes. Other soft, grunting calls. Voice very similar in all populations.

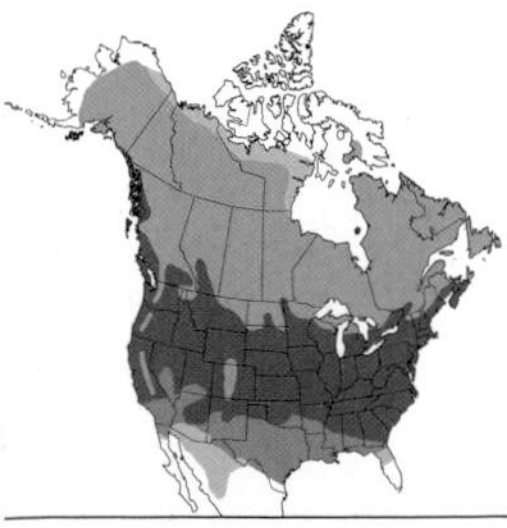

Very common and widespread, found on or near any body of water from urban parks to tundra wetlands. Forms large flocks Aug–May, often seen grazing on lawns, golf courses, and farmland. Southern breeders nearly sedentary, northern birds highly migratory.

Brant

Branta bernicla

L 25" WS 42" WT 3.1 lb (1,400 g) ♂>♀

A relatively small, coastal goose, overall blackish with white rear end. Usually shows thin white necklace. Note small black bill, black legs. Relatively pointed wings more angled in flight than in other geese. Flight fast and agile; flock tends to fly lower than Canada Geese in uneven U formation or lines.

VOICE: a soft, gargling *rrot* or *cronk* and a hard *cut-cut* in flight. Flock chorus a constant, low, murmuring, gargling sound with little variation. All populations identical.

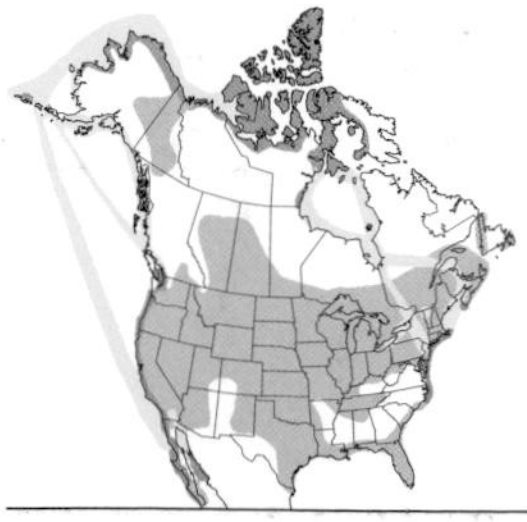

Locally common at favored wintering and staging areas, elsewhere uncommon to rare. Rarely mixes with other geese. Winters in large flocks almost exclusively on shallow saltwater bays and estuaries, sometimes foraging on short grass fields. Nests on coastal tundra around saltwater estuaries.

Subspecies differ mainly in belly and flank color, slightly in necklace size and back color. No differences in voice, shape, etc. Note that juveniles of all forms have unpatterned gray flanks and belly, with white tips on wing coverts. Adults of all forms except Dark-bellied have mostly white flanks. Black nests in western Arctic, winters on Pacific coast. Gray-bellied nests on Prince Patrick Island, winters mainly around Puget Sound. Dark-bellied nests in northeastern Arctic, winters in Europe (a very rare visitor to our Atlantic coast). Pale-bellied nests in eastern Arctic, winters along Atlantic coast.

Barnacle Goose

Branta leucopsis

L 27" WS 50" WT 3.7 lb (1,700 g) ♂>♀

About the size of Cackling Goose. Black, gray, and white overall, with no brown in plumage. Among Canada Geese look for black breast and white flanks; on close inspection boldly barred upperside and white face are apparent. In flight, note pale gray wing coverts.

VOICE: Hoarse, monosyllabic bark *henk*; generally higher than Snow Goose; sometimes rapidly repeated and resembling shrill yapping of small dog. Wings produce creaking noise.

Rare but increasing visitor to Northeast Oct–Apr, Most records are of single birds with flocks of Canada Geese. Occasional hybrids × Canada or × Cackling Goose are seen.

Barnacle × Cackling Goose hybrid

Emperor Goose

Chen canagica

L 26" WS 47" WT 6.1 lb (2,800 g)

A relatively small, stocky goose; very round-bodied, with short neck and broad wings, dark gray body, and white head; no other goose has dark tail coverts contrasting with white tail. Also note yellow to orange legs.

VOICE: High-pitched, rapid, triple- or double-note tinny *honk kla-ha* like Snow Goose; also high, clear, trumpeting *tedidi*. Low grunting from ground.

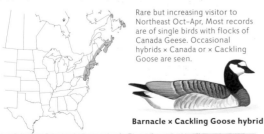

Locally common in winter range, elsewhere uncommon to rare. Winters on salt water, mainly on sheltered bays and lagoons, but also along rocky coasts. Nests on wet coastal tundra.

Snow Goose

Chen caerulescens

LESSER: L 28" WS 53" WT 5.3 lb (2,420 g) ♂>♀
GREATER: L 31" WS 56" WT 7.4 lb (3,400 g) ♂>♀

Two color morphs. White morph told from all-white domestic geese by pink bill and legs, dark primaries; see Ross's Goose. Dark morph distinguished from other dark geese by very pale wing coverts, pink bill and legs, and by adult's white head.

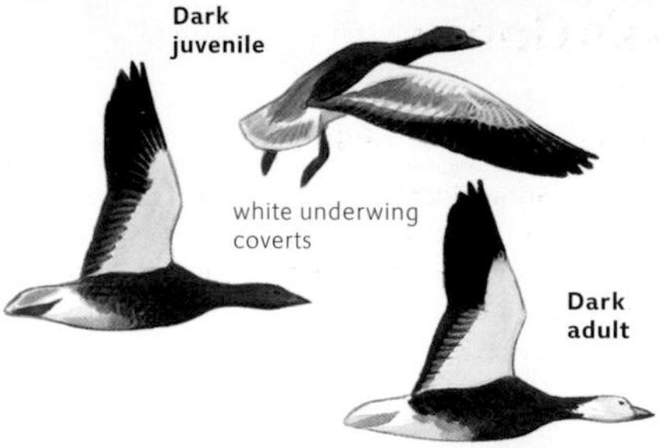

VOICE: Common call a harsh, monosyllabic, descending *whouk* or higher *heenk*; harsher and more raucous than other geese, recalling Great Blue Heron. Flock chorus slow-paced with single honks on varied pitches; whole pitch range greater than other geese. Also lower, grunting *hu-hu-hur* from foraging birds.

A complete range of intermediate birds occurs between white and dark morph.

Locally common at a few key wintering areas, elsewhere uncommon to rare. In winter forms large flocks that forage in agricultural fields and coastal saltmarshes, roosts on sheltered water. Nests in large colonies on tundra.

Two subspecies: Lesser is widespread, Greater nests in eastern Canada and winters in the mid-Atlantic states, where Lesser is generally uncommon. Dark morph is rare in Greater and in populations wintering in California and New Mexico, but the majority of Gulf Coast wintering population is dark.

Ross's Goose

Chen rossii

L 23" WS 45" WT 2.7 lb (1,250 g)

Very similar to Snow Goose, but smaller with relatively small bill, round head, short neck; white immature plumage unlike dusky gray immature Snow Goose. Reliably distinguished from Snow only by details of bill. Hybrids x Snow Goose are rare but regular, complicating identification.

VOICE: Generally quiet, less vocal than Snow, flock sounds overall higher-pitched and more rapid than Snow Goose; calls a high sharp *keek* or squealing *keek-keek*, analogous to *heenk* of Snow Goose, and a grunting low *kowk*.

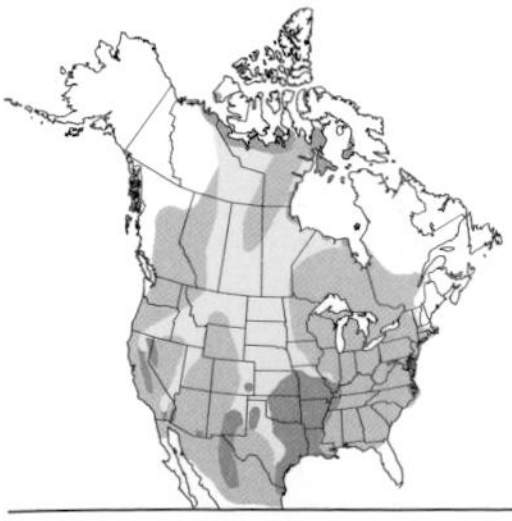

Locally common and forming large pure flocks in central California, elsewhere found in small numbers mixed with Snow Geese; uncommon east to Louisiana, rare on Atlantic coast. Habits and habitat like Snow Goose. Dark morph rare throughout.

Geese Head and Bill Shapes

Ross's Goose
small with rounded head and stubby bill; little or no "grin patch"; border at base of bill straight and vertical; bluish on base of bill; dark morph rare

Ross's × Lesser Snow Goose hybrid
intermediate in size and bill structure

Lesser Snow Goose
larger than Ross's with more wedge-shaped head; obvious black "grin patch"; strongly curved border at base of bill

Greater Snow Goose
averages 20 percent longer-billed than Lesser; looks longer-faced with head even more strongly wedge-shaped

Mute Swan

Cygnus olor

L 60" WS 75" WT 22 lb (10 kg) ♂>♀

Larger than any goose, long necked, and all white; similar to other swans but note long, pointed tail and orange bill with black knob at base. Juvenile can be white or gray-brown. Wings produce loud throbbing hum in flight.

VOICE: Not mute; gives a variety of calls. In aggression an explosive, exhaling *kheorrrr* with rumbling end; sometimes a clear, bugling *kloorrr* reminiscent of Tundra Swan, also hissing and snorting. Immature gives higher gurgling note. Wings produce loud, resonant, throbbing hum in flight, unlike other swans.

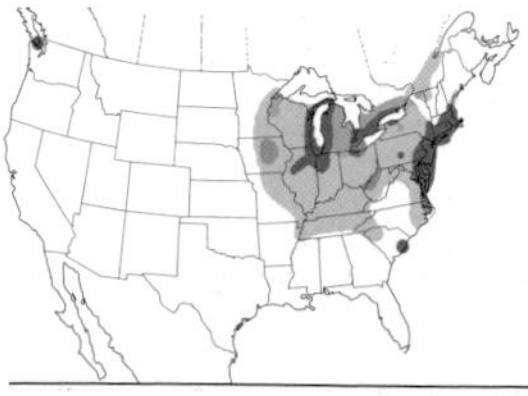

Uncommon but conspicuous. Usually seen in territorial pairs or family groups on shallow ponds and lakes with plenty of submerged and emergent vegetation. When small ponds freeze in winter flocks gather on unfrozen water including sheltered coastal bays. Occasional escapes from captivity can appear continent-wide.

Trumpeter Swan

Cygnus buccinator

L 60" WS 80" WT 23 lb (10.5 kg) ♂>♀

Very similar to Tundra Swan, slightly larger, with longer bill. Best distinguished by details of bill shape and by voice. Immature retains gray-brown plumage into first summer (Tundra becomes white by mid-winter).

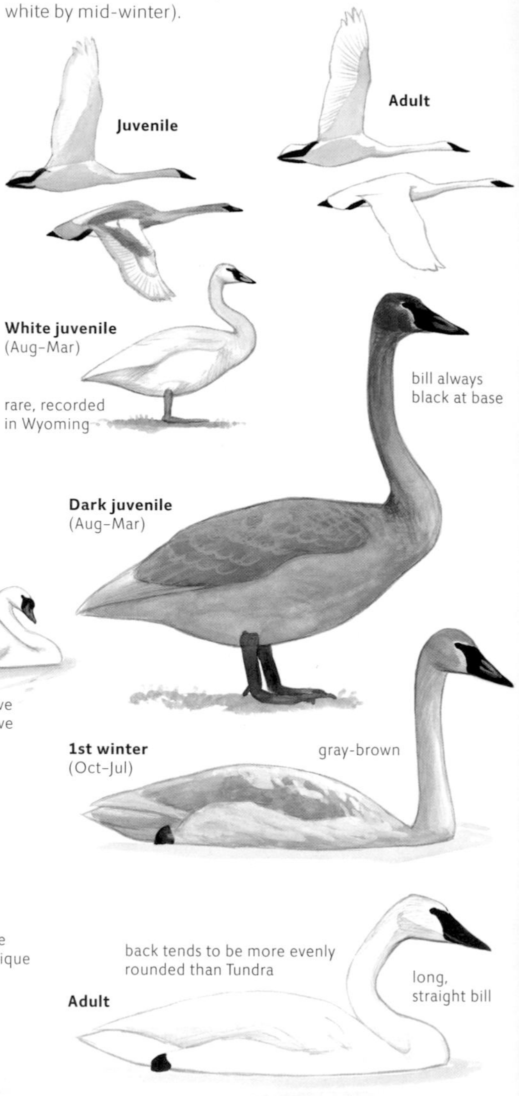

VOICE: Less vocal and voice much lower-pitched than Tundra Swan. Gentle nasal honking, slightly hoarse *hurp* or *hur di di*, like the honk of a European taxi; lower-pitched and less urgent than Canada Goose. Immature gives higher-pitched toy-trumpet-like calls, changing to hoarse version of adult calls during 1st winter.

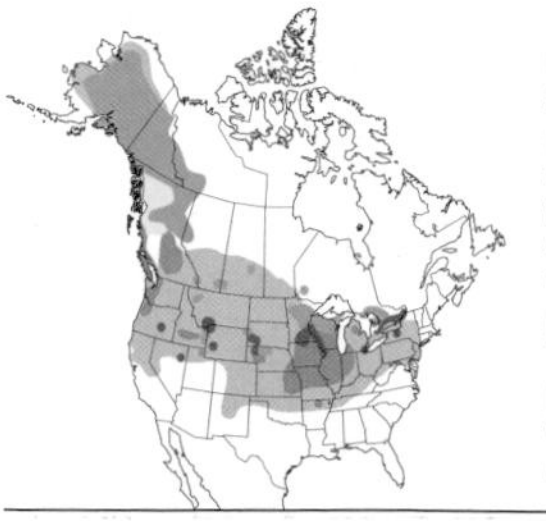

Uncommon and very local, but increasing in the East with a successful reintroduction program. In summer usually seen as pairs or family groups on shallow marshy ponds and lakes surrounded by trees, and year-round tolerates smaller ponds and more trees than Tundra Swan. Migratory populations gather in flocks and winter on favored lakes and agricultural fields.

Tundra Swan

Cygnus columbianus

L 49" WS 75" WT 13.7 lb (6,200 g) ♂>♀

Our smallest swan, but still larger and relatively longer-necked than any goose. Distinguished from Mute Swan by pink to black bill without knob at base, shorter tail, straighter neck. See very similar but larger Trumpeter Swan. Flies in long lines and V formation like geese.

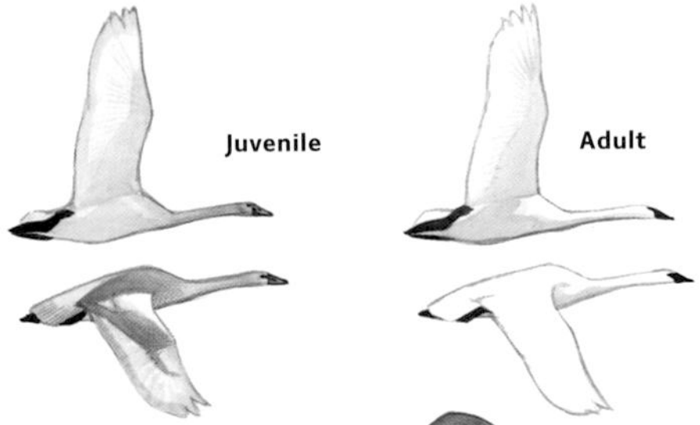

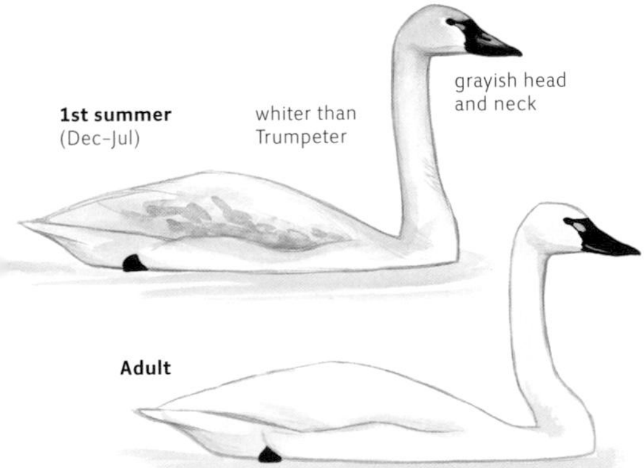

VOICE: A melancholy, clear, singing *kloo* or *kwoo* with hooting or barking quality. Distant flock sounds like baying hounds, rather goose-like, resting flock gives gentle, musical murmuring. Immature calls wheezier, becoming adultlike by 2nd year.

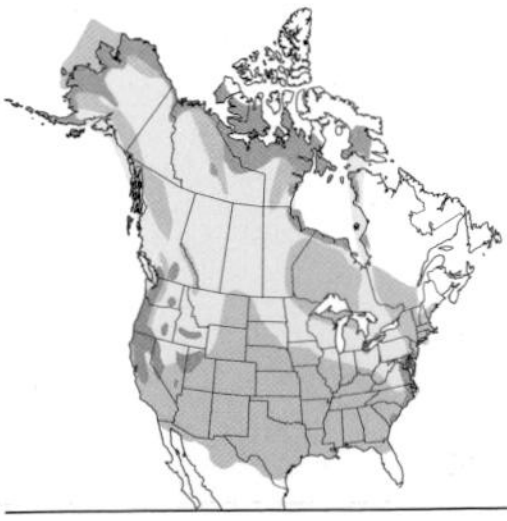

Locally common at traditional staging and wintering areas, uncommon to rare elsewhere. Nests on marshy tundra lakes. In winter forms large flocks and forages on agricultural fields by day, roosts on open water at night.

Identification of Swans

Trumpeter

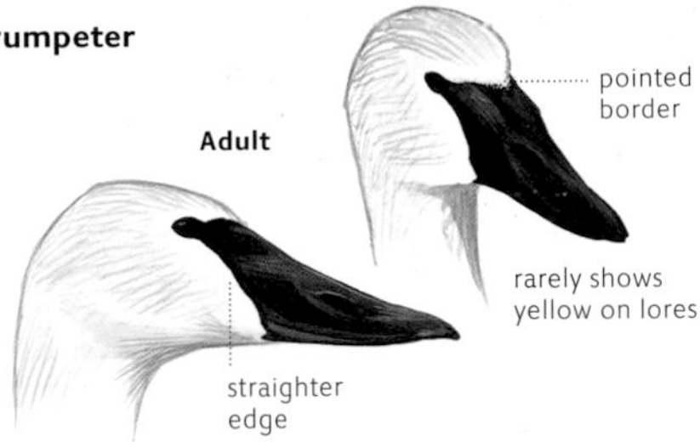

Tundra (American or Whistling)

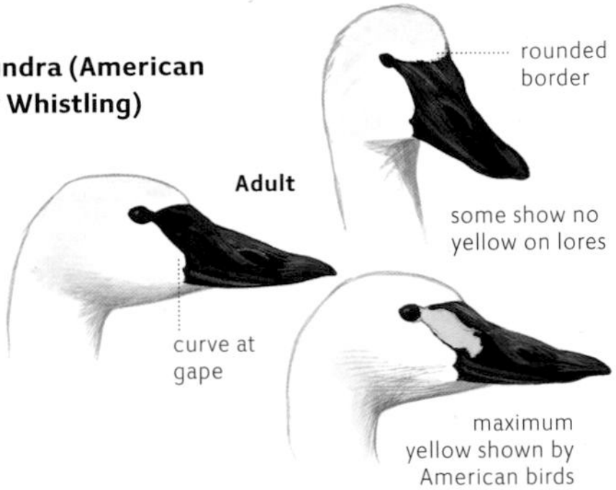

Tundra (Eurasian or Bewick's)

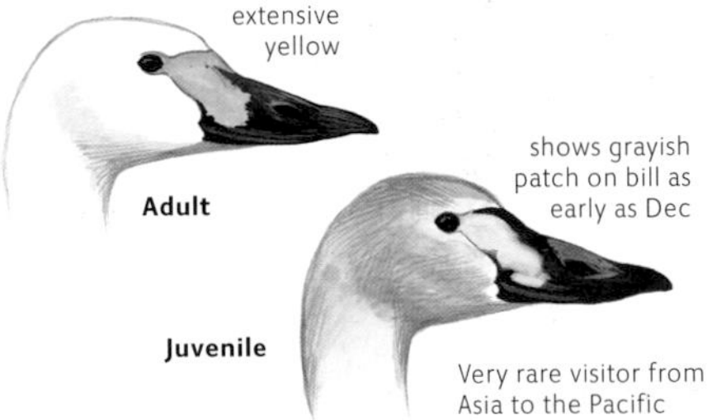

Very rare visitor from Asia to the Pacific coast, seen mostly among large flocks of Tundra Swans wintering in California.

Whooper Swan

Cygnus cygnus

L 59" WS 85" WT 18.3 lb (8,300 g) ♂>♀

A large swan with extensive pale yellow covering base of bill. Most similar to Tundra (Bewick's) Swan, but note larger size, longer bill, yellow extending farther forward on bill.

Small numbers winter in the Aleutian Islands east to Adak. Very rare visitor to mainland Alaska, and a few records south to California are accepted as natural vagrants. Occasional records farther east are presumed or known escapes.

Black-bellied Whistling-Duck

Dendrocygna autumnalis

L 21" WS 30" WT 1.8 lb (830 g)

A relatively large and long-necked duck with bright rose-pink bill and white wing patches. Whistling calls heard constantly in flight.

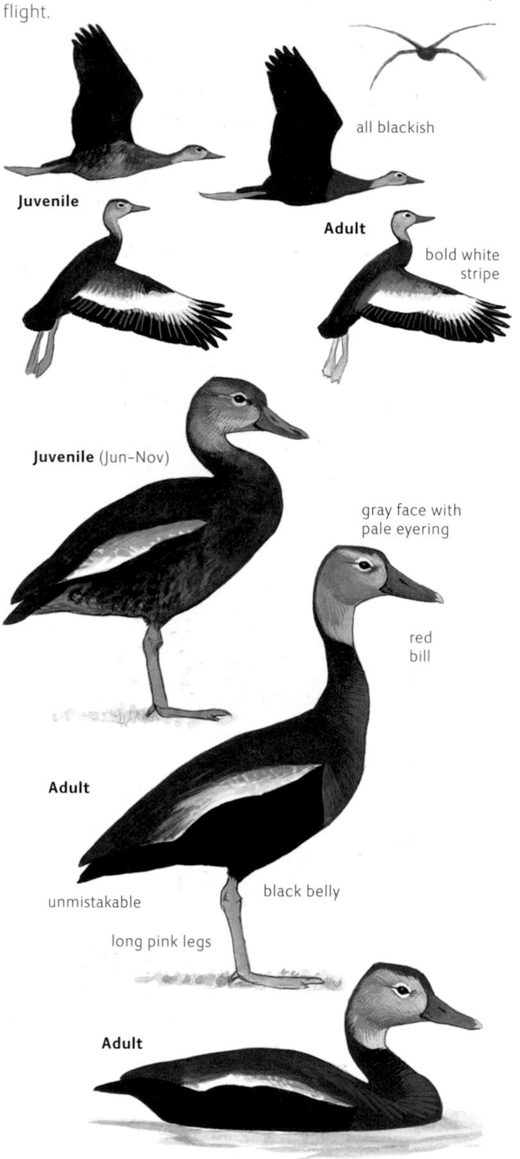

VOICE: Wheezy but sharp whistle, softer and more musical than Fulvous and typically five or six syllables *pit pit pit WEEE do deew.* Also high, weak *yip* singly or in series when flushed.

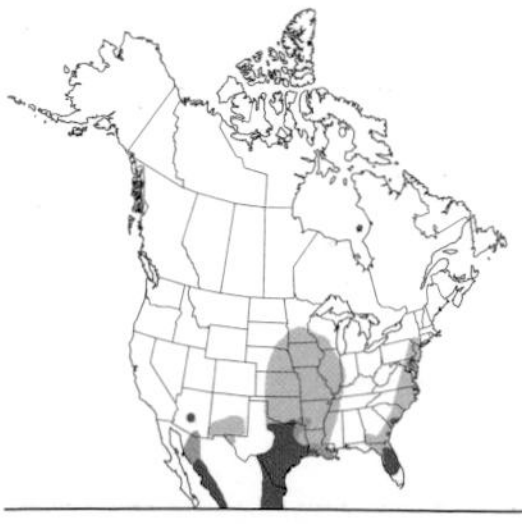

Common. Found in flocks in shallow ponds, Locally common. Look for it in shallow ponds or flooded fields, often in flocks resting on sloping grassy or brushy margins or on large tree branches.

Fulvous Whistling-Duck

Dendrocygna bicolor

L 19" WS 26" WT 1.5 lb (670 g)

Overall warm tan brown color with darker back, all-dark wings, dark bill and legs; white band on uppertail coverts. Other species (e.g. female Northern Pintail) can be stained rusty on face and breast, resembling the color of Fulvous Whistling-Duck.

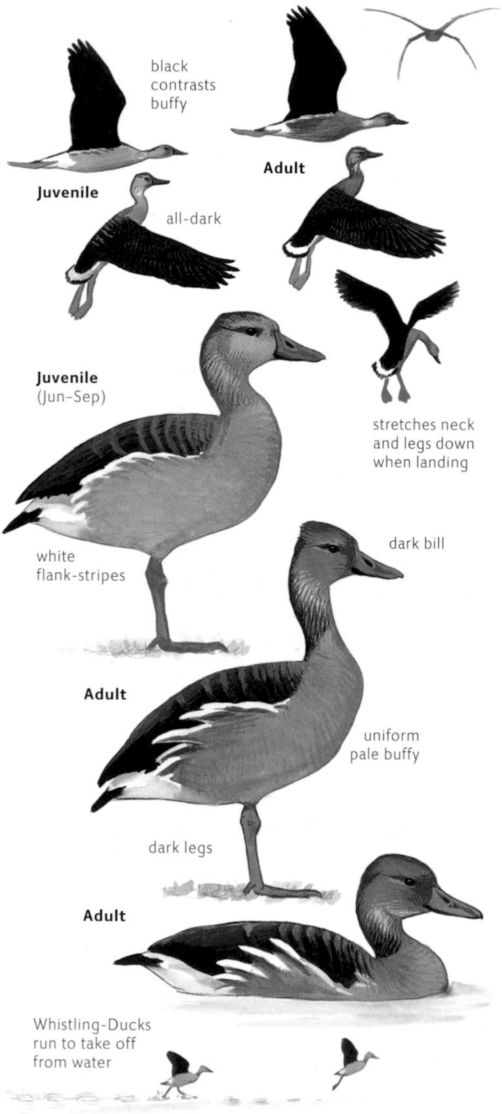

VOICE: A thin, squeaky whistle *pi-piTEEEEW* or *pitheeew.* Also a soft, conversational *cup-cup-cup.* Male calls distinctly higher-pitched than female.

Locally common. Found in small to large flocks in grassy marshes or shallow ponds with emergent vegetation and ringed with reeds or brush, especially flooded rice fields. Often active at night and in twilight, whistling constantly in flight.

Muscovy Duck

Cairina moschata

♂: L 31" WS 48" WT 6.6 lb (3,000 g)
♀: L 25" WS 38" WT 3.3 lb (1,500 g)

A large bulky duck, with disproportionately short legs and long tail. Domestic birds are heavier than wild ones, with partly to entirely white head and body, bright red face, and yellow legs (see page 16).

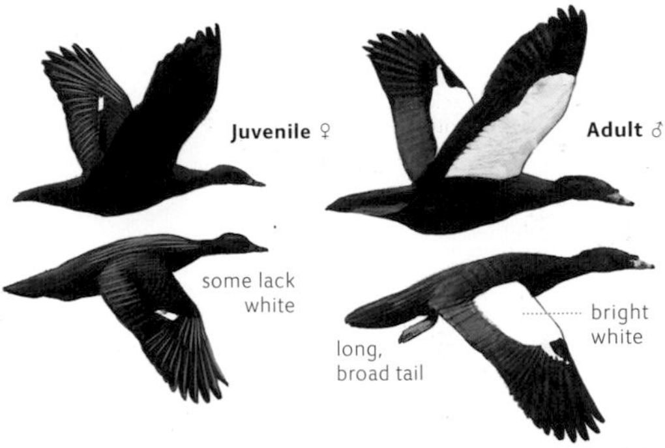

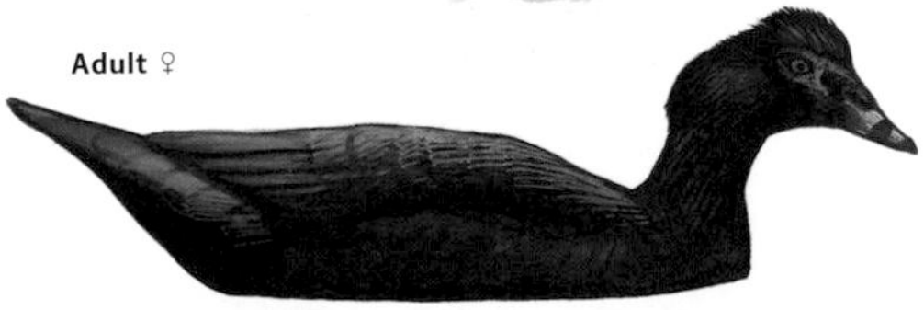

VOICE: Generally quiet. Female gives a soft quack. Male gives rhythmic puffs in display or when alarmed, also hisses.

Widespread in urban parks and farmyards in a variety of domestic forms. Escaped from captivity and breeding in the wild in Florida and perhaps other southern states. A few individuals along the Rio Grande in southern Texas are presumed wild visitors from native range in Mexico.

Wood Duck

Aix sponsa

L 18.5" WS 30" WT 1.3 lb (600 g) ♂>♀

Smaller than Mallard, dark, with contrasting white belly and long dark tail. Breeding male distinctive. Both sexes have rather small bill, drooping crest, and long broad tail usually raised when swimming.

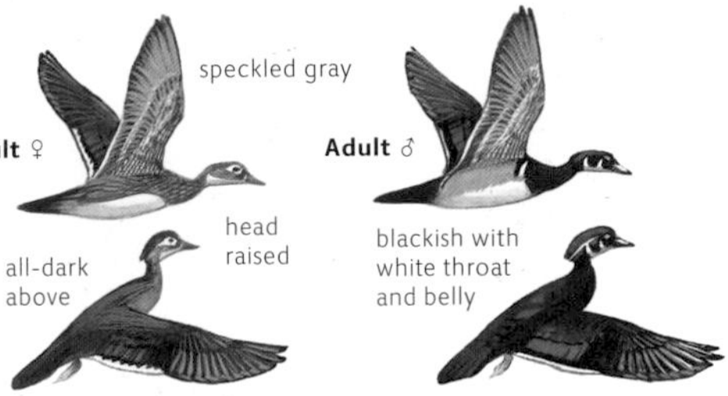

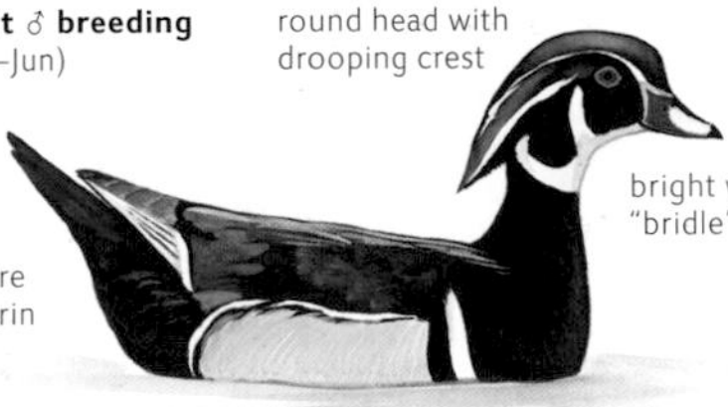

VOICE: Mainly thin, squeaky whistles. Female a penetrating squeal *ooEEK ooEEK*... also a raucous quack, rarely heard. Male a thin, high, drawn-out *jeweeep* or *sweeoooo*, *kip kip kip*. Male wing whistle similar to Mallard.

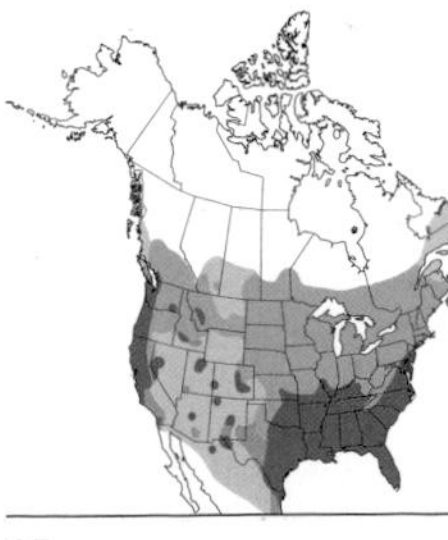

Uncommon on sheltered ponds, rivers, swamps, or wherever there is standing water among trees, generally stays near emergent vegetation. Nests in tree cavities or boxes. Usually in pairs or small groups. Feeds mainly on acorns and other seeds.

Domestic Geese

The common geese seen at city parks and farm ponds are Graylag and Swan Geese, as well as hybrids with Canada Goose. Variation in size, shape, and color is extensive, and domestic birds are often pot-bellied and partly or mostly white.

Greylag × Swan Goose hybrid adult

Swan Goose

Domestic Graylag × Swan Goose hybrid adult

Swan Goose × Canada Goose hybrid adult

Graylag Goose × Canada Goose

Hybrid adult

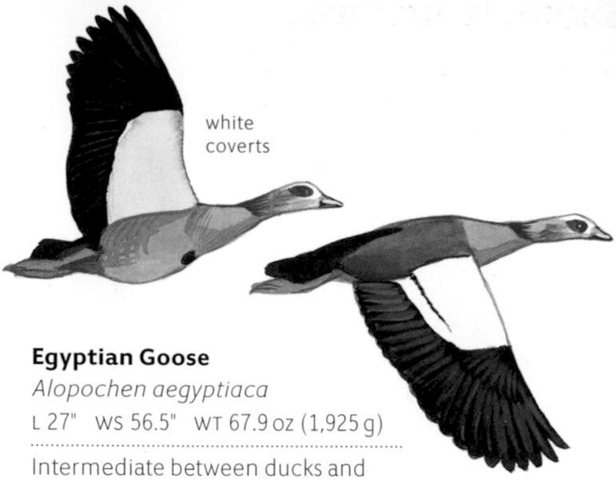

Egyptian Goose
Alopochen aegyptiaca
L 27" WS 56.5" WT 67.9 oz (1,925 g)

Intermediate between ducks and geese in size and shape, with relatively long legs, varied brownish color, bold dark eye patch, white wing coverts.

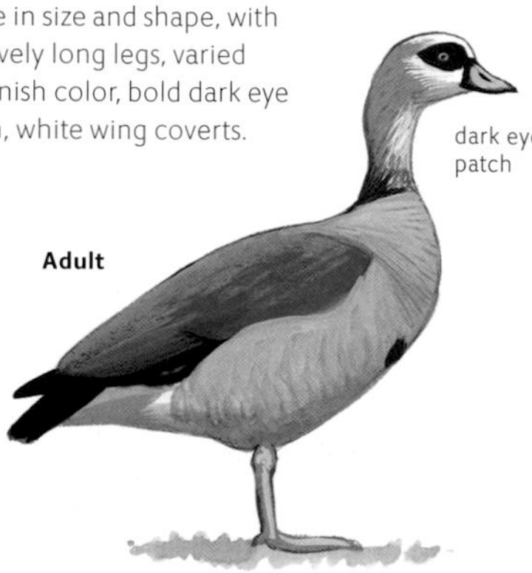

Adult

Found around open shallow water such as ponds on golf courses or city parks. Native to Africa, small numbers are breeding in the wild and apparently established in Florida, southern California, and Texas, and escapes can be seen anywhere.

Domestic Ducks

Muscovy Duck and Mallard are the common ducks of city parks. Muscovy has established feral populations in many southern states. Both species occur in a bewildering variety of colors and shapes, and hybrids also occur.

Muscovy Duck

Mallard Domestic variants

Common Shelduck
Tadorna tadorna
L 24.5" WS 44" WT 39.7 oz (1,125 g)

Native to Eurasia. Some in the Northeast could be wild visitors.

Adult

Ruddy Shelduck
Tadorna ferruginea
L 25.6" WS 52" WT 45.2 oz (1,291 g)

Native to central Asia.

Adult

Exotic Waterfowl

Many exotic waterfowl are held in zoos and private collections and occasionally escape. The following are some of the more numerous species, and could be encountered anywhere, but many other species are also possible. See page 25 for more exotic ducks.

Black Swan

Cygnus atratus

L 49.6" WS 75" WT 15.4 lb (7,000 g)

Native to Australia.

Adult

Bar-headed Goose

Anser indicus

L 29.1" WS 56" WT 4.7 lbs (2,150 g)

Native to Asia.

Adult

Ringed Teal

Callonetta leucophrys

L 12.8" WS 20" WT 8.8 oz (250 g)

Native to South America.

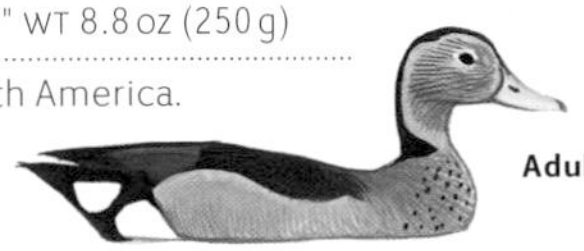

Adult

White-cheeked Pintail

Anas bahamensis

L 16.9" WS 27" WT 17.6 oz (499 g)

Native to tropical America. Some individuals in the southeastern US could be wild visitors.

Adult

Mandarin Duck

Aix galericulata

L 17.7" WS 27" WT 18.9 oz (536 g)

Native to Asia. A small population is established in Sonoma County, California, and individuals can be encountered anywhere.

Adult ♂

Adult ♀

Hybrid Mallard Identification

THE "BROWN MALLARDS": The following two pages show the "Mallard group"—four forms (in three closely related species). All are large dabbling ducks, with relatively long bill, plain brownish plumage, white underwing coverts, etc.

American Black Duck is darkest, then Mottled, then "Mexican" Mallard, then female "Northern" Mallard, and within each form males are darker than females. Sexes can be distinguished by bill color—clear yellow or yellow-green on males, and drab olive or dusky orange on females—and this can be a helpful clue for identification.

Other features to focus on include throat pattern, white bars on the speculum, and range.

"Northern" Mallard hybridizes with all three brown forms, and male hybrids show mixed features, female hybrids are difficult or impossible to identify.

Mexican × Northern Mallard intergrade adult ♂ breeding

Virtually no pure Mexican Mallards occur in North America.

American Black Duck × Mallard hybrid adult ♂ breeding

Seen regularly. Mottled Duck × Mallard is also regularly recorded.

Drake-plumaged ♀ Northern Mallard

Beware that occasional female Northern Mallards show male-like plumage, but small size and typical female bill color

Downy Young

Young waterfowl leave the nest within hours of hatching and follow the adult female. Each species has a distinctive plumage pattern, and six are shown here. Some species, in particular Hooded Merganser and Redhead, often lay eggs in the nests of other species, and downy young of these species are occasionally seen among a brood of some other species.

Black-bellied Whistling Duck

Wood Duck

Mallard

Redhead

Common Goldeneye

Hooded Merganser

Mallard

Anas platyrhynchos

L 23" WS 35" WT 2.4 lb (1,100 g) ♂>♀

Male breeding plumage distinctive, with yellow bill, green head, and pale body. Female more difficult to identify; note large size, prominent dark line through eye, brownish belly, orange legs, and orange and black bill. The many domestic variations may cause confusion (see page 16).

Mexican

Very similar to female Northern but darker overall, with mostly dark tail and undertail coverts, male (with all yellow bill) darker than female. Most in the US show evidence of introgression with Northern.

VOICE: Female gives familiar, loud quacking calls; also deep, reedy laughing; similar calls by other dabblers are shorter, harsher, often higher. Also single loud quacks in a variety of situations (e.g., when flushed a series of single rising quacks *brehk, brehk*...). Male gives a similar short, rasping *quehp*. Displaying male gives short whistle *piu* similar to teals but weaker. Wings whistle faintly in flight.

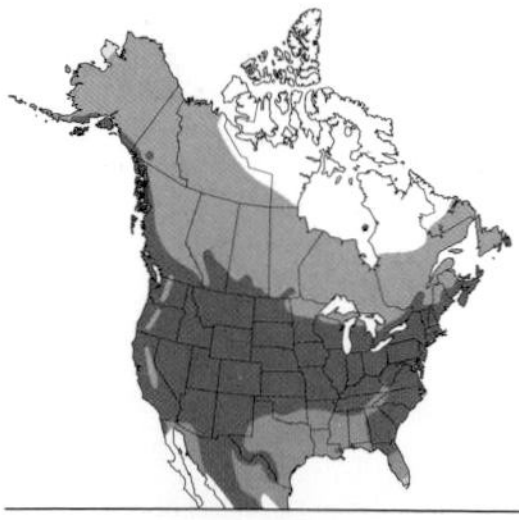

Common and widespread on shallow water from coastal lagoons to city parks; in most areas the most frequently seen species of duck. Nests on the ground in concealing vegetation, usually near water. Usually in small groups or pairs, but large numbers gather at favored ponds. Feeds mainly on seeds taken from the water.

Uncommon and local on marshy ponds and rivers near the Mexican border from Arizona to southern Texas, with rare records north to Colorado and Kansas. Usually seen in pairs or small groups, often mixed with "Northern" Mallards or intergrades.

Mexican

Mottled Duck

Anas fulvigula

L 22" WS 30" WT 2.2 lb (1,000 g) ♂>♀

Darker than female Mallard; slightly paler than Black Duck with unmarked warm buffy throat. Range barely overlaps with American Black Duck, but Mallard is spreading into Mottled Duck's range and hybridization is increasing (see page 17).

VOICE: Voice like Mallard, but female quack may be a little weaker and softer on average.

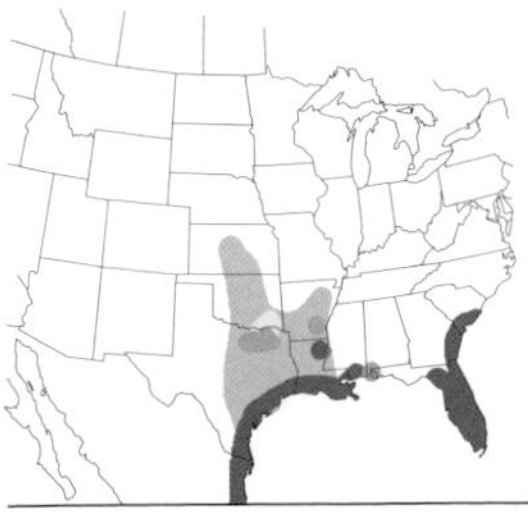

Uncommon and local. Found primarily in saltmarshes and on marshy ponds. Habits similar to Mallard and American Black Duck.

American Black Duck

Anas rubripes

L 23" WS 35" WT 2.6 lb (1,200 g) ♂>♀

Very similar in appearance to Mallard, but darker overall, especially male. Female told from Mallard by darker color, dark tail, dark olive bill. Often hybridizes with Mallard; hybrids show intermediate characteristics (see page 17).

VOICE: Like Mallard, but female quack averages slightly lower-pitched.

Common in Atlantic coast saltmarshes, uncommon to rare inland. Usually found with Mallard. Habits similar to Mallard, but diet has higher proportion of animal prey.

Gadwall

Anas strepera

L 20" WS 33" WT 2 lb (910 g) ♂>♀

A rather stocky duck with rounded head and subdued plumage pattern. Male distinctive; grayish with black tail coverts. Female resembles female Mallard, but with smaller bill and usually showing small white triangle on secondaries.

VOICE: Female quack similar to Mallard but coarser, more nasal. Male in courtship gives low, nasal burp *mepp*; often combined with quiet, high squeak *tiMEPP*.

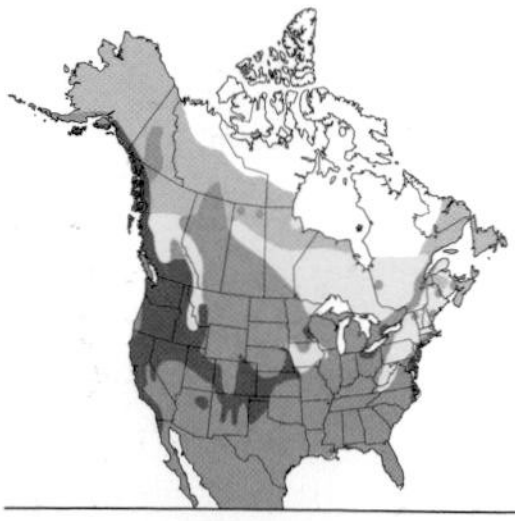

Common, usually in small numbers. Found on shallow fresh water and sometimes sheltered salt water; often with American Wigeon. Nests on the ground near water. Swims in pairs or small groups, dabbling for aquatic plants.

Mallard × Gadwall hybrid

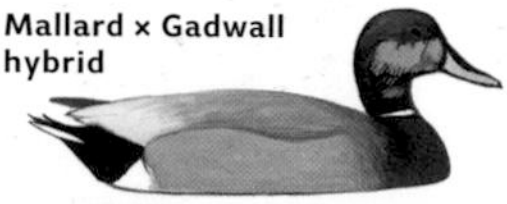

Northern Pintail

Anas acuta

L 21" (♂ to 25") WS 34" WT 1.8 lb (800 g) ♂>♀

Slender, elegant, long-necked and narrow-winged. Breeding male unmistakable, with white breast, brown head, and long tail. Female buffy-brown with very plain head and gray bill.

VOICE: Female quack quieter, hoarser than Mallard. Courting male gives high, wiry, drawn-out *zoeeeeeaa* and short, mellow whistle *toop* or *prudud*; lower-pitched and more melodious than Green-winged Teal, often doubled.

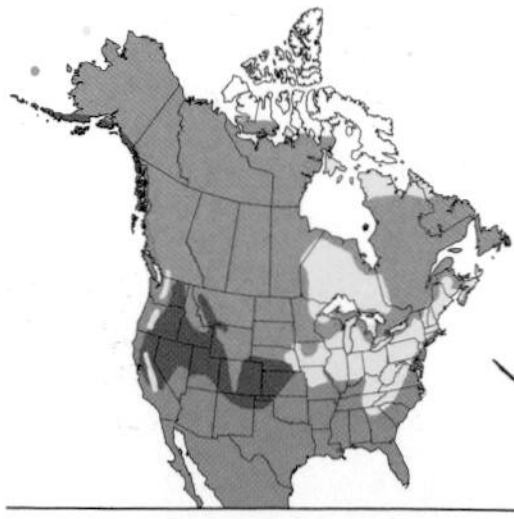

Common in West, uncommon in East. Found on shallow ponds and marshes usually with some emergent vegetation. Often in large flocks. Sometimes grazes on open fields.

Mallard × Pintail hybrid

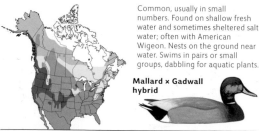

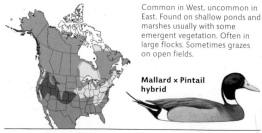

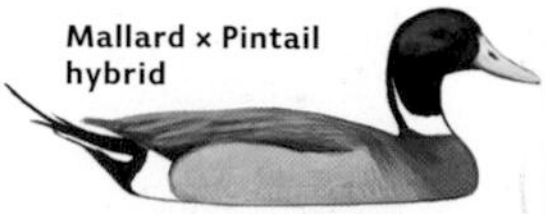

American Wigeon

Anas americana

L 20" WS 32" WT 1.6 lb (720 g) ♂>♀

Small gray bill, usually angled down, and round head distinctive. Breeding male has white hip-patch, white or buffy forehead, and dark eye-patch. Female has plain gray-brown head with dark smudge around eye.

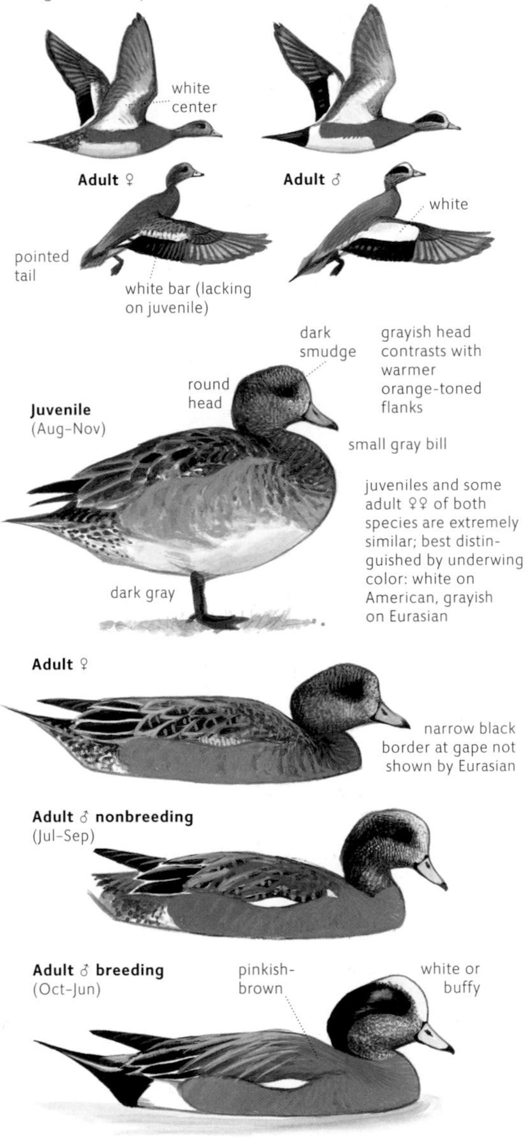

VOICE: Female quack low, harsh, growling *rred* or *warr warr warr* similar to *Aythya* ducks. Male a distinctive airy whistle of two or three syllables *wi-WIW-weew* or *Wiwhew*.

Common on fresh water and sheltered salt water. Often in large flocks, picking plants from waters' surface; also often grazes on fields.

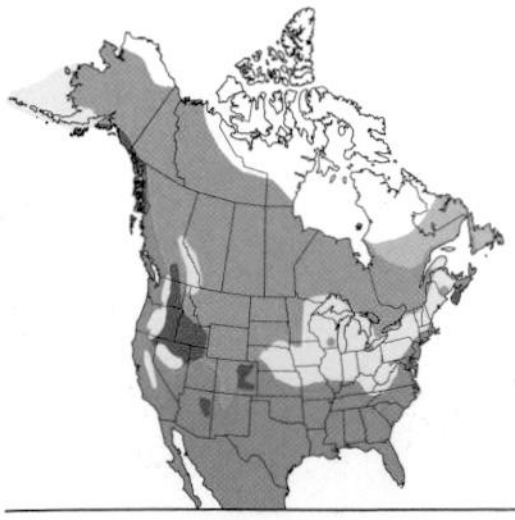

Eurasian Wigeon

Anas penelope

L 20" WS 32" WT 1.5 lb (690 g) ♂>♀

Breeding male differs from American by dark rufous head and breast contrasting with paler gray body. Female very difficult to distinguish from American; generally has warmer brown head, slightly grayer body, and gray underwing coverts.

VOICE: Female quack may be even harsher than American. Male whistle distinctive, similar to American but higher, stronger, more vibrant descending single note *hwEEEEEr*.

Rare visitor from Eurasia; usually found singly among flocks of American Wigeon. Habits similar to American.

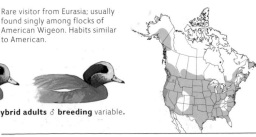

American × Eurasian Wigeon hybrid adults ♂ breeding variable.

Blue-winged Teal

Anas discors

L 15.5"   WS 23"   WT 13 oz (380 g)   ♂>♀

Relatively small and slender with long bill. Male breeding plumage distinctive with white crescent on gray head, white hip patch. Female told from Green-winged by longer bill, white eye-arcs. See Cinnamon Teal. In flight, pale blue wing coverts obvious.

VOICE: Female quack coarse, high, less nasal than Green-winged; higher than Northern Shoveler. Courting male gives a rather thin, high whistle *pwis* or *peeew*; sometimes a nasal *paay* like Northern Shoveler

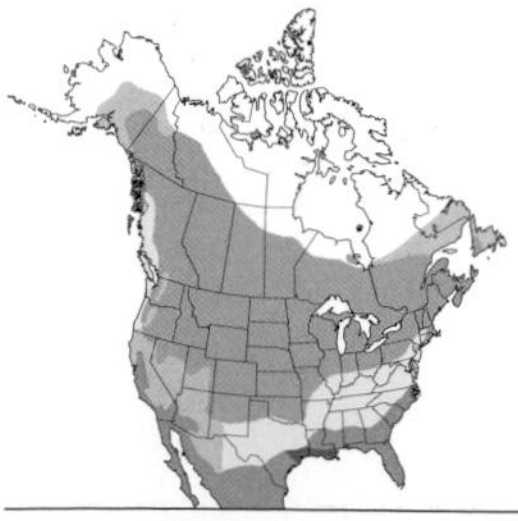

Common on shallow marshy ponds and mudflats. In small groups usually close to vegetation. Feeds mostly by swimming in shallow muddy water with bill outstretched, picking seeds and plant material from surface.

Cinnamon Teal

Anas cyanoptera

L 16"   WS 22"   WT 14 oz (400 g)   ♂>♀

Breeding male unmistakable, with dark reddish-brown plumage. Female very similar to female Blue-winged; distinguished only on average by slightly larger bill and overall warmer brown and less distinctly patterned plumage.

VOICE: Female quack like Blue-winged. Male in display gives a dry, chattering or rattling *gredek gredek*...; vaguely reminiscent of Northern Shoveler, unlike the whistle of Blue-winged.

Fairly common; found on shallow ponds in or near marshy vegetation. Usually in small groups or pairs, often mixed with Blue-winged Teal. Feeding habits similar to Blue-winged.

Blue-winged × Cinnamon Teal hybrid adult ♂ breeding. Seen occasionally.

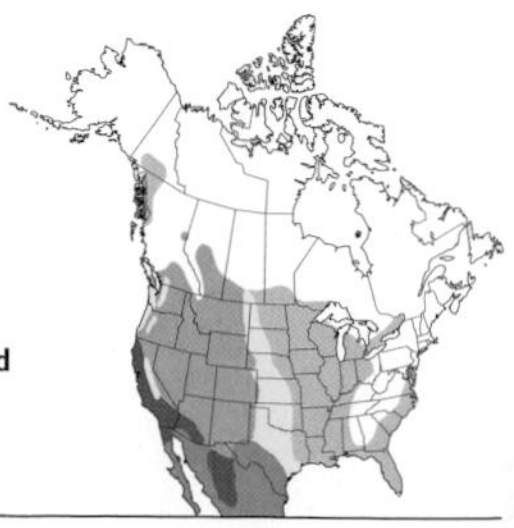

Garganey

Anas querquedula

L 15.5" WS 24" WT 13 oz (375 g) ♂>♀

Breeding male distinctive, with curving white stripe on brown head, pale gray flanks. Female and juvenile difficult to distinguish from other female teal; note dark mark around eye and dark line from gape across cheek, gray legs, and pale gray upperwing.

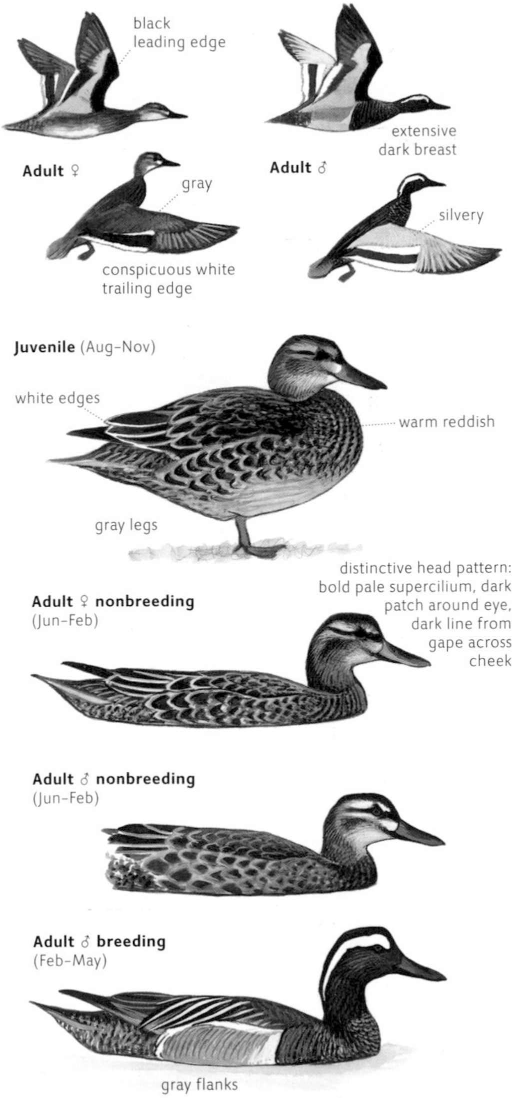

VOICE: Female gives a feeble, shrill croak like Green-winged Teal; some calls like Blue-winged Teal or Northern Shoveler. Male in display gives a drawn-out, dry clicking like winding of a fishing reel, similar to Cinnamon Teal.

Very rare visitor, regular in western Alaska, most records south and east are of males in spring, when they are easily identified. Found on small marshy or weedy ponds, usually with Blue-winged or Cinnamon Teal.

Northern Shoveler

Anas clypeata

L 19" WS 30" WT 1.3 lb (610 g) ♂>♀

Very long, broad bill distinctive on all birds. Breeding male has striking plumage pattern: white breast with rufous flanks and belly. Female similar to female Mallard, but with relatively obvious pale markings, especially broad pale edges on tertials. In flight, note broad white bar on coverts of both sexes.

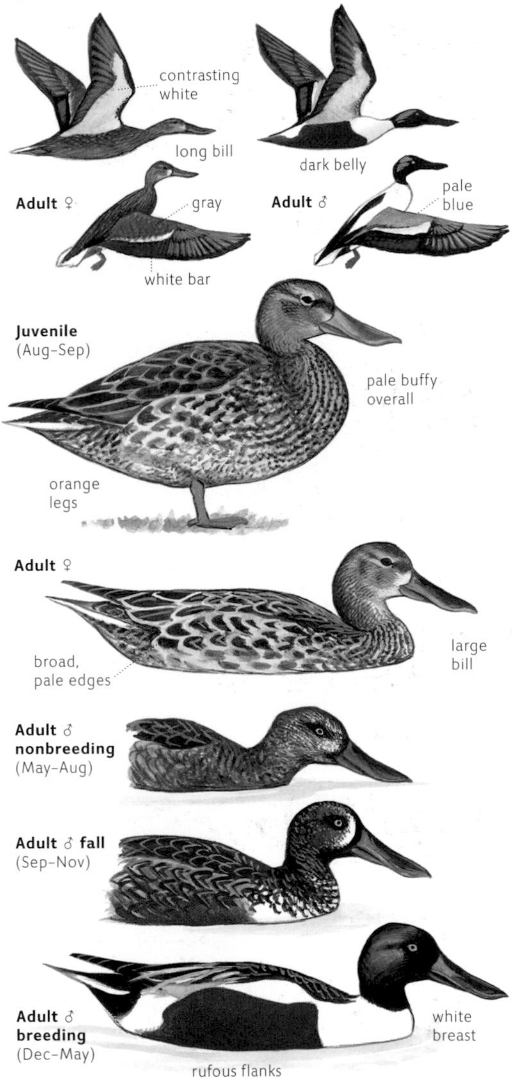

VOICE: Female quack deep and hoarse *kwarsh* and short *gack gack ga ga ga*. Male in courtship gives nasal, unmusical *thuk-thUK* and in fall a loud, nasal *paay*. Male's wings produce rattling sound on takeoff.

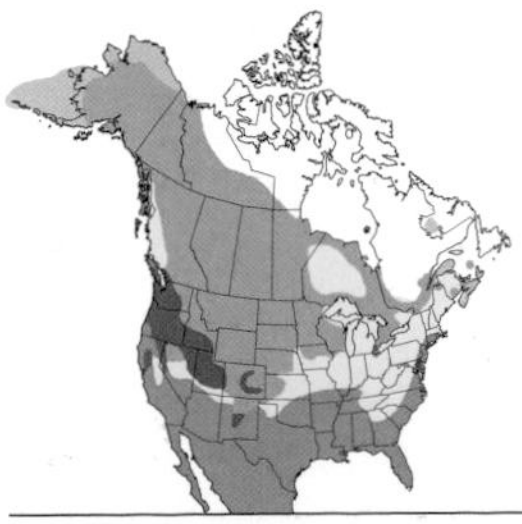

Locally common; typically seen in small flocks on shallow weedy or grassy ponds. Feeds on plankton and some seeds, gathered from water's surface by straining with bill.

Green-winged Teal

Anas crecca

L 14" WS 23" WT 12 oz (350 g) ♂>♀

Our smallest duck; stocky and short-bodied, with relatively small, slender bill. Breeding male distinctive, with dark head and white bar on side of breast. Female very similar to Blue-winged Teal, but note smaller size, smaller bill, darker and more patterned face, and obvious pale buffy streak on tail coverts.

Eurasian (Common Teal)

Eurasian population differs from American only in plumage details as noted, although birds from east Asia (found in western North America) average slightly larger overall.

American

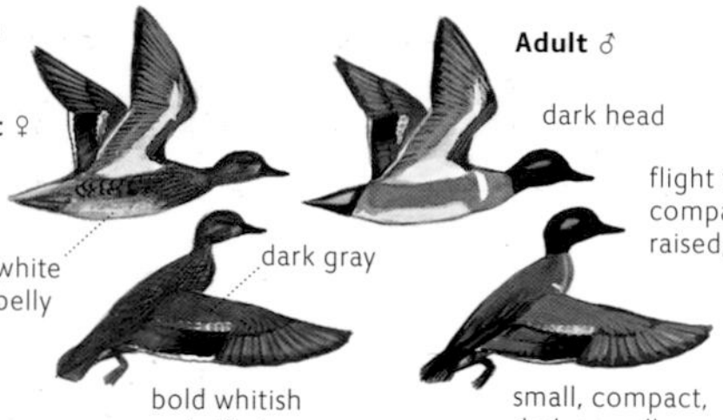

flight fast, twisting; in compact bunches; head raised; wingbeats deep

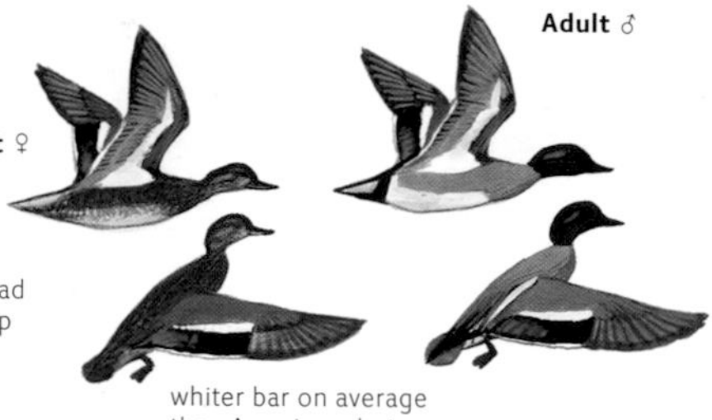

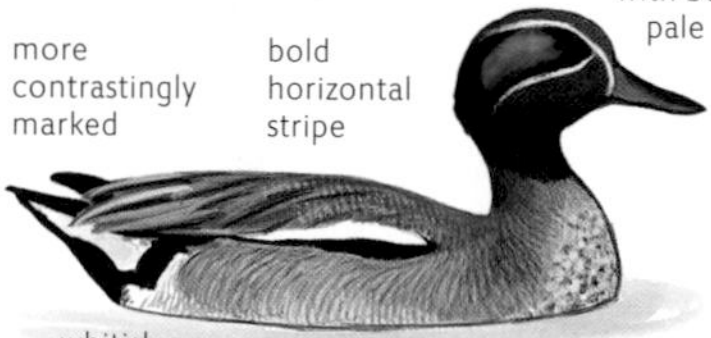

American × Eurasian intergrade adult ♂ breeding
Combines white bar on breast and white stripe on scapulars.

VOICE: Female voice shriller, feebler than other ducks: high, nasal, scratchy *SKEEE we we we*; quality like snipes. Courting male gives a shrill, ringing whistle *kreed* or *krick* like Spring Peeper, sometimes in series of short phrases *te tiu ti*, etc.; males in fall give hoarser whistle.

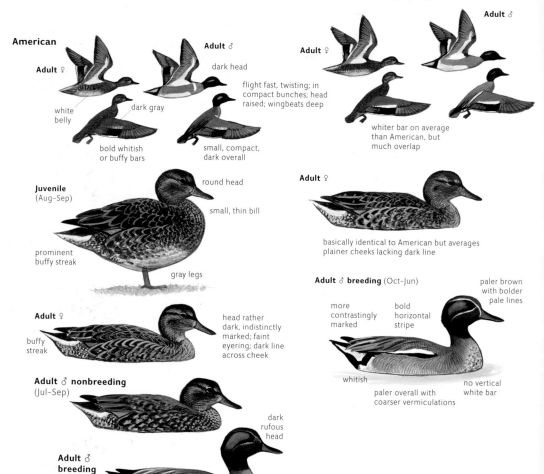

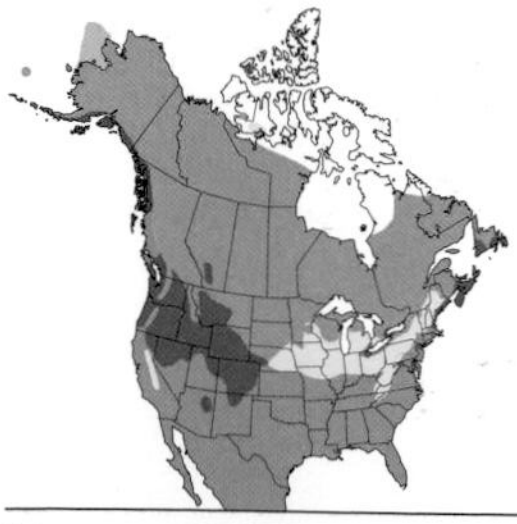

Common. Found on shallow ponds, marshes, flooded fields. In small groups that often congregate in large flocks. Forages mainly by dabbling at surface of water or walking on mudflats skimming surface with bill.

The Eurasian form of Green-winged Teal, also known as Common Teal, is a rare but regular visitor to both coasts. Considered a separate species by many authorities, males differ in details of plumage, but not in shape, voice, or behavior. Females differ in wing pattern and possibly in face pattern, but are not reliably distinguishable. Intergrade males, showing mixed features, are only slightly less frequent than pure Eurasian.

Rare and Exotic Ducks

Falcated Duck
Anas falcata

L 19.7" WS 31" WT 1.4 lb (650 g)

Wigeon-like in shape, size and habits, male has slender black bill, long "mane" of brown feathers, long curving (falcate) tertials. Female is plain brownish.

Very rare visitor from Eurasia, most records are from far western Alaska, with a few records south along Pacific coast to California.

Baikal Teal
Anas formosa

L 16.1" WS 26" WT 15.5 oz (439 g)

Male has striking head pattern, thin vertical white bar on side of breast and rear flanks. Note that rare hybrids of other species (e.g. American Wigeon × Green-winged Teal) can show a very similar head pattern.

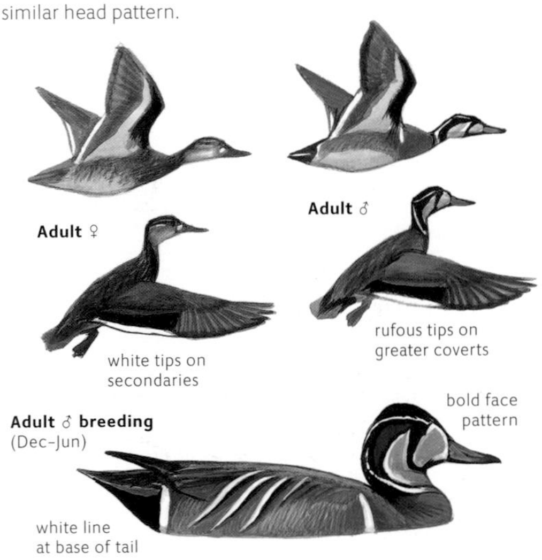

Very rare visitor from Asia, most records are from the western Aleutians, with a few farther south to California and Arizona.

Common Pochard
Aythya ferina

L 18.1" WS 30" WT 2 lb (907 g)

Male has whitish back and sloping forehead like Canvasback, reddish head like Redhead, and black bill with blue-gray "saddle." Distinguished from less-rare Canvasback × Redhead hybrid by details of bill color and forehead color.

Native to Eurasia, recorded regularly in far western Alaska, and several times south to California, with a few records elsewhere suspected of being escapes from captivity.

Hybrid Redhead × Canvasback

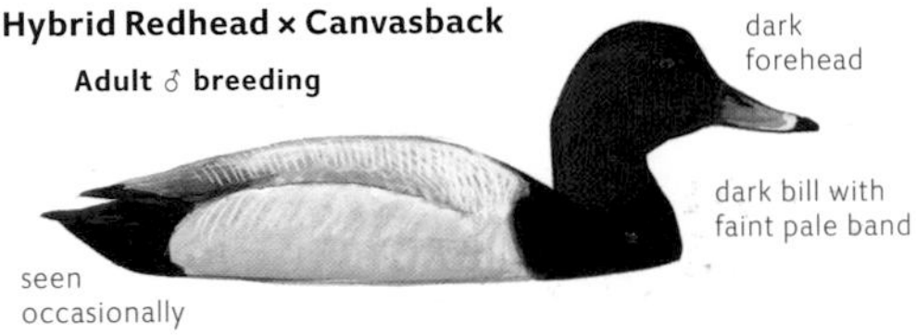

Rosy-billed Pochard
Netta peposaca

L 21.7" WS 28" WT 2.2 lb (1,000 g)

Native to South America, occasionally escaped from captivity

Red-crested Pochard
Netta rufina

L 21.3" WS 29" WT 2.5 lb (1,134 g)

Native to Europe. Occasionally escaped from captivity.

Canvasback

Aythya valisineria

L 21" WS 29" WT 2.7 lb (1,220 g) ♂>♀

One of our largest diving ducks; very long, tapered black bill and sloped forehead distinctive in all plumages.

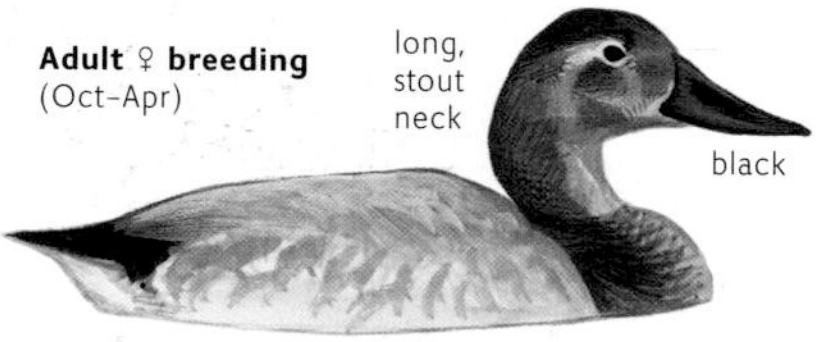

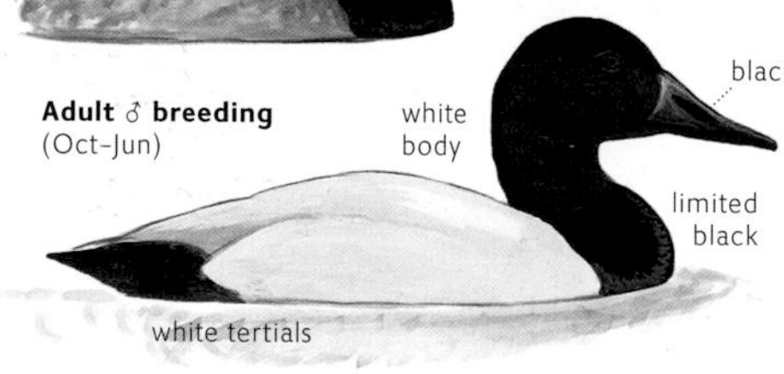

VOICE: Female gives a low, rough, growling *grrrt grrrt*... like other *Aythya* ducks; also a repeated *kuck*. Male in display gives eerie hooting *go-hWOOO-o-o-o* with weird, squeaky overtones.

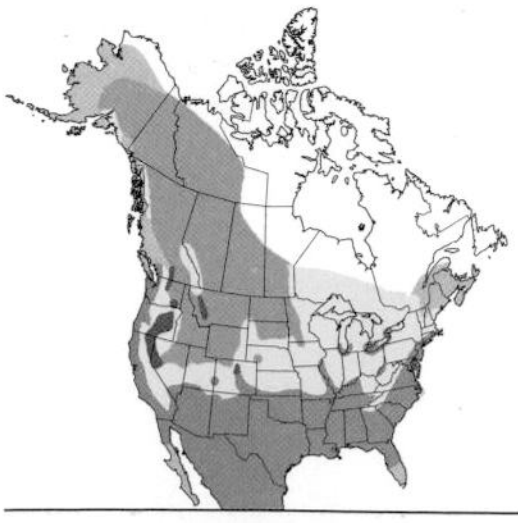

Common locally. Nests in marshy ponds, winters on open lakes and bays, often in large rafts with other *Aythya* ducks.

Redhead

Aythya americana

L 19" WS 29" WT 2.3 lb (1,050 g) ♂>♀

Note round head and blue-gray bill in all plumages. Breeding male has bright rufous head and gray body. Female pale warm brown overall.

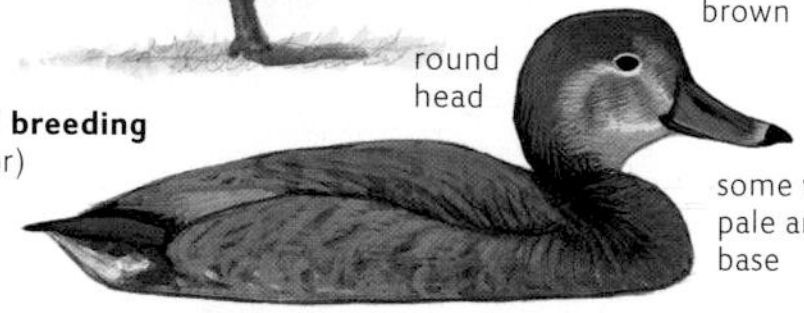

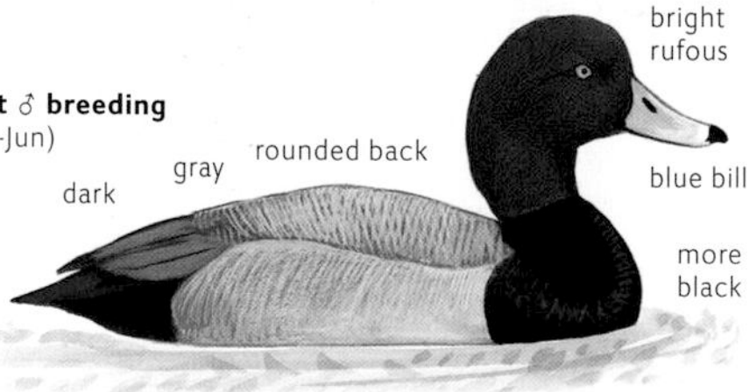

VOICE: Female gives a rather soft, low, nasal *grehp* or harsher *squak*. Courting male gives distinctive and far-carrying cat-like, nasal *waow*.

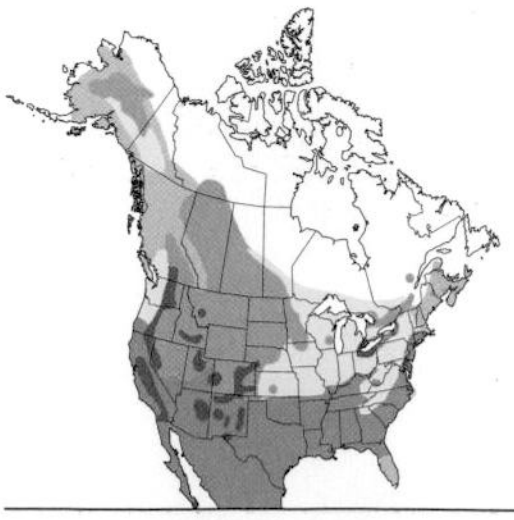

Common locally. Nests in ponds with open water and dense reedy vegetation. Winters on open water of lakes and bays, often mixed with other *Aythya* ducks in large flocks. Tolerates shallower water than most other diving ducks, and most numerous winter in shallow saltwater lagoons along the Texas coast.

Ring-necked Duck

Aythya collaris

L 17" WS 25" WT 1.5 lb (700 g) ♂>♀

Best identified by tall head with sharp peak on rear crown. Breeding male distinctive, with black back and white bar on side. Female has faint pale bar and white eyering. In flight, note gray wingstripe.

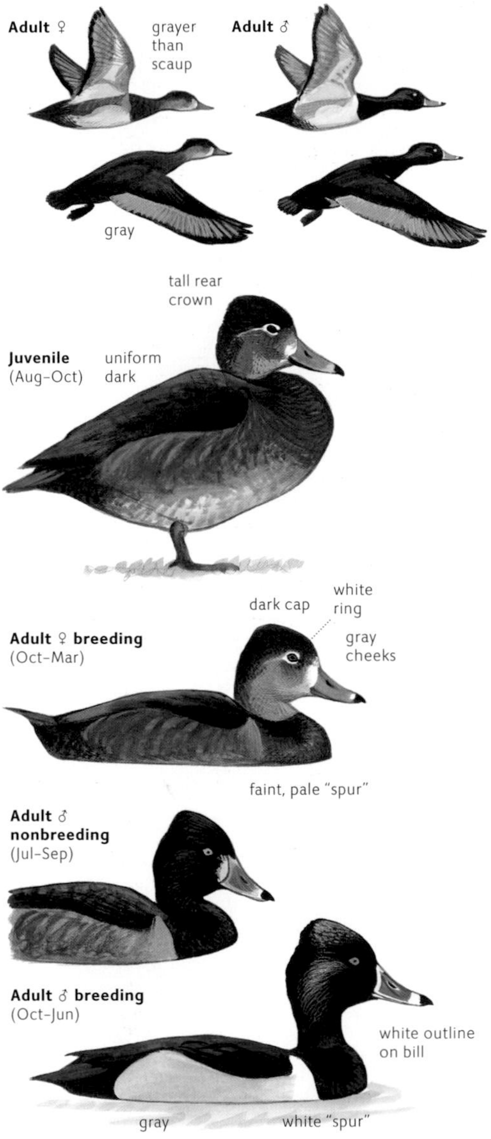

VOICE: Female gives a purring or rough growl *kerp kerp*.... Male usually silent; during display gives a low-pitched, hissing whistle like a person blowing through a tube.

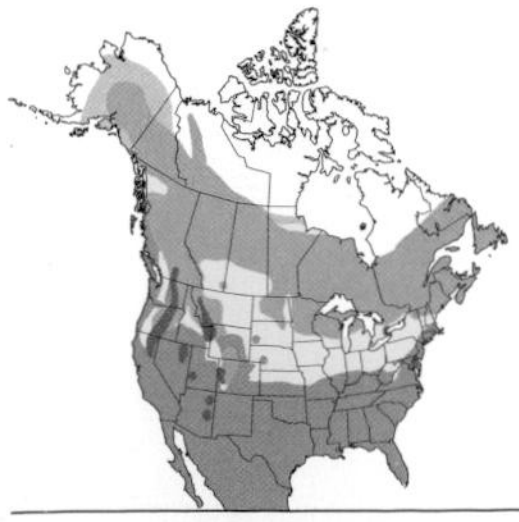

Common. Nests on ponds with emergent vegetation. Winters on ponds and rivers, often near or among trees, on smaller and more enclosed ponds than other *Aythya*, rarely on salt water

Tufted Duck

Aythya fuligula

L 17" WS 26" WT 1.6 lb (740 g) ♂>♀

Male stands out among scaup by black back, round head, long tuft; female very similar to scaup but has darker brown back and breast, paler flanks.

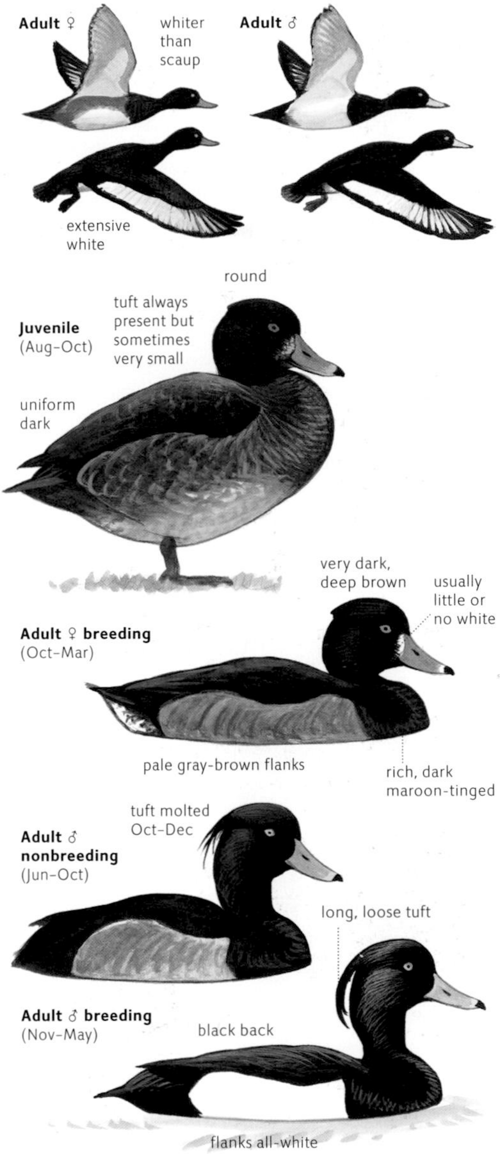

VOICE: Female gives a soft, growling *kerrb*. Courting male produces a rapid, whistled giggle *WHA-wa-whew*; also a hoarse *wheeoo* and high peeping during display.

Rare visitor. Usually found singly among flocks of Lesser Scaup or Ring-necked Duck on ponds, lakes, or bays.

Greater Scaup

Aythya marila

L 18" WS 28" WT 2.3 lb (1,050 g) ♂>♀

Male has whitish body with black front and rear (like Lesser Scaup). Female dark brown head usually with contrasting white around base of bill, slightly paler on body.

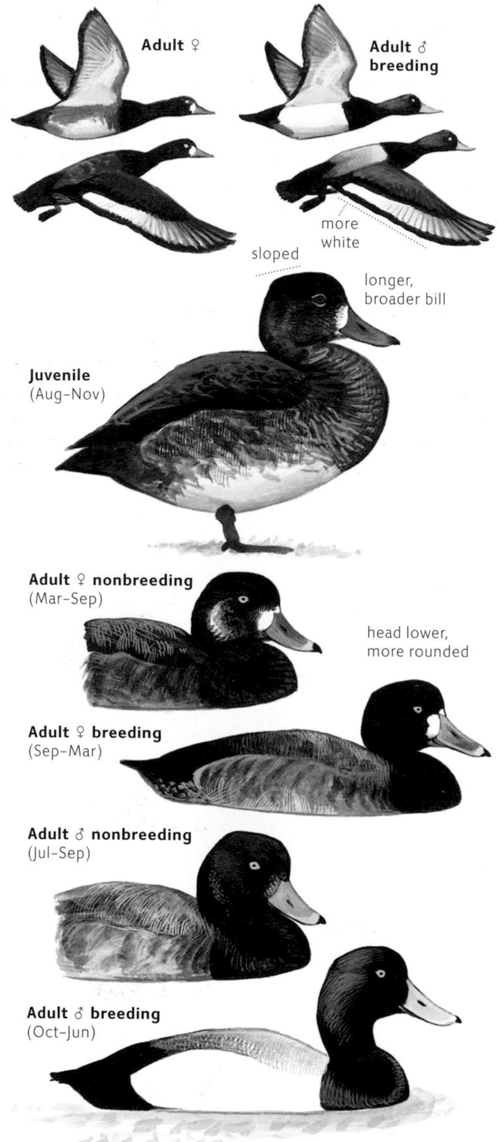

VOICE: Female gives very rough, hoarse *karr karr*...; more rasping than other *Aythya* ducks; also softer, muffled *garrp garrp*.... Male in display often silent; sometimes produces a soft, hollow, bubbling hoot *blup BIDIVooo*.

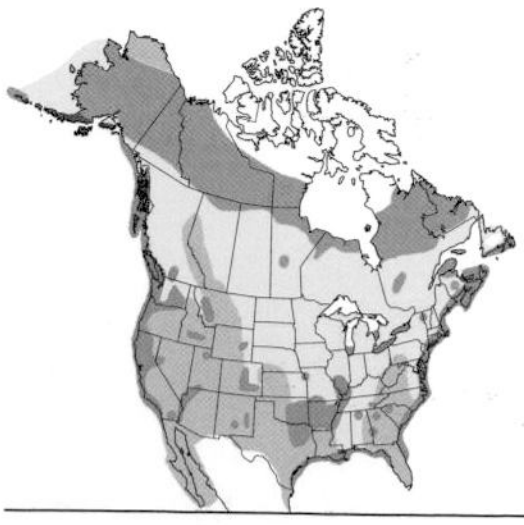

Common. Nests on ponds and lakes. Winters on lakes and bays, favors salt water and larger lakes than Lesser Scaup.

Lesser Scaup

Aythya affinis

L 16.5" WS 25" WT 1.8 lb (830 g) ♂>♀

Extremely similar to Greater Scaup, distinguished (with difficulty) by subtle differences in head shape. Also averages slightly smaller, and is the expected species on most interior lakes and ponds.

VOICE: Like Greater but slightly higher-pitched. Male display includes a husky whistle, higher than Greater and less bubbling.

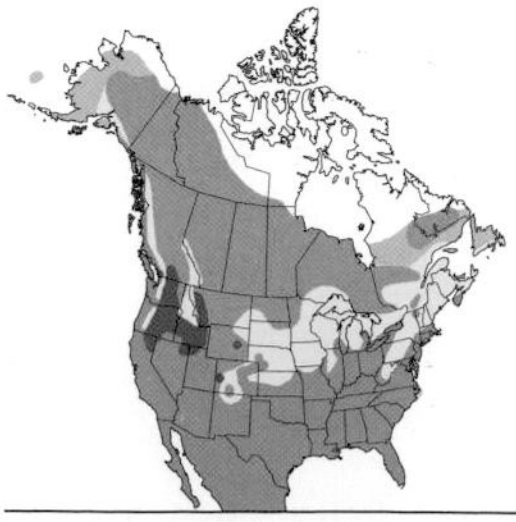

Common and widespread. Nests on ponds and lakes. Winters on ponds, lakes, and protected bays. Favors fresh water and smaller lakes than Greater Scaup, but the two species can often be found together.

Identification of Scaups

The head shape of both species of scaup varies depending on activity, from active diving (top) to relaxed (bottom). Differences between the species are most apparent when relaxed and largely disappear when birds are active. (Watch for similar changes in the head shape of other ducks, e.g., goldeneyes.) Note the more rounded nape contour and the peak of the head farther forward on Greater.

Greater Scaup **Lesser Scaup**

Lesser has a taller and narrower head (most evident when relaxed), with narrower, more straight-sided bill. Black nail at tip of bill averages smaller on Lesser, but there is much overlap.

Greater Scaup **Lesser Scaup**

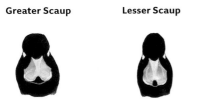

Differences in head shape are easily seen on sleeping birds in fully relaxed state. Scaup are often more easily identified when sleeping than when awake.

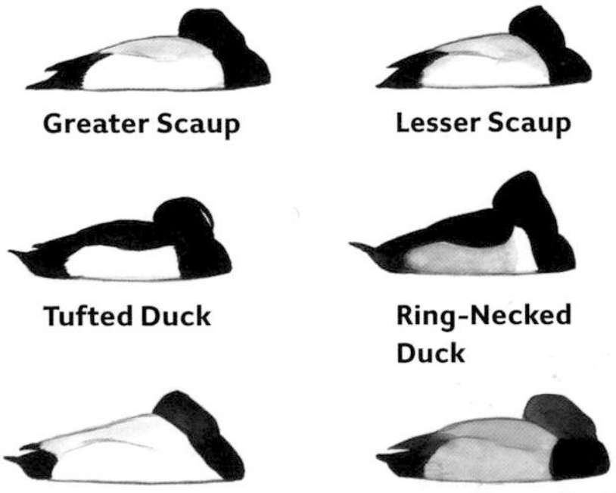

Greater Scaup **Lesser Scaup**

Tufted Duck **Ring-Necked Duck**

Canvasback **Redhead**

Hybrid *Aythya*

Hybrid Tufted × Scaup

Adult ♂ breeding

Hybrids between Tufted Duck and Scaup are seen occasionally. These differ from pure Tufted Duck in having the back dark gray and faintly patterned (not solid black), the tuft short and bushy, and intermediate extent of black on the bill tip. Hybrids of Ring-necked Duck × Scaup, and Tufted Duck × Ring-necked Duck, and other combinations, also occur. Female hybrids are barely distinguishable.

Identification of Eiders

The exact shape and structure of the bill can be an important identification clue on eiders. Note especially the pattern of feathering around the base of the bill. The four populations of Common Eider shown here differ only on average, with many intermediate individuals. Females are shown; the bill shapes of adult male Common and King Eiders are quite different.

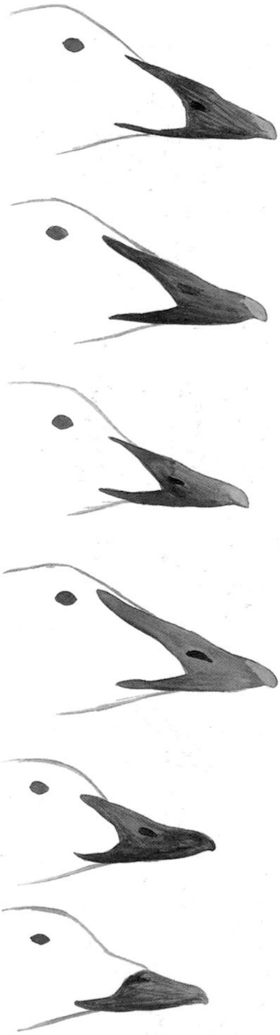

Common Eider West Arctic

Common Eider Hudson Bay

Common Eider East Arctic

Common Eider Atlantic

King Eider

Spectacled Eider

Common Eider

Somateria mollissima

L 24" WS 38" WT 4.7 lb (2,150 g) ♂>♀

Our largest duck, male mostly white with black belly, female all dark brown. All have large bill with long wedge-shaped head.

Common. Nests on tundra ponds close to salt water. Winters exclusively on open salt water, especially along rocky coasts and forms large flocks in favored areas. Often rests on rocks.

VOICE: Female gives hoarse, guttural croaking or groaning sounds, from single *grog* to rapid clucking series *kokokokok*. Courting male gives a very low, hollow, ghostly moan *oh-OOOOOooo*.

King Eider

Somateria spectabilis

L 22" WS 35" WT 3.7 lb (1,670 g) ♂>♀

Male differs from Common Eider in black back, blue crown, orange bill knob. Female by tawny color, shorter bill and less wedge-shaped head.

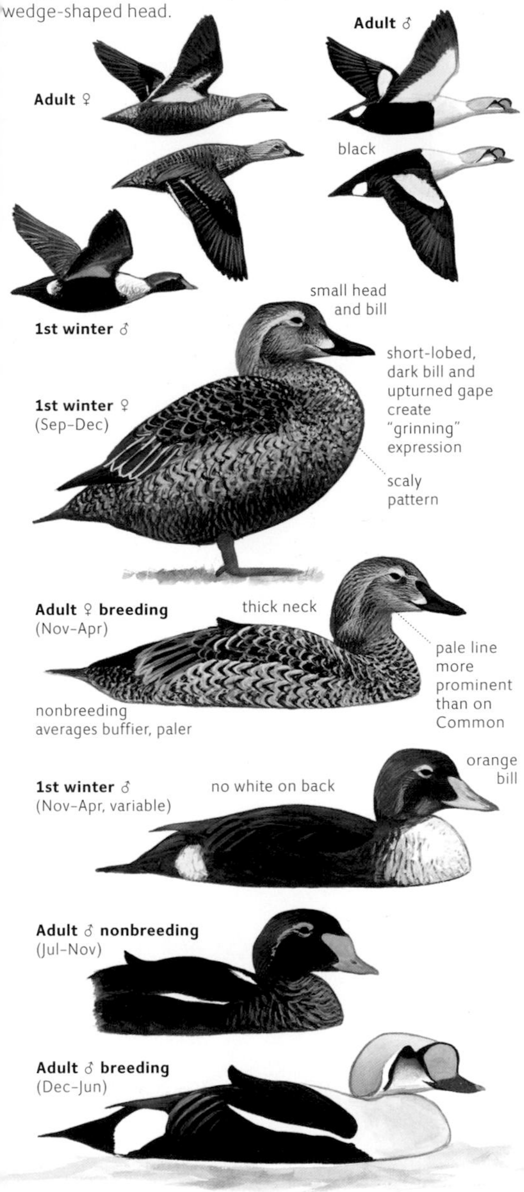

VOICE: Female gives low, wooden *gogogogogo*...; deeper than Common Eider. Courting male gives a low, hollow, quavering moan in crescendoing series *broo broooo brOOOOO broo.*

Common. Nests on tundra ponds. Winters on open salt water, especially along rocky coasts. Rare south of Canada, and usually found with Common Eider or with scoters, sometimes rests on rocks.

Spectacled Eider

Somateria fischeri

L 21" WS 33" WT 3.4 lb (1,570 g) ♂>♀

Male smaller than Common Eider with dark forehead, gray breast. Female told from other eiders by dark forehead, pale eye-patch, small bill.

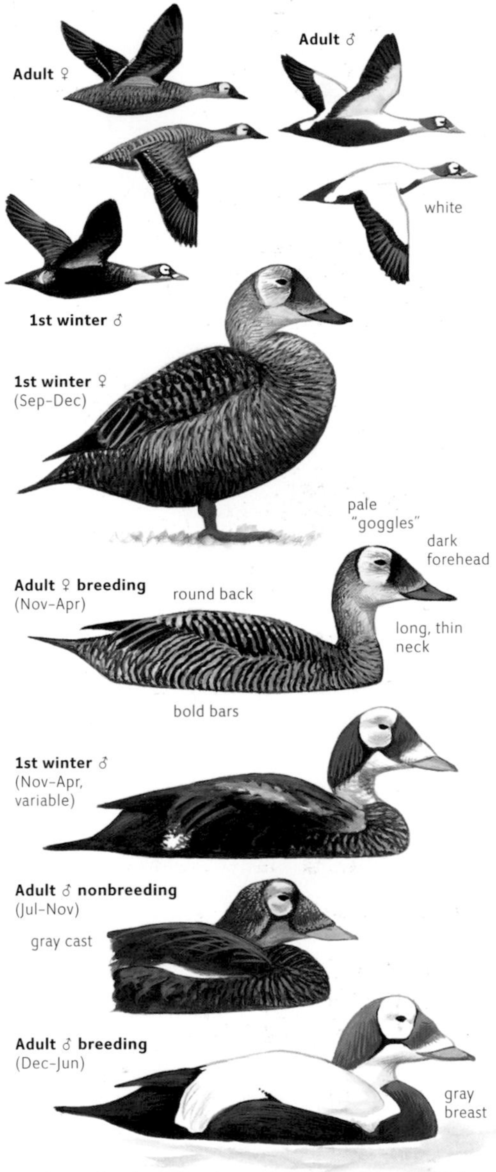

VOICE: Generally silent. Female gives *gogogo*... like other eiders; croaking and clucking sounds at nest. Male in display gives faint *ho-HOO*; weaker than similar call of Common Eider.

Uncommon. Nests on tundra ponds. Winters on open salt water, mainly at openings in pack ice in Bering Sea. North American population under 20,000 individuals.

Harlequin Duck

Histrionicus histrionicus

L 16.5" WS 26" WT 1.3 lb (600 g) ♂>♀

A small dark duck, male dark gray with intricate pattern of small white markings; female dark gray-brown, told from scoters by smaller size, short bill, small white spots on face

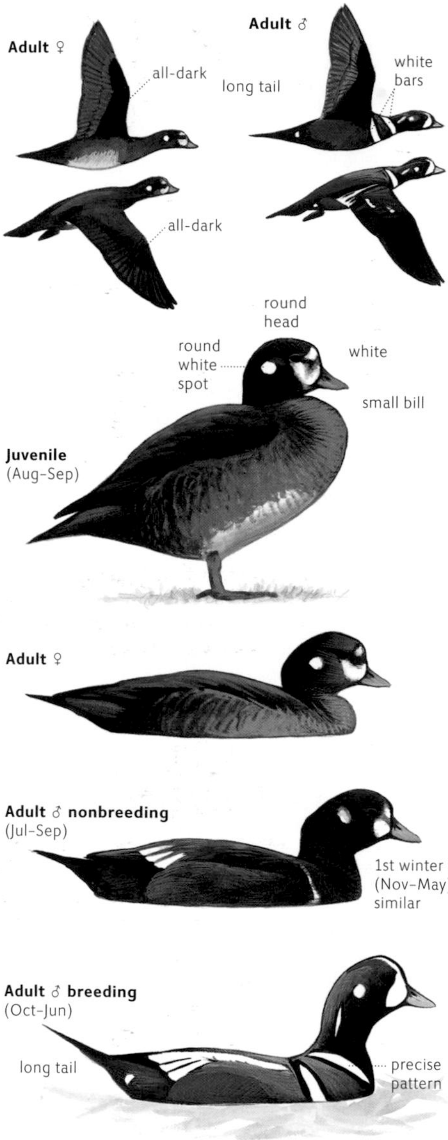

VOICE: Female gives an agitated, nasal *ekekekekek*...; also smooth quacking. Male gives a high, squeaky whistle *tiiv*, sometimes rapidly repeated; reminiscent of Spotted Sandpiper alarm note.

Uncommon and local. Nests along fast-flowing rocky rivers. Winters in small groups on salt water on rocky shorelines with strong wave action often loosely associated with scoters or eiders.

Steller's Eider

Polysticta stelleri

L 17" WS 27" WT 1.9 lb (860 g) ♂>♀

Male mostly white with dark patch around eye, black collar, tawny belly. Female dark brown overall with stout gray bill.

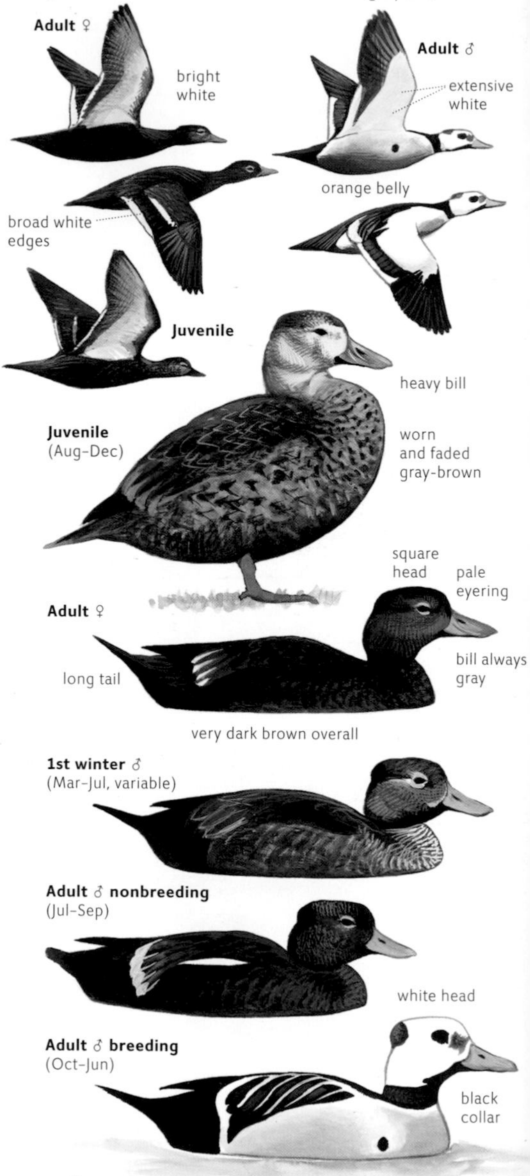

VOICE: Generally quiet. Female gives rapid, guttural calls and loud *cooay* in winter flocks; lacks *gog* call of *Somateria* eiders. Male essentially silent. Wings produce whistle in flight, louder than goldeneyes.

Uncommon. Nests on tundra ponds. Winters mostly on shallow salt-water bays and lagoons. Look for it in small groups or pairs swimming near shore or flying low over water. Does not mix with other eiders. Declining, North American population only a few thousand individuals

Long-tailed Duck

Clangula hyemalis

L 16.5" (adult ♂ to 21")   WS 28"   WT 1.6 lb (740 g)   ♂>♀

A relatively small and very stocky sea duck, with round head, short bill, and round body. Complex seasonal plumage changes but always shows dark breast, all-dark wings, white belly, and some white on head.

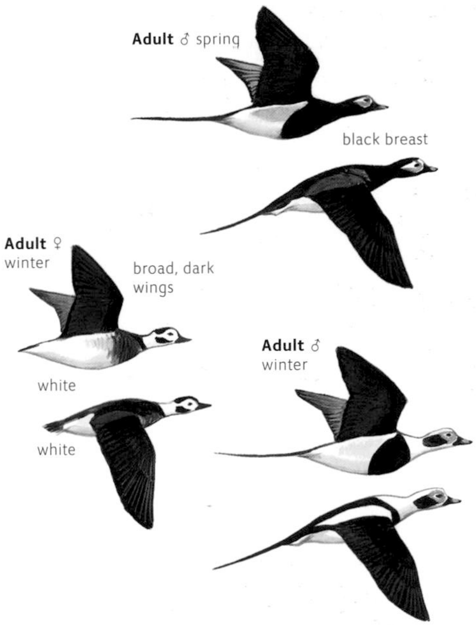

VOICE: Very vocal Feb–Jun. Female calls vary: soft grunting/quacking *urk urk* or *kak kak kak kak*. Male gives a melodious yodeling with clear, baying quality *upup OW OweLEP*.

Common locally on shallow open ocean especially along sandy shorelines. Nests on tundra ponds. Generally in small groups; rarely mixes with other ducks. Dives for crustaceans and some mollusks and plants.

Diving Motions

Diving motions of birds can be useful for identification at great distance.

Most diving ducks (and other diving birds) use only their feet for underwater propulsion and dive with wings closed tightly against the body.

Some ducks use their wings and feet for propulsion underwater and dive with a noticeable wing flick: Long-tailed Duck, Harlequin Duck, scoters (except Black), and eiders.

Alcids use only their wings for propulsion underwater and dive by simply tipping their bodies forward with their wings open and feet up and held loosely.

Black Scoter

Melanitta nigra

L 19" WS 28" WT 2.1 lb (950 g) ♂>♀

The smallest, most compact scoter, with relatively small bill and rounded head. Female has pale cheek contrasting with dark crown.

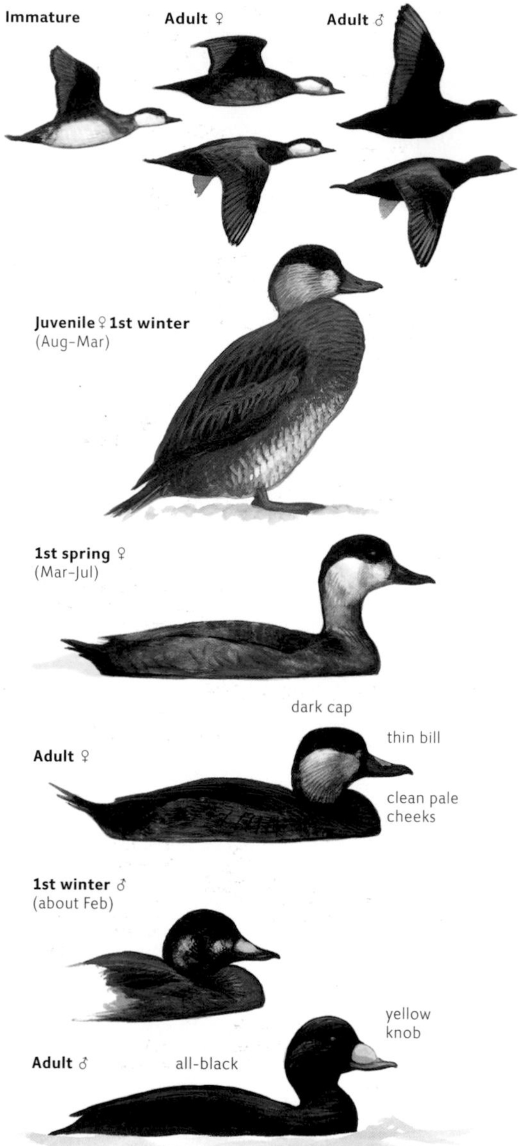

VOICE: Most vocal scoter. Female calls low, hoarse, growling *kraaaa.* Male gives slurred, mellow, piping *peeeew* and plaintive whistle *cree.* Adult male's wings produce quiet, shrill whistling.

Common on open salt water; uncommon to rare inland. Nests among large tussocks of grass near tundra ponds. In winter found on nearshore ocean waters, favors rocky shorelines.

bill small and upturned with vertical border at base

Surf Scoter

Melanitta perspicillata

L 20" WS 30" WT 2.1 lb (950 g) ♂>♀

Stocky and dark, with relatively large head and triangular bill. Male's white head spots distinctive. Female told by heavy bill, vertical white spot at base of bill.

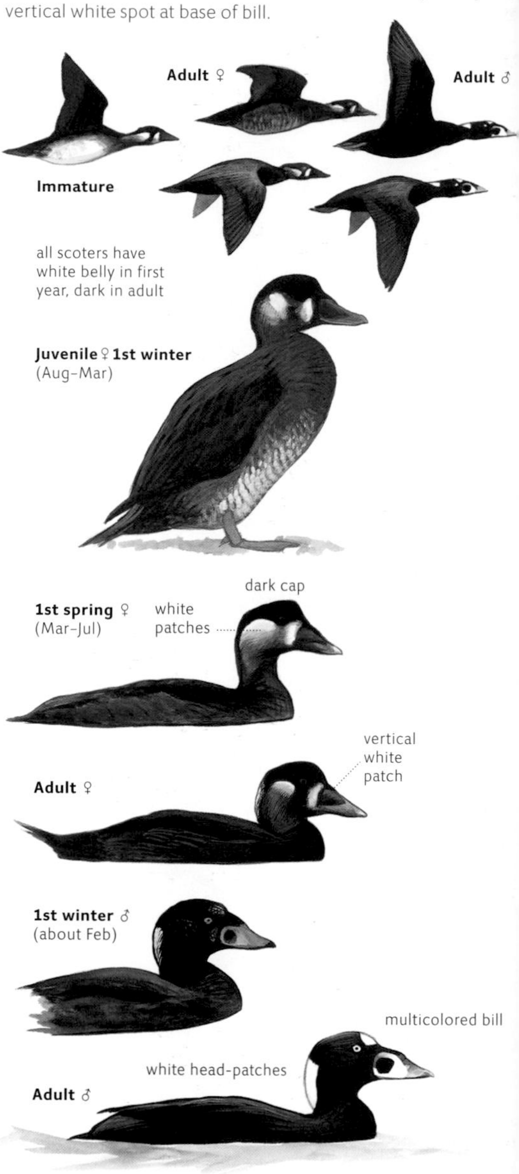

VOICE: Usually silent. Female gives a croaking *krrrraak krrraak.* Male produces a low, clear whistle or liquid, gurgling *puk puk.* Adult male's wings produce a low, hollow whistling.

Common on both coasts, rare inland. Nests on ponds and lakes within spruce forest. Winters on open salt water, especially along rocky shorelines. More common along Pacific coast

bill large and wedge-shaped, feathering extends forward along top

White-winged Scoter

Melanitta fusca

L 21" WS 34" WT 3.7 lb (1,670 g) ♂>♀

A very large and dark duck showing white on secondaries at rest (usually) and in flight. Slightly thicker-necked and with more wedge-shaped head than other scoters.

VOICE: Usually silent; calls brief, crude. Female and male give harsh, guttural croak/quack. Male in display gives thin whistle *wher-er*. Male's wings produce quiet, whistling sound in certain flight displays.

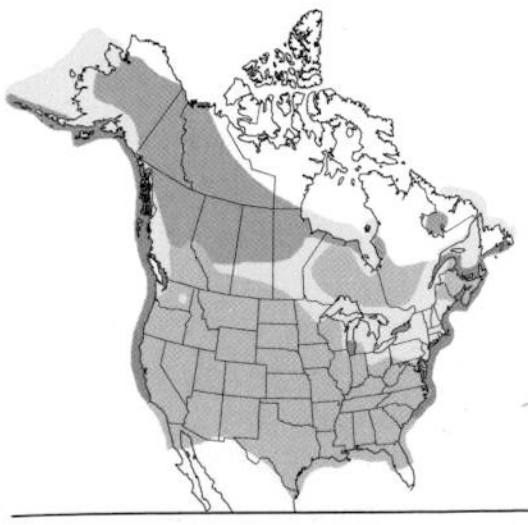

Uncommon. Nests on large lakes in boreal forest. Winters on open salt water, favoring sandy shorelines.

bill large with U-shaped border of feathers covering base

Smew

Mergellus albellus

L 16" WS 24.5" WT 22.5 oz (640 g)

A very distinctive small merganser, male mostly white, female gray with rufous and white head.

VOICE: Usually silent.

Very rare visitor to western Alaska with a few records farther south to California and east to Rhode Island. Some records may involve escapes from captivity.

Goldeneye Hybrids

These two hybrid combinations are seen occasionally.

Common × Barrow's Goldeneye

Hybrid adult ♂ breeding

Intermediate shape and pattern with maroon head gloss.

Hooded Merganser × Goldeneye

Hybrid adult ♂ breeding

White-winged Scoter

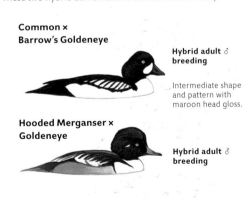

Stejneger's (Siberian) ♂

Siberian males have blackish flanks (not brownish), slightly different bill shape and pattern. Recorded a few times in far western Alaska

Barrow's Goldeneye

Bucephala islandica

L 18" WS 28" WT 2.1 lb (950 g) ♂>♀

Very similar to Common Goldeneye. Male distinguished by head shape and plumage patterns; female by head shape and bill color.

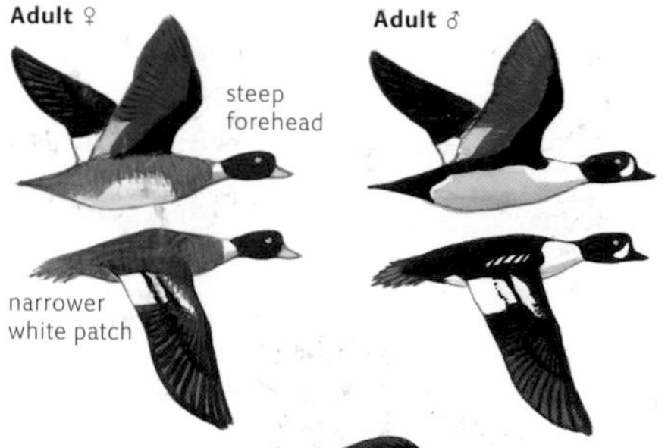

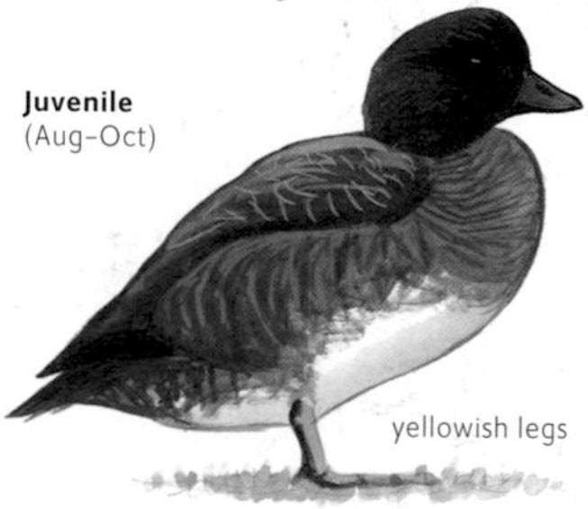

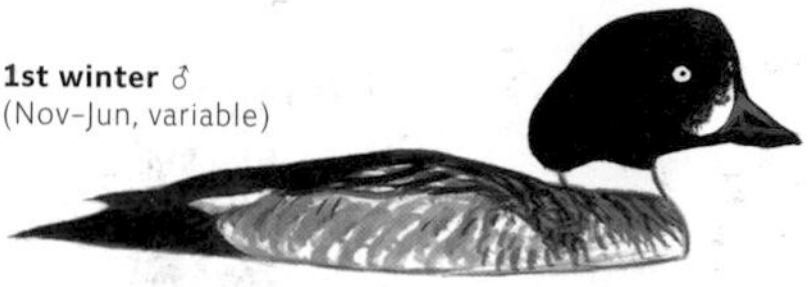

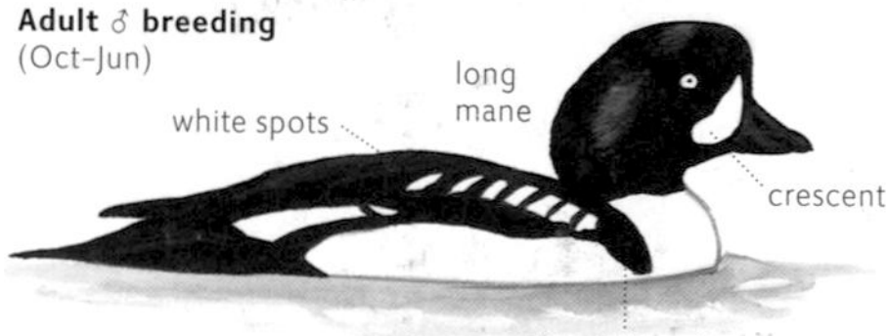

VOICE: Female call like Common but possibly a little higher-pitched. Male in display gives weak, grunting *kaKAA*, unlike Common. Wings produce low whistle in flight; possibly quieter and less metallic than Common.

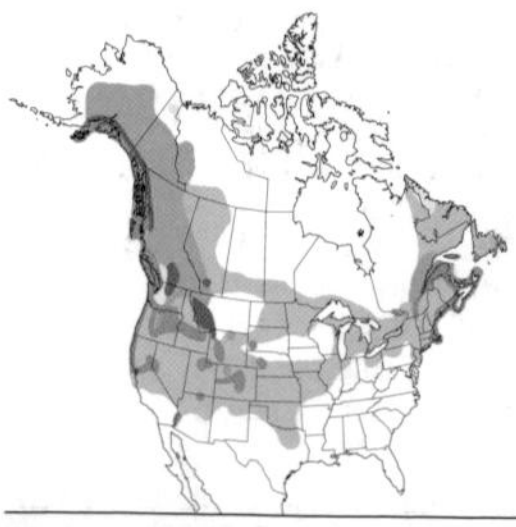

Common locally in northwest, uncommon to rare elsewhere. Winters on open rivers, lakes, and bays, usually mixed with Common Goldeneye.

Common Goldeneye

Bucephala clangula

L 18.5" WS 26" WT 1.9 lb (850 g) ♂>♀

Male mostly white, with black head and round white spot on face. Female gray with brown head. See very similar Barrow's.

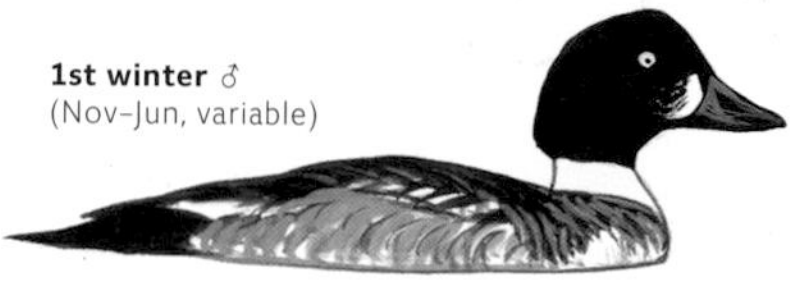

VOICE: Female gives low, grating *arr arr*... like *Aythya* ducks, often in flight. Male in display gives rasping, buzzy whistle *jip JEEEEV* and low, hollow rattle. Wings produce low, metallic whistle in flight; loudest in adult male.

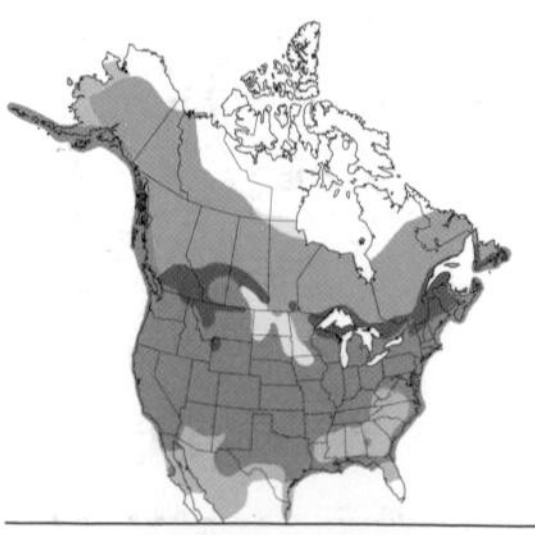

Common. Winters on open lakes, rivers, and bays in small flocks that sometimes gather in larger numbers. Nests in tree cavities around shallow marshy lakes and beaver ponds. Dives mainly for animal prey such as crustaceans, fish, insects, and mollusks.

Bufflehead

ucephala albeola

13.5" WS 21" WT 13 oz (380 g) ♂>♀

early our smallest duck. Breeding male strikingly white with ttle black. Female gray-brown with obvious oval white patch on heek.

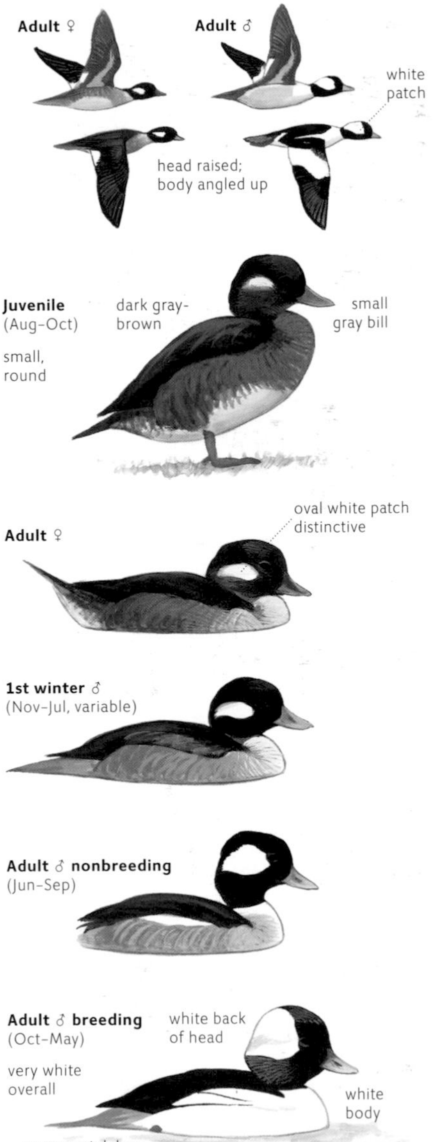

VOICE: Usually silent. Female gives a low, hollow *prrk prrk*… similar to goldeneyes but weaker, softer. Male produces a squealing or growling call.

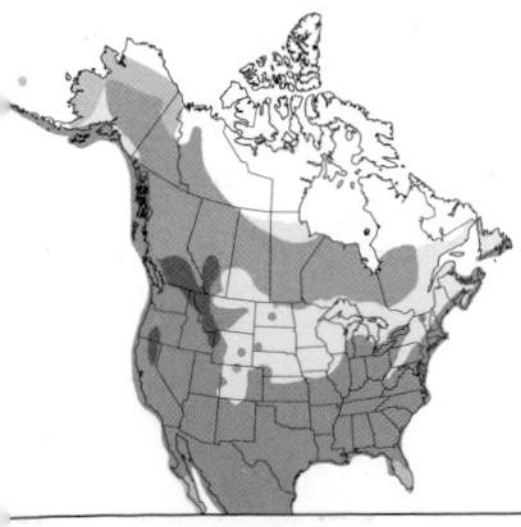

Common. Winters on open lakes, harbors, and bays, especially coastal bays and tidal creeks. Nests in tree cavities near ponds and rivers. Usually in small flocks that sometimes gather in larger numbers. Dives for mollusks, crustaceans, and insect larvae.

Hooded Merganser

Lophodytes cucullatus

L 18" WS 24" WT 1.4 lb (620 g) ♂>♀

Small and long-bodied; often raises tail when swimming. All have hammerhead crest; mostly white on male, frosted brown on female.

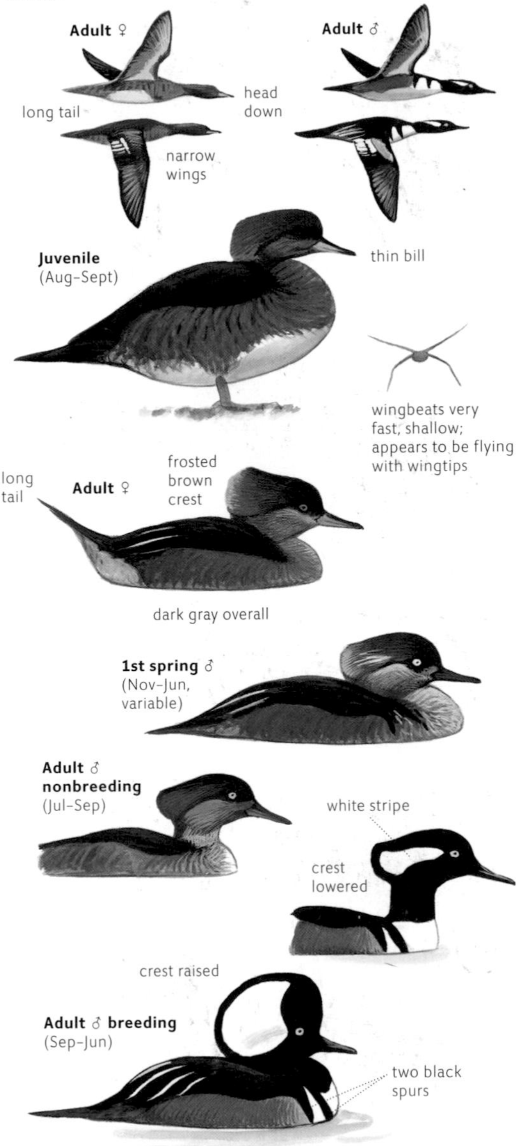

VOICE: Female gives a soft croak *wrrep*; sometimes *ca ca ca ca ca*… in flight. Male in display gives a low, purring croak, descending and slowing *pah-hwaaaaaa*. Wings produce high, cricket-like trill in flight; loudest in adult male.

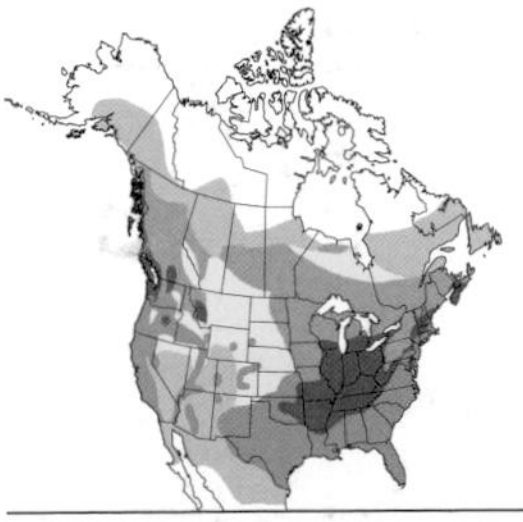

Uncommon. Winters in small flocks on relatively small and sheltered open water—wooded ponds, swamps, tidal creeks—often swimming among standing trees. Nests in tree cavities near ponds, rivers, wooded swamps.

Common Merganser

Mergus merganser

L 25" WS 34" WT 3.4 lb (1,530 g) ♂>♀

Similar to Red-breasted Merganser, but heavier and with thicker bill. Male mostly white. Female has darker brown head than Red-breasted, with sharply contrasting white chin and neck.

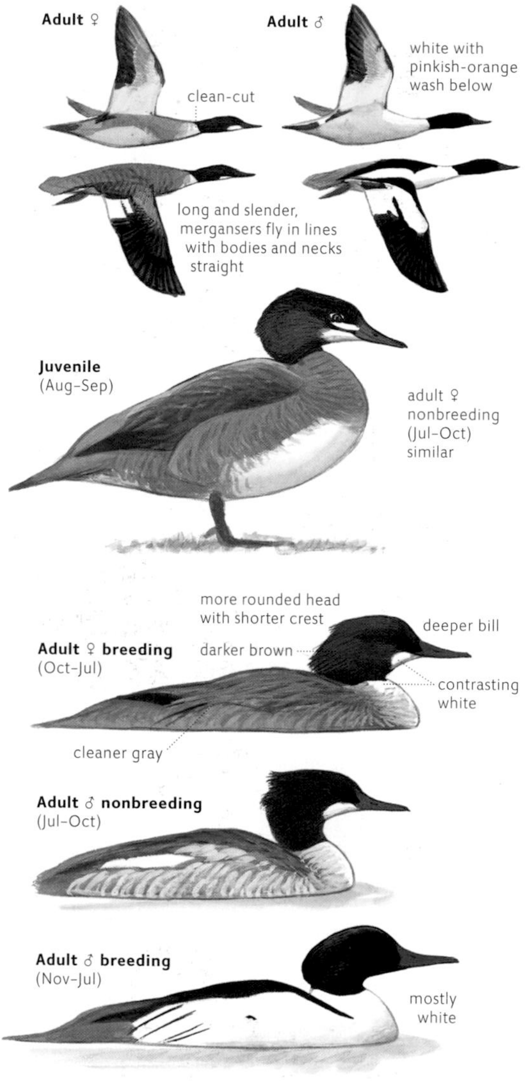

VOICE: Female call a deep, harsh, croaking *kar-r-r* in flight, sometimes accelerating to cackling *kokokokok*. Male also gives hoarse croaking notes; in display faint twanging or bell-like single notes.

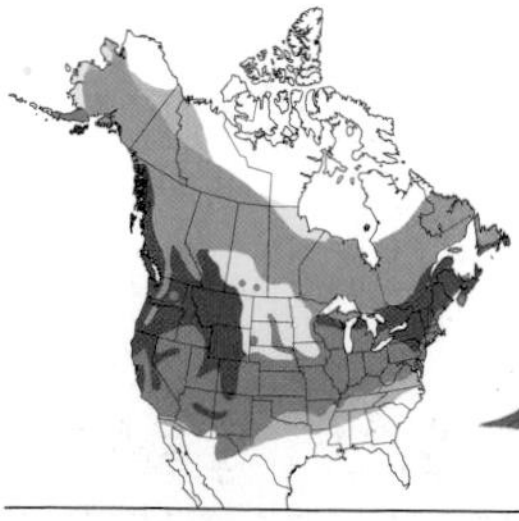

Common locally. Found on deep clear lakes and rivers; rarely on salt water. Nests in tree cavities and on the ground near large lakes. In winter forms small groups that may gather into large numbers at favored sites.

Red-breasted Merganser

Mergus serrator

L 23" WS 30" WT 2.3 lb (1,060 g) ♂>♀

Slender and long-bodied; shaggy crest and thin red bill distinctive. Female distinguished from Common Merganser by diffuse border between brownish head and pale chin and neck.

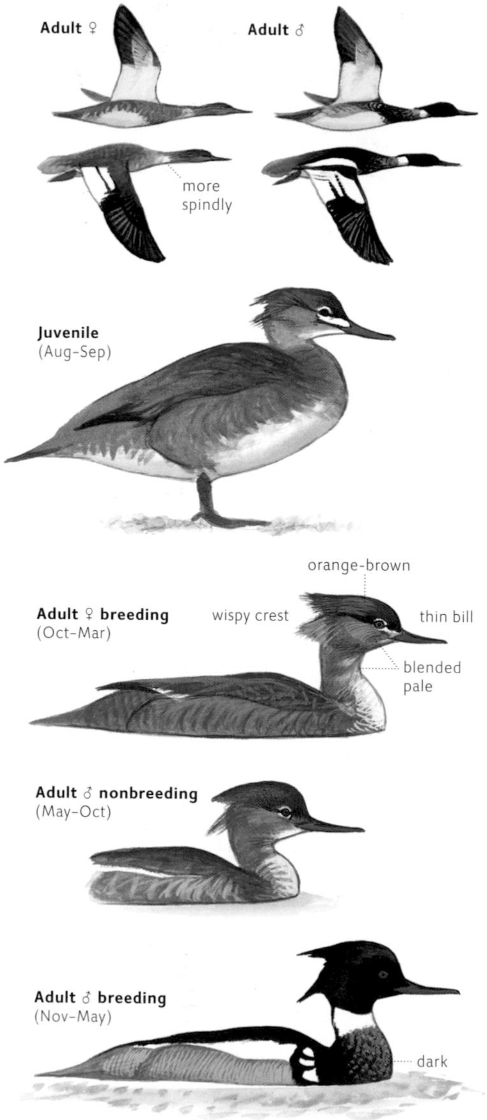

VOICE: Female gives *prek prek...* in flight, similar to Common Merganser but slightly higher-pitched. Male mostly silent; in display a purring or scraping *ja-aah* and wheezy, metallic *yeow*.

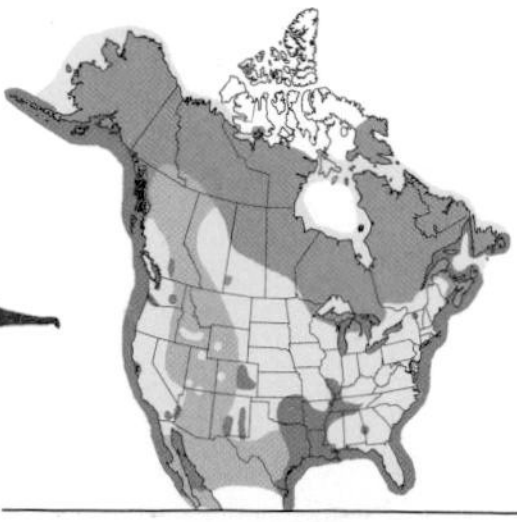

Common. In winter found on salt water in lagoons and bays and along sheltered coasts; less common inland on lakes. Nests on the ground in a variety of settings close to water. Found in small groups. Feeds on small fish in shallow water; sometimes hunts cooperatively in groups.

Masked Duck

Nomonyx dominicus

L 13.5" WS 17" WT 13 oz (380 g) ♂>♀

Smaller than Ruddy Duck, with heavier bill. Breeding male unmistakable. Female distinguished from Ruddy Duck by intricately patterned body and striped head.

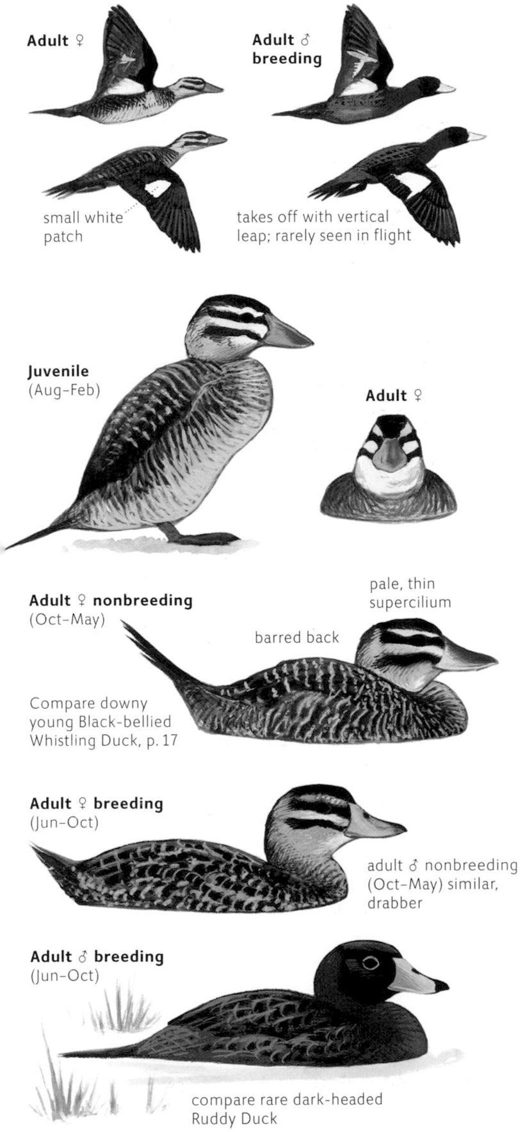

VOICE: Generally silent. Female gives a short hissing noise. Male in display gives a long series *kirri kirroo kirri kirroo kirroo kirroo kirrrr*; also quiet puffing sounds and a low bark.

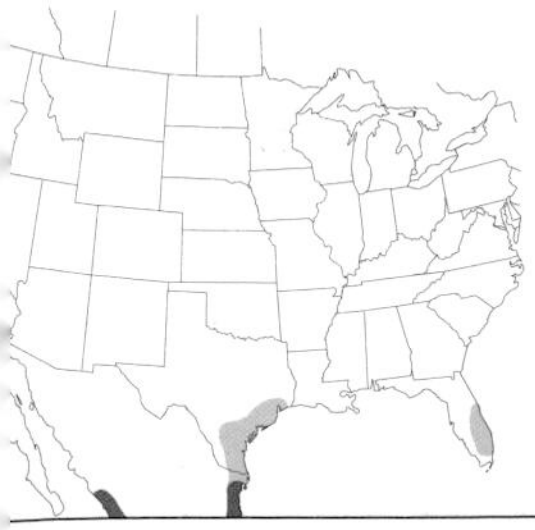

Rare visitor, averaging a couple of records each year, mostly in southern Texas. Found in shallow grassy or weedy ponds, where it stays mostly hidden and rarely ventures far from cover. Never joins flocks of Ruddy Ducks on open water.

Ruddy Duck

Oxyura jamaicensis

L 15" WS 18.5" WT 1.2 lb (560 g) ♂>♀

Small and compact, with relatively large head and bill, sloping forehead, and long tail often raised. Male has obvious white cheeks at all seasons, female dark gray-brown with dark stripe across cheek.

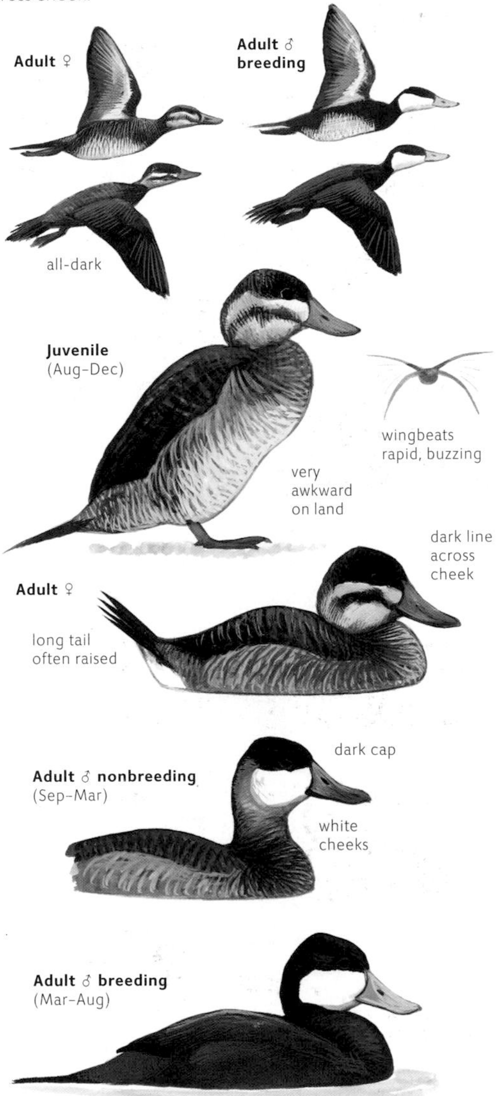

VOICE: Female gives a low, nasal *raanh*; also a high, sharp squeak and thin *tweet*. Male essentially silent except during display: muffled popping series *jif jif jif jif ji ji ji ji jijijijijijwirrrrr*; also series of staccato pops produced by feet running on surface of water.

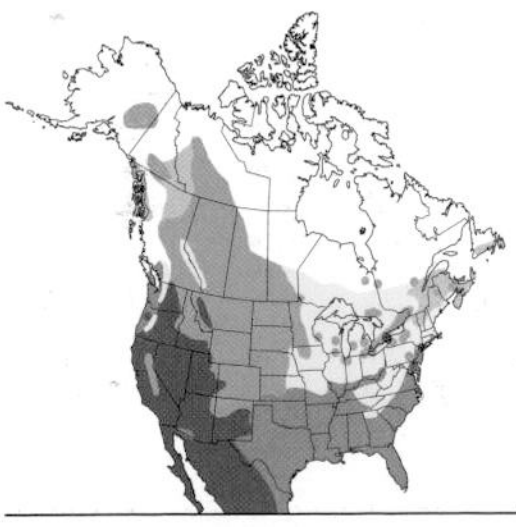

Common, especially in West. Found on open water of ponds, lakes, and sheltered bays in tightly clustered flocks, sometimes mixed with other diving ducks. Nests in marshy vegetation in shallow ponds.

Dark-headed variant ♂

Upland Game Birds

FAMILIES: CRACIDAE, ODONTOPHORIDAE, PHASIANIDAE

20 native species and 4 with established introduced populations. Several additional species have local feral populations, and many others are seen as occasional escapes. All in the family Phasianidae except Plain Chachalaca (Cracidae) and six species of quail (Odontophoridae). All are chicken-like ground birds that forage mainly on the ground for plant matter and some insects. They are usually quite secretive, crouching in dense cover. Note explosive takeof and direct flight alternating bursts of rapid wingbeats with sailin glides on stiffly downcurved wings. In all species, downy youn develop functional flight feathers within days of hatching and ca fly when only one-third of their full-grown size. Beware of mistak ing these for smaller species. Adult females are shown.

Genus *Ortalis*

Plain Chachalaca, page 42

Genus *Oreortyx*

Mountain Quail, page 45

Genus *Colinus*

Northern Bobwhite, page 43

Genus *Cyrtonyx*

Montezuma Quail, page 42

Genus *Bonasa*

Ruffed Grouse, page 52

Genus *Callipepla*

Scaled Quail, page 45

California Quail, page 44

Gambel's Quail, page 44

Genus *Dendragapus*

Dusky Grouse, page 49

Sooty Grouse, page 49

Genus *Alectoris*

Chukar, page 46

Genus *Perdix*

Gray Partridge, page 46

Genus *Phasianus*

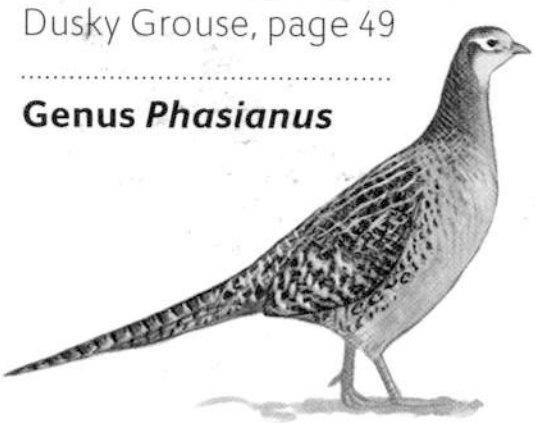

Ring-necked Pheasant, page 47

Genus *Lagopus*

Willow Ptarmigan, page 51

Rock Ptarmigan, page 50

White-tailed Ptarmigan, page 51

Genus *Falcipennis*

Spruce Grouse, page 48

Genus *Tympanuchus*

Sharp-tailed Grouse, page 52

Greater Prairie-Chicken, page 53

Lesser Prairie-Chicken, page 53

Genus *Centrocercus*

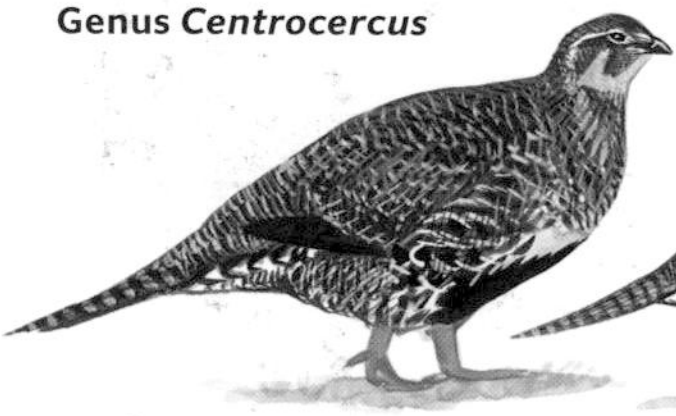

Greater Sage-Grouse, page 54

Gunnison Sage-Grouse, page 54

Genus *Meleagris*

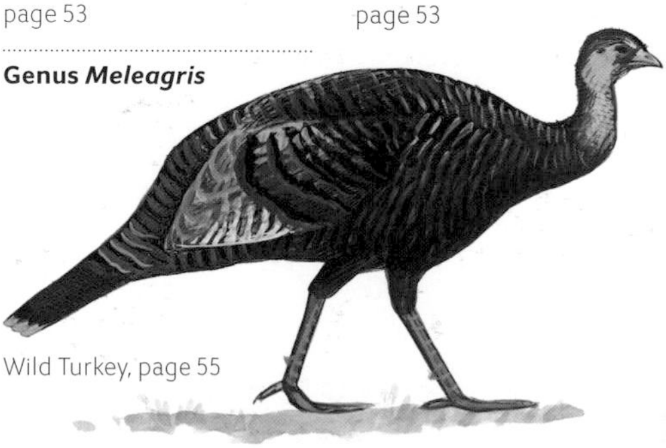

Wild Turkey, page 55

Himalayan Snowcock

Tetraogallus himalayensis

L 26" WS 42" WT 6 lb (2,722 g)

Very large, stocky, and relatively short-tailed; gray-brown with striped face; mostly white wings are striking in flight. Much larger and heavier than Dusky Grouse, and no other large grouse overlaps in range and habitat.

VOICE: A curlew-like whistle in display; low clucking and shrill cackles.

Native to Himalayas. Introduced and with a very small population established in the Ruby Mountains, Nevada. Found on rocky slopes near or above treeline, foraging on the ground.

Red Junglefowl

Gallus gallus

L 23.2" WS 26" WT 2 lb (907 g)

Ancestor of the familiar domestic chicken, with many domestic varieties, but most feral birds have simple rust and black color. Easily distinguished from all native grouse by color pattern, long legs, long tail held upright, habitat.

VOICE: Males give raucous crowing "cock-a-doodle-doo"; both sexes (but especially females) give a variety of clucking sounds.

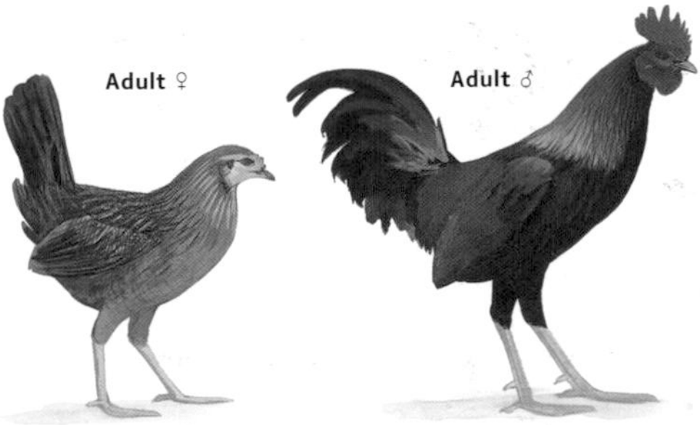

Native to Southeast Asia; a feral population is established in the Florida Keys, and this common barnyard species can be seen and heard almost anywhere near human habitation.

Helmeted Guineafowl

Numida meleagris

L 23.2" WS 26" WT 3 lb (1,361 g)

Large round body, with bare head that appears very small, and almost no tail. Natural plumage is dark gray with small white spots, but many domestic birds are all or partly white.

VOICE: Raucous notes in rhythmic series.

Widely kept in parks and farmyards. No known feral populations, but often seen roaming freely around farms and parks.

Golden Pheasant

Chrysolophus pictus

L 34.4" WS 22" WT 1.4 lb (635 g)

Gold and red male is unmistakable. Female similar to Ring-necked but darker, with strongly barred breast.

Native to Asia. Occasionally escapes from captivity, no feral population established.

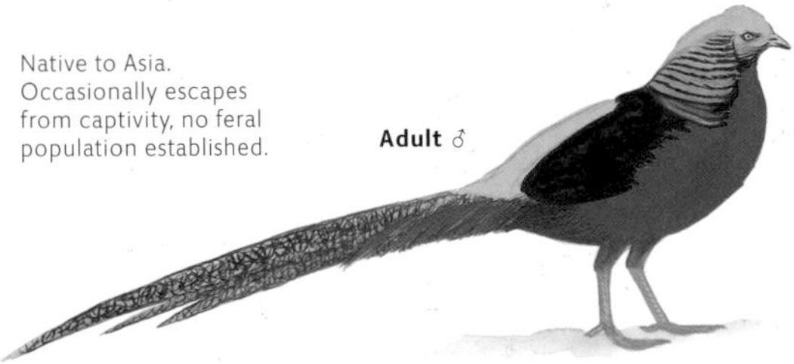

Indian Peafowl

Pavo cristatus

L 63" WS 48" WT 9.4 lb (4,264 g)

Extremely large, with long broad tail and small head; unmistakable. Many domestic birds are all or partly white. Technically, the male's "tail" is very long uppertail coverts.

VOICE: Male gives very loud, wailing cries.

Native to Southeast Asia. A popular ornamental bird at parks and estates, free-living at many locations and in some, such as southern California, may have established feral populations.

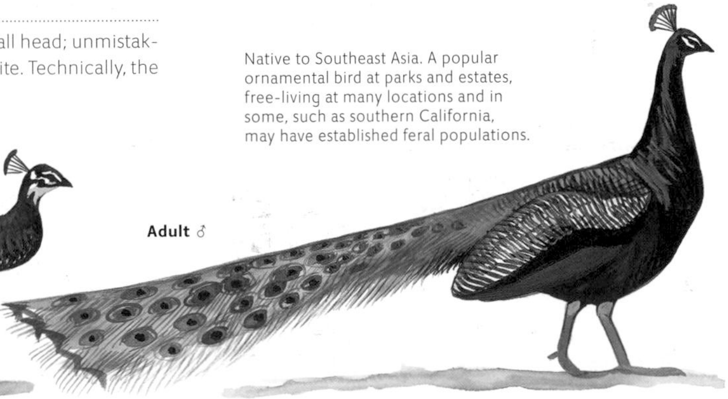

Plain Chachalaca

Ortalis vetula

L 22" WS 26" WT 1.2 lb (550 g) ♂>♀

Chicken-size, but long-necked with long broad tail. Overall unpatterned drab brown color and oversized dark tail with white corners distinctive. Flies with heavy wingbeats and short sailing glides across clearings or to escape danger. Bare skin on side of throat is most obvious and most reddish on displaying males.

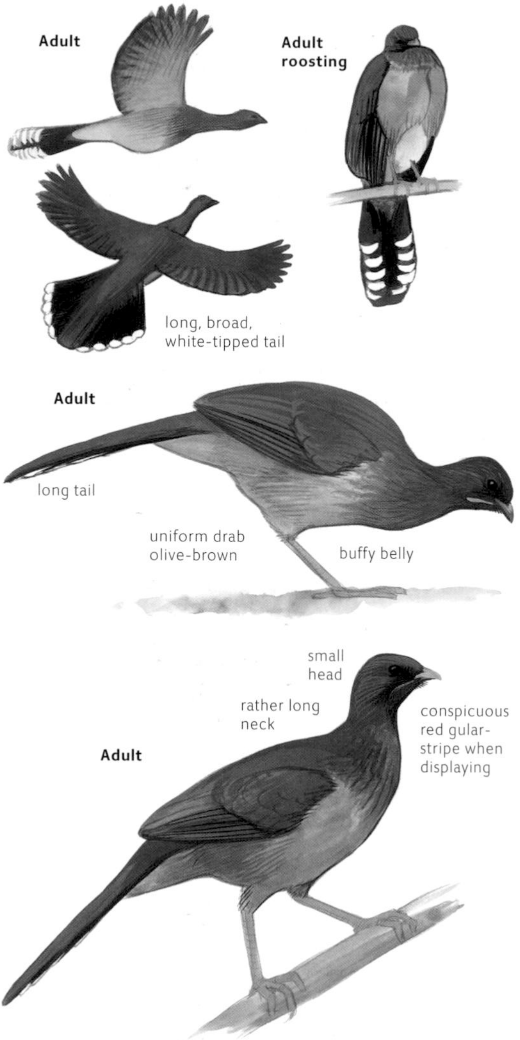

VOICE: Incredibly raucous, loud calls given by many birds in synchronized chorus, especially at dawn. Lead bird low-pitched *GRA da da* joined by others giving squeakier, higher-pitched *REEK a der*. A muffled, toneless *krrrr* given when nervous, leading into raucous clattering *KLOK aTOK aTOK aTOK* as birds take flight.

Common within limited range. Usually seen in small social groups of five to 10 birds that move along the ground or through branches in riparian woods and mesquite brushlands. A small population is also introduced and established on Sapelo Island, Georgia.

Montezuma Quail

Cyrtonyx montezumae

L 8.75" WS 15" WT 6 oz (180 g) ♂>♀

Distinguished from other quail by very round shape with shor neck, round head, and very short tail. Intricate clown-like heac pattern is unique. When flushed, note round body and very shor tail, darker overall than other quail in its range, and usually th only quail in its range and habitat.

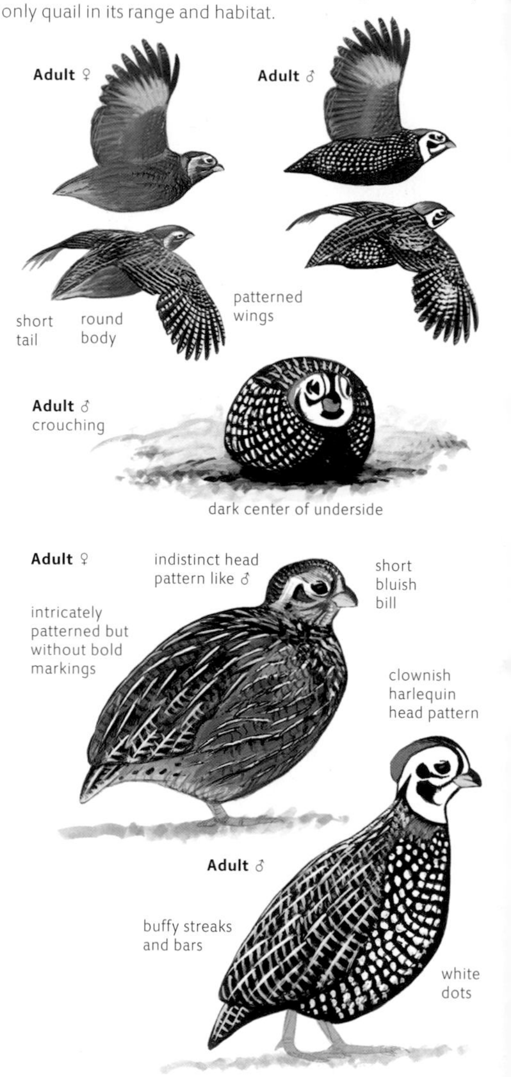

VOICE: Male song an eerie, melancholy, vibrant, descending whistle *vwirrrrr*; also gives piping *querp* or louder *querp-queeeep*. Other calls include low *wep wep*...and soft, low *pwew* assembly call a low, whistled series of six to nine notes descending in pitch. Flush with loud, popping wing noise.

Uncommon and local on arid grassy slopes with scattered oaks and yuccas, generally at higher elevation than Scaled Quail. Found in pairs or small family groups, not gathering in larger coveys like other quail. Occasionally emerges to forage at sides of roads or trails, otherwise on the ground hidden in grass, where it sits still to avoid detection, flushing only at very close approach.

Northern Bobwhite

Colinus virginianus

9.75" WS 13" WT 6 oz (170 g)

Small, short-tailed, and short-legged; intricately patterned in rufous, black, and white. In flight, note gray wings and tail. The only quail in most of its range. Where range overlaps with Gambel's and Scaled Quail, smaller size, shorter tail, and rich brown pattern on body easily distinguishes Bobwhite.

Eastern

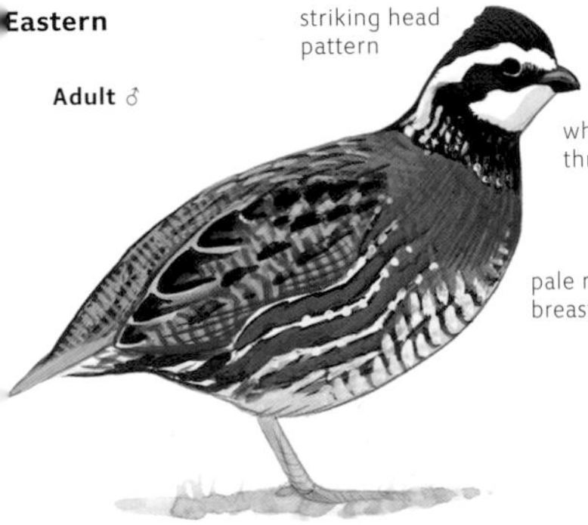

VOICE: Male song a strong, clear whistle *pup-WAAAYK* or *bob-WHITE* (often imitated by starlings, mockingbirds, etc.); also a loud, harsh *quaysh* or *quEEEak*. Covey calls include *hoy*, *hoypoo*, and *koilee*; contact calls a soft *took* and *pitoo*. Ground predators elicit soft, musical *tirree*, changing to *ick-ick-ick* or *toil-ick-ick-ick*; avian predators elicit throaty *errrk*. Male in southern Texas gives harsher *bob-WHIISH* than northern birds.

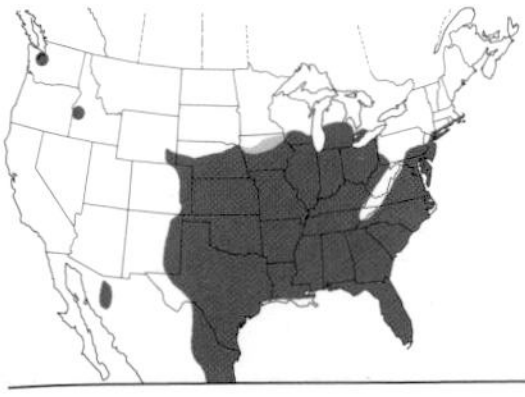

Uncommon, local, and declining. Found in mixed weedy and brushy habitat such as old fields and pastures with hedgerows or brushy edges. Males sing from low exposed perch such as fence post, otherwise found on the ground in small flocks, which flush all together at close range in an explosion of noisy wingbeats.

Masked (Southwestern)

Masked is found in the arid grasslands of western Mexico and very locally in southeastern Arizona. Males are distinctive with rufous belly and black face, but females are indistinguishable from other subspecies, and voice is similar.

Great Plains and Texas

Florida

Regional variation is considerable but clinal, most obvious in males. Northern birds average larger than Southern. Western birds average paler and grayer than eastern. Florida population is distinctive, small and dark with extensive black on breast and flanks.

California Quail

Callipepla californica

L 10" WS 14" WT 6 oz (180 g)

The only quail in most of its range. Very similar to Gambel's Quail, but range barely overlaps. Distinguished by strongly scaled belly and nape, paler olive-gray on flanks, and lack of blackish patch on belly.

VOICE: Location call a repeated, nasal *put way doo* similar to Gambel's, but lower-pitched, usually three syllables, and final note longer and descending, individually variable. Male song a single note *caaw*. Other calls clucking and sputtering.

Common. Found in coastal chaparral-type habitat of dense brush and weedy thickets. Often seen along roadsides or field edges in small coveys. Usually on the ground, in flocks, often disappearing into dense chaparral brush. Males sing from exposed low perch such as fencepost.

Gambel's Quail

Callipepla gambelii

L 10" WS 14" WT 6 oz (180 g)

Very similar to California Quail, but range barely overlaps. Dis tinguished from Scaled by stronger pattern on head, weake pattern on breast and flanks; dark curved topknot, also usuall occupies less grassy habitat than Scaled.

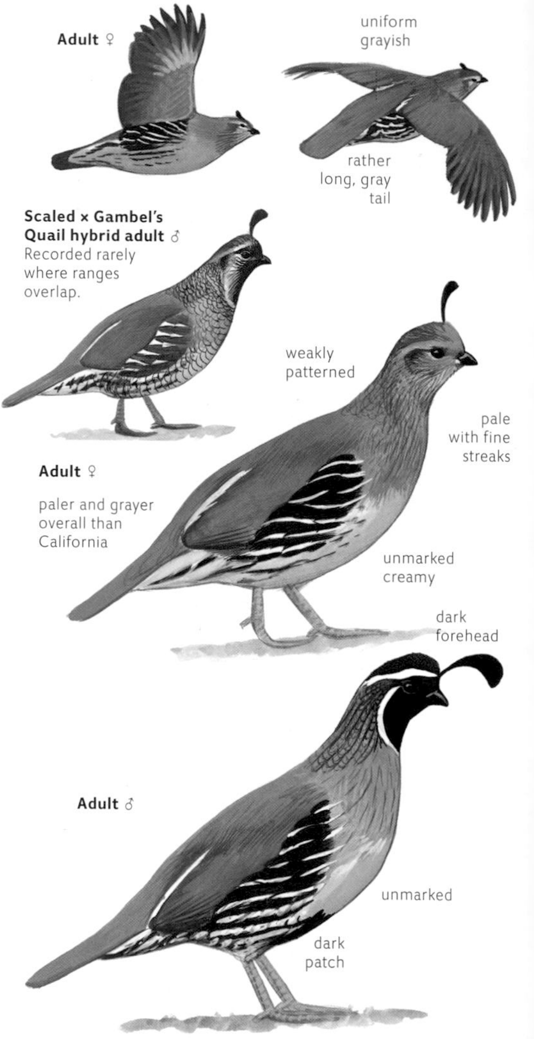

VOICE: Location call a repeated, nasal *pup waay pop pop*, very similar to California Quail. Male song a nasal descending *caaw*, like a drawn out syllable of location call. Other calls include a descending or moaning *where* or *uweeea*; high, sharp *spik* notes; hoarse, trumpeting *krrt*; soft clucking notes. Some calls quite distinct from California.

Common. Found in dry semidesert with tall shrubs; also in adjacent agricultural areas. Habits like California Quail.

Scaled Quail

Callipepla squamata

L 10" WS 14" WT 6 oz (180 g)

Size and shape similar to Gambel's Quail, but more uniform gray and scaly overall, with pale tuft on crown. Also note relatively unpatterned face. In flight appears all-gray.

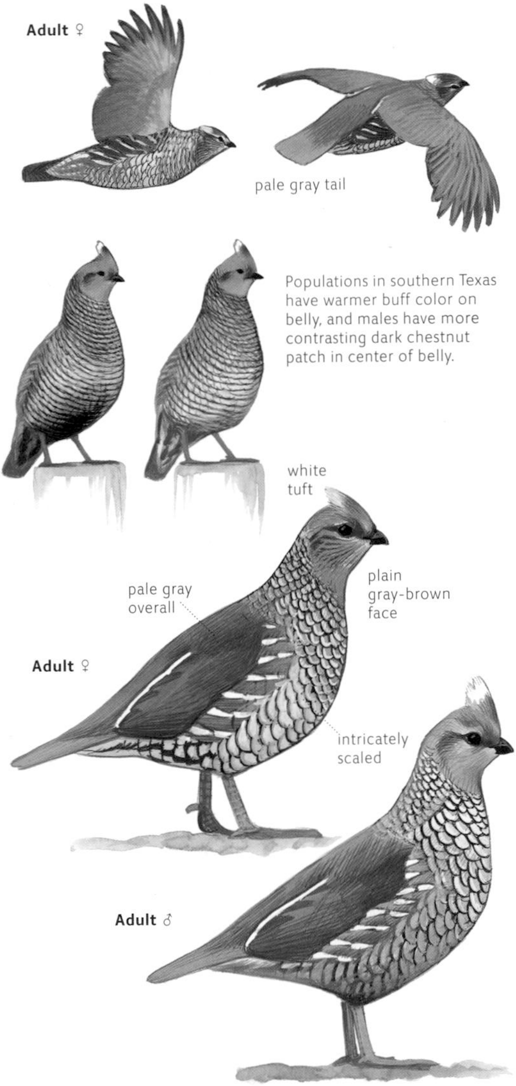

VOICE: Male song a high, raucous outburst *QUEESH* or *BEISH*; location call by both sexes a slow, rhythmic, coarse clucking *kek kut, kek kut*...or *kek kyurr, kek kyurr*....(or *Pe-cos*). When flushed a sharp *cheep* or *chipee*. Other calls include high, sharp chip or trill of sharp notes reminiscent of small songbirds.

Uncommon. Found in arid grasslands mixed with taller brush. Usually in small coveys, often easy to see as males perch on prominent singing posts and coveys walk between patches of cover.

Mountain Quail

Oreortyx pictus

L 11" WS 16" WT 8 oz (220 g)

Distinguished from other quail by slender straight head plumes, slightly larger size, bold white bars (not streaks) on chestnut flanks leaving a band of unmarked brown with a thin white border at edge of flanks. Also note that range and habitat overlaps very little with other quail.

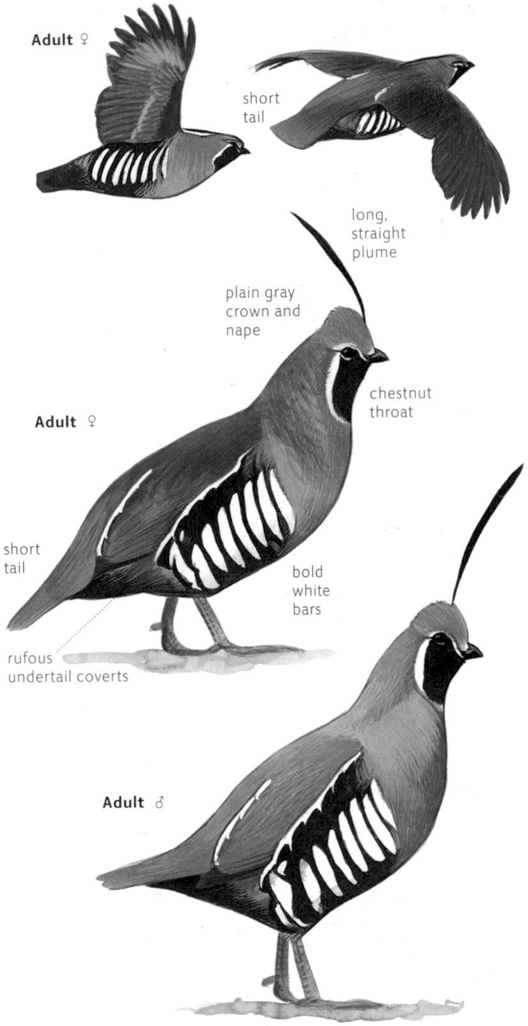

VOICE: Male song loud, raucous, ringing, two-syllable *QUEark*, like Scaled Quail but lower, descending (and no range overlap); delivered about every 6 seconds. Loud sharp whistled *cle-cle-cle*... or *kow kow kow*... series reunites covey. Alarm call a creaking *cree-auk*. Other soft clucking and trilled notes given by both sexes.

Uncommon in dense brushy chaparral on dry mountain slopes; usually the only quail within its range and habitat. Reclusive, males sing from prominent rocks or tree stumps, but most often seen crossing roads or trails early in the day.

Chukar

Alectoris chukar

L 14" WS 20" WT 1.3 lb (590 g) ♂>♀

Chunky and medium-size, slightly larger than quail; pale gray overall, with bold black bars on flanks and black border on throat. Red bill and legs conspicuous. Rusty outer tail feathers conspicuous when fanned on takeoff or before landing.

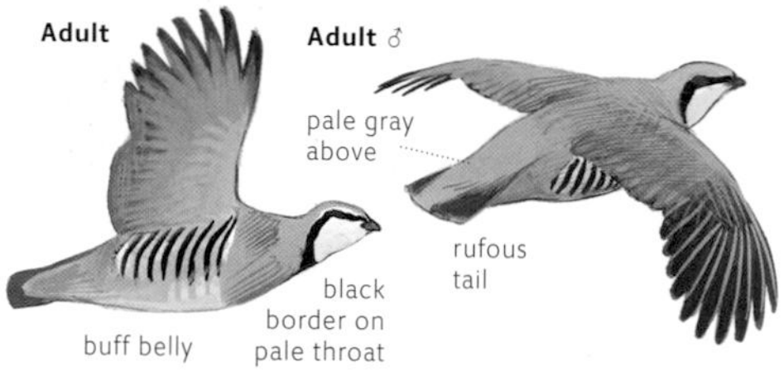

VOICE: A series of nasal, clipped clucks *kakakaka-kakakachukAR chuKAR chuKAR.* In flight quiet clucking sounds and loud, harsh note followed by softer notes *PITCH-oo-whidoo.*

Introduced, native to Middle East. Uncommon and local on dry rocky canyon slopes with widely scattered grass and brush, usually in coveys of up to about 40 birds. Released or escaped individuals are seen occasionally continent-wide.

Gray Partridge

Perdix perdix

L 12.5" WS 19" WT 14 oz (390 g)

A chunky, medium-size game bird; larger than a quail and smaller than a grouse. Best identified by overall gray-brown color, cinnamon bars on flanks, and orange-rufous outer tail feathers that are often flicked open, conspicuous when taking off or just before landing.

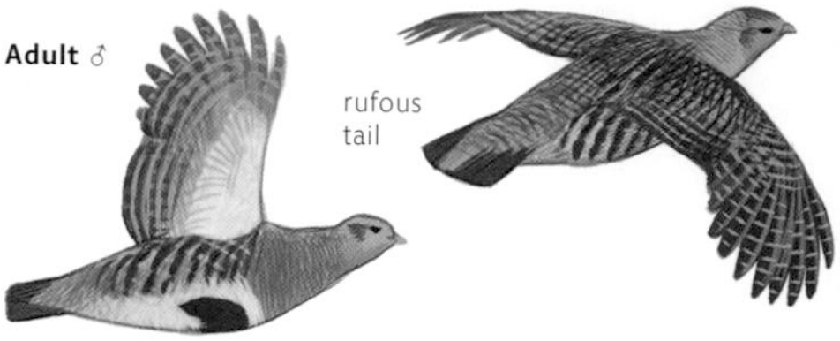

VOICE: Male gives repeated scratchy *Kishrrr* or *kshEEErik*; quality reminiscent of guineafowl. Both sexes give *keeeah* and clucking notes. Covey flushes with wing noise and loud, high, creaking *keep, keep*...in cackling chorus.

Native to Europe. Introduced and uncommon on disturbed prairie and agricultural land. Usually seen in groups of up to 15, foraging on the ground; when flushed all take off together and usually fly in unison.

Ring-necked Pheasant

Phasianus colchicus

L 21" (adult ♂ to 35") WS 31" WT 2.5 lb (1,150 g) ♂>♀

Chicken-size, with long pointed tail. Male unmistakable. Female pale sandy-brown overall; best identified by pointed tail and relatively pale, unpatterned head. In flight note pale barred flight feathers.

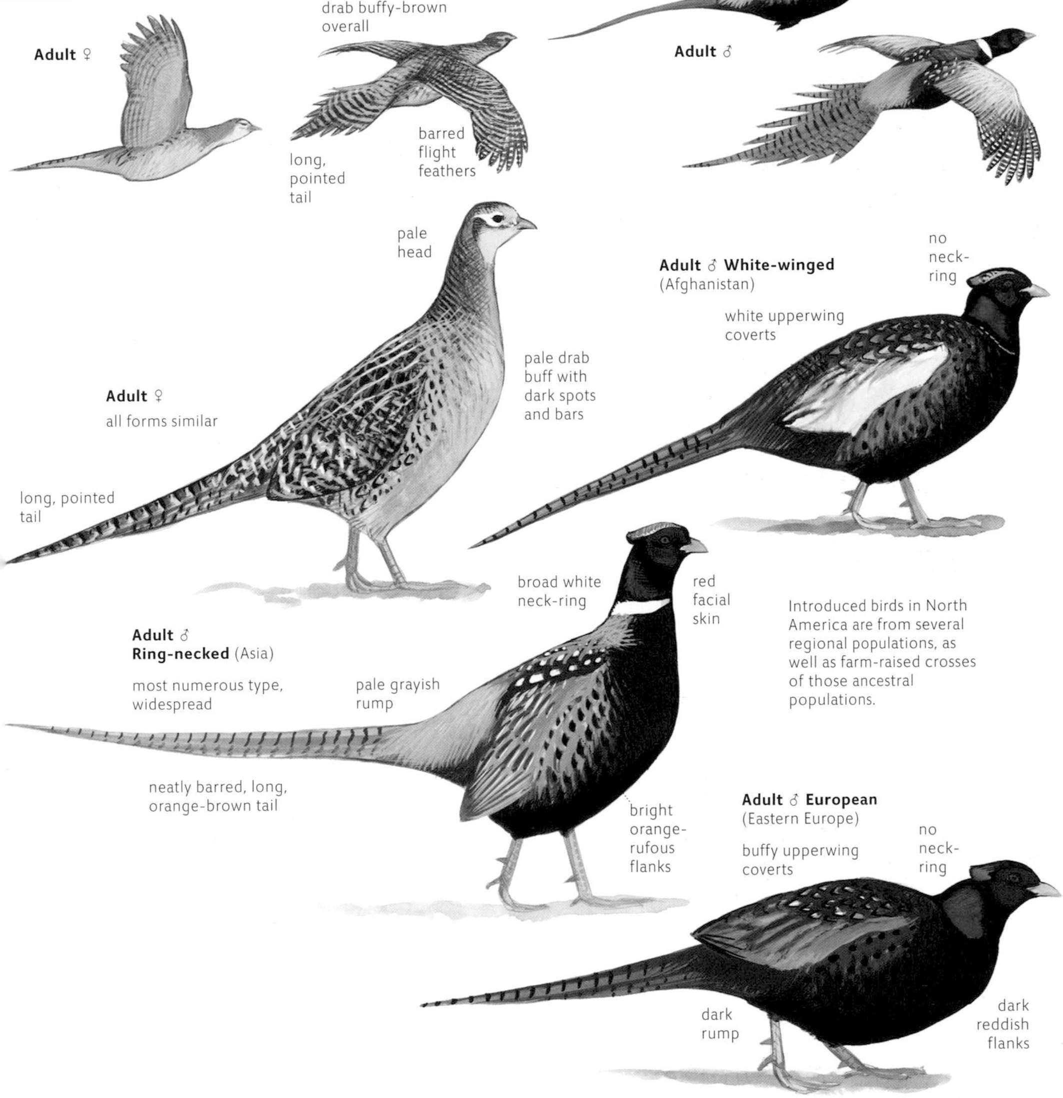

Introduced birds in North America are from several regional populations, as well as farm-raised crosses of those ancestral populations.

VOICE: Male display a loud, harsh, hacking *uurk-iik* followed by short burst of noisy wingbeats. When flushed males give loud, harsh cackle; females sometimes a quieter cackle. Also a variety of low, clucking notes, some very similar to domestic hen.

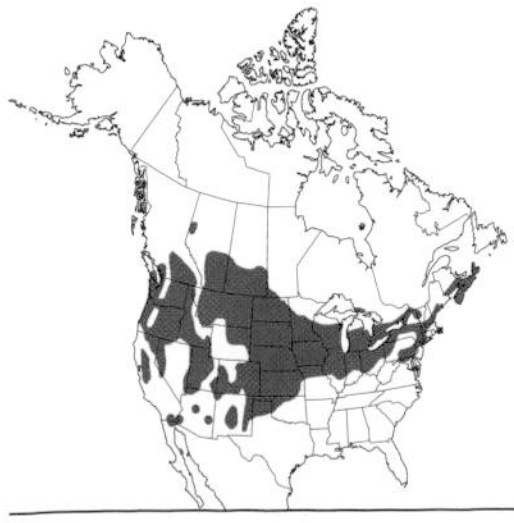

Native to Asia. Widely introduced and locally common but declining in many areas. Found in expansive open grassy or weedy areas with some taller cover such as hedgerows, especially agricultural areas either dry or irrigated. Usually solitary; in winter forms small single-sex flocks.

Japanese Green Pheasant

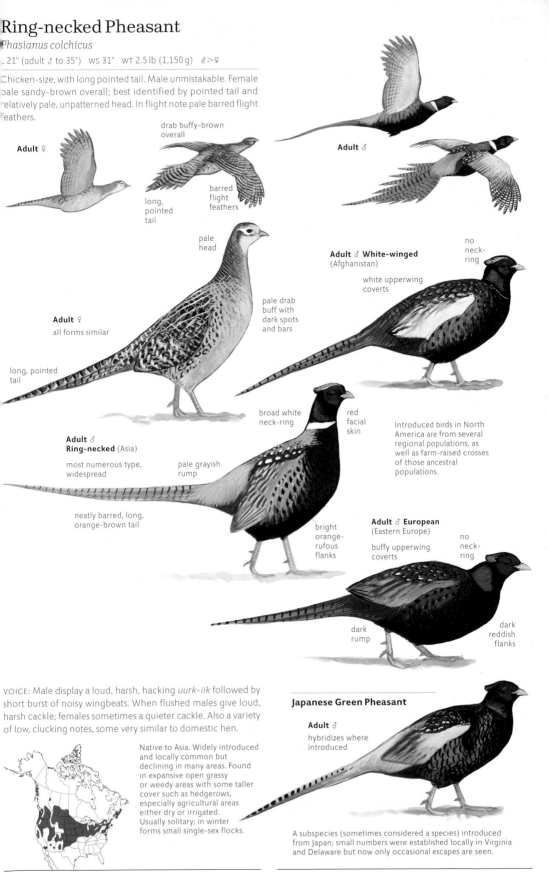

A subspecies (sometimes considered a species) introduced from Japan; small numbers were established locally in Virginia and Delaware but now only occasional escapes are seen.

Spruce Grouse

Falcipennis canadensis

L 16" WS 22" WT 1 lb (460 g) ♂>♀

Male distinctive, with black chest, white-spotted underparts, and mostly black tail. Female distinguished from Ruffed Grouse by short dark tail and barred underparts; from Dusky and Sooty Grouse by short tail and barred breast.

VOICE: Both sexes give guttural notes and clucks. Display of male differs between populations: Taiga display involves gliding from 10 to 20 feet up in tree, turning body vertically and hovering to ground; Franklin's display involves gliding 50 to 100 yards from perch and producing two loud wing claps (like gunshots) before settling on ground.

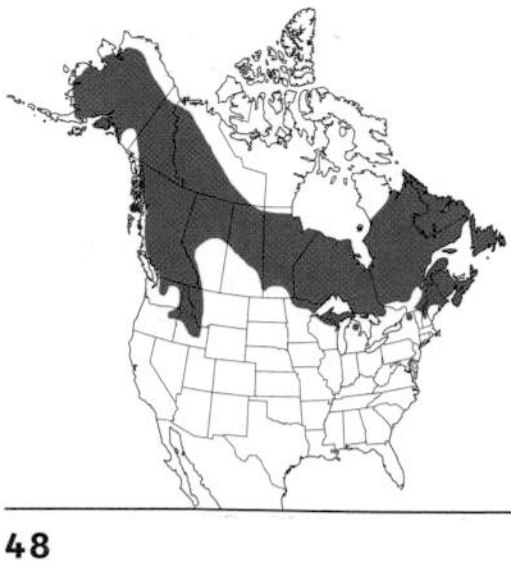

Uncommon. Found in dense spruce forests with mossy ground. Usually solitary and unwary. Found on the ground or sitting quietly in spruce tree eating buds or needles.

Two subspecies groups differ significantly in male plumage and display, but females are not reliably distinguishable. These populations may represent two species, but intergrade locally where range overlaps across central British Columbia. A third population on islands off southeast Alaska, is similar to Franklin's but with narrower white tips on tail coverts (on males).

Sooty Grouse

Dendragapus fuliginosus

L 20" WS 26" WT 2.3 lb (1,050 g) ♂>♀

Very similar to Dusky Grouse, until recently considered the same species. Male Sooty averages darker overall; where range overlaps most are distinguished by light gray tips of tail feathers (northern populations of Dusky lack gray tips), other minor differences detailed below. Females are not reliably distinguishable by sight.

VOICE: Male advertising call a series of very low, pulsing hoots *whoof whoof whoof whoof whoof whoof*, rising then falling slightly (compare Great Gray Owl), higher and louder than Dusky. Sooty audible at a quarter mile, Dusky only at a few hundred feet. In display male gives single low hoot. Both sexes give a low *gr gr gr gr* and soft clucking or barking sounds.

Uncommon. Found in open coniferous forests. Solitary. Displaying males tend to be in trees (Dusky on ground) but this probably has more to do with the available perches in chosen habitat rather than any real behavioral difference.

Dusky Grouse

Dendragapus obscurus

L 20" WS 26" WT 2.3 lb (1,050 g) ♂>♀

A very large grouse; larger than Spruce Grouse, with relatively longer tail. All plumages fairly uniform grayish or gray-brown without strong contrast. Female speckled and spotted overall, without clear barred pattern of other grouse. Male more uniform dark gray, with blackish tail.

VOICE: Male advertising call similar to Sooty but lower and quieter, audible only at close range; in display gives single low hoot like Sooty. Both sexes give a low *gr gr gr gr* and soft clucking or barking sounds like Sooty.

Uncommon. Found in open mixed forests of conifers and aspen with some clearings and shrubs in understory, including dry slopes with only scattered small trees, bordering prairie. In winter wanders to other habitats, mainly in Douglas-Fir and Lodgepole Pine forests.

Rock Ptarmigan

Lagopus muta

L 14" WS 23" WT 15 oz (420 g) ♂>♀

Plumages similar to Willow, but spring male mostly white into June, with only scattered gray-brown feathers on head and neck. Female and nonbreeding plumages distinguished from Willow by smaller bill and different habitat; nonbreeding adult male by black eyeline. Distinguished from White-tailed Ptarmigan by range and habitat, larger size, and black outer tail feathers.

VOICE: Male in display gives guttural, croaking rattle followed by quiet hiss *krrrr-Karrrrr, wsshhh* very unlike Willow. Displays mainly on ground: sliding on breast, rolling, leaping. Flight display ending with croaking call may be more common in Aleutian populations. Both sexes give clucking notes.

Uncommon in barren rocky tundra, avoiding the low thickets favored by Willow Ptarmigan, although it moves into willow thickets in winter. Habits similar to Willow Ptarmigan, but gathers in larger groups in winter.

Aleutian populations average larger with heavier bill than continental birds. They are found in grassy lowlands rather than rocky tundra, and differ markedly in breeding plumage. Display may also differ.

All ptarmigan have a complex pattern of three molts each year, and the timing and extent of these differs between males and females, as well as regionally. Molts in warmer climates are more extensive and earlier in spring, later in fall. A partial molt in spring produces the "courtship plumage" shown above, a second molt produces the mostly dark summer plumage, and a third molt (complete) in fall produces the all-white winter plumage.

Willow Ptarmigan

Lagopus lagopus

L 15" WS 24" WT 1.2 lb (550 g) ♂>♀

Like other ptarmigan all-white in winter and brown with white flight feathers in summer. Spring male distinctive, with rich rufous head and neck contrasting sharply with white body. Female and winter plumage similar to Rock Ptarmigan; note larger bill and different habitat.

VOICE: Male in display gives comical, nasal barking calls in series *goBEK goBEK goBEK, poDAYdo poDAYdo*...and a smoothly accelerating laugh. Female gives barking *dyow*; both sexes give clucking notes.

Uncommon. Found mainly on low flat tundra in and near stunted willows near tree line. Solitary in summer; forms small groups in winter.

White-tailed Ptarmigan

Lagopus leucura

L 12.5" WS 22" WT 13 oz (360 g) ♂>♀

Wings mostly white all year (like all ptarmigan, but unlike other grouse). Told from other ptarmigan by white tail (which can be hard to confirm), smaller size, details of plumage color, range and habitat.

VOICE: Male in display gives rapid clucking *pik pik pik pik piKEEA* and low, hoarse *pwirrr* while alternating fast and slow strutting; no flight display. Both sexes give clucking notes.

Uncommon and local. Found in alpine tundra, especially where exposed rocks and wet mossy ground mix. Solitary in summer, forms small flocks in winter. Males winter above treeline in stunted willow thickets, females often winter at or below treeline in taller and denser willow thickets such as along streams.

Ruffed Grouse

Bonasa umbellus

L 17" WS 22" WT 1.3 lb (580 g) ♂>♀

A relatively slender and long-tailed grouse. Identified by dark tail-band and dark bars on flanks. Rufous and gray morphs and intermediates occur throughout range.

VOICE: Male display is non-vocal, a series of accelerating, muffled thumps, produced by beating wings with increasing speed while standing, like a distant motor starting (this low-pitched drumming may be felt more than heard). Both sexes give clucking notes and higher squeal when alarmed; flush with explosive burst of wingbeats.

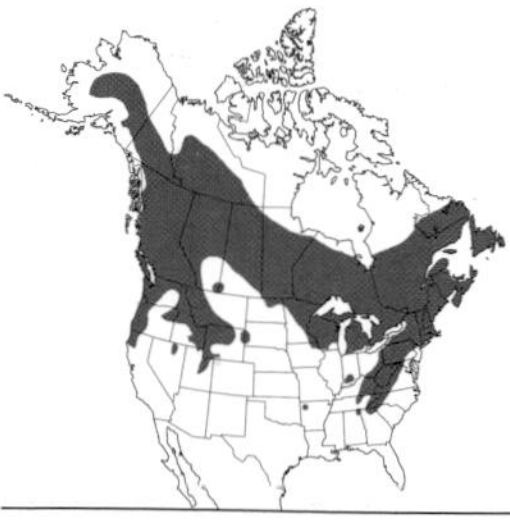

Uncommon, found in deciduous or mixed wooded areas with a diverse mosaic of clearings, dense brush, and trees, often near wet areas or second-growth as those support denser undergrowth, most species of grouse favor more open areas.

Sharp-tailed Grouse

Tympanuchus phasianellus

L 17" WS 25" WT 1.9 lb (880 g) ♂>♀

Smaller than Ring-necked Pheasant, the pale, pointed tail is obvious in flight and unlike all other grouse. Closely related to prairie-chickens, but patterned in spots instead of bars, tail longer, pointed and mostly whitish.

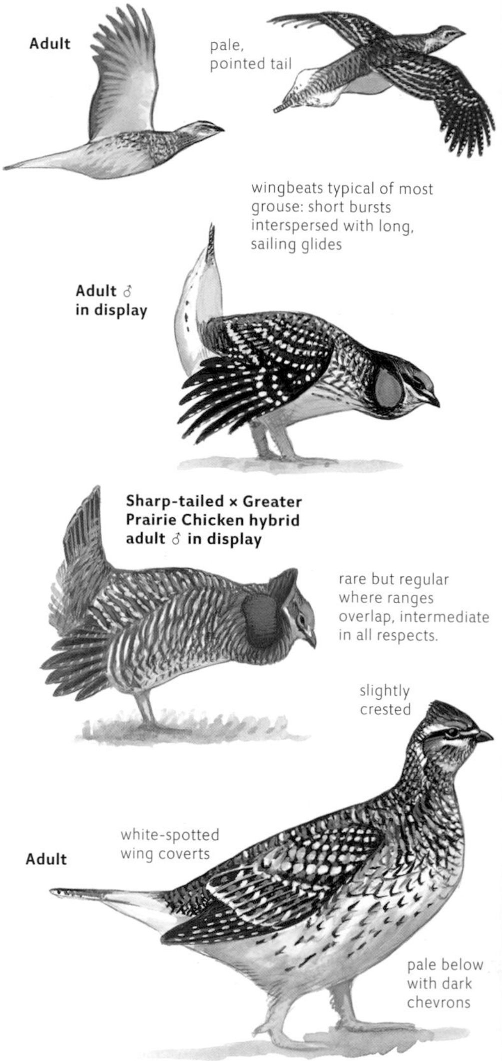

VOICE: Male display includes weird, unearthly hoots *yooowm, gyowdowdyom, gloooowm*...; higher-pitched, more varied than prairie-chickens; hoots interspersed with quacking *wek* and soft, dry chatter produced by stamping feet on ground. Both sexes give clucking notes. When flushed a cackling *kek-kek-kek-kek*.

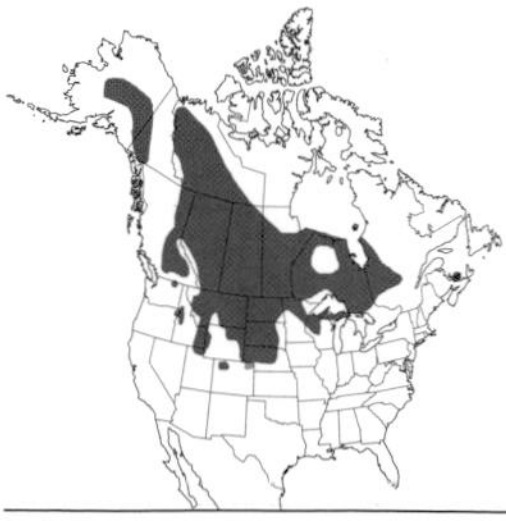

Uncommon and local. Found in grassy habitat where prairie and woodland mix, in open grasslands near patches of trees, such as aspen parklands. Usually in small groups, which flush in ones and twos rather than all together.

Greater Prairie-Chicken

Tympanuchus cupido

L 17" WS 28" WT 2 lb (900 g) ♂>♀

Densely barred dark brown, with short rounded dark tail. Told from Sharp-tailed Grouse by shorter all-dark tail, barred belly, lack of white spots on upperwing coverts. See Lesser Prairie-Chicken.

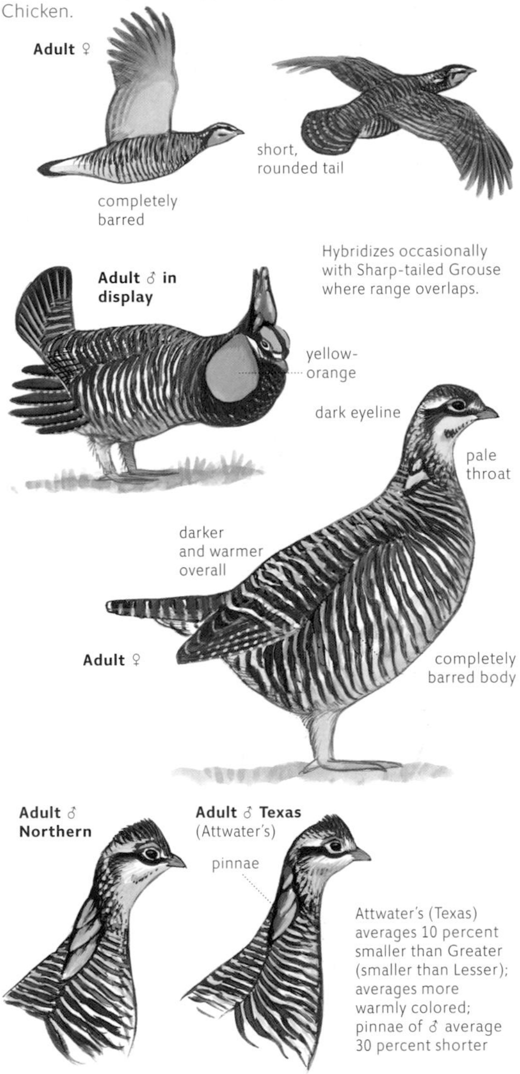

VOICE: Male display includes long, low, hooting moan *oooa-hooooooom* about two seconds long (like air blown across the top of a bottle); also high, wild clucking sounds, *hoaa* notes, foot stamping, and a *pwoik* in presence of female. Both sexes give clucking notes. Usually silent when flushed (except for loud wing noise) unlike Sharp-tailed Grouse.

Rare, local, and declining. Found in tallgrass prairie where diverse grasses mix with low shrubs and weeds. Gathers in flocks in winter (flush individually, not as a group), and at that season more likely to wander into agricultural land for food.

Lesser Prairie-Chicken

Tympanuchus pallidicinctus

L 16" WS 25" WT 1.6 lb (750 g) ♂>♀

Very similar to Greater, slightly smaller and paler overall but best distinguished by range and by male courtship display. Hybrids are known from the very small area of overlap in Kansas.

VOICE: All calls higher-pitched and shorter than Greater. Male display a bubbling, hooting *wamp wamp wodum wodum* and wild clucking in descending series; also a sharp *pike* in presence of female. Both sexes give clucking notes.

Rare, local, and declining in arid grasslands with low shrubs—sandsage, shin oak, etc. In winter forms flocks and may be found in agricultural land as well as native prairie.

Greater Sage-Grouse

Centrocercus urophasianus

♂ L 28" WS 38" WT 6.3 lb (2,900 g)
♀ L 22" WS 33" WT 3.3 lb (1,500 g)

Very large, with long pointed tail. Dark gray overall, with pale breast and black belly, in flight the white underwing coverts contrast sharply with dark flanks. Dusky Grouse has darker, rounded tail, darker underwing; Ring-necked Pheasant is more slender and differently colored.

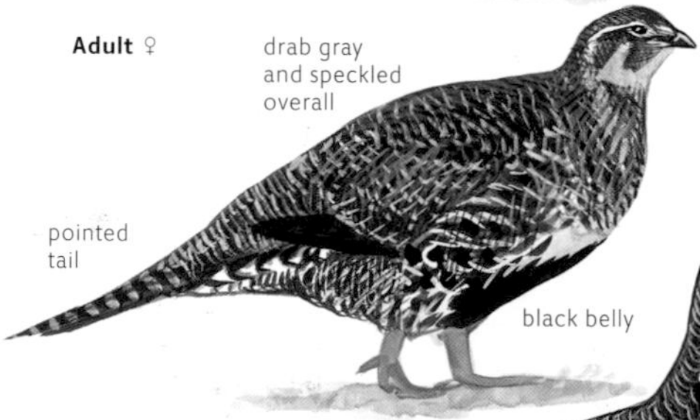

VOICE: Display of male includes two swishing sounds as wings brush against body, then two weird hooting, popping sounds *oo-WIdoo-WIdoo-wup*. Also gives guttural clucking and cackling notes. Flushes with noisy wingbeats.

Uncommon and local. Found in expansive sagebrush plains. Usually in flocks, up to hundreds together in winter.

Gunnison Sage-Grouse

Centrocercus minimus

♂ L 22" WS 30" WT 4.6 lb (2,100 g)
♀ L 18" WS 26" WT 2.4 lb (1,100 g)

Very similar to Greater, but about two-thirds as large, and range does not overlap. Male has paler whitish bars on tail. In display male has much more prominent black filoplumes, different movements and vocalizations. Females difficult to identify, differ only in size and range.

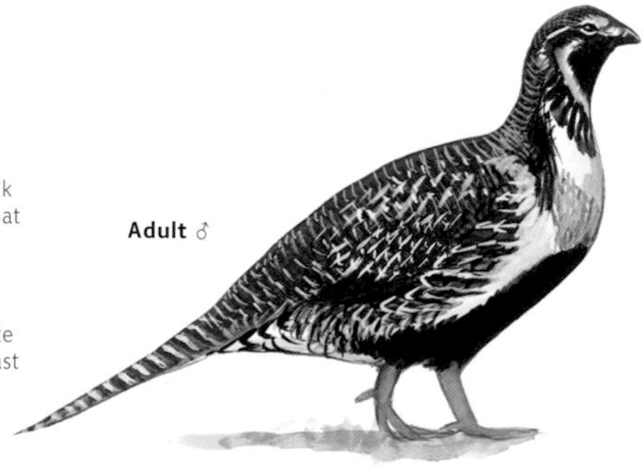

VOICE: Display of male, with nine lower-pitched hooting or popping sounds; three faint wing-swish sounds in middle of display (very little wing movement); the whole display low-pitched and monotonic. Also constantly raises thicker filoplumes above head, and its display often ends with a tail-shaking motion (with the tail still raised) absent in Greater. Year-round both sexes give guttural clucking and cackling notes.

Uncommon and local within its very limited range. Found in expansive sagebrush plains. World population of a few thousand individuals and declining.

Wild Turkey

Meleagris gallopavo

♂ L 46" WS 64" WT 16.2 lb (7,400 g)
♀ L 37" WS 50" WT 9.2 lb (4,200 g)

Huge size, dark color (appearing blackish), long legs, and tiny head make this species unmistakable. Note that young can fly when only about a week old, barely larger than quail.

VOICE: Male in display gives familiar descending gobble. Female gives loud, sharp *tuk* and slightly longer, whining *yike*…repeated in slow series. Both sexes give a variety of other soft clucks and rolling calls.

Overall color varies regionally. Widespread Eastern subspecies have dark reddish-brown tips on all feathers, broadest on the rump. Wild Turkeys in the arid Southwest have these feathers tipped pale whitish. Birds in Texas are intermediate, and a few Eastern birds have pale buffy feather tips similar to Southwestern birds.

Common and increasing. Found in open woodlands (especially oaks) with clearings or agricultural fields, including suburban neighborhoods. Most often seen walking along forest roads or foraging in open fields, often in flocks of up to 40 or more. Roosts in trees at night.

Loons and Grebes

FAMILIES: GAVIIDAE AND PODICIPEDIDAE

LOONS: 5 species in 1 genus. Loons are medium to large birds that eat mainly small fish captured by diving and pursuing underwater. All nest on banks of lakes or ponds and winter on large bodies of open water. Compared to ducks, loons are longer-bodied, with legs set far back on the body, smoothly-curved neck, and straight, dagger-like bill. Distinguishing species of loons requires careful attention to overall size, head and neck pattern, and bill shape. All species may hold their bill angled up, but certain species do so habitually. Nonbreeding adults are shown.

Genus *Gavia*

Red-throated Loon, page 59

Pacific Loon, page 58

Arctic Loon, page 58

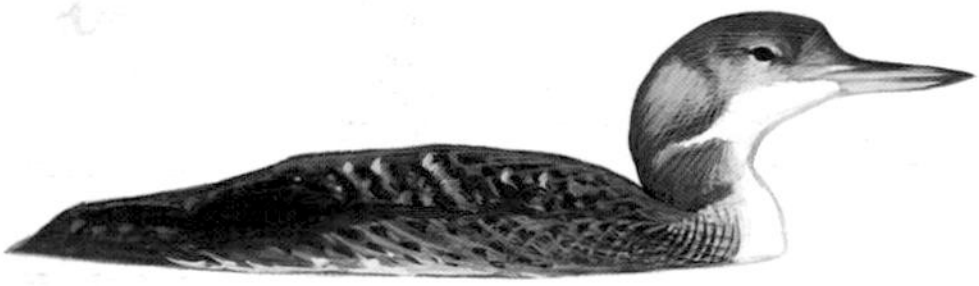

Common Loon, page 57

Yellow-billed Loon, page 57

GREBES: 7 species in 4 genera. Superficially similar to loons grebes are smaller, with lobed toes and relatively longer necks Their insignificant tail feathers are invisible among fluffy tai coverts. Grebes fly less often and less strongly than loons. They forage by diving for small aquatic animals. All build floating nest of matted plant material on shallow ponds or lakes, and winter on open water. Seasonal changes are relatively small (only the species in the genus *Podiceps* acquire a very different breeding plumage) and age differences are minimal. All have striped heads briefly in juvenal plumage, but after that it is generally impossible to determine the age of a grebe in the field. The fundamenta differences between loons and grebes have long been recognized, and recent DNA studies confirm that these two families are not closely related. Research suggests that Loons are more closely related to shearwaters, cormorants, herons, etc., while grebes are more closely related to flamingos and pigeons! (See Bird Families, page 256).

Compare loons, cormorants, ducks (especially mergansers), and alcids. Nonbreeding adults are shown.

Genus *Tachybaptus*

Least Grebe, page 63

Genus *Podilymbus*

Pied-billed Grebe, page 63

Genus *Podiceps*

Horned Grebe, page 60

Eared Grebe, page 60

Red-necked Grebe, page 61

Genus *Aechmophorus*

Western Grebe, page 62

Clark's Grebe, page 62

Common Loon

Gavia immer

L 32" WS 46" WT 9 lb (4,100 g) ♂>♀

The second-largest loon (only Yellow-billed is larger), much heavier and heavier-billed than Pacific and Red-throated, although size is not always apparent. In nonbreeding plumages note jagged border between light and dark on neck, white around eye. In breeding plumage note blackish head and bill, white checkering on back.

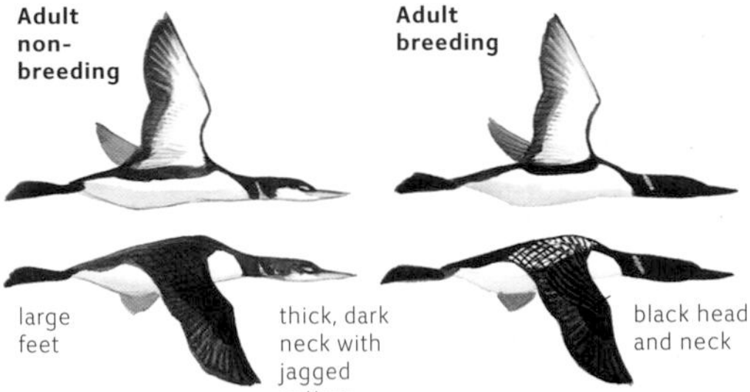

Juvenile (Aug–Feb)

1st summer (Mar–Sep)

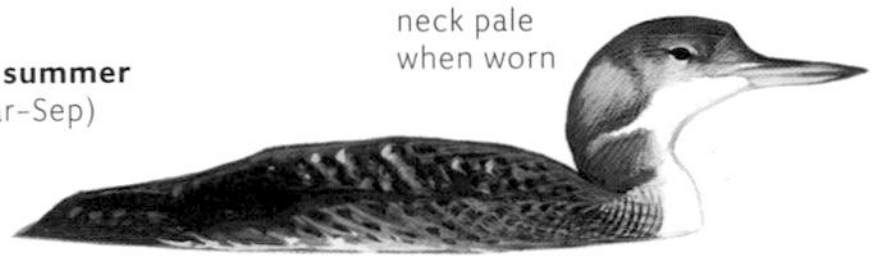

Adult nonbreeding (Sep–Mar)

Adult breeding (Mar–Oct)

VOICE: Low, melancholy yodeling or wailing cries. Tremolo of five to ten notes on even pitch *hahahahahaha* heard year-round, often in flight; sometimes a short *kuk* or *gek* in flight. In summer an undulating *whe-ooo quee* and rising wail *hoooo-lii*.

Generally uncommon. Nests on deep clear forest lakes large enough to provide fish for a family of loons. Winters on open water of large clear lakes, open ocean and bays. On most fresh water and along the Gulf Coast this is by far the most commonly seen loon species.

Yellow-billed Loon

Gavia adamsii

L 35" WS 49" WT 11.8 lb (5,400 g) ♂>♀

Very similar to Common Loon but larger, breeding plumage with more white on back, nonbreeding and juvenile plumages paler overall. Best distinguished from Common by details of bill: culmen straight so bill appears upturned, and tip of bill entirely pale whitish.

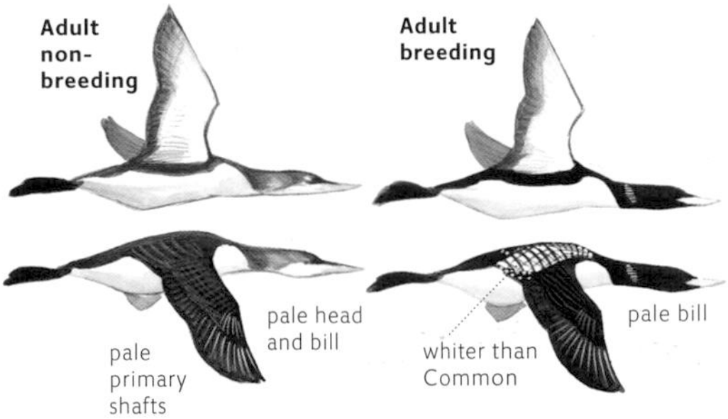

Juvenile (Aug–Apr)

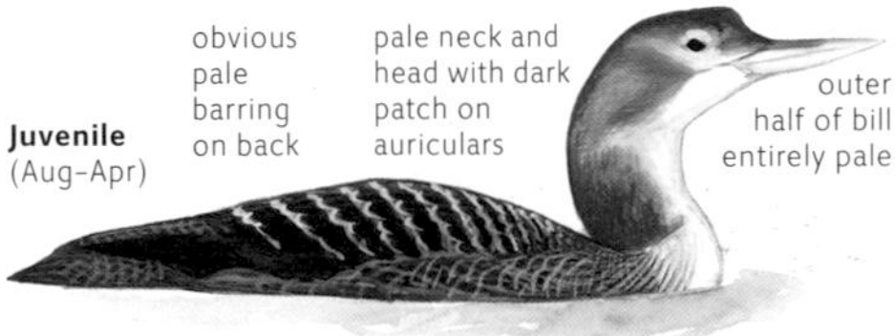

1st summer (May–Sep)

Adult nonbreeding (Sep–Mar)

Adult breeding (Mar–Oct)

VOICE: Voice similar to Common but lower-pitched, harsher, delivered more slowly.

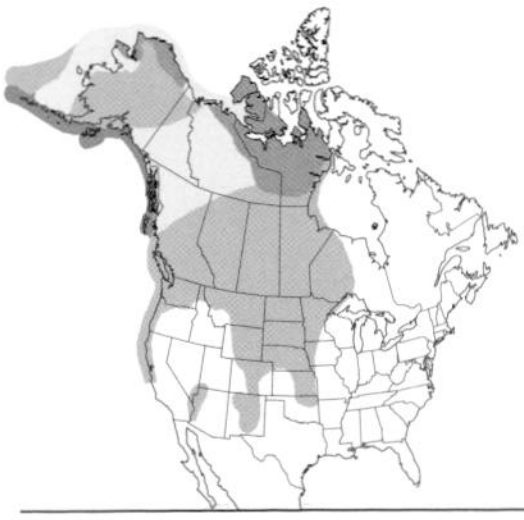

Uncommon to rare, and always in very small numbers. Expected only in Arctic Canada and on the coasts of Alaska and British Columbia, very rare elsewhere. Nests on relatively large and deep tundra lakes. In winter found on open salt water: ocean, bays and inlets; very rarely inland on deep clear lakes.

Arctic Loon

Gavia arctica

L 27" WS 40" WT 5.7 lb (2,600 g) ♂>♀

Nearly identical to Pacific Loon, averages slightly larger with heavier bill; best distinguished by white patch on flanks, which varies with posture but is always present. Common and Red-throated occasionally show whitish flanks, and any loon may lean to expose white belly, often when preening.

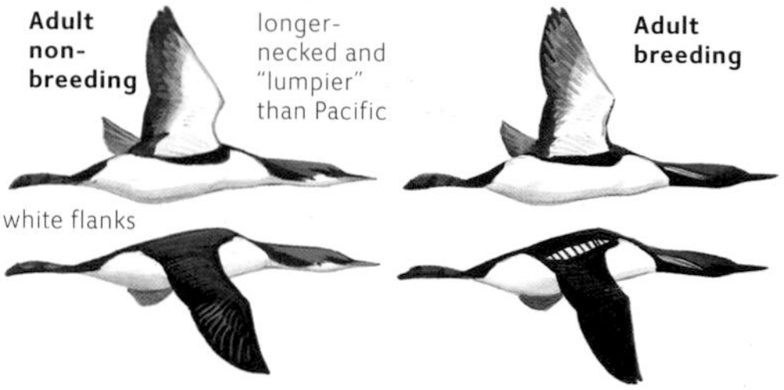

VOICE: Calls mainly on breeding grounds; similar to Pacific but deeper. Long call a slurred *owiiil-ka owiiil-ka owiiil-ka*; lower-pitched, less strident, simpler than Pacific. Also a raven-like *kraaw* and a muffled *aahaa* like Canada Goose.

Rare and local breeder on tundra ponds near coast in western Alaska. Very rare visitor (mainly juveniles) to open ocean and bays along Pacific coast, even rarer inland on large lakes.

Pacific Loon

Gavia pacifica

L 25" WS 36" WT 3.7 lb (1,700 g) ♂>♀

Smaller than Common Loon, slightly stockier than Red-throated. In nonbreeding plumage note grayish nape with unbroken straight border, all dark around eye. In breeding plumage note pale gray nape, black throat, and white patches on back.

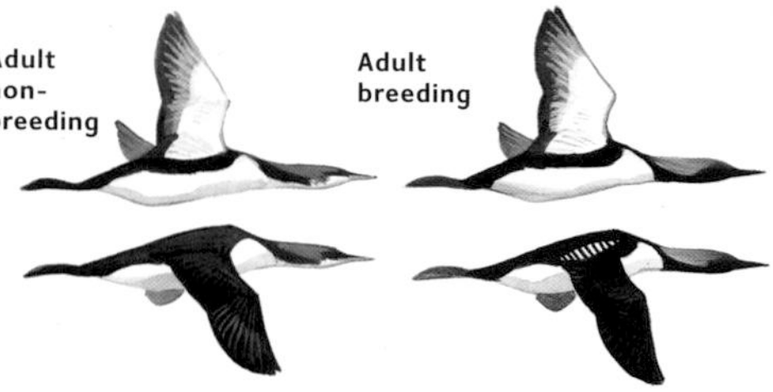

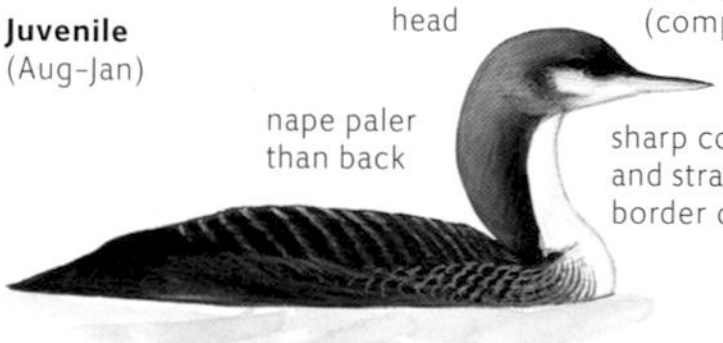

VOICE: Calls mainly on breeding grounds, giving plaintive, yodelling *o-lo-lee*. Long call a mournful but rather high and strident *ooaLEE-kow, ooaLEE-kow, ooaLEE-kow*. Also a raven-like *kowk* and other growls and croaks.

Uncommon to locally common. Nests on relatively large and deep tundra ponds. In winter and migration found on deep, clear, open salt water, often farther from land than other loons. Outside of breeding range very rare on fresh water.

Red-throated Loon

Gavia stellata

L 25" WS 36" WT 3.1 lb (1,400 g) ♂>♀

Our smallest and most slender loon, distinguished from grebes by longer body and relatively short neck. Slender bill usually held angled up. In breeding plumage note all-dark neck and back, in nonbreeding note usually extensive white on face and neck.

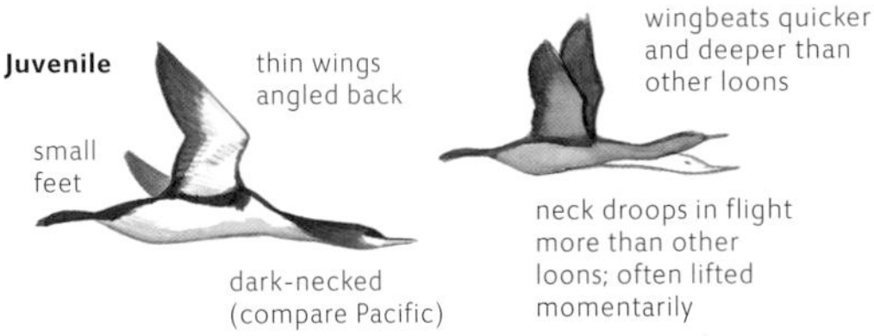

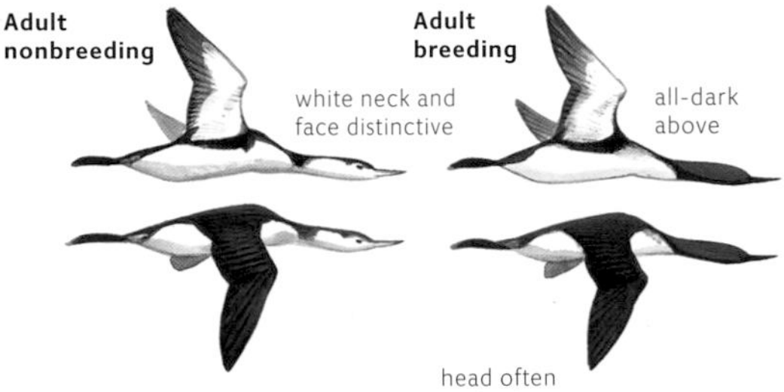

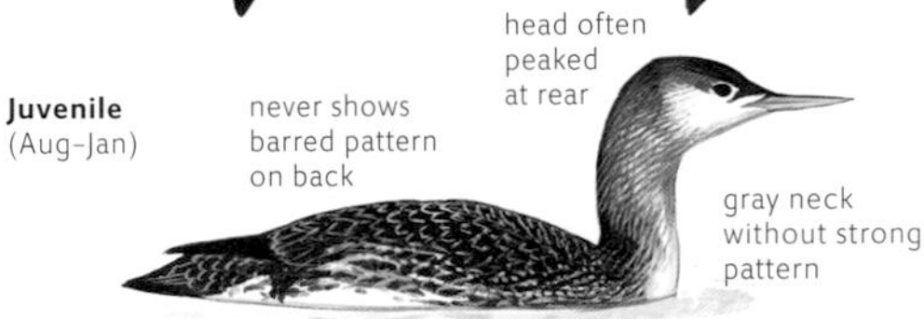

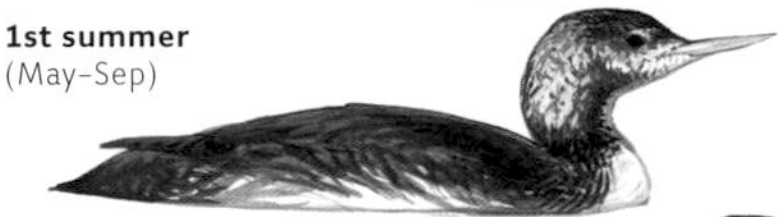

VOICE: Drawn-out, gull-like wailing or shrieking *aarOOoa, aarOOoa*…; less rhythmic than Red-necked Grebe. Adult flying over nesting territories gives short, nasal quacks *bek, bek, bek*…, sometimes building to cackling *kark kark kark karkarak karkarak*.

Uncommon to locally common. Nests on relatively small and shallow tundra ponds. In winter and migration found on deep, clear, unobstructed open water, primarily nearshore ocean waters and less commonly on bays, often in loose flocks. Common on the Great Lakes but generally rare on all other fresh water.

Loon Habits

Among many things that set loons apart from ducks is their large feet with legs set very far back on the body. This makes their feet very prominent in flight, and nearly useless for landing and walking.

Loons migrate singly or in loose, widely-spaced groups, never in organized flocks. They often fly high, up to 100 feet over water or hundreds of feet over land, with strong, steady, continuous wingbeats.

Loons have very large feet (and short tails) so that the rear end of their silhouette in flight is formed by the feet trailing behind.

When landing on water, loons make a very gradual approach and slide to a stop, with wings raised.

Loons walk rarely and with difficulty, usually coming out of the water only when nesting, and just sliding on their belly as they push with their feet

Eared Grebe

Podiceps nigricollis

L 13" WS 16" WT 11 oz (300 g)

Small grebe with delicate proportions, similar to Horned but note smaller head, thinner neck, upturned bill, usually more rounded back. Some individual Horned can be very similar to Eared, especially when molting to breeding plumage in Mar–Apr.

VOICE: Essentially silent except on breeding grounds. Generally quieter and less harsh than Horned. Song a high, rising, squeaky whistle *ooEEK* or *ooEEKa* repeated; reminiscent of Sora but shorter, weaker. Other calls on breeding grounds mainly shrill, chittering series usually ending in upslur *hik*.

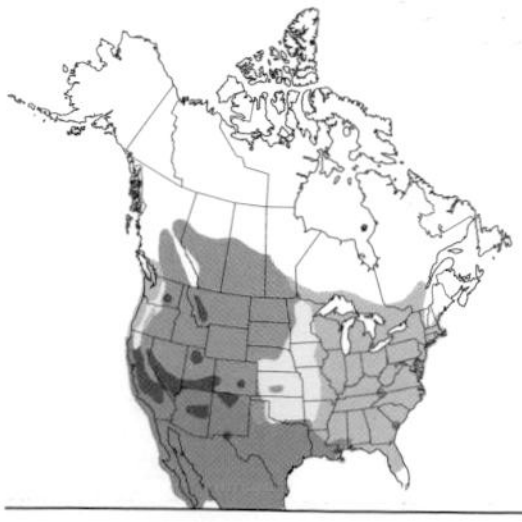

Common. Breeds on shallow ponds and lakes with emergent reeds. Forages and winters on open water, from shallow small ponds to open ocean.

Horned Grebe

Podiceps auritus

L 14" WS 18" WT 1 lb (450 g)

Small grebe, similar to Eared but differs in head shape, bill shape and head pattern. Some Eared are confusingly similar, especially juveniles in fall which can appear white-cheeked like Horned.

VOICE: Essentially silent except on breeding grounds. Song trilling, usually in duet, with rising and falling pulses of sound: squeaky, giggling, nasal. Drier chatter in alarm. Common call in summer a whining, nasal *way-urrr* or *ja-orrrr* descending, ending in throaty rattle, repeated. High, thin notes sometimes heard in winter.

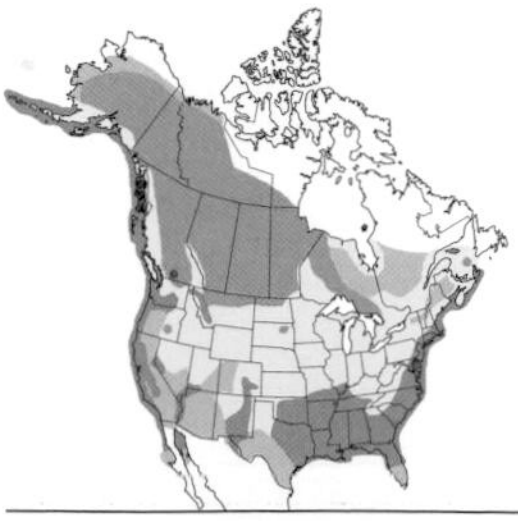

Breeds on ponds with some emergent vegetation. Winters on open water on larger lakes or open ocean, rarely on small ponds.

Red-necked Grebe

Podiceps grisegena

L 18" WS 24" WT 2.2 lb (1,000 g)

A relatively large grebe; distinguished from Horned Grebe by larger size, larger yellowish bill often angled slightly down, and (in winter) dull brownish neck with pale ear-patch.

Adult nonbreeding

Adult breeding

Juvenile (Jul–Oct)

1st winter (Sep–Mar)

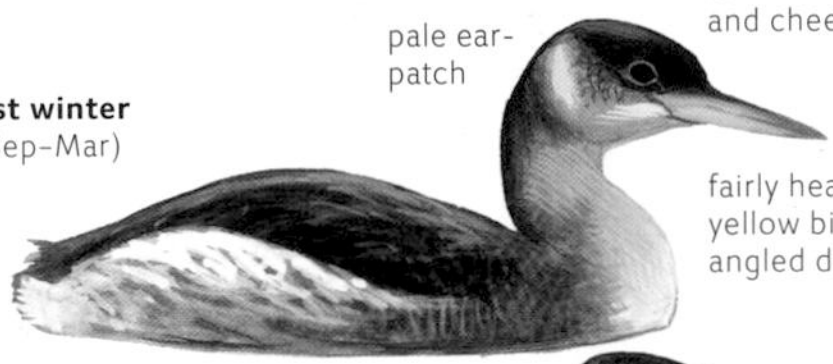

Adult nonbreeding (Sep–Mar)

Adult breeding (Feb–Aug)

VOICE: Essentially silent except on breeding grounds. Song a loud series, beginning with nasal, gull-like quality, then braying chatter with quavering end. Male calls higher-pitched than female. Also a loud, nasal quacking series *ga-ga-ga-ga*.... Low, grunting notes sometimes heard in winter.

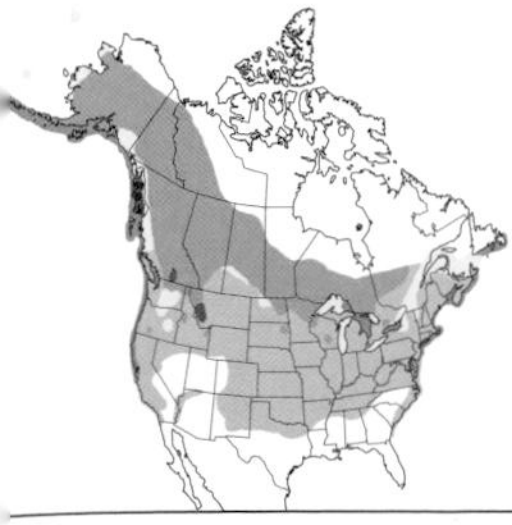

Uncommon. Breeds on ponds with some emergent marshy vegetation. Winters on open water on larger lakes or (mostly) open ocean, rarely seen on small ponds.

Identification of Grebes

The three grebes in the genus *Podiceps* can be difficult to tell apart in nonbreeding plumage. Red-necked is significantly larger, but size is hard to judge at a distance on open water. The pattern and shape of the head from behind provides useful clues. Horned generally shows the most white, and Eared has a relatively small and rounded head.

During molt from nonbreeding to breeding plumage both species go through confusing transitional stage in which head pattern can resemble the other species. Use extra caution and rely more on head and bill shape at this season.

Grebes sleep with neck laid on back and bill pointed forward.

Young grebes (like loons) ride on their parents' backs.

Grebes walk only rarely and only a few steps at a time.

Western Grebe

Aechmophorus occidentalis

L 25" WS 24" WT 3.3 lb (1,500 g)

Our largest grebe, with long graceful neck and long straight bill. Clean white neck and narrow dark stripe down back of neck distinctive. See Clark's Grebe.

VOICE: Gives high, creaking, far-carrying calls. Most common year-round call a two-part *kreed-kreet* (Clark's gives long single-syllable call). Other calls similar to Clark's. Courting female gives long series of begging notes: *krDEE krDEE*.... Various other calls on breeding grounds include a high, thin whistle.

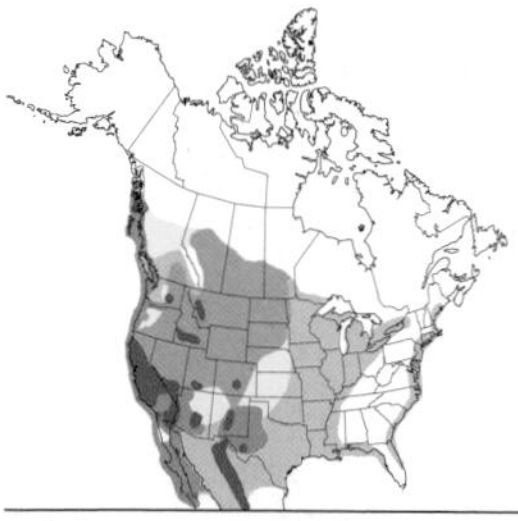

Uncommon but found locally in large numbers. Nests on shallow lakes with emergent vegetation, forages and winters on deeper open water of nearshore ocean, bays, and large lakes.

Clark's Grebe

Aechmophorus clarkii

L 25" WS 24" WT 3.1 lb (1,400 g)

Structurally identical to Western but averages fractionall smaller. Candidates are most easily picked out among Western by lighter color—paler flanks and more white on neck and face—then confirmed by cleaner yellow-orange bill color. Occasiona intermediates cannot be identified with certainty.

VOICE: Similar to Western, but common year-round call a drawn-out single syllable *kreeeed* or *kreee-eed* (Western gives two-parted call). Courting female gives high, scratchy *kweec kweea*...or *weeweeweewee*....

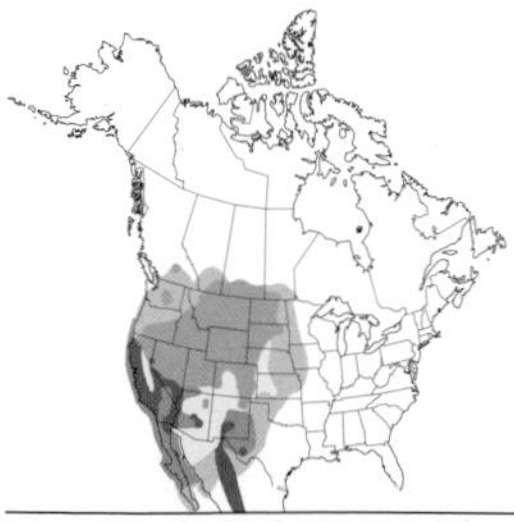

Uncommon. Habitat and habits similar to Western Grebe, and usually found together in mixed flocks. Clark's more common southwards, but even there is usually outnumbered by Western.

east Grebe

achybaptus dominicus

9.5" WS 11" WT 4 oz (115 g)

ur smallest grebe, much smaller than Pied-billed and overall arker and grayer, with relatively small dark bill and bright yellow ye. Usually hiding in vegetation.

Adult nonbreeding **Adult breeding**

Juvenile
(Jun–Oct)

Adult nonbreeding
(Sep–Mar)

Adult breeding
(Feb–Oct)

OICE: Quite vocal year round. Loud, ringing, nasal *beep* or *teeen* nd nasal, whining *verr*. Greeting call a rapid, nasal, buzzy chat-er descending *vvvvvvvvvvvvv* similar to Pied-billed call but igher-pitched, much more rapid and buzzing. No sound anal-gous to song of Pied-billed.

Uncommon and local, found on still water of shallow ponds with reeds, lily pads, and other plants, rarely venturing far from emergent vegetation. Tolerates smaller ponds, shallower water, and thicker vegetation than any other grebe, but habitat overlaps broadly with Pied-billed.

Pied-billed Grebe

Podilymbus podiceps

L 13" WS 16" WT 1 lb (450 g) ♂>♀

The most widespread grebe, found on small marshy ponds and waterways across most of the continent. Stout bill and overall brownish color is distinctive; never shows contrasting dark and light pattern on head and neck.

Adult non-breeding **Adult breeding**

Juvenile
(Apr–Oct)

Adult nonbreeding
(Sep–Mar)

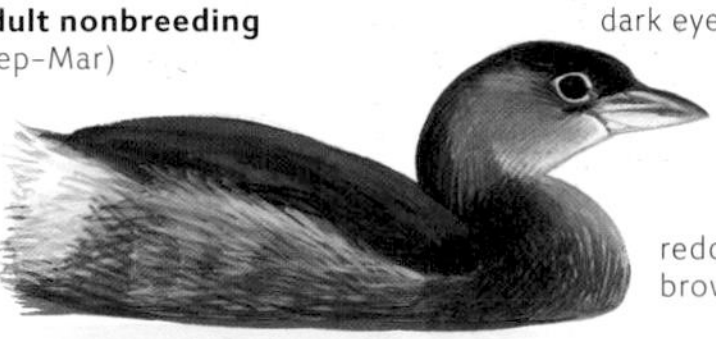

Stockier overall than Horned and Eared Grebes, with thicker bill, brownish neck, breast, and flanks.

Adult breeding
(Feb–Sep)

VOICE: Essentially silent except on breeding grounds. Song far-carrying, vibrant, throaty barks *ge ge gadum gadum gadum gaum gaom gwaaaaaow gwaaaaaaow gaom* given by male and female. Female also gives low grunting notes. Greeting call a drawn-out, nasal chatter slightly descending and fading away.

Uncommon and local. Nests in shallow ponds and marshes with emergent vegetation, in winter found on a wider range of water, including open ponds and sheltered salt water. In many inland areas the most frequently seen species of grebe.

Albatrosses, Petrels, and Shearwaters

FAMILIES: DIOMEDEIDAE, PROCELLARIIDAE

31 species in 6 genera in 2 families (7 rare species not shown here). Albatrosses in Diomedeidae, shearwaters and petrels in Procellariidae. All are aerial seabirds with relatively long, narrow, and stiff wings, usually found far offshore over open ocean, coming to land only to nest. All fly with stiff wing-beats and long, arcing glides. They feed on small animals and carrion found at or near the water's surface, sometimes making shallow dives underwater, and gathering in large numbers where food is abundant. All nest on islands, and most are only visitors to North America, nesting elsewhere. Adults are shown.

Genus *Phoebastria*

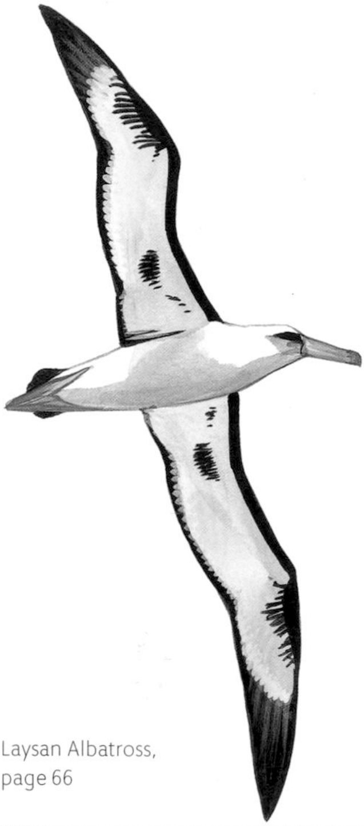

Laysan Albatross, page 66

Black-footed Albatross, page 66

Short-tailed Albatross, page 67

Genus *Pterodroma*

Herald Petrel, page 74

Bermuda Petrel, page 73

Fea's Petrel, page 73

Black-capped Petrel, page 72

Murphy's Petrel, page 71

Mottled Petrel, page 70

Hawaiian Petrel, page 71

Cook's Petrel, page 70

;enus *Puffinus*

ink-footed
hearwater, page 77

Flesh-footed
Shearwater, page 78

Great Shearwater,
page 76

Wedge-tailed
Shearwater, page 82

Buller's Shearwater,
page 77

ooty Shearwater,
age 79

Short-tailed
Shearwater, page 79

Manx Shearwater,
page 80

Black-vented
Shearwater, page 80

Audubon's Shearwater,
page 81

;enus *Fulmarus*

Jorthern Fulmar,
age 69

Genus *Calonectris*

Streaked Shearwater,
page 82

Cory's Shearwater,
page 76

Black-footed Albatross

Phoebastria nigripes

L 31" WS 83" WT 7 lb (3,200 g) ♂>♀

Extremely large, comparable to Brown Pelican but with very long, slender wings held stiff and straight in flight, shearwater-like. Distinguished from other albatrosses by dark color with white face and usually whitish tail coverts.

VOICE: Noisy in groups at sea: groaning or squealing noises, also bill-snapping. Courtship display at nest site involves melancholy groaning and bill-snapping in a simple dance.

Uncommon visitor from nesting grounds in central Pacific to open ocean, mainly from December to July. Usually solitary, but may gather in feeding groups. Feeds mainly on squid.

Laysan Albatross

Phoebastria immutabilis

L 30" WS 79" WT 6.8 lb (3,100 g) ♂>♀

Habits and size similar to Black-footed Albatross, but with clea white head and underparts. May recall Western Gull, but not different flight style, much longer wings with dark borders o underwing, dark tail, and large pinkish bill. Dusky smudge aroun eye is surprisingly obvious.

VOICE: Voice quieter, higher, and less harsh and nasal than Black footed; otherwise similar. Courtship dance at nest site mor elaborate and with slower tempo than Black-footed.

Uncommon but increasing visito from nesting grounds in central Pacific (and recently nesting off Mexico) to cold open ocean waters far offshore; most numerous off Alaska.

Short-tailed Albatross

hoebastria albatrus

34" WS 89" WT 12.8 lb (5,800 g)

arger than other albatrosses, with relatively broad wings and assive pink bill. Juvenile dark brown overall, with contrasting ink bill. Soon develops white patches on upperwing, and gradally acquires mostly white adult plumage over 12 or more years.

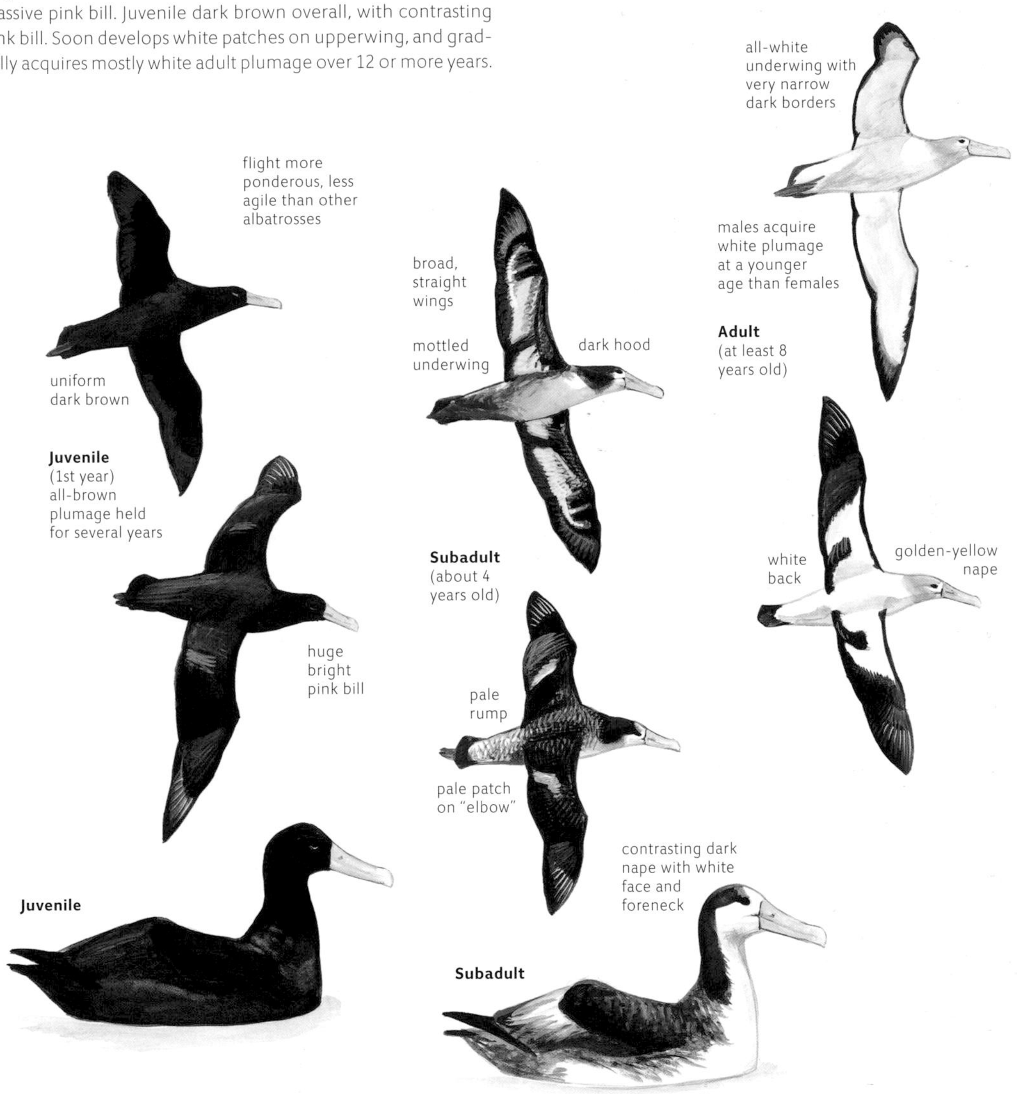

OICE: Generally silent at sea. Groaning sounds in display.

Found on cold open ocean waters. Uncommon visitor from nesting islands off Japan to open ocean near Aleutian Islands; very rare visitor to open ocean farther south. World's population only about 2,000 individuals, increasing.

Yellow-nosed Albatross

Thalassarche chlororhynchos

L 30" WS 79" WT 4.8 lb (2,200 g)

Relatively small and slender for an albatross, with mostly black bill and narrow and neat dark borders on mostly white underwing. Can be confused with Great Black-backed Gull or Northern Gannet, but note short dark tail, dark eye patch, wing shape and bill shape.

VOICE: Usually silent at sea.

Very rare visitor from southern oceans to Atlantic coast. Found on open ocean and normally pelagic, but many records are of birds seen from shore, a few even resting on land or some distance inland.

Black-browed Albatross

Thalassarche melanophris

L 33" WS 86" WT 8.7 lb (3.950 g)

A small albatross, similar to Yellow-nosed but slightly large thicker-necked, and bulkier, with broader and messier dark bo ders on underwing. Adult has orange bill, immature dusky.

VOICE: Usually silent at sea.

Very rare visitor from Southern Hemisphere to open ocean off Atlantic coast. Several confirme records from North Carolina to Newfoundland, once in Northwest Territories.

Shy Albatross

Thalassarche cauta

L 37" WS 95" WT 8.6 lb (3,900 g)

A large and heavy albatross, larger than Laysan, with almos entirely clean white underwing and greenish-gray bill. Also not paler gray back and tail than Laysan.

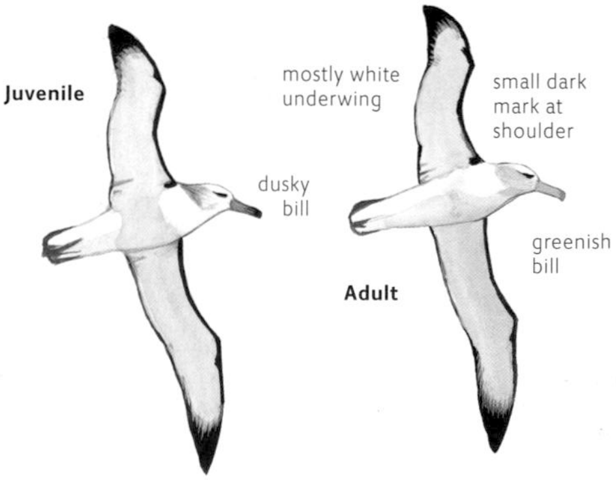

VOICE: Usually silent at sea.

Very rare visitor to open ocean off Pacific Coast. North America records are of three different subspecies, sometimes considered three species.

Northern Fulmar

ılmarus glacialis

18" WS 42" WT 1.5 lb (680 g) ♂>♀

ocky and thick-necked compared to shearwaters, with stout ·llowish bill. Plumage varies from mostly white with gray upper-ing to all-gray. Light morph superficially gull-like but note iff-winged flight, pale inner primaries, dark smudge around /e, and short gray tail.

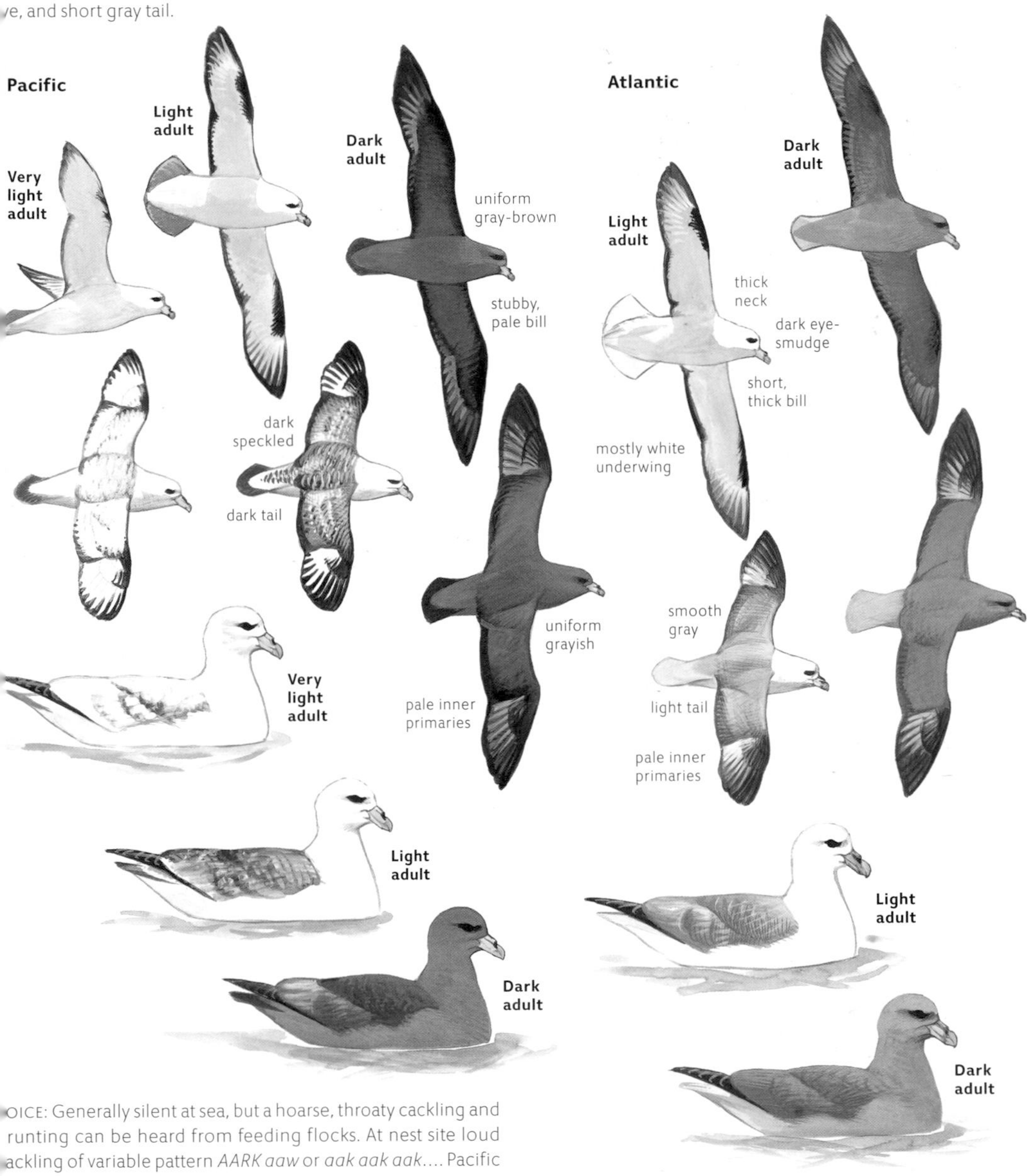

OICE: Generally silent at sea, but a hoarse, throaty cackling and runting can be heard from feeding flocks. At nest site loud ackling of variable pattern *AARK aaw* or *aak aak aak*.... Pacific pparently similar to Atlantic.

Common in far northern waters; uncommon to rare and irregular southward. Found on open ocean, usually on cold water. Nests on narrow ledges on sea cliffs, often alongside kittiwakes and murres. Often gathers in large groups around food sources and roosts in flocks on the water. Feeds mainly on fish and carrion and often follows fishing boats.

Variable, with two color morphs (and intermediates) and distinct Atlantic and Pacific populations. Pacific birds are similar to Atlantic in shape and habits but average a little thinner-billed. Best distinguished by tail color: always contrastingly dark on Pacific but pale gray (like the rump) on Atlantic. Pacific also has lighter and darker extremes than Atlantic, and the upperparts are usually more mottled (less smoothly colored). Percentage of dark and light morph birds varies between colonies, dark generally most numerous in Aleutian and Labrador breeders.

Cook's Petrel

Pterodroma cookii

L 13" WS 31" WT 6.5 oz (185 g)

Small and dainty for a *Pterodroma* petrel, with relatively quick, agile, and erratic flight. Pale gray back contrasting with darker wing coverts and almost entirely clean white underside distinctive.

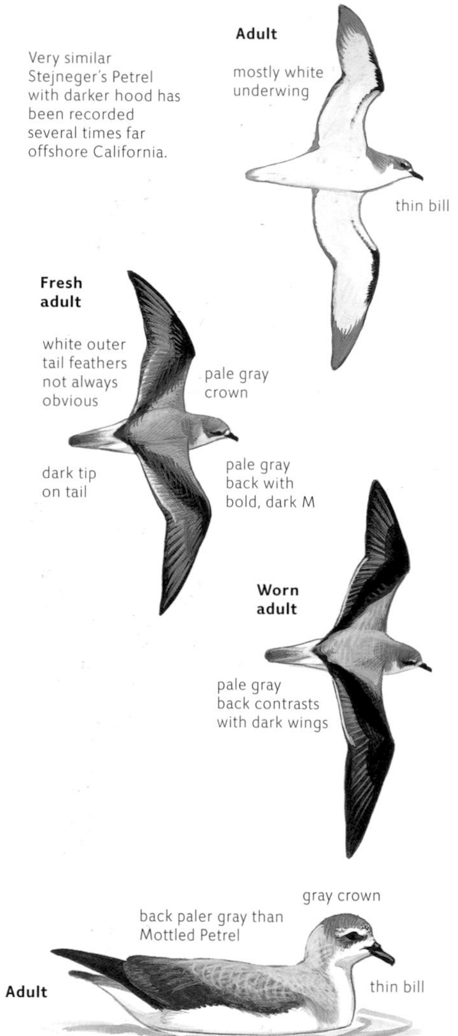

VOICE: Generally silent at sea.

Rare visitor from nesting grounds around New Zealand to warmer ocean waters far offshore; not recorded annually, but sometimes seen in numbers when present. Usually solitary. Feeds mainly on squid.

Mottled Petrel

Pterodroma inexpectata

L 14" WS 34" WT 11 oz (320 g)

A very strong flier, with high arcing flight typical of *Pterodron* petrels. Distinguished from all other Pacific tubenoses by da gray belly contrasting with white vent and extensively whi underwing, and by bold black bar on underwing coverts.

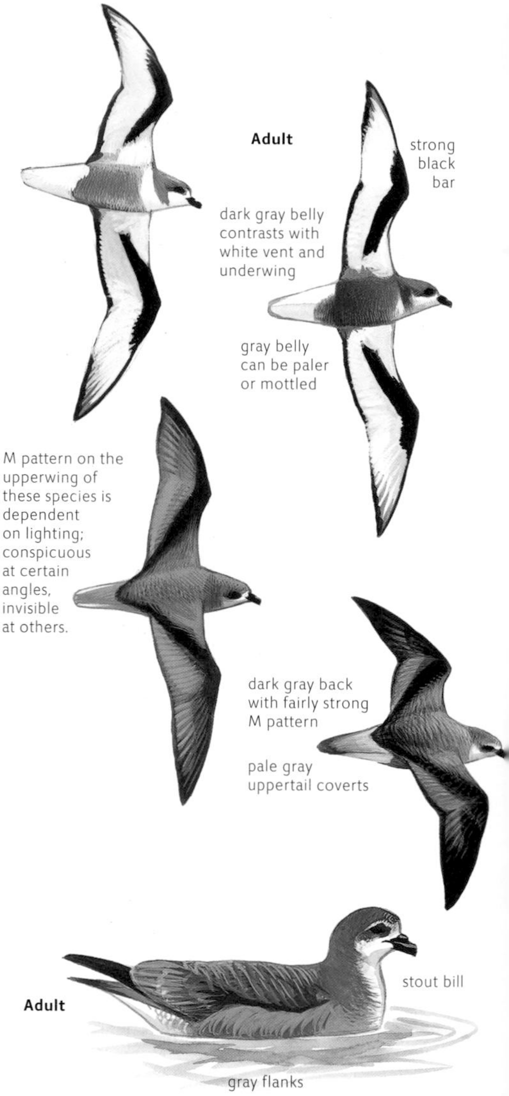

VOICE: Generally silent at sea.

Rare visitor from nesting ground around New Zealand to cold ocean waters far offshore, mainl in Alaska. Usually solitary; does not mix with other tubenoses. Feeds on small fish and squid.

/lurphy's Petrel

'terodroma ultima

14" WS 35" WT 13 oz (360 g)

l-dark plumage similar to shearwaters and jaegers, but differs flight style (angled wings and high arcs), short dark bill, faint pattern above, pale chin, and mostly dark underwing with pale ash on underside of primaries.

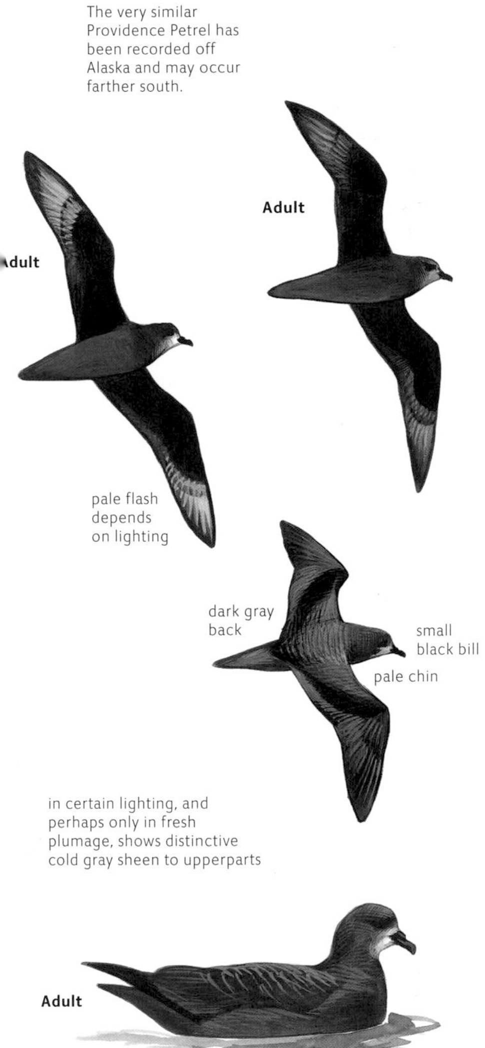

OICE: Generally silent at sea.

Rare and irregular visitor from nesting grounds in South Pacific to cold ocean waters far offshore, mainly from May to June. Usually solitary. Feeds on squid and fish.

Hawaiian Petrel

Pterodroma sandwichensis

L 15" WS 39" WT 15.5 oz (435 g)

Long-winged and slender, with long pointed tail. More slender than shearwaters, with fast flight and high arcing path. Also note uniform dark upperside, clean white underparts, mostly white underwing with bold dark bar on leading edge, and dark hood.

VOICE: Generally silent at sea.

Rare visitor from nesting grounds in Hawaii to open ocean far offshore California and Oregon. Usually solitary.

Black-capped Petrel

Pterodroma hasitata

L 16" WS 37" WT 16.5 oz (470 g)

Most similar to Great Shearwater, but with much more obvious white uppertail coverts (often looking white-tailed), white collar (usually obvious), short bill, and dark bar on underwing. Like other *Pterodroma* species, holds wings more angled than shearwaters and arcs high above the water even in light winds.

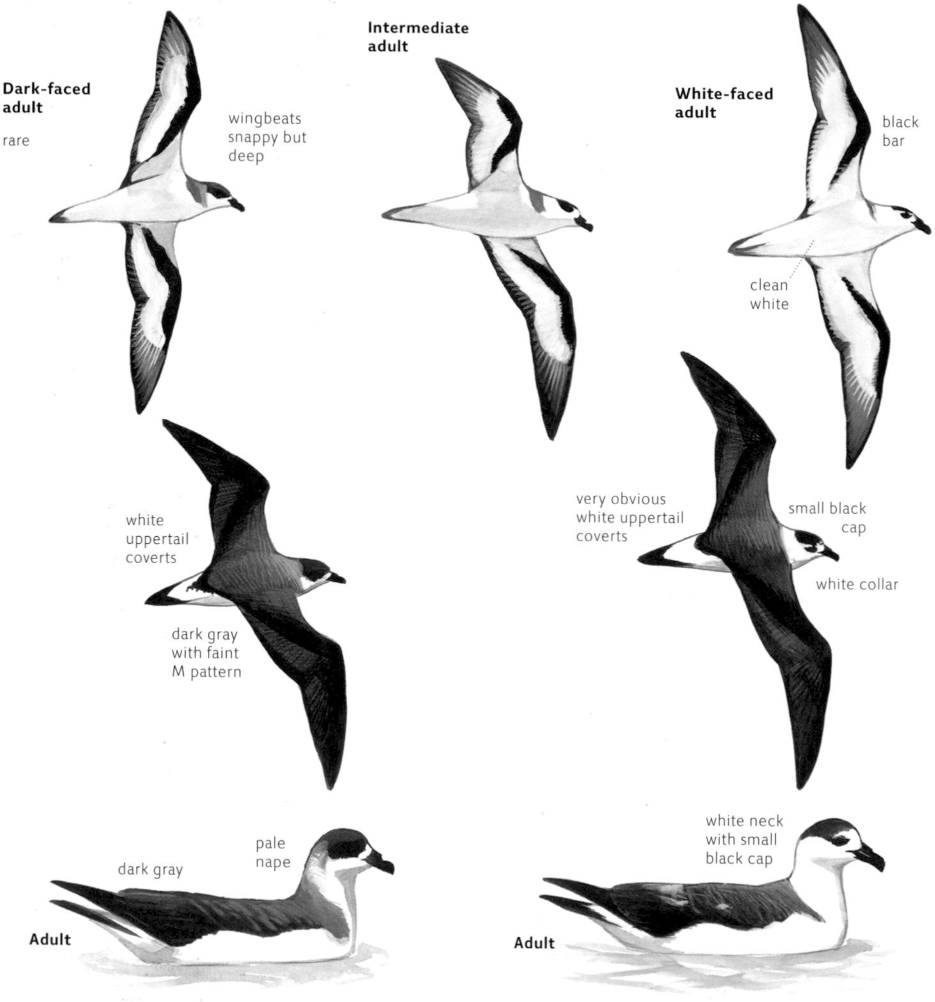

VOICE: Generally silent at sea.

Uncommon visitor from nesting grounds in Caribbean to warm Gulf Stream waters. Often seen flying singly, but forms groups to roost and often to feed.

At least two forms presumably representing different breeding populations differ subtly in head pattern, bill size and underwing pattern. White-faced birds are slightly thicker-billed, with limited dark on crown and neck, and average less black on underwing coverts. Dark-faced birds have smaller bill and more extensive dark on face, neck, and underwing coverts. White-faced is less numerous overall, but predominates in spring, while dark-faced birds are more numerous and predominate from July to December.

ʒermuda Petrel

terodroma cahow

14" WS 35" WT 8.5 oz (240 g)

ery similar to Black-capped Petrel, but averages slightly smaller, ·ith smaller bill and relatively longer wings, leading to more uoyant flight. Also note dark nape and narrower white rump and (with dark-tipped uppertail coverts), averages more dark n underwing. Best distinguished by combination of plumage nd shape.

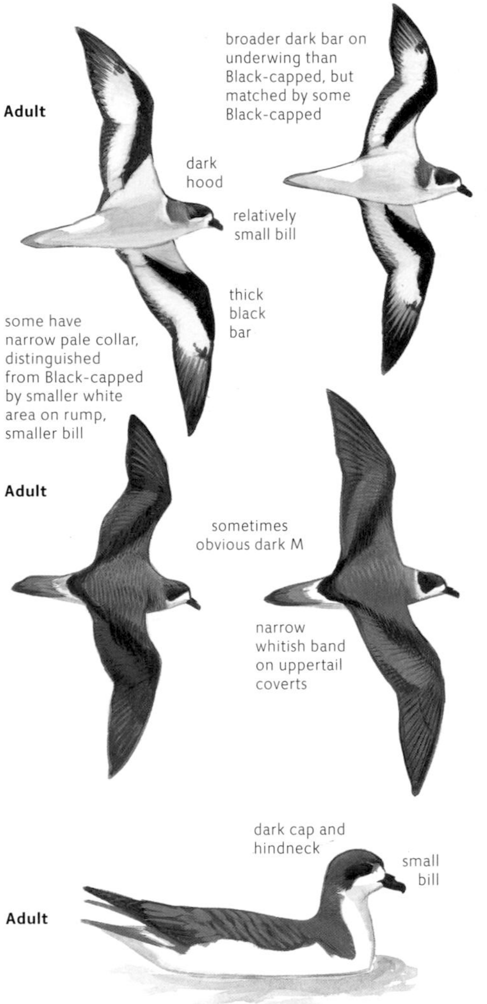

OICE: Generally silent at sea.

Very rare visitor from Bermuda to warm waters off mid Atlantic coast, recorded annually in recent years. Current world population estimated to be about 200 individuals.

Fea's Petrel

Pterodroma feae

L 15" WS 35" WT 11 oz (310 g)

Smaller than Black-capped Petrel, with mostly dark underwing, broad gray smudges on sides of neck, paler gray head and back and, especially, uniform pale gray uppertail coverts and tail.

The very similar Zino's Petrel is slightly smaller, with relatively thinner bill, and averages more white on underwing coverts. It has been recorded at least once off North Carolina.

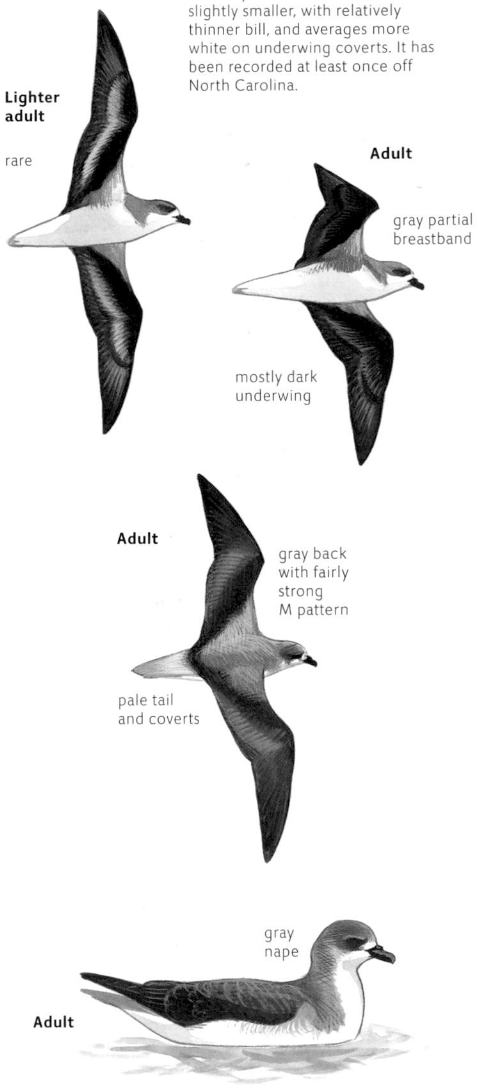

VOICE: Generally silent at sea.

Rare visitor from nesting grounds in eastern Atlantic; several recorded annually on warm Gulf Stream waters off southeastern states. Usually solitary; does not mix with other seabirds.

Herald Petrel

Pterodroma arminjoniana

L 15" WS 37" WT 12 oz (330 g)

Dark morph (more frequent) similar to Sooty Shearwater, but more slender, longer-tailed, and shorter-billed, with different underwing pattern and looser wingbeats. Light and dark morphs recall jaegers, but with different flight style.

North American records at least 75% dark morph, 20% light morph, and very few intermediate.

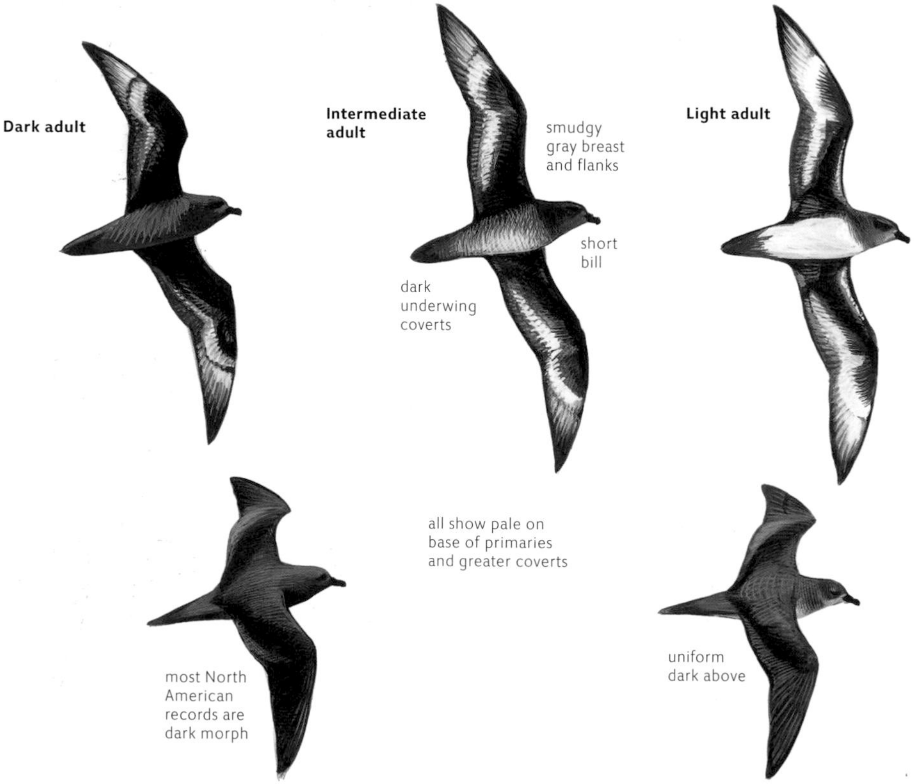

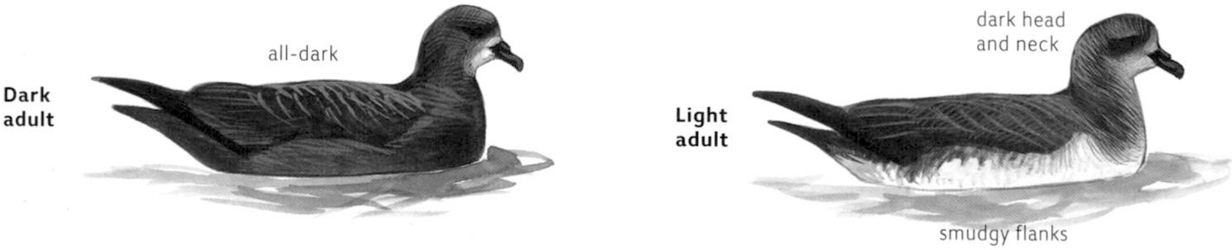

VOICE: Generally silent at sea.

Rare visitor from nesting grounds in South Atlantic; several recorded annually on warm Gulf Stream waters off southeastern states. Usually solitary, but joins feeding flocks of other seabirds.

ɜreat-winged Petrel

'terodroma macroptera

17" WS 43" WT 19 oz (540 g)

, relatively stocky all-dark petrel, larger and relatively broader-/inged than Murphy's, with more buoyant flight; note very stout lack bill and pale face including pale forehead.

OICE: Generally silent at sea.

Very rare visitor recorded several times on open ocean off California.

Vhite-chinned Petrel

'rocellaria aequinoctialis

21" WS 54" WT 43 oz (1,215 g)

, large and bulky petrel with thick neck, large head, and heavy ill, white chin spot is small and inconspicuous. Similar to Sooty hearwater but much heavier and broader-winged, with mostly ark underwing; told from Murphy's Petrel by larger size, large nd pale bill, broader wings.

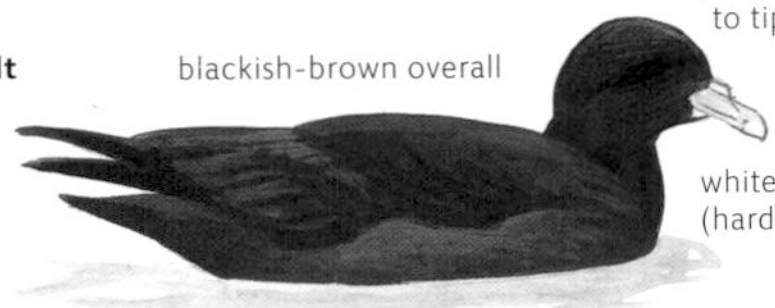

The very similar Parkinson's Petrel has been recorded off California, and two additional similar species should be considered.

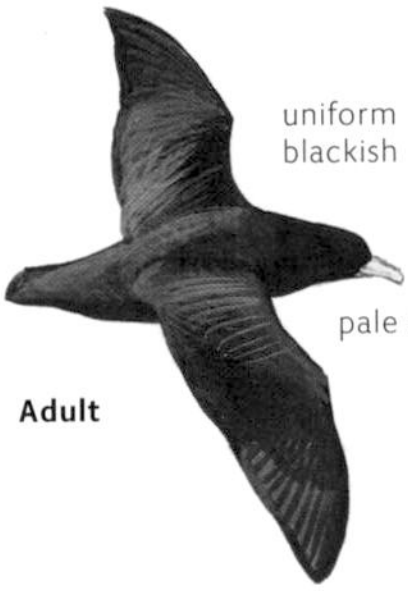

OICE: Generally silent at sea.

Very rare visitor to offshore waters, recorded Texas, Maine, North Carolina, and California.

Bulwer's Petrel

ɜulweria bulwerii

. 11" WS 26" WT 3.8 oz (105 g)

ntermediate between storm-petrels and shearwaters in size, with elatively longer wings and tail than either. All-dark plumage, /ery long, pointed tail, and buoyant, weaving flight that may ecall nighthawks.

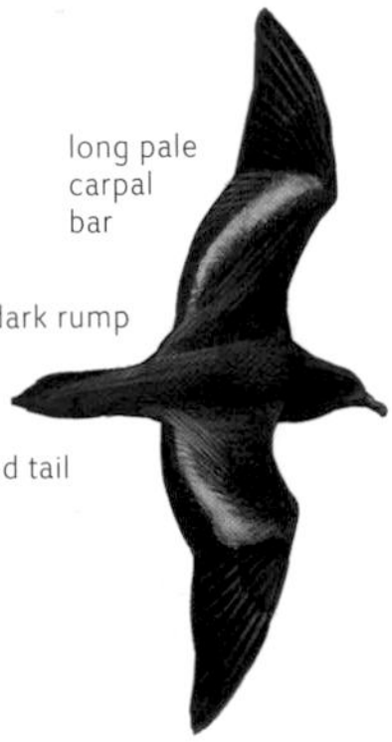

/OICE: Generally silent at sea.

Very rare visitor to warm waters offshore, with only two definite records: North Carolina and California.

Cory's Shearwater

Calonectris diomedea

L 19" WS 46" WT 1.8 lb (800 g) ♂>♀

Our largest shearwater; relatively broad-winged and slow-flying, with wings arched and crooked. Sandy-brown upperside, clean white underwing, mostly dark head, and heavy yellow bill distinctive.

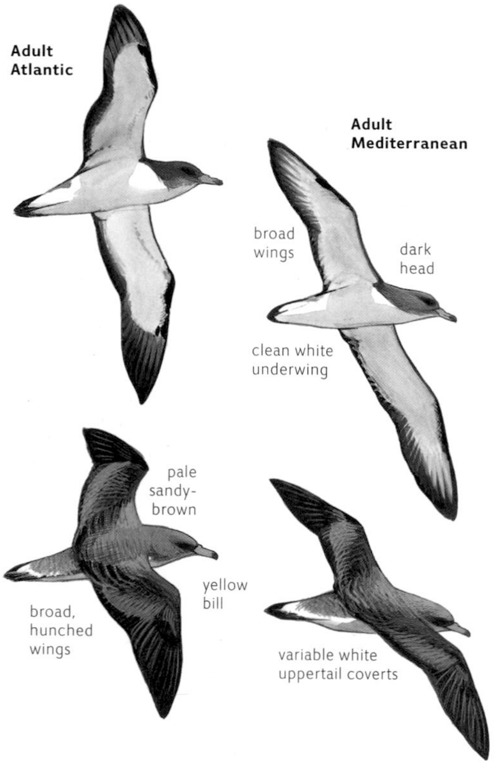

Two populations, both disperse to North America and are often considered separate species. Mediterranean breeders (Scopoli's Shearwater) best distinguished from Atlantic by more extensive white on inner webs of primaries (visible from below), also average smaller overall with smaller bill but much overlap.

VOICE: Generally silent at sea. In feeding groups a raucous, nasal, descending bleating similar to Great.

Common on warm open ocean waters mainly from Jun–Oct; visiting from nesting grounds in eastern Atlantic and Mediterranean. Often solitary, but gathers where food is abundant in loose flocks mixed with other shearwaters, rests on the water in flocks. Feeds mainly on fish and squid.

Great Shearwater

Puffinus gravis

L 18" WS 44" WT 1.8 lb (840 g)

Nearly as large as Cory's Shearwater, but more slender and holds wings straighter, with dark pattern on underwing and well defined dark cap. Distinguished from Manx Shearwater by larger size and by white on nape and uppertail coverts.

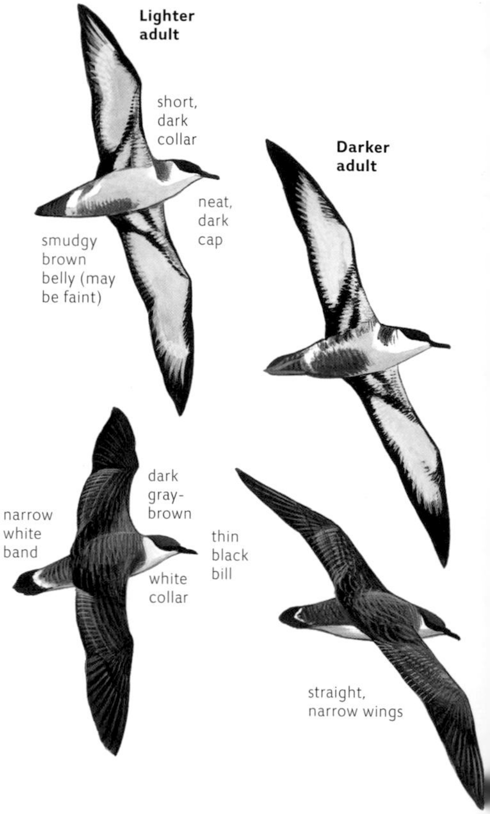

VOICE: Generally silent at sea, but vocal in feeding groups, often bleating *waaan* like a lamb.

Common visitor from nesting grounds in South Atlantic. Found on cold ocean waters mainly from May–Jul, many lingering to October in north; irregularly common and often in large loose groups; Rests in flocks. Feeds on fish and squid.

Buller's Shearwater

Puffinus bulleri

L 17" WS 40" WT 14 oz (400 g)

A relatively slender, long-tailed shearwater with very elegant markings. Pure white below, mostly white underwing, bold M pattern on upperside, and neat dark cap. Also note grayish bill.

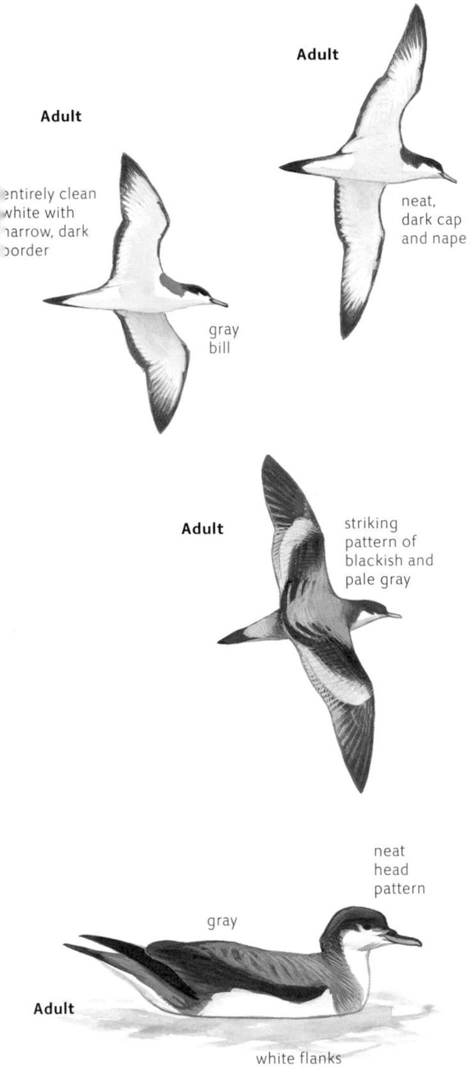

VOICE: Generally silent at sea.

Uncommon visitor from nesting grounds around New Zealand. Found on cold open ocean, passing through quickly from Sep–Oct. Often travels in small groups and joins other shearwaters in feeding and resting flocks. Feeds mainly on squid and fish.

Pink-footed Shearwater

Puffinus creatopus

L 18" WS 44" WT 1.6 lb (720 g)

The largest and heaviest pale-bellied shearwater regularly seen off our west coast. Plumage is similar to the much smaller Black-vented Shearwater, but note pale bill, much larger size, slower wingbeats, broader wings.

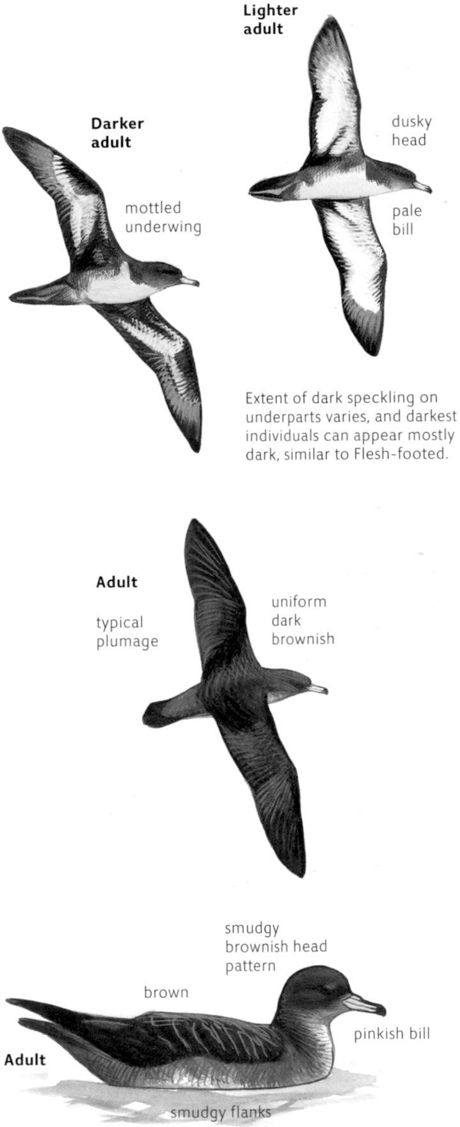

VOICE: Generally silent at sea. Vocal in feeding groups, giving a nasal, descending whinny.

Common visitor to open ocean from nesting grounds on islands off Chile. Found on cold open ocean waters (mainly Apr–Oct). Usually solitary or in relatively small numbers among other shearwaters. Feeds on squid and fish.

Molt in Seabirds

Molt in seabirds can create confusion as missing and growing feathers alter color pattern, wing shape, and tail shape, and altered wing shape can even result in different wingbeats, wing posture, and flight style. Make a habit of watching for signs of molt, such as gaps or irregularities along the tips of the wing feathers, and use a bit more caution identifying those birds by flight style or overall shape.

Molting Great Shearwater
The large group of missing greater coverts reveals white bases of secondaries; the primaries are replaced in sequence from inner to outer, here showing new and growing inner primaries and old outer primaries. This timing is typical of a Southern Hemisphere breeding species in about July; Northern Hemisphere breeding species would not reach this stage until September to October.

In species that breed in the Southern Hemisphere, such as Great Shearwater and Sooty Shearwater, with young fledging around February, the molt occurs mostly between June and August on the Northern Hemisphere "wintering grounds." Some nonbreeders begin molt in the Southern Hemisphere before migration, and arrive in May to June with molt already underway.

Northern Hemisphere breeders such as Manx Shearwater and Cory's Shearwater, with young fledging around August, molt primaries from about September to February, with nonbreeders molting a little earlier, about August to January. In June to July, when Great Shearwaters are in the midst of molt, no Manx Shearwaters have started to molt, and by September, when some Manx are beginning to molt, the Greats have nearly finished.

The same principle allows separation of winter- and summer-breeding populations of species such as Band-rumped Storm-Petrel. Birds nesting in the Northern Hemisphere winter follow a molt schedule similar to Great and Sooty Shearwaters, while populations nesting in the Northern Hemisphere summer match Manx and Cory's Shearwaters.

Winter-breeding Band-rumped Storm-Petrels (Grant's) do not have to work around a long migration as Southern Hemisphere species do, and begin molting when young fledge in February, arriving off the US in May to June with primary molt more than half completed, finishing by August. Summer breeding populations (Madeiran) begin molt in September (possibly as early as July in nonbreeders) and complete in February to May.

Flesh-footed Shearwater

Puffinus carneipes

L 18" WS 44" WT 1.4 lb (620 g)

Very closely related to Pink-footed Shearwater, differs mainly in all-dark plumage. Distinguished from Sooty by bulkier proportions, pale pinkish bill, all-dark underwing coverts.

pale silvery remiges (dependent on lighting)
Adult
pale pinkish bill
uniform dark underwing coverts

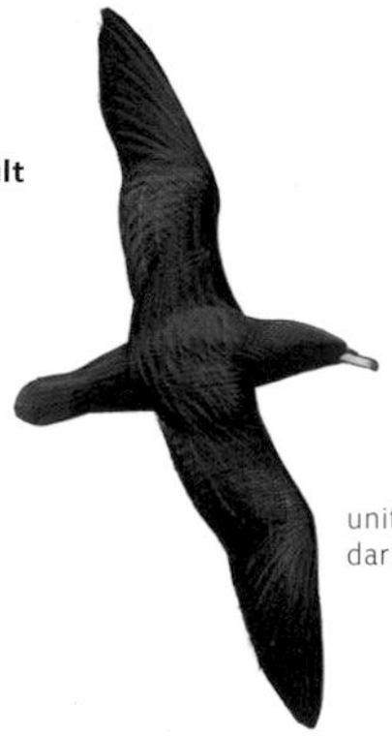
Adult
uniform dark

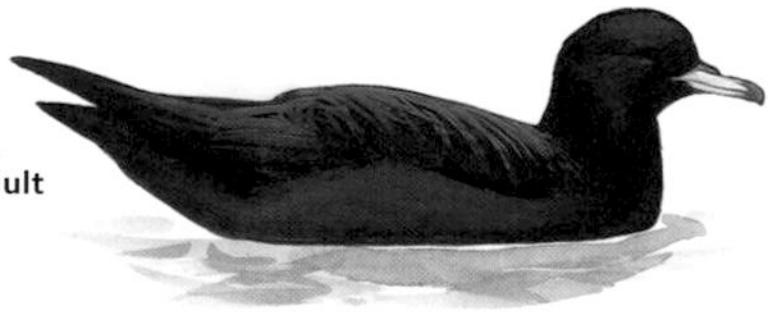
Adult

VOICE: Generally silent at sea.

Rare but regular visitor from nesting grounds around New Zealand. Found over relatively cold open ocean, mainly (Jul–Oct), Usually seen singly, mixed with other species of shearwaters. Feeds on squid and fish.

Short-tailed Shearwater

Puffinus tenuirostris

16.5" WS 38" WT 1.4 lb (635 g)

Extremely similar to Sooty Shearwater, distinguished (with care) by smaller bill, rounder head, and more uniform underwing color, palest on secondary coverts.

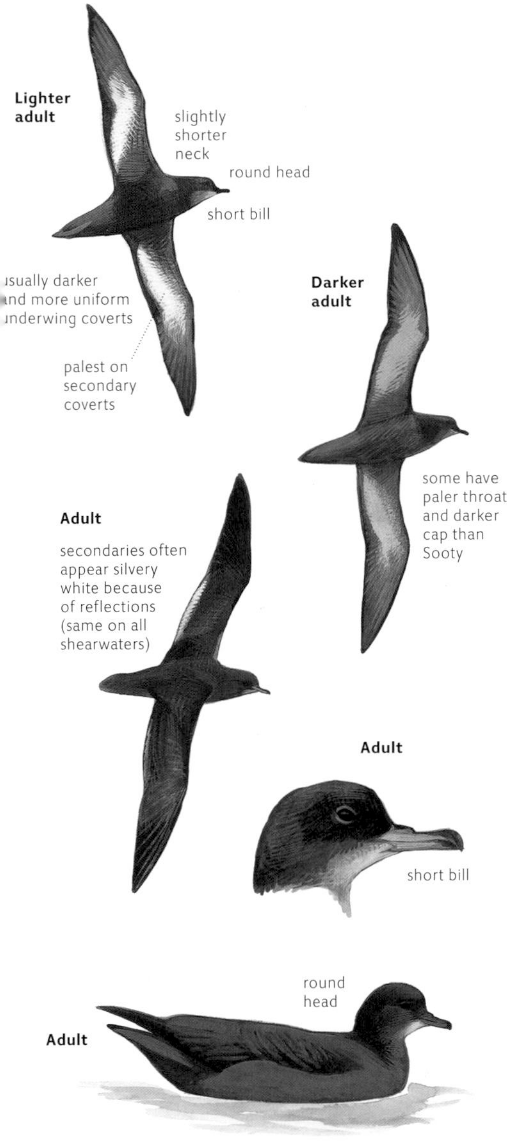

VOICE: Generally silent at sea. Calls at nest higher, more screeching than Sooty.

Common visitor from nesting grounds around Tasmania to open ocean off Alaska; peak numbers Jul–Aug, uncommon to rare southwards, mainly Oct–Mar. Greatly outnumbered by Sooty off California except in winter. May gather in flocks numbering in the millions off Alaska. Feeds on squid and fish.

Sooty Shearwater

Puffinus griseus

L 17.5" WS 40" WT 2 lb (900 g)

The only all-dark shearwater in the Atlantic, and the most common dark shearwater in the Pacific south of Alaska. Slender proportions, with narrow and very pointed wings, slender body, and strong flexible wingbeats. Body all-dark but flashes silvery white patches on the underwing especially on primary coverts. Very similar to Short-tailed Shearwater.

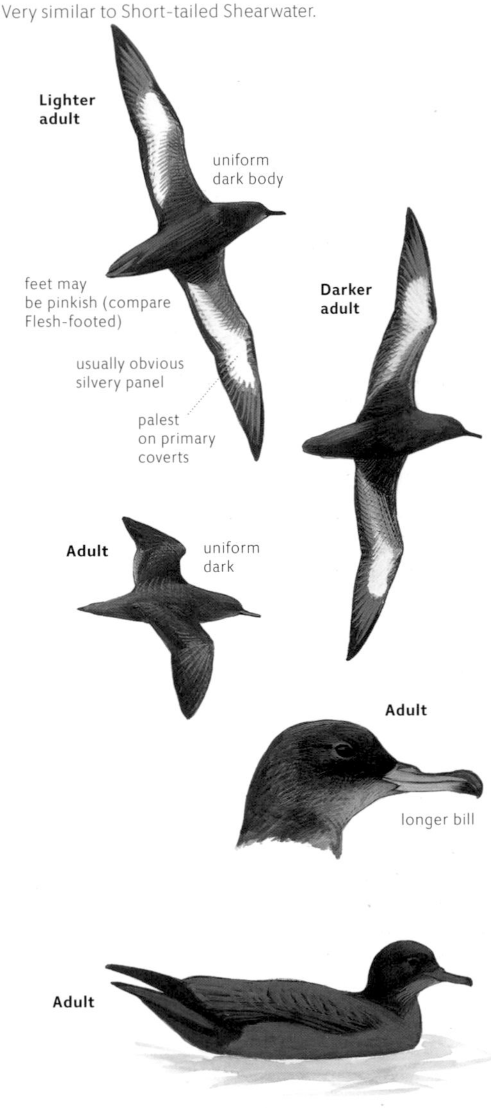

VOICE: Generally silent at sea. In feeding group occasionally gives raucous cry: a relatively high, nasal *aaaa*.

Common visitor from nesting grounds in Southern Hemisphere. In most places the most likely species of shearwater to be seen from land. Most numerous May–Jun in eastern North America, with some lingering to Oct. Peak numbers Aug–Oct off Pacific coast, where it may form flocks of many thousands and often mixes with other shearwaters. Feeds mainly on squid and fish.

Black-vented Shearwater

Puffinus opisthomelas

L 14.5" WS 32" WT 15 oz (425 g)

The smallest regularly-occurring Pacific shearwater. Plumage pattern similar to Pink-footed Shearwater, with indistinct border between dark upperside and pale underside, but note smaller size, quicker wingbeats, and dark bill. Compare rare Manx Shearwater. Flight generally low and direct, with quick wingbeats.

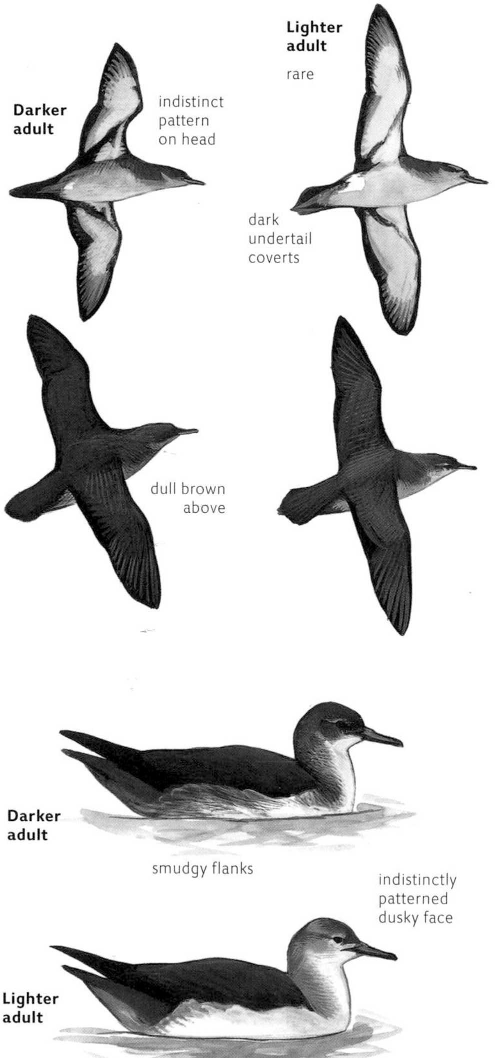

VOICE: Generally silent at sea. Voice at nest site undescribed.

Common visitor from nesting grounds off western Mexico to warm open ocean waters within a few miles of land, usually visible from land, mainly Oct–Mar. Solitary or in groups; often joins other shearwaters. Feeds mainly on squid and fish.

Manx Shearwater

Puffinus puffinus

L 13.5" WS 31.5" WT 1.1 lb (500 g)

A small shearwater, with quick wingbeats; proportions similar t Great Shearwater. All-dark above and white below, with broa dark smudge on sides of neck. Note thin pale crescent behin dark cheek and all-white undertail coverts and flanks.

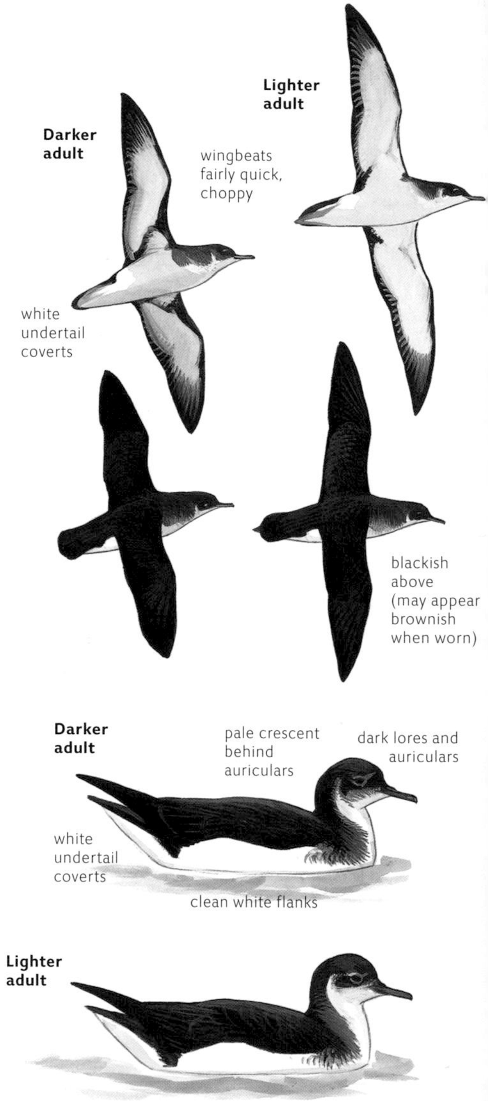

VOICE: Generally silent at sea. At nest site at night a cackling ther crooning series *cack cack cack carrr ho*. From ground a squall-ing sound rising to a crescendo of crowing when excited.

Uncommon visitor from nesting grounds in North Atlantic to cold offshore waters, rare but increasing in Pacific. Nests in burrows on grassy slopes. Often solitary, or in pairs or small groups, mixing with other species of shearwaters. Feeds mainly on fish.

Audubon's Shearwater

Puffinus lherminieri

L 12.5" WS 27" WT 7 oz (200 g)

Our smallest expected shearwater; relatively short-winged and long-tailed, with rapid wingbeats and flight often close to the water (not arcing high). Flight and plumage may recall large alcids. Best distinguished from Manx Shearwater by proportions (relatively shorter wings and longer tail) and by white markings around eye.

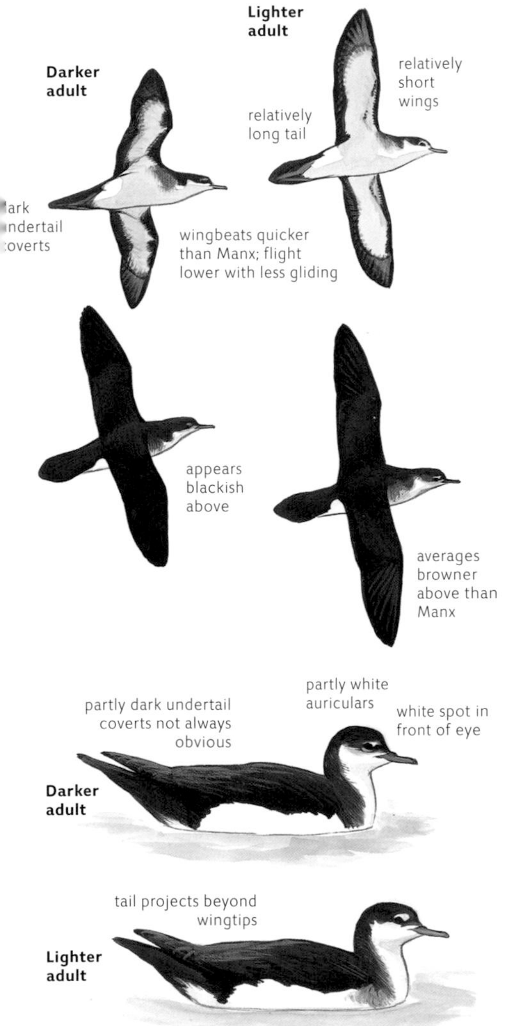

VOICE: Generally silent at sea. Sometimes gives a high, nasal whining.

Locally common visitor from nesting grounds in Caribbean. Found over warm pelagic waters, especially around floating mats of sargassum at edge of Gulf Stream. Solitary or in small groups. Feeds mainly on fish and squid.

Barolo Shearwater

Puffinus baroli

L 10.5" WS 23" WT 6 oz (170 g)

A very small shearwater, similar to Audubon's and Manx but even smaller, with extensive white above eye, relatively short wings and tail, small bill, and mostly white undertail coverts. White tips on upperwing coverts form narrow pale bands and are distinctive when present.

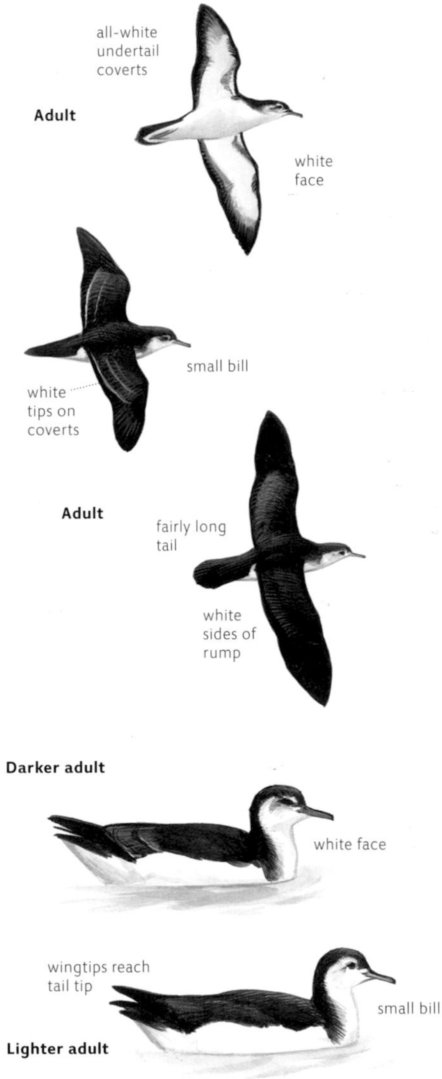

VOICE: Generally silent at sea.

Very rare visitor from nesting grounds in eastern Atlantic, recorded several times far offshore from Massachusetts to Newfoundland.

Streaked Shearwater

Calonectris leucomelas

L 19" WS 42" WT 1.1 lb (500 g)

A relatively large and bulky shearwater, with broad wings and lazy, languid flight. White head, pale back, gleaming white underparts, and dark carpal patch on underwing create distinctive appearance even at a distance.

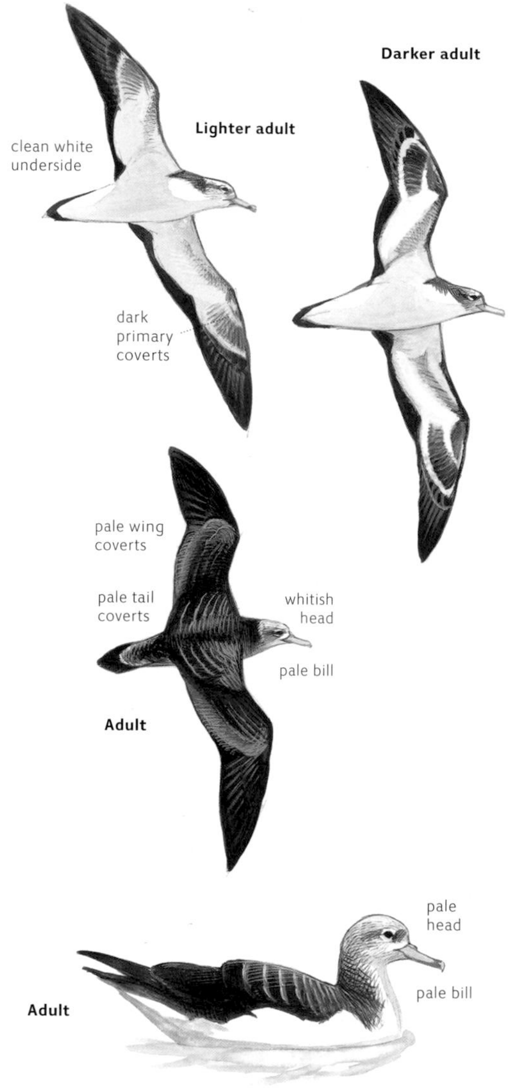

VOICE: Generally silent at sea.

Very rare visitor from western Pacific. Recorded almost annually off central California mostly in Sep–Oct. Once off Washington and once in Wyoming.

Wedge-tailed Shearwater

Puffinus pacificus

L 18" WS 41" WT 14 oz (400 g)

Slightly smaller and much lighter weight than Pink-footed Shear water, with broad wings, long pointed tail, and buoyant flight. O dark morph note mostly dark underwing, long, pointed tai grayish bill. Light told from Pink-footed by wing shape and b gray bill.

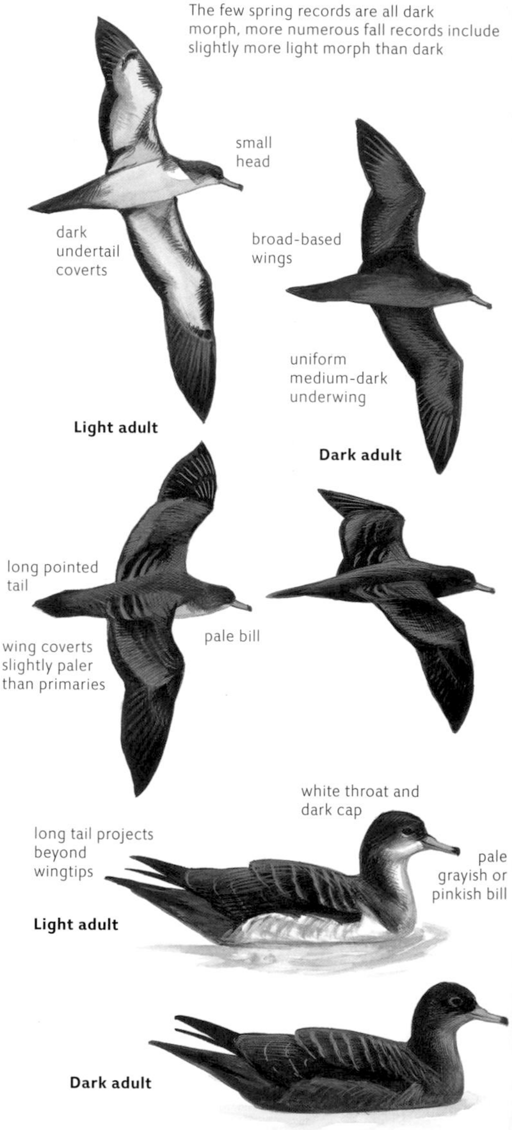

VOICE: Generally silent at sea.

Very rare visitor from subtropical Pacific. Recorded a few times in summer and fall from warm waters off California.

Storm-Petrels

ᴬAMILY: HYDROBATIDAE

2 species in 3 genera (3 rare species are not shown on this page). ll are small oceanic birds found mainly far offshore over open cean, and coming to land only at night when nesting. Sort into wo main groups: Southern Hemisphere breeders with relatively ong legs and short wings (e.g. Wilson's) and Northern Hemiphere breeders with relatively long wings and short legs. All orage on plankton and other small prey captured at the water's urface, often while flying and "foot-pattering." Flight actions are mportant for identification but depend on many factors, includng wind conditions and the motivation of the bird. Foraging light, traveling flight, and the less often seen "escape flight" (as vhen alarmed by an approaching boat) all differ. In judging tail hape, note that the outer tail feathers of all species flex upward n flight, creating or enhancing the impression of a notched tail, nd that long toes of some species can also look like a notched ail. Adults are shown.

Molting Wilson's Storm Petrel

Timing of primary molt can be a useful clue for identification. This Wilson's Storm-Petrel, with 5 or 6 old, worn and faded outer primaries, is typical of adults seen around June in the North Atlantic. Timing differs between species, e.g. adult Leach's molts Nov–Apr on its Southern Hemisphere wintering grounds.

Genus *Oceanites*

Wilson's Storm-Petrel, page 84

Genus *Pelagodroma*

White-faced Storm-Petrel, page 89

Genus *Hydrobates*

European Storm-Petrel, page 88

Genus *Oceanodroma*

Fork-tailed Storm-Petrel, page 87

Leach's Storm-Petrel, page 85

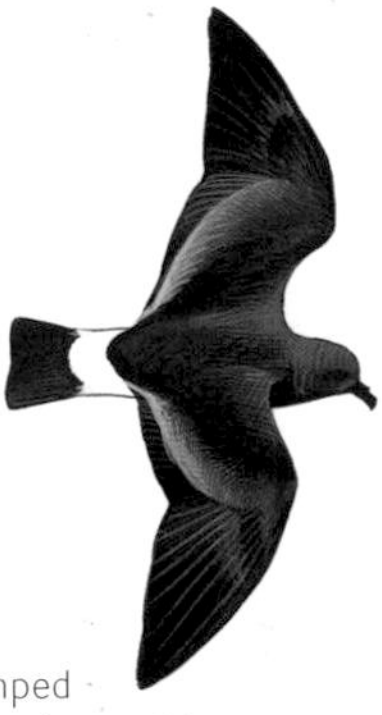

Band-rumped Storm-Petrel, page 84

Black Storm-Petrel, page 86

Ashy Storm-Petrel, page 86

Least Storm-Petrel, page 87

Wilson's Storm-Petrel

Oceanites oceanicus

L 7" WS 15" WT 1.3 oz (37 g)

Differs fundamentally from other dark storm-petrels in structure (shorter arm, longer legs) and shallower wingbeats with wings held stiffly and angled down, not arched. Plumage averages darker than most other storm-petrels, and white of rump wraps around extensively on sides.

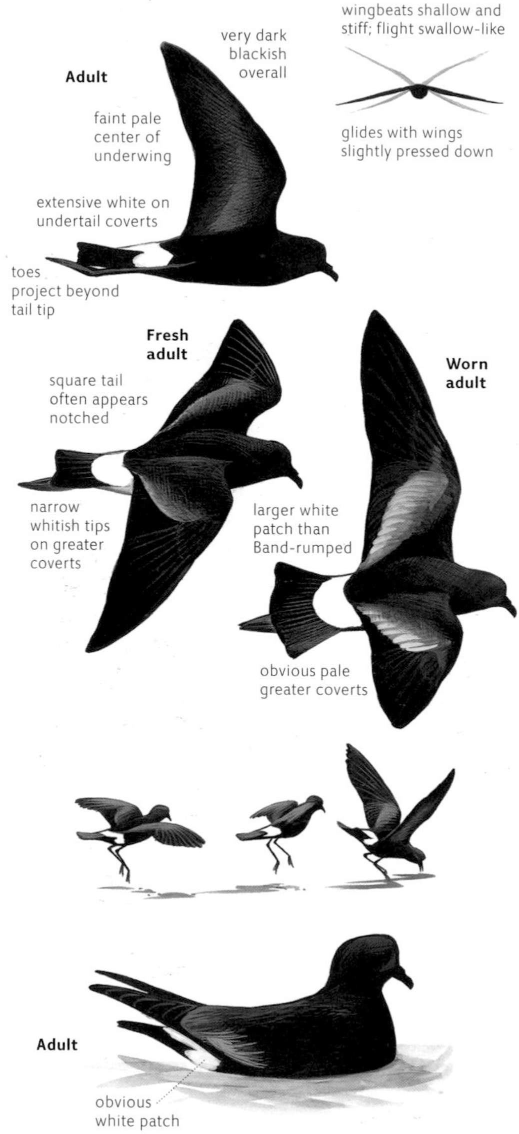

VOICE: Generally silent at sea. At close range a peeping or chattering sound may occasionally be heard from feeding groups.

Common visitor to the North Atlantic from nesting grounds in sub-Antarctic. Found over cold open ocean waters, mainly May–Aug, often visible from land; forms loose groups. Rare in North Pacific mainly Aug–Oct. Feeds on plankton.

Band-rumped Storm-Petrel

Oceanodroma castro

L 8" WS 18.5" WT 1.7 oz (48 g)

Larger than Wilson's, with long arched wings, notched tail, an short legs like Leach's; distinguished from Leach's by relativel shorter wings and less bounding flight; shorter tail; weaker car pal-bar not reaching leading edge of wing; and, especially, mor white wrapping around sides of rump.

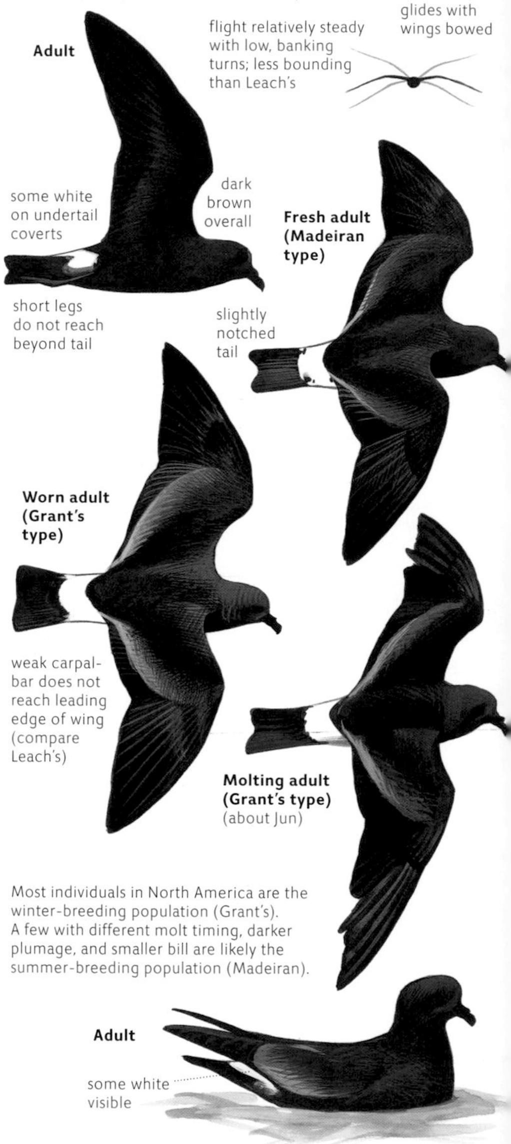

VOICE: Generally silent at sea. At nest site a squeak like a finge across a wet pane of glass, followed by low purring.

Uncommon to rare visitor from breeding grounds in eastern North Atlantic. Found on warm Gulf Stream waters far off southeastern states, mainly May–Sep. Habits similar to other storm-petrels.

Leach's Storm-Petrel

Oceanodroma leucorhoa

L 8" WS 18" WT 1.5 oz (42 g)

Distinguished from other storm-petrels by longer, more angled, arched wings; deep wingbeats and bounding flight (may recall nighthawks); longer, strongly notched tail (sometimes difficult to see); and short legs.

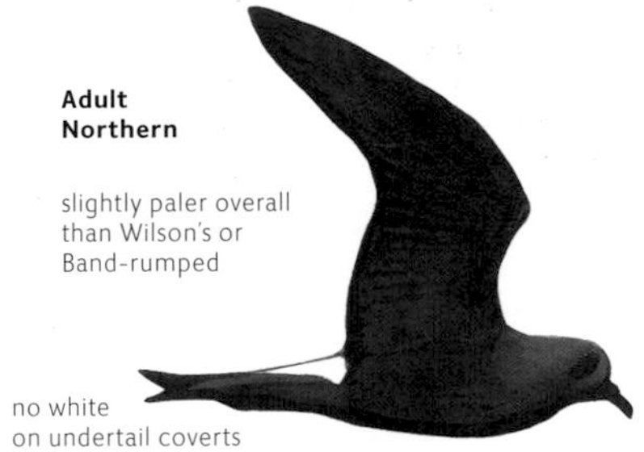

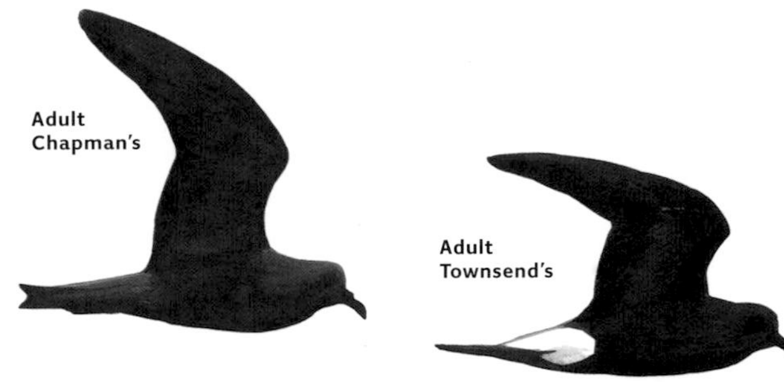

VOICE: Generally silent at sea. At nest site a musical purring three to four seconds long, rising slightly in pitch and interspersed with inhaled *whee*. In flight over colony a chuckling chatter *pwa putaDEEPto-ditado*. Winter breeding population of Guadalupe Island (not proven to occur in North America) has slightly different vocalizations.

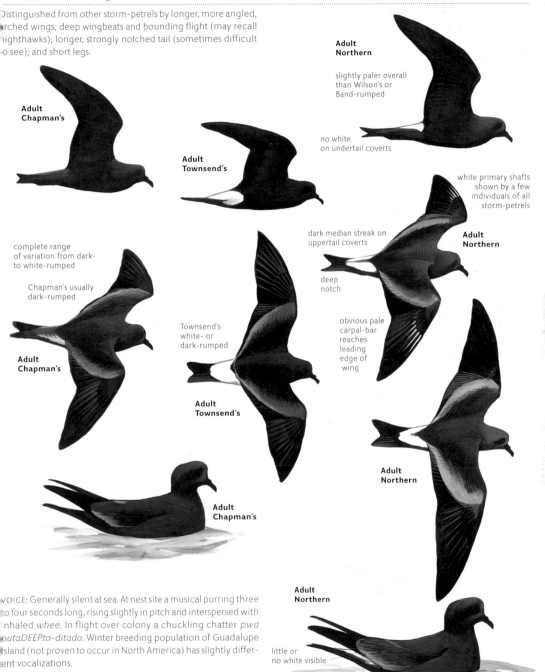

Common to uncommon on open ocean, usually over deep water; rarely seen from land, except dark-rumped populations off southern California found over shallow near-shore waters. Nests in burrows or crevices on islands; visits nest only at night. Usually solitary, but may form groups or mix with other storm-petrels at roosting and feeding areas. Feeds on plankton.

All Atlantic birds, and Pacific birds south to central California are large and white-rumped. Off southern California two additional forms occur, both smaller with shorter wings and tails. Chapman's occurs regularly, often close to land, and is usually very dark-rumped. Townsend's less numerous, smallest, can be white-rumped or dark-rumped; white-rump is brighter and more extensive than northern birds; dark-rumped birds very difficult to distinguish from Chapman's, but averages more pale on sides of rump.

Black Storm-Petrel

Oceanodroma melania

L 8.7" WS 21" WT 2.1 oz (59 g)

The largest storm-petrel in North America; noticeably larger and longer-winged than other species, with deep languid wingbeats.

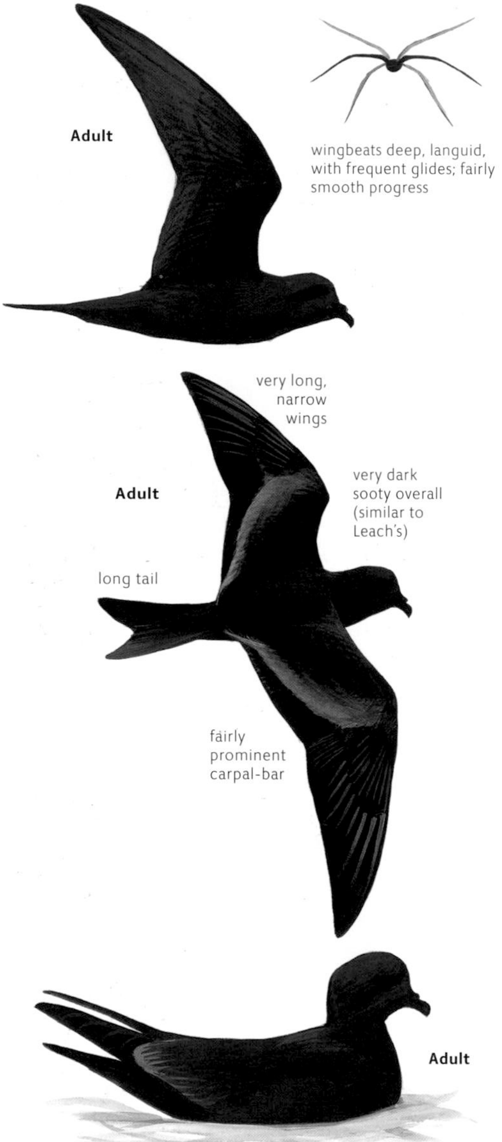

VOICE: Generally silent at sea. At nest site a purring or rattling 10 to 12 seconds long. In flight over colony a loud, screechy chattering *pukaree puck-puckaroo*.

Common on warm open ocean waters. Nests in underground crevices on islands. Usually solitary at sea, but roosts in flocks on the water. Feeds on fish, squid, and a variety of plankton, including fish eggs, shrimp, and larval crustaceans.

Ashy Storm-Petrel

Oceanodroma homochroa

L 7.5" WS 17" WT 1.3 oz (37 g)

Smaller than Black Storm-Petrel, with distinctively shallow wingbeats, pale gray sides of rump, and pale underwing coverts. Distinguished from Leach's Storm-Petrel by stiffer and shallower wingbeats, grayish sides of rump, pale underwing, relatively narrow wings and long tail.

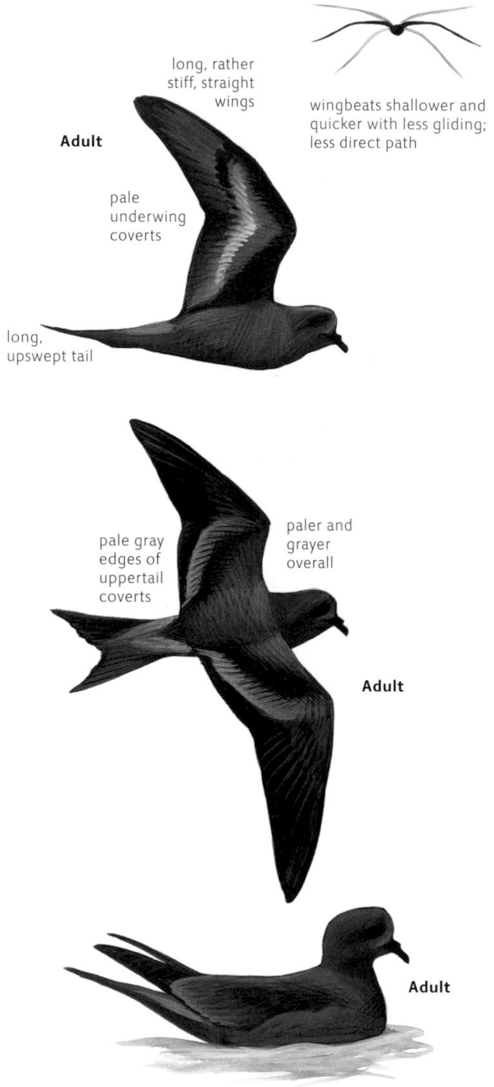

VOICE: Generally silent at sea. At nest site a purring like Leach's rising and falling with inhaled gasp. In flight over colony a harsh squeaky chuckling.

Common locally. Found on warm open ocean waters. Nests in underground crevices and burrows on islands. Usually solitary at sea, but roosts in flocks on the water. Feeds on plankton.

Least Storm-Petrel

Oceanodroma microsoma

5.75" WS 13.5" WT 0.7 oz (20 g)

he smallest storm-petrel off Pacific coast; smaller than Ashy and uch smaller than Black Storm-Petrel. Note very short, wedge-haped tail. Look for combination of all-dark color, very small ze, short tail, and deep wingbeats.

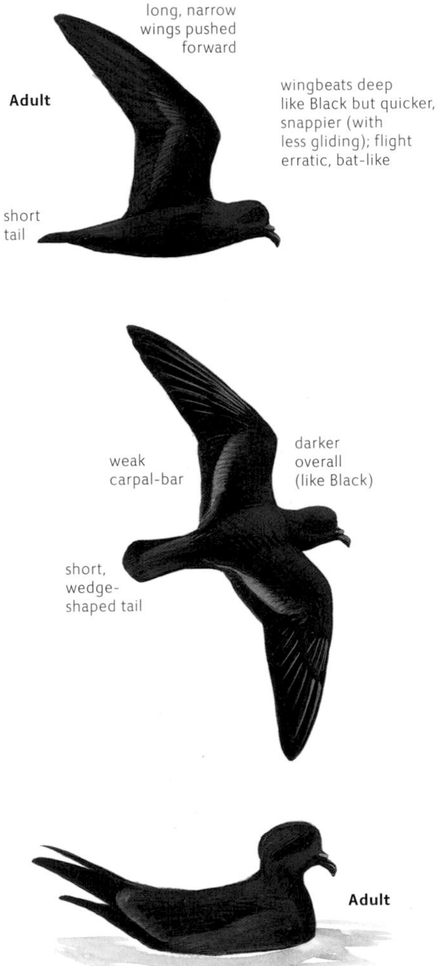

VOICE: Generally silent at sea. At nest site a soft, rapid purring call, three to four seconds long with occasional inhaled notes. In flight over colony a harsh, accelerating chatter like Black but weaker.

Uncommon and irregular visitor from nesting grounds off western Mexico to warm offshore ocean waters, mainly Aug–Oct. Usually solitary at sea, but may join other storm-petrels in roosting groups. Feeds on plankton.

Fork-tailed Storm-Petrel

Oceanodroma furcata

L 8.75" WS 19" WT 2.1 oz (60 g)

Pale gray plumage unlike any other storm-petrel; note dark underwing coverts, pale carpal-bar on upperwing, and dark eye-smudge on pale gray head. Superficially similar to phalaropes but different flight with much gliding, no pale wingstripe.

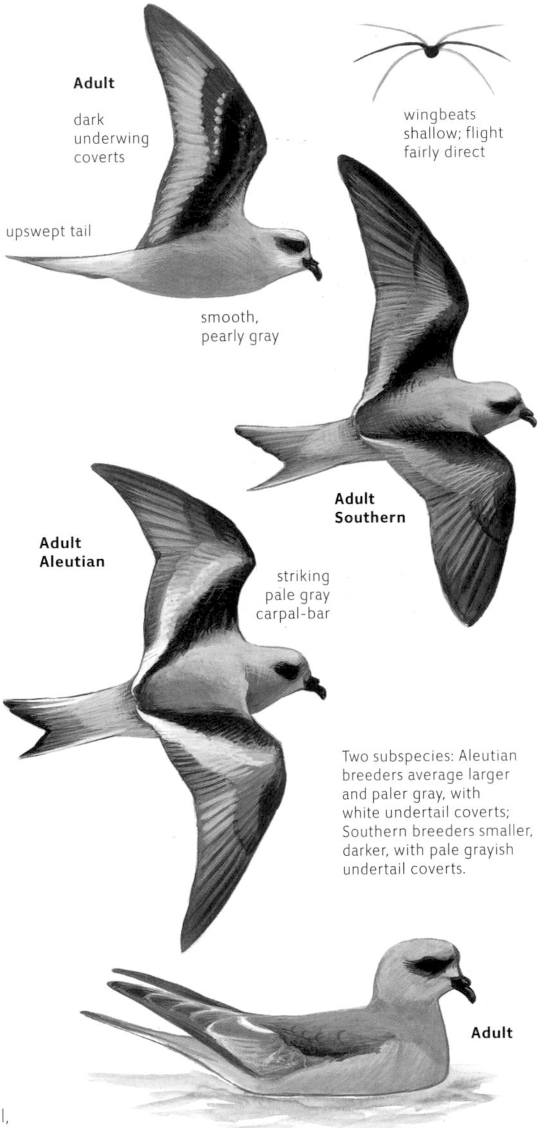

VOICE: Generally silent at sea. At nest site a soft twittering or high, complaining, rasping notes *skveeeee skwe skwe*.

Common. Found over cool open ocean waters. Nests in colonies in underground burrows and crevices on islands. Usually solitary, but may gather in small groups and join other storm-petrels when resting on the water. Feeds on plankton.

Swinhoe's Storm-Petrel

Oceanodroma monorhis

L 7.5" WS 18.5" WT 1.5 oz (42 g)

The only all-dark storm-petrel recorded in the Atlantic, but some Leach's have very little white on the rump. Look for uniform dark rump (same color as back) with no trace of paler edges, pale at base of primary shafts, and somewhat more erratic, bounding flight.

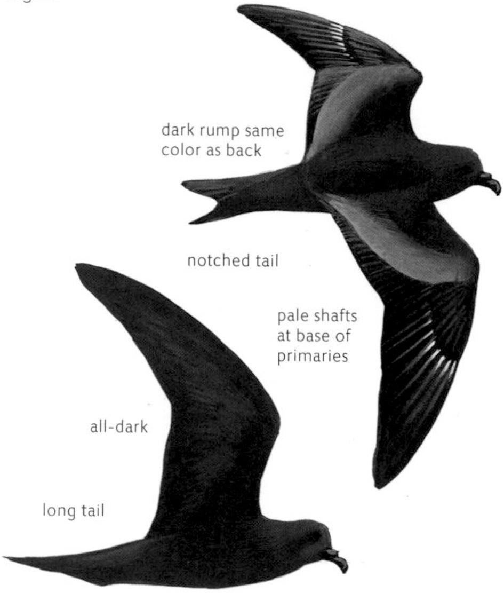

VOICE: Generally silent at sea.

Very rare visitor to warm offshore waters, recorded several times off North Carolina.

Wedge-rumped Storm-Petrel

Oceanodroma tethys

L 6" WS 14" WT 0.9 oz (25 g)

A very small storm-petrel with extensive white on the rump, even appearing to have a white tail with dark corners. Distinguished from smaller subspecies of Leach's (Townsend's) by smaller size, more extensive white rump, thicker bill, shallower tail fork.

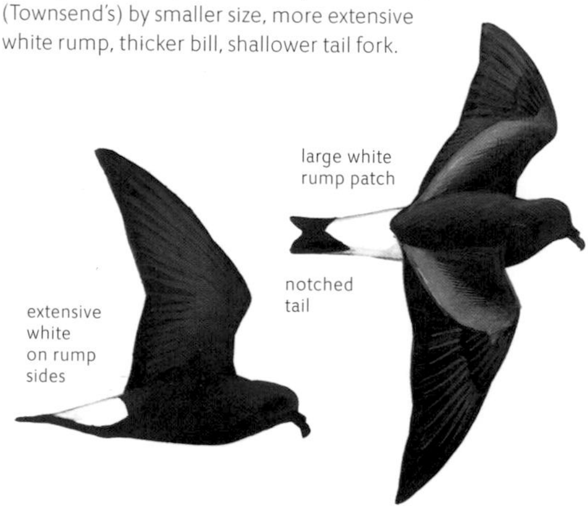

VOICE: Generally silent at sea.

Very rare visitor to open ocean far offshore, recorded several times off California.

European Storm-Petrel

Hydrobates pelagicus

L 6.3" WS 14" WT 1 oz (28 g)

Very small and dark, with rectangular white rump patch; longe arm than Wilson's, fluttering, bat-like flight. White band o underwing coverts is diagnostic but not easy to see and can b confused with a similar pale patch on Wilson's.

VOICE: Generally silent at sea.

Very rare visitor from breeding grounds in northeast Atlantic, a few recorded each spring off North Carolina in recent years, all in late May–early June.

White-faced Storm-Petrel

elagodroma marina

7.75" WS 16" WT 1.9 oz (54 g)

Vhite head and underparts with gray-brown upperside distinctive. Note rather broad, paddle-shaped wings and very long legs. lumage pattern reminiscent of phalaropes, but flight utterly ifferent.

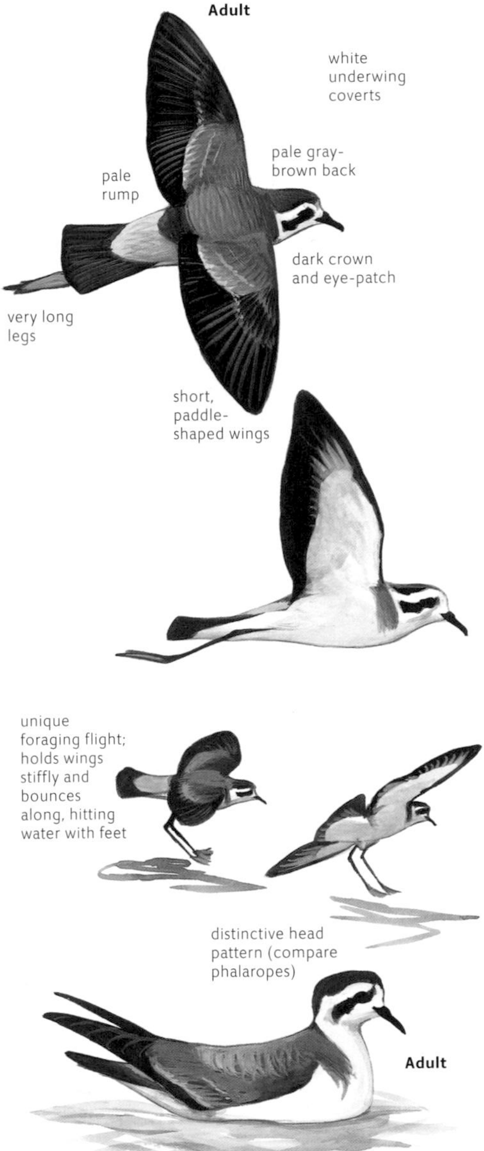

VOICE: Generally silent at sea.

Very rare visitor from breeding grounds in eastern Atlantic; several recorded annually at Gulf Stream eddies far offshore Massachusetts to North Carolina, mainly Aug–Sep. Usually solitary; does not mix with other storm-petrels. Feeds on plankton.

Black-bellied Storm-Petrel

Fregetta tropica

L 8" WS 17" WT 1.9 oz (54 g)

A relatively stocky, broad-winged, and short-tailed storm petrel. The combination of shape, flight style, and extensive white on flanks is distinctive (but beware rare partial albino Wilson's or other species with white blotches on underparts).

VOICE: Generally silent at sea.

Very rare visitor from breeding grounds in the southern oceans. Recorded several times in warm waters off North Carolina.

Pelicans, Boobies, and Frigatebirds

FAMILIES: PELECANIDAE, SULIDAE, FREGATIDAE

10 species in 3 families occur regularly. All are large fish-eating seabirds with all four toes joined by webbing (a trait also shared by tropicbirds, cormorants, and anhinga). Pelicans are huge, with a very long bill supporting an expandable pouch used to scoop fish and water, then allow water to drain away. Boobies and gannet have long narrow wings and catch fish underwater by diving from the air in powerful, streamlined head-first dive. Frigatebirds are extremely graceful aerial species, mostly black, usually seen soaring buoyantly high above beaches and open ocean, swooping down to pluck prey from the water's surface or to steal fish captured by other seabirds.

Genus *Pelecanus*

American White Pelican, page 90

Brown Pelican, page 91

Genus *Fregata*

Magnificent Frigatebird, page 96

Genus *Morus*

Northern Gannet, page 95

Genus *Sula*

Masked Booby, page 93

Blue-footed Booby, page 93

Brown Booby, page 92

Red-footed Booby, page 94

American White Pelican

Pelecanus erythrorhynchos

L 62" WS 108" WT 16.4 lb (7,500 g) ♂>♀

A very large and ponderous bird. White plumage, black flight feathers, and huge yellow bill and pouch distinctive.

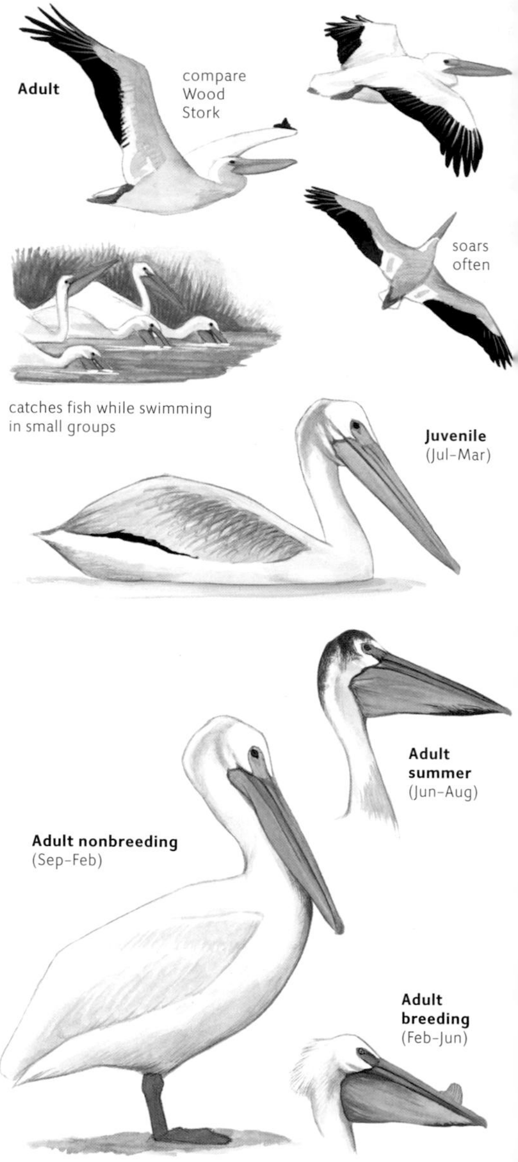

VOICE: Generally silent away from breeding grounds. At nest utters quiet, low grunting or croaking sounds. Young give whining grunt or weak, sneezy bark.

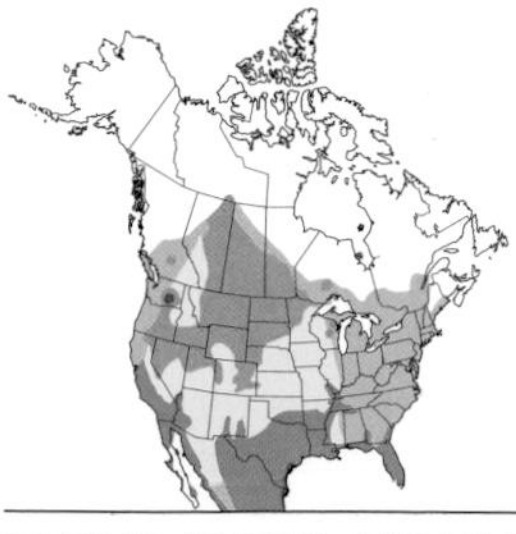

Locally common. Nests on islands in large lakes, forages in shallow protected water, fresh or salt. Often in flocks and often seen soaring high overhead. Feeds on small fish; groups forage cooperatively, driving fish ahead and plunging in bills simultaneously.

ʒrown Pelican

elecanus occidentalis

51" WS 79" WT 8.2 lb (3,740 g)) ♂>♀

arge size, gray-brown or silvery color, and large bill and pouch nmistakable. Flight graceful; often seen flying in lines, rising as ach bird flaps and then gliding down, coasting just inches above ıe water.

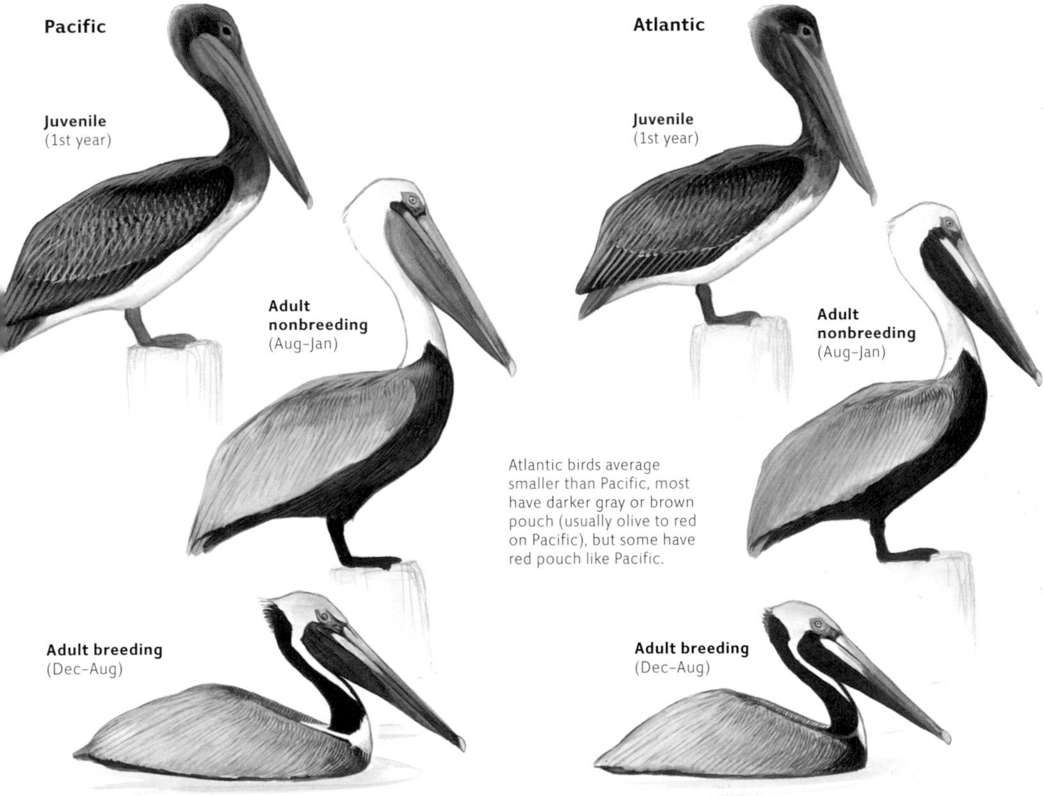

ᐯOICE: Generally silent after fledging. Young in nest give groan-ng or screeching calls.

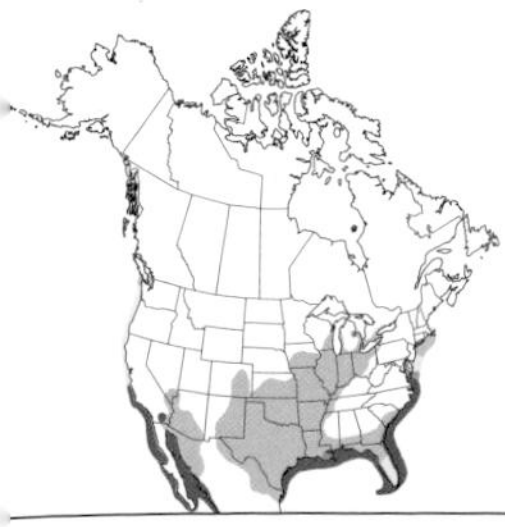

Common along coasts, foraging in shallow waters of ocean, bays, and lagoons. Nests in colonies on small protected islands. Roosts in large groups on sandbars and pilings. Feeds on small fish captured in spectacular plunge-dive.

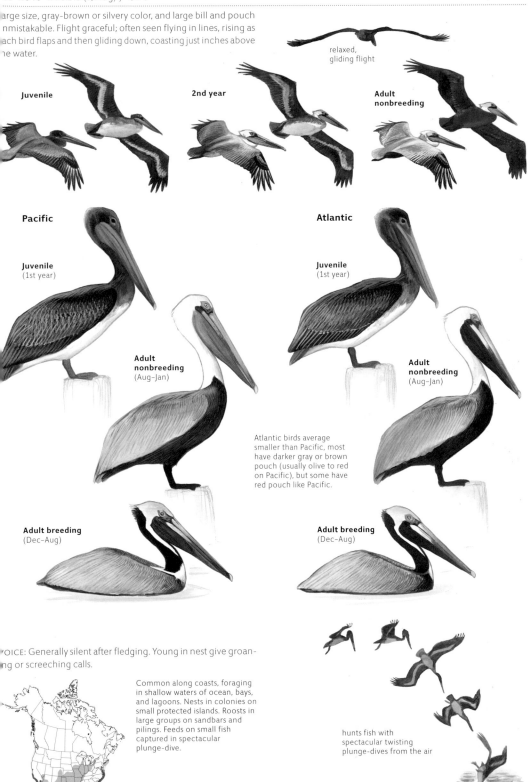

Brown Booby

Sula leucogaster

L 30" WS 57" WT 2.4 lb (1,100 g) ♀>♂

wingbeats slightly deeper than Northern Gannet

Smaller than Blue-footed or Masked Boobies. On all plumages note uniform dark upperside lacking white uppertail coverts of other boobies, and yellowish feet. Contrasting, clean white belly of adult distinctive.

Adult

Juvenile

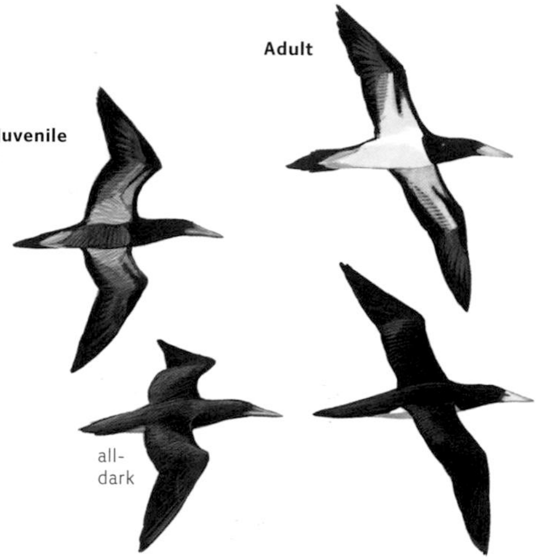

all-dark

Brown Booby often flies low, fanning tail when turning and executing very low-angle plunge-dives.

Adult ♀ Pacific

♀ has yellow face, like bill color; ♂ dark slaty face

Adult ♂ Pacific

Adult males of eastern Pacific population have very pale, frosty gray head.

Juvenile (1st year)

uniform dull brown

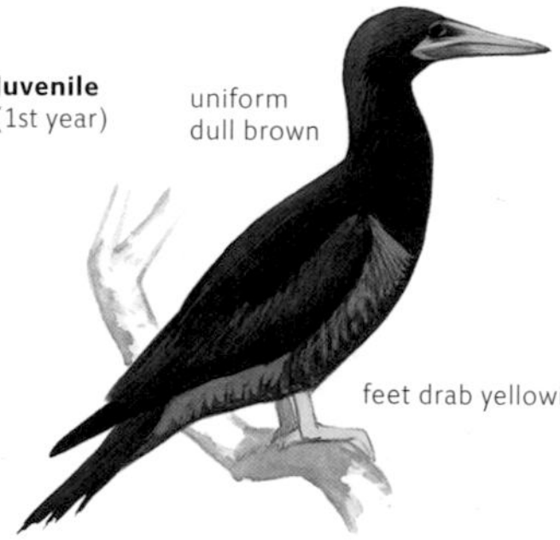

feet drab yellowish

Adult ♀ Atlantic

Juvenile (1st year)

Adult ♂ Atlantic

dark brown and clean white

bright yellow feet

VOICE: Generally silent away from breeding colony. Immatures and adult female give strident honking and grunting; male gives quieter, high, whistled *schweee*; in aggression a nasal croaking *gaan gaan*.

Rare but regular visitor to warm ocean waters in California and southern Florida, rarely seen from land. Solitary at sea; roosts on buoys, channel markers, and trees, sometimes in loose groups. Feeds on small fish captured in shallow plunge-dive.

Two subspecies reliably distinguished in adult male plumage: Pacific male has head frosted white (vs. uniform dark brown like back on Atlantic males), also darker gray bill and may average darker iris. These differences also apparent to a lesser extent on females.

Masked Booby

ula dactylatra

32" WS 62" WT 3.3 lb (1,550 g) ♀>♂

naller than Gannet, but larger than other boobies; immature stinguished by clean white underside, white collar and mostly hite underwing, adult from gannet by dark face, black second-ies and coverts.

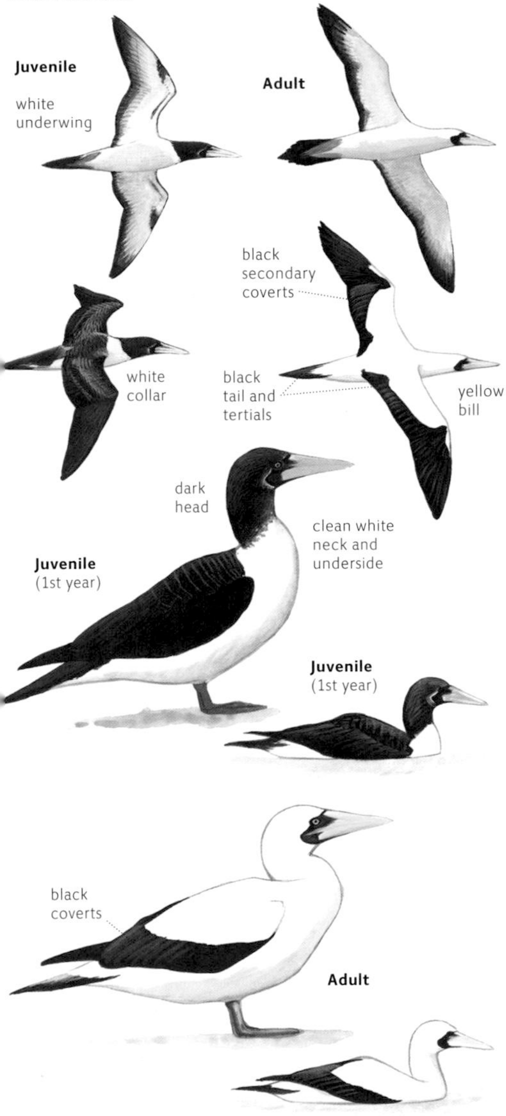

OICE: Generally silent away from breeding colony. Immatures nd adult female give loud honking or braying; adult male gives iping or wheezy whistle.

Rare visitor to warm ocean waters in southern states, rarely seen from land. Regular only around the Dry Tortugas, Florida, where it has nested. Solitary at sea; roosts in groups on sandbars. Feeds on small fish captured in plunge-dive.

Blue-footed Booby

Sula nebouxii

L 32" WS 62" WT 3.4 lb (1,550 g) ♀>♂

Large with relatively long and slender bill and long legs. Distinguished from other boobies by combination of mostly dark underwing, light-colored tail and lower back, bluish feet, and gray bill.

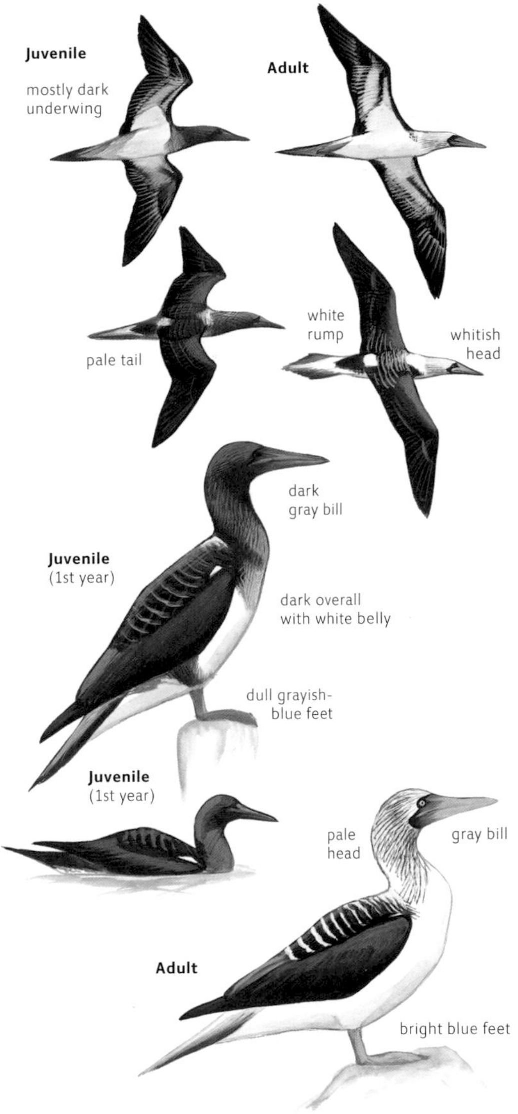

VOICE: Generally silent away from breeding colony. Juvenile and adult female give resonant trumpeting or honking; adult male gives weak, plaintive whistles; both calls slightly deeper than Brown Booby.

Very rare visitor from nesting grounds in western Mexico; most records from Salton Sea, California, Aug–Nov. Solitary at sea; roosts in groups on rocks. Feeds on fish captured in plunge-dive.

Red-footed Booby

Sula sula

L 28" WS 60" WT 2.2 lb (1,000 g) ♀>♂

The smallest booby, with slender bill. Juvenile always dark, with all-dark underwing. Adult light or dark; usually with dark tail in Pacific (many Atlantic adults have white tail). White morph has dark secondaries but white tertials (unlike Masked Booby). Red feet obvious on adult; juvenile has dull orange feet.

wingbeats deeper and bouncier than other boobies; flight buoyant with wings swept back

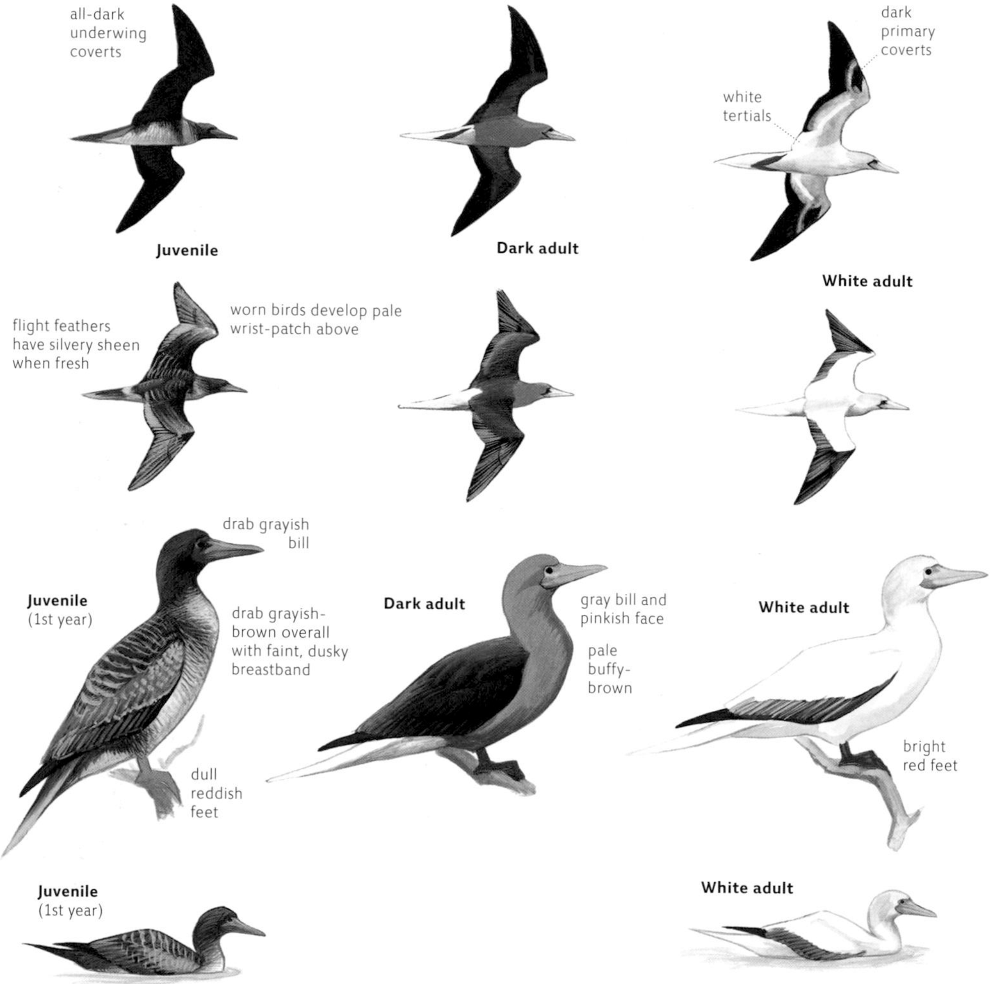

VOICE: Generally silent away from breeding colony. Both sexes give guttural screeching or rattling squawks

Very rare visitor from tropical oceans to warm ocean waters; not recorded annually. Solitary at sea; roosts in groups in trees. Forages mainly at night for small fish.

Most North American records are of juveniles, which are always dusky brownish with paler underparts. Adults are polymorphic. Pacific adults range from white to dark brown on head and body, usually with dark tail. Caribbean adults range from white to medium brown on body (most are brown), nearly always with white tail. Adult plumage is acquired in about three years.

Northern Gannet

orus bassanus

37" WS 72" WT 6.6 lb (3,000 g)

early pelican-size and with similar slow wingbeats and ponderous flight, but much more slender with sharply-pointed wings, il, and bill. Bright white color and black primaries of adult distinctive, gradually acquired over four years. Subadults stinguished from gulls by larger size, rakish shape.

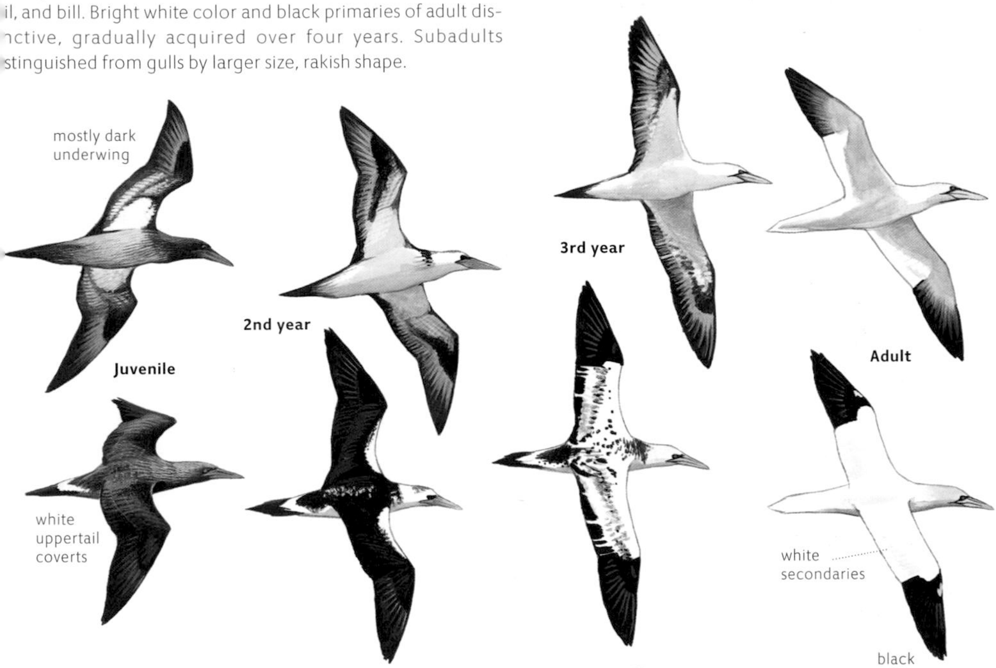

gannets and boobies catch fish in spectacular plunge-dives, piercing the water with almost no splash

OICE: Generally silent away from breeding grounds. Groups at a give grating, descending, growling *urrrrah*. In colony gives oud, harsh, gurgling call.

Common on open ocean often close to land; almost never over land. Nests in colonies on a few cliff tops in Canada. Often in small groups, traveling in low arcing flight or foraging by patrolling high above the water. Forms large feeding groups. Feeds on fish captured in spectacular plunge-dive.

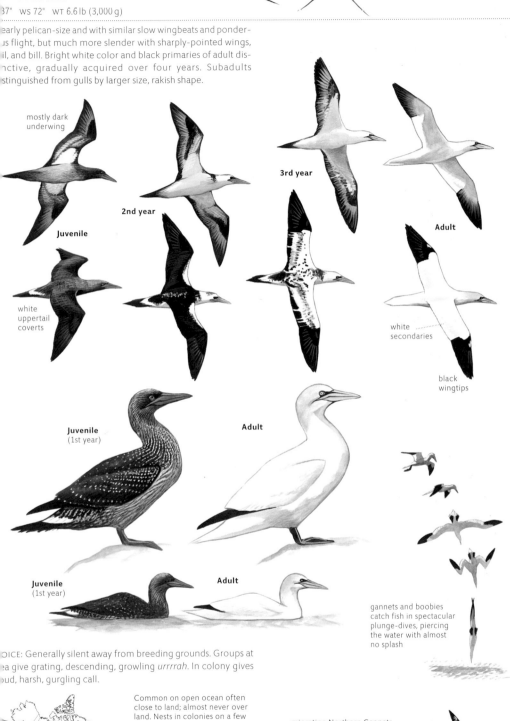

migrating Northern Gannets travel in short lines

Magnificent Frigatebird

Fregata magnificens

L 40" WS 90" WT 3.3 lb (1,500 g) ♀>♂

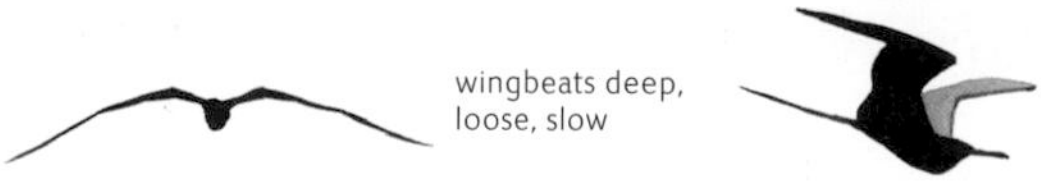

Sinister-looking, angular, and black, with long, sharply-pointed wings, and long forked tail usually held closed. Extremely buoyant in flight, soars and glides effortlessly for hours.

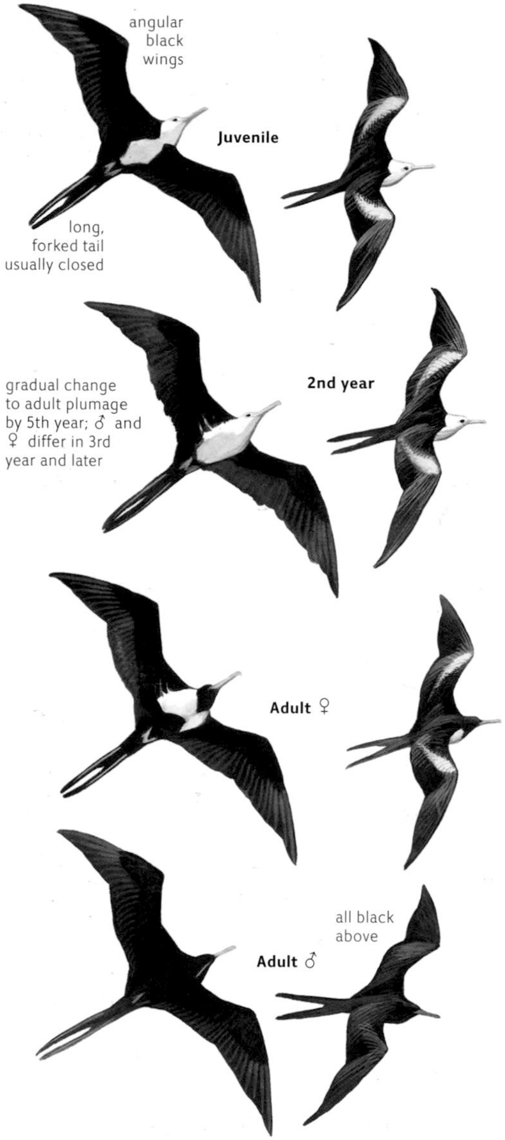

VOICE: Generally silent away from colony. Displaying male gives rapid bill-clattering and resonant, knocking sounds; female and young give high, wheezy sounds. Short, wheezy or grating calls in interactions: *wik wik wikikik*, etc.

Uncommon but conspicuous along beaches and over warm ocean in southern Florida and (during late summer) Texas. Rare visitor elsewhere. Roosts in trees or on navigational towers. Forages by sailing high in the air watching for fish or for other seabirds carrying fish, which it steals in aerial chase. Usually solitary, but roosts and nests in groups.

;reat Frigatebird

'egata minor

37" WS 85" WT 3.1 lb (1,400 g)

ery similar to Magnificent Frigatebird, adult male has pale car-
al bar above (all-black on Magnificent); adult female has gray
roat and red orbital ring. Subadult stages very difficult to dis-
nguish.

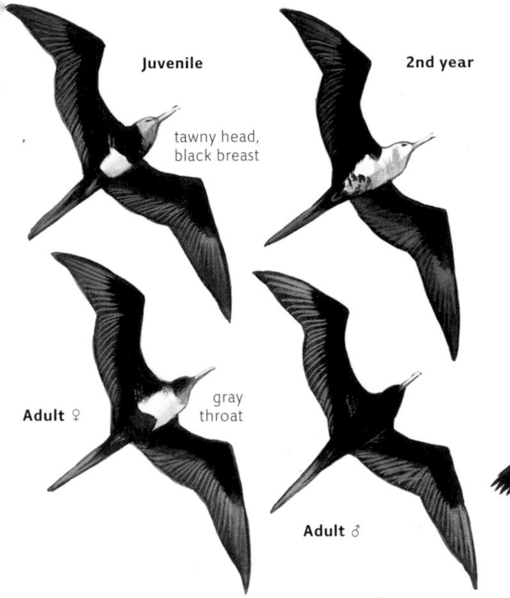

OICE: Generally silent away
om nesting colony.

Very rare visitor from tropical oceans, recorded several times in California, once in Oklahoma.

.esser Frigatebird

'egata ariel

30" WS 71" WT 1.7 lb (760 g)

ightly smaller than other frigatebirds, with white axillaries
all plumages. Juvenile has tawny head and black breast like
reat (unlike Magnificent).

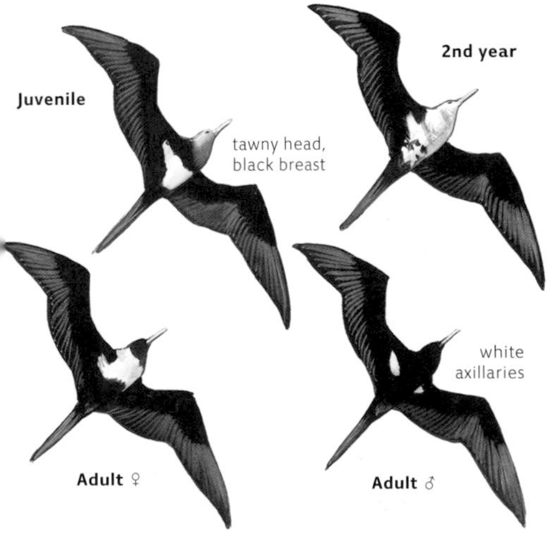

OICE: Generally silent away
om nesting colony.

Very rare visitor from tropical oceans, recorded several times on both coasts and interior.

Cormorants

FAMILY: PHALACROCORACIDAE, ANHINGIDAE

7 species in 2 families; generally blackish waterbirds that eat fish captured in underwater pursuit. Anhinga differs from cormorants in pointed bill, longer tail, silvery upperwing coverts, and habit of swimming almost completely submerged. Cormorants are distinguished from loons and other similar swimming birds by hooked bill tip, relatively long neck and tail, bare skin around chin often colored red to yellow. The six species of cormorants segregate largely by range and by habitat. The three eastern cormorants (top row) are often found on fresh water, fly with neck kinked, often fly high over land. The three Pacific species are found exclusively on salt water, rarely fly over land, and fly with neck straight. Juveniles are shown.

Genus *Phalacrocorax*

Great Cormorant, page 100

Double-crested Cormorant, page 101

Neotropic Cormorant, page 100

Brandt's Cormorant, page 98

Red-faced Cormorant, page 99

Pelagic Cormorant, page 99

Genus *Anhinga*

Anhinga, page 98

These two species often soar, and are best distinguished by Anhinga's triangular tail and long pointed bill. (Double-crested Cormorant has hourglass-shaped tail end.)

Anhinga

Anhinga anhinga

L 35" WS 45" WT 2.7 lb (1,250 g)

Swims completely submerged with only head exposed, spears fish with pointed bill. Long, fan-shaped tail, long pointed wings, long neck tapering to very sharp bill, and whitish upperwing coverts distinctive.

VOICE: Quite vocal when perched. A descending series of mechanical croaks, almost clicking: *krr kr krr kr kr krrrr krr*; also low, nasal, frog-like grunts.

Uncommon. Found in wooded swamps and along ponds and canals. Roosts in trees over water. May form small groups; migrates in flocks that soar like hawks.

Brandt's Cormorant

Phalacrocorax penicillatus

L 34" WS 48" WT 4.6 lb (2,100 g)

Larger than Pelagic Cormorant, with thicker neck and bi shorter tail, pale throat-patch, more graceful neck and mo upright stance. Also more gregarious, often feeding and flyir in flocks.

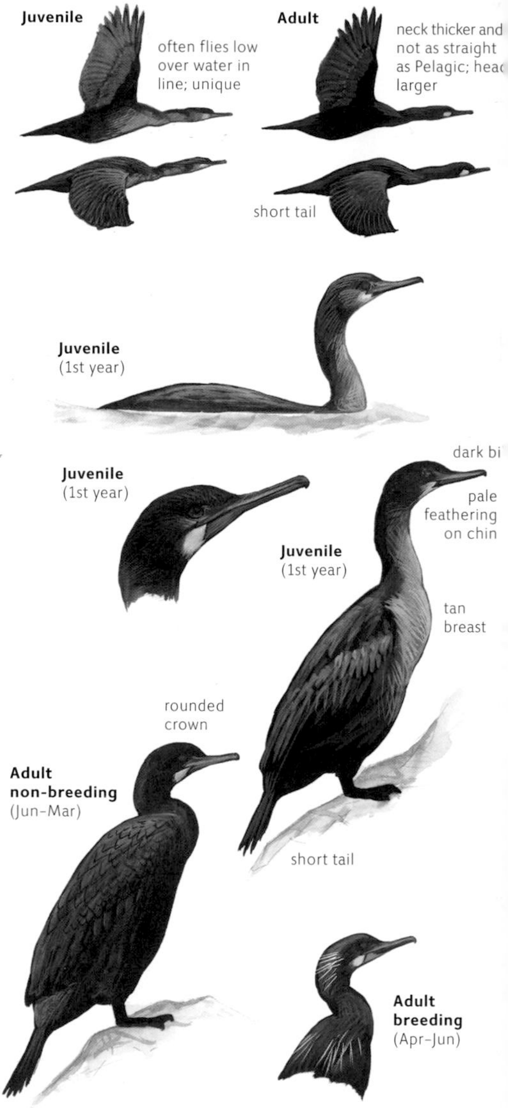

VOICE: Usually silent away from nest site. Hoarse guttural croa or growl given in aggression.

Common on ocean waters, foraging mainly just beyond sur but often far from land. Nests in colonies on flat or gently slopin coastal rocks. Often seen roosting in large groups on low rocks or flying in lines; may gather in large feeding flocks (Pelagic Cormorant solitary when feeding, may roost in sma groups). Dives for relatively large fish.

'elagic Cormorant

'halacrocorax pelagicus

20" WS 39" WT 3.9 lb (1,800 g)

ways all-dark (except white hip patch and neck plumes of 'eeding adult) with dark-red face, smaller and more slender an other cormorants, with extremely thin bill. Flies with very in neck held straight.

OICE: Usually silent away from nest site. Varied low grunts, painıl groaning, also hissing.

Common along rocky coasts. Nests and roosts on narrow ledges on sheer cliffs. Forages on ocean usually close to rocks. Typically seen singly, occasionally flies in lines of several birds, or joins small groups of flying Brandt's Cormorants.

Red-faced Cormorant

Phalacrocorax urile

L 29" WS 46" WT 4.6 lb (2,100 g)

Stockier than Pelagic Cormorant, with thicker neck and usually thicker bill; also note pale yellowish bill and, on adult, brighter red facial skin extending across forehead.

VOICE: Usually silent away from nest site. Low, droning, guttural croak.

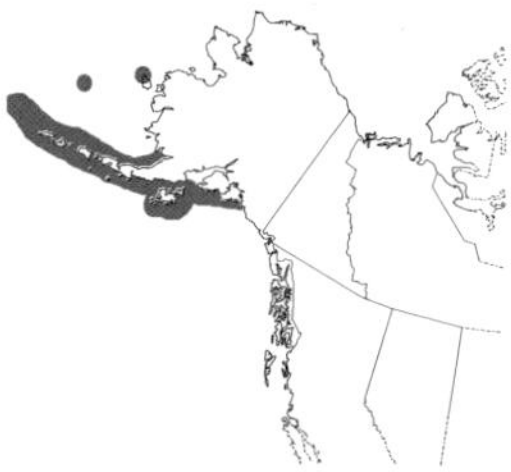

Common locally along rocky coasts. Usually solitary when foraging; gathers in groups on narrow cliff ledges when nesting and roosting. Dives for fish.

Great Cormorant

Phalacrocorax carbo

L 36" WS 63" WT 7.2 lb (3,300 g)

Our largest cormorant. Distinguished from Double-crested by larger size, yellowish chin-patch, grayish bill, and white throat, and by immature's clean white belly contrasting with brown neck.

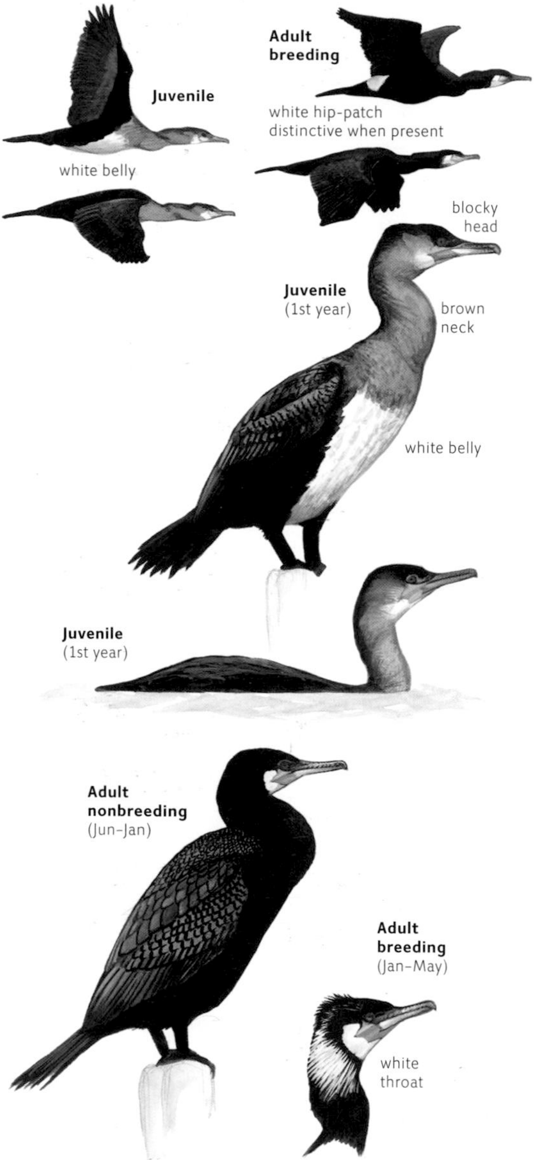

VOICE: Usually silent away from nest site. Guttural, grunting notes; loud, raucous *tock gock gock*; subdued, hoarse *fhi fhi fhi*.

Common within limited range. Found along rocky coastlines. Habits similar to other cormorants, but rarely seen on fresh water; often mixes with Double-crested.

Neotropic Cormorant

Phalacrocorax brasilianus

L 25" WS 40" WT 2.6 lb (1,200 g) ♂>♀

Best distinguished from Double-crested by smaller size wit relatively long tail and short bill. At close range note mostly gra bill and dark lores (with limited orange color), and V-shape border on chin-patch. Juvenile averages darker on breast tha Double-crested but varies from blackish to pale brown.

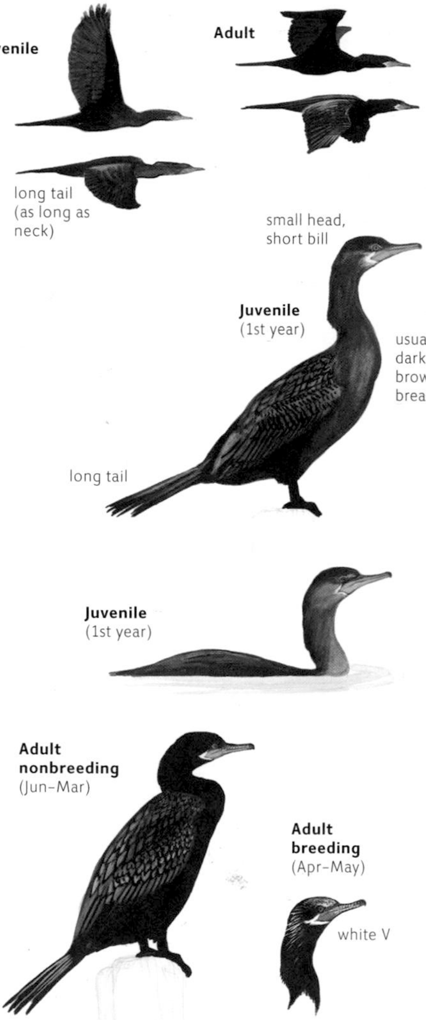

VOICE: Usually silent away from nest site. Low, short, frog-lik grunts and baritone croaking.

Common within limited range. Found on lakes, ponds, and river less numerous on protected saltwater bays. Habits similar to larger Double-crested and mixe with that species in all habitats, but Neotropic predominates on smaller and shallower fresh wate and is greatly outnumbered by Double-crested on salt water.

)ouble-crested Cormorant

ıalacrocorax auritus

ɜ3" WS 52" WT 3.7 lb (1,700 g) ♂>♀

many inland locations this is the only cormorant, but can be fficult to distinguish from Neotropic or Great Cormorant ıere range overlaps. Those three species all fly with kinked ɛck, unlike Pacific coast cormorants.

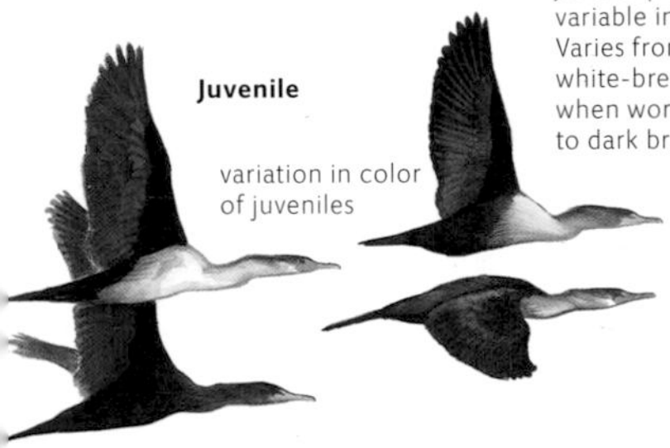

ɔICE: Usually silent away from nest site. Hoarse, bullfrog-like ·unting; clear-spoken *yaaa yaa ya*.

Common. Nests on islands or in dead trees standing in water. Forages in open water, including shallow water of ponds or tidal creeks to deep water of large lakes or open ocean. Often in flocks, often seen flying in lines or high overhead in ragged V formation.

Identification of Cormorants

Cormorants fly in untidy, wavering lines, generally less organized than geese, silent, and individuals often glide (geese never do)

Their feathers are not water repellent, so cormorants and Anhinga enter water only to feed and bathe, then spend much time on exposed perches with wings spread to dry

These three species are often seen in flight high over water or land (Pacific species usually low over water). Subtle differences in size and proportions allow identification, and average differences in immature color pattern are also helpful.

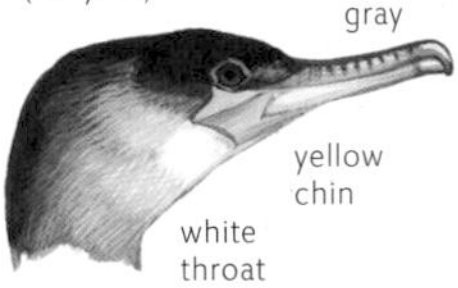

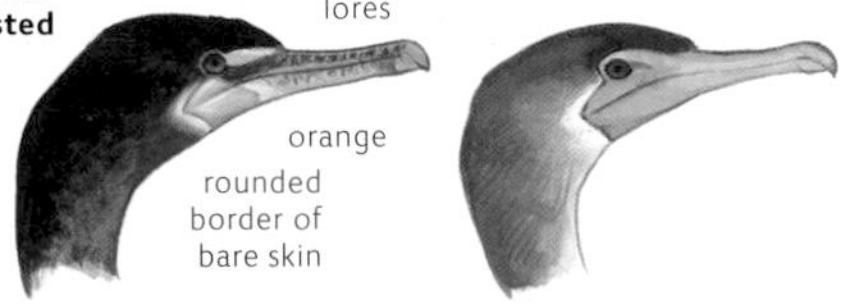

Tropicbirds

FAMILY: PHAETHONTIDAE

3 species in 1 genus. Superficially tern-like, with mostly white plumage and red or yellow bill, but with relatively small wings, shallow and rapid wingbeats. Adults have very long and slender central tail feathers, but these are not always obvious, and beware terns and other species trailing fishing line or other debris. The three species are best distinguished by details of upperwing pattern. Juveniles are shown below.

Genus *Phaethon*

Red-tailed Tropicbird, page 102

Red-billed Tropicbird, page 103

White-tailed Tropicbird, page 103

Red-tailed Tropicbird

Phaethon rubricauda

L 18" (adult to 31") WS 44" WT 2 lb (900 g)

Heavier than other tropicbirds, with relatively broad wings. At ages note primaries almost all white, and very little black on he or back. Reddish tail streamers of adult are not obvious, and oth tropicbirds can show off-white streamers.

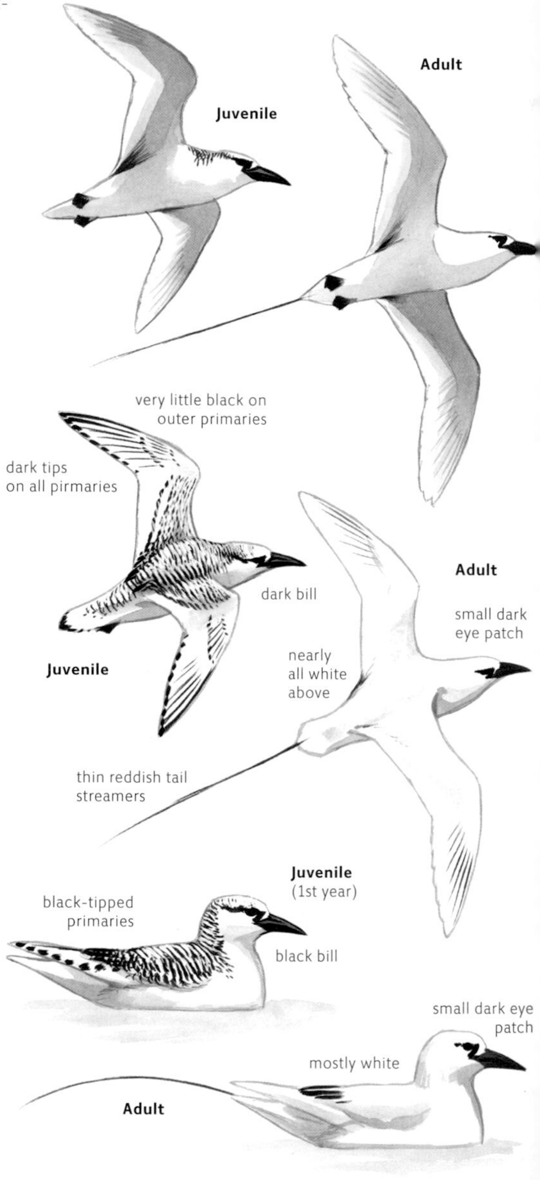

VOICE: Generally silent at sea

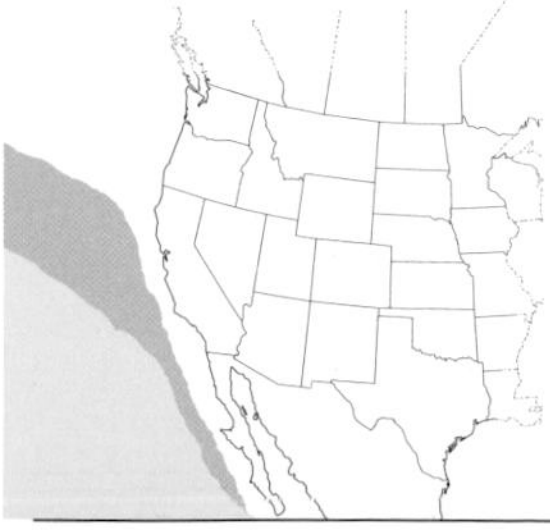

Very rare visitor from central Pacific to warm waters far offshore California. Like other tropicbirds generally seen flyin steadily and high over water, an often approaches and circles over boats.

ed-billed Tropicbird

naethon aethereus

18" (adult to 34") WS 44" WT 1.6 lb (750 g)

stinctly larger than White-tailed, with heavier bill. Note dark e-stripe continuing around back of head, barred wing coverts nd back. Best distinguished from immature White-tailed by ack on primary coverts.

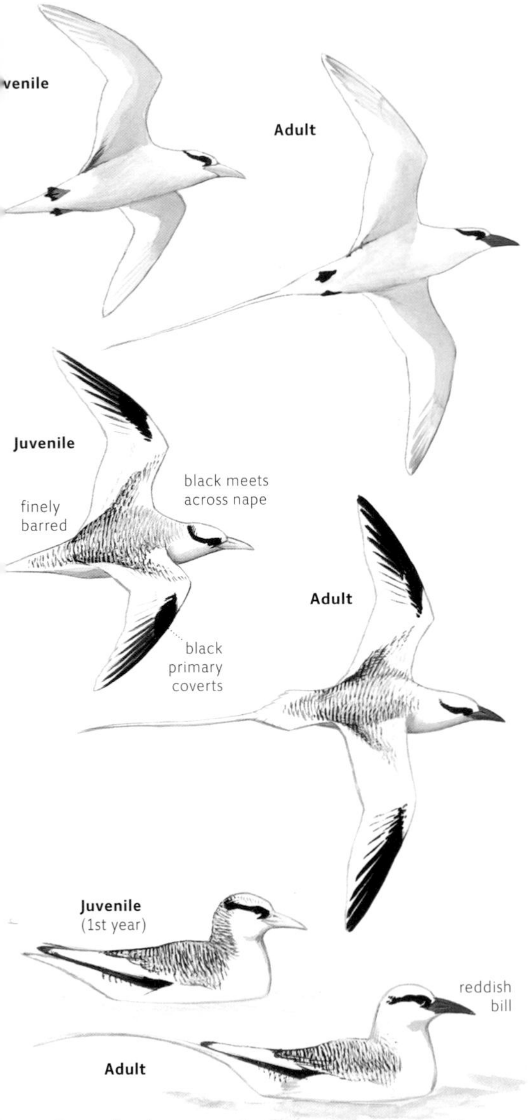

OICE: Generally silent at sea. Sometimes gives shrill grating or ttling notes. In display, shrill, harsh screams, recalling Common rn in quality.

Rare visitor from nesting grounds in tropical oceans to warm ocean waters far offshore. Most often seen resting on the water or in level flight about 50 feet above surface. Solitary; does not mix with other seabirds. Feeds on small fish captured in plunge-dive.

White-tailed Tropicbird

Phaethon lepturus

L 15" (adult to 29") WS 37" WT 11 oz (300 g)

Flies with steady, shallow, pigeon-like wingbeats. Royal Tern can be mistaken for tropicbirds, but has longer wings and deeper, more fluid wingbeats. Long tail streamers of adult tropicbirds are distinctive (but beware terns or gulls trailing grass or fishing line).

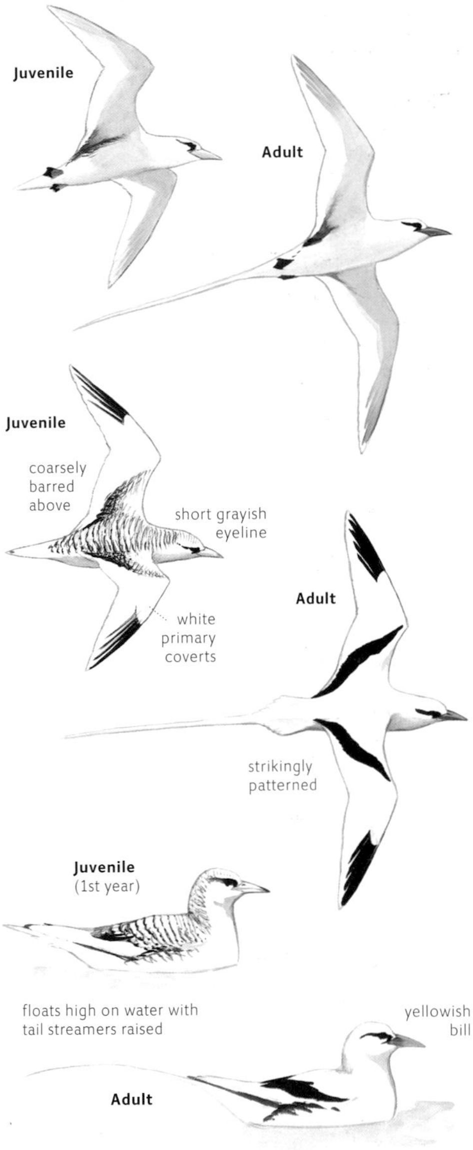

VOICE: Generally silent at sea. Voice harsh, squeaky, and ternlike.

Rare visitor from Caribbean to warm open ocean waters far offshore. Habits like other tropicbirds; nests in crevice high on cliff above ocean, otherwise seen perched on land only when exhausted, for example when blown inland by a tropical storm.

Wading Birds

FAMILIES: ARDEIDAE, THRESKIORNITHIDAE, CICONIIDAE, PHOENICOPTERIDAE

24 species in 4 families, 14 genera (several rare or escaped species not shown here). Bitterns, herons, and egrets, in family Ardeidae, have pointed bills and seize prey in lightning-quick forward strike, fly with neck coiled (all others fly with neck straight). Ibises and Roseate Spoonbill, in Threskiornithidae, have downcurved or spatulate bill, forage by walking with head down, probing or sweeping bill in water and mud. Wood Stork and Jabiru, in Cic niidae, are very large with massive bills, move slowly throu shallow water with varied foraging methods, often soar. Flami gos, in Phoenicopteridae, are incredibly tall and slender, fora by sweeping angled bill along surface of shallow water.

Genus *Ardea*

Great Blue Heron, page 108

Great Egret, page 109

Genus *Egretta*

Little Egret, page 110

Snowy Egret, page 109

Little Blue Heron, page 111

Tricolored Heron, page 111

Reddish Egret, page 112

Genus *Bubulcus*

Cattle Egret, page 114

enus *Botaurus*

nerican Bittern, page 107

Genus *Ixobrychus*

Least Bittern, page 107

Genus *Butorides*

Green Heron, page 114

Genus *Nycticorax*

Black-crowned Night-Heron, page 115

enus *Nyctanassa*

ellow-crowned Night-Heron, ige 115

Genus *Eudocimus*

White Ibis, page 116

Genus *Plegadis*

Glossy Ibis, page 117

White-faced Ibis, page 117

enus *Platalea*

oseate Spoonbill, page 118

Genus *Phoenicopterus*

American Flamingo, page 119

Genus *Mycteria*

Wood Stork, page 106

Wood Stork

Mycteria americana

L 40" WS 61" WT 5.3 lb (2,400 g) ♂>♀

White body, black flight feathers, and heavy bill distinctive. Larger than most other wading birds, with very methodical and slow foraging movements, standing still or walking slowly, often in lines or small groups. Often seen soaring in groups, sometimes with vultures.

VOICE: Usually silent after 1st year except for hissing and bill-clattering in nest displays. Young bird occasionally gives nasal, barking *nyah nyah nyah*.

Uncommon and local in muddy ponds where receding water levels concentrate fish, even very small and isolated ponds can attract storks. Nests and roosts in colonies in trees. Forages singly or in small groups, holding bill open in water and seizing fish.

Jabiru

Jabiru mycteria

L 55" WS 96" WT 14 lb (6,500 g) ♂>♀

Huge, much larger than Wood Stork, with massive black bi body and wings entirely whitish. Most records in North Ameri involve immatures, with dusky smudges on wing coverts.

VOICE: Generally silent.

Very rare visitor from Mexico in late summer, recorded as far north and east as Oklahoma an Mississippi. Look for it in shallo ponds or wet grassy meadows, o soaring.

merican Bittern

otaurus lentiginosus

28" WS 42" WT 1.5 lb (700 g)

rge and heavy-bodied with relatively long neck tapering to ointed bill. Distinguished from night-herons by richer brown olor, bold whisker stripe, longer bill, tapered neck, secretive abits, and dark flight feathers.

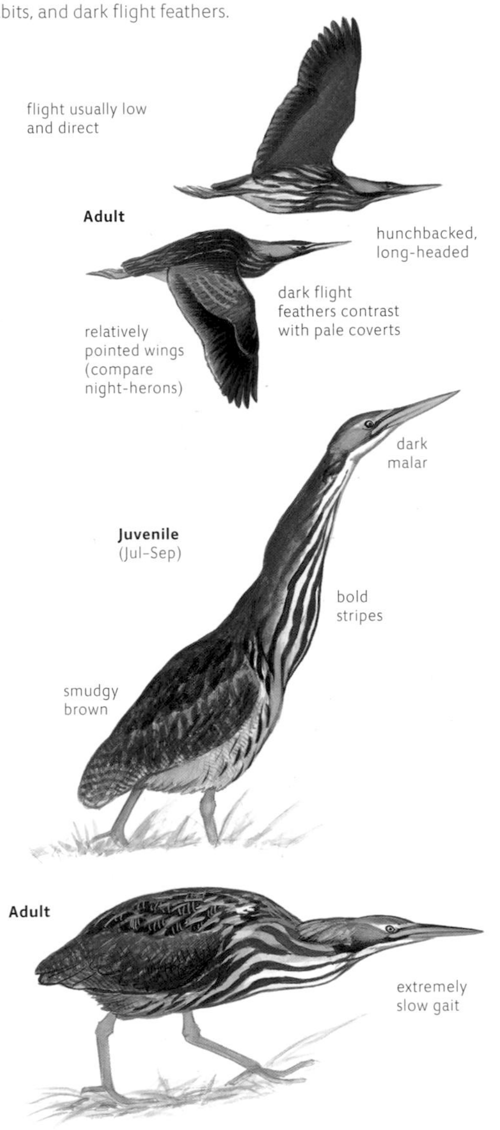

OICE: Song a deep, gulping, pounding *BLOONK-Adoonk* epeated. When flushed a rapid, throaty *kok kok kok*. In flight a oud, hoarse, nasal *squark* intermediate between night-herons nd Mallard.

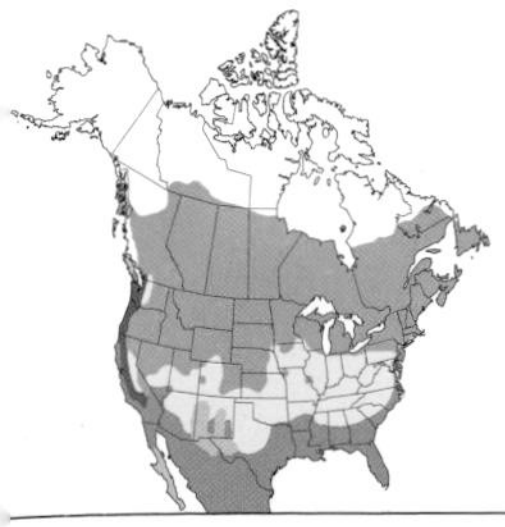

Uncommon and very inconspicuous. Nests and forages in expansive marshes or meadows with dense grassy or reedy vegetation, often with shrubs. Occasionally seen in other grassy habitats.

Least Bittern

Ixobrychus exilis

L 13" WS 17" WT 2.8 oz (80 g)

Our smallest heron; small size and pale buff color distinctive. Long neck can be extended but is usually held coiled as bird crouches in dense reeds.

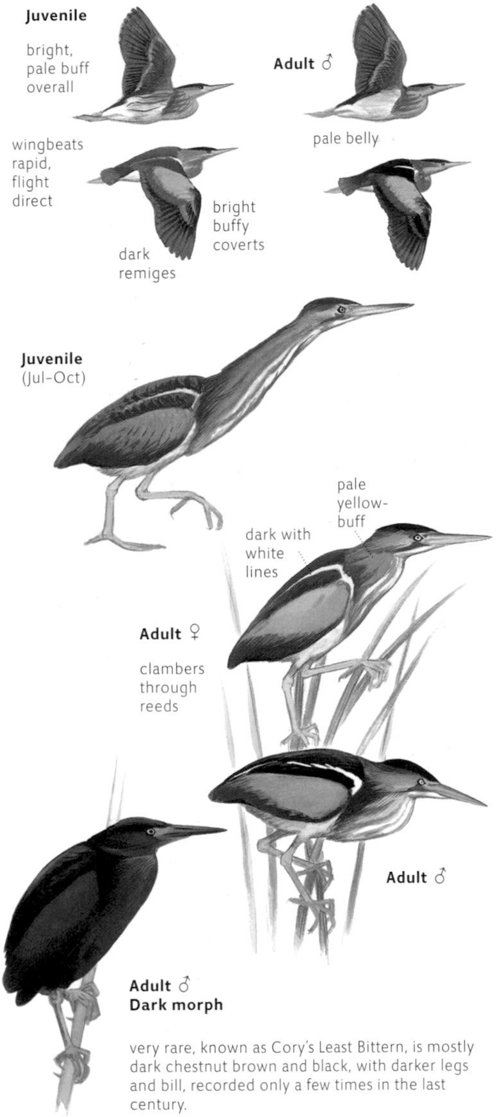

VOICE: Song a low, cooing *poo-poo-poo* descending, like Black-billed Cuckoo but lower-pitched with wooden, not whistled, quality. Common call (year-round) loud, harsh, quacking, rail-like *rick-rick-rick-rick*. In flight short, flat *kuk* or *gik*.

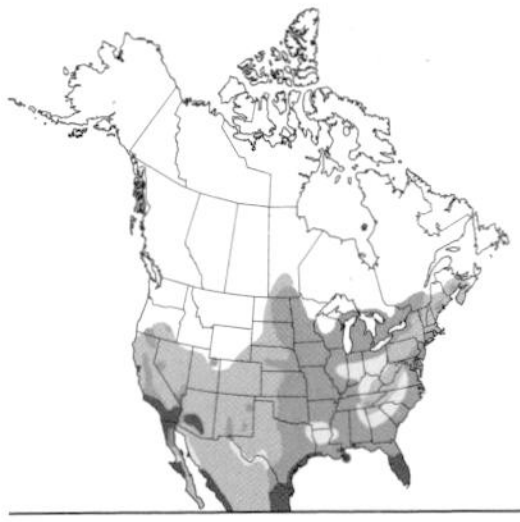

Uncommon and very inconspicuous. Nests and forages (for small fish) in marshes with dense reedy vegetation. Look for it where dense reeds meet open water.

Great Blue Heron

Ardea herodias

L 54" WS 72" WT 6.4 lb (2,900 g) ♂>♀

differs from Sandhill Crane by coiled neck, dark morph has more patterned plumage

Our largest and heaviest heron, grayish overall (except white in southern Florida). Compare Sandhill Crane.

VOICE: Flight call a very deep, hoarse, trumpeting *fraaahnk* or *braak*; often heard at night. In aggression a slow series *fraank fraank fraank taaaaw taaaaw*; last notes lower, croaking. Voice of all forms similar.

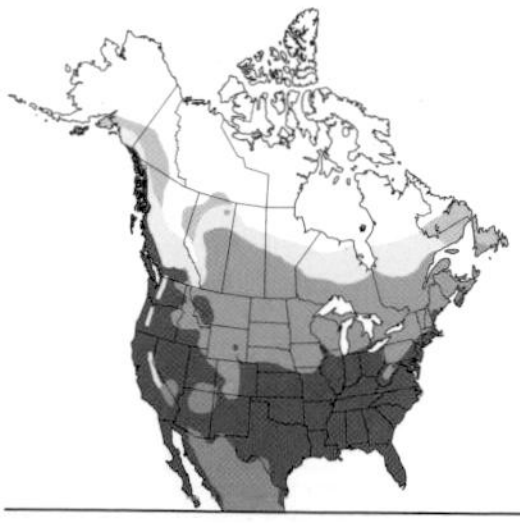

Common. Nests in colonies in trees standing in water. Forages in any open grassy to shallow water habitat. Usually solitary, but roosts and migrates in groups. Often more active at night.

Great Egret

rdea alba

39" WS 51" WT 1.9 lb (870 g) ♂>♀

 most areas the largest white heron, much taller than Snowy gret and relatively long-necked. Note yellow bill and black legs nd feet.

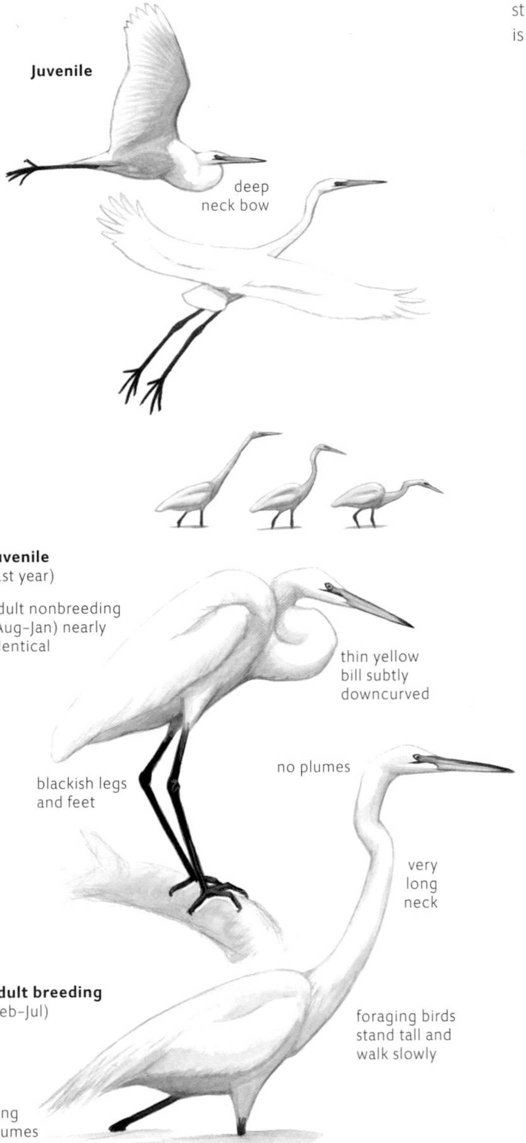

OICE: Very deep, low, gravelly *kroow*, grating unmusical *karrrr*, nd other low croaks; fading at end; lower and coarser than Great lue Heron without trumpeting quality.

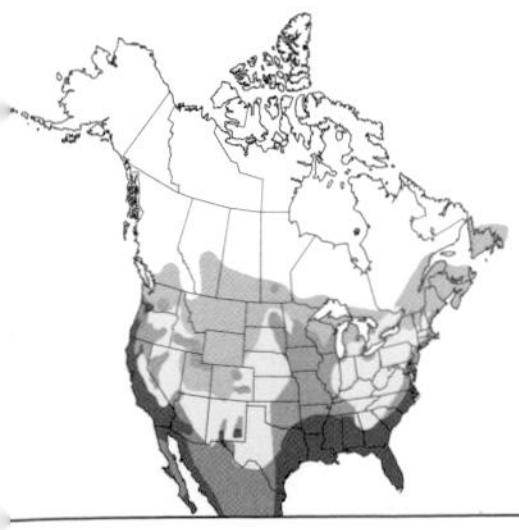

Common. Nests in trees on islands or standing in water. Forages in shallow water with or without emergent vegetation, sometimes in grassy fields. Often in loose groups where prey is abundant.

Snowy Egret

Egretta thula

L 24" WS 36" WT 13 oz (360 g)

Always all-white. Smaller than Great Egret and larger than Cattle Egret, with dark bill and yellow feet, often very active when foraging. Can be extremely similar to immature Little Blue Heron, study foraging posture and details of bill and leg color. Little Egret is also very similar, but rare.

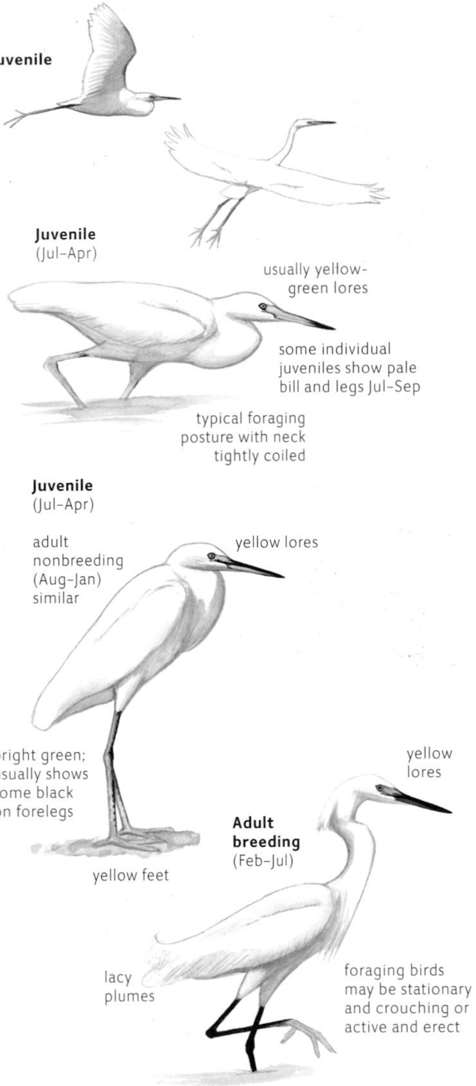

VOICE: Hoarse, rasping *raarr* or nasal *hraaa* very similar to Little Blue Heron; higher and more nasal than Great Egret. In flight occasionally a hoarse cough *charf*.

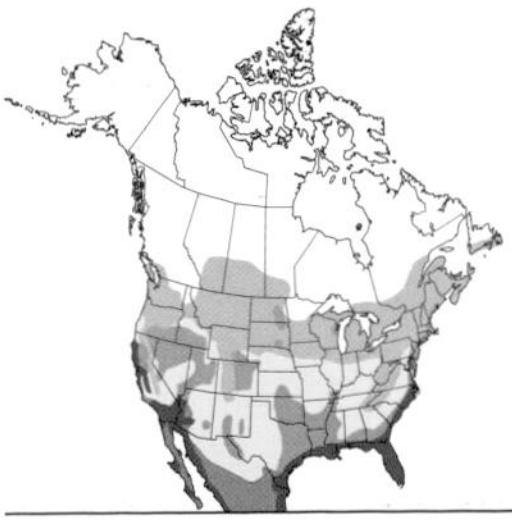

Common. Nests in trees on islands or standing in water. Forages in shallow water with or without emergent vegetation, sometimes in grassy fields. Often in loose groups where prey is abundant.

Little Egret

Egretta garzetta

L 24" WS 36" WT 16 oz (450 g)

Very rare visitor from Eurasia, nearly identical to Snowy Egret in appearance and behavior. Adults are readily distinguished by long head plumes (but these may be missing Aug–Dec), immatures average slightly larger than Snowy, with more black on legs

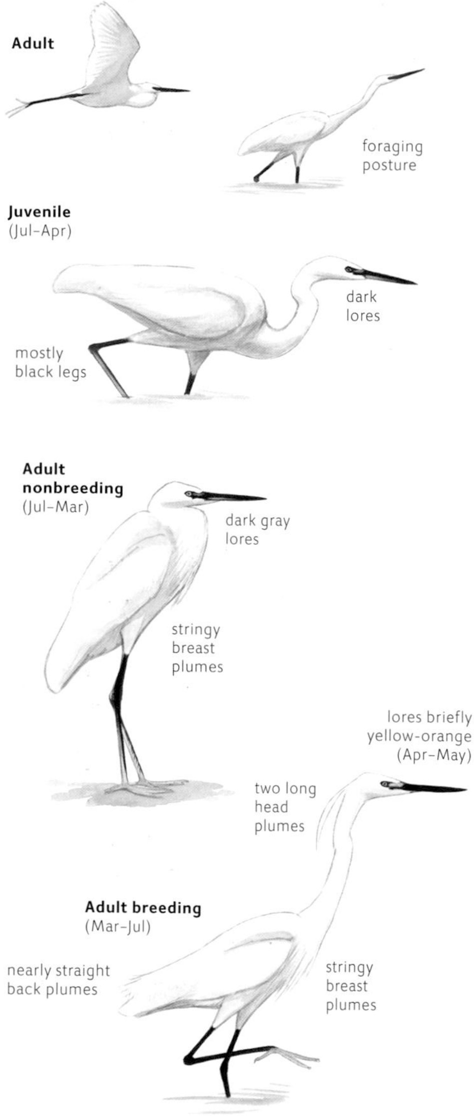

VOICE: Presumably similar to Snowy Egret.

Very rare visitor to Atlantic coast with records from Newfoundland to Virginia, once in Alaska. Found in marshy ponds and tidal creeks with Snowy Egret.

Western Reef-Heron

Egretta gularis

L 25" WS 36" WT 16 oz (450 g)

Closely related to Little Egret, but plumage mostly dark. Distir guished from adult Little Blue Heron by more slende proportions, active foraging habits, white chin, and yellow fee

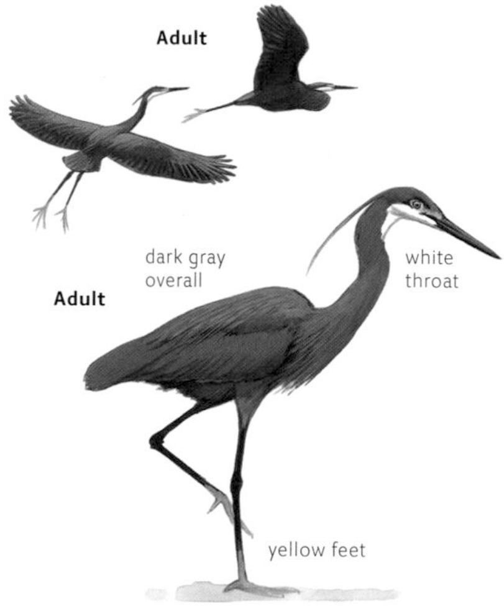

VOICE: Presumably similar to Snowy Egret.

Very rare visitor from Africa to Atlantic coast, only a few record from New Jersey to Newfoundland, Apr–Sep. Found in shallow salt water on tidal flats, beaches, and tidal creeks.

Snowy Egret vs. Little Egret

Some juvenile Snowy Egrets have dark gray lores at fledgin (Jul–Sep), like Little Egret, changing within days or weeks t greenish. Very young birds of these species may be indistinguish able. Little Egret has, on average, a longer and less tapered bil feathering extending farther forward on forehead and chin, an more black on legs, but all of these features overlap.

Snowy Egret vs. Little Blue Heron

Some juvenile Little Blue Herons have greenish lores like Snow Egret. Little Blue Heron has a subtly rounder head, thicker-basec bill, larger eye with more "staring" expression, and pale greenisł legs with no black, but best distinguished by foraging posture.

ittle Blue Heron

gretta caerulea

24" WS 30" WT 12 oz (340 g)

small heron, adults always dark blue-gray, immatures always hite and extremely similar to Snowy Egret, best distinguished y foraging posture with neck extended forward, greenish legs nd feet, slightly thicker bill with pale base.

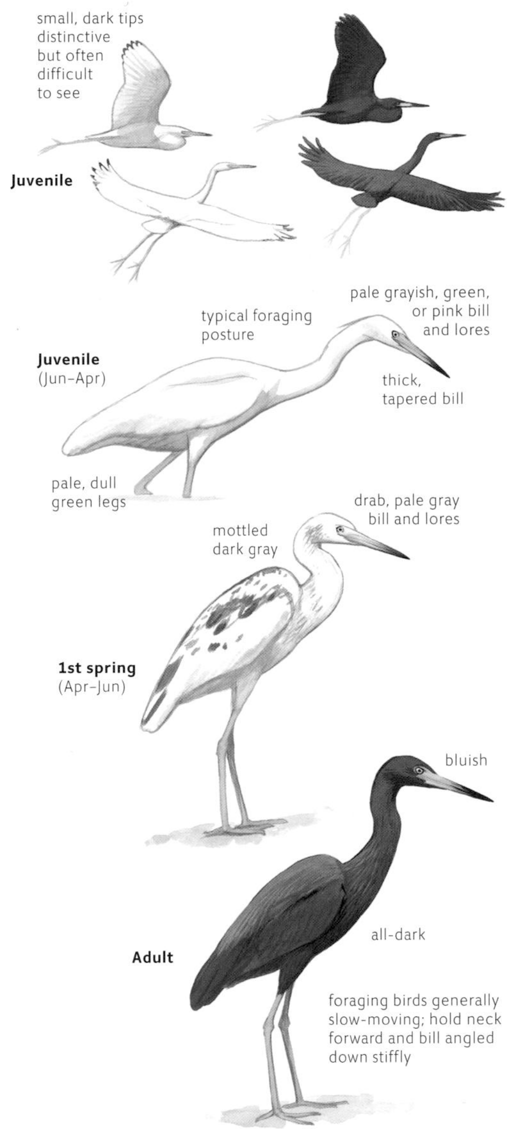

OICE: Various hoarse squawks, fairly high *raaaaa raaa*; some-mes with trumpeting or squealing quality.

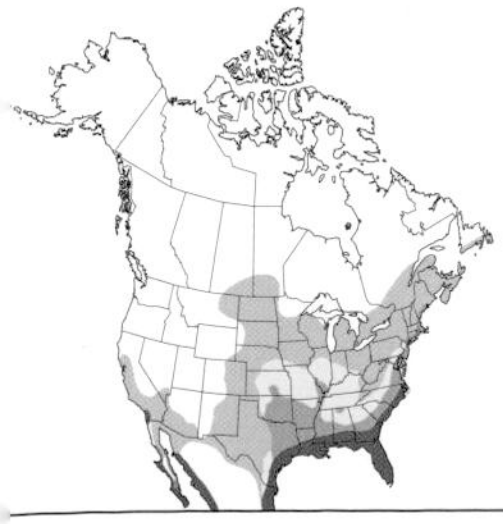

Common. Nests in trees on islands or standing in water. Forages in shallow water usually with emergent vegetation, sometimes in grassy fields. Generally prefers more fresh water and edges of grassy pools than Snowy Egret, and usually solitary (vs. Snowy often gathers in dense groups where prey is abundant; in that case look for Little Blue Heron away from the group.)

Tricolored Heron

Egretta tricolor

L 26" WS 36" WT 13 oz (380 g)

A small heron with very long and slender neck and bill, always has clean and contrasting white belly and white underwing coverts, unlike other grayish herons. Foraging habits like Snowy Egret but even more active, often running after fish.

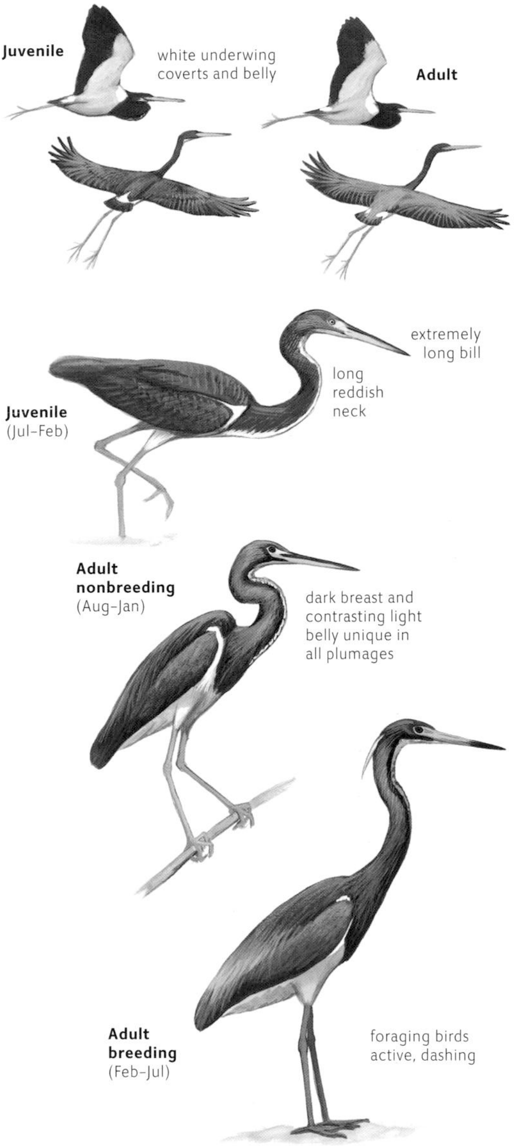

VOICE: Soft, nasal moaning; usually lacking scratchy or rasping quality of most other herons and egrets. Similar to ibises.

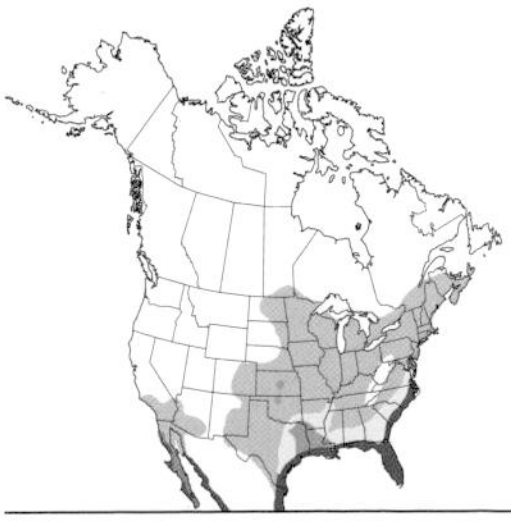

Uncommon. Nests in trees on islands or in standing water. Forages in shallow water with or without emergent vegetation. Usually solitary, but sometimes joins other herons where prey is abundant.

Reddish Egret

Egretta rufescens

L 30"   WS 46"   WT 1 lb (450 g)

Either all-white or all-dark, a relatively large heron with long neck and long gray legs. Bill and lores dark (except pink with dark tip on breeding adults), and adults have shaggy plumes all over head and neck. Very active "dancing" behavior is distinctive. Color of bill and legs offer best clues for identifying white morph.

Distinctive animated actions of foraging Reddish Egret

White Morph

Dark Morph

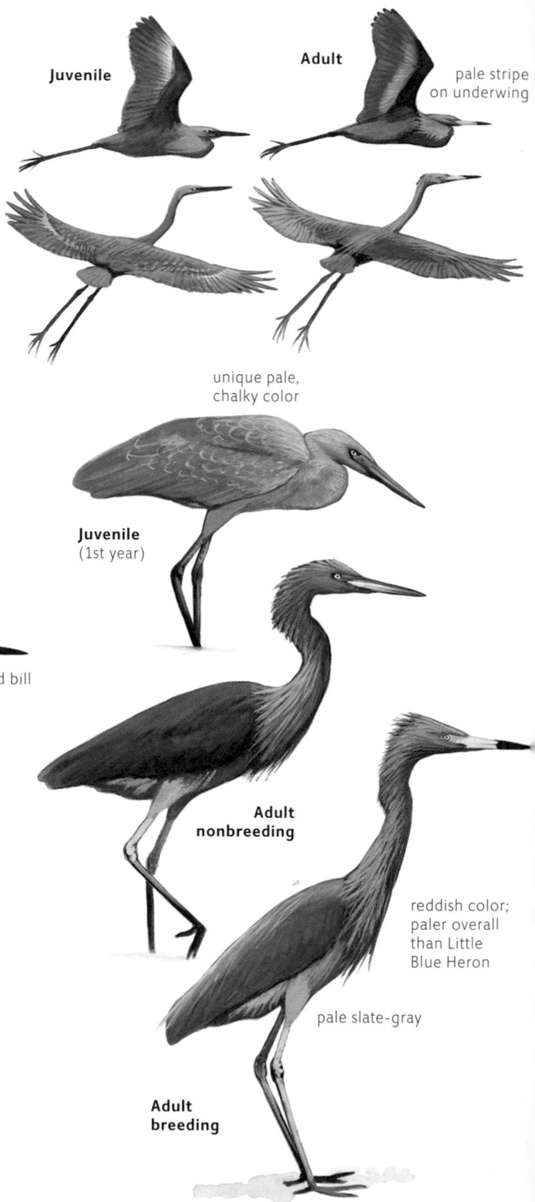

VOICE: Calls relatively infrequent, soft groan and short grunt, not rasping; similar to Tricolored Heron but lower-pitched.

Uncommon. Nests in mixed species colonies on islands or in low shrubs in standing water. Forages in expanses of shallow water such as on broad tidal flats. Almost always solitary and not associating closely with other herons. Total US population estimated under 5,000 birds.

Two morphs with few intermediates. In Gulf of Mexico, white morph birds represent 2 to 7 percent of the total. Pacific population is all-dark.

ray Heron

rdea cinerea

37″ WS 52″ WT 3.3 lb (1,500 g)

ery similar to Great Blue Heron, and identification in North merica is challenging. A Gray Heron should be noticeable by naller size, shorter legs and neck, and cleaner gray and white lumage. But Great Blue Herons can approach those same features, so confirming the ID will depend on careful examination f details.

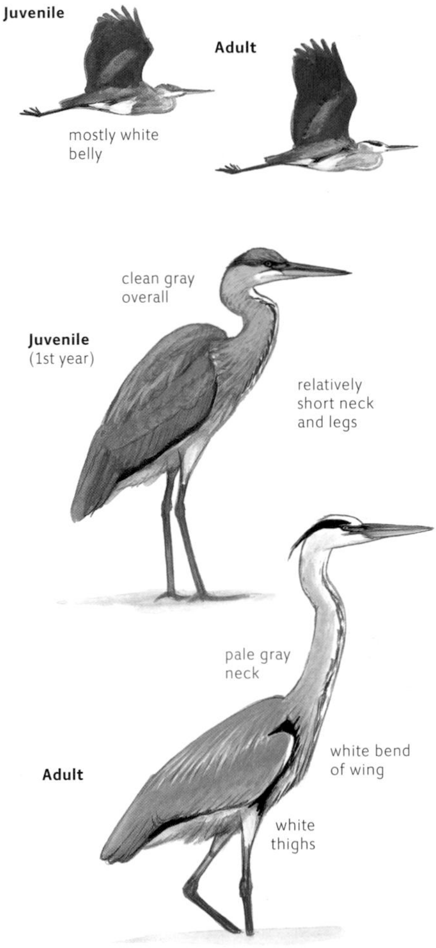

OICE: Similar to Great Blue Heron but averages higher-itched.

Very rare visitor from Eurasia to North America with only a few records in Alaska and Newfoundland.

Courtship Colors

In early spring, during courtship at the beginning of the nesting cycle, adult herons and egrets develop much brighter bill and lore colors. Colors reach peak intensity for a few days, and should be expected only near an active nesting colony. Through most of the breeding season bill and lore colors are as shown in the adult breeding illustrations in the species accounts. Colors of legs and feet also become brighter in some species.

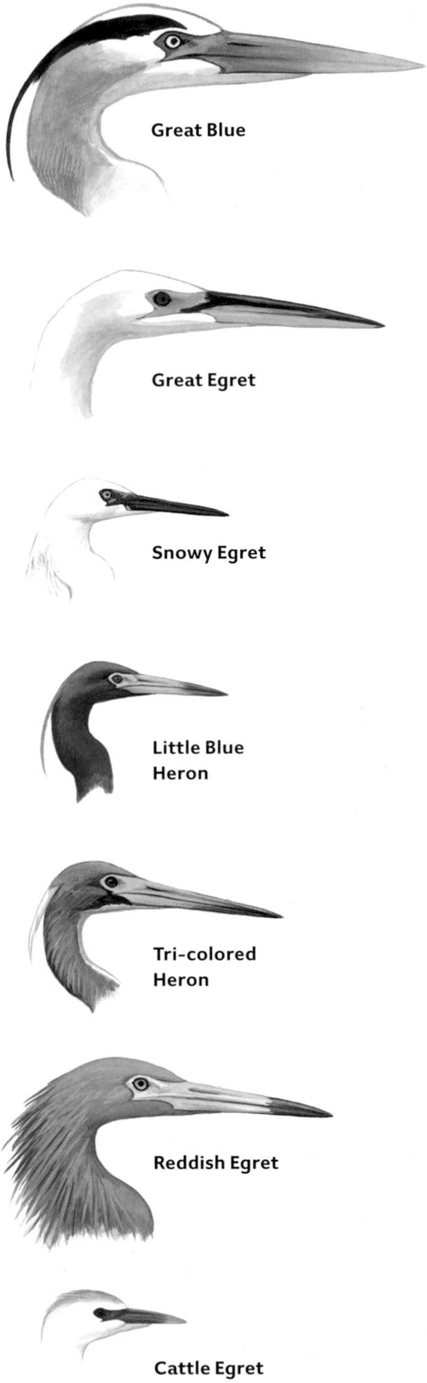

Cattle Egret

Bubulcus ibis

L 20" WS 36" WT 12 oz (340 g)

Smallest of the white herons, slightly smaller than Snowy Egret with much shorter legs and bill, thicker neck. Combination of small size, white plumage, and all-dark legs distinctive,

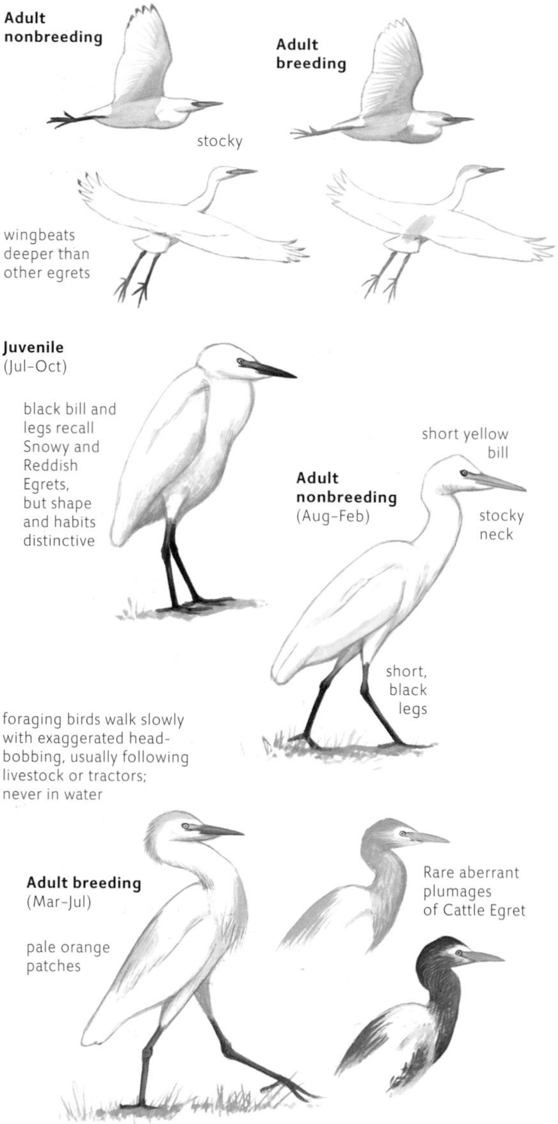

VOICE: Short croaks or quacks on breeding grounds; generally silent elsewhere. Most common year-round call a subdued, nasal quack *brek* or *rick rak*; occasionally a short, soft moan.

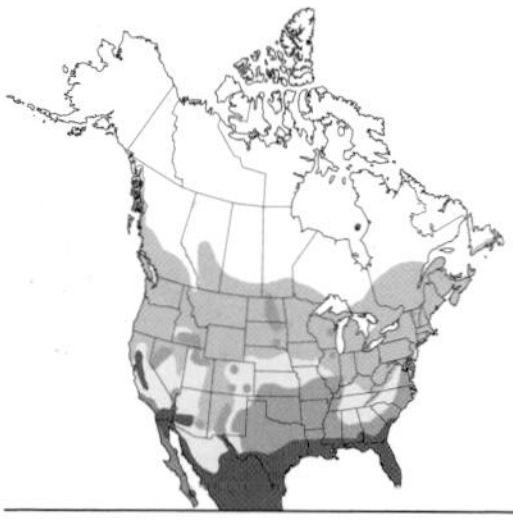

Common locally. Nests in trees or shrubs with other herons. Forages in upland fields with short grass, often following livestock or tractors to catch insects disturbed. Usually in loose groups, often flying in short lines.

Green Heron

Butorides virescens

L 18" WS 26" WT 7 oz (210 g)

A very small and stocky heron. Very dark overall without contras ing markings. Note stocky shape, short broad wings, and sho legs.

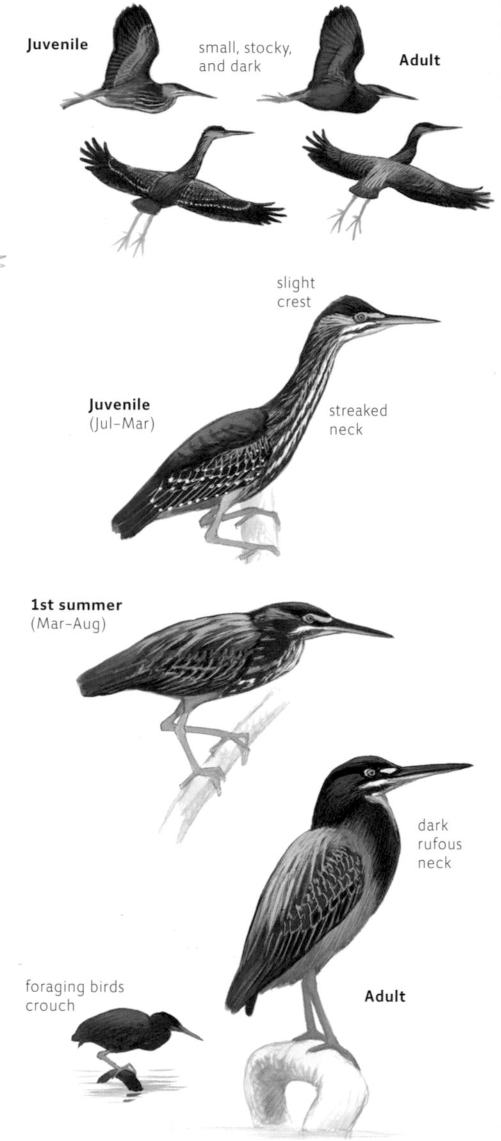

VOICE: Common call in flight an explosive, sharp, swallowe *SKEEW* or *skeow*. When nervous an irregular series of low, knock ing notes *kuk kuk kuk kuk*....

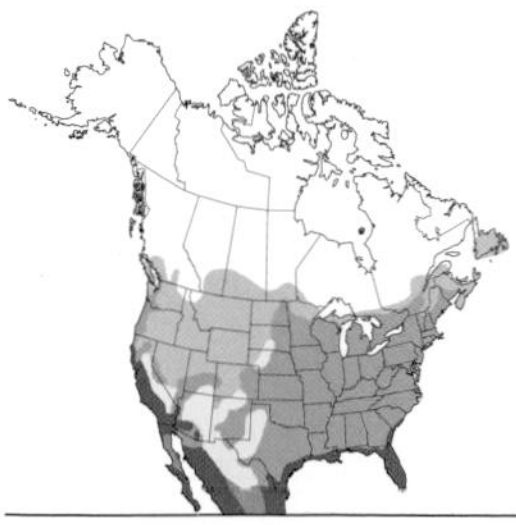

Uncommon and inconspicuous. Nests and roosts singly in dense trees or shrubs over water. Forages most typically along tree-lined or tree-covered streams and ponds, but also seer in more open grassy or weedy marsh or pond-edge habitats

lack-crowned Night-Heron

ycticorax nycticorax

25" WS 44" WT 1.9 lb (870 g)

stocky, large-headed heron that rarely extends its neck. Imma-ure can be difficult to distinguish from Yellow-crowned ight-Heron; note paler color, large white spots on upperwing, nd more tapered bill with yellowish base (also compare Amer-can Bittern).

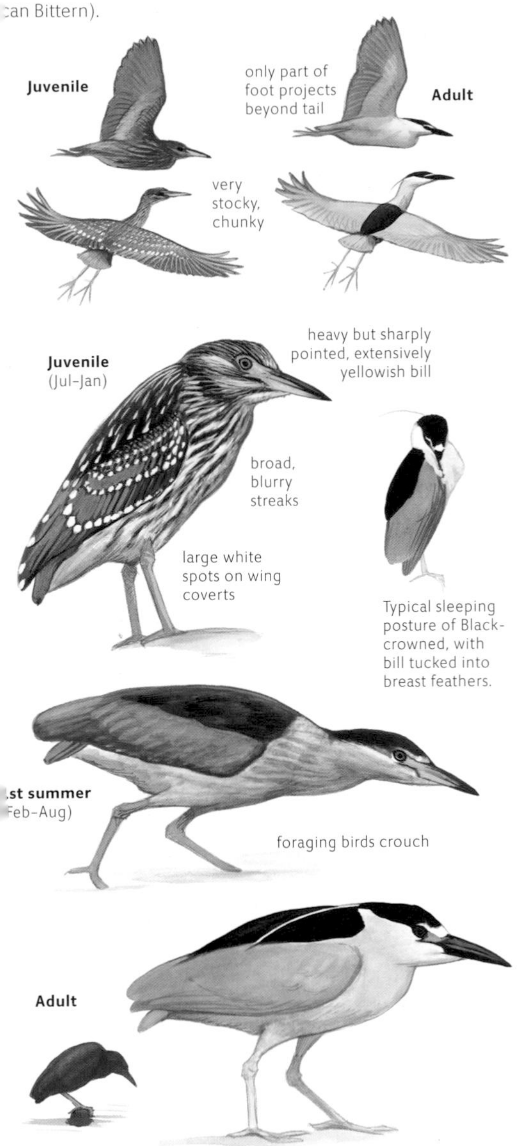

OICE: Common call in flight a flat, barking *quok* or *quark*. Other alls of similar quality given in nesting colony.

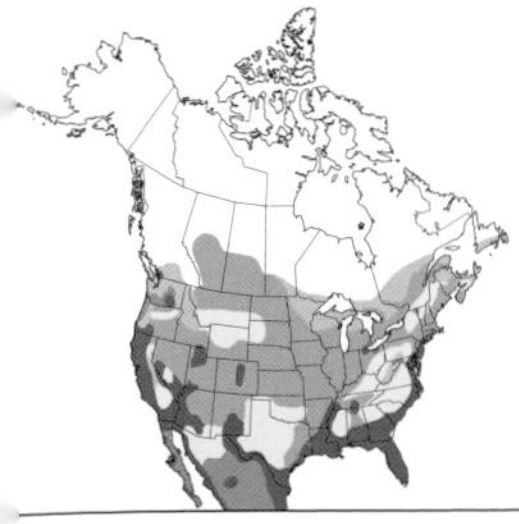

Common but inconspicuous. Nests colonially in trees or shrubs on islands or over water, often with other herons. Roosts in trees, shrubs, or reeds dense enough to provide cover during day. Forages mainly at twilight and at night in shallow water at edges of ponds or streams. Usually solitary.

Yellow-crowned Night-Heron

Nyctanassa violacea

L 24" WS 42" WT 1.5 lb (690 g)

Longer-necked (neck often extended) and longer-legged than Black-crowned Night-Heron, with stouter bill. Immature differs from Black-crowned by much smaller white spots on upperwing coverts, darker gray flight feathers, and mostly dark bill.

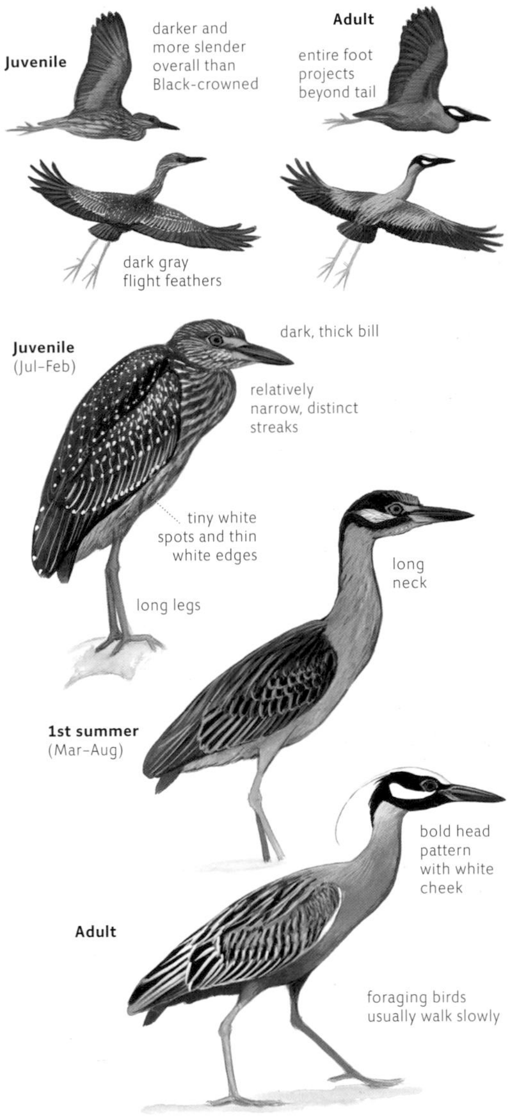

VOICE: Common call in flight a high, squawking bark *kowk* or *kaow* similar to Black-crowned but higher, more crow-like; approaching Green Heron but deeper and less sharp.

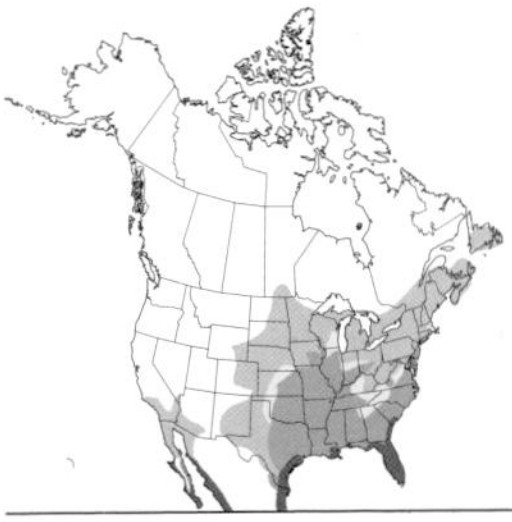

Common but inconspicuous. Nests colonially hidden in trees or shrubs usually on islands or over water, often with other herons. Roosts in trees, shrubs, or reeds dense enough to provide cover during day. Forages mainly on crabs at twilight and at night in shallow water at edges of ponds or streams. Usually solitary.

White Ibis

Eudocimus albus

L 24" WS 42" WT 1.5 lb (690 g)

Larger and heavier than other ibises. Adult all white with black wingtips, and red-orange bill and legs. Immature mostly brown, distinguished from Glossy by orange bill and legs, white rump and underparts. Adult from egrets by bill color and shape and by different foraging habits and flight.

VOICE: In flight a harsh, nasal *urnk urnk urnk* lower than other ibises.

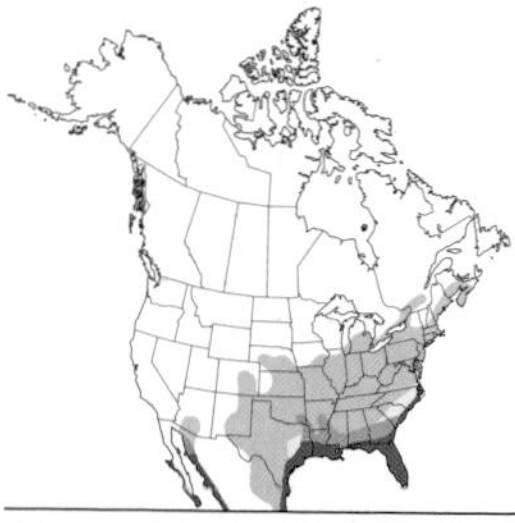

Common. Forages by probing in mud in shallow water, either in wide shallow pools, grassy marshes or meadows, often on wet lawns. Nests and roosts colonially in shrubs or small trees.

Scarlet Ibis

Eudocimus ruber

L 25" WS 38" WT 2 lb (900 g) ♂>♀

Identical to White Ibis except red plumage of adult. Hybridize with White Ibis, producing pink offspring.

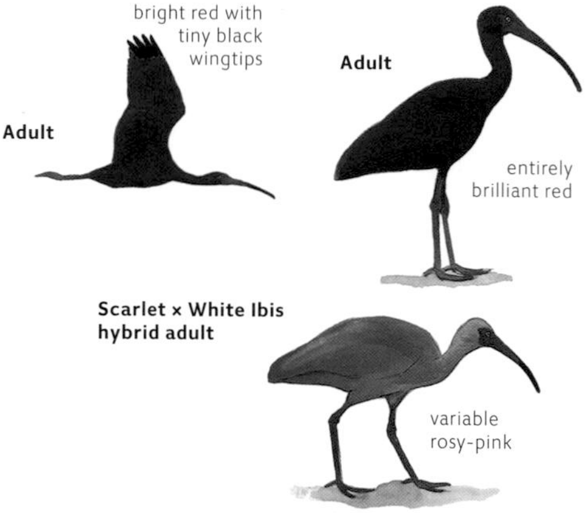

VOICE: Similar to White Ibis.

Occasional records in Southern Florida and elsewhere are presumably escaped from captivity. Native to South America. Habitat and habits similar to White Ibis.

Identification of Dark Ibises

Variation in face pattern of breeding adults is shown here; note that the face pattern of White-faced Ibis can nearly match Glossy but most are easily separated by face color and eye color. Apparent hybrids (such as the intermediate bird shown at bottom) are rare but regular, and not all are identifiable.

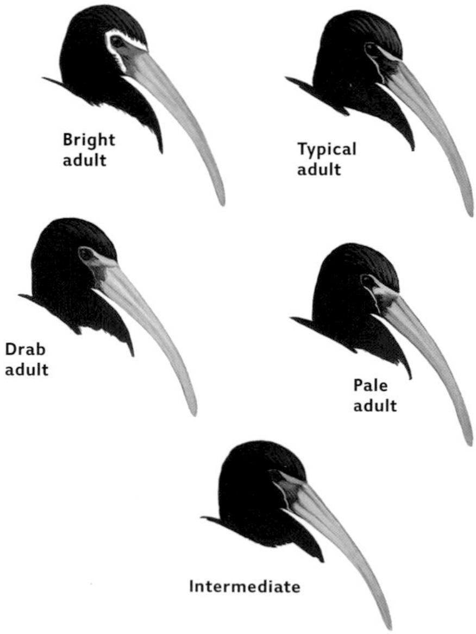

White-faced Ibis

Plegadis chihi

L 23" WS 36" WT 1.3 lb (610 g) ♂>♀

Adult has red iris and facial red skin; with white feathering around face in breeding season. Immature may be indistinguishable from Glossy Ibis; some are paler overall, with paler and more golden-green iridescence on upperparts.

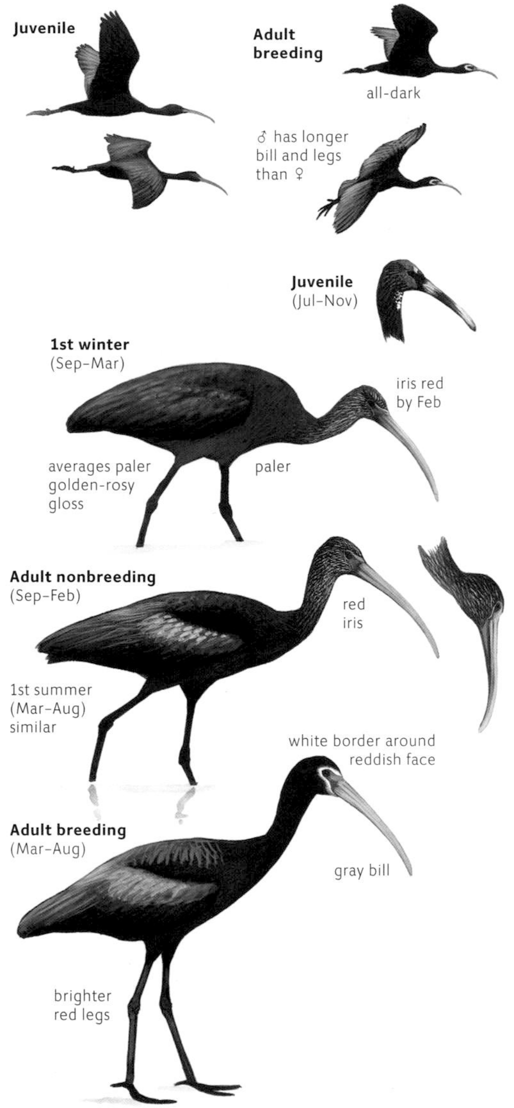

VOICE: Apparently identical to Glossy Ibis: when flushed a nasal, moaning *urnn urnn urnn* or a rapid series of nasal quacks *waa waa waa waa*.... Birds in feeding flock give soft, nasal, often doubled grunt *wehp-ehp.*

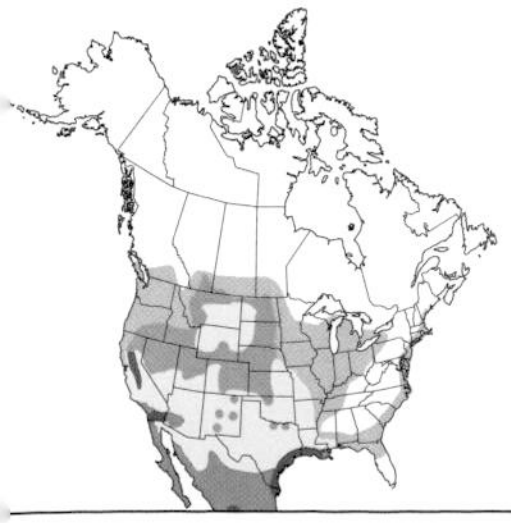

Locally common. Forages by probing in mud in shallow water, either in wide shallow pools, grassy marshes or meadows. Nests and roosts colonially in large reed beds, or in shrubs or small trees.

Glossy Ibis

Plegadis falcinellus

L 23" WS 36" WT 1.2 lb (550 g) ♂>♀

Adult has dark blue-black face with pale borders, dark eye, and lacks white feathering around face in breeding season. Immature may be indistinguishable from White-faced Ibis.

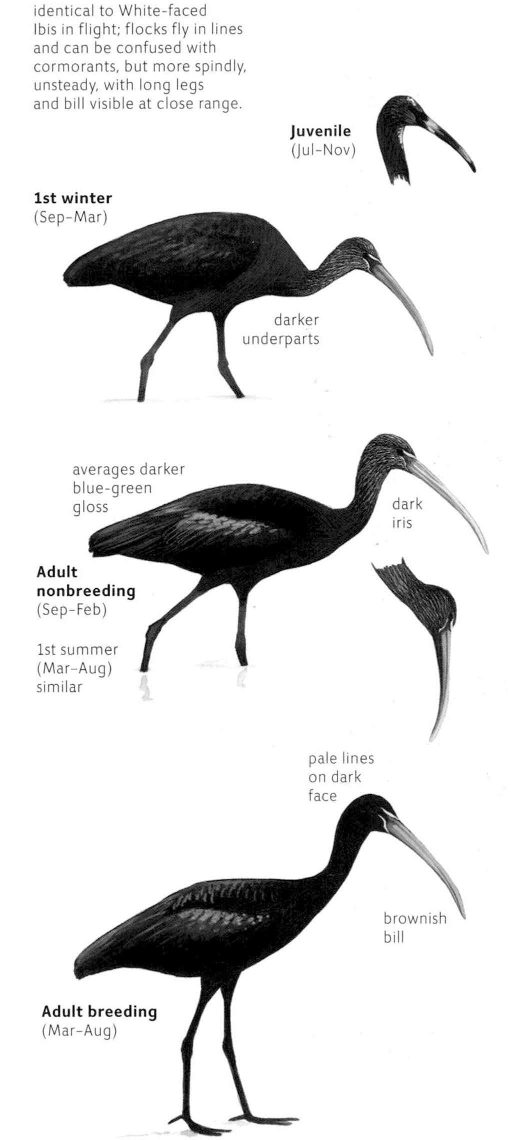

VOICE: Like White-faced Ibis; higher and softer than White Ibis.

Uncommon. Found near coast but forages mainly in shallow muddy pools of fresh or brackish water with fringes and patches of grass and other emergent vegetation; not in very open, tidal, or sandy habitats. Habits like White-faced Ibis, and the two species mix freely whenever they meet.

Roseate Spoonbill

Platalea ajaja

L 32" WS 50" WT 3.3 lb (1,500 g)

Pale pink color and spoon-shaped bill distinctive. Immatures nearly white, adults develop scarlet wing coverts and orange tail. Larger than ibises and flies with slower wingbeats and little gliding.

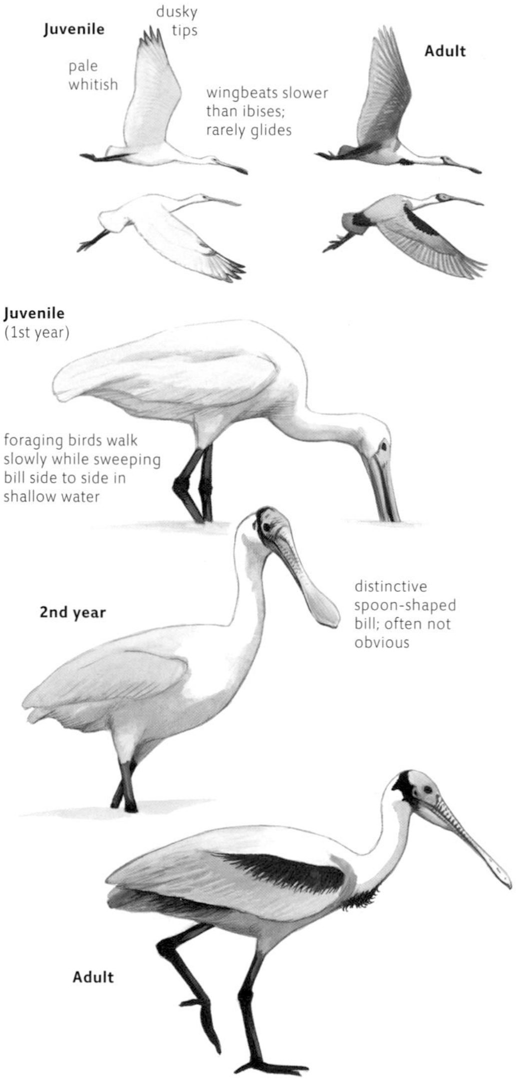

VOICE: Low, ibis-like grunting *huh-huh-huh-huh* without change in pitch or volume. Also a fairly rapid, dry, rasping *rrek-ek-ek-ek-ek-ek*, much lower, rougher, faster than ibises.

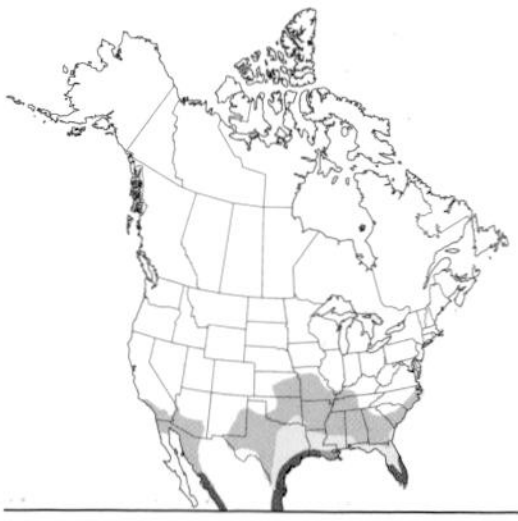

Uncommon. Forages in expanses of shallow water with muddy substrate. Often with ibises, but spoonbills tend to be in slightly deeper water and move more rapidly. Nests and roosts colonially in shrubs or small trees.

Sacred Ibis

Threskiornis aethiopicus

L 29" WS 45" WT 3 lb (1375 g)

Larger than White Ibis, with black head and bill, black legs, lac black plumes on back, and black tips on all flight feathers.

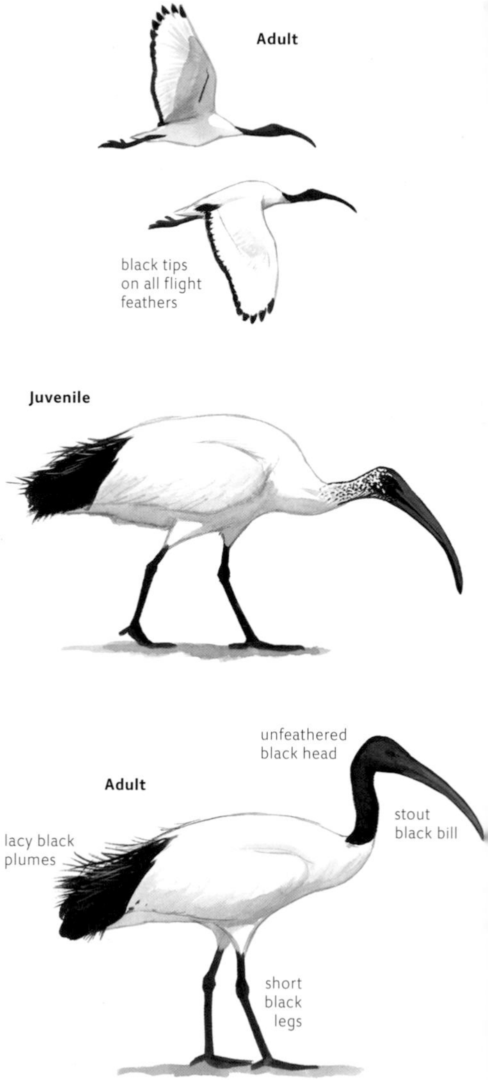

VOICE: Silent except for occasional grunting or croaking sounds.

Native to Africa. A few individual escaped and breeding in the wild in Florida, but active eradication efforts there make it unlikely the species will become established.

\merican Flamingo

hoenicopterus ruber

46" WS 60" WT 5.6 lb (2,550 g) ♂>♀

xtremely tall and slender; pink color like spoonbill but easily istinguished by much longer legs and neck and by bill shape. hree other species of flamingos are seen rarely in North Amer- :a as escapes from captivity. Flies with very long neck extended nd drooping, long legs trailing.

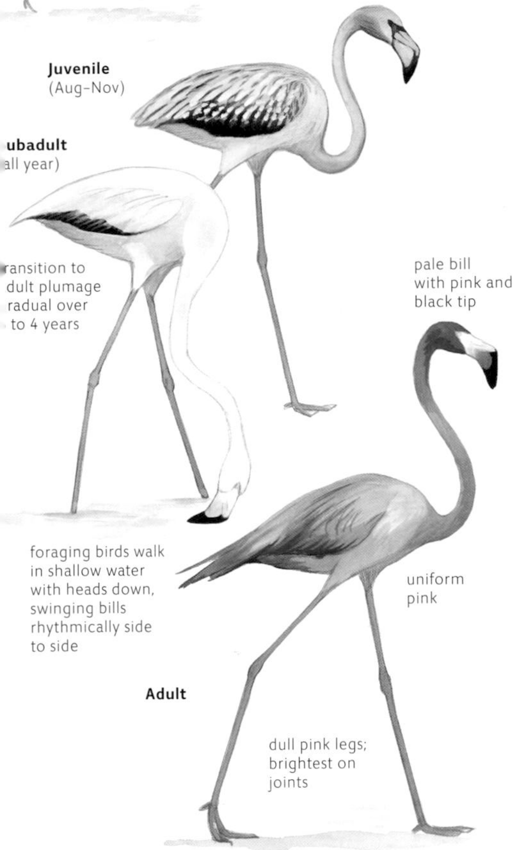

VOICE: In flight honk like barnyard geese but deeper. Flock gives ow, conversational gabble while feeding. Courting pair gives *eep eep cak cak eep eep cak cak*...; higher *eep* notes by female, ower *cak* notes by male.

Rare visitor from Caribbean to extreme southern Florida, mostly in winter. Very rare records elsewhere mostly escaped captives (and often other species of flamingos). Found on expansive mudflats, foraging in shallow water by swinging bill rhythmically from side to side and filtering small organisms out of the mud.

Greater Flamingo

Phoenicopterus roseus

L 46" WS 60" WT 6.3 lb (2,900 g) ♂>♀

Distinguished from American by bright pink bill (not whitish) with very small black tip, mostly whitish body plumage, and uniformly bright pink legs.

Native to Eurasia; occasionally escapes from captivity in North America.

Chilean Flamingo

Phoenicopterus chilensis

L 44" WS 57" WT 5 lb (2,300 g) ♂>♀

Distinguished from other flamingos by extensive black at bill tip, pale whitish body with pink neck, and grayish legs with contrasting pink ankle joints and feet.

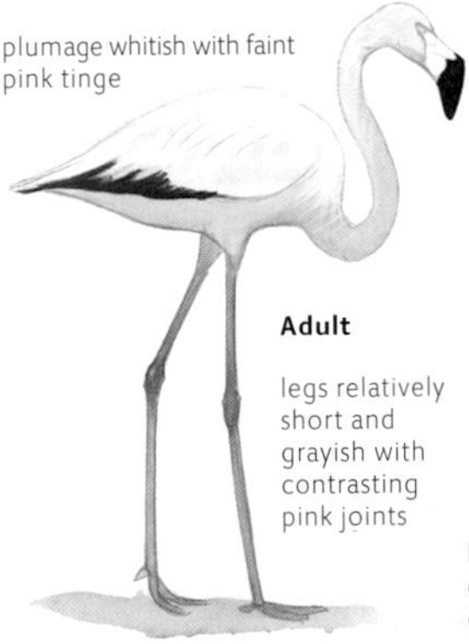

Native to South America; occasionally escapes from captivity in North America.

Lesser Flamingo

Phoeniconaias minor

L 37" WS 48" WT 4.2 lb (1,900 g) ♂>♀

Distinguished from other flamingos by smaller size, very dark bill and facial skin.

Averages smaller overall with shorter bill; bill much darker; legs bright red; plumage strongly washed pink.

Native to Africa and India; occasionally escapes from captivity in North America.

Hawks and Vultures

FAMILIES: ACCIPITRIDAE, CATHARTIDAE, PANDIONIDAE

28 species in 3 families (2 rare species not shown here), 16 genera, all in family Accipitridae (hawks and eagles) except Osprey in Pandionidae and vultures and condor in Cathartidae. The falcons, family Falconidae, formerly thought to be related to hawks, now begin on page 323. As predators, raptors are always relatively scarce, and are often seen at a distance in flight. Identification often depends on subtle differences in wing shape and body proportions, which are related to foraging style; specie with similar shapes tend to have similar habits. Soaring (circlin on updrafts without flapping) is a common behavior for mos hawks and vultures, and also some unrelated species includin falcons, pelicans, frigatebirds, Anhinga, storks, cranes, gulls, an ravens. Juveniles are shown (except vultures).

Genus *Coragyps*

Black Vulture, page 123

Genus *Cathartes*

Turkey Vulture, page 123

Genus *Aquila*

Golden Eagle, page 142

Genus *Pandion*

Osprey, page 125

Genus *Haliaeetus*

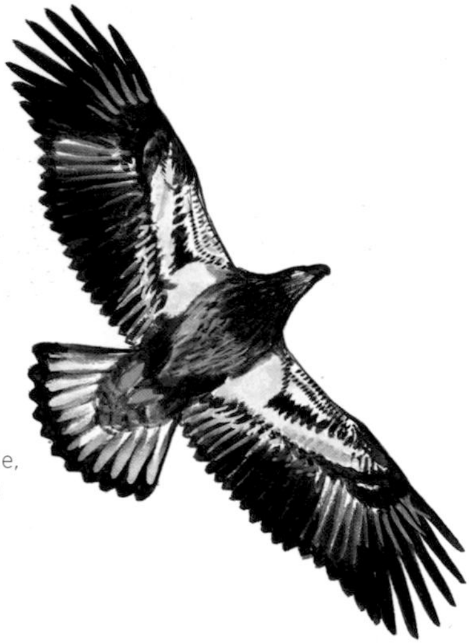

Bald Eagle, page 143

Genus *Chondrohierax*

Hook-billed Kite, page 125

Genus *Elanoides*

Swallow-tailed Kite, page 127

Genus *Elanus*

White-tailed Kite, page 126

Genus *Rostrhamus*

Snail Kite, page 126

Genus *Ictinia*

Mississippi Kite, page 127

Genus *Circus*

Northern Harrier, page 124

enus *Buteo*

ed-shouldered Hawk, age 133

Broad-winged Hawk, page 135

Gray Hawk, page 132

Short-tailed Hawk, page 134

wainson's Hawk, page 136

White-tailed Hawk, page 137

Zone-tailed Hawk, page 131

Red-tailed Hawk, page 138

erruginous Hawk, page 140

Rough-legged Hawk, page 141

Genus *Parabuteo*

Harris's Hawk, page 130

Genus *Buteogallus*

Common Black-Hawk, page 130

Genus *Accipiter*

Sharp-shinned Hawk, page 128

Cooper's Hawk, page 128

Northern Goshawk, page 129

California Condor

Gymnogyps californianus

L 46" WS 109" WT 23 lb (10.5 kg)

One of our largest birds, size alone often enough for identification; in flight can be mistaken for a small airplane. Also note broad wings with splayed "fingers" on wingtips, short tail, and white on under- and upperwing coverts.

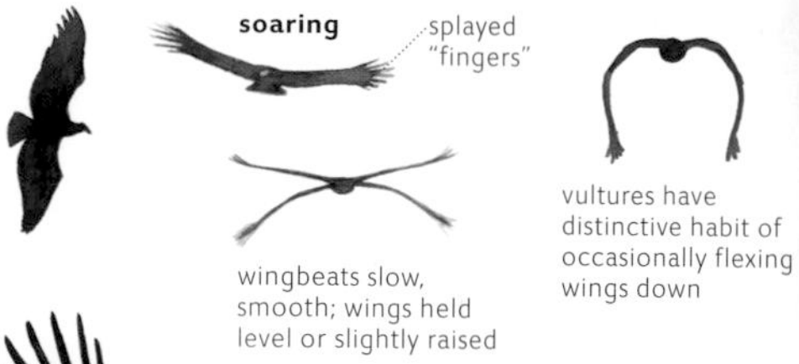

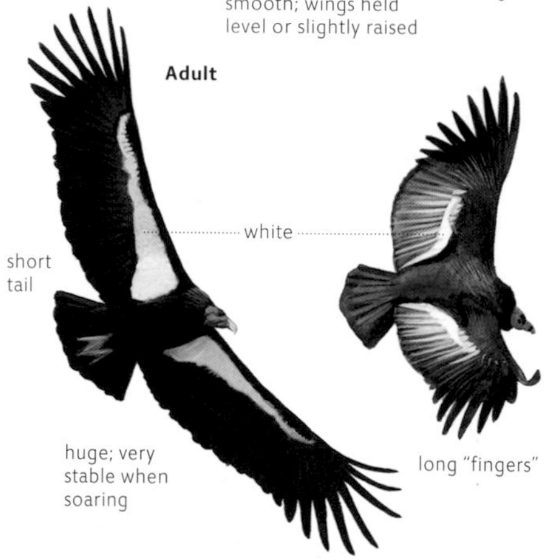

The huge size of the California Condor is usually evident. Here it is shown with other raptors to scale.

VOICE: Usually silent; limited to hissing and grunting. Wing flaps produce loud swishing noise.

Rare and extremely local. Only a few hundred individuals survive. Most often seen in flight over foothills and ridges in open savannah-like habitat. Roosts on cliffs or large trees.

Turkey Vulture

athartes aura

26" WS 67" WT 4 lb (1,830 g)

dentified as a vulture by large size, dark color, and small naked ead; differs from Black Vulture in brownish plumage and longer ail; in flight note two-toned wings raised in dihedral and habit f rocking gently from side to side while flying.

OICE: Usually silent; limited to soft hissing, clucking, and whining.

Common. Most often seen in flight, especially over mixed habitat of woods and open areas, rarely over water. Feeds on carrion found on the ground, often along roads. Roosts in groups on buildings, trees, and towers.

Black Vulture

Coragyps atratus

L 25" WS 59" WT 4.4 lb (2,000 g)

Similar to Turkey Vulture, but with shorter wings pale only near tips, and much shorter tail, also quicker wingbeats, wings held flatter and lack of rocking motion in flight. Dark gray head is matched by juvenile Turkey Vulture.

VOICE: Usually silent; limited to soft hissing and barking.

Common. Habits and habitat like Turkey Vulture, but often flies higher, and usually in groups. A low-flying solitary Black Vulture is more likely to be a Turkey Vulture.

Northern Harrier

Circus cyaneus

L 18" WS 43" WT 15 oz (420 g) ♀>♂

A very slender, long-winged, and long-tailed hawk. White uppertail coverts obvious and distinctive. At close range note owl-like facial disc. Extremely buoyant in flight, with wings always raised in strong dihedral.

VOICE: Two call types. Piercing, insistent whistle *eeeya* or high, very thin *sseeeew* given mainly by female and young. Dry clucking or barking series *chef chef chef*... to rapid staccato *kekekekeke*... given mainly by male.

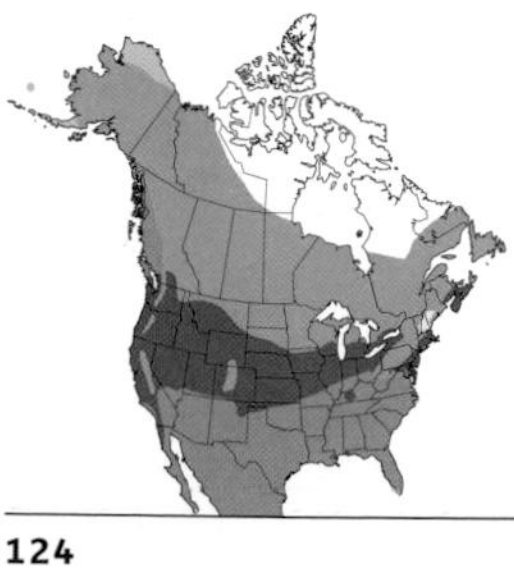

Uncommon. Usually seen flying just above vegetation over open fields, grasslands, or marshes. Perches on the ground or on fence posts, lower than most hawks. Generally solitary.

Harrier Flight Shapes

Like other raptors, harriers present a wide array of differer shapes, depending on their mode of flight. Northern Harrier ca be particularly confusing in this respect since most observers ar accustomed to seeing it low over fields or marshes; its unex pected appearance high overhead baffles many observers. Fiv different shapes here show the range of variation between soa ing and gliding flight.

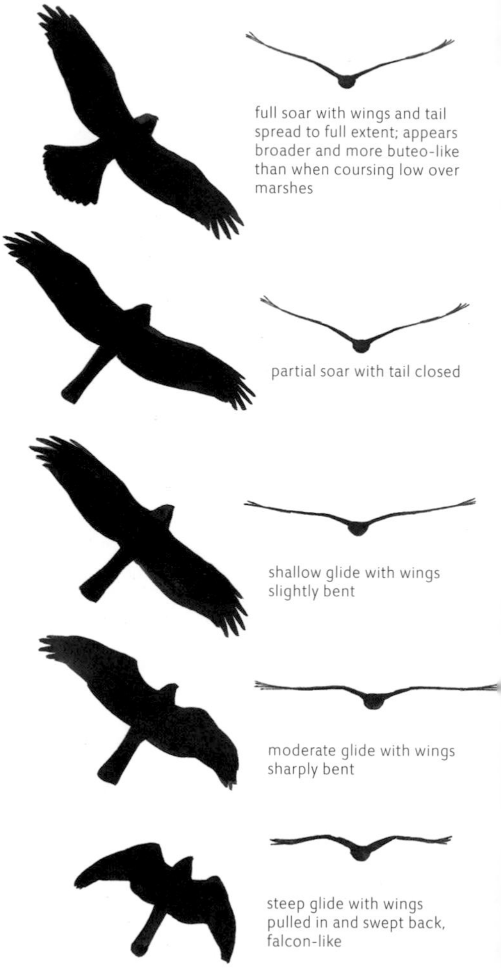

Osprey Caribbean adult

Ospreys in southern Florida average less black on head than northern birds, some individuals have head nearly all-white.

)sprey

andion haliaetus

23" WS 63" WT 3.5 lb (1,600 g)

ɔng crooked wings (held slightly arched) and white underparts, ull-like. Note all-dark upperside, including rump and tail, and ark patches on underwing. At close range dark eye-stripe and hite cap distinctive.

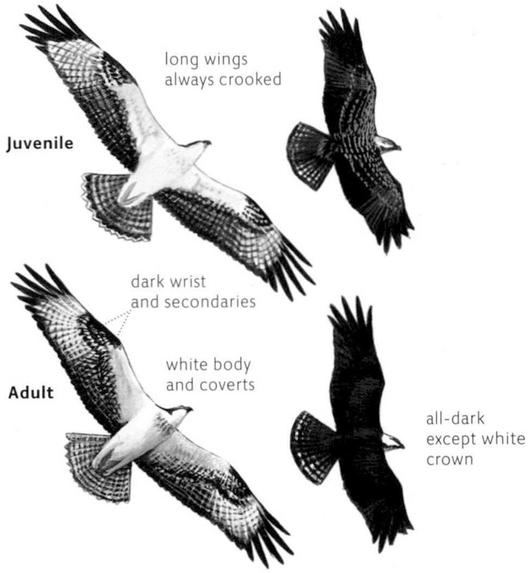

OICE: Quite vocal; all calls short, shrill whistles *tewp, tewp, tewp, eelee, teelee, tewp*; commonly single loud, shrill, slightly slurred vhistle *teeeeaa*. Juvenile similar to adult.

Locally common around saltwater bays and estuaries, uncommon to rare on lakes and rivers. Look for it flying high above the water, patrolling and hovering as it searches for fish below. Perches and nests on isolated trees, poles, towers, channel markers, etc. Mostly solitary, but vocal and interactive in small groups.

Hook-billed Kite

Chondrohierax uncinatus

L 18" WS 36" WT 10 oz (283 g)

Shape somewhat accipiter-like, but with extremely broad, paddle-shaped wings and much less agile in flight. Pale staring eye is conspicuous and distinctive in close view.

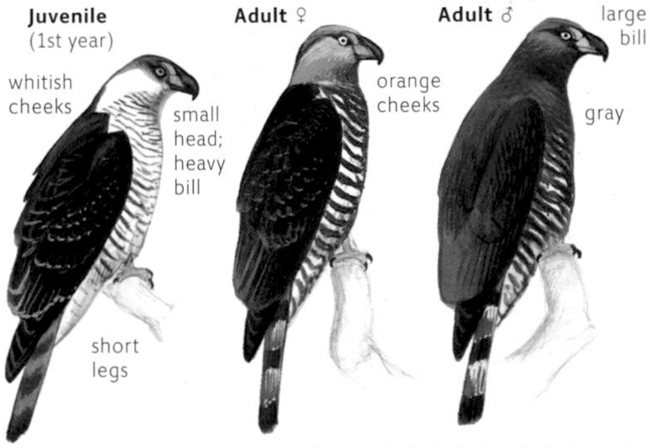

VOICE: A clucking or rattling chatter *kekekekeke-kekekekeke*; highest in middle, quality like Northern Flicker but more rapid; also a soft, conversational *huey*. Juvenile call similar to adult.

Rare, only a few individuals resident in brushy woods along lower Rio Grande in Texas. Usually seen in direct flight between roosting and feeding areas, just above woods or brushland, Forages in trees for snails, but rarely seen perched. A rare dark morph, blackish overall, has been recorded in Texas.

White-tailed Kite

Elanus leucurus

L 15" WS 39" WT 12 oz (340 g)

A small hawk; slender and long-tailed. Very white overall, with conspicuous white tail and contrasting black shoulders. Confused as much with gulls and terns as with hawks.

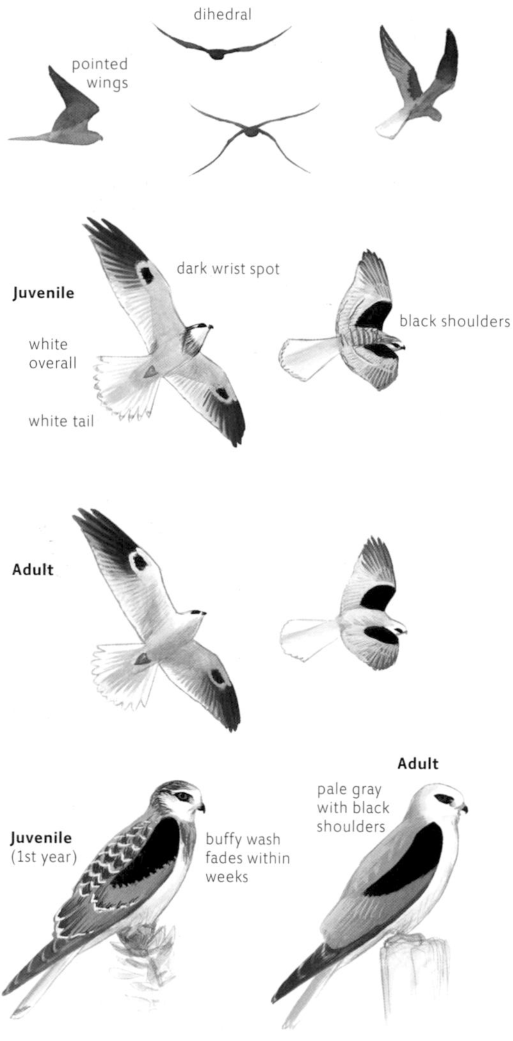

VOICE: Varied: mellow whistle/yelp *eerk, eerk...*; high, thin, rising whistle followed by low, dry, harsh notes *sweeekrrkrr*. In aggression a low, grating *karrrrr*. Juvenile gives an Osprey-like whistle *teewp*.

Uncommon. Found in open grassy areas with scattered shrubs. Hunts from the air, often hovering gracefully then dropping straight down. Perches on prominent posts or treetops.

Snail Kite

Rostrhamus sociabilis

L 17" WS 42" WT 15 oz (420 g)

Flies with floppy wingbeats and slow floating flight across mars settling gently on water to pick up apple snails. White aroun base of tail could cause confusion with Northern Harrier, bu broad arched wings, short tail, and floppy flight distinctive.

VOICE: Usually silent. Gives a harsh, cackling *ka-ka-ka-ka-ka...* or grating *krrkrr...*; harsh *ker-wuck* repeated. Juvenile gives harsh, coarse scream.

Rare. Fewer than 1,000 individual in US. Found in expansive sawgrass marshes, often in loosely-associated family groups flying low over grass and open water in search of snails, perching in nearby shrubs and trees.

ʌississippi Kite

ʼtinia mississippiensis

14" WS 31" WT 10 oz (280 g)

ender and buoyant in flight, with long narrow wings and square il. Falcon-like in shape, but flight more graceful. Short outer-ost primary is distinctive

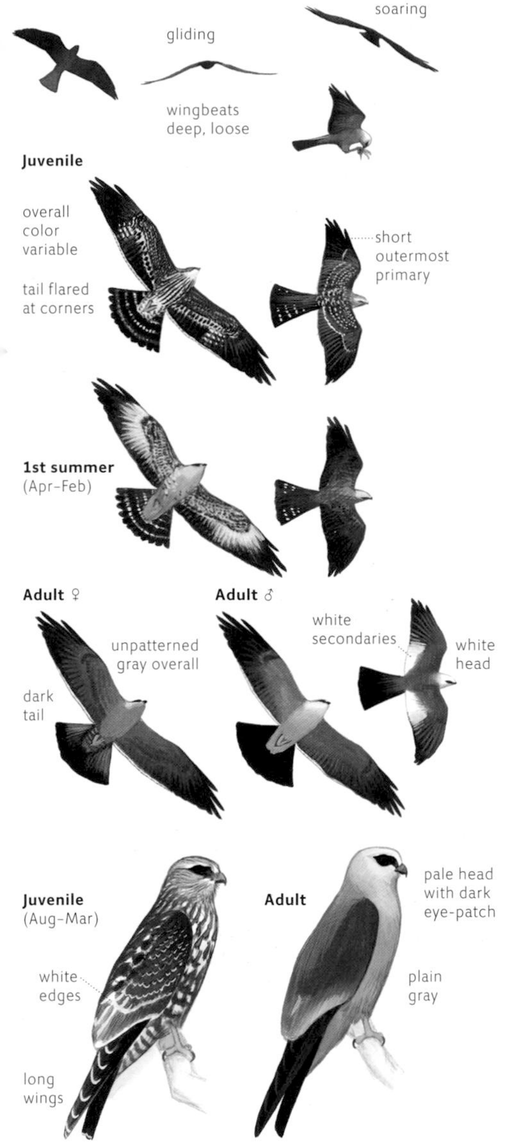

ɔICE: A high, thin whistle *pe-teew* or *pee-teeeer*, similar to ʳoad-winged Hawk but descending; sometimes a more excited *ee-tititi*. Juvenile call similar to adult.

Uncommon and local. Found where large trees are intermixed with open areas, such as golf courses, riparian corridors, etc. Forages high in the air, mainly catching large insects in flight. Forms small flocks during migration.

Swallow-tailed Kite

Elanoides forficatus

L 22" WS 51" WT 15 oz (420 g)

Extremely graceful, with long pointed wings and long forked tail. Striking black and white plumage and large size may recall Osprey, but shape is unique.

VOICE: Short, weak, high whistles *hu-kli-kli-kli*; high, clear, rising *eeep* or *eeeip* repeated. Juvenile call similar to adult.

Uncommon and local. Found in forested areas near rivers or swamps. Rarely seen perched, look for it flying gracefully above the forest, capturing flying insects or swooping to pluck lizards or birds from treetops. US population estimated at about 1,000 nesting pairs.

Sharp-shinned Hawk

Accipiter striatus

L 11" WS 23" WT 5 oz (140 g) ♀>♂

Our smallest accipiter; relatively small-headed and short-tailed. Very similar to Cooper's Hawk; best identified (with experience) by shape and quick snappy wingbeats, but plumage details also offer useful clues.

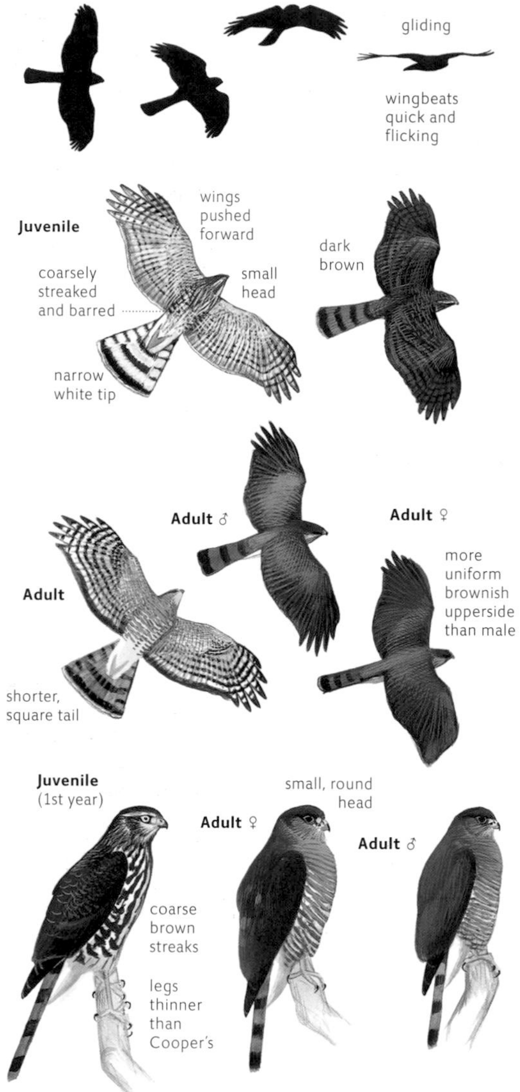

VOICE: Adult at nest gives a series of short, sharp notes *kiw kiw kiw*...; sometimes a high, thin *keeeeeep*. Juvenile gives a clear, light chip *tewp* in fall; juvenile begging call a high-pitched *eee*.

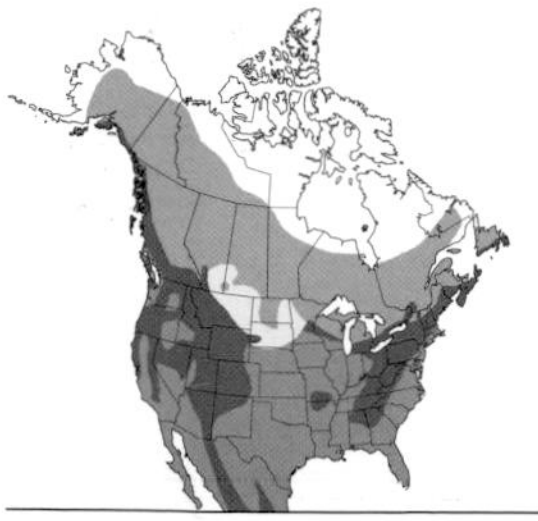

Uncommon but widespread. Nests in mixed forest. Forages mainly along forest edges, hedgerows, suburban yards. Hunts small birds captured by surprise in flight. Usually chooses inconspicuous perch, seen mainly in flight below treetop level, but sometimes soars higher, especially when migrating.

Cooper's Hawk

Accipiter cooperii

L 16.5" WS 31" WT 1 lb (450 g) ♀>♂ E>W

A medium-size accipiter; always larger than Sharp-shinne Hawk, but size difficult to judge. Relatively the most slender ar longest-tailed accipiter, with rather long head and neck. Be identified (with experience) by shape and stiff wingbeats.

VOICE: Adult at nest gives a series of flat, nasal, barking notes *pe pek*...; also long *keeee* (male) or *whaaaaa* (female); male als gives a single *kik*. Juvenile begs with repeated squeaky whist *kleeer* (sometimes low and nasal, like sapsuckers' call). Juveni also gives high, rapid *kih kih kih*.

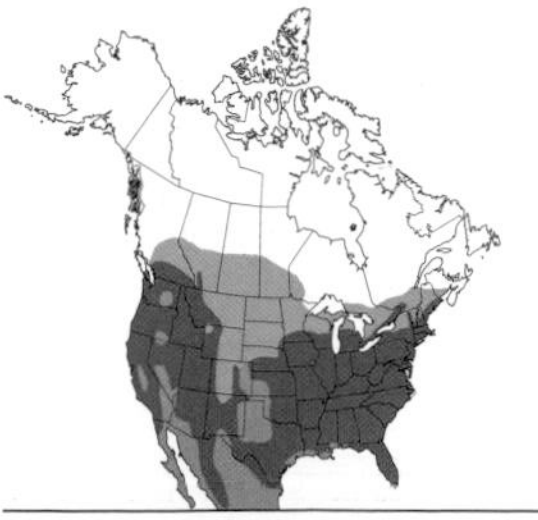

Uncommon but widespread. Habits and habitat like Sharp-shinned, but slightly mor likely to be seen on an open perch such as fencepost. Nests farther south and accepts more broadleaf trees for nesting, such as suburban habitat and riparian corridors.

ɅNorthern Goshawk

ccipiter gentilis

21" WS 41" WT 2.1 lb (950 g) ♀>♂

ur largest and bulkiest accipiter, approaching buteo-like proportions; note broad head, body, and tail and relatively pointed ings. Adult plumage distinctive, with pale underparts and bold ce pattern. Immature best identified (with experience) by overl bulkiness and wing shape.

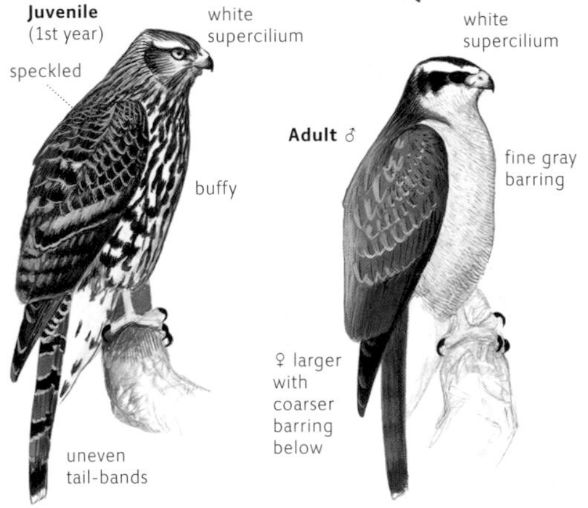

ɔOICE: Adult at nest gives a loud cackling *kye kye kye*...; wilderɔunding, slower, and higher-pitched than Cooper's. Display call f adult a wailing, gull-like *KREE-ah* repeated regularly. Male ives wooden *guck* near nest. Juvenile begging call a plaintive ɔream *kree-ah*.

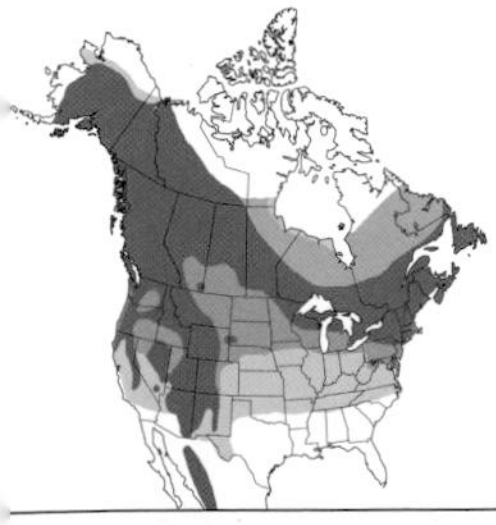

Rare. Found in very small numbers in mixed forests where clearings, wetlands, or other features provide open areas. Usually seen within forest or at edges, occasionally soaring above treetops. Hunts mainly from perch for grouse, rabbits, and squirrels, launching explosive attack once prey is sighted.

Accipiter Identification

Accipiters are characterized by relatively short rounded wings and long tails. They are bird eating hawks of woodlands and suburbs (often hunting at bird feeders), with very agile flight, short bursts of speed, and quick turns through foliage. At other times they soar high, and can also be seen hunting open areas.

Other species can be confused with accipiters, especially Red-shouldered Hawk, Gray Hawk, and Harris's Hawk.

Identification of accipiters is always challenging, particularly because they are usually seen briefly and unexpectedly. Adult Northern Goshawk has distinctive plumage, others differ only in details of color and pattern, and differences in shape and size are variable, subtle, and overlapping.

Determining age is critical to using plumage colors and patterns for identification. All immature accipiters are brownish above (usually patterned with paler spots and feather edges) and streaked below, with brownish head and more or less obvious pale eyebrow. Adults are smooth bluish-gray to brownish-gray above with underparts barred reddish-brown (Cooper's and Sharp-shinned) or vermiculated gray (Northern Goshawk).

Identifying a perched accipiter:

- Focus on head shape and nape pattern.
- On immatures check streaking on the belly.

Identifying a flying accipiter:

- Focus on tail length and shape.
- Also look at head size and the shape of the leading edge of the wing.

With experience you'll be able to use other, more subtle, features like wingbeats, but even expert birders can be stumped by accipiters. Use multiple clues and don't expect to identify every one.

dark cap of adult Sharp-shinned continues down nape, Cooper's has pale nape (enhancing square-headed appearance); adult male Cooper's has gray cheeks (orange on all others)

all perched accipiters can show bright white spots on back when feathers are fluffed

adult Cooper's Hawk and Northern Goshawk in breeding season often fly with white undertail coverts conspicuously flared

Common Black-Hawk

Buteogallus anthracinus

L 21" WS 46" WT 2.1 lb (950 g)

On adult note dark color, yellow face, and white-banded tail. Extremely broad wings distinctive in flight. Can be difficult to distinguish from Zone-tailed when perched, note yellow lores, long tertials.

VOICE: Adult gives a series of high, sharp whistles/screams *kle KLEE KLEE klee kle kle kle kle kle,* increasing abruptly then trailing off in intensity. Juvenile gives rapid, high whistles.

Rare and local. Found in mature cottonwood forest along permanent streams, where it hunts mainly from a perch for frogs, snakes, and rodents. Migrants in Arizona hunt in Pecan groves. US population fewer than 300 nesting pairs.

Harris's Hawk

Parabuteo unicinctus

L 20" WS 42" WT 2 lb (900 g) ♀>♂

Structurally intermediate between buteos and accipiters: bulk body and broad wings and tail, but wings relatively short an rounded and tail long. Note overall dark color with white aroun base of tail and broad white tip on tail.

VOICE: Very raucous, harsh *raaaak* lasting up to three second from nest a low, grating, toneless *keh keh keh keh keh keh ke keh.* Juvenile gives slow, squealing *skeeei skeeei*...or *kweeeurr*

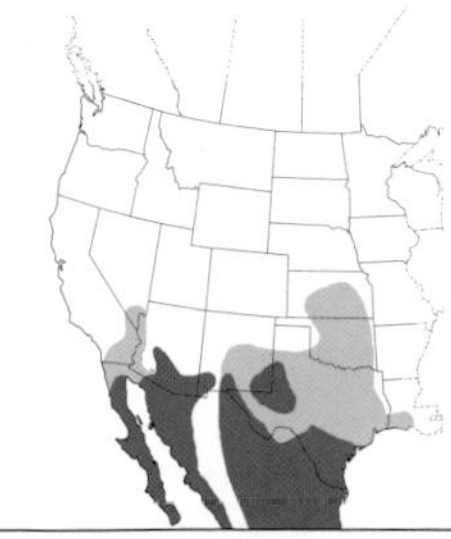

Uncommon. Found in mesquite woodland or saguaro-mesquite desert. Nests in relatively large trees. Social; often seen in pairs or trios perched close together on trees and posts; and often hunts cooperatively, chasing pre through cover. Feeds mainly on small mammals captured in brushy woods.

;one-tailed Hawk

ıteo albonotatus

:0" WS 51" WT 1.8 lb (810 g) ♀>♂

flight shape, size, and color extremely similar to Turkey Vulture ınd may use this to surprise its prey). Check for slightly "cleaner" ıes, dark feathered head, yellow cere and feet, light tail bands. ›rched birds similar to Common Black-Hawk but note smaller llow area on face, longer wings, shorter legs.

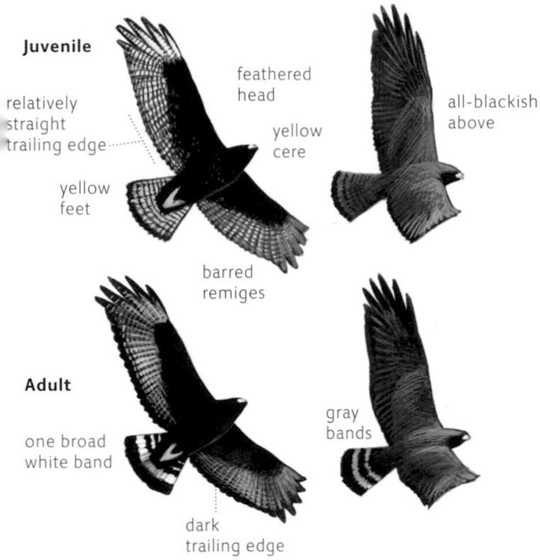

ɔICE: High scream *koweeeeeur* clearer and lower than Red-ıiled; nearly level but variable. Juvenile call a high, clear, nasal *eeeeer*; also high, squealing series *hwee hwee hwee ɯEEEeeeer.*

Rare and local in foothill canyons with permanent streams and open woodland. Usually seen in flight, often with Turkey Vultures, hunting open areas for bird and mammal prey. US population only about 300 nesting pairs.

Buteo Identification

Buteos as a group average large size, with long rounded wings and short tails. They are relatively slow and stable in flight, and usually hunt by patiently watching from a perch or from the air, dropping suddenly on ground-dwelling prey.

Distinguishing the species of buteos is complicated by wide variation in plumage; many species have dark and light morphs, and the fact that they are often seen in flight at a distance.

On flying buteos focus on wing and tail proportions, color pattern on the underwing, and tail pattern.

On perched buteos the color pattern on breast and belly will narrow the possibilities, also focus on where the wingtips fall relative to the tail tip, and tail pattern.

Dark morph buteos present special challenges, but can be identified by structural clues and by the pattern of primaries, secondaries and tail feathers (which are the same as on light morph birds).

Getting to know Red-tailed Hawk very well is one of the keys to buteo identification, as this is the default buteo in most regions, often seen along roadsides, farmland, playing fields, city parks—any open habitat near trees.

Perched light morph Red-tailed Hawks of all ages typically show a dark "belly-band" and a V-shaped pattern of white speckling on the back, both of which are excellent clues for identification.

The smaller buteos—including Red-shouldered, Broad-winged, and Gray Hawks—often perch on utility wires, but larger species such as Red-tailed Hawk rarely do.

Rough-legged Hawk often perches on small twigs at the top of trees or shrubs; other buteos choose larger branches. Rough-legged, Red-tailed, and a few other large species typically perch in the open on treetops, fenceposts, hay bales, etc., while other species (including Broad-winged, Red-shouldered, Zone-tailed, and others) usually choose less conspicuous perches within trees and are seen in the open less frequently.

Roadside Hawk

Buteo magnirostris

L 14.6" WS 28" WT 9.7 oz (275 g) ♀>♂

A small buteo, with relatively short wings and long tail reminiscent of accipiters. Note obvious pale band on uppertail coverts, relatively large head and bill.

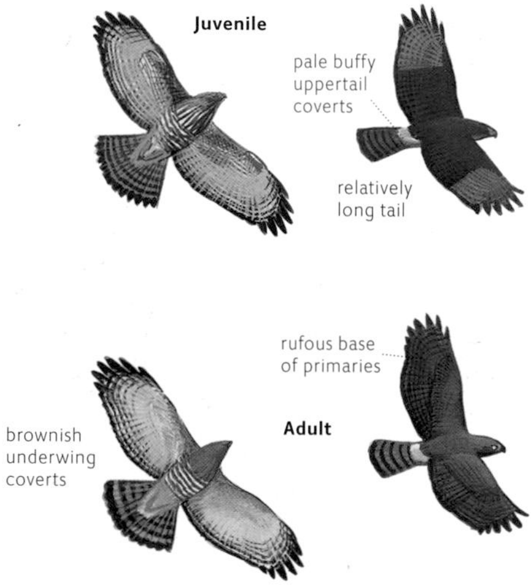

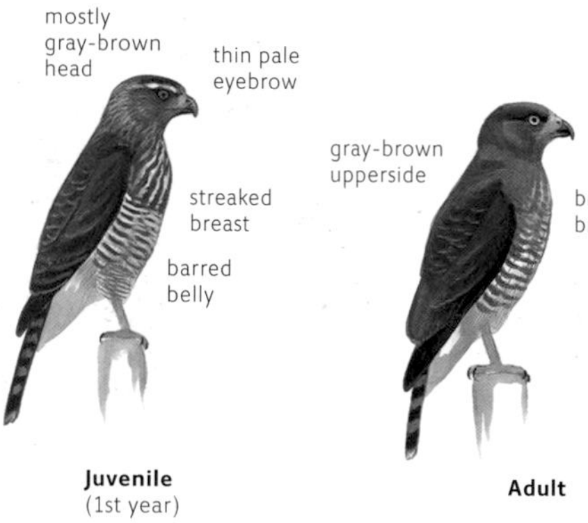

VOICE: Vagrants are usually silent. Common call in normal range a prolonged whining scream *reeeeeah*. Display call a series of short laughing *bek* notes.

Very rare visitor from Mexico to southern Texas, recorded only a few times. Found around openings and brushy edges of forest, such as clearings, ponds, and roadsides. Habits similar to Red-shouldered Hawk.

Gray Hawk

Buteo plagiatus

L 17" WS 34" WT 1.2 lb (540 g) ♀>♂

A small buteo with some characteristics of accipiters, relative short wings and long tail. Adult's pale gray color is distinctive. C juvenile note bold facial stripes, white band across rump.

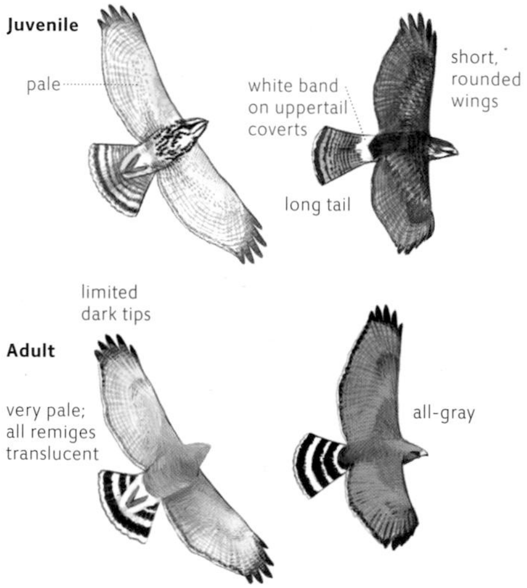

VOICE: Display call a series of long, plaintive whistles *hoooooweє hoooooweeo*...or higher *pidiweeeeeeerh pidiweeeeeeerh*. Juvenile call a high, squeaky *KEE-errrrrrrrrr*; adult gives simil scream when alarmed.

Rare and very local, regular at only a few locations near the Mexican border in mature trees along permanent streams. Usually seen flying through or just above trees, or perched on relatively low branches or posts. US population only about 100 nesting pairs.

ed-shouldered Hawk

teo lineatus

7" WS 40" WT 1.4 lb (630 g) ♀>♂

ather compact, accipiter-like buteo, very vocal. Shorter-nged and longer-tailed than other widespread buteos. All ımages show distinctive pale translucent crescent across ter primaries (by far the most important field mark to watch on a flying bird).

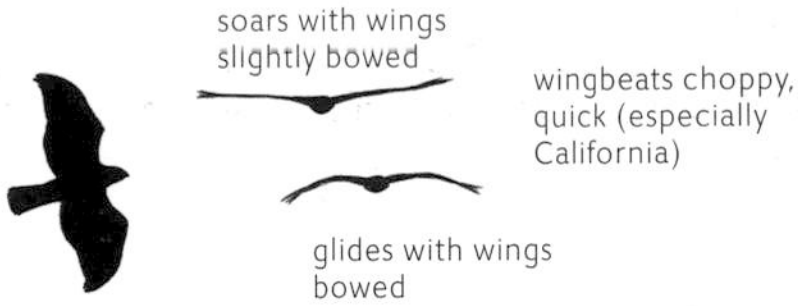

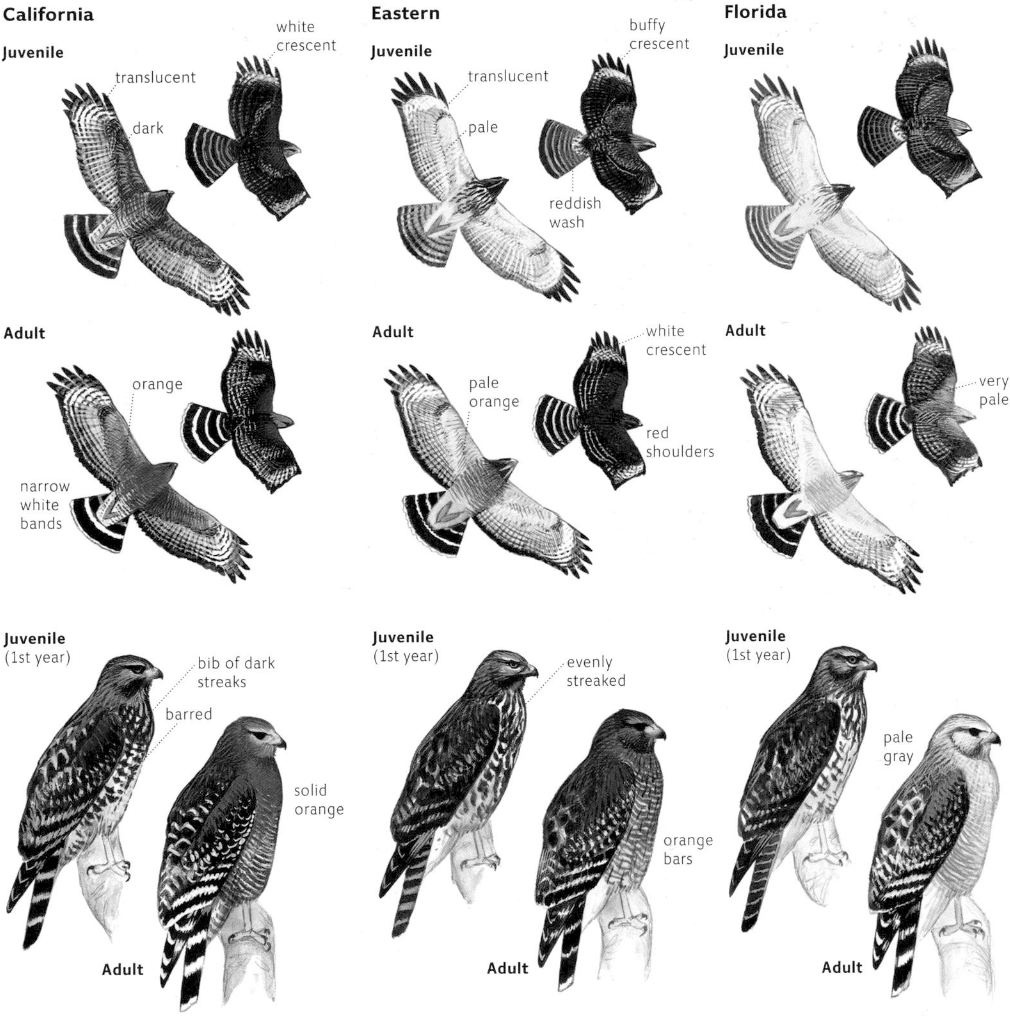

ICE: Very vocal, with distinctive, far-carrying calls. Adult ter-orial call a high, clear, squealing *keeyuur keeyuur*...repeated eadily; often imitated by Blue Jay and Steller's Jay. Also a single slowly repeated high, sharp *kilt*. Juvenile similar to adult. Calls California may be a bit shorter, higher, and sharper than East-n and Florida but very similar.

Uncommon in wooded areas with clearings and water. Often seen perched in trees or on poles or wires at edges of clearings.

Eastern adults richly-colored, except very pale in Florida; juvenile brownish overall and streaked below, similar to juvenile Broad-winged Hawk. California adult is more richly colored with solid orange breast lacking dark streaks, tibia feathers darker orange than belly, and fewer and broader white tail-bands. Juvenile is more like adult than eastern juveniles, with black and white wings and tail, and dark rufous underwing coverts.

Short-tailed Hawk

Buteo brachyurus

L 16" WS 37" WT 15 oz (420 g) ♀>♂

A small buteo with relatively long wings; long primaries curve up more in flight than on Broad-winged Hawk, and reach tail tip when perched. In flight from below, on all plumages note secondaries darker gray than outer primaries. Wing coloration (dark flight feathers) suggests Swainson's, but note wing shape and more extensive white on outer primaries.

Light juvenile

Dark juvenile

Light adult

Dark adult

Light juvenile
(1st year)

Light adult

Dark juvenile
(1st year)

occurs in distinct light and dark morphs with no intermediates; light morph with clean white underside and underwing covert distinctive. Florida population up to 80 percent dark morph

Dark adult

VOICE: High, clear *keeeea* long and drawn-out, sometimes quavering; also high *keee* or *kleee*, often in slow series; nearly as high as Broad-winged but with harsher, less thin quality. Juvenile similar to adult.

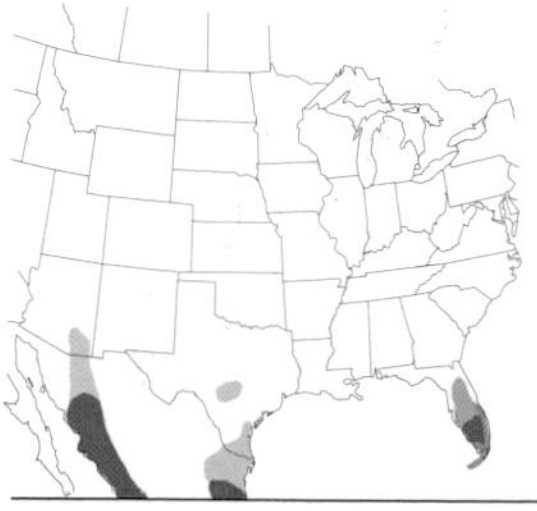

Rare, local, and inconspicuous. Rarely seen perched, look for it flying high above wooded areas, kiting and circling with upswept wingtips as it hunts for small birds in the treetops. US population under 100 nesting pairs

›road-winged Hawk

ıteo platypterus

ı5" WS 34" WT 14 oz (390 g) ♀>♂

small and inconspicuous forest buteo, with relatively pointed, ·aight-edged wings. In flight, note uniform dark upperside and ›le underwing with dark border. Adult has single broad white ›nd across tail and very pale underwing with broad dark border. venile has pale secondaries, and finely barred tail with broader bterminal tail-band. Occurs in distinct light and dark morphs th no intermediates.

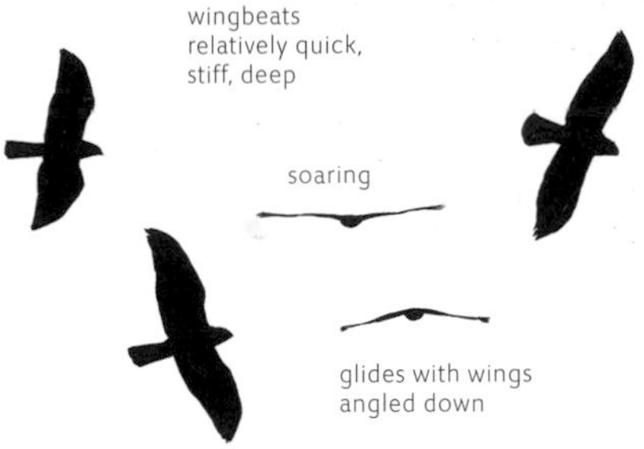

Light juvenile

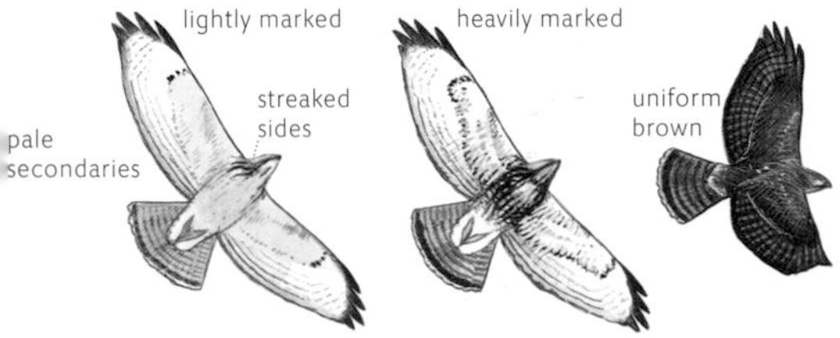

Dark juvenile

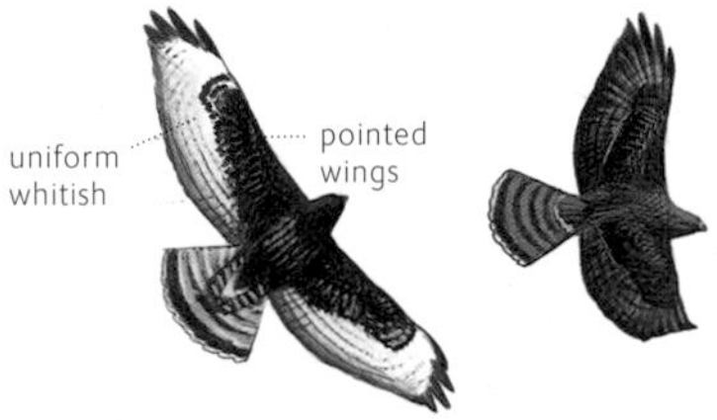

Light adult

Dark adult

juvenile differs from Red-shouldered by unbarred secondaries, mostly unmarked coverts, relatively unmarked center of breast, short legs

dark morph rare (less than 1% of total population); breeds only on western edge of range; migrants very rare east of Great Plains

ɔICE: A piercing, thin, high whistle *teeteeeeee* on one pitch; ale higher than female but no other significant call known. venile similar to adult. Compare Mississippi Kite.

Uncommon and usually inconspicuous, found in wooded areas where it is most often seen in flight through the trees, occasionally soaring high above. Usually perches in trees within forest, less often on an exposed branch, post, or wire. During migration gathers in large numbers along favored routes.

Molting juvenile

For a brief period May–Jun (age 10–12 months) molting outer primaries create translucent crescent superficially similar to Red-shouldered.

Swainson's Hawk

Buteo swainsoni

L 19" WS 51" WT 1.9 lb (855 g) ♀>♂

A lanky buteo, slender and long-winged. More slender than Red-tailed Hawk, with more pointed wingtips, uniform dark upperside, dark chest, and dark gray flight feathers.

VOICE: Long, high scream like Red-tailed but higher, clearer, weaker; may be short *cheeeeew* or long, drawn-out *kwee-aaaaaah*; sometimes a series of whistled notes *pi tip, pi tip....* Juvenile gives soft, plaintive whistle or mew.

Occurs in range of color morphs: light, rufous, and dark, with many intermediates. Combined dark and intermediate morphs less than 10 percent of most populations, more in far West.

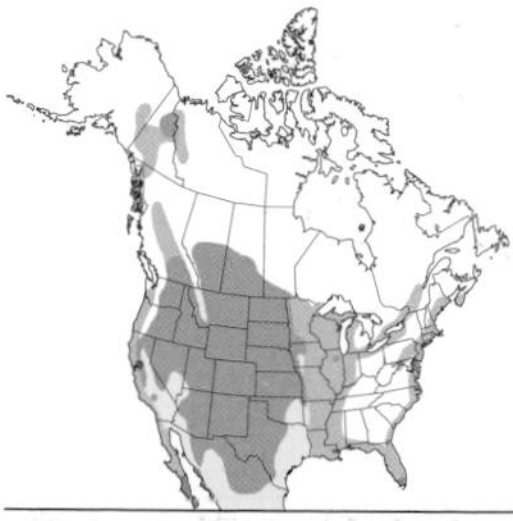

Uncommon in prairies and farmland. Nests in isolated trees. Usually solitary, but migrates in large flocks along favored routes at migration points. Hunts from the air or a perch or while walking on the ground. Feeds mainly on small mammals and reptiles in summer; grasshoppers and other invertebrates the rest of the year.

Swainson's, Broad-winged, and White-tailed Hawks are the only buteos with four notched primaries (other buteos have five). These "fingers" can usually be seen on soaring birds and fewer notched feathers creates a more pointed wingtip.

Vhite-tailed Hawk

uteo albicaudatus

20" WS 51" WT 2.3 lb (1,035 g) ♀>♂

 large buteo; stocky body but long wings with narrow and
ɔinted tips. All have dark flight feathers below. Adult plumage
riking, clean white and gray. On juvenile note dark color with
ȷht patch on breast and pale tail.

Light juvenile

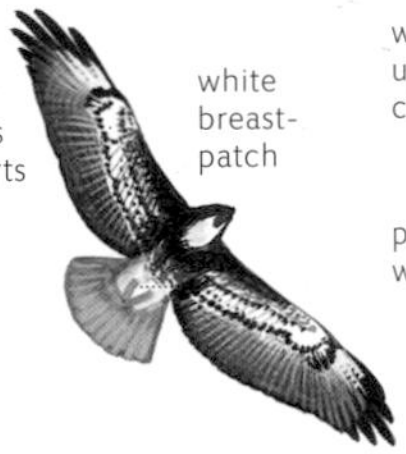

2nd year

all juveniles and 2nd year birds differ from Swainson's by light tail, white breast, dark leading edge of wing; all other buteos have light remiges

Dark juvenile

Adult

Juvenile (1st year)

2nd year

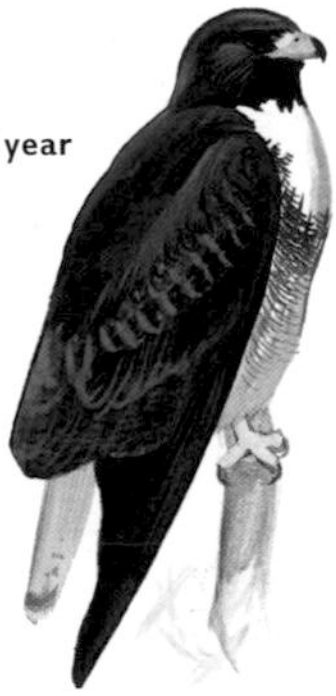

Adult

OICE: Rather high, laughing *reeeEEE ke-HAK ke-HAK keHAK* 'emale call differently accented); also a high, harsh *kareeeev,* ıeezy *tseef.* Juvenile (through 1st winter) begs with high, mew-ıg squeal *meeeii.*

Uncommon in open savannas. Nests in low trees or shrubs. Solitary. Hunts mostly from the air while kiting or hovering, but also from perch; feeds on variety of prey, mainly small birds and mammals. Attracted to fires.

White-tailed Hawk is the only buteo with an easily recognizable second-year plumage, all other buteos molt directly from juvenal plumage to an adultlike body plumage when about one-year-old. Those species can still be aged as second-year by details of wing feathers. Not all primaries and secondaries are replaced in that molt, and the contrast of retained juvenal feathers (longer, more pointed, and with smaller dark subterminal band than adult) alongside new adult-like feathers identified second-year buteos.

Red-tailed Hawk

Buteo jamaicensis

L 19" WS 49" WT 2.4 lb (1,080 g) ♀>♂

Numerous and conspicuous, this is the default buteo in most areas. Stocky and broad, with rounded wings and short tail. Reddish tail of adult distinctive. Light morph has pale breast contrasting with brown head and streaked belly-band, white-speckled V on back, and dark mark on leading edge of underwing.

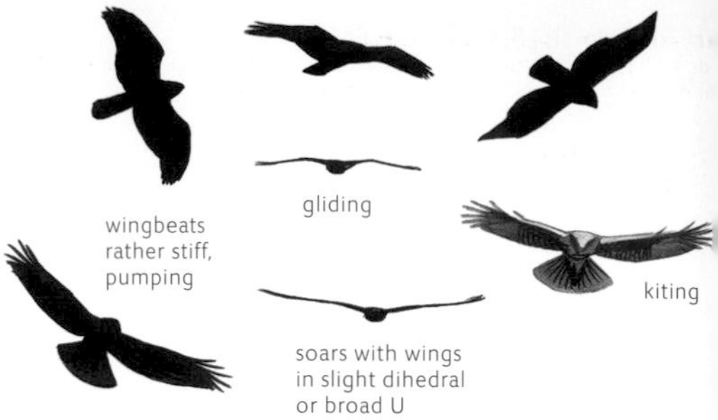

Western

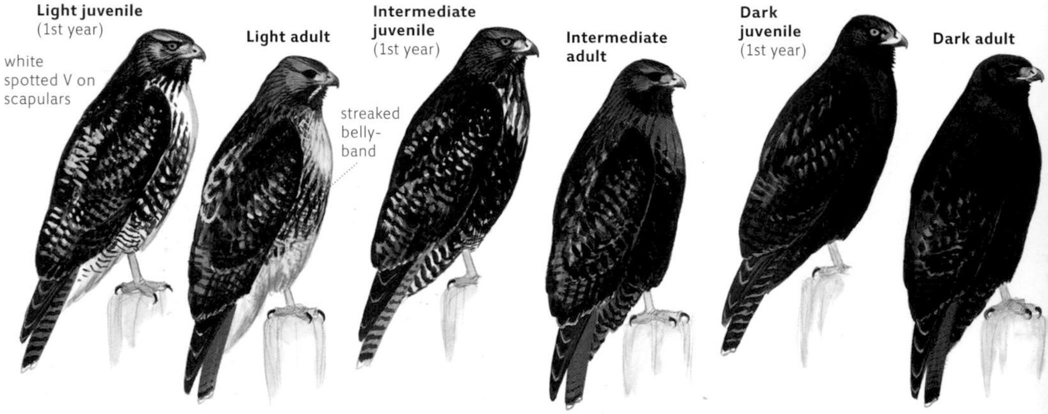

VOICE: A distant-sounding, rasping, scraping scream, falling in pitch and intensity *cheeeeeeewv*; also a shorter scream that may be repeated in series. Juvenile begs with measured, insistent whistles *pweee, pweee....*

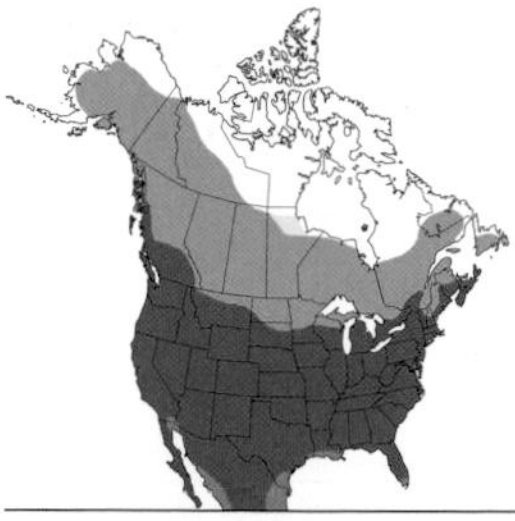

Generally uncommon but conspicuous and widespread. Often seen perched in trees and on posts along roadsides or in open fields.

Western population is the most variable population of Red-tailed Hawk, with complete range from light to dark plumage. Dark and intermediate morphs account for 10–20 percent of Western population with scattered records east to Atlantic states.

rtial albino Red-tailed Hawk

ds with mostly white plumage occur rarely throughout the species' ıge; beware confusion with Gyrfalcon or Snowy Owl.

Adult

uthwestern

oportions like Western; belly faintly streaked or clean, tail often pale.

Juvenile

Adult

stern

ghtly shorter-winged than Western; white throat, well-defined lly-band and whitish breast. Populations in Florida and eastern nada show heavier markings like Western.

Juvenile

Adult

rider's

is scarce, pale prairie variant is always outnumbered by normal ht-morph birds. Underparts white with faint or no belly-band, tail ıitish, much white on upperparts, often white-headed. Compare rruginous Hawk.

Juvenile

Adult

The subspecies shown above are typical of their respective regions but blend with adjacent subspecies where ranges meet. Large areas (e.g., much of western Canada) are occupied by intergrades.

Harlan's

Plumage blackish and white, lacking brown tones. Intergrades with Western Red-tailed where ranges overlap.

Dark juvenile

Light adult

light morph less than 1 percent of population

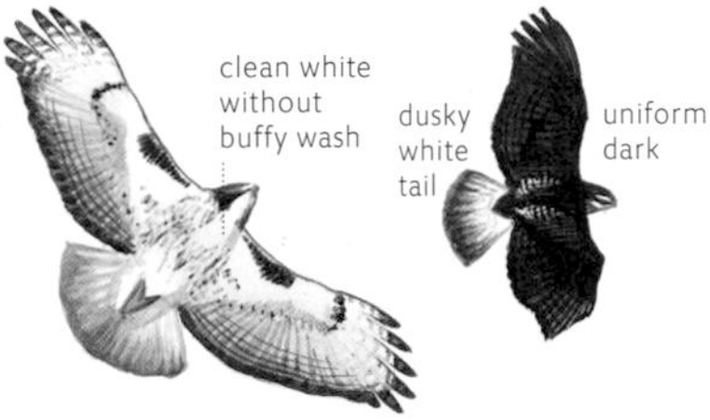

Dark adult

Dark juvenile (1st year)

Dark adult

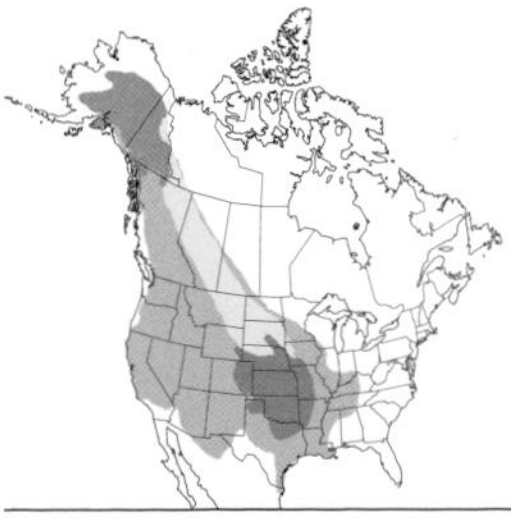

Generally uncommon, nesting in central Alaska and wintering mainly from Nebraska and Iowa to Texas and Arkansas, with small numbers west to California, but very rare eastward.

Harlan's

Ferruginous Hawk

Buteo regalis

L 23" WS 56" WT 3.5 lb (1,600 g) ♀>♂

Our largest buteo; note broad head and long, tapered wings. Light morph told from Red-tailed Hawk by clean snowy-white underparts, mostly dark back, dark leg-feathering, and pale tail. Always has very pale flight feathers with small dark tips.

Light juvenile

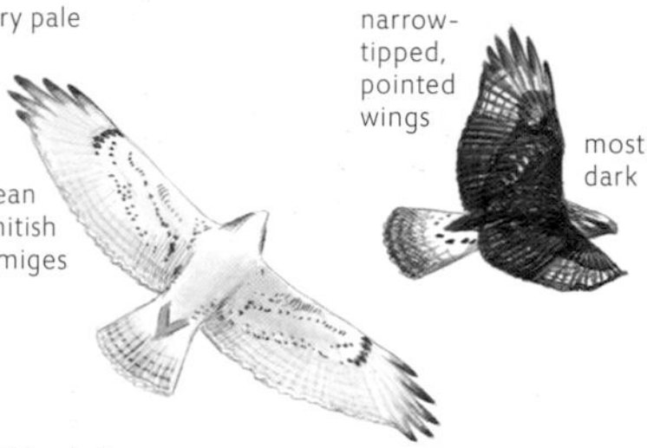

Dark juvenile

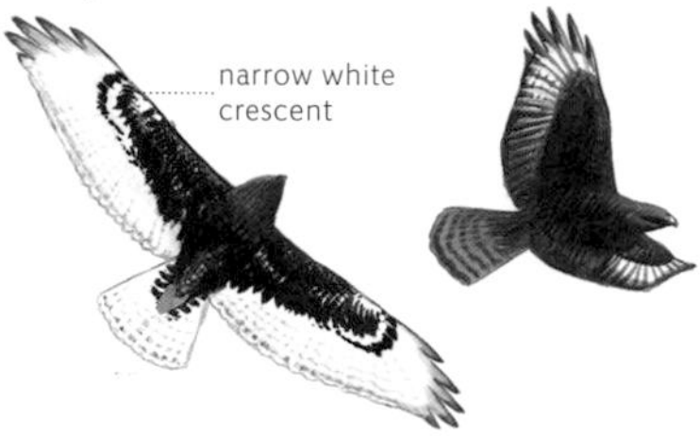

Light adult

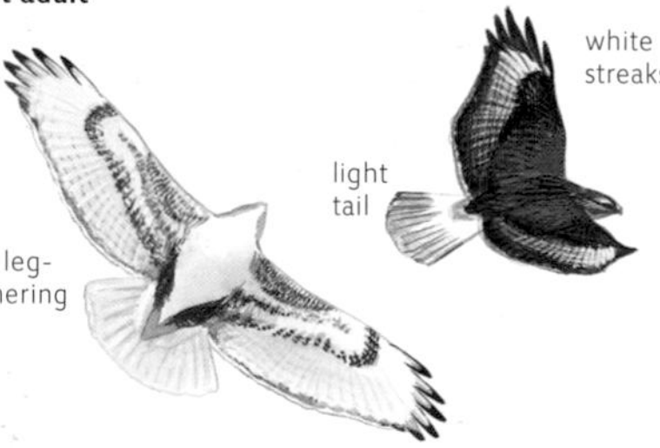

Dark adult

Light juvenile (1st year)

Light adult

Dark juvenile (1st year)

Dark adult

Occurs in distinct light and dark color morphs with few intermediates; dark morph less than 10 percent of population

VOICE: A melancholy whistle *k-hiiiiiiiw* or *geeer*; lower-pitched and less harsh than other buteos. Juvenile gives high scream.

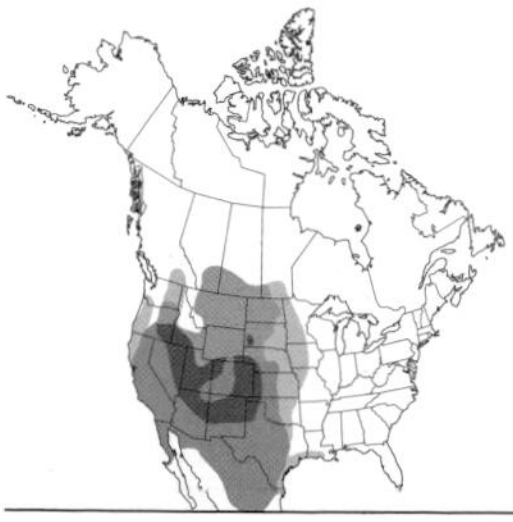

Rare and local in arid grasslands and adjacent farmland. Nests in isolated trees or on rock outcrops. Perches on posts or often on the ground. Solitary. Hunts from the air or from a perch, mainly for small mammals.

On average our largest buteo, this species has a broad head and distinctively tapered and narrow-tipped wings. Very clean white underside of primaries and secondaries with tiny dark tips is not matched by any other buteo. White inner webs of primaries and sometimes secondaries can be seen on upperside of spread wing, a pattern matched only by Rough-legged Hawk and some paler Red-tailed Hawks.

Rough-legged Hawk

uteo lagopus

21" WS 53" WT 2.2 lb (990 g) ♀>♂

large and rather slender buteo, with long, broad-tipped wings nd long tail. Light morph boldly patterned below: dark belly and rist-patch with very pale flight feathers; brownish or grayish bove, with white base of tail.

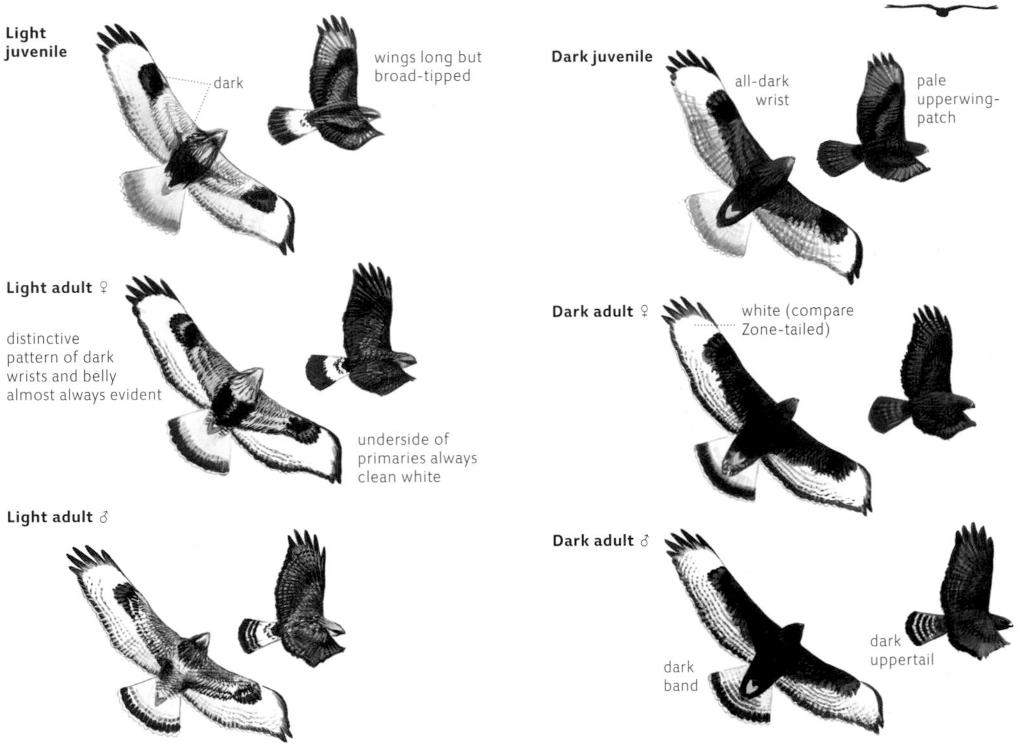

OICE: A simple, high squeal *keert* or *keeeeeer*, less rasping than ed-tailed. Juvenile begs with thin, rising *skwee, skwee*....

Uncommon to rare on open tundra, farmland, prairie, marshes, and other expansive open habitats. Look for it perched on fenceposts or isolated trees in open areas, or flying and hovering over open ground. Nests on cliff ledges in tundra. Solitary. Feeds mainly on small rodents.

Occurs in distinct light and dark color morphs with few intermediates; dark morph 10 percent in west, up to 40 percent in east

Fluffy plumage, small bill, and small feet are all adaptations to Arctic life, giving distinctive appearance.

This species has the most different male and female plumages of all our buteos; the sex of most adults can be determined readily by plumage patterns, but intermediates do occur and variation is complex.

Golden Eagle

Aquila chrysaetos

L 30" WS 79" WT 10 lb (4,575 g) ♀>♂

Very large and dark, with buteo-like proportions and wing posture. All-dark body and underwing coverts, without white markings shown by subadult Bald Eagles. Told from Turkey Vulture by larger size, larger head, and steady flight.

VOICE: Rather weak, high yelping. Adult gives two-syllable *kee-yep* or *chiup* in slow, measured series. Juvenile begs with piercing, insistent *sssееееееe-chk* or chittering *kikikikikikiki*-yelp.

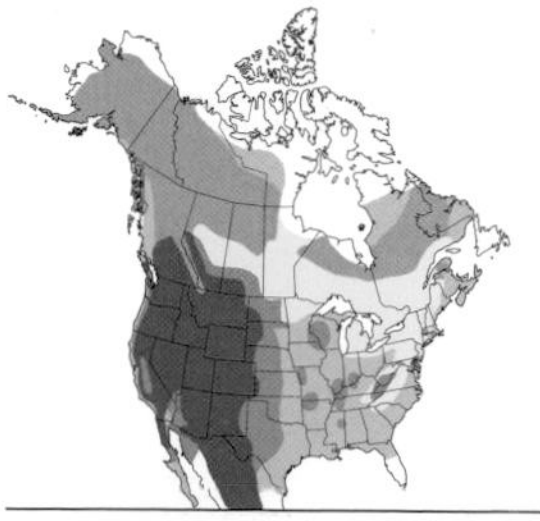

Rare but widespread, in grasslands, deserts, and other open country relatively far from people, especially in mountainous areas. Nests on cliff ledges or less often in tall trees. Solitary. Hunts mainly mammals, especially squirrels and rabbits, from prominent perch or from the air, contouring over hillsides or soaring high above. Often seen perched on utility poles or rocky crags.

Golden Eagle could be mistaken for Turkey Vulture or dark buteos, but note much larger size, relatively long wings, and stability in flight. Most similar to Bald Eagle, although the two species are not very closely related, and usually readily distinguished by wing and tail patterns. Occasional Golden Eagles have white markings on the underwing coverts suggesting immature Bald Eagle. Check color pattern on upperside, proportions of bill and head, and details of wing shape and pattern.

Bald Eagle

aliaeetus leucocephalus

31" WS 80" WT 9.5 lb (4,325 g) ♀>♂

ery large and dark, with plank-like wings and relatively large ead and bill. Adult distinctive, with white head and tail. Juvenile ark with white on underwing coverts, whitish streaks on tail, and ale belly.

OICE: Call rather weak, flat, chirping whistles, stuttering, vari-ble. Immature calls generally harsher, more shrill than adult ntil three to four years old.

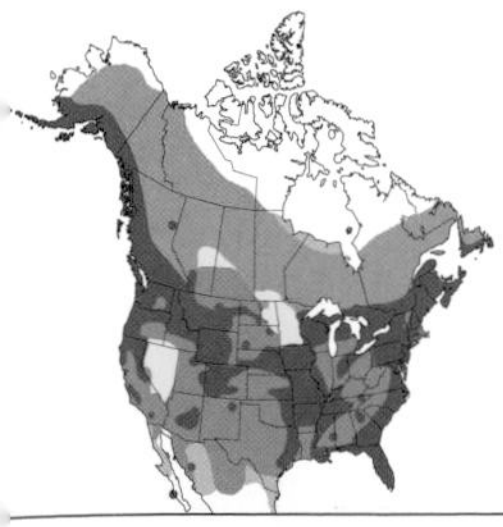

Rare to locally uncommon. Usually found near water: ocean bays, lakes, and rivers with abundant fish and waterfowl for prey. Catches live prey but also scavenges or steals food. Look for it perched on prominent large trees, or on the ground along shoreline.

Southern breeders smaller, average 10 percent shorter-winged and 20 percent shorter-tailed than northern breeders; differences are broadly clinal. Southern juveniles fledge in Mar; northern juveniles are not independent until Aug. Southern juveniles wander north to Canada May–Sep and can sometimes be distinguished from the northern by date and relatively worn plumage. Juveniles of both Bald and Golden Eagles have broader wings and longer tails than adults. 2nd year birds retain longer juvenile feathers among shorter new ones.

Rails, Coots, Cranes, and Limpkin

FAMILY: RALLIDAE, ARAMIDAE, GRUIDAE

14 species in 9 genera (1 rare species not shown here). Most are in family Rallidae, but cranes in Gruidae and Limpkin in Aramidae. All inhabit marshy wetlands, where they feed on plants and small animals, although the feeding habits of genera differ greatly. Cranes are tall and conspicuous, traveling in flocks and flying in goose-like V formation (compare herons and geese Limpkin is solitary and secretive. Rails are very secretive ar solitary in grassy marshes, while gallinules are often seen in th open, and coots are gregarious and bold, swimming in larg flocks duck-like. Adults are shown.

Genus *Coturnicops*

Yellow Rail, page 153

Genus *Laterallus*

Black Rail, page 153

Genus *Porzana*

Sora, page 152

Genus *Fulica*

American Coot, page 147

Genus *Rallus*

Clapper Rail, page 150

King Rail, page 151

Virginia Rail, page 152

Genus *Gallinula*

Common Gallinule, page 147

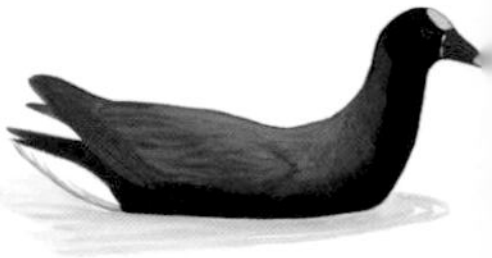

Purple Gallinule, page 146

Genus *Aramus*

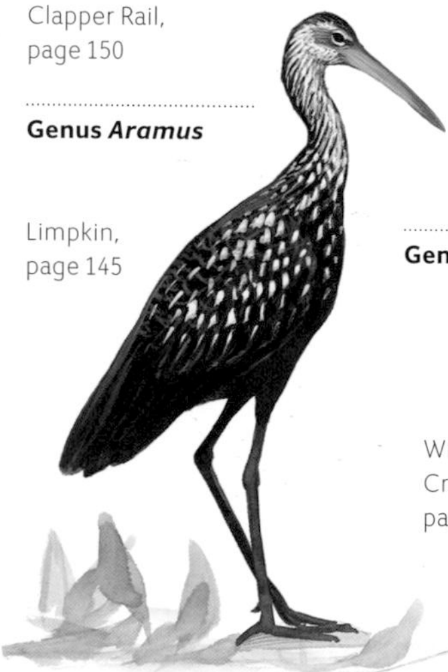

Limpkin, page 145

Genus *Grus*

Whooping Crane, page 148

Sandhill Crane, page 149

Genus *Porphyrio*

Purple Swamphen, page 146

dentification of Coots

oots are often confused with ducks, as they are typically seen vimming on open ponds mixed with ducks and geese. Their ark color and white bill immediately identifies them as coots, ut other fundamental differences that distinguish them from ucks include lobed toes (not webbed), more rounded back, very ort tail, stout conical bill, and whining and clucking calls. They ive underwater for aquatic plants and sometimes resurface tail rst, dragging vegetation up in their bill.

Downy young American Coot

owny young coots and gallinules are mostly blackish, like rails ut unlike ducks. American Coot has red bill and some reddish own on head and neck.

American Coot variant adult ♂

American Coot typical adult ♂

)n adult coots and gallinules the forehead is unfeathered, merg- ng with the bill in an expanded frontal shield. This is largest and ost showy in breeding season, and slightly more exaggerated males than females.

Typical American Coots have the frontal shield small, flat, and ark reddish. On some individuals throughout North America, resumably males, the frontal shield is expanded, bulging, and ostly whitish. This matches the appearance of Caribbean Coot *F. caribaea*), but that species is unlikely to occur here, and all re assumed to be simply variants of American Coot.

Crane Behavior

airs of cranes engage in elaborate "dancing" rituals during ourtship in early spring. These involve synchronized dipping, owing, leaping, head-tossing, and flapping while giving wild ugling cries.

Limpkin

Aramus guarauna

L 26" WS 40" WT 2.4 lb (1,100 g)

An unusual bird, like a very large rail (or a small crane) dark brown overall with variable white spots, long legs and long bill. Larger than ibises, with nearly straight bill, and usually seen near reeds, shrubs, or overhanging trees.

VOICE: Loud, anguished, wild-sounding scream/wail, clear with some rattling overtones, higher-pitched than cranes; *kwEEEeeer* or *klAAAar* with quieter wooden clicking *t-t-t-t-t-tklAAAaar.* Unmistakable.

Uncommon, local, and inconspicuous in wooded and brushy swamps; usually seen walking slowly along edges of ponds or streams searching through shallow water for snails and other mollusks. Usually solitary.

Purple Swamphen

Porphyrio porphyrio

L 16" WS 35" WT 1.5 lb (700 g)

Superficially similar to Purple Gallinule but much larger and heavier, with thick red bill, dark red legs. Most Florida birds have pale grayish head, unlike Purple Gallinule.

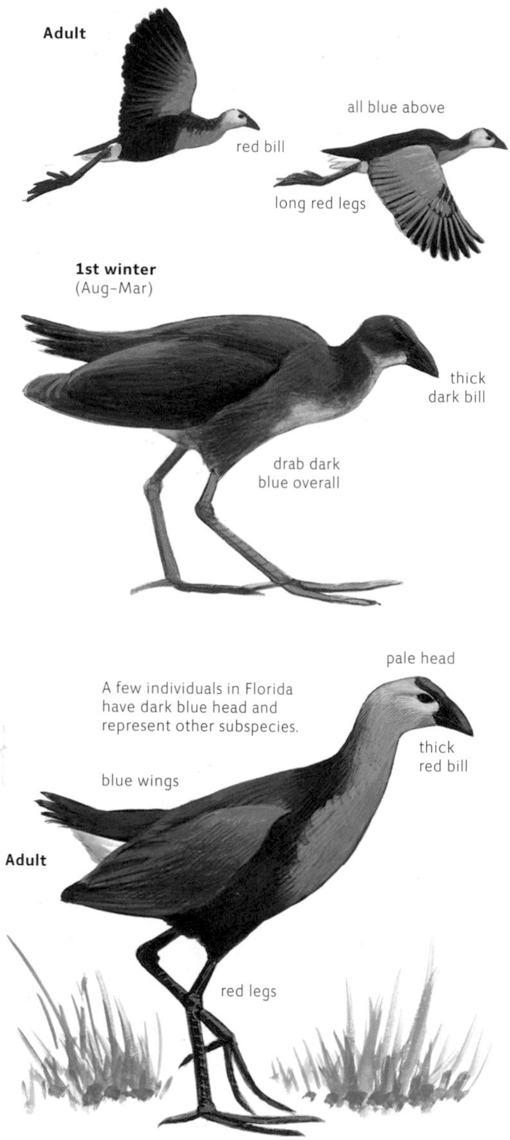

VOICE: Extremely varied. Loud, abrupt, shrill creaking calls like some calls of coot or gallinule, and much lower honking notes reminiscent a very low-pitched Laughing Gull.

Native to Eurasia. A small feral population is introduced in Florida and established in wet areas with dense reeds. Feeds mainly on plant material such as reed stalks.

Purple Gallinule

Porphyrio martinica

L 13" WS 22" WT 8 oz (235 g) ♂>♀

Adult unmistakable, with yellow legs and green, purple, an blue feathers. Juvenile also unique, with pale brown and gree plumage.

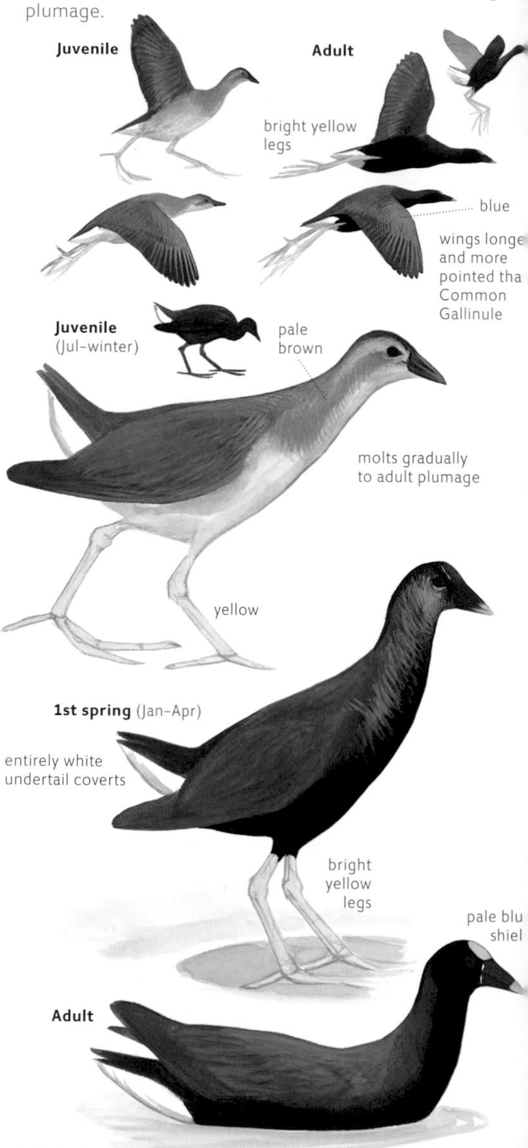

VOICE: Varied, but never whining notes. Long rhythmic serie distinctive: rather slow, nasal, honking *pep pep pep pePA pePAA*...or *to to to terp to terp to to terp*.... Also a deep gruntin series similar to King Rail and a variety of sharp, single notes: lov *pep*; short, high *kit*; and high, sharp *kidk*.

Uncommon and local in ponds and marshes with abundant lily pads and other emergent vegeta tion and brushy edges. Walks around pond edges or on vegetation, often climbing into bushes or low trees; rarely swims Solitary.

Common Gallinule

Gallinula galeata

14" WS 21" WT 11 oz (312 g) ♂>♀

ike coots, unlike other swimming birds, gallinules bob their heads s they swim. Duck-like when swimming, but note long legs and ed bill, rail-like when walking. White stripe on flanks distinctive.

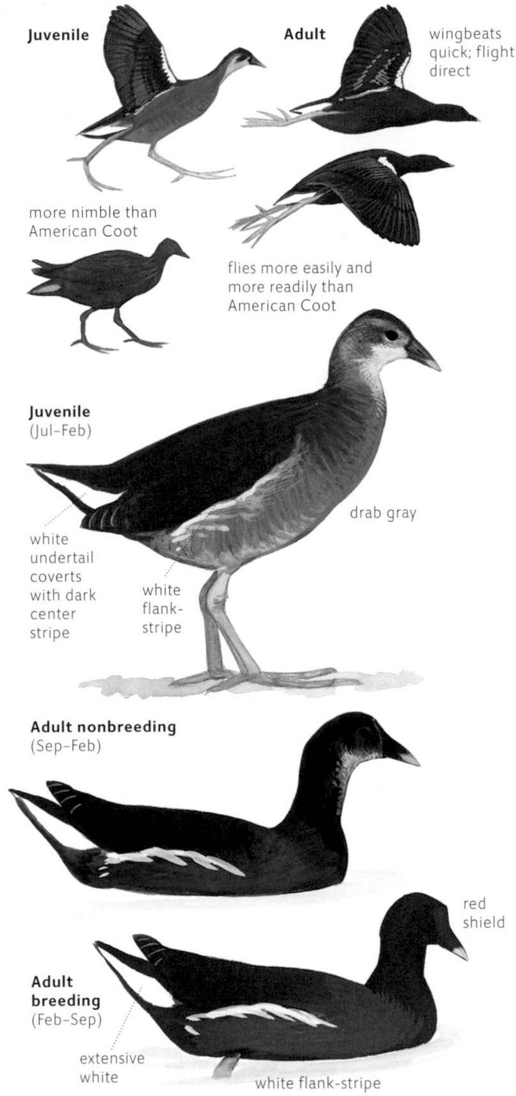

VOICE: Varied; a slowing series of clucks ending with distinctive ong, whining notes *pep pep pep pehr pehr peehr peeehr pehr*. Quality varies from low and nasal to higher and creaking. Other whining, creaking notes; also varied short clucks from low, heavy *kulp* to high, sharp *keek*.

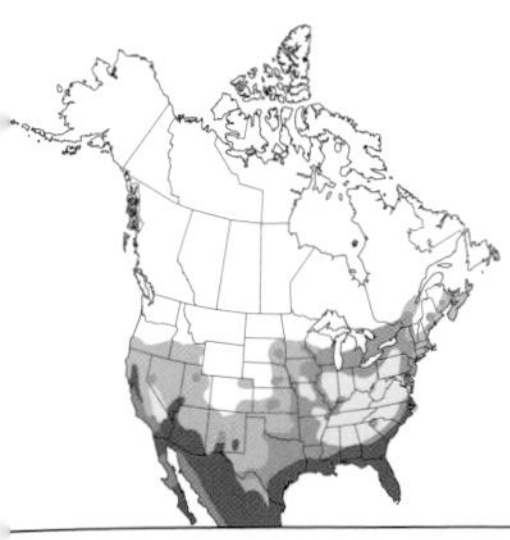

Prefers water with abundant emergent vegetation, swimming along the edges of marshy pools or around lily pads, never swims in deep open water. Generally solitary or in small loose groups, never large flocks like American Coot.

American Coot

Fulica americana

L 15.5" WS 24" WT 1.4 lb (650 g) ♂>♀

Duck-like, but blackish with stubby white bill, round head and body. Swims buoyantly with bill angled down, often diving for plants; also walks easily. Often forms large compact flocks.

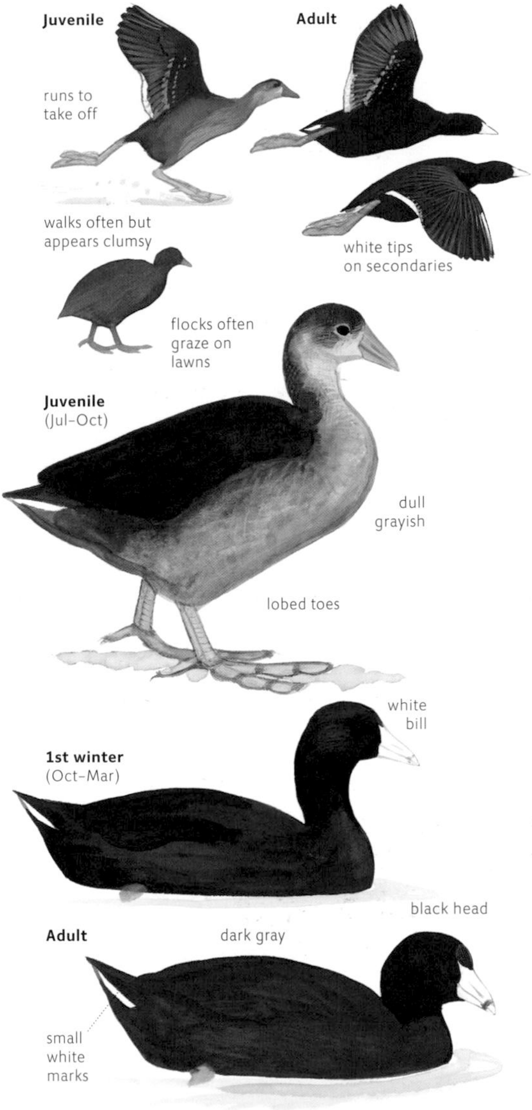

VOICE: A variety of short, clucking notes: most with hollow, trumpeting quality and coarse, rattling undertone; usually lower than Common Gallinule. Most commonly a short single note *krrp* or *prik*, often strung together in long series *priKI priKI*.... Voice of female low and nasal, male high and clear.

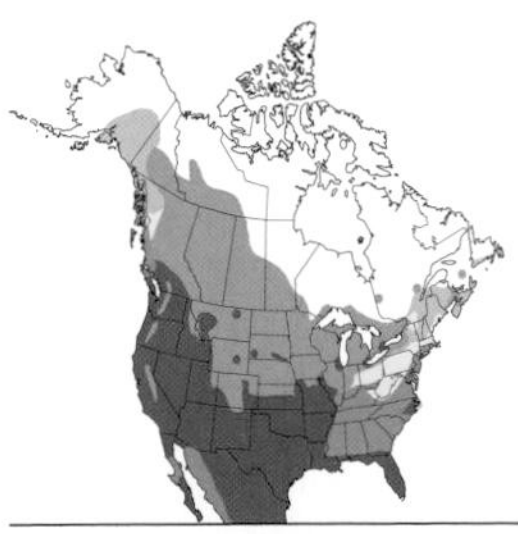

Found on shallow ponds and lakes with submerged vegetation near the surface often near stands of reeds. In winter often on small ponds, such as on golf courses, foraging both by diving for submerged plants and by grazing on lawns.

Whooping Crane

Grus americana

L 52" WS 87" WT 15 lb (6,850 g)

Our tallest bird; large size and characteristic crane-like shape and habits combined with white plumage and black wingtips distinctive. Courtship involves elaborate dancing displays.

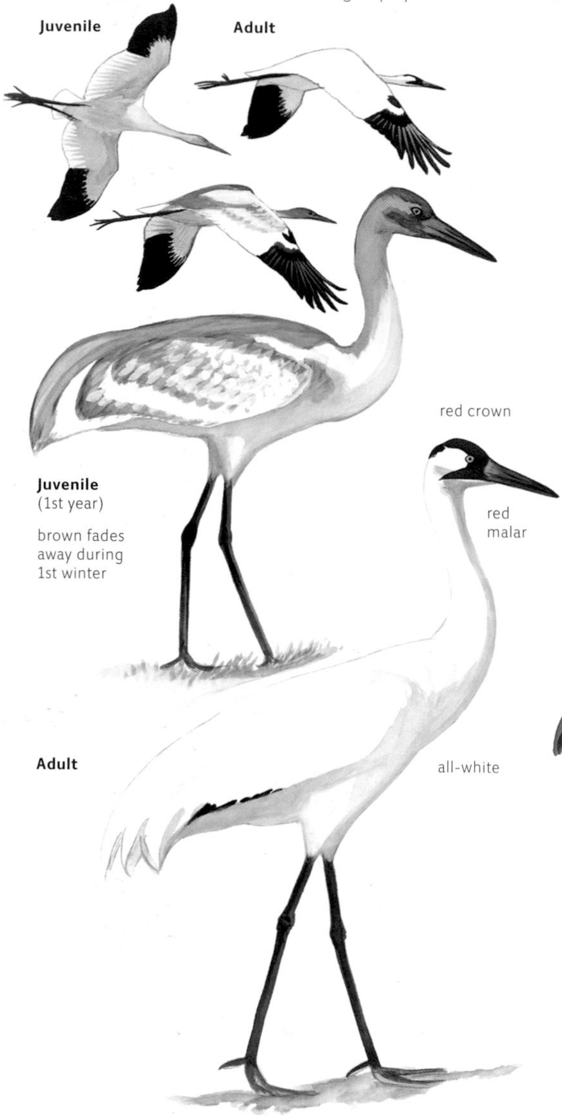

VOICE: A loud, clear, bugling *bKAAAH*, clearer, higher-pitched, and longer than Sandhill. Juvenile gives a high, whistling *ddeer* or persistent, high, clear whistle *swee, swee*.... In dancing display female calls shorter, more rapid, and slightly higher-pitched than male.

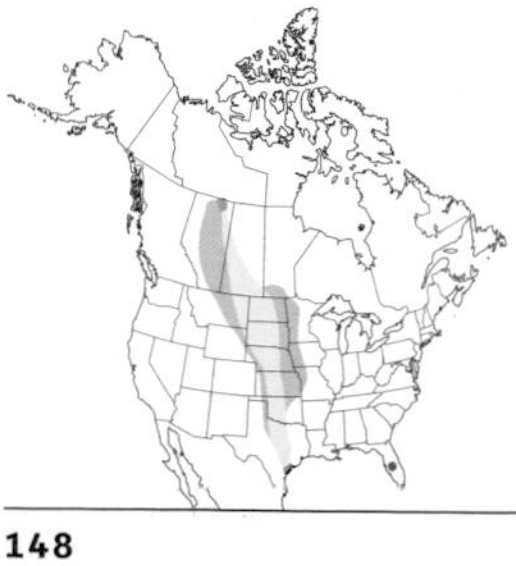

One of our rarest birds, with only about 200 individuals in the wild. Nests in expansive freshwater marshes near tree line. Winters in coastal saltmarshes. Travels in small family groups year-round, often two adults with one or two juveniles.

Common Crane

Grus grus

L 45" WS 82" WT 12.1 lb (4,588 g)

Similar to Sandhill Crane, but plumage always gray, never brown ish; adult has black neck with white on nape, all ages shov contrasting dark flight feathers.

VOICE: Flight calls very similar to Sandhill Crane, may average slightly higher, shorter, and more musical.

Very rare visitor from Eurasia. Recently about one record per year in North America, usually single birds with migrating flocks of Sandhill Cranes, some paired with Sandhill Crane to produce hybrid offspring.

Sandhill Crane

Grus canadensis

GREATER: L 46" WS 77" WT 10.6 lb (4,850 g) ♂>♀
LESSER: L 41" WS 73" WT 7.3 lb (3,350 g) ♂>♀

A tall gray bird easily distinguished from Great Blue Heron by voice, flocking behavior, straighter neck (never coiled), stiffer wingbeats, and rusty color in summer. Performs elaborate dancing displays in courtship.

cranes fly with slow, rolling downbeat and quick upbeat

Lesser (Northern)

Greater (Southern)

VOICE: A loud, resonant, wooden rattle or a rolling bugle, slightly descending, but some variation. Juvenile until at least April (age ten months) gives high, squeaky or trilled *tweer*. Dancing display of adults accompanied by complex duet, female giving more rapid, higher-pitched calls than male.

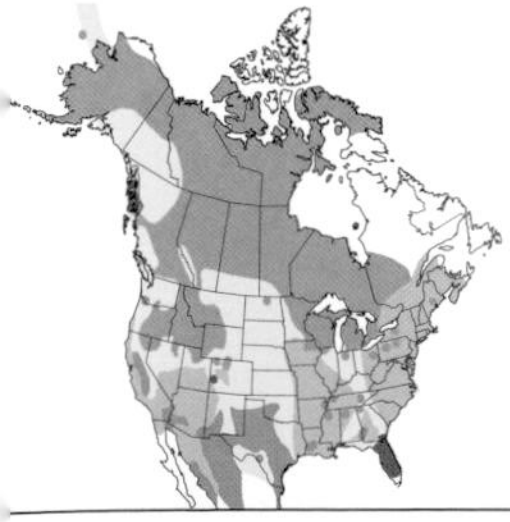

In migration and winter forages in large flocks on open grassland and grain fields, roosting at night in shallow water. Nests in expansive wet bogs and marshes, such as old beaver ponds, where grassy marsh and meadow habitat is interspersed with shrubs.

Lesser breeds in far north; it is small, short-billed, and short-legged, with relatively long wings. Greater breeds from central Canada southward; it is larger and up to 50 percent longer-billed than Lesser, with little overlap, and is relatively short-winged. Greater also has paler primaries with whitish shafts. Extremes are easily identified, but much of central Canada is occupied by an intermediate population. Most of the intermediate birds winter midcontinent, and in those areas many individuals are difficult to assign to subspecies.

Clapper Rail

Rallus longirostris

L 14.5" WS 19" WT 10 oz (290 g) ♂>♀

Much larger and longer-billed than Virginia Rail. Atlantic subspecies very drab grayish overall, Gulf Coast and California birds more like Virginia or King Rail but somewhat less richly colored, with less distinct dark streaks on back.

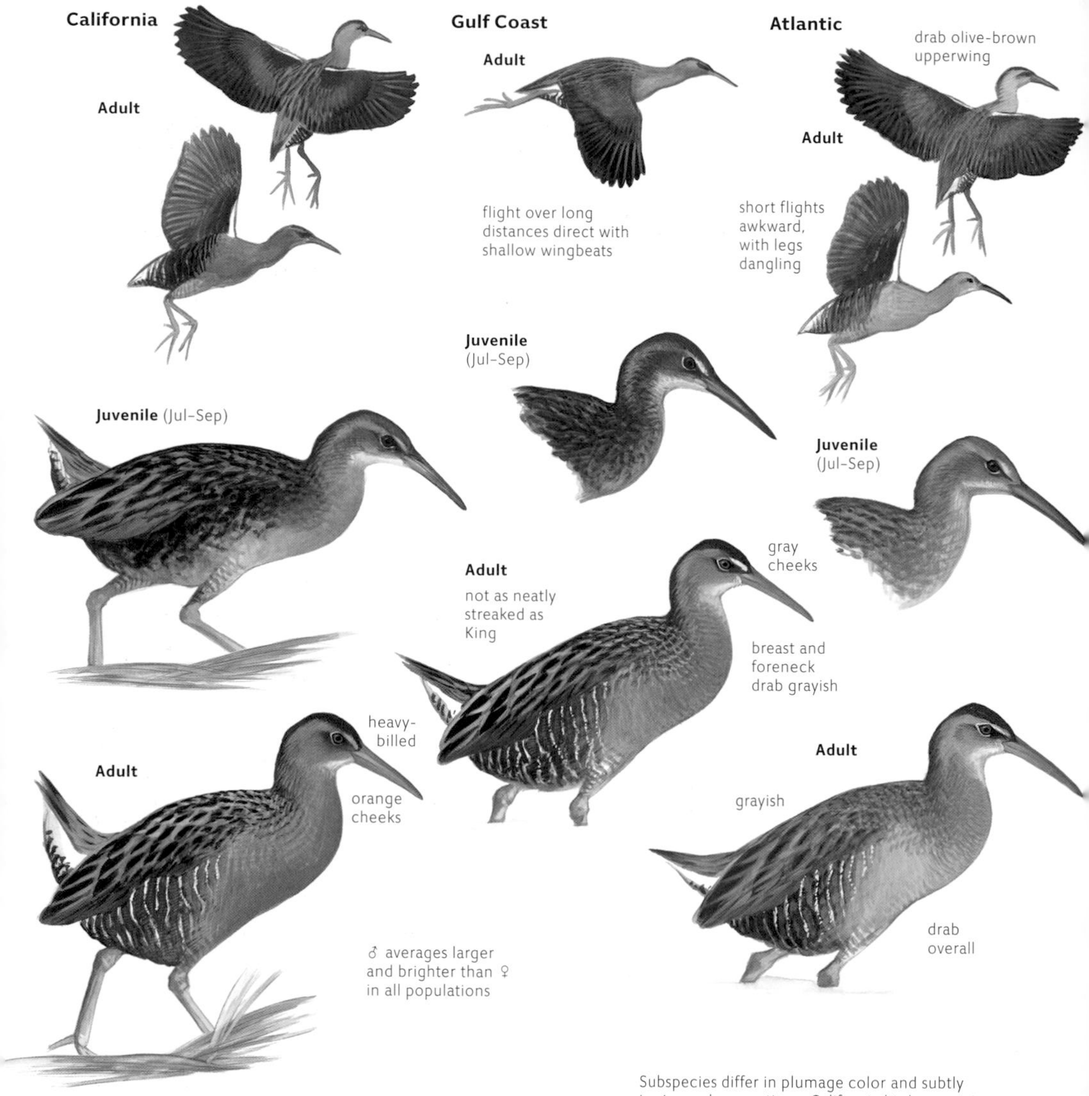

VOICE: Atlantic and Gulf Coast birds give clappering series (4–5 notes/sec); grunting series (4–7 notes/ sec). California gives clappering series (5–6 notes/sec); grunting call undescribed, may be rare. See King Rail.

Common locally in grassy, saltwater and brackish marshes. Generally solitary, but occurs at high densities in some marshes. Often seen in the open or crossing tidal creeks.

Subspecies differ in plumage color and subtly in size and proportions. California birds, sometimes considered a separate species, are relatively large and brightly colored like King, with orange cheeks unlike other Clappers. Atlantic coast birds are smaller and much drabber, without blackish or rufous markings. Gulf Coast birds are similar to Atlantic but more richly colored. King Rail hybridizes with Clapper Rail on Atlantic coast; these hybrids appear very similar to Gulf Coast Clapper Rails. King presumably also hybridizes with Clapper Rail along Gulf Coast, but such hybrids would be almost impossible to detect given the similarity of male Clapper to female King.

‹ing Rail

allus elegans

15" WS 20" WT 13 oz (360 g) ♂>♀

ırge and long-billed like Clapper; differs only slightly in voice ıd in details of plumage, which is darker and more richly col-ıred; especially note blackish centers of feathers on upperparts, eep orange or rufous breast and upperwing coverts, and bold ırs on flanks.

OICE: Shares basic vocal repertoire with Clapper Rail but with a ower tempo. Clappering series and grunting series (usually 2 otes/sec). All calls average deeper and more resonant than Clapper, but quality overlaps extensively; slower tempo is more eliable for identification.

Uncommon and local in grassy or reedy, fresh and brackish marshes. Habits similar to Clapper Rail, but occurs at much lower density.

Habits of Rails

Shy and secretive, rails are found in dense, marshy vegetation. Calls are usually the best clue to their presence, but patiently watching the edges of marshy ponds may yield views. They may be seen running rapidly across openings in the marsh, and they often swim short distances rather than fly. Rails' bodies are laterally compressed, which allows them to escape into dense grass or reeds.

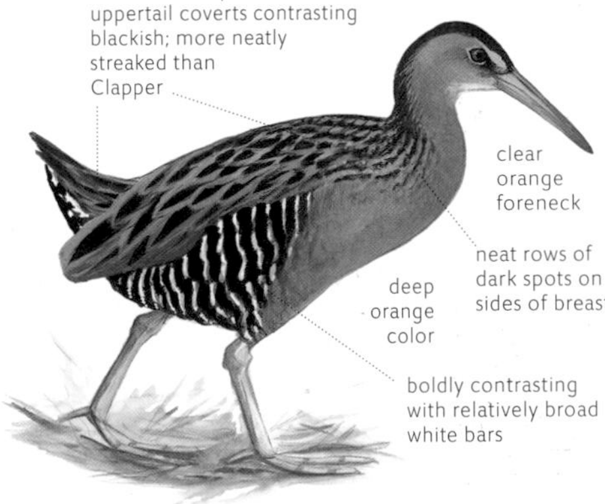

King and Clapper Rail share basic vocal repertoire. Clappering call a series of unmusical *kek* notes, slower at beginning and end; the tempo of the fastest portion is useful for species identification. Grunting series is of steady tempo but falling pitch. Individuals may give faster or slower calls depending on mood, but such departures are usually brief. Long, consistent bouts of typical calls can be reliably identified. Males of both species give single hard *ket* notes repeated monotonously (about 2 notes/sec). Females of both species give dry, clattering *ket ket karrrrr*. Other calls include rapid, high *kek* notes and a raucous squawk like a startled chicken.

Virginia Rail

Rallus limicola

L 9.5" WS 13" WT 3 oz (85 g) ♂>♀

Small, stocky, and dark, with short thick neck, fairly long bill, and overall dark reddish-brown color. In flight note longer bill and broader wings than Sora, without pale trailing edge on wings.

Juvenile
(Jul–Sep)

Adult

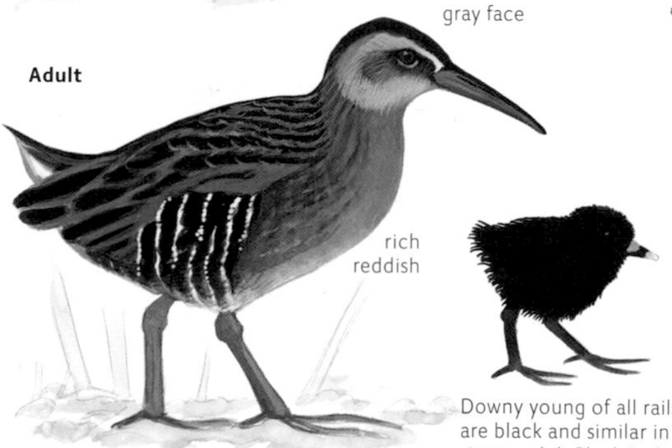

Downy young of all rails are black and similar in size to adult Black Rail.

VOICE: Wheezy, pig-like grunting series *wep wep wep wepwep-wepwepwepppprrr* descending and usually accelerating; higher-pitched than Clapper Rail. Female gives sharp, metallic notes followed by rich churring *chi chi chi chi treerrr*. Male gives a hard, mechanical *gik gik gik gidik gidik gidik gidik*; also squealing or clattering, hard skew, *kweek* or *kikik ik-ik* in dispute or in alarm; most calls harsher or more nasal than Sora.

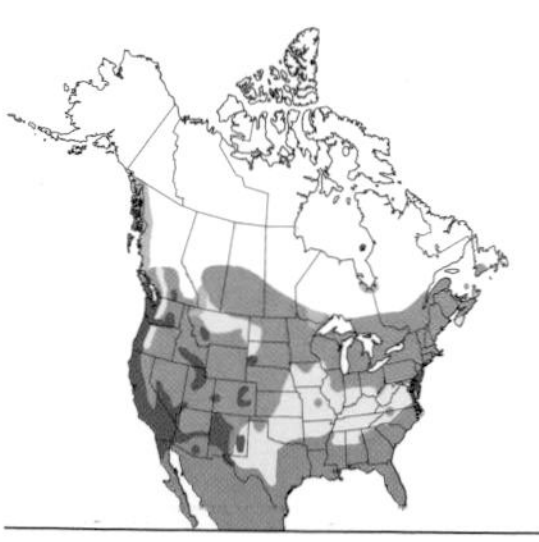

Found mainly in marshes with tall reedy vegetation like cattails; also wet meadows, rank growth at pond margins, etc. Birds attempting to winter far to the north are forced into any unfrozen water, such as woodland streams, outflow pipes, etc.

Sora

Porzana carolina

L 8.75" WS 14" WT 2.6 oz (75 g) ♂>♀

Small and stocky, quail-size, with stout bill. A little more slend overall than Virginia Rail, with shorter greenish bill, paler and le reddish color, and clear buffy undertail coverts. In flight no narrower wings with pale trailing edge.

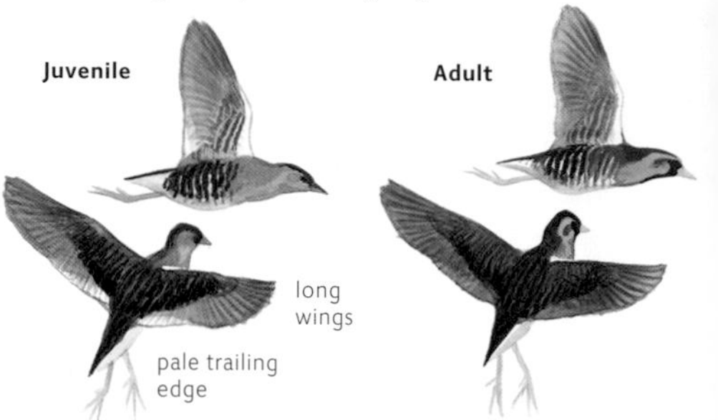

Juvenile
(Jul–Nov)

Adult nonbreeding
(Aug–Feb)

Adult breeding
(Mar–Aug)

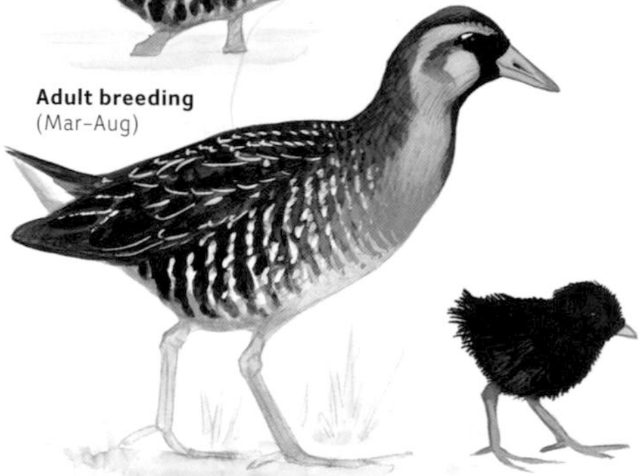

VOICE: A long, high, squealing whinny, descending and slowin at end *ko-WEEeee-e-e-e-e-e, ee, ee* given by both sexes; high clear, sharp, whistled *kooEE*. In alarm a surprisingly loud, shar *keek* and a variety of other notes; some fairly mellow, some wit plaintive quality (*keeeoo*). Also a sharp, staccato *kiu* like Virgini Rail, but other calls usually clearer, more whistled.

Found in any wet marshy habitat especially cattail marshes and grassy marshes. Migrants may occur in any wet or damp grassy or weedy habitat. Most often seen at openings in marshy habitat, along the edges of marshy vegetation, or walking or mud or shallow water near vegetation.

ellow Rail

oturnicops noveboracensis

7.25" WS 11" WT 1.8 oz (50 g)

ery small and stocky. Yellow streaks on back with fine white ossbars distinctive. In flight note very small size, white second-ies. Beware confusion with juvenile Sora, which is larger, fferently patterned, and much more likely to be seen in the pen.

Adult

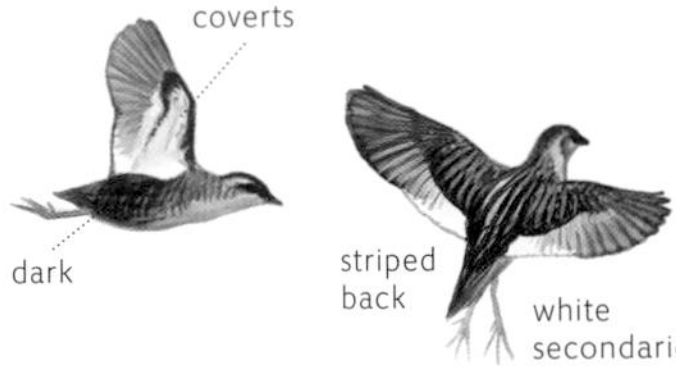

Juvenile
(Aug–Feb)

Adult breeding
(Mar–Aug)

OICE: Territorial song of male (given at night) a mechanical licking sound (easily imitated by tapping two pebbles together) sually in distinctive rhythm *tic-tic tictictic, tic-tic tictictic...*; licking sound nearly duplicated by Cricket Frog but rhythm nique to Yellow Rail. Other calls include descending cackle of bout ten notes; three to four notes that sound like distant knock-ng on door; quiet croaking; also soft wheezing or clucking notes.

Rare and local, probably not truly rare but extremely secretive and very rarely seen. Found in damp grassy meadows or sedge marshes, the same habitat often occupied by Sedge Wren and LeConte's Sparrow. Shuns sunlight and never walks out into exposed view (unlike Sora).

Black Rail

Laterallus jamaicensis

L 6" WS 9" WT 1.1 oz (30 g)

Tiny and all-dark (beware that downy young of all rails are black). Note white speckling on back, rufous nape, short pointed tail, black bill, and red eye.

Adult

Juvenile
(Aug–Feb)

Adult ♂

VOICE: Territorial call of male (given mainly at night) a rich, nasal *keekeedrrr* or *deedeedunk*; typically two high notes followed immediately by one lower solid note, but number of notes variable; when agitated a growling *krr-krr-krr*. Female (rarely) gives a quick, low, cooing *croocroocroo* of two or three notes similar to Least Bittern. Adult also gives soft, nasal barking *churt* or scolding notes *ink-ink-ink-ink*.

Rare and local in grassy, fresh and brackish marshes. As secretive as a mouse: virtually never seen in the open; if forced out of sheltering grasses, dashes quickly across open space and back into cover. Solitary.

Shorebirds

FAMILIES: CHARADRIIDAE, HAEMATOPODIDAE, RECURVIROSTRIDAE, JACANIDAE, SCOLOPACIDAE

76 species in 19 genera (rare species not included here). Most are true sandpipers in family Scolopacidae. Plovers are in family Charadriidae, oystercatchers in Haematopodidae, avocets and stilts in Recurvirostridae, and rare Northern Jacana in family Jacanidae. All are small to medium-size with relatively thin bills and long legs; almost all frequent open habitats, especially mudflats and shorelines, where they forage on small aquatic insect worms, and other animals by picking or probing. Habitat choic and foraging actions often provide identification clues; bill shap and body proportions are also very important for identificatio Nonbreeding adults are shown.

Genus *Pluvialis*

Black-bellied Plover, page 156

American Golden-Plover, page 157

Pacific Golden-Plover, page 157

Genus *Actitis*

Spotted Sandpiper, page 170

Genus *Charadrius*

Snowy Plover, page 158

Wilson's Plover, page 158

Semipalmated Plover, page 159

Piping Plover, page 159

Killdeer, page 160

Mountain Plover, page 160

Genus *Arenaria*

Ruddy Turnstone, page 176

Black Turnstone, page 176

Genus *Calidris*

Surfbird, page 179

Red Knot, page 186

Sanderling, page 183

Semipalmated Sandpiper, page 182

Western Sandpiper, page 182

Least Sandpiper, page 183

White-rumped Sandpiper, page 185

Baird's Sandpiper, page 185

Pectoral Sandpiper, page 184

Purple Sandpiper, page 179

Rock Sandpiper, page 178

Dunlin, page 187

Stilt Sandpiper, page 188

Buff-breasted Sandpiper, page 188

Ruff, page 189

enus *Limosa*

udsonian Godwit, page 175

Marbled Godwit, page 175

Genus *Scolopax*

American Woodcock, page 193

enus *Phalaropus*

Vilson's Phalarope, page 194

Red-necked Phalarope, page 195

Red Phalarope, page 195

Genus *Gallinago*

Wilson's Snipe, page 192

enus *Limnodromus*

hort-billed Dowitcher, age 191

Long-billed Dowitcher, page 190

Genus *Recurvirostra*

Black-necked Stilt, page 165

Genus *Himantopus*

American Avocet, page 165

enus *Haematopus*

merican Oystercatcher, age 164

Black Oystercatcher, page 164

enus *Numenius*

ong-billed Curlew, page 173

Whimbrel, page 172

Genus *Tringa*

Solitary Sandpiper, page 167

Wandering Tattler, page 177

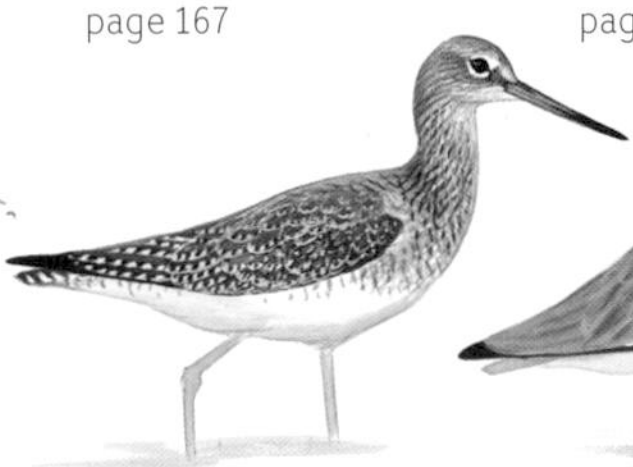

Greater Yellowlegs, page 166

Willet, page 168

Lesser Yellowlegs, page 166

Genus *Bartramia*

Upland Sandpiper, page 171

Black-bellied Plover

Pluvialis squatarola

L 11.5" WS 29" WT 8 oz (240 g)

Our largest plover, with stout bill. In flight, note white tail and wingstripe; all plumages have black armpit on white underwing, and breeding plumage has all-black belly.

VOICE: Flight call distinctive: high, clear whistles; melancholy and gently slurred *PLEEooee* or *peeooEEE*; variations with first or last syllable highest. Flight display song a melodious, ringing *kudiloo* or *trillii* repeated.

Common coastally; uncommon to rare inland. Found on open mudflats and beaches and in fields. Often forms small flocks when roosting or flying, but foraging birds move independently and are often seen singly. Nests on relatively dry tundra ridges and knolls with sparse vegetation.

European Golden-Plover

Pluvialis apricaria

L 10.5" WS 26" WT 7 oz (210 g)

Very similar to American Golden-Plover, but stockier with shorte wings and shorter bill. In breeding plumage, shows less black c underside, never black on undertail coverts. Best distinguishe in flight by clean white underwing coverts.

VOICE: Flight call similar to American and Pacific, but lower and simpler: a plaintive, liquid *tlui*; soft, piping *wheep wheep*; and short, single notes *teep*. Flight song *perPEEoo* repeated.

Very rare visitor from Europe; nc recorded annually. Virtually all records from Newfoundland in late Apr–May; some records of small flocks. Habitat similar to American Golden-Plover.

ʼacific Golden-Plover

ʼuvialis fulva

10.25" WS 24" WT 4.6 oz (130 g)

early identical to American Golden-Plover, averages slightly ore yellowish color and larger spots above; slightly longer-gged and longer-billed, details of wing and tail proportions elpful.

ᴐICE: Flight call a rising *quit, koWIT*, or *kowidl*; lower, simpler ıan American; sharply accented on rising second syllable like emipalmated Plover. Flight song relatively slow, slurred whistles ith long pauses between phrases *tuee tooEEEE; tuee tooEEEE....*

Uncommon. Habitat and habits very similar to American Golden-Plover, and the two species often occur together on migration in plowed farm fields, pastures, and drier mudflats, but when nesting Pacific is usually found on wetter lowland tundra with more vegetation.

American Golden-Plover

Pluvialis dominica

L 10.5" WS 26" WT 5 oz (145 g)

Slightly smaller than Black-bellied Plover, with relatively small head and bill. Best distinguished in flight by dark rump and tail and uniform gray underwing coverts. Golden spots on upperside not distinctive, often fade to nearly white, and juvenile Black-bellied can appear quite yellowish.

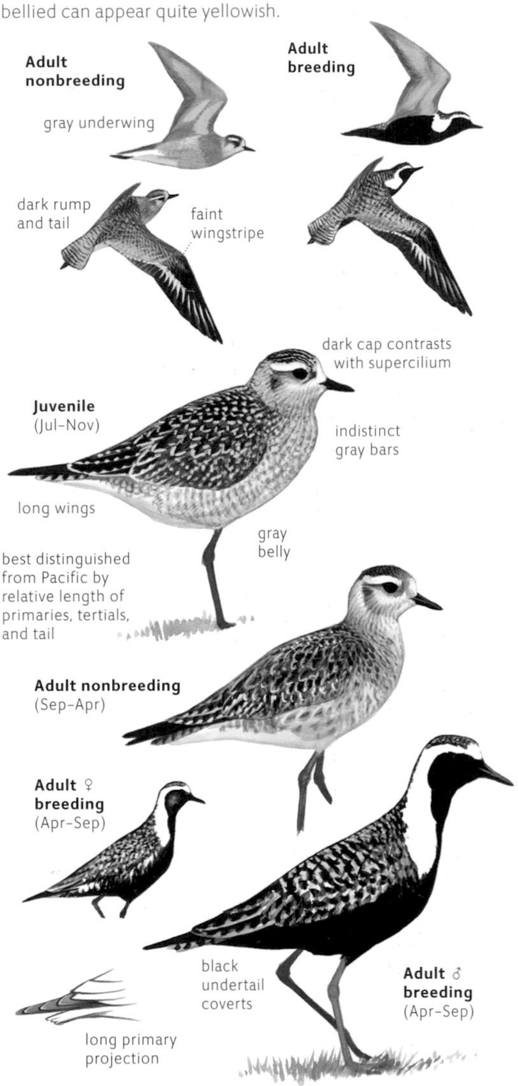

VOICE: Flight call a sad-sounding, urgent *queedle*; higher than Black-bellied with little pitch change. Varies from high, sharp *quit* to *koweeawi* but always with vibrant, urgent tone. Flight song endlessly repeated *koweedl* or *tlueek* notes or a repeated *wit wit weee wit wit weee....*

Uncommon on dry mudflats and in shortgrass fields and pastures. Usually in small flocks, often mixed with Black-bellied Plover. Nests on well-drained rocky tundra slopes and knolls with lichens and sparse low vegetation.

Snowy Plover

Charadrius nivosus

L 6.25" WS 17" WT 1.4 oz (40 g)

Relatively short-winged and long-legged. Note pale gray-brown upperparts, drab gray legs, thin bill, and smudgy dark eyeline.

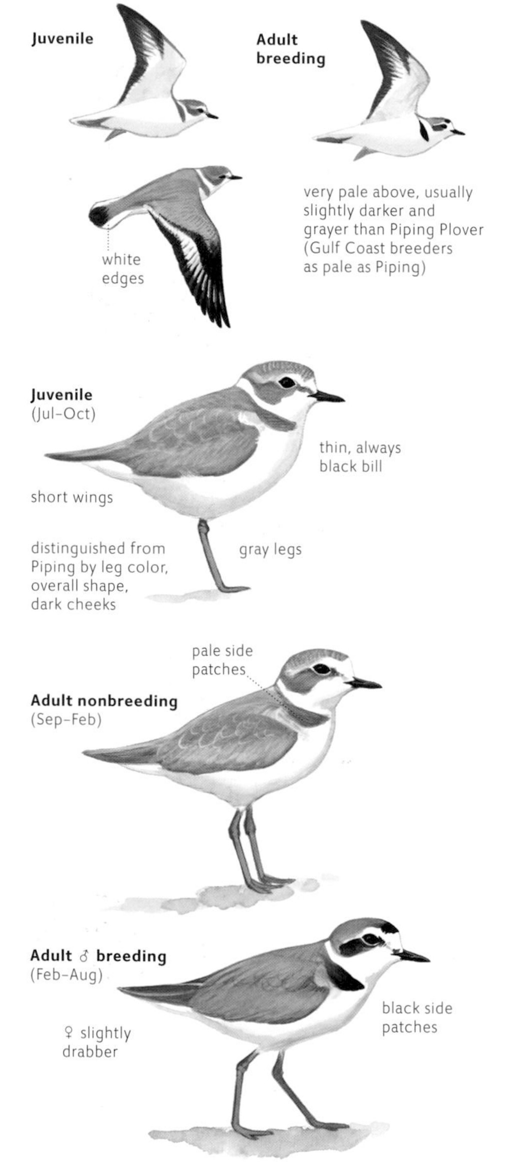

VOICE: Flight call *koorWIJ*; more nasal, husky, and complex than the whistle of Semipalmated; also rather hard *quip* or slightly rough *krip* or *quirr*. Display song (from ground) a repeated whistled *tuEEoo*.

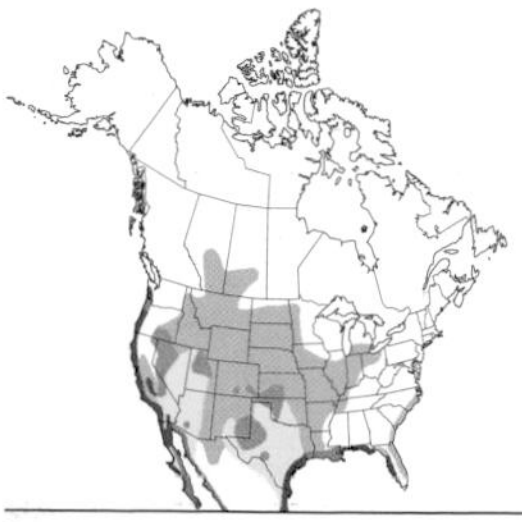

Uncommon and local on sandy beaches and at some shallow inland lakes, favoring drier areas of sand or dried mud far from water. Solitary when foraging; small numbers may roost together. North American population about 25,000.

Wilson's Plover

Charadrius wilsonia

L 7.75" WS 19" WT 2.1 oz (60 g)

Similar to Semipalmated Plover but larger, with very large blac bill. Upperparts slightly paler than Semipalmated. Note pinkisł gray legs, broad white forehead, appears front-heavy with sho wings and large head and bill.

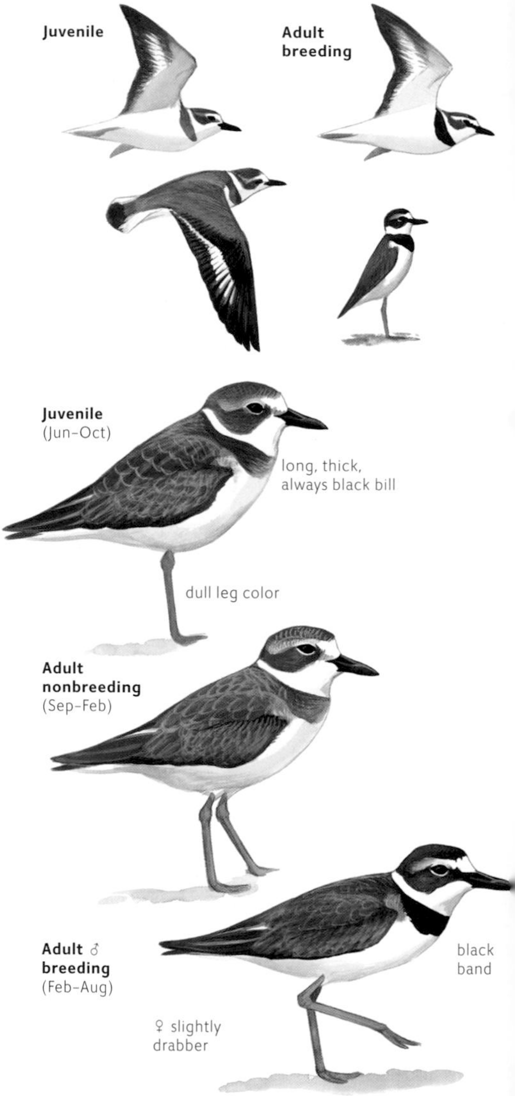

VOICE: Flight call loud, sharp, high, and liquid *quit* or *quee* When flushed a high, hard *dik* or *kid*; also a higher, squeakie *keest*. Grating or rattling, rasping notes in flight display and whe agitated *jrrrrrid jrrrrrid....*

Uncommon and local on sandy beaches and drier sandflats or mudflats usually well away from water's edge. Solitary when foraging; small numbers may roost together. North American population under 10,000.

iping Plover

haradrius melodus

7.25" WS 19" WT 1.9 oz (55 g)

relatively stocky, small plover. Note very pale upperparts, large ırk eye isolated on pale face, and orange legs.

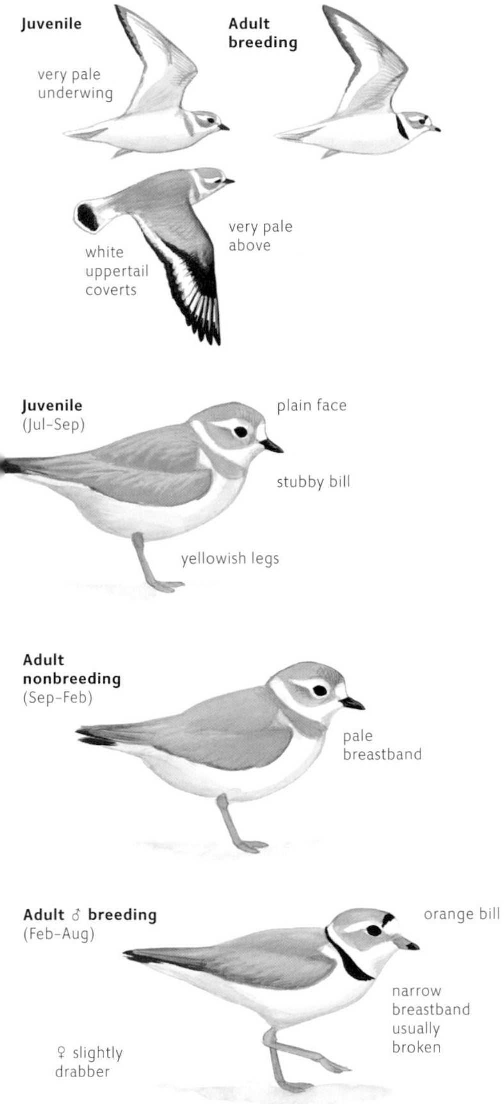

OICE: Clear, mellow whistles *peep, peeto,* or *peep-lo;* low-tched and gentle. When agitated endless series of low, soft histles *pehp, pehp, pehp, pehp....* Display song a steadily peated, whistled *pooeep pooeep...* and variations.

Uncommon (world population under 10,000 birds) and very local on open sandy beaches. Usually seen on dry sand at or above high tide line, roosting and foraging among jetsam on higher levels of beach. Occasionally wanders down to water's edge at low tide. Sometimes gathers in small groups but mostly solitary and does not join other shorebirds.

Semipalmated Plover

Charadrius semipalmatus

L 7.25" WS 19" WT 1.6 oz (45 g)

The most numerous small plover, relatively slender with dark brown upperside. Also note dark cheek, orange legs, and orange bill base. Distinguished from Killdeer by smaller size, single dark breastband, orange legs.

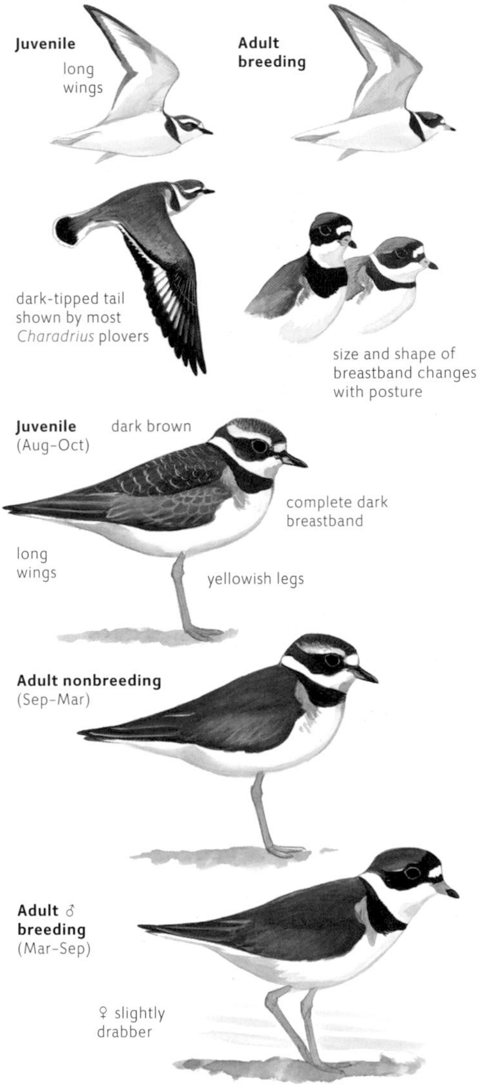

VOICE: Flight call a short, husky whistle *chuWEE* or *kweet* and variations. In aggression a low, husky *kwiip.* Threat call a rapid, descending series *wyeep wyeep yeep yip yipyiyiyiyiyi.* Flight song a repeated husky whistled *too-ee, too-ee...* and various other calls.

Common on open mudflats and beaches. Roosts and flies in flocks, often with other small shorebirds; foraging birds widely spaced or only loosely associated with others. Nests on tundra and shingle beaches.

Mountain Plover

Charadrius montanus

L 9" WS 23" WT 3.7 oz (105 g) ♀>♂

Relatively short-tailed and long-legged. Distinctively plain and pale brown in all plumages. Can be confused with nonbreeding American Golden-Plover, but note clean white belly and underwing, and pale legs.

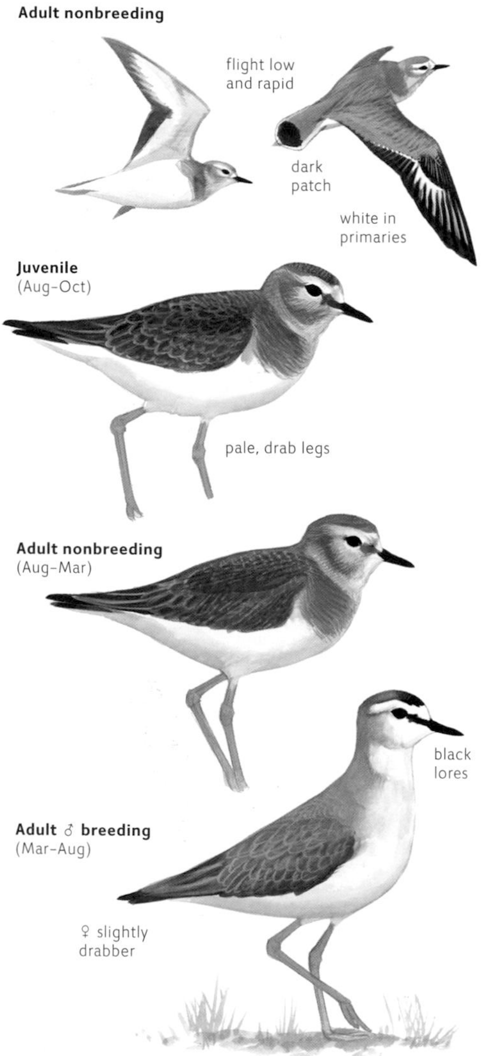

VOICE: Flight call a coarse *grrrt*; lower single note *dirp* like Western Meadowlark; also a coarse, grating *ji ji ji ji ji*. In courtship a low, soft moan very similar to distant cow mooing. Flight song a rapidly repeated, wild, harsh whistle *we we we we we*....

Uncommon and local. Nests on shortgrass prairies often in the very short-cropped grass around prairie-dog towns. Winters on dry barren ground, smooth dirt fields, and shortgrass prairies. Found in small flocks in winter. World population about 20,000.

Killdeer

Charadrius vociferus

L 10.5" WS 24" WT 3.3 oz (95 g) ♀>♂

A relatively tall, slender plover, with unusually long tail. Da double breastband distinctive. In flight, note rufous rump ar tail. Loud and insistent calls announce presence. Downy juveni has single breastband, like small plovers.

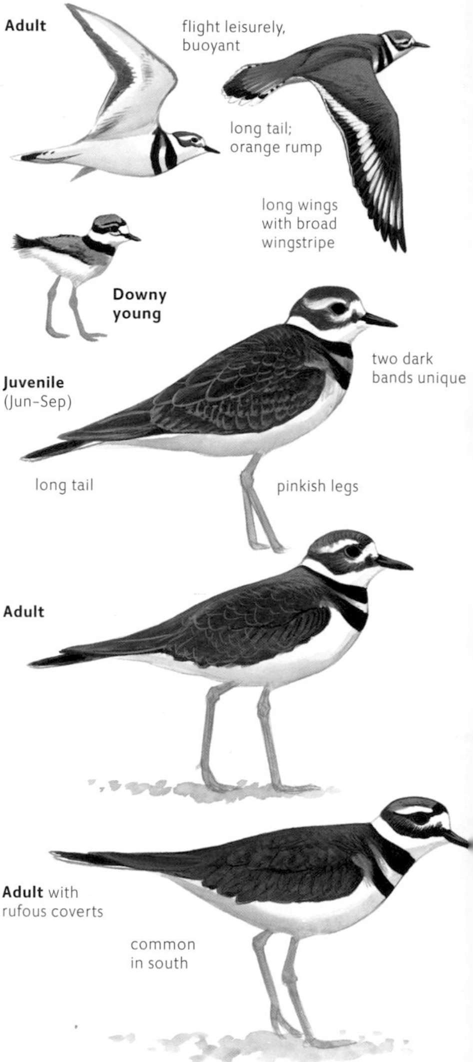

VOICE: Calls high, strident piping, often drawn-out *dee deeeyee, tyeeeeeee deew deew, Tewddew* (or "kill deer"). Whe agitated a strident, clear *teeeee di di* repeated; high, rapid tr *tttttttttt* when very nervous. Display song a rapid, high, rollin *didideeerr didideeerr*... with high, thin quality like other calls.

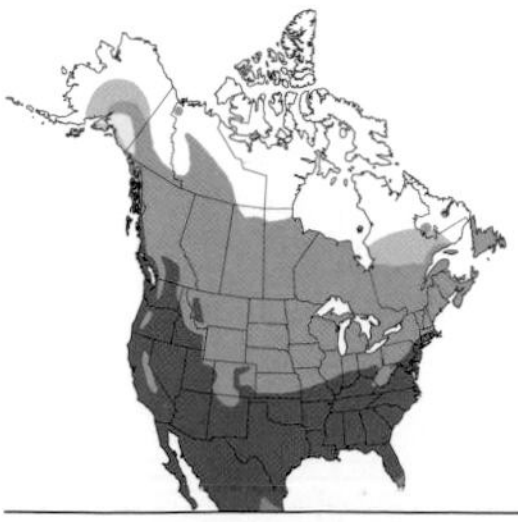

Common and widespread on an open ground, often not near water. Nests and forages on gravel parking lots, shortgrass fields, and similar open areas. Often in small groups; never forms large flocks and usually does not associate closely with other shorebirds.

ᐯorthern Lapwing

anellus vanellus

12.5" WS 33" WT 7 oz (210 g)

bout as large as Black-bellied Plover, but unmistakable with riking black and white plumage, black breast, and wispy crest. ight pigeon-like; note very broad, rounded wingtips, black pperside and breast, white rump.

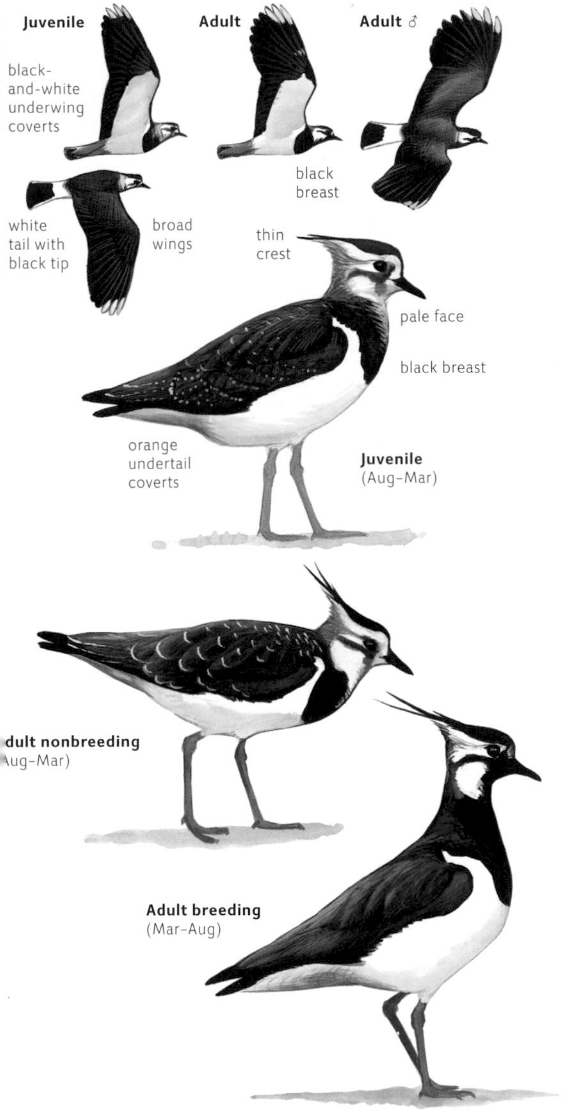

OICE: High, thin whistles with quality of kitten mewing *chee* or *eewi*, and short, sharp *peet*. Song given in roller-coaster flight *irr willucho weep weep weep ee yo weep*.

Very rare visitor from Europe; not recorded annually. Most records of solitary birds Dec–Jan, on the coast from Newfoundland to New York. Found on agricultural fields, pastures, mudflats.

Northern Jacana

Jacana spinosa

L 9.5" WS 20" WT 3.3 oz (95 g) ♀>♂

In the family Jacanidae, combines features of sandpipers and rails, with extraordinarily long toes. Adult dark overall with yellow bill; juvenile bicolored; all ages reveal startling pale yellow flight feathers when wings are spread.

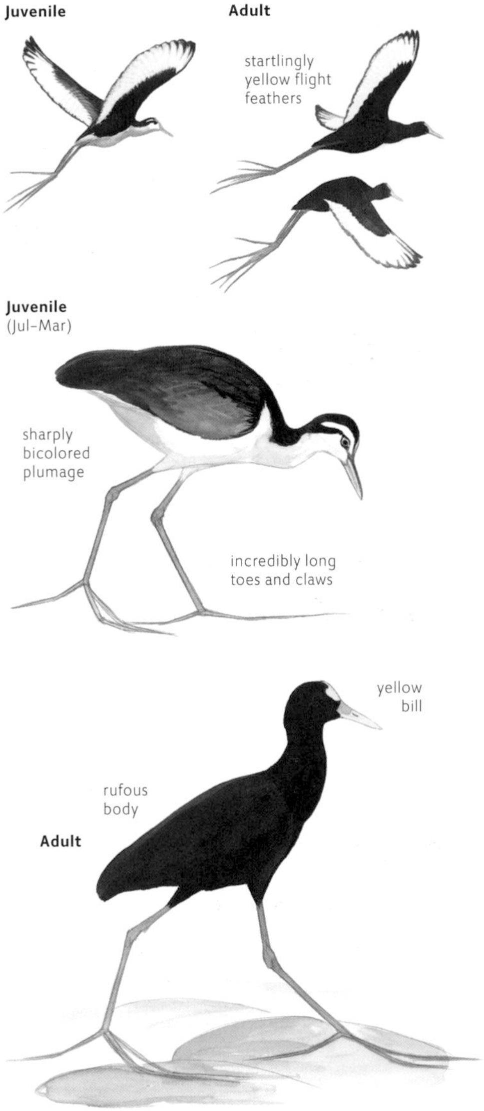

VOICE: High, harsh squawking (usually in flight) *scraa scraa scraa*...; usually with sharp *keek* notes interspersed.

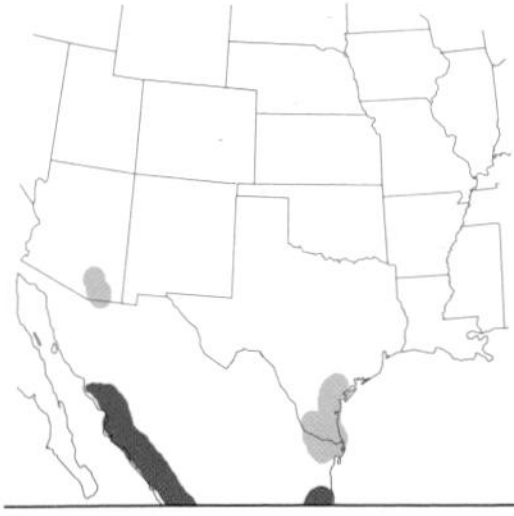

Very rare visitor from Mexico to southern Texas, where it has nested. Found on ponds with dense emergent vegetation such as lily pads. Solitary. Walks on floating vegetation, delicately picking food.

Common Ringed Plover

Charadrius hiaticula

L 7.5" WS 20" WT 2.1 oz (60 g)

Virtually identical to Semipalmated Plover; slightly paler back, more white over eye, or more contrasting dark cheeks can help locate a candidate, but voice is by far the most reliable distinguishing feature.

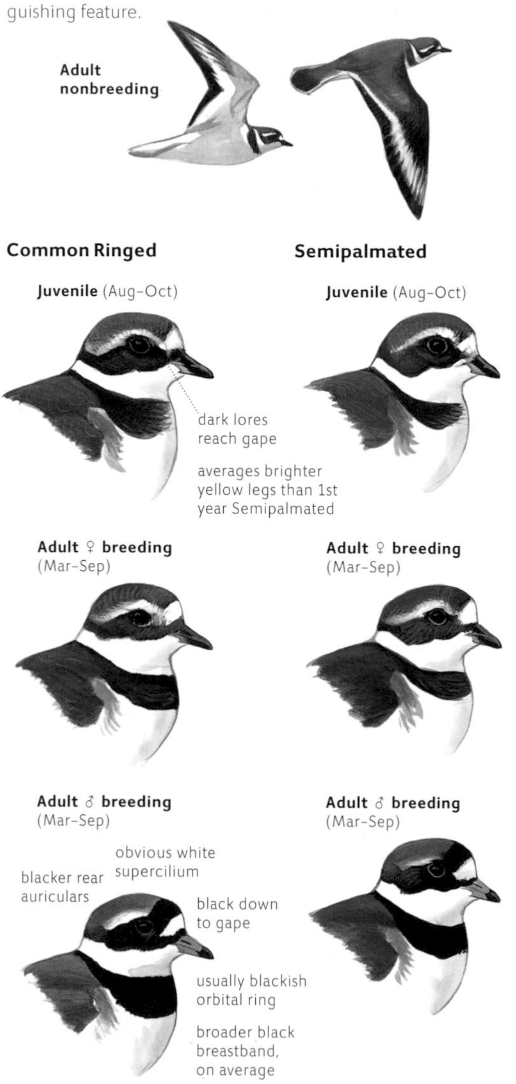

VOICE: Common call a soft *tooe* or *toolip*; lower and more wooden than Semipalmated (quality more like Piping) with emphasis on first rather than second syllable. Threat call a rapid, low, mellow *towidi towidi towidi*... reminiscent of Lesser Yellowlegs. Display song *TOO-widee-TOO-widee*....

Small numbers breed on Baffin Island, migrating to Europe. Rare visitor south to Newfoundland and very rare farther south on Atlantic coast, also a rare visitor from Siberia to northwestern Alaska (and once in California). Habitat and habits similar to Semipalmated Plover.

Lesser Sand-Plover

Charadrius mongolus

L 7.5" WS 22" WT 2.4 oz (68 g)

Similar to Semipalmated Plover but slightly larger, with longe grayish legs, longer black bill, and lacks obvious whitish colla Breeding plumage distinctive, with rufous breast.

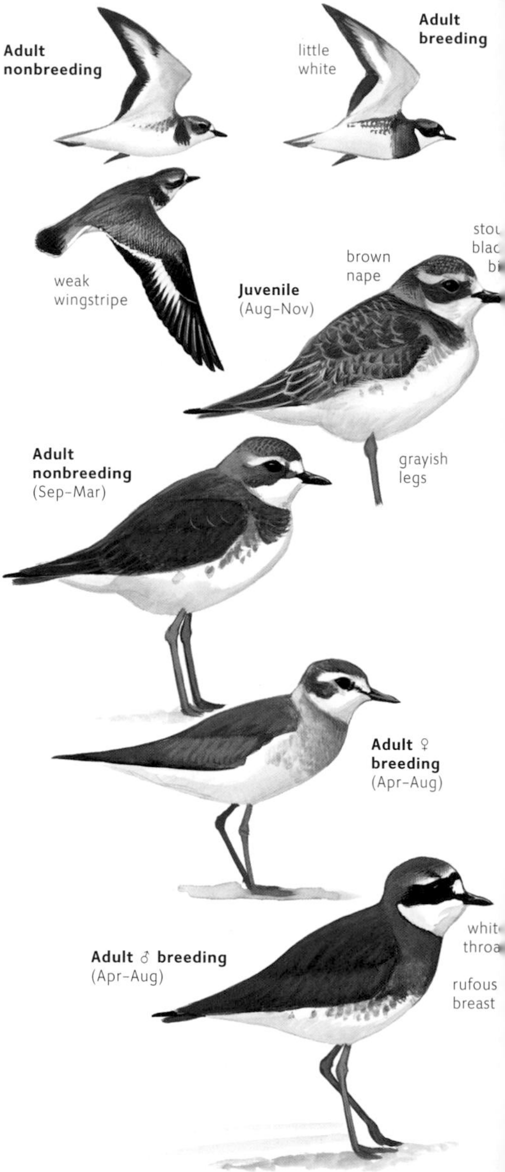

VOICE: Flight call a short, hard rattle *drrit* or *ddddd* slightly risin reminiscent of longspurs or turnstones.

Rare visitor from Siberia; recorded annually in Alaska, wit a few records farther south, mainly from Jul–Sep and usually of single birds with Semipalmated Plover. Habits and habitat like Semipalmated Plover.

urasian Dotterel

aradrius morinellus

.25" WS 23" WT 3.9 oz (110 g) ♀>♂

ımp and small-billed, with yellow legs, long pale eyebrows that eet at back of head, and plain gray wings without obvious pat-n. Breeding adult distinctive, with dark rufous belly. Juvenile ffy overall, with faint pale breastband.

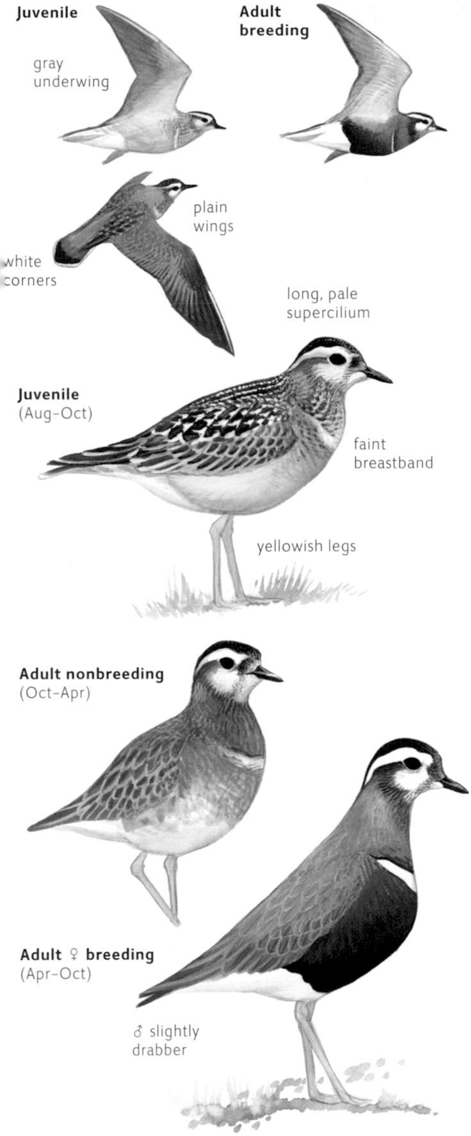

ɔICE: Most common call a ınlin-like *keerrr* when king off; also a soft *put put* other soft or grating ɔtes. Flight song a high, pidly repeated *pwit pwit vit....*

Rare visitor from Siberia; not recorded annually in western Alaska; has been suspected of breeding on barren gravel ridgetops. A few records farther south to California, mainly in Sep. Found on barren dry ground such as sand dunes, plowed fields, and rocky tundra ridges. Usually solitary or in pairs.

Greater Sand-Plover

Charadrius leschenaultii

L 8.7" WS 23" WT 3.1 oz (88 g) ♀>♂

Very similar to Lesser Sand-Plover, differing mainly in slightly larger size and in proportions: longer and heavier bill, longer legs, with toes projecting just beyond tail tip in flight. All features overlap and identification can be very challenging.

VOICE: Similar to Lesser Sand-Plover.

Very rare visitor from Asia, with records in California and Florida. Habits and habitat similar to Semipalmated Plover.

Eurasian Oystercatcher

Haematopus ostralegus

L 16.7" WS 31" WT 19.4 oz (550 g) ♀>♂

Similar to American Oystercatcher, but with black back, nonbreeding plumages show small white collar. In flight note extensively white rump, longer white wingstripe.

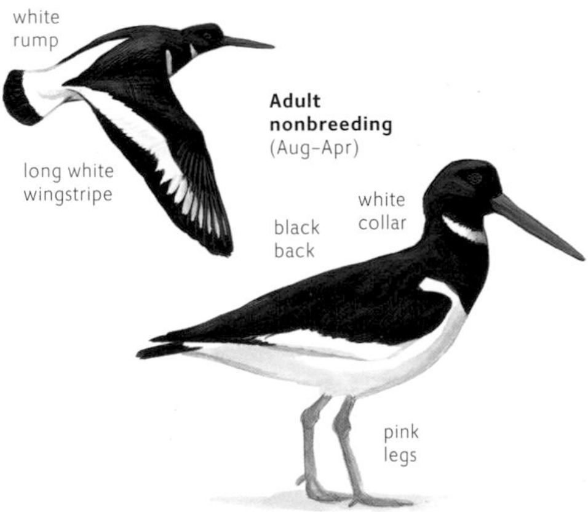

VOICE: Similar to American.

Very rare visitor from Eurasia, recorded several times in Newfoundland and once in Alaska. Less restricted to tidal sandbar habitat than American Oystercatcher, often foraging on mudflats, even in brackish or fresh water, and on grassy fields.

Black Oystercatcher

Haematopus bachmani

L 17.5" WS 32" WT 1.4 lb (650 g) ♀>♂

Relatively broad-winged and slow-flying for a shorebird. Overall blackish color, pale legs, and bright orange-red bill distinctive.

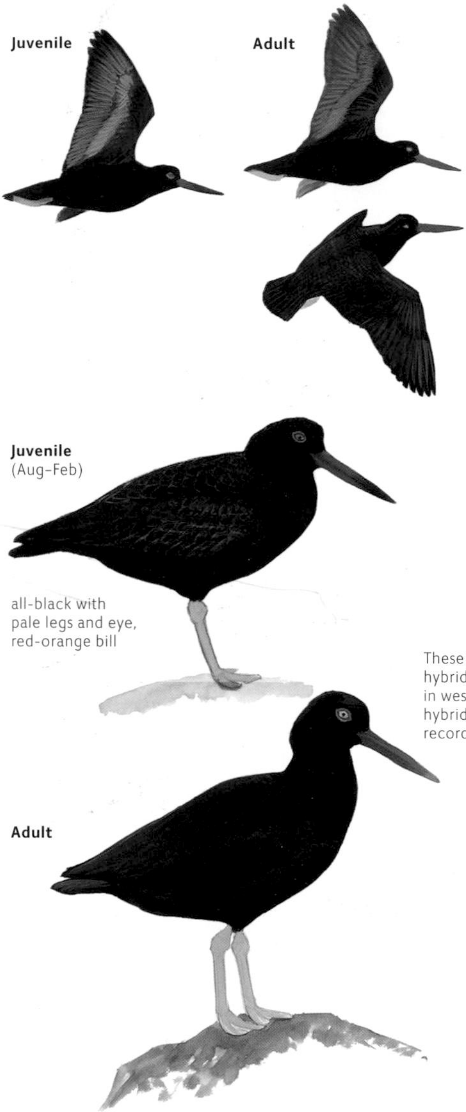

These two species hybridize occasionally in western Mexico; hybrids have been recorded in California.

VOICE: loud whistled yelps or high, clear, piping whistles *queep, weeyo,* etc. In display a long accelerating series *queep queep quee deedeedeedeedeededededeeddddddrrr* rising and then descending. Alarm a clear *kleep, kleep, klidik-klideeew*; falcon alarm a rapid *whidididideew.*

Sparsely but widely distributed along coast. Look for singles or small groups on rocks at ocean's edge, only loosely associated with other shorebirds. Nests on rocks and pebbles just above high-tide line.

American Oystercatcher

Haematopus palliatus

L 17.5" WS 32" WT 1.4 lb (630 g) ♀>♂

A crow-sized shorebird, boldly patterned in black and white wi black head and bright red-orange bill. More strongly black a white than any other shorebird.

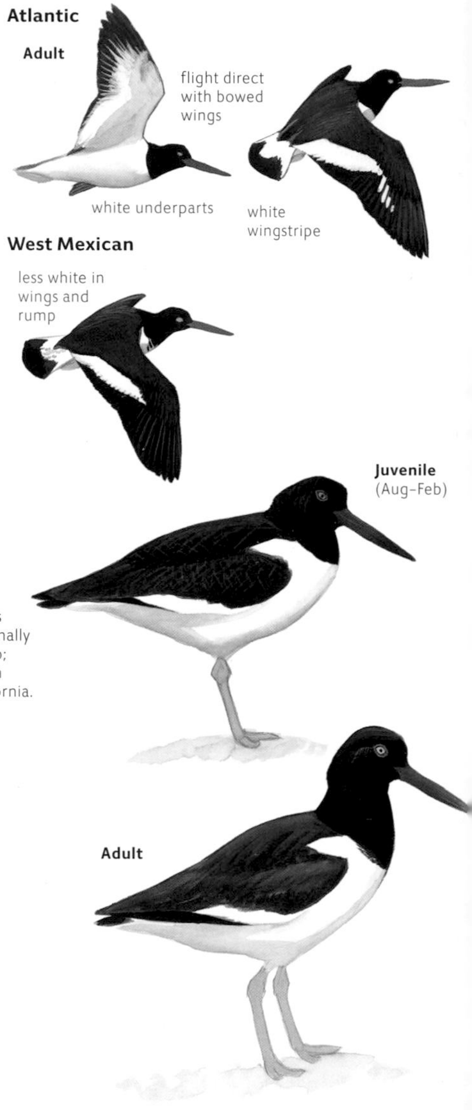

VOICE: Voice apparently identical to Black Oystercatcher.

Uncommon to locally common gathering in large groups at roosting sites with other large shorebirds; forages on sand flat and shell bars at low tide. Nests on sandbars and dunes just abo high-tide line.

lack-necked Stilt

mantopus mexicanus

ı" WS 29" WT 6 oz (160 g)

ılks delicately, tilting forward to pick food with needle-like bill. ender shape, black and white plumage, and extraordinarily ıg red legs unmistakable.

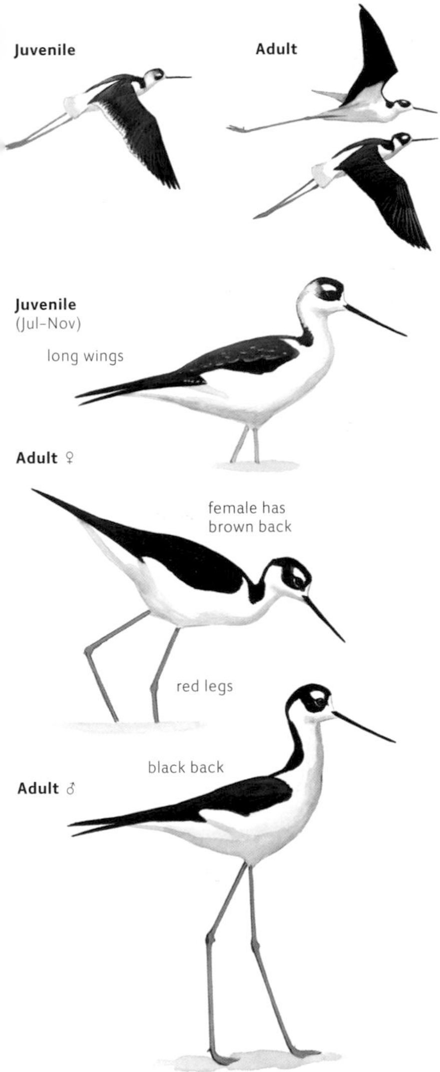

ICE: Common call a loud, sharp *taawh* or *pleek* often repeated cessantly; ranges from short *kik kik*... to complaining *keef ef*... and harsher *wreek wreek*...; lower and louder than American Avocet. Juvenile often gives a high, sharp *peek* or *pidi* nilar to Long-billed Dowitcher.

Common but somewhat local on shallow still ponds with muddy bottoms and grassy edges. Usually in pairs or small groups. Nests along edges of shallow weedy ponds.

American Avocet

Recurvirostra americana

L 18" WS 31" WT 11 oz (315 g)

A large sandpiper, note mostly white plumage with black stripes on back, unmarked pale head and neck, very slender, upturned bill, and gray legs. In flight relatively broad-winged.

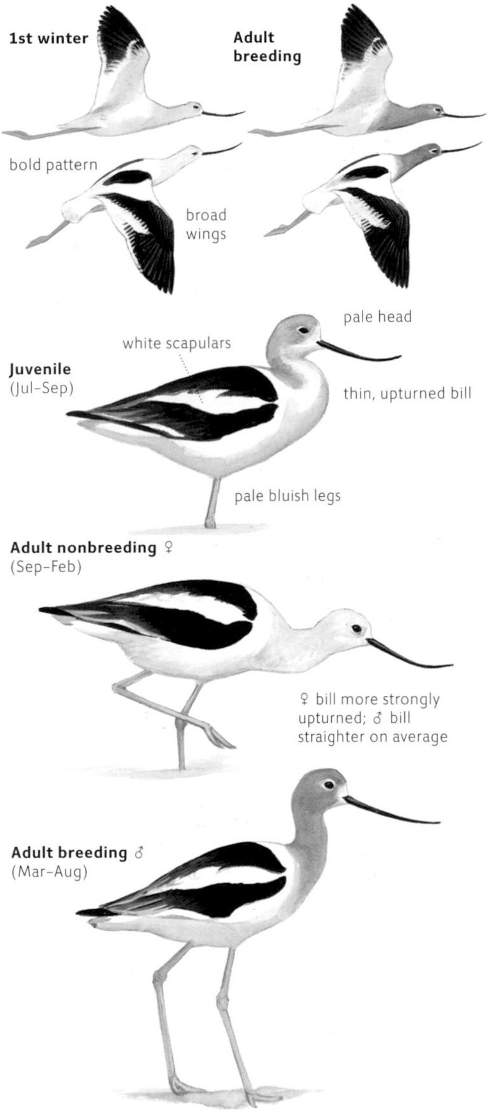

VOICE: Less vocal than Black-necked Stilt. Mainly single note calls ranging from high, sharp *kweep* to lower *pwik* or sharp *pleek*; all higher than similar calls of Black-necked Stilt. Some calls reminiscent of yellowlegs.

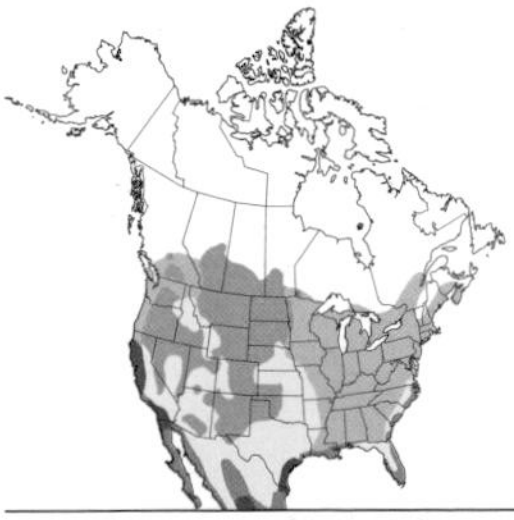

Common locally in shallow water, where groups of several to hundreds march purposefully with heads down, sweeping bills from side to side through the water. Rests in compact flocks. Nests along edges of shallow weedy ponds or lakes.

Greater Yellowlegs

Tringa melanoleuca

L 14" WS 28" WT 6 oz (160 g)

Foraging actions varied, from gracefully walking and picking invertebrates to actively chasing fish in shallow water. Larger than Lesser Yellowlegs, with longer bill (noticeably longer than head). Molting primaries is a common sight Aug–Sep.

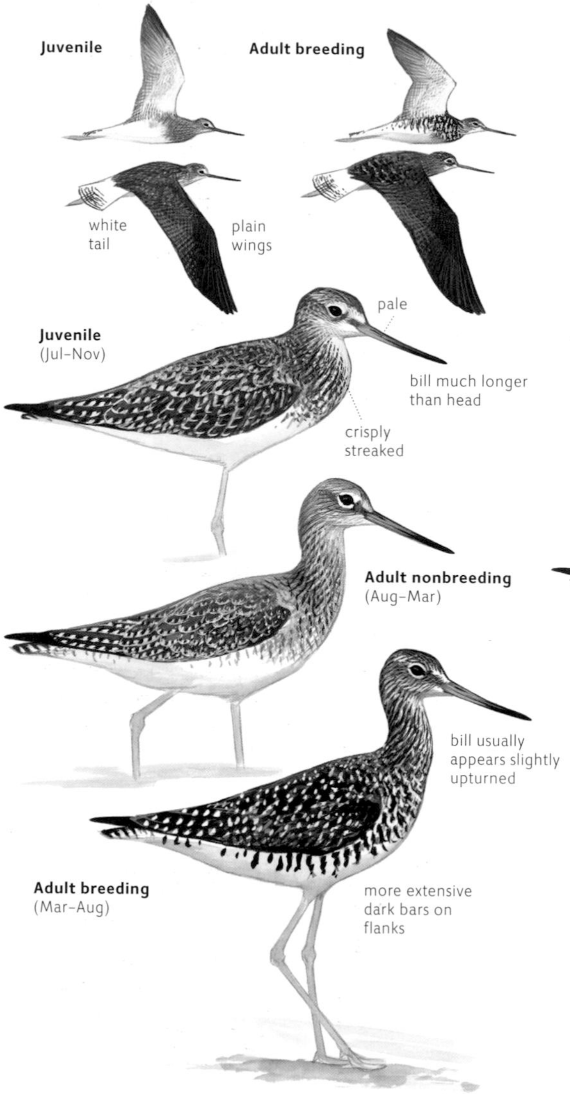

VOICE: Flight call three or four loud ringing notes *deew deew deew;* higher-pitched than Lesser with strident overtones. In agitation an endlessly repeated single note *tew, tew*.... Feeding bird gives soft, single notes. Display song a melodious, rolling *kleewee kleewee*....

Common in any shallow-water or mudflat habitat. Nests around ponds in patchy boreal spruce forests. Solitary or in loose groups.

Lesser Yellowlegs

Tringa flavipes

L 10.5" WS 24" WT 2.8 oz (80 g)

Very similar to Greater Yellowlegs, best distinguished by smal size, relatively shorter bill (not much longer than head), and c

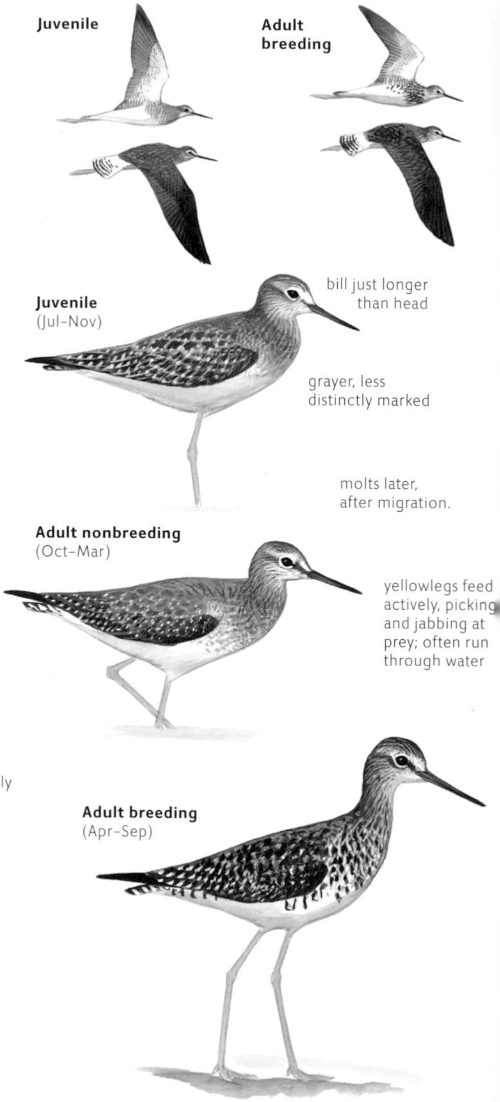

VOICE: Flight call of one or two short whistles *tip,* or *too-t* typically shorter, flatter and softer than Greater. In agitation repeated *tiw, tiw*.... Alarm a rising, trilled *kleet*. Threat a low, ro ing trill. Display song a rapid, rolling *towidyawid, towidyawid* lower-pitched than flight call.

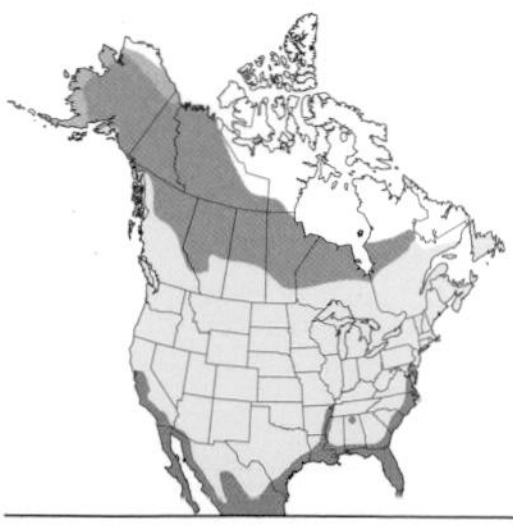

Common. Habitat similar to Greater Yellowlegs, but may prefer slightly shallower, more grassy, and less open sites. Nest farther north than Greater, in slightly drier areas. Solitary or i loose groups, often mixing wit Greater.

olitary Sandpiper

inga solitaria

.5" WS 22" WT 1.8 oz (50 g)

sembles yellowlegs, but much smaller, relatively short-legged, rker above, with all-dark underwing and dark center of rump d tail. Told from Spotted Sandpiper by white eyering, darker perparts with pale spots, all-dark wings.

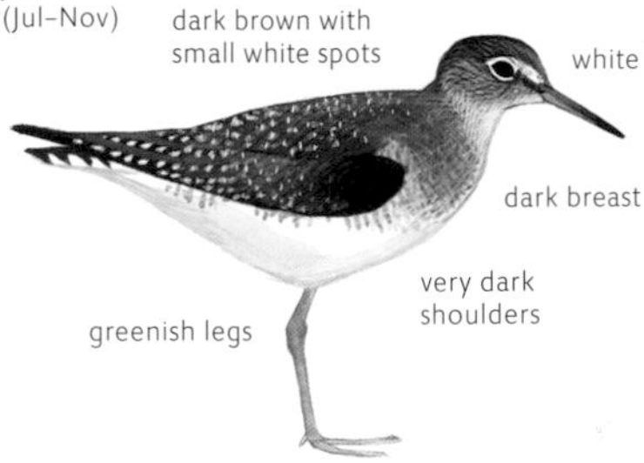

ICE: Flight call a clear, high, rising whistle *peet-WEET* or *peet eet weet*; higher and more urgent than Spotted Sandpiper. arm from ground a very hard, sharp *plik* repeated. Display song eries of short phrases reminiscent of flight call but more comex, with startling, bell-like quality.

Uncommon but widespread on small freshwater mudflats and at edges of brushy ponds or ditches, where few other shorebirds occur. Nests in boreal forests by marshy ponds, using songbird nest from previous year in spruce or fir tree. Usually solitary, but sometimes forms small loose groups.

Identification of Yellowlegs

The two species of yellowlegs (like other sandpipers of the genus *Tringa*) are tall elegant shorebirds. Alert, noisy, and active, they walk briskly or even run in shallow water, and bob their head and body emphatically when alarmed. Their long yellow legs, long neck, and habits distinguish them from all other shorebirds, but distinguishing them from each other is more difficult.

Overall size, relative bill length, and calls offer the best clues. Bill length can be judged relative to the length of the head—bill usually distinctly longer than the head on Greater, and about equal to the head on Lesser.

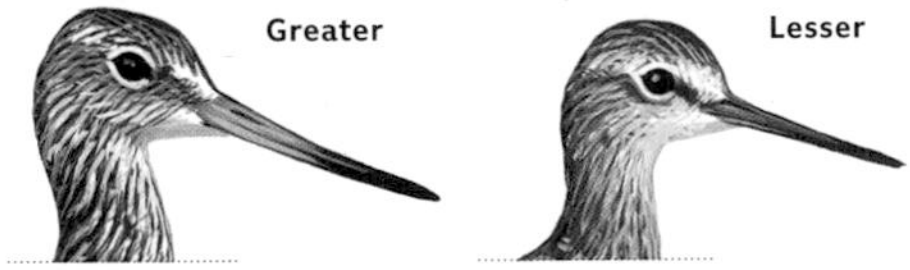

Molt timing also offers a clue. Greater Yellowlegs with molting primaries is a common sight Aug–Sep. Lesser molts later, on the wintering grounds.

Common Greenshank

Tringa nebularia

L 12.8" WS 27" WT 7.2 oz (204 g)

Very similar to Greater Yellowlegs in all respects. Best distinguished in flight by white rump extending in long wedge to lower back, cleaner white underwing coverts. Also note greenish legs, paler and finely streaked head and neck.

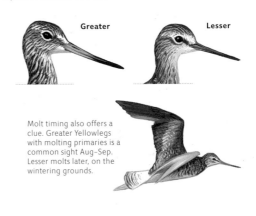

VOICE: Flight call three or four clear whistled notes *tew tew tew*, very similar to Lesser and Greater Yellowlegs.

Rare visitor from Eurasia, recorded regularly in spring in far western Alaska, with a few records south and east to Florida and Newfoundland

Willet

Tringa semipalmata

L 15" WS 26" WT 8 oz (215 g)

Walks purposefully, picking or probing the ground for prey. Always drab-plumaged; larger and stockier than yellowlegs, with heavy bill and thick gray legs. Broad white wingstripe is striking in flight. Also note relatively broad, rounded wings.

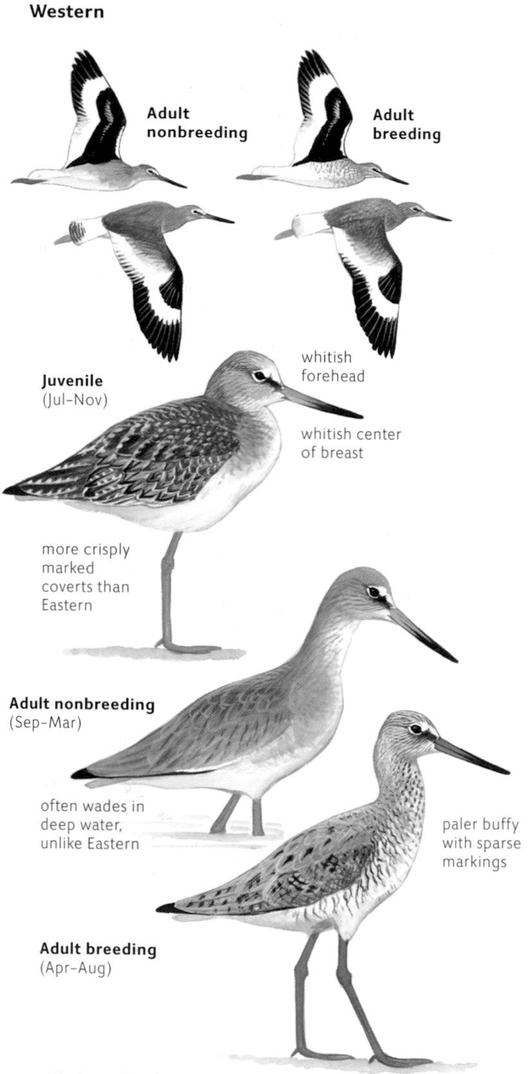

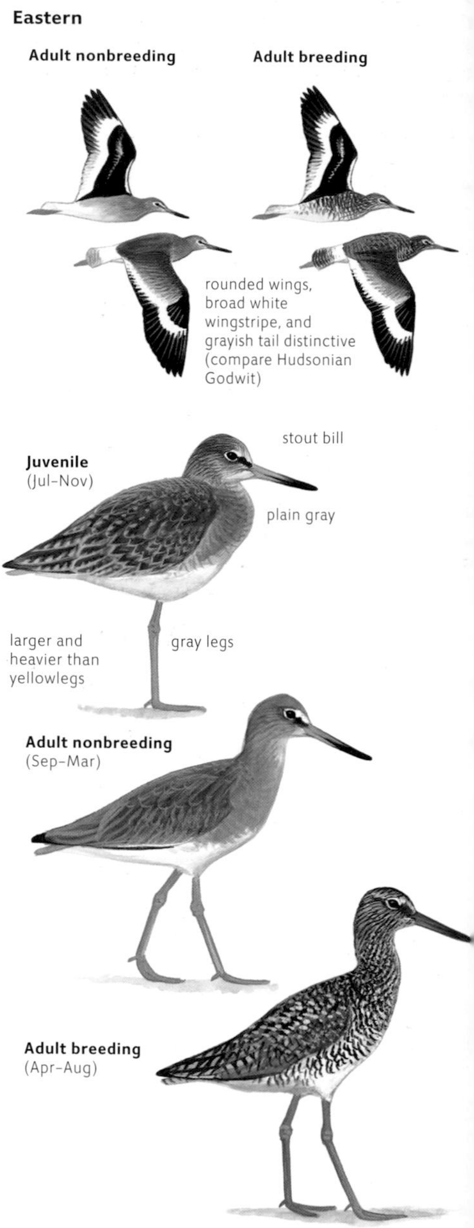

VOICE: Flight call a clear, loud, ringing *kyaah yah* or *kleee lii* or simply a descending *haaaa*. When flushed *kleeliilii* and variations; in alarm a monotonously repeated *wik, wik*... or more intense *kliK, kliK*.... Territorial song a rolling, clear chant *pilly WILL WILLET* repeated steadily.

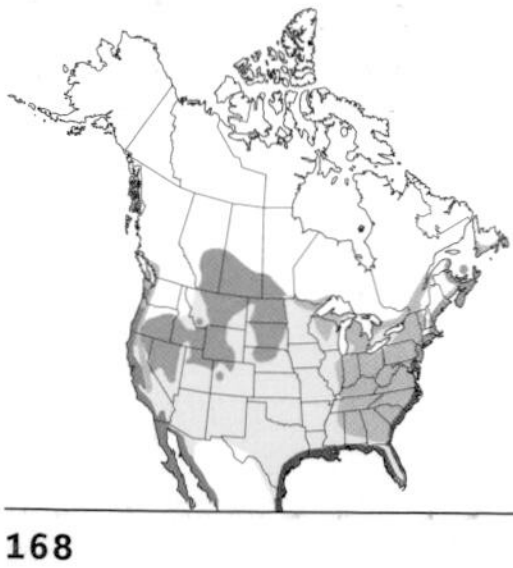

Common. In winter and migration found on open beaches and mudflats. Nests in grassy marshes and meadows. Often solitary when foraging, but gathers in flocks when migrating and roosting, associating with godwits and other large shorebirds.

Eastern birds strictly coastal all year; Western breeds inland and migrates to both coasts. Western birds are 10 percent larger with 15 percent longer bill and legs and little overlap; bill is relatively slender and often slightly upturned; wingtips may project slightly farther beyond tail tip. Western's more godwit-like feeding habits are related to its structure. It is paler overall and its white wingstripe averages broader. Voices of subspecies similar but all calls of Western birds average lower-pitched than Eastern. Western call on flushing a more raucous rolled *krrri lii liit*; territorial song of Western birds lower-pitched and longer with relatively long *will* note.

potted Redshank

inga erythropus

.2.5" WS 25" WT 6 oz (165 g)

ɔmmon foraging method involves walking in belly-deep water th neck outstretched and bill pointed down, jabbing at prey. stinguished from yellowlegs by reddish legs, thin bill with slight oop at tip, and white rump. Mostly dark breeding plumage stinctive.

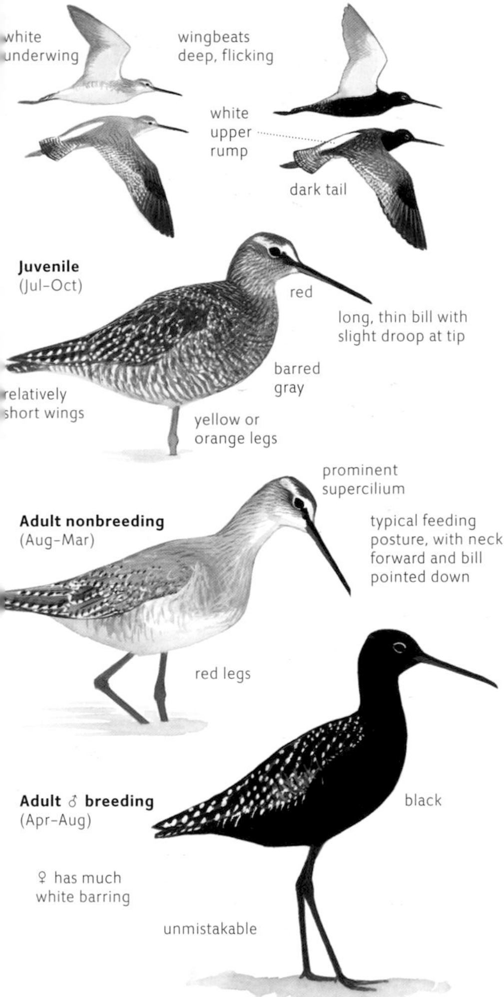

ɔICE: Flight call a husky *ıuwit* or *kawit;* very similar Semipalmated Plover but ırder, stronger, deeper. On ıing sometimes gives a ıuckling *chu, chu* or loud, ghtly husky *kweep kip kip.* splay song varied; includes rill, grinding *krr-WEEa* ırases and short bouts of arp notes.

Very rare visitor from Eurasia; not recorded annually, mainly from Apr–Oct. Usually found with yellowlegs in shallow freshwater pools, but often forages in deeper water than yellowlegs.

Wood Sandpiper

Tringa glareola

L 8" WS 22" WT 2.5 oz (71 g)

Size and shape most like Solitary Sandpiper, color more like Lesser Yellowlegs; note bolder spots above, broad pale eyebrow and contrasting dark cap, white rump, and pale underwing coverts. Repeatedly bobs rear when agitated.

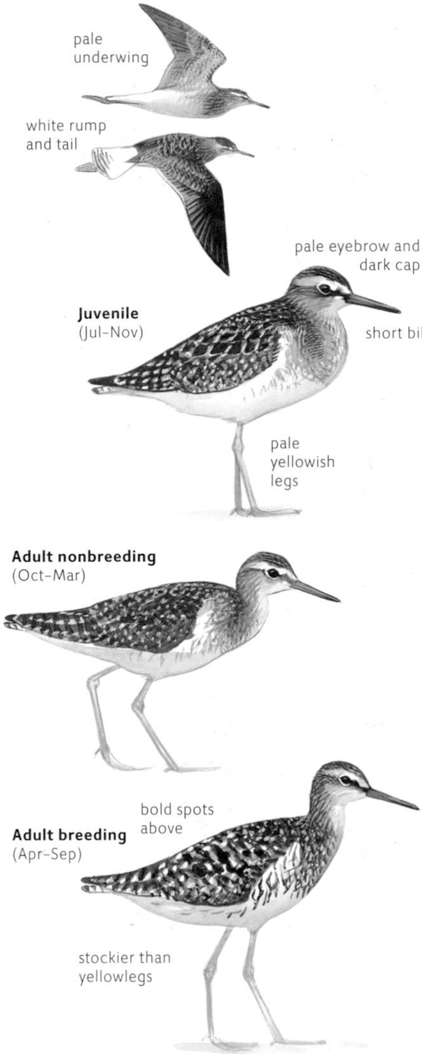

VOICE: Flight call a series of several high, slightly husky whistled notes all on one pitch *teef teef teef;* lower-pitched and less strident than Solitary Sandpiper, higher and more drawn-out than call of Lesser Yellowlegs. Alarm a sharper *kew* repeated. Song short whistles in rapid rolling chant *pidilo-pidilo-pidilo....*

Rare visitor from Eurasia to far western Alaska in small numbers each spring, has nested. Very rare visitor elsewhere with a few scattered records on Pacific coast south to California and on Atlantic coast from Rhode Island to Delaware; in shallow grassy or weedy pools, not associating closely with other shorebirds.

Spotted Sandpiper

Actitis macularius

L 7.5" WS 15" WT 1.4 oz (40 g)

Constantly teeters, raising and lowering rear body. Crouching appearance enhanced by short legs and neck. Relatively dark and unpatterned above, with dark patch on side of breast.

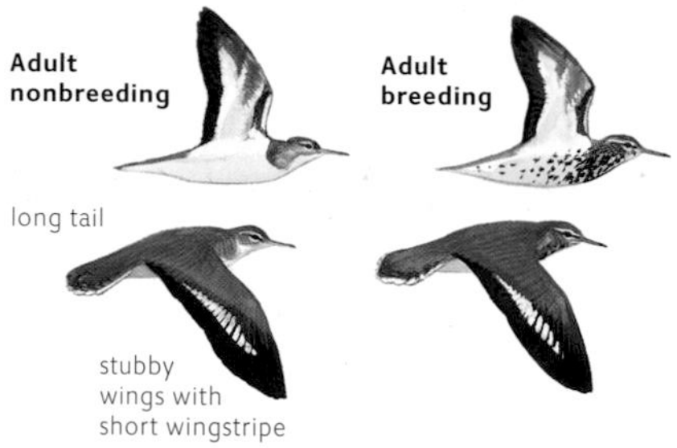

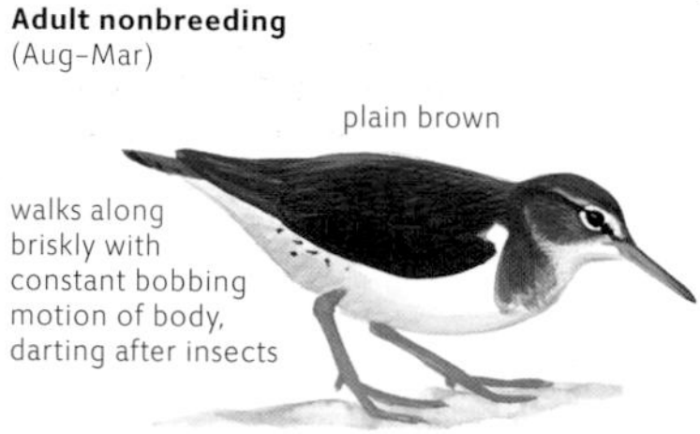

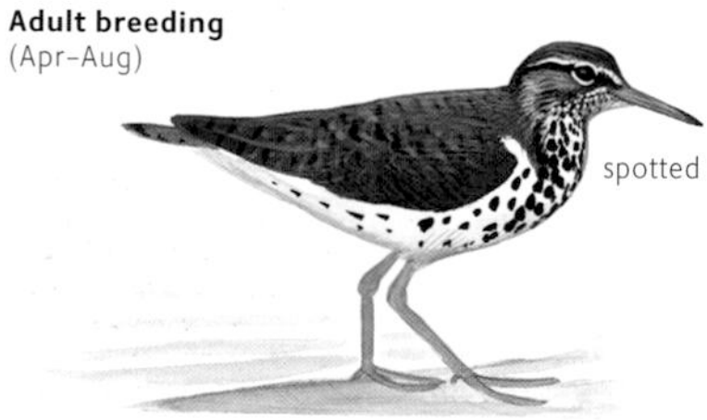

VOICE: Flight call a high, clear, whistled *twii twii* or a descending series *peet weet weet;* lazier and lower-pitched than Solitary; often a single note *peet* or repeated whistle *pweet, pweet....* Display song a rolling, clear, whistled *tototowee, tototowee....*

Uncommon but widespread on ponds and streams, particularly on rocky shores and steep banks. Usually solitary and usually on fresh water. Nests on dirt or gravel with clumps of grassy or weedy vegetation, such as on sandbars, riverbanks, field edges.

Terek Sandpiper

Xenus cinereus

L 9.1" WS 24" WT 2.9 oz (82 g)

A distinctive species, slightly larger than Spotted Sandpiper, w relatively short legs and long upswept bill, also note flat bac large square head. In all plumages relatively plain grayish overa with white trailing edge of wings.

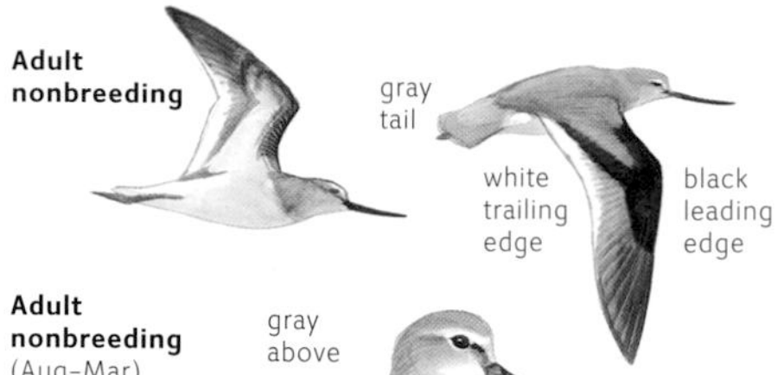

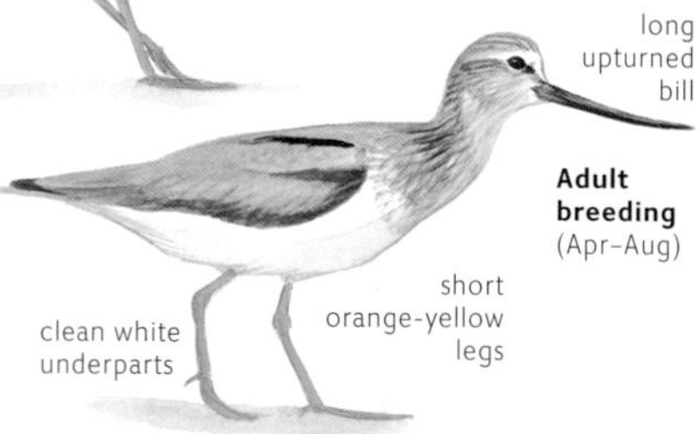

VOICE: Flight call a rapid series of clear piping whistles, nearly a trill *lidididid,* very similar to Wandering Tattler, but notes all on one pitch.

Rare visitor from Eurasia to western Alaska, with a few records south to California, onc to Massachusetts. Often choose open, exposed beaches or flats, foraging either by picking methodically at surface of mud or making quick dashes after insects, less often wading.

Common Sandpiper

Actitis hypoleucos

L 7.9" WS 15.7" WT 1.8 oz (51 g)

Very similar to Spotted Sandpiper in all respects, most eas distinguished by lack of spots in breeding plumage; at oth seasons note relatively longer tail, drabber bill and leg color.

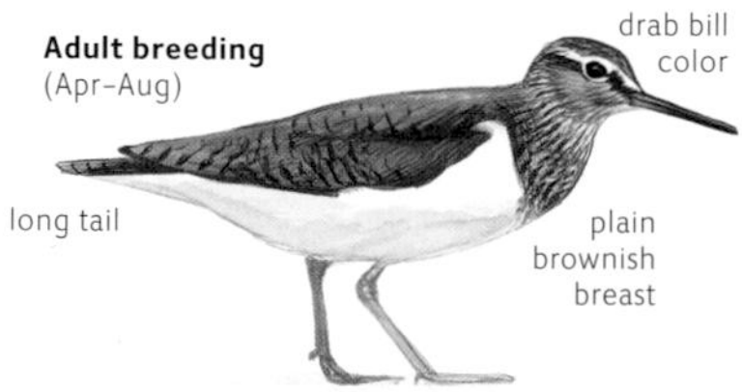

VOICE: Flight call a descending series of strident whistles, higher-pitched than Spotted, more like Solitary *peet-weet weet.*

Rare but regular visitor from Eurasia to western Alaska, where it has nested. No confirmed records south or east of Alaska. Habits and habitat like Spotted Sandpiper.

'pland Sandpiper

rtramia longicauda

2" WS 26" WT 6 oz (170 g) ♀>♂

ated to curlews. Shape distinctive, with thin neck, small head, d long tail. Note yellow legs, short yellowish bill, and dove-like ɔression on pale face. In flight, note pale inner wing and dark maries.

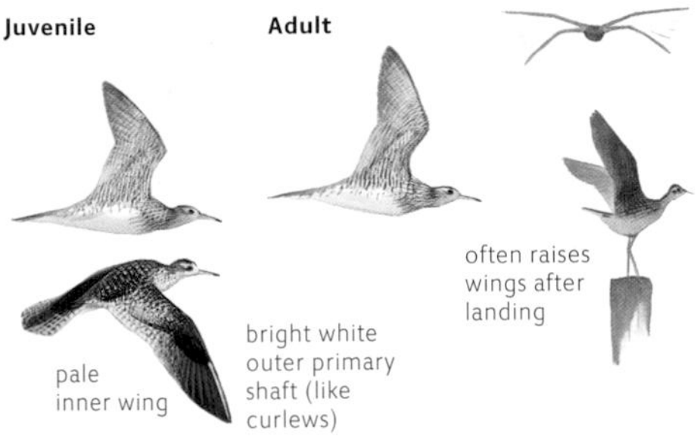

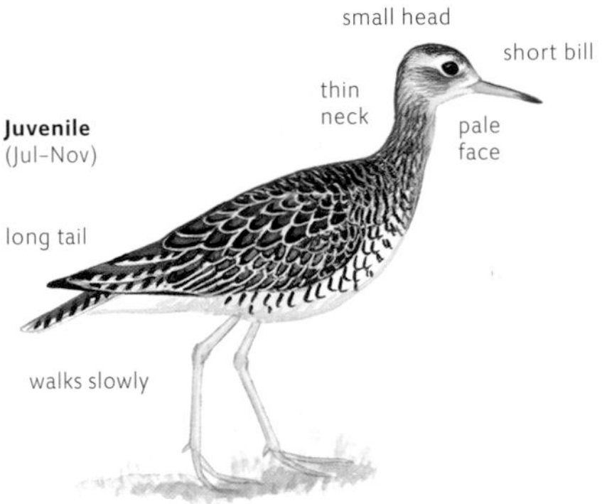

ɔICE: Flight call a low, strong, liquid *qui-di-di-du;* last note wer and weaker; distinctively loud and clear. Alarm a nasal, owling *grrgrrgrrgrrgrr*. Flight song a weird, unearthly, bubbling histle, slowly rising then falling *bububuLEE-hLEEyooooooo*.

Uncommon and local. Nests on grassy fields where grass is about 4 to 8 inches high with patches of open ground. Migrants choose similar habitat: short-grass playing fields, recently mowed hay fields. Solitary or in small groups.

Little Curlew

Numenius minutus

L 12.6" WS 27" WT 6.2 oz (175 g)

A very small and delicate curlew with relatively short bill and long wings. Very similar to Eskimo Curlew (which is presumed extinct) but with pale lores, less strongly marked flanks, and buffy (not cinnamon) underwing coverts.

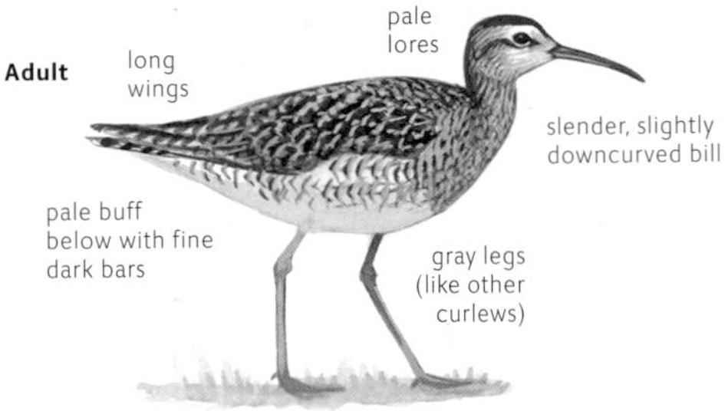

VOICE: Flight call a melodic, two-syllabled, upslurred whistle *pwee-wheet*.

Very rare visitor from Asia, recorded four times in California (fall), twice in Alaska (spring), once in Washington (spring). Favors open fields, short grass, dunes, the same habitat as American Golden-Plover.

Bill Length in Curlews

Bill length is variable in all curlews, females are longer-billed than males, and some juveniles are particularly short-billed. Longest-billed female Whimbrel overlaps with shortest (juvenile) Long-billed Curlew, and must be distinguished by other body proportions, plumage, and voice.

Shortest-billed juvenile Whimbrel overlaps with longest-billed Eskimo and Little Curlews. Whimbrel always has much stouter and thicker bill than the smaller species, even when short, and should also be distinguishable by larger body size, stockier overall proportions, and bold head stripes.

Whimbrel

Numenius phaeopus

L 17.5" WS 32" WT 14 oz (390 g) ♀>♂

A large sandpiper, with relatively short legs and long downcurved bill. Uniformly speckled gray-brown overall, with dark crown-stripes the only contrasting markings. In flight, note large size, strong wingbeats, plain color, and very pointed wings.

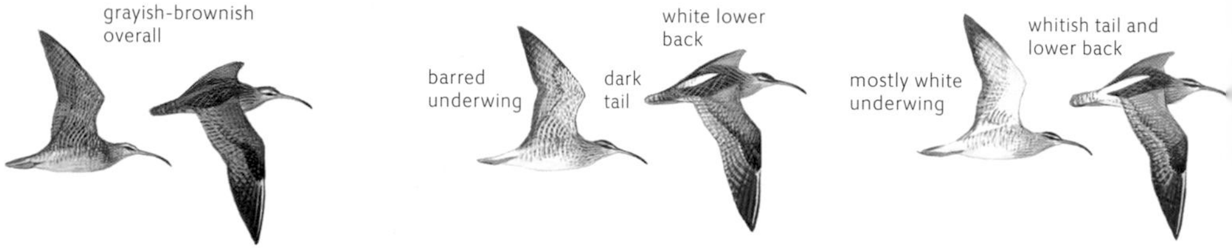

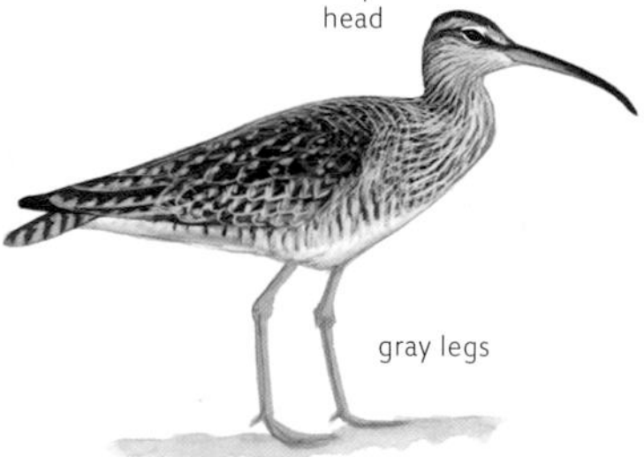

At a distance all curlews can be distinguished from godwits by pale gray legs and more graceful foraging actions

VOICE: Flight call a rapid, forceful, liquid *quiquiquiquiqui* with no change in pitch. Chase call, given in flight, a series of short, varied whistles and trilled notes. Display song a low, clear whistle *oook* repeated, followed by long trill and often ending with several emphatic high whistled phrases; also a low, vibrato whistle followed by slightly higher trill given frequently on nesting grounds (e.g., when landing).

Uncommon and local. During migration and winter found on grassy mudflats, tidal flats, beaches and coastal rocks; always rare inland. Foraging birds solitary or in loose groups; may form small flocks that fly in lines; generally does not mix with other species. Nests on mossy tundra.

Three subspecies occur, differing only in plumage color. American population (the expected form throughout North American range) is overall brownish including the underwing coverts and rump. Siberian (rare but regular in western Alaska, with a few records farther south on Pacific coast) has heavily barred white underwing and rump with white lower back. European (rare visitor to Northeast coast) has mostly white underwing, tail, and lower back.

ristle-thighed Curlew

umenius tahitiensis

.7" WS 32" WT 1.1 lb (490 g) ♀>♂

bits and appearance very similar to Whimbrel, but note striking pale buffy rump and tail, obvious in flight.

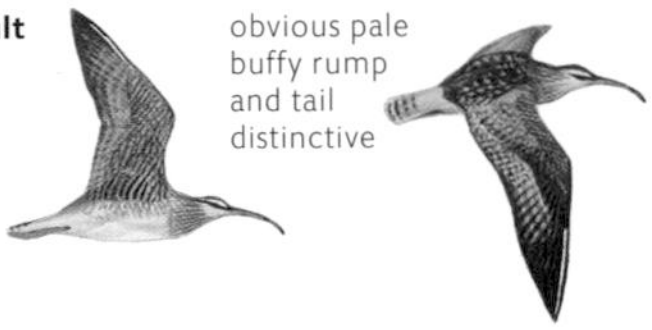

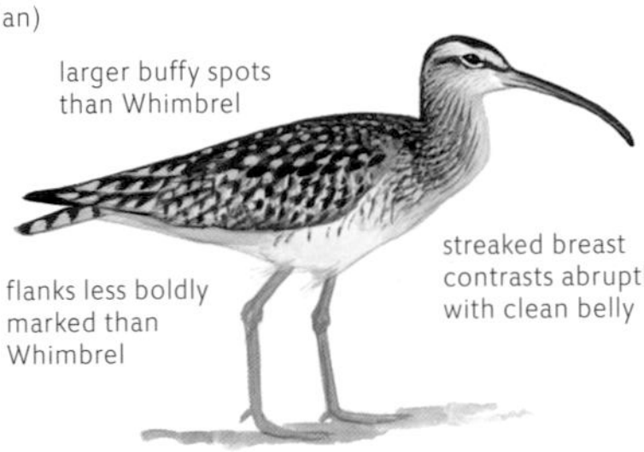

OICE: Flight call a clear, insistent whistle of even emphasis *eeoip;* remarkably like human attention whistle; lower than lack-bellied Plover and quite different in character. Display ight soaring; begins with about five *wiiteew* notes, then epeated, complex *pidl WHIDyooooo.*

Rare and local. Migrants found on beaches, or marshes. Winters on South Pacific islands. Usually in small groups. Nests on remote hilltops with grassy tundra in Alaska. World population about 10,000.

Long-billed Curlew

Numenius americanus

L 23" WS 35" WT 1.3 lb (590 g) ♀>♂

Extremely large, with extraordinarily long bill. Told from Whimbrel by larger size, longer bill, warmer color, cinnamon wings. From Marbled Godwit by larger size, downcurved bill, paler legs. broader wings and stockier body.

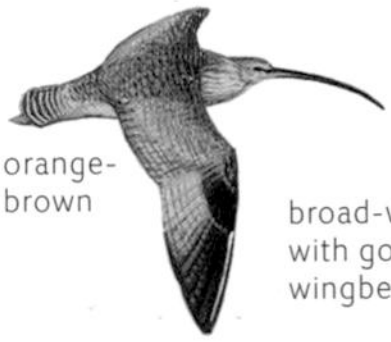

VOICE: Flight calls clear whistles, most commonly a short, rising *coooLI* with sharp rise at end. Also *kwid wid wid wid,* a loud, whistled *wrrreeep,* and variations. Song of low, rich, whistled notes building to long, slurred whistle *pr pr pr pr pr prrreeeep prrrreeeeeerrr.*

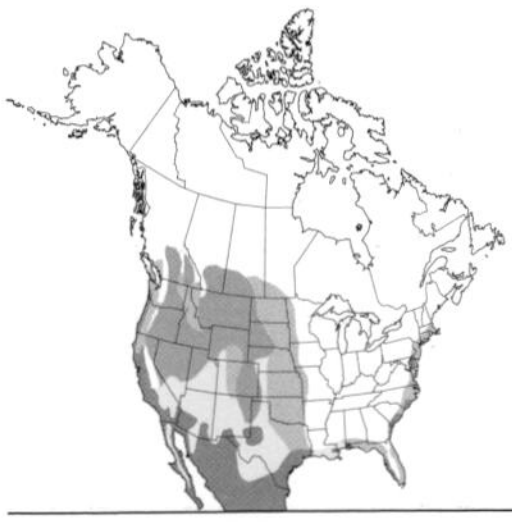

Uncommon. Winters in marshes and fields and on lawns and beaches. Foraging birds usually solitary; may form loose flocks when flying and roosting. Uses prodigious bill to probe deep into mud and sand, bringing prey up to surface. Nests on relatively arid short-grass prairie not necessarily near water, foraging mainly on grasshoppers.

Bar-tailed Godwit

Limosa lapponica

L 16" WS 30" WT 12 oz (340 g) ♀>♂

Distinguished from other godwits by relatively short legs, white or barred underwing coverts, and white or barred rump. In breeding plumage, note clean rufous underparts. Two subspecies, European and Asian, differ in rump and underwing color.

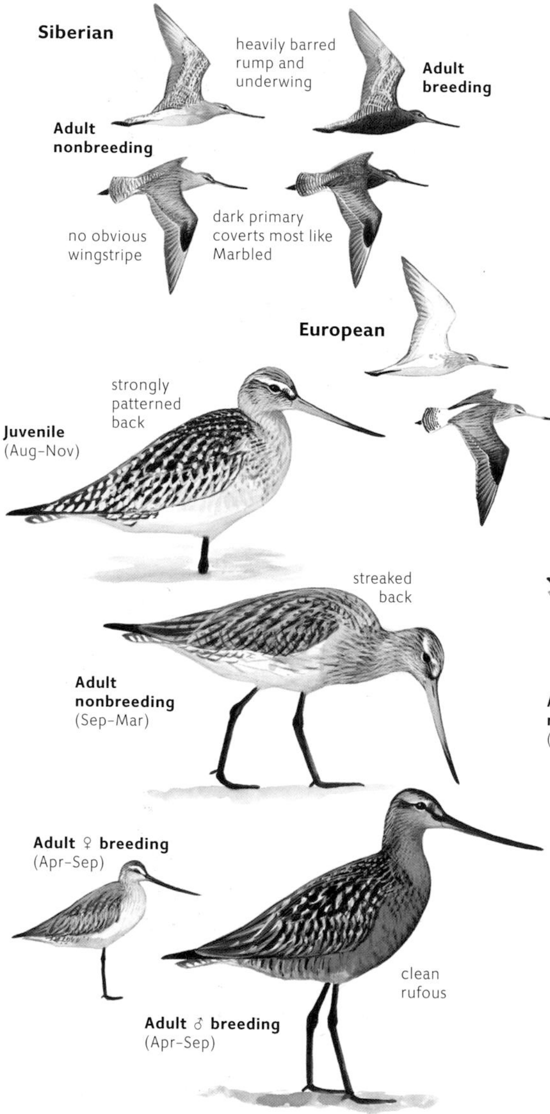

VOICE: Calls creaking, nasal. Flight call *kee kee kaa* or *kooy kooy kooy* very similar to Hudsonian but slightly lower-pitched and with more even pitch. Display song more melodic and varied in rhythm and pitch than other godwits.

Uncommon breeder on low tundra in western Alaska; very rare visitor elsewhere in North America. Most records south of Alaska involve single birds, sometimes mixed with flocks of other large shorebirds. Found almost exclusively on coastal sandflats and mudflats. Nests in dwarf-shrub and sedge habitat on tundra.

Black-tailed Godwit

Limosa limosa

L 16.5" WS 29" WT 11 oz (300 g) ♀>♂

Long-billed and long-legged, with nearly straight bill. Disti guished from Hudsonian Godwit by mostly white underwir note orange neck in breeding plumage.

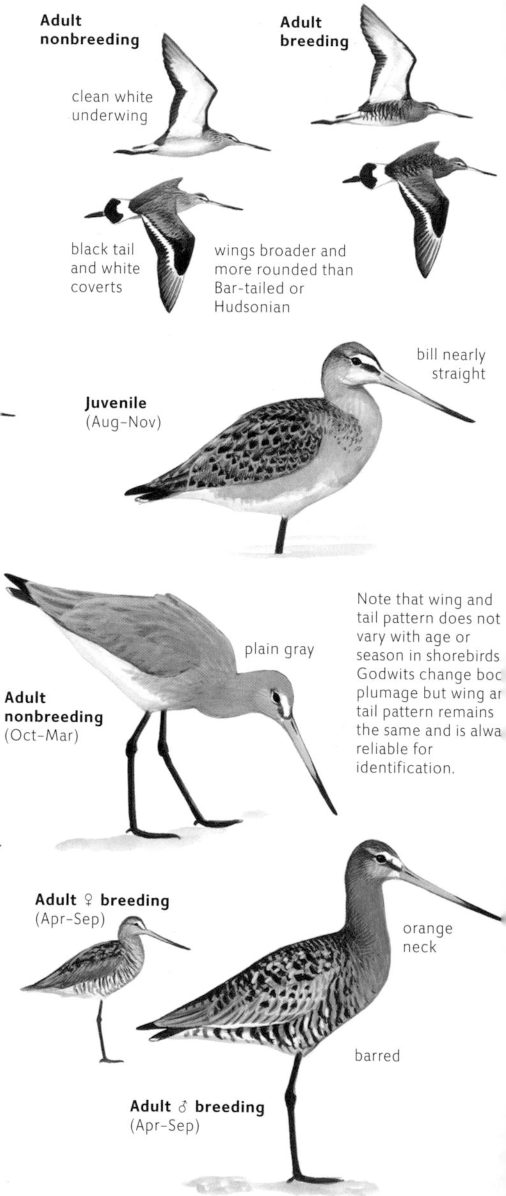

Note that wing and tail pattern does not vary with age or season in shorebirds Godwits change boc plumage but wing ar tail pattern remains the same and is alwa reliable for identification.

VOICE: Calls rather strident, nasal, mewing; most commonly *weeka weeka weeka;* variety of other short notes includes soft *kaa* with mewing quality and sharper *kip*. Display song of repeated two- or three-syllable phrases.

Very rare visitor from Eurasia, recorded less than annually along Atlantic coast and in extreme western Alaska. Most records of single birds, sometimes mixed with other large shorebirds. Generally prefers more grassy habitats tha Bar-tailed Godwit; even found ir freshwater ponds.

ludsonian Godwit

mosa haemastica

l5.5" WS 29" WT 11 oz (300 g) ♀>♂

e smallest godwit, with slightly upturned bill; slightly larger an Greater Yellowlegs, but significantly heavier. Pale eyebrow ominent in all plumages. In flight, note very dark wings with rrow white stripe, white rump, and black tail.

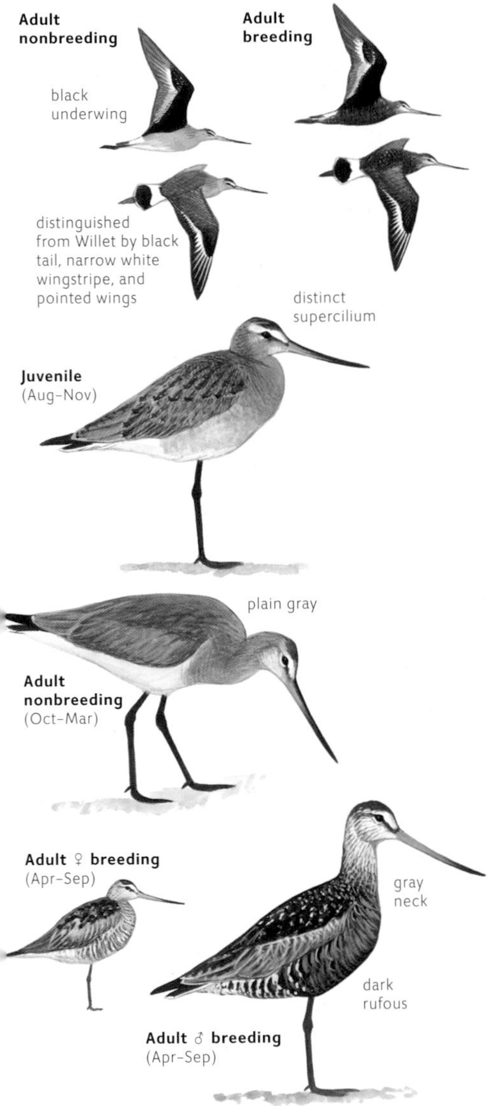

OICE: Flight call high falsetto *kwidWID* or *kweh-weh* and varia-ons, each syllable rising; higher-pitched than other godwits; so high *week* or *kwee* like Black-necked Stilt but softer. Display ong a repeated three-syllable phrase.

Uncommon and local. Forages on expansive mudflats and in shallow water. Nests around ponds within spruce woods. Usually in small flocks that fly in lines. Nests on wet grassy tundra near treeline.

Marbled Godwit

Limosa fedoa

L 18" WS 30" WT 13 oz (370 g) ♀>♂

A large sandpiper, with very long, slightly upturned bill. Overall buffy color without contrasting markings distinctive. In flight, note overall cinnamon color.

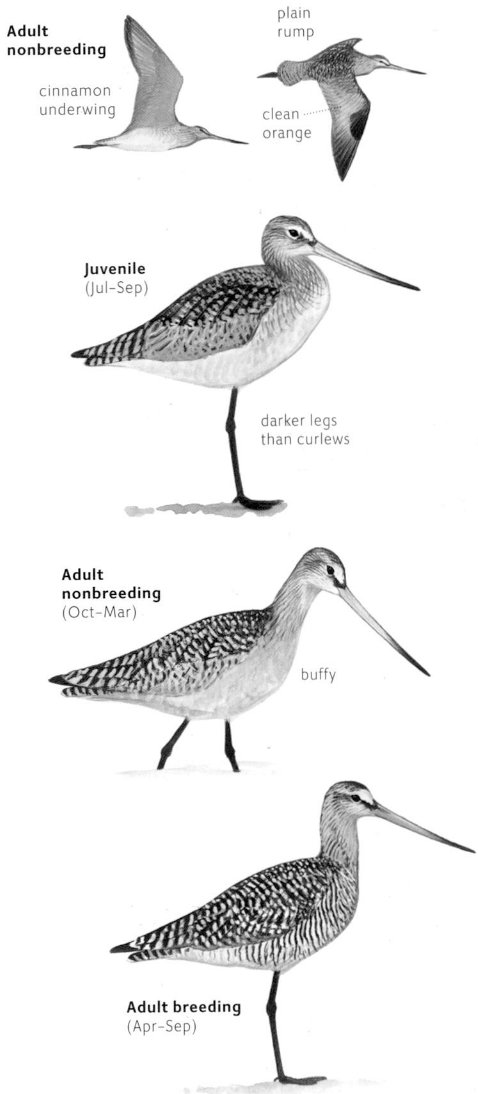

VOICE: In flight tentative, hoarse, trumpeting *kweh* or *kaaWEK* like Laughing Gull; lower-pitched than other godwits. Rolling series with nasal quality *kowEto kowEto*... and soft *ked ked*... combined into long, alternating series in territorial chase.

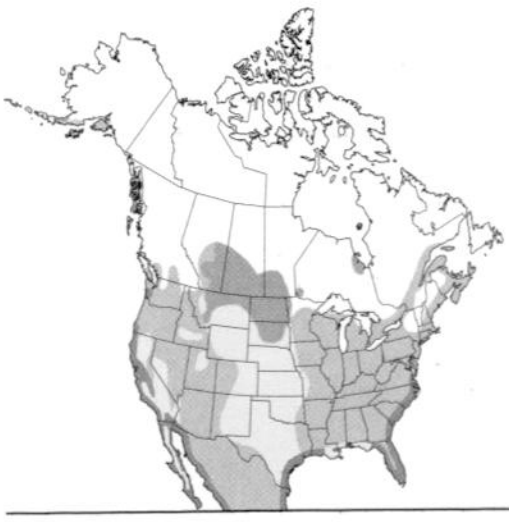

Uncommon on expansive mudflats and sandflats and on beaches. Nests around prairie ponds. May form flocks; often associates loosely with other shorebirds.

Black Turnstone

Arenaria melanocephala

L 9.25" WS 21" WT 4.2 oz (120 g) ♀>♂

Slightly bulkier and much darker overall than Ruddy Turnstone, with darker legs usually obvious; also note uniform dark head, breast, and upperparts, lacking pale patches on throat and breast-sides, and with no rufous color on upperside.

VOICE: Flight call a shrill, high chatter *keerrt* similar to high call of Belted Kingfisher; higher and more clicking than Ruddy. Threat a long, low rattle with slight pitch changes *krkrkrkrkrkrkr...;* also clear, nasal *WEEpa WEEpa WEEpa....* Song a long trill or rattle, changing pitch a few times.

Common. Winters along rocky ocean shores. Habits and appearance similar to Ruddy, and the two species are often seen together, but Black is essentially never seen away from salt water. Nests in marshy coastal tundra.

Ruddy Turnstone

Arenaria interpres

L 9.5" WS 21" WT 3.9 oz (110 g) ♀>♂

Relatively short-legged, with short, pointed bill. Short oran legs and calico plumage pattern with dark breast and white be distinctive.

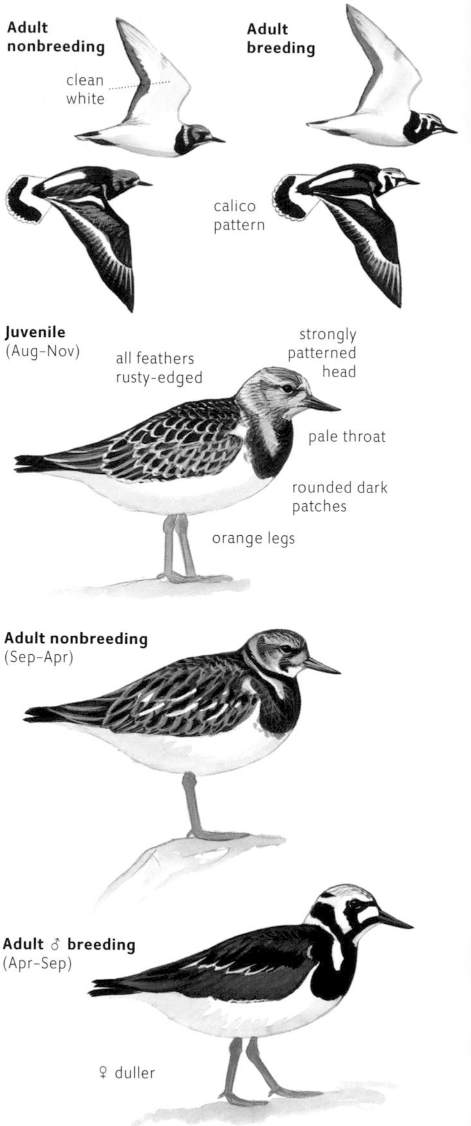

VOICE: Flight call a relatively low, mellow, bouncing rattle; mo nasal, harder than Short-billed Dowitcher. Also a single, shar *klew* or hard, nasal *gaerrt*. Threat a long, low rattle *k-k-k-k-k-k-k*. Flight song a long, rolling rattle.

Common along coasts but generally rare inland. Winters along rocky or sandy shorelines Usually in small flocks. Uses sho upturned bill to flip over rocks and debris in search of food. Nests on sparsely-vegetated tundra near marshes, streams, and ponds.

Vandering Tattler

'inga incana

.1" WS 26" WT 3.9 oz (110 g)

verall dark gray color with unmarked wings and tail, relatively ort yellow legs, and long wings distinctive. Bobs while walking, e Spotted Sandpiper.

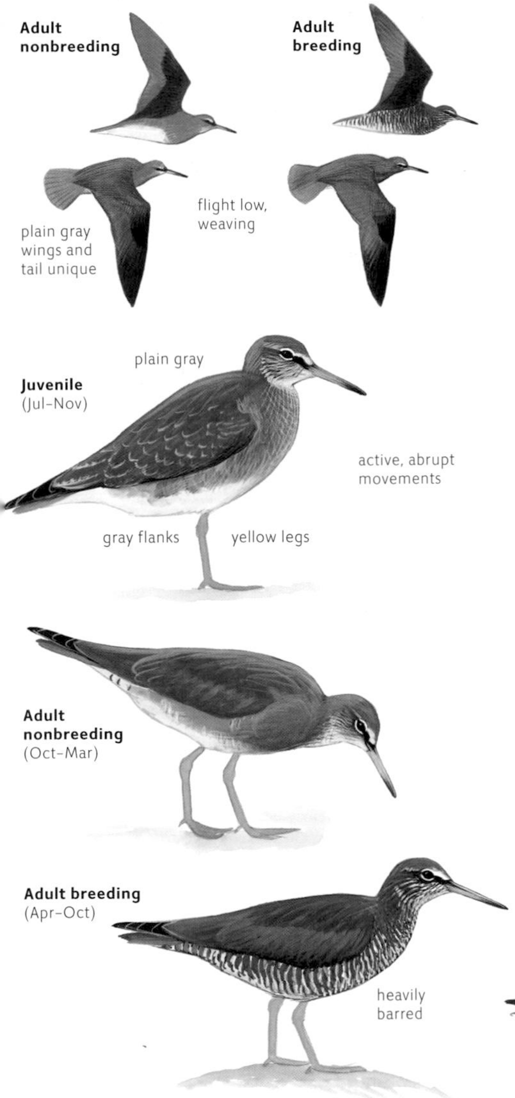

OICE: Flight call a high, clear piping of about five notes in ightly descending series, a rapid, almost unbroken trill *kwidi-idid*; also a sharp, high *klee-ik*. Flight song a ringing, whistled *eedle-deedle-deedle-dee*.

Uncommon. Winters along rocky ocean shores, rarely seen on mudflats. Usually solitary, but may associate with feeding or roosting flocks of Surfbirds and turnstones on coastal rocks. Nests along rocky streams and ponds in mountainous areas.

Gray-tailed Tattler

Tringa brevipes

L 10" WS 25" WT 3.1 oz (88 g)

Very similar to Wandering Tattler in all respects, slightly paler overall, breeding plumage with finer bars on underparts not extending to belly.

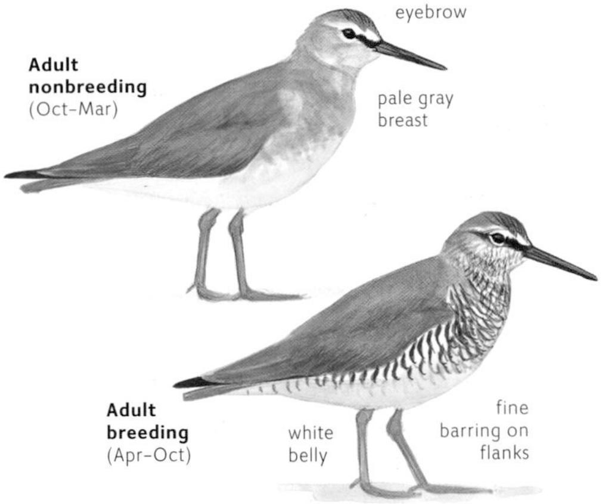

VOICE: Flight call pitch and quality similar to Wandering Tattler, but only one or two drawn-out rising whistled notes *kewee* or *kewee kwee*, unlike the rapid five-note series of Wandering.

Rare but regular visitor from Asia to western Alaska, mainly in fall, with a few records south to California, once in Massachusetts. Often on mud or sand flats, unlike Wandering Tattler.

Great Knot

Calidris tenuirostris

L 10.8" WS 24" WT 6.3 oz (179 g)

Similar to Red Knot but much larger and relatively longer-billed; lacks wingstripe

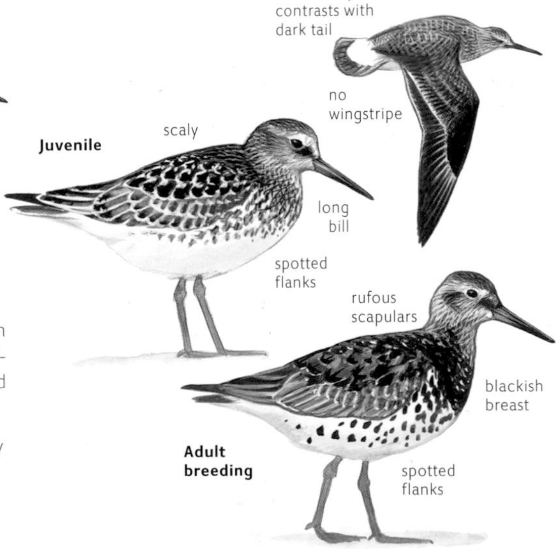

VOICE: Rarely heard.

Very rare visitor to northwestern Alaska, mainly in spring, with a few fall records along Pacific coast south to California, once in West Virginia.

Rock Sandpiper

Calidris ptilocnemis

L 9" WS 17" WT 2.5 oz (70 g) ♀>♂

Stocky and dark, with round body and short drab greenish legs. Especially dark on head and breast; also note heavily streaked flanks.

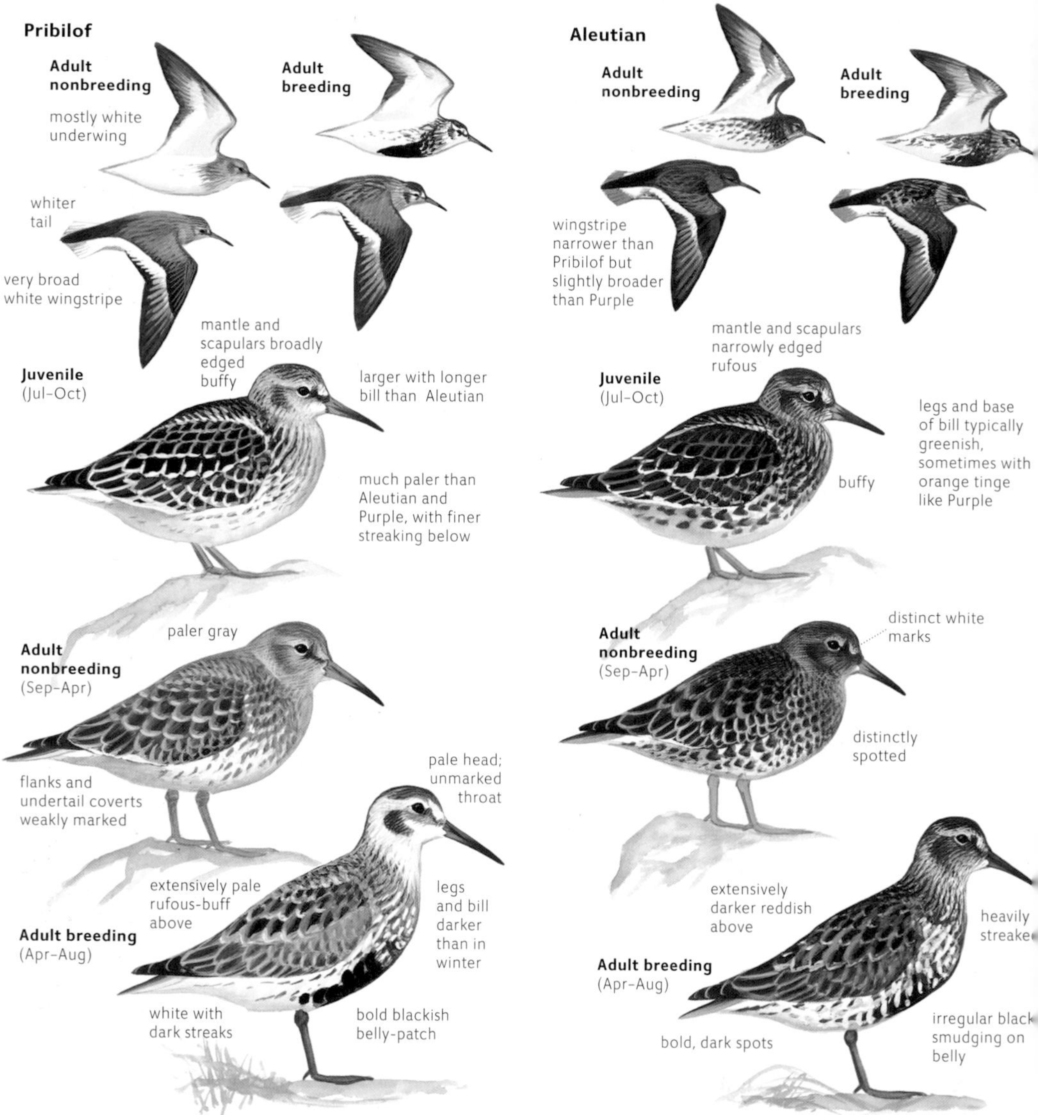

VOICE: Voice not known to differ from Purple Sandpiper, though flight calls may be slightly higher-pitched. Pribilof population's flight call a coarse, husky *cherk* sometimes more whistled; flight song of low, growling trills, a series of about ten rising *grreee* notes followed by similar low *grrdee* notes.

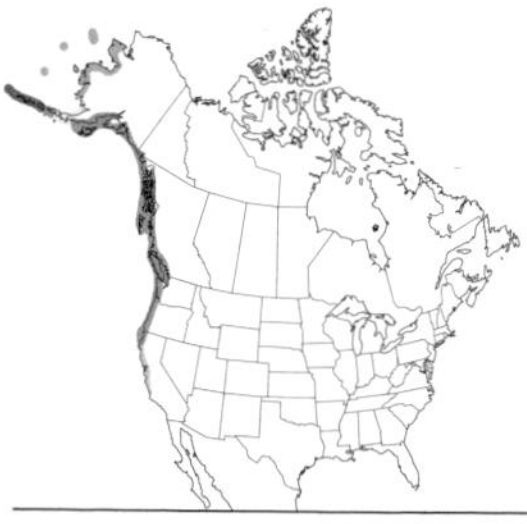

Uncommon. Winters along rocky shores; winter range extends farther north than any other sandpiper. Often in small flocks mixed with turnstones. Picks prey from rocks. Nests on moist tundra with low heath or sedge vegetation and nearby shallow ponds.

Most of breeding range and winter range occupied by relatively small dark individuals. A distinctive population nests on Pribilof Islands. Pribilof and Aleutian populations represent extremes of variation. Many intermediates may not be safely identified. Pribilof population winters south to southeastern Alaska and is distinctively large and pale in all plumages. Aleutian birds are resident, small and dark. Birds nesting along Bering Sea coast (not shown) winter south to California; they average slightly larger and paler than Aleutian, and their breeding plumage is patterned more like Pribilof but darker.

'urple Sandpiper

alidris maritima

9" WS 17" WT 2.5 oz (70 g) ♀>♂

tocky and dark, with round body and short, dull orange legs; specially dark on head and breast. Note heavily streaked flanks.

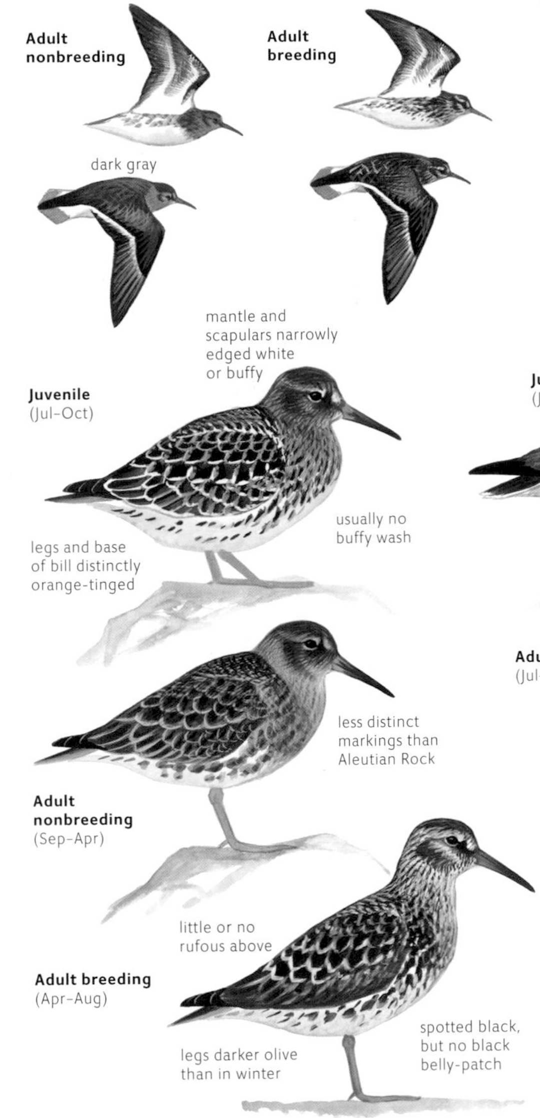

'OICE: Flight call a scratchy, low *keesh* or sharper *kwititit-kwit.* 'eeding birds give soft, scratchy notes. Chase call a higher-ɔitched *kif-kif-kif*, not very aggressive. Display song a Dunlin-like vheezing and trilling with several changes in rhythm; also rapid augh *pupupupupu*....

Common locally. Found almost exclusively on wave-washed rocks. Usually in small flocks; almost never found on beaches or mudflats with other sandpipers. Winters further north than other sandpipers. Nests on barren rocky or mossy tundra.

Surfbird

Calidris virgata

L 10" WS 26" WT 7 oz (190 g) ♀>♂

Larger than turnstones and other rock-loving shorebirds. Stocky and stout-billed, with short yellow legs, mostly unpatterned head, and striking black and white tail pattern.

VOICE: Contact calls soft and inconspicuous. Flight call a soft *iif iif iff*.... Feeding flock gives constant chatter of high, nasal squeaks. Flight song nasal, buzzy notes in series *kwii kwii kwii kwirr kwirr kwirr kwirr skrii skrii skrii kikrrri kikrrri kikrrri kikrrri*; similar phrases given in other situations, e.g., alarm.

Uncommon. Winters along rocky shorelines, nearly always in small flocks and often associated with turnstones. Picks prey from large wave-washed rocks at the ocean edge. Nests on barren gravel ridge tops in mountainous areas.

Red-necked Stint

Calidris ruficollis

L 6.25" WS 14" WT 0.88 oz (25 g) ♀>♂

Size and shape similar to Semipalmated Sandpiper, but with slightly shorter legs and longer wings. Beware confusion with brightly colored individuals of other peeps, especially fresh juvenile Least Sandpipers in late July, and with adult breeding Sanderling. Also compare Little Stint.

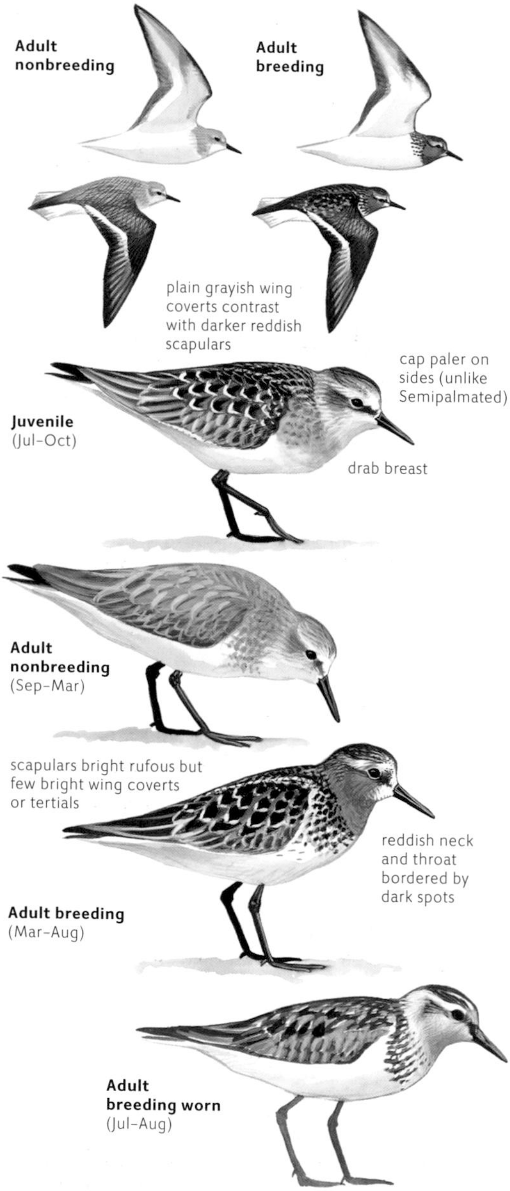

VOICE: Flight call a high, scraping *quiit* most like Western Sandpiper but huskier; variable and may overlap with other species. Display song a long, steady repetition of short *yek* notes or low, almost whistled *huee huee huee*... series.

Rare visitor from Siberia and Asia; small numbers occur in western Alaska and rarely farther south. Recorded annually along Pacific coast and along Atlantic coast from Maine to North Carolina; extremely rare elsewhere. Found on open mudflats with Semipalmated or Western Sandpipers.

Little Stint

Calidris minuta

L 6" WS 14" WT 0.84 oz (24 g) ♀>♂

Size and shape very similar to Semipalmated Sandpiper. Plumag can be very similar to Red-necked Stint, but breeding adult ha whitish throat with reddish color covering spots on breast.

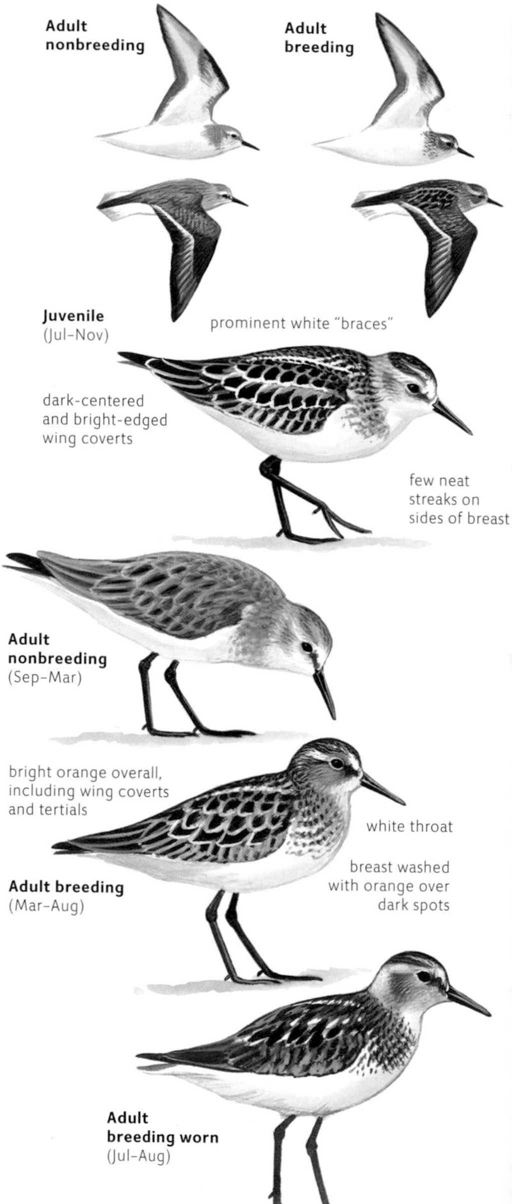

VOICE: Flight call a thin, short *tit* or *stit*; other variations similar to Red-necked. Threat a high, weak, level *tee tee tee tee*. Display song includes series of high, thin, squeaky notes *tsee-tsee-tsee*...; recalls distant Arctic Tern courtship.

Rare visitor from Eurasia; recorded annually in western Alaska and along the Pacific coast and Northeast coast, with scattered records elsewhere in North America. Found on open mudflats with other peeps, showing a slight preference for fresh water and more muddy and grassy edges.

Temminck's Stint

Calidris temminckii

6.2" WS 14" WT 0.85 oz (24 g) ♀>♂

imilar to other peeps but with relatively long tail, plain upperarts, indistinct head pattern with thin white eyering, and attling call.

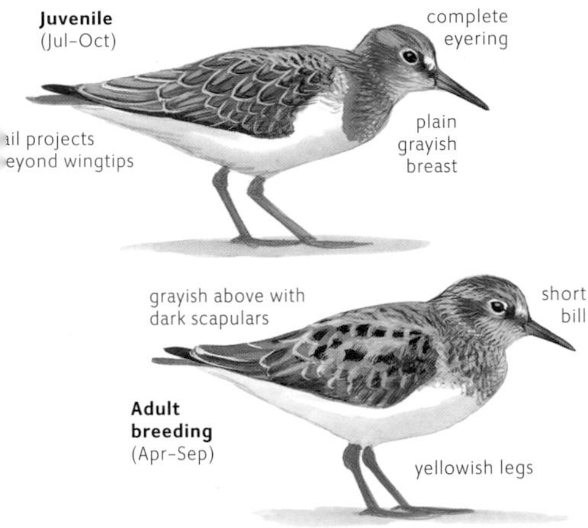

OICE: Flight call a dry rattle.

Rare visitor to western Alaska, with one record south to British Columbia. Habitat like Least Sandpiper, grassy and muddy pools, but generally solitary, not flocking with other "peeps."

Long-toed Stint

Calidris subminuta

6" WS 14" WT 0.8 oz (23 g) ♀>♂

ery similar to Least Sandpiper, and must be identified with xtreme care. Averages slightly larger, longer-necked, and tends o stand taller; long toes can be obvious. Best distinguished by face attern, with structure, voice, and behavior as additional clues.

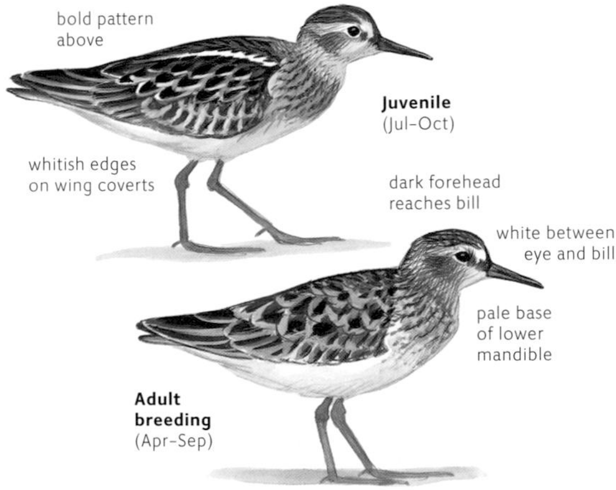

VOICE: Flight call a musical rill like Least Sandpiper but ower-pitched and not rising.

Rare visitor to western Alaska, mainly in spring, with a few records along Pacific coast south to California. Habitat like Least Sandpiper, but vagrants have been solitary, not joining flocks of Least Sandpipers.

Identification of Small Sandpipers

Often seen as distant mixed flocks, the small sandpipers known as "peeps" (Least, Semipalmated, and Western) are always challenging to identify. One of the keys to understanding the subtle field marks and variations in these species is distinguishing juvenile from adult and breeding from nonbreeding plumage. The illustrations below show the plumages of a single Western Sandpiper through its first two years of life, with some changes resulting from molt and others from wear.

Fresh juvenile
July, bright buff and cinnamon tones; crisp pale fringes on back feathers create a scaled pattern

Worn juvenile
within weeks, by late Aug, cinnamon color has faded to gray and white fringes have worn away

First winter
molt to all-gray winter plumage similar to adult

First summer
variable, individuals molt in spring and acquire more or less breeding plumage

Adult winter
molt to all-gray

Fresh adult breeding
molt in spring, fresh feathers are bright rufous and tipped whitish

Worn adult breeding
by July white tips have worn away, leaving a much darker overall appearance

Western Sandpiper

Calidris mauri

L 6.5" WS 14" WT 0.91 oz (26 g) ♀>♂

Slightly larger than Least and Semipalmated Sandpipers, with slightly longer and more drooping bill. Breeding and juvenile plumages show rufous highlights on scapulars and head. Smaller than Dunlin, in nonbreeding plumage paler above with whitish breast.

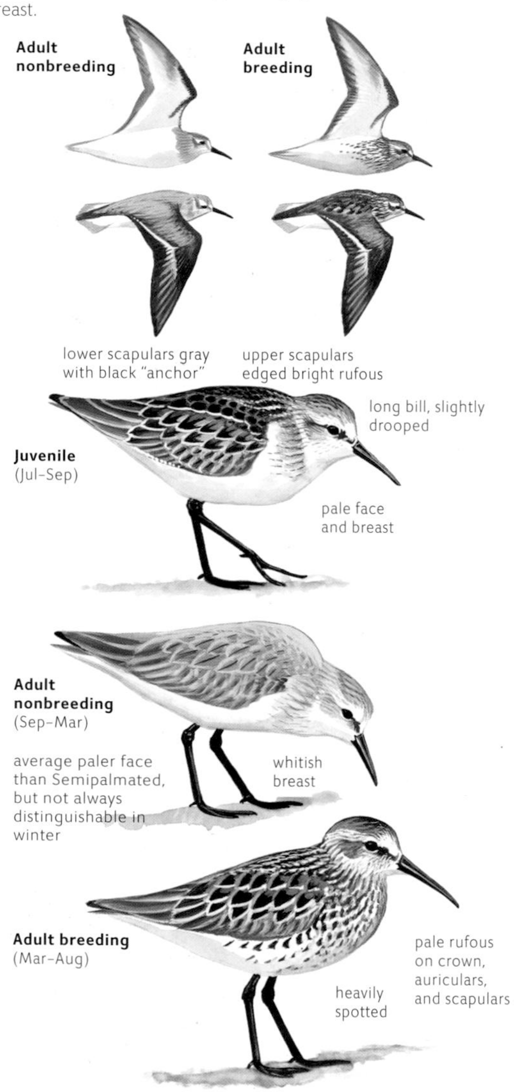

VOICE: Call typically a thin, high, harsh *cheet*. Feeding flock emits a constant twitter of quiet, scratchy notes. In high aggression, gives peevish, weak, rising *twee twee twee twee*. Display song scratchy, weak, thin; shorter and simpler than Semipalmated's song, with higher, weaker quality.

Common on mudflats and sandy beaches. Often in large flocks, mixes freely with Semipalmated Sandpiper or Dunlin.

Semipalmated Sandpiper

Calidris pusilla

L 6.25" WS 14" WT 0.88 oz (25 g) ♀>♂

Larger than Least Sandpiper; characterized by relatively shor blunt-tipped bill, dark legs, and usually plainer gray-brown plum age. Variable in color; some show strong rufous tones similar t Western. In all plumages very similar to Western and some ca be extremely difficult to distinguish.

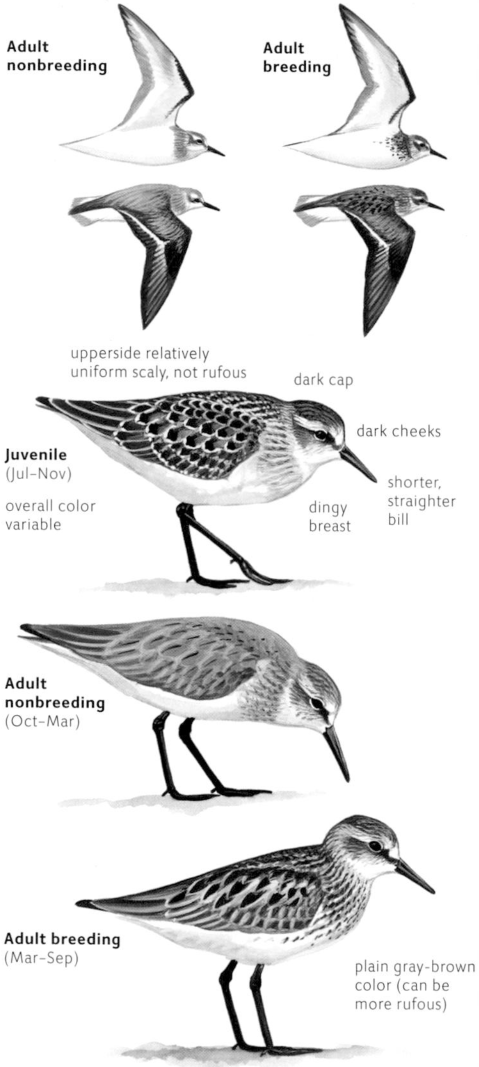

VOICE: Typical flight call a short, husky *chrup* or *chrf*; also give sharp, thin *cheet* similar to Western. Feeding flock very voca with rapid, giggling/arguing *twee do do do do*. Display song a continuous rolling trill of varying pitches *grrridigrrrridi....*

Common to uncommon on mudflats, often in large flocks. Common on grassy tundra breeding grounds. Winters in South America; migrants often mixed with Western or Least Sandpipers.

Least Sandpiper

alidris minutilla

6" WS 13" WT 0.7 oz (20 g)

he smallest peep, characterized by greenish-yellow legs, rouching posture, short fine-tipped bill. In all plumages slightly arker and browner than other peeps,

OICE: Typical flight call a high, trilled *prreep*; gentle, rising, and ιusical. Flock in wheeling flight gives short, weak, high notes in apid twittering chorus. Threat rapid, giggling, high *didididi.* isplay song similar to dowitchers: *b-b-b-trree-trree-trree*; each hrase rising but the series level.

Common; the most widely distributed peep, but generally in small groups, not large cohesive flocks. Found on virtually any bit of exposed mud or grassy mudflat both coastally and inland, choosing smaller and muddier pools with more vegetation than other peeps, less often on open mudflats, rarely on sandy beaches.

Sanderling

Calidris alba

L 8" WS 17" WT 2.1 oz (60 g) ♀>♂

Size close to Dunlin, larger than peeps, with short bill and different behavior. Most often seen in very pale gray nonbreeding plumage, paler than any other shorebird. In flight, note broad white wingstripe and dark leading edge of wing.

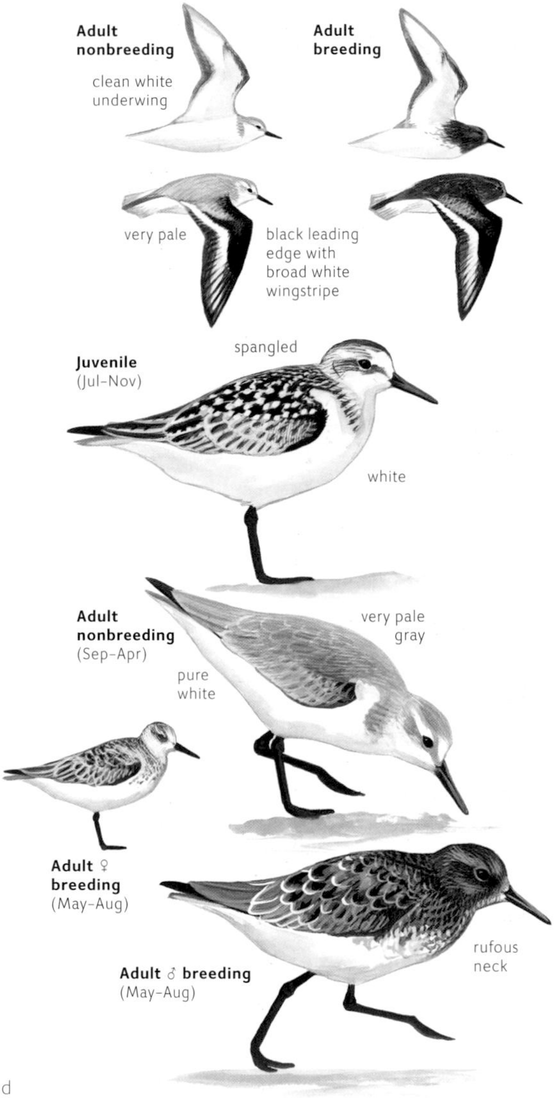

VOICE: Common call a short, hard *klit* or *kwit*. Feeding flock gives high, scratchy *tiv* calls. Threat a high, thin, relatively slow *twee twee twee....* Flight song complex: short bursts of churring and trilling with croaking or hissing sounds.

Common. In nonbreeding season found almost exclusively along sandy beaches with some wave action; small flocks run rapidly up and down beach, chasing waves and frenetically probing sand for invertebrate prey. Nests mainly on gravel ridges and slopes near wet tundra.

Pectoral Sandpiper

Calidris melanotos

L 8.75" WS 18" WT 2.6 oz (73 g) ♂>♀

Resembles a very large Least Sandpiper. Note greenish legs, overall brownish color, and especially dense streaking on breast that ends abruptly at white belly. In flight, note size, long wings, and very weak pale wingstripe. Male larger than female, so size variation within flock is considerable.

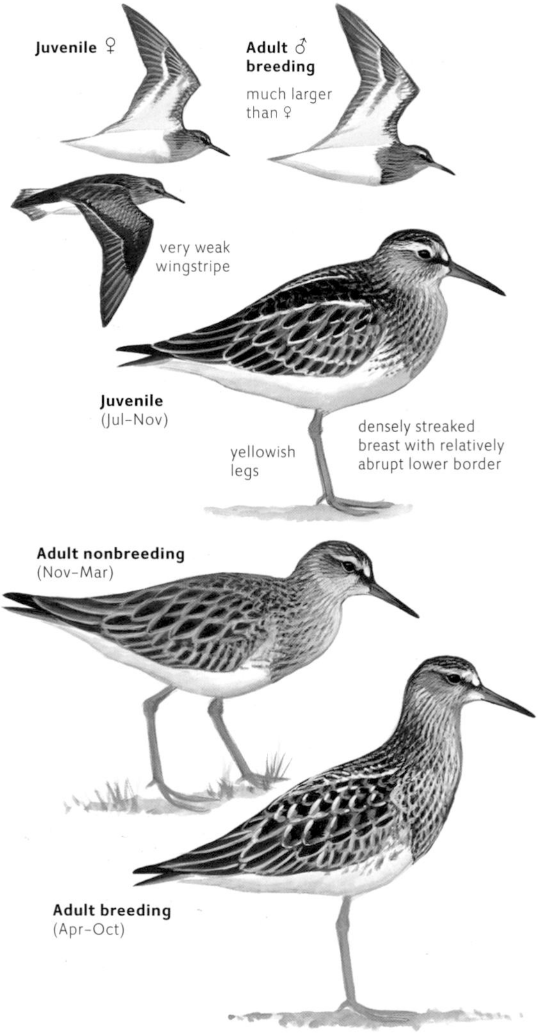

VOICE: Flight call a rather low, rich, reedy, harsh trill *drrup* or *jrrff*. Threat a low, soft *goit goit goit*. Display song in low flight a remarkable, rapid, foghorn-like hooting *ooah ooah*... continuing 10 to 15 seconds; more complex harsh and hooting calls from ground.

Common to uncommon in small groups on weedy or grassy mudflats and in flooded fields. Nests on grassy lowland tundra.

Sharp-tailed Sandpiper

Calidris acuminata

L 8.5" WS 18" WT 2.4 oz (68 g) ♂>♀

Size and shape similar to Pectoral, but juvenile has buffy breas with few streaks and without sharp lower border; also note wel defined rufous cap and thin but distinct white eyering. Male larger than female.

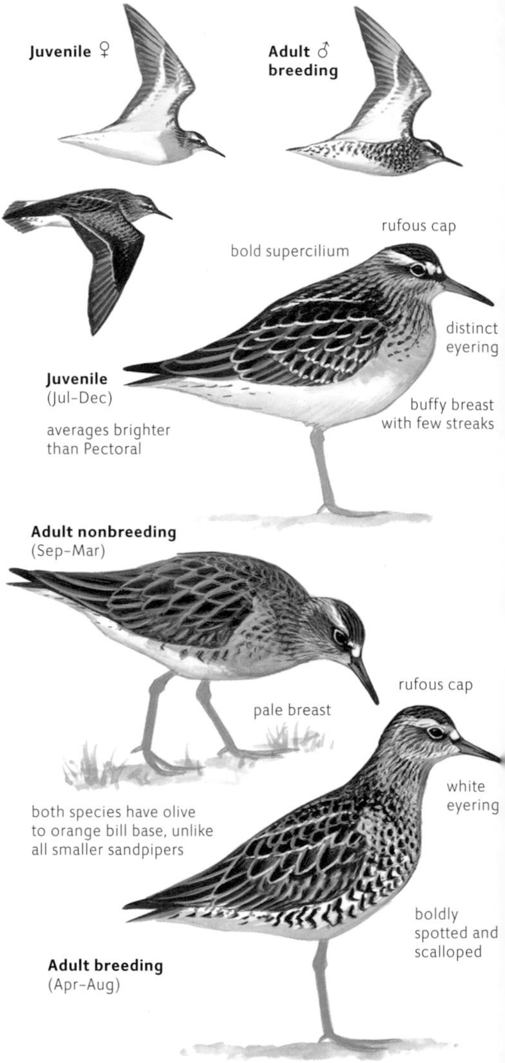

VOICE: Flight call a soft *weep* or *chewt;* higher, more musical tha Pectoral; also high trills, often in short, twittering sequence. Fligh song includes long, rhythmic trill and short, low-pitched *hoot*.

Rare visitor from Asia; recorded annually in small numbers. Most records are of single juveniles with Pectoral Sandpipers from Sep–Nov, usually along muddy and grassy edges of small freshwater pools.

aird's Sandpiper
alidris bairdii
7.5" WS 17" WT 1.3 oz (38 g) ♀>♂

Iearly as large as White-rumped Sandpiper; long wings give lender tapered appearance. In all plumages relatively pale-eaded with buffy tones. Juvenile has neat scaly pattern on pperside.

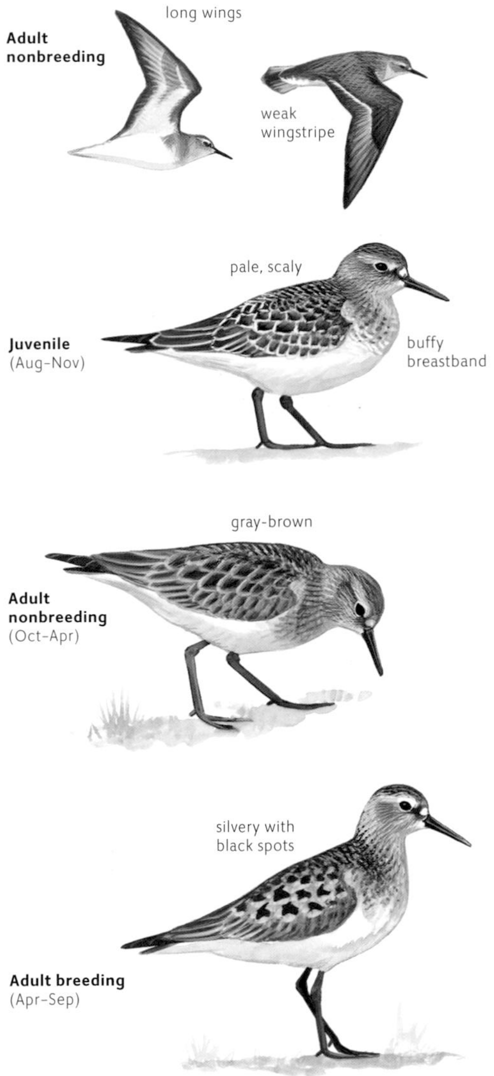

OICE: Flight call a rough *kreep;* reminiscent of Pectoral Sand-iper but higher-pitched, drier, more distinctly trilled. Flight ong a series of buzzy, rising notes and long, level rattling.

Uncommon on damp upper edges of mudflats and sometimes in shortgrass fields. Usually near other peeps, but forages on drier substrate, rarely in water or on wet flats. Nests on barren gravel ridges with lichens and a few low plants in coastal tundra.

White-rumped Sandpiper
Calidris fuscicollis
L 7.5" WS 17" WT 1.5 oz (42 g) ♀>♂

Larger than other peeps and longer-winged; body appears long and tapered. White rump unique among regularly-occurring small sandpipers. Breeding plumage has streaks extending back along flanks. Nonbreeding plumage shows prominent white eyebrow.

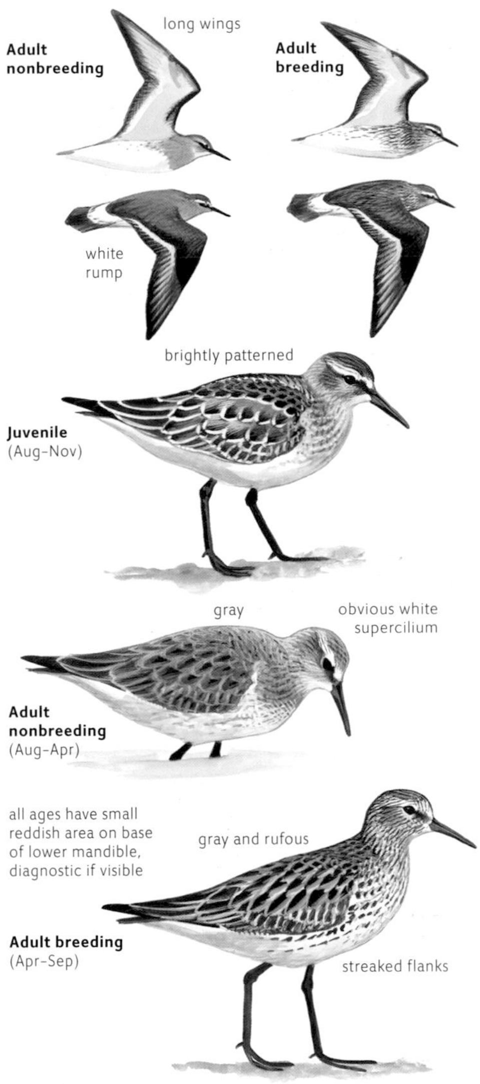

VOICE: Flight call distinctive, a very high, thin, mouse-like squeak *tzeek* or *tseet.* Threat a very high, insect-like rattle *t-k-k-k-k-k.* Flight song extremely high, mechanical, and insect-like.

Common to uncommon on mudflats. Usually mixed with other peeps, such as Semipalmated Sandpiper, but often feeds in slightly deeper water.

Red Knot

Calidris canutus

L 10.5" WS 23" WT 4.7 oz (135 g) ♀>♂

Large and sturdy, with relatively short legs and bill. Salmon and gray breeding plumage distinctive. On nonbreeding plumage, note barred flanks. In flight, note weak wingstripe, gray rump and tail, and very long wings.

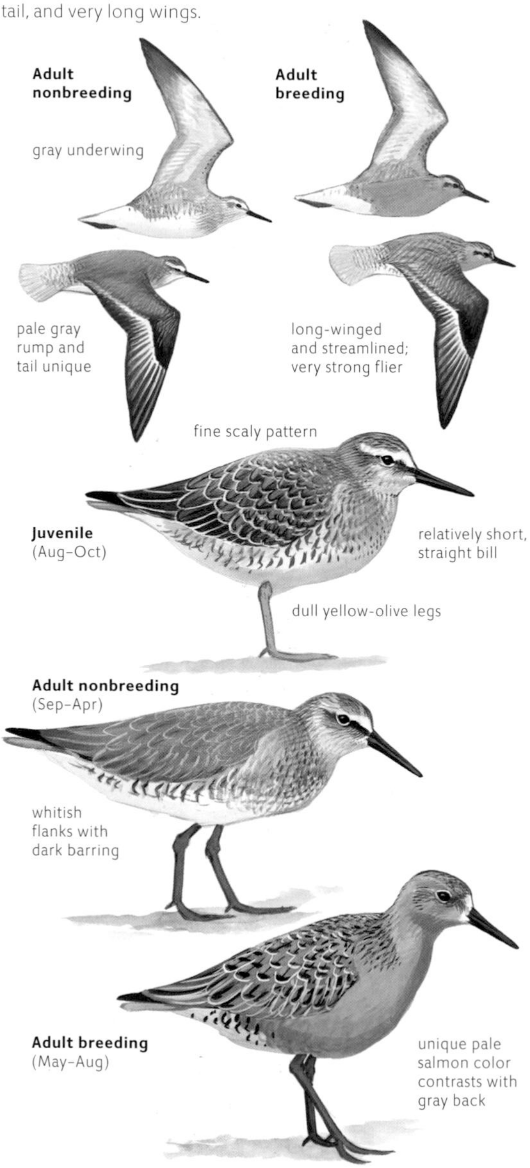

VOICE: Usually silent; sometimes a low, husky *wett-wet*. Threat call a godwit-like *kowet-kowet*. Display song a low, husky, whistled *kwa-wee, kh-where, kh-where*....

Common locally on migration and in winter. Winters along sandy beaches, usually in flocks; often with Ruddy Turnstone and Black-bellied Plover. Walks and picks food methodically. Nests on tundra near marshy areas.

Curlew Sandpiper

Calidris ferruginea

L 8.5" WS 18" WT 2.1 oz (60 g) ♀>♂

More slender and elegant than Dunlin, with longer wings, leg and neck, white rump. Similar to Stilt Sandpiper, which has lon ger greenish legs and lacks white wingstripe. Breeding plumag striking with dark rufous head and underparts.

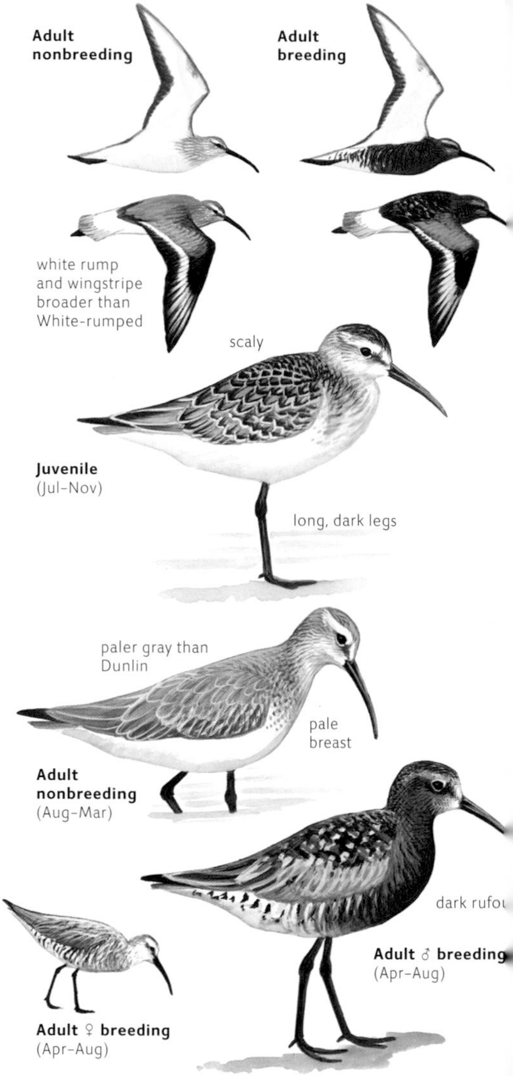

VOICE: Flight call a soft, musical *chirrup;* more liquid than Dun lin, recalls Pectoral Sandpiper. Feeding flock gives constar chorus of low, soft notes. Song complex: chatter followed b harsh two-note phrases ending with several clear, rising *whaaa* notes; fundamentally similar to Stilt Sandpiper.

Rare visitor from Eurasia; a few recorded annually along Atlantic coast, less frequent elsewhere. Most records are of single birds, often with Dunlin on mudflats and shallow ponds, less often or sandy beaches.

Dunlin

Calidris alpina

L 8.5" WS 17" WT 2.1 oz (60 g)

Larger than peeps, with relatively stocky body, short legs and long, slightly drooping bill. Both juvenile and breeding adult have distinctive black belly. Nonbreeding plumage drab gray-brown with weakly marked head; best identified by size, shape, and habits.

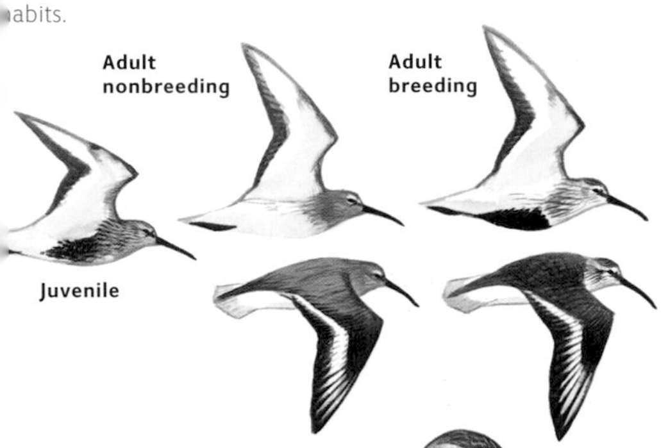

Juvenile
(Jul–Sep)

rarely seen south of breeding grounds

Adult nonbreeding
(Aug–Mar)

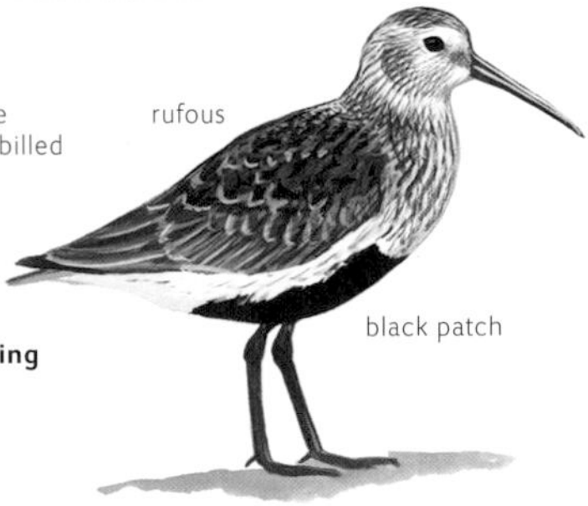

females average slightly longer-billed than males

Adult ♀ breeding Atlantic
(Apr–Aug)

VOICE: Flight call distinctive, buzzy, rasping *pjeev*. Threat call a low, hoarse *gwrr-drr-drr-drr*. Display song (often heard on migration) a series of rolling, harsh trills *jrrre jrrre jrrrrijijijijijiji jrrr jrrr jrrr*.

Common along coasts on sandy beaches and on mudflats; uncommon inland on mudflats and lakeshores. Often in large dense flocks that move slowly, probing busily in mud or shallow water. Nests on moist to wet tundra with grassy tussocks and many small shallow pools.

Adult ♂ breeding Greenland
(Apr–Oct)

Greenland breeding populations (recorded several times on Atlantic coast) are smaller and shorter-billed than American, drabber in breeding plumage, and paler-breasted and grayer in nonbreeding; they molt during or after fall migration.

Adult ♂ breeding Pacific
(Apr–Aug)

Pacific birds average paler on breast and head than Atlantic, brighter rufous on edge of scapulars, and lack fine streaks in rear flanks, but all subspecies are variable; females are more heavily streaked and drabber than males, and immatures even more so.

Dunlin × White-rumped Sandpiper

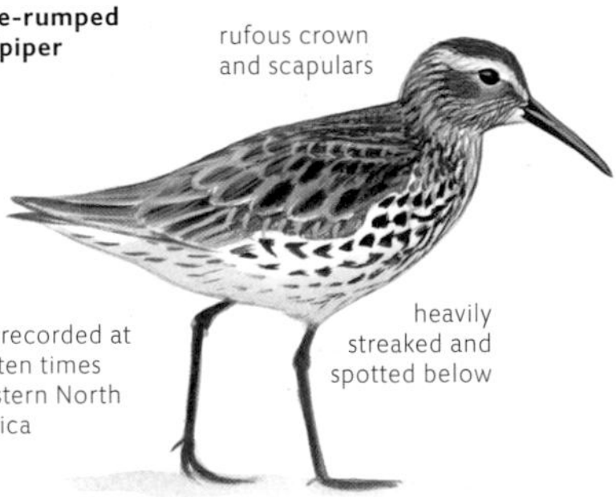

Rare, recorded at least ten times in eastern North America

Buff-breasted Sandpiper

Calidris subruficollis

L 8.25" WS 18" WT 2.2 oz (63 g) ♂>♀

A medium-sized, slender sandpiper with delicate features, walks with high-stepping gait and picks food from the ground. Unmarked pale buffy face, broad pale eyering, and yellow legs distinctive. In flight, note plain upperside and mostly white underwing coverts.

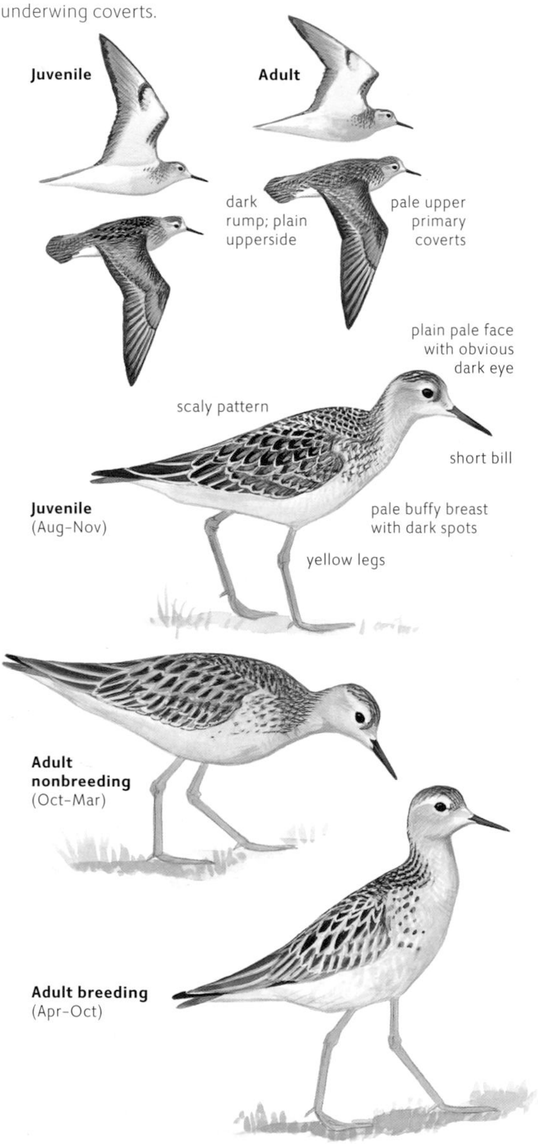

VOICE: Flight call a quiet *greet* similar to Pectoral and Baird's Sandpipers but dry, rattling; also a short *chup* and quiet *tik*. Display includes rapid clicking noises

Uncommon and local, usually in small groups. Forages in dry shortgrass habitats, such as sod farms or drying edges of ponds. Nests on tundra; males display at leks.

Stilt Sandpiper

Calidris himantopus

L 8.5" WS 18" WT 2 oz (58 g) ♀>♂

Shape and habits blend features of yellowlegs and dowitchers. Note rather long, drooping bill and long greenish-yellow legs. In flight, note plain wings and white rump. When feeding, long legs and short bill force it to lean farther forward than dowitche

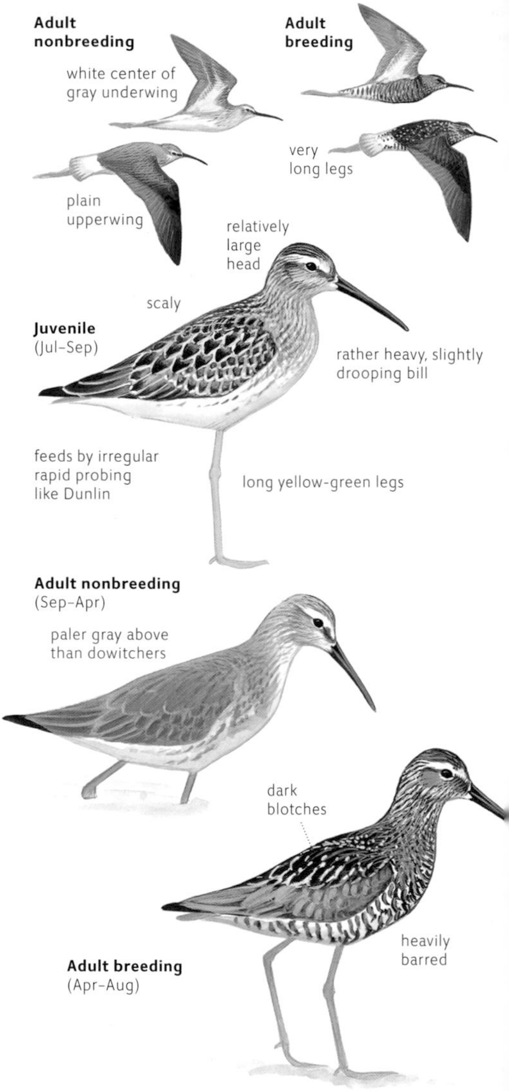

VOICE: Flight call a low, soft, muffled, husky *toof* or *jeew*, a quie background noise among Lesser Yellowlegs; also a sharpe wheezy *keewf* or a clearer, godwit-like *koooWI*. Display song remarkable series of nasal, dry, buzzy trills (quality like Surfbird

Uncommon on open shallow muddy ponds such as flooded fields, salt pans, etc., almost always wading in water to forage. Forages and roosts in small flock in water, often alongside dowitchers. Nests in wet tundra meadows, often with low shrubb willows or birches.

Ruff

Calidris pugnax

♂ L 11" WS 23" WT 6.4 oz (181 g)
♀ L 9" WS 19" WT 3.9 oz (110 g)

As tall as yellowlegs, but with shorter bill and more methodical actions. Legs greenish to orange; bill often orange. Male much larger than female, with bizarre and variable ruff in breeding plumage. Other plumages drab gray or with blotchy blackish patterns. In flight, note white U-shape on uppertail coverts, and weak wingstripe.

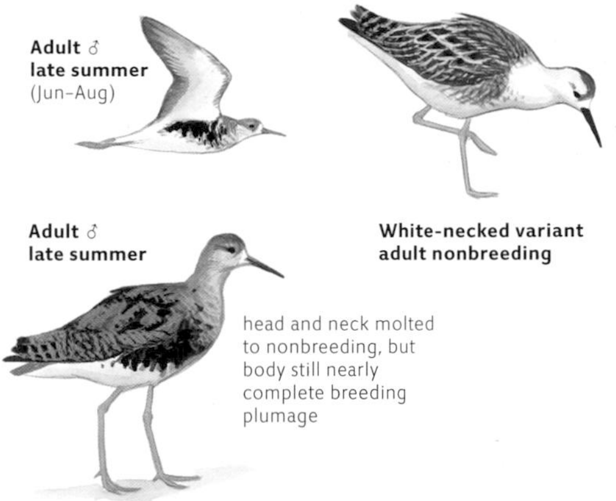

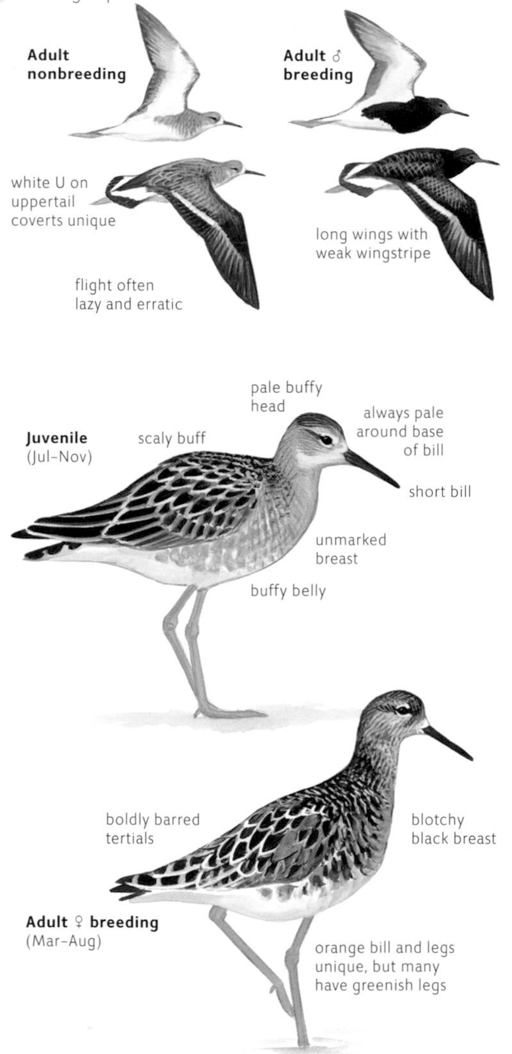

Adult ♂ breeding in display posture, showing plumage variation.

VOICE: Essentially silent. Flight call (rarely heard) a muffled croak or low grunt; sometimes a shrill, rising *hoo-ee*.

Rare visitor from Eurasia. Found mainly in shallow water along grassy edges of muddy ponds. Usually single birds that forage independently, but often associate with yellowlegs or Pectoral Sandpipers.

Color of male breeding plumage matches status. Males with black or brown ruffs defend small territories at a lek. Satellite males (about 16% of the population) have white ruffs, average slightly smaller, and remain outside of the lek. Cryptic "faeder" males (<1% of population) are close to females in appearance and size, and this allows them to move freely within the lek.

Dowitcher Identification

In late summer adult dowitchers of both species have worn breeding plumage with extensive orange and little barring below. Close examination reveals that Long-billed Dowitcher has short dark bars on the sides of the breast, Short-billed Dowitcher has dark spots on the side of the breast. Overall shape can be helpful (the fuller chest of Long-billed Dowitcher) but voice is the most reliable distinguishing feature.

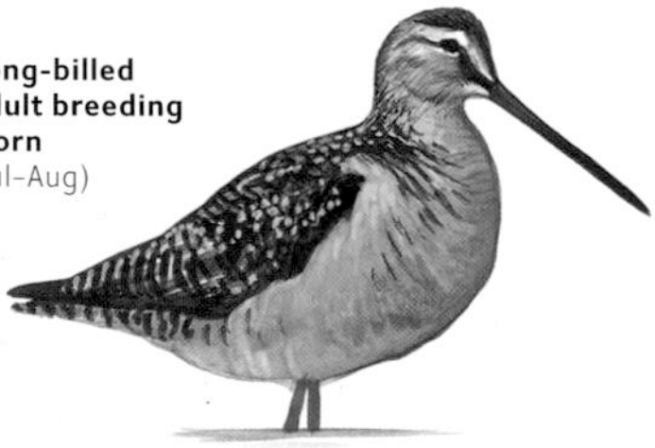

When feeding, Stilt Sandpiper's (left) longer legs and shorter bill force it to lean farther forward than dowitchers (right).

When foraging, dowitchers typically gather in tight flocks and wade in shallow water moving slowly and methodically with their bills probing up and down steadily in "sewing-machine" motion. Other shorebirds usually spend more time walking and move more quickly and erratically. Stilt Sandpipers often associate with dowitchers, but have longer legs and shorter bouts of probing. Dunlin have foraging actions similar to dowitchers but are smaller and move more quickly with much quicker and shorter up-and-down head motions.

Long-billed Dowitcher

Limnodromus scolopaceus

L 11.5" WS 19" WT 4 oz (115 g) ♀>♂

Very similar to Short-billed Dowitcher, averages longer billed bu much overlap; most easily distinguished by call. In all plumage slightly bulkier than Short-billed, with deep chest and mor rounded back. Identification by details of plumage depends o knowledge of age and molt.

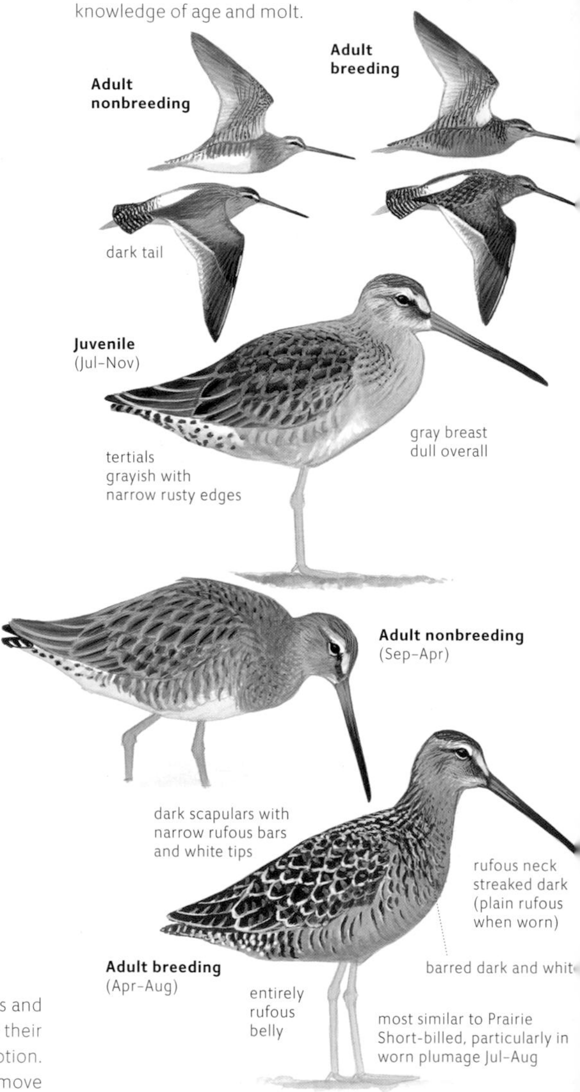

VOICE: Flight call a high, sharp *keek* or *pweek*, sometime repeated in accelerating quick, sharp series *kik-kik-kik-kik*. Floc gives constant, soft chatter while feeding (Short-billed flock quiet). Display song rapid and buzzy; sharper and higher tha Short-billed.

Common, often in large flocks. Forages in shallow muddy pools, more frequent than Short-billed on freshwater ponds, much less frequent on tidal mud flats. Migrates somewhat later in fall than Short-billed, and more expected inland in winter. Nests on grassy tundra.

Short-billed Dowitcher

imnodromus griseus

11" WS 19" WT 3.9 oz (110 g) ♀>♂

stocky, long-billed, short-necked sandpiper, very similar to
ɔng-billed. Dowitchers feed by rapid, rhythmic, vertical, sew-
ıg-machine-like probing. Often seen with yellowlegs (much
ıller and more elegant in shape and actions) or Dunlin (smaller
ith shorter bill and dark legs).

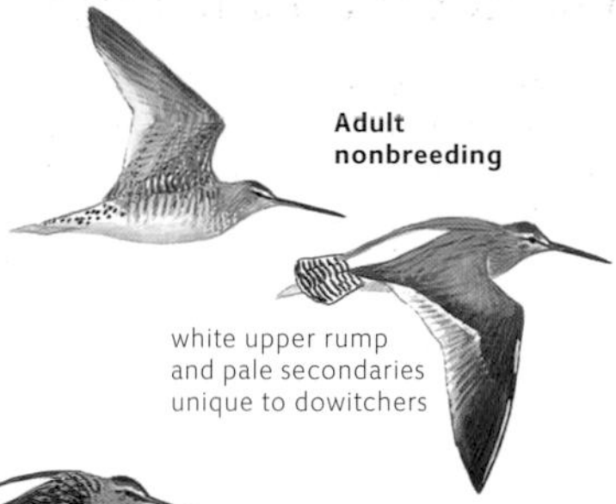
Adult nonbreeding
white upper rump
and pale secondaries
unique to dowitchers

Adult breeding Pacific/Atlantic
Adult breeding Central

Juvenile Pacific
(Jul–Nov)
tertials dark
with contrasting
rufous-buff edges
and bars

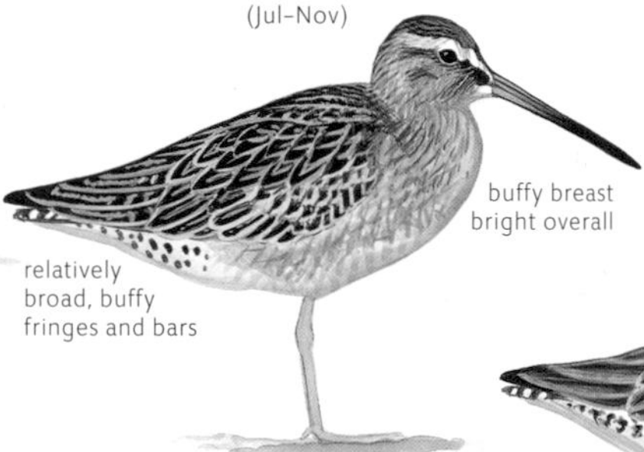
Juvenile Central/Atlantic
(Jul–Nov)
buffy breast
bright overall
relatively
broad, buffy
fringes and bars

Short-billed molts remiges only on wintering grounds, while Long-billed often molts during migration (Jul–Sep)

Adult nonbreeding
(Sep–Apr)
all populations similar
flat back
dark,
barred flanks
unlike other
sandpipers
generally pale
with streaked
face, speckled
breast, and paler flanks
than Long-billed

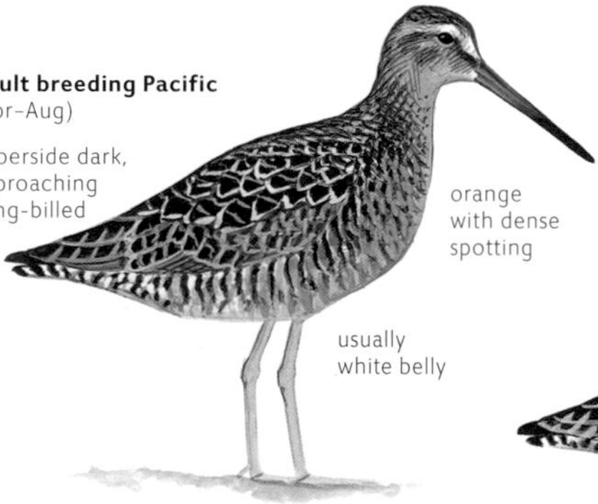
Adult breeding Pacific
(Apr–Aug)
upperside dark,
approaching
Long-billed
orange
with dense
spotting
usually
white belly

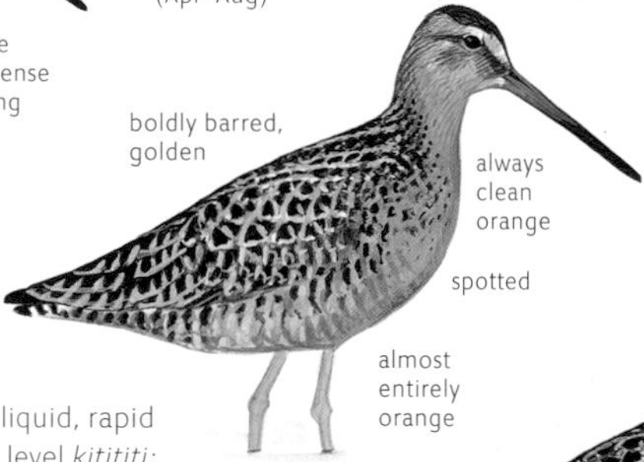
Adult breeding Central
(Apr–Aug)
boldly barred,
golden
always
clean
orange
spotted
almost
entirely
orange

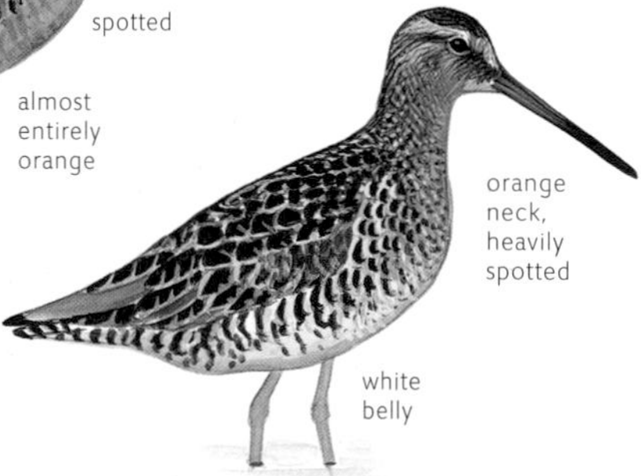
Adult breeding Atlantic
(Apr–Aug)
orange
neck,
heavily
spotted
white
belly

ɔICE: All calls lower than Long-billed. Flight call liquid, rapid
ɘwtutu or *tlututu*, slightly descending; also harder, level *kitititi*;
ways faster than Lesser Yellowlegs; cadence recalls Ruddy Turn-
one. Feeding flock is normally silent, but occasionally a single
ı is heard (feeding Long-billed flock gives constant chatter of
ɔft, high notes). Display song similar to Long-billed Dowitcher
ıd Least Sandpiper: *gididi drreee drrooo*, each phrase slightly
ɜing but the series falling (Least Sandpiper series is level).

Common, often seen in large flocks. Forages in shallow muddy pools, usually forming tight flocks all with heads down busily probing in mud; most frequent on muddy or sandy tidal flats, less often in freshwater ponds (compared to Long-billed). In winter almost exclusively coastal. Nests on grassy tundra.

Three populations largely separated by range, but some overlap and some intergrades. In breeding plumage, Pacific and Atlantic populations are very similar, with white belly, heavily spotted neck and breast, and narrow reddish fringes on upperparts (Pacific averages darker overall). Central population is brighter, with little or no white on belly, limited dark spotting on breast, and broader buff fringes on upperparts. Juvenile Pacific averages darker on the breast, with narrower and darker rufous fringes above than Central and Atlantic.

Wilson's Snipe

Gallinago delicata

L 10.5"   WS 18"   WT 3.7 oz (105 g)   ♀>♂

Dark brownish overall, with bold cream-colored stripes on back. Very stocky and long-billed, easily distinguished from dowitchers by more crouching posture, shorter legs, and darker color. In flight display (known as winnowing), outer tail feathers produce a low pulsing whistle.

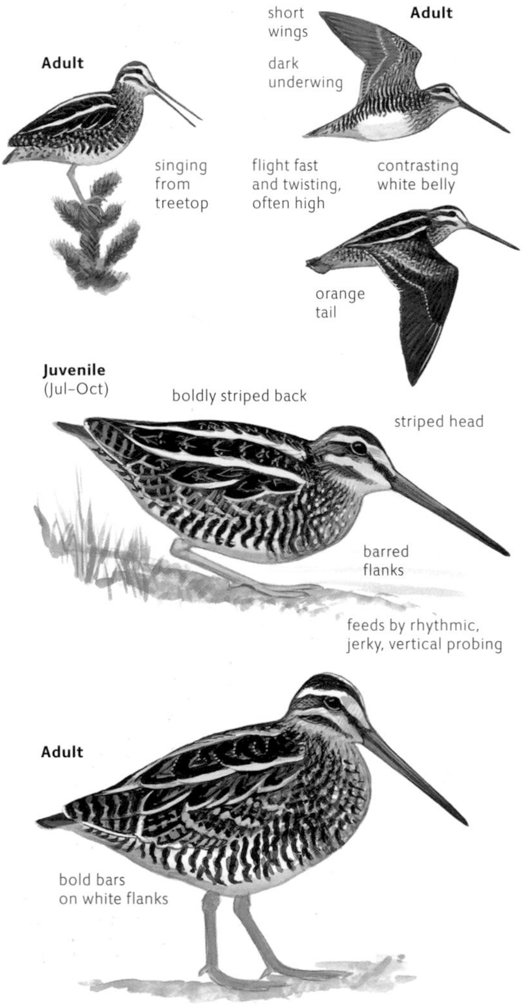

VOICE: Flight call a dry, harsh, scraping *scresh* or *kesh*. Display song from perch a loud *TIKa TIKa TIKa*... or *kit kit kit*.... In winnowing flight display outer tail feathers produce hollow, low whistle *huhuhuhuhuhuhuhuhuhu* very similar to Boreal Owl's song.

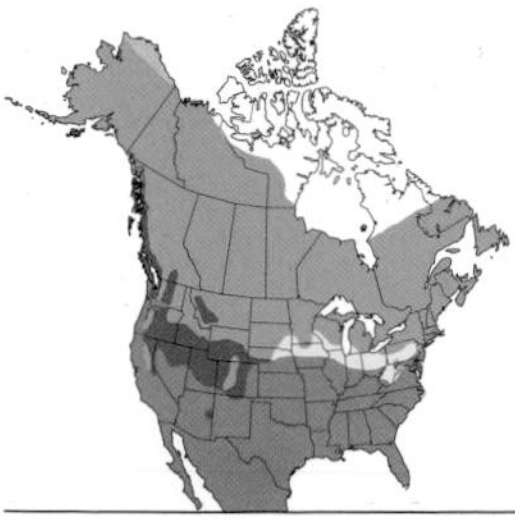

Uncommon and inconspicuous along grassy edges of freshwater ponds or among muddy stubble in flooded fields. Often seen flying high over these habitats then plunging down to land in cover. May form loose groups of up to ten or more, but essentially solitary and does not mix with other shorebirds.

Common Snipe

Gallinago gallinago

L 10.5"   WS 18"   WT 3.7 oz (105 g)   ♀>♂

Until recently considered the same species as Wilson's Snipe; ver similar in all respects. Best distinguished by white bar on unde wing coverts (visible only in flight), also averages broader whit tips on secondaries, and other minor differences in plumage.

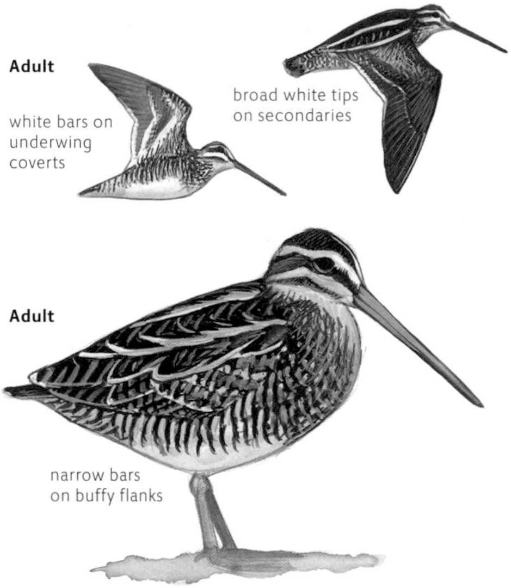

VOICE: Calls identical to Wilson's. Winnowing display produced by tail is lower-pitched with gruff, buzzing, nasal quality, reminiscent of bleating sheep; unlike the musical hooting sound of Wilson's.

Rare but regular visitor from Eurasia to islands of far western Alaska, where it has nested. Very rare elsewhere in North Americ in winter, recorded in California and Newfoundland.

Jack Snipe

ymnocryptes minimus

7.3" WS 15.5" WT 1.9 oz (54 g) ♀>♂

imilar to Wilson's Snipe but smaller and shorter-billed. More ecretive than other snipe; and tends to stay hidden, flushing only t close range, then flying low for a short distance and dropping ack into cover.

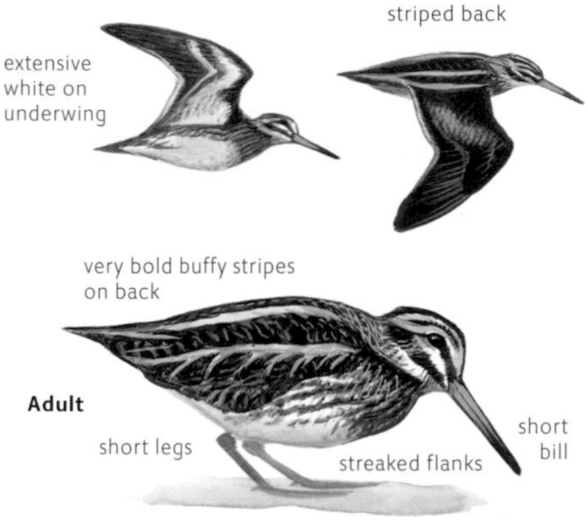

OICE: Usually silent when ushed, sometimes a harsh, uiet *kesh*.

Very rare visitor from Eurasia, recorded a few times along Pacific coast from Alaska to California, and once in Newfoundland, most records in late fall and winter. Habitat similar to Wilson's Snipe, but usually stays hidden in grassy vegetation.

Aerial Displays of Snipes and Woodcocks

Vinnowing flight display of Common Snipe is performed day or ght by birds circling high in the air, then suddenly diving down t high speed. The rush of air past the outspread outer tail feath-rs produces a low, pulsing, whistling sound. Flight display of merican Woodcock is performed at night over open, brushy elds. The movement of air over the narrow outer primaries roduces a high twitter. It begins with steady twittering as the rd rises, becoming well-spaced bursts of twittering while bird rcles at top of climb. Finally, as the bird plunges toward the round with sharp changes in direction, a series of louder and ore varied chirps results: *tewp tilp tiptooptip....*

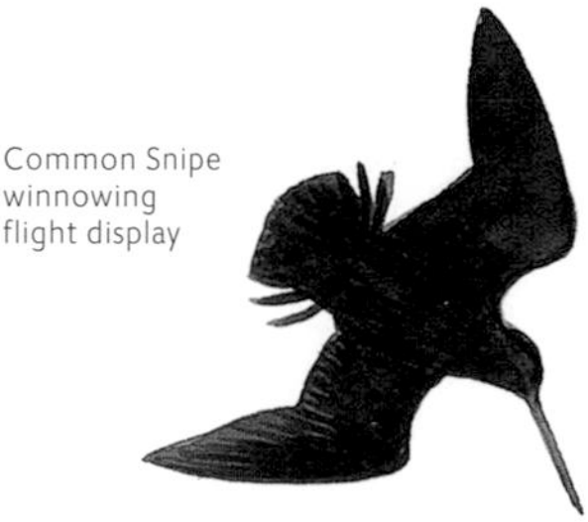

American Woodcock

Scolopax minor

L 11" WS 18" WT 7 oz (200 g) ♀>♂

Round body, long bill, large head, and uniform buffy underparts distinctive. Color and habits very different from snipe. When flushed the whistling wings and long bill are obvious and unique.

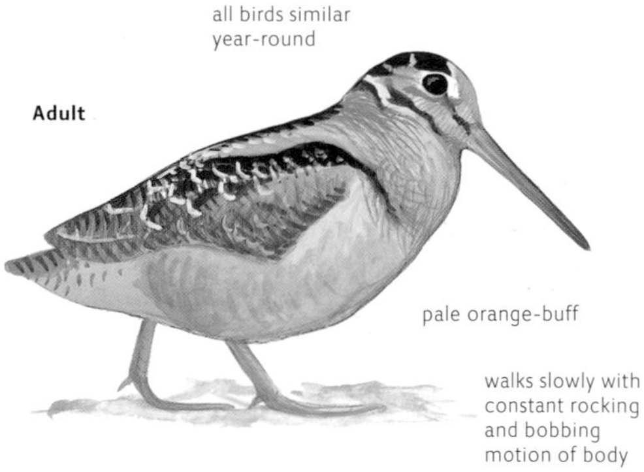

VOICE: Flight call absent. Wings produce a high twittering on takeoff and when making sharp turns in flight; higher-pitched and clearer than Mourning Dove. Displaying bird on ground gives explosive, very nasal *beent* similar to Common Nighthawk but level, less harsh.

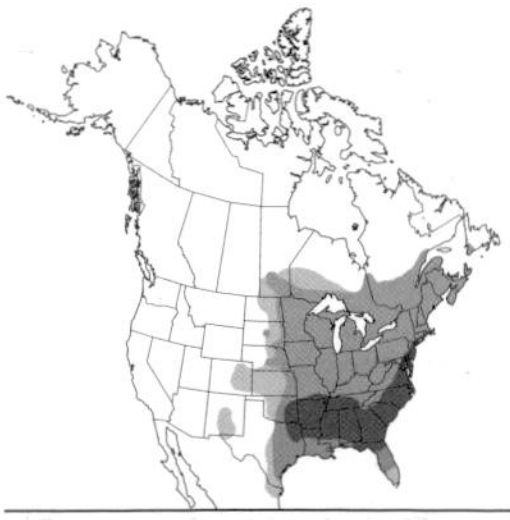

Uncommon and secretive on damp ground under dense cover within woods, where it is rarely seen except when flushed at close range. Displaying birds emerge into open grassy fields at dusk in spring. Secretive and solitary; rarely seen in daylight and never mixes with other shorebirds.

Wilson's Phalarope

Phalaropus tricolor

L 9.25" WS 17" WT 2.1 oz (60 g) ♀>♂

A very active sandpiper, with quick jerky actions. Elegant and dainty, with slender pointed bill. Juvenile and nonbreeding adult have very white face and underparts, cleaner than similar species, and pale yellow legs (dark on breeding birds).

VOICE: Flight call a low, muffled, nasal grunting or moaning *wemf, vint,* or *vimp;* soft but distinctive, reminiscent of muffled Black Skimmer.

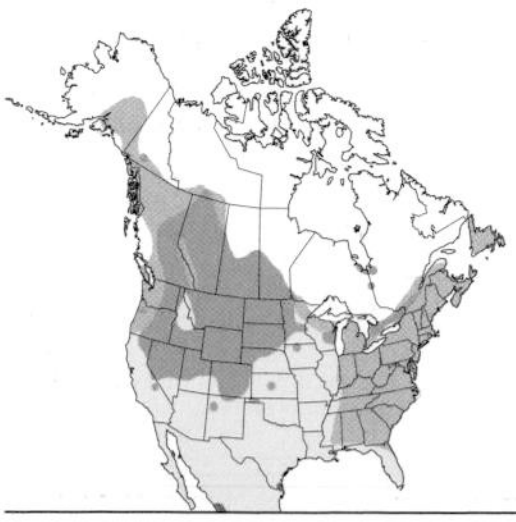

Common on shallow ponds within grassy marshes. Solitary or in small groups. Picks minute prey from water's surface while walking around muddy pond edges or swimming in shallow water.

Identifying Phalaropes

The three species in this genus have lobed toes and other distinc tive features and at times have been considered a separat family. Females, which are larger and more brightly colored tha males, leave after egg-laying, and the males incubate the egg and raise the young alone.

Phalaropes feed by picking minute food items from the su face of the water. They are very active and nervous, darting an jabbing constantly. They often swim (other sandpipers do so on occasionally) and when swimming often spin rapidly in tigh circles to create an upwelling that raises food items to the surfac

Red-necked Phalarope spinning

Stilt Sandpiper and Lesser Yellowlegs are superficially similar t Wilson's Phalarope in nonbreeding plumage but have very di ferent foraging actions.

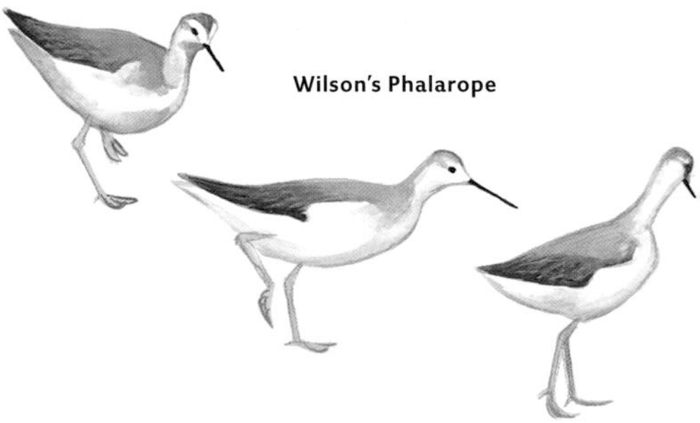

Wilson's Phalarope is very active, with nervous picking action often leaning, crouching, and darting; it is also brighter whi and pale gray (other species are drabber).

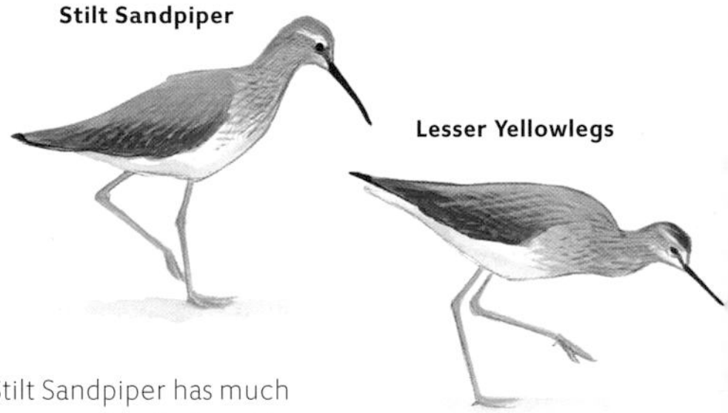

Stilt Sandpiper has much more methodical probing actions.

Lesser Yellowlegs is often quite active, dashing and jabbing at prey, but is more elegant and graceful and les erratic than Wilson's Phalarope.

ed Phalarope

halaropus fulicarius

.5" WS 17" WT 1.9 oz (55 g) ♀>♂

rger and longer-winged than Red-necked Phalarope, with icker bill. Rufous breeding plumage with white cheek distinc-e. Nonbreeding plumage paler than Red-necked.

OICE: Flight call a distinct, high *piik;* higher and clearer than ed-necked; like flat Long-billed Dowitcher; also a softer *dreet.* n breeding grounds a high, rising *sweeep,* a bit tinny and nasal start but finishing high and clear; also often a high, rough, uzzy *jeeer* and *pity pity pity pity.*

Uncommon. Nests in grass near high-Arctic tundra ponds. Migrates and winters in small flocks on open ocean; rarely seen from land or on inland lakes and ponds. Forages mainly by picking insect larvae and other tiny prey from water's surface while swimming.

Red-necked Phalarope

Phalaropus lobatus

L 7.75" WS 15" WT 1.2 oz (35 g) ♀>♂

The smallest phalarope, smaller and more compact than Wilson's, with needle-like bill and short dark legs. Breeding plumage shows distinctive white throat.

VOICE: Flight call a short, hard *kett;* sharper, higher, and harder than Sanderling; flatter than Red; at times strongly reminiscent of icterids (e.g., Brewer's Blackbird). Various other buzzy calls given on breeding grounds.

Common locally. Nests in grass near tundra ponds. Migrates and winters in small flocks on open ocean along lines of floating weeds and debris; generally uncommon to rare inland, but massive numbers gather at certain alkaline lakes in the west in fall. Forages mainly by picking insect larvae and other tiny prey delicately from water's surface while swimming.

Gulls

FAMILY: LARIDAE

27 species in 8 genera (1 rare species not shown here). Terns and skimmers (see page 224) are also in the family Laridae. Generally conspicuous and gregarious species found in open areas, usually near or on water, and often attracted to dumps, dams, parking lots, fishing piers, and other manmade concentrations of food. Genus *Larus*, with 16 species, includes the confusingly similar large white-headed species, which are omnivorous and mix freely where food is abundant. Smaller species, most with dark hoods in breeding plumage, are more specialized, sometim found in large flocks but not mixing freely with large specie Many species are difficult to identify and some hybridize fr quently. Transition from brownish juvenal plumage gray-and-white adult plumage takes two to four years, and inte mediate stages add to the identification challenges. Most nest colonies, on open ground. Nonbreeding adults are shown.

Genus *Rissa*

Black-legged Kittiwake, page 199

Red-legged Kittiwake, page 199

Genus *Pagophila*

Ivory Gull, page 220

Genus *Xema*

Sabine's Gull, page 198

Genus *Chroicocephalus*

Bonaparte's Gull, page 201

Black-headed Gull, page 201

Genus *Hydrocoloeus*

Little Gull, page 200

Genus *Rhodostethia*

Ross's Gull, page 198

Genus *Leucophaeus*

Laughing Gull, page 203

Franklin's Gull, page 202

ack-tailed Gull,
ge 220
Heermann's Gull,
page 219
Mew Gull,
page 204
Ring-billed Gull,
page 206
alifornia Gull,
age 207
Herring Gull,
page 208
Lesser Black-
backed Gull,
page 217
Great Black-
backed Gull,
page 218
ayer's Gull,
ge 210
Iceland Gull,
page 211
Glaucous Gull,
page 212
Glaucous-
winged Gull,
page 213
aty-backed Gull,
ge 214
Western Gull,
page 215
Yellow-
footed Gull,
page 216
Kelp Gull,
page 221

Ross's Gull

Rhodostethia rosea

L 13.5" WS 33" WT 6 oz (180 g)

A small gull, with very small bill, relatively long wings, and wedge-shaped tail. Juvenile has bold M pattern above. On adult note uniform pale gray upperwing and gray underwing. The pink suffusion shown by most is related to diet and can be shown (faintly or intensely) by individuals of other small gull species.

VOICE: High and melodious. Some calls with hollow barking quality like Black-legged Kittiwake; soft, mellow barking *p-dew* or *prrew* and tern-like *kik-kik-kik-kik-kik*. Generally silent in winter.

Rare visitor in northern Alaska and one or two nesting pairs occur annually at Churchill, Manitoba; only a few scattered records elsewhere, most often among flocks of Bonaparte's Gulls.

Sabine's Gull

Xema sabini

L 13.5" WS 33" WT 6 oz (180 g) ♂>♀

A small, delicate, tern-like gull. Dark gray or brownish abo[ve.] Boldly patterned upperwing is distinctive, similar pattern is a[lso] shown by juvenile Kittiwakes (and other species) but no[te] Sabine's darker head and back, smaller size.

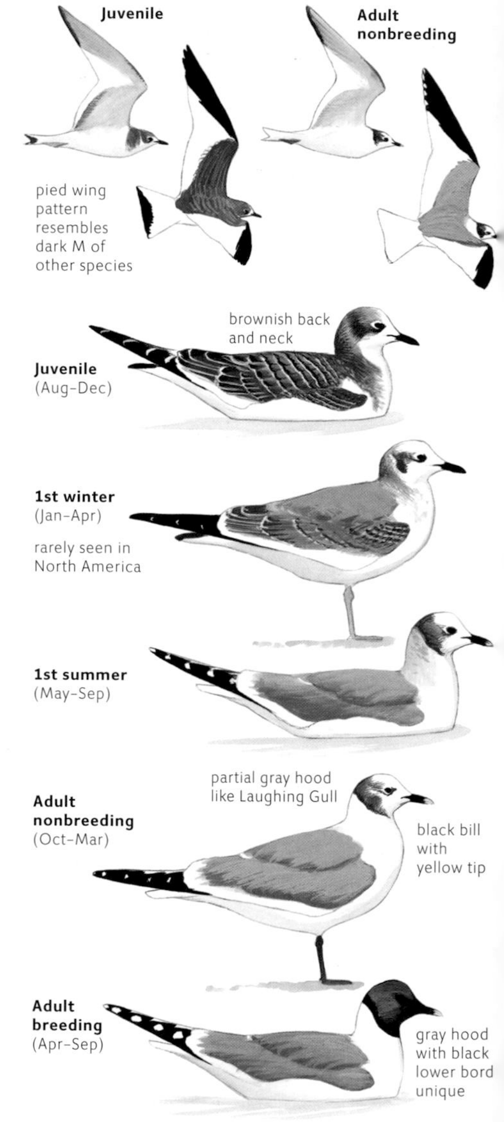

VOICE: Grating, buzzy or trilling tern-like *kyeer, kyeer...* or gra[t]ing *krrr*. Juvenile call a high, clear trill *dedededededer* sligh[tly] descending at end.

Uncommon. Nests on tundra ponds. Migrates mainly off Pacific coast, rarely seen from land; and migrants rarely appea[r] on inland ponds, lakes or rivers. Solitary or in small groups, not mixed with other gulls.

lack-legged Kittiwake

ssa tridactyla

17" WS 36" WT 14 oz (400 g) ♂>♀

small gull, with relatively long and narrow wings, quick and stiff ngbeats. Juvenile has bold M pattern above, dark collar. Adult le gray above, with paler primaries and neat black triangle on ingtips, unmarked yellow bill, and black legs.

ICE: Hollow, nasal quality like Heermann's Gull. Rhythmic, peated *kitti-weeeik*...given frequently at or near nest site, often chorus. Generally silent in winter.

Common. Almost exclusively seen on open salt water. Nests in large colonies on steep cliffs above ocean, often with murres and fulmars; scavenges and often attracted to fish-processing plants or boats. The only gull commonly seen plunge-diving head-first like terns.

Red-legged Kittiwake

Rissa brevirostris

L 15" WS 33" WT 13 oz (380 g) ♂>♀

A small gull similar to Black-legged; note much shorter bill, shorter red legs, darker gray upperside, and gray underwing. Immature has white triangle on wings similar to Black-legged and to Sabine's Gull, lacks dark tail band.

VOICE: Much higher than Black-legged; a high, falsetto squeal *suWEEEEr* repeated. Generally silent in winter.

Common within limited range. Nests in colonies on cliff ledges, usually on taller cliffs and narrower ledges than Black-legged. Usually in small groups. Forages for larval fish mainly at night over open ocean.

Identification of Gulls

Gull identification represents one of the most challenging and subjective puzzles in birding and should be approached only with patient and methodical study. A casual or impatient approach will not be rewarded.

Identification problems among the large gulls are compounded by wide variation in virtually every aspect of their appearance.

Males are larger than females, with larger and deeper bills. Juveniles have thinner bills than adults that gradually thicken over several years.

Head shape varies between individuals and also changes with age, weather conditions, and position. Apparent head shape can also be influenced dramatically by facial markings and bill color and size.

Variation in plumage falls into three main categories.

- AGE VARIATION The long period of immaturity (up to four years) of large gulls means that some species go through as many as eight different plumage stages, becoming gradually more adultlike.
- WEAR AND FADING After each molt the feathers gradually change due to abrasion and bleaching until they are finally replaced by the next molt.
- INDIVIDUAL VARIATION Even when comparing birds of the same age and stage of molt, there is tremendous variation in the color and pattern of the plumage, as well as the colors of bill, legs, and eyes.

These two Herring Gulls are the same age and plumage stage, at the end of their 1st winter, approaching one year old. The upper bird is much paler, presumably because its feathers contain less melanin, which has made them more susceptible to gradual wear and fading. These effects are particularly obvious in spring and summer. The differences in overall darkness of these two individuals would have been less obvious a few months earlier, when the feathers were newer, and the plumage of adults shows very little individual variation at any season.

Little Gull

Hydrocoloeus minutus

L 11" WS 24" WT 4.2 oz (120 g)

Our smallest gull, with relatively broad, rounded wings. Juven has bold M pattern above, dark secondaries, and smudgy ca On adult note pale gray upperwing (without black tips) and da gray underwing with broad white border.

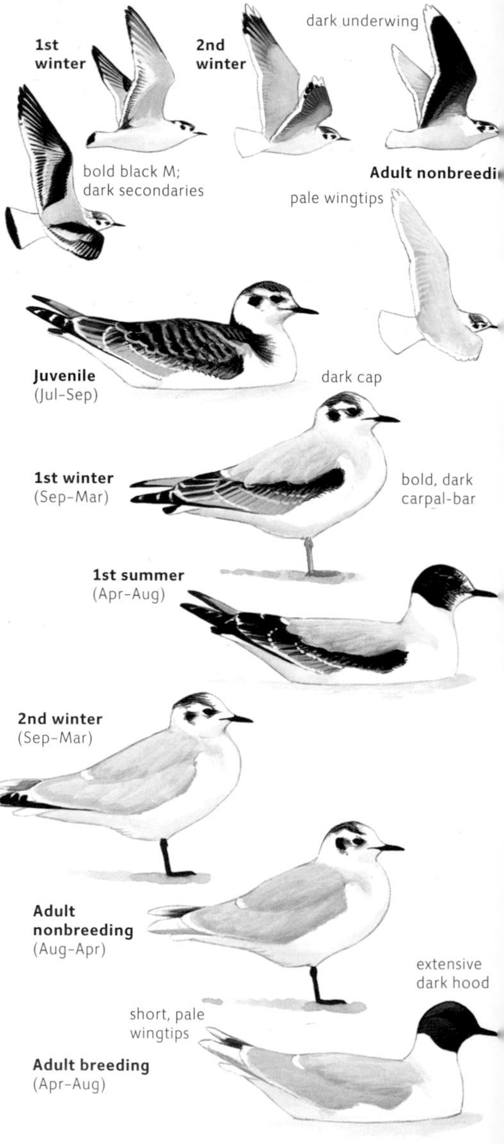

VOICE: Grating or clear, nasal, most often a short *kek* like Bla Tern. In display flight a clear, nasal *teew* and long series *tew te tikik tikik tikik tikeew tikeew....*

Rare, but gathers in small numbers at a few sites in easter Great Lakes and Chesapeake Ba area; usually around expansive open water, even far offshore o open ocean; often joins flocks o Bonaparte's Gulls, or roosting with terns. Has nested in grassy marshes in Canada.

Bonaparte's Gull

hroicocephalus philadelphia

13.5" WS 33" WT 7 oz (190 g)

ur smallest commonly seen gull; differs from terns by broader ings, shorter bill, and details of plumage, often swims and rarely overs. Note thin black bill, pink to red legs, and white outer rimaries.

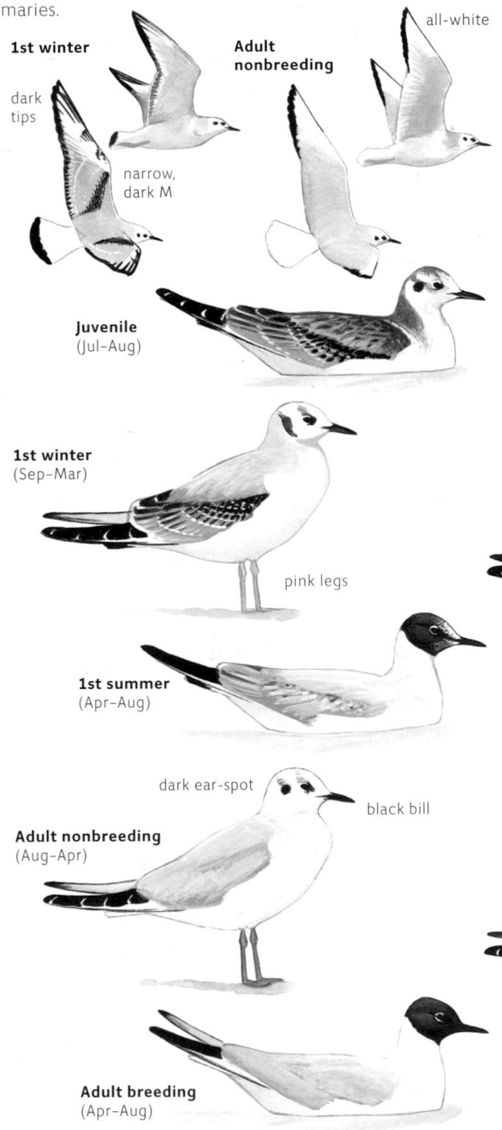

OICE: Low, wooden, grating or rasping *gerrrr* or *reeek*; tern-like ut lower-pitched than most terns. Also clear *kew* notes from lock. Juvenile gives high, somewhat nasal squeal *peeeur*.

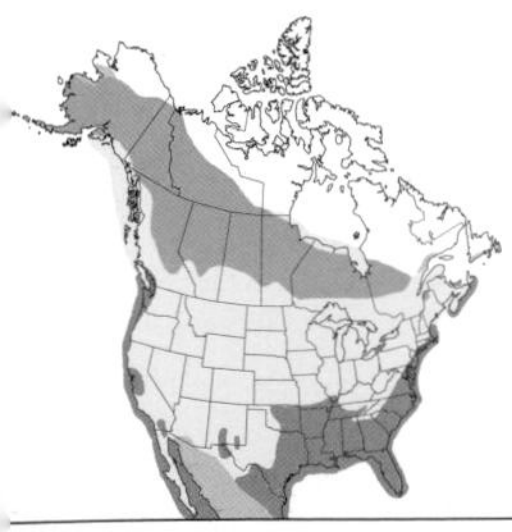

Common locally on lakes, rivers, and ocean, especially where strong currents stir the water; always around water and usually does not mix with other gulls. Gathers in flocks (sometimes hundreds) where food is abundant. Forages for small fish and crustaceans primarily by picking prey from water's surface.

Black-headed Gull

Chroicocephalus ridibundus

L 16" WS 40" WT 9 oz (270 g)

Similar in all plumages to Bonaparte's, but slightly larger, with heavier reddish bill and dark underside of primaries. Also note paler gray upperside, dark red legs, less extensive and more brownish hood in breeding plumage.

VOICE: Harsh grating similar to Bonaparte's but lower, with richer quality like Laughing Gull.

Uncommon and local around inlets and harbors along Canadian Atlantic coast; rare to very rare elsewhere. Usually seen singly among flocks of Bonaparte's Gulls, but sometimes joins Ring-billed or Laughing Gulls and may forage or roost with them in open fields (which Bonaparte's does very rarely).

Franklin's Gull

Leucophaeus pipixcan

L 14.5" WS 36" WT 10 oz (280 g) ♂>♀

Similar overall to Laughing Gull; slightly smaller with shorter wings and bill. Also distinguished by mostly white underwing, smaller black tips on primaries. In nonbreeding plumages note very neat half hood and white nape.

1st winter

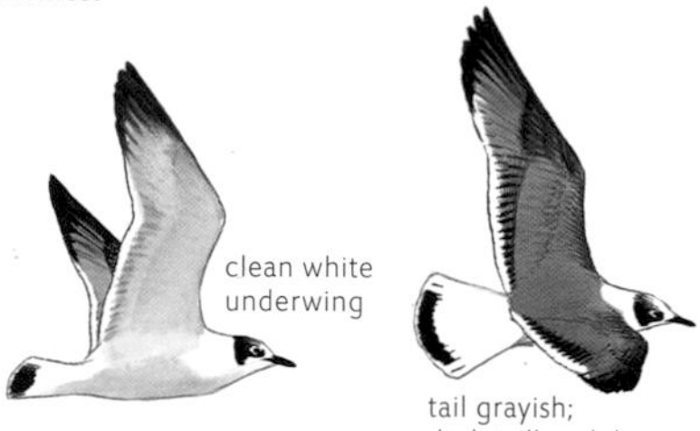

1st summer

Adult nonbreeding

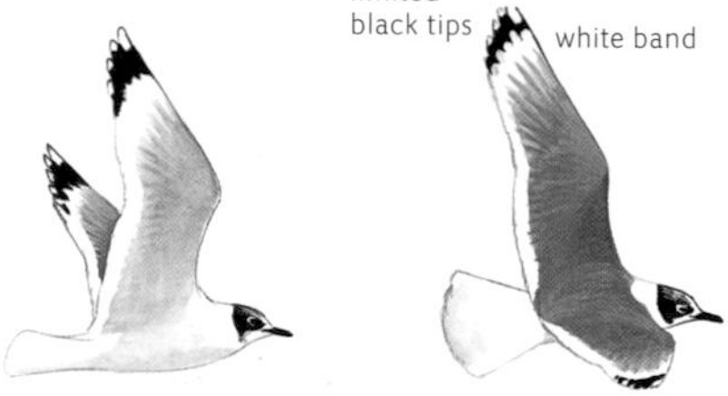

VOICE: Voice nasal and laughing but hollow-sounding and less penetrating than Laughing Gull. Common call a short, hollow *kowii* or *queel*. Long call descending and accelerating, each note rising; the series with much greater pitch change than Laughing Gull.

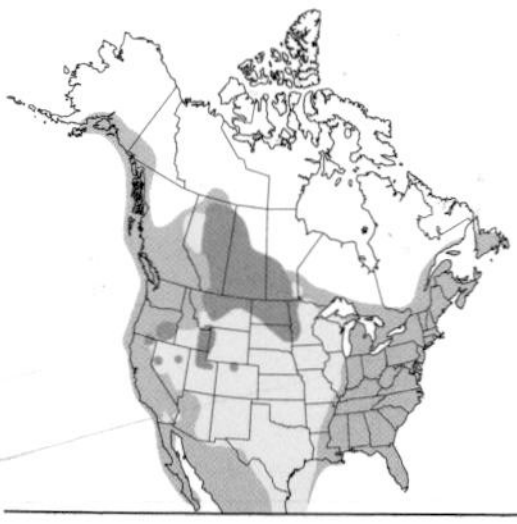

Uncommon on beaches, lakes, and farmland. Often in small flocks.

Juvenile
(Aug–Sep)

1st winter
(Sep–Apr)

1st summer
(Apr–Aug)

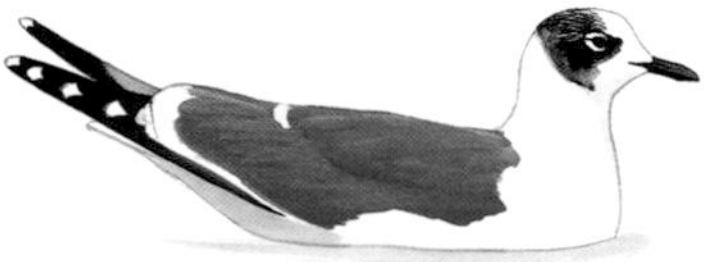

2nd winter
(Sep–Apr)

Adult nonbreeding
(Aug–Mar)

Adult breeding
(Apr–Aug)

This is the only gull that undergoes two complete molts each year. Thus, the plumage is always fresh (other gulls molt in late summer and appear worn in the following summer); compare 1st summer with same age Laughing Gull.

aughing Gull

ucophaeus atricilla

6.5" WS 40" WT 11 oz (320 g) ♂>♀

latively small (although the largest of the dark-hooded gulls), th very long wings and long drooping bill. Note dark gray perside blending onto extensive black wingtips, dark reddish blackish bill and legs.

1st winter

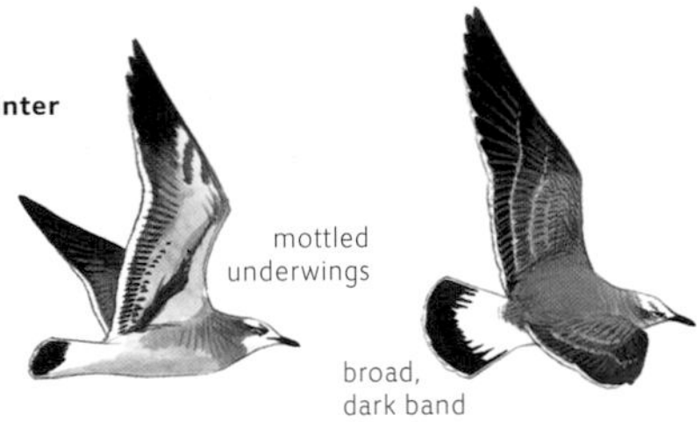

2nd winter

Adult nonbreeding

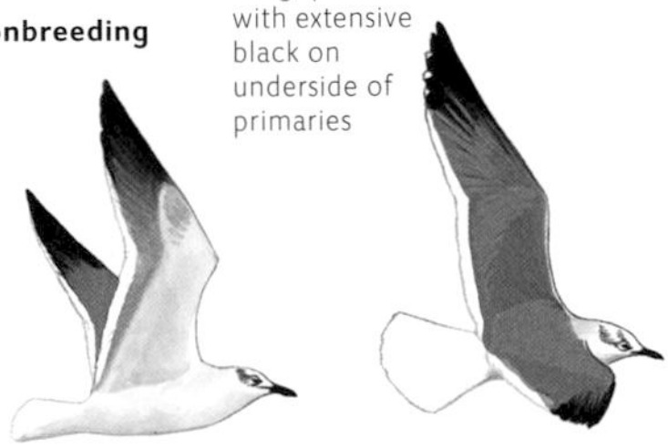

OICE: Nasal laughing; common call of adults a two-syllable ugh *kiiwa* or *kahwi*. Long call a laughing series, an iconic sound f southeastern shores, rapid then slowing.

Common and conspicuous; the most frequently seen gull along southern Atlantic and Gulf coast beaches, also on the Salton Sea, California, in summer and fall; rare elsewhere. Found mainly on sandy, saltwater beaches and marshes, often with other gulls. An avid taker of handouts; may gather in large numbers at parks and parking lots where food is offered.

Juvenile
(Aug–Nov)

1st winter
(Oct–Mar)

1st summer
(Mar–Aug)

2nd winter
(Sep–Mar)

Adult nonbreeding
(Sep–Mar)

Adult breeding
(Mar–Sep)

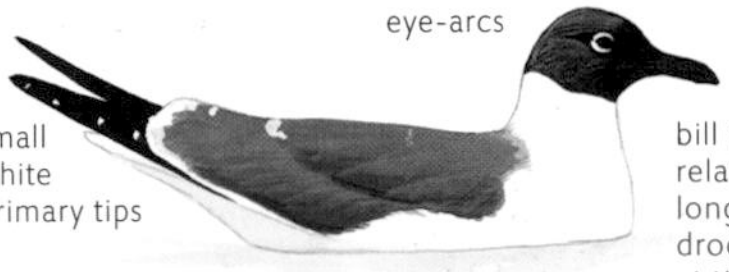

Mew Gull

Larus canus

L 16" WS 43" WT 15 oz (420 g) ♂>♀

Similar in all plumages to Ring-billed Gull. Averages smaller, with shorter legs and smaller bill. On adult note dark iris, slightly darker gray upperside, and unmarked or only faintly ringed yellow bill. First-year birds gray-brown overall with mostly dark tail.

Short-billed (American)

1st winter

2nd winter

Adult nonbreeding

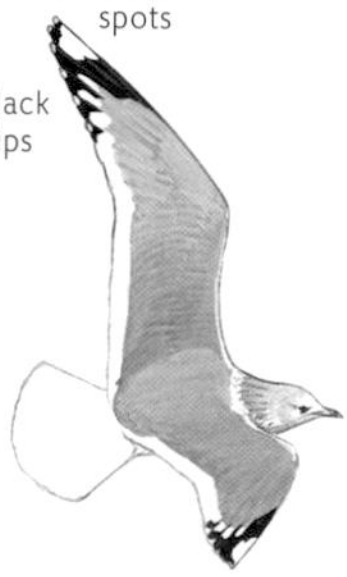

VOICE: Voice high and squealing with strident nasal quality, less harsh and more mewing than Ring-billed. Long call falsetto, ending with rapid series of short notes.

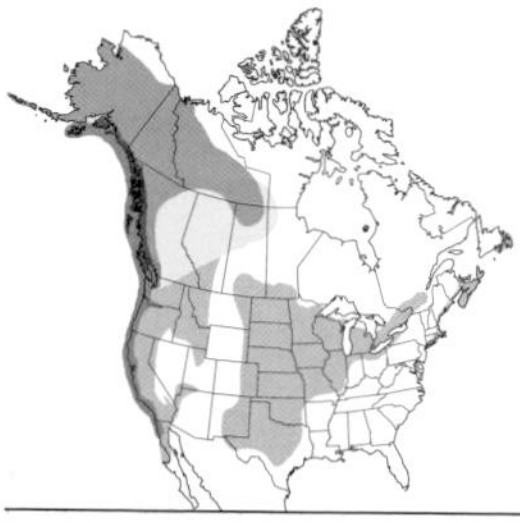

Common. Nests at ponds and rivers in boreal forests. Winters mainly along ocean beaches and coastal ponds.

Juvenile (Jul–Sep)

1st winter (Oct–Apr)

1st summer (Apr–Aug)

2nd winter (Aug–Apr)

Adult nonbreeding (Sep–Apr)

Adult breeding (Apr–Sep)

ommon (European)

subspecies of Mew Gull. Adults differ in having more black on e wingtips, slightly paler gray upperside, more distinct head reaking. Immature has cleaner and more contrasting pattern an Short-billed, white tail with well-defined band, mostly whit- n underwing.

1st winter

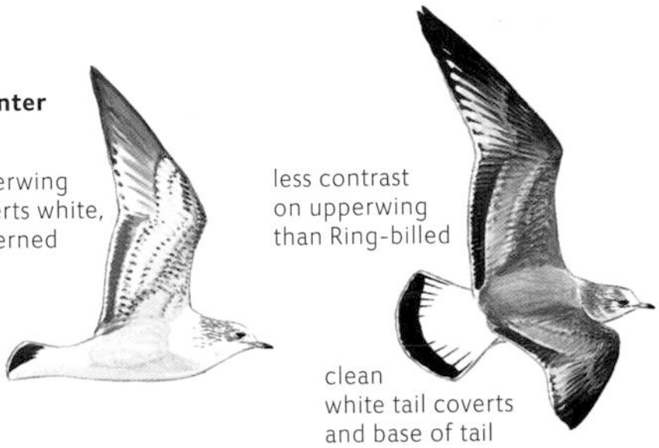

2nd winter

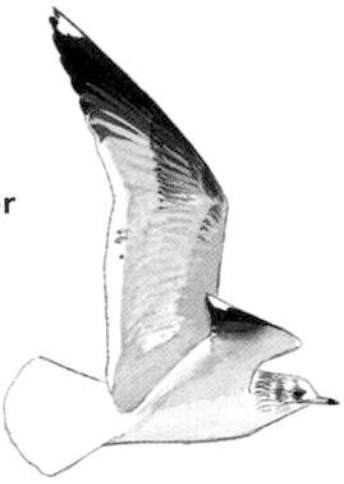

Adult nonbreeding

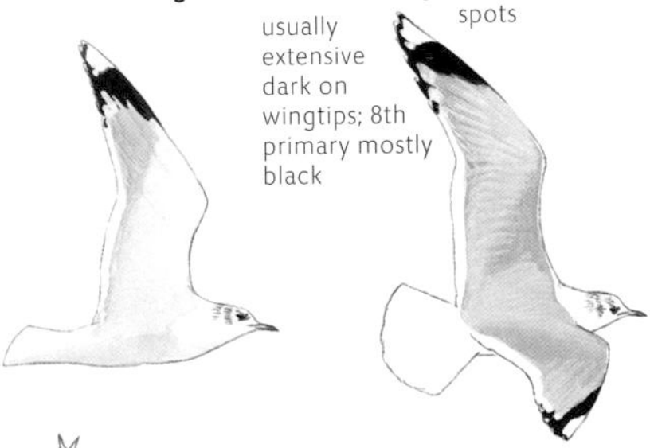

VOICE: Voice very similar to American, but may average lower-pitched and with more hollow sound; more mewing than Ring-billed, higher than Herring. Long call begins with high squeals, then rapid phrases about 4 per second.

Very rare visitor from Eurasia. A few are seen annually in coastal ponds and bays from Newfoundland to Massachusetts, mainly Nov–Mar. Mew Gulls from Siberia are recorded annually in far western Alaska mostly in spring and summer, with a few probable winter records south and east to Massachusetts.

Juvenile
(Jul–Sep)

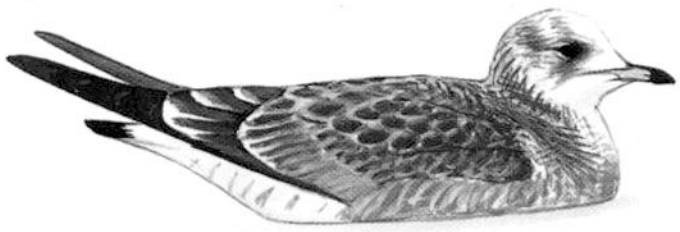

1st winter
(Sep–Apr)

1st summer
(Apr–Aug)

2nd winter
(Aug–Apr)

Adult nonbreeding
(Sep–Apr)

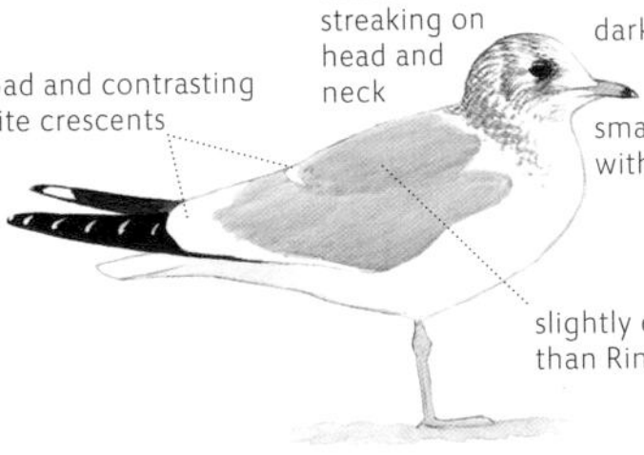

Adult breeding
(Apr–Sep)

Siberian population (called Kamchatka Gull) differs subtly from European, slightly larger and heavier, immature more heavily marked, but identification is very difficult.

Ring-billed Gull

Larus delawarensis

L 17.5" WS 48" WT 1.1 lb (520 g) ♂>♀

The smallest of the common white-headed gulls, with relatively short bill and long slender wings. On adult note very pale gray back without contrasting white tertial crescent, dark-ringed yellow bill, yellow legs, and sharply contrasting black wingtips. Juvenile has mostly white underside and rump with dark tail-band.

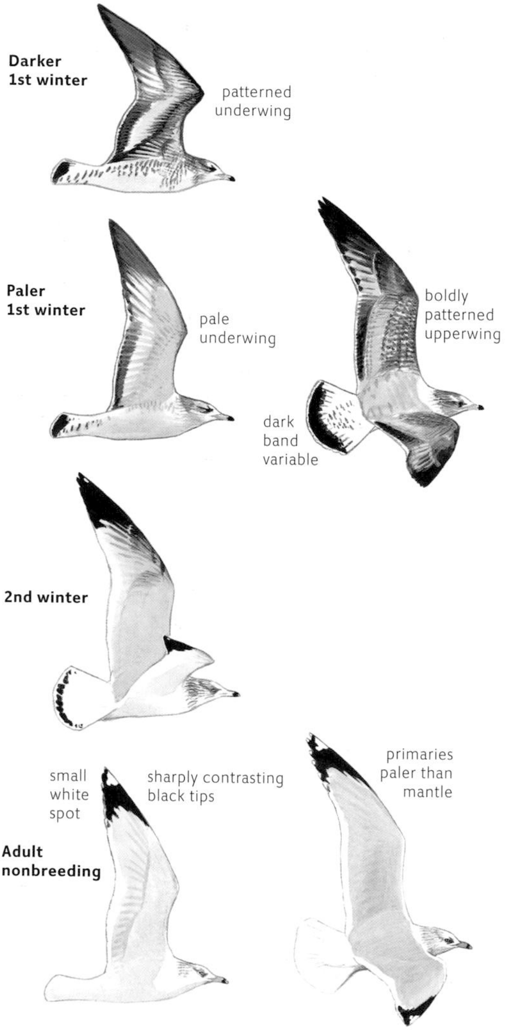

VOICE: High and rather hoarse with wheezy, scratchy quality; higher than California Gull; higher, harsher, less nasal than Mew. Long call level and rather slow; begins with long, high squeals, then rapid phrases (about 3 per second); ends with long, slurred notes. Also a high *kuleeeeuk* repeated, and *kleeeea* or *k-heeer*. Flight call a high, thin *keeel*.

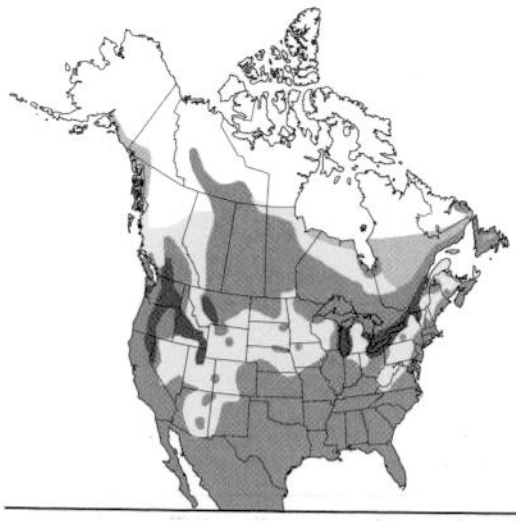

Common; our most widely seen gull. Found on all bodies of water from small lakes and rivers to ocean; often also seen foraging on agricultural land or loitering around restaurants, parking lots, and city parks looking for handouts. In small groups where food is plentiful.

;alifornia Gull

ırus californicus

21" WS 54" WT 1.3 lb (610 g) ♂>♀

termediate in size between Ring-billed and Herring Gulls with latively longer wings; adult has darker gray back with extensive ack on wingtips, dark eye, greenish-yellow legs. First-year birds ostly brownish like Herring or Western, distinguished with care / shape, details of plumage and bill color.

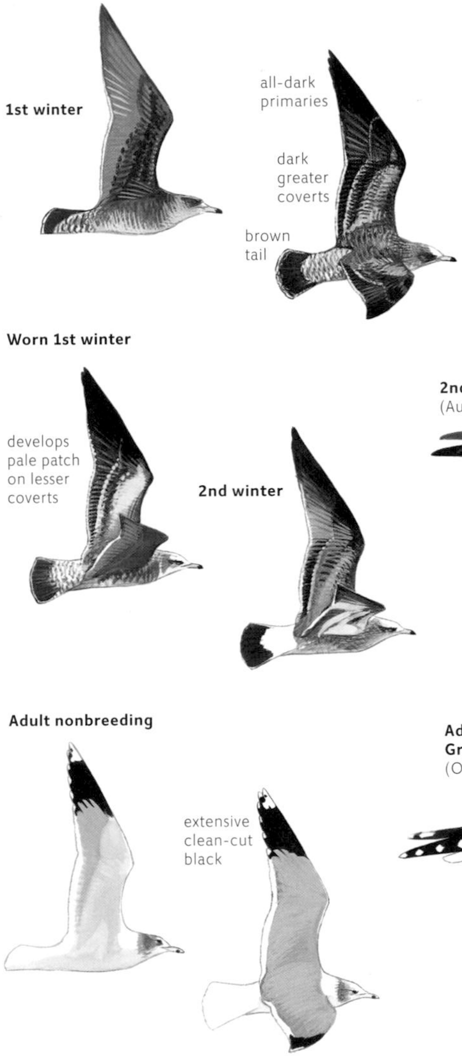

'OICE: Hoarse, scratchy, somewhat deeper than Ring-billed; arsher and higher than Herring; never clear tones. One com- non single note *gaaal* similar to Great Black-backed. Long call apid, high, wheezy.

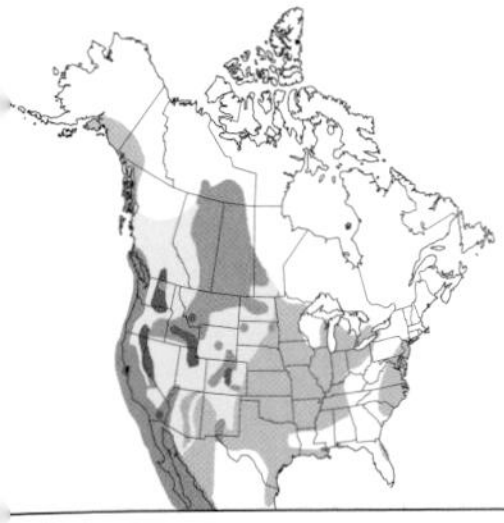

Common on ponds, lakes, and coastlines, often with Ring-billed or Herring Gulls. Habitat and habits similar to Ring-billed.

Northern Great Plains breeders average larger than Great Basin, and adults average paler-mantled, approaching Herring Gull. These differences are broadly overlapping, however, and not safely identifiable.

Herring Gull

Larus argentatus

L 25" WS 58" WT 2.5 lb (1,150 g) ♂>♀

In most areas (except on Pacific coast), the most numerous large gull; on adult note pale gray upperparts and black wingtips. First-year birds mostly brownish. Hybridizes extensively with Glaucous-winged and Glaucous Gulls where range overlaps.

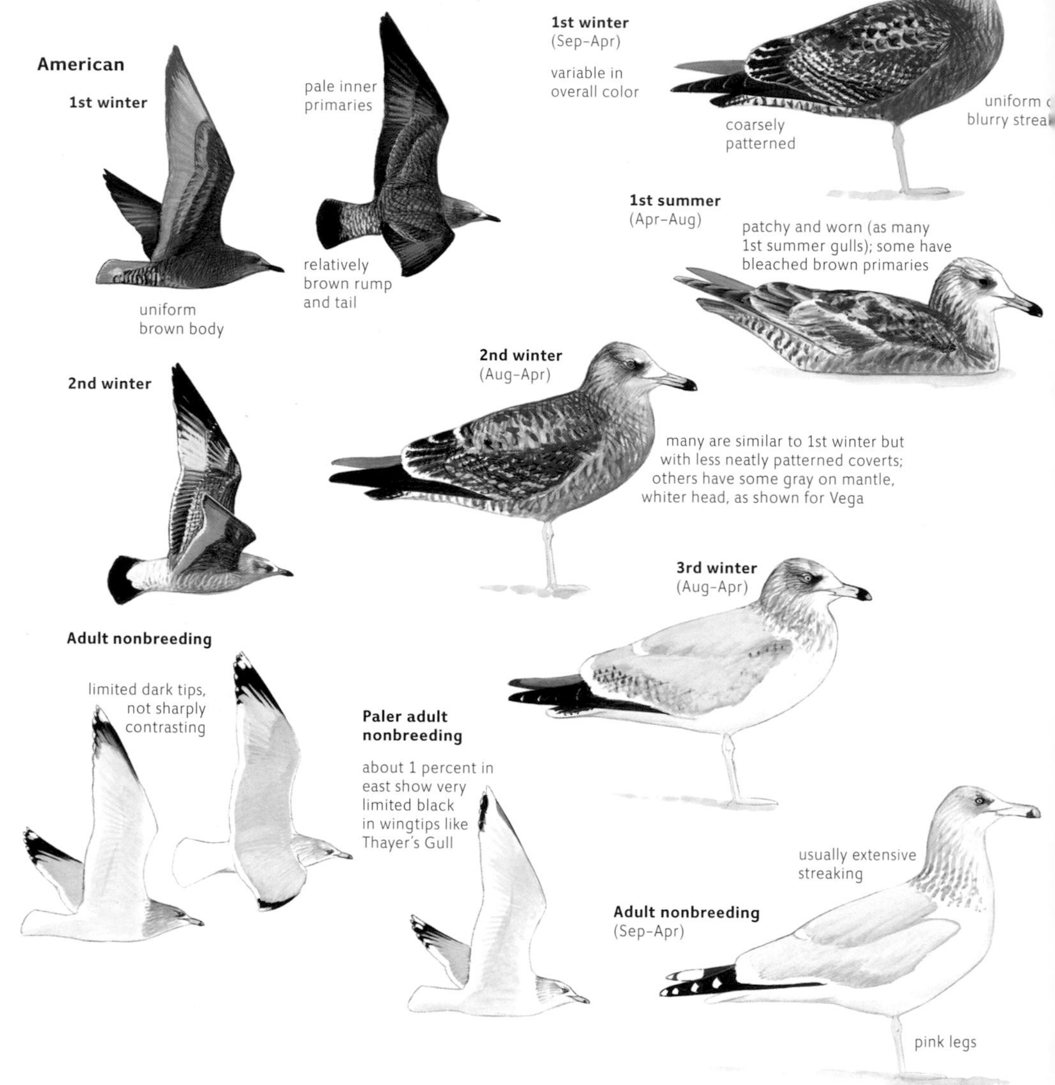

VOICE: Clear, flat bugling. Long call higher, clearer, and more two-syllabled than Western; also single notes *klooh, klaaw,* and short, low, hollow *kaaw.* Flight call trumpeting, lower than Ring-billed.

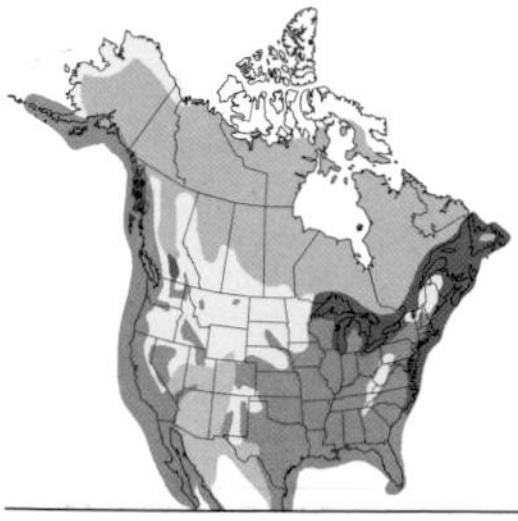

Common along coast; common to uncommon at larger bodies of water inland; only locally common in west. Often roosts on beaches, in parking lots, open fields, and similar areas. Like other large, white-headed gulls gathers in large groups where food is abundant. Forages for fish and other animal prey, largely scavenged, primarily on the water and at garbage dumps.

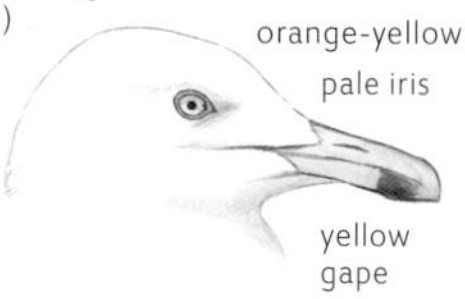

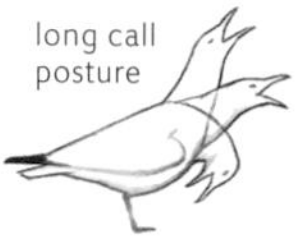

Vega (Siberian)

Siberian subspecies (called Vega Gull) is a rare but regular visitor to western Alaska. Adults have dark iris and slightly darker mantle than American, immatures average paler than American with contrasting pale rump

Juvenile
(Jul–Sep)

1st winter

paler gray inner primaries

broad, dark tail-band

mostly white rump and tail coverts

1st winter
(Sep–Apr)

barred scapulars

checkered coverts

streaked

averages paler and more crisply marked than American with distinct tail-band, but some American birds have similar markings

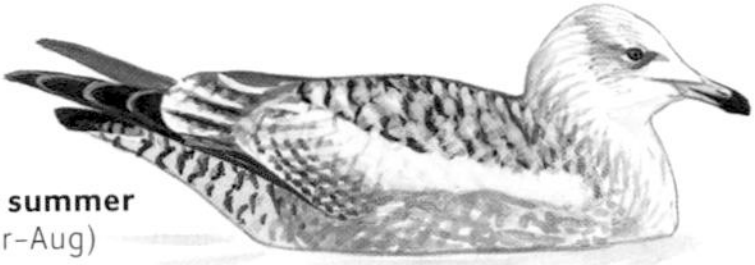

1st summer
(Apr–Aug)

often very pale with dark outer primaries and tail

2nd winter

2nd winter
(Aug–Apr)

barred coverts

Adult nonbreeding
(Sep–Apr)

distinguished from American by darker mantle and different bare-parts colors

Adult nonbreeding

white subterminal spots usually obvious

Adult breeding
(Apr–Sep)

orange-red orbital ring

brownish-yellow iris

yellow gape

European

A rare visitor to Newfoundland and perhaps farther south on our Atlantic coast. Averages smaller and sleeker (less bulky) than American birds, but much overlap. Adults indistinguishable from American.

1st winter

like Vega overall; paler and more crisply marked than American, with even narrower dark tail-band, but all these features matched by some American birds

1st winter
(Sep–May)

VOICE: Some short call notes hoarser than American. Long call posture like American.

Thayer's Gull

Larus thayeri

L 23" WS 55" WT 2.2 lb (1,000 g) ♂>♀

Identification is complex and must be based on a combination of details of size, shape, plumage, etc. Averages slightly smaller and paler than Herring Gull, slightly larger and darker than Iceland Gull; matched in many features by Herring × Glaucous-winged hybrids.

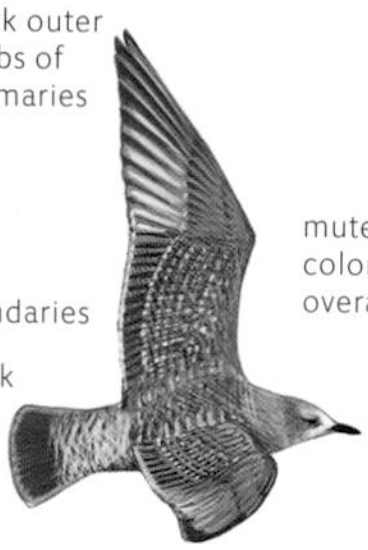

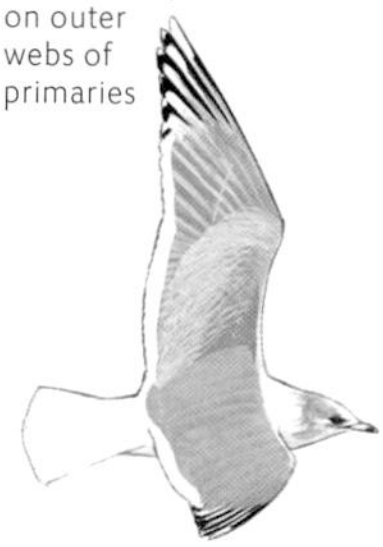

VOICE: Lower and flatter than Herring. Long call muffled, lower than Herring Gull; notes simple and flat, monotone. Rarely heard in winter.

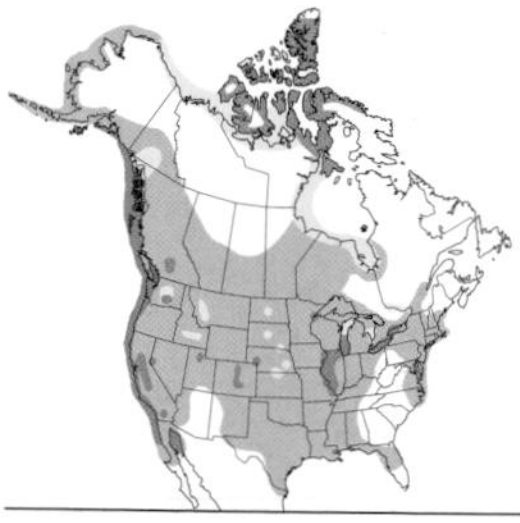

Uncommon on breeding ground, and along Pacific coast, rare elsewhere. Habitat and habits similar to Herring Gull. Usually in small numbers mixed with other large gulls, but gathers in large numbers at favored locations.

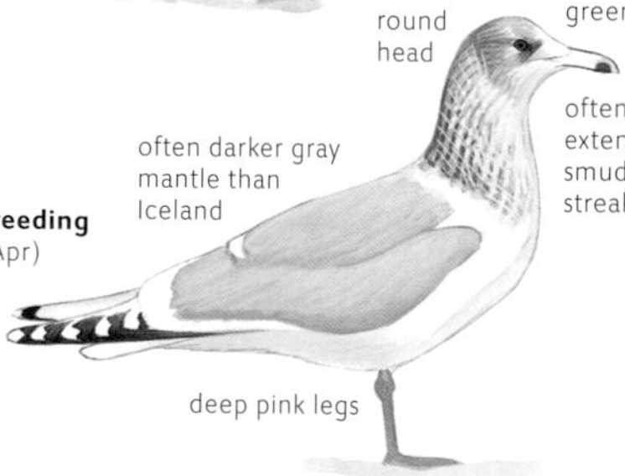

Iceland Gull

Larus glaucoides

L 22" WS 54" WT 1.8 lb (820 g) ♂>♀

Overall similar to Herring Gull, but averages slightly smaller and more dainty, with mostly white wingtips. Wingtip color ranges from all-white to extensively dark (blackish on adult, brown on immature); darker individuals overlap with Thayer's Gull.

VOICE: Similar to Herring but said to be more shrill; presumably similar to Thayer's. Rarely heard in winter.

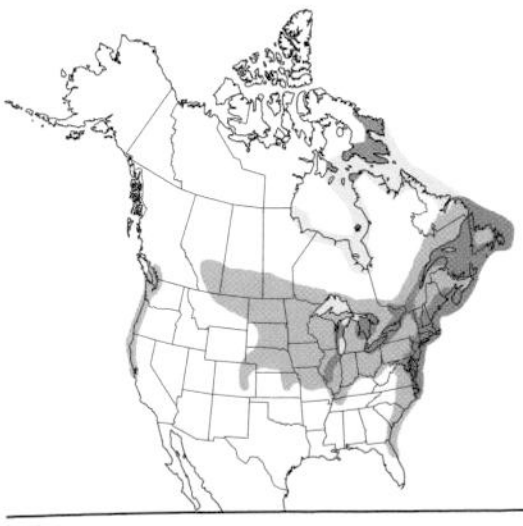

Nests in Arctic. Common in winter around Newfoundland and Gulf of Saint Lawrence; rare south of Lake Erie and Massachusetts, very rare visitor elsewhere. Gathers in large groups where food is abundant and generally found with other large gulls.

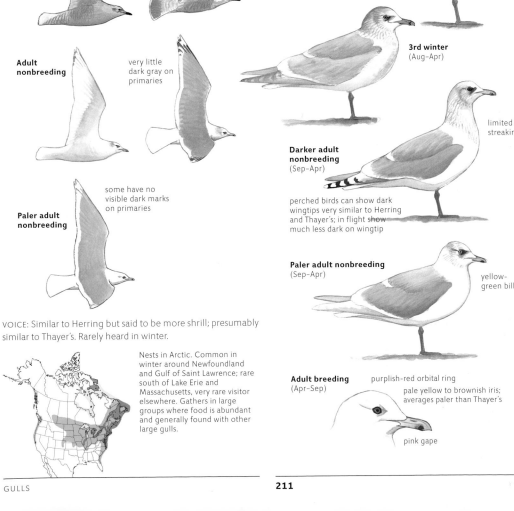

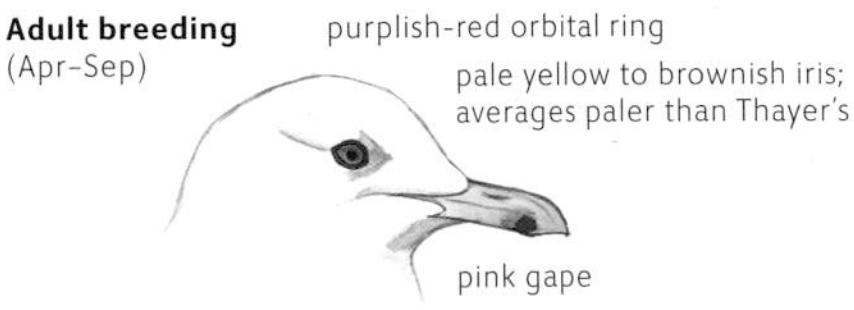

Glaucous Gull

Larus hyperboreus

L 27" WS 60" WT 3.1 lb (1,400 g) ♂>♀

One of our largest gulls, with heavy bill. Always pale: juvenile pale brown or whitish overall, with wingtips paler than body and with sharply bicolored bill. Adult paler gray above than Herring, with white wingtips. Hybridizes extensively with Herring and Glaucous-winged Gulls where range overlaps.

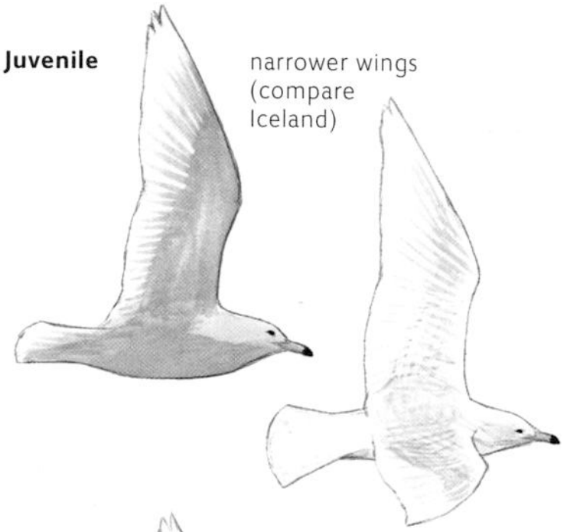

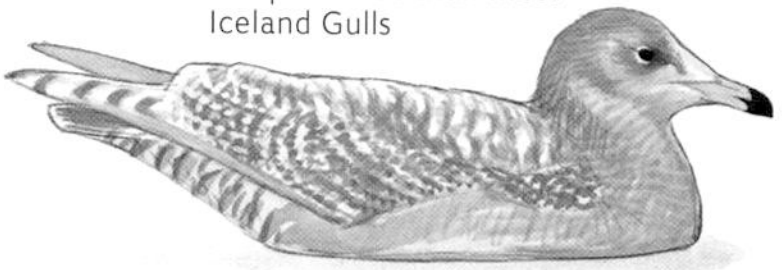

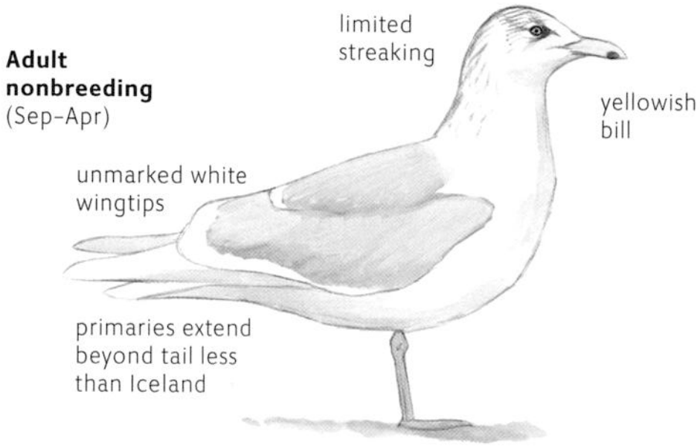

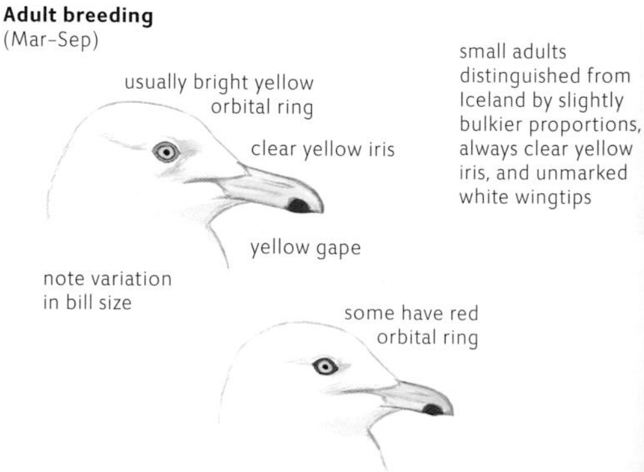

VOICE: Long call similar to Herring but slightly hoarser; lower-pitched than Iceland. Call note *k-lęee* sharply two-part, with clucking first syllable, distinctive; higher and weaker than Herring.

Common in northern and western Alaska; uncommon to rare elsewhere. Habitat and habits similar to Herring Gull. Usually in small numbers among other large gulls.

Glaucous-winged Gull

Larus glaucescens

L 26" WS 58" WT 2.2 lb (1,000 g) ♂>♀

Size and shape varies from stocky and large-billed like Western Gull to more slender and delicate like Herring Gull. Note relatively unpatterned and uniform plumage, with wingtips about the same color as back. Hybridizes extensively with Western, Herring, and Glaucous Gulls where range overlaps.

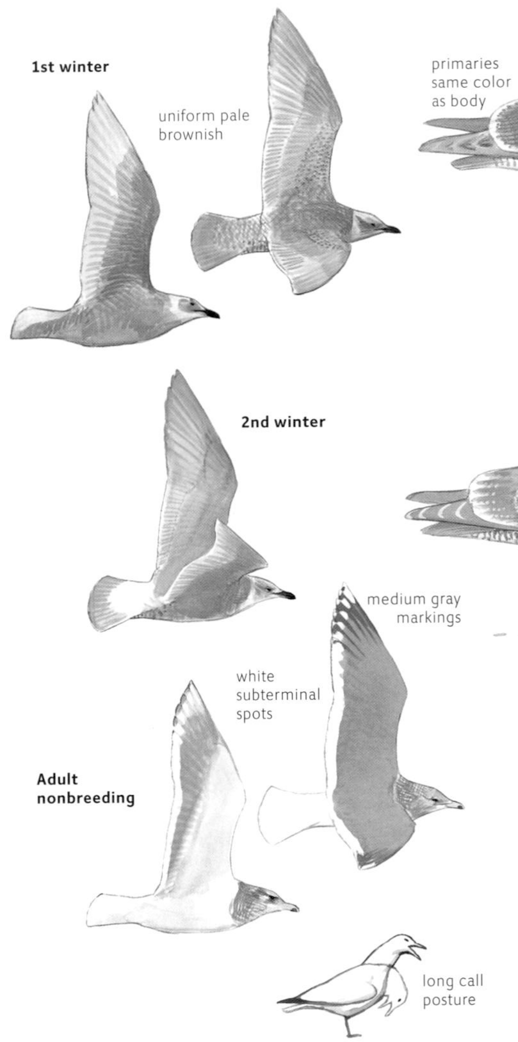

VOICE: Slightly flatter, more hollow-sounding than Western. Long call of Washington populations much like Western but slightly higher-pitched with longer notes. Long call of Siberian populations lower, slower (more like Slaty-backed).

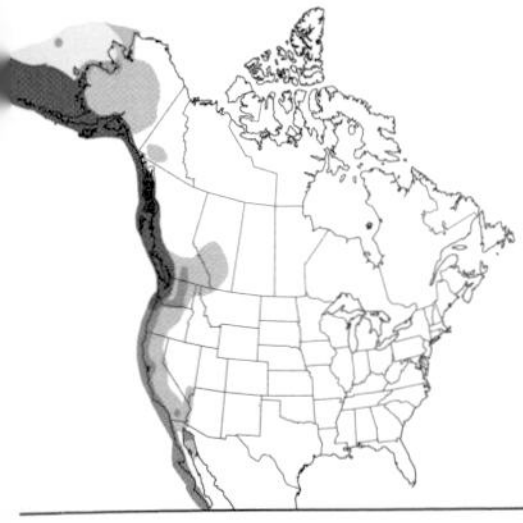

Common in any open coastal habitat, from ocean and mudflats to open fields. Habits similar to Herring Gull.

Slaty-backed Gull

Larus schistisagus

L 25" WS 58" WT 3 lb (1,350 g) ♂>♀

From second winter, shows dark gray (slaty) back, darker than any other gull species usually seen in Alaska. Note white subterminal spots and limited black on wingtips; pale iris; and, in winter, extensive head streaking, especially around eye and nape.

Along with individuals without white subterminal spots on primaries, distinguished from Western by streaked head (in winter), black only on outer webs of primaries, paler underside of primaries, slightly more slender structure. Hybridizes with Glaucous-winged Gull in Siberia; adult hybrids are paler-mantled but otherwise difficult to distinguish from pure birds.

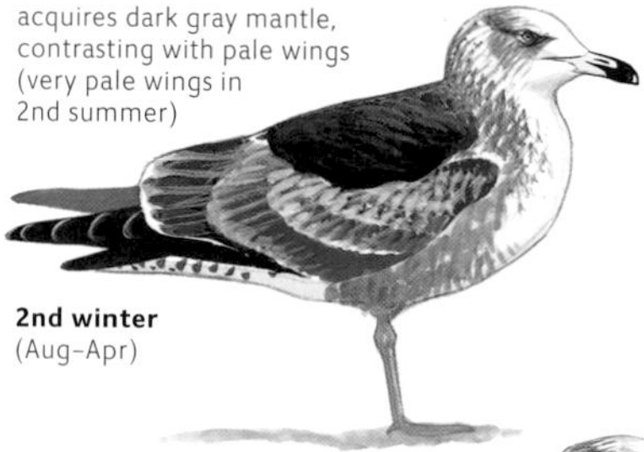

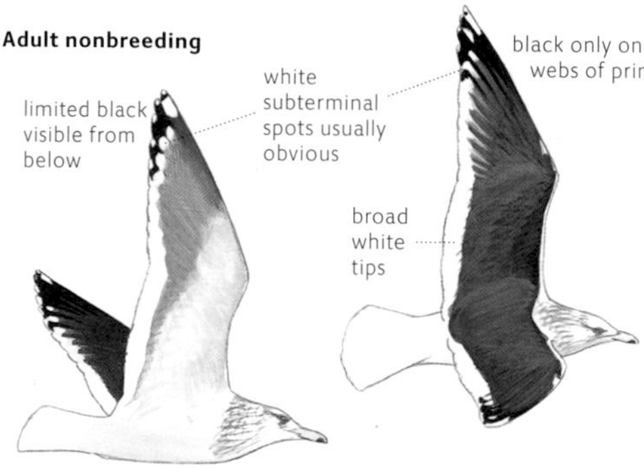

VOICE: Virtually identical to Siberian populations of Glaucous-winged. Long call lower and slower overall than Western.

Uncommon and local in western Alaska; rare elsewhere but increasing and now found in small numbers each winter along Pacific coast south to California. Very rare east to Newfoundland and south to Florida and Texas. Habitat and habits similar to Herring Gull.

Western Gull

arus occidentalis

25" WS 58" WT 2.2 lb (1,000 g) ♂>♀

arge and stocky, with heavy bill that is typically thick-tipped and ightly drooping. Note rounded head with peak over eye and oped rear crown. The only expected dark-backed gull in most f its range. Hybridizes extensively with Glaucous-winged Gull ⁄here range overlaps.

Juvenile (Aug–Oct)

uniform dark, sooty brown

1st winter (Sep–May)

pale base of lower mandible

1st winter

all-dark tail and paler rump

1st summer (Apr–Sep)

2nd winter (Aug–Mar)

2nd winter

3rd winter (Aug–Mar)

Adult nonbreeding

extensive but poorly defined black

Adult nonbreeding Northern (Sep–Mar)

broad tertials and drooping "skirt"

Adults in spring can develop rich yellow tinge on legs, compare Yellow-footed Gull.

Northern birds paler-mantled than Southern with average darker eye, more head-streaking in winter. Variation is clinal, with step in central California.

long call posture

Adult nonbreeding Southern (Sep–Mar)

virtually unstreaked

'OICE: Generally similar to Herring but notes a little more ·lipped, less drawn-out, lower, and less clear. Long call fairly flat, imple, and low. Adult in flight over roost gives deep, gruff *ooowa* ›r *kwaow* unlike high, clear notes of Herring.

Common in any open coastal habitat, but in most areas extremely rare more than a mile from salt water. Habits similar to Herring Gull.

Adult breeding (Feb–Sep)

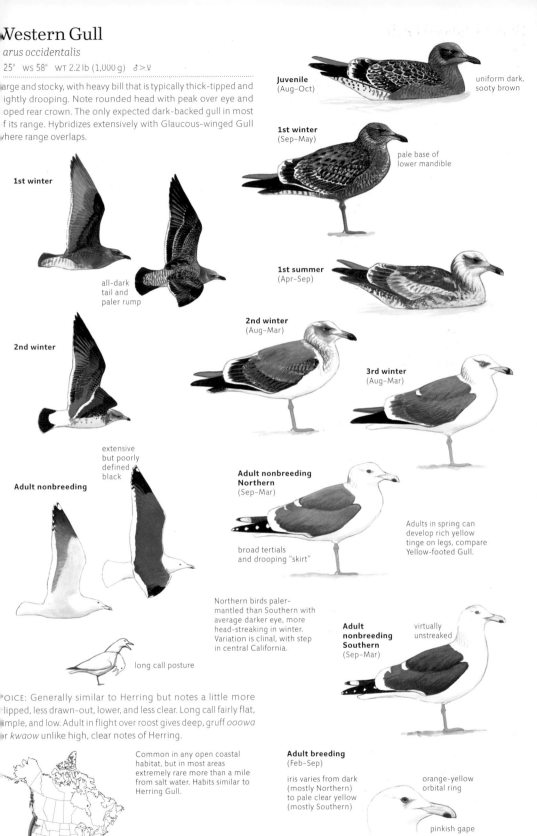

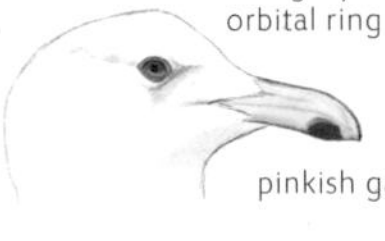

iris varies from dark (mostly Northern) to pale clear yellow (mostly Southern)

orange-yellow orbital ring

pinkish gape

Yellow-footed Gull

Larus livens

L 27" WS 60" WT 2.8 lb (1,260 g)

Appearance similar to Western Gull, but with heavier bill and relatively longer neck. Best distinguished from Western by bright yellow legs, obvious by second winter (but beware some breeding Western Gulls develop yellow legs). Juvenile distinguished by clean white belly and mostly white rump. Acquires adult plumage by third winter, one year earlier than Western.

Juvenile
(Jun-Sep)

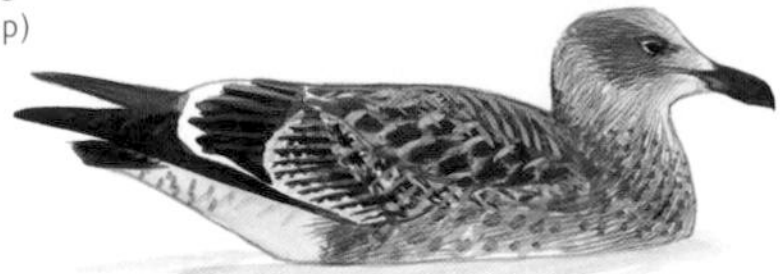

Juvenile

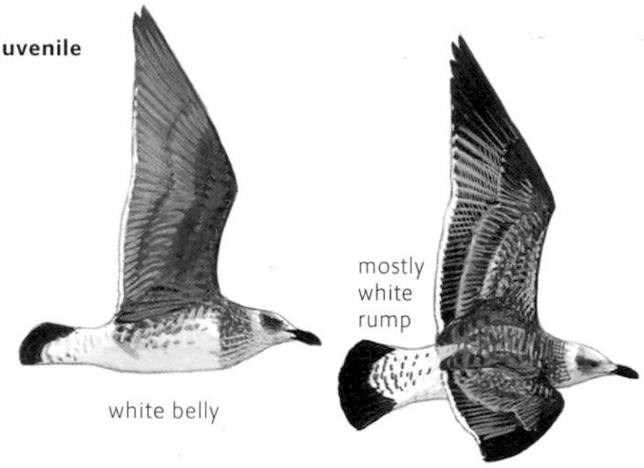

1st winter
(Oct-Jan)

1st summer
(Feb-Jul)

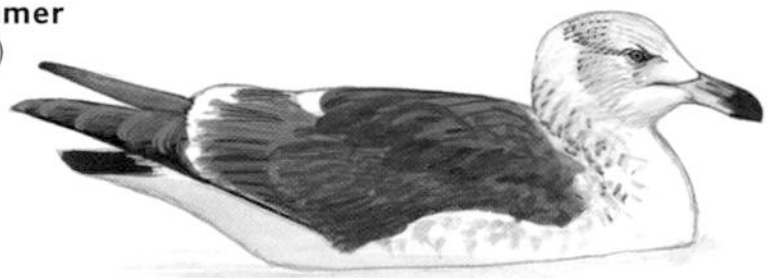

2nd winter

Dark-mantled like Southern Western Gull, but with bright yellow legs; distinguished from Lesser Black-backed by much heavier structure, lack of head-streaking, yellow orbital ring; from Kelp Gull by different head shape, brighter legs, yellow orbital ring, slightly paler back.

2nd winter
(Aug-Dec)

Adult nonbreeding

Adult nonbreeding
(Aug-Dec)

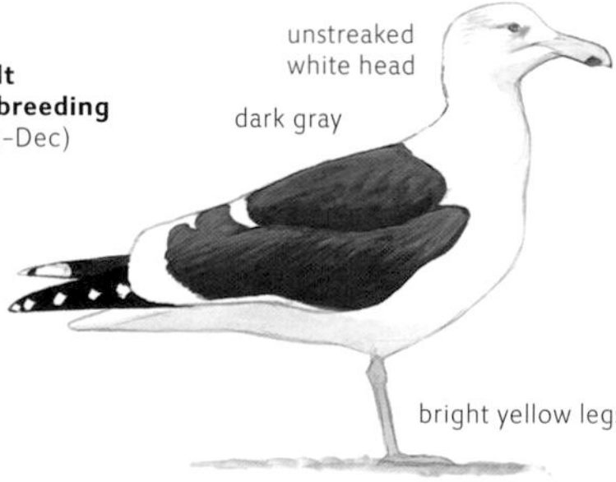

VOICE: Controlled and low with nasal, talking quality, not strident or squeaking; much lower and more nasal than Western. Long call of rapid one-syllable notes; differs from Western in lower pitch and lack of rough first segment of each phrase.

Uncommon visitor from nesting areas in Gulf of California to Salton Sea, California (where Western Gull is rare), especially from Jun-Oct; recorded on the coast of southern California only a few times. Habits similar to Western Gull.

Adult breeding
(Jan-Jul)

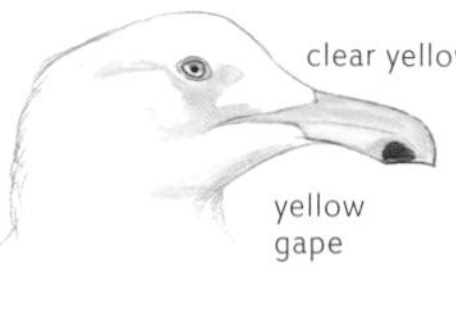

ꞁesser Black-backed Gull

arus fuscus

21" WS 54" WT 1.8 lb (800 g) ♂>♀

ꞁightly smaller and more slender than Herring Gull, with relatively long, slender wings. Adult darker gray above than Herring, sually slightly paler than Great Black-backed Gull; note yellow ꞁgs and (nonbreeding) streaked head. First-year birds have arker wings and paler body than Herring.

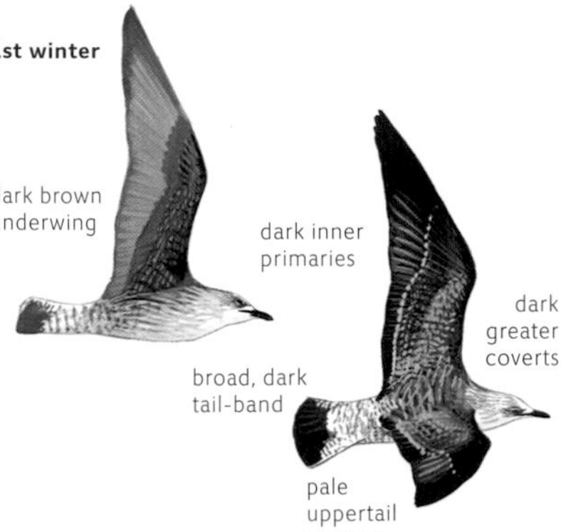

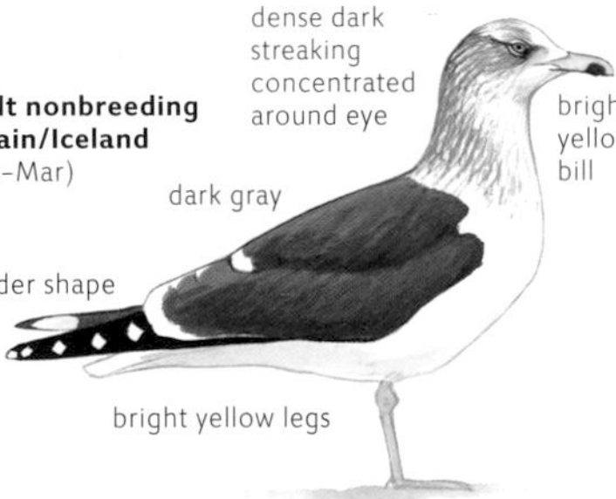

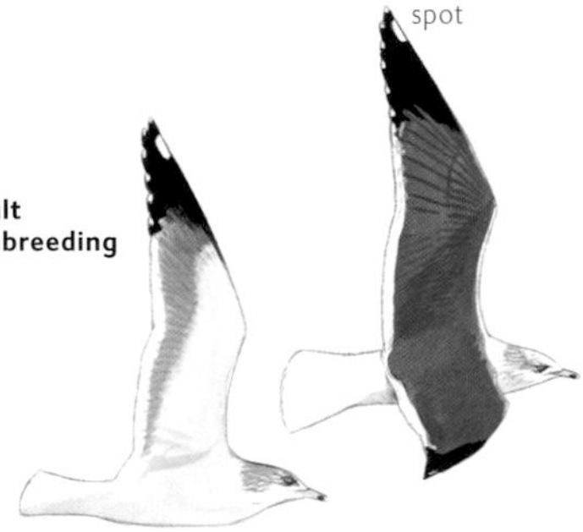

OICE: Guttural; slightly deeper and more nasal than Herring, ut not as gruff as Great Black-backed. Long call begins with ɔng, low note, then rapid, short, rising notes (about 4 per second), low and gruff. Some short calls nasal, hollow-sounding.

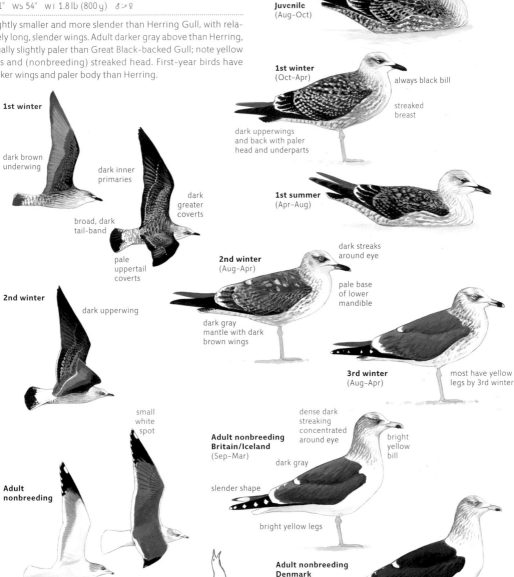

Rare but increasing winter visitor from Europe; locally fairly common at a few locations in Atlantic states. Found with other large gulls. Habitat and habits similar to Herring Gull.

Nearly all North American records are of the paler-mantled Britain/Iceland population. A few records apparently refer to the darker Denmark population, which has same mantle color as Great Black-backed and relatively longer wings.

Great Black-backed Gull

Larus marinus

L 30" WS 65" WT 3.6 lb (1,650 g) ♂>♀

Our largest gull, bulky with heavy bill and broad wings. First-year birds have mostly white head and underparts contrasting with large black bill and speckled brown upperside; also note mostly white rump and tail, with narrow dark tail-band. Adult very dark above, with pale pink legs, usually dusky iris, and large white spots on wingtips.

Juvenile
(Aug–Oct)

1st winter
(Oct–Mar)

1st winter

1st summer
(Mar–Sep)

2nd winter

2nd winter
(Aug–Mar)

3rd winter
(Aug–Mar)

Adult nonbreeding

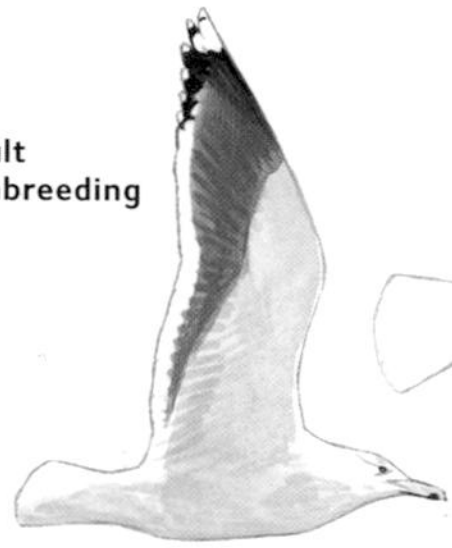

Adult nonbreeding
(Sep–Feb)

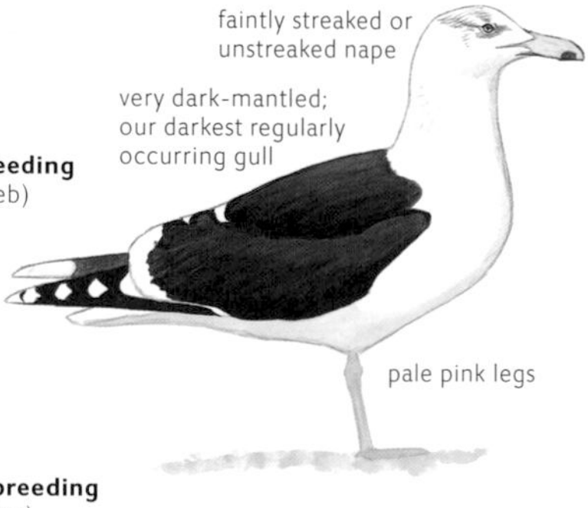

Adult breeding
(Mar–Sep)

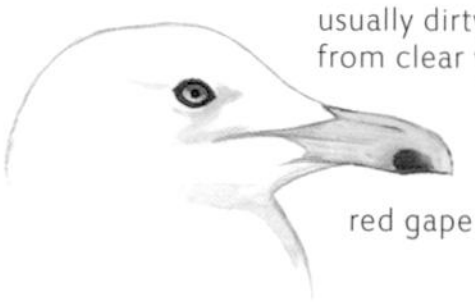

VOICE: Very deep, swallowed, hoarse; much lower than Herring. Long call slow and hoarse phrases (less than 3 per second); phrases poorly defined, each one fading. Common flight calls deep, gruff *gowl* or *gawp*.

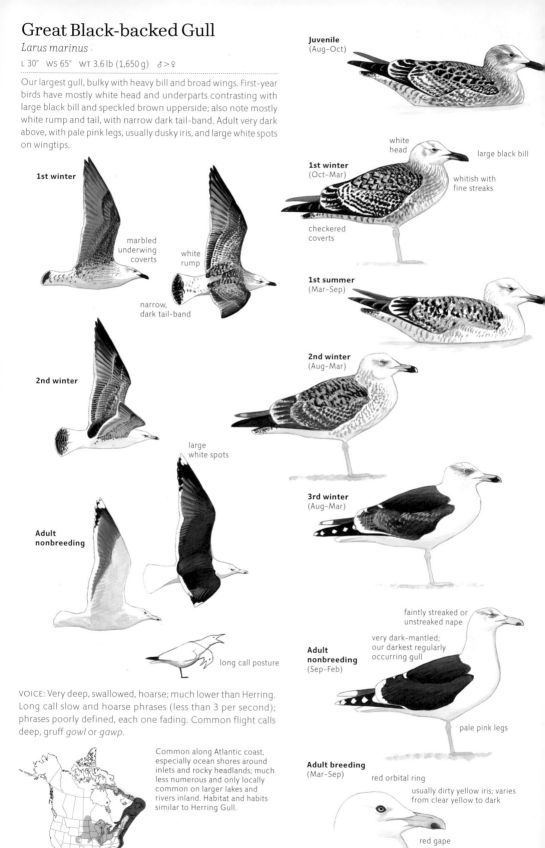

Common along Atlantic coast, especially ocean shores around inlets and rocky headlands; much less numerous and only locally common on larger lakes and rivers inland. Habitat and habits similar to Herring Gull.

Heermann's Gull

Larus heermanni

. 19" WS 51" WT 1.1 lb (500 g) ♂>♀

Jnpatterned dark plumage distinctive. Similar in size to Ring-billed Gull, but stockier, with very broad wings; also note red bill and black legs. Often mistaken for a jaeger due to its dark plumage, pointed wings, and habit of chasing other seabirds to steal food.

VOICE: Low, hollow trumpeting. Common call a nasal *yoww*; also a short *yek* and series *ye ye ye ye*.... Long call like other large gulls but quality strikingly different.

Common along coast, mainly on ocean beaches and open ocean near shore. Often in large groups.

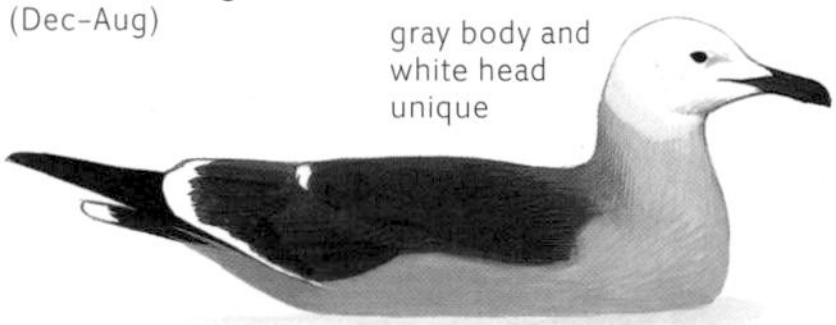

Ivory Gull

Pagophila eburnea

L 17" WS 37" WT 1.4 lb (630 g) ♂>♀

A fairly small, stocky gull; note small bill, short black legs, and relatively broad, pointed wings. All plumages mainly ivory-white; First-year birds have dark spots on tips of all wing and tail feathers and dark-smudged face. Compare albinos of other gull species, and white Rock Pigeon.

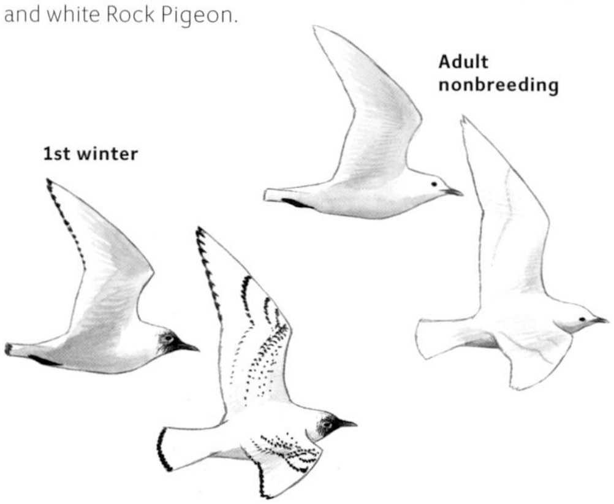

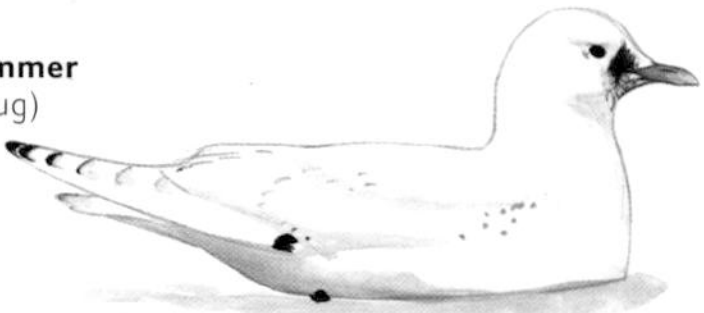

VOICE: Somewhat grating and tern-like: a mewing, high whistle, strongly descending and sometimes slightly trilled *wheeew* or *preeo*.

Rare. Found almost exclusively in extreme far north among pack ice; rarely ventures farther south. Solitary or in small flocks. Does not mix with other gulls, always near water, and most often feeding on carrion.

Black-tailed Gull

Larus crassirostris

L 19" WS 49" WT 1.2 lb (530 g)

Adult dark gray above and slender overall, similar to Laughing or Lesser Black-backed Gull, intermediate in size between those two. Note mostly black tail, red and black tip on bill, yellowish legs, and all-black wingtips.

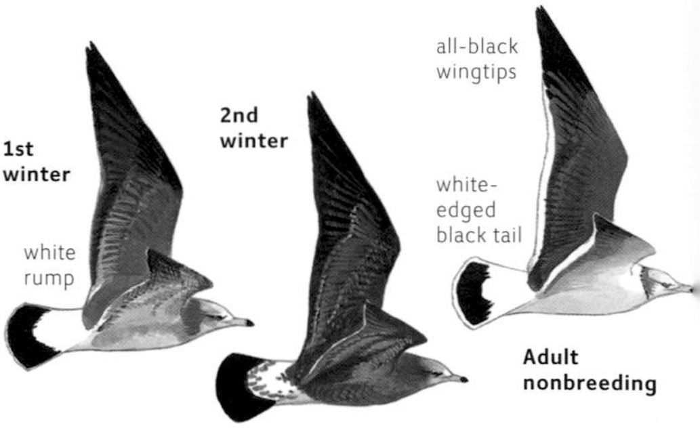

VOICE: Nasal, rasping mewing reminiscent of tomcat, higher than Herring Gull.

Very rare visitor from Asia; records widely scattered both coastal and inland, from Alaska to Newfoundland and south to California, Texas, and North Carolina. Usually found with concentrations of other gulls.

elp Gull

arus dominicanus

24" WS 55" WT 2.3 lb (1,035 g)

imilar in size to Herring Gull, but slightly slimmer with relatively ırge bill. First-year birds extremely difficult to distinguish from rst-year Lesser Black-backed. Adult slightly darker above than ;reat Black-backed, with greenish-yellow legs and pale iris.

OICE: Similar to Herring ;ull but hoarser. Call note a wo- or three-syllable, taccato, repeated *kwee-ah*. ong call noisier and higher han Yellow-footed, lower han Western.

Very rare visitor from South America; has nested in Louisiana, hybridizing there with Herring Gull. Few records elsewhere, but found north to Maryland, Indiana, Ontario, and Colorado. Habitat and habits similar to Herring.

Yellow-legged Gull

Larus michahellis

L 25" WS 58" WT 2.5 lb (1,125 g)

Adult told from Herring Gull by slightly darker gray back, bright yellow legs, limited dark streaking on back of head (none on neck), and other details, but beware that these features can be matched by individual Herring Gulls and Herring × Lesser Black-backed hybrids.

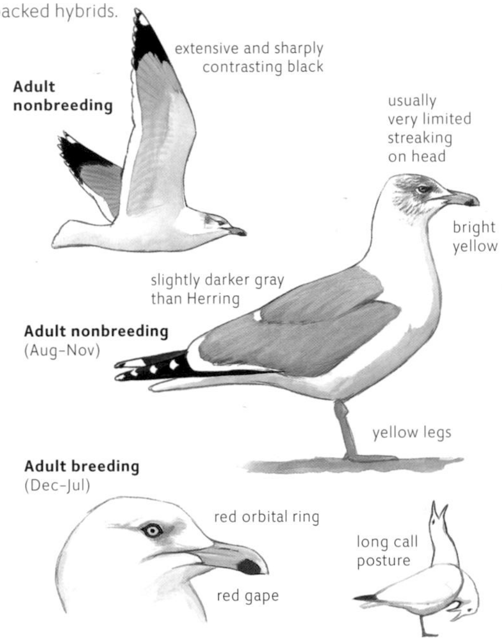

VOICE: Calls similar to Lesser Black-backed, deeper and gruffer than Herring. Long call display also similar to Lesser Black-backed, with bill raised beyond vertical.

Extremely rare visitor from Europe; only a few accepted records in North America, mostly from eastern Newfoundland in fall and winter. Habitat and habits similar to Herring.

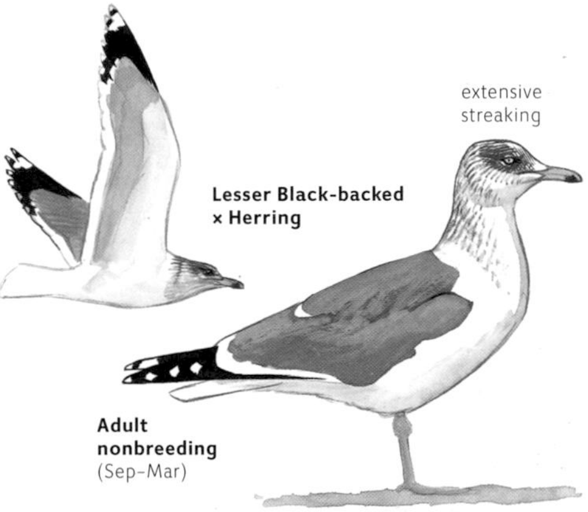

Distinguishing Yellow-legged Gull from this hybrid combination is very difficult. Hybrids generally have extensive dusky streaking on the neck, and retain streaking longer into early spring (Yellow-legged usually white-headed by Dec). Hybrids usually show a darker gray back, drabber pinkish-yellow legs, and subtle differences in wingtip pattern, but these are variable features that require careful assessment.

Identification of Hybrid Gulls

All large gulls are very closely related; some hybridize frequently. Identification of hybrid gulls is difficult and often conjectural. Most hybrids are intermediate between parent species, but individual variation and backcrosses produce a continuum of variation. In cases of frequent hybridization, such as those illustrated on these pages, it is often necessary to use vague titles such as "Herring tending toward Glaucous-winged." Other combina tions occur rarely (e.g., Herring × Great Black-backed) and a fe are unrecorded (Herring × Iceland). Small gulls hybridize muc less frequently, but Black-headed Gull occasionally crosses wit Laughing and Ring-billed Gulls.

Glaucous-winged × Herring Gull

Fairly common from Alaska to California; mainly coastal. Variable in plumage and structure. Head shape may be blocky like Glaucous-winged or slender and angular like Herring. Most are larger, bulkier, and heavier-billed than Thayer's, but some may be indistinguishable from Thayer's.

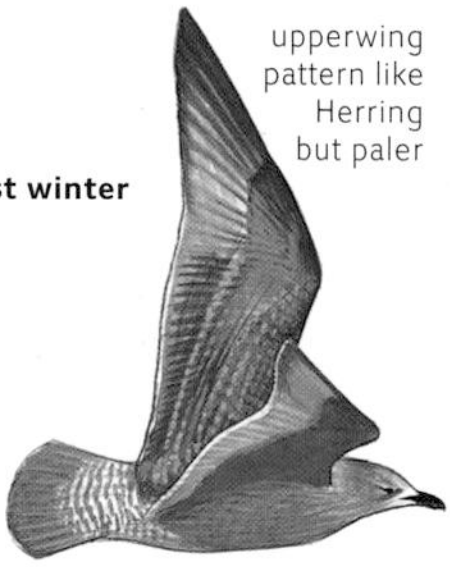

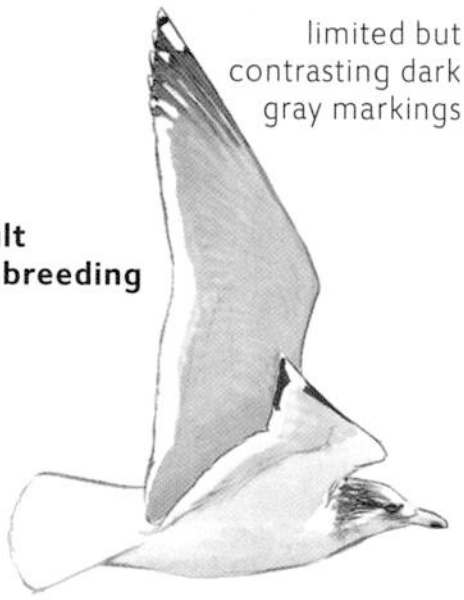

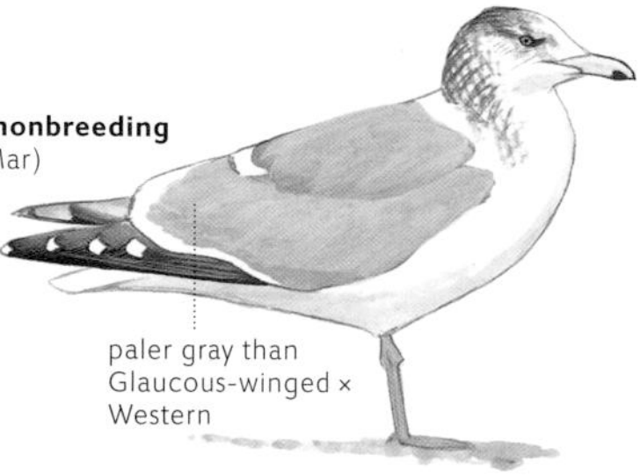

Glaucous-winged × Western Gull

Common coastally from British Columbia to California. Som populations in Washington are mostly hybrids and backcrosse with very few pure birds. Always bulky, broad-winged, and heavy billed like the parent species.

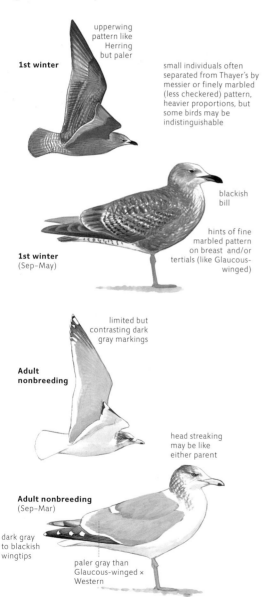

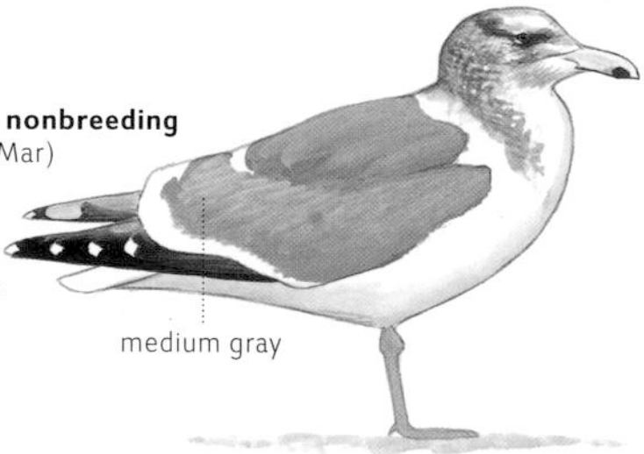

■laucous × Herring Gull

ɿirly common in some Arctic breeding areas; less numerous ɿrther south wherever Glaucous occurs. Generally Glaucous-ke with some intermediate darker plumage characteristics. ever as bulky or broad-winged as most Glaucous-winged.

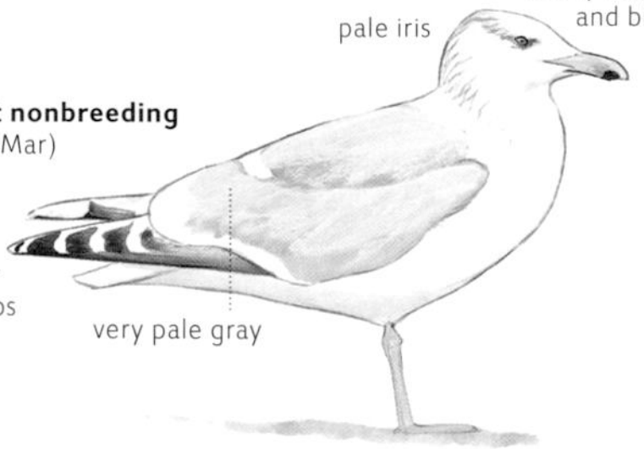

Glaucous-winged × Glaucous Gull

Fairly common locally in western Alaska; less numerous farther south on Pacific coast but probably overlooked. Very difficult to distinguish from pale Glaucous-winged. Always large and large-billed.

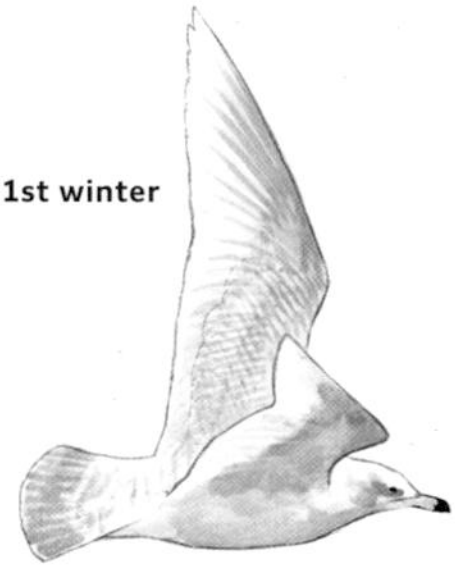

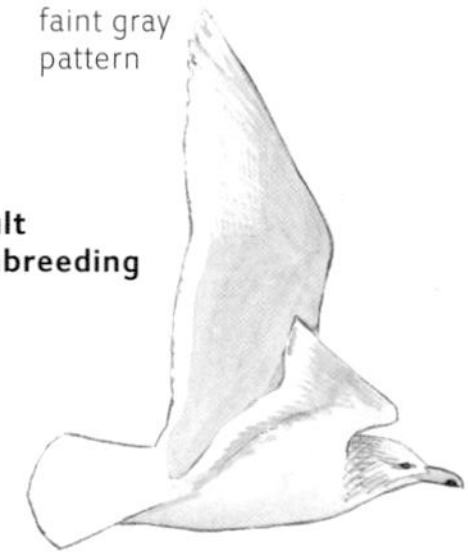

ɿlbinism in Gulls

ɔerrant plumages such as albinism (pale or white plumage) ust always be considered when identifying pale gulls. Check ll shape and color and eye color. Albinistic birds may be indis-ɿguishable from Herring × Glaucous or other hybrids and can ɿuse confusion with other pale species, such as Thayer's and eland. The presence of unusual plumage contrasts (such as a ɛry pale-winged bird with dark brown tail) is a strong indication albinism.

Terns and Skimmer

FAMILY: LARIDAE

18 species in 9 genera (1 rare species not shown here). Terns are in the same family as gulls, Laridae, but in a separate subfamily, Sterninae. Generally smaller than gulls, with straight, pointed bill, relatively long and narrow wings, short legs, and most species have black cap in breeding plumage. Most species feed exclusively on small fish captured by plunge-diving head-first from the air into the water, often beginning in a hovering position. A few species swoop and snatch prey from the water's surface. Black Skimmer has unique foraging style of flying with lower mandib slicing the water's surface. Foraging groups of terns may be larg and dense where prey is concentrated; all species also roost i large groups on the shoreline, often near (but not mixed witl roosting flocks of gulls. All species nest colonially at the water edge, usually on islands; most species lay eggs in a scrape on th ground, noddies build stick nest in low bush. Nonbreeding adul are shown.

Genus *Anous*

Brown Noddy, page 235

Black Noddy, page 235

Genus *Onychoprion*

Sooty Tern, page 233

Bridled Tern, page 233

Aleutian Tern, page 232

Genus *Thalasseus*

Royal Tern, page 226

Sandwich Tern, page 227

Elegant Tern, page 226

Genus *Hydroprogne*

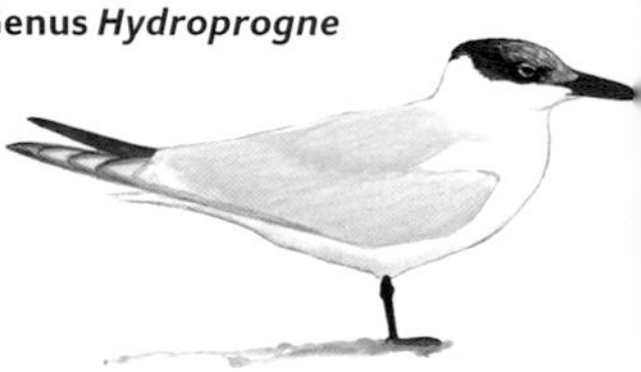

Caspian Tern, page 225

Genus *Sterna*

Roseate Tern, page 231

Common Tern, page 228

Arctic Tern, page 229

Forster's Tern, page 230

Genus *Sternula*

Least Tern, page 231

Genus *Gelochelidon*

Gull-billed Tern, page 232

Genus *Chlidonias*

Black Tern, page 234

Genus *Rynchops*

Black Skimmer, page 236

Skuas and Jaegers

FAMILY: STERCORARIIDAE

5 species in 1 genus. Closely related to gulls, these species are primarily oceanic, coming to land only when nesting. On nesting grounds they are predatory, feeding on lemmings, small birds, and other animal prey. At other seasons they acquire much of the food by piracy, stealing from other seabirds. Juveniles are show

Genus *Stercorarius*

Great Skua, page 241

South Polar Skua, page 240

Pomarine Jaeger, page 239

Parasitic Jaeger, page 238

Long-tailed Jaeger, page 237

ꞆaspianTern

Iydroprogne caspia

21" WS 50" WT 1.4 lb (660 g) ♂>♀

)ur largest tern; about the size of Ring-billed Gull. Large size and ıassive red bill distinctive. Distinguished from Royal Tern by roader wings with dark under primaries, heavier bill, and usually ark forehead.

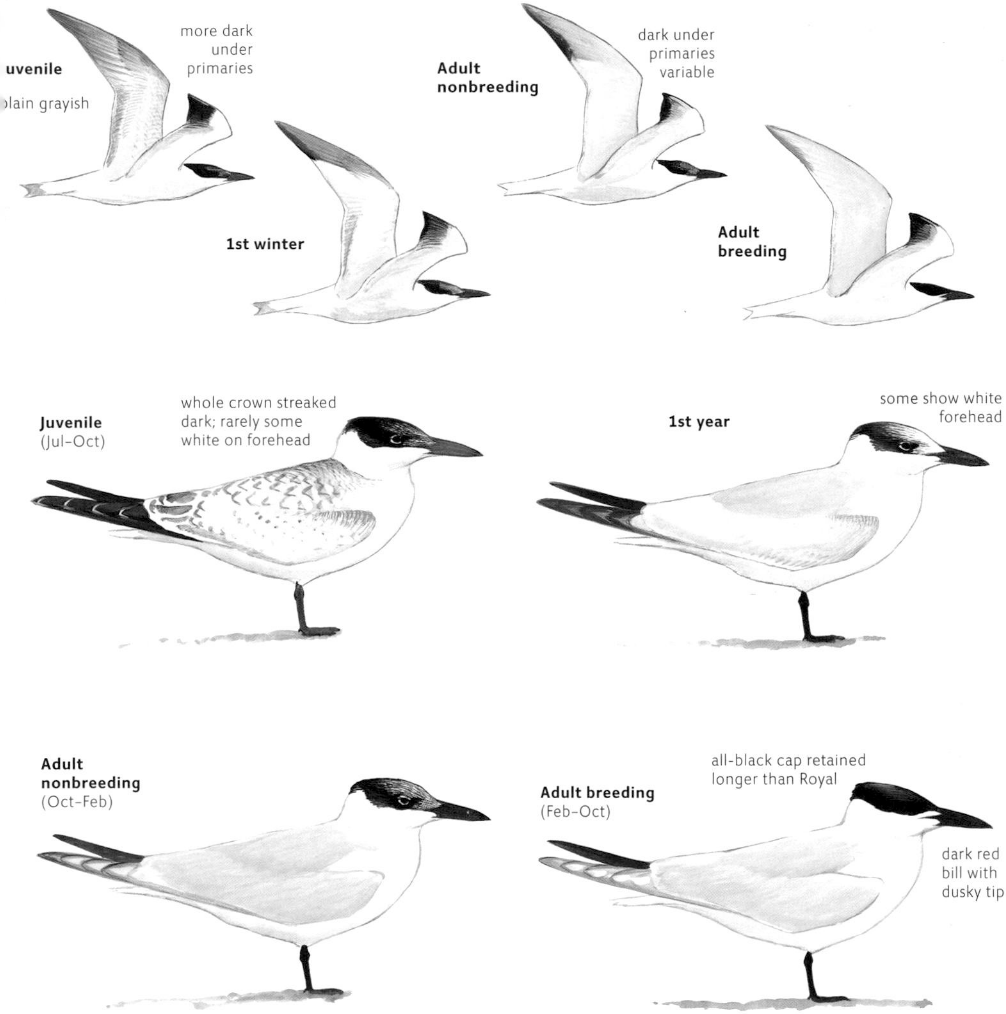

'OICE: Call a deep, very harsh, heron-like scream *aaayayaum*. ıvenile call a high, thin, wheezy whistle *sweeeea* or *fweeeee-er* heard Jul–Apr).

Common locally. Nests in small colonies on sand islands, roosts on beaches and sandbars. Patrols over open water along sandy ocean beaches, lagoons, lakes, and rivers in search of fish. Usually solitary when foraging.

Royal Tern

Thalasseus maximus

L 20" WS 41" WT 1 lb (470 g) ♂>♀

A very large tern, with heavy orange-red bill. Nearly as large as Caspian Tern, but more slender with thinner wings and bill; also note mostly white underwing and, for most of the year, white forehead. Larger and heavier than Elegant Tern.

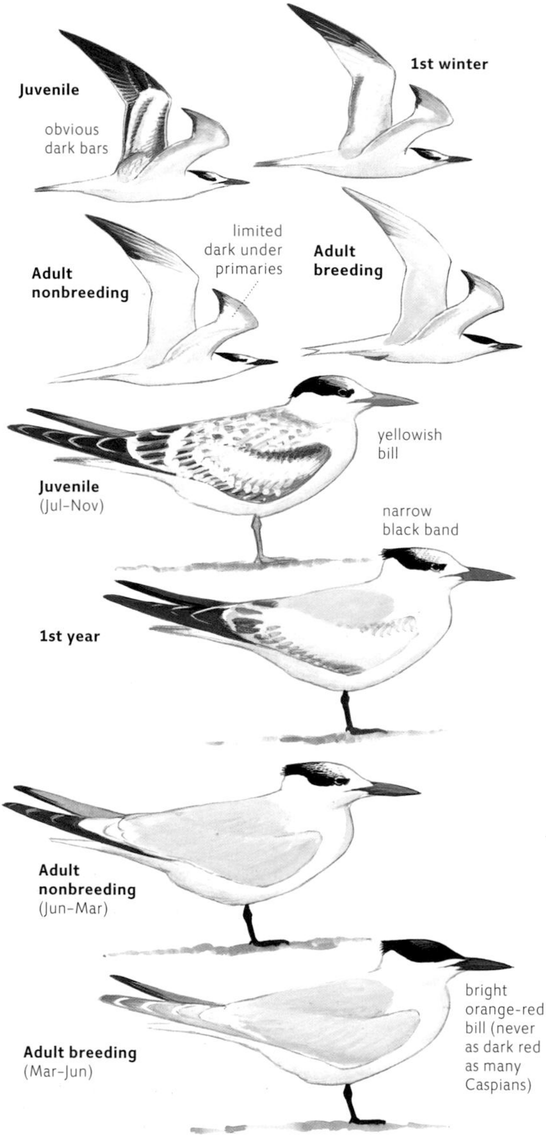

VOICE: Call loud, throaty, rolling *kerrra* or lower *koorrrick*; lower, longer than Sandwich; also soft *youm* or *yeek*. Juvenile begging call two to three piping whistles; also high, ringing *kerreeep*.

Common; found on open salt water along sandy beaches and especially around oceanfront bays and inlets. Nests in large colonies on isolated sand islands. Patrols over water and plunge-dives for fish.

Elegant Tern

Thalasseus elegans

L 17" WS 34" WT 9 oz (260 g) ♂>♀

Habits and appearance very similar to Royal Tern, but smaller wit relatively larger head and shorter tail; best distinguished by lon ger, thinner, downcurved bill; longer drooping crest; and mor extensive black crown in nonbreeding plumage.

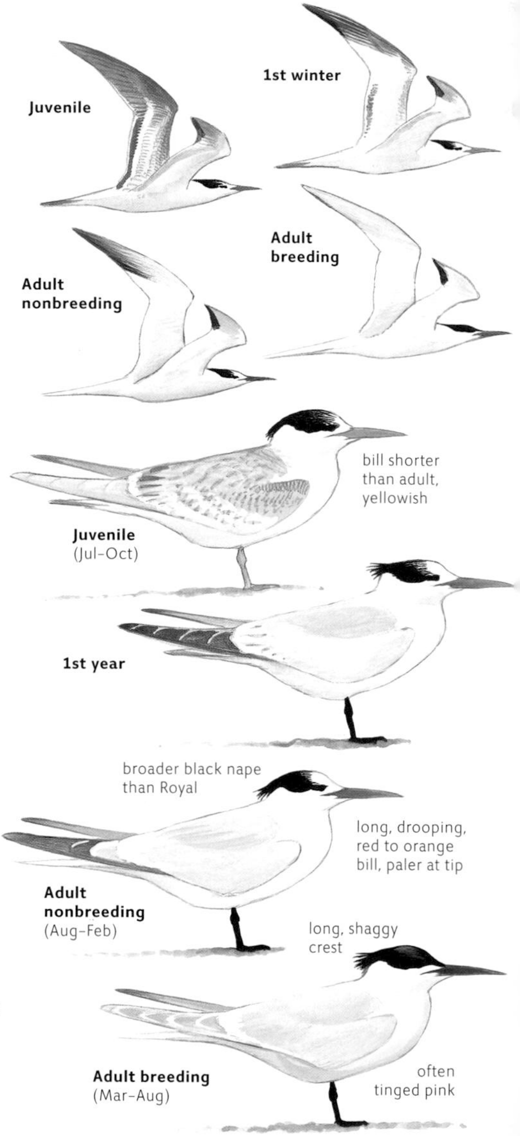

VOICE: Call a short, low, loud *keerik*; very similar to Sandwich bu deeper. Juvenile gives insistent, thin, whistled *sip sip sip*... simila to Royal.

Common on open coastal waters, especially on inlets of bays and lagoons and along sandy beaches. Recorded a few times along Gulf and Atlantic coasts, north to Massachusetts.

andwich Tern

alasseus sandvicensis

5" WS 34" WT 7 oz (210 g) ♂>♀

ermediate in size between Royal and Common Terns; relatively nder-winged and long-billed. Pale tip of black bill distinctive. stinguished from Common and Forster's Terns by larger size, atively slender wings, long bill, paler gray upperparts, and aggy crest.

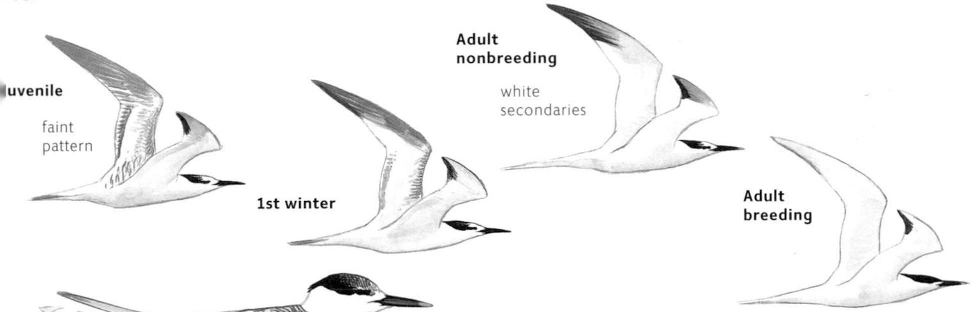

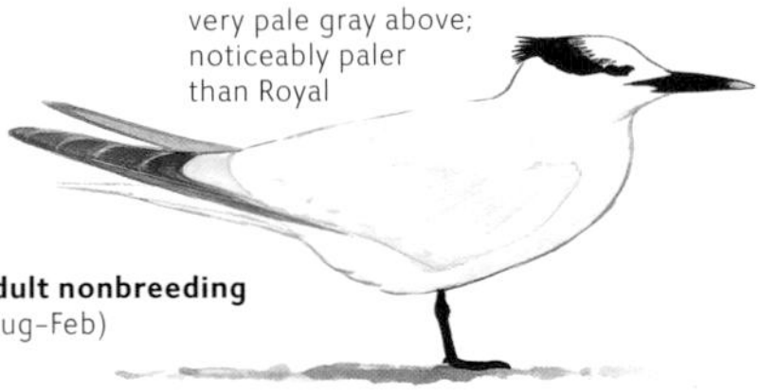

ICE: Call *keerik* or *krjik*; higher, more abrupt, and more grating an Royal (reminiscent of Roseate Tern). Juvenile begging call igh *see see see* like Royal; also a high, ringing *kreep*.

Common on open salt water. Usually found with Royal Tern and has similar habitat and habits. Recorded a few times inland to Great Lakes and north to Newfoundland, several times in California.

Bill Details

Royal Tern

Elegant Tern

Sandwich Tern

Sandwich Tern (Cayenne)

Common Tern

Sterna hirundo

L 12" WS 30" WT 4.2 oz (120 g)

Very similar to Forster's and Arctic Terns. Told from Forster's by slightly slimmer proportions, darker gray upperwing; in breeding plumage gray belly; in nonbreeding plumages dark hindcrown, dark carpal-bar. Told from Arctic by details of proportions and wing pattern.

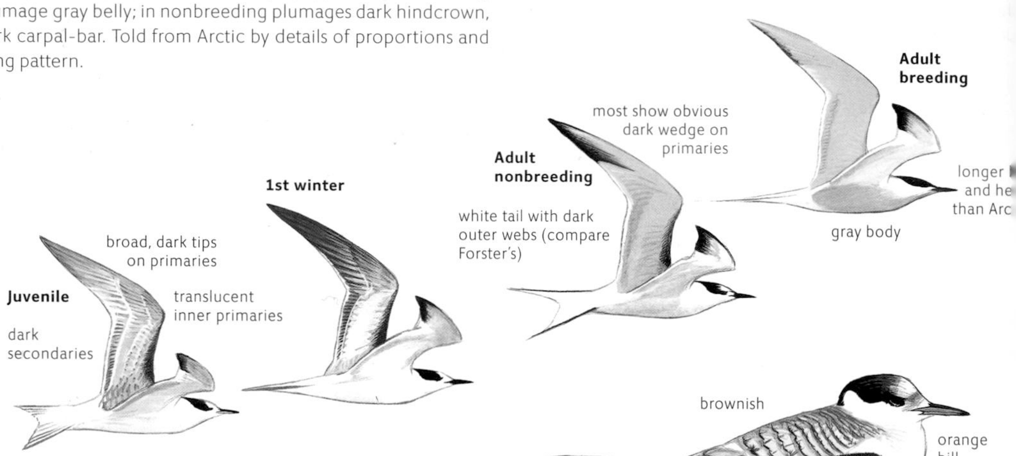

VOICE: All calls like Arctic and Forster's but richer, more musical, with slurred, two-syllable pattern. Common call a rich, high *keeeyurr* with obvious drop in pitch; a high *kit* or *tyik* and longer, deeper *kiiw*. Adult courtship call a begging *kerri kerri kerri....* Attack call a hacking, dry *k-k-k-k*.

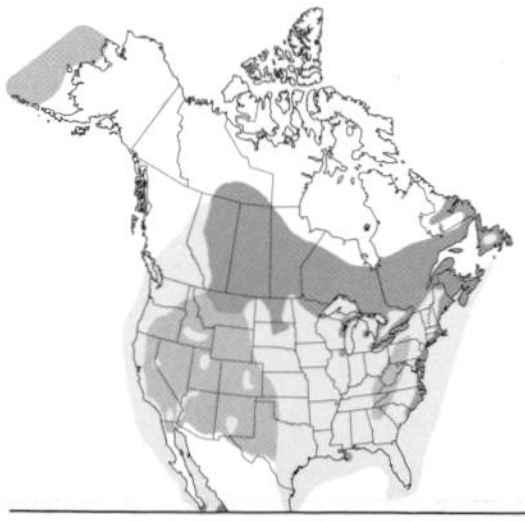

Common along ocean beaches and lagoons; less common on large lakes inland. Usually outnumbers Forster's on open ocean (Forster's is more numerous on shallow bays and lagoons). Nests in large colonies on sandy or rocky islands. Feeds on small fish captured in plunge-dive.

Records in western Alaska are of the Siberian population, which is identical to American populations except breeding adults have darker gray body plumage (more like Arctic Tern), all-black bill (but a small percentage have red bill base), and dark reddish-brown legs.

.rctic Tern

ɜrna paradisaea

2"   WS 31"   WT 3.9 oz (110 g)

ghtly smaller than Common Tern, with shorter bill, rounder ad, much shorter legs, and narrower wings. All plumages show le gray and white primaries with small dark tips. In breeding ımage note long tail, all-red bill.

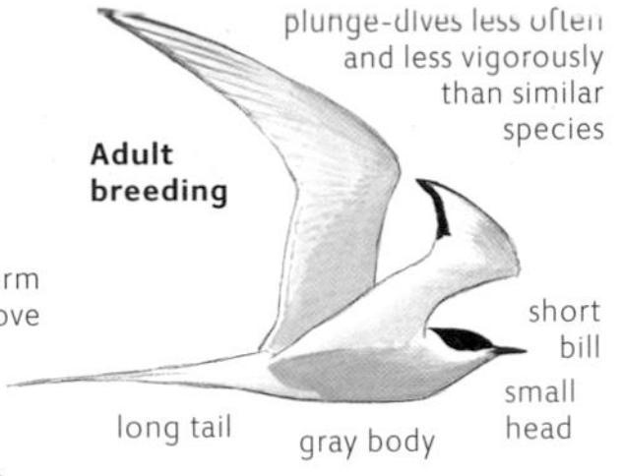

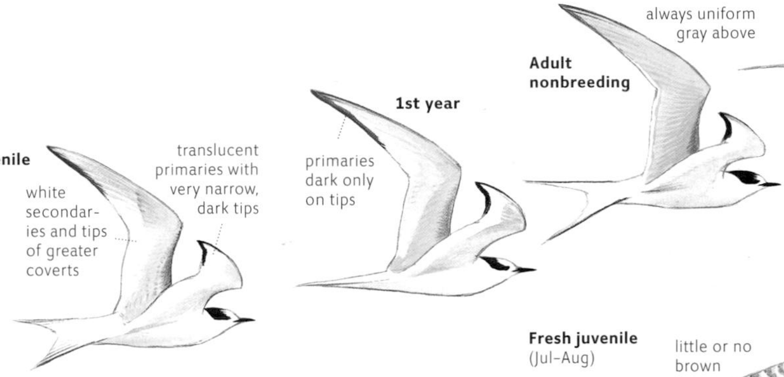

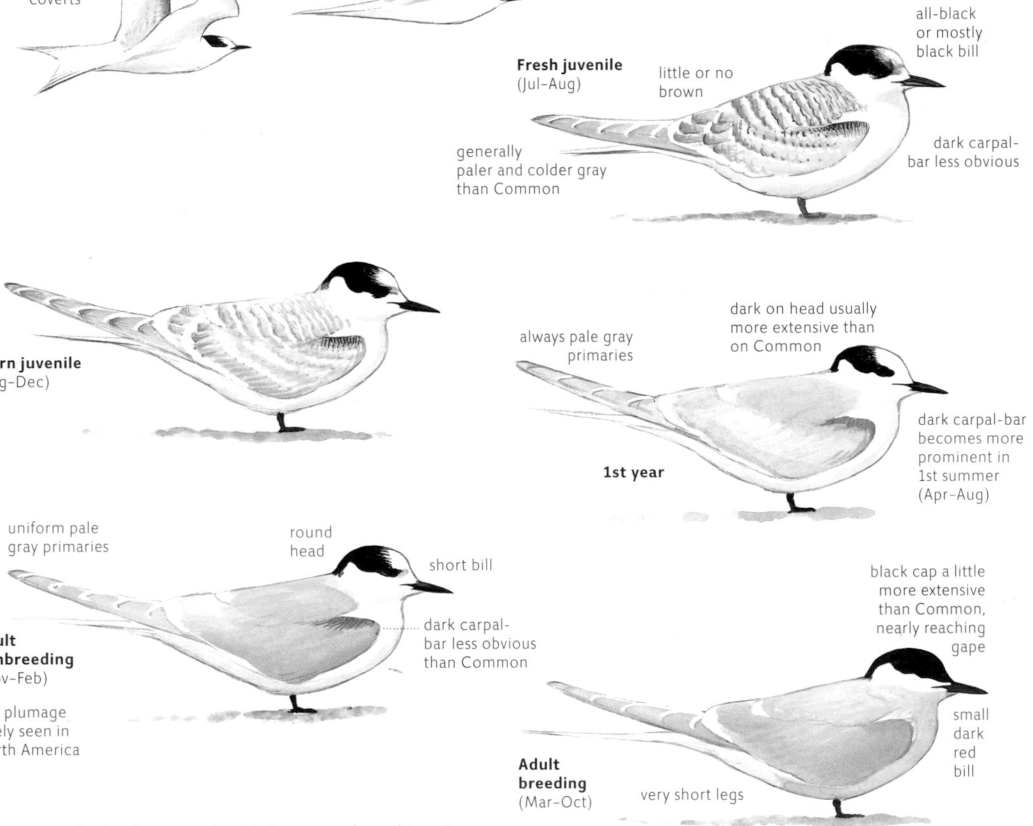

ɔICE: All calls like Common but higher, squeakier, drier. Com-on calls rather harsh and buzzy *keeeyurr*; often clipped with arper pitch change than Common; also *kit* or *keek* notes like ·ng-billed Dowitcher. Agitated bird gives rhythmic series *titik-ri titikerri*.... Attack a dry, nasal *raaaz*.

Common within breeding range; uncommon migrant in small numbers over open ocean far offshore. Nests in colonies on small islands, often mixed with Common Tern. Habits similar to Common Tern.

All juvenile terns give high, squeaky renditions of adult calls, gradually acquiring adult voice over a year or more. Even 1st-summer (one-year-old) terns usually sound higher and squeakier than adults.

Forster's Tern

Sterna forsteri

L 13" WS 31" WT 6 oz (160 g)

Slightly bulkier and broader-winged than Common Tern. Nonbreeding plumages have white crown with isolated black eye-patch and usually paler upperwing than Common, especially flashing silver-white on primary coverts. Breeding plumage has all-white underparts.

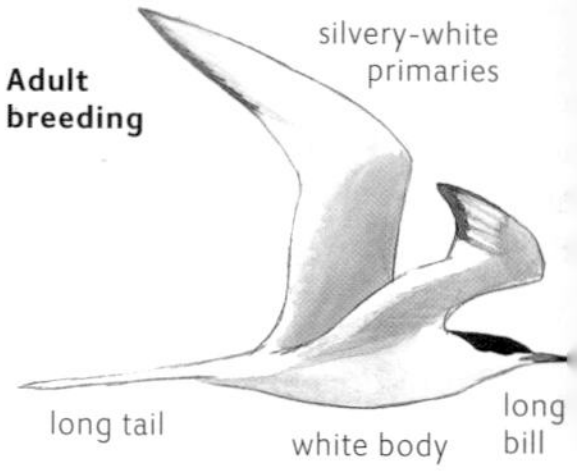

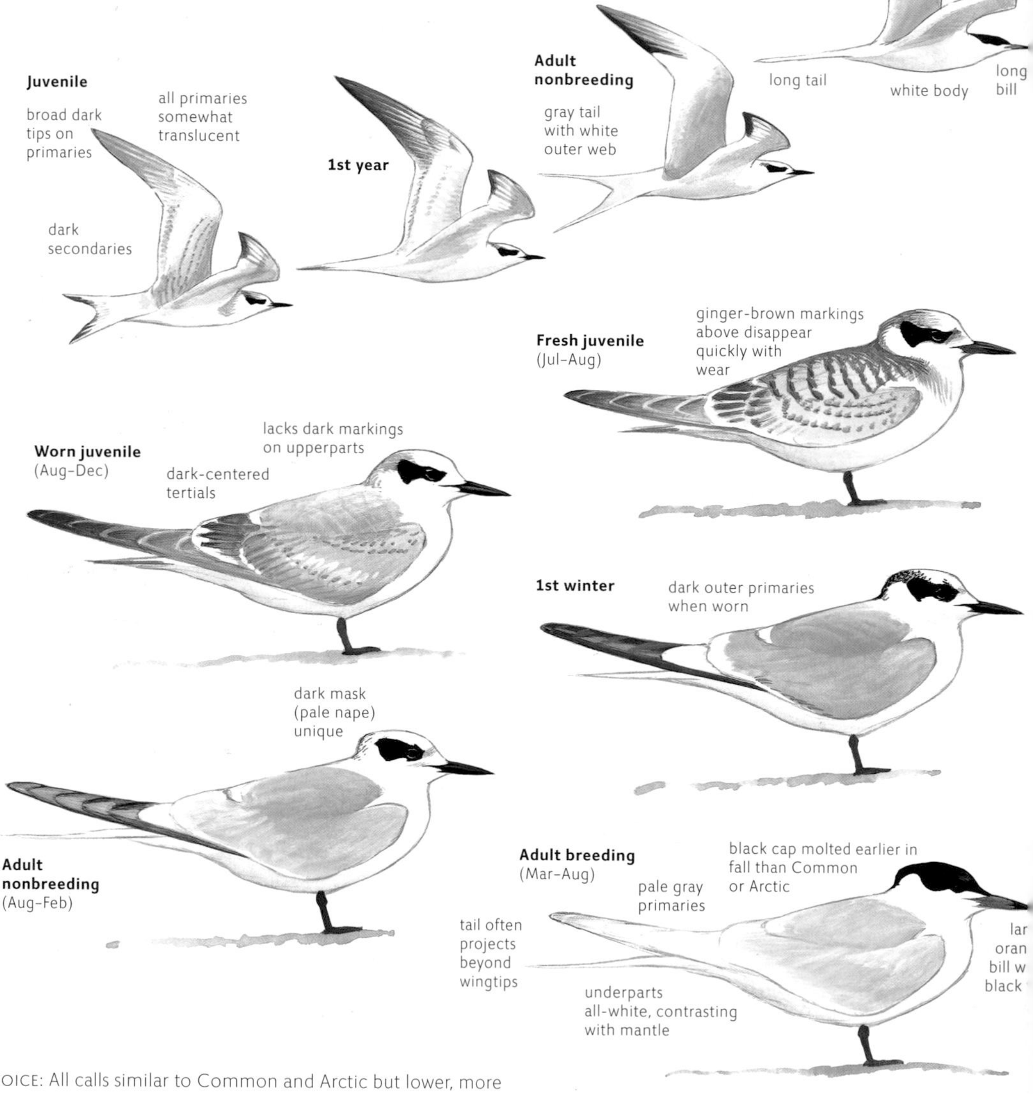

VOICE: All calls similar to Common and Arctic but lower, more wooden and rasping, one syllable. Common call a simple, descending *kerrr*; lower and more wooden-sounding than Common; also *kit* or *kuit*. Adult courtship begging *kerr kerr kerr*. Attack a very low, rasping *zaaaar*.

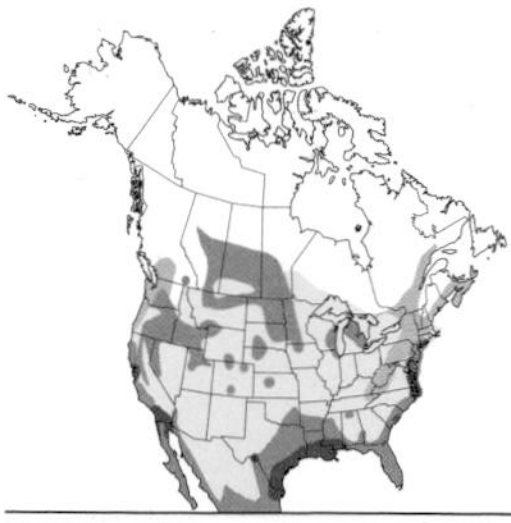

Common on open water and in marshes; outnumbers Common Tern inland and around sheltered bays, marshes, and tidal creeks (Common favors open water of large lakes or ocean). Nests in small colonies on marshy islands. Habits similar to Common.

The upperwing patterns of terns illustrated in this book show the typical fresh condition (paler primaries) on adult breeding birds and the typical worn condition (darker primaries) on adult nonbreeding birds. In reality, fresh-looking primaries are not strictly related to breeding plumage and can be seen at any time of year. The extent of dark color on the upperside of the primaries is a useful identification clue for some species. Worn 1st year birds of most species develop mostly dark remiges above. The notable exception is Arctic Tern, which always has pale gray and translucent primaries, regardless of age or wear.

Roseate Tern

erna dougallii

12.5" WS 29" WT 3.9 oz (110 g)

milar to Common Tern, but with relatively shorter and more ender wings, quicker wingbeats. In breeding plumage note ostly black bill, white underparts with faint rosy tinge, only two ark outer primaries, and very long white tail. Wingbeats fast and iff like Least Tern.

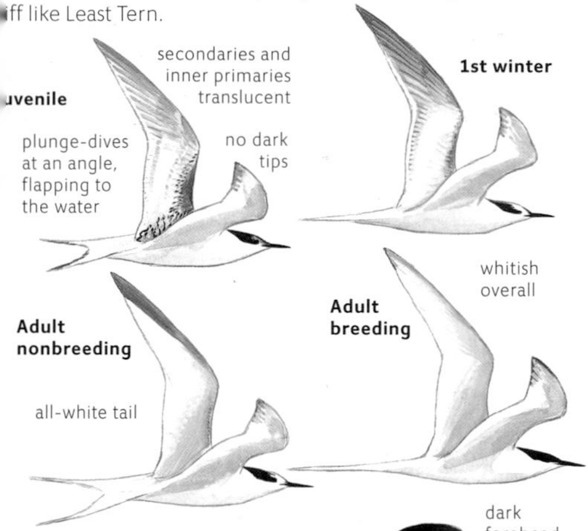

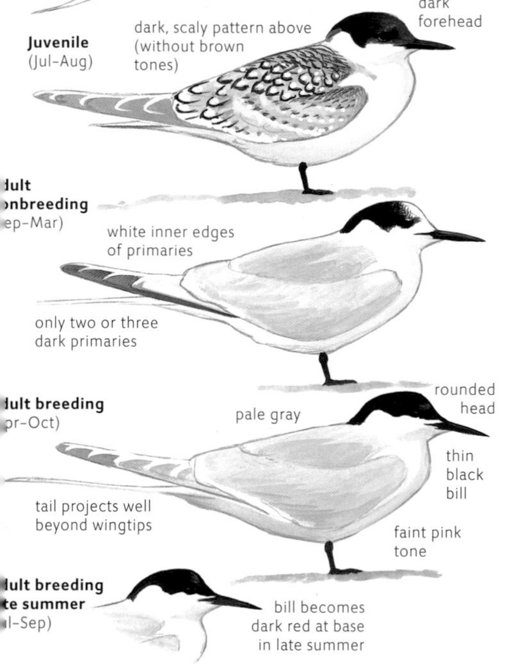

ICE: Distinctive; abrupt, harsh, two-syllable calls: a rather ft and plover-like *CHIvik* or *chewVI* or a more insect-like *ivvik*. Alarm a high, clear *keer*. Attack a very harsh, rasping *rrraaaaach*. Juvenile gives higher, more trilled *dreewid*.

Common locally around nesting colonies on small rocky islands; seldom seen elsewhere except at staging areas on beaches of Cape Cod, Massachusetts in Aug–Sep. Mixes freely with other terns.

Least Tern

Sternula antillarum

L 9" WS 20" WT 1.5 oz (42 g)

Much smaller than all other terns, with quicker wingbeats, slender wings, short tail, and relatively large bill. In breeding plumage note white forehead, two dark outer primaries, yellow bill and legs.

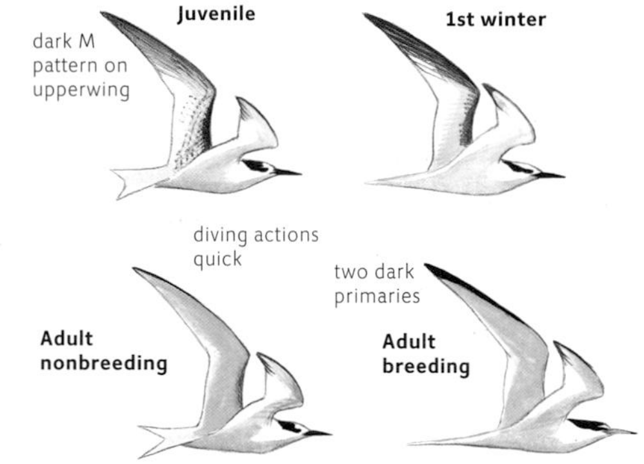

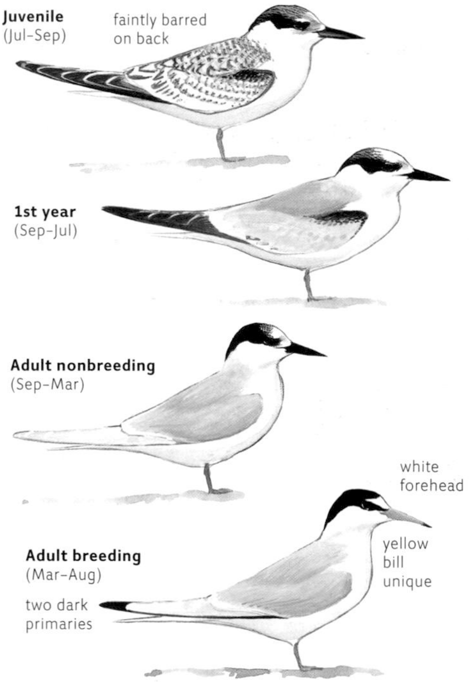

VOICE: Common calls include a rapid, shrill, sharp *piDEEK-adik* or *keDEEK*. Also a weak, nasal *whididi*; high, sharp squeaks *kweek* or *kwik*. Alarm a sharp, rising *zreek*.

Common locally around nesting colonies; uncommon to rare elsewhere. Nests on sand dunes just above high tide line among scattered debris and grass as well as flat rooftops near water. Habits similar to Common Tern and other terns; mixes freely with other terns at foraging and roosting sites.

Gull-billed Tern

Gelochelidon nilotica

L 14" WS 34" WT 6 oz (170 g)

Slightly larger than Forster's Tern, with long wings and thick black bill, short tail, relatively long black legs. Nonbreeding and immature plumages have mostly white head.

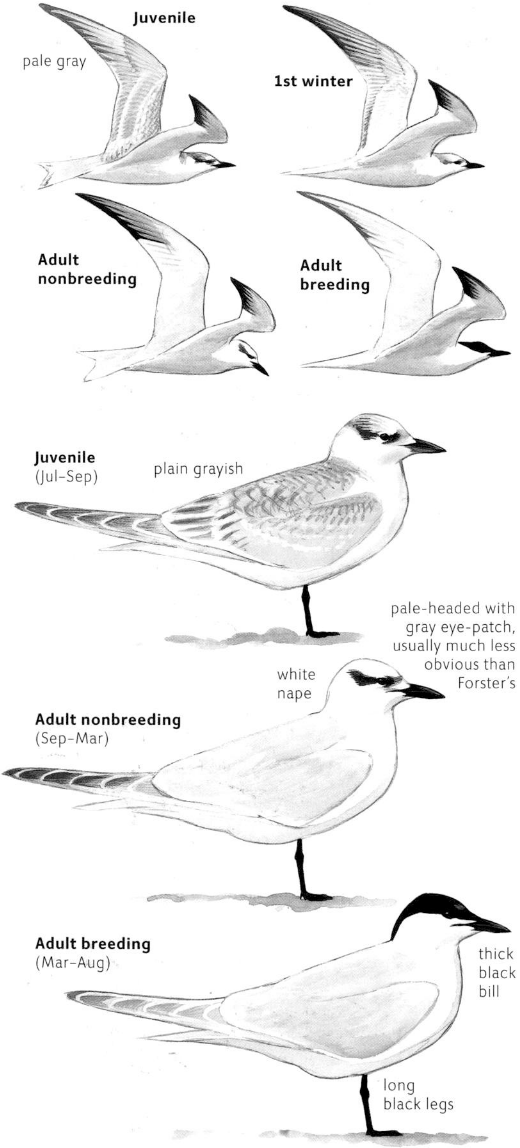

VOICE: Common call a distinctive, nasal yapping *kayWEK*; higher and sharper than Black Skimmer. Alarm a rattling, rasping *aach*. Juvenile begs with high, thin, two-syllable whistle *see-lee*.

Uncommon; usually seen flying singly or in loose groups over sand flats or grassy marshes with patches of open mud, swooping to snatch crabs, insects, and other prey from the ground or from water's surface, never plunging into water. Nests in small colonies on sand. Usually does not mix with other terns.

Aleutian Tern

Onychoprion aleuticus

L 12" WS 29" WT 3.9 oz (110 g)

Appears short-necked in flight. Juvenile has bright cinnamc edges on upperparts and cinnamon wash on breast. Adult dark gray above and below than Common and Arctic Terns, with da bar on secondaries in all plumages and white forehead in bree ing plumage.

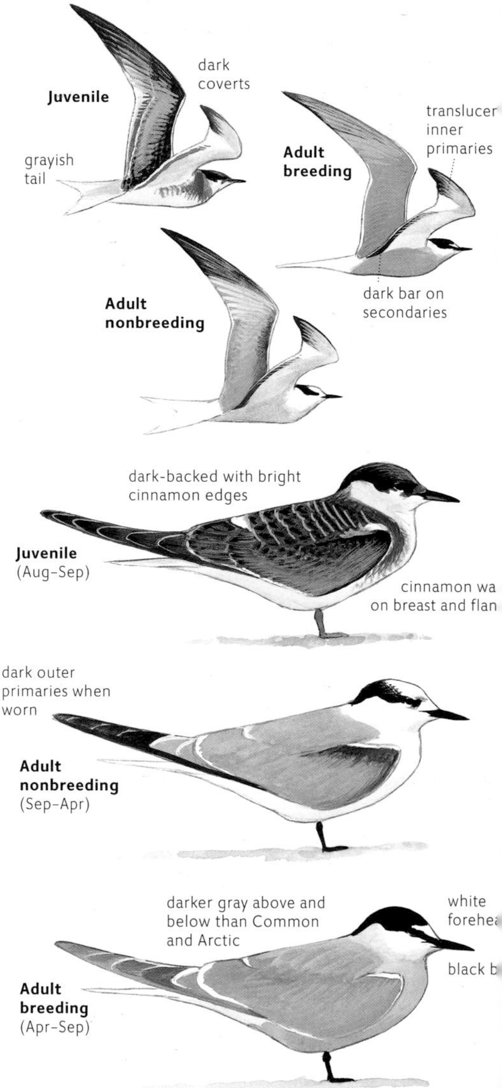

VOICE: Common call a soft, whistled *whidid* or *whididid* slight descending; reminiscent of shorebird or even House Sparro rather than tern. Slightly harsher *whirrr* near nest.

Uncommon and local. Nests in colonies on small islands in coastal lagoons and ponds. Forages for small fish in lagoons and over open ocean. Apparently never plunge-dives.

ꞵridled Tern

nychoprion anaethetus

15" WS 30" WT 3.5 oz (100 g)

ightly larger than Common Tern, with very dark gray upperside. istinguished from Sooty by more slender wings, more graceful ight, extensive white in longer tail, and habit of perching on oating debris.

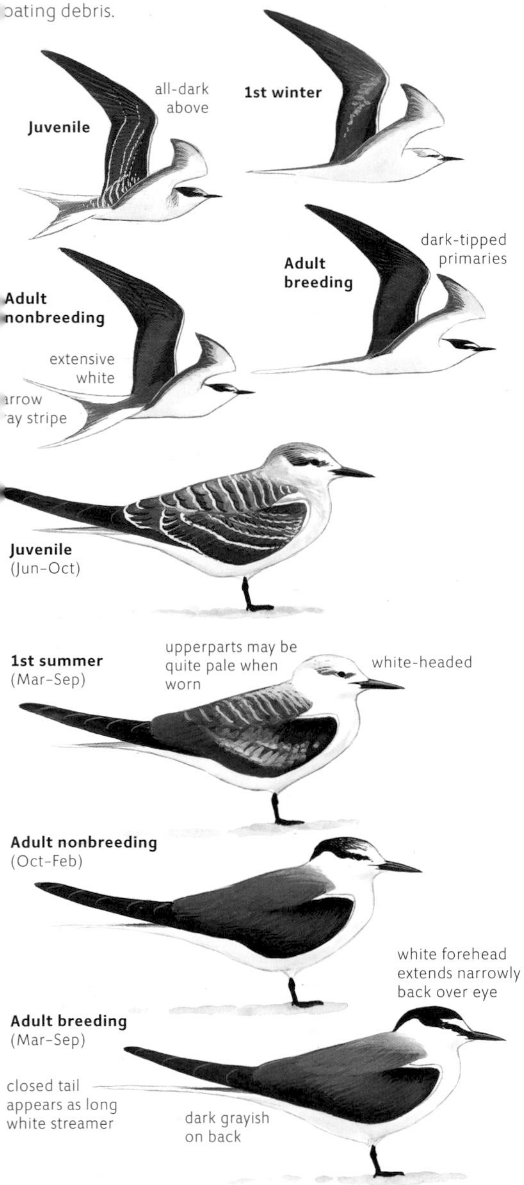

ᴐICE: Common call a rather soft, mellow whistle: a rising, nasal *eeeep*; quality like Sooty but higher, softer.

Uncommon on warm water far offshore. Very rare inland after tropical storms. Usually seen in groups of two to five along lines of floating weeds, often perching on boards or other debris. Feeds on small fish snatched from water's surface.

Sooty Tern

Onychoprion fuscatus

L 16" WS 32" WT 6 oz (180 g)

Distinctly larger, broader-winged, and more powerful than Common Tern. Note blackish upperside of adult from crown to tail, with sharply contrasting clean white underside. Immature is dark overall with whitish underwing and belly.

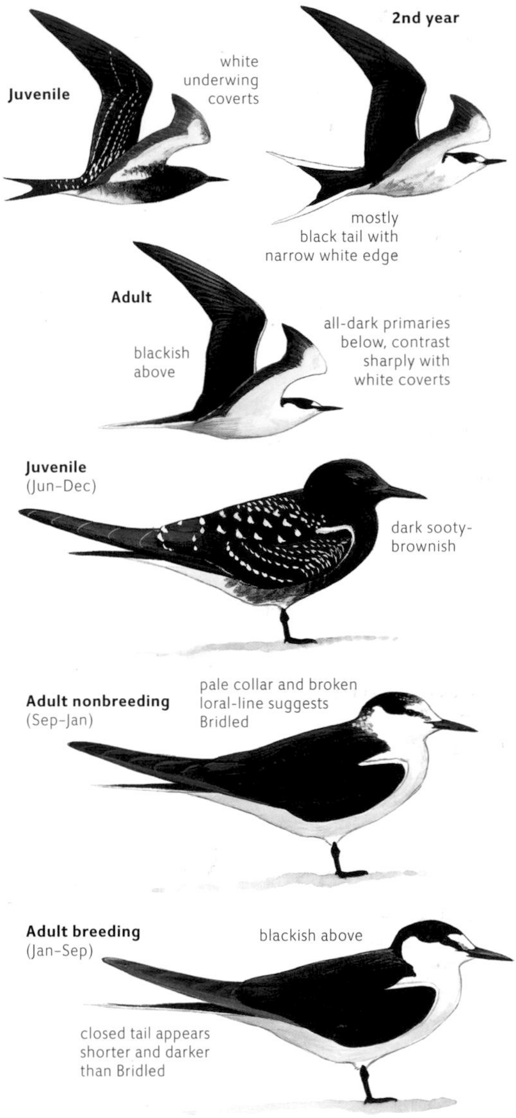

VOICE: Calls nasal, laughing *ka-weddy-weddy* or rapid *ka-WEEda-WED*; higher than Gull-billed. Threat a nasal, rasping moan *draaaaa*. Juvenile gives high, shrill *wheeer*.

Common off Florida Keys and at nesting colony on Dry Tortugas, Florida; uncommon to rare elsewhere. Very rare inland after tropical storms. Briefly forms flocks where prey is concentrated. Virtually never perches on floating debris (unlike Bridled). Feeds on small fish snatched from water's surface.

Black Tern

Chlidonias niger

L 9.75" WS 24" WT 2.2 oz (62 g)

In most areas the only tern with dark gray wings and back. Also the smallest tern in most areas, just slightly larger than Least Tern, with much broader wings, short tail, and buoyant flight. Breeding plumage unique with blackish body, gray wings; other plumages dark gray above, including gray underwing.

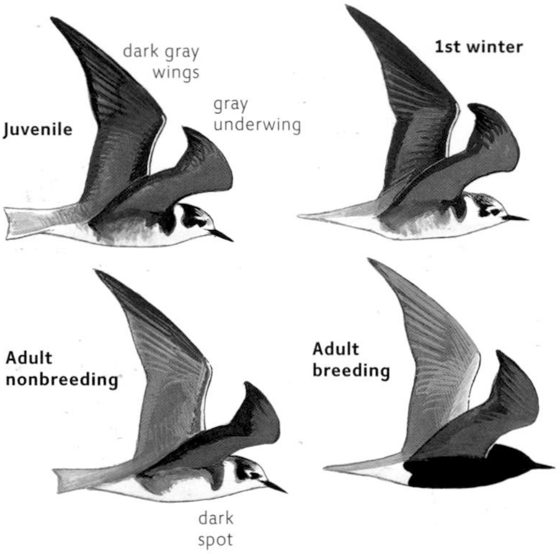

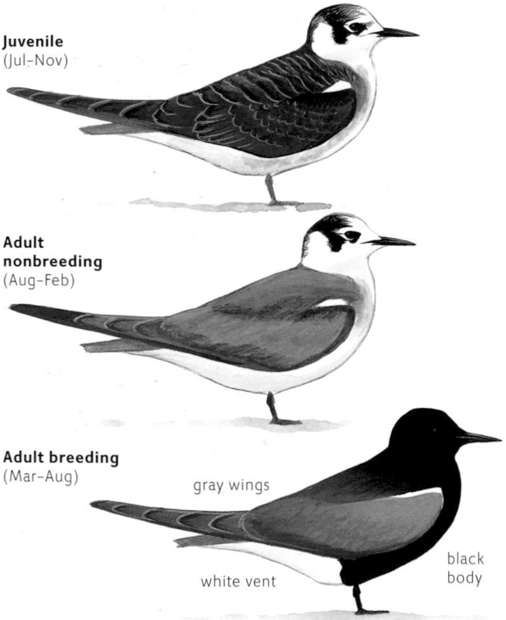

VOICE: Call a harsh, sharp, scraping, complaining *keff* or *keef* and higher *kyip*; reminiscent of Black-necked Stilt but higher.

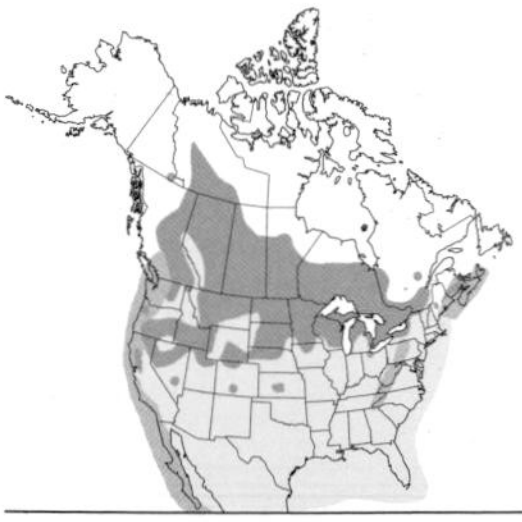

Uncommon and local. Nests on marshy ponds with emergent vegetation; migrants can be seen over any water from marshes to open ocean and often roost with other terns on sandbars. Usually in small groups. Feeds on small fish, insects, and other aquatic prey captured in flight; rarely plunge-dives.

White-winged Tern

Chlidonias leucopterus

L 9.5" WS 23" WT 2.2 oz (63 g)

Similar to Black Tern. Breeding plumage distinctive with brig white upperwing coverts, black underwing. Nonbreeding plum age much paler than Black, with whitish underwing and rum and white flanks; lacks dark spur on side of breast.

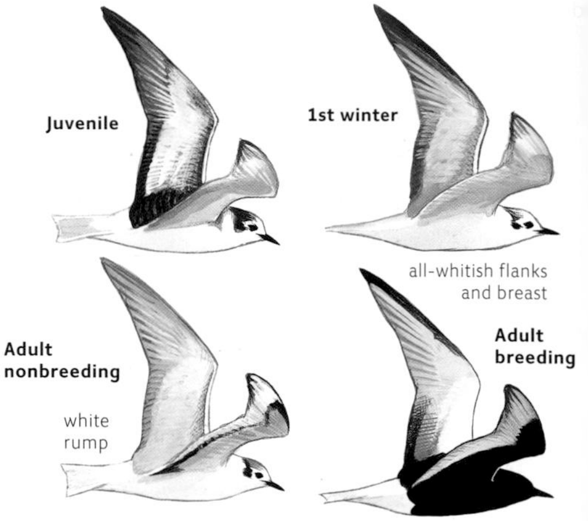

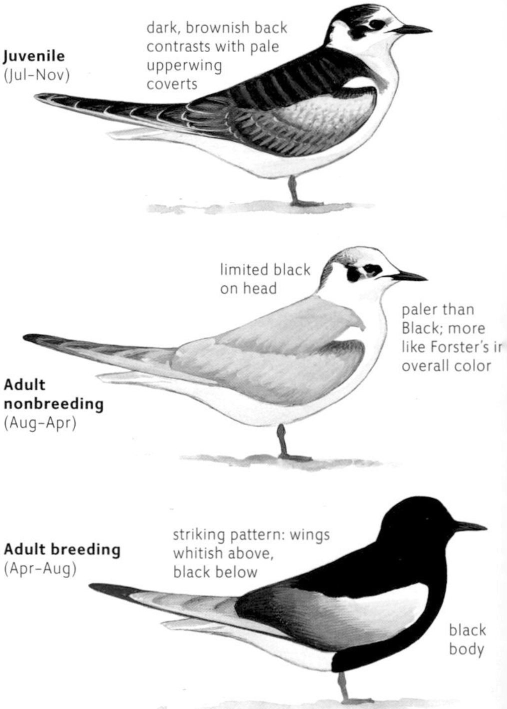

VOICE: Calls short *kesch* and *kek*; deeper and harsher than Black Tern.

Very rare visitor from Eurasia to lakes and marshes, habitat and habits like Black Tern. Most records in Northeast from Ontario and Maine to Virginia, i May-Aug, also recorded in Alas and California.

ȝlack Noddy

.nous minutus

13.5" WS 30" WT 3.9 oz (110 g)

abits and appearance very similar to Brown Noddy, but darker ɳd grayer (less brown) overall, with paler gray tail; also smaller, ᵛith thinner neck and relatively longer, more slender, and ːraighter bill. Immature has strongly contrasting white cap; dult's cap blends to gray neck.

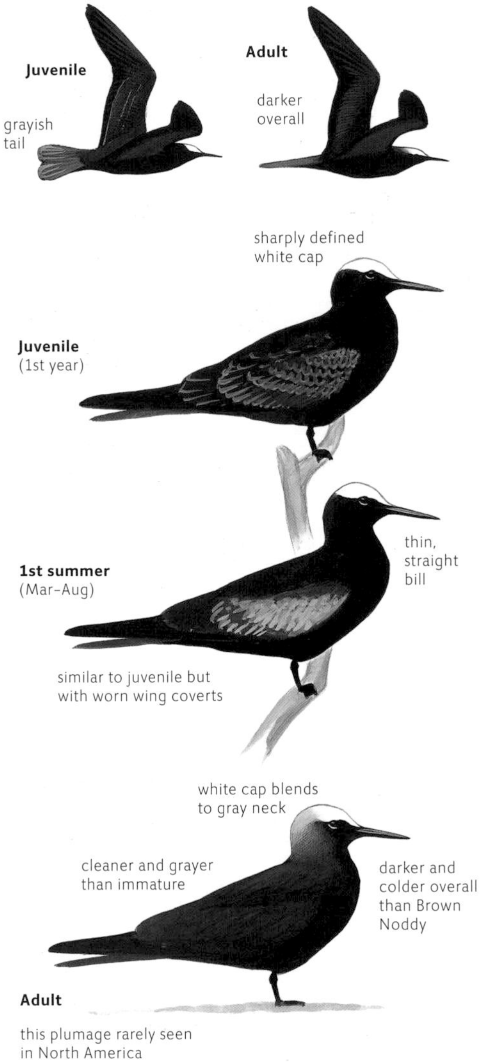

OICE: All calls similar to Brown Noddy but higher and sharper. ˌ low, grating, stuttering *krkrkrkr.* Call of 1st year birds a high, ɦin, piping *suwee.*

Very rare visitor from tropical Atlantic; several are recorded annually on Dry Tortugas, Florida, mostly immatures lingering around nesting colony of Brown Noddies.

Brown Noddy

Anous stolidus

L 15.5" WS 32" WT 7 oz (200 g)

A medium-size tern, larger than Common Tern, with relatively broad and stiff wings and broad tail. All-dark plumage and white forehead is shared only by rare Black Noddy (but compare juvenile Sooty Tern, storm-petrels, jaegers, etc.).

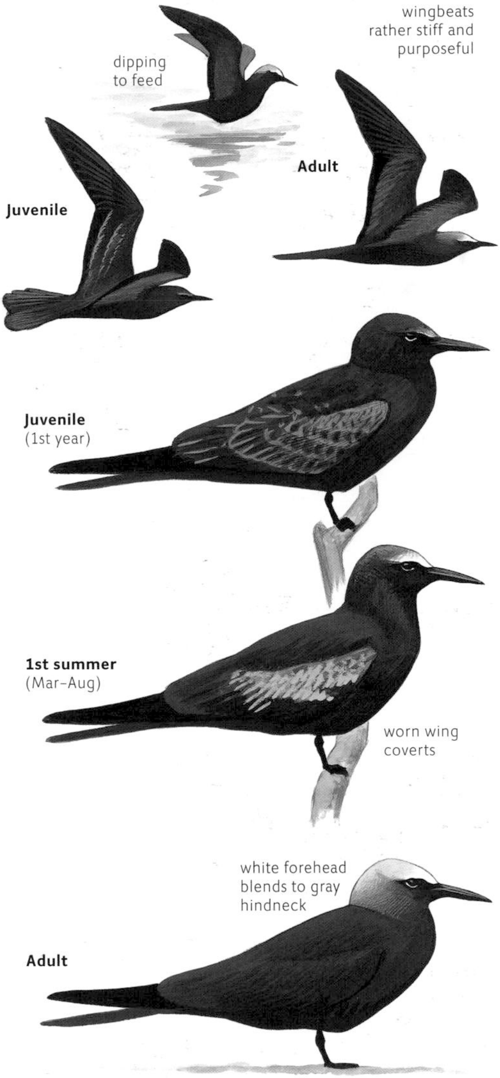

VOICE: All calls of adults low, grunting, grating, or buzzing; a deep, rattling *garrr* and a rising croak *brraak.* In courtship flight a grating, low *nek nek nek nek nek nekrrr.* 1st year birds give high whistle *tsooeek* and high, clear squawk *kweeer.*

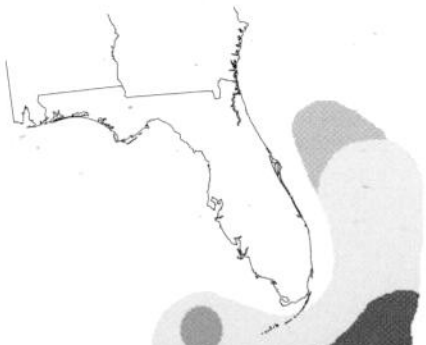

Common around nesting colony on Dry Tortugas, Florida; rare over warm pelagic waters away from Florida Keys. Very rarely seen from land (from Texas to Maine), mainly after hurricanes. Forages for small fish over warm water far offshore, often with Sooty Tern. Snatches fish from water's surface by swooping or while fluttering distinctively at surface.

Black Skimmer

Rynchops niger

L 18" WS 44" WT 11 oz (300 g) ♂>♀

Very long wings, simple black and white plumage pattern, and long red and black bill are distinctive; also note unique bill shape, with lower mandible much longer than upper. Flight graceful and buoyant, with slow beats of long wings.

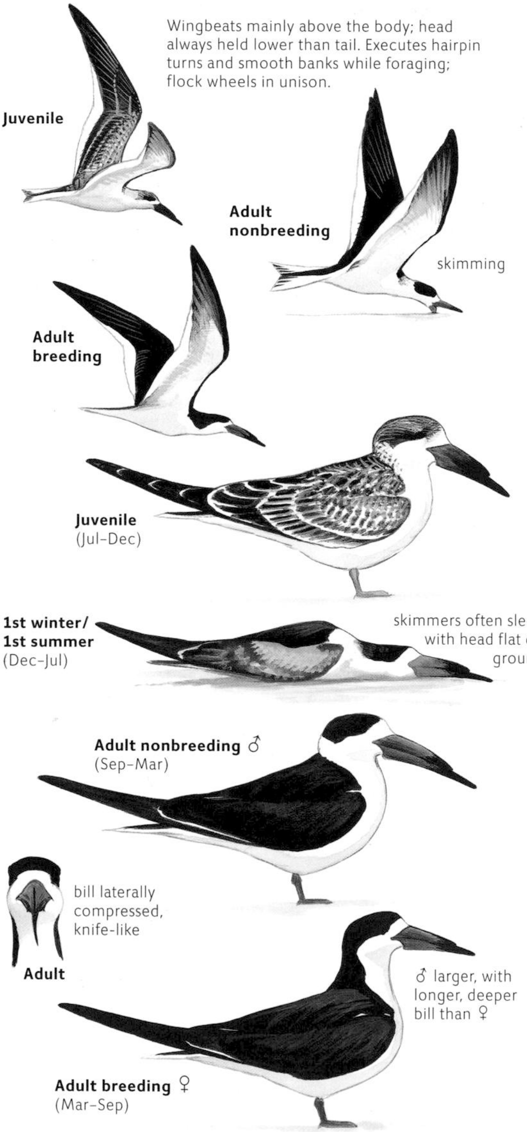

VOICE: Call a distinctive, hollow, soft, nasal barking *yep* or *yip*. Alarm a longer *Aaaaw*; quality like Long-tailed Duck. Juvenile call a higher-pitched *iip*, more squawky than adult.

Common locally around sheltered bays, inlets, and lagoons. Nests and roosts in large groups on low sandbars, usually alongside but not mixing with gulls and terns. Forages for small fish mainly at night in shallow sheltered water, often flying in small groups.

Jaeger Identification

Adults have distinctive tail shape and plumage differences, bu subadults are variable and confusing, and many brief or distar sightings must be left unidentified. Overall size and shape ar often helpful for identification, but impressions can be mislead ing. Barred underwing coverts indicate an immature, all-dar coverts an adult (except some dark morph immatures have al dark coverts). On all ages it is most helpful to focus on tail shap (even juveniles have distinctively shaped central tail feathers) an the pattern and structure of head and bill.

Jaeger Bill Shapes

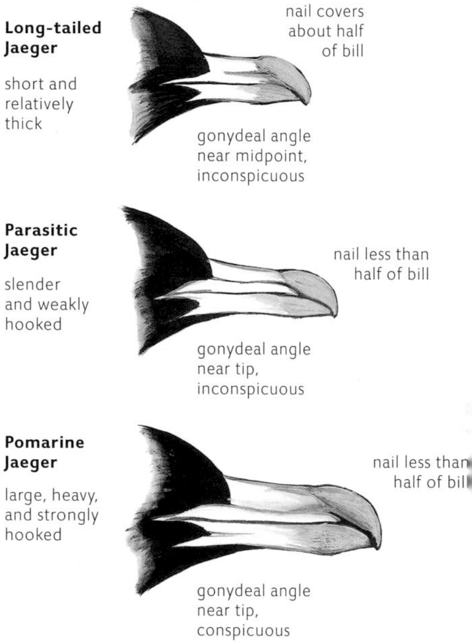

Typical flight views of Parasitic Jaeger

The actions of a jaeger or skua chasing its victim are distinctive. Certain gulls (especially Laughing and Heermann's) engage in the same pirating behavior, but usually appear less powerful and angular than jaegers.

Typical Wingbeats of Jaegers

In normal flight Long-tailed Jaeger has wingbeats mainly above horizontal

Pomarine and Parasitic have wingbeats more even.

In pursuit flight when attacking another bird all jaegers flap with very deep wingbeats.

Long-tailed Jaeger

tercorarius longicaudus

15" (adult breeding to 23") WS 43" WT 11 oz (300 g) ♀>♂

he smallest and most elegant jaeger, with relatively slender vings and long "hand," relatively long tail (even without tail treamers), and short stout bill. About the size of Laughing Gull, hases other birds less than Parasitic. Central tail feathers blunt-pped on juvenile, pointed on older birds.

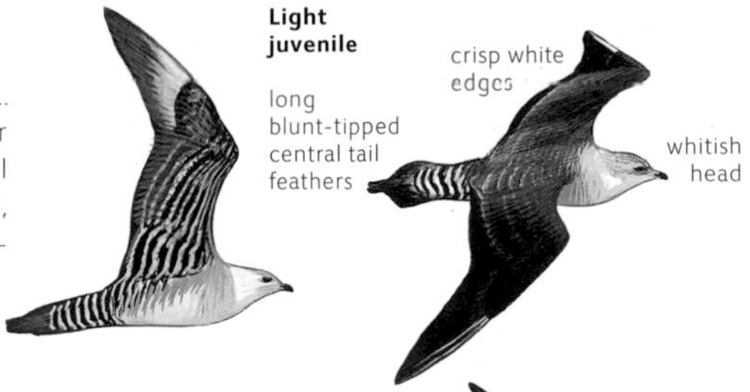

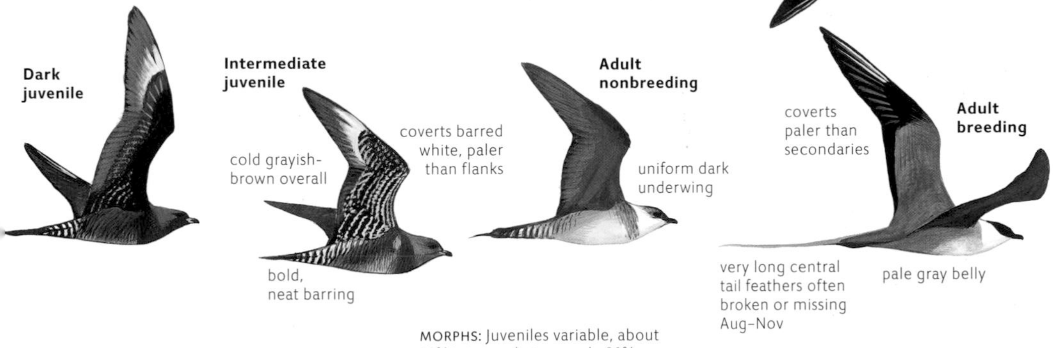

MORPHS: Juveniles variable, about 50% intermediate morph, 30% dark, and 20% light. Adults are always light, with no dark morph.

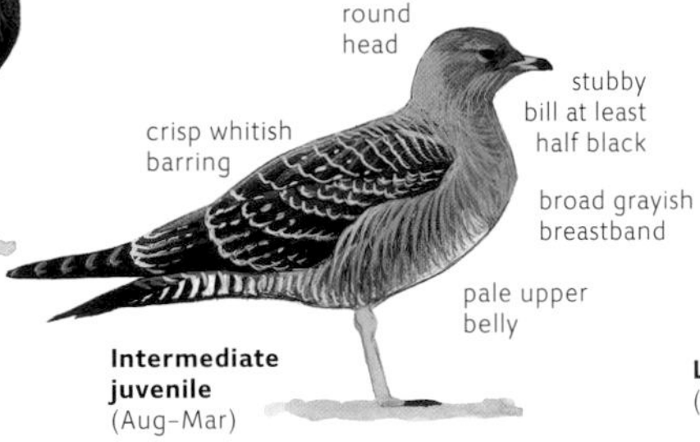

Juvenile overall color generally grayish, with crisp, pale feather edges, unpatterned grayish breast, and bold barring on tail coverts and underwing coverts.

Always shows subtle contrast between dark secondaries and paler upperwing coverts.

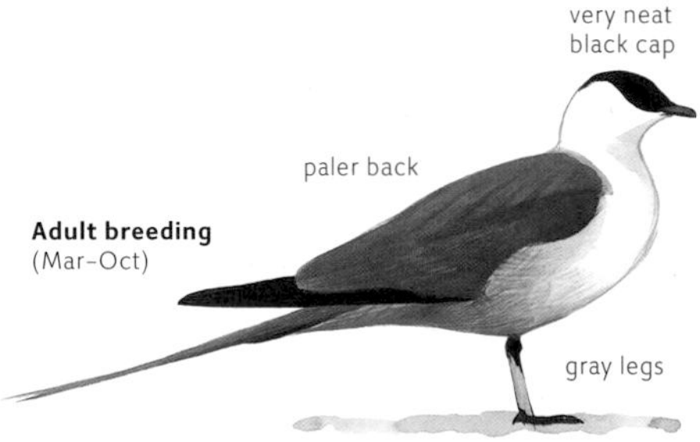

'OICE: Generally silent away from breeding grounds. All calls ıigher than other jaegers. Common call a short, trilled *krri*; also ı nasal *kee-ur* like Red-shouldered Hawk, and sharp *kl-dew*. Long :all a rattling followed by long, plaintive *feeeeoo* notes.

Uncommon to rare. Nests on Arctic tundra. Migrates on open ocean far offshore, winters mainly well south of our area; almost never seen from land (but migrants appear rarely at inland ponds and lakes in fall). Usually solitary. Diet includes lemmings and berries on tundra, and fish and other animal prey at sea; sometimes steals food from other seabirds.

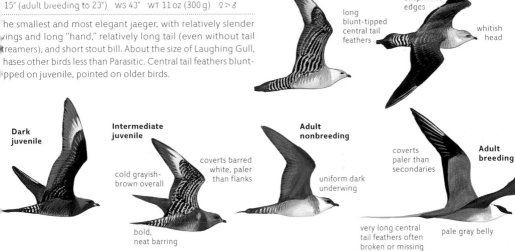

Parasitic Jaeger

Stercorarius parasiticus

L 16.5" (adult breeding to 20") WS 46" WT 1 lb (470 g) ♀>♂

The medium-size jaeger, slender but powerful, with intermediate wing and tail proportions and relatively slender bill. Slightly smaller overall than Ring-billed Gull, chases terns and small gulls frequently and relentlessly to steal food (more than other jaegers). Central tail feathers always pointed.

Juvenile overall color generally warm buff or cinnamon, with pale and faintly streaked nape, tail coverts not as strongly barred as other jaegers.

MORPHS: Juveniles variable, about 75% are intermediate morph, 15% light, and 10% dark. Adults vary from light to dark (with few intermediates), varies regionally, over 75% dark morph on Aleutian Islands to over 95% light on islands of Arctic Canada.

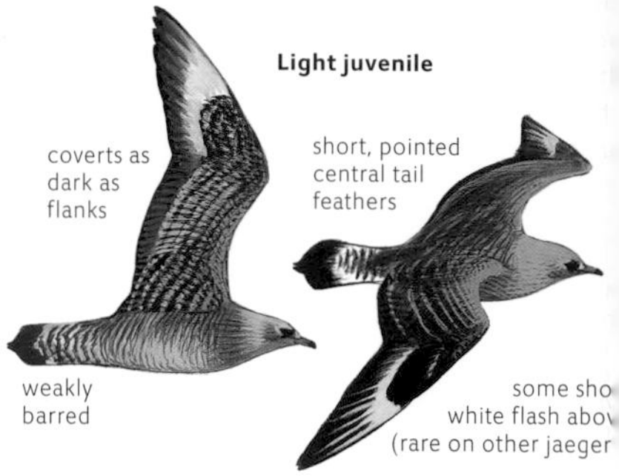

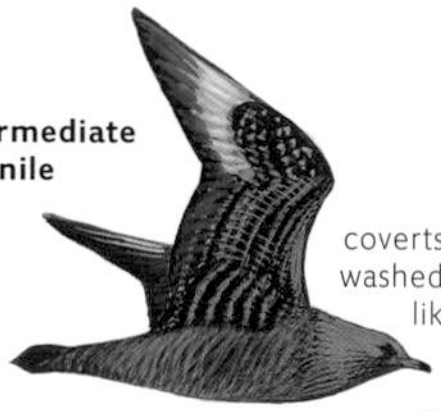

VOICE: Generally silent away from breeding grounds. Pitch intermediate between other jaegers. Common call a nasal *KEwet, KEwet*... similar to Gull-billed Tern, a little higher than Pomarine Jaeger; also short barking *gek* notes. Long call includes nasal whining or crowing *feee-leerrrr.*

Uncommon. Nests on Arctic tundra. Winters on open ocean mostly within a few miles of land and this species is the jaeger most frequently seen from land. Usually solitary. Feeds mainly on fish stolen from other seabirds.

Pomarine Jaeger

Stercorarius pomarinus

L 18.5" (adult breeding to 23") WS 52" WT 1.5 lb (700 g) ♀>♂

The largest and bulkiest jaeger, with relatively broad wings and heavy bill, slightly larger (but appearing much more powerful) than Ring-billed Gull. Often chases other birds to steal food, but chases are generally brief and direct. Central tail feathers always broad and rounded.

Juvenile overall color generally plain brownish, without pale feather edges and with plain dark head and nape. Combination of dark head and pale uppertail coverts not shown by Parasitic; pale bill contrasting with dark head also differs from Parasitic.

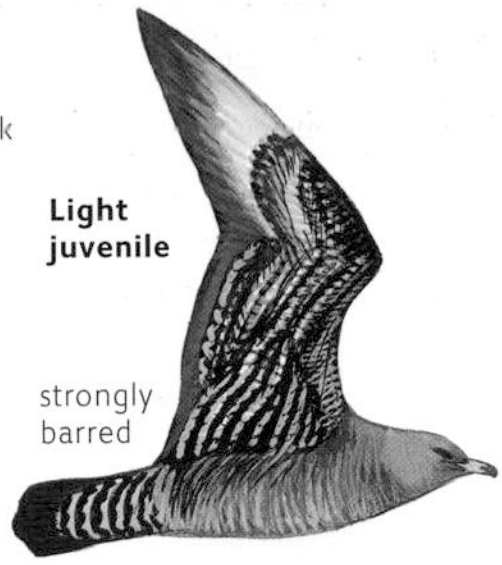

MORPHS: Juveniles less variable than other jaegers, about 90% intermediate morph, 9% dark, 1% light. Adults are 90% light morph with about 10% dark morph and very few intermediates.

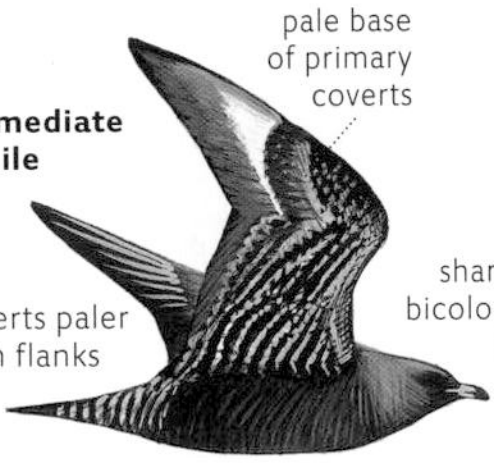

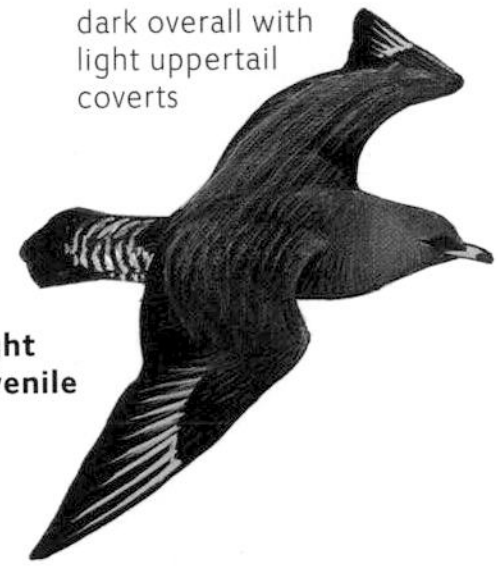

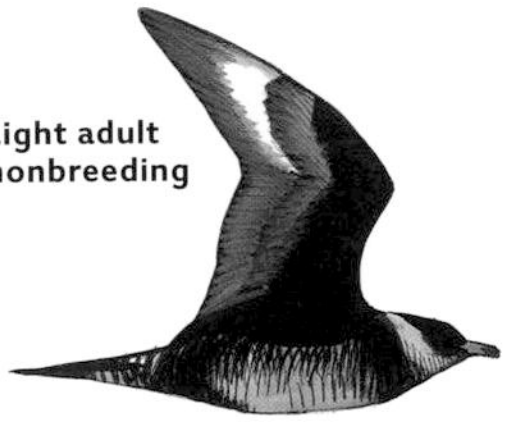

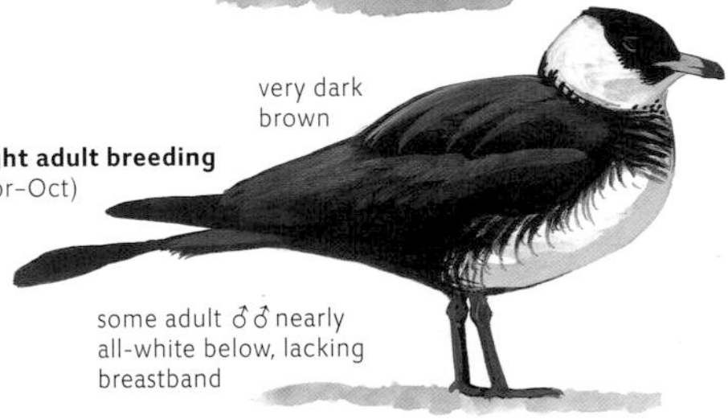

VOICE: Occasionally gives barking calls in winter. On breeding grounds a yelping *vee-veef*, also a short, barking *geck* or complaining and wavering *ehwewewewe*. Long call a series of rising, nasal *week* notes, unlike the other rising and falling notes of other jaegers.

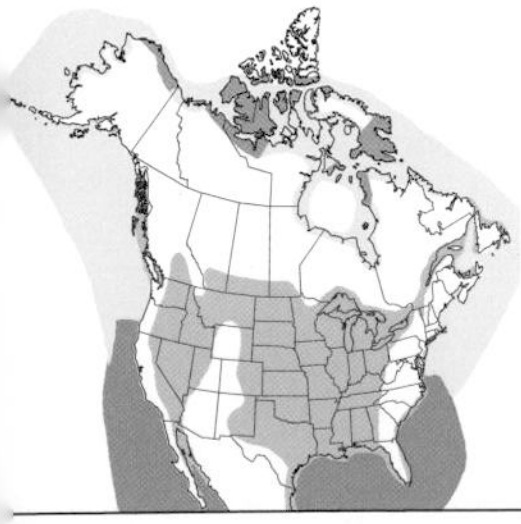

Uncommon. Nests on Arctic tundra. Winters on open ocean and only occasionally seen from land. Solitary. On tundra feeds mainly on lemmings; at sea eats mostly fish, often stolen from other seabirds. The least frequent jaeger inland in most of the West.

South Polar Skua

Stercorarius maccormicki

L 21" WS 52" WT 2.5 lb (1,150 g) ♀>♂

Very large and stocky, with broad wings, short tail, and heavy bill; darker, bulkier, and more powerful than immature gulls. Obviously larger, broader-winged, and shorter-tailed than Pomarine Jaeger, with more obvious white wing-patch.

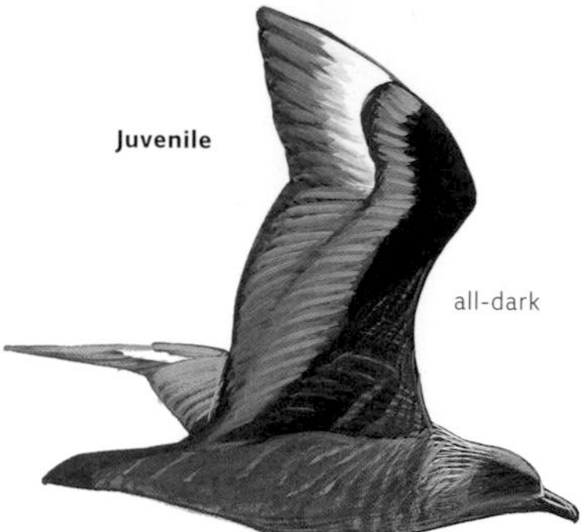

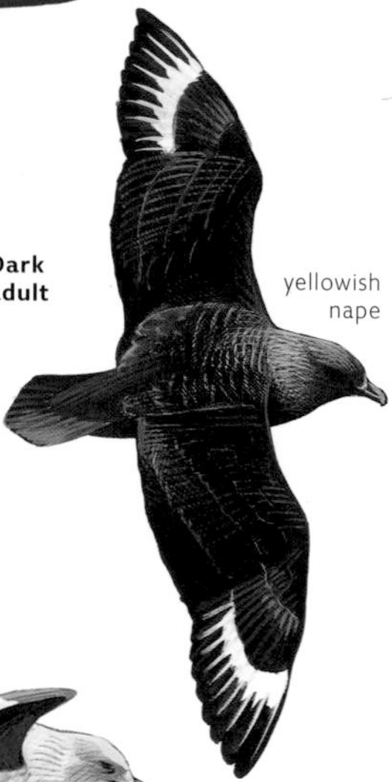

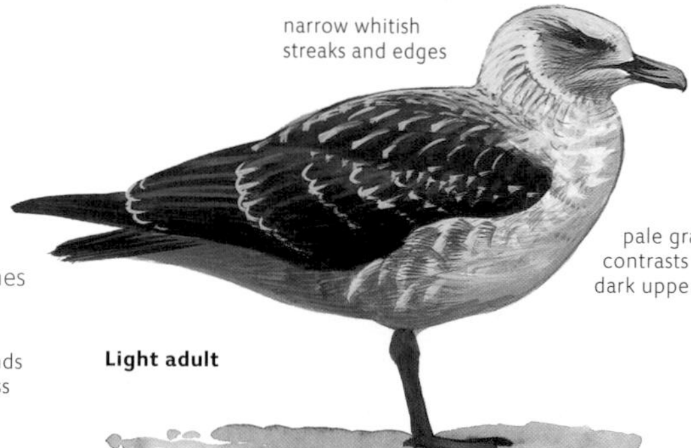

VOICE: Generally silent away from breeding grounds. Sometimes gives a weak, nasal, gull-like *haaasi*.

Rare visitor from nesting grounds in Antarctic; small numbers pass through eastern and western North America on open ocean far offshore, mainly from May–Oct. Solitary. Feeding habits similar to jaegers: eats mostly fish at sea and often steals fish from other seabirds.

Great Skua

Stercorarius skua

L 23" WS 55" WT 3.2 lb (1,450 g) ♀>♂

Very similar to South Polar Skua, but with usually obvious cinnamon tones on underparts and pale spangles on upperparts, averages larger with heavier bill. South Polar has colder gray-brown color without pale spots, but some immature Great Skuas can be very similar.

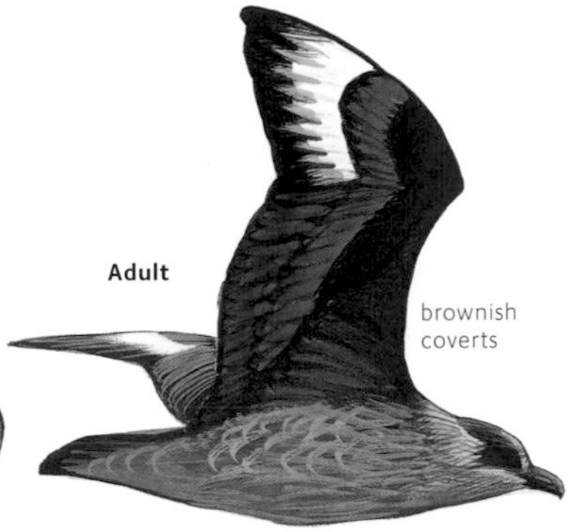

VOICE: Generally silent away from breeding grounds. Short call a *hek*. Long call a series of short, nasal *pyeh* notes

Rare visitor from nesting grounds in Iceland and Europe; small numbers found on open ocean far offshore mainly from Aug–Mar. Solitary. Feeds mainly on fish, often stolen from other seabirds.

Alcids

FAMILY: ALCIDAE

22 species in 10 genera. All are stocky, oceanic birds, very rarely seen on fresh water and coming to land only when nesting. Most species nest colonially in crevices or burrows, (murres nest on cliff ledges) on remote islands. Juveniles of most species leave the nest before fully grown and complete development at sea accompanied by one parent. Some species form feeding aggregations where prey is abundant, but are not strongly gregarious, and other species (e.g., most murrelets) are almost always seen as singles or pairs. All species pursue prey by diving, using thei wings to "fly" underwater, and their feet only for steering. Th relatively small wings needed for underwater propulsion are les efficient in air, takeoff can be difficult, and all species have ver direct flight with continuous buzzing wingbeats. Specialized bi shapes are related to feeding habits; some have elaborate bi ornaments or head plumes in breeding season. Compare loons grebes, and ducks. Nonbreeding adults are shown.

Genus *Alle*

Dovekie, page 245

Genus *Uria*

Common Murre, page 244

Thick-billed Murre, page 244

Genus *Alca*

Razorbill, page 245

Genus *Cepphus*

Black Guillemot, page 243

Pigeon Guillemot, page 243

Genus *Ptychoramphus*

Cassin's Auklet, page 249

Genus *Brachyramphus*

Long-billed Murrelet, page 249

Marbled Murrelet, page 248

Kittlitz's Murrelet, page 248

Genus *Synthliboramphus*

Scripps's Murrelet, page 246

Guadalupe Murrelet, page 246

Craveri's Murrelet, page 247

Ancient Murrelet, page 247

Genus *Aethia*

Parakeet Auklet, page 250

Least Auklet, page 250

Whiskered Auklet, page 251

Crested Auklet, page 251

Genus *Cerorhinca*

Rhinoceros Auklet, page 252

Genus *Fratercula*

Atlantic Puffin, page 253

Horned Puffin, page 253

Tufted Puffin, page 252

Pigeon Guillemot

epphus columba

13.5" WS 23" WT 1.1 lb (490 g)

maller than murres and round-bodied, with relatively broad, ounded wings. Breeding plumage unique: all-black with white ving coverts forming conspicuous patch even when perched. Note distinctive wing pattern on juvenile and nonbreeding lumages. Very similar to Black Guillemot.

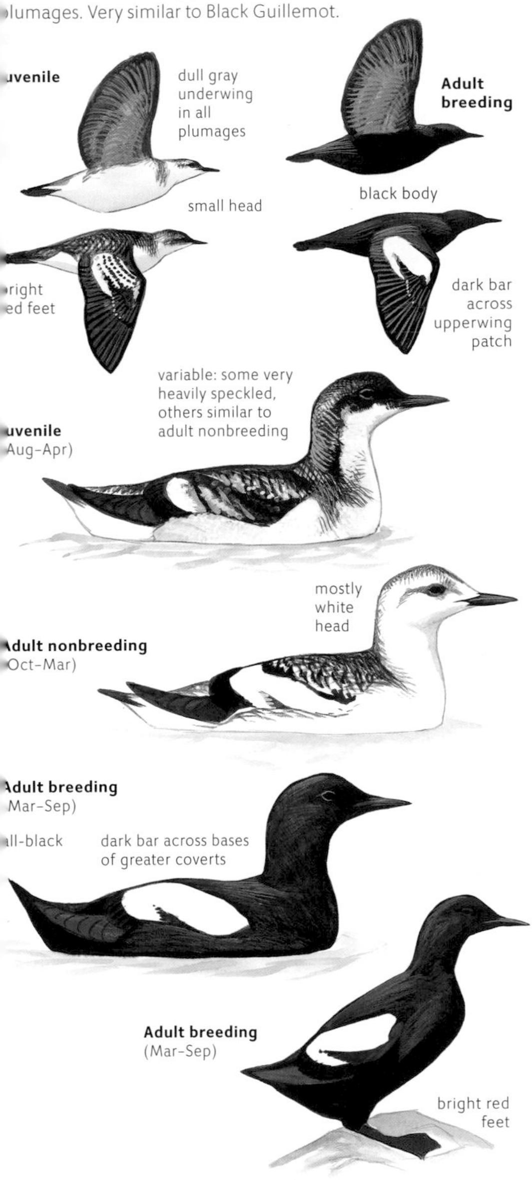

VOICE: Similar to Black Guillemot; no differences known.

Uncommon along rocky shorelines, seldom far from land. Often in pairs or small groups. Nests in crevices and among boulders at ocean's edge, even under piers or buildings. Feeds on small fish captured underwater.

Black Guillemot

Cepphus grylle

L 13" WS 21" WT 15 oz (430 g)

Habits and appearance very similar to Pigeon Guillemot, but range barely overlaps; slightly smaller and smaller-billed overall, lacking dark bar across white upperwing patch. Best distinguished in flight by clean white underwing in all plumages. Juvenile and nonbreeding plumages average paler and whiter than Pigeon Guillemot.

VOICE: Calls frequently near breeding areas: extremely high-pitched, thin, squeaky, or piping whistles; a drawn-out screaming whistle *see-oo* or *swweeeeeer* up to two seconds long. Also short *sit sit sit*... notes often leading into series, rising and accelerating then falling at end. Courting group sits on rocks or water and calls while displaying red mouth-lining.

Uncommon and local along rocky shorelines. Habits like Pigeon Guillemot.

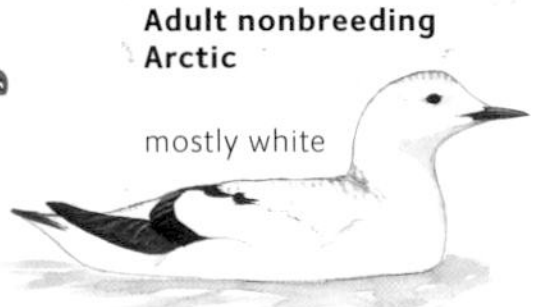

Common Murre

Uria aalge

L 17.5" WS 26" WT 2.2 lb (990 g)

Very similar to Thick-billed Murre; both football-shaped with clean black and white plumage, but Common has browner upperparts and longer, thinner, straighter bill; more dusky markings on flanks, more white on head in nonbreeding plumage.

Thick-billed Murre

Uria lomvia

L 18" WS 28" WT 2.1 lb (970 g)

Very similar to Common Murre; upperparts blacker and bi shorter, thicker, and more downcurved. In nonbreeding plumag head still mostly dark, with white only on throat. Overall a littl heavier with broader wings, cleaner white flanks and underwing

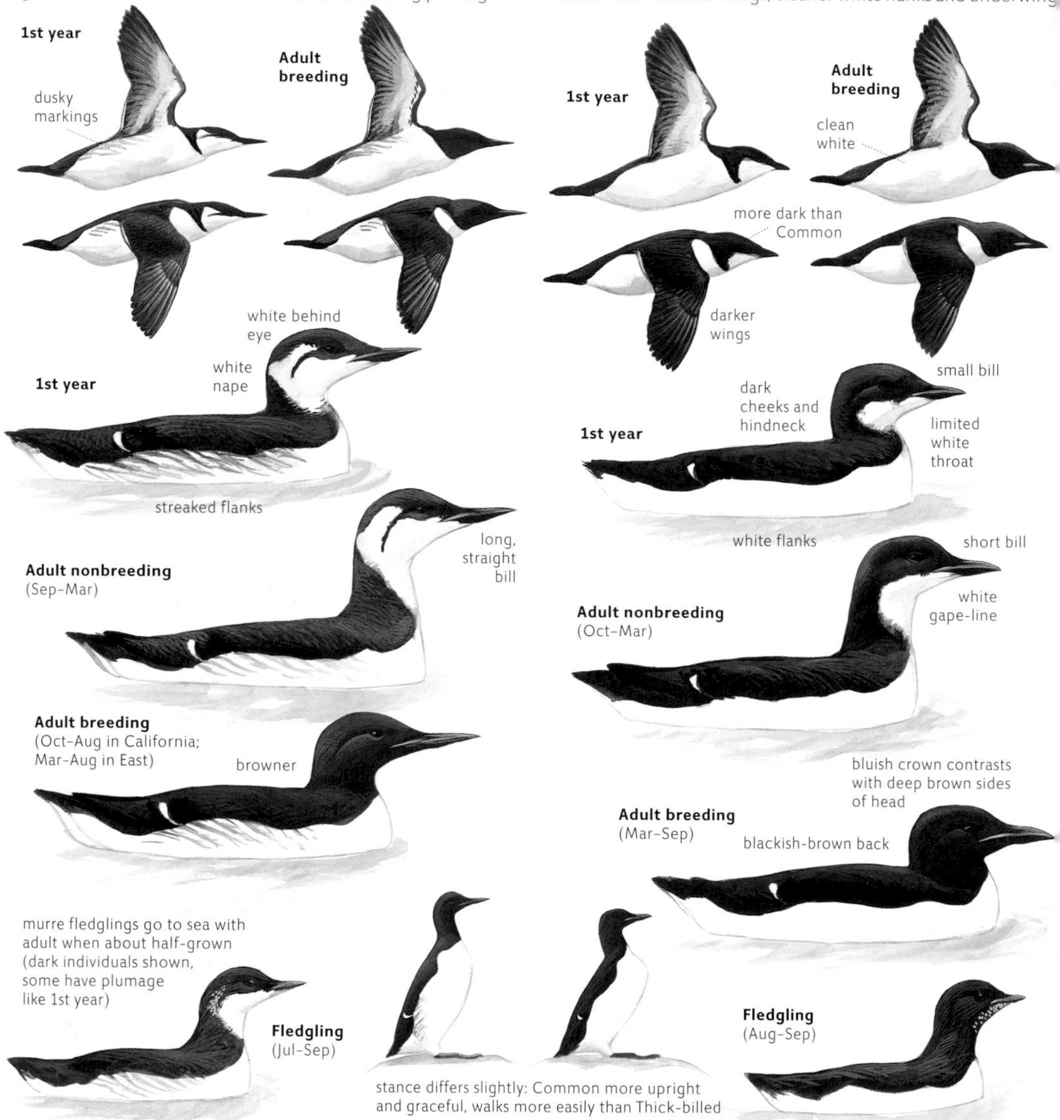

VOICE: Adults silent at sea. Call in nesting colony a moaning, nasal *aarrrr*; higher-pitched and more pleasant-sounding than Thick-billed. Juvenile at sea begs with insistent, short, low whistle *pidoo-WHIdoo* while following adult.

Common at nesting colonies on rocky cliffs rising out of the ocean, laying eggs on open rock ledges. Winters on open ocean over shallow banks far offshore and often seen from land. Solitary or in small groups. Feeds on small fish captured underwater.

VOICE: Adults silent at sea. Call a roaring, groaning, angry-sounding *aoorrr*. Juvenile at sea gives whistle call like Common.

Common at nesting colonies on rocky cliffs above ocean, where i chooses narrow ledges with overhanging rocks (Common chooses broader and more open ledges or plateaus). Winters on open ocean mostly far offshore, but often seen from land. Solitary or in small groups. Feeds on small fish captured underwater.

Razorbill

Alca torda

L 17" WS 26" WT 1.6 lb (720 g)

Similar to murres in size and plumage. Note deep bill (with vertical white line on adult) and long tail, giving posterior end a tapered appearance; tail often raised while swimming. In nonbreeding plumage note some white behind eye, but not as much as on Common Murre.

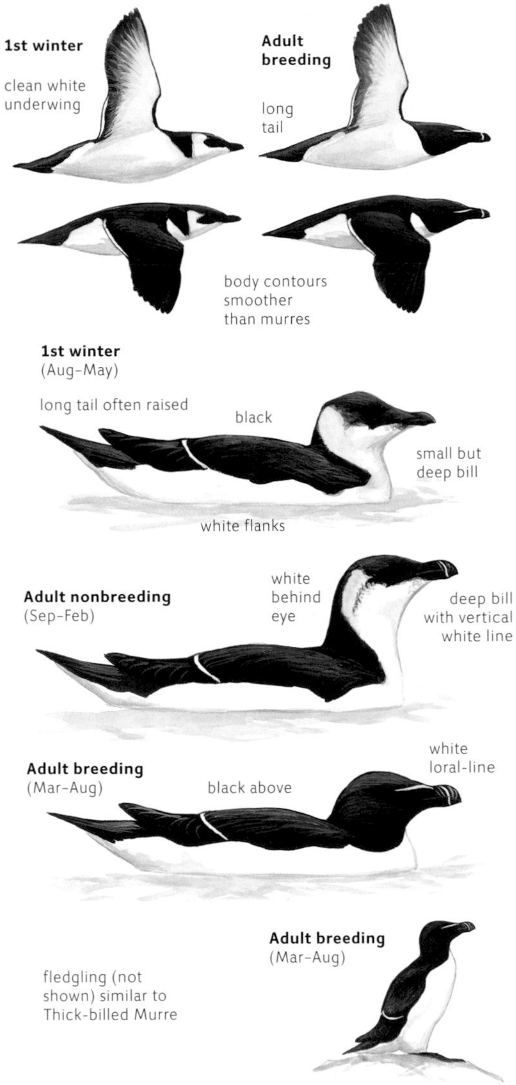

VOICE: Adults silent at sea. Call a grunting *urrr* with dry, rattling quality; more mechanical than murres, with melancholy sound. Juvenile at sea begs with whistled call similar to murres.

Uncommon. Nests in crevices on cliffs or among boulders on small rocky islands with other alcids. Winters on open ocean close to shore; often seen from land, singly or in small groups that swim and fly in lines. Feeds on small fish captured underwater.

Dovekie

Alle alle

L 8.25" WS 15" WT 6 oz (160 g)

Small, neckless, with clean black-and-white plumage. Similar to murres in plumage, but much smaller, with very stubby bill, chunky body, and relatively long wings. In comparison to murrelets and other small alcids shows more sharply contrasting black and white plumage.

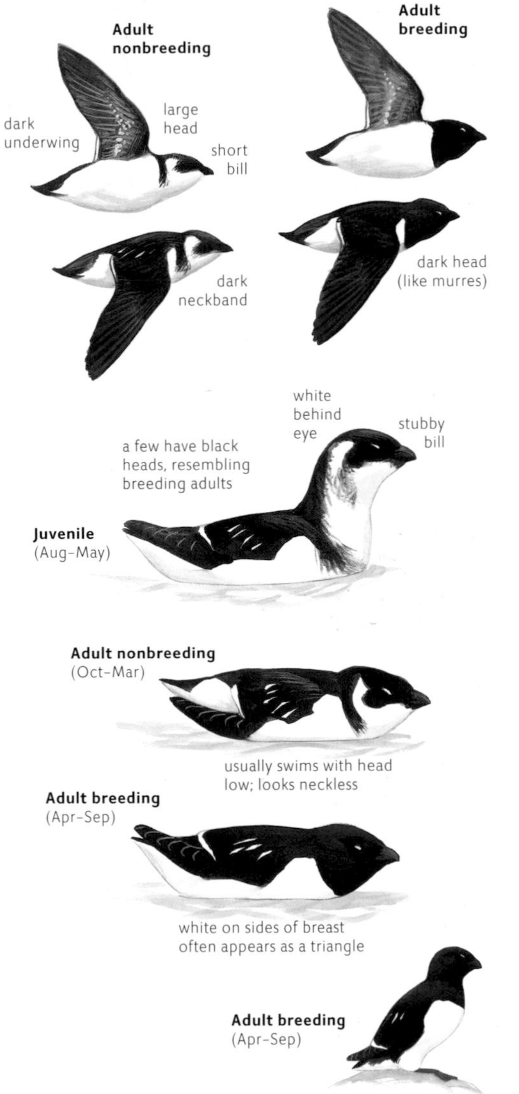

VOICE: Silent at sea. Call a high, screaming trill, rising and falling, lasting one to three seconds. Juvenile at sea gives shrill peeping calls.

Common visitor from nesting colonies in Greenland and northern Europe; most numerous and seen from land in Canada, irregular farther South. Also nests in very small numbers in colonies of other auklets in northern Bering Sea. Winters on open ocean, usually far offshore. Solitary or in small groups. Feeds on plankton captured underwater.

Scripps's Murrelet

Synthliboramphus scrippsi

L 9.8" WS 15" WT 6 oz (170 g)

A small alcid, relatively slender-bodied and thin-billed, with striking black and white pattern. Similar to Craveri's Murrelet; best distinguished in flight by clean white underwing, less dark on flanks, at close range by less black on face with white extending slightly up in front of eye.

VOICE: Call a series of three to eight high-pitched peeping notes *seep seep seep seep* unlike Craveri's. Juvenile gives a thin *peer*.

Uncommon on open ocean on relatively warm water, often very far from shore. Nests in crevices on rocky islands. Usually in pairs. Feeds on small fish and plankton. Rare postbreeding visitor north to Washington, never recorded inland.

Guadalupe Murrelet

Synthliboramphus hypoleucus

L 9.8" WS 15" WT 6 oz (170 g)

Very similar to Scripps's Murrelet, and until recently considered a subspecies. Best distinguished by face pattern, white extending up in front of and above eye, white above gape, broader white crescent below eye, and paler grayish auriculars. Also averages slightly longer bill and different voice.

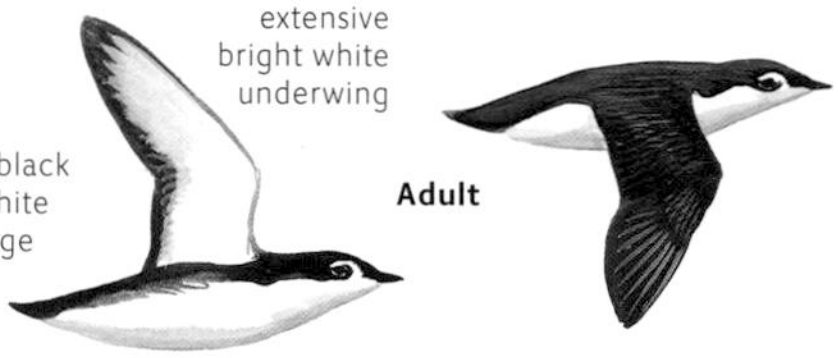

VOICE: Call a rattle similar to Craveri's, unlike the high peeping of Scripps's.

Rare visitor from nesting grounds off western Mexico (a few pairs may nest on Channel Islands off southern California), recorded irregularly on open ocean north to Monterey, very rarely north to Washington. Habits like Scripps's.

ʗraveri's Murrelet

ynthliboramphus craveri

9.8" WS 15" WT 5 oz (150 g)

ery similar to Scripps's Murrelet; best distinguished in flight by rayish (not white) underwing. On water note more extensive lack on face extending to chin, more extensive dark on flanks, nd more brownish (not grayish) tinge on dark upperside.

fledgling (not shown) similar to Scripps's; identified by face pattern and accompanying adult

ʼOICE: Poorly known. Adult at sea gives a high, cricket-like rattle ɔr trill *sreeeeer.* Other calls undescribed.

Rare visitor from nesting colonies on islands off Mexico. Found on open ocean on relatively warm water far offshore. Nests in crevices on rocky islands. Usually seen in pairs. Feeds on small fish and plankton.

Ancient Murrelet

Synthliboramphus antiquus

L 10" WS 17" WT 7 oz (205 g)

Strikingly patterned, with uniform gray back, black face, and white neck. White underwing coverts usually obvious in flight. Distinguished from other small alcids by bright white underparts, from murres by small size, white collar.

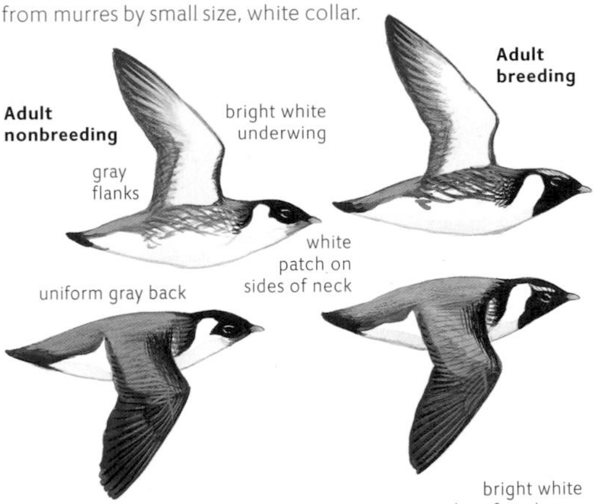

VOICE: Adult at sea gives a short, whistled *teep.* Variety of calls at nest site include a short *chirrup* and varied chips, trills, and rasping sounds.

Uncommon on open ocean, usually on relatively cold water along rocky coasts near shore. Very rare visitor to inland lakes and Atlantic coast. Nests in colonies in burrows on offshore islands. Often seen in pairs or small groups at sea, sometimes forming flocks of up to 50. Feeds on small fish and plankton.

Kittlitz's Murrelet

Brachyramphus brevirostris

L 9.5" WS 17" WT 8 oz (220 g)

Very similar to Marbled Murrelet but note shorter bill. Breeding plumage shows white undertail coverts and paler, more golden-brown color overall; nonbreeding plumage shows mostly white face (including above eye) and narrow dark crown-stripe.

flight of all murrelets low, rapid, with very fast beats of long, pointed wings

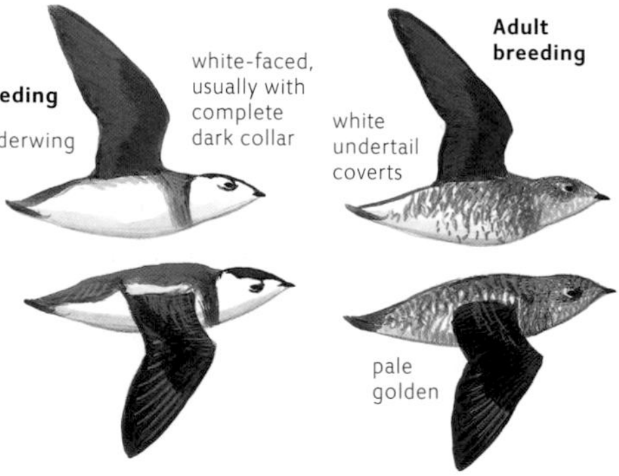

VOICE: Less vocal than Marbled. Call a quiet, low groan *urrrhhn*, similar to rarely heard groan call of Marbled. Also a short quack *urgh*.

Uncommon on open ocean near shore. Habits like Marbled (e.g. usually in pairs or small groups on the water), but always nests on the ground (never on a tree branch).

Marbled Murrelet

Brachyramphus marmoratus

L 9.75" WS 16" WT 8 oz (220 g)

Breeding plumage entirely dark brownish; nonbreeding plum age black and white with obvious white collar and partial dar breastband.

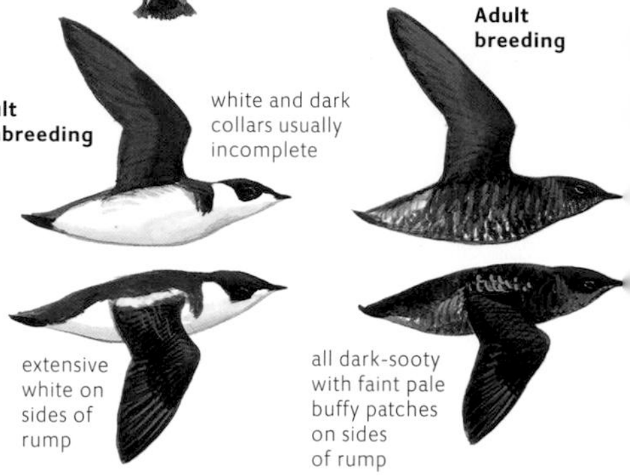

overall extent of white changes dramatically when active or diving (wings exposed) or relaxed (wings concealed by white flanks)

VOICE: Quite vocal in flight near nest site as well as at sea. Fligh call a squealing, gull-like but high-pitched, slightly descendin series *kleeer kleeeer kleeeer*.... Also a rarely heard groan *urrrrr* and a high, clear *quip*.

Uncommon on open ocean along rocky coasts. Often seen ir pairs or small groups. Feeds on small fish and plankton. Nests up to 45 miles (70 km) inland on the ground or high on a mossy tree branch.

ʟong-billed Murrelet

rachyramphus perdix

10" WS 17" WT 10 oz (290 g)

milar to Marbled, note all-dark nape (Marbled has white or pale ollar). In breeding plumage dark brown body contrasts with ʰhitish throat. Bill averages slightly longer, and overall size larger, ıan Marbled, but this is generally not helpful in the field.

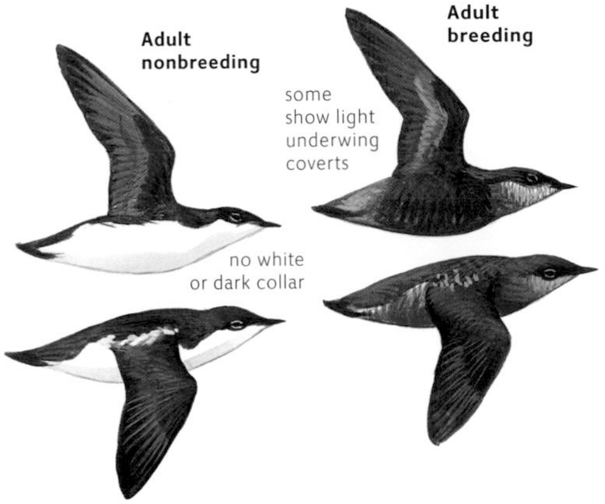

ᵛOICE: Undescribed. Usually silent at sea.

Rare visitor from Siberia to open ocean, very rarely on inland lakes and to Atlantic coast, where Marbled Murrelet has never been recorded. Feeds on small fish and shrimp captured underwater. Until recently considered a subspecies of Marbled.

Cassin's Auklet

Ptychoramphus aleuticus

L 9" WS 15" WT 6 oz (185 g)

Small and chunky, with large head and short triangular bill. Grayish overall; white eye of adult surprisingly conspicuous at a distance. Distinguished from other small alcids by overall gray color, pale belly, stout triangular bill.

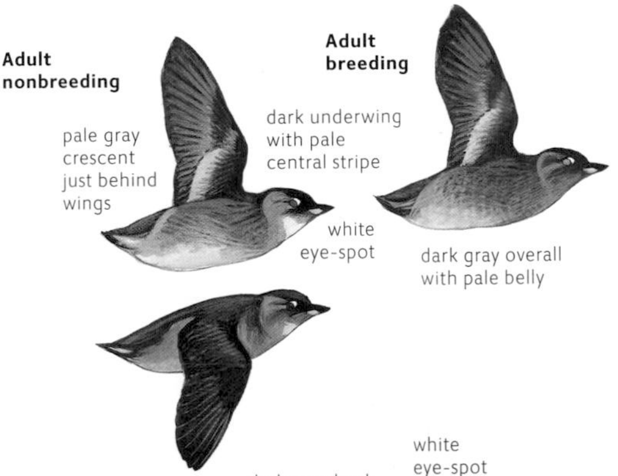

VOICE: Silent at sea. Adults in colony at night give chorus of hoarse, rhythmic calls *RREP-nerreer* with many variations (similar to calls of Parakeet Auklet); also a short, harsh *skreer.*

Common on open ocean usually far from shore, but can be seen from land regularly. Nests in colonies in burrows or crevices on islands. At sea may be solitary or in large feeding groups. Feeds mainly on plankton.

Parakeet Auklet

Aethia psittacula

L 10" WS 18" WT 11 oz (315 g)

Stocky and round-headed, with upturned, nearly circular orange bill and relatively broad, rounded wings. Combination of uniform dark gray upperside and flanks with extensive bright white on underside is unique among alcids.

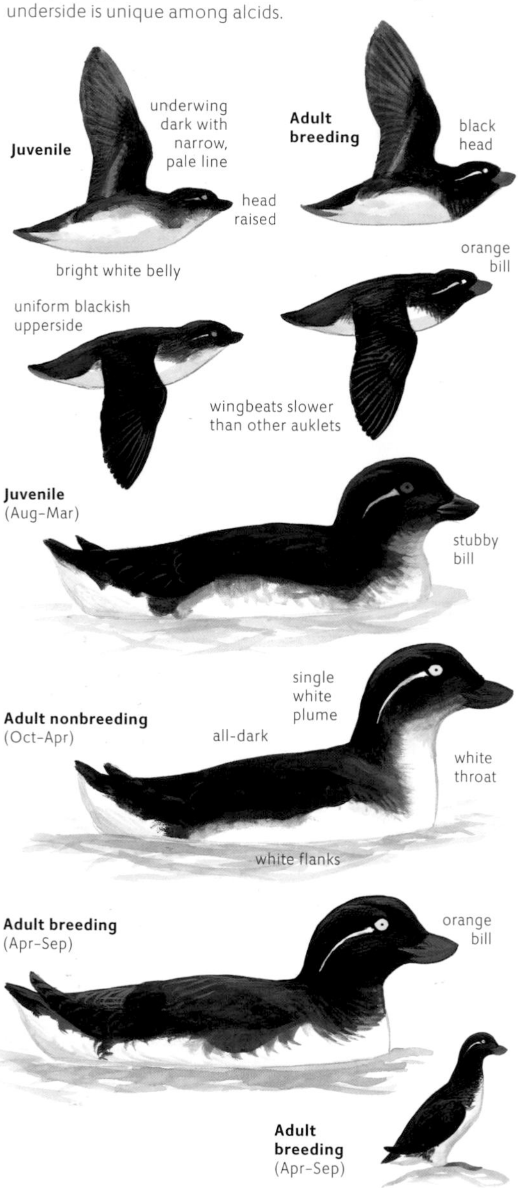

VOICE: Silent at sea. Adults in colony give rhythmic, hoarse calls similar to Cassin's Auklet; also a quavering, descending squeal.

Common on open ocean. Nests in colonies in crevices or burrows on islands. Often solitary at sea, never in large groups. Feeds on jellyfish, as well as various plankton.

Least Auklet

Aethia pusilla

L 6.25" WS 12" WT 3 oz (85 g)

Our smallest alcid; tiny and compact, with small bill. Note pal scapular-stripe and pale stripe on underwing coverts in all plum ages. Breeding plumage shows speckled belly and white throa nonbreeding plumage shows entirely white underparts.

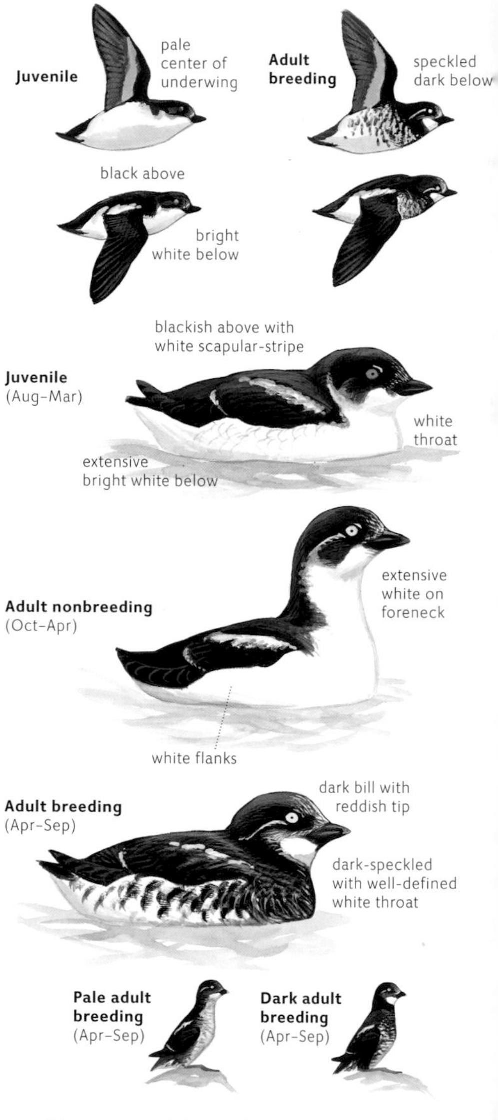

VOICE: Silent at sea. Adults in colony vocal; varied repertoire no similar to other auklets; mainly a high, grating or churring trill o chatter in pulsing series reminiscent of White-throated Swift.

Common on open ocean. Nests in large colonies in crevices among boulder fields. Usually in small flocks at sea. Feeds on plankton.

Vhiskered Auklet

ethia pygmaea

/.75" WS 14" WT 4.2 oz (120 g)

mall and small-billed, with thin dark crest. Only slightly larger an Least Auklet. Distinguished from other gray auklets by small ze, small bill, pale gray rump, thin white plumes on face, and ale gray belly.

OICE: Silent at sea. Adults in and near colony give plaintive, high, itten-like *meew* and rapid series of sharp, two-syllable notes *eedeer beedeer beedeer beedeer bideer bideer bideer bidi idi bidi bidee.*

Common locally on open ocean where powerful tidal currents flow between islands. Nests in colonies in crevices on rocky islands. Often in large and very compact, dense flocks at sea. Feeds on plankton and small fish.

Crested Auklet

Aethia cristatella

L 10.5" WS 17" WT 10 oz (285 g)

Heavy-bodied and dark gray overall, with rather long slender wings, large head, and deep bill. More uniform gray than other auklets in all plumages; adult's stubby orange bill and shaggy crest conspicuous.

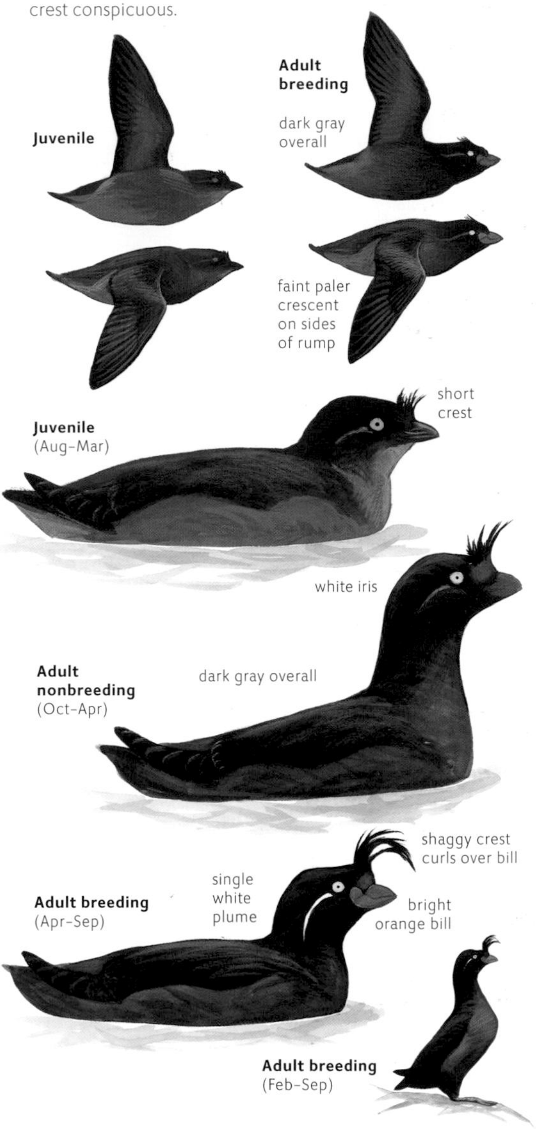

VOICE: Silent at sea. Adults in and near colony give a variety of low barking and hooting calls; a nasal, barking *kyow* like a small dog is frequently heard. Also a short, accelerating series of honking notes.

Common on open ocean. Nests in colonies in crevices and burrows on rocky islands. Often in large flocks at sea. Feeds on plankton and small fish.

Rhinoceros Auklet

Cerorhinca monocerata

L 15" WS 22" WT 1.1 lb (520 g)

Our largest auklet, but distinctly smaller than murres. Note relatively long, stout, yellowish bill; wedge-shaped head; and overall gray color with fairly extensive white belly.

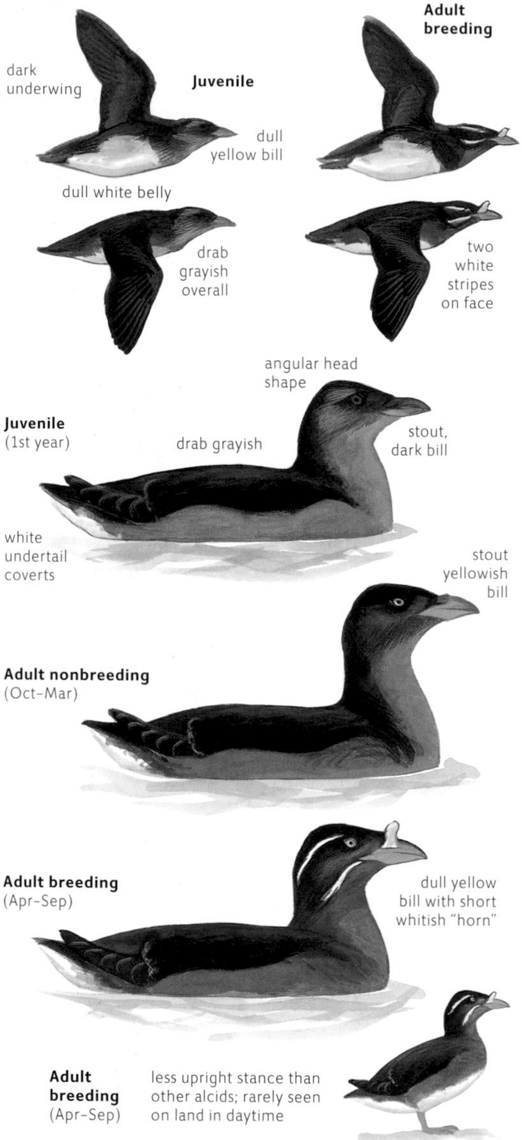

VOICE: Generally silent at sea. Adult near colony at night gives a building and fading series of about ten low, mooing notes; also short barks and single groaning calls.

Uncommon or locally common on open ocean. Nests in colonies in burrows or deep crevices on islands. Nearly always solitary at sea. Feeds almost exclusively on small fish.

Tufted Puffin

Fratercula cirrhata

L 15" WS 25" WT 1.7 lb (780 g)

Our largest puffin. Note blackish overall color (some juvenile have pale belly), large orange bill, orange feet, and broad wing tips. Juveniles distinguished from Rhinoceros Auklet by deeper bill, pale cheeks, broader wings, orange feet.

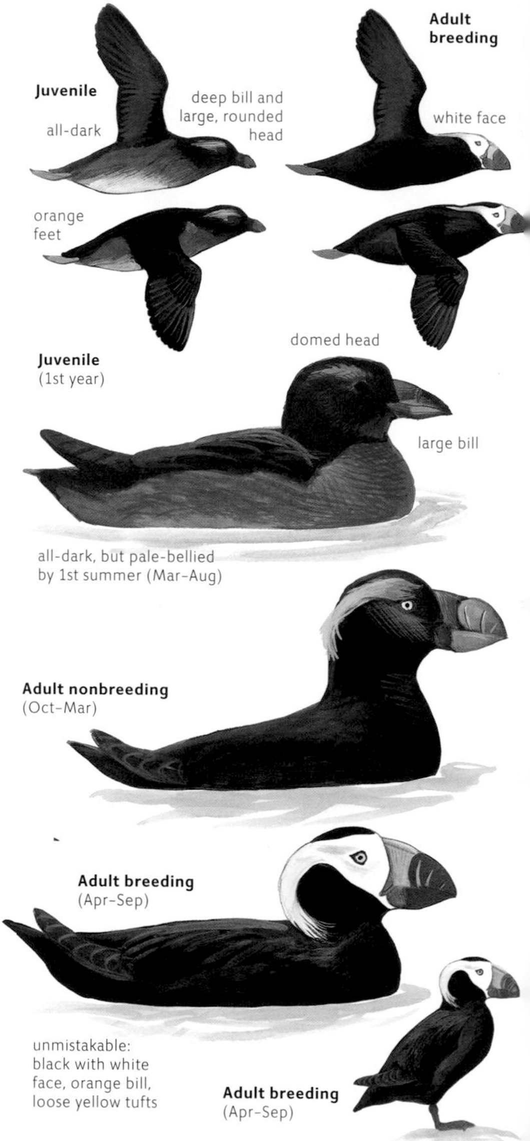

VOICE: Silent at sea. Adult in colony not very vocal; only a low rumbling, groaning sound heard.

Uncommon on open ocean. Nests in mixed colonies with other alcids, in burrows or deep crevices on islands. Nearly always solitary at sea. Feeds almost exclusively on fish.

Iorned Puffin

ratercula corniculata

L 15" WS 23" WT 1.4 lb (620 g)

own-like face, deep bill, broad wingtips, and orange feet typi-
l of puffins, unlike other alcids. Note mostly white underparts
ontrasting cleanly with dark neck and all-dark underwing.

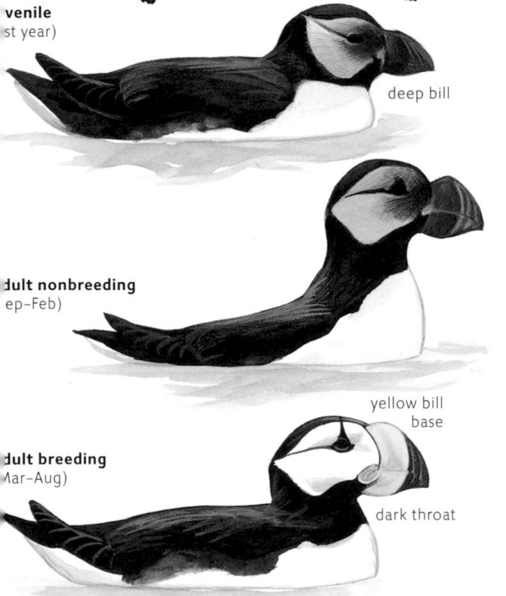

Range does not overlap with Atlantic Puffin, but readily distinguished in all plumages by larger size, dark throat.

OICE: Silent at sea. Adults in colony give low groaning or growl-
ng calls, often in rhythmic pattern unlike the continuous
oaning of Atlantic Puffin.

Uncommon on open ocean. Nests in colonies in burrows or crevices on islands. Nearly always solitary at sea. Feeds almost exclusively on fish.

Atlantic Puffin

Fratercula arctica

L 12.5" WS 21" WT 13 oz (380 g)

Smaller than murres. Deep, triangular, orange-tipped bill distinctive. Also note pale face, dark collar, and orange feet. Distinguished from other Atlantic alcids by white belly, deep bill, orange feet, dusky underwing coverts, relatively broad wings, less tapered head

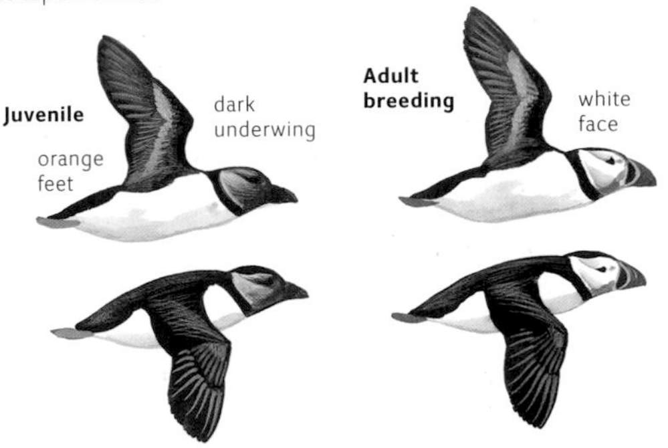

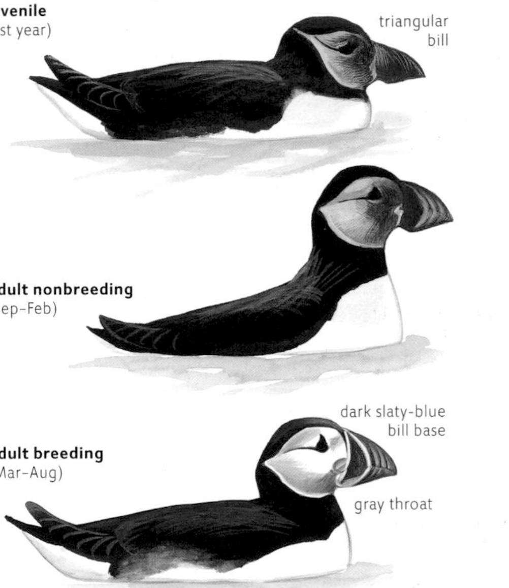

VOICE: Silent at sea. Adults in colony give variations on a low, unmusical moaning or bellowing with slight pitch changes, like the sound of a distant chain saw.

Common at nesting colonies in Canada and on isolated islands off Maine. Nests in burrows or in deep crevices among boulders. Forages for small fish and winters on open ocean; rarely seen from land. Usually solitary; often seen flying higher above the water than other alcids.

Pigeons and Doves

FAMILY: COLUMBIDAE

17 species in 7 genera (4 rare species not shown here). All have short, blunt bill, relatively small head, short legs; walk with mincing steps and bobbing head. All forage on fruit and seeds picked from the ground (doves and Rock Pigeon) or in treetops (other pigeons). Flight is strong and direct with flicking wingbeats. Doves are usually found in small groups, pigeons may form larg flocks and travel long distances. All give low cooing calls. Nest all species is a flimsy platform of a few twigs, placed on a hor zontal branch. Can be confused with falcons in flight, or wi upland gamebirds. Adults are shown.

Genus *Patagioenas*

White-crowned Pigeon, page 257

Red-billed Pigeon, page 257

Band-tailed Pigeon, page 256

Genus *Zenaida*

White-winged Dove, page 258

Mourning Dove, page 258

Genus *Columba*

Rock Pigeon, page 255

Genus *Streptopelia*

Eurasian Collared-Dove, page 260

African Collared-Dove, page 260

Spotted Dove, page 261

Genus *Columbina*

Inca Dove, page 262

Common Ground-Dove, page 263

Ruddy Ground-Dove, page 263

Genus *Leptotila*

White-tipped Dove, page 261

:ock Pigeon

ɔlumba livia

ɩ2.5" WS 28" WT 9 oz (270 g)

tremely variable in plumage after centuries of domestication. ost are gray with white rump (like ancestral wild form). Wings nger and more pointed than other pigeons. Sleek and powerl in flight, falcon-like but with less regular rhythm of wingbeats. ›ur main plumage variations; all share pink legs and white cere.

ɔICE: Song a low, muffled *whoo, hoo-witoo-hoo* with bowing ıd strutting display; at nest site a simpler *hu-hu-hurrr*, repeated. 'ings produce soft whistle on takeoff; on display takeoff a slapng *T-T-T-T-t-t-t-t*.

Common and widespread; introduced from Europe and escaped from domestication to become one of the most familiar birds in North America and the common city- and farm-dwelling pigeon. Usually in small flocks. Nests and roosts on narrow ledges on buildings, bridges, and cliffs. Forages mainly on the ground for seeds.

Band-tailed Pigeon

Patagioenas fasciata

L 14.5" WS 26" WT 13 oz (360 g) ♂>♀

Larger and more elongate than Rock Pigeon, with longer tail and more rounded wings. Pale gray and lavender overall, including gray underwing coverts; tail broadly tipped with very pale gray. Bright white collar, and yellow bill and legs unique among pigeons.

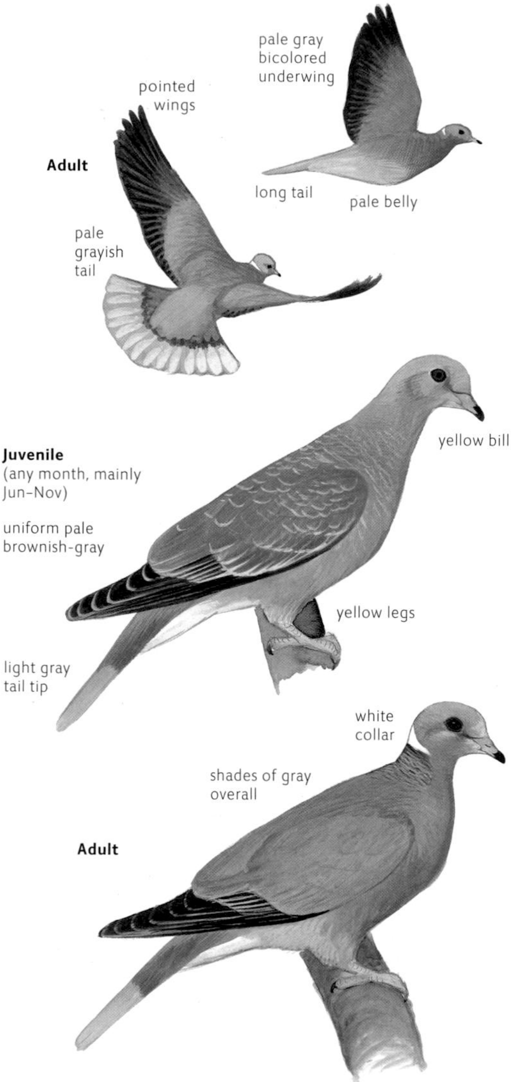

VOICE: Song a deep, somewhat hoarse, repeated, owl-like hooting *hu whoo, hu whoo*.... Also a nasal, grating *raaaaaan*. Wings produce loud clapping on takeoff and muffled but far-carrying swoosh in fast descent.

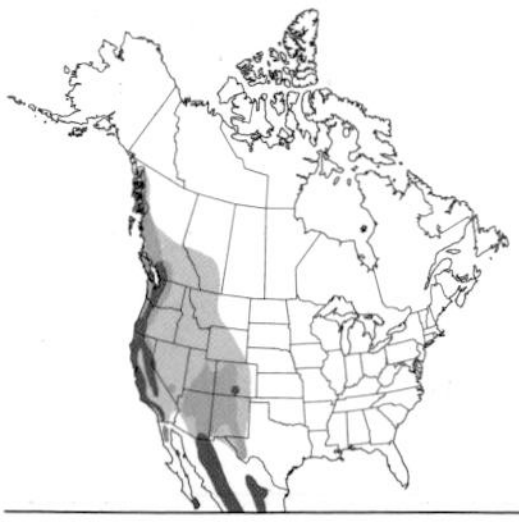

Common in oak and conifer-oak woods, usually in small flocks. Feeds on fruit and seeds, including acorns. Large numbers may gather to forage in agricultural fields near woods; flocks are also attracted to mineral-rich soils around certain springs.

Bird Families

In North America there is a clear distinction between pigeo (large, conspicuous, often flying high, mostly gray plumage) ar doves (smaller, more secretive, often flying low, mostly brow plumage), but the two names have no technical definition ar are used interchangeably in other parts of the world.

The native pigeons of North America (Band-tailed, Re billed, and White-crowned) were considered part of a worldwic genus *Columba* until recently, along with the introduced Ro Pigeon and many other Old World pigeons. Taxonomists stud ing the appearance and behavior of these birds could fir differences that distinguished most New World and Old Wor pigeons, but no single feature that would separate all, and so were retained in the same genus.

New techniques for DNA analysis reveal clearly that the tv regional groups have been diverging independently for millio of years, and they were recently split into separate genera, b still no visible or audible difference is known that will definitive identify a pigeon to genus.

The higher level relationships of the pigeons and doves a much less clear. As with the genera, DNA analysis has reveale deep relationships that would never be imagined by taxonomis studying the birds' appearance. Recent studies suggest th pigeons are related to tropicbirds, grebes, and flamingos, amor many other surprises. Research is ongoing to verify these result but the family tree outlined here might represent the futu arrangement of these groups.

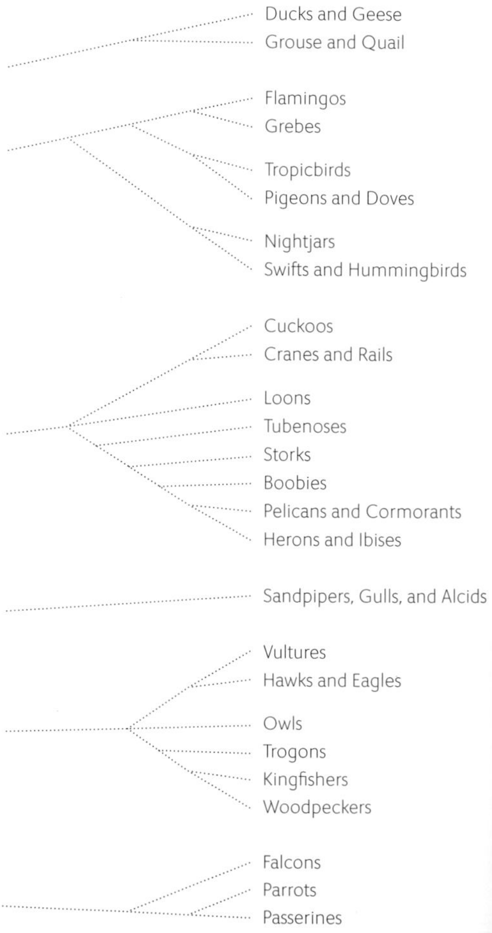

ℇed-billed Pigeon

ıtagioenas flavirostris

ℓ4.5" WS 24" WT 11 oz (315 g)

ınger-necked than Rock Pigeon, with shorter and more rounded ıngs. Adult dark grayish overall, with slightly paler underwing ›verts and reddish wash on neck. Note reddish bill with yellow ›. Range does not overlap with White-crowned Pigeon.

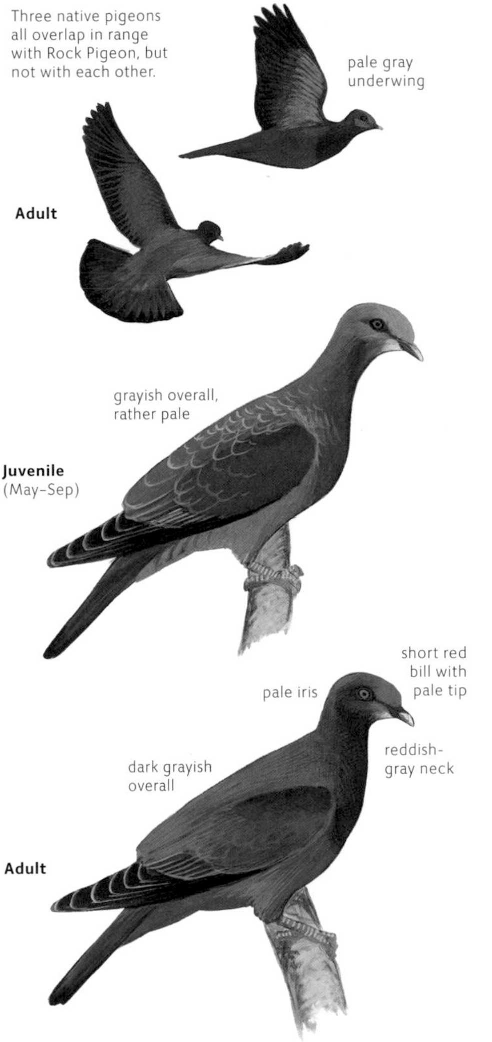

ɔICE: Song a hoarse cooing *hup-hupA-hwooooo*; often in long ɛries introduced by single, low, hoarse note *Hhooo*; *hwooo ·wooooo hup hupA hwoooo, hup hupA hwooooo....* When armed, a few loud wing claps on takeoff.

Uncommon in riparian woods mainly from Apr–Sep. Shy and wary, perching and foraging for fruit within foliage of treetops; most often seen in flight in small groups.

White-crowned Pigeon

Patagioenas leucocephala

L 13.5" WS 24" WT 10 oz (290 g)

Relatively longer-necked and longer-tailed than Rock Pigeon, with shorter and more rounded wings. Note uniform dark gray color, including underwing coverts and rump, with white to grayish-white crown and reddish bill.

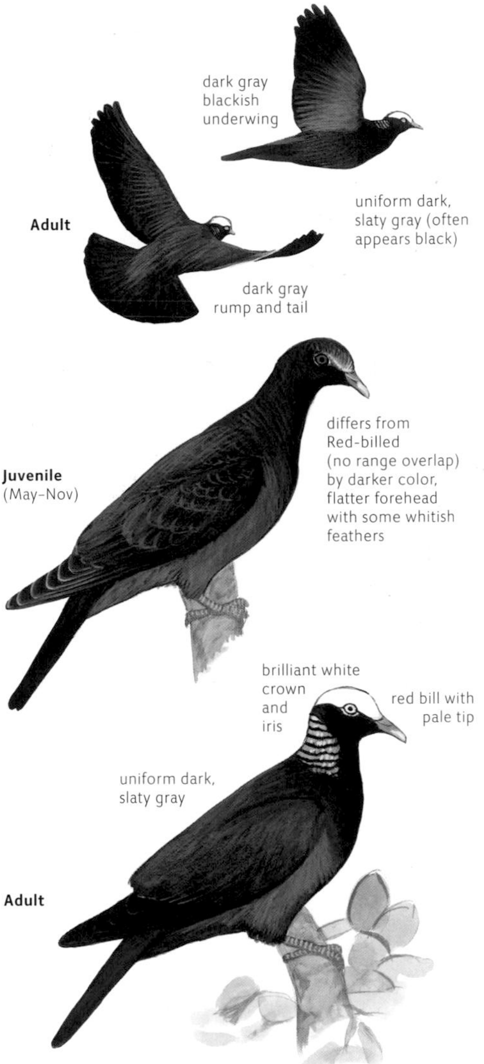

VOICE: Song slow, well spaced, fairly high-pitched like Mourning Dove: *woopop wooooo, woopop wooooo....* Rhythmic high, hoarse *hudi-ho-HOOOO* repeated. A few wing claps given when taking off in alarm or aggression.

Uncommon; found mainly in mature mangroves and in hammock forests, but small flocks disperse widely each day to feed on fruiting trees, including large shade trees in suburban habitats. Shy and wary, usually perching and foraging for fruit within foliage of treetops; most often seen in flight in small groups.

White-winged Dove

Zenaida asiatica

L 11.5" WS 19" WT 5 oz (150 g)

Larger and heavier than Mourning Dove, with broader wings and short rounded tail; very strong, pigeon-like flight. White upperwing coverts form narrow edge on folded wing and striking band in flight, contrasting with very dark primaries.

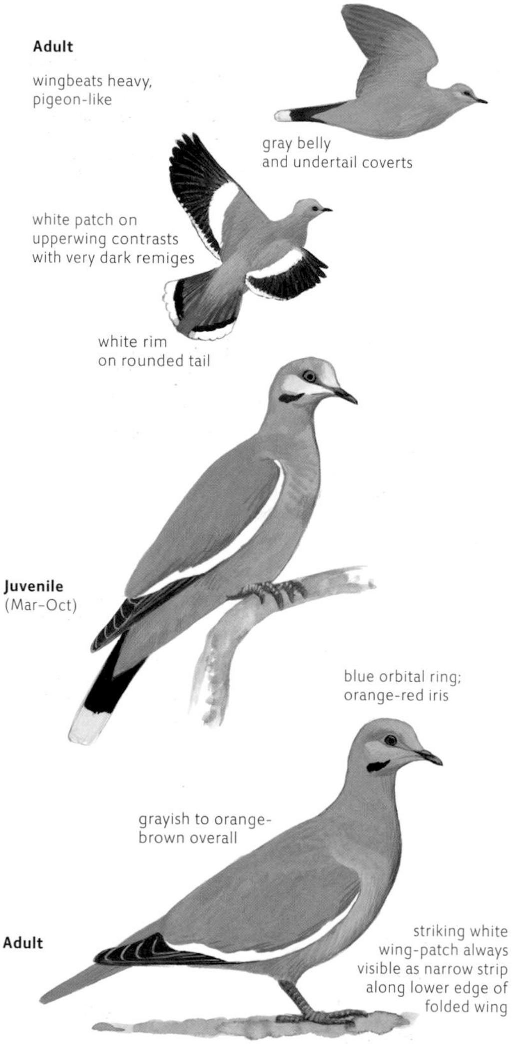

VOICE: Song a rhythmic hooting *hhhHEPEP pou pooooo* ("who cooks for you"), reminiscent of Barred or Spotted Owl, and a slow, measured series *pep pair pooa paair pooa paair pooa*.... Wings produce weak whistle on takeoff; lower and softer than Mourning Dove.

Common and increasing in brushlands and suburban areas with trees. Habits similar to Mourning Dove, but often flies higher.

Mourning Dove

Zenaida macroura

L 12" WS 18" WT 4.2 oz (120 g) ♂>♀

Our most slender dove, with long pointed tail and relatively na row, pointed wings. Subtly colored in brown and gray; genera warm brown with buffy undertail coverts and black spots on wi coverts.

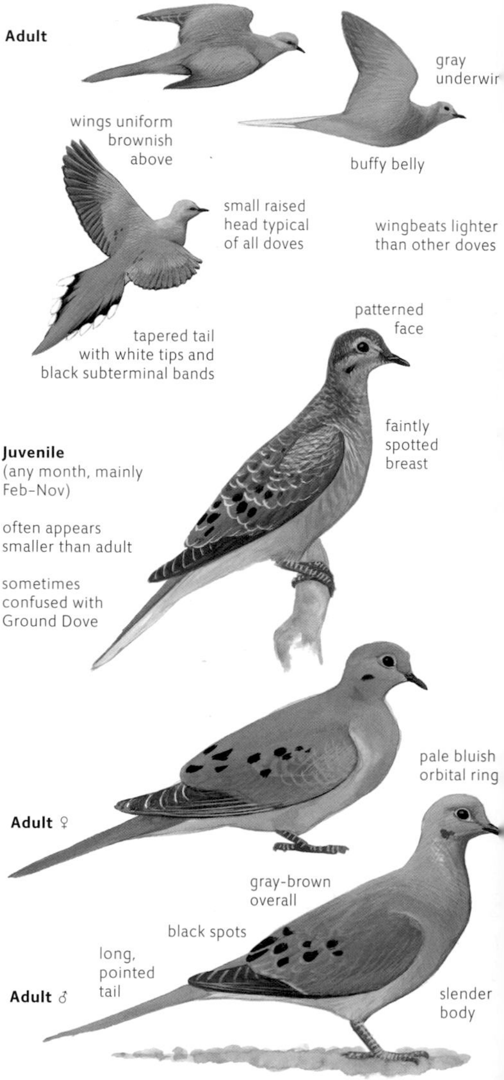

VOICE: Song a mournful hooting *ooAAH cooo coo coo*; ofte mistaken for an owl by inexperienced listeners. Much minc individual variation in pattern. Sometimes a strong, sing *pooooo*. Wings produce light, airy whistling on takeoff.

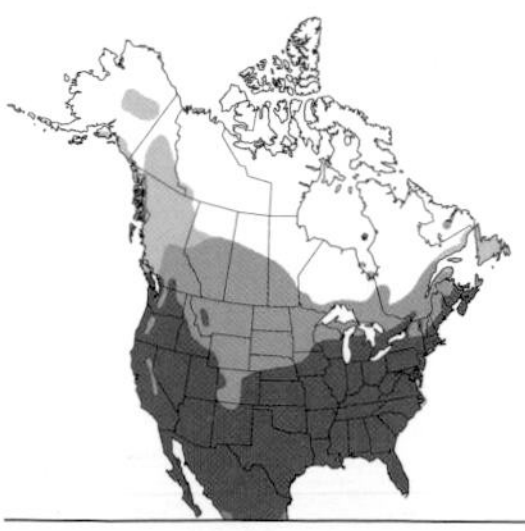

Common and widespread in many suburban and agricultural habitats with mix of open grour and brushy cover, usually not in forest. Usually in small groups. Forages mainly on the ground fo seeds. Frequently seen on overhead wires or on the groun on road edges, lawns and at birc feeders.

ᵢenaida Dove

ᵣnaida aurita

.1.5" WS 10.5" WT 5.3 oz (150 g)

ghtly larger and stockier than Mourning Dove, with short unded tail and obvious white trailing edge on wings. On aver- e body plumage is darker and redder than Mourning Dove, ıt some overlap.

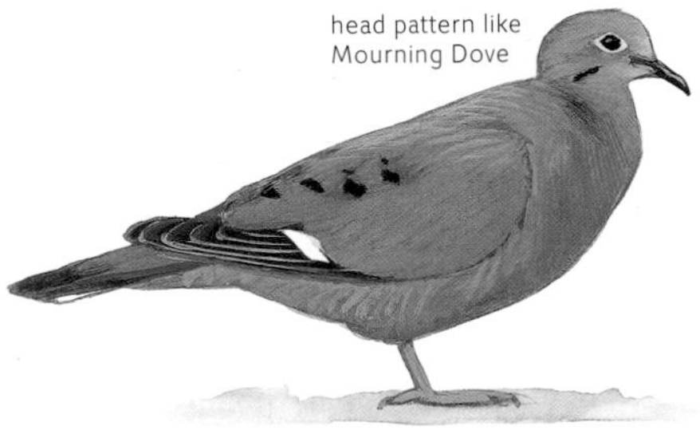

VOICE: Song very similar to Mourning Dove.

Very rare visitor from the West Indies to southeastern Florida, recorded a few times in recent decades in native hardwood hammock habitat or suburban neighborhoods with many trees and shrubs, not in more open suburban habitats where Mourning Dove is common.

)riental Turtle-Dove

:reptopelia orientalis

13" WS 21.5" WT 7.75 oz (220 g)

ırger, stockier, and overall darker than Mourning Dove, with ›unded tail, gray head, patterned wing coverts. Similar to Spot- d Dove but slightly larger, with bold scaly pattern above; rump ıler gray.

VOICE: Song low pigeon-like cooing with hoarse growling quality, repeated rhythmically as two longer notes then two shorter: *kroo-KRoo, ko-koo, kroo-KRoo, ko-koo....*

Very rare visitor from Asia, recorded about ten times from Alaska and Yukon to British Columbia and California, southerly records mainly in fall and winter, in woods edges or visiting bird feeders.

:uropean Turtle-Dove

treptopelia turtur

11" WS 19.5" WT 5.3 oz (150 g)

ıze similar to Mourning Dove, with relatively shorter and ›unded tail, white belly, orange and black upperwing coverts, ıd white lines on neck. Very similar to Oriental Turtle-Dove, but ote white belly, more white on neck, brighter orange edges on ·ing coverts.

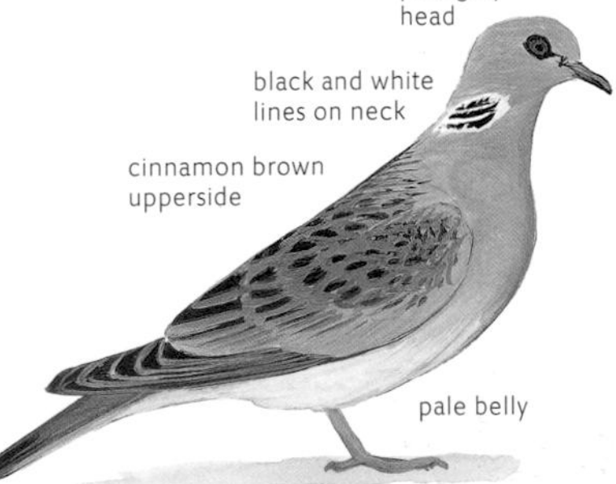

VOICE: Song a low, wooden, rolling or purring sound, repeated with only short pauses *kurrrrr, kurrrrr, kurrrrr...;* sometimes two-syllabled *Ku-kurrrrr.* On takeoff wings clatter and produce faint breezy whistle.

Extremely rare visitor from Europe, recorded only twice (spring records in Florida and Massachusetts).

Eurasian Collared-Dove

Streptopelia decaocto

L 13" WS 22" WT 7 oz (200 g)

Larger and paler than Mourning Dove, with long broad tail, heavier flight, habit of lifting tail after landing. Most similar to African Collared-Dove, larger and usually darker overall; note gray undertail coverts, darker flight feathers.

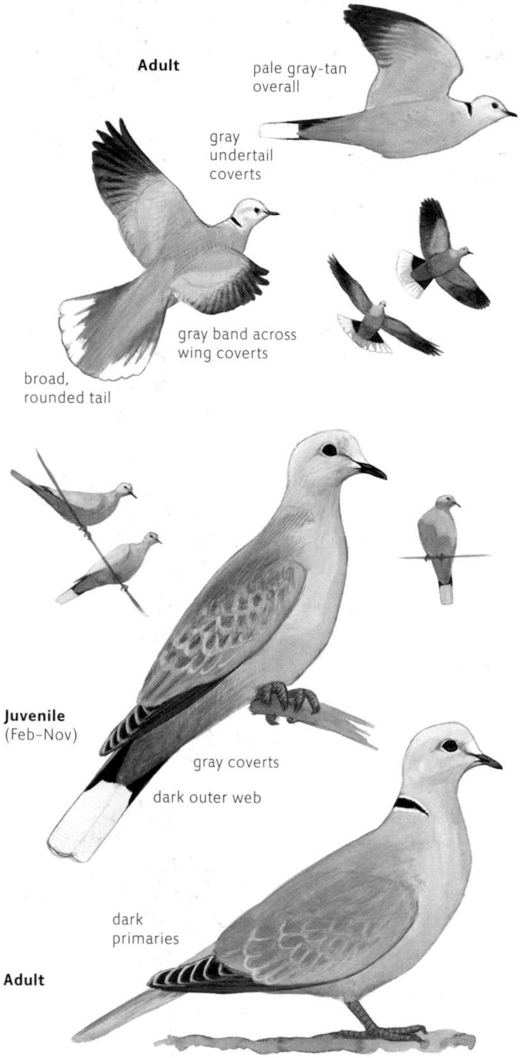

VOICE: Song a rhythmic, three-syllable hooting *coo COOO cup* steadily repeated; slightly lower-pitched than Mourning Dove. Display call shorter *COO COO co;* also a rather harsh, nasal *krreeew* during display flight, reminiscent of Catbird but low and hollow. No wing whistle in flight.

Common; introduced from Europe and increasing rapidly. Found mainly in suburban habitats with relatively little vegetation, but also in more natural settings. Usually solitary or in small groups, seen perched on wires, rooftops, poles. Forages mainly on the ground for seeds.

African Collared-Dove

Streptopelia roseogrisea

L 11" WS 20" WT 6 oz (160 g)

Very similar to Eurasian Collared-Dove; smaller and more slend overall, usually with paler color (often nearly white, but variabl best distinguished by white undertail coverts with white out web of outer tail feathers, and pale outer primaries. Also disti guished by voice.

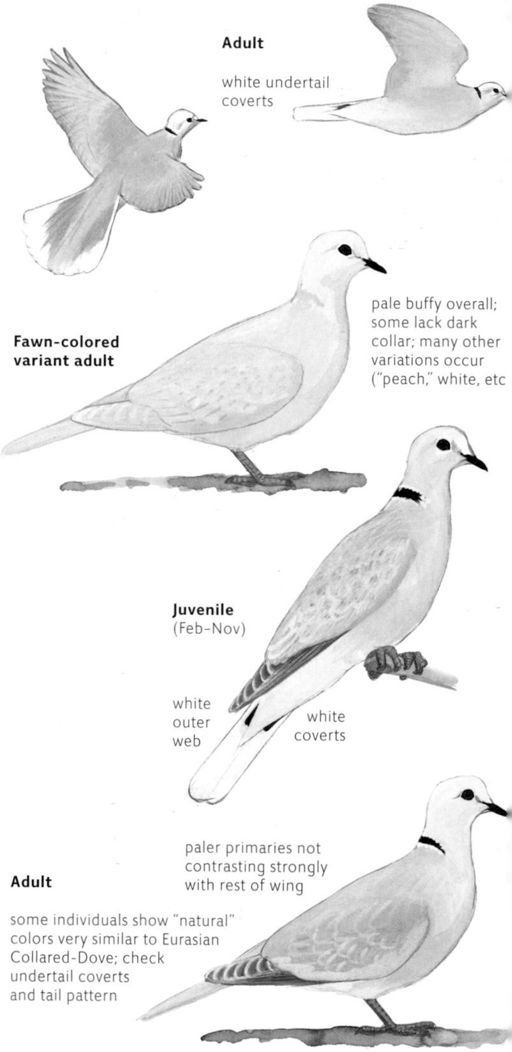

VOICE: Song a soft, rolling, cooing *cooeh-crrrrooa* not repeated quickly; also a soft, nasal laugh given from perch: a low, hollow *hodo-hoo-hoo-hoo-hoo.*

Rare and local. The domestic form of African Collared-Dove, known as Ringed Turtle-Dove, is commonly kept and occasionall escaped from pet stores or dove breeders throughout North America. Feral (but apparently not self-sustaining) populations found in some southern cities, especially in Florida and California.

potted Dove

eptopelia chinensis

2" WS 21" WT 6 oz (170 g)

ghtly larger and heavier than Mourning Dove. Relatively dark own above and reddish-brown below; note very dark underng coverts and long rounded tail with broad white corners.

ICE: Song a forceful *poo pooorr* or *coo CRRRRRoo cup*; mide note very rough, rolling; final note separate; higher and uder than Mourning Dove. Wings do not whistle in flight.

Uncommon and declining in suburban habitats with large trees in southern California. Introduced from Asia. Closely related to Eurasian Collared-Dove and similar in habits.

White-tipped Dove

Leptotila verreauxi

L 11.5" WS 18" WT 5 oz (150 g)

Very stocky, with short rounded tail and rounded wings. Note pale grayish underparts and very pale face. Rufous underwing coverts and pale belly are conspicuous in flight.

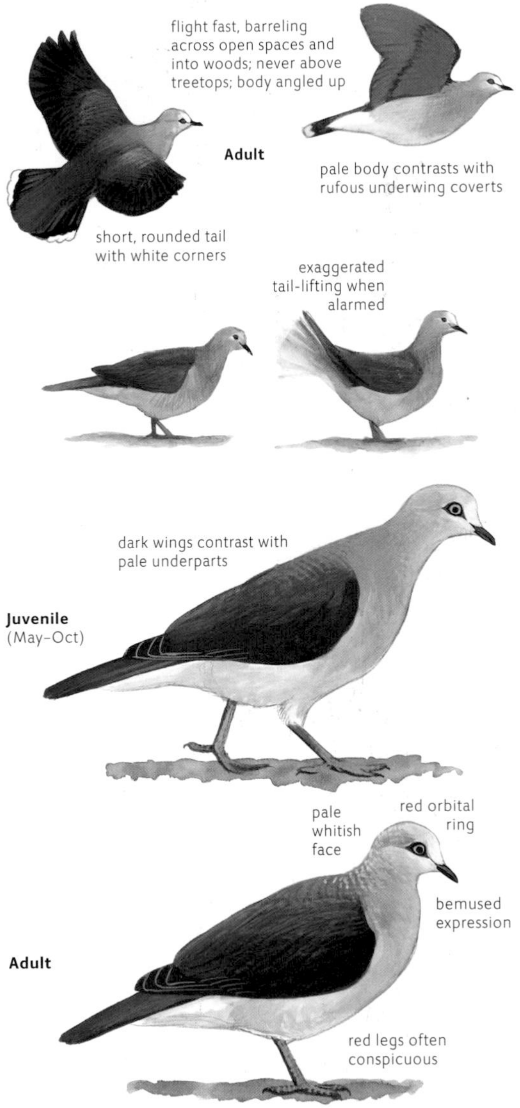

VOICE: Song a low, hollow moaning or cooing *oh-oohoooooooo*, like the sound made when one blows across the top of a bottle. Wings produce high, thin, sharp whistle (higher than Mourning Dove) on takeoff.

Common locally in dense riparian woods along Rio Grande. Solitary and secretive, walking on forest floor and seldom coming into view; most often seen at feeders or flying across forest openings. Forages on the ground for seeds and fruit.

Key West Quail-Dove

Geotrygon chrysia

L 12" WS 19" WT 6 oz (170 g)

Very stocky and rotund, with heavy body and short tail. Overall reddish-brown color with whitish underparts and striking white stripe across cheek distinctive.

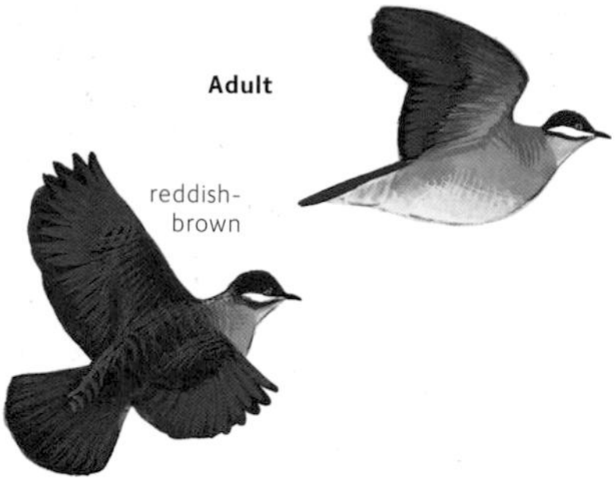

VOICE: Song a low, moaning *ooooo* slightly descending, or *ooowoo* with second part accented and slightly higher; very similar to White-tipped Dove. Though not loud, call is far-carrying. Calls from low branches (4–6 feet high) within forest.

Very rare visitor from Bahamas or Cuba to dense brushy hammock forests in southern Florida. Solitary. Feeds on fruit and seeds. Difficult to see in dense vegetation, usually walking briskly on the ground on short legs.

Inca Dove

Columbina inca

L 8.25" WS 11" WT 1.6 oz (47 g)

Very small and slender, with long square tail. Pale gray color, sc plumage pattern, rufous primaries, and white sides on long distinctive. Easily distinguished from Ground Doves by long t rattling wing noise.

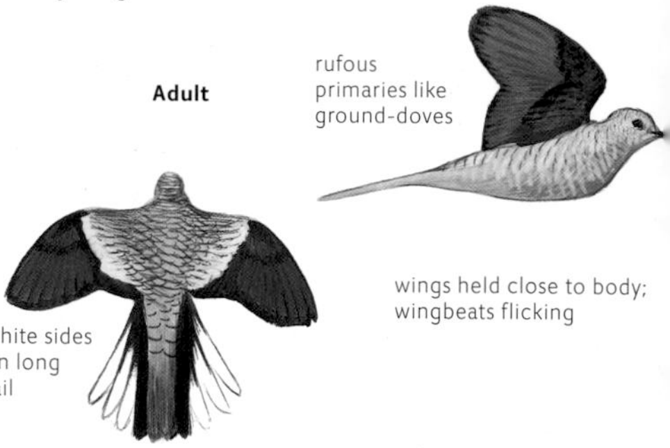

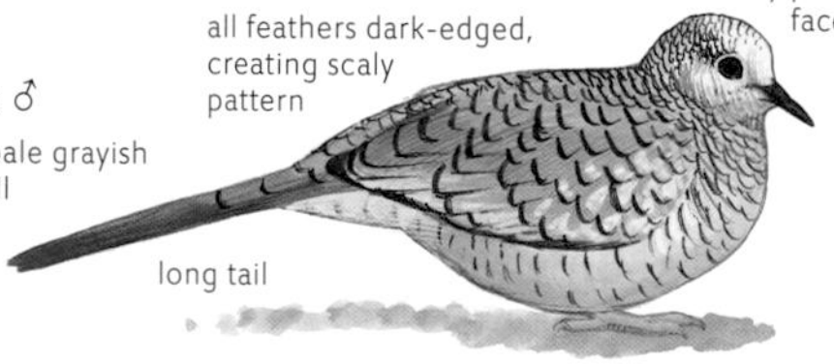

VOICE: Song a fairly high, forceful cooing *POO-pup* ("no hope repeated monotonously, less than once per second; also a mo intense, hoarse growling *krooor* with soft, burry quality. Win produce distinctive quiet, dry rattle on takeoff.

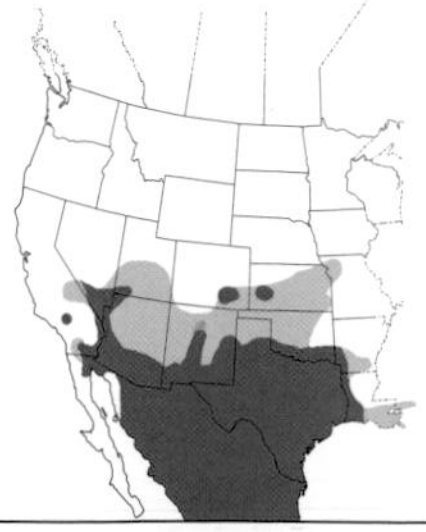

Common and increasing in suburban habitats where trees and shrubs surround lawns, usually near buildings, especial attracted to well-watered lawns and shrubs in arid landscape. Almost always in small groups c three to ten. Forages on the ground for seeds.

.uddy Ground-Dove

lumbina talpacoti

.75" WS 11" WT 1.4 oz (40 g)

ry similar to Common; averages slightly larger and longer-led, males are obviously bright rufous with gray crown, but ight male Common can cause confusion. Best distinguished unpatterned breast and nape and spotted scapulars.

Western birds are paler and grayer, and Eastern richer rufous, such that males of the pale Western populations resemble females of the brighter Eastern.

)ICE: Song slightly lower, faster, more complex, than Common ound-Dove: *pidooip* repeated rapidly about twice per second.

Rare but increasing visitor from Mexico. Prefers wetter habitat than Common Ground-Dove, such as irrigated lawns. Often in small groups associating with Inca Dove.

Common Ground-Dove

Columbina passerina

L 6.5" WS 10.5" WT 1.1 oz (30 g)

Very small and stocky, with flicking wingbeats and low direct flight. Distinguished from Inca Dove by short tail and unpatterned plumage with dark spots on wing coverts.

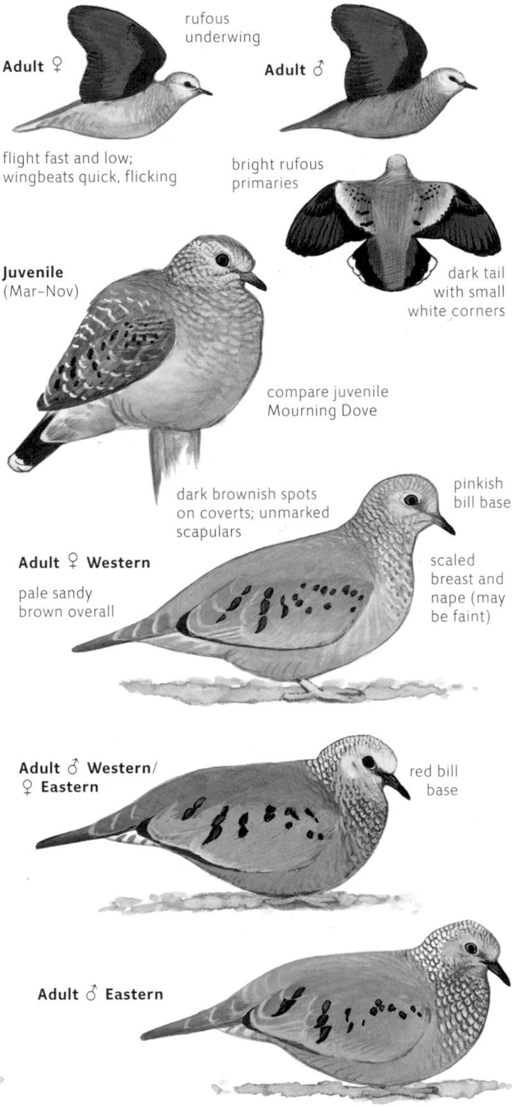

VOICE: Song a series of simple, rising coos *hoooip, hoooip, hoooip*... repeated slowly about once per second; a little higher and clearer than Mourning Dove. Wings produce faint rattle on takeoff; not as clicking as Inca Dove.

Uncommon among brushy and weedy vegetation in dry sandy soil. Roosts in low branches. Usually in pairs or small groups. Forages on the ground for seeds.

Cuckoos

FAMILY: CUCULIDAE

7 species in 4 genera (1 rare species not shown here). Cuckoos forage in dense foliage of shrubs and trees, plucking caterpillars, large insects, lizards, and other prey; flight is smooth and flowing like kingbirds, calls are guttural cooing. Anis are found in weedy or brushy fields, feeding on insects and some fruit; usually very inconspicuous, they perch prominently at times; flight is weak and flopping with sailing glides; calls are squeaky whistles. Greater Roadrunner is found in desert, brushland, and open woods, feeding on lizards, snakes, insects, and other animal prey; runs smoothly and rapidly, flight is infrequent and mainly low glides. Nest of all species is a sloppy or flimsy platform of twigs in bush or low tree. Adults are shown (except juvenile cuckoos).

Genus *Coccyzus*

Yellow-billed Cuckoo, page 265

Mangrove Cuckoo, page 264

Black-billed Cuckoo, page 265

Genus *Geococcyx*

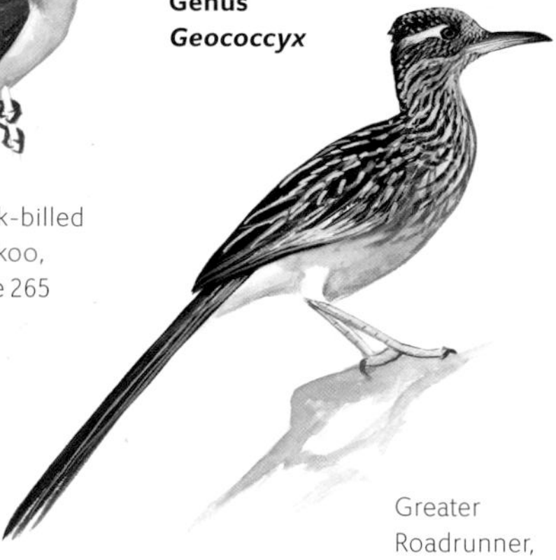

Greater Roadrunner, page 266

Genus *Crotophaga*

Smooth-billed Ani, page 267

Groove-billed Ani, page 267

Mangrove Cuckoo

Coccyzus minor

L 12" WS 17" WT 2.3 oz (65 g)

Differs from other cuckoos in thicker bill, shorter and mo rounded wings, buffy underparts. Most similar to Yellow-bill Cuckoo, but note lack of rufous in primaries, less white on out web of tail, grayer upperparts, and more prominent dark mas

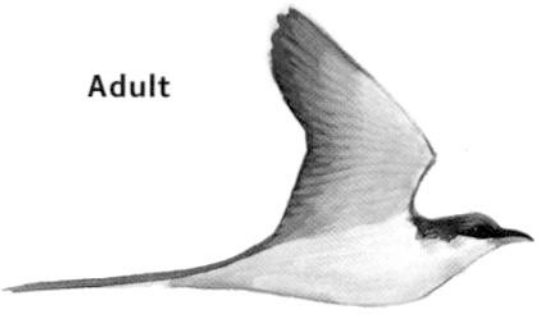

VOICE: Very nasal, unmusical *aan aan aan aan aan aan a urmm urmm*, slightly accelerating; last two notes lower ar longer.

Uncommon and inconspicuous in mature mangroves and dens hammock woodlands. Solitary. Like other cuckoos, moves furtively through dense foliage search of insects, lizards, and other prey.

'ellow-billed Cuckoo

ccyzus americanus

.2" WS 18" WT 2.3 oz (65 g)

‹e other cuckoos, more slender than Mourning Dove, with ng tail and sinuous movements. Note mostly yellow bill (gray r a few weeks on very young birds), clean white underparts, fous primaries, and large contrasting white tail spots.

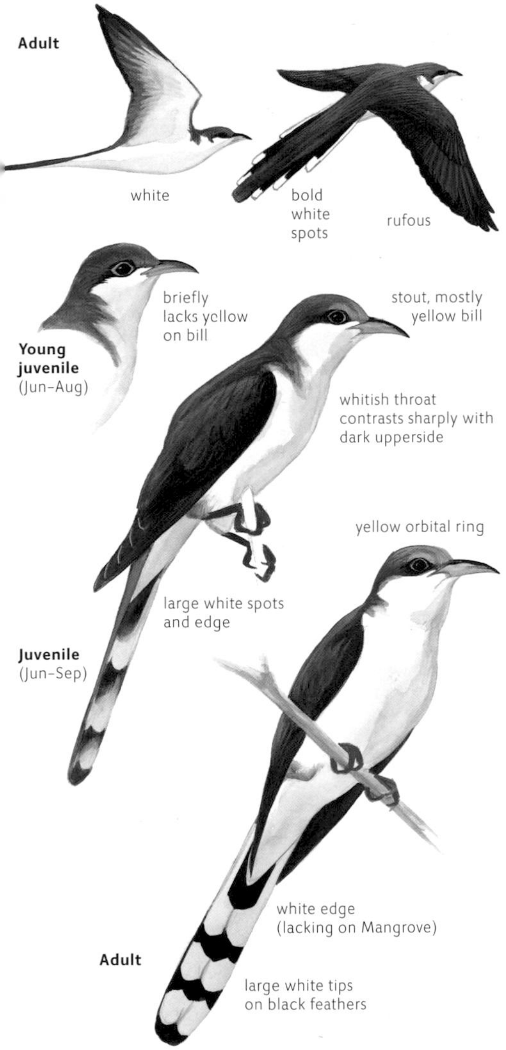

)ICE: Song a guttural, hard, knocking series *ku-ku-ku-ku-ku- -kddowl-kddowl...*; sometimes single *kddowl* notes. Also deep, vallowed, dove-like cooing *cloom* repeated with long pauses; slow cooing series descending and weakening *too too too too to to*; a fairly rapid series of single, hard *tok* notes on one pitch *k tok tok tok....*

Uncommon in woods, often in wet areas such as riparian willow groves or in mid-levels of trees at field edges with dense tangle of undergrowth. Solitary. Often sits for long periods watching, then moves smoothly through branches to feed on caterpillars and other prey gleaned from twigs and leaves.

Black-billed Cuckoo

Coccyzus erythropthalmus

L 12" WS 17.5" WT 1.8 oz (52 g)

More slender than other cuckoos, with thinner bill. Distinguished from Yellow-billed Cuckoo by dark bill, little or no rufous in primaries, gray-brown tail with small white spots, and less contrasting plumage overall.

VOICE: Song a hollow, whistled *po po po* repeated; sometimes a long, rapid series gradually falling into triplet pattern (see Least Bittern). A rolling call of *kddow* notes higher-pitched, quicker, and not as guttural as Yellow-billed; beginning rapidly and ending with decelerating *cloo* notes. Also a rapid, hard, descending *k-k-k-k* or two-part, descending *kru-dru*.

Uncommon and local in woods, often in wet openings with willows and dense undergrowth. Solitary. Feeds mainly on caterpillars.

Common Cuckoo

Cuculus canorus

L 13" WS 23" WT 4.1 oz (120 g)

Slender and long-tailed with pointed wings; flight is strongly reminiscent of small hawk or falcon. Adult male and some females are gray overall, juveniles and some adult females are rufous with dark bars.

VOICE: Visitors to North America are generally silent. Song a repeated two-syllabled *cu-coo, cu-coo*... like the familiar cuckoo clock.

Very rare visitor from Eurasia to western Alaska, mainly in Jun–Jul, also recorded once each in Quebec, Massachusetts, and California. Found in thickets and brushy understory of woods.

Greater Roadrunner

Geococcyx californianus

L 23" WS 22" WT 13 oz (380 g)

Distinctive; large and long-tailed, with shaggy streaked appea ance and short ragged crest often raised. Running action smoc and strong; standing bird adopts variety of comical poses.

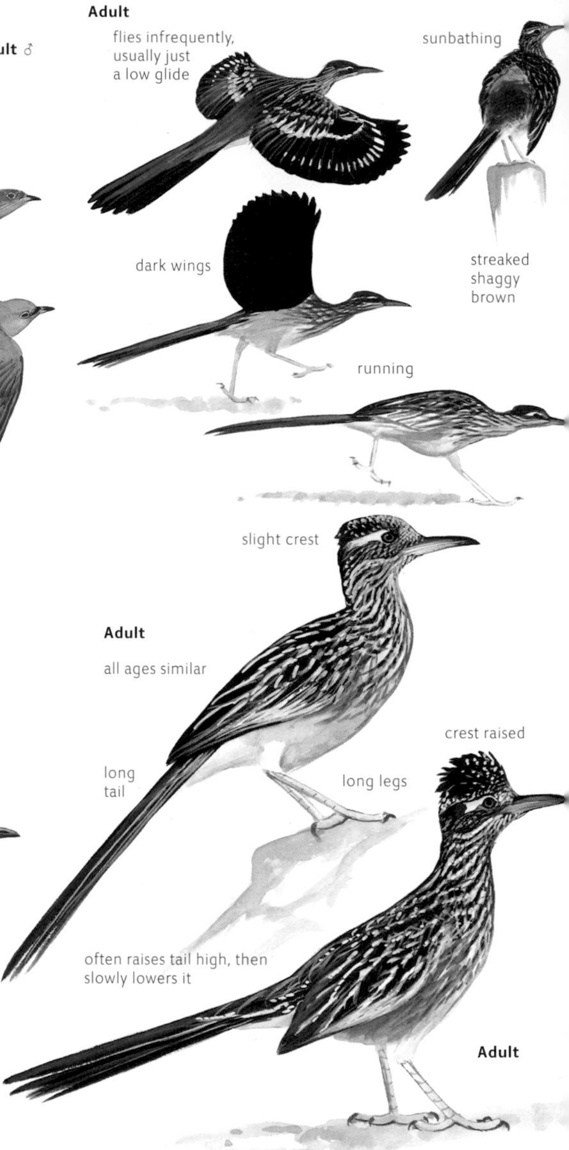

VOICE: Song a slow, descending series of about six resonar low-pitched coos: *cooo cooo cooo cooo coo coo*; weaker at er Also a low, hollow, wooden clatter or rattle *trrrt* produced by b

Uncommon in dry areas where patches of brush and other low vegetation are separated by op ground, often seen along road edges. Walks or stands patientl watching for prey, then runs briefly to capture preferred pre of large insects, lizards and snakes. Often perches on fence posts or rocks. Usually solitary, but occasionally seen in pairs.

ɪroove-billed Ani

ʼotophaga sulcirostris

.3.5" WS 17" WT 3 oz (85 g) ♂>♀

ıis often look disheveled, with wings drooping and tail spread. black plumage and long tail can be confused with grackles, ıt note deep bill, awkward posture, floppy flight low and slow, th sailing glides. Very similar to Smooth-billed Ani but range ɾely overlaps.

ɔICE: Common call a sharp, high whistle with slurred, whining ıding *PEET-uaay* or *PEE-ho*. Often simply a sharp, inhaled *PEEt*; ɔo a sharp, hollow *pep, pep*... or low, grating *krr krr*....

Local and uncommon summer visitor from Mexico to brushy and grassy habitats. Usually seen in small groups clambering and gleaning on low vegetation. Feeds on insects, other animal prey, seeds, and fruit. Look for small groups in brushy tangles and edges.

Smooth-billed Ani

Crotophaga ani

L 14.5" WS 18.5" WT 3.7 oz (105 g) ♂>♀

Very difficult to distinguish from Groove-billed Ani but usually no range overlap. Best identified by voice and by details of bill shape: bill smooth or only weakly grooved, and lower mandible has more pronounced gonydeal angle. Some individuals have sharp rise at base of culmen, unlike any Groove-billed.

VOICE: Common call a whining, metallic, ascending whistle *queee-ik*; also a thin, descending *teeew*. Other whistles and clucks less melodious than Groove-Billed.

Rare and declining in wet weedy areas and brushy patches. Formerly an uncommon and local resident occurring in small groups, now a rare visitor recorded only a few times each year, mostly as single individuals. Habits like Groove-billed Ani.

Owls

FAMILIES: TYTONIDAE, STRIGIDAE

19 species in 11 genera; all in family Strigidae except Barn Owl in Tytonidae. Owls are mainly nocturnal predators, distinctive with large eyes facing forwards and surrounded by facial disc. All have hooked bills and needle-sharp talons, used for catching animal prey. Forage mainly by perching and listening or watching for prey, and different species specialize in different prey: many smaller species are insectivores, larger species take mainly mammals, pygmy-owls mainly small birds. Owls do not build nes most species nest in cavities or crevices and readily accept ma made structures such as birdhouses; some use old nests of oth birds or squirrels. Young owls leave the nest while still downy a before fully flighted. Voice of most species is whistling, screec ing or barking; only a few species hoot, and cooing sounds doves are frequently mistaken for owls. Adults are shown.

Genus *Tyto*

Barn Owl, page 270

Genus *Asio*

Long-eared Owl, page 270

Short-eared Owl, page 271

Genus *Surnia*

Northern Hawk Owl, page 28

Genus *Psiloscops*

Flammulated Owl, page 277

Genus *Megascops*

Whiskered Screech-Owl, page 278

Western Screech-Owl, page 278

Eastern Screech-Owl, page 279

Genus *Aegolius*

Boreal Owl, page 275

Northern Saw-whet Owl, page 275

Genus *Athene*

Burrowing Owl, page 276

Genus *Micrathene*

Elf Owl, page 277

Genus *Glaucidium*

Northern Pygmy-Owl, page 280

Ferruginous Pygmy-Owl, page 281

enus *Bubo*

Great Horned Owl, page 272

Snowy Owl, page 273

enus *Strix*

otted Owl, page 274

Barred Owl, page 274

reat Gray Owl, page 273

Owling

As birders, it is good practice to disturb birds as little as possible, and several factors make owls unusually sensitive to disturbance. Most species are nocturnal, and during the day they seek out sheltered roost sites where they can rest, unnoticed. This may be in tree cavities, or concealed within dense vegetation such as tangles of vines and twigs. Most species have cryptic coloration and also adopt "cryptic poses" using ornamental plumage like ear tufts to blend with bark and branches.

Once discovered, an owl will be harassed by crows, jays, chickadees, and other species, and may even be forced to fly in search of a new roosting site.

Many birding information networks now have a policy of not broadcasting the location of roosting owls, as the popularity of these birds, and the resulting traffic of visiting birders can cause enough disturbance to force the owl to move.

To look for a roosting owl during the day:

- Follow the sounds of scolding crows, jays, chickadees, etc.
- Look for whitewash and pellets on the ground, which will indicate a habitual roost site.
- Scan the densest tangles of branches.

If you see a roosting owl, chances are it is already alarmed by your presence. Move away quietly. For an owl, holding still in its camouflage posture is a sign of agitation, and you should resist the temptation to get closer or try for a clearer view, as any additional stress could cause the owl to fly. Carefully consider the situation before telling other birders. Can the owl be seen from a safe distance without disturbing it?

Eastern Screech Owl

camouflage posture in daytime

Great Horned Owl

Juvenile

Branching Owls

Before young owls are fully-grown they leave the nest and follow the adults through the forest towards prime feeding areas. These young owls cannot fly, but use their partly grown wings to help them clamber up trees, through branches, and glide from tree to tree. If you find one of these young owls, it will try to make itself look more imposing by fanning its wings, swaying, and snapping its bill to make a loud popping sound.

Beware of a defending adult, and move away to watch from a distance where the bird does not feel threatened.

Barn Owl

Tyto alba

L 16" WS 42" WT 1 lb (460 g) ♀>♂

Medium-size, short-tailed, and long-legged. Pale overall with fawn-colored upperside, white to buff underside; dark eyes and white face surrounded by heart-shaped border.

VOICE: Heard only at night. Common call year-round simply a long hissing shriek *csssssshhH*. Female call averages softer than male and juvenile averages less hoarse, but overall little variation.

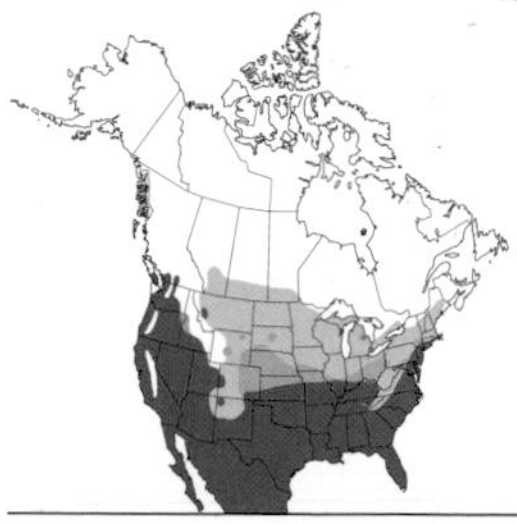

Uncommon. Nests in old barns and other structures. Strictly nocturnal: roosts during day in barns, caves, or less often in dense trees; at night flies over open grassland, agricultural fields, marshes, and brushy areas in search of small mammals. Usually solitary.

Long-eared Owl

Asio otus

flying away

L 15" WS 36" WT 9 oz (260 g) ♀>♂

Medium-size and slender. Coloration similar to Great Horne
Owl but much smaller, with more strongly streaked patte
below. Long "ears" are tufts of feathers that can be raised ar
lowered at will.

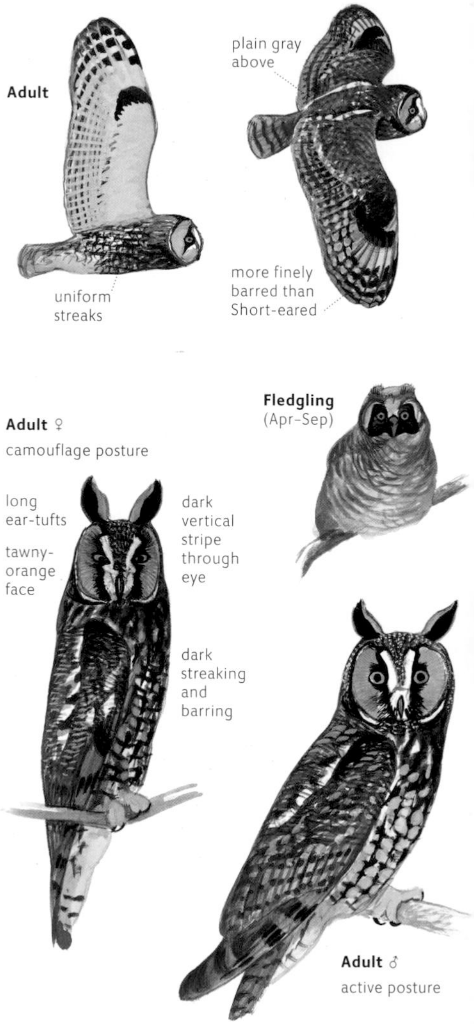

VOICE: Heard only at night. Male gives a low, soft hoot *woo*
about every three seconds. Female call higher and softer *sheoc*
Alarm (both sexes) variable soft, nasal barks *bwah bwah bwa*
also a quiet moan and squealing or mewing calls. Juvenile giv
a high, squeaky *wee-ee* like rusty hinge. Both sexes wing-cla
during display flight, producing a sound like cracking whip give
singly at irregular intervals.

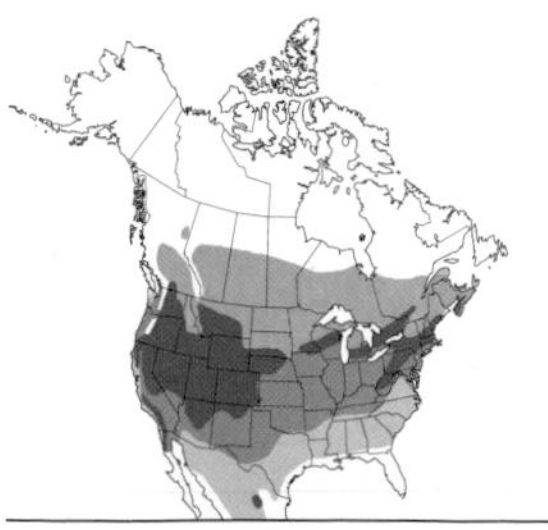

Rare. Most often seen at winter roost sites, where one to severa birds routinely spend the day hidden in dense foliage, especially in willow thickets or evergreen trees. Nests in dense conifer groves adjacent to open fields and wetlands. Forages at night, flying slowly along hedgerows, brushy fields, forest edges.

hort-eared Owl

sio flammeus

15" WS 38" WT 12 oz (350 g) ♀>♂

edium-size and relatively slender. Pale buffy overall, with eaked underside; note dark triangle around each eye. Foraging bits and habitat similar to Northern Harrier and the two species ten occur together. Short-eared Owl has floppier, somewhat ratic flight, pale patch on primaries, shorter tail, and larger head.

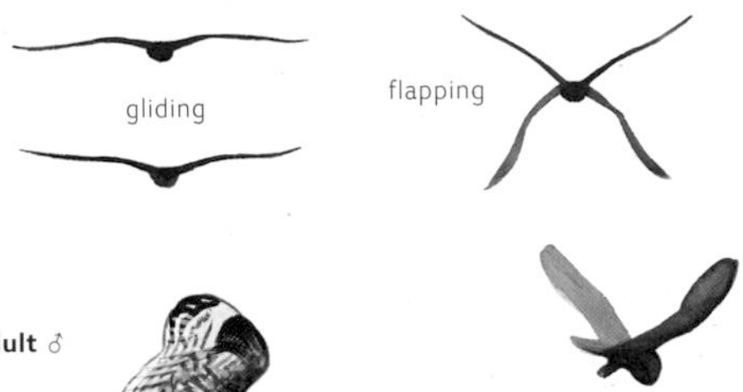

Northern

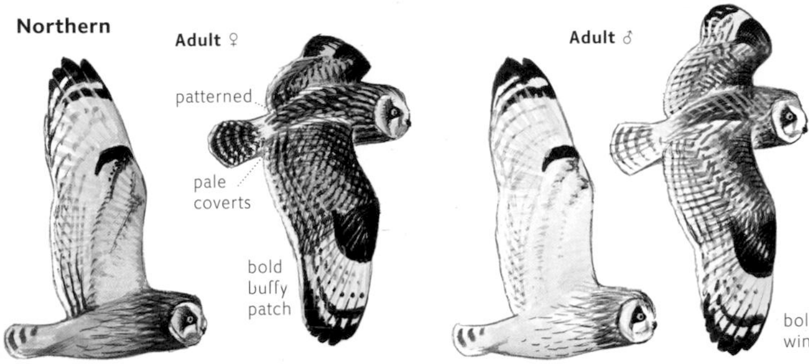

♀ averages darker and buffier than ♂ (as in other owls)

Caribbean

Caribbean Short-eared Owl is a rare visitor from the West Indies to Florida, mainly Apr–Jul. Most records are from the Florida Keys, where Northern subspecies is very rare or unrecorded. Averages smaller than Northern. Caribbean has a bolder frame of dark feathers around the face, finer streaks below, and the mantle and scapulars are dark with bold buffy spots, rather than the more patterned upperside of Northern. Check also for dark uppertail coverts (usually conspicuously pale on Northern); it is also possible that remiges average more neatly barred, though this may be age-related.

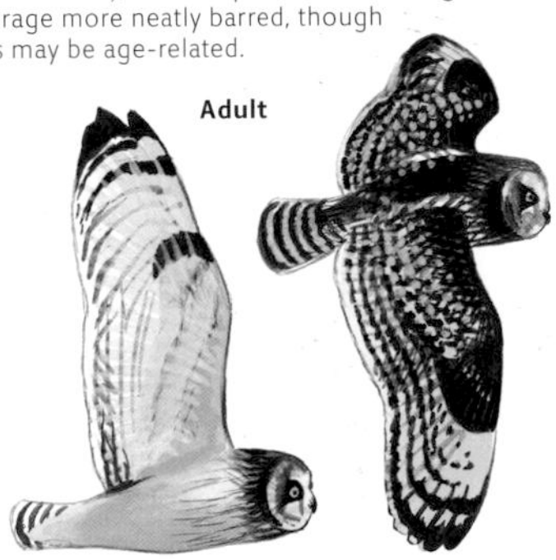

OICE: Silent except in nesting season (at night). Male gives uffled *poo poo poo*... (5–6/sec in series 2 sec long). Alarm oth sexes) high, nasal barks and wheezy notes *cheef cheef* and *heewaaay*. Wing-clapping sounds like cracking whip, given in apid, rattling series. Voice of Caribbean birds not known to dif- er from Northern.

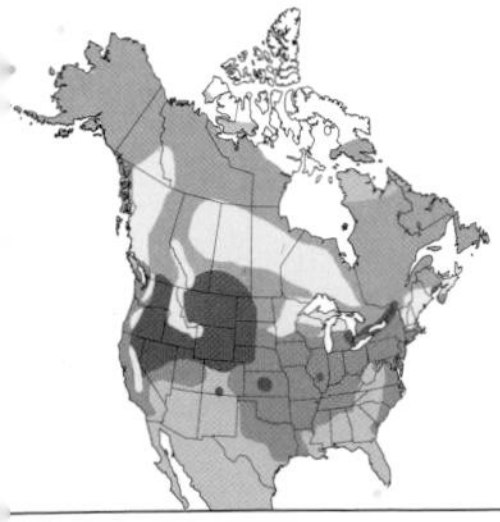

Rare to locally uncommon. Look for it flying low over expansive open fields, grasslands, or marshes, essentially the nocturnal counterpart of Northern Harrier. Often seen in daylight, especially early and late. Roosts on the ground among weeds and grass (for example, sand dunes). Usually solitary, but small numbers may roost together in winter.

Great Horned Owl

Bubo virginianus

L 22" WS 44" WT 3.1 lb (1,400 g) ♀>♂

Large (size similar to Red-tailed Hawk) and bulky; broad ear tufts create cat-like head shape. Note overall grayish upperside, barred underside, and tawny-orange face.

wingbeats steady, stiff, mostly below horizontal

Adult West Taiga

Adult Pacific

Adult Eastern

Adult West Taiga
some birds so pale as to resemble immature Snowy, but note ear-tufts and gray face

Adult Pacific

very dark overall

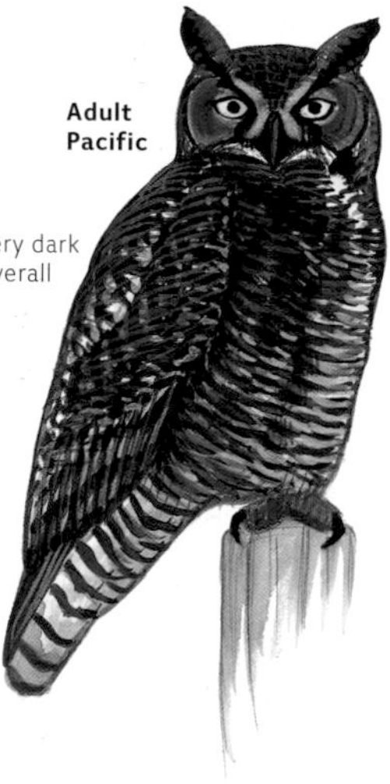

Fledgling Eastern (Mar–Aug)

color of fledgling is geographically variable, paralleling variation in adults

Adult Eastern

large ear-tufts

Adult Eastern

tawny-orange face

densely barred below

Adult Southwest

gray to rusty face

VOICE: Heard mainly at night. Song a deep, muffled hooting in rhythmic series *hoo hoodoo hoooo hoo* or longer *ho hoo hoo hoododo hooooo hoo*; only slightly deeper than Mourning Dove. Female voice higher-pitched than male; courting female answers male with low, nasal, barking *guwaay*. Juvenile begs with high, wheezy, scratchy, or hoarse bark *reeeek* or *sheew* or *cheeoip*; variable, usually shorter and less rasping than Barn Owl but some very similar.

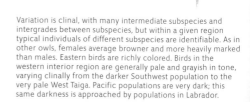

Uncommon but widespread. Nocturnal: roosts during day in trees, on sheltered cliff ledges, or in other secluded spots; at night forages in any habitat—open woods, fields, marshes, desert—for small mammals up to size of rabbits and skunks. Usually solitary.

Variation is clinal, with many intermediate subspecies and intergrades between subspecies, but within a given region typical individuals of different subspecies are identifiable. As in other owls, females average browner and more heavily marked than males. Eastern birds are richly colored. Birds in the western interior region are generally pale and grayish in tone, varying clinally from the darker Southwest population to the very pale West Taiga. Pacific populations are very dark; this same darkness is approached by populations in Labrador.

;nowy Owl

ıbo scandiacus

23" WS 52" WT 4 lb (1,830 g) ♀>♂

ne of our largest owls and relatively sleek, with small head and ɔ "ear" tufts. Mostly white in all plumages; face and underwing ɔverts always pure white. Some very pale Great Horned Owls ıperficially similar.

ɔICE: Call a high-pitched, drawn-out scream heard occasionally ı territorial disputes year-round. Song of male a deep, muffled oot *brooo* repeated. Barking (female) or quacking (male) ɔunds in alarm; also a slurred whistle.

Uncommon to rare and irregular. Nests on open tundra. In winter found on open fields, marshes, coastal sand dunes, etc., where it perches on the ground or fence posts even in daylight. Healthy birds are mainly nocturnal, like other owls, but individuals seen far to the south of normal range are often stressed for food, and thus active in daylight. Solitary.

Great Gray Owl

Strix nebulosa

L 27" WS 52" WT 2.4 lb (1,080 g) ♀>♂

Our largest owl by length but not by weight, long-tailed, fluffy, and large-headed. Note uniform soft gray colors, imposing gaze with yellow eyes, and black-and-white bow-tie pattern.

VOICE: Courtship call (heard only at night) of male five to ten very deep, muffled, pumping hoots, slightly lower and weaker at end; higher-pitched than grouse and longer. Female answers with emphatic, mellow whistle *iihWEW*; also a low, hooting *hooahp*. Juvenile gives muted, trumpet-like *bweek*.

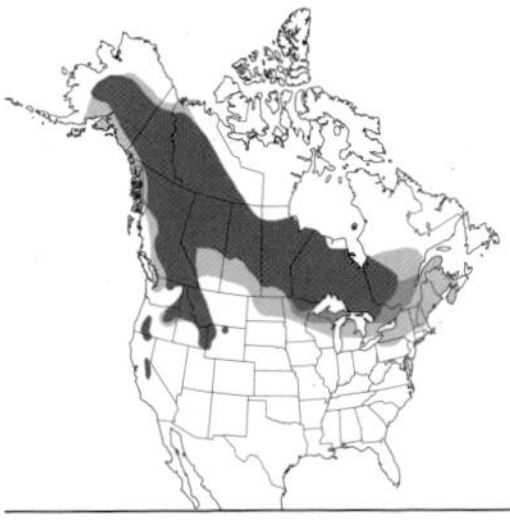

Rare. Found in coniferous forest, especially around forest clearings such as mountain meadows; roosts in trees in the day, hunts edges of fields and meadows at night, perching quietly on low fence posts or branches and listening for prey. Birds seen south of normal range in winter frequent forest edges, hedgerows, etc., and are often active in daylight early or late.

Spotted Owl

Strix occidentalis

L 17.5" WS 40" WT 1.3 lb (610 g) ♀>♂

A relatively stocky, medium-size owl; note dark eyes and overall brown color with white spots. Similar to Barred Owl, but smaller, darker overall and with different pattern of dark bars and pale spots on belly. Hybridizes occasionally where range overlaps.

VOICE: Heard mainly at night. Strong resonant hooting/barking with distinctive rhythm *whup, hoo-hoo, hooooo* or longer series; slightly higher-pitched than Barred Owl, notes more monotone with longer pauses. Also a rising, nasal whistle *toweeeeeeip* given by female and a more hissing, rasping *kssssshhip* by begging juvenile.

Uncommon to rare and declining. Coastal subspecies in dense shaded old-growth forest of conifers such as redwood and Douglas-fir; Interior birds associated with oaks and conifers in shaded canyons especially close to rocky cliffs. Nocturnal: roosts during day in trees; at night hunts from perch for small mammals. Solitary or in pairs.

Barred Owl

Strix varia

L 21" WS 42" WT 1.6 lb (720 g) ♀>♂

A relatively stocky, medium sized owl with distinctive dark eye Similar to Spotted in brown color and dark eyes, but easily di tinguished by paler overall color, streaked belly.

VOICE: Clear-voiced, expressive, hooting/barking *hoo ho ho-ho, hoo hoo ho-hoooooaw* ("who cooks for you, who cook for you all") ending with descending and rolling *hoooaaaw* note Often given during daylight. In chorus a tremendous variety c barking, cackling, and gurgling notes. Juvenile begs with risin hiss *kssssssshhip* like Spotted.

Uncommon in mature forest, particularly hardwood swamps—bald cypress in the Southeast and red maple in the north—but any forest with tall trees and relatively open understory is suitable. Usually solitary. Mostly nocturnal: hunts mainly for small mammals.

Boreal Owl

egolius funereus

10" WS 21" WT 4.7 oz (135 g) ♀>♂

abits like Northern Saw-whet Owl and similarly fluffy and large-eaded, but larger, with more or less spotted pattern above and elow. Note dark V through eyes creating "angry" expression.

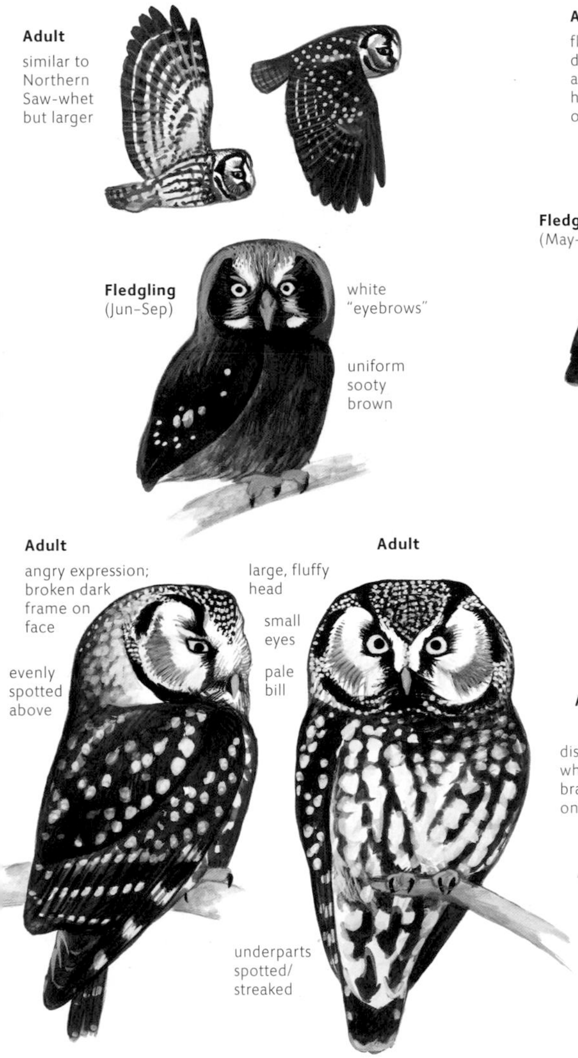

OICE: Song (heard only at night) a rapid series of low, whistled ›ots *po po po po po po po po po*, slightly louder, clearer at end, wo seconds long; similar to Wilson's Snipe winnowing but does ot fade at end. Also a low, nasal *hoooA* falling at end and a hort, sharp *skiew*. Juvenile gives a rather high, clear chirp and ccasionally a short chatter.

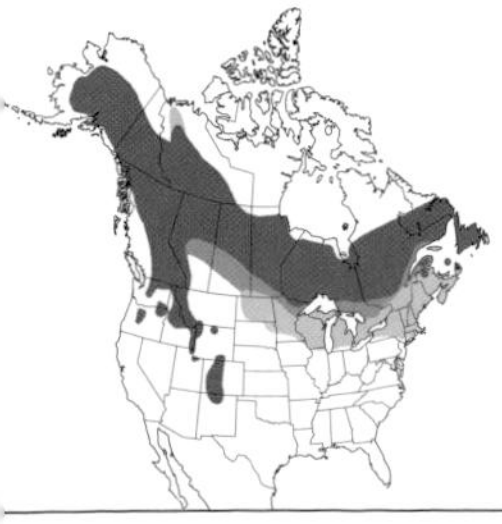

Uncommon and sparsely distributed in mixed aspen and spruce-fir forest; seldom seen. Nests in cavities in mature aspen or poplar groves within coniferous forest, but most foraging and roosting occurs in conifers. Rarely disperses southward or downslope in winter, then found in forest edges or hedgerows, wherever prey is available.

Northern Saw-whet Owl

Aegolius acadicus

L 8" WS 17" WT 2.8 oz (80 g) ♀>♂

The smallest northern owl; fluffy and large-headed. Note prominent white V on face, broad brown streaks on underparts, and white spots on scapulars.

VOICE: Song (heard only at night) repeated, low, whistled toots (about 2/sec) *poo poo poo...* or *toit toit toit...* very similar to Northern Pygmy-Owl but with regular rhythm. Also wheezy, rising, cat-like screech *shweeee*; soft, nasal barks *keew* or *pew* very similar to Elf Owl; whining, soft whistle *eeeooi*.

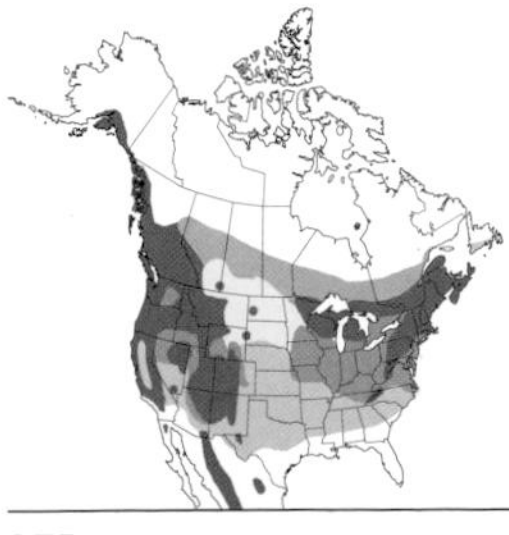

Uncommon in wooded areas, mostly mature and diverse mixed coniferous and deciduous woods, often near wet areas. Nests in tree cavities. Nocturnal: roosts during day in dense foliage; hunts at night for small mammals. Solitary.

Burrowing Owl

Athene cunicularia

L 9.5" WS 21" WT 5 oz (155 g)

A small owl with long legs and distinctive habit of perching in the open. Agitated birds have quick bobbing "deep knee-bend" action. Note barred underparts, spotted upperside, white throat and arched white eyebrow.

VOICE: Heard mainly at night. Male gives high, nasal, trumpeting *coo-cooo* call on one pitch; female answers with short, clear *eeep* or harsher, rasping *ksshh*. Short, sharp, husky barking series often with a rasping scream *kweee-ch-ch-ch-ch* heard year-round. Other smacking or warbling calls rarely heard. Juvenile begs with short, harsh, rasping calls.

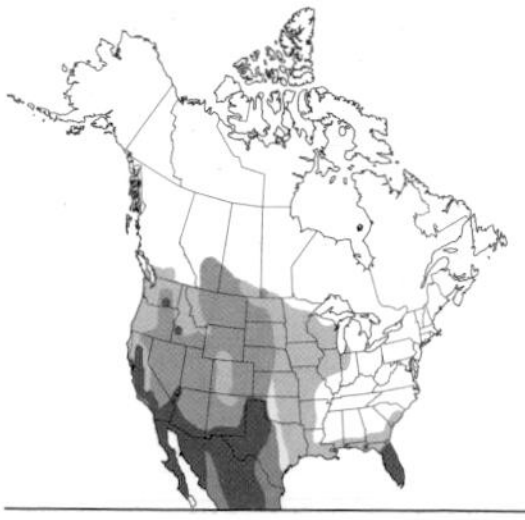

Uncommon and local. Found in expansive, nearly flat open areas such as prairie grassland, with a few scattered shrubs. Relies on mammals to create its burrows for roosting and nesting, so found especially around Prairie Dog towns or where ground squirrels are common. Hunts small rodents at night, but often seen perched on the ground or fence posts in daylight. Usually solitary.

Two populations with no range overlap (except vagrants of both populations along Atlantic coast). Florida adults are darker brown overall; details are shown above. Florida birds have feathering on tarsi reduced to sparse hair-like shafts (Western more feathered but variable and can become worn by spring). Spotted underwing coverts of Florida birds reliable but difficult to see in the field.

Elf Owl

Micrathene whitneyi

L 5.75" WS 13" WT 1.4oz (40g)

Our smallest owl; distinguished from other owls by size alone. Note overall speckled gray and brown pattern, blurry streaks below, and arched white eyebrows.

VOICE: Heard only at night. Common call a fairly sharp, high bark *pew* or *peew*; slightly squeaky series with steady rhythm *pe pe pe pe pe pe pe pe pe* highest in middle. Also soft, high, quiet, whistled *meeeew* descending.

Uncommon and local in open dry woodlands and patches of dense brushy vegetation along desert washes, often associated with large cacti such as Saguaro. Nests and roosts in cavities such as old woodpecker holes. Nocturnal: forages at night mainly for insects. Usually solitary.

Flammulated Owl

Psiloscops flammeolus

L 6.75" WS 16" WT 2.1oz (60g)

The only small owl with dark eyes. Similar to screech-owls, but smaller, with short rounded ear tufts; vermiculated gray overall, with streaks of black and rufous.

VOICE: Heard only at night. Song a low, soft hoot *poot* or *pooip*, sometimes *podo poot*; repeated once every two to three seconds; lower-pitched than screech-owls. Female call dissimilar; higher-pitched with quavering, whining quality.

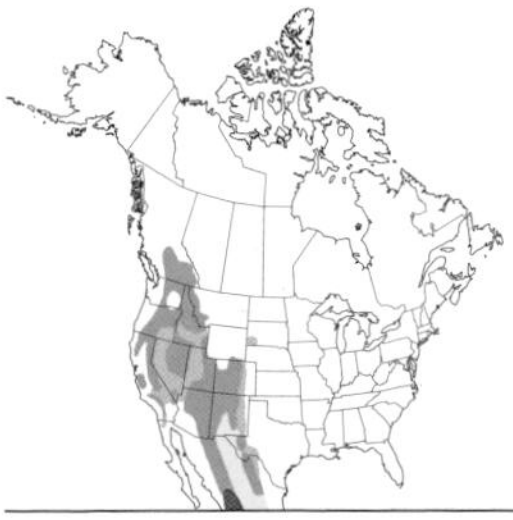

Uncommon and difficult to detect in mixed oak and conifer woods, often associated with Ponderosa Pine. Nocturnal; roosts during day in cavities or in dense foliage high in trees; at night forages in woods mainly for insects, often remaining hidden in foliage. Usually solitary.

Whiskered Screech-Owl

Megascops trichopsis

L 7.25" WS 17.5" WT 3.2 oz (90 g) ♀>♂

Very similar to Western Screech-Owl but little overlap in habitat; best distinguished by voice. Slightly smaller than Western, with relatively smaller feet and coarser markings above and below giving a somewhat more spotted appearance.

Adult

found in oaks, usually at higher elevation than Western, but much overlap

Fledgling
(May–Aug)

more coarsely barred than other screech-owls; best distinguished from Western by size, foot size, bill color

Adult

more coarsely marked above than Western

only gray morph occurs in US (red morph in Mexico)

eyes tinged orange

Adult

bill color varies from dark gray to yellowish, usually paler than Western

short, thick dark streaks below look somewhat spotted

relatively small feet

overlaps with Mexican population of Western

VOICE: Heard mainly at night. Common call a steady series of four to eight evenly spaced notes *po po po po po po* higher in middle, slightly slower at end. Syncopated, rhythmic *pidu po po, pidu po po, pidu po po, po* (pitch same as Western Screech-Owl). Female voice slightly higher than male. Also a descending soft whistle higher than other calls.

Fairly common locally within limited range. Found in oak woods, generally at higher elevations than Western Screech-Owl. Habits similar to Western.

Western Screech-Owl

Megascops kennicottii

L 8.5" WS 20" WT 5 oz (150 g) ♀>♂

Similar to Eastern Screech-Owl and best distinguished by voic Small and stocky; relatively short-tailed, broad-winged, an large-headed. Variable in color from brownish to plain grayis but always intricately patterned.

Fledgling
(May–Aug)

similar to gray morph Eastern but with dark bill

Brown adult Pacific

Gray adult Pacific

Adult Great Plains

Adult Mojave

Adult Mexican

dark bill

prominent dark streaks and weak cross bars

VOICE: Heard mainly at night. Primary song an accelerating serie of short whistles (bouncing-ball song) *pwep pwep pwep pwe pwepwepwepepepep* slightly lower at end. Tremolo song a two part whistled trill *dddd-ddddddr* slightly falling at end; notes little more distinct than Eastern. Other calls barking and chuck ling; similar to Eastern.

Uncommon but widespread in open woods such as riparian cottonwoods, open oak woodlands, or pinyon-juniper woodlands. Strictly nocturnal. Habits like Eastern Screech-Owl.

astern Screech-Owl

egascops asio

3.5" WS 20" WT 6 oz (180 g) ♀>♂

much of its range the only expected small owl. Very similar to estern Screech-Owl. Best distinguished by voice. Small and ocky, relatively short-tailed, broad winged, and large headed. riable in color from bright rufous to plain grayish, but always tricately patterned.

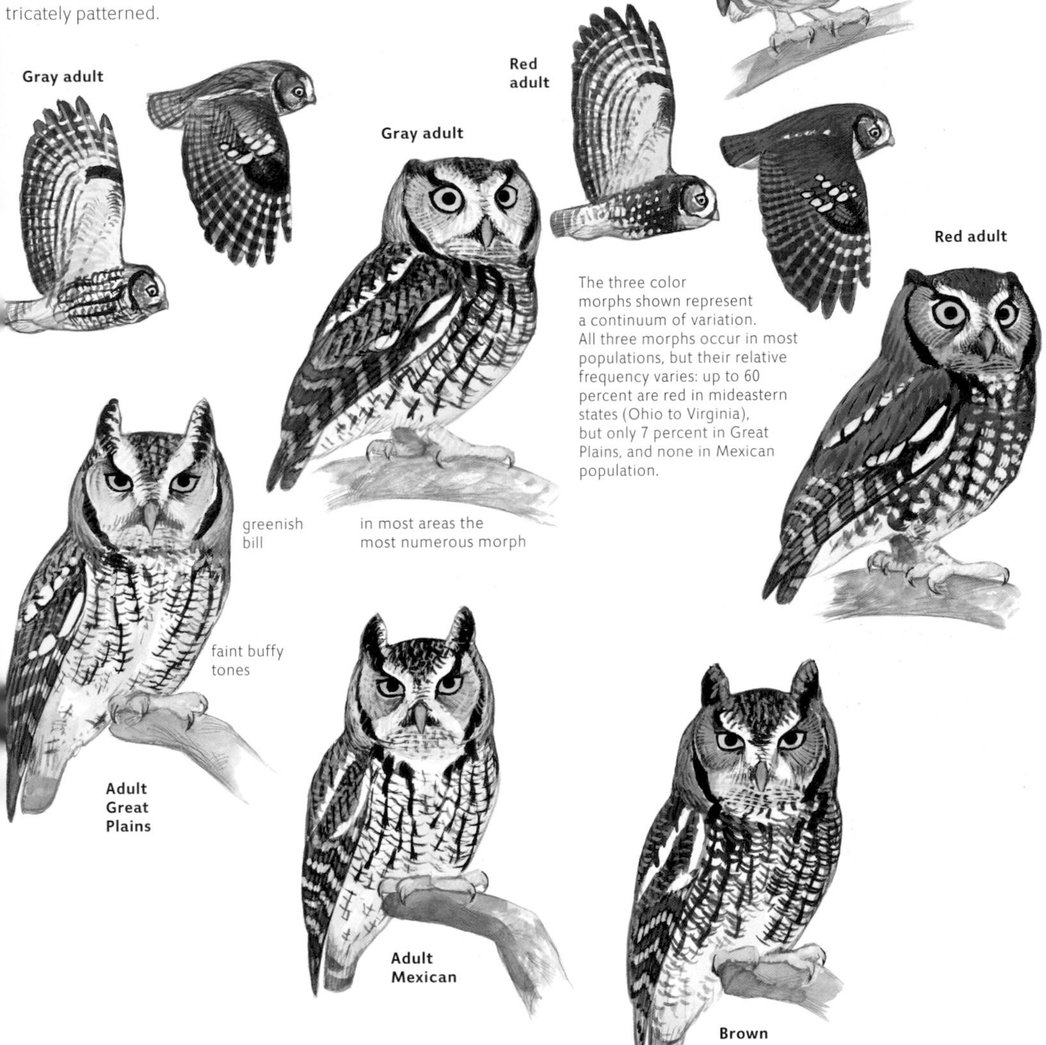

OICE: Heard mainly at night. Primary song a strongly descend- g whinny with husky falsetto quality reminiscent of whinnying orse. Tremolo song a long, whistled trill on one pitch, up to ree seconds long. All calls given by both sexes, but female voice ightly higher-pitched than male. Juvenile begs with short, harsh asp usually falling in pitch. Other calls infrequently heard clude soft bark and short chuckle.

Most Eastern Screech-Owls are much browner than Western, but the two species look nearly identical where their ranges meet (compare Great Plains and Mexican populations of both); hybrids have been recorded, and not all individuals can be safely identified.

Regional variation in Eastern Screech-Owls is clinal. In general, northern birds are larger, paler, and fluffier than southern. Mexican birds (in southern Texas) are most distinctive, always gray with markings tending toward Western Screech-Owl and slightly different voice: whinny call short, weak, and infrequently given; tremolo call more rapid, wooden-sounding, with more distinct notes and uneven tempo.

Uncommon but widespread in open wooded habitats. Strictly nocturnal: roosts during day in old woodpecker holes and other cavities or next to tree trunk; perches at night on low limbs in orchards, open woods and along forest edges in search of insects and rodents. Usually solitary.

Northern Pygmy-Owl

Glaucidium gnoma

L 6.75"   WS 12"   WT 2.5 oz (70 g)   ♀>♂

A very small owl active in daylight, with relatively long tail that is often jerked sideways when perched. Note bold dark streaks on underparts, false eye-spots on nape, and dark head with short white eyebrow.

VOICE: Song (often heard during day) of monotonously repeated single or double toots; pattern and rate of delivery varies regionally. Always lower-pitched and slower than Ferruginous Pygmy-Owl; most populations slower than Northern Saw-whet, but Mexican populations virtually identical. Sometimes begin with a low, descending series of rapid toots followed by normal series *popopopopo, too-too too*.... All populations give a very high rattle or trill *tsisisisisisisi*.

Uncommon in broken forests of mixed oaks and conifers with clearings and patches of brush or chaparral. Active in daylight, perching within trees on horizontal branches to hunt for small birds. Solitary.

Two or three regional populations differ slightly in overall color and voice, with Pacific birds darkest and brownest. They give very slow single toots (1 note every 2 or more sec). Interior West birds grayer overall, and the pale spots on breast and crown are often broadened into short bars. They give mainly single toots with some paired notes, resulting in irregular rhythm *too, too-too, too, too, too-too, too*... (about 1 note every 1.4 sec or 1 pair every 2 sec). Mexican population (not shown), found in southeastern Arizona, averages smaller and darker than Interior West birds; Mexican birds give mainly paired notes more rapidly (about 1 pair every sec).

'erruginous Pygmy-Owl

laucidium brasilianum

6.75" WS 12" WT oz (70 g) ♀>♂

abits and appearance very similar to Northern Pygmy-Owl, but nge and habitat do not normally overlap. Average color is more ange-brown than Northern, especially on tail; occurs adjacent grayest population of Northern.

OICE: Song (often heard during day) rapidly repeated whistled otes (about 3/sec), each note slightly rising *pwip pwip pwip...*; ften a few higher, weaker notes at beginning; usually higher-itched than Northern Pygmy-Owl but variable in pitch and uality. When agitated may give a sharper, more barking whistle.

Rare and local in patchy live-oak, mesquite, and riparian woods in Texas; in Saguaro and mesquite in Arizona. Active in daylight; perches in trees on horizontal branches and hunts small birds. Usually solitary.

Northern Hawk Owl

Surnia ulula

L 16" WS 28" WT 11 oz (320 g) ♀>♂

Size similar to Cooper's Hawk, but with large head of owl, very round fluffy body. Flight accipiter-like: low and fast with quick stiff wingbeats; swoops up to perch and often hovers over open ground. Note uniformly barred underside and black frame on whitish face.

VOICE: Courtship call (heard mainly at night) a series of popping whistles up to six seconds long *popopopopo...*; reminiscent of Boreal Owl but higher, sharper, and longer. Female and juvenile give weak, screeching *tshooolP*. Also a thin, rising whistle *feeeee*. Alarm a shrill, chirping *quiquiquiqui*.

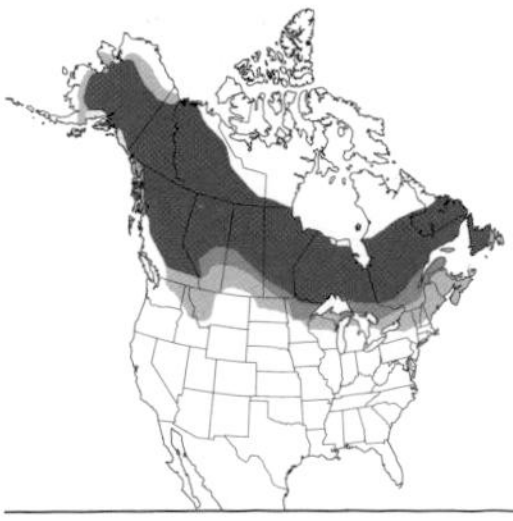

Rare in open spruce woods and around bogs or burned areas with widely scattered tall trees and open ground. Active in daylight, perching shrike-like on the highest treetops or poles to watch for small mammals. Solitary.

Nightjars

FAMILY: CAPRIMULGIDAE

9 species in 4 genera. All are active mainly at night, foraging on insects captured in flight; all have tiny bill but very large mouth. Roost during the day on the ground or on low horizontal branches, relying on very cryptic coloration for camouflage. All species except nighthawks have a relatively large head and rounded wingtips; they forage exclusively at night by perching on the ground or on low branches in open woods or shrublands, or clearings, watching for passing insects, and give lou repeated, whistled song. Nighthawks have long pointed win and relatively smaller heads, and are sometimes seen in daylig flying over open areas or woods to catch insects in flight; muc less vocal than other nightjars. Nest is a simple scrape on leafy c gravelly ground. Nestlings are capable of flight when just ten da old, much smaller than adult size. Adults are shown.

Genus *Chordeiles*

Lesser Nighthawk, page 286

Common Nighthawk, page 287

Antillean Nighthawk, page 286

Genus *Antrostomus*

Chuck-will's-widow, page 284

Buff-collared Nightjar, page 284

Eastern Whip-poor-will, page 285

Mexican Whip-poor-will, page 285

Genus *Nyctidromus*

Common Pauraque, page 283

Genus *Phalaenoptilus*

Common Poorwill, page 283

Eyeshine

The nightjars (excluding nighthawks) are strictly nocturnal and very difficult to see. They become active after dark and move into clearings where they hunt for flying insects. They often sit on quiet back roads, and their large eyes strongly reflect any light shone on them, most strongly when the beam of light originates near the observer's eyes. This "eyeshine" is an excellent way to locate nightjars (and other nocturnal animals).

;ommon Pauraque

yctidromus albicollis

11" WS 24" WT 1.8 oz (52 g)

›nger-tailed and with more rounded wings than nighthawks, aking only short flights from the ground. Combination of pale ır across wingtip and extensive white in tail unique.

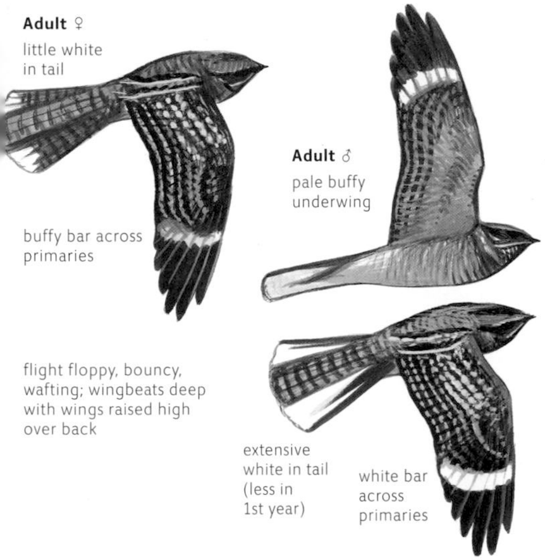

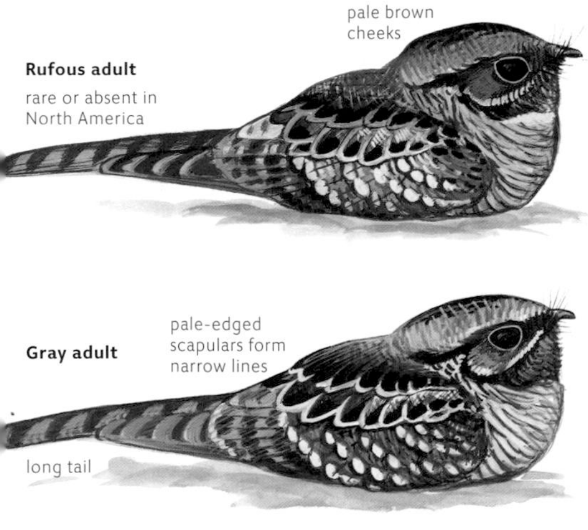

ɔICE: Song (heard only at night) a sharp, wild, buzzy whistle ɔWIzheeeeeer; year-round call more slurred, builds slowly to ·escendo *po po po po... po pup purrEEyeeeeeeeerrrr*. Also a ·mple, husky, buzzy *urrREEErrr*. When flushed gives a low, soft, ȷuid *quup*.

Common locally. Roosts during day on the ground under overhanging low branches within brushy woods. Active at night; moving out to field edges and roads, sitting on the ground and flying up to catch passing insects, rarely flying higher than 10 feet. Solitary or in pairs.

Common Poorwill

Phalaenoptilus nuttallii

L 7.75" WS 17" WT 1.8 oz (50 g)

Our smallest nightjar, with short tail and relatively short, rounded wings. Grayish overall. Range and habitat usually do not overlap other nightjars, and very short tail is distinctive, but beware (in summer) young fledglings of other species can fly before fully grown.

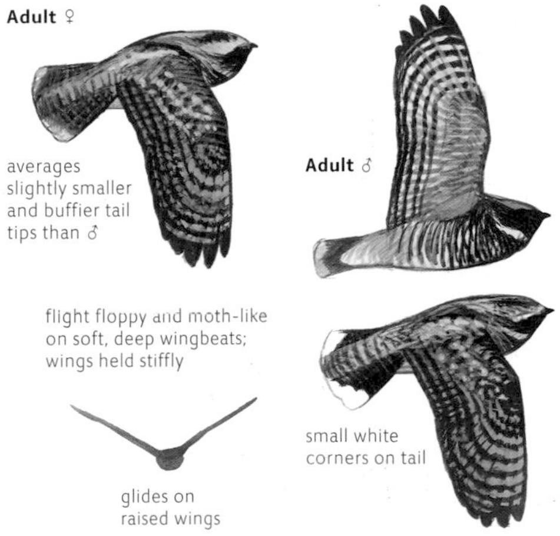

VOICE: Song (heard only at night) a gentle, low whistle *poowJEE-wup*; from distance sounds like simple, soft whistle *poor will*. Also a rough, low *wep* or *gwep* slightly rising, lower than any part of normal call, and soft clucks.

Common locally. Roosts during day on dry gravelly or rocky ground, in the shade of scattered bushes, often associated with chamise, scrub-oak, sagebrush or pinyon-juniper. Active at night; moving to roadsides, trails or other clearings; sits in open and flies up to catch flying insects. Solitary.

Buff-collared Nightjar

Caprimulgus ridgwayi

L 8.75" WS 18" WT 1.7 oz (48 g)

Larger and longer-tailed than Common Poorwill, smaller and shorter-tailed than whip-poor-wills. Also compare tail pattern. Other species can show a buffy collar, but not as clean-cut and obvious as on this species.

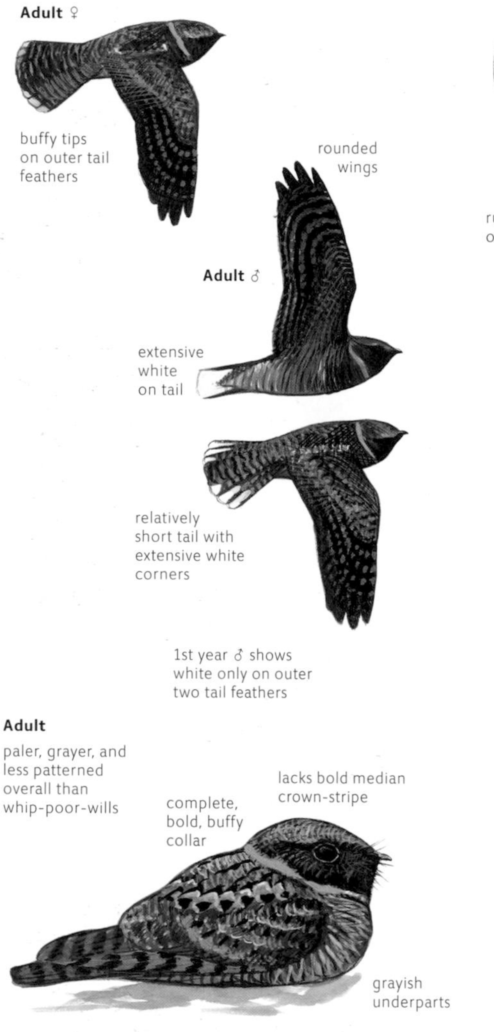

VOICE: Song (heard only at night) of rapid, sharp, dry chips in crescendoing series *tup to to tu tu ti ti trridip* accelerating and rising to flourish at end; pattern similar to dawn song of Cassin's Kingbird and Vermilion Flycatcher but notes sharper and drier. Also rapid, dry, clicking notes.

Rare and local in dry, brushy, desert washes. Shares habitat with Common Poorwill, while Mexican Whip-poor-will is found at higher elevations among trees. Habits similar to whip-poor-wills. Often hunts from perch at top of small bush

Chuck-will's-widow

Caprimulgus carolinensis

L 12" WS 26" WT 4.2 oz (120 g)

Our largest nightjar, with relatively long, pointed wings and da reddish color. Usually identified by song, rarely seen in dayligh Distinguished from Eastern Whip-poor-will by larger size, diffe ent tail pattern.

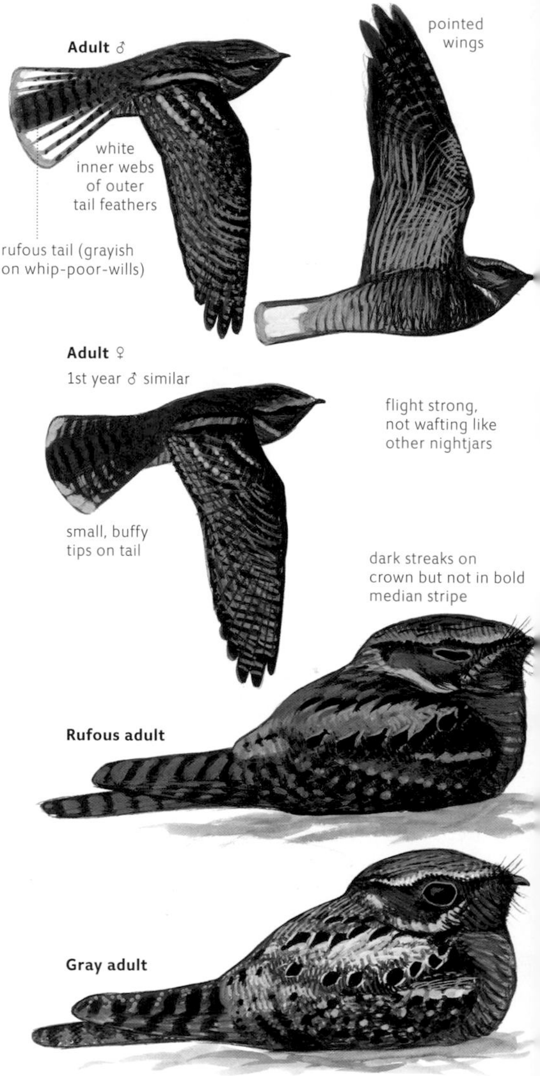

VOICE: Song (heard only at night) a loud, emphatic whistle *CH wido WIDO* or *CHUCK-wills-WIDOW*, repeated; also low, nasa frog-like croaking or growling *wukrr wukrr-wukrr...* and har tongue-clicking cluck. When flushed often gives several muffle low, gruff barks *grof, grof, grof.*

Common locally. Nests on the ground in pine and deciduous woods. Active at night; found in clearings, perching or flying low in search of insects. Roosts during day on the ground or on low branch. Solitary.

⁄lexican Whip-poor-will

aprimulgus arizonae

.75" WS 19" WT 1.9 oz (54 g)

ery similar to Eastern Whip-poor-will and until recently consid-ed the same species. Distinguished from Common Poorwill by oice and habitat, larger size and longer tail. Told from Buff-ollared Nightjar by voice and habitat, longer tail, bold dark own stripe, and less obvious buffy collar.

Adult ♂

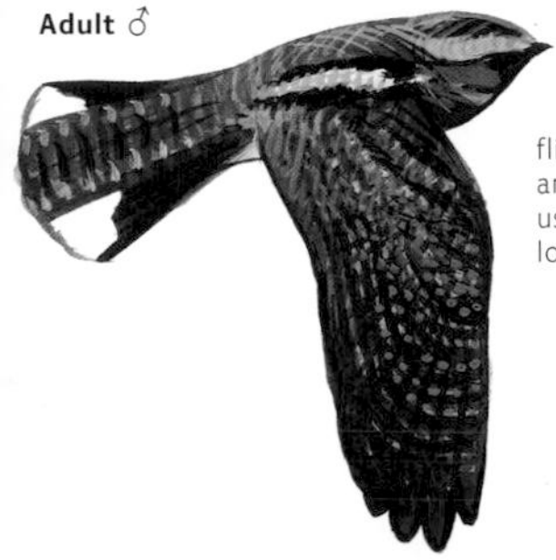

flight buoyant and floppy; usually roosts on low branches

Juvenile (Jun–Sep)

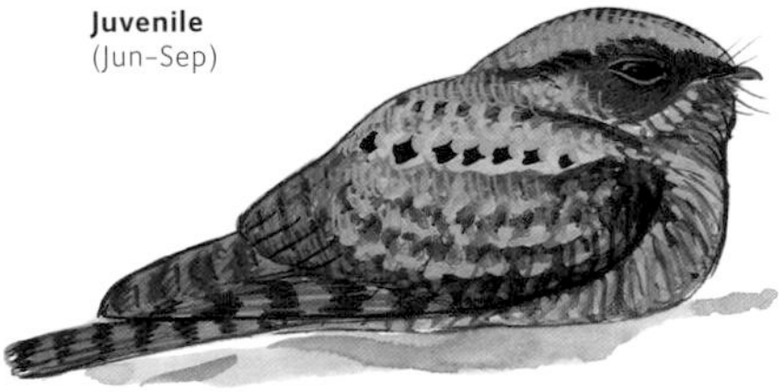

faint buffy collar

gray scapulars

Adult

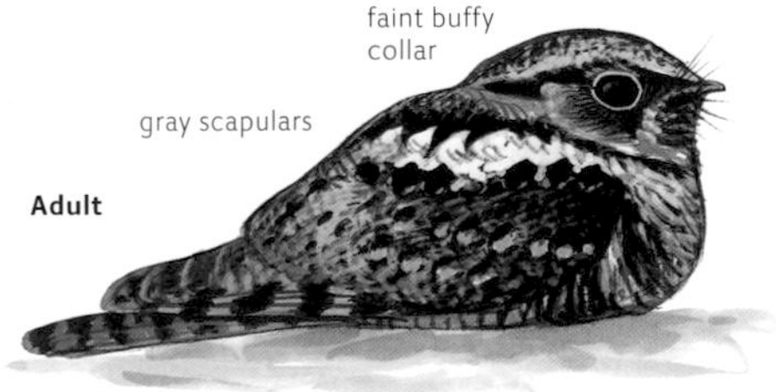

OICE: Song (heard only at night) a rolling, trilled *g-prrip prrEE*; wer, rougher, and with different rhythm than Eastern. When ushed gives rather deep, muffled, rising *gwirp* like start of song.

Common locally. Nests on the ground in open mixed pine and oak woods. Habits like Eastern Whip-poor-will.

Eastern Whip-poor-will

Caprimulgus vociferus

L 9.75" WS 19" WT 1.9 oz (54 g)

Rather dark brown and gray overall with cryptic pattern; note pale corners on tail. Distinguished from Chuck-will's-widow by voice and by smaller size, more rounded wings, and details of plumage. Very closely related to Mexican Whip-poor-will but range does not normally overlap.

Adult ♀

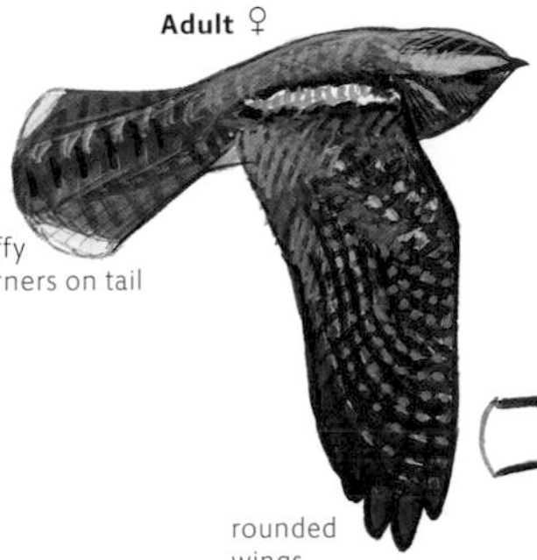

buffy corners on tail

rounded wings

Adult ♂

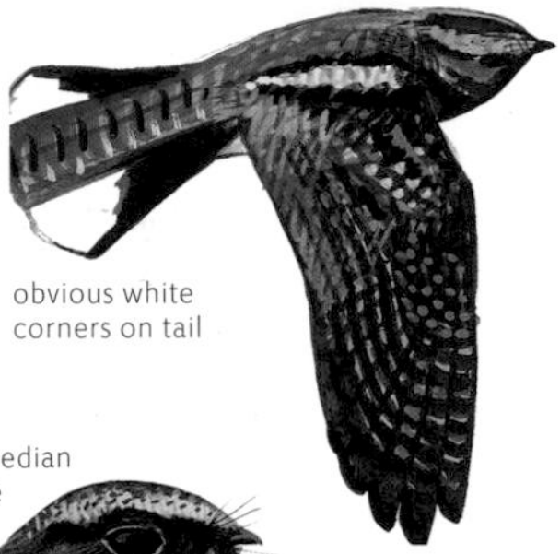

obvious white corners on tail

juveniles of all nightjars pale buffy or sandy-colored and faintly patterned; they can fly before fully grown, resembling shape of Common Poorwill, but are rarely seen

bold, dark median crown-stripe

obvious pale gray braces on back

Rufous adult

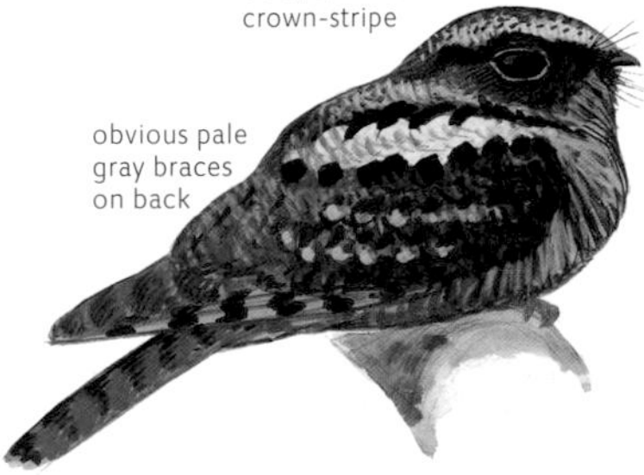

Averages slightly smaller and darker than Mexican, and adult males have more white in tail, but these characteristics overlap and are difficult to judge.

Gray adult

VOICE: Song (heard only at night) a loud, clear, emphatic whistle *WHIP puwiw WEEW* ("whip poor will"); also gives a single, liquid *pwip* like Swainson's Thrush.

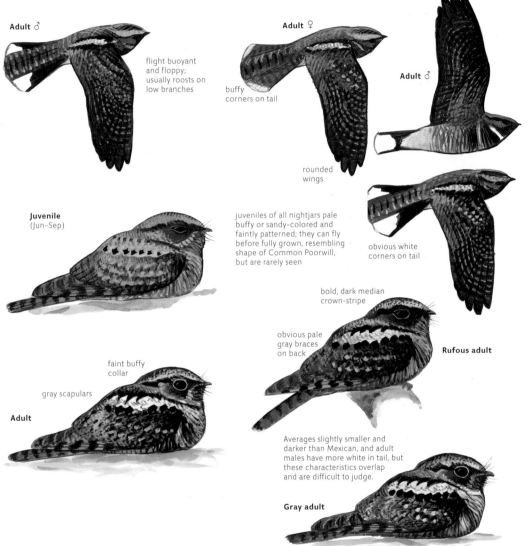

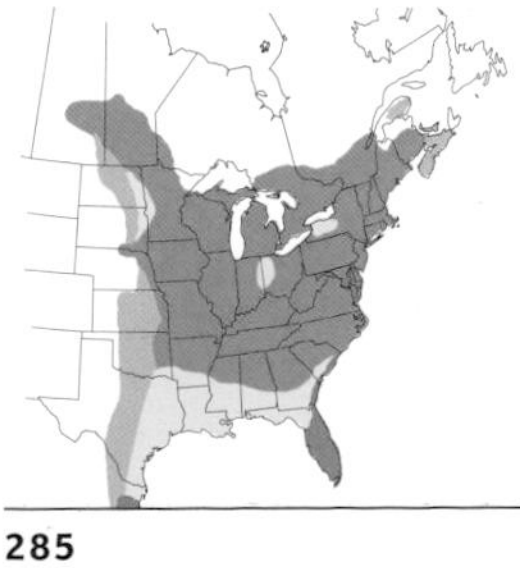

Common locally but declining. Nests on the ground in open mixed pine and deciduous woods. Rarely seen in daytime as it roosts on low horizontal branch or on the ground. Active at night, sitting on the ground or on exposed perch and flying out to catch passing insects. Solitary.

Lesser Nighthawk

Chordeiles acutipennis

L 9" WS 22" WT 1.8 oz (50 g)

Slightly smaller than Common Nighthawk; best distinguished by details of wing pattern. Lacks diving display of other nighthawks. Molts on breeding grounds Jun–Sep.

VOICE: Song up to ten seconds long: a low, whistled trill on one pitch like tremolo of Eastern Screech-Owl but longer, notes more distinct. Also a nasal laughing or bleating *mememeng*. Lacks diving display; silent in flight.

Common in low, arid, open brushlands or deserts. Forages mainly at night, over any habitat, but especially coursing low over ponds or along desert washes in search of insects. Roosts during day on loose gravel on the ground, often in the shade of a small shrub; sometimes on a low branch. Usually solitary.

Antillean Nighthawk

Chordeiles gundlachii

L 8.5" WS 21" WT 1.8 oz (50 g)

Very difficult to distinguish from Common Nighthawk except b voice. Slightly smaller than Common and more richly colore on average, with buffy belly but whitish breast. (Commons wi buffy belly also have buffy color covering breast.)

VOICE: Male gives two to six very rapid, staccato syllables in dece erating series *pitti-pit-pit* or *killy-ka-dik*; quality hard ar sputtering, unlike the somewhat more nasal buzz of Commo Display dive of male produces humming *whoosh* similar to Com mon but much quieter, usually inaudible, and slight higher-pitched.

Uncommon Apr–Aug within limited range in Florida Keys, recorded only a few times nortl of that range. Habitat and habit similar to Common and often found together where range overlaps, especially in spring an fall as Common Nighthawks migrate through range of Antillean.

Common Nighthawk

hordeiles minor

.5" WS 24" WT 2.2 oz (62 g)

ng pointed wings held angled and raised, bounding flight, and hite bar across primaries distinctive. Not related to true hawks. ry similar to Lesser and especially Antillean Nighthawks.

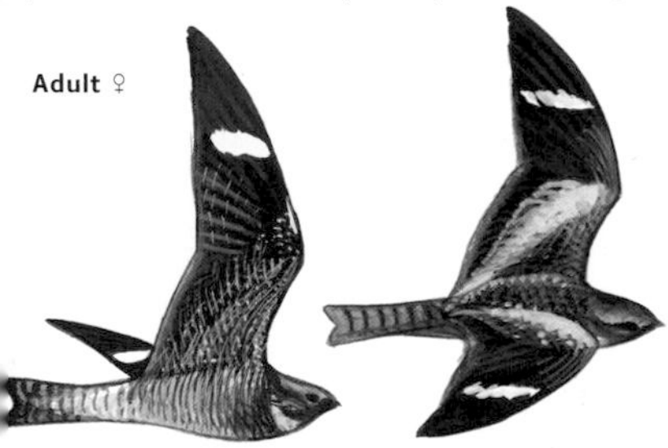

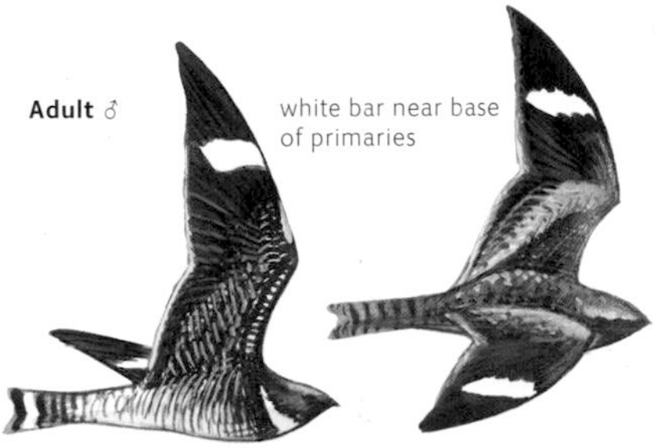

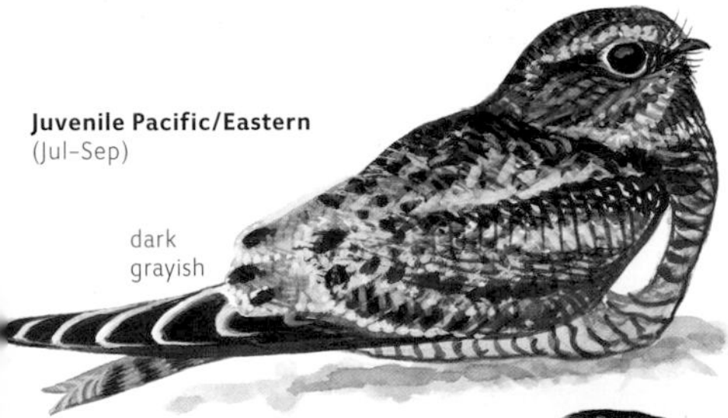

Color varies regionally, but all variation is clinal (most conspicuous in juveniles, as illustrated).

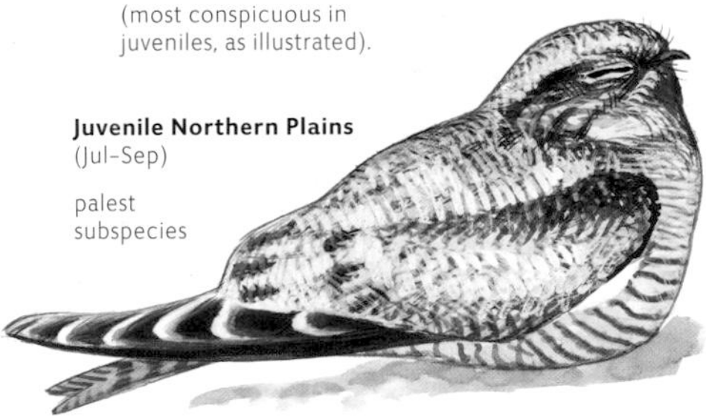

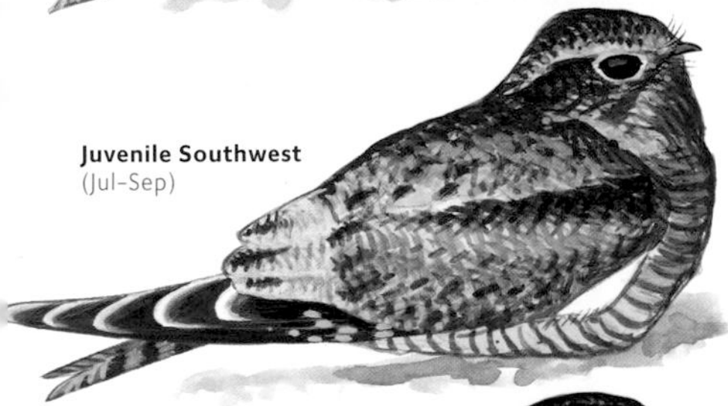

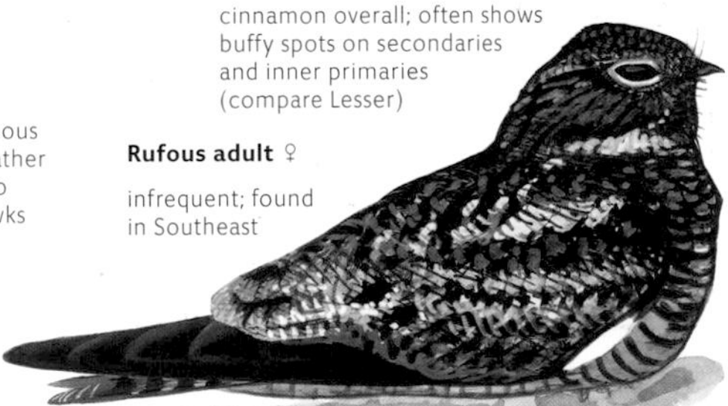

OICE: Territorial male gives rasping, nasal, descending buzz *EEErzh*; given in flight with several quick stutter flaps. Occasionly gives rapid, clucking *quit-quit-quit-quit* in chase; not as accato as Antillean. At nest site female gives low clucks and urring or hissing growls. Display dive of male produces huming, whooshing *Hoooov* at bottom of dive.

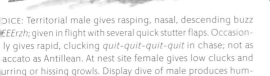

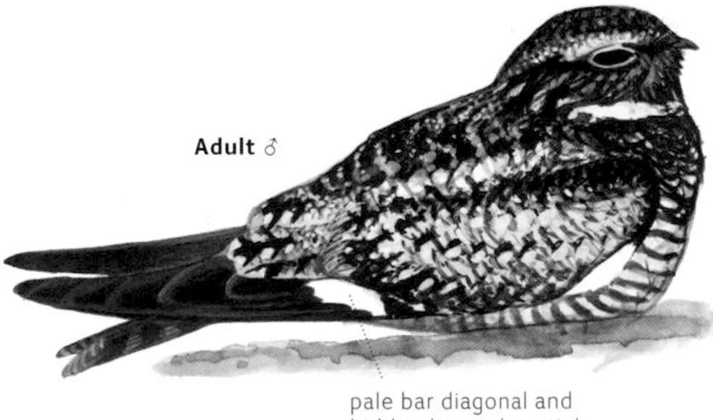

Common around clearings, fields, ponds, and other open areas, as well as over towns and cities. Nests on bare ground and on gravel rooftops. Roosts on the ground or perched lengthwise on a branch. Mostly active at night, but often seen flying in daylight. Flies relatively high in search of insects. Usually solitary, but often forms loose groups when foraging or migrating.

Swifts

FAMILY: APODIDAE

7 species in 5 genera (3 rare species not shown here). Swifts are related to hummingbirds, and have similar wing shape but very different flight style, bill shape, and size. Entirely aerial insectivores, they are seen only in flight, perching only at nesting or roosting sites on concealed vertical walls in cliff crevices, hollow trees, and chimneys. Normally forage at high altitude for tiny insects and spiders; most conspicuous early or late in the day, or during inclement weather when forced to forage at low altitude, often over wetlands with swallows. Superficially similar to swallows, but have different wing shape with short arm; flight is fast and direct with relatively stiff and rapid wingbeats and abrupt turns, unlike the more buoyant and graceful flight of swallows. Adults are shown (except juvenile Black).

Genus *Cypseloides*

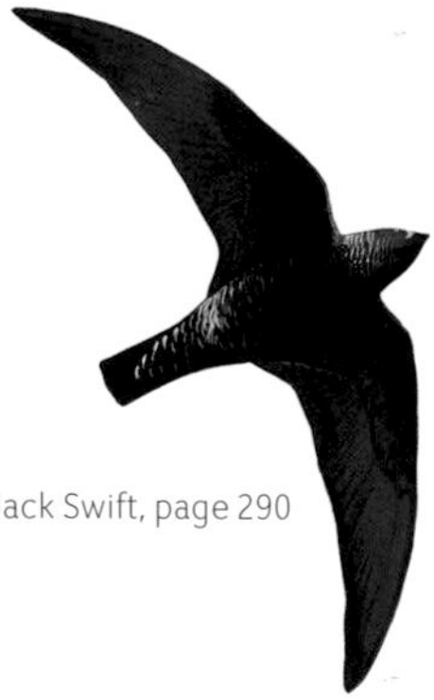

Black Swift, page 290

Genus *Aeronautes*

White-throated Swift, page 290

Genus *Chaetura*

Chimney Swift, page 291

Vaux's Swift, page 291

White-collared Swift

Streptoprocne zonaris

L 8.2" WS 19.7" WT 3.8 oz (249 g)

A very large swift, nearly nighthawk-size, broad-winged ar overall blackish. Larger than Black Swift with relatively long wings. The white collar is obvious in good light, but can be ha to see.

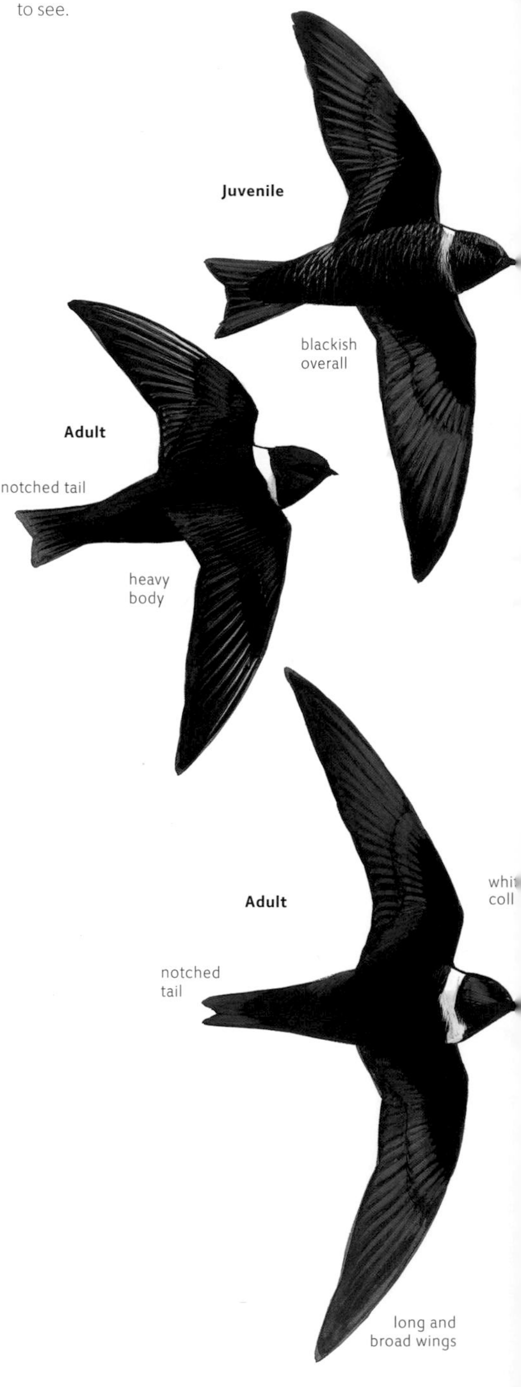

VOICE: Vagrants are usually silent. Calls are high chattering, reminiscent of parakeets.

Very rare visitor from Mexico an the Caribbean with eight widely scattered records—California, Michigan, Texas (several), and Florida.

;ommon Swift

ɔus apus

;.4" WS 17.7" WT 1.4 oz (40 g)

large, long-winged and long-tailed swift. Dark overall like ıimney Swift, but larger and much more slender. Long forked ıl is usually held tightly closed in a point.

ɔICE: Vagrants are usually ıent. Typical call a very high, ırill scream, unlike the sharp ıipping sounds of most New 'orld swifts, most similar to 'hite-throated Swift.

Very rare visitor from Eurasia, recorded several times in Alaska, northeastern North America, and California.

Fork-tailed Swift

Apus pacificus

L 6.9" WS 18.5" WT 1.6 oz

Large and slender like Common Swift, but slightly larger with longer wings and tail, obvious white band across rump.

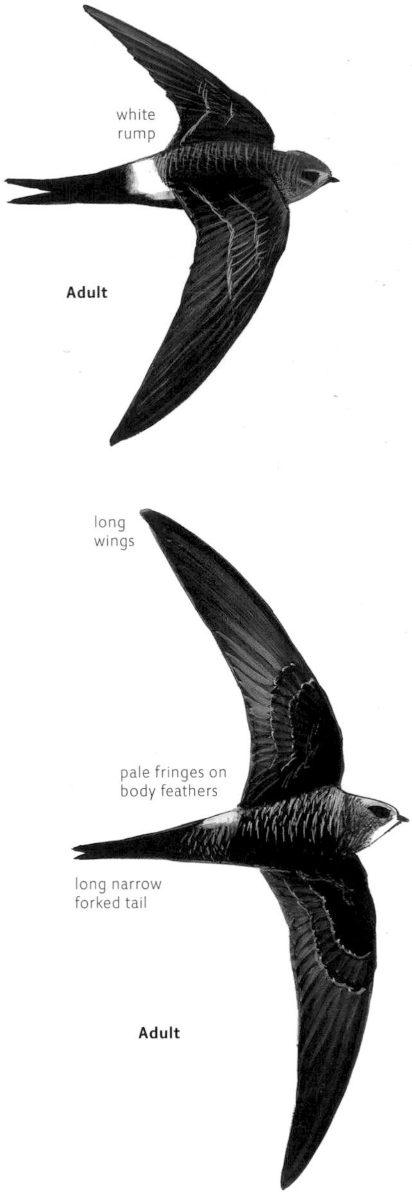

VOICE: Vagrants are usually silent. Typical call a shrill scream, similar to Common Swift but harsher.

Very rare visitor from Asia, recorded several times in coastal Alaska

Black Swift

Cypseloides niger

L 7.25" WS 18" WT 1.6 oz (45 g)

Our largest regularly-occurring swift; all-dark, with relatively broad-based wings and broad square tail. Flight appears slow and erratic for a swift, but still easily distinguished from swallows by quicker and stiffer wingbeats.

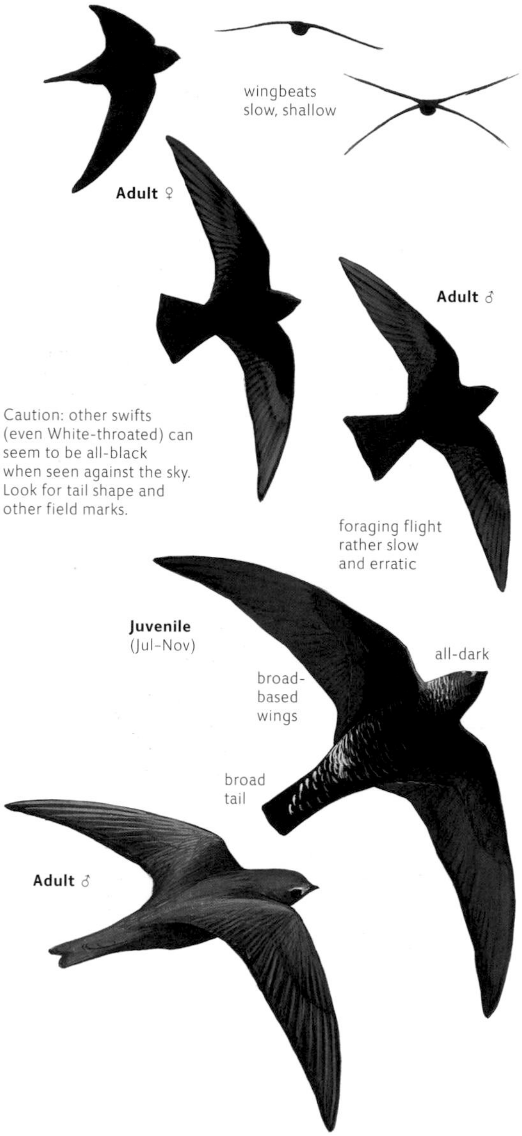

VOICE: Low, flat, twittering chips, often a rapid series of chips slowing at end; lower-pitched than Chimney Swift. Individual notes reminiscent of higher, clearer notes of Red Crossbill.

Uncommon and very local; seen mainly around nest sites on damp coastal cliffs or cliff ledges behind waterfalls. Foraging birds fly at high altitude over any habitat, traveling long distances each day and rarely seen. Often in small groups.

White-throated Swift

Aeronautes saxatalis

L 6.5" WS 15" WT 1.1 oz (32 g)

Larger than Vaux's Swift, with relatively longer wings and long more pointed tail. The only expected swift in our area with wh patches on plumage; note clean white on throat, sides of rum and tips of secondaries.

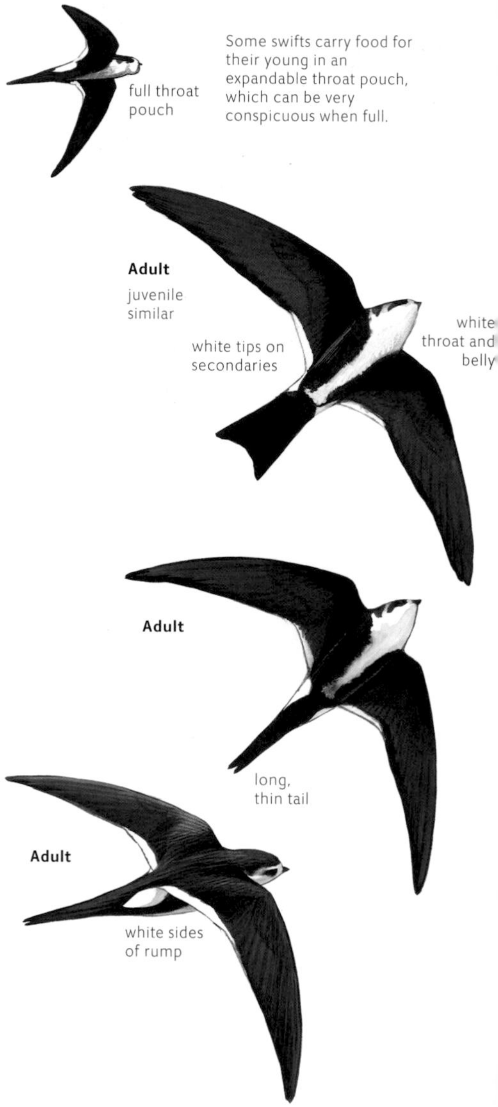

VOICE: Common call a long, descending series of scraping not *ki ki ki kir kir kiir kiir kirsh krrsh, krrsh*; begins high and twitterin ends lower and rasping; often given in chorus by small, wheelir flocks.

Common locally; seen mainly high in air and around nest sites on rocky cliffs. Usually in small groups.

;himney Swift

ıaetura pelagica

·.25" WS 14" WT 0.81 oz (23 g)

ırk gray overall with no contrasting markings, short tail and ry stiff-winged flight. Habits and appearance nearly identical Vaux's Swift. Slightly larger and longer-winged than Vaux's, th less contrasting pale throat and rump.

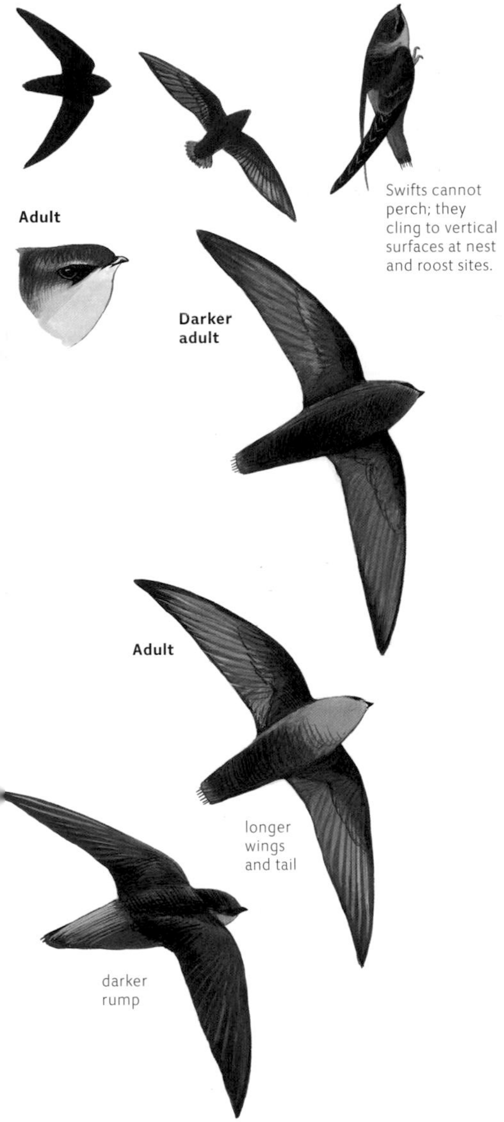

ɔICE: Single high, hard chips run together into rapid, uneven, vittering, chattering series: a rolling, descending twitter. Single ıips often heard are similar to some warblers.

Common and widespread. Nest almost exclusively in chimneys, so most frequently seen around towns and cities. Usually in small groups, but gathers by the thousands at favored roost sites.

Vaux's Swift

Chaetura vauxi

L 4.75" WS 12" WT 0.6 oz (17 g)

Much smaller and shorter-tailed than White-throated Swift. Nearly identical to Chimney (but little range overlap); slightly smaller, with shorter wings and usually paler throat and rump.

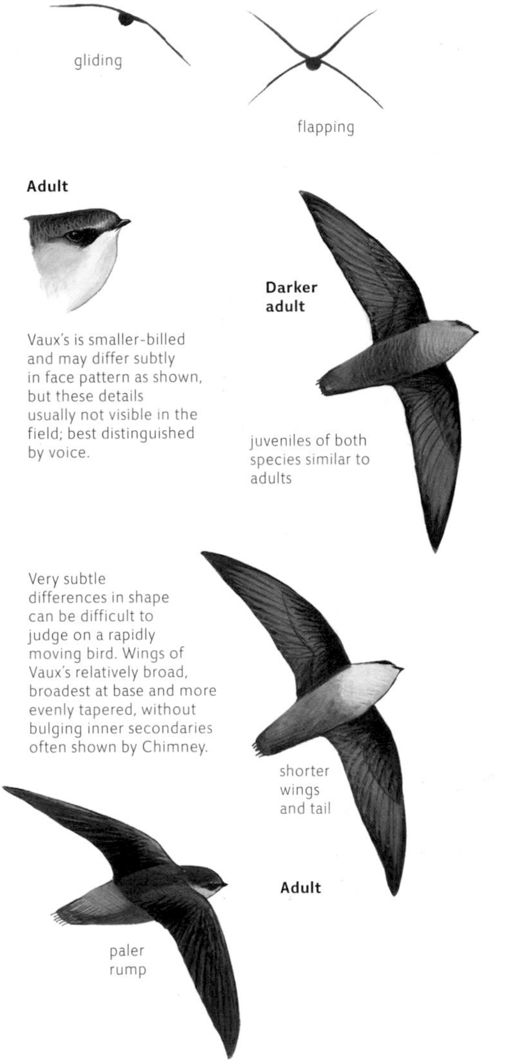

VOICE: Sharp chips higher than Chimney Swift; full call is several sharp chips followed by buzzy trills *tip tip tip tipto tipto tzeeeerip.* Chipping notes higher and final buzz much higher and finer than Chimney; reminiscent of Eastern Kingbird.

Uncommon. Nests in large hollow trees. Usually in small groups seen flying high over any habitat, but especially stream valleys, towns; lower in cool or rainy weather.

Hummingbirds

FAMILY: TROCHILIDAE

19 species in 12 genera. With their small size, long bills, and hovering flight, hummingbirds are truly distinctive. Subtle differences in habitat preferences are of little use for identification, as all species congregate at feeders or patches of flowers. All feed primarily on nectar from flowers, but also take many tiny insects, either picked from vegetation or captured in flight. N is a small cup built of spider webs and decorated with liche placed on a horizontal branch or in an upright fork. Adult fema are shown.

Genus *Colibri*

Green Violetear, page 303

Genus *Anthracothorax*

Green-breasted Mango, page 303

Genus *Eugenes*

Magnificent Hummingbird, page 302

Genus *Heliomaster*

Plain-capped Starthroat, page 300

Genus *Lampornis*

Blue-throated Hummingbird, page 302

Genus *Calothorax*

Lucifer Hummingbird, page 298

Genus *Archilochus*

Ruby-throated Hummingbird, page 295

Black-chinned Hummingbird, page 295

Genus *Selasphorus*

Broad-tailed Hummingbird, page 296

Rufous Hummingbird, page 297

Allen's Hummingbird, page 297

Calliope Hummingbird, page 296

Genus *Calypte*

Anna's Hummingbird, page 294

Costa's Hummingbird, page 294

Genus *Cynanthus*

Broad-billed Hummingbird, page 299

Genus *Hylocharis*

White-eared Hummingbird, page 299

Genus *Amazilia*

Berylline Hummingbird, page 301

Buff-bellied Hummingbird, page 301

Violet-crowned Hummingbird, page 300

ummingbird Wing Shapes

tails of primary shape can be useful for distinguishing some ilar species of hummingbirds. For example the genus *Archi-hus* has much narrower inner primaries than outer primaries. mmingbirds in the genera *Calypte* and *Selasphorus* have just htly narrower inner primaries than outer.

Shape of primaries provides one of the best features for dis-guishing very similar female Black-chinned and Ruby-throated mmingbirds.

Calypte

Anna's and Costa's

Inner primaries on *Selasphorus* and *Calypte* hummingbirds are not as narrow as on *Archilochus*.

Archilochus

Black-chinned

very narrow

tips broad and curved

Ruby-throated

very narrow with hooked tips

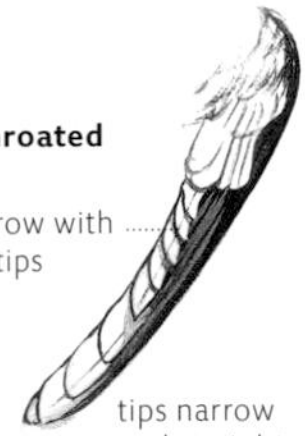

tips narrow and straight

Selasphorus

Broad-tailed

adult ♂

narrow tips produce trill in flight

Anna's Hummingbird
Juvenile ♂
dusted with pollen

ımmingbirds feeding on flower nectar often become dusted stained with pollen on their throats and/or foreheads (differ-t species of flowers deposit pollen in different locations). This eates a yellow to orange patch that can be quite conspicuous t should not cause identification problems, beyond momen-ry surprise, as no hummingbird species is normally yellow on e head.

Hummingbird Displays

Many male hummingbirds perform spectacular dive display for a female. All species also use short back-and-forth "shuttle" flight for close courtship.

All dive displays typically begin with male hovering close above female, then climbing nearly vertically (while looking down at the female) then diving steeply and accelerating rapidly, pulling out of the dive near the female. Some variation in behavior, but general shape and pattern of dives is distinctive, as well as associated sounds.

Display paths are diagrammed here: fainter lines indicate a slow upward movement, stronger lines indicate a rapid descent.

Lucifer Hummingbird
Steep dive usually ending with a short shuttle display.

Anna's Hummingbird
Steep dive ending with a sharp *tewk* sound near the bottom; often repeated several times.

Costa's Hummingbird
Series of steep dives each following quickly after the other without a pause at the bottom, producing a long shrill whistle during descent.

Calliope Hummingbird
Steep dive with muffled hum at bottom; and slow ascent with several pauses.

Ruby-throated Hummingbird
Series of short U-shaped dives with a short high rattle, relatively short and deep arcs.

Black-chinned Hummingbird
Series of short U-shaped dives with a soft stuttering sound; relatively wide and shallow.

Broad-tailed
Steep dive with short buzzy rattle at bottom; the male's wing trill is audible especially at the top of the ascent.

Rufous Hummingbird
Series of steep dives all following a similar path; each ending with a stuttering hum.

Allen's
Steep dive similar to Rufous but usually only a single dive, and humming sound at end of dive is not stuttering; often begins with shuttle display.

Anna's Hummingbird

Calypte anna

L 4" WS 5.25" WT 0.15 oz (4.3 g)

Larger and sturdier than Black-chinned Hummingbird, with relatively short bill. Slightly grayer and drabber overall and less clean-looking than female Black-chinned. Note male's red throat and crown and female's red central throat-patch.

VOICE: Call a very high, sharp *stit.* Chase call a rapid, dry chatter *zrrr jika jika jika jika jika.* Song from perch scratchy, thin, and dry *sturee sturee sturee, scrrrr, zveeee, street street.* Male dive display ends with explosive buzz/squeak *tewk* very similar to some Ground Squirrel alarm calls.

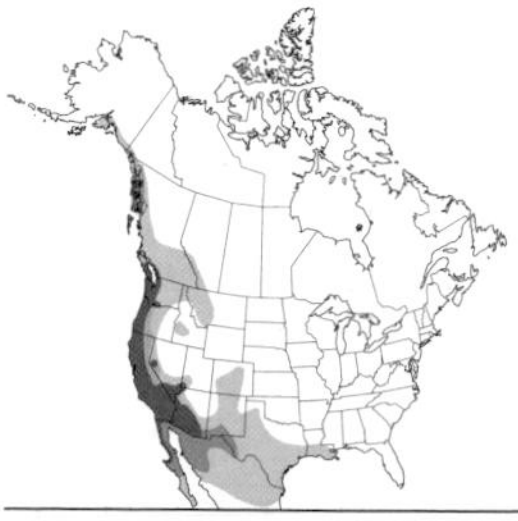

The most common hummingbird in well-watered suburban plantings and flower gardens within the oak/chaparral habitat of the Pacific coast. Range has recently expanded northwards, now common to British Columbia.

Costa's Hummingbird

Calypte costae

L 3.5" WS 4.75" WT 0.11 oz (3.1 g) ♀>♂

One of our smallest birds; tiny and short-tailed. Male has purp crown and long flared throat feathers. Female very similar Black-chinned Hummingbird, but note short tail, wingtips reac ing tail tip, pale underparts, and pale eyebrow.

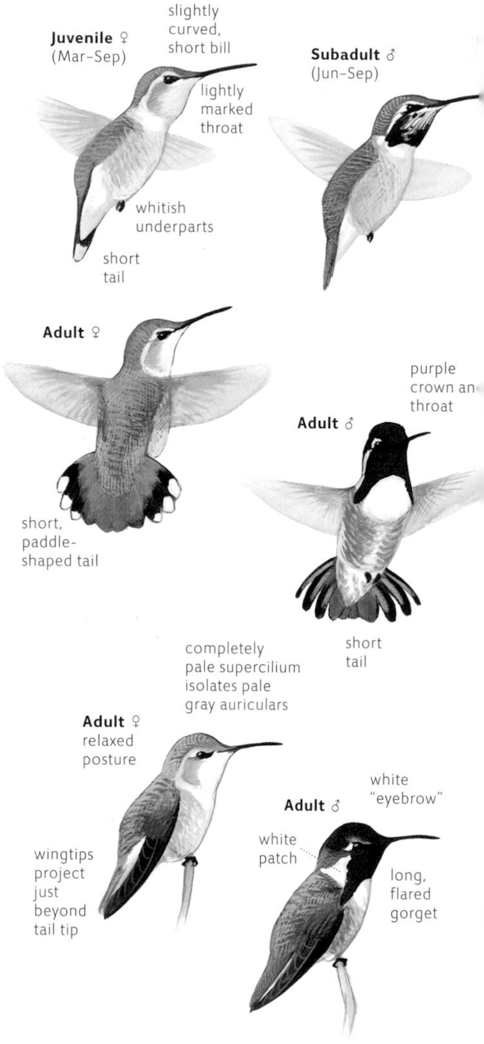

VOICE: Call a light, dry *tink* somewhat like cardinals. Chase ca very sharp, high twitter *stirrr, stirrr* or rapid series of *tink* no followed by lower, scratchy squeal. Song an extremely thin, hi buzz *szeeeee-eeeeeeeeeew* rising then falling. Male dive displ produces continuous shrill whistle.

Common in low desert habitat, where it nests from Jan–Mar following winter rains; uncommon in varied habitats a other seasons, including coasta chaparral, flower gardens, and even high mountain meadows.

lack-chinned Hummingbird

chilochus alexandri

.75" WS 4.75" WT 0.12 oz (3.3 g) ♀>♂

e western counterpart of Ruby-throated Hummingbird. Male's k head and purple throat distinctive. Female similar to Anna's d Costa's, but relatively long-billed and thin-necked, with ter calls.

ICE: Call a soft, husky *tiup* or *tiv* or *tipip*. Chase call sharp, ıttering, cascading *spirrr spididddr* and variations. Male dive splay produces soft, stuttering *didididit*. Wings of adult male oduce soft low whistle in flight. All sounds similar to Ruby-oated.

Common in dry, lowland riparian and oak woods. In East a very rare fall and winter visitor to feeders and flower gardens. Extremely rare in winter in the West, but regular (in very small numbers) then along the Gulf Coast.

Ruby-throated Hummingbird

Archilochus colubris

L 3.75" WS 4.5" WT 0.11 oz (3.2 g) ♀>♂

Very similar to Black-chinned, but range barely overlaps. Note brighter golden-green upperside, usually green crown, male's distinctive red throat and black chin, and female's shorter bill and narrower straighter tips of primaries. Habit of holding tail still while hovering also helpful.

VOICE: Call soft and husky like Black-chinned but may average slightly sharper and higher. A high, rattling *t t t t* is given during dive display; adult male's wings produce faint high buzz in flight.

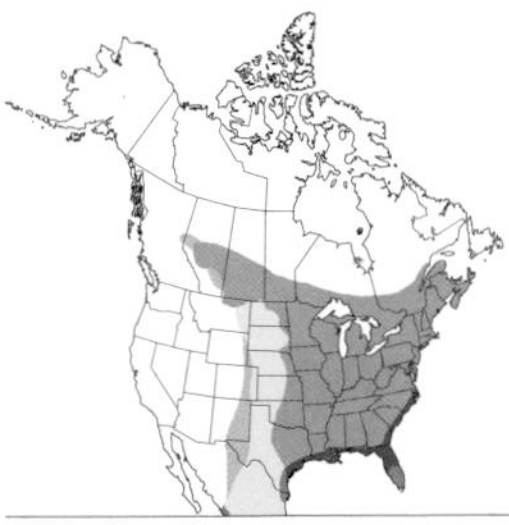

Common in wooded areas and edges of woods, especially at flowers or hummingbird feeders. Often nests near water. The only hummingbird common in the East, but nearly all migrate to Central America, and by November other species (such as Rufous) are equally likely in most of the East.

Calliope Hummingbird

Stellula calliope

L 3.25" WS 4.25" WT 0.1 oz (2.7 g) ♀>♂

Our smallest hummingbird; tiny and short-tailed, with relatively short, thin bill. Male has distinctive streaked rosy gorget. Female similar to *Selasphorus* hummingbirds, but smaller and shorter-tailed, with wingtips reaching tail tip, thin white line above gape.

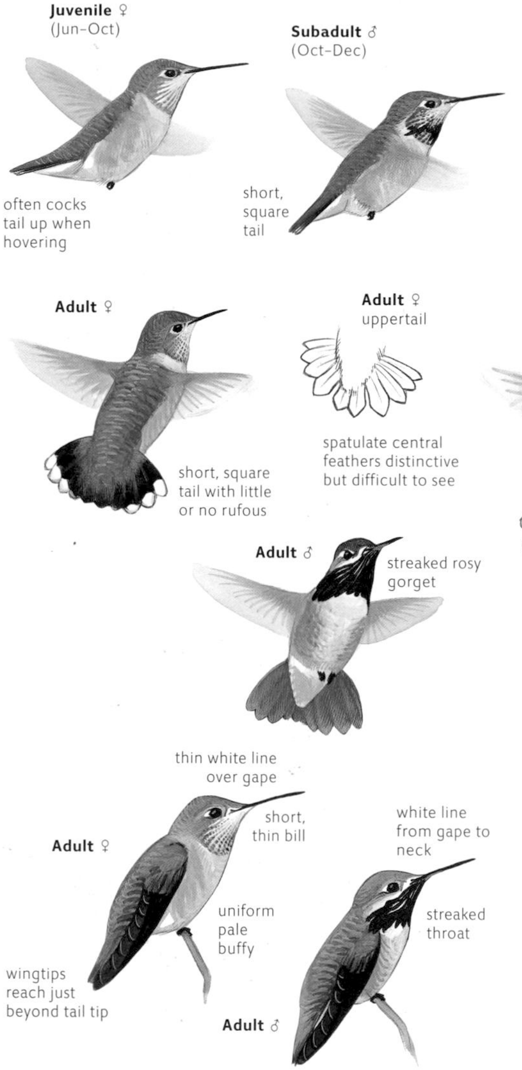

VOICE: Call a quiet, very high, musical chip. Chase call alternates rattle and buzz *tototo zeee tototo zeeee*.... Song a very high, thin whistle *tseeeee-ew*. Male dive display produces short, high, muffled *pvrrr*.

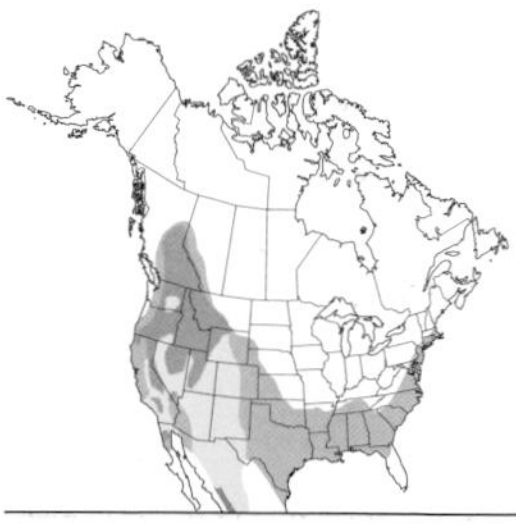

Uncommon and local in riparian thickets and meadow edges within montane coniferous forests.

Broad-tailed Hummingbird

Selasphorus platycercus

L 4" WS 5.25" WT 0.13 oz (3.6 g) ♀>♂

Slightly larger than Rufous Hummingbird. Orange-buff flank female similar to Rufous and Calliope, but tail longer and broa with rufous only on outer feathers.

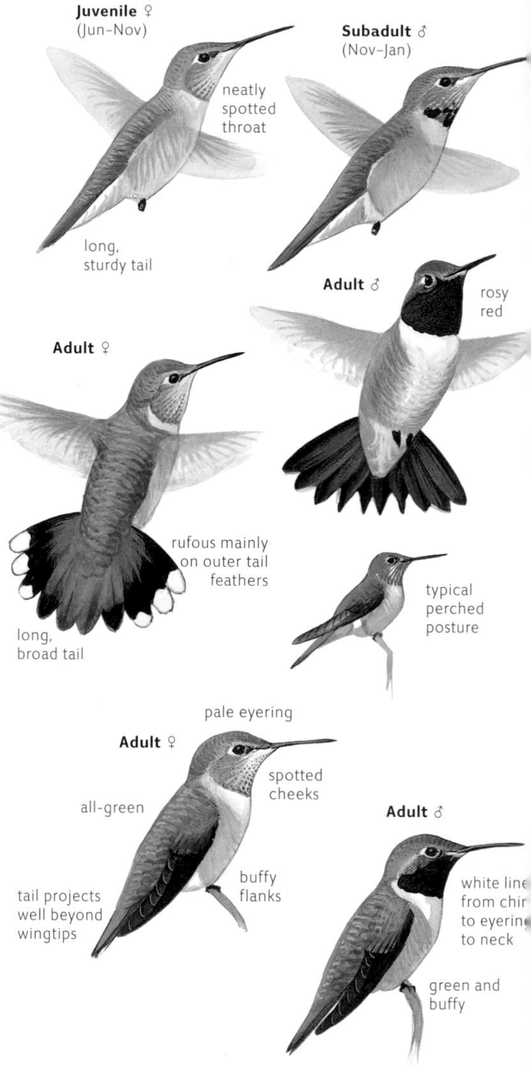

VOICE: Call a sharp, high, chip similar to Rufous but sligh higher. Chase call *tiputi tiputi*... like Rufous but lower, variab Male dive display produces loud wing buzz. Wings of adult m produce high trill in flight like Cedar Waxwing call; lower a more musical than Rufous and Allen's.

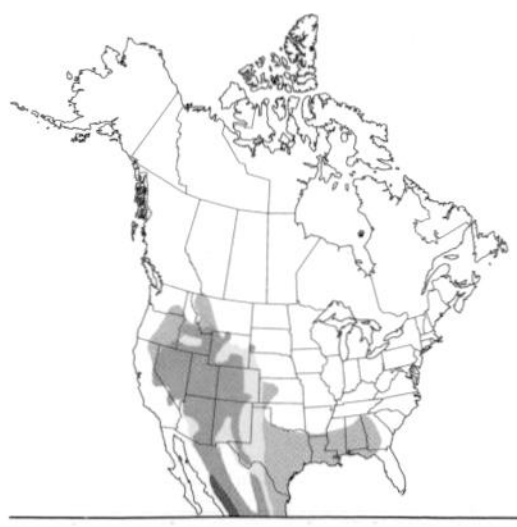

Common in montane dry coniferous woods with openin such as aspen groves, meadow and riparian thickets of willow Common at feeders in any adjacent habitat.

ufous Hummingbird

lasphorus rufus

.75" WS 4.5" WT 0.12 oz (3.4 g) ♀>♂

mall and compact hummingbird with relatively short wings. :ensive orange-rufous color in most plumages distinguishes s species from all except Allen's. Drab females with limited 'ous are very similar to Broad-tailed and Calliope.

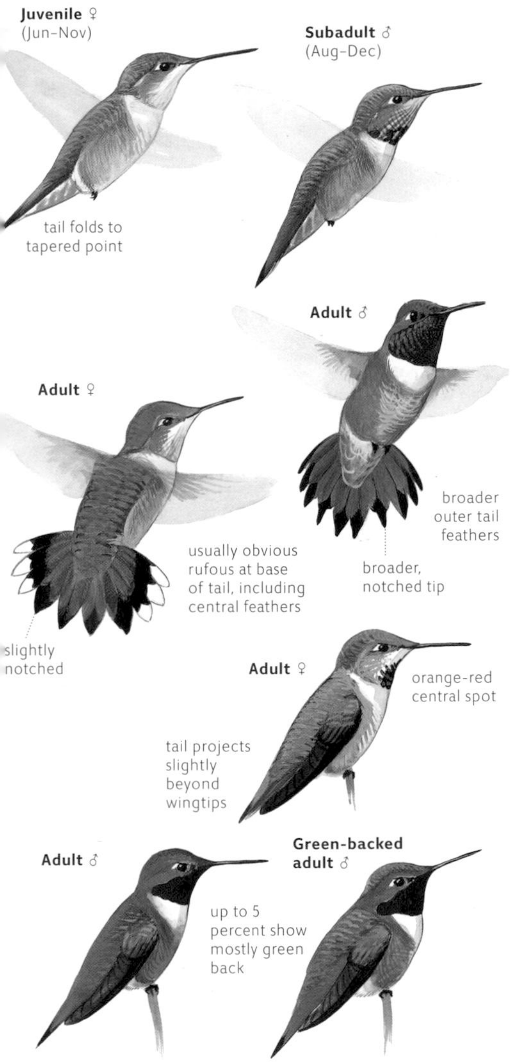

ɔICE: Call a high, hard chip *tyuk.* Chase call a sharp buzz fol-ᴊwed by three-syllable phrases *tzzew tzupity tzupity tzup*; also *ee chew chew chew* or *zeelk zeelk.* Male dive display produces ɪttering, humming *vi vi vi virrr.* Wings of adult male produce gh, buzzy trill like Allen's, higher than Broad-tailed.

Common. Nests in open coniferous forests and riparian woods. Migrants common in many habitats, especially mountain meadows. Rare but increasing visitor to flower gardens and feeders in East.

Allen's Hummingbird

Selasphorus sasin

L 3.75" WS 4.25" WT 0.11 oz (3 g) ♀>♂

Virtually identical to Rufous Hummingbird in all respects. Adult male has greenish back (but beware that some male Rufous also show greenish back). Identifiable with certainty only in the hand by measurements of tail feathers.

VOICE: Calls apparently all like Rufous. Male dive display produces high, humming whistle, not stuttering like Rufous. Wings of adult male produce high, buzzy trill, faster and higher than Broad-tailed; like Rufous but may average slightly higher-pitched.

Common in coastal chaparral and low riparian woods. Habits essentially identical to Rufous Hummingbird, males of both species are aggressively territorial at feeders, attempting to drive away all other hummingbirds.

Lucifer Hummingbird

Calothorax lucifer

L 3.5" WS 4" WT 0.11 oz (3.1 g) ♀>♂

A small hummingbird with strongly curved bill. Male has forked tail and long magenta gorget. Female and immature washed with buffy below and show distinctive buffy stripe above and behind gray cheek.

The strongly curved bill is an excellent field mark, but beware that many other species have slightly curved bills, particularly Black-chinned, and that the curvature can be exaggerated at certain angles, such as head-on views.

VOICE: Call a dry, sharp, twittering chip, often doubled. Chase call louder, sharper chips in series. Shuttle display of male produces fluttering rattle like shuffling of a deck of cards; a quieter fluttering sound produced at end of dive display.

Uncommon and very local in brushy deserts with large numbers of agave plants; small population nests at Big Bend National Park, Texas, and a few pairs in extreme southeastern Arizona. Occasionally wanders into adjacent foothill oak habitat for feeders and flower patches.

Hybrid Hummingbirds

Hummingbirds hybridize more frequently than many other sp cies of birds (although hybrids are still rare). Anna's × Cost Hummingbird is probably the most frequent hybrid combir tion, and should be expected occasionally where those spec overlap in southern California. Other combinations are rare extremely rare, but many combinations have been recorded, a the possibility of a hybrid should be considered when a bird j doesn't fit the standard species.

Anna's × Costa's Hummingbird hybrid adult ♂

Anna's × Rufous Hummingbird hybrid adult ♂

Identification problems are more often the result of variation hummingbird colors rather than hybrids. The iridescent colc of hummingbirds are structural, not pigment—created by lig reflecting off of tiny air bubbles in the surface of the feather. O angles of light can create a very different impression for tl observer. The glittering green back shown by most species c have blue or golden highlights depending largely on lighting

The normally brilliant red or orange throats of male hur mingbirds such as Ruby-throated, Rufous, or Anna's often appe golden-green at some angles, flat black at others. The mc brilliant color is seen on the throat when the bird faces direc towards the observer.

There is variation in the brilliant iridescent throat colors male hummingbirds. For example, normally red colors can va from orange to magenta. There are even several records of ma Ruby-throated Hummingbirds with brilliant, iridescent gree throats.

road-billed Hummingbird

nanthus latirostris

" WS 5.75" WT 0.1 oz (2.9 g)

ıall, but relatively stocky and broad-tailed. Male mostly dark, :h bright red bill and pale undertail coverts. On female note ıgy gray underparts, pale stripes bordering dark cheek, and 1 on lower mandible.

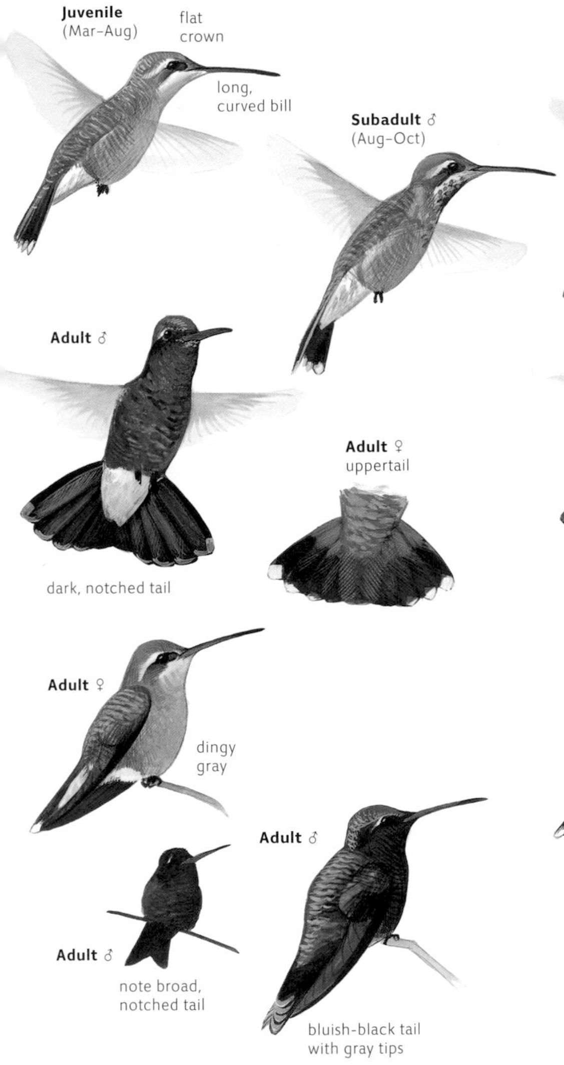

ɔICE: Call a dry *tek* or *tetek* like Ruby-crowned Kinglet but arper; also high, sharp *seek* and thin *tseeew*. Chase call stac-ıto chips followed by dry laughing: *tsik tsitik tilk-ilk-ilk-ilk*. ɔng high, tinkling with buzzy end *situ ti ti ti ti ti ti zreet zreet eet*. Male dive display produces a high *zing*.

Common locally in riparian woods and low elevation wooded canyons; occasionally visits feeders and flower gardens.

White-eared Hummingbird

Hylocharis leucotis

L 3.75" WS 5.75" WT 0.12 oz (3.3 g)

Similar to Broad-billed Hummingbird, but with rounder head and shorter straighter bill. All plumages show striking broad white eyebrow, contrasting abruptly with dark cheeks. Female and immature spotted green on underparts.

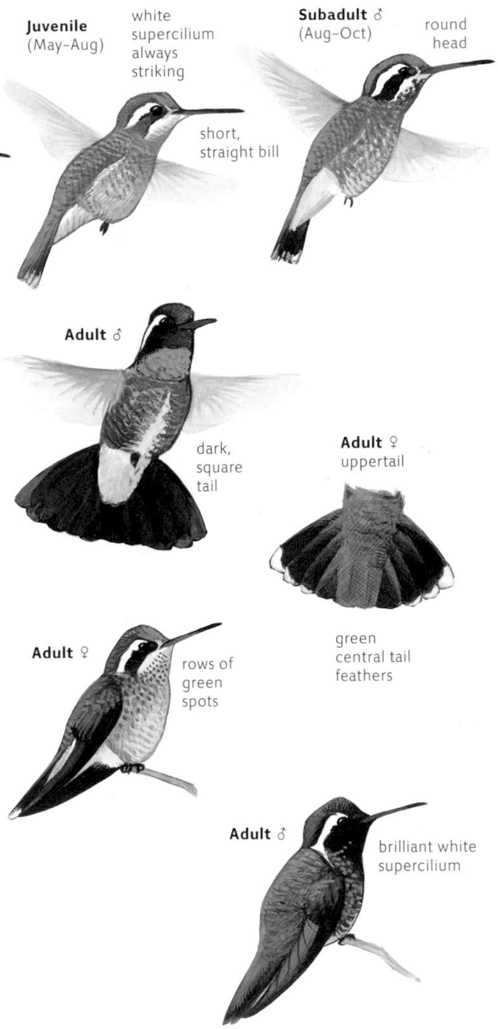

VOICE: Call a high, flat chip similar to Magnificent Humming-bird; often doubled or tripled with the series slightly descending (rising in Magnificent). Chase call a very rapid series of about five sharp, high chips. Song a series of high chips alternating with short, staccato rattles.

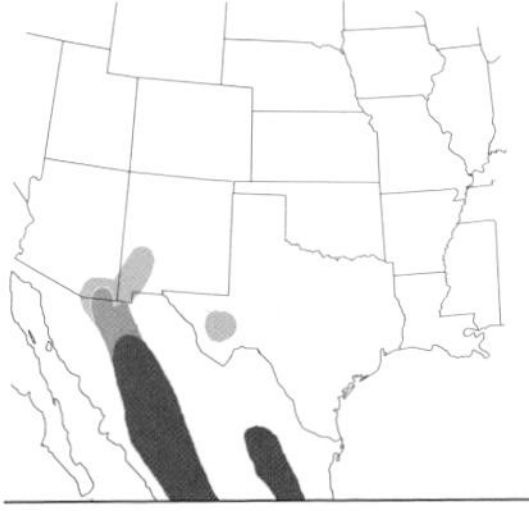

Rare visitor from Mexico to high mountain canyons and pine-oak forests. Only a few recorded each year from Mar–Oct; seen mostly at feeders, but has also nested in our region.

Plain-capped Starthroat

Heliomaster constantii

L 5" WS 7" WT 0.26 oz (7.3 g) ♀>♂

As large as Magnificent Hummingbird, with extremely long bill. Drab dusky olive and gray overall, obvious white stripes on head together with white on flanks and rump distinctive.

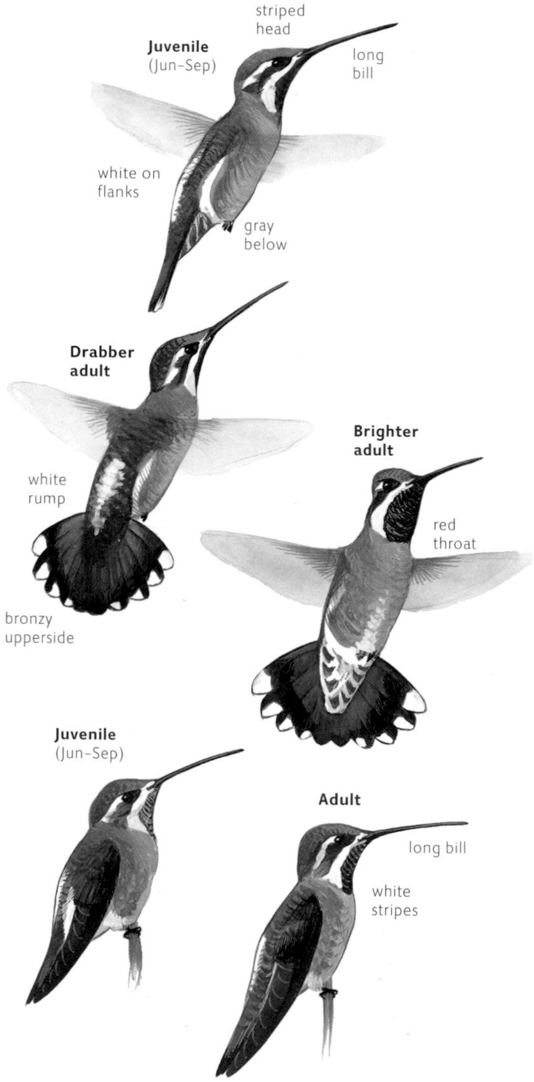

VOICE: Call strong, sharp chips like Magnificent Hummingbird but huskier. Song a series of sharp chips with occasional two-syllable chips *chip chip chip pichip chip....*

Rare visitor from Mexico to oak woodlands or sycamores along streams; one or two recorded annually from May–Sep, mostly at feeders.

Violet-crowned Hummingbird

Amazilia violiceps

L 4.5" WS 6" WT 0.19 oz (5.5 g) ♂>♀

Large and rather long-bodied. Plain brownish upperparts w clean white underparts unique among hummingbirds in weste North America; also note red bill and plain brownish tail.

VOICE: Call a rather dry *tak* or *chap*; sometimes a drier *tek* li Broad-billed. Chase call a squeaky, laughing series *kweesh tw twik twik wik wik*. Song a series of very high, thin, descendir notes *seew seew seew seew seew*.

Rare and local. Nests in cottonwoods and sycamores at few locations along permanent streams in foothills.

·erylline Hummingbird

nazilia beryllina

.25" WS 5.75" WT 0.16 oz (4.6 g) ♂>♀

ɔmbination of entirely dark green head, breast, and back with fous tail and curved red bill shared only with Buff-bellied Hum- ingbird (range does not overlap). Rufous-tinged wings are vious and diagnostic.

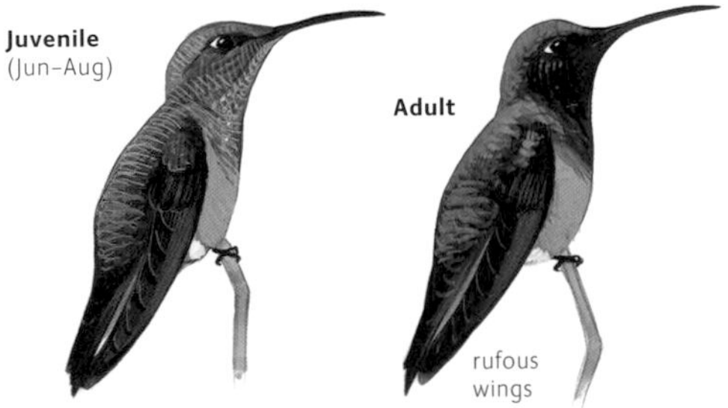

ɔICE: Call a short, buzzy rattle *trrrk*. Chase call a high, thin *seek* ke Blue-throated Hummingbird. Song scratchy and squeaky *rr-toPIKoPIKoPIKo* and variations.

Rare visitor from Mexico to mountain canyons and oak forests. A few recorded annually mainly from Jun–Sep; seen mostly at feeders, but has also nested.

Buff-bellied Hummingbird

Amazilia yucatanensis

L 4.25" WS 5.75" WT 0.13 oz (3.8 g) ♂>♀

Much larger than Ruby-throated Hummingbird; dark greenish overall with rusty-brown tail and grayish buff belly. Red bill most conspicuous on adult male.

VOICE: Call a very high, sharp, metallic *tack*; nearly always given in short series of two to four notes; weaker single notes when perched. Chase call a dry, sharp buzz *jjjjjjj*; a piercing *seek-seek* when chasing predator.

Uncommon in oak woods and suburbs within limited range. Most often seen at feeders or flower gardens, rare north of southern Texas.

Blue-throated Hummingbird

Lampornis clemenciae

L 5" WS 8" WT 0.27 oz (7.6 g) ♂>♀

Noticeably larger than most other hummingbirds and with slower wingbeats than all others. Relatively long, broad tail with large white corners often fanned and very conspicuous. Also note plain gray underparts, bronze rump, and white stripes on head.

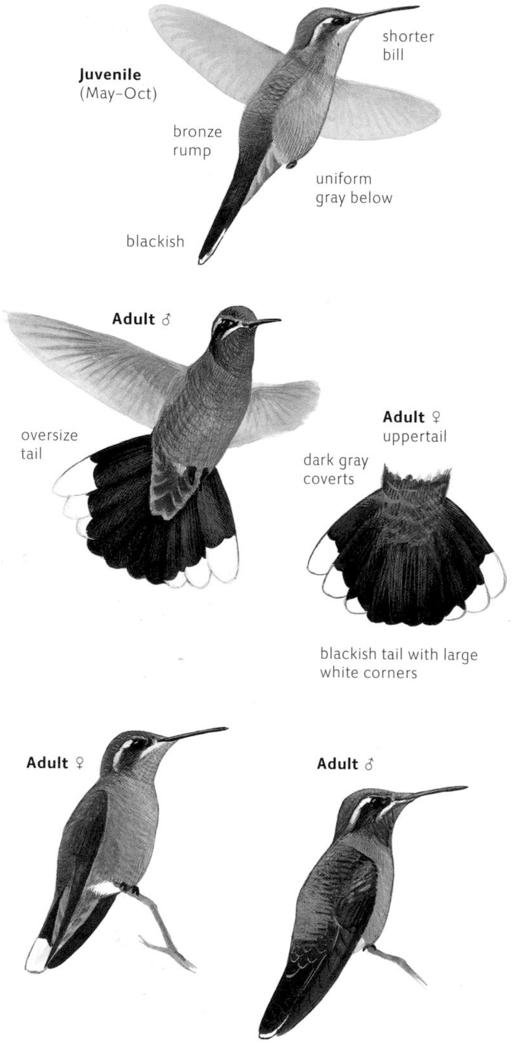

VOICE: Call a penetrating, high, clear *seek* (given repeatedly by perched male). Chase call not well-developed; sometimes a loud, popping chip or *seek krrkr*. Both calls similar to some Magnificent Hummingbird vocalizations. Song a quiet, mechanical rattle with hissing quality *situtee trrrrrrrrr* repeated.

Uncommon in moist shady mountain canyons with mixed forests of mature trees. Wanders to adjacent habitats in search of flowers and feeders.

Magnificent Hummingbird

Eugenes fulgens

L 5.25" WS 7.5" WT 0.25 oz (7 g) ♂>♀

Nearly as large as Blue-throated Hummingbird, but relativ slender, with very long bill and slightly smaller tail; tail all-da on male and with small white corners on female and immatu Male often appears all-blackish; female and immature have gra ish underparts with green spots.

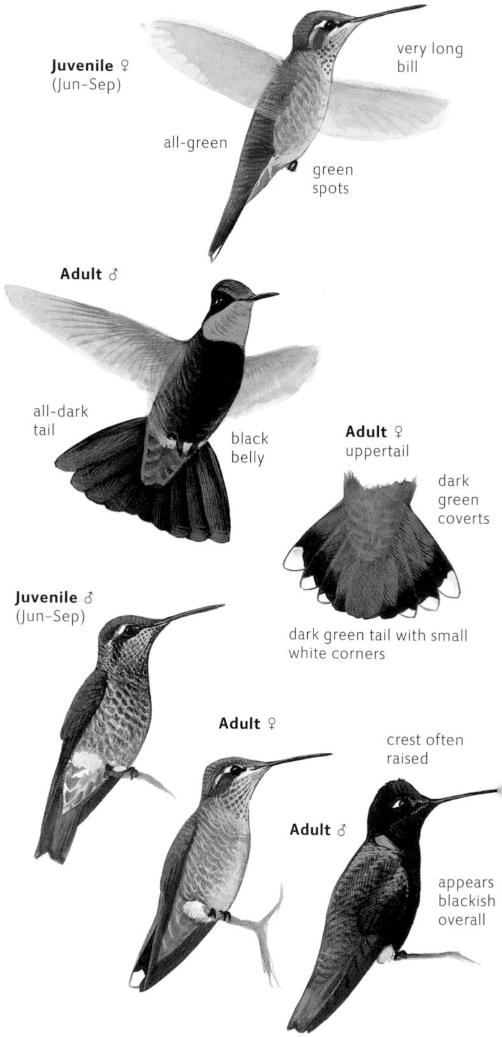

VOICE: Call a sharp chip; varies from high chip like Anna's to lo solid chip or flat, squeaky *tiip*. Chase call variable: a rapid, laugh ing series *twik twik wik wik wik wik ik ik ikikikikikik* rising; a stead whining *twee kwee kwee kwee kwee kwee*; and a crackling *ch krr krr*.

Uncommon in montane pine-oak forests. Found in a wid range of habitats wherever flowers and feeders are present.

reen Violetear

libri thalassinus

.5" WS 7" WT 0.21 oz (5.9 g) ♂>♀

rge and dark greenish overall, with curved bill and square tail. ze and nearly uniform dark color distinctive; note dark band ross tail and blue-violet ear-patch.

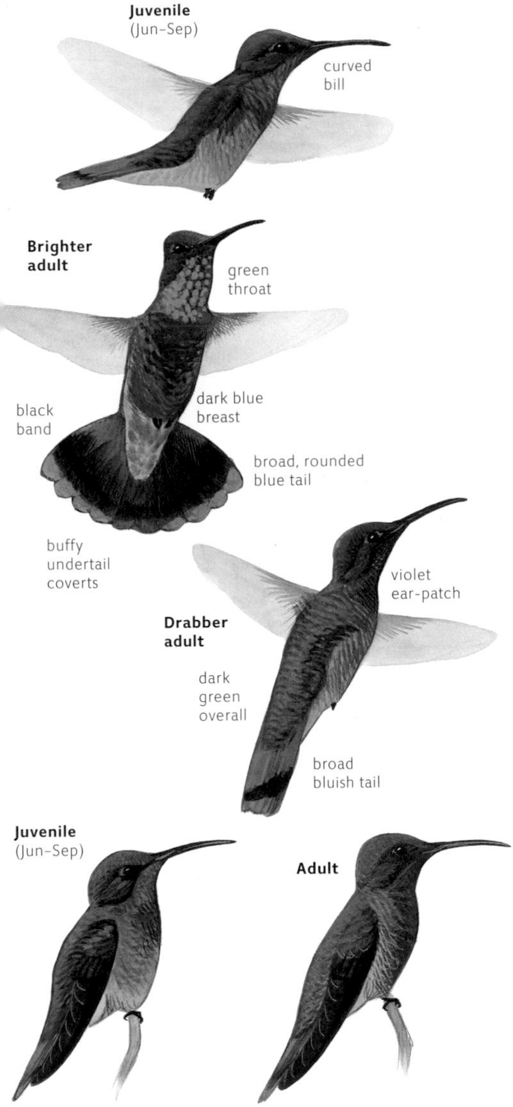

OICE: Call a rapid rattling series of 2–4 very sharp, dry chips, ccasionally a single very short chip, quality dry and unmusical, eminiscent of Violet-crowned Hummingbird. Song a series of ry, metallic chips *chitik-chitik, chitik-chitik*... with irregular hythm.

Very rare visitor from Mexico to feeders and flower gardens; most records from central and southern Texas, from Apr–Jul. Recorded to California, Alberta, Wisconsin, and New Jersey.

Green-breasted Mango

Anthracothorax prevostii

L 4.7" WS 6.5" WT 0.2 oz

A large dark hummingbird with strongly curved bill. Female and immature have distinctive white stripe on each side of underparts. Adult male most similar to Green Violetear, but bill more strongly curved, tail reddish purple.

VOICE: Calls infrequently, a hard *chewp* with strongly descending inflection, longer and sweeter than other hummingbirds, reminiscent of Yellow Warbler. Other high squeaky calls in aggression and display.

Very rare visitor from Mexico, with most records in Texas in fall and winter, but also recorded north to Wisconsin and east to North Carolina.

Trogons and Kingfishers

FAMILIES: TROGONIDAE, ALCEDINIDAE

2 species in 2 genera. A very distinctive family distantly related to woodpeckers. Trogons are peculiar tropical birds, usually seen perched quietly on branches within forest canopy, titling head and suddenly flying out to pluck insects or fruit from twigs or leaves. Both species often perch with rump projecting and tail straight down. Nest in tree cavities. Juveniles are shown.

Genus *Euptilotis*

Eared Quetzal, page 305

Genus *Trogon*

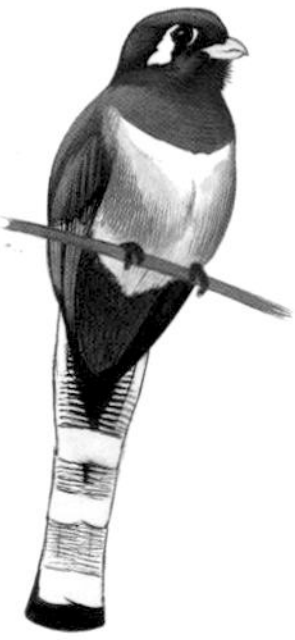

Elegant Trogon, page 305

3 species in 2 genera. Kingfishers are fish-eating birds found c
sheltered waters. They use branches, posts, or wires over water
a vantage point to watch for fish, or hover over water to look f
prey, then plunge head-first into water. Belted and Ringed a
large, conspicuous and loud; they are found around open wat
perch conspicuously, fly high, and hover frequently. Green is sma
inconspicuous, and quiet; it is found along sheltered creeks ar
pools, perches on low twigs, flies low, and rarely hovers. All speci
nest in holes excavated in dirt banks. Adult females are shown.

Genus *Megaceryle*

Ringed Kingfisher, page 307

Belted Kingfisher, page 306

Genus *Chloroceryle*

Green Kingfisher, page 307

ared Quetzal

ptilotis neoxenus

.4" WS 24" WT unknown

rger than Elegant Trogon and quite different in shape and
ice; note smaller head, small dark bill, very broad tail, and all-
rk head and breast with all-red belly.

OICE: Call a high, squealing *kweeeeeeee-chk* strongly rising
ith sharp cluck at end. Series of high whistles rising in pitch and
ntensity *fwee fwee fwee... fwEErk fwEErk*. Toneless descending
ackle given in flight *chikikikik* similar to some calls of Steller's
ay, but harder.

Very rare visitor from Mexico to Arizona, where it has nested; especially in Huachuca Mountains. Nests in tree cavities in pine-oak forests up to edge of spruce-fir zone, generally at higher elevations than Elegant Trogon and less strongly associated with water. Solitary. Feeds on insects and berries.

Elegant Trogon

Trogon elegans

L 12.5" WS 16" WT 2.5 oz (70 g)

Distinctive shape, posture, and color make this species easily identifiable. Note stout yellow bill, red belly, and long, square-tipped tail.

VOICE: Common call of both sexes soft, hoarse, croaking *brr brr brr brr...* or stronger *bwarr bwarr bwarr...*; sometimes a hoarse, spitting *weck weck weck...* (in series of 5–15 notes). Alarm call a rapid, hoarse *bekekekekekek*. Also gives low, soft, hooting or clucking notes.

Rare and local along canyon streams. Nests in tree cavities, usually in sycamores. Solitary. Forages in trees for insects and berries, perching quietly with rump bulging and tail straight down while watching for prey.

Belted Kingfisher

Megaceryle alcyon

L 13" WS 20" WT 5 oz (150 g)

Dark blue-gray above and mostly white below, with prominent white collar, shaggy crest, and large bill. In flight note irregular rowing wingbeats and white patch at base of primaries.

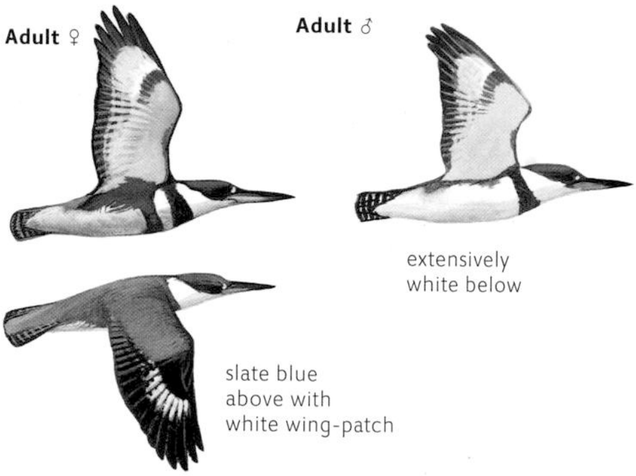

VOICE: Common territorial call a long, uneven rattle most similar to Hairy Woodpecker rattle but harsher, unsteady, clattering. Also a higher, shorter, more musical, rapid trill *tirrrrr.*

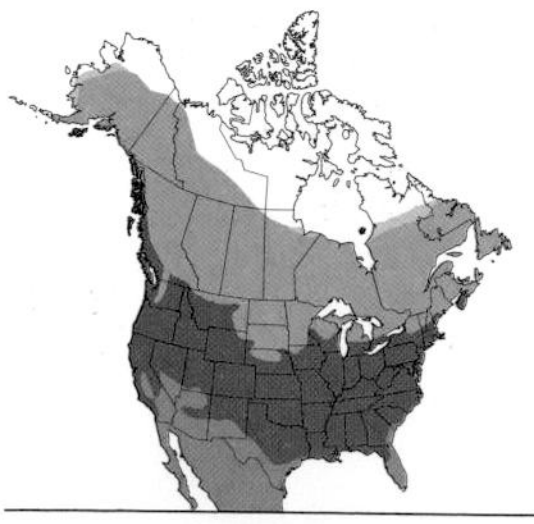

Uncommon but widespread around any sheltered open water with small fish. Often sits on prominent exposed perch such as a wire, branch, pole, or dock usually 5–20 feet over water and often hovers over water to locate small fish before diving headfirst.

Kingfisher Behavior

All three North American kingfishers capture fish by plungi head-first into the water. Belted and Ringed Kingfishers oft hover above the water while looking for prey, and dive direc from the air. Green Kingfisher stays lower, usually hunting fro twigs at the water's edge, and rarely hovers.

Kingfishers have a very distinctive pattern of wingbeats, unlik any other group of birds. Many species (such as ducks and sanc pipers) use continuous wingbeats with no gliding. Most lan birds use a regular pattern of short bursts of quick wingbeats an intervals of gliding, which leads to an undulating path as the climb when flapping and descend when gliding. This undulatin flight is especially pronounced in finches and woodpeckers.

In kingfishers, wingbeats are intermittent and irregular, with out a regular rhythm of flapping and gliding. Short bursts of tw or three quicker wingbeats (a sort of "stutter flap") are separate by intervals with several slower wingbeats. The overall path of th flight is level, without undulations.

?inged Kingfisher

'egaceryle torquata

16" WS 25" WT 11 oz (315 g)

uch larger than Belted Kingfisher, with massive bill. Dark blue-ray upperside and white collar like Belted, but easily stinguished by entirely rufous belly.

/OICE: Common call in flight a loud, double-note *ktok* similar o (but deeper than) some calls of Great-tailed Grackle. Rattle call a loud, very hard machine-gun rattle all on one pitch *ke ke ke ke ke ke ke ke...*; much slower and lower-pitched than Belted Kingfisher rattle.

Uncommon along Rio Grande and nearby ponds and lakes. Habits similar to Belted Kingfisher, perches on exposed snags or wires over water, flies higher than Belted, usually found around relatively large bodies of water.

Green Kingfisher

Chloroceryle americana

L 8.75" WS 11" WT 1.3 oz (36 g)

Blackish above with green gloss, with obvious broad white collar and very long, dark bill. In flight note flicking wingbeats and white outer tail feathers.

VOICE: Common call a dry, quiet, staccato clicking like tapping pebbles together, usually single or double, given from a perch. In flight a short harsh descending buzz *tsheeersh*. Other calls during interactions include a high, squealing *tseelp* and grating sounds.

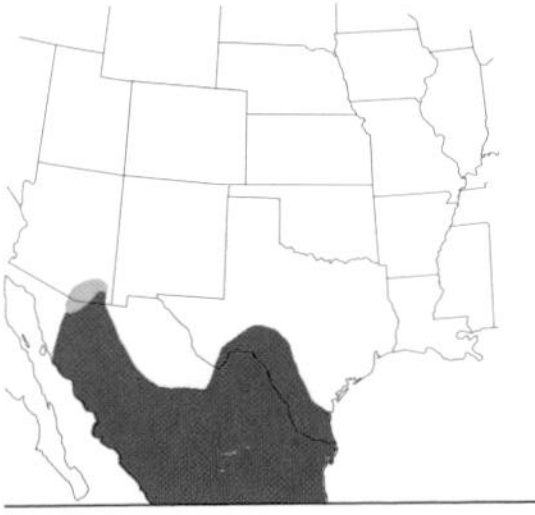

Uncommon along clear streams or ponds. More retiring and much smaller than Belted Kingfisher. Prefers very clear shallow water shaded by overhanging trees or shrubs, perches on branches or posts low over water at edge of ponds or streams, often within a tangle of branches. Flies just over water's surface and rarely hovers.

Woodpeckers

FAMILY: PICIDAE

22 species in 5 genera. All woodpeckers climb tree trunks with stiff tails used as props while the bird's feet cling to the bark. All have sturdy, chisel-like bills used to peel bark and excavate wood to uncover insects and larvae. All species excavate nest and roost cavities in trees, the size and shape of the hole varies slightly between species, and abandoned woodpecker holes are then used by dozens of other species of birds and mammals. Mo species are found singly, a few form loose groups. Flight of mo species is deeply undulating. Only Brown Creeper has simila tree-clinging posture (but is much smaller); nuthatches use di ferent climbing method. Adult females are shown.

Genus *Melanerpes*

Lewis's Woodpecker, page 309

Red-headed Woodpecker, page 310

Acorn Woodpecker, page 309

Gila Woodpecker, page 310

Golden-fronted Woodpecker, page 311

Red-bellied Woodpecker, page 311

Genus *Colaptes*

Northern Flicker, page 320

Gilded Flicker, page 321

Genus *Sphyrapicus*

Williamson's Sapsucker, page 312

Yellow-bellied Sapsucker, page 313

Red-naped Sapsucker, page 313

Red-breasted Sapsucker, page 312

Genus *Picoides*

Ladder-backed Woodpecker, page 314

Nuttall's Woodpecker, page 314

Downy Woodpecker, page 316

Hairy Woodpecker, page 317

Arizona Woodpecker, page 315

Red-cockaded Woodpecker, page 319

White-headed Woodpecker, page 315

American Three-toed Woodpecker, page 318

Black-backed Woodpecker, page 319

Genus *Dryocopus*

Pileated Woodpecker, page 322

Lewis's Woodpecker

Melanerpes lewis

L 10.75" WS 21" WT 4 oz (115 g)

A large, long-winged woodpecker; very dark overall. Blackish plumage with pink belly and pale gray collar unique. Flight crow-like, with rowing wingbeats.

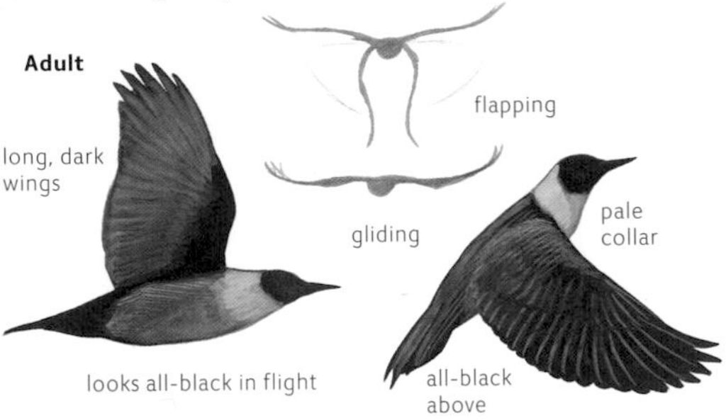

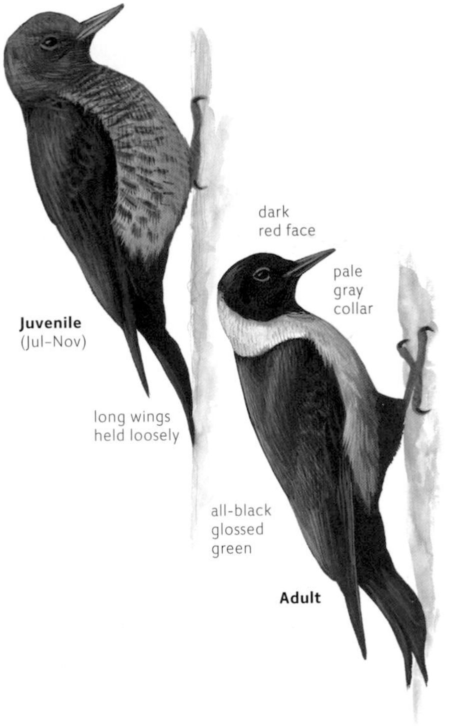

VOICE: Contact call a weak, sneezy *teef* or *kitsif*; also a high, squeaky, descending *rik rik rik*. Dry, rattling chase series reminiscent of European Starling. Song a series of short, harsh *chr* calls. Drum short, weak, medium speed; followed by several individual taps.

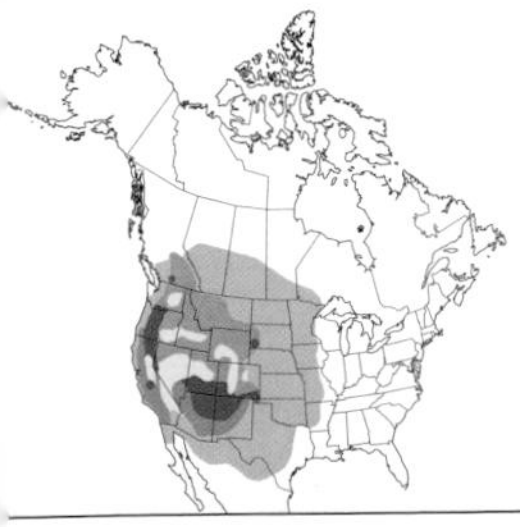

Uncommon and local in dry open pine forests and other habitat with scattered trees, such as orchards or riparian cottonwoods. Somewhat social and may form small loose groups. Makes frequent short flights from perch to capture flying insects.

Acorn Woodpecker

Melanerpes formicivorus

L 9" WS 17.5" WT 2.8 oz (80 g)

A striking black-and-white woodpecker, bright white rump and white at base of primaries contrasts boldly with otherwise jet-black upperside. Also note clownish face pattern, red crown, white iris.

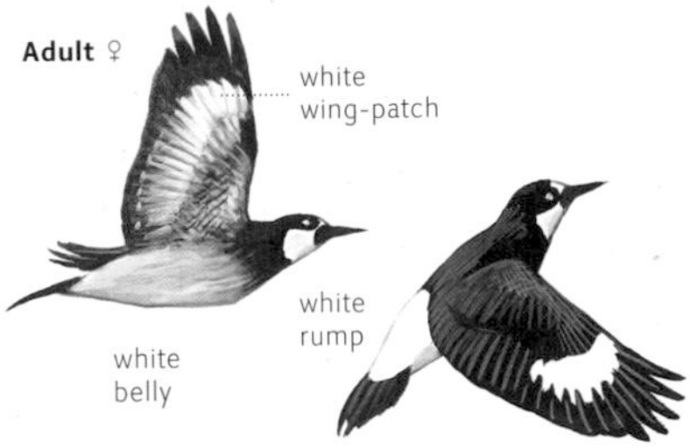

VOICE: Very vocal. Raucous laughing; most common call a nasal *wheka wheka*... or *RACK-up RACK-up*.... Also nasal, vibrant, trilled *ddddrri-drr* or high, burry *ddrreeerr*; nasal *waayk, waaayk* or rising *quaay*. Close contact a high squeaky *Ik a Ik a Ik a*. Drum short, slow, accelerating, variable.

Common and conspicuous in open oak woods or mixed woods with oaks. Noisy and gregarious, with complex social system, distinctive voice, and striking plumage. Gathers acorns to store in holes drilled in certain trees or poles.

Red-headed Woodpecker

Melanerpes erythrocephalus

L 9.25" WS 17" WT 2.5 oz (72 g)

Adult unmistakable, with red head and striking white rump and secondaries. Juvenile less striking, but still distinctive, with plain brown head, white secondaries with dark bars, and whitish underparts.

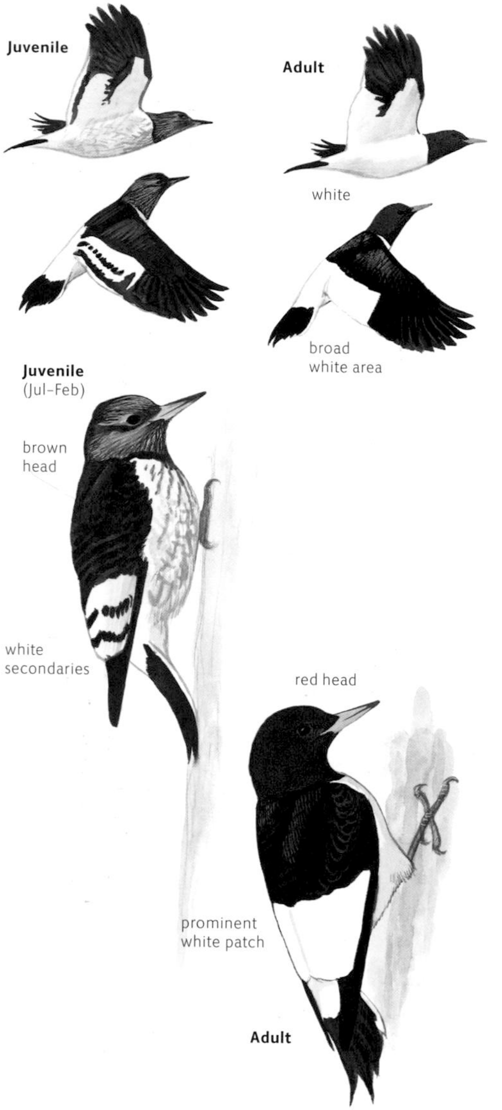

VOICE: Contact call a wheezy *queeah* or *queerp*; weaker and less vibrant than Red-bellied, variable. In flight a low, harsh *chug* like Red-bellied. Close contact call a gentle, dry rattle *krrrrrr*. Drum short, weak, fairly slow.

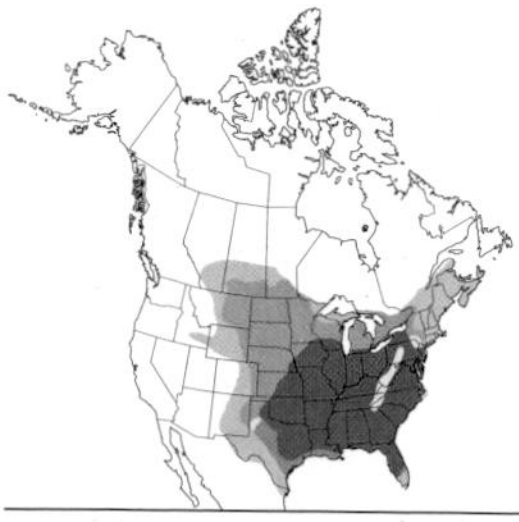

Common locally in open park-like woodlands, including clear-cut areas where a few trees are left standing, or riparian edges. Often seen perched on exposed snags, flying out to catch insects in mid-air.

Gila Woodpecker

Melanerpes uropygialis

L 9.25" WS 16" WT 2.3 oz (66 g) ♂>♀

A medium-size woodpecker, with plain brown head and under parts and uniformly barred back and wings. Note small whit wing-patch and barred rump not contrasting with back. Simila to Golden-fronted but no range overlap.

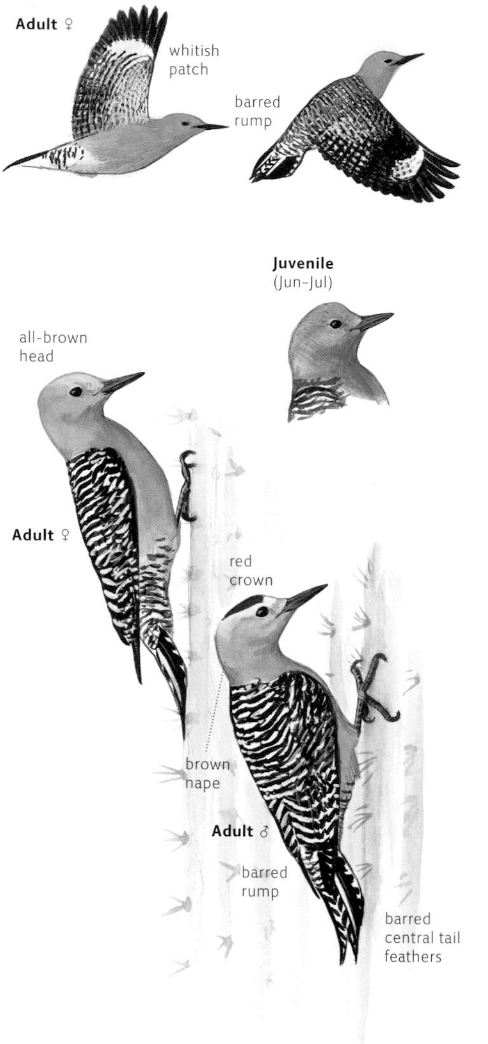

VOICE: Contact call a loud, harsh *quirrr* slightly rising like Red-bellied. Also raucous, laughing, well-spaced *geet geet geet geet geet* and high, nasal, squeaky *kee-u kee-u kee-u*... higher and clearer than comparable calls of Red-bellied and Golden-fronted. Drum long and steady.

Common in low-elevation deserts with woody plants large enough to provide nest sites, including areas with saguaro cactus or cottonwoods.

ɜolden-fronted Woodpecker

ʼelanerpes aurifrons

ɟ.5" WS 17" WT 2.9 oz (82 g) ♂>♀

milar in all respects to Red-bellied Woodpecker, but with ɔlden yellow (not red) nape and forehead, and, on male, isoted red crown-patch. Also note all-white rump and all-black il.

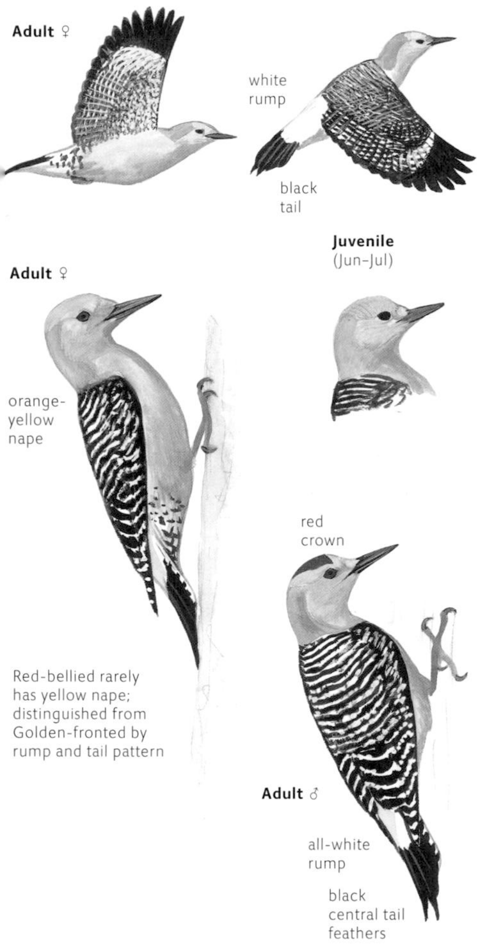

ɔICE: Contact call a loud, harsh, level *kirrr*; all calls harder and ırsher than Red-bellied and Gila. Also slow, strained *kih-wrr h-wrr...*; *tig tig...* calls harder than Red-bellied; a grating *krrr rr....* Drum medium speed, short, much slower than Ladderıcked.

Common in mature deciduous woods, including mesquite and other relatively dry and brushy habitat. Where range overlaps with Red-bellied, Golden-fronted generally occupies drier and more scrubby woodlands. Hybridizes regularly with Red-bellied where range overlaps in Texas and southwestern Oklahoma.

Red-bellied Woodpecker

Melanerpes carolinus

L 9.25" WS 16" WT 2.2 oz (63 g) ♂>♀

Note uniformly barred back, brown underparts, and red nape. Faint wash of red on belly rarely visible, but bright red nape always conspicuous.

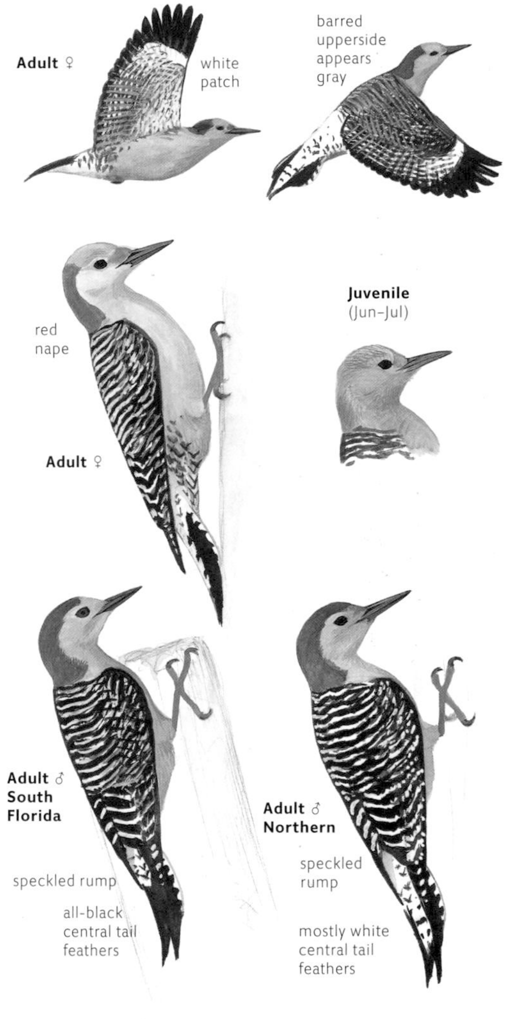

Populations in Florida Keys have tail feathers all black, male has gray-brown forehead without red, average slightly smaller overall. In most of Florida birds are variable and intermediate.

VOICE: Contact call a loud, harsh, but rich *quirrr* slightly rising; in flight a single, low *chug*. Also a harsh *chig-chig*, a series of *chig* notes delivered slowly, or a rapid, chuckling series descending. Drum medium speed and length with steady tempo.

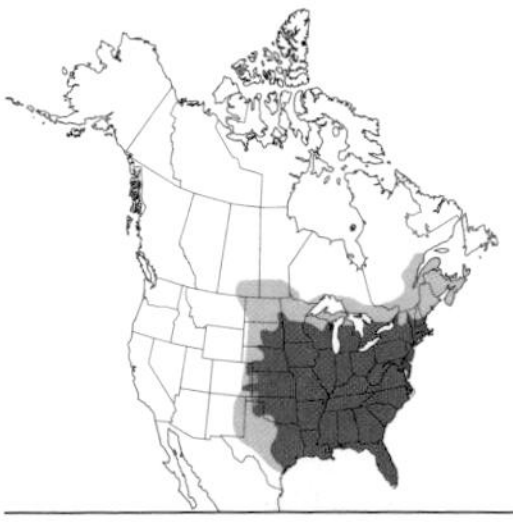

Common in mature deciduous woods or mixed pine forest in southern states. Especially numerous in wet lowland forest such as cypress swamps, pine savanna. Range is expanding rapidly to the north, where it inhabits suburban woodlots and visits bird feeders.

Williamson's Sapsucker

Sphyrapicus thyroideus

L 9" WS 17" WT 1.8 oz (50 g)

Habits and size similar to Red-naped and other sapsuckers, but plumage very different. Male striking, with black back, white rump and wing-patch, and white stripes on face; female has densely barred body and plain brown head.

VOICE: Contact call a strong, clear *queeah*; less mewing than other sapsuckers. Initial burst of drumming faster than other sapsuckers, followed by loud taps with fading vibration; pauses between these latter bursts longer than other sapsuckers.

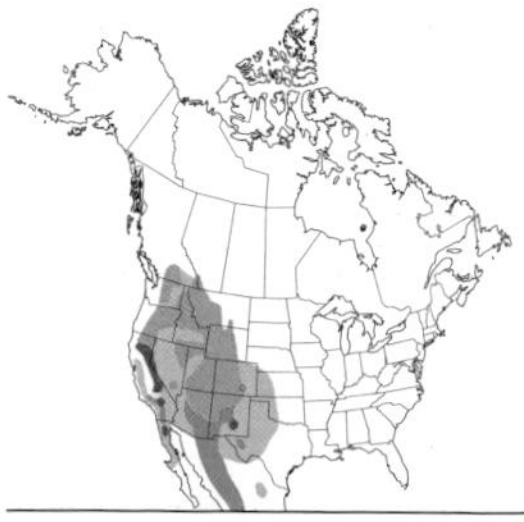

Uncommon. Nests mainly in Ponderosa Pine forests, rarely in other coniferous forest; nests often placed in aspen trees within or adjacent to Ponderosa Pine. Wanders to lower elevations and other forest types in winter.

Red-breasted Sapsucker

Sphyrapicus ruber

L 8.5" WS 16" WT 1.8 oz (50 g)

Identified by mostly red head with little black and by more exter sive black on back and tertials with limited white barring. Fema has less red and more white than male.

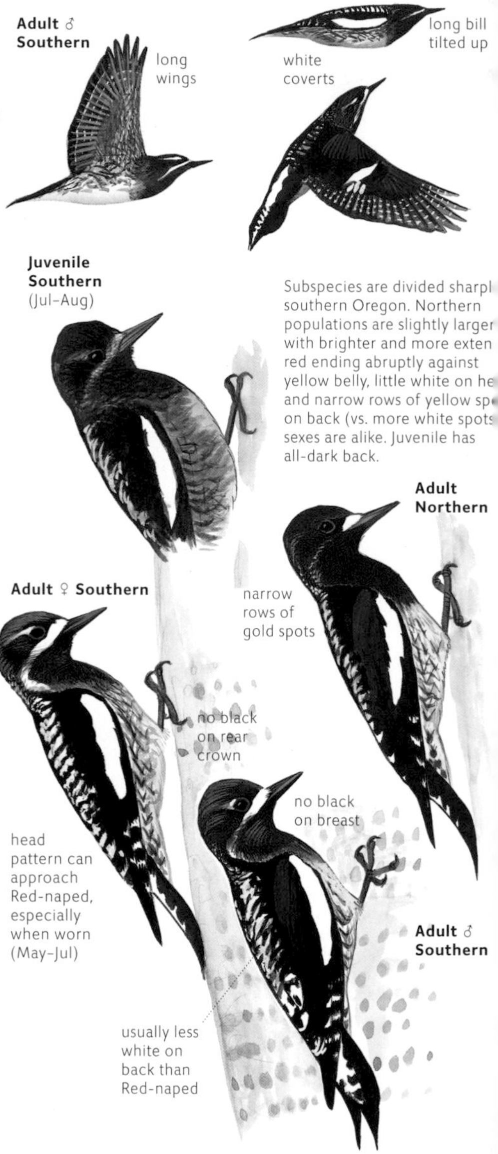

VOICE: All calls and drum similar to Yellow-bellied and Rec naped. Contact call may average lower and hoarser tha Yellow-bellied, but variable. All sapsuckers are quiet in winter.

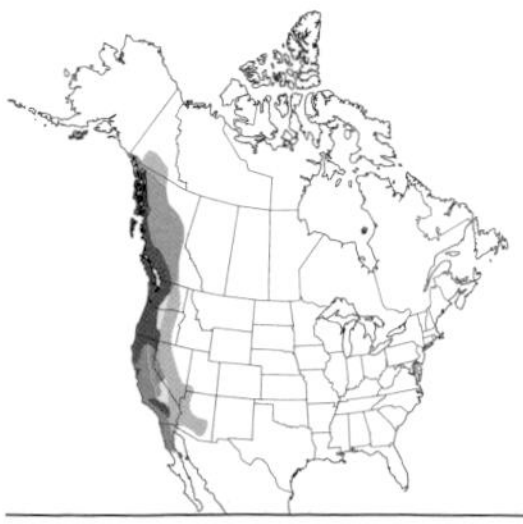

Common in mixed coniferous and deciduous forests. Habits and appearance very similar to Red-naped and Yellow-bellie Sapsuckers and hybridizes with both species where range overlaps.

‹ed-naped Sapsucker

›hyrapicus nuchalis

3.5" WS 16" WT 1.8 oz (50 g)

›ry similar to Yellow-bellied and Red-breasted Sapsuckers, ffering only in extent of red on head, pattern of black and hite on head and back. All three species hybridize where nges overlap.

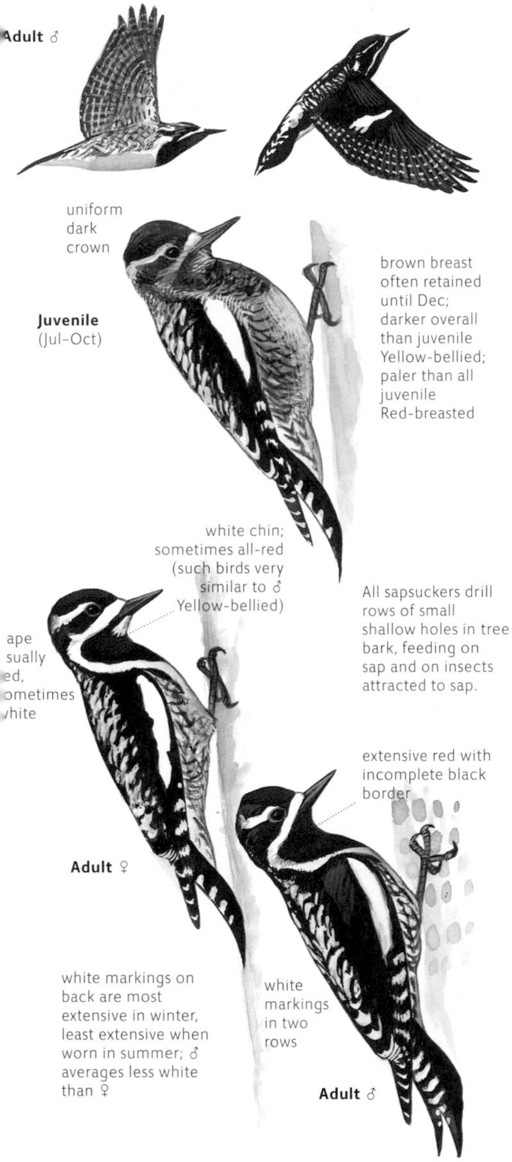

›ICE: Essentially identical to other sapsuckers.

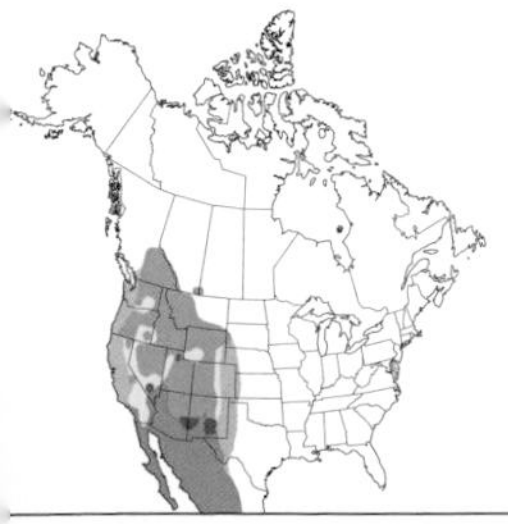

Uncommon in aspen groves in mountains, also in coniferous forests where aspens are intermixed. In migration and winter found in lowland riparian (e.g. willow and cottonwood) and suburban habitats (where mainly in conifers).

Yellow-bellied Sapsucker

Sphyrapicus varius

L 8.5" WS 16" WT 1.8 oz (50 g)

Bold white upperwing coverts, mottled back and flanks, and white stripe from bill to belly distinguish sapsuckers from other woodpeckers. Nearly identical to Red-naped but with less red and black, more white on head. Juvenile retains brown plumage through winter.

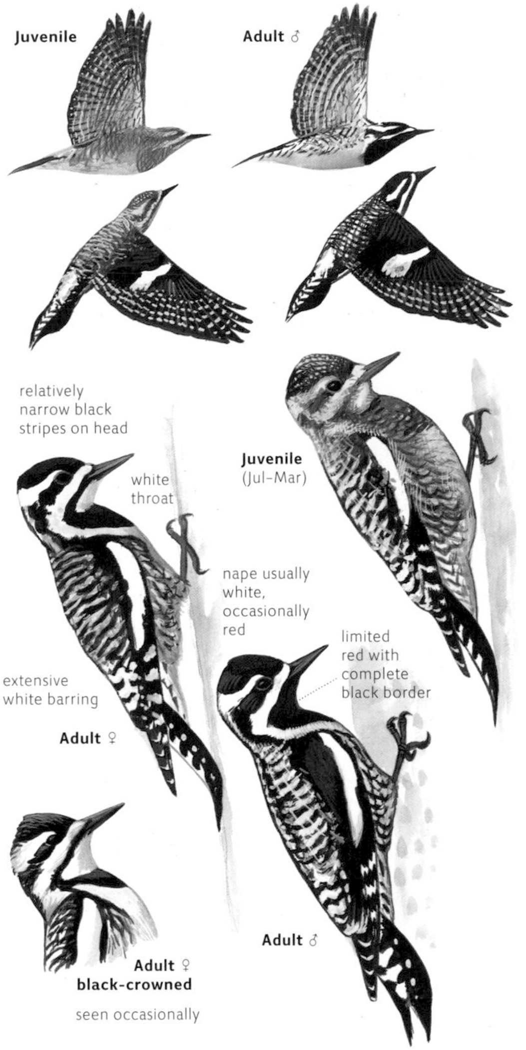

VOICE: Contact call a nasal squealing or mewing *neeah*; on territory an emphatic *QUEEah*. Close contact call *wik-a-wik-a...* series, hoarse and uneven. In flight sometimes gives nasal *geert*. Drum a burst of about five rapid taps followed by gradual slowing with occasional double taps.

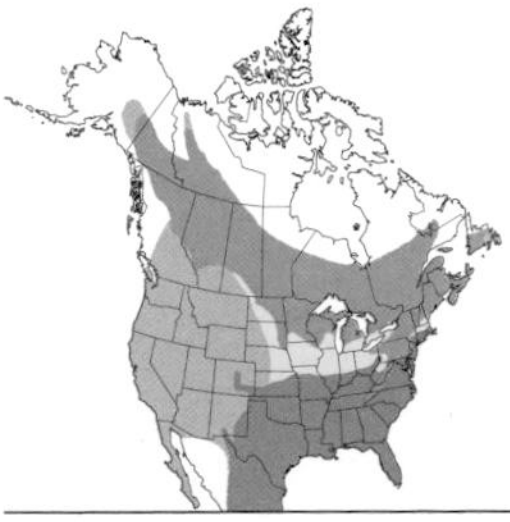

Uncommon. Nests in mixed woodlands. Less active than other woodpeckers, perching and tapping quietly. Drills small shallow holes in tree bark (note characteristic pattern of small holes lined-up in rows), feeding on sap and insects attracted to sap. The only sapsucker found in the East.

Nuttall's Woodpecker

Picoides nuttallii

L 7.5" WS 13" WT 1.3 oz (38 g)

Slightly larger than Downy Woodpecker, with barred back and spotted breast. Very similar to Ladder-backed, but range overlaps only slightly; note darker face and solid black upper back.

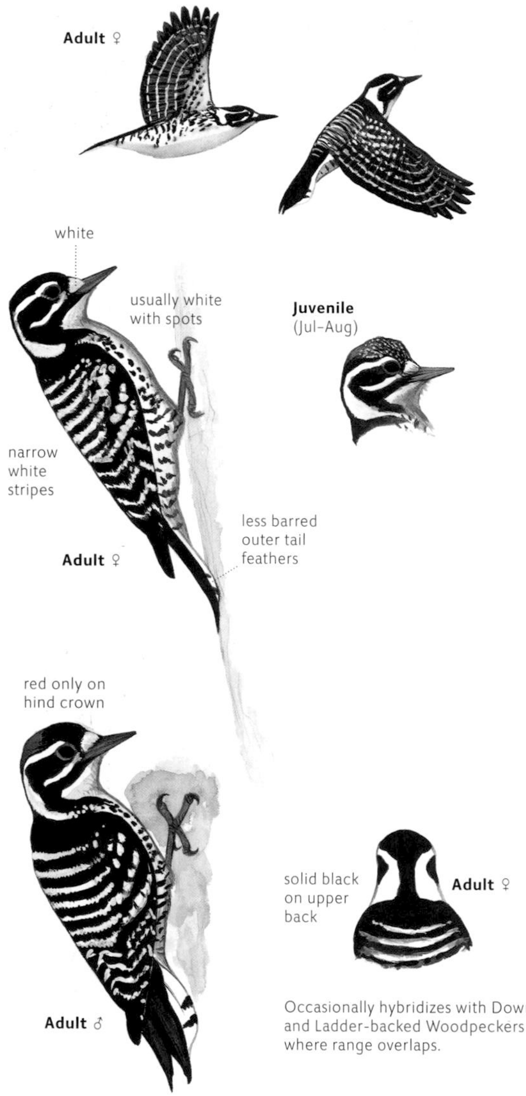

Occasionally hybridizes with Downy and Ladder-backed Woodpeckers where range overlaps.

VOICE: Contact call a sharp, rising, two- or three-note *pitik* (occasionally a single-note *pik*); quality like Hairy. Rattle call level and steady *pitikikik*.... Drum steady, medium speed, relatively long; noticeably faster than Downy.

Uncommon in oak and riparian woods. Where range overlaps with Downy Woodpecker, this species is more numerous and found in a wider range of habitat, including oak savanna, while Downy is restricted to riparian willows.

Ladder-backed Woodpecker

Picoides scalaris

L 7.25" WS 13" WT 1.1 oz (30 g)

Size and shape similar to Downy Woodpecker, but range ar habitat barely overlaps. Note dark triangle enclosing whi cheek, boldly barred back and spotted flanks of both sexes, ar red crown on male.

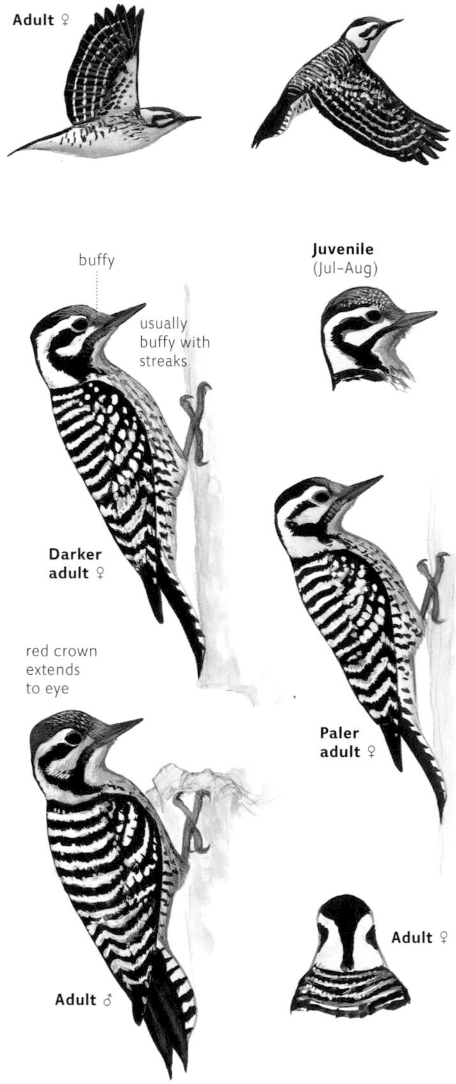

VOICE: Contact call a sharp *pwik* slightly lower-pitched and mo musical than Downy. Rattle call always ends with low, gratir notes *kweekweekweekweekweechrchr*. Drum very rapid buz ing, medium length (about 1 second long); shorter than Nuttall much faster than Downy.

Common in mature mesquite and other arid wooded habitats including desert scrub, often foraging on small shrubs and yuccas. Compared to similar species (Downy, Nuttall's) this species is usually found in drier habitat with smaller trees and shrubs.

Arizona Woodpecker

icoides arizonae

7.5" WS 14" WT 1.6 oz (47 g) ♂>♀

maller than Hairy Woodpecker and relatively short-tailed. Our nly brown and white woodpecker; stocky and dark, with large hite neck-patch and entirely brown back.

OICE: Contact call a sharp *keech* similar to Hairy but higher, istinctly long and squeaky. Rattle call of grating notes, descendg *keechrchrchrchr.* Drum long, 3–4 per minute.

Uncommon in foothill oak woods in mountains of southeastern Arizona. Habits similar to Hairy Woodpecker.

White-headed Woodpecker

Picoides albolarvatus

L 9.25" WS 16" WT 2.1 oz (61 g) ♂>♀

Plumage striking and unlike any other North American bird: body entirely black and head entirely white. White bases of primaries show as white streak on folded wing.

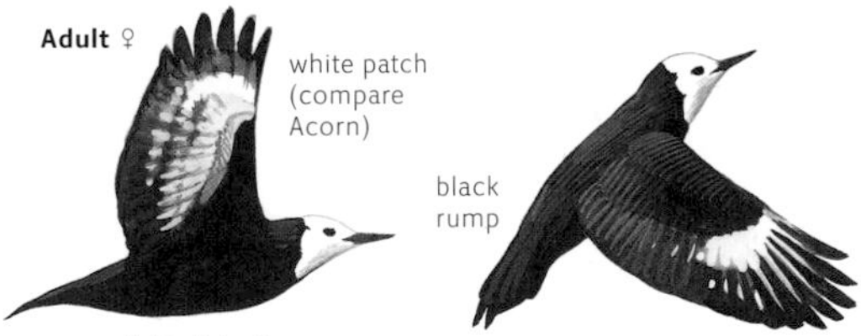

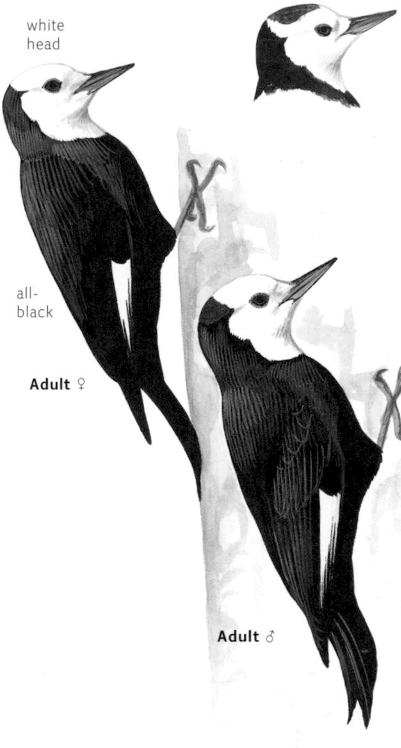

VOICE: Contact call a sharp, two- or three-note *pitik* very similar to Nuttall's but higher, more metallic and usually descending. Rattle call an extended contact call *peekikikikikkikik.* Drum medium speed, fairly long, with increasing or decreasing tempo.

Uncommon in mature coniferous woods with openings such as meadows. Habits similar to Hairy Woodpecker.

Downy Woodpecker

Picoides pubescens

L 6.75" WS 12" WT 0.95 oz (27 g) ♂>♀

Our smallest woodpecker, with very short bill. Distinguished from all woodpeckers except Hairy by white patch on back and mostly unmarked whitish flanks. Differs from Hairy by smaller size, relatively small bill, foraging habitat, and voice.

When judging bill size, note that males have larger bills than females, so male Downy and female Hairy approach each other in bill size. Determining sex (by red on head) can aid identification.

Adult ♀ Eastern

Adult ♀ Pacific

Juvenile ♂ (Jul–Aug)

Adult ♀ Eastern

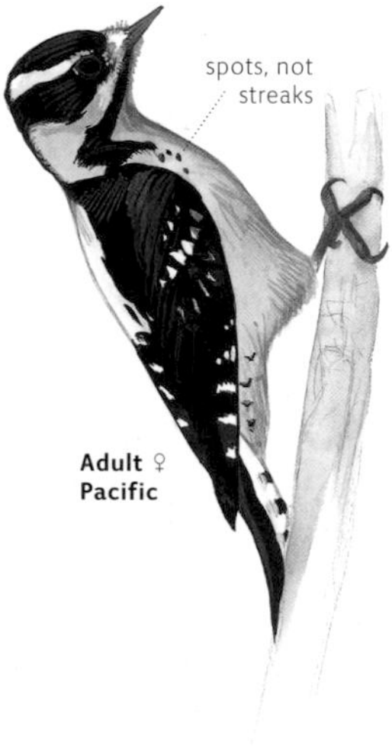

Adult ♀ Pacific

average more white on wing coverts than Hairy, but some overlap

Downy usually has pale eyebrow continuous around back of head. Hairy usually has pale band broken by black.

Adult ♀ Interior West

Adult ♂ Eastern

Geographic variation mostly parallels variation in Hairy, but western birds usually show some white on wing coverts and tertials (vs. usually all-black on western Hairy).

VOICE: Contact call a short, gentle, flat *pik*. Rattle call slow, similar in quality to *pik* note; beginning slow and squeaky, ending lower, faster; varies from *kikikikiki...* to slow, squeaky *twi twi twi...*; variable *chik* or *kweek* notes in spring. Rattle call of Pacific and Interior West birds often level or lacks low ending: a shorter, cutoff *kikikikikikikik*. Drum almost slow enough to count, short; usually several drums in fairly rapid sequence (9–16 per minute, only a few seconds pause between drums).

Common in any wooded habitat, especially riparian and deciduous woods with patches of smaller trees or brush. More local and restricted to riparian habitats in West. Often forages along twigs and weed stems.

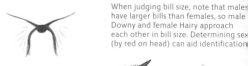

Downy × Nuttall's Woodpecker hybrid ♀

Rare but regular where range overlaps in California.

-lairy Woodpecker

icoides villosus

9.25" WS 15" WT 2.3 oz (66 g) ♂>♀

ırge and strong, with relatively large bill. Distinguished from all oodpeckers except Downy by white patch on back and mostly ımarked whitish flanks. Differs from Downy by size and propor-ons, usually unmarked white outer tail feathers, foraging ıbitat, and voice.

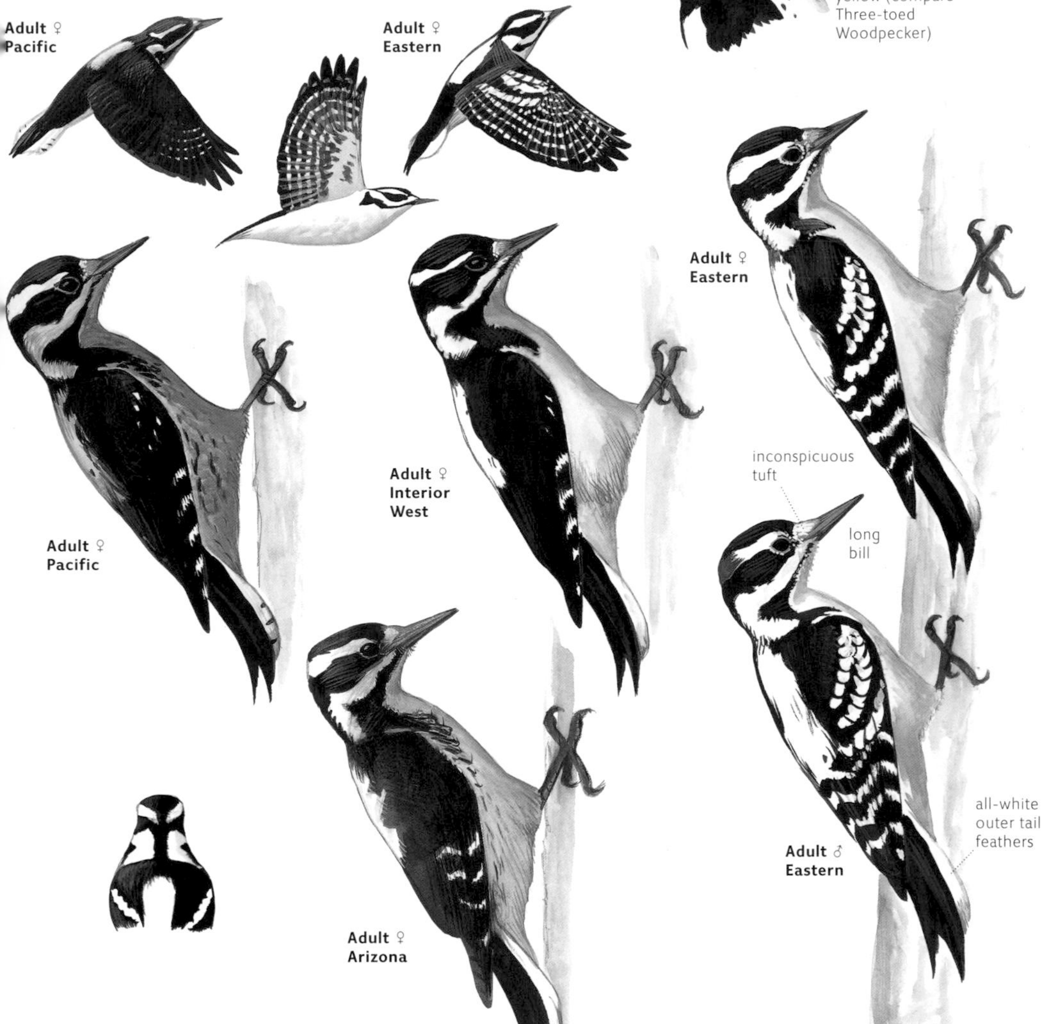

ɔICE: Contact call a sharp, strong *peek* or *peech,* sharper, ·uder, and higher-pitched than Downy. When agitated a stut-ring *peek rr krr.* Rattle call sharp and high like contact call: a pid series of notes on one pitch; variable high *kweek...* notes spring. Drum very fast buzzing, fairly long, 4–9 per minute.

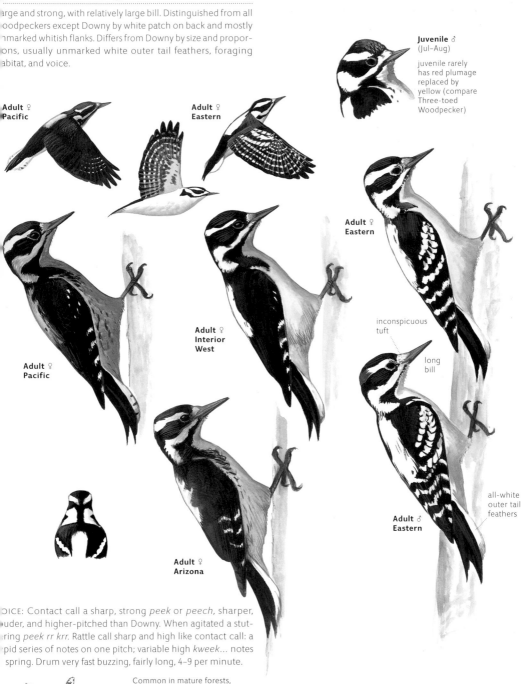

Common in mature forests, where it forages on trunks and major limbs of large trees. In West found in any mature, open forest, unlike Downy which is restricted to riparian habitat.

Geographic variation parallels Downy: Eastern birds are pale with large white spots on wing coverts, western birds are darker with mostly black wings (blacker than most Downy Woodpeckers). Darkest in Pacific Northwest; paler in Rocky Mountains. Birds in southeast Arizona relatively dark, with extensive dark on sides of breast and brownish (not black) back.

American Three-toed Woodpecker

Picoides dorsalis

L 8.75" WS 15" WT 2.3 oz (65 g)

Dark overall, with limited white on back, white head-stripes, and dense barring on flanks. A relatively large-headed and short-tailed woodpecker. Intermediate in size between Downy and Hairy Woodpeckers, darker overall.

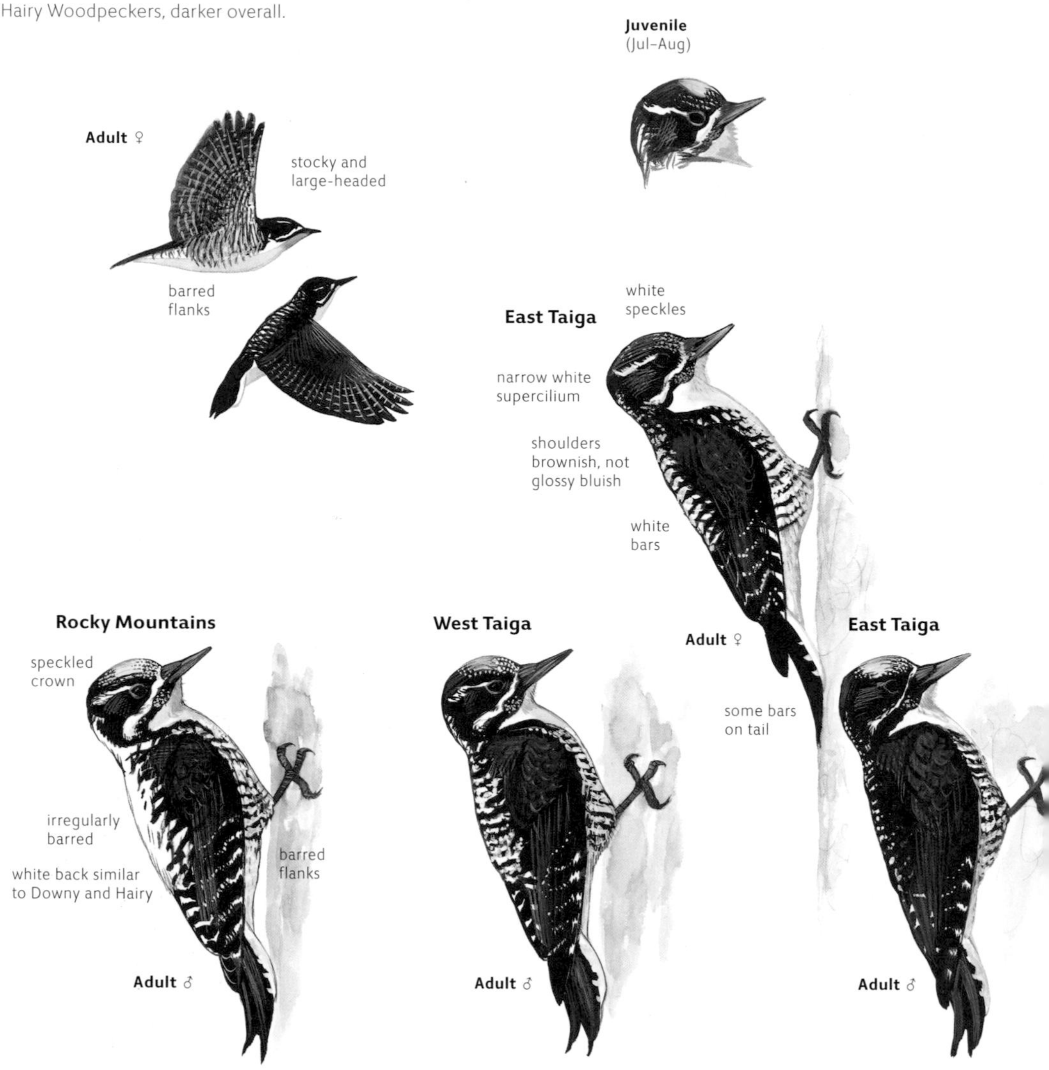

VOICE: Contact call a relatively low, flat *pwik*; similar to Downy but deeper, more hollow or wooden; weaker than Black-backed. Rattle call short, shrill *kli kli kli kli kli*; varied *twiit twiit*, etc. Drum short, slow, speeding up and trailing off at end, 2 per minute; slower and shorter than Black-backed.

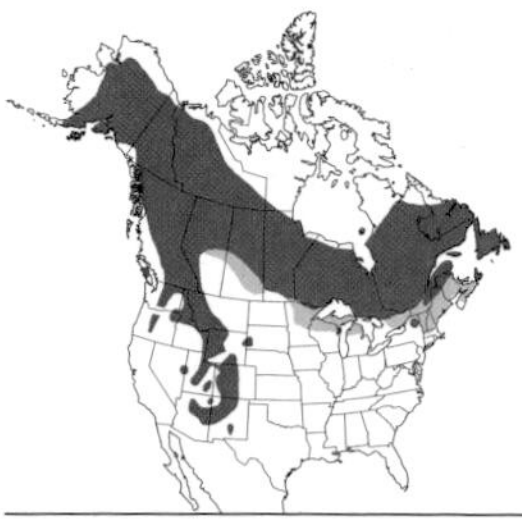

Uncommon and inconspicuous in mature spruce-fir forest. Generally uncommon and very sparsely distributed, but occurs in higher densities where large numbers of recently dead trees are found (e.g., after forest fires).

Geographic variation is slight: blackest in East, slightly more white in western Canada, and distinctly more white in the Rocky Mountains. All are variable, however, and change is clinal.

This species and Black-backed flake off outer bark scales to uncover insect larvae rather than excavating wood. Bark-flaking behavior produces a quiet scraping and tapping sound and results in large patches of flaked bark on favored trees.

Black-backed Woodpecker

icoides arcticus

9.5" WS 16" WT 2.5 oz (70 g)

ppears relatively large-headed and short-tailed. Very dark over-l, with entirely black back, mostly black head, and dense barring n flanks.

ICE: Contact call reminiscent of Hairy but much deeper, hol-w, wooden. Rattle call a peculiar grating, then a rasping snarl. pparently no *wicka* type notes as in other *Picoides*. Drum fairly ng, slow but distinctly accelerating.

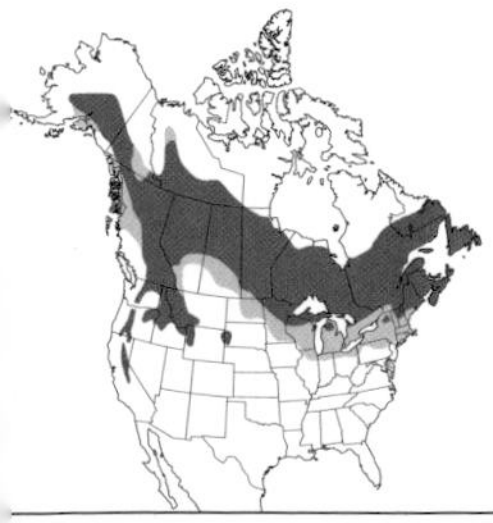

Uncommon and inconspicuous. Habits like Three-toed, but found in a wider variety of forest habitat including deciduous trees mixed with spruce and fir. More numerous than Three-toed in East and in Sierras; Three-toed is more numerous in northern Rocky Mountains.

Red-cockaded Woodpecker

Picoides borealis

L 8.5" WS 14" WT 1.5 oz (44 g)

Larger and relatively longer-tailed than Downy Woodpecker. Clean white cheek patch striking; also note barred back and flanks.

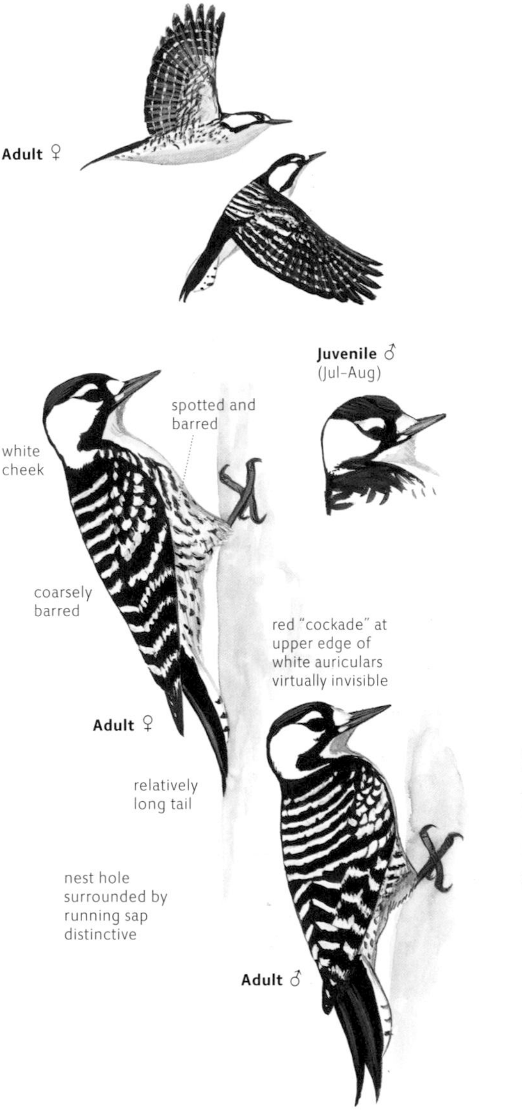

VOICE: Rather vocal in groups. Contact call a unique sharp, nasal, buzzy/squeaky *shirrp* or *shrrit*. Rattle call infrequent *shirrp-chrchrchrchr*.... Varied notes low and short to high clear *wica wica*.... Drum infrequent and rather quiet.

Rare and very local. Found in mature Longleaf Pine savannas, which exist now only in scattered patches of managed forest where nest trees are marked by researchers. Usually in small groups. World population about 14,000 individuals.

Northern Flicker

Colaptes auratus

L 12.5"   WS 20"   WT 4.6 oz (130 g)

A large woodpecker, with long, slightly downcurved bill. Striking plumage always distinctive: brownish with barred back, spotted belly, mostly gray or brown head, and black breastband. White rump and reddish or yellow underwing startling in flight.

Red-Shafted (Western)

Yellow-Shafted (Taiga/Eastern)

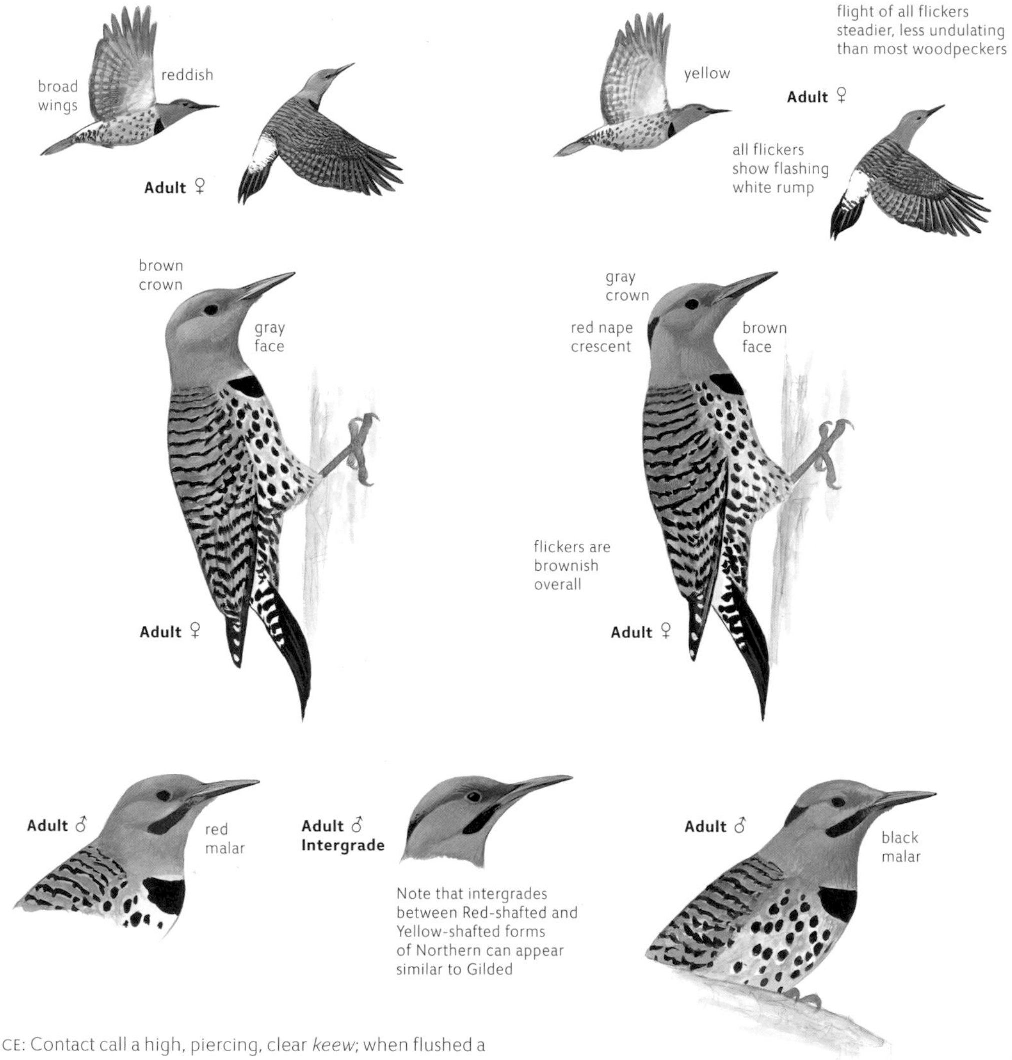

VOICE: Contact call a high, piercing, clear *keew;* when flushed a soft, muffled *bwirr.* Close contact a soft, lilting *wik-a-wik-a-wik-a....* Song a long, strong, relatively low-pitched series *kwikwikwikwi...* continued steadily for up to 15 seconds. Drum averages moderate to fast speed, but variable.

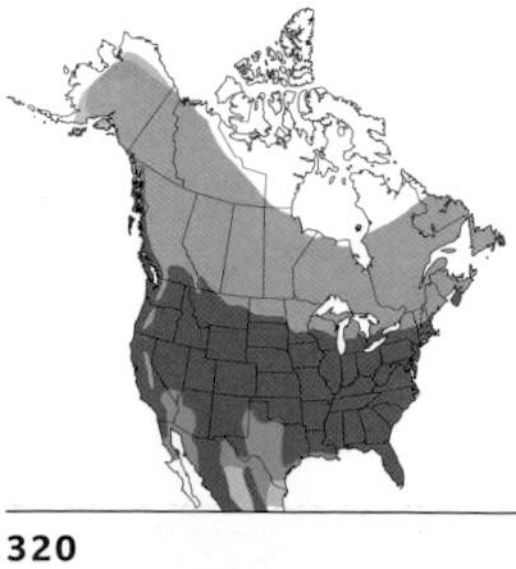

Common and widespread in wooded areas with openings such as forest edges, golf courses, riparian cottonwoods. Unlike other woodpeckers, this species forages largely on the ground, feeding on ants.

"Red-shafted" flickers, nesting in the western US and southwestern Canada, have orange-red on the underside of wings and tail, brown crown, and gray face. They always lack red on the nape, and males have a red malar stripe. "Yellow-shafted" flickers, nesting in Alaska, most of Canada, and throughout the eastern US, have yellow underside of wings and tail, gray crown and nape, and brown face. They always have a red crescent on the nape, and males have a black (not red) malar stripe. Intergrades are common across a broad area from Oklahoma to British Columbia, and can show any combination of features.

ɜilded Flicker

olaptes chrysoides

11" WS 18" WT 3.9 oz (111 g)

abits and appearance virtually identical to Northern Flicker; verages slightly smaller. Head pattern similar to Red-shafted rm of Northern, but with yellow on underside of wings and tail ke Yellow-shafted; more black on tail than either.

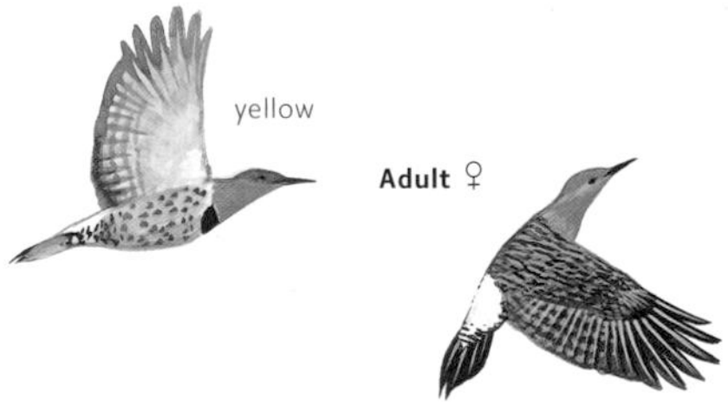

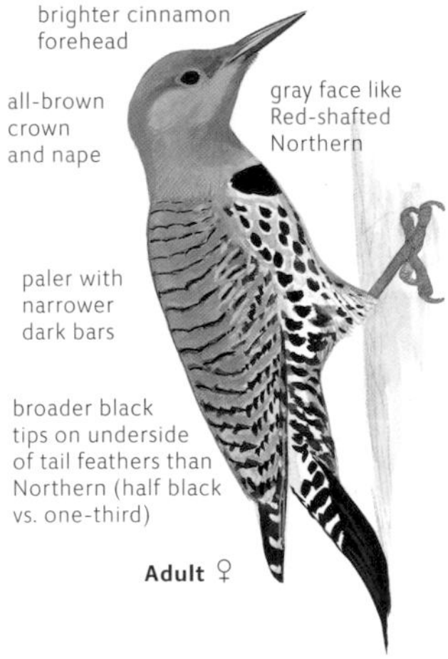

OICE: Essentially identical to Northern Flicker but calls may verage slightly higher-pitched.

Common within limited range in riparian woods and saguaro deserts. Habits like Northern Flicker, and hybridizes where range overlaps.

Woodpecker Drumming

All woodpeckers use drumming instead of a vocal song to proclaim territory and attract mates. The drumming sound is produced by tapping the bill rapidly and sharply against a tree trunk, pole, or other drumming post selected for its resonating quality. Within a region, species can often be identified by their drumming, listen especially for the length and speed of the drumming as well as changes in tempo, frequency, or intensity. Most species give a continuous burst of sound, but sapsuckers have a distinctive rhythmic pattern. Pileated Woodpecker usually sounds bigger than Downy and other small woodpeckers, but the quality of the sound is strongly influenced by the resonance of the drumming post and should not be relied on for identification.

Shown below are typical drumming rhythms for selected species over a 3-second interval. Listen for the differences in overall length, tempo, and fading in or out.

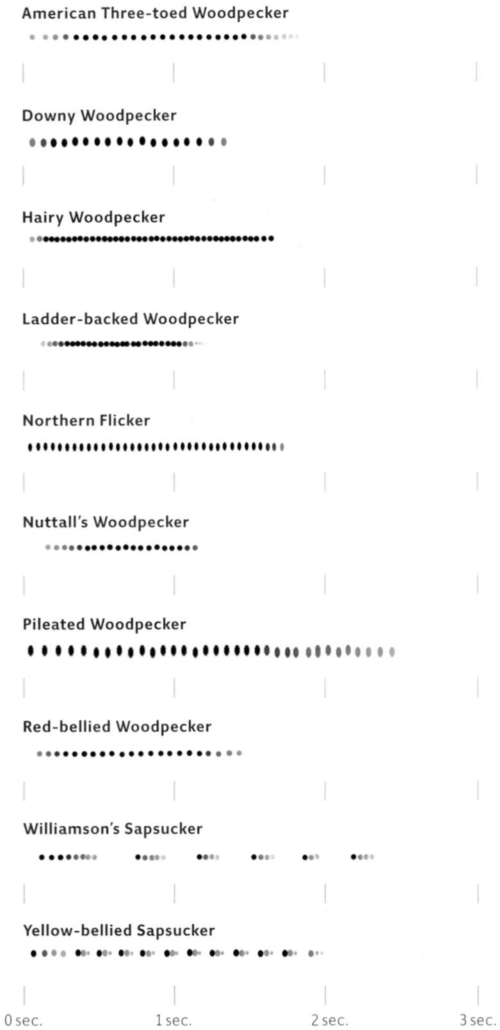

Pileated Woodpecker

Dryocopus pileatus

L 16.5" WS 29" WT 10 oz (290 g) ♂>♀

Unmistakable: crow-size, with long neck, red crest, and obvious white wing-patch. Flight fairly direct (not as undulating as other woodpeckers) with deep, irregular, rowing wingbeats.

VOICE: Contact call single, loud, deep, resonant *wek* or *kuk* notes, often given in flight. Main territorial call higher-pitched *kuk kuk keekeekeekeekeekeekeekuk kuk,* like flickers but slower, with irregular rhythm and deeper, wilder sound. Drum slow, powerful, accelerating and trailing off at end; short or up to three seconds long, only one or two per minute.

Uncommon but widespread; sparsely distributed in mature hardwood and coniferous forests. Favorite foods include carpenter ants, so often forages low on dead trees or even on fallen logs; creates distinctive oval or rectangular holes.

alcons

AMILY: FALCONIDAE

species in 2 genera (2 rare species not shown here). Caracara a distinctive, long-legged scavenger with rounded wingtips. other species are fast-flying predators with pointed wings and eamlined body shape. Formerly thought to be related to hawks amily Accipitridae), but DNA shows that falcons (and parrots) share a common ancestor with songbirds. Prey is mainly birds, captured in flight; only a few species (especially Peregrine Falcon) perform high speed stoop from high altitude; kestrels hover and drop on prey, mainly rodents. Falcons do not build nests, using pre-existing tree cavity or cliff ledge. Juveniles are shown.

enus *Falco*

nerican Kestrel,
ge 326

Merlin,
page 327

Aplomado
Falcon,
page 329

Gyrfalcon,
page 325

regrine Falcon,
ge 324

Prairie Falcon,
page 326

Genus *Caracara*

Crested Caracara,
page 329

Peregrine Falcon

Falco peregrinus

L 16" WS 41" WT 1.6 lb (720 g) ♀>♂

Sleek and powerful, with very pointed wings and relatively short tail. Prominent dark mustache; also note uniformly patterned underwing. Told from Merlin by larger size, relatively long wings, bold face pattern.

VOICE: Alarm a slow, scolding *rehk rehk rehk...*; harsh, raucous, each note rising. In aggression a hard, mechanical *wiSHEP koCHE koCHE koCHEcheche.* Juvenile call similar to adult.

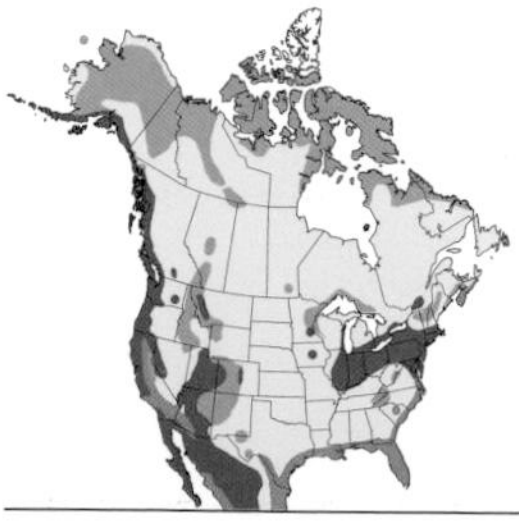

Uncommon in open areas, especially near water. Nests on cliff ledges or on buildings or bridges in cities. Solitary. Hunts from perch or from high in the air, stooping on prey at very high speed; sometimes stoops repeatedly at same prey in series of U-shaped dives. Feeds mainly on small or medium-size birds.

Tundra breeding population is widespread in migration and winter. The Interior West population is similar to Pacific but with unmarked, pale breast and cheeks. Pacific is mainly resident coastally from Washington to Alaska. Recent reintroductions in many areas involve captive-bred form similar to Pacific, and Pacific-like birds are now found across North America.

;yrfalcon

lco rusticolus

:2" WS 47" WT 3.1 lb (1,400 g) ♀>♂

ɹr largest and most powerful falcon, overall color ranges from rk brownish-gray to almost pure white. Told from Peregrine longer tail, flight feathers paler than underwing coverts, weak ·ad pattern. Can also be confused with Northern Goshawk.

ɔICE: Alarm a hoarse, nasal *KYHa KYHa KYHa…*; a deep, hoarse *vah kwah kwah…* lower, gruffer, with more trumpeting quality an Peregrine. Juvenile begs with wailing call and gives hoarse ries like adult.

Rare. Found in expansive open spaces such as tundra, marshes, and farmland. Nests on cliff ledges. Solitary. Hunts from perch or in low-level flight, using ground contours for cover; less often from high in the air. Feeds on birds, especially ptarmigan and other rather large birds, and some small mammals.

Dark birds can be confused with Northern Goshawk, but Gyrfalcon favors open country (never in woods); Goshawk favors wooded areas (occasionally briefly in the open)

Overall color varies from dark to white. Gray (intermediate) morph is widespread. White morph is scarce and nests mainly in northern Greenland; darkest morph nests mainly in Labrador.

Prairie Falcon

Falco mexicanus

L 16" WS 40" WT 1.6 lb (720 g) ♀>♂

Slightly longer-tailed and blunter-winged than Peregrine Falcon. Overall pale color with very dark "wingpits" distinctive. Also note pale gray-brown tail, and mostly white underparts.

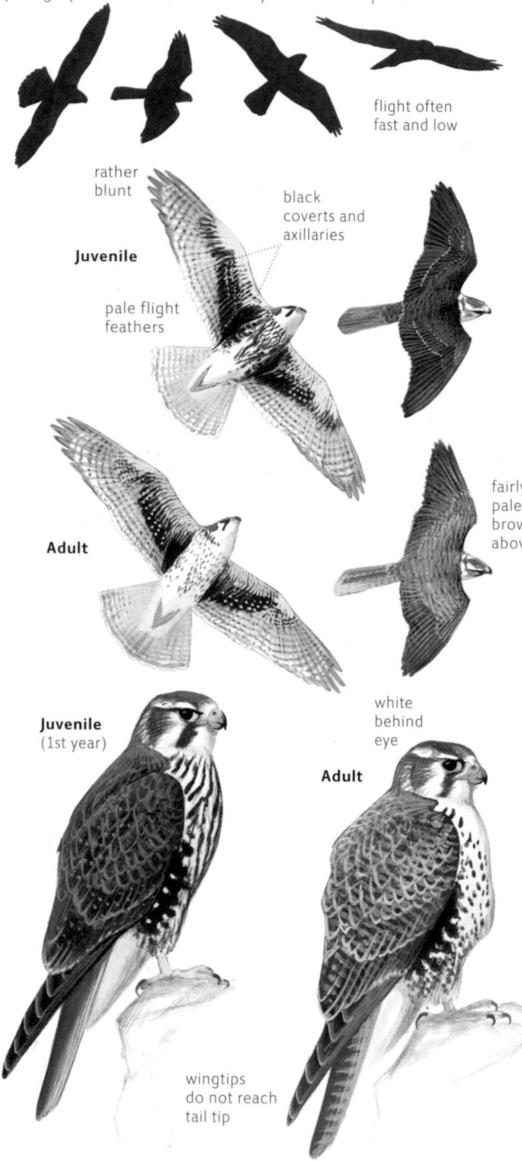

VOICE: Generally similar to Peregrine. Alarm an angry, harsh *ree-kree-kree-kree*... similar to Peregrine but higher; also a high, rising scream *keeeee* repeated.

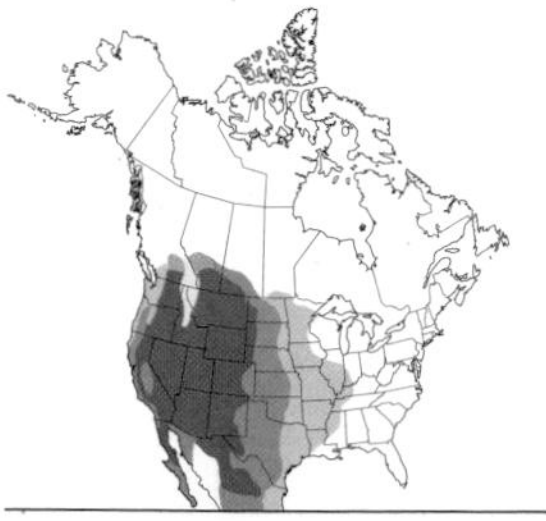

Uncommon in open deserts, grasslands, and agricultural land. Nests on cliff ledges. Solitary. Hunts from perch, from low contouring flight, or from high in the air. Feeds mainly on small mammals such as ground squirrels, but also takes many birds and some insects.

American Kestrel

Falco sparverius

L 9" WS 22" WT 4.1 oz (117 g) ♀>♂

Our smallest falcon. Small and slender, with boldly pattern head, rufous back and tail, and relatively pale underparts; oft hovers in search of prey, unlike other small falcons or hawks.

VOICE: Common call a shrill, clear screaming *kli kli kli kli kli kli kli* or *killy killy killy*, higher and weaker than other rapto Juvenile call similar to adult.

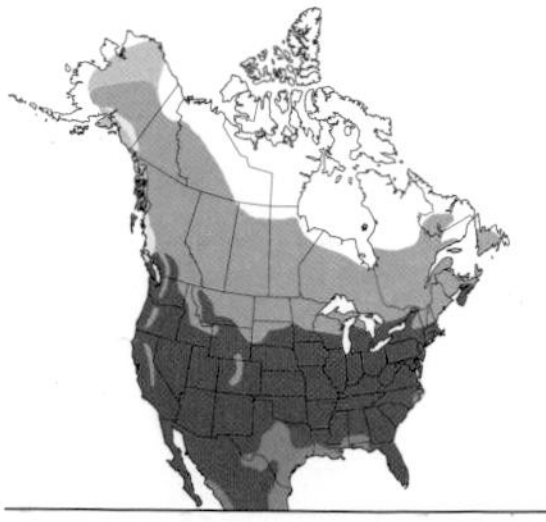

Uncommon in many open habitats from desert grasslands to meadows to brushy fields; often seen on roadside wires or fenceposts, pumping its tail. Nests in tree cavities, birdhouse or crevices in buildings. Solitary Hunts mainly for insects and small mammals from perch or by hovering and dropping straight down.

Merlin

lco columbarius

L0" WS 24" WT 6.5 oz (190 g) ♀>♂

nall, compact, and powerful, with angular shape, relatively ort, broad, and very pointed wings, and usually fast and direct ght. Also note entirely gray to brownish back and wings and ack tail with narrow white bands.

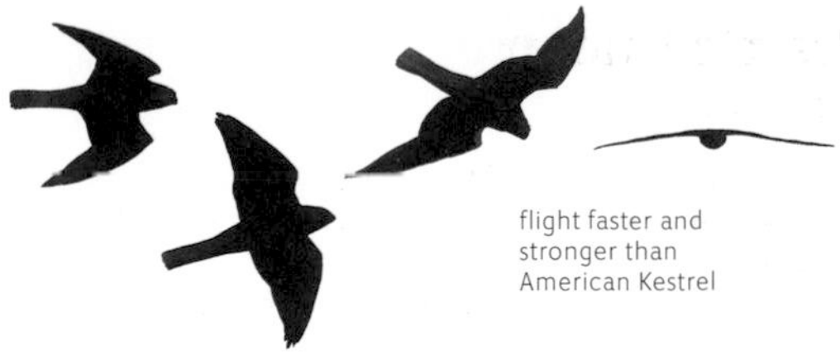

flight faster and stronger than American Kestrel

ICE: Alarm call rapid, accelerating series of strident notes, sing then falling, with quality like Killdeer but harsher *twi twit-itwitititititititit*; female lower, slower, harsher than male; also ngle hard *peek* reminiscent of Black-headed Grosbeak. Juvenile ll similar to adult.

Uncommon in open habitats. Nests in trees in forests with open areas. Solitary. Pugnacious; often harasses much larger birds. An active and energetic hunter: spots prey from perch or during low fast flight, closes with incredible speed, and attacks with abrupt turns, often from below. Feeds almost entirely on small birds; also takes dragonflies in midair.

Three different populations differ in overall color, reflected in the prominence of the dark mustache, width of dark tail-bands, extent of flank and underwing covert markings, and back color. Typical individuals are readily identified, but intermediates occur where ranges meet. Occasional Taiga birds from the East apparently resemble Black. Taiga population is widespread, wintering continent-wide but especially along both coasts. Prairie birds winter from Canada to Mexico, occasionally west to California. Black population winters along Pacific coast south to California, rarely east to Texas.

Eurasian Kestrel

Falco tinnunculus

L 13.5" WS 30" WT 7.2 oz (204 g)

Similar to American Kestrel but larger and relatively longer-tailed; best distinguished by single dark mustache line on face, male has blue-gray tail, female usually paler rufous above than American, with contrasting dark outer wing.

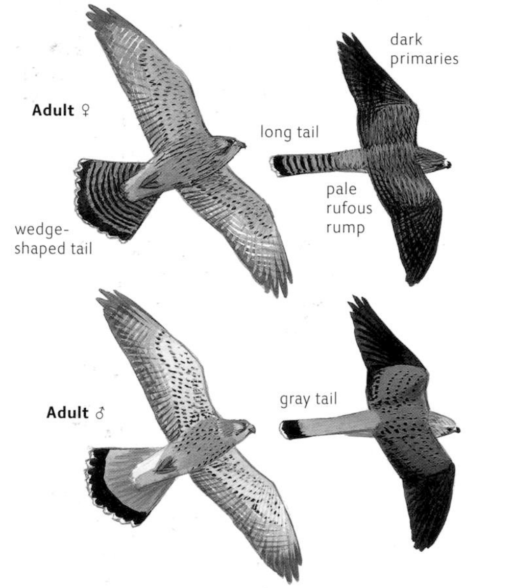

VOICE: Generally silent away from breeding grounds. Typical call a series of short squeals, similar to American Kestrel but slightly lower-pitched and each note single-syllabled *kwee-kwee-kwee....*

Very rare visitor from Eurasia to both coasts, recorded in Alaska, California, Massachusetts, New Jersey, and Florida. Habitat and habits like American Kestrel.

Eurasian Hobby

Falco subbuteo

L 12.7" WS 33" WT 8.5 oz (241 g)

A small, compact falcon with relatively long wings and short ta Most similar to Peregrine, but smaller. Note uniform dark gr upperside, boldly-marked white cheeks with single black mou tache mark, and (on adult) cinnamon buff undertail coverts.

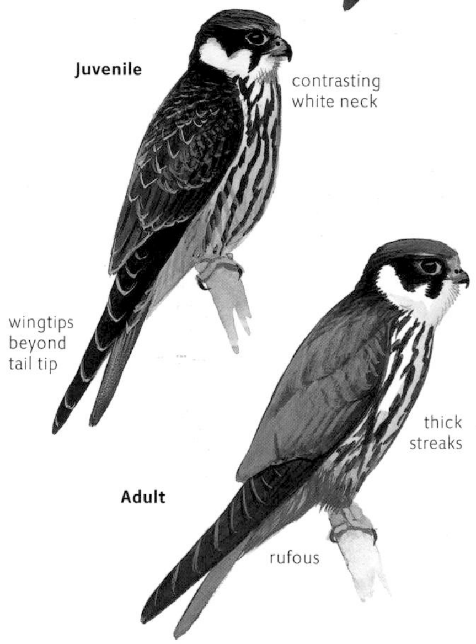

VOICE: Generally silent away from breeding grounds. Typical calls a high strident *kwee-kwee-kwee...* series or a weaker *kew-kew-kew....*

Very rare visitor from Eurasia to far western Alaska with several records farther south and east including Washington, Newfoundland, and Massachusetts. Feeds almost entirely on small birds, including swallows, captured in midair.

plomado Falcon

lco femoralis

16" WS 35" WT 12 oz (335 g) ♀>♂

ightly smaller than Peregrine, more slender with distinctly nger tail. Note striking pattern: dark gray upperside, black anks, cinnamon undertail coverts, and bold pale eyebrow. In ght all-gray upperside with white trailing edge on wings.

OICE: Alarm a harsh series *kek kek kek*... faster and higher than rairie or Peregrine; also a sharp chip; female gives plaintive ailing. Juvenile begs with high-pitched chittering.

Rare and local. Found in desert grasslands and coastal prairies with scattered yuccas or bushes. Formerly (pre-1910) nested in desert grasslands from Arizona to Texas. It is now being reintroduced in southern Texas. Nests in yuccas or low trees. Often in pairs, which hunt cooperatively, pursuing prey in level flight. Feeds mainly on small birds, but also takes many insects.

Crested Caracara

Caracara cheriway

L 23" WS 49" WT 2.2 lb (990 g)

Very un-falcon-like, with long legs and neck, rounded wingtips, scavenging behavior like raven. In flight, note shape and dark plumage with white neck, tail, and wingtips. Perched note long legs, very dark body, pale neck.

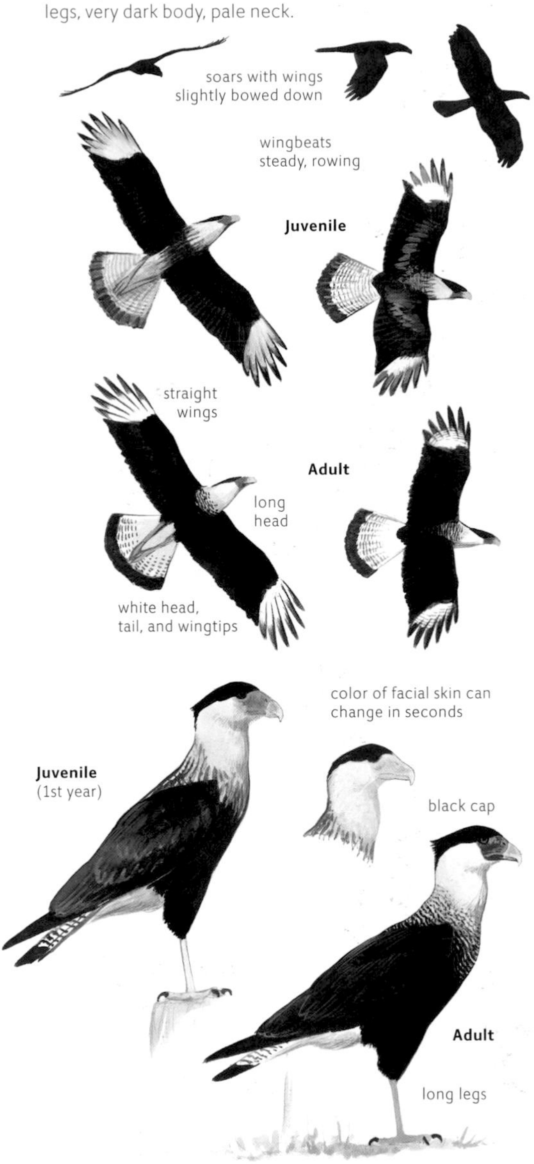

VOICE: Usually silent; low, toneless, croaking *grrrrk* or more complex call of several syllables. Juvenile begs with hoarse or wheezy scream.

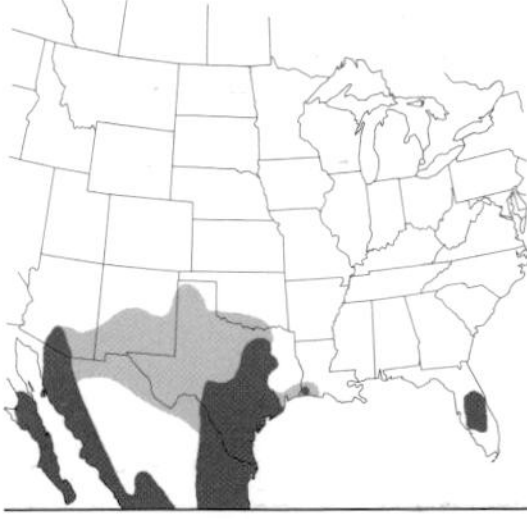

Uncommon and local in semi-open lowland habitats from mesquite brushlands to prairies and farmland. Nests in low bushes or palms. Usually in pairs. Feeds on carrion and some lizards and mammals. Flies relatively low in search of prey, often patrolling along roads in early morning to find carrion before vultures fly.

Parrots

FAMILY: PSITTACIDAE, CACATUIDAE

With nearly 30 species seen regularly, and dozens more species possible as local escapes, the identification of parrots in North America can be a challenge. The species shown here include the most frequently encountered, but in southern cities you should expect to find others. Look at overall shape (especially tail shape) and size to narrow down the group; the most numerous are two South American genera *Amazona* (with short tail and broad rounded wings) and *Aratinga* (with long pointed tail ar narrow wings).

Parrots are generally located by their loud raucous c screeching calls, but can be surprisingly difficult to see whe perched in foliage, and are most easily seen in flight. They gath at communal roosts in the evening, often with many specie mixed together.

Sulphur-crested Cockatoo

Cacatua galerita

L 19.5" WS 40" WT 30 oz (850 g)

A large and stocky parrot with broad wings and short tail. Essentially all white.

Native to Australia. Occasionally seen as escapes in southern cities.

VOICE: Extremely loud, harsh, and toneless.

Cockatiel

Nymphicus hollandicus

L 12.5" WS 20" WT 3.2 oz (90 g)

Slender, with long tail and pointed wings. Overall pale gray with pale cheeks, thin crest, pale upperwing coverts.

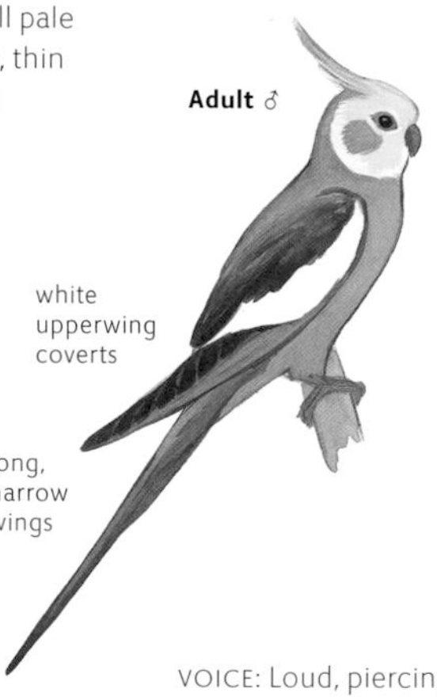

Native to Australia. Occasionally seen as escapes, mostly in southern cities, but no established populations.

VOICE: Loud, piercing, clea whistles *queeep* or *queeo*

Budgerigar

Melopsittacus undulatus

L 7" WS 12" WT 1 oz (29 g)

Small and slender, with long, pointed tail and wings. Most are greenish, like the ancestral wild form, but escapes can be blue, yellow, or white.

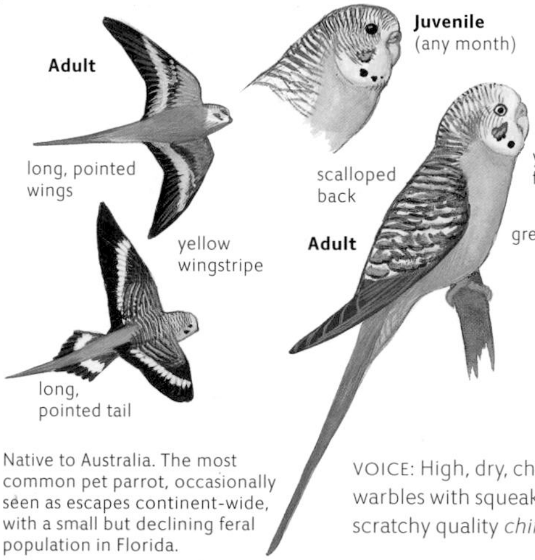

Native to Australia. The most common pet parrot, occasionally seen as escapes continent-wide, with a small but declining feral population in Florida.

VOICE: High, dry, chattering warbles with squeaky and scratchy quality *chirrup*.

Rosy-faced Lovebird

Agapornis roseicollis

L 6.5" WS 13" WT 1.9 oz (54 g)

A very small parrot with relatively large head and short pointed tail.

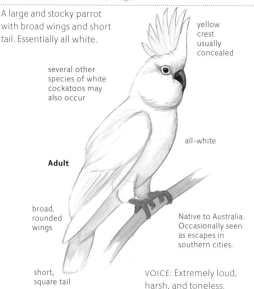

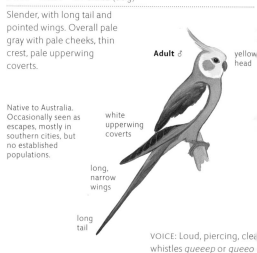

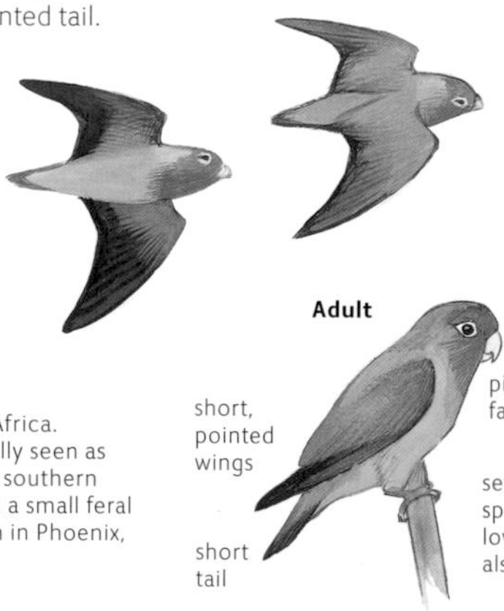

Native to Africa. Occasionally seen as escapes in southern cities, with a small feral population in Phoenix, Arizona.

VOICE: A short sharp *skree*

lack-hooded Parakeet

andayus nenday

2" WS 23" WT 4.5 oz (128 g)

large and long-tailed parakeet, pale yellow-green with dark ce and dark wings.

Native to South America. Occasionally seen as escapes in southern cities, with a substantial and increasing feral population in Florida and California.

VOICE: Higher than Blue-crowned; a more screeching *graad graad....*

Rose-ringed Parakeet

Psittacula krameri

L 16" WS 18.5" WT 4.1 oz (117 g)

Superficially similar to *Aratinga* parakeets, but larger, with distinctive fluttering wingbeats and very long, slender tail.

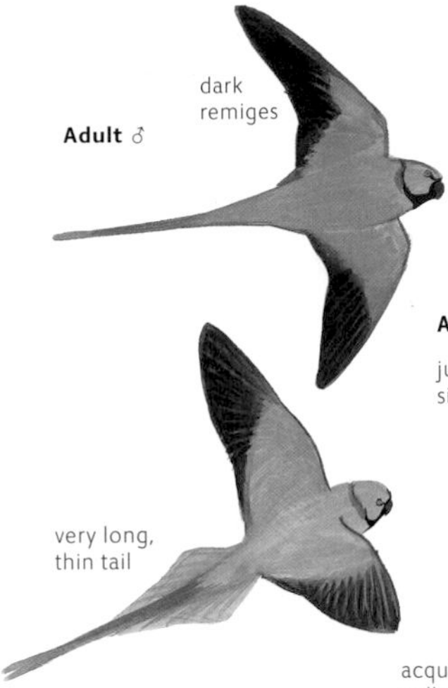

Native to Asia. Occasionally escaped, with small established populations in southern California and in Florida.

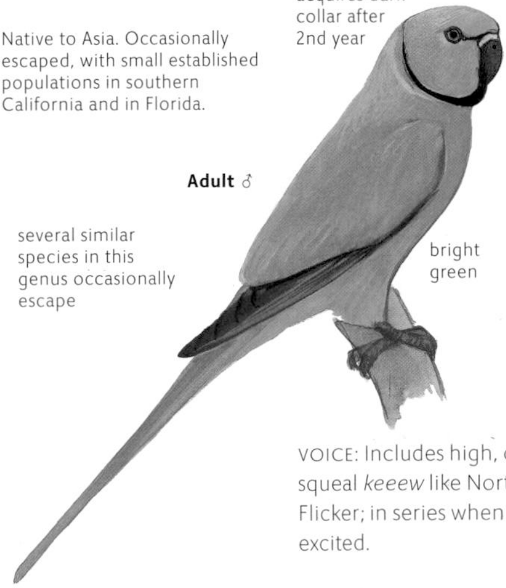

VOICE: Includes high, clear squeal *keeew* like Northern Flicker; in series when excited.

onk Parakeet

yiopsitta monachus

11.5" WS 19" WT 3.5 oz (100 g)

milar to *Aratinga* parakeets, but note paler green color and ay face. Builds a stick nest; all other species nest in tree cavities.

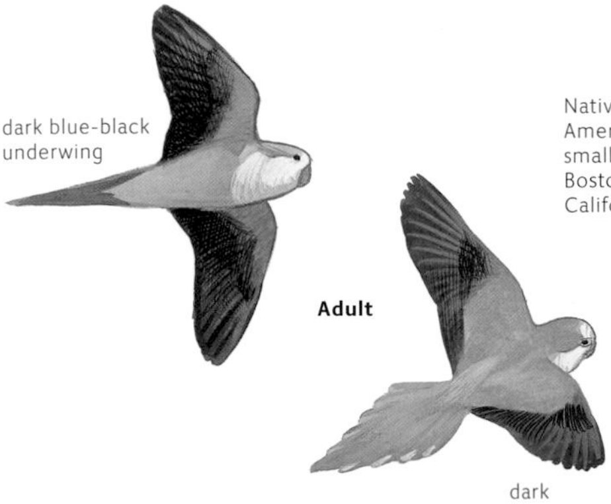

Native to southern South America, escaped and with small feral populations from Boston and Chicago to California, Texas, and Florida.

VOICE: A varied, throaty rattling *graaa* and higher, rasping *skveet*; both slightly rising.

Blue-crowned Parakeet

Aratinga acuticaudata

L 14.5"   WS 24"   WT 7 oz (200 g)

Fairly large and long-tailed; note blue forehead and bicolored bill.

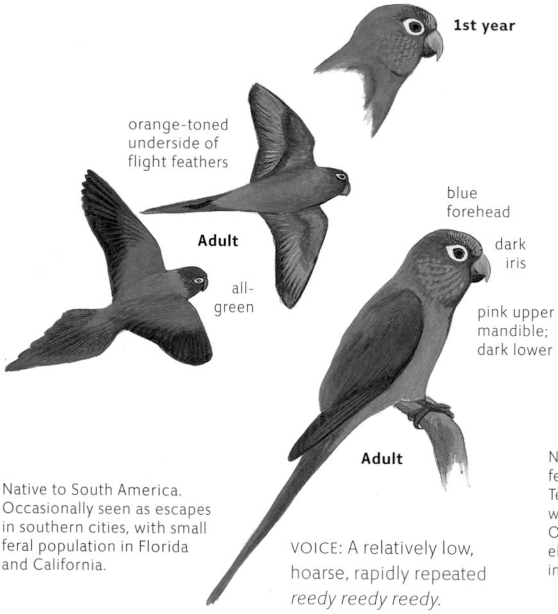

Native to South America. Occasionally seen as escapes in southern cities, with small feral population in Florida and California.

VOICE: A relatively low, hoarse, rapidly repeated *reedy reedy reedy.*

Green Parakeet

Aratinga holochlora

L 13"   WS 21"   WT 8 oz (230 g)

Stocky and relatively short-tailed; uniform bright green all ove with pale bill.

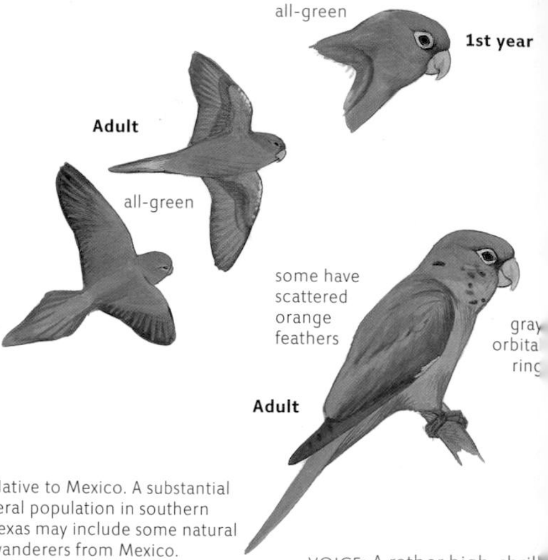

Native to Mexico. A substantial feral population in southern Texas may include some natural wanderers from Mexico. Occasionally escapes seen elsewhere, especially in Florida.

VOICE: A rather high, shrill raucous chattering.

Mitred Parakeet

Aratinga mitrata

L 15"   WS 25"   WT 7 oz (205 g)

Largest of the *Aratinga* parakeets, bright green with red patches on face.

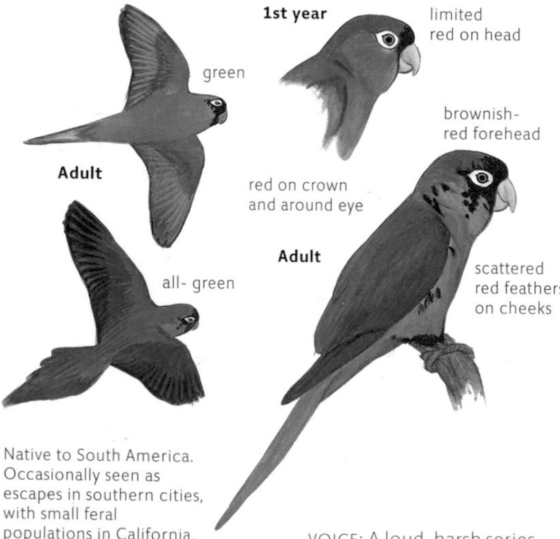

Native to South America. Occasionally seen as escapes in southern cities, with small feral populations in California, Florida, and New York.

VOICE: A loud, harsh series and rising *kreeep.*

Red-masked Parakeet

Aratinga erythrogenys

L 13.5"   WS 22"   WT 6 oz (170 g)

Very similar to Mitred Parakeet, but smaller, with more red c head and red underwing coverts.

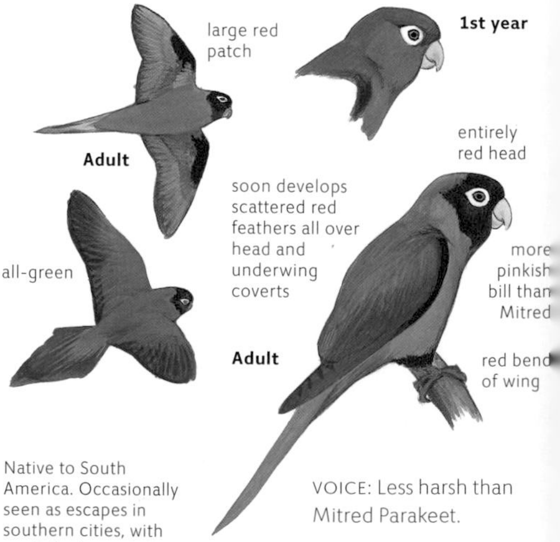

Native to South America. Occasionally seen as escapes in southern cities, with small feral populations in California and Florida.

VOICE: Less harsh than Mitred Parakeet.

)usky-headed Parakeet

ratinga weddellii

11" WS 18" WT 3.9 oz (110 g)

small and reltively short-tailed *ratinga* parakeet; note gray head ith large white eye-patch, ellowish belly.

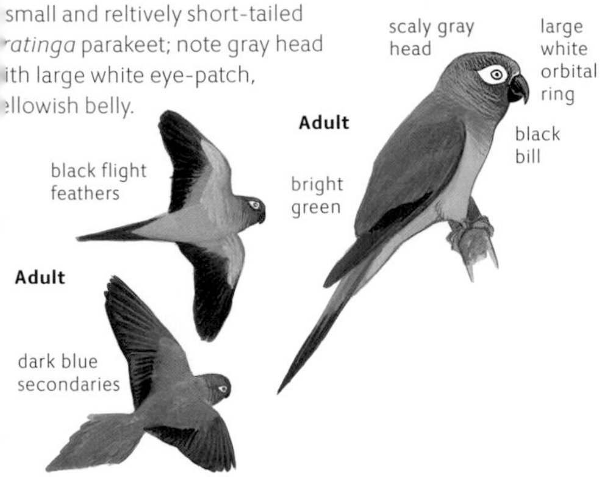

ative to South America. ccasionally seen as escapes in uthern cities, with a small feral opulation in Florida.

VOICE: Rather nasal; less grating than similar species.

Vhite-eyed Parakeet

ratinga leucophthalma

13.5" WS 22" WT 6 oz (170 g)

ery similar to Mitred and Red-masked Parakeets.

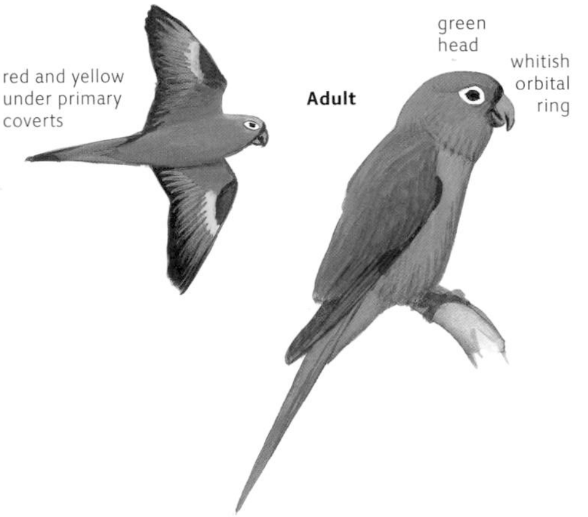

Native to South America. Occasionally seen as escapes in southern cities, with small feral populations in California and Florida.

VOICE: Similar to Mitred Parakeet.

Virtually any of the world's parrot species could be seen as escapes in North America. A few species not shown here but seen multiple times in recent years include Senegal Parrot, Patagonian Conure, Green-cheeked Parakeet, Orange-fronted Parakeet, Orange-chinned Parakeet, Red-shouldered Macaw, and several species of large macaws. Parrots are long-lived, and a single release or series of releases, in a suitable climate, can create a local population that persists for many years without ever reproducing in the wild.

White-winged Parakeet

Brotogeris versicolurus

L 8.75" WS 15" WT 2.1 oz (60 g)

Noticeably smaller than *Aratinga* parakeets, with relatively small pointed wings and medium-length pointed tails. Wing beats fluttery.

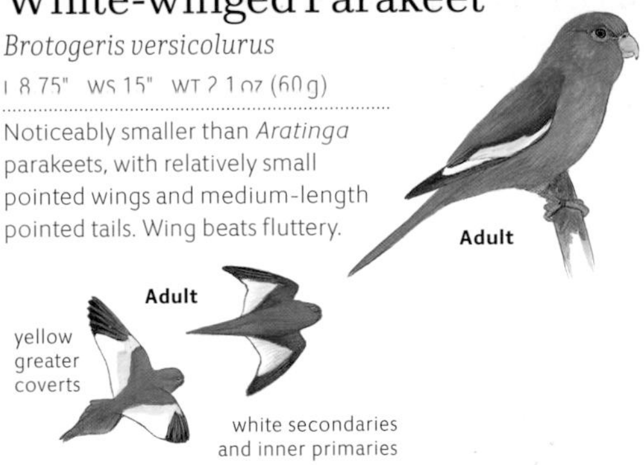

Native to South America. Escaped and feral in small numbers in southern Florida and California

VOICE: Similar to Yellow-chevroned but mellower.

Yellow-chevroned Parakeet

Brotogeris chiriri

L 8.75" WS 15" WT 2.1 oz (60 g)

Very similar to White-winged, but more numerous. With dark flight feathers, subtle differences in head and body colors.

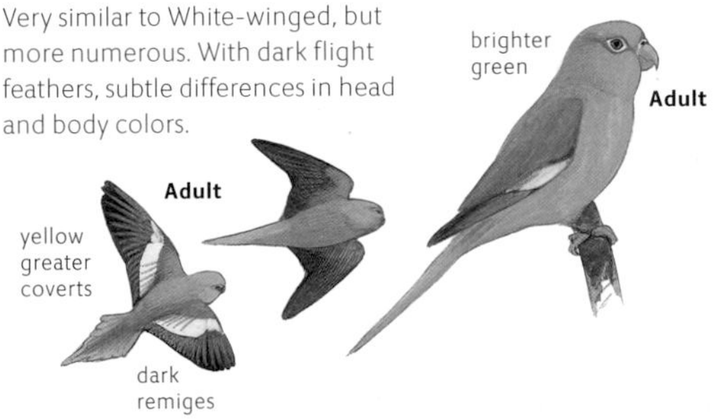

Native to South America. Escaped and feral in small numbers in southern Florida and California

VOICE: A rather high, scratchy *krere-krere*.

Thick-billed Parrot

Rhynchopsitta pachyrhyncha

L 15" WS 32" WT 15 oz (440 g)

Large and stocky, with relatively short tail, note black bill.

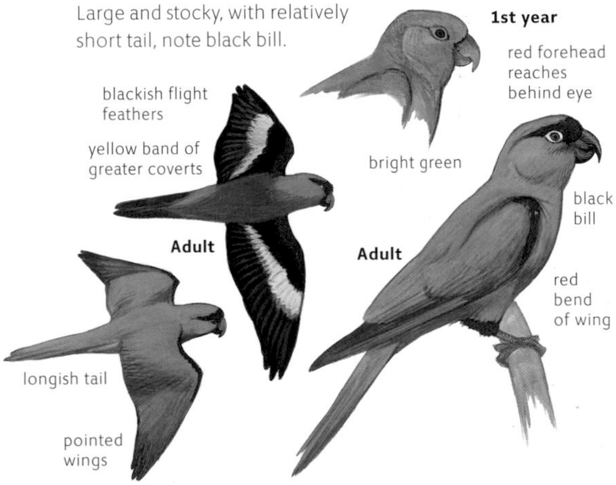

Formerly a very rare visitor to Arizona from montane pine forests of Mexico; now rare in Mexico, last recorded in the wild in Arizona in 1938. Reintroduction efforts in Arizona in the 1980s failed.

VOICE: All calls talky, clear, not grating; may recall flock of Snow Geese.

Red-crowned Parrot

Amazona viridigenalis

L 12" WS 25" WT 11 oz (300 g)

One of several species of large parrots collectively known as Amazons. Relatively large, with blocky bodies and heads, broad wings, and short square tails. Wingbeats shallow; wings not raised above body in flight. Flying birds often organize into short lines.

Native to eastern Mexico. Established feral populations in California, Florida and Texas (at least some Texas birds may be natural vagrants from Mexico). Ironically, feral population in the US now thought to exceed the native population in Mexico.

VOICE: Common call a high squeal *weeeoo* followed by harsh, pounding *daak daak daak daak*.

Lilac-crowned Parrot

Amazona finschi

L 12.5" WS 24" WT 11 oz (300g)

Very similar to Red-crowned, but with paler green cheeks, smaller and darker maroon forehead patch.

Native to Mexico. Occasionally escapes in southern cities with a small feral population in southern California.

VOICE: Similar to Red-crowned.

Yellow-headed Parrot

Amazona oratrix

L 14" WS 28" WT 1 lb (450 g)

Larger than Red-crowned, with extensive pale yellow on head, very pale bill.

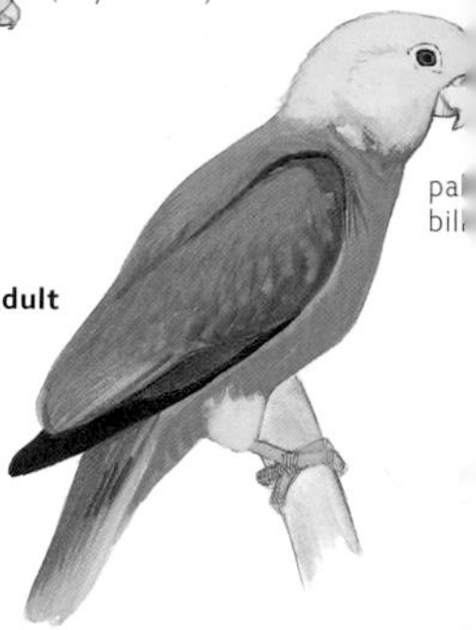

VOICE: Generally mellow and human-voiced *herra* or long, descending *yadadadadada*.

Native to Mexico. Occasionally seen as escapes in southern cities.

White-fronted Parrot

Amazona albifrons

L 9.5" WS 23" WT 7 oz (200 g)

Smaller than most Amazons, with dark blue wings and (on male) red primary coverts; small white forehead patch.

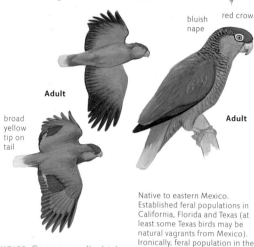

VOICE: Varied, includes some high, squeaky notes.

Native to Mexico and Central America. Occasionally seen as escapes in southern cities.

ꞮRed-lored Parrot

mazona autumnalis

12.5"

milar to Red-crowned but ith yellow below eye, darker ill.

ative to Mexico and Central merica. Occasionally seen as scapes in southern cities.

Orange-winged Parrot

Amazona amazonica

L 12.5"

Similar to Blue-fronted, but smaller with yellow forehead, blue line through eye.

Native to South America. Occasionally seen as escapes in southern cities, with small numbers in Florida.

Mealy Parrot

mazona farinosa

15.5"

argest Amazon, plain drab reenish color with bold vhite orbital ring.

Jative to South America. Occasionally seen as escapes in outhern cities.

Blue-fronted Parrot

Amazona aestiva

L 15"

Larger than most Amazons, with yellow face and small blue patch on forehead.

Native to South America. Occasionally seen as escapes in southern cities.

Yellow-crowned Parrot

Amazona ochrocephala

L 14"

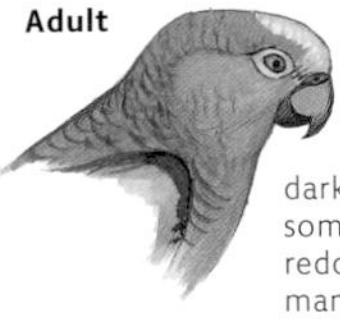

Native to Mexico. Rarely seen as escapes in southern cities.

Yellow-naped Parrot

Amazona auropalliata

L 14"

This species and Yellow-crowned very closely related to Yellow-headed, differ in darker bill and less yellow on head.

Native to Mexico. Rarely seen as escapes in southern cities

Chestnut-fronted Macaw

Ara severus

L 18" WS 30" WT 15 oz (430 g)

Much larger than *Aratinga* parakeets, with bare white face, black bill, mostly reddish underwing.

Native to South America. Occasonally seen as escapes in southern cities, with a small feral population in southeastern Florida.

VOICE: Very loud, harsh screeching; sometimes like braying donkey.

Blue-and-yellow Macaw

Ara ararauna

L 34" WS 50" WT 42 oz (1191 g)

Extremely large and long tailed, with brilliant blue and yellow color.

Native to South America. Rarely escaped from captivity.

Tyrant Flycatchers and Becard

FAMILIES: TYRANNIDAE, TITYRIDAE

44 species in 15 genera (9 rare species not shown here); all except Rose-throated Becard in family Tyrannidae. Most species sort out into well-defined groups based on genera; plumage and structure within some genera are so similar that voice is the primary field mark. Nest structure and placement can offer useful clues. Most have drab colors and short, rather broad-based and flattened bills. Small species are found mainly in brushy woods. Larger species are more conspicuous and aggressive, some foun in open areas. Nearly all feed mainly on insects captured in fligh flight is strong, buoyant, and agile, with quick turns and abrup movements as birds pursue flying insects. Aerial flycatching hab its are also shared by warblers, kinglets, gnatcatchers, waxwing hummingbirds and others. Juveniles are shown.

Genus *Sayornis*

Black Phoebe, page 348

Eastern Phoebe, page 348

Say's Phoebe, page 349

Genus *Pyrocephalus*

Vermilion Flycatcher, page 34

Genus *Contopus*

Olive-sided Flycatcher, page 338

Greater Pewee, page 338

Western Wood-Pewee, page 339

Eastern Wood-Pewee, page 339

Genus *Myiarchus*

Dusky-capped Flycatcher, page 350

Ash-throated Flycatcher, page 350

Great Crested Flycatcher, page 351

Brown-crested Flycatcher, page 351

Genus *Pitangus*

Great Kiskadee, page 359

Genus *Myiodynastes*

Sulphur-bellied Flycatcher, page 359

Genus *Camptostoma*

Northern Beardless-Tyrannulet, page 360

enus *Empidonax*

ellow-bellied Flycatcher, age 343

Acadian Flycatcher, page 343

Alder Flycatcher, page 345

Willow Flycatcher, page 344

east Flycatcher, page 345

Hammond's Flycatcher, page 346

Gray Flycatcher, page 347

Dusky Flycatcher, page 346

'acific-slope Flycatcher, age 342

Cordilleran Flycatcher, page 342

Buff-breasted Flycatcher, page 347

;enus *Tyrannus*

'ropical Kingbird, page 356

Couch's Kingbird, page 356

Cassin's Kingbird, page 357

Thick-billed Kingbird, page 354

Vestern Kingbird, page 357

Eastern Kingbird, page 355

Gray Kingbird, page 355

Scissor-tailed Flycatcher, page 353

Genus *Pachyramphus*

Rose-throated Becard, page 361

Greater Pewee

Contopus pertinax

L 8" WS 13" WT 0.95 oz (27 g)

Similar to Olive-sided Flycatcher but relatively longer-tailed, with thin pointed crest, plain grayish underparts including throat and belly, lacking darker vest. Larger than wood-pewees, with larger bill and pointed crest.

VOICE: Song a lazy, clear whistle *soo saay sooweeoo* or (José Maria) usually introduced by several gentle *hoo didip* phrases; that phrase often repeated without song. Call a soft, low *pip* or *pip-pip-pip*; higher, softer, and slower than Olive-sided.

Uncommon in montane pine forests and shaded canyons, spending most of its time high in the trees watching for flying insects, which it catches in short flights like Olive-sided Flycatcher. Rarely seen away from known nesting sites, then found in open woods or parks with scattered trees.

Olive-sided Flycatcher

Contopus cooperi

L 7.5" WS 13" WT 1.1 oz (32 g)

Large, with relatively large head and short tail. Note dark hea and flanks sharply contrasting with white throat and belly. Hab of perching on very conspicuous treetop or wire is a good iden tification clue.

VOICE: Song a sharp, penetrating whistle *whip WEEDEEER* o "quick, three beers" (Pacific birds sing subtly different "wha peeves yoou" with equal emphasis on all syllables). Call a low hard *pep pep pep* with variations from soft, rapid *piw-piw-piw* to harder *pew, pew* to single note *pep* or *quip*.

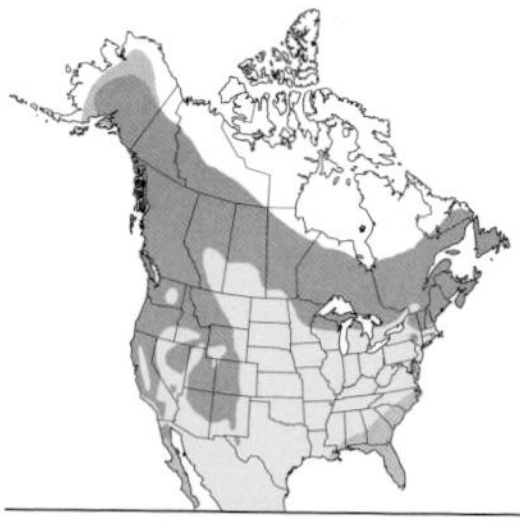

Uncommon. Nests in spruce-fir forests where trees of varied heights are mixed with clearings. Solitary. Always chooses most conspicuous perch available such as a dead tree in the middle of a clearing from which to watch for flying insects; perches upright, watching intently all around.

Western Wood-Pewee

ontopus sordidulus

6.25" WS 10.5" WT 0.46 oz (13 g)

istinguished from *Empidonax* flycatchers by habits (including ck of tail-flicking), larger size, and longer wings; also by darker ead, dusky breast, and dusky smudges on undertail coverts. early identical to Eastern Wood-Pewee, best distinguished by ong.

OICE: Song a burry, nasal whistle *DREE-yurr* or *breerrr* or *rreeee* with distinctive rough quality. Also gives a variety of clear, whistled phrases virtually identical to Eastern: a thin, clear *peee-didip* or *pee-ee* or *peeaa*. Call a flat, dry *plit*; varied sputtering notes in aggression.

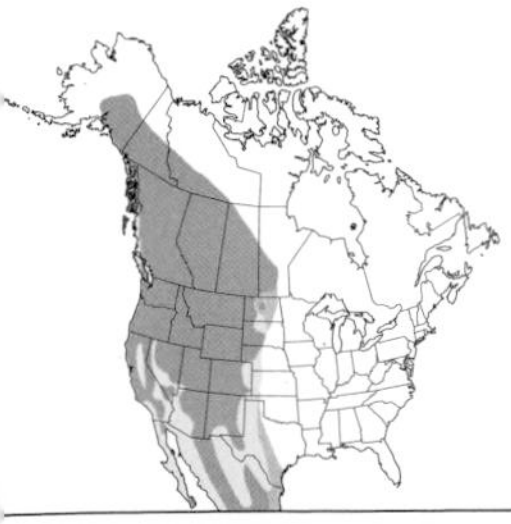

Common in mature deciduous and mixed forests and forest edges. Solitary. Forages for small flying insects from conspicuous perch in middle to upper branches of trees. Nest of both wood-pewees a neat cup decorated with lichens on horizontal branch.

Eastern Wood-Pewee

Contopus virens

L 6.25" WS 10" WT 0.49 oz (14 g)

Nearly identical to Western Wood-Pewee, and safely distinguished only by song. Eastern averages paler below, and more greenish above, with bolder white wingbars, but all features overlap with Western.

Both wood-pewees are distinguished from all *Empidonax* flycatchers by longer wings, long primary projection, and relatively short legs; also by dark face with weak partial eyering, dusky "vest," and grayish smudges on undertail coverts. Wood-pewees do not flick their tails when perched; foraging birds choose a prominent, high perch and return to it repeatedly.

VOICE: Song plaintive, slurred, high, clear whistles *PEEaweee* and *peeyoooo*; also short, upslurred *pawee* (given by migrants), downslurred *peeaaa*, and others. Dawn song alternates regular song phrases with *peee-didip*. Call a flat, dry chip *plit*; sputtering notes in aggression.

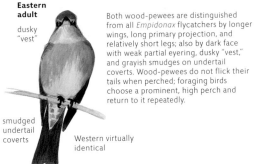

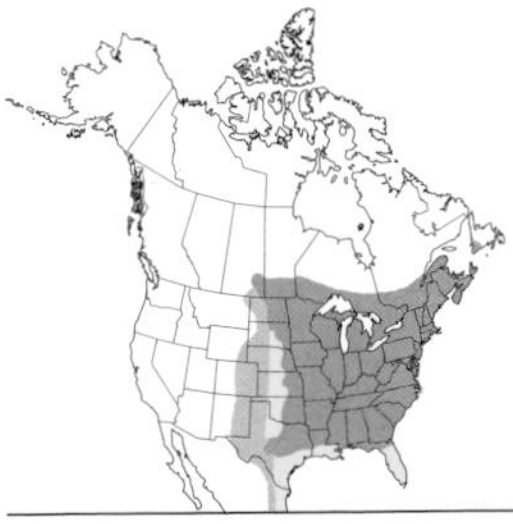

Common in mature deciduous forests and forest edges. Solitary. Forages for small flying insects from comspicuous perch in middle of upper branches of trees.

Cuban Pewee

Contopus caribaeus

L 6.3" WS 9.5" WT 0.5 oz (14 g)

Small size, short primary projection, and pale breast suggest *Empidonax* rather than pewee, but differs in drab wingbars, more prominent crest, bold white arc behind eye, limited tail flicking, and voice.

VOICE: Very high, clear whistles *twee-teeh-teeh* which may be somewhat reminiscent of the flight call of Solitary Sandpiper, and other short weak notes equally high and weak; also a long drawn-out, descending *tweeeeeer.*

Very rare visitor from West Indies. Recorded several times in wooded areas in extreme southeastern Florida.

Tufted Flycatcher

Mitrephanes phaeocercus

L 4.8" WS 7" WT 0.3 oz (9 g)

Distinctive. Very small, with rich cinnamon-buff breast an[d] always shows thin pointed crest. Differs from Buff-breasted Fly-catcher in pointed crest, weaker wingbars, more extensiv[e] cinnamon color below.

VOICE: Call an emphatic, high, rising *churrree churrree* sometimes single or tripled; a rough and sputtering buzz at beginning, then clear and high at end; also a *bik* call similar to Hammond's Flycatcher, and a high clear penetrating *tseeew* repeated every few seconds in breeding season.

Very rare visitor from Mexico, with a few records just north of the border in Texas, Arizona, and Nevada, mainly in fall and winter.

Identification of *Empidonax*

The main source of variation in the appearance of *Empidonax* flycatchers involves the wear and fading of the plumage. Fresh plumage has richer colors, more buff, yellow, and olive tones, with broader and buff-colored wingbars. Worn plumage is drabber, more grayish overall, with narrower wingbars faded to whitish. All species have a complete molt of all feathers in the late summer or fall, and a partial molt of body feathers in early spring. Plumage is—on average—freshest in fall and winter and most worn and faded in mid-summer, but at any season it is possible to see individuals of the same species with contrastingly different plumage wear.

Dusky Flycatcher

The primaries and secondaries are molted only once a year—either in late summer before migration or early winter after migration. New fledglings in late summer have fresh wing feathers, in contrast to the worn feathers of unmolted adults.

Early molting species: Adults of Acadian, Hammond's, and Buff-breasted flycatchers molt primaries and secondaries on the breeding grounds before fall migration, so that all fall migrants have similarly fresh wing feathers.

Late molting species: Adults of Pacific-slope, Cordilleran, Yellow-bellied, Willow, Alder, Least, Dusky, and Gray flycatchers molt after fall migration, so that worn adults and fresh-plumaged juveniles seen together during fall migration appear slightly different.

This provides several identification clues:

- Any *Empidonax* molting wing feathers in July to September must be one of the early-molting species.
- Any *Empidonax* with very worn wing feathers during fall migration must a late-molting species.
- Birds with fresh wing feathers in fall could be immatures of any species, or older birds of the early-molting group.

Call notes are a very useful aid to identifying *Empidonax* flycatchers away from the breeding grounds. Five species sound similar with sharp, dry *whit* calls: Willow Flycatcher, Least Flycatcher, Dusky Flycatcher, Gray Flycatcher, Buff-breasted Flycatcher. Other species: Pacific-slope and Cordilleran flycatchers give very high thin *tseet*; Yellow-bellied Flycatcher a clear whistled *tuwee* or a sharp *pyew*; Acadian Flycatcher a sharp *pweest*; Alder Flycatcher a low flat *pip*; Hammond's Flycatcher a sharp *peek*.

Other species with wingbars and eyering that can be confused with *Empidonax* include Hutton's Vireo and Ruby-crowned Kinglet, distinguished by more constant movement through foliage, horizontal posture, longer legs, different bill shape, and calls.

Comparing *Empidonax* and Other Flycatchers

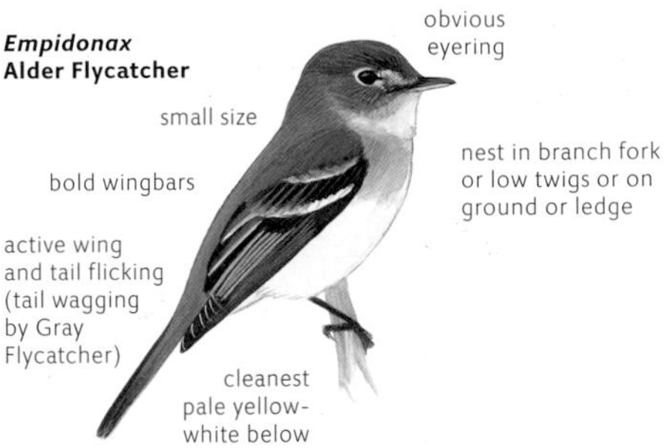

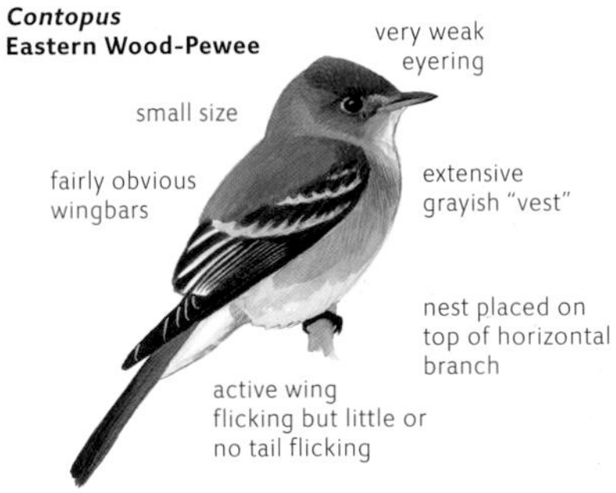

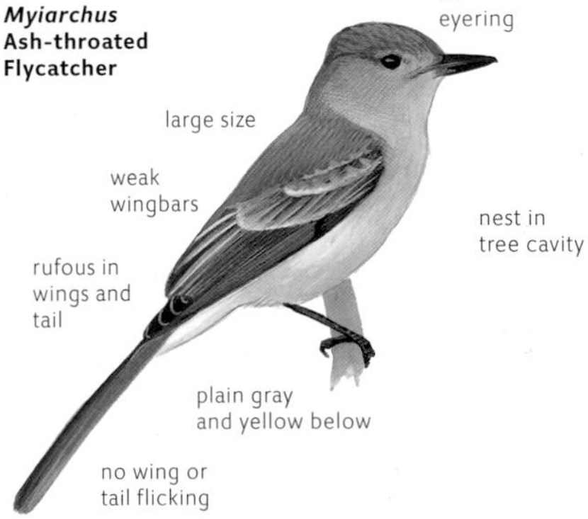

Pacific-slope Flycatcher

Empidonax difficilis

L 5.5" WS 8" WT 0.39 oz (11 g)

Usually yellowish overall, with oval eyering pointed behind eye. Yellow-olive color (especially on throat) distinguishes from all other *Empidonax* except Yellow-bellied and Cordilleran Flycatchers; those are best distinguished by range and voice.

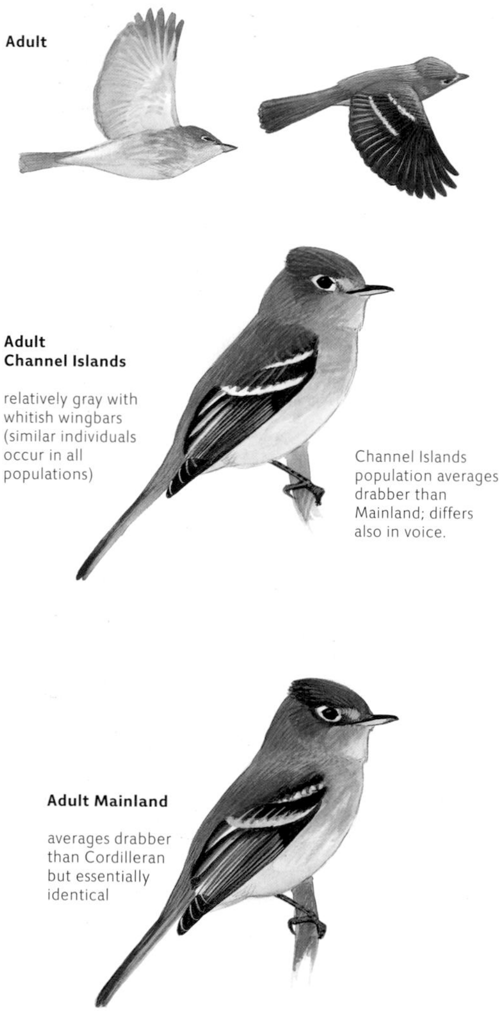

VOICE: Song higher-pitched and thinner than other *Empidonax*. Three phrases repeated in sequence: a high, sharp *tsip*; a thin, high, slurred *klseewii*; and an explosive *ptik*. Female call a very high, short *tseet*. Male position note a very high, thin whistle; Mainland gives a slurred *tseeweep*. Channel Islands a rising *tsweep*. Juvenile call a husky squeak *wiveet*.

Common in moist shaded coniferous and mixed forests. Often seen along streams, where it forages mostly low in shady undergrowth. Nest on ledge or steep bank.

Cordilleran Flycatcher

Empidonax occidentalis

L 5.5" WS 8" WT 0.39 oz (11 g)

Essentially identical to Pacific-slope Flycatcher; distinguishable only by details of song and call of male. Differences in song and call are most distinct at southern end of range, the two species blend together in northern areas.

VOICE: Barely differs from Pacific-slope. In song, *ptik* note has first syllable higher than second (reverse of Pacific-slope). Male position note variable; usually a two-part *tee-seet*, but other birds give rising *tsweep* or slurred *tseeweep* like Pacific Slope. Other calls very similar.

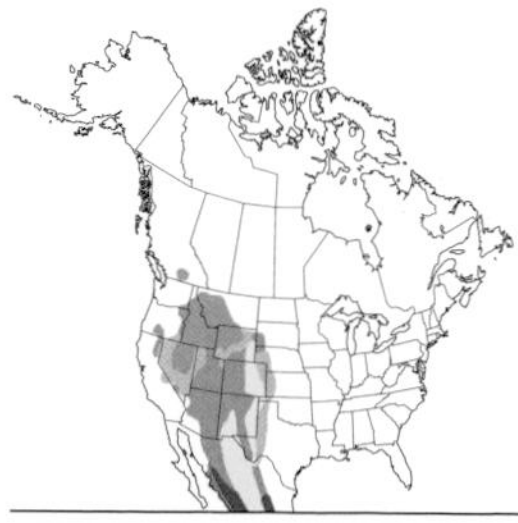

Common in shaded coniferous forests, especially along streams in narrow canyons. Nests on ledge or steep bank.

Yellow-bellied Flycatcher

mpidonax flaviventris

5.5" WS 8" WT 0.4 oz (11.5 g)

elatively small, small-billed, and long-winged. Usually yellow-ish-olive overall; no other flycatcher except Pacific-slope and Cordilleran shows yellowish throat. Best identified by song.

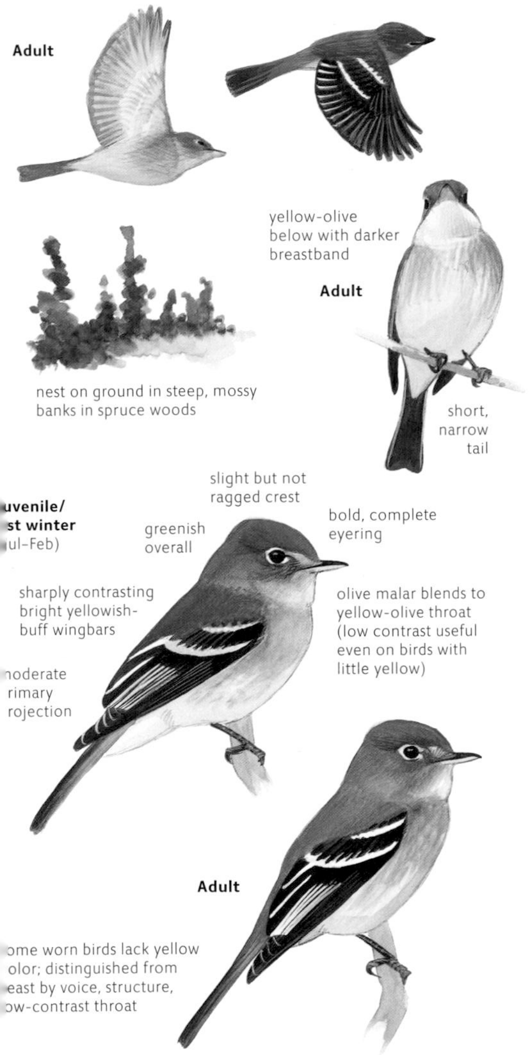

VOICE: Song a hoarse *chebunk* or *cheberk* very similar to Least Flycatcher but lower, buzzier, softer, without strong emphasis. Call a short, clear, rising whistle *tuwee* reminiscent of wood-pewees. Also shorter versions *pwee* or *peee*; a sharp, descending *pyew* (given by migrants); somewhat plaintive, long *peehk*; a harp monotone *wsee*.

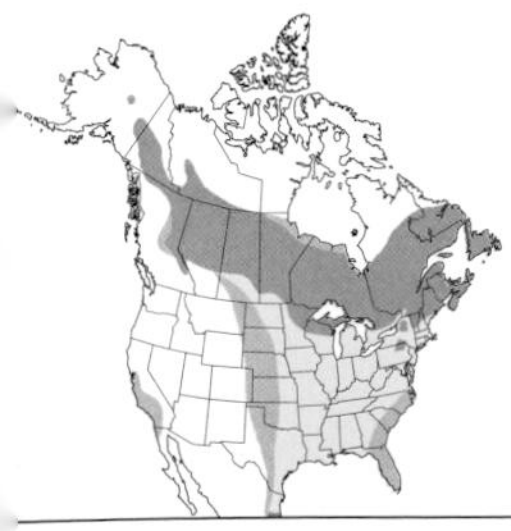

Uncommon in dense spruce woods; migrants found in low understory and brushy vegetation, more often within woods rather than at edges. Habits similar to Least Flycatcher.

Acadian Flycatcher

Empidonax virescens

L 5.75" WS 9" WT 0.46 oz (13 g)

Relatively large, long-billed, and long-winged. Note overall greenish color, clean pale breast and throat, very pale malar, thin but distinct and complete eyering. Best identified by song.

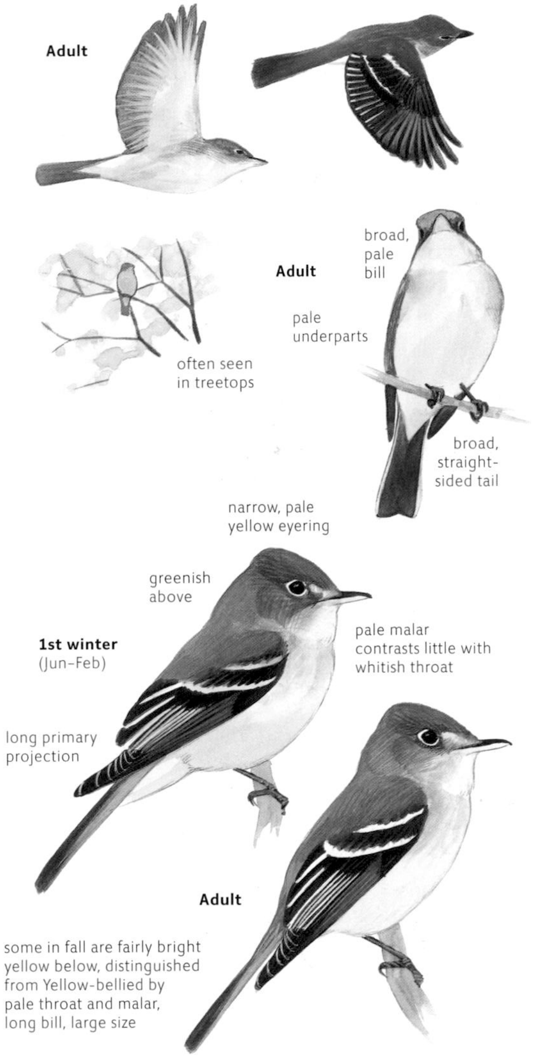

VOICE: Song an explosive, loud, high, rising *spit-a-KEET*; recalls some phrases of Red-eyed Vireo but always more explosive. Call a loud, flat *peek* and strong, squeaky *pweest*; sharper and higher than other *Empidonax* flycatchers. Common call on nesting grounds a relatively long, low *wheeeew*; also a slow, whistled *pwipwipwipwipwipwipwi*.

Common in mature lowland forests; generally in larger trees than other *Empidonax* flycatchers, and stays within forest canopy rather than edges or brushy vegetation. Solitary. Forages for flying insects. Nests in a shallow cup in fork of horizontal branch.

Willow Flycatcher

Empidonax traillii

L 5.75" WS 8.5" WT 0.47 oz (13.5 g)

This species and Alder Flycatcher, our largest *Empidonax* flycatchers. Relatively long-billed and flat-headed, with weak eyering. Can be mistaken for a pewee, but note habits, shorter wings, and clean whitish undertail coverts.

Adult

Eastern

broad, pale bill

Adult

broad, straight-sided tail

Western

dark brownish with indistinct brown wing markings

1st winter (Jul–Feb)

short to moderate primary projection

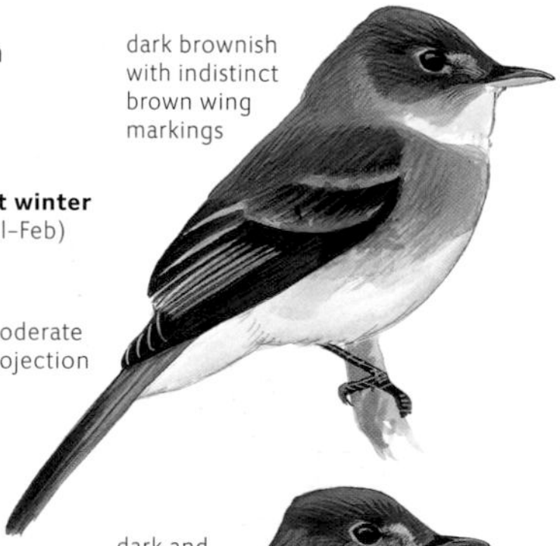

dark and brownish

Adult

little or no eyering

narrow, pale loral-line

1st winter (Jul–Feb)

brownish-olive with contrasting white throat

fairly contrasting wing markings

moderate primary projection

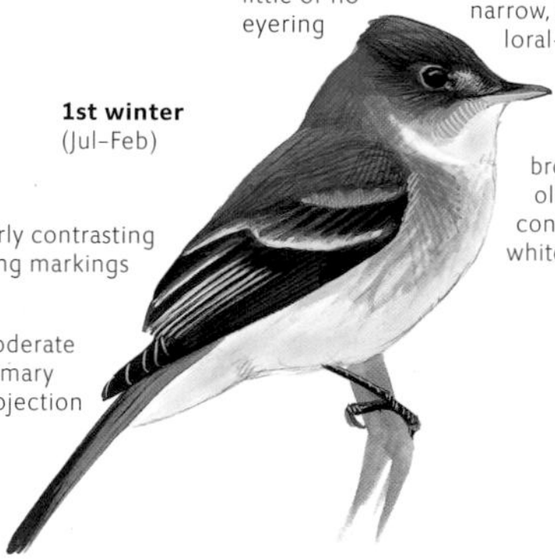

slightly paler than Western, more gray and olive

Adult

VOICE: Song a harsh, burry *RITZbew* or *RRRITZbeyew* or *rrrEEP-yew,* often alternating among these variations: sometimes merely a strong *rrrrrIP,* rough and low with emphatic rising ending (similar calls of Alder are not sharply rising); also *zweeoo, churr,* and *churr-weeoo,* suggesting Alder. Call a thick, liquid *whit* unlike Alder; averages lower and fuller-sounding than similar calls of other *Empidonax.*

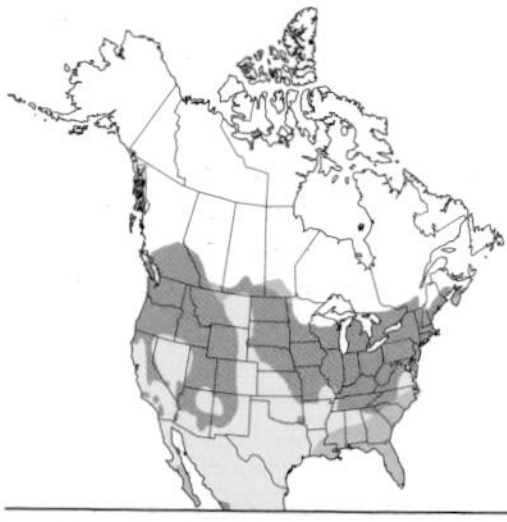

Common in low brushy vegetation in wet areas, in the West especially in riparian willow thickets. Solitary. Forages for small flying insects along brushy areas and sings from bush tops. Nest a neat cup close to the ground among vertical stems.

\lder Flycatcher

'mpidonax alnorum

5.75" WS 8.5" WT 0.47 oz (13.5 g)

ogether with Willow Flycatcher our largest *Empidonax* fly-atcher. Essentially identical to Willow in appearance; best dentified by song. Willow and Alder were formerly considered single species (Traill's Flycatcher).

'OICE: Song a harsh, burry *rreeBEEa*; sometimes merely an scending *rrreep* or *rrreeea* similar to Willow but not as sharp or ising. Also low, clear, whistled *pew* and *peewi*. Call a flat *pip* eminiscent of single note of Olive-sided, unlike the *whit* of Villow and other *Empidonax*.

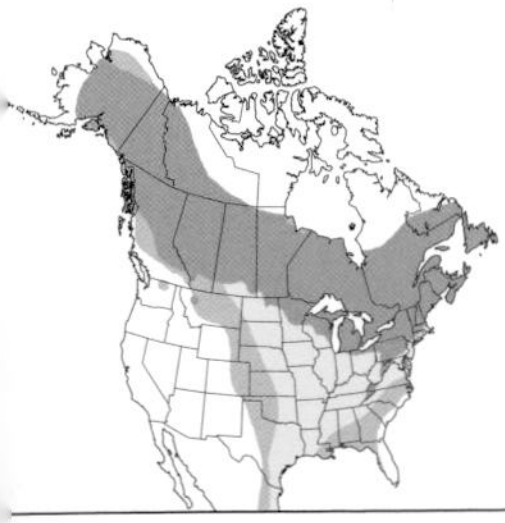

Common locally in low brushy vegetation in wet areas, such as alder thickets surrounding bogs or marshes. Solitary. Forages for small flying insects in low brush and hedgerows. Nest like Willow but shaggier.

Least Flycatcher

Empidonax minimus

L 5.25" WS 7.75" WT 0.36 oz (10.3 g)

Relatively small, with short bill and wings, bold complete eyering, and overall pale grayish-olive color. In comparison to Hammond's and Dusky, Least shows slightly more contrasting and colorful plumage. Best identified by song.

VOICE: Song an emphatic, dry *CHEbek* or *cheBIK* repeated rapidly. Call a sharp, dry *pwit* or *pit*; shorter, drier, and sharper than Willow. Also gives high *wees wees wees*... interspersed with *pit* notes.

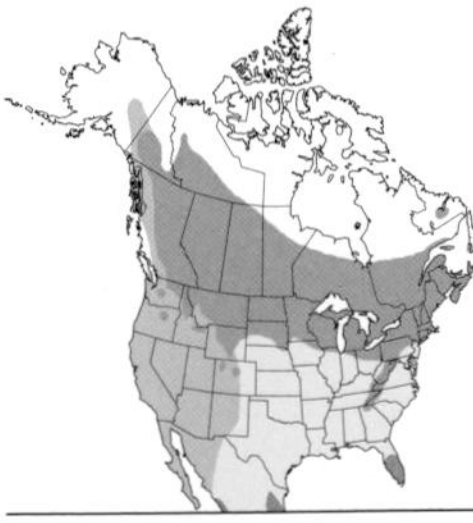

Common in mature deciduous forests with brushy understory; migrants found in brushy woodland edges and thickets. Solitary. Forages for small flying insects in low to middle levels of trees and at forest edges. Nest a neat cup usually on horizontal branch.

Hammond's Flycatcher

Empidonax hammondii

L 5.5" WS 8.75" WT 0.35 oz (10 g)

A small *Empidonax* with short bill, long wings, and short tail. Very similar to Dusky Flycatcher, but differs slightly in proportions, and significantly in habitat choice. Like all *Empidonax* flycatchers, best distinguished by voice.

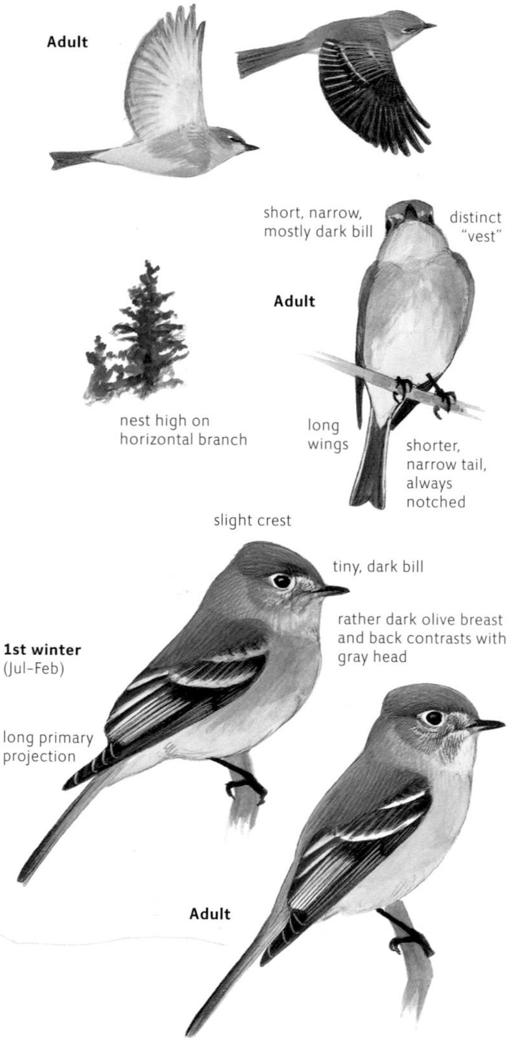

VOICE: Song of three phrases (usually in this sequence): *tsi-pik* suggesting Least; *swi-vrk* high then scratchy; *grr-vik* lower and rougher than any phrase of Dusky. Phrases more strongly two-syllabled than Dusky; never includes high, clear notes. Call a sharp *peek*; sharper and higher than Alder. Also a low, whistled *weew*.

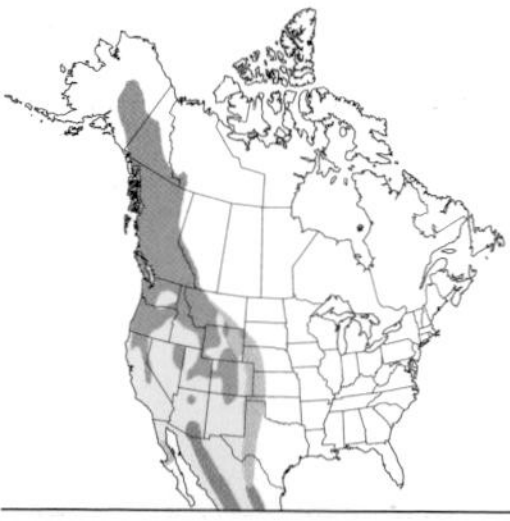

Common in coniferous or mixed forests, usually perching high in tall trees. Generally chooses mature coniferous forest, and forages high in trees and within the canopy. Dusky chooses brushy second-growth and edges of clearings, forages lower and along edges.

Dusky Flycatcher

Empidonax oberholseri

L 5.75" WS 8.25" WT 0.36 oz (10.3 g)

Gray-olive overall, small-billed, and short-winged. Similar t Least and Hammond's Flycatchers, but averages slightly longer tailed, rounder-headed; plumage averages less contrasting tha Least; best distinguished by voice.

VOICE: Song of three phrases (usually in this sequence): a short high, quick *sibip*; a rough, nasal *quwerrrp*; a clear, high *psuwee* similar to Pacific-slope Flycatcher position note. Call a soft, dr *whit* or fuller *twip* similar to Willow, Least, and Gray. Male ca a soft, plaintive whistle *deew* or *dew-hidi* repeated at regula intervals.

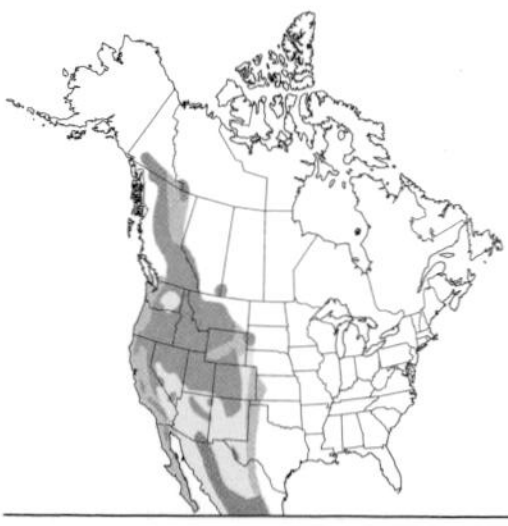

Common in brushy patches within open forests or in forest clearings. Usually solitary. Forages for small flying insects.

Gray Flycatcher

mpidonax wrightii

6" WS 8.75" WT 0.44 oz (12.5 g)

, relatively large and long-tailed *Empidonax* flycatcher. Note ong, narrow bill with yellowish lower mandible and small, harply defined dark tip. Phoebe-like habit of wagging tail geny downward distinctive.

OICE: Song of two phrases: mostly a rough, emphatic *jr-vrip* (or milar rough *jhrrr*) with a high, whistled *tidoo* interspersed; the wo phrases are so different they seem unrelated. Also gives brupt *chivip* and low, whistled *weew*. Common call a sharp, dry *hit* similar to Dusky, Least, and Willow Flycatchers.

Common in sagebrush and similar arid brushy habitats of well-spaced low shrubs. In migration and winter also found in brushy thickets and edges, but usually low and in more open habitat than other *Empidonax*.

Buff-breasted Flycatcher

Empidonax fulvifrons

L 5" WS 7.5" WT 0.28 oz (8 g)

Our smallest *Empidonax* flycatcher, with very short bill and short tail. Rich orange-buff color on breast and neck distinctive, although it can be faded in summer; the only similar species is the very rare Tufted Flycatcher.

VOICE: Song of two similar phrases: sharp and musical, often alternating *PIdew, piDEW, PIdew, piDEW*...; also gives rolling *prrrew* or *pijrr*. Common call a hard, dry *pit* sharper and higher than the *whit* calls of other *Empidonax*.

Uncommon and very local in dry open pine forests with patches of bushes; mainly in Huachuca Mountains of Arizona. Nest on horizontal branch near trunk of tree.

Black Phoebe

Sayornis nigricans

L 7" WS 11" WT 0.67 oz (19 g)

A medium-size flycatcher, active and conspicuous, with long tail strongly wagged (similar to Eastern Phoebe); blackish head and dark gray back and breast with contrasting white belly unlike any other flycatcher, may recall juncos.

VOICE: Song of two high, thin, whistled phrases usually alternated *sisee, sitsew, sisee, sitsew...*; also gives high, thin, whistled *tseew* year-round. Common call a simple, high, clear chip like Eastern but flatter and more whistled.

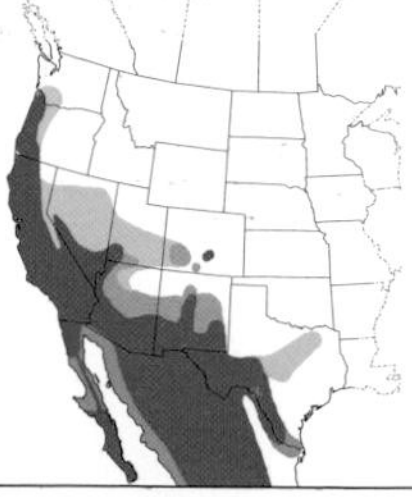

Uncommon along streams and ponds with marshy vegetation and overhanging trees. Always on low open perch; even perches on rocks in streams. Nest of all phoebes a mud cup under eaves, bridges, and the like.

Eastern Phoebe

Sayornis phoebe

L 7" WS 10.5" WT 0.7 oz (20 g)

Larger than pewees and *Empidonax* flycatchers, with dark face weak wingbars, dark tail constantly wagged. Juvenile in fall show buff wingbars and weak eyering, distinguished from other fly catchers by size, dark head, tail-wagging.

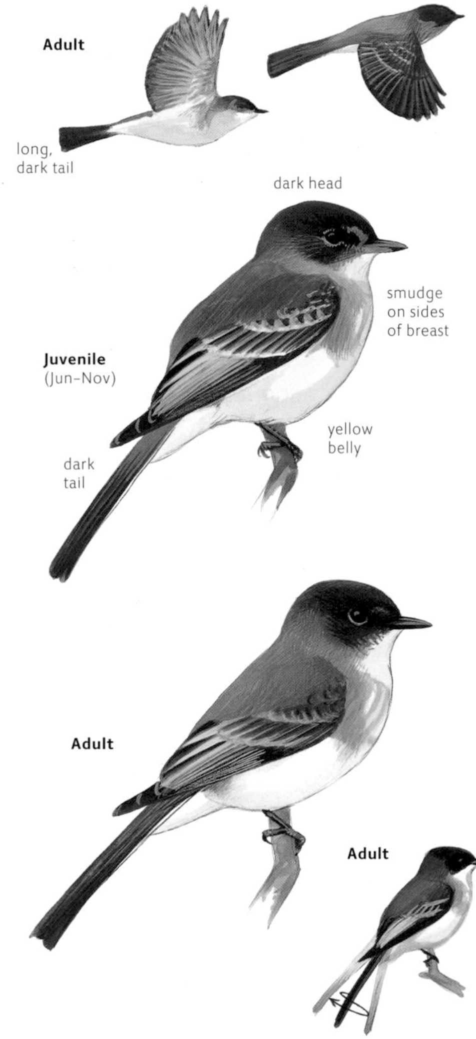

VOICE: Song of two rough, whistled phrases usually alternate *seeeriddip, seebrrr, seeeriddip, seebrrr...*; also gives clear, whis tled *weew* or *tiboo* and abrupt *wijik* year-round. Common call distinctive simple *chip*: high, clear, and descending (sharpe than Black).

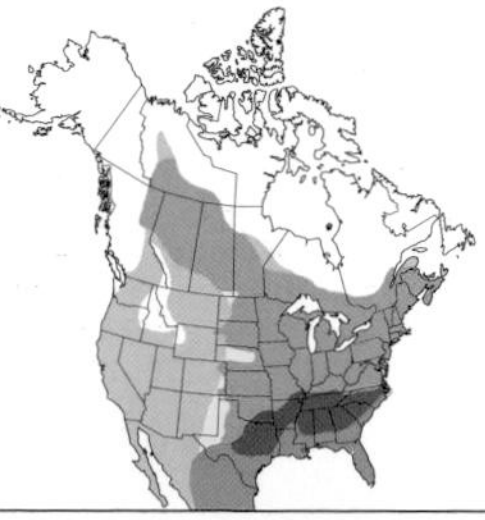

Uncommon in open woods and woodland edges. Nests almost exclusively on manmade structures such as bridges and buildings, and often found along streams or near houses. Solitary. Forages mainly for smal flying insects from low perch; may perch on open fences, rock branches, and the like. When perched constantly wags tail in characteristic motion.

Say's Phoebe

ayornis saya

7.5" WS 13" WT 0.74 oz (21 g)

Overall pale gray and rust colored. Slightly more slender shape than other phoebes. Note pale gray wings, pale rufous belly. In flight sharp-edged black tail is conspicuous. Black tail can suggest Western Kingbird, but note smaller size and tail-wagging behavior.

VOICE: Song of two relatively low, whistled phrases, usually alternated *pidiweew, pidireep, pidiweew, pidireep....* Common call low, plaintive, clear whistle *pdeeer* or *tueeee*.

Uncommon in expansive open areas such as prairies, tundra, farmland, and playing fields. In summer associated with nesting sites in crevices on cliffs, rock outcrops, buildings, bridges, etc. Habits similar to other phoebes, but found in more open settings, sometimes hovers when foraging over open fields.

Vermilion Flycatcher

Pyrocephalus rubinus

L 6" WS 10" WT 0.51 oz (14.5 g)

Adult male unmistakable with brilliant red body and crown. On female and immature note small size, dark mask, short dark tail, streaked breast, and pinkish or yellowish belly. Dips and spreads tail like a phoebe.

VOICE: Song of high, sharp, flat notes and a higher, rolling trill, falling at end *pit pit pit pidddrrrreedrr*; given from prominent perch or in flopping song flight with breast expanded. Common call a high, sharp *pees*.

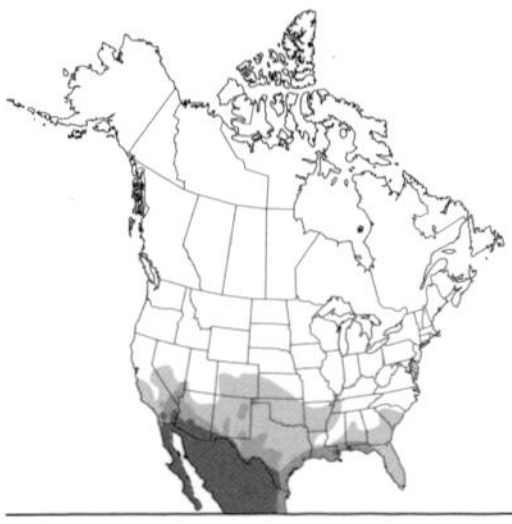

Uncommon and local around wet oases in desert habitat and in diverse mixes of trees, brush, and grassy openings near open water. Forages for small flying insects from open perch, such as a small tree or fence. Solitary.

Dusky-capped Flycatcher

Myiarchus tuberculifer

L 7.25" WS 10" WT 0.7 oz (20 g)

Smaller than Ash-throated Flycatcher; distinguished by relatively thin bill, slightly brighter yellow and darker gray underside, browner head and especially limited rufous in tail.

VOICE: Song *pidi pew pew pedrrrrrrrrr* with long trill or an impatient, rhythmic *pididi-peeeeeer* or *"I'm over here"*; higher-pitched than other *Myiarchus*. Call prolonged, soft whistles: a melancholy *hweeeeeew* or *pwE-Deeeeeeeeew*. Also a sharp *whit* like other *Myiarchus* flycatchers and a rapid, complaining *treedr treerdr teeer teer*.

Common locally in shady oak woods in mountain canyons, generally found at slightly higher elevation and in more heavily wooded habitat than Ash-throated. Habits similar to Ash-throated Flycatcher.

Ash-throated Flycatcher

Myiarchus cinerascens

L 8.5" WS 12" WT 0.95 oz (27 g)

Slightly smaller than Great Crested and Brown-crested Flycatchers, larger than Dusky-capped. Note very pale color of underparts (but other species in late summer can be equally pale), distinctive dark tip on undertail of adult, and voice.

VOICE: Song a series of repeated phrases *kibrr, kibrr...*, or short musical *kaBRIK*. All calls have low, flat quality; mostly short abrupt, two-syllable phrases; always higher-pitched and lighter than Brown-crested; never raucous level notes. Common call sharp *bik*; also *ki-brrnk-brr*. In fall and winter simply a soft *prrt*.

Common in brushy and lightly wooded habitats, generally at low elevations and ranges into drier habitats than other *Myiarchus* in the Southwest. Somewhat secretive; perches on twigs within thickets and trees, rarely in open. Solitary or in pairs. Feeds on insects and berries.

3rown-crested Flycatcher

1yiarchus tyrannulus

8.75" WS 13" WT 1.5 oz (44 g)

abits and appearance similar to Ash-throated Flycatcher; best istinguished by larger bill, tail pattern, and voice. Distinguished om very similar Great Crested by paler underparts, weak pale dges on tertials, and voice.

ʘICE: Song of low, rolling, alternating phrases *prEErrr-prdrdrrr, ·rrp-didider*... lower than Ash-throated; low *trrrp*, strong *kreep* ɜss rough and more nasal than Great Crested. Loud, raucous, escending *keeerp* often in series (e.g., *keeerp keeerp keeerp reeek brit brit bik*...). Low, strong, liquid *whip* often followed by rolling *brrrg*.

Common locally. Nesting habitat overlaps broadly with Ash-throated and Dusky-capped, but Brown-crested is usually only in areas with larger trees (such as in riparian woods with cottonwoods, sycamores, or willows) or desert washes where saguaro cactus provide old woodpecker holes for nesting.

Great Crested Flycatcher

Myiarchus crinitus

L 8.75" WS 13" WT 1.2 oz (34 g)

The most richly colored of the *Myiarchus* flycatchers, with gray head, olive back, rufous wings and tail, and bright yellow belly; also note contrasting blackish and white tertials. Yellow belly extends farther forward than on other *Myiarchus*, merging with gray breast in olive patch on sides of breast.

VOICE: Song of alternating phrases *quitta, queeto, quitta*.... Distinctive call a strong, clear, rising *wheeeep* or *queEEEEP.* Other calls include a very rough, level *KRREEEP*; lower, softer *krrrriip*; low, sharp, dry *kwip*; all often combined into excited series, e.g., *KRREEEP, KRREEP, kwip-kwip-kwip-kwip-kwip kweep kweep, krrriip*.

Common in mature deciduous forests. Solitary or in pairs. Forages for insects and berries from perch within middle to upper levels of trees, rarely in open. A large and somewhat secretive flycatcher.

Nutting's Flycatcher

Myiarchus nuttingi

L 8" WS 11.5" WT 0.75 oz (21g)

Very similar to Ash-throated Flycatcher. Slightly smaller and stockier with brighter yellow belly, rufous edges on secondaries, and lacks dark tips on tail feathers, but these features matched by juvenile Ash-throated. Best distinguished by voice.

VOICE: Generally clearer and sharper than Ash-throated, most distinctive is a clear rising whistle *weeep*.

Very rare visitor from Mexico, recorded in Arizona, New Mexico, and southern California. Found in brushy thickets of mesquite or other low trees or shrubs, often with Ash-throated Flycatcher.

La Sagra's Flycatcher

Myiarchus sagrae

L 7.25" WS 10.5" WT 0.63 oz (18 g)

Habits and appearance similar to other *Myiarchus*, smalle with little rufous in tail or wings and nearly white undersid Distinguished from Ash-throated Flycatcher by bushy crest, lon thin bill, lack of rufous in tail, and voice. Superficially like Easter Phoebe.

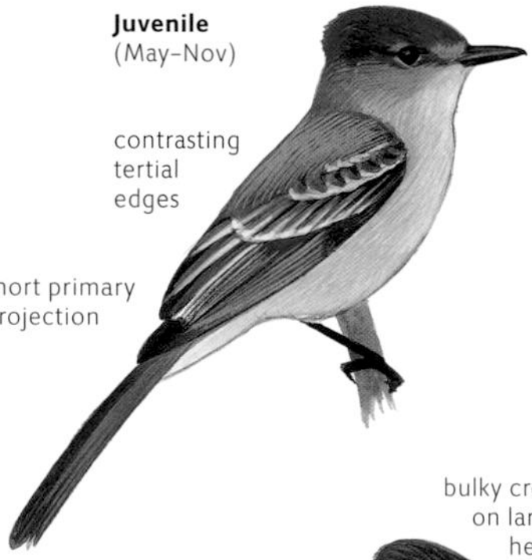

VOICE: Call a high, rising, clear, whistle *weeek* or *weeeit* lik high-pitched Great Crested; a series becoming lower an rougher *weeek weeep kreeep kreep krip*; a clear *whidip* an sharp, buzzy *keee* or *keew*. Also *weep wida-weer* like Dusky capped but higher and more rapid.

Very rare visitor from Bahamas t brushy woods of native hardwood trees and shrubs in southeastern Florida; one or tw recorded annually.

›cissor-tailed Flycatcher

'yrannus forficatus

10" (adult ♂ to 15") WS 15" WT 1.5 oz (43 g)

ery pale gray head and back with long black and white tail dis-nctive. Even immatures have noticeably long tail with extensive ›hite edges, otherwise similar to Western Kingbird, but note aler gray head and back.

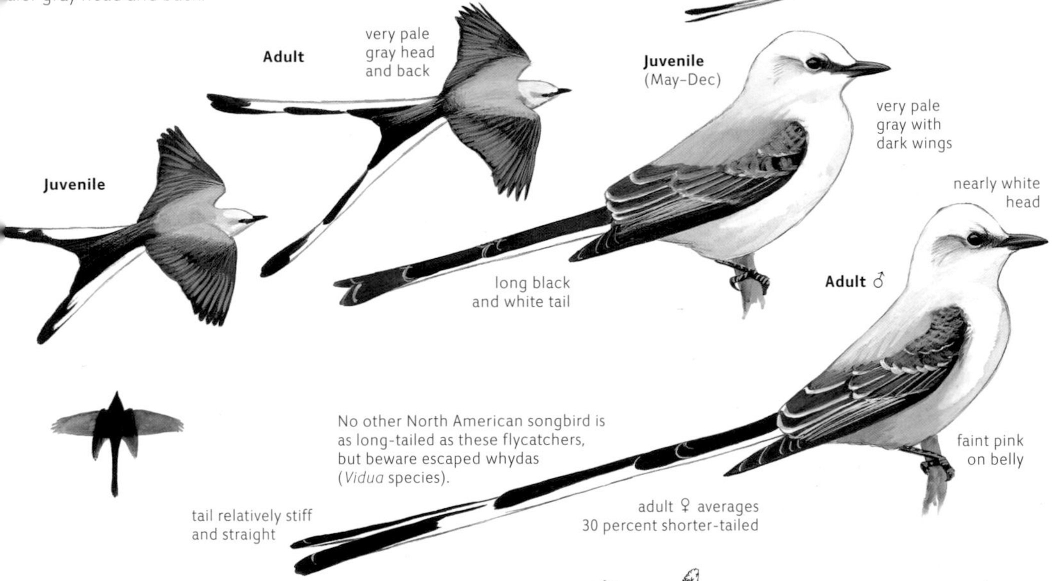

OICE: Song lower-pitched and flatter than Western Kingbird, ›ith slightly different pattern *pidik pek pik pik pidEEK.* Dawn ›ng *pup pup pup pup pup perek.* Common call a relatively low, at *pik*; lower than Western Kingbird. Also *pik-prrr* or *kopik* or igher, sharper *kid.*

Common in prairies and open fields with scattered trees and bushes, often along fences or overhead wires. Often in pairs or small groups; migrants travel during day in small flocks. Feeds on flying insects, often just above the ground.

›ork-tailed Flycatcher

'yrannus savana

10" (adult ♂ to 16") WS 14" WT 1 oz (29 g)

›maller than kingbirds or Scissor-tailed Flycatcher, with very long ll-black tail. Overall plumage similar to Eastern Kingbird but paler ray on the back, with mostly white breast and underwing coverts.

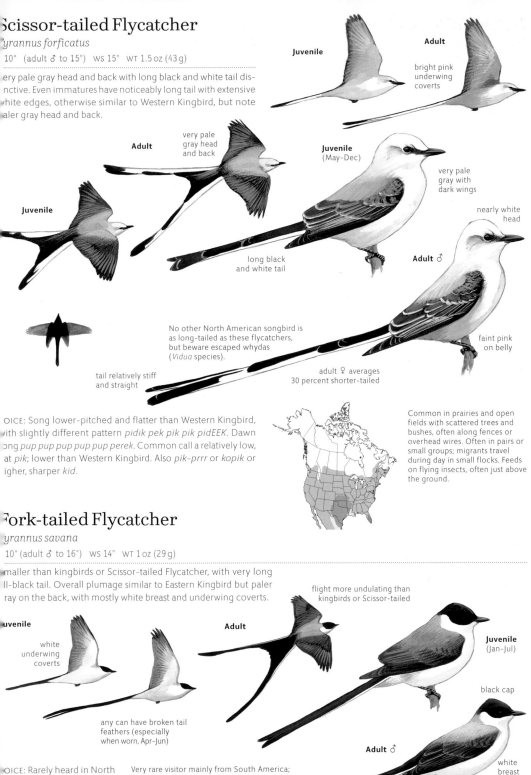

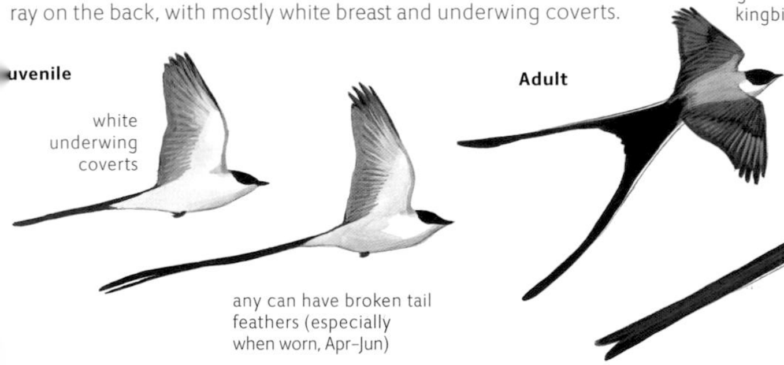

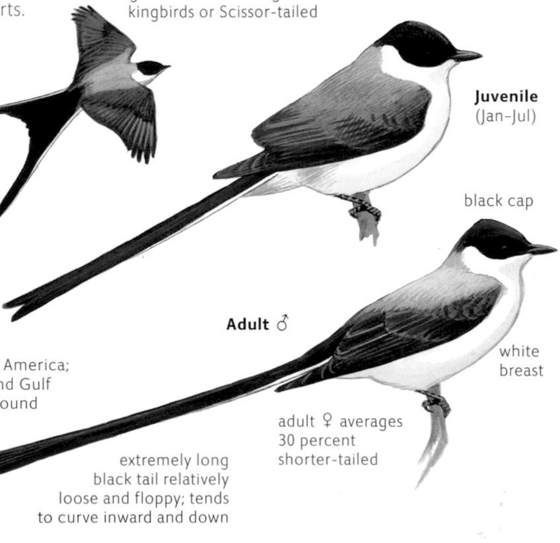

OICE: Rarely heard in North ›merica. Call of very high, ard chips and twittering, taccato, and hummingbird-›ke *tik tik krkrkr....*

Very rare visitor mainly from South America; recorded annually along Atlantic and Gulf coasts, and very rarely elsewhere. Found in any open habitat such as fields, dunes, and marshes. Habits kingbird-like.

Thick-billed Kingbird

Tyrannus crassirostris

L 5" WS 7.25" WT 0.46 oz (13 g)

Large and thick-necked, with very large bill. Uniform dark gray-brown above, with all-dark tail and dark mask; pale yellow or whitish below.

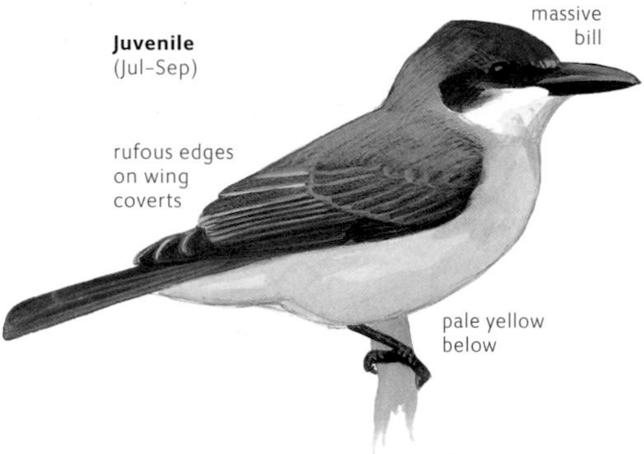

VOICE: Call a buzzy or dry, metallic *tzwee-eerrr* or *to to to zweeerrr.* Also a dry, clicking *ket ket-ket ket...* or a buzzy, nasal *kotoREEEF,* sharply ascending, squealed, burry, with emphatic questioning end.

Uncommon and local at a few locations in southeastern Arizona. Found in lowland cottonwoods along permanent streams, where it perches on exposed twigs. Winter vagrants in California are found in city parks with scattered large trees.

Loggerhead Kingbird

Tyrannus caudifasciatus

L 9" WS 14" WT 1.5 oz (43g)

Similar to Eastern Kingbird, but with longer bill and large bush crest, more brownish color above and clean white breast. Fro Gray Kingbird by thinner bill, blackish crown, pale tip on tail. Calls also differ.

VOICE: Call a sharp single note *kip,* similar to some other kingbirds; a squeaky, sputtering rattle *squeer* slightly descending, often combined with low *put* notes in series *squeer-pu-pu-pu-squeer.*

Very rare visitor from West Indie with several records in extreme southern Florida. Records to dat are of the Cuban subspecies. No as conspicuous as other kingbirds, perching low and often foraging within foliage.

;ray Kingbird

yrannus dominicensis

)" WS 14" WT 1.5 oz (44 g)

ıperficially similar to Eastern Kingbird, but larger and paler; ith longer bill, longer notched tail, and broader wings; also ɔte paler gray crown, whitish underwing coverts, and lack of hite tip on tail.

ᴏɪᴄᴇ: Common call a high, sharp, twittering chatter *tik-teeerr* r *preeerr-krrr*; similar to Tropical Kingbird but fuller and ɔugher.

Common locally among mangroves and other dense vegetation near coast. Often seen perched on wires, treetops, and other prominent lookouts. Habits similar to Western Kingbird.

Eastern Kingbird

Tyrannus tyrannus

L 8.5" WS 15" WT 1.4 oz (40 g)

A relatively large and conspicuous flycatcher, easily identified by plain gray back and clean white underparts, sharp contrast between dark cap and white throat, and bold white tip on black tail.

VOICE: Song of sharp, electric, rasping or sputtering notes in series ending with emphatic, descending buzz *kdik kdik kdik PIKa PIKa PIKa kzeeeer*; elements often given separately. Most frequently heard call a sharp, buzzy *kzeer.* Dawn song a high, rapid, electric rattling building to crescendo *kiu kitttttttttttttiu ditide.*

Common in semi-open habitats with mix of grassy fields and trees, such as orchards or marsh edges; often near water. Usually solitary or in small family groups; migrants may gather in hundreds. Perches on any prominent open perch from which to watch for passing insects.

Tropical Kingbird

Tyrannus melancholicus

L 9.25" WS 14.5" WT 1.4 oz (40 g)

Habits and appearance similar to Western Kingbird; larger and longer-billed with brighter yellow belly and breast, greener back, and notched brownish tail without white sides (some worn Westerns lack white tail sides). Virtually identical to Couch's Kingbird and reliably distinguished only by voice.

VOICE: Song has sharp, metallic quality; rapid, rising, tinkling *tr tr twididīd twididid twididi*. Common call of all populations a high, sputtering twitter of sharp, metallic notes *twitrrr* or *tzitz-itzitzi*; never a single note like Couch's.

Uncommon and local in its limited range. Found in open areas with some large trees, such as golf courses and city parks; often near water. Often in pairs or small groups. Vagrants are found in open areas where they perch on fences, treetops, or wires.

Couch's Kingbird

Tyrannus couchii

L 9.25" WS 15.5" WT 1.5 oz (43 g)

Slightly larger and more brightly colored than Western; be distinguished by longer bill, brighter yellow belly, and browni (not black) tail without white outer edge. Nearly identical Tropical Kingbird.

VOICE: Song of high, thin, nasal squeals *towi towi toWITItoo* *gewit gewit geWEETyo*. Dawn song a series of long, slow phras *pleerrr pleerrr pleerrr plity plity plity plity chew*. Common ca a high, sharp *dik* (like Western) and an insect-like, trailing bu *kweeeerz* or *dik dik dikweeeerz*; also a sharp, dry *ch-eek*.

Uncommon in open areas with trees and often with dense brus such as river floodplains. Often in pairs or small groups. Habits similar to Western Kingbird.

assin's Kingbird

rannus vociferans

" WS 16" WT 1.6 oz (46 g)

bits and overall appearance similar to Western Kingbird; best stinguished by darker gray head and breast setting off small hite throat-patch and by dark tail with indistinct pale tip. ote also that upperwing coverts are paler than back (darker Western).

ICE: Song of clear, nasal notes and hoarse *churr* notes rising crescendo *teew, teew, teew tewdi tidadidew.* Call a husky *HI-Vrrrr.* Also clearer, nasal *keew* notes in series; rough, burry, scending *ch-queeer*; and nasal *gdeerr-gdeerr...* in rapid series up to ten notes, lower and more nasal than Western.

Locally common where tall trees are mixed with large open grassy spaces. In nesting season can be found at higher elevation than other kingbirds, in oak savanna, clearings in Ponderosa Pine forest, as well as tall cottonwoods along streams.

Western Kingbird

Tyrannus verticalis

L 8.75" WS 15.5" WT 1.4 oz (40 g)

Appearance quite different from Eastern Kingbird: longer wings and tail, more robin-like flight, and very different plumage. Note yellow belly, pale gray breast and head, and black tail with narrow white sides.

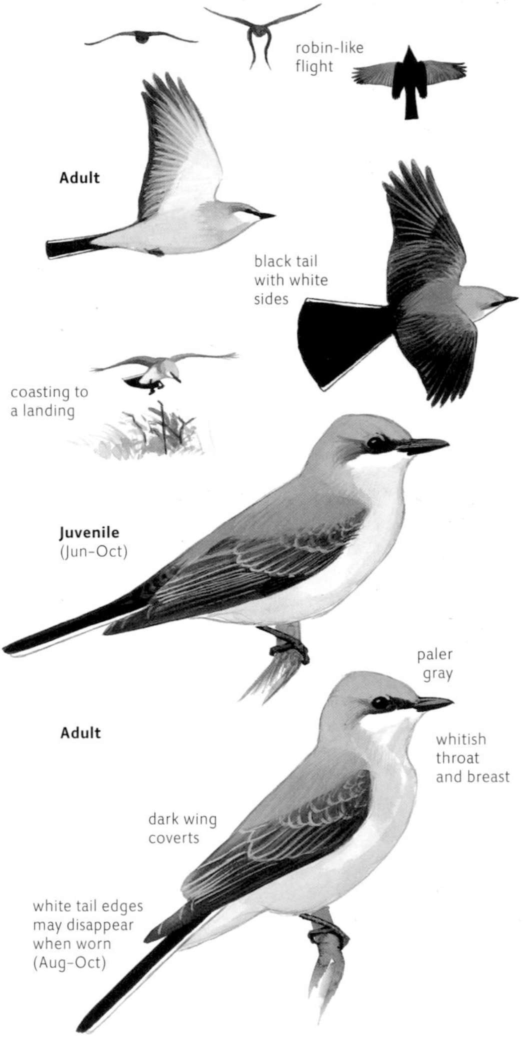

VOICE: Song high, hard, squeaky *pidik pik pidik PEEKado.* Calls lower-pitched than Eastern; higher and clearer than Cassin's. Calls include a rapid, rising series of shrill, sputtering notes *widik pik widi pik pik pik*; very hard, sharp *kit* often in accelerating rising series or long, sputtering series; also a lower series *kdew kdew kdew kdew.*

Common in open habitats with scattered trees or hedgerows, such as prairies and farmland, where it perches conspicuously on wires or exposed twigs, flying out to catch passing insects. Solitary or in small family groups.

Piratic Flycatcher

Legatus leucophaius

L 5.8" WS 9.2" WT 0.9 oz (26g)

Phoebe-sized, with bold head pattern. Smaller than Variegated Flycatcher, with shorter and slightly downcurved bill, reduced white edges on wing coverts, no streaking on back, less rufous on uppertail coverts, and slightly broader dark mask.

VOICE: Vagrants are generally silent.

Very rare visitor from Central and South America, recorded mainly in fall from Texas and New Mexico, one spring record in Florida. Found in open woods and edges, foraging for insects and fruit at middle to upper levels of trees.

Variegated Flycatcher

Empidonomus varius

L 7.1" WS 10.5" WT 0.9 oz (26g)

Bold head pattern like Sulphur-bellied Flycatcher, but sligh smaller, with small bill, weak and blurry streaking on underpar mostly dark tail with rufous edges. Also compare smaller a plainer Piratic.

VOICE: Vagrants are generally silent.

Very rare visitor from South America with records mostly in late fall in Maine, Tennessee, Florida, Washington, and Ontario. Found in open woods and edges, foraging for insects and fruit at middle to upper levels of trees.

ulphur-bellied Flycatcher

yiodynastes luteiventris

.5" WS 14.5" WT 1.6 oz (46 g)

rge, large-headed, and short-tailed. Bold plumage pattern ique: striped head (similar to female Black-headed Grosbeak), atly streaked underparts, and bright rufous tail.

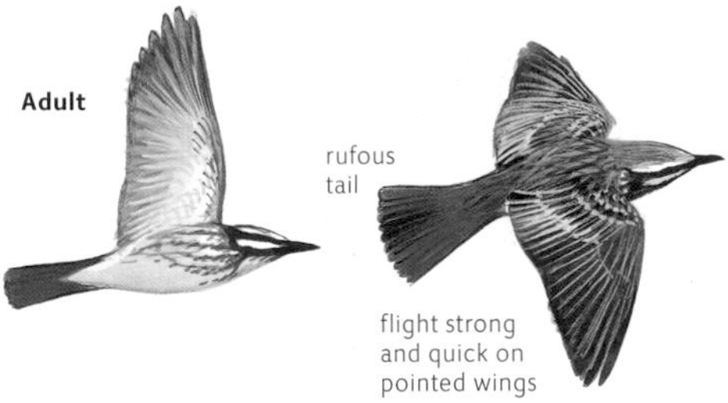

ICE: Common calls high, squeaky, whistled phrases, penetrat- g squeaks like a child's toy *tooWI drrdip* or *wee-dee-yoo*; a rceful *seedeeyee*; a lower, nasal *kweeda kweeda*...; a rather w, hollow, nasal *ket* often in series, suddenly exploding into ud, high squealing.

Uncommon in sycamores along mountain canyons; usually stays within canopy of tall trees. Solitary or in pairs. Feeds on insects and berries. Nests in tree cavity.

Great Kiskadee

Pitangus sulphuratus

L 9.75" WS 15" WT 2.1 oz (60 g)

Relatively large and stocky, with short tail and wings. Striped head, yellow belly, and rufous wings and tail unmistakable. Loud calls and habit of perching conspicuously in small groups in treetops make this species easy to find.

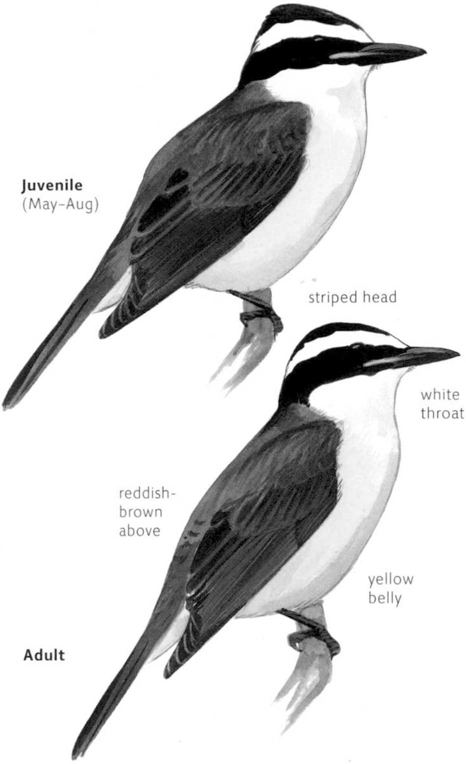

VOICE: Song a raucous, high, repeated *KREEtaperr* ("kiskadee" or "Christopher"); often repeated in chorus with several individuals. All calls loud, clear, raucous, repeated; most common a rather long, level squeal *weee*; also a high, rough *grrt*.

Common in dense woodlands near water. Solitary or in small groups. Perches in trees, sometimes on lower open branches over water. Feeds on insects, berries, fish, lizards, and other prey. Nest, in a tree, is a large ball of twigs and grasses with side entrance.

Northern Beardless-Tyrannulet

Camptostoma imberbe

L 4.5" WS 7" WT 0.26 oz (7.5 g)

Very small. Distinguished from slightly larger *Empidonax* flycatchers by short, blunt-tipped bill, bushy crest. drab pale wingbars, pale eyebrow, and by different habits (more gleaning of prey and less flycatching).

VOICE: Song lazy, clear, piping whistles in descending series usually introduced by short, rising series *piti pi pi PEEE dee dee* or slow series *peeh peeeh peeh peeh peeh peeh peeh* gradually descending. Common call a clear, piping *peeehk*; also a descending *peeewk* and a chuckling *piklkhlk*.

Uncommon. Found in lowland riparian woods, especially dense stands of mesquite mixed with cottonwoods, sycamores, or oaks. Solitary, but may join foraging flocks of kinglets and warblers. Feeds on small insects and larvae gleaned from leaves. Nest globular with side entrance in tree.

White-crested Elaenia

Elaenia albiceps

L 5.7" WS 8.5" WT 0.57 oz (16g)

Differs from *Empidonax* flycatchers in slightly larger size, re tively long and bushy crest. White stripe on center of crov usually obvious, but can be almost entirely concealed when cre is flattened.

VOICE: Call a relatively low clear whistle *wheew*, lower and more relaxed than any North American *Empidonax* call.

Very rare visitor from South America, only one definite record in Texas, but other recor in Illinois, Rhode Island, and Florida could have been this species or the very similar Small-billed Elaenia.

ꞁose-throated Becard

ıchyramphus aglaiae

ꞁ.25" WS 12" WT 1.1 oz (30 g)

stocky, large-headed, short-tailed, and stout-billed songbird. ark cap, plain wings, and unpatterned body distinctive. Not osely related to other flycatchers.

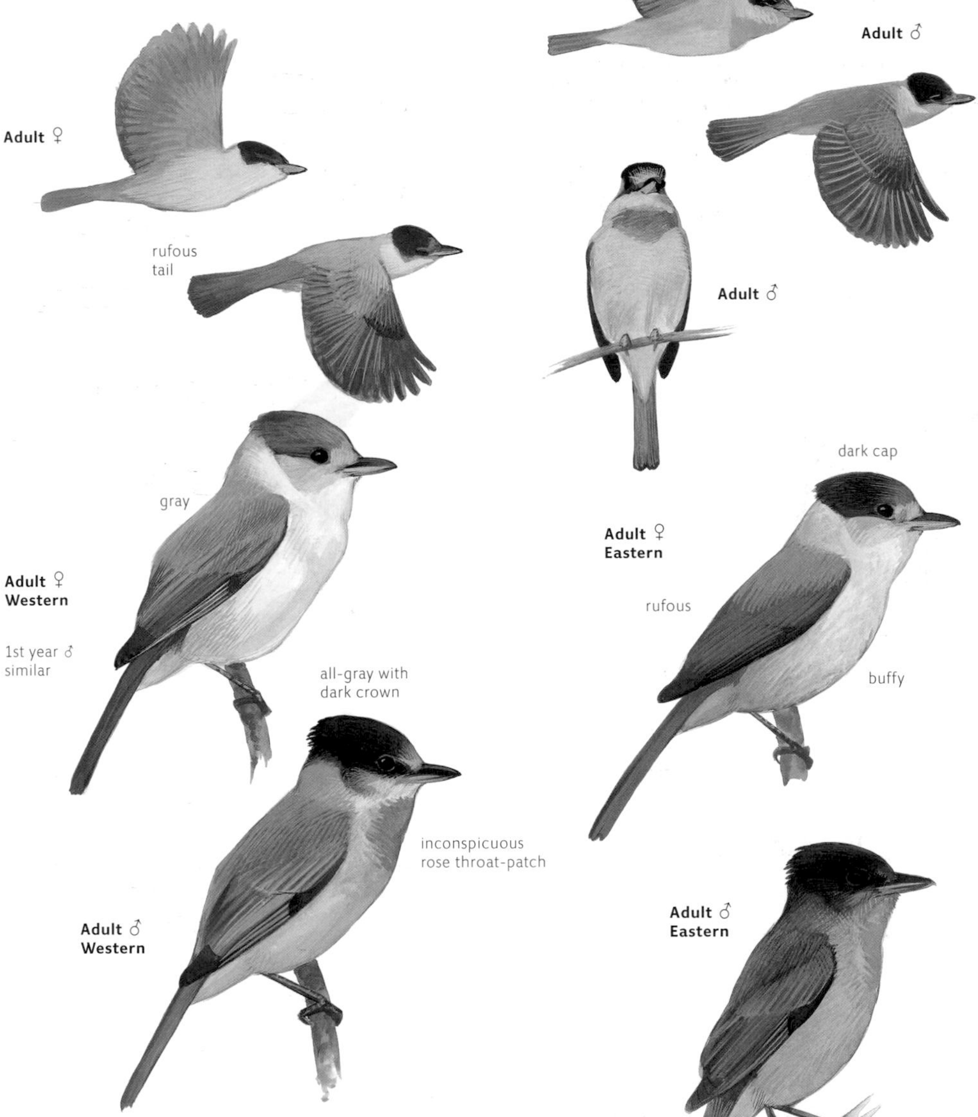

OICE: Common call a very high, thin whistle *sweetsoo* or *eelee*. Also a high, squeaky, nasal chatter often ending with ıin, descending note *kit-kiddleit-ti-ti-teew*; a very high, ringing *eeeeew*.

Rare in tall riparian sycamores, cottonwoods, and mature woods, where it moves through upper branches of trees. Solitary, but may join loose flocks of warblers and kinglets. Forages mainly on fruit gleaned from trees. Nest is large and globular, suspended near tip of long branch.

Birds of the west Mexican subspecies (found in Arizona), smaller and paler overall than east Mexican birds (found in Texas). Note strong contrast between dark cap and pale nape on Western male, and nearly white underparts.

Shrikes

FAMILY: LANIIDAE

3 species in 1 genus. Shrikes are found singly in open brushy fields and hedgerows. They are predatory songbirds with strong, hooked bills, feeding on prey ranging from insects to small animals (including lizards, rodents, and small birds). When foraging, all species perch conspicuously on wires, posts, or treetops. Captured prey may be hung on thorns or twigs for later consumption. Compare Northern Mockingbird, Blue Jay, and American Kestrel. Adults are shown.

Genus *Lanius*

Loggerhead Shrike, page 363

Northern Shrike, page 363

Brown Shrike, page 362

Brown Shrike

Lanius cristatus

L 7.3" WS 10.6" WT 1.3 oz (37g)

Smaller than Loggerhead Shrike, and distinctly brown in a
plumages. Can be mistaken for brown juvenile Northern, b
smaller, with all-brown tail, wings not much darker than back ar
always lacks white patch at base of primaries.

VOICE: Generally silent in winter.

Very rare visitor from Asia to western Alaska (several records) California (four fall and winter records), and Nova Scotia (one fall record). Habits similar to Loggerhead Shrike, but somewhat less bold, perching lower and staying close to dens hedgerows and thickets.

,oggerhead Shrike

anius ludovicianus

9" WS 12" WT 1.7 oz (48 g)

ery similar to Northern Shrike, but smaller with relatively smaller ll; note especially broader black mask. Distinguished from ockingbird by larger head, quicker wingbeats and more direct ght, smaller white wing patch.

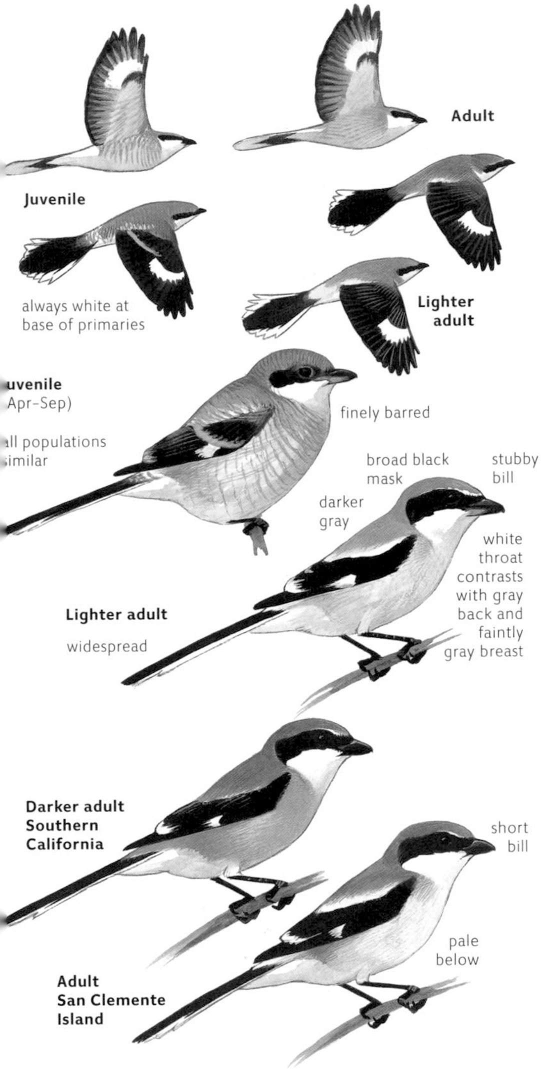

OICE: Song generally sharp, precise, mechanical two-syllable hrases *krrDI* or *JEEuk* (etc.) with the phrase often repeated over nd over at short intervals. Calls include a harsh, scolding *jaaa,* rating *teeen raad raad raad raad raad,* and other variations milar to song phrases.

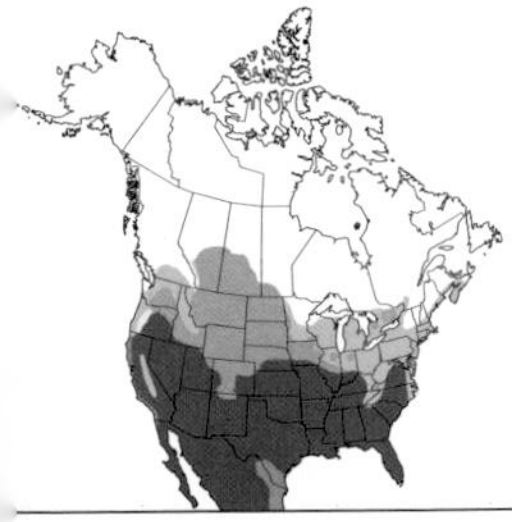

Uncommon and declining in open pastures and prairies with scattered bushes, hedgerows, and trees. Solitary. Feeds on grasshoppers and other insects, small birds, and rodents.

Northern Shrike

Lanius excubitor

L 10" WS 14.5" WT 2.3 oz (65 g)

Very similar to Loggerhead Shrike, but larger; note narrower dark mask with white markings around eye. Immature has much smaller white wing patch than Loggerhead, often strongly brownish color, very weak dark mask.

VOICE: Song of repeated short phrases like Loggerhead, but lower-pitched with mellower, more liquid quality *kdldi* or *plid-plid.* Can sound soft and thrasher-like. Calls include a very nasal, complaining *fay fay...,* a harsher *reed reed reed...,* and a very harsh, dry *shraaaa.*

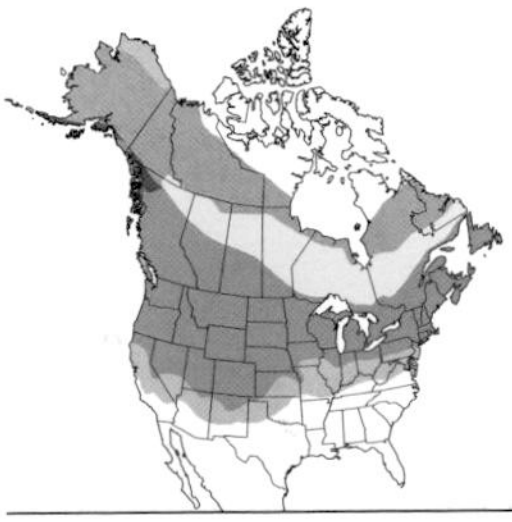

Uncommon to rare. Nests in open spruce woods. Winters in open habitats with scattered bushes or trees. Solitary. Perches on high exposed branch or wire when foraging for small birds or rodents.

Vireos

FAMILY: VIREONIDAE

15 species in 1 genus. Vireos are small, relatively stocky songbirds with blunt-tipped, hooked bills and short, strong legs. Plumage of most species is drab with subtle markings. Found in dense foliage of bushes or trees, vireos feed on fruit, insects and larvae gleaned from twigs and leaves. They habitually fly to a perch and sit still for many seconds (warblers rarely pause longer than or or two seconds). Nest is a neat, compact cup suspended in horizontal fork. Song of most species is simple whistled phrase repeated; several small brush-dwelling species have more complex chattering song. Immatures are shown.

Genus *Vireo*

Bell's Vireo, page 365

Black-capped Vireo, page 366

White-eyed Vireo, page 367

Thick-billed Vireo, page 367

Yellow-throated Vireo, page 369

Plumbeous Vireo, page 368

Cassin's Vireo, page 368

Blue-headed Vireo, page 369

Warbling Vireo, page 370

Philadelphia Vireo, page 370

Hutton's Vireo, page 372

Gray Vireo, page 366

Red-eyed Vireo, page 371

Yellow-green Vireo, page 372

Black-whiskered Vireo, page 371

Bell's Vireo

Vireo bellii

4.75" WS 7" WT 0.3 oz (8.5 g)

very small, active, and plain vireo. Facial markings include faint spectacles and faint eyeline. Smaller and drabber than White-eyed Vireo, with weaker face pattern and only a single white wingbar, also note different tail movements.

Two populations are normally distinguishable in the field, but intermediate birds are found in west Texas. Eastern birds are brighter yellow and olive and shorter-tailed than Western. Eastern birds bob tail like Palm Warbler; Western birds flip longer tail up and sideways like gnatcatchers. Eastern may have brighter blue legs than Western. Western population is subdivided into Arizona and California (latter also called Least: smaller with virtually no yellow tones). There is no known difference in voice.

VOICE: Song husky, chatty, musical *chewede jechewide cheedle eeew*; slight emphasis on ending but basically flat; variations in details, but quality distinctive. Call fairly high, soft, nasal *biiv biiv...* or *chee chee*. Similar call also given singly: nasal, rising *mreee* repeated.

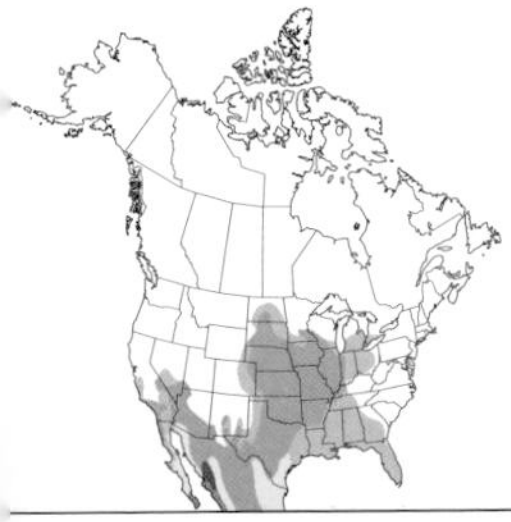

Uncommon and declining. Found especially in willow and mesquite thickets near water; stays low within vegetation. Usually solitary. Feeds on insects and larvae gleaned from leaves.

Gray Vireo

Vireo vicinior

L 5.5" WS 8" WT 0.46 oz (13 g)

Plain gray overall. Often confused with Plumbeous Vireo, but with thin whitish eyering on plain face, only faint wingbars, longer tail often raised and flicked, and different habitat. Also compare smaller Bell's.

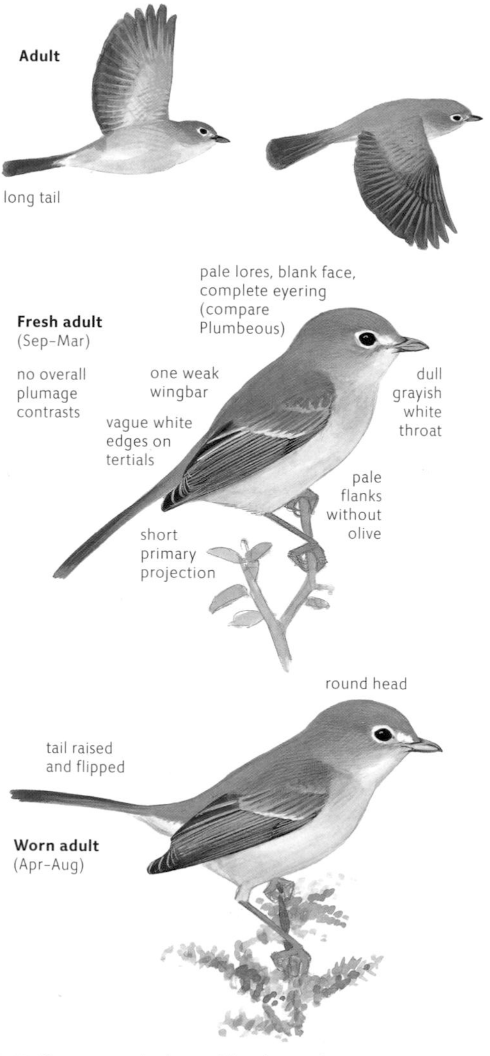

VOICE: Song very similar to Plumbeous but phrases average shorter, simpler, delivered more rapidly (average one phrase every 1.5 sec); less varied *tiree pwideer dew tiree pwideer dew....* Call a low, harsh *charrr* similar to Bewick's Wren; also a short, harsh *chik,* a rapid series of popping whistles, and a harsh descending series.

Uncommon and local on rocky arid hillsides with widely scattered junipers, oaks, or mesquite and patches of bare ground in between; stays within foliage of low shrubs. Solitary. Feeds on insects and larvae gleaned from leaves.

Black-capped Vireo

Vireo atricapilla

L 4.5" WS 7" WT 0.3 oz (8.5 g)

A small active vireo, related to White-eyed but with striking hea pattern, white lores and eyering contrasting cleanly with da gray or black hood. Also note greenish upperside, obvious wing bars, yellowish flanks, and short tail.

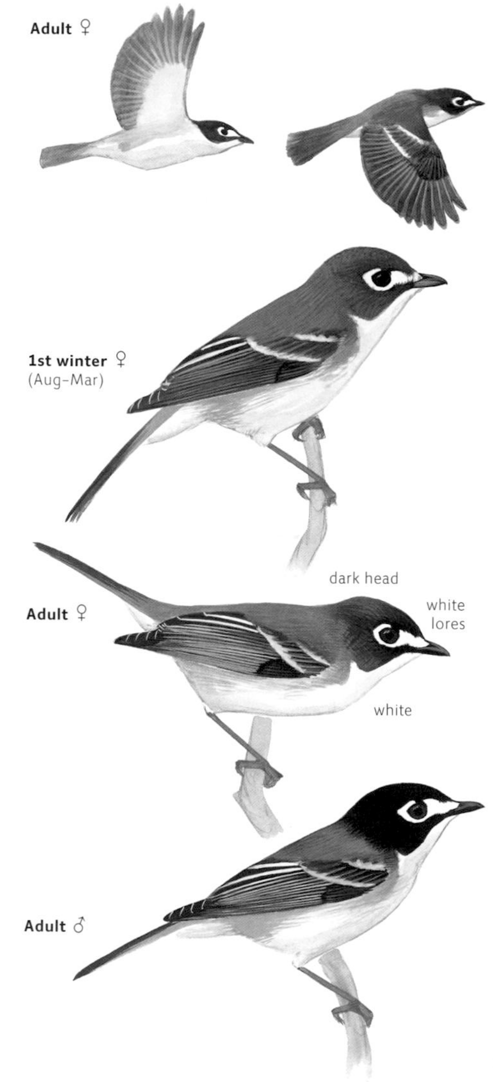

VOICE: Song of well-spaced complex phrases *grrtzeepididic prididzeegrrt...,* with husky chattering quality. Call a long, harsh rising *zhreee* similar to Bewick's Wren scold; many variations. Also a short, dry *tidik* like Ruby-crowned Kinglet.

Uncommon and local. Found on dry hillsides where patches of dense shrubby oaks and junipers are interspersed with open spaces. Usually solitary or in pairs. Feeds on insects and larvae gleaned from leaves.

White-eyed Vireo

ireo griseus

5" WS 7.5" WT 0.4 oz (11.5 g)

small, active, drab grayish-olive bird, often confused with king- ets and warblers. Note bright yellow spectacles on grayish head, ale wingbars, yellowish flanks, and thick bill. White iris of adult distinctive.

VOICE: Song rapid, nasal, harsh; typically begins and/or ends with sharp *chik* notes and includes at least one long, whining note: *tik-a-purrreeer-chik* or *chik-errrr-topikerreerr-chik*; many variations. Call a harsh *meerr* level or slightly descending; commonly a level series *rikrikrikrik rik rik rik rik*; sometimes a single rising *rik*.

Common in dense foliage of shrubs and vines within or along edges of woods; stays low and usually hidden in vegetation. Usually solitary. Feeds mainly on insects and larvae gleaned from leaves. A small secretive vireo; rather stocky and active, often flicks wings open.

Thick-billed Vireo

Vireo crassirostris

L 5" WS 7.25" WT 0.46 oz (13 g)

Very similar to immature White-eyed Vireo (with dark eye); note fairly uniform buffy-yellow underparts, olive head, and different head pattern: eyering white (not yellow), spectacles broken above eye.

VOICE: Song very similar to White-eyed; averages harsher (but given variation in White-eyed this is of limited use). Call a slow series of harsh scold notes; notes longer and series slower than White-eyed.

Very rare visitor from Bahamas; recorded a few times in dense brushy habitat in southern Florida. Habits similar to White-eyed Vireo.

Plumbeous Vireo

Vireo plumbeus

L 5.75" WS 10" WT 0.63 oz (18 g)

Habits, shape, and overall plumage pattern very similar to Blue-headed and Cassin's Vireos; differs only in near total lack of yellow pigment in plumage. Similar to Gray, but with bold white spectacles and wingbars, shorter tail, usually different habitat.

VOICE: Song of rough, burry phrases similar to Yellow-throated and Gray; lower-pitched, rougher, and with slightly shorter phrases on average than Blue-headed or Cassin's; averages one phrase every three seconds. Call a harsh series like Yellow-throated, often ending with rising *zink* note.

Common in dry open pine woods, especially Ponderosa Pine, usually at middle to upper levels of trees. Occasionally seen in low shrubs, especially on migration. Habits like Blue-headed and Cassin's Vireos.

Cassin's Vireo

Vireo cassinii

L 5.5" WS 9.5" WT 0.56 oz (16 g)

Olive and gray with bold white wingbars and spectacles. Brighte individuals are very similar to drab Blue-headed Vireo (but wit less contrast between cheeks and throat), drabbest are ver similar to Plumbeous. Also similar to Hutton's, but larger, wit clean white spectacles and underparts.

VOICE: Song similar to Plumbeous but averages slightly higher-pitched and with more clear phrases, tending toward Blue-headed; averages lower-pitched and more burry than Blue-headed (but much overlap). Calls like Plumbeous and Blue-headed.

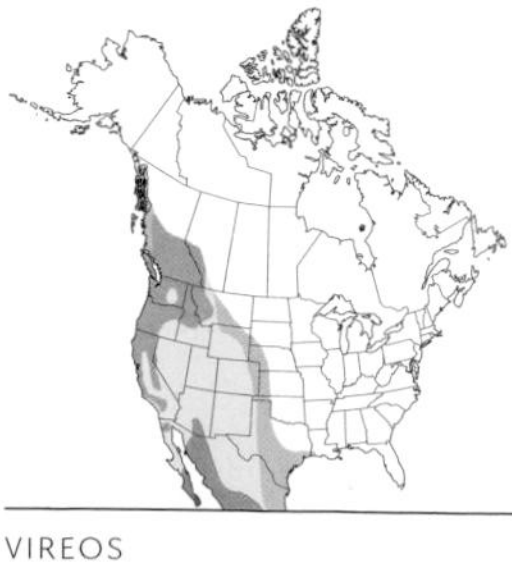

Common in coniferous and mixed woodlands. Habits like Blue-headed and Plumbeous.

Blue-headed Vireo

Vireo solitarius

L 5.5" WS 9.5" WT 0.56 oz (16 g)

Closely related to Cassin's, Plumbeous, and Yellow-throated Vireos. All are relatively large-headed and short-billed with similar voice. Note dark gray head sharply contrasting with white spectacles and throat, bright yellowish flanks, and clean white belly.

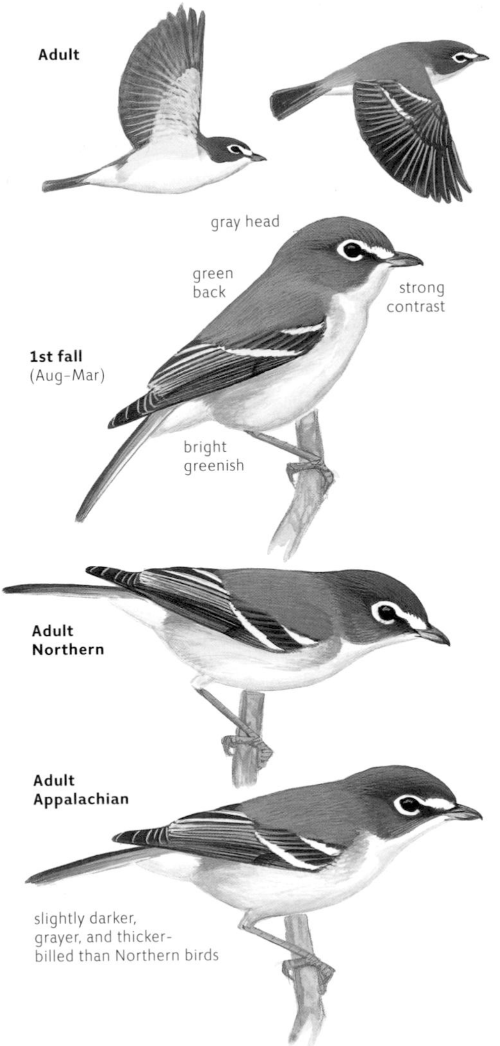

VOICE: Song of high, clear, sweet phrases with slurred notes *see you, cheerio, be-seein-u, so-long, seeya...*; sometimes gives burry phrases, especially in western portion of range. Averages one phrase every 2.5 seconds, slightly slower than Red-eyed. Call a strongly descending series of short harsh notes, like Cassin's and Plumbeous Vireos.

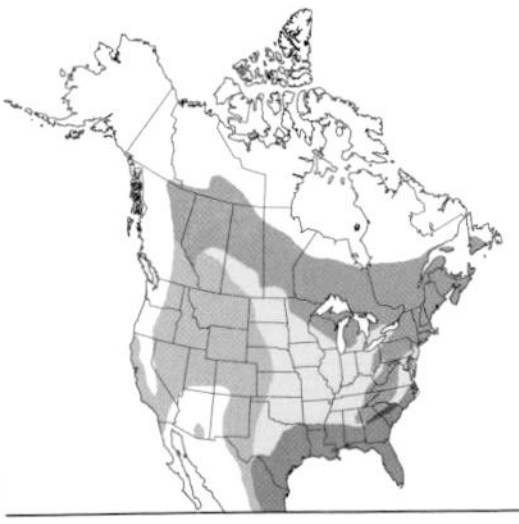

Common to uncommon in mixed forests, usually at middle to upper levels of trees. Solitary. Feeds on insects and larvae gleaned from leaves.

Yellow-throated Vireo

Vireo flavifrons

L 5.5" WS 9.5" WT 0.63 oz (18 g)

A stocky and slow-moving vireo, with bright yellow spectacles and throat. Shape, size, and voice similar to Blue-headed Vireo, but colored very differently. Plumage superficially similar to male Pine Warbler, but note different shape.

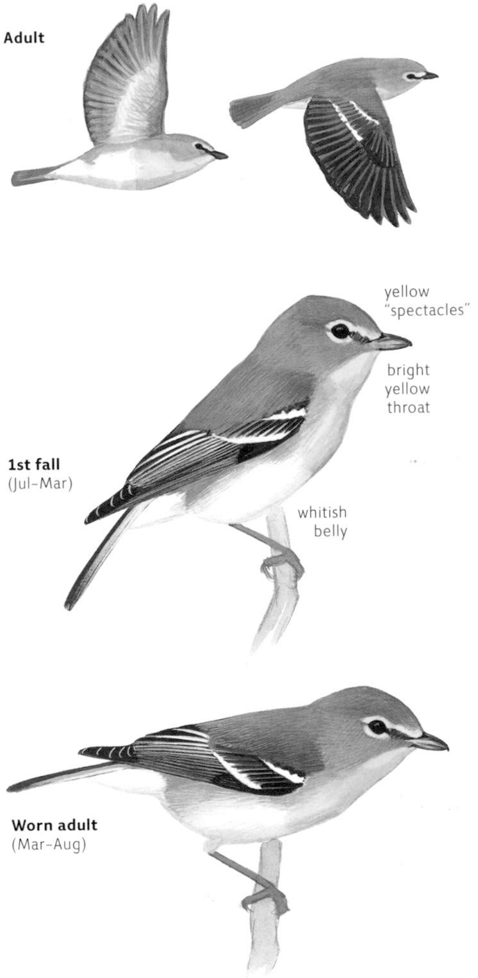

VOICE: Song of short two- or three-syllable phrases, most slurred with a burry quality *rrreeyoo, rreeooee, three-eight...*; averages one phrase every three seconds; distinct from all except Plumbeous. Call of rapid, harsh notes falling or steady *ship shep shep shep shep shep shep shep* very similar to Blue-headed.

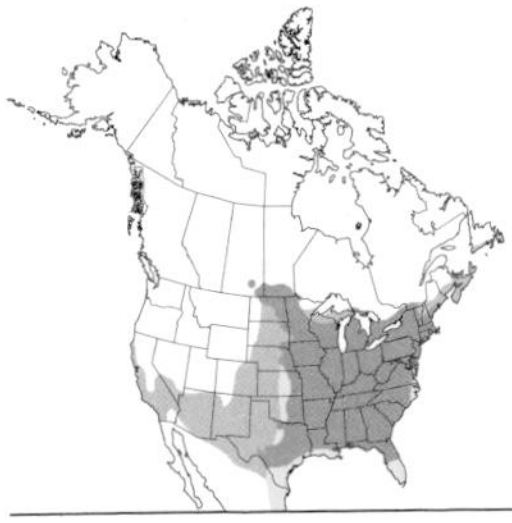

Uncommon in mature deciduous forests, usually at middle to upper level of trees. Solitary. Feeds on insects and larvae gleaned from leaves.

Warbling Vireo

Vireo gilvus

L 5.5" WS 8.5" WT 0.42 oz (12 g)

Plain, pale, and grayish overall, with hints of olive or yellow on back and flanks. Smaller than Red-eyed Vireo, with weaker pale eyebrow and much weaker dark line through eye. Very similar to Philadelphia; note paler lores, less yellow on throat.

Western population differs from Eastern, may represent two species. Song of Western averages slightly; choppier, and buzzier, often ending with a descending buzzy note. Eastern song, clearer with singsong rhythm; usually ends with emphatic high note.

Western averages very slightly smaller bill, darker crown, and drabber flanks than Eastern.

VOICE: Song a rapid, run-on warble *viderveedeeviderveedeevi-derVEET* without pause and with distinctive, husky vireo quality. Call a harsh, nasal mewing *meeerish* (slightly two-syllabled) or a swelling *meeezh*; also a short, dry *git* or *gwit*; a high, slightly nasal chatter rising then falling.

Common in large trees usually near water, especially cottonwoods and aspens. Usually solitary. Feeds on insects and larvae gleaned from leaves at middle to upper levels of foliage.

Philadelphia Vireo

Vireo philadelphicus

L 5.25" WS 8" WT 0.42 oz (12 g)

Smaller and shorter-billed than Red-eyed Vireo, with yellov throat. Very similar to Warbling; note brightest yellow on cente of throat and slightly more contrasting face pattern with broade dark lores.

VOICE: Song very similar to Red-eyed Vireo but higher-pitched on average, weaker and choppier, with longer pauses averaging one phrase every three seconds (but Red-eyed can match these characteristics). Call nasal and soft like Red-eyed but three to five short notes in slightly descending series *weeej weeezh weeezh weeezh*.

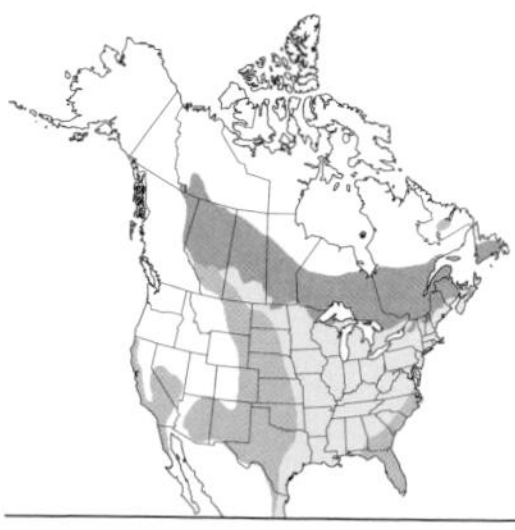

Uncommon in deciduous forests, usually at middle to upper levels of trees within foliage. Usually solitary. Feeds mainly on caterpillars and insects.

Red-eyed Vireo

Vireo olivaceus

L 6" WS 10" WT 0.6 oz (17 g)

A large and elongate vireo, plain greenish above and whitish below, lacking wingbars. The long bill, long pale eyebrow and dark eyeline, and relatively straight horizontal posture all help to accentuate long-headed look.

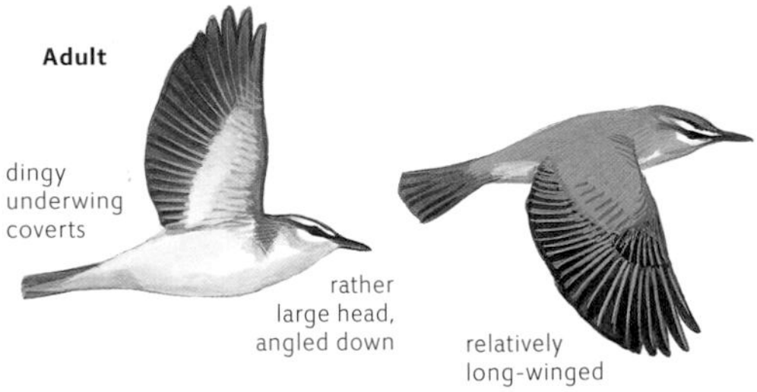

VOICE: Song of simple, hurried, whistled phrases *here-I-am, in-the-tree, look-up, at-the-top*...; averages one phrase every two seconds. Call a nasal, fairly soft, mewing *meerf* or *zherr*; a longer, more whining, descending *rreeea* when agitated; also chatter and *gwit* note similar to Black-whiskered, but rarely heard.

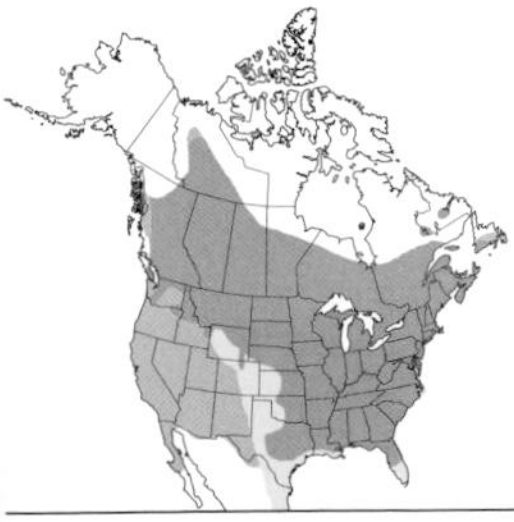

Common and widespread. Nests in broadleaf trees in forests; migrants found in any wooded habitat. Usually at middle to upper levels of trees within foliage. Usually solitary. Feeds on insects, larvae, and berries gleaned from trees.

Black-whiskered Vireo

Vireo altiloquus

L 6.25" WS 10" WT 0.63 oz (18 g)

Habits and appearance similar to Red-eyed Vireo; best distinguished by larger bill, slightly more brownish overall color, and drab striped pattern on head. Thin dark lateral throat-stripe (whisker) is distinctive, but difficult to see.

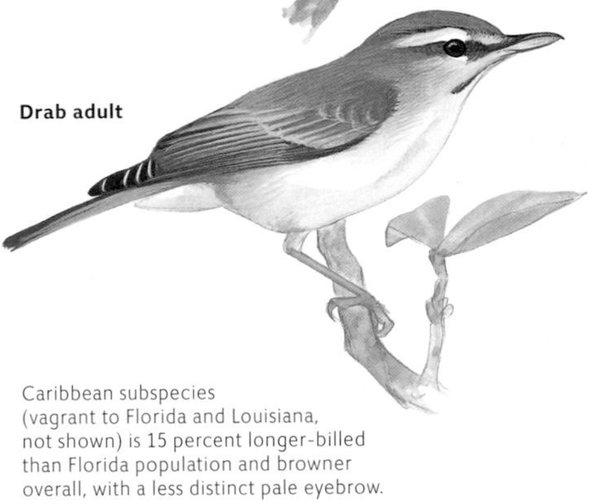

VOICE: Song of three- or four-syllable phrases *chip john phillip, chiip phillip..., chillip phillip*; sharper with more abrupt pitch changes than Red-eyed, averaging one phrase every two seconds. Call varies from nasal mew like Red-eyed to soft, low *vwirr* like Veery; also a rapid, nasal chatter and a short, low *gwit*.

Common in limited range among mangroves and adjacent broadleaf trees.

Yellow-green Vireo

Vireo flavoviridis

L 6.25" WS 10" WT 0.63 oz (18 g)

Habits and appearance similar to Red-eyed Vireo; best distinguished by larger bill, drabber head pattern (with less contrasting pale eyebrow), and overall suffusion of yellow, especially on sides of neck and breast.

VOICE: Song similar to Red-eyed but shorter, less musical, and delivered more rapidly, averaging one phrase every 1.5 seconds; overall impression very similar to House Sparrow chirping. Calls similar to Red-eyed.

Rare visitor to deciduous woods and mature riparian woods, usually near water. Has nested in Texas. Vagrants in California and elsewhere found in dense leafy foliage of trees and shrubs.

Hutton's Vireo

Vireo huttoni

L 5" WS 8" WT 0.39 oz (11 g)

Small, stocky, and active. Strikingly similar to Ruby-crowne Kinglet in plumage and in wing-flicking habit, but heavier, wit thicker bill, thicker and grayish legs, broader tail, and lack of dar bar across base of secondaries. Also note different calls.

Two populations. Mexican birds average about 10 percent larger and are grayer with paler cheeks, more distinct whitish eyering and wingbars, while Pacific birds are washed with yellow-olive.

VOICE: Song of simple whistled phrases repeated every one to two seconds, each phrase repeated many times before switching to another phrase *trrweer, trrweer, trrweer,... tsuwiif, tsuwiif,....* Call a nasal, rising *reeee dee de* or laughing *rrrreeeee-dee-dee-dee-dee*. Also a high, harsh mewing *shhhhhrii shhhri shhr shhr* and a dry *pik* like Hermit Thrush.

Common in oak woods and mixed oak-conifer forest, where it moves through trees with flocks of other small songbirds, foraging mainly at lower to mid levels of foliage.

Crows and Jays

FAMILY: CORVIDAE

23 species in 9 genera (4 rare species are not shown here). Most ays are long-tailed with bright blue plumage; crows and ravens are larger and all-black. All species are gregarious, usually seen n small groups. They are omnivorous but feed mainly on seeds and nuts; some species' diets are very specialized. Flight is generally strong and buoyant with rowing wingbeats. Nest is a bulky cup of sticks in a tree; ravens usually build nests on cliff ledge or tower. All are noisy and bold, but can be very inconspicuous when nesting. All species mob predators, and owls and hawks can sometimes be found by following agitated jays and crows. Adults are shown.

Genus *Perisoreus*

Gray Jay, page 381

Genus *Cyanocitta*

Steller's Jay, page 374

Blue Jay, page 375

Genus *Cyanocorax*

Green Jay, page 379

Genus *Aphelocoma*

Florida Scrub-Jay, page 377

Island Scrub-Jay, page 377

Western Scrub-Jay, page 376

Mexican Jay, page 378

Genus *Psilorhinus*

Brown Jay, page 379

Genus *Gymnorhinus*

Pinyon Jay, page 380

Genus *Pica*

Black-billed Magpie, page 382

Yellow-billed Magpie, page 382

Genus *Corvus*

American Crow, page 384

Northwestern Crow, page 384

Fish Crow, page 385

Tamaulipas Crow, page 385

Chihuahuan Raven, page 383

Common Raven, page 383

Genus *Nucifraga*

Clark's Nutcracker, page 380

Steller's Jay

Cyanocitta stelleri

L 11.5" WS 19" WT 3.7 oz (105 g)

Broad-winged, with long dark crest. Closely related to Blue Jay, and shares similar shape and habits, but mostly dark, lacking white in wings and tail. Differs from Western Scrub-Jay in crest, dark throat and belly, black bars on wing and tail feathers, shorter tail and different flight.

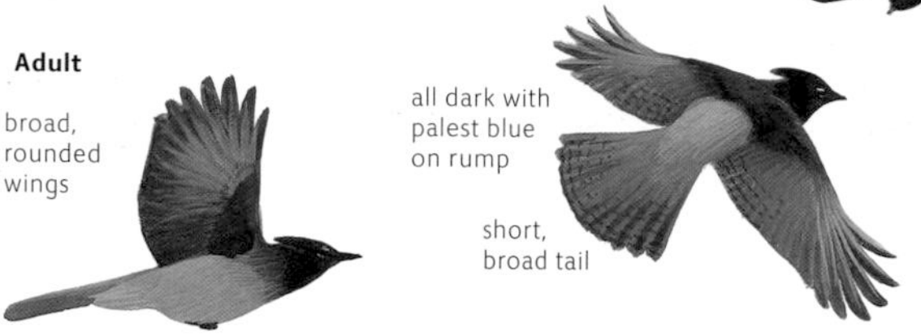

flight of Steller's and Blue Jays steady on rowing wingbeats; wingbeats almost entirely below horizontal, with short, sailing or swooping glides.

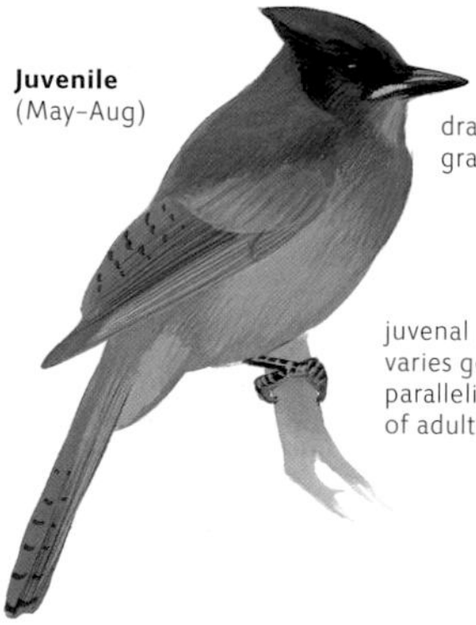

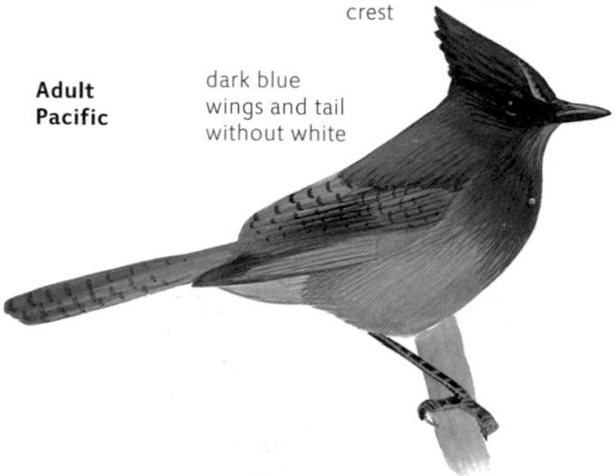

VOICE: Varied; most common a very harsh, unmusical, descending *shaaaaar*; also a rapid, popping *shek shek shek shek* or *chook chook chook chook*; generally lower, harsher than scrub-jays. Also a clear, whistled *whidoo* and quiet, melodious thrasher-like song.

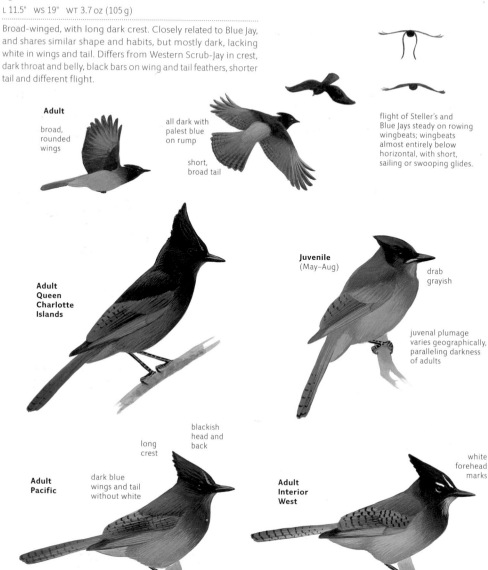

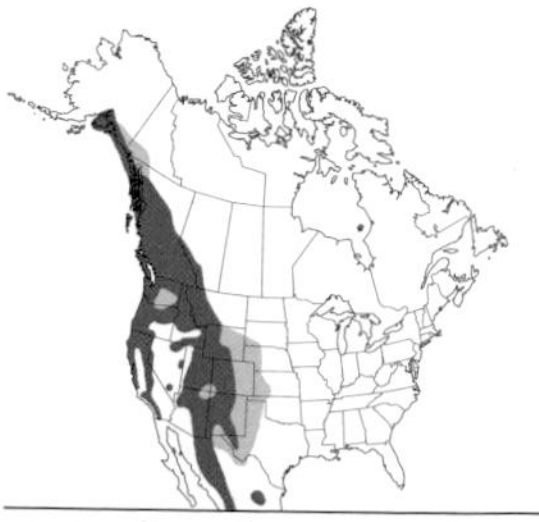

Common in mature and thick coniferous woods, usually in the shade of dense forest. Can be seen in adjacent oak and other woods, especially in winter, and at all seasons forages commonly on the ground. Usually in small groups and relatively high in trees. Feeds on variety of seeds, fruit, and insects. Often visits bird feeders.

Geographic variation is complex and clinal, involving overall color and white or blue facial markings. Darkest overall in Pacific Northwest (especially Queen Charlotte Islands), palest in south. All interior birds have pale marks around eye, where coastal birds are all-dark. Populations from southern Wyoming southwards differ distinctly from all others: pale blue and gray, lacking brown tint and with contrasting black head; prominent whitish markings on face, and relatively long crest. All of these characteristics are more pronounced to the south.

Blue Jay

Cyanocitta cristata

. 11" WS 16" WT 3 oz (85 g)

Jnmistakable, the only North American bird with bright blue and white on wings and tail. Also note blue crest, pale underparts, and dark necklace.

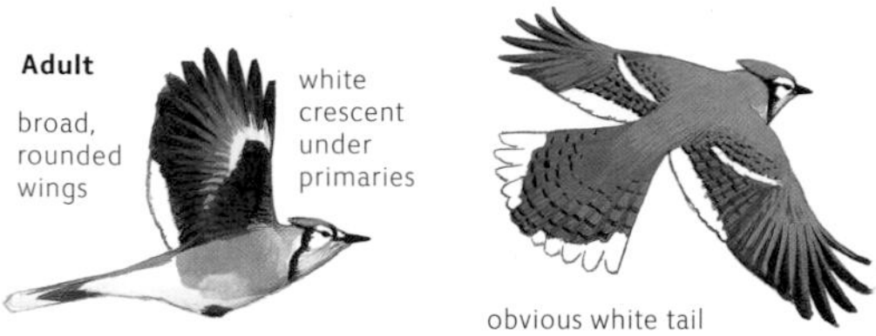

VOICE: Varied; most common a shrill, harsh, descending scream *jaaaay*. Many other calls include clear, whistled phrases *toolili*; quiet, clicking rattle; and rarely a quiet, thrasher-like song. Short, harsh *shkrrr* when attacking predator. Expert mimic, especially of raptors.

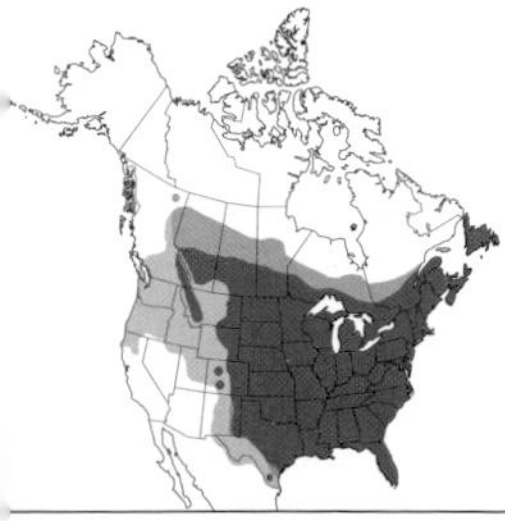

Common and widespread in woods, often relatively high in trees. Travels through woods in small groups or pairs. Feeds on a variety of insects as well as acorns and other seeds. Often visits bird feeders.

Black-throated Magpie-Jay

Calocitta colliei

L 27" WS 25" WT 7 oz (198 g)

Shares some features with Blue Jay, but much larger, with magpie-like long tail and no white on wings or tail.

VOICE: High strident rolling calls such as *skraaa* or a quieter *krriit*, more reminiscent of parrots than other jays.

Native to Mexico, recently escaped and breeding in very small numbers around San Diego, California.

Western Scrub-Jay

Aphelocoma californica

L 11.5" WS 15.5" WT 3 oz (85 g)

Relatively longer-tailed than Steller's Jay, whitish below, and lacks crest. Rather dark blue overall, with grayish or whitish underparts and dark cheek.

flight more undulating than Steller's Jay; wingbeats quicker and stiffer

VOICE: Generally harsh and angry-sounding. Most common call a harsh, rising *shreeeeenk* with some musical quality; often in rapid series *wenk wenk wenk wenk* or *kkew kkew kkew*...; also a harsh, pounding *sheeyuk sheeyuk*... in long series. Other calls include a low clucking *chudduk* or clicking sounds.

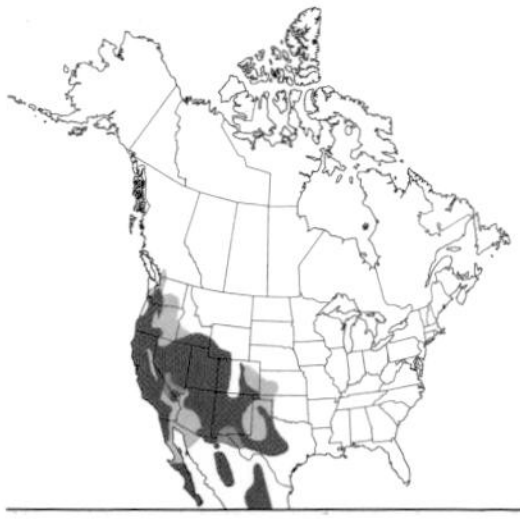

Common in coastal regions, closely tied to oaks in brushy areas and open woods, foraging largely on the ground; uncommon inland in dry open oak-juniper woods. Usually travels in small groups, flying low from tree to tree with quick wingbeats and stiff-winged, sailing glides. Feeds on variety of seeds, fruit, and insects, but mainly acorns. Often visits bird feeders.

Two well-differentiated populations, with very little range overlap, are distinguishable by plumage; few intermediates. Pacific population is thicker-billed and darker overall, with richer colors; contrasting blue breastband and other details. Pacific birds are generally bold and conspicuous; Interior birds are shy and inconspicuous, and more sparsely distributed. Voice of Interior population averages lower, hoarser, less electric-sounding than Pacific.

Island Scrub-Jay

Aphelocoma insularis

L 13" WS 17" WT 4.1 oz (116 g)

Habits and appearance very similar to Pacific populations of Western Scrub-Jay, but no range overlap. Distinguished by larger size, relatively large bill, long legs, blacker cheek, and darker blue color including blue on the undertail coverts.

VOICE: Very similar to Pacific population of Western Scrub-Jay, but averages slightly louder and harsher.

Common within limited range, on Santa Cruz Island off southern California, with a total population estimated at only 2,500. Found in open oak woods and brushy patches, the same habitat occupied by Western Scrub-Jay on the adjacent mainland.

Florida Scrub-Jay

Aphelocoma coerulescens

L 11" WS 13.5" WT 2.8 oz (80 g)

Very similar to Western Scrub-Jay, but range does not overlap; differs in whitish forehead, paler back, and pale gray-brown underparts. Told from Blue Jay by plain blue wings and tail, lack of crest, longer tail, calls and habits also differ.

VOICE: Similar to Western Scrub-Jay but distinctly lower, harsher, flatter, less rising *kreesh*; also a low, husky *kereep* and other variations. All calls unlike the higher, more expressive voice of Blue Jay.

Uncommon, local, and declining in dry palmetto and oak scrub. Usually in small groups; seen perched on tops of low trees or in low flight between trees. World population estimated at about 10,000 birds.

Mexican Jay

Aphelocoma wollweberi

L 11.5" WS 19.5" WT 4.4 oz (125 g)

Slightly larger than scrub-jays, with broader wings and shorter, broader tail. Relatively plain blue and gray above and plain gray below with paler throat; lacks contrasting dark necklace of scrub-jays, and more gregarious.

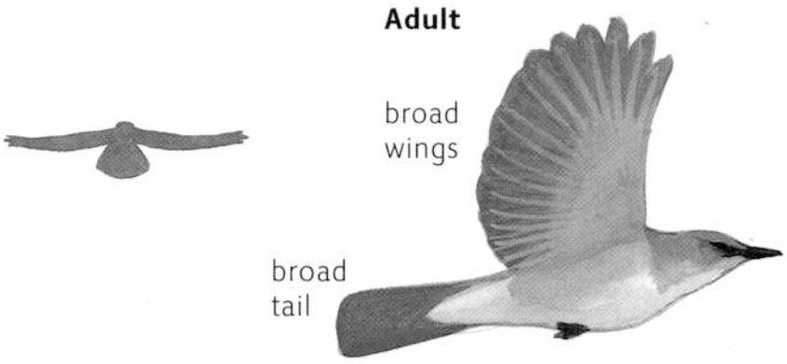

VOICE: Somewhat less varied than other jays; common call a rather soft, musical, rising *zhenk* or *wink*; repeated in series especially in flight *wink wink wink*...; notes softer and shorter than scrub-jays and with almost finch-like quality. Also gives a soft thrasher-like song.

Common in montane pine-oak woods within its limited range in western North America. Always in small groups, even flocks up to about 25 birds. Feeds on acorns and variety of other food. Occasionally visits bird feeders.

Texas and Arizona populations differ in plumage, size, voice, and habits, but with some intermediate populations in Mexico. Texas birds are ten percent smaller, darker overall, and brighter blue with a whitish throat (vs. grayish like breast); juvenile acquires black bill just after fledging (Arizona juvenile is pale-billed for two years); pairs are territorial (vs. cooperative flocks of six to ten birds year-round); eggs are speckled (vs. unmarked). Texas calls average slightly shorter, harsher, and less musical than Arizona: *jink jink jink*.... Texas birds also give a mechanical rattle (female only); lacking in Arizona populations.

Brown Jay

Psilorhinus morio

. 16.5" WS 26" WT 7 oz (200 g) ♂>♀

Much larger than other jays, with big dark tail and very, broad wings. Overall dark brown plumage all-dark above with pale belly obvious at a distance. Flies with very deep flopping wingbeats, short swooping glides.

OICE: Common call an intense, clear bugling *keerg* or *paow* ften repeated in series and given in chorus by flock; much like ed-shouldered Hawk but higher, more raucous, without pslur at beginning. Also gives a popping rattle audible at close ange.

Rare and very local in riparian woods; a few pairs formerly nested in Texas, along the Rio Grande, but in recent years just an occasional visitor. Travels in noisy family groups of three to eight, moving through treetops and often flying long distances. Feeds on variety of seeds, fruit, and insects. Occasionally visits bird feeders.

Green Jay

Cyanocorax yncas

L 10.5" WS 13.5" WT 3.5 oz (100 g)

Size and proportions similar to other jays, but pale yellow and green color with mostly black head and breast is unique. Also note yellow sides on tail, obvious when flying away.

VOICE: Varied; common call a rapid series of four to five notes; harsh, electric *jeek jeek jeek jeek* and high, mechanical *slikslikslikslik*. Also a variety of quiet, squeaky or buzzy croaks: nasal *been*; high, nasal *nnneeek-neek* or *grreen-rreen*; drawn-out clicking *ree urrrrrrrr it*.

Common in dense brushy woods within limited range, often low in dense foliage. Watch for it flying across small openings with bursts of quick wingbeats. Usually in small groups, calling frequently. Feeds on variety of fruit, seeds, and insects. Often visits bird feeders.

Pinyon Jay

Gymnorhinus cyanocephalus

L 10.5" WS 19" WT 3.5 oz (100 g)

An unusual jay with short tail and long tapered bill. Overall dusty blue color without contrasting markings distinctive. Flight direct, with rapid stiff wingbeats; flies relatively high for long distances.

VOICE: Calls rather nasal and soft for a jay; quality reminiscent of California and Gambel's Quail. Calls constantly in flight: a soft, conversational series *hoi hoi hoi*... or single *hoya*; also a series of harsher rising notes *kwee kwee kwee*... and loud, clear, nasal *waoow*.

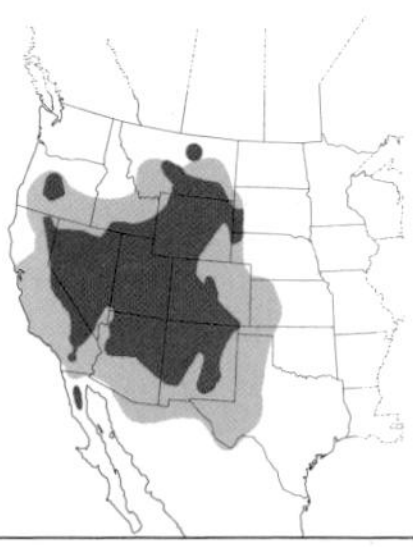

Uncommon and irregular in pinyon pine forests and open pinyon-juniper habitat, often ranges into adjacent sagebrush or pine or oak forest when foraging. The most gregarious jay, nearly always in noisy flocks of a few birds to several hundred. Feeds largely on seeds of pines, especially pinyon pine.

Clark's Nutcracker

Nucifraga columbiana

L 12" WS 24" WT 4.6 oz (130 g)

Larger than jays, with short tail and long bill. Pale gray color and black and white wings and tail distinctive. Often first detected by raucous calls, perched on treetop or flying high overhead; note black-and-white wings and gray body, short tail mostly white.

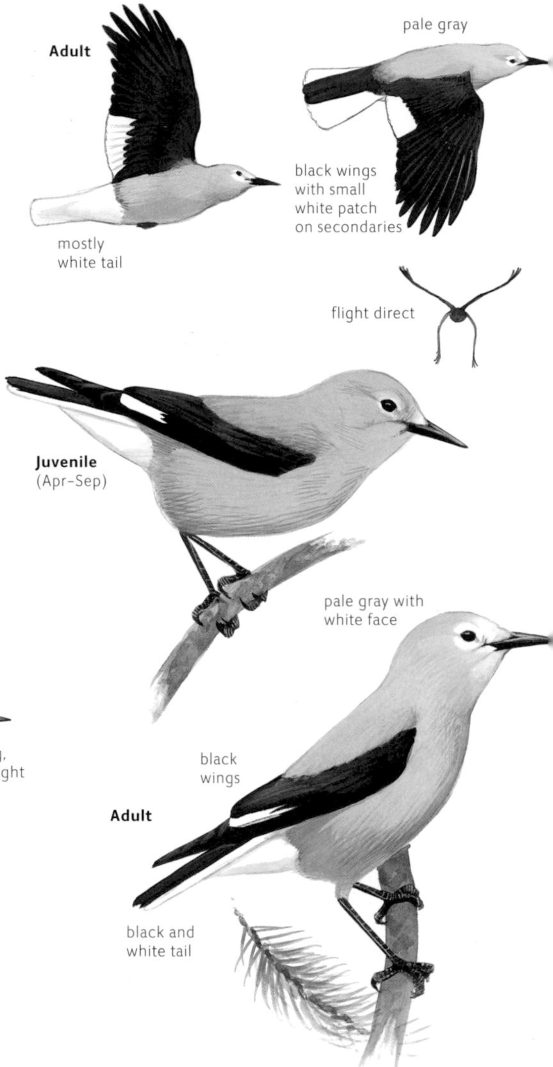

VOICE: Varied; common call a long, harsh, slightly rising *shraaaaaaa*; hollow and slightly buzzing. Other calls include higher, descending *taaar*; a very hard, slow rattle *k-k-k-k-k*; strong yelping *keeeew*; a clear, nasal *waaat* reminiscent of Pinyon Jay; and a high, clear, electric *deeen*.

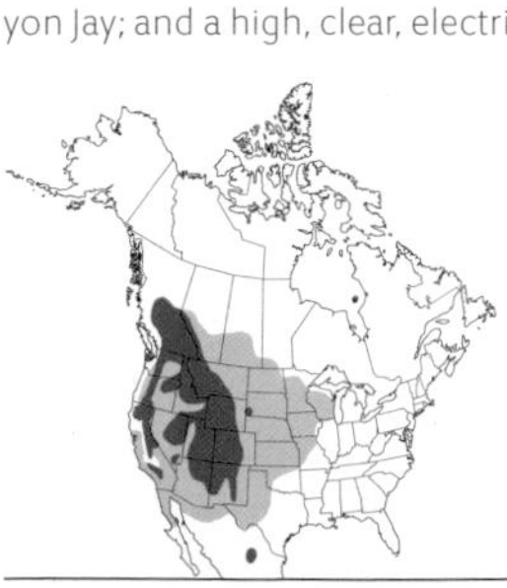

Uncommon in mature mixed coniferous forests in mountains, usually near treeline or around open meadows or rocky areas. Usually in small groups flying around mountain slopes or perching on conspicuous trees or rocks. Feeds mainly on seeds of pines, but omnivorous, often traveling long distances up or down slope to find food.

Gray Jay

Perisoreus canadensis

11.5" WS 18" WT 2.5 oz (70 g)

arge and fluffy, with muted gray colors overall. Reminiscent of large chickadee, with dark crown and short bill. Flight relatively low, with burst of soft flopping wingbeats and a sailing glide.

OICE: Calls varied, generally soft, whistled or husky notes in hort series. Short, mellow whistled *weeoo*, and soft husky *chef hef chef chef* are most common and distinctive calls. Also nown to mimic other species such as Blue Jay and various rap-ors.

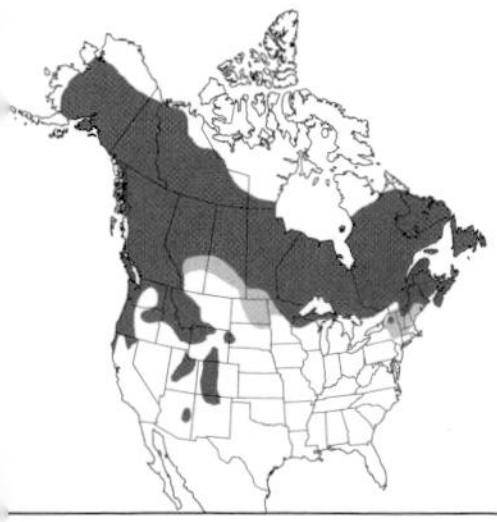

Uncommon and sparsely distributed in coniferous woods, usually in areas with low trees or dense undergrowth such as around bogs or clearings. Travels in small groups year-round. Feeds on a variety of seeds, fruit, and insects as well as carrion, and quickly becomes habituated to human food sources such as picnic areas. Occasionally visits bird feeders.

Complex geographic variation with three distinguishable populations, but all are connected by intermediate populations. Pacific populations are most distinct—darker overall but with whitish belly, extensively dark head, lacking white on tail and secondaries, and have brownish-tinged back with pale shaft streaks. In Rocky Mountains, especially southwards, birds have very limited black on head. There may also be consistent variations in juvenal plumage. Any regional differences in voice are overshadowed by the tremendous variation of calls given by individuals.

Black-billed Magpie

Pica hudsonia

L 19" WS 25" WT 6 oz (175 g) ♂>♀

Magpies are easily identified by large size, bold black and white plumage pattern, and long black tail. Flight level, with relatively steady, rowing wingbeats and long tail reminiscent of large grackles; often flies high for long distances.

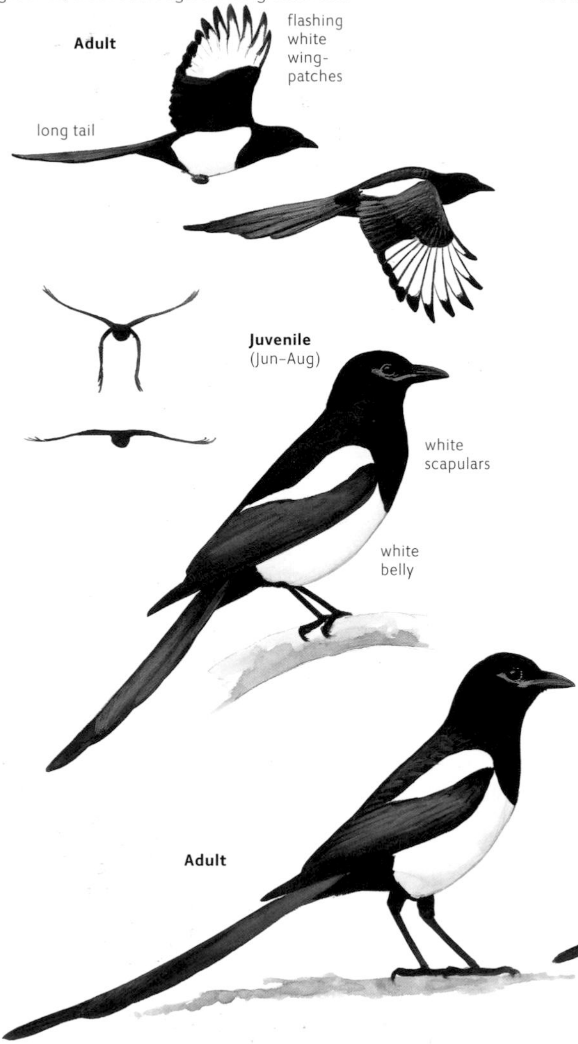

VOICE: Call a nasal, rising *jeeeek*; harsher, lower *rek rek rek rek* or *weg weg weg weg weg*; rapid *shek shek shek shek*, three to five notes. Also higher, long, nasal *gway gway* or *gwaaaaay*; rather nasal, hard, querulous *ennk*.

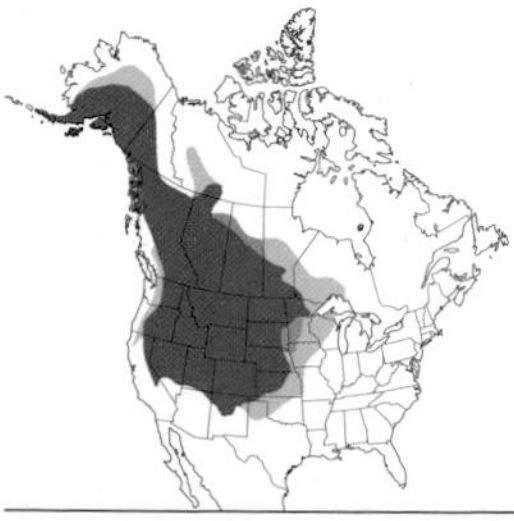

Common in prairies and parklands with scattered trees. Usually in small groups. Often seen on fenceposts or along roadsides. Feeds on a variety of seeds and animal prey, foraging mostly on the ground.

Yellow-billed Magpie

Pica nuttalli

L 16.5" WS 24" WT 5 oz (155 g) ♂>♀

Habits and appearance (aside from bill color) virtually identica to Black-billed Magpie, but range does not overlap. Differs fror Black-billed in smaller size, yellow bill, and yellow bare skir around eye.

VOICE: Nearly identical to Black-billed; chatter call reportedl higher-pitched and clearer than Black-billed.

Common locally in oak savanna resident and almost never recorded outside its very limited range.

Chihuahuan Raven

:orvus cryptoleucus

. 19.5" WS 44" WT 1.2 lb (530 g) ♂>♀

ntermediate between American Crow and Common Raven. Told rom crows by heavier bill, longer wings and tail, different call. From Common Raven by subtle differences in habitat, voice, and tructure; white neck rarely visible.

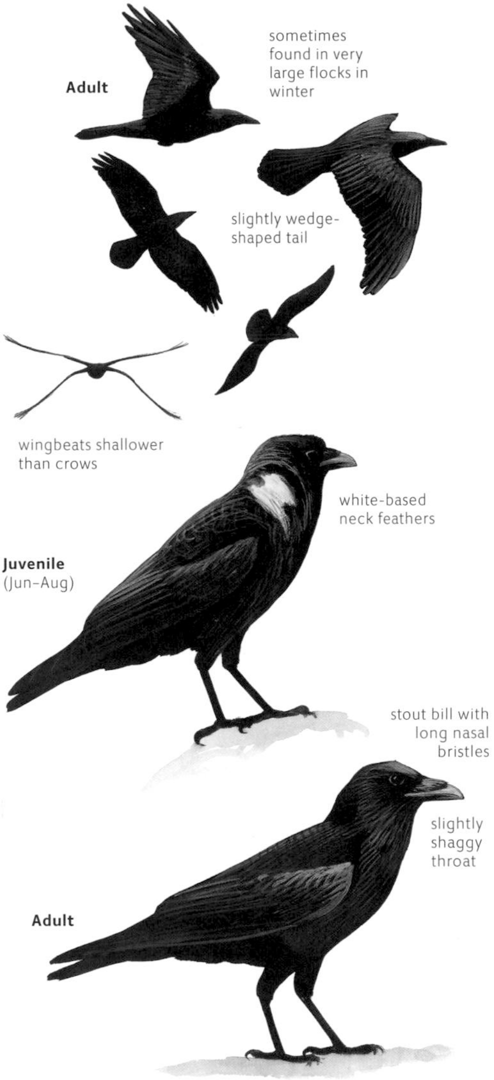

VOICE: Less varied than Common. Typically fairly high, crow-like croak; usually a slightly rising *graak*, but this can be matched by Common. Apparently also gives *kre* and growling sounds similar o Common.

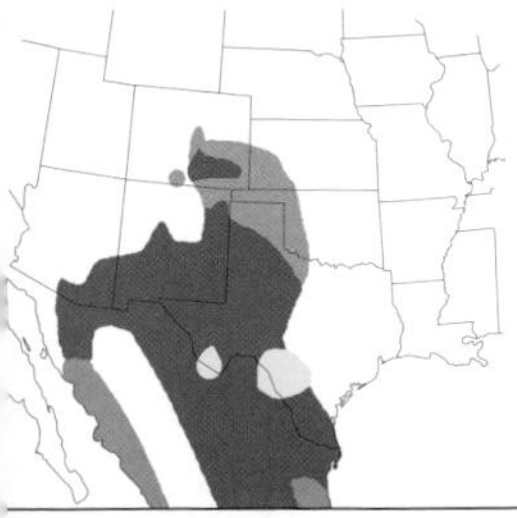

Common locally in arid grasslands and brushlands; usually not in mountainous areas. Usually in pairs or small groups; may gather in flocks of hundreds at roost sites and where food is abundant. Diet diverse, as in crows.

Common Raven

Corvus corax

L 24" WS 53" WT 2.6 lb (1,200 g) ♂>♀

Larger than crows, with longer and narrower wings; long, wedge-shaped tail; long bill, and usually deeper, croaking calls. Often soars or glides for long periods on flat wings (crows flap more often).

VOICE: Incredibly varied: from low, deep baritone croaks to high, bell-like and twanging notes. Long, hoarse *kraaah*; lower, hollow *brrronk*; and deep, resonant *prruk* are typical. Some calls like Chihuahuan. Juvenile calls higher, more squawking than adult.

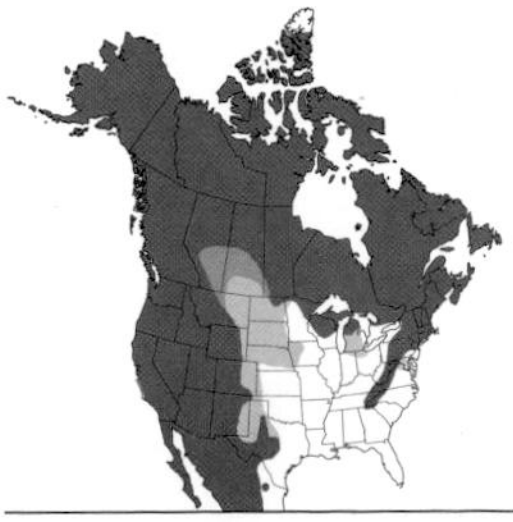

Uncommon. Habitat varied, from tundra to coniferous forests to arid brushlands, often in mountainous regions. Usually in pairs or small groups; rarely mixes with crows (in fact, is often attacked vigorously by them). Diet diverse.

American Crow

Corvus brachyrhynchos

L 17.5" WS 39" WT 1 lb (450 g)

The most widespread large, all-black bird; distinguished from other crows and ravens by structure and voice. Flight direct, with steady rowing wingbeats.

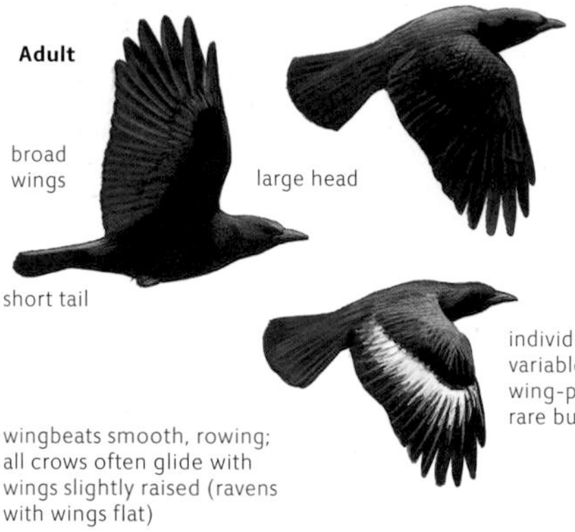

VOICE: Call varied; typically the familiar full-voiced, descending, *carrr* or *caaw* with great variety of inflection and pitch. Also a rapid, hollow rattle *tatatato*; sometimes soft call *prrrk* similar to Common Raven. Juvenile gives higher, hoarser, nasal calls; usually also longer calls (e.g., *carrrr*).

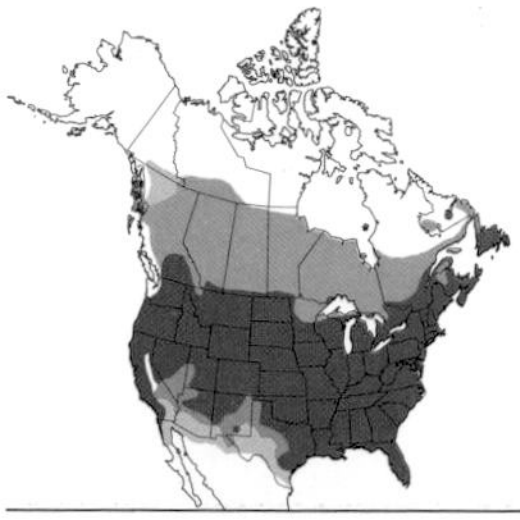

Common and widespread; certainly one of the best-known birds in North America. Usually in small groups (occasionally large flocks) in all open habitats from beaches and farmland to suburbs and open woods. Forms large communal roosts at night where thousands of individuals gather in trees. Diet extremely diverse, from small animals, bird eggs, fish, snails, and insects to seeds, fruits, and human refuse.

Northwestern Crow

Corvus caurinus

L 16" WS 34" WT 13 oz (380 g)

Habits and appearance essentially identical to American Crov averages slightly smaller and shorter-tailed, but not distinguish able by appearance, and doubtfully identifiable by voice.

VOICE: Some calls a little lower, hoarser, and more rapid tha American Crow, but differences are subtle and seem to chang clinally towards the south and east, with intermediate bird around Puget Sound in Washington and further slight difference evident south to California and east to the Rocky Mountains.

Common in coastal coniferous forests and along beaches and shorelines from Kodiak Island, Alaska, to southern British Columbia. Exact limits of range uncertain.

Hooded Crow

Corvus cornix

L 18" WS 38.5" WT 18 oz (510 g)

Size and shape similar to American Crow but with contrasting pale gray neck and breast.

VOICE: Higher and harsher than American Crow, very similar to some calls of immature American Crow.

Very rare visitor from Europe, one recent record in New York and New Jersey could be a natural vagrant, or that and the few other records may involve escapes from captivity

House Crow

Corvus splendens

L 15.7" WS 33" WT 10.8 oz (306 g)

Smaller than American Crow but with relatively large bill, dusky gray cheeks, neck, and breast.

VOICE: Harsh and raucous, and on a steady pitch, not descending like American Crow.

Very rare visitor from Asia, presumably arriving via ship A few have nested in Florida and could occur elsewhere.

ｉamaulipas Crow

Corvus imparatus

14.5" WS 30" WT 8 oz (240 g)

›maller than other crows, with relatively long tail and wings and ›uoyant flight. Glossy blue-black overall. In its range the only imilar species are Chihuahuan Raven (much larger with heavy ›ill) and Great-tailed Grackle (smaller with long tail and pale eye).

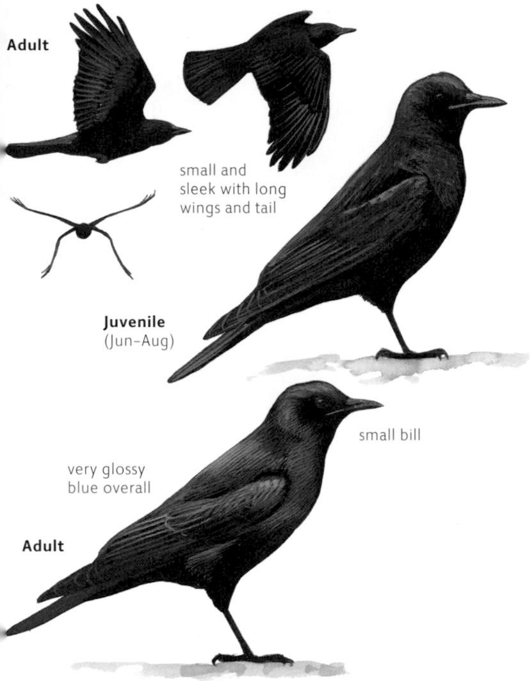

'OICE: Call a very rough, nasal, flat burp/croak *brraarp* usually ı series: *brraap brraap brraap.*

Rare visitor from Mexico to extreme southern Texas; found only in immediate vicinity of Brownsville. Formally numerous but now only recorded a few times each year.

Eurasian Jackdaw

Corvus monedula

14.5" WS 29" WT 7 oz (200 g)

maller than crows, with short ill, gray nape, pale eye.

'OICE: Call a sharp, ard *jeck* reminiscent f *chig* note of ed-bellied Voodpecker.

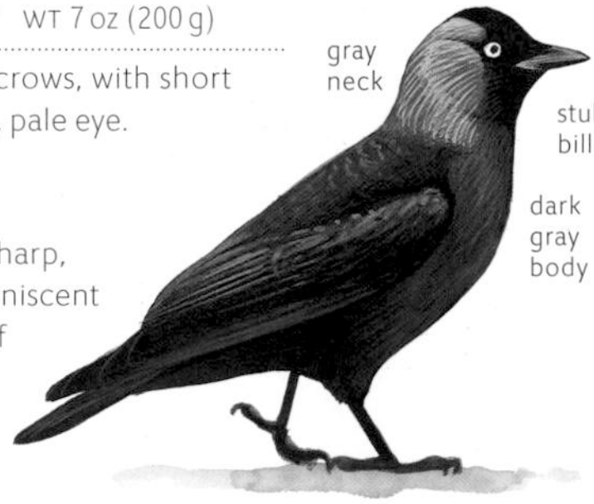

Very rare visitor from Europe, recorded a few times in the Northeast, usually with American Crows.

Fish Crow

Corvus ossifragus

L 15" WS 36" WT 10 oz (280 g)

Appearance very similar to American, but averages slightly smaller. Sometimes identifiable by relatively smaller head, longer wings and tail, or shorter legs, but reliably distinguished only by voice.

VOICE: Call a simple, short, nasal *cah* or *cah-ah*; less often a hoarser *cahrr* like some high-pitched calls of American. Juvenile gives high *keeer* like American but higher, clearer. Throaty rattle *grrrr* higher and more nasal than American.

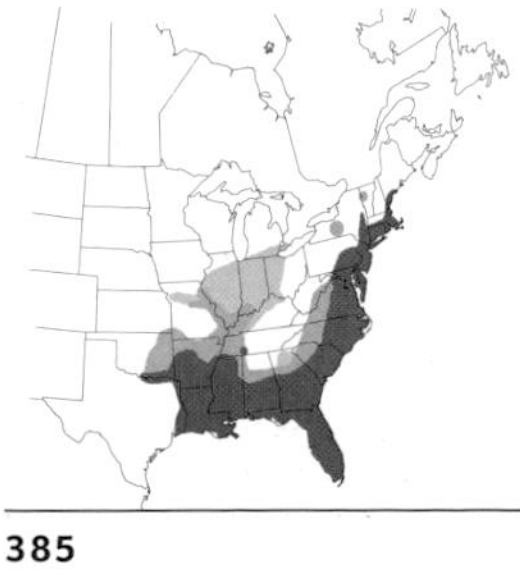

Common along coast; uncommon inland. Found in wide variety of open habitats usually near water. Forms small groups (occasionally large flocks); may mix with American Crow. Diet similar to American.

Larks

FAMILY: ALAUDIDAE

2 species in 2 genera. These cryptically-colored birds are found in large farm fields and other open areas, often in large flocks in nonbreeding season; nearly always perch on the ground or on a low post or wire. They have short legs and run or walk with a shuffling gait. Flight is generally smooth and undulating; often sweeping low across open ground. Larks are renowned for the song given by males in flight. Nest is a grassy cup placed on the ground at the base of a shrub or grass tuft. Adult males are shown.

Genus *Alauda*

Sky Lark, page 386

Genus *Eremophila*

Horned Lark, page 387

Sky Lark

Alauda arvensis

L 7.25" WS 13" WT 1.4 oz (40 g)

Slightly larger than Horned Lark or sparrows, with relatively broa chest and short tail. Note warm brown color, streaks above an below, slight crest, and white trailing edge on wings.

Adult

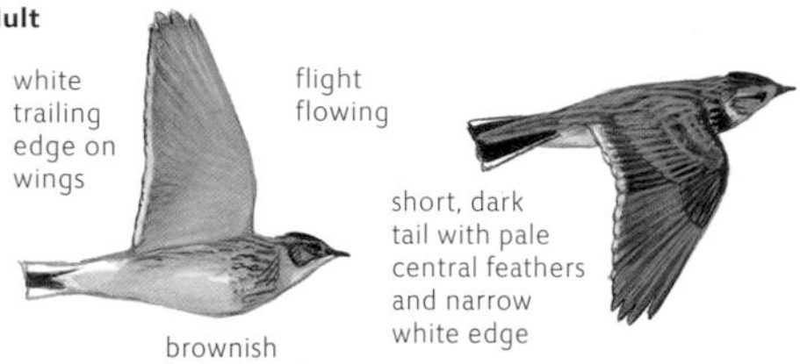

Juvenile (Jun–Aug)

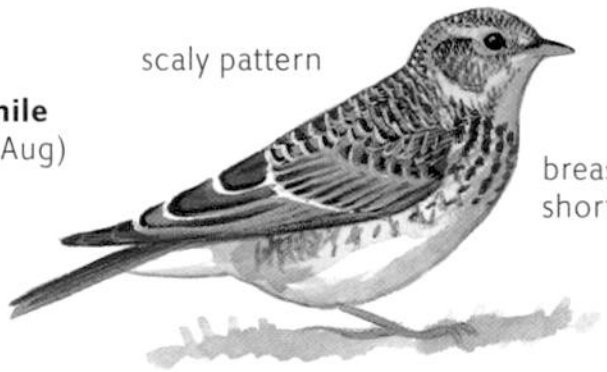

Fresh adult (Sep–Mar)

Worn adult (Apr–Aug)

VOICE: Song a spectacular varied warble of high, liquid, rolling notes in long series; often including mimicry of other species; given from high in the air. Flight call a low, rolling chortle *drirdrirk.*

Uncommon and local. Introduced from Europe and ver small numbers are established o farmland around Victoria, British Columbia, Canada. Usually seen in small groups, foraging on the ground for seeds and insects. Vagants from Siberia occur regularly in western Alaska and very rarely south to California.

Horned Lark

remophila alpestris

7.25" WS 12" WT 1.1 oz (32 g)

lightly larger than sparrows, longspurs, or pipits, with short legs nd crouching posture, relatively pale sandy-brown plumage, old face pattern with dark mask, and dark breastband. In flight, ote long wings and dark square tail. Flight smooth and flowing; ften very low to the ground.

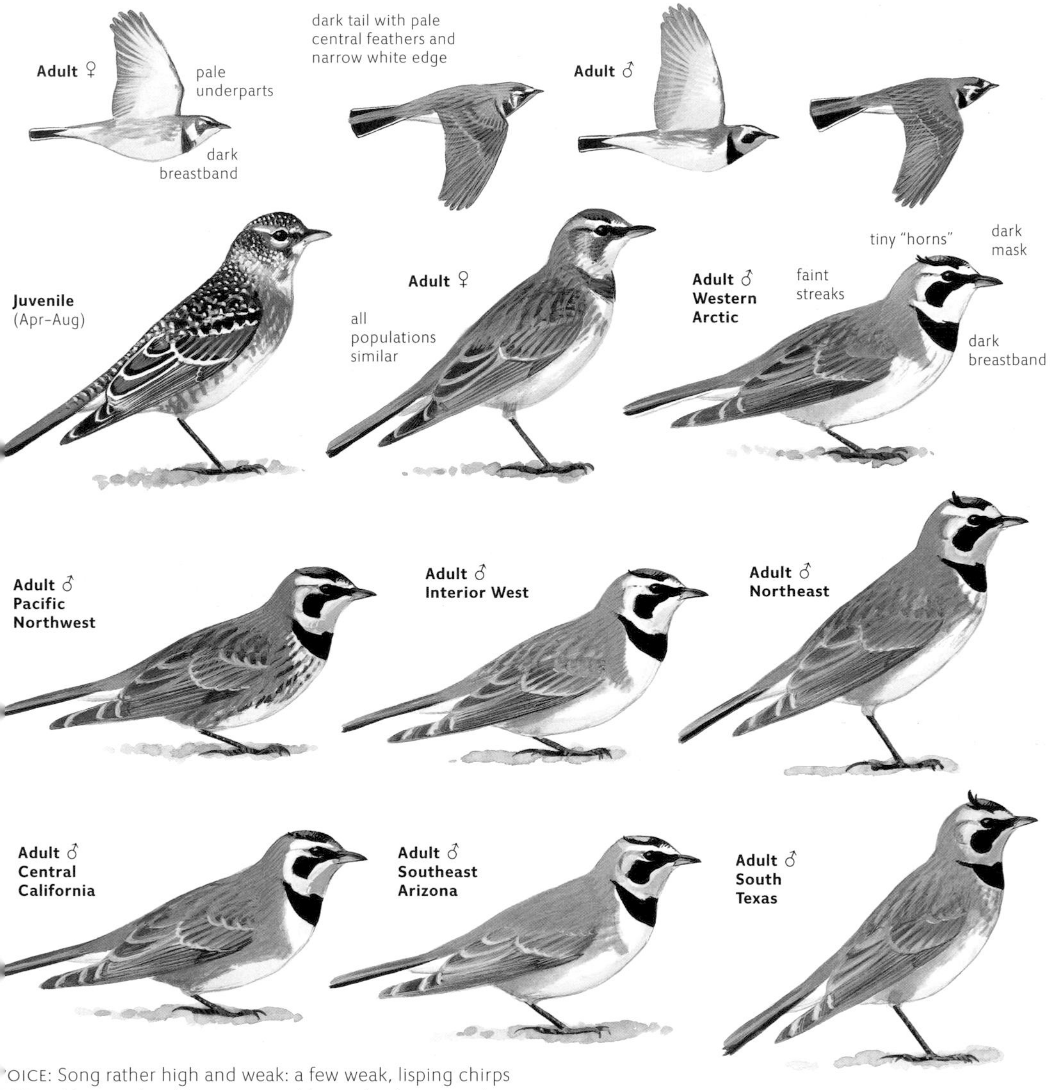

OICE: Song rather high and weak: a few weak, lisping chirps ollowed by a rapid, tinkling, rising warble *reeek trik treet tritilititi reet*. Flight calls high, rather soft, weak, lisping *see-tu* or *see-titi* nd other variations weaker than American Pipit; some buzzy otes unlike pipits.

Common in expansive open areas with barren or only sparsely vegetated ground such as beaches, plowed fields, and even parking lots or airport runways. Usually in flocks (often hundreds together), often mixed with longspurs or Snow Bunting. Feeds on seeds and insects gleaned from low vegetation or the ground.

Geographic variation, which is complex and clinal with many intermediates, mostly involves plumage color and is most evident in summer males. Look for differences in back color, intensity of yellow on the face, and streaking on the breast. Broad patterns of variation are evident. Pacific populations have a dark reddish nape, heavy streaking, and bright yellow face. Interior West populations are pale and grayish, less streaked, culminating in the very pale Southeast Arizona birds. Northeast and Arctic populations are darker, with pinkish nape and moderate streaks.

Swallows

FAMILY: HIRUNDINIDAE

12 species in 7 genera (4 very rare species are not shown here). Highly aerial, with pointed wings, notched tails, and smooth, flowing flight, all swallows feed almost exclusively on insects captured in sustained flight. Often seen perched on twigs or wires or on the ground, sometimes in groups of hundreds or thousands. Nesting habits vary by species and between genera. Two genera, including Cliff, Cave, and Barn Swallows, build mud nests on cliffs or buildings; Tree, Violet-green, and Purple Martin nest in cavities of birdhouses; Bank and Northern Rough-winged nest in burrows in sand banks. Adult females are shown.

Genus *Stelgidopteryx*

Northern Rough-winged Swallow, page 391

Genus *Progne*

Purple Martin, page 389

Genus *Tachycineta*

Tree Swallow, page 390

Violet-green Swallow, page 390

Genus *Riparia*

Bank Swallow, page 391

Genus *Hirundo*

Barn Swallow, page 393

Genus *Petrochelidon*

Cliff Swallow, page 392

Cave Swallow, page 392

Molting Swallows

Tree, Violet-green, Northern Rough-winged, and Cave Swallows molt primaries in late summer through fall. All other swallows molt exclusively on the wintering grounds outside North America. Thus, any swallow molting primaries in North America is likely to be one of these four species.

Active molt in primaries (several outermost feathers old) is typical of Tree Swallow around September.

›urple Martin

rogne subis

8" WS 18" WT 2 oz (56 g)

›ur largest swallow; relatively long-winged and long-headed. dult male uniform bluish-black; can be confused with European :arling. All plumages blackish above, with dark head and breast, ›ry dark underwing coverts, and streaked or speckled belly.

:astern

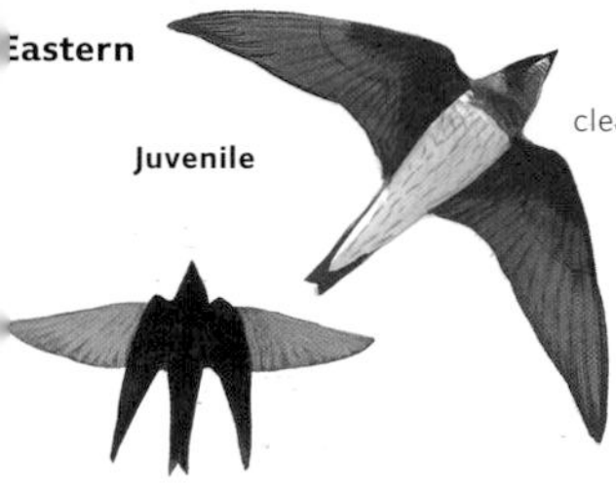

Western

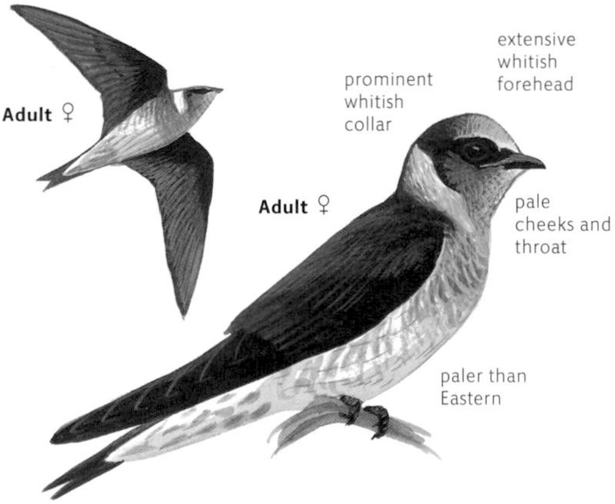

›ICE: Male song a low-pitched, rich, liquid gurgling; female ›ng a mixture of chortle calls and downslurred whistles. Calls ›elodious, rich, low whistles; most common call a rich, descend-›g *cherr*; also a more complex chortle. Variety of other calls ›clude a harsh, buzzy *geerrt* in alarm; a dry, rattling *skrrr* when :tually stooping on a predator; and a hard *gip* given by juvenile.

Common locally. In many areas nests almost exclusively in manmade martin houses placed in the open near water; also nests in tree cavities and saguaro cactus. Usually in small groups. Forages over any open habitat; foraging flight often higher than other swallows.

Western birds are similar to Eastern: some populations average smaller; adult males are identical in plumage; western female and immature average paler than eastern. Voice also differs: *cherr* and chortle calls slightly higher-pitched and often repeated in rapid series; male song longer (2–6 sec vs. 1.5–3 sec) with longer notes and usually at least three grating phrases interspersed (Eastern gives grating phrases only at end, if at all); female song includes more whistles than chortle calls (reverse in Eastern female).

Tree Swallow

Tachycineta bicolor

L 5.75" WS 14.5" WT 0.7 oz (20 g)

Relatively broad-winged, with notched tail. Always dark above and white below: adult metallic blue-green above, juvenile brownish; all show sharp contrast between dark cheek and white throat.

VOICE: Song of clear, sweet whistles *twit-weet twit-weet liliweet twit-weet*...; also a clear *tsuwi tsuw* repeated. Call high, liquid chirping or twittering. Alarm a harsh chatter. Noise from large autumn flocks mainly thin, scratchy *tzeev* notes.

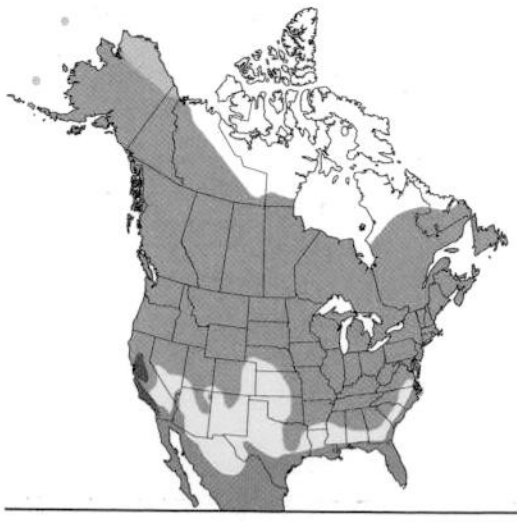

Common. Nests singly in birdhouses or tree cavities in open fields or over water. Often seen in large flocks, perching on wires or in bushes or reeds. Forages over fields or water for berries and insects.

Violet-green Swallow

Tachycineta thalassina

L 5.25" WS 13.5" WT 0.49 oz (14 g)

Similar to Tree Swallow, but smaller with shorter tail and narrowe wings. Wings extend beyond tip of short tail when folded. In fligh note that white color of underside wraps onto sides of rump.

VOICE: Song a creaking *teer twee, tsip-tsip-tsip*.... Call chirpin like Tree Swallow but sharper, harder *chilp* or *chip-lip*; almo swift-like without twittering quality of Tree. Also clear, descenc ing notes in alarm *tseer* or *teewp* repeated.

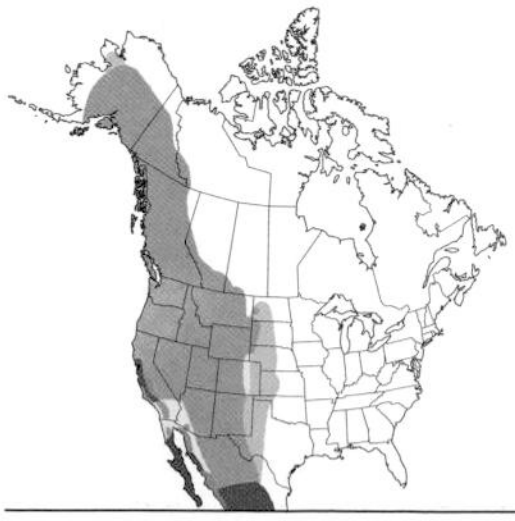

Common. Nests in tree cavities and cliff crevices in open areas. Usually in small groups. Forages over open areas, usually near water.

ɜank Swallow

ɪparia riparia

5.25" WS 13" WT 0.47 oz (13.5 g)

ur smallest swallow, with relatively long, notched tail. Pale ·own above, with rump distinctly paler than back; note dark ·eastband (juvenile Tree Swallow can have dusky breastband, ɹt not as contrasting as Bank).

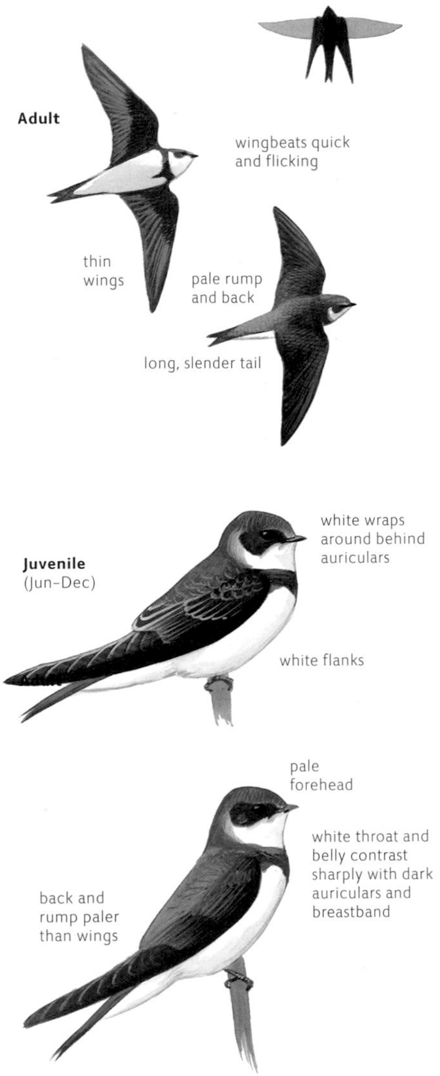

OICE: Song *wit wit dreee drr drr drr* repeated. Call a short, dry, cratchy *chirr* or *shrrit*; often repeated in rapid, run-on series. larm at colony a longer, descending buzz and a thin, mewing *eeew* like Cliff but higher and longer.

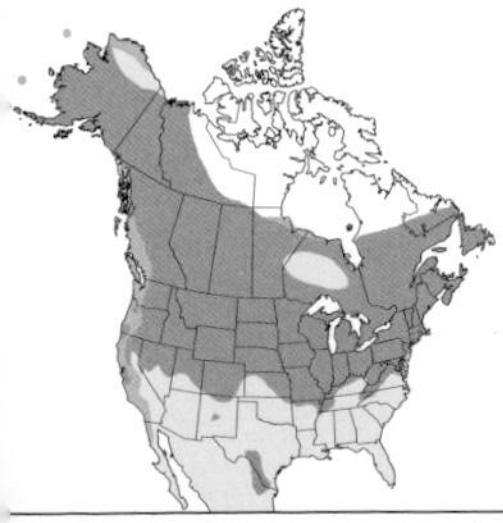

Common. Nests in colonies, excavating tunnels into vertical sandbanks. Forages over any habitat but especially marshes, meadows, and water.

Northern Rough-winged Swallow

Stelgidopteryx serripennis

L 5.5" WS 14" WT 0.56 oz (16 g)

A very drab brownish swallow, uniform brownish above, pale below with gray-brown throat. Broad-winged and short-tailed like Tree Swallow. Striking white undertail coverts often visible from above in flight.

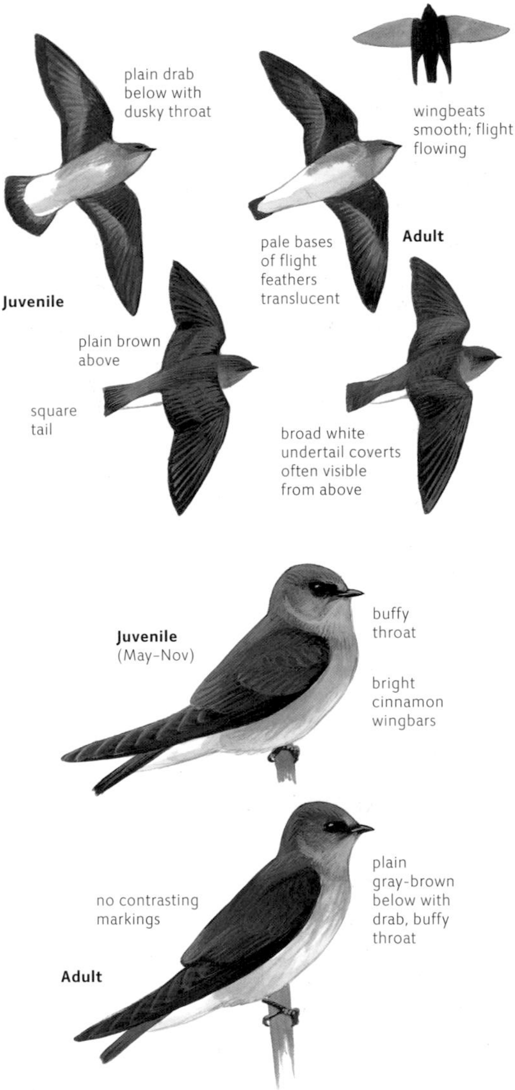

VOICE: Song simply a steady repetition of rough, rising notes like call *frrip frrip frrip*.... Call a low, coarse *prriit* slightly rising; lower and softer than Bank.

Uncommon. Nests singly in holes in sandbanks or in crevices or pipes in cliffs or walls, often perching on nearby twigs. Forages over any open area, especially ponds and rivers.

Cliff Swallow

Petrochelidon pyrrhonota

L 5.5" WS 13.5" WT 0.74 oz (21 g)

A relatively stocky swallow with short square tail. Note pale buffy rump and dark throat, and on most adults bright white forehead. Distinguished from Cave Swallow by all-dark head contrasting with pale collar (vs. dark cap only).

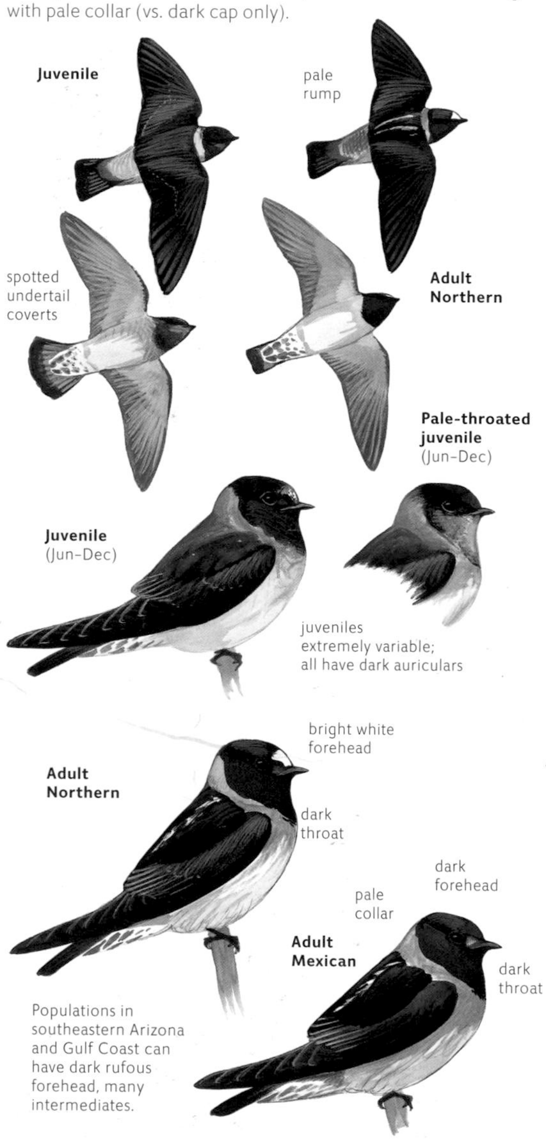

VOICE: Song thin, strained, with drawn-out creaking and rattling sounds; shorter and simpler than Barn Swallow. Call a low, soft, husky *verr* or lower, drier, rolled *vrrrt*. Alarm at colony a soft, low *veew*.

Common. Nests mostly on manmade structures such as under bridges or house eaves; also under overhanging ledges on rocky cliffs. Forages over fields and ponds. Nest a gourd-shaped mud cone with small entrance hole; built in tightly packed clusters in large colonies.

Cave Swallow

Petrochelidon fulva

L 5.5" WS 13" WT 0.53 oz (15 g)

Very similar to Cliff Swallow, with pale rump, but note relative pale throat and dark forehead creating a dark-capped appea ance unlike any other swallow. Underside similar t Rough-winged, but note contrasting dark cap, different call.

VOICE: Song similar to Cliff Swallow. Common flight call a sof rising *pwid* like Barn Swallow but softer, clearer. Alarm a sharpe higher *jeewy* higher than Cliff; also a very high, thin *teeer*.

Common. Nest a partially enclosed half-bowl of mud on vertical wall well inside sheltering cave or culvert; nests in colonies, often alongside Clif or Barn Swallows. Forages over fields and ponds.

Barn Swallow

Hirundo rustica

L 6.75" WS 15" WT 0.67 oz (19 g)

Our most graceful swallow, with long forked tail and long pointed wings; flight is more flowing, with smoother wing strokes than other swallows. Note blue-black upperside, whitish to orange underside, and dark rufous throat.

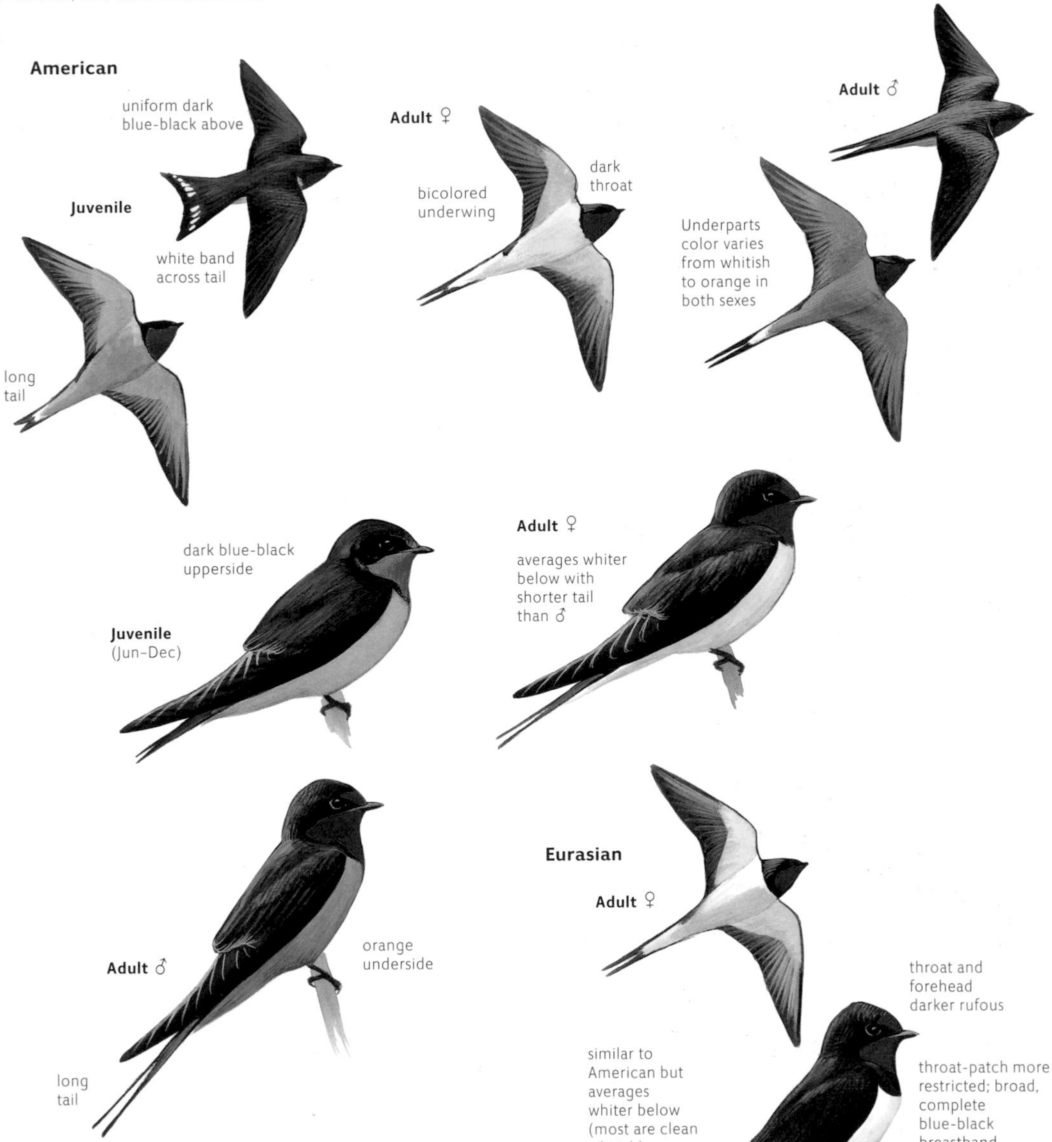

VOICE: Song lilting with husky, squeaky quality like call notes; irregularly interrupted by dry, creaking rattles. Call a short, husky *vit* or *vit-vit*; alarm a stronger and sharper *vit-VEET* or *si-VLIT*.

Common. Nests almost exclusively on structures, such as under bridges or house eaves; also on cliff ledges. Forages over fields and ponds. Nest a partial bowl of mud; large numbers may congregate at prime nesting sites.

Eurasian birds have been recorded regularly in western Alaska, several times south to Washington. Although best distinguished by throat pattern, suspects might be picked out by whitish belly and long tail. Note that shortest-tailed Eurasian juveniles are similar to adult female American; adult female Eurasian has tail length like adult male American; and adult male Eurasian is longer-tailed than any American.

Brown-chested Martin

Progne tapera

L 6.9" WS 17" WT 1.1 oz (31 g)

Size and shape similar to Purple Martin, and identification requires very good views. Look for contrast of pale throat with dark breastband. Lacks pale forehead and pale collar of female Purple Martin, but these features are surprisingly hard to confirm.

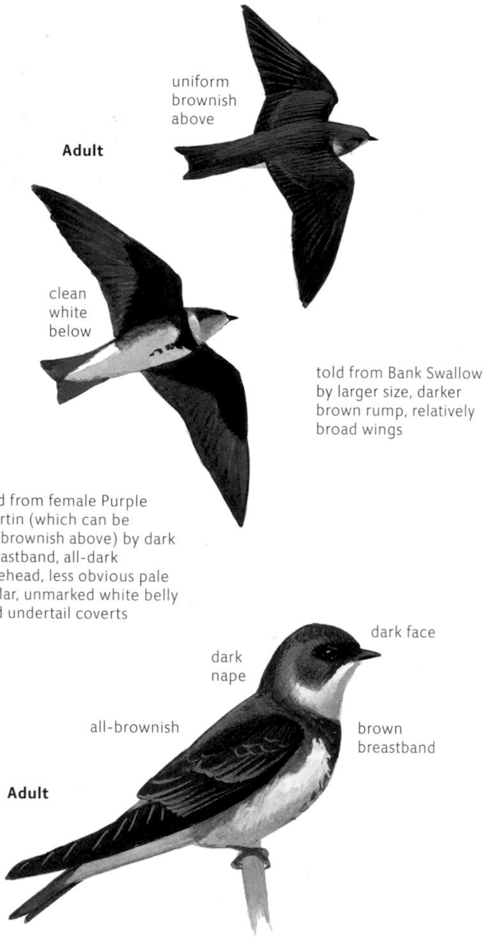

VOICE: Vagrants are generally silent. Call a two-note electric buzz *dzi-dzit*. Song much higher than Purple Martin, with tinkling quality.

Very rare visitor from South America, recorded a few times along Atlantic coast north to Massachusetts (mostly late fall, once in Jun), once in Arizona (Feb).

Common House-Martin

Delichon urbicum

L 5" WS 11" WT 0.7 oz (20 g)

Similar to Tree Swallow, but with bright white rump, longer an more deeply-forked tail, narrower wings, and paler underwin coverts.

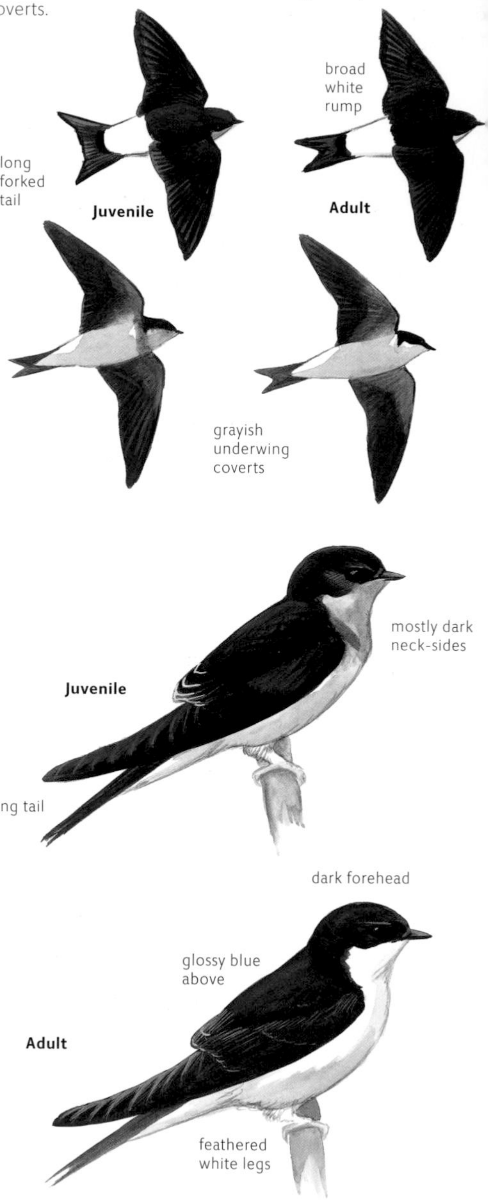

VOICE: Vagrants generally quiet. Call a thin sputtering rattle *prrt*, similar to Bank Swallow but sharper, with notes more distinct. Song a jumble of lower-pitched throaty rattles.

Very rare visitor from Eurasia, recorded only a few times in Alaska, once in St. Pierre and Miquelon off Newfoundland.

ahama Swallow

chycineta cyaneoviridis

.75" WS 14" WT 0.6 oz (17 g)

milar to Tree Swallow, but has long forked tail like Barn Swallow. all plumages note white cheek contrasting sharply with dark own, and white underwing coverts unlike all other North merican swallows.

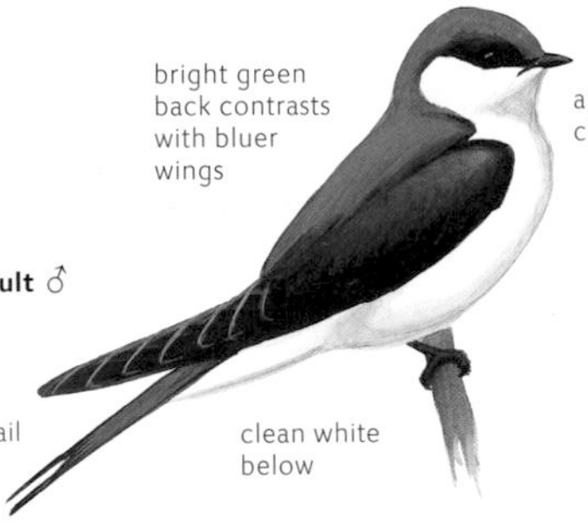

OICE: Song liquid gurgling eminiscent of Tree Swallow ut softer; also a high, laintive *seew-seew-seew-eew* given by pair. Calls *hilpilp* and *killf* softer and hirpier than Tree Swallow.

Very rare visitor from Bahamas to southern Florida mainly from Mar–Jul; most records from Big Pine Key, where open pine forests are similar to nesting habitat in the Bahamas. Forages over variety of open habitats. Formerly more regular, with records nearly annual, but unrecorded since 1992.

Mangrove Swallow

Tachycineta albilinea

L 5.1" WS 11.9" WT 0.53 oz (15 g)

Slimmer than Tree Swallow, with bright white rump, whitish underwing coverts, and thin white eyebrow. Very similar to rare Common House Martin, note smaller white rump, white eyebrow.

VOICE: Call a high twittering trill, level or rising *trrip* or *trreep*. Song a varied mix of twittering calls.

Very rare visitor from Central America, recorded once in Florida. Habits generally like Tree Swallow.

Chickadees and Allies

FAMILIES: PARIDAE, REMIZIDAE, AEGITHALIDAE

14 species in 4 genera all in family Paridae except Verdin (Remizidae) and Bushtit (Aegithalidae). Generally small and drab (but often contrastingly patterned), these species all have short legs and short, strong bills. Found in small flocks, all species are fairly social and inquisitive, often joining other small woodland songbirds in mixed-species flocks. They feed on insects and seeds gleaned from twigs and bark, and are frequent visitors to bi feeders. Flight is rather slow, hesitant and undulating. Chickade and titmice nest in tree cavities or birdhouses. Verdin's nest is spherical ball of twigs; Bushtit's nest a long pendulous hangi basket. Adults are shown.

Genus *Poecile*

Carolina Chickadee, page 398

Black-capped Chickadee, page 399

Mountain Chickadee, page 398

Mexican Chickadee, page 40

Chestnut-backed Chickadee, page 400

Boreal Chickadee, page 401

Gray-headed Chickadee, page 401

Genus *Baeolophus*

Bridled Titmouse, page 404

Oak Titmouse, page 402

Juniper Titmouse, page 402

Tufted Titmouse, page 403

Black-crested Titmouse, page 403

Genus *Auriparus*

Verdin, page 404

Genus *Psaltriparus*

Bushtit, page 405

hickadee and Titmouse dentification

herever they occur, chickadees and titmice are resident in oodland habitat, and roam around their territories in pairs or nall groups. These groups often form the nucleus of mixed-ecies foraging flocks, attracting nuthatches, warblers, vireos, nglets, and others to join the resident chickadees.

As some of the most vocal, bold, and inquisitive birds in the rest, chickadees and titmice are relatively easy to find and are nong the most aggressive at mobbing predators. They are quick respond to imitations of owl calls and to imitations of mobbing alls ("pishing"), and an agitated chickadee or titmouse will draw many other species of small birds. Simply walking through the oods might reveal the sounds of a few chickadees. An astute rder who pauses and tries to attract and see chickadees will sually find other species with them.

Strong legs and feet allow chickadees and titmice to hang upside down when foraging, a habit shared by only a few other birds.

Titmice and chickadees often hold a food item in their feet while pounding it with their bill. This behavior is shared only by jays.

Eurasian Blue Tit

Cyanistes caeruleus

L 4.4" WS 7.1" WT 0.4 oz (11 g)

Smaller than chickadees and relatively short-tailed. Bluish wings and tail, dark eyeline, pale yellow belly distinctive.

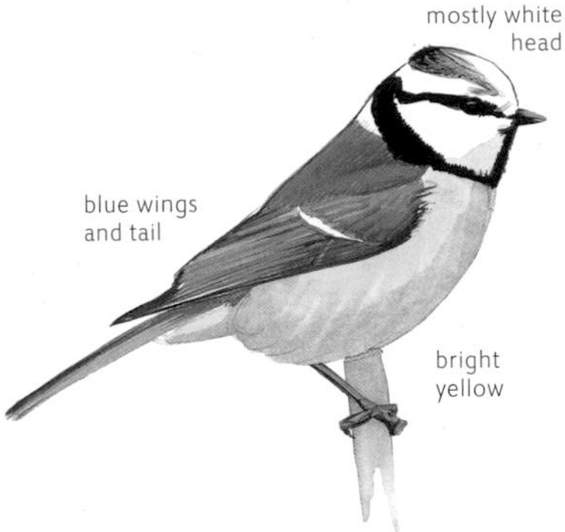

VOICE: Calls include piercing high clear notes and a low husky chatter, often combined in titmouse-like series *tseea tseea ch-ch-ch-ch-ch-ch.*

Native to Eurasia. Occasionally escapes from captivity in North America.

Great Tit

Parus major

L 5.5" WS 9.4" WT 0.6 oz (17 g)

Larger than chickadees, and relatively short-tailed. Bright white cheeks and yellow underparts with black central stripe distinctive.

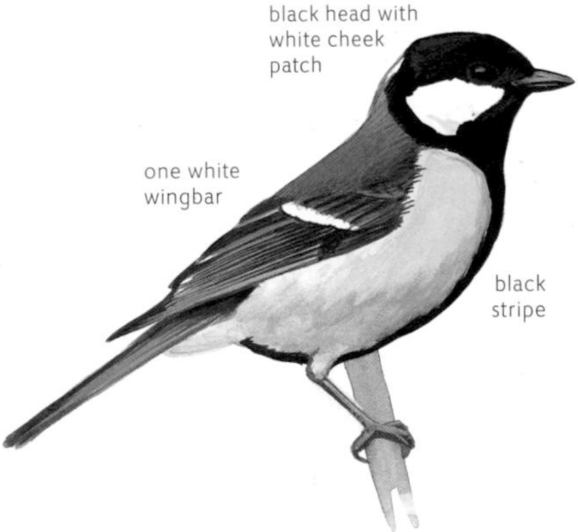

VOICE: Calls include high thin notes often followed by husky chatter or musical phrases *tsee tsee ch-ch-ch-ch or tsee tsee tsewdi.* Song a rhythmic series of two-parted whistled phrases *teew-ti teew-ti teew-ti.*

Native to Eurasia. Occasionally escapes from captivity in North America, and has nested in the wild.

Mountain Chickadee

Poecile gambeli

L 5.25" WS 8.5" WT 0.39 oz (11 g)

Habits and appearance very similar to Black-capped Chickadee; note plain grayish wings lacking white edges, white eyebrow, slightly thinner bill and shorter tail, and tendency to choose drier and more coniferous forest.

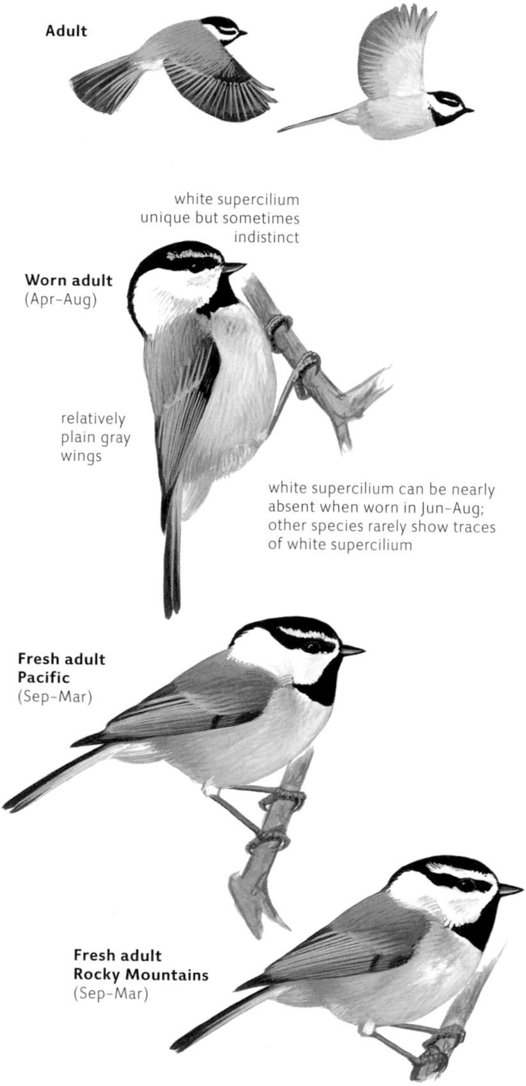

VOICE: Song of clear, high whistles like Black-capped but with three to six syllables; pattern variable with many local dialects. Call *chika dzee dzee* similar to Black-capped but *dzee* notes slightly harsher, slower, and often descending. Other calls like Black-capped.

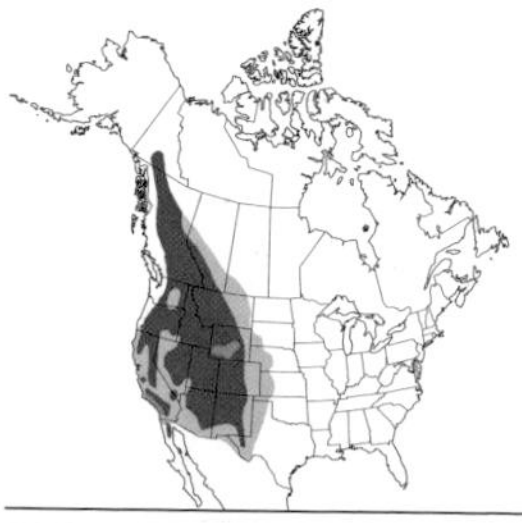

Common in montane aspen and conifer forests, usually on dry slopes in open forest or where large trees are interspersed with clearings. Often visits bird feeders.

Carolina Chickadee

Poecile carolinensis

L 4.75" WS 7.5" WT 0.37 oz (10.5 g)

Nearly identical to Black-capped Chickadee, but range overla only in a narrow band; averages smaller, shorter-tailed, smalle headed, and less brightly colored. Best distinguished by rang and by combination of plumage, shape, and voice.

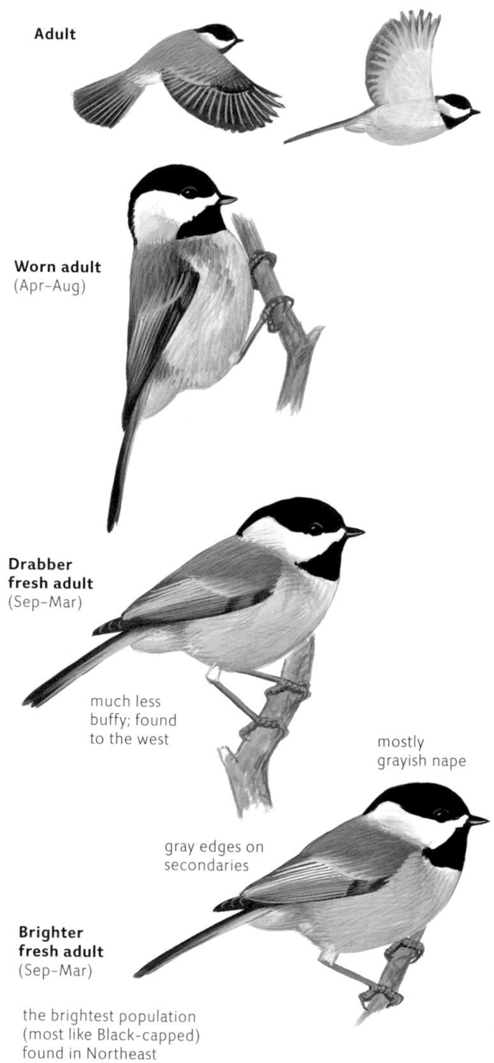

VOICE: Song three to five notes on different pitches *see bee se bay* and other variations; like Black-capped but higher-pitched all notes clearly separated by pauses. Call *chikadeedeedeede* higher-pitched and more rapid than Black-capped (5–7 *de* notes/sec vs. 3–4 in Black-capped).

Common and widespread in any wooded habitat. Habits similar to Black-capped Chickadee. Often visits bird feeders.

3lack-capped Chickadee

oecile atricapillus

5.25" WS 8" WT 0.39 oz (11 g)

ack cap and throat with white cheek and buffy flanks instantly entifies a chickadee. Very similar to Carolina Chickadee. Simiar to Mountain Chickadee but note white edging on wings, ore contrasting colors overall.

Adult

like other chickadees, flight slow, deeply undulating, and usually flies only short distances between trees

Fresh adult
Pacific
(Sep–Mar)

Fresh adult
Rocky Mountains
(Sep–Mar)

OICE: Song of most populations a simple, high, pure whistle *eebee*, with second note lower than first and relative pitch of wo notes constant; sometimes sounds three-noted, second part roken by slight falter but no real temporal break *fee beeyee*. ommon and familiar call *chikadee dee dee dee*. Contact call sharp *chik* or *tsik* slightly harsh, often leading into *chik-a-dee* all. Gargle call a complex, descending jumble of short notes and larm a very high, thin series *teeteeteeteetee*; both similar in all hickadees. Voice of Pacific population differs.

Common and widespread in any wooded habitat. Like all chickadees forms small groups; often joined by other woodland songbirds. Feeds on seeds, insects, and spiders gleaned from twigs. Often visits bird feeders.

Black-capped and Carolina Chickadee Identification

Black-capped and Carolina Chickadees are extremely similar in all respects. Overall, Black-capped is brighter, more colorful, and more contrastingly marked than Carolina; it is larger, fluffier, larger-headed, and longer-tailed, with darker tail and wings that have brighter white edges; its cheek-patch is entirely white (Carolina blends to pale gray at rear), and it has a greenish back and buffy flanks (Carolina is duller grayish); its song is lower-pitched and its call slower. All these features are relative and subject to variation, but in combination they should serve to identify most birds. Hybrids are recorded in the narrow zone of overlap where some individuals appear intermediate and many are not safely identifiable. Song is learned, so not very helpful for identification, as individual birds can learn the "wrong" song type.

Worn adult
Eastern
(Apr–Aug)

Fresh adult
Eastern
(Sep–Mar)

Regional populations of Black-capped Chickadee vary subtly in plumage and size; most variation is clinal, with many intermediates. Eastern birds are relatively bright and contrastingly marked; Rocky Mountain populations are larger and paler with broad white edges on flight feathers. Pacific populations are most distinctive, being small and dark, with drab grayish-olive edges on flight feathers and different voice. Variable songs also given by birds on islands off Massachusetts.

Chestnut-backed Chickadee

Poecile rufescens

L 4.75" WS 7.5" WT 0.34 oz (9.7 g)

Relatively small and short-tailed. Quite dark for a chickadee, with dark reddish-brown back and dark gray wings and tail; most populations have dark reddish-brown flanks. Note relatively narrow white cheek-patch.

VOICE: Lacks whistled song; an accelerating series of chips may function as song. Calls generally higher than other chickadees. Typical call high, buzzy notes with lower nasal, husky notes: *tsidi-tsidi-tsidi-cheer-cheer* or weaker *tsity ti jee jee*. Some high buzzy notes similar to warbler flight calls.

Common in moist woods of oaks, pines, and Douglas-Fir. Habits similar to Black-capped Chickadee. Occasionally visits bird feeders.

Mexican Chickadee

Poecile sclateri

L 5" WS 8.25" WT 0.39 oz (11 g)

Habits and appearance similar to Black-capped and Mountai Chickadees, but range does not overlap. Note extensive blac bib, dark gray flanks about the same color as back, and gray win with narrow wingbar.

VOICE: Song a series of short, abrupt phrases *peeta peeta peet* may recall Oak and Juniper Titmice. Characteristic call note high, buzzy *sschleeeer* level or slightly descending; also hig buzzy trills preceding slow, hissing notes *tzee tzee tzee shhh shhhh* and a low, husky trill *dididididi*.

Common in high-elevation pine and fir forests. Found in the US only in the Chiricahua Mountain of Arizona and the nearby Animas Mountains of New Mexico, where it is normally the only chickadee present.

Gray-headed Chickadee

Poecile cinctus

L 5.5" WS 8.5" WT 0.44 oz (12.5 g)

A relatively large and long-tailed chickadee, with very fluffy plumage. Told from Boreal Chickadee by white sides of neck and white-edged wing feathers. From Black-capped by gray-brown cap about the same color as back, reduced black bib.

Adult

VOICE: Short, high buzzing phrases in whistled song. Call *tsiti ti jeew jeew jeew* or simply *jeew jeew jeew*, lower than Boreal.

Uncommon to rare in stunted birch, alder, and spruce woods along rivers near tree line. Usually in pairs or family groups. Feeds on seeds, insects, and spiders.

Boreal Chickadee

Poecile hudsonicus

L 5.5" WS 8.25" WT 0.35 oz (10 g)

Compared to Black-capped Chickadee darker overall, with plain gray wings and tail, gray nape leaving only small white cheek-patch, and relatively dark brownish flanks. Also slightly shorter-tailed, and nearly always in conifers.

Adult

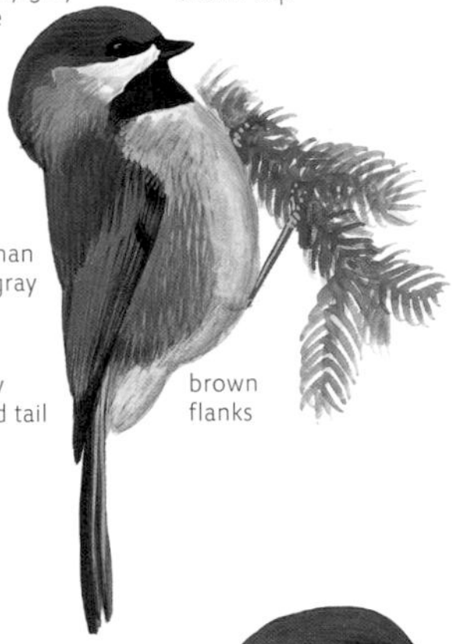

VOICE: Song a simple, clear trill with short introductory note *p-twee-titititititititi*. Call a labored, wheezy, nasal *tsi-jaaaay* or *tsi ti jaaaay jaaay* ("yesterdaaay"); *chit* or *tchidk* call (harsher, more staccato than Black-capped) very common year-round in many variations.

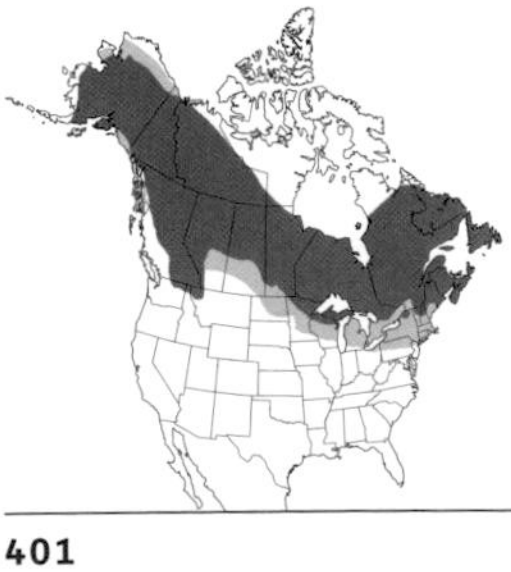

Uncommon in dense spruce-fir woods. In small groups; less social and more secretive than other chickadees. Feeds on seeds, insects, and spiders gleaned from twigs. Rarely visits bird feeders.

Oak Titmouse

Baeolophus inornatus

L 5.75" WS 9" WT 0.6 oz (17 g)

Nearly identical to Juniper Titmouse, but range overlaps only on the Modoc Plateau of northeastern California. Averages slightly darker and browner, best distinguished by range and voice.

VOICE: Song of strong, whistled, repeated phrases *tjiboo...* or paired *tuwituwi...* and other variations. Also a rapid, popping trill. Call a few high, thin notes followed by a single harsh scold *si si si chrr*; the same pattern often given in clear, whistled notes *pi pi pi peeew*.

Common in open oak woods or nearby trees. Usually in pairs or family groups. Feeds on seeds and insects. Occasionally visits bird feeders.

Juniper Titmouse

Baeolophus ridgwayi

L 5.75" WS 9" WT 0.6 oz (17 g)

Nearly identical to Oak Titmouse. Very plain; drab gray-brow overall with short bushy crest, no contrasting markings. Drabb and shorter-crested than Black-crested Titmouse, larger an drabber than Bridled and found in more arid habitat.

VOICE: Song averages lower-pitched and more rapid than Oa with less whistled quality; phrases often in groups of three, result ing in a pulsing rattle on one pitch *jijiji jijiji jijiji....* Also gives rapid, popping trill faster than Oak. Call a rapid *sisisi-ch-ch-ch ch* or *si-ch-ch-ch* and other variations.

Uncommon in arid oak-juniper or pinyon-juniper woods. In pair or small groups. Feeds on seeds and insects. Occasionally visits bird feeders.

Black-crested Titmouse

Baeolophus atricristatus

L 6.5" WS 9.75" WT 0.75 oz (21.5 g)

Very similar to Tufted Titmouse and sometimes considered the same species, differs in pale forehead and blackish crest. Hybrids are common in narrow zone of range overlap, showing mix of features and often chestnut forehead.

These two species hybridize frequently in a narrow zone where ranges meet in central Texas. Such birds combine characteristics of both populations; they typically have dark gray crest and chestnut brown forehead, but a complete range occurs.

VOICE: Song typically consists of five to seven slurred phrases delivered rapidly *peew peew peew peew peew,* averages slightly faster and less clearly two-syllabled than in Tufted. Call a series of angry, nasal, rising notes often preceded by very high, thin notes: *ti ti ti sii sii zhree zhree zhree.*

Common in oak woods. In pairs or small groups, often in mixed flocks with warblers, kinglets, and other small songbirds. Feeds on insects and seeds. Often visits bird feeders.

Tufted Titmouse

Baeolophus bicolor

L 6.5" WS 9.75" WT 0.75 oz (21.5 g)

Uniform pale gray above, with short, pointed crest, short bill, black forehead, and orange-buff flanks. Chickadee-like in shape and actions, but larger, with distinct crest, relatively short tail, and different plumage pattern.

VOICE: Song a low, clear whistle *peter peter peter peter* usually strongly two-syllabled. Call angry, nasal, rising notes often preceded by very high, thin notes *ti ti ti sii sii zhree zhree zhree.* Also simple whistled notes such as *see-toit* and high, sharp *tsip* like Field Sparrow.

Common and widespread in mature deciduous woods. In pairs or small groups, often in mixed flocks with chickadees and other forest songbirds. Feeds on insects and seeds. Range has expanded northwards in recent decades. Often visits bird feeders.

Bridled Titmouse

Baeolophus wollweberi

L 5.25" WS 8" WT 0.37 oz (10.5 g)

Our smallest titmouse, chickadee-size but shorter-tailed. Black throat and bridled face pattern with short crest unique. Range and habitat barely overlaps with other titmice or chickadees.

VOICE: Song a rapid series of six to eight low, clear, whistled phrases with popping quality: *pidi pidi pidi pidi pidi* or *pipipipi....* Call simply a rapid series of low, harsh notes in chickadee-like chatter *ji ji ji ji ji* or *jedededededed*; sometimes preceded by high, clear notes *tsi tsi tsi jedededededed*.

Common in oak woods at middle elevations, especially along streams where oaks mix with junipers and sycamores. In winter some move to cottonwood/willow thickets along streams at lower elevations. Usually in small groups, often mixed with other small songbirds. Feeds on insects and seeds. Occasionally visits bird feeders.

Verdin

Auriparus flaviceps

L 4.5" WS 6.5" WT 0.24 oz (6.8 g)

Very small, with sharply pointed, triangular bill and relatively short tail. Pale grayish color and yellow head distinctive; also note dark lores. Juvenile plain gray like Bushtit but different shape and habits, pale bill base.

VOICE: Call a high, piercing *tseewf*; also a lower-pitched, strong whistle *tee too too* or *tee too tee tee*. Contact call a flat, hard *kit* or *tsik* often repeated rapidly, three to four times per second; also high, sharp, slightly nasal *kleeu*.

Common and conspicuous in brushy deserts and suburbs. Nearly always solitary, moving briskly and acrobatically through brush in search of small insects, fruit, and nectar. Often perches on exposed twigs and flies long distances from shrub to shrub. Visits hummingbird feeders, not seed feeders.

Bushtit

Psaltriparus minimus

L 4.5" WS 6" WT 0.19 oz (5.3 g)

Extremely small. Combination of tiny size, round body, and overall gray-brown color, with stubby bill and long tail distinctive. Some (mostly juvenile males of interior population) show partial to full black ear patch.

Pacific

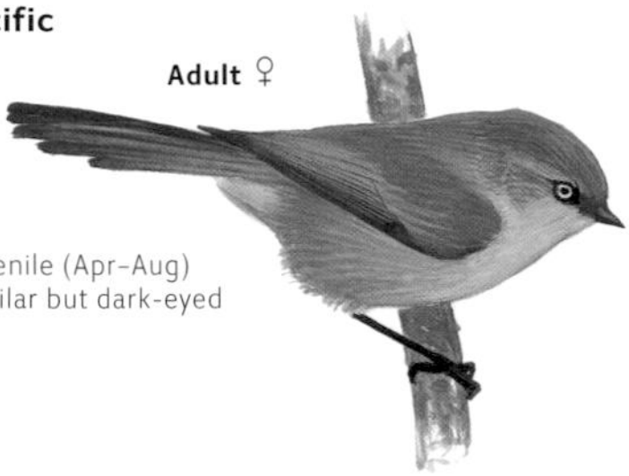

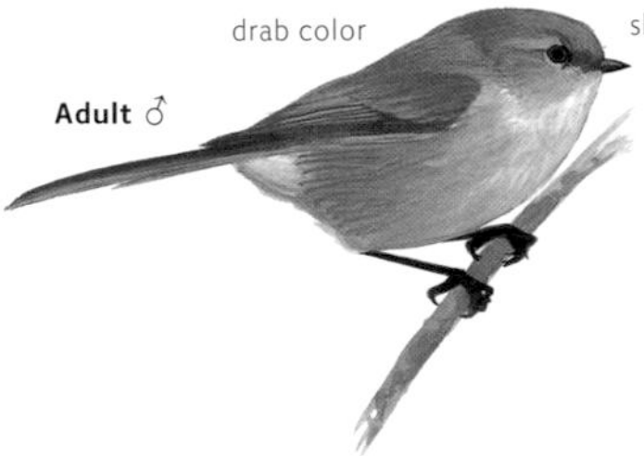

Interior

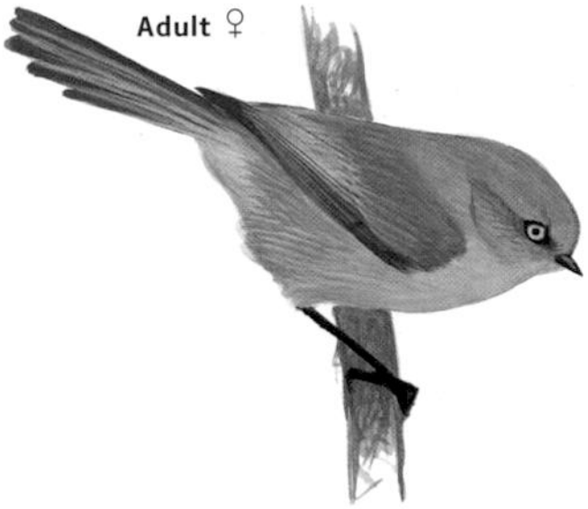

Although similar in all other respects, Interior and Pacific populations differ consistently in plumage; intermediates occur in narrow zone where ranges meet. Interior birds are paler and grayer overall, with gray crown matching back color; Pacific birds are browner overall, with dark brownish crown that contrasts with back. In addition, Pacific birds may have more obvious dark lores (Interior have paler lores).

complete range of variation between typical and black-eared forms

Black-eared form is uncommon in southwestern mountains; seen mainly in Texas but recorded rarely (juveniles only) to Colorado and eastern California. Shown mostly by juvenile male, occasionally by juvenile female or adult male. Fully black-eared adults are rare in North America, common in Mexico.

VOICE: Calls include a sharp, scraping buzz ending with several high, clear chips *skrrti ti ti*; dry *spik* notes like Lazuli Bunting; and high, thin, scraping *tseeez tzee tzee tzee...* in long, slightly descending series. All calls given constantly by flock, producing sharp, buzzing, sparkling chatter: a mixture of thin, scratchy notes and high, clear chips. Aerial predator alarm very high, clear, tinkling, descending *tsidididididi*; usually given by several birds at once as flock seeks cover.

Common in brushy chapparal habitat or open woods especially oak and juniper. Always in flocks that buzz with activity, moving rapidly through trees and chattering constantly in high-pitched notes. Flock swarms into trees and bushes in search of aphids and other tiny insects, then flies single file to next tree. Not attracted to bird feeders.

Nuthatches and Creeper

FAMILIES: SITTIDAE, CERTHIIDAE

5 species in 2 genera. Nuthatches in family Sittidae; creeper in Certhiidae. Nuthatches are short-tailed and long-billed with a distinctive tree-climbing method; using strong legs to support their body, they can climb head down, gleaning insects from bark crevices. Black-and-white Warbler uses the same style of climbing. The cryptically-colored Brown Creeper uses its long tail as a prop (as woodpeckers do) to climb trees head-up and glean food from bark. Nuthatches nest in cavities, Brown Creeper nests behind a sheet of loose bark. Flight of all species is undulating. All species often join flocks of chickadees and other woodland songbirds during nonbreeding season. Adults are shown.

Genus *Sitta*

Red-breasted Nuthatch, page 406

White-breasted Nuthatch, page 407

Pygmy Nuthatch, page 408

Brown-headed Nuthatch, page 408

Genus *Certhia*

Brown Creeper, page 409

Red-breasted Nuthatch

Sitta canadensis

L 4.5" WS 8.5" WT 0.35 oz (10 g)

One of the smallest songbirds, very short-tailed, blue-gray above and orange below. Much smaller than White-breasted Nuthatch, with dark eyeline, white eyebrow, and pale orange underparts, and more often perches on twigs rather than clinging to bark.

Adult ♂

nests in coniferous forests

Adult ♀

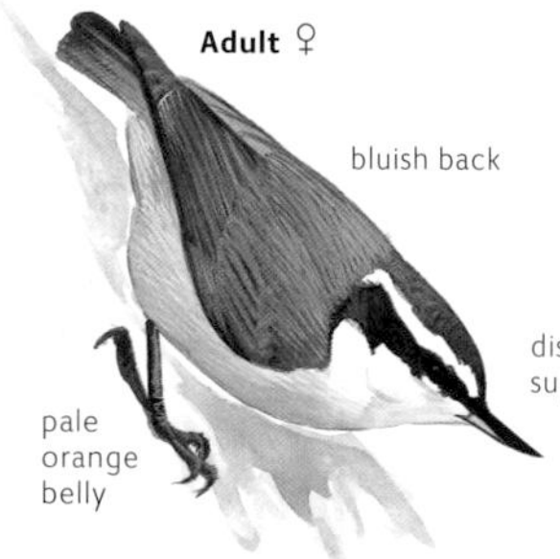

Adult ♂

VOICE: Song a monotonous series of clear, nasal, rising calls repeated slowly *eeeen eeeen eeeen*.... Call a weak, nasal *ink* or *yenk*; when agitated a longer, rising *iiink* and lower, hoarser series *iik iik iik*.... Calls generally shorter, more nasal than White-breasted.

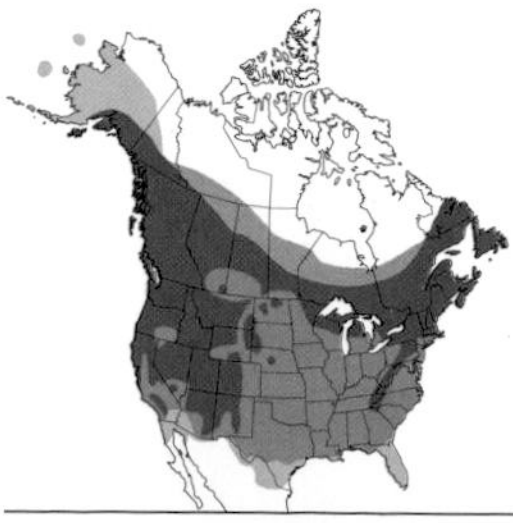

Common but somewhat irregular in coniferous and mixed woods. Subject to occasional irruptions when large numbers move south and into lowlands. Generally solitary, but may be found in pairs or small groups, and often joins mixed flocks of small songbirds. Feeds on insects and seeds. Often perches on twigs. Often visits bird feeders.

White-breasted Nuthatch

Sitta carolinensis

L 5.75" WS 11" WT 0.74 oz (21 g)

Our largest nuthatch; note extensively white head, with white face surrounding eye, narrow dark crown and long bill. Short, broad tail with pale corners, black, gray and white overall, usually seen clinging to bark sideways or head-down.

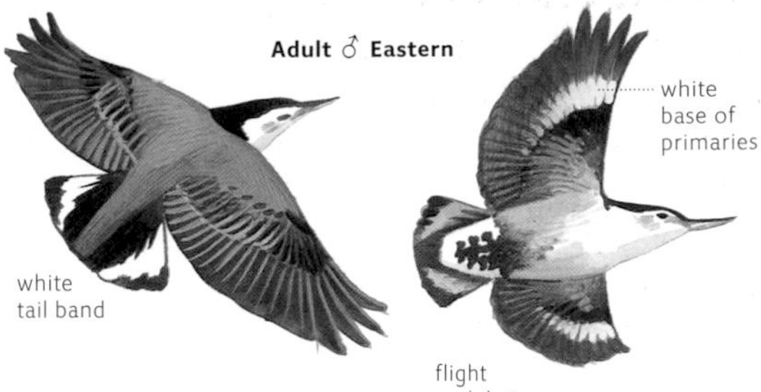

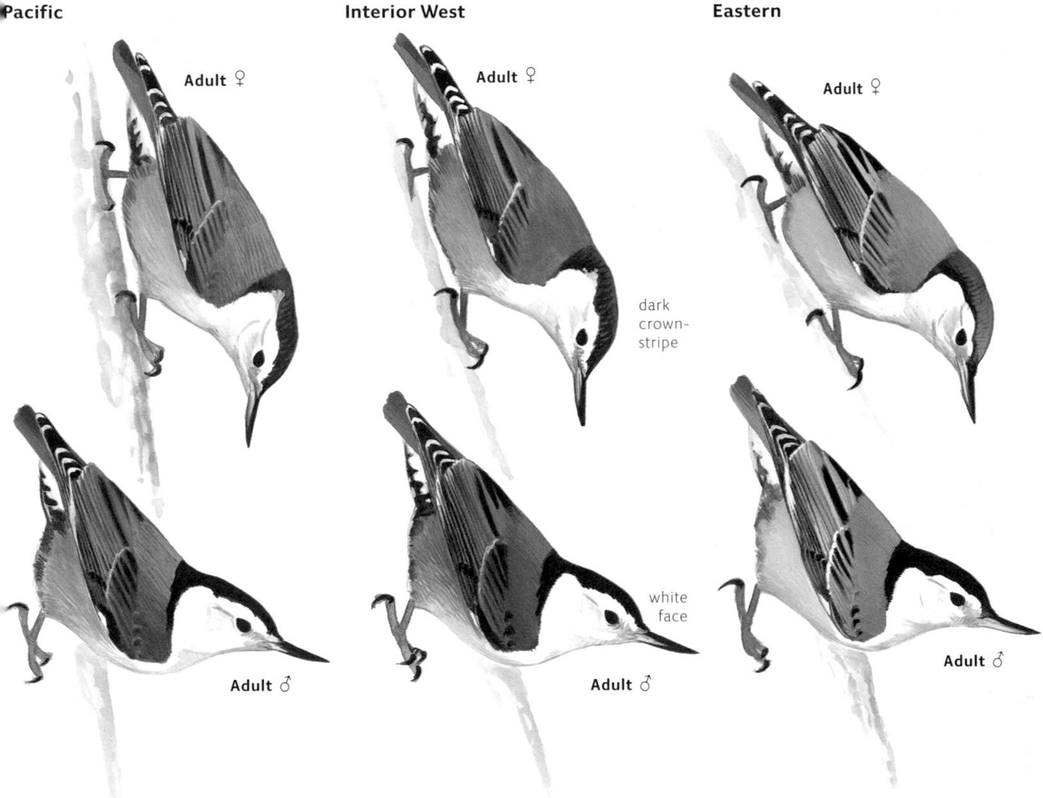

VOICE: Song of all populations a series of soft, slightly nasal, whistled notes on one pitch *whi-whi-whi-whi-whi-whi-whi*. All give short, nasal contact calls that vary regionally and soft, high *ink* notes while foraging.

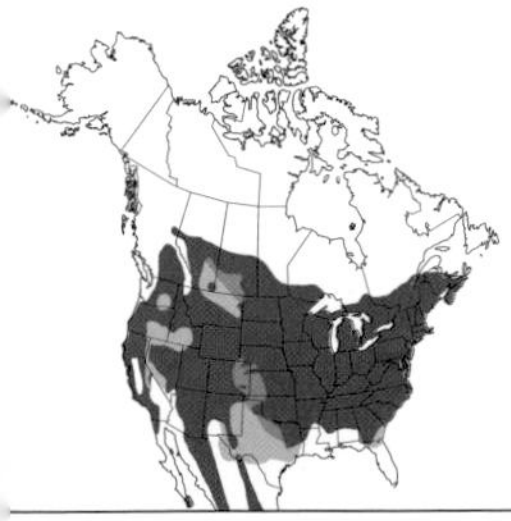

Common in mature deciduous and mixed woods. Usually solitary or in pairs, but may be loosely associated with mixed flocks of small songbirds. Feeds on insects and seeds gleaned from bark. Usually seen on trunks or major limbs of trees, often climbing head-down in characteristic nuthatch fashion. Often visits bird feeders.

Three populations differ in voice (see below) and subtly in plumage and shape. Eastern has thickest bill and palest gray back with sharply contrasting black marks on tertials and coverts. Eastern also has buffier flanks and a broader dark crown-stripe than either of the western populations. Pacific and Interior West both have thinner bills, darker gray backs with less contrasting dark marks on tertials and coverts, and narrower dark crown-stripes than Eastern. Interior West differs from Pacific in darker back and much darker flanks.

Eastern call a low and hoarse nasal *yank* (lower and stronger than Red-breasted Nuthatch call), and song relatively low and hoarse, each phrase rising *pway pway pway*.... Interior West call higher, short, repeated in rapid series *yidididid*, or in pairs in series *yidi yidi yidi*..., song higher than Eastern. Pacific call highest, a thin nasal descending *eearrn* or *beerrf*, song highest and usually with two-syllabled phrases *tuey tuey tuey*.... DNA analysis reveals a fourth group distinct from the Interior West population, in the Sierra Nevada and northern Rocky Mountains, which apparently has calls and song with rapid pattern like Interior West, but higher-pitched like Pacific.

Pygmy Nuthatch

Sitta pygmaea

L 4.25" WS 7.75" WT 0.37 oz (10.5 g)

Tiny, short-tailed with relatively large head and bill. Distinguished from Red- and White-breasted Nuthatches by small size, dark crown reaching eye, pale buffy and gray underparts, and voice. Very similar to Brown-headed Nuthatch but no range overlap.

Adult

Worn adult
(Apr–Aug)

Fresh adult
(Sep–Mar)

VOICE: Call high, clear, hard peeping usually given in chorus by flock: amazing variety of loud chips, most frequently *bip-bip-bip* and many higher squeaks and chips; loud *kip* on variable pitch; rather flat, high, strident *peet* or *peeta*. In flight weaker *imp imp*.

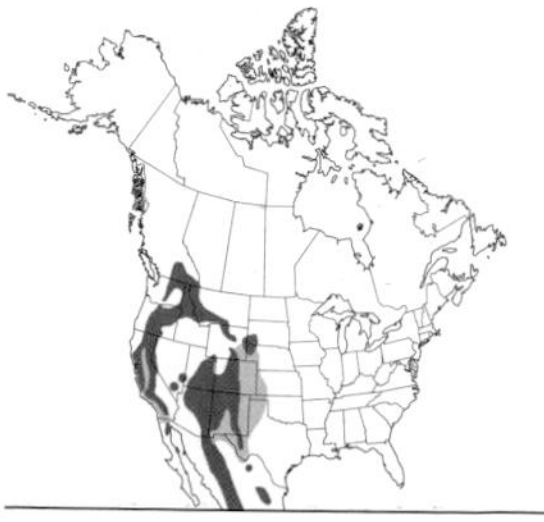

Common in coniferous forests, especially pines. A small and active nuthatch, usually found in noisy and active flocks mixed with other songbirds, such as chickadees, kinglets, and warblers. Feeds on insects and seeds. Occasionally visits bird feeders.

Brown-headed Nuthatch

Sitta pusilla

L 4.5" WS 7.75" WT 0.35 oz (10 g)

Smaller than Red- or White-breasted Nuthatch, with relativel large head and bill. Also note brownish crown extending to eye pale buffy and grayish underparts, and voice.

Adult

Worn adult
(Apr–Aug)

Fresh adult
(Sep–Mar)

VOICE: Call a high, sharp, spunky, slightly nasal *KEWde* usually followed by lower, hard, nasal notes *KEWdodododo teew*; also a hard *pik*.

Common in pine forests. Usually in small groups, active and noisy as they move briskly through treetops in search of food; often mixes with chickadees and other woodland flocking birds. Feeds on insects and seeds. Often visits bird feeders.

3rown Creeper

:erthia americana

5.25" WS 7.75" WT 0.29 oz (8.4 g)

'ery small and delicate, with long tail and slender, curved bill. ılways seen clinging tightly to bark of large trees, braced with ail. Creeping habits, cryptic brown pattern, and whitish underɔarts distinctive.

There is much subtle variation in plumage within this species, but regional variation is overshadowed by the presence of reddish, brown, and gray morphs within many populations. In general, Western populations are relatively small, dark, and long-billed; Eastern populations are slightly larger, paler, buffier, and shorter-billed. Mexican population resident from southeastern Arizona into Mexico is darker, particularly on the gray belly and chestnut rump, and averages shorter- and rounder-winged and shorter-billed than other western birds. Regional differences in voice, some of which are described, deserve more study.

'OICE: Song a very high, thin series of accelerating, cascading ıotes. Call a very high, thin, quavering *seee* or *sreee* similar to Golden-crowned Kinglet, but single, with relaxed, liquid quality; veaker than waxwings. In flight much weaker, extremely high ıotes *tit, titip*, or *zip* sharper than Golden-crowned Kinglet.

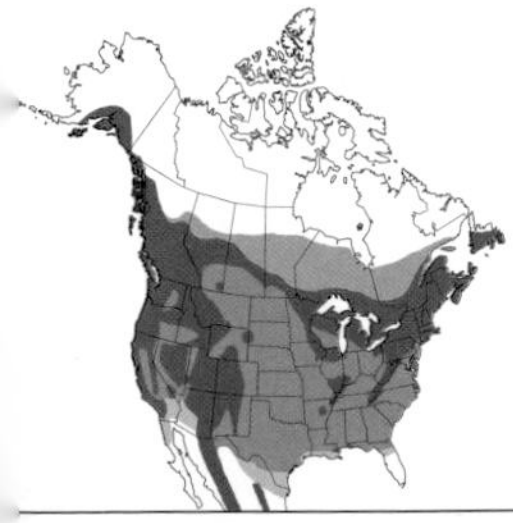

Uncommon in mature woods, particularly in wet, shaded areas where it creeps on trunks and large limbs. Generally solitary, but often mixes with flocks of other small songbirds such as kinglets and chickadees. Feeds on insects and spiders gleaned from bark. Not attracted to bird feeders.

Eastern song begins with two long, high notes followed by irregular jumble ending on a low note *seee sooo sideeda sidio*; Western song usually ends on a high note *seee sitsweeda sowit-see* ("trees trees pretty little trees"), overall clearer, simpler with less jumbled sound than East. Mexican song begins with one or two widely spaced trilled notes, then a descending jumble with slightly "looser" and slower tempo than East, ending on a low note.

Wrens

FAMILY: TROGLODYTIDAE

11 species in 7 genera (1 rare species not shown here). Wrens are mostly small, brown, and active but secretive; they creep through tangled and dense vegetation, foraging for insects and fruit, often with their tails raised above their backs. They are often detected by their loud and rollicking songs and calls, surprisingly forceful from such small birds, and seeming out of character for species that are otherwise inconspicuous. Their flight is quick and erratic on short, rounded wings. All have narrow heads and long, slender bills, an adaptation for probing crevices. Most nest in cavities, including birdhouses; Cactus, Sedge, and Marsh Wrens build globular nests of sticks or reeds. Adults are shown.

Genus *Salpinctes*

Rock Wren, page 417

Genus *Catherpes*

Canyon Wren, page 417

Genus *Troglodytes*

Pacific Wren, page 413

Winter Wren, page 413

House Wren, page 412

Genus *Thryomanes*

Bewick's Wren, page 411

Genus *Cistothorus*

Sedge Wren, page 414

Marsh Wren, page 415

Genus *Thryothorus*

Carolina Wren, page 410

Genus *Campylorhynchus*

Cactus Wren, page 416

Carolina Wren

Thryothorus ludovicianus

L 5.5" WS 7.5" WT 0.74 oz (21 g)

Reddish-brown overall; note bold, white eyebrow. Stockier tha Bewick's Wren, with shorter tail, brighter reddish-brown upper side, orange-buff flanks, and pale legs.

Adult

Juvenile
(Apr–Sep)

Adult Northern

Adult Mexican

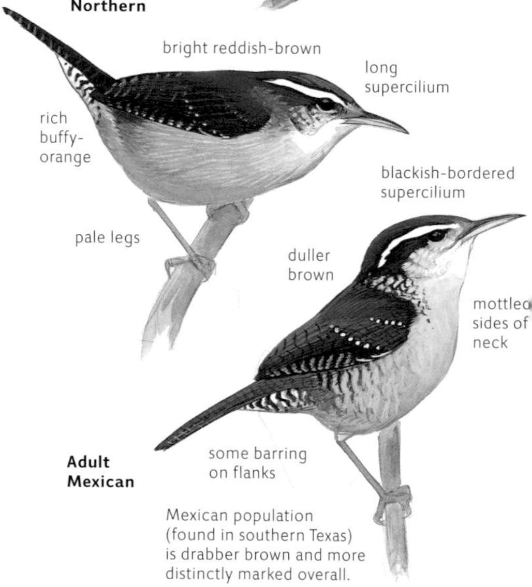

Mexican population (found in southern Texas) is drabber brown and more distinctly marked overall.

VOICE: Song a rolling chant of rich phrases *pidaro pidaro pidarc* or *TWEE pudo TWEE pudo TWEEP* and other variations. A long buzzing chatter sometimes given by female while male sings Varied calls generally richer than other wrens: a harsh, complain-ing *zhwee zhwee zhwee*...; a descending, musical trill; a low, solic *dip* or *didip*.

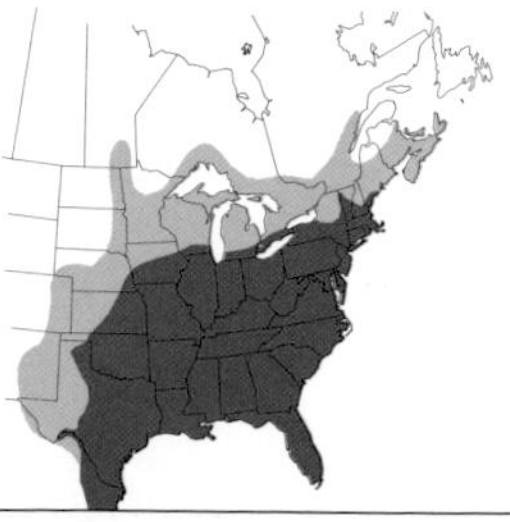

Common in dense brushy tangles within woods or edges, including suburban yards with dense shrubbery. Occasionally visits bird feeders.

Bewick's Wren

Thryomanes bewickii

L 5.25" WS 7" WT 0.35 oz (10 g)

Larger and more slender than House Wren, with longer tail often flicked and fanned showing white corners. Also note plain brown upperparts, gray flanks, white breast, and bold white eyebrow. More slender than Carolina Wren and not as colorful.

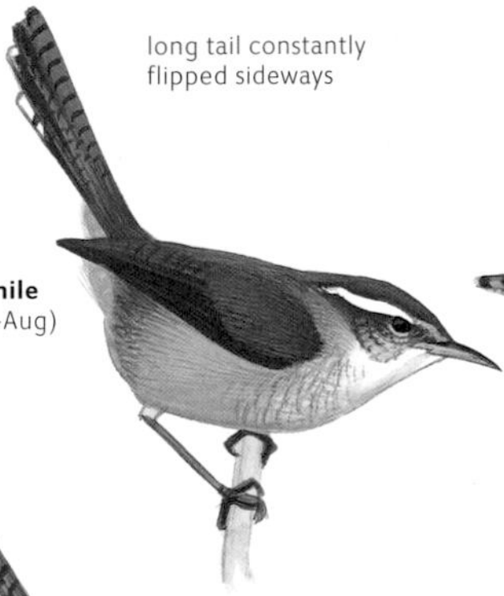

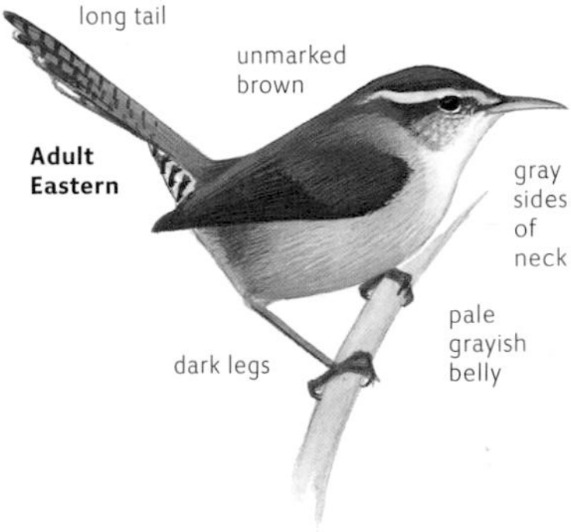

Regional variation in plumage color is subtle and clinal, but the three extremes are quite distinctive. Interestingly, song variation corresponds to plumage variation: simplest songs are given by the grayest birds (Southwest) and much more complex songs are given by the most reddish birds (Eastern).

VOICE: Extremely varied. Song varies regionally but always has thin, rising buzzes and slow trills, with distinctive quality and overall descending pitch: Arizona birds simple (e.g., *tuk zweee-drrrrrrr*); Pacific birds more complex (e.g., *t-t zree drr-dree tutututututu*); Eastern birds most complex, recalling Song Sparrow with high, clear notes and musical trills (e.g., *zrink zrink oozeeee delzeedle-eedle-ooh tsetetetetetete*). Call generally dry, harsh, and unmusical, but also varied soft notes. Scold of drawn-out harsh notes *shreeee, zheeeeer* or *jree jree*... and other variations; also a soft, dry *chrrr*, harsh *jik*, soft *wijo*, or sharp *spik*. Most distinctive is a high, rising *zrink* with characteristic zippy quality, often incorporated into song.

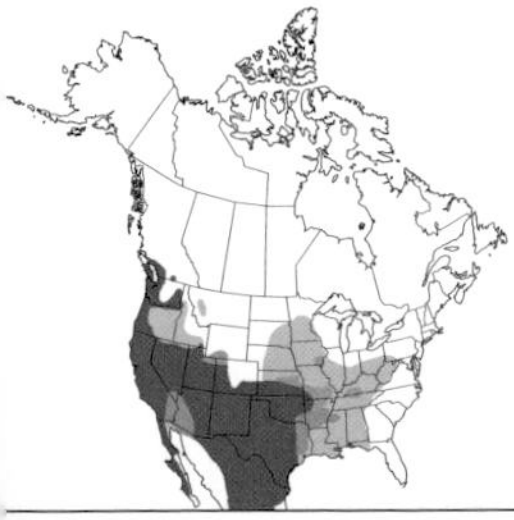

Common in many dense brushy habitats from mesquite thickets to chapparal to riparian willows to openings in coniferous forests. Rarely visits bird feeders. Eastern population, east of central Texas, rare and declining.

Sinaloa Wren

Thryothorus sinaloa

L 5.2" WS 7" WT 0.62 oz (18 g)

A small, brownish wren with bold white eyebrow similar to Bewick's Wren but stockier, with shorter tail. Note bright rufous tail, contrasting with rest of upperparts, and black-and-white streaks on cheeks and sides of neck.

VOICE: Song of varied, loud whistles, some low and rich like orioles, others higher with very quick upslur or downslur. Also short harsh scold notes and a ratcheting dry rattle.

Very rare visitor from Mexico, with several records in southern Arizona. Secretive in dense riparian thickets.

House Wren

Troglodytes aedon

L 4.75" WS 6" WT 0.39 oz (11 g)

The familiar small, grayish-brown wren of gardens and hedgerows, often known by its loud, bubbling song. Small and relatively slender. Drab gray-brown overall, with weak eyebrow and eyering and pale underparts.

Adult

Rufous juvenile Western
(May–Aug)

some bright reddish as shown; most are similar to adult and to Eastern juvenile

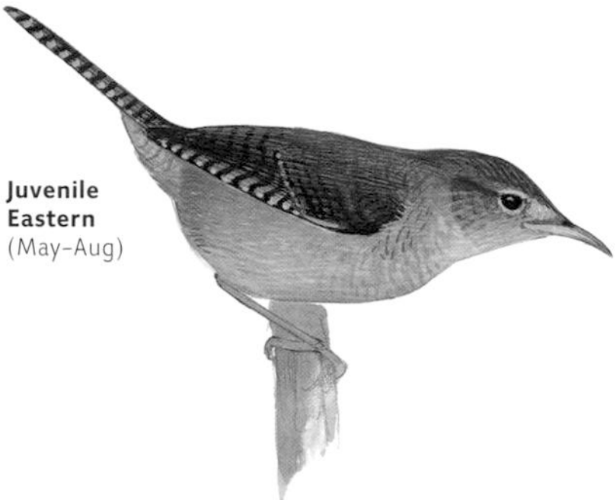

Juvenile Eastern
(May–Aug)

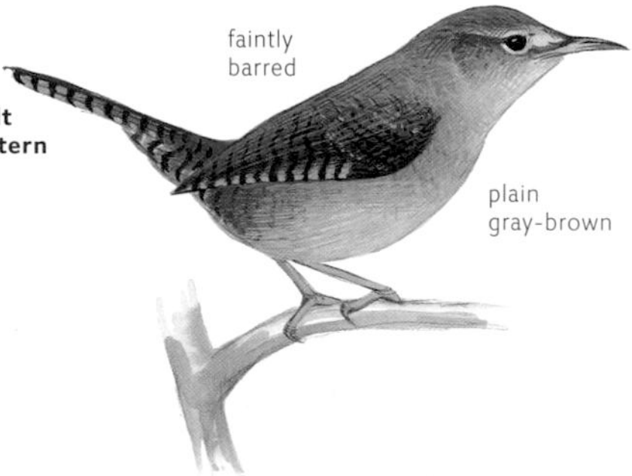

Adult Western

faintly barred

plain gray-brown

Gray adult Eastern

Adult Mexican
(Brown-throated)

Brown adult Eastern

weak supercilium

drab gray-brown

VOICE: Extremely varied. Song a rapid, rolling series of rattles and trills culminating in a descending series of bubbling liquid trills. Calls generally low, dry, short, often rather soft notes. One call commonly heard in West, but never in East: a rolling, trilled, musical *dirrd*; other possible differences overshadowed by variation. Scold a nasal whining or mewing *merrrrr* reminiscent of gnatcatchers or Gray Catbird; also long, harsh rising *sshhhhp*; series of harsh *chrrf* notes generally weaker and more hissing than Bewick's, much softer and drier than Carolina. Also low *ch* or *chek* notes given singly.

Common in dense brushy patches, overgrown gardens, and hedgerows, where its loud bubbling song is a familiar sound. Not attracted to bird feeders.

Western birds average grayer than Eastern with more contrasting reddish brown tail coverts and flanks, but there is extensive overlap. Some Western juveniles are strikingly rufous. One Western call is unique. Brown-throated population of southeastern Arizona (mainly Huachuca and Santa Rita Mountains) averages warmer buff on underparts, with more barring on flanks and more prominent buffy supercilium, but Arizona populations are variable and intermediate between northern birds and the true Brown-throated populations in Mexico. Brown-throated song is slightly sweeter and more complex than Eastern; calls are like other western populations.

Pacific Wren

Troglodytes pacificus

L 4" WS 5.5" WT 0.62 oz (9 g)

Tiny and dark, this species was recently split from Winter Wren and is extremely similar but with little range overlap. Averages slightly darker and more rufous overall, best identified by voice.

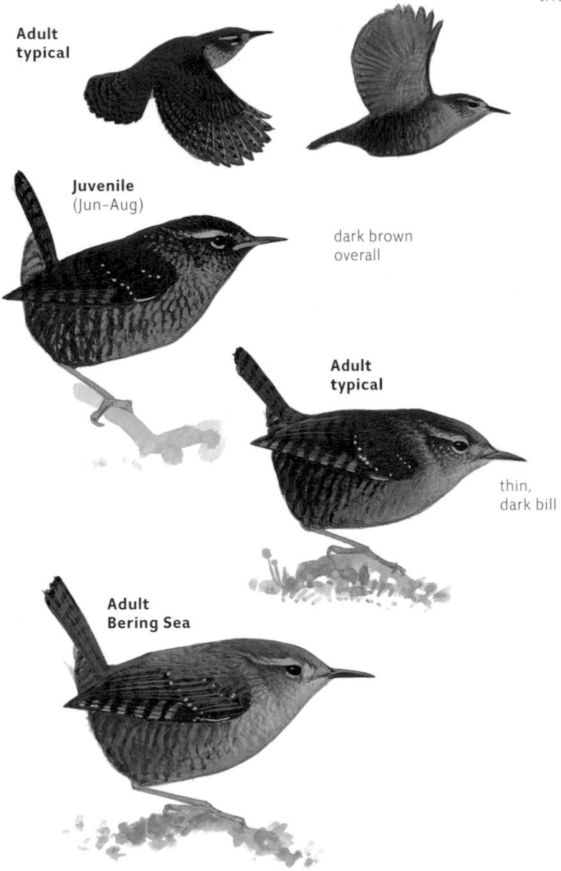

Populations resident on western Alaskan islands average up to 20% larger overall and are paler brown. Calls are longer and squeakier. Song is more varied, buzzy lower-pitched.

VOICE: Song long and complex, a remarkable continuous warble of very high, tinkling trills and thin buzzes, very similar to Winter Wren. Call short, usually doubled, sharp *chat-chat*, similar to Winter but higher and sharper (like Wilson's Warbler). Agitated birds give a rapid, high, staccato series, a similar call is given by Winter Wren.

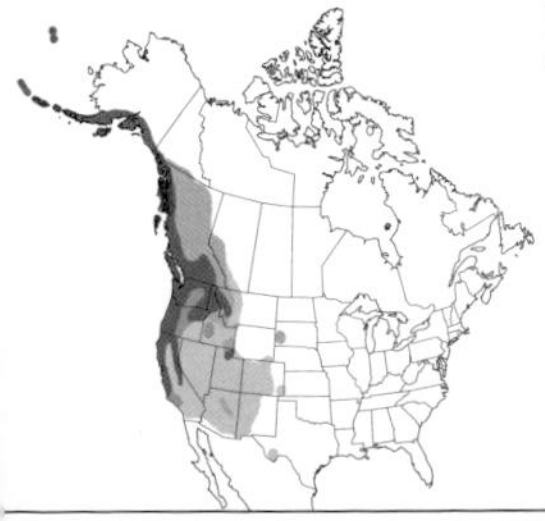

Like Winter Wren, uncommon in damp shaded areas.

Winter Wren

Troglodytes hiemalis

L 4" WS 5.5" WT 0.62 oz (9 g)

This and Pacific Wren our smallest and darkest wrens, very short tail often raised vertically over back. Note short pale eyebrow, dark brownish belly. Very similar to Pacific (formerly considered the same species); best distinguished by voice.

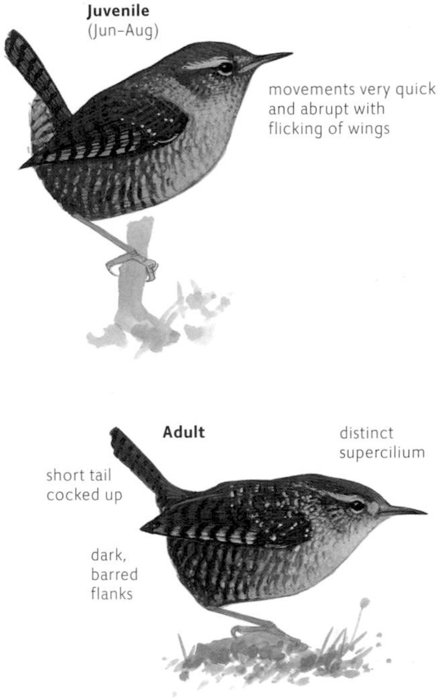

Songs of both Winter and Pacific Wrens are long, complex series of very high trills and warbles. There are subtle differences in quality, mainly due to the faster tempo of the trills of Pacific, which leads to a more mechanical and rattling sound. In addition to sound quality, listening for variety in a singing performance can help identify the species. Each male Winter sings a repertoire of only two songs, while each male Pacific switches between thirty or more songs.

VOICE: Song a remarkable long series of very high tinkling trills and warbles, like Pacific Wren but trills slower, more musical, with relatively soft quavering sound (vs. faster trills, more mechanical, buzzing). Call a short hard note, usually doubled *jip-jip*, distinctly lower and softer than Pacific, with quality close to Song Sparrow but harder.

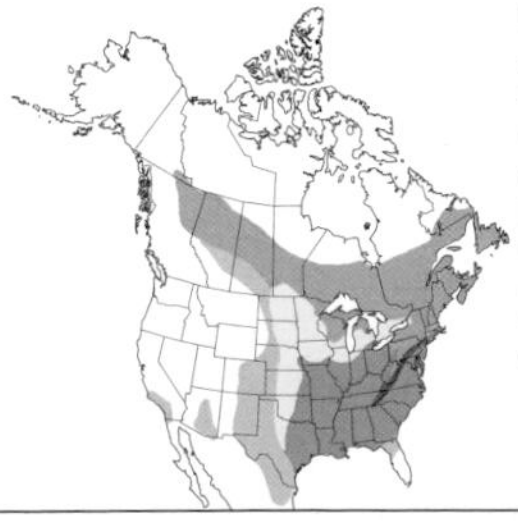

Uncommon in damp shaded areas, such as along streams in coniferous woods, where it climbs around fallen logs and overturned stumps, or in brush piles, working in and out of crevices and through tangles of branches. Generally solitary and does not join mixed flocks of other woodland songbirds.

Sedge Wren

Cistothorus platensis

L 4.5" WS 5.5" WT 0.32 oz (9 g)

Very small, with short bill and tail, and overall pale sandy-buff or cinnamon color. Smaller and paler than Marsh Wren, with boldly barred wings and streaked crown. Compare LeConte's Sparrow, often found in same habitat and similarly small and pale buff.

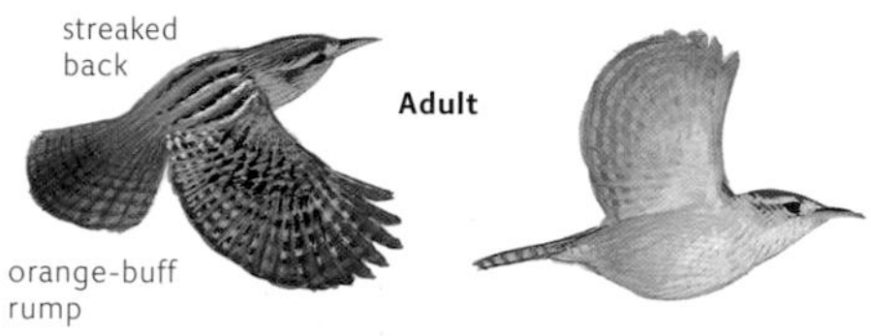

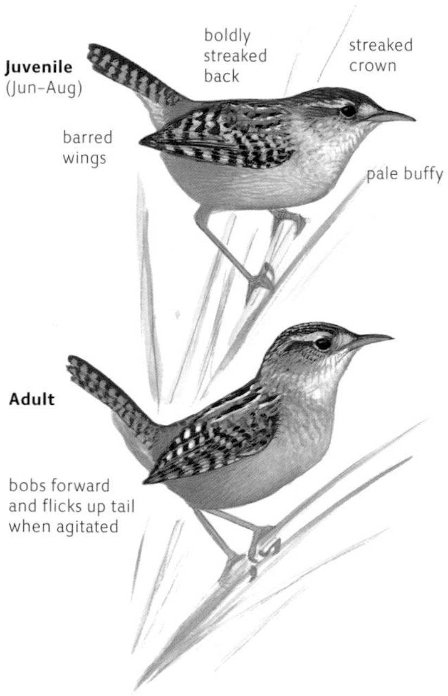

VOICE: Song very sharp, staccato chips followed by a more rapid series *chap chap chatatatat* or *chap chap ch jee jee*; compare more rapid staccato rattle of Common Yellowthroat. Common call a very sharp, staccato, bouncing *chadt*; less intense *chep* sometimes in series like House Wren. Scold a quiet, nasal, low, buzz *krrt*.

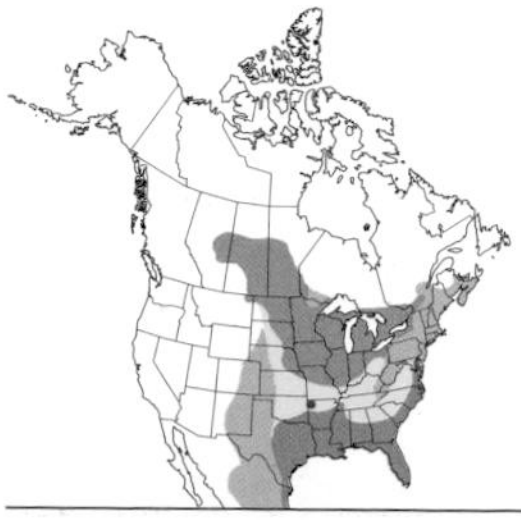

Uncommon and local in sedge marshes and damp grassy meadows with scattered shrubs. Creeps furtively through grass; secretive and difficult to see except when singing. Habitat generally does not overlap with Marsh Wren, Sedge Wren being found at the drier upper edge of the marsh, or in weedy fields.

Sedge Wren and Grass Sparrows

Sedge Wren is one of several species of small, secretive birds that spend most of their time hiding in dense, tall grass. Identifying these species is difficult because they usually flush very close underfoot and fly away quickly before dropping back into cover. Knowing what to expect and being prepared will allow identification of many of these little brown blurs, even if the views are not very satisfying. Pay close attention to tail shape, tail pattern, back and rump pattern, and overall color. Although difficult to do, the best hope of identifying many of these birds is to get a glimpse of the face pattern. Habitat can also provide clues; however remember that not all can be identified. Note that any species of sparrow can act secretive, and many species will flush out of dense grass and drop quickly back into cover; the following species do so habitually.

SEDGE WREN Found in damp, dense grass with scattered bushes; the smallest and weakest flying of the group; short, rounded tail and wings; often gives sharp calls when flushed.

LE CONTE'S SPARROW Often shares habitat with Sedge Wren; quite similar in overall appearance but more boldly patterned, with longer tail; more brightly marked than other grass sparrows.

HENSLOW'S SPARROW Found in more weedy habitat with open ground; often flies to bushes or into woods when flushed; small and dark.

GRASSHOPPER SPARROW Found in drier habitat but can occur together with Le Conte's; larger and relatively plain gray-buff but very difficult to distinguish from Le Conte's and Baird's in normal flight views.

BAIRD'S SPARROW Found in dry grassland, often with Grasshopper; very difficult to distinguish from Grasshopper and Savannah.

NELSON'S AND SALTMARSH SPARROWS Normally found in wet, marshy grasses, but migrants can occur in drier habitat; small and dark with bright orange face.

SAVANNAH SPARROW Relatively longer-winged and shorter-tailed with stronger and more bounding flight than other grass sparrows, but this is difficult to judge; note clean white belly revealed in higher flight.

In southwestern grasslands, watch for the heavier and longer-tailed Cassin's and Botteri's Sparrows. The similar Bachman's is found in southeastern pinewoods. Other species that flush out of dense grass include Sprague's Pipit and some longspurs, but those species typically climb high into the air and fly long distances.

Marsh Wren

:istothorus palustris

. 5" WS 6" WT 0.39 oz (11 g)

\ small and rather dark wren, with usually rufous-brown wings ınd tail, and prominent white eyebrow. Also note striped back, ›rominent barring only on tail. Found almost exclusively in tall 'eeds of wet marshes

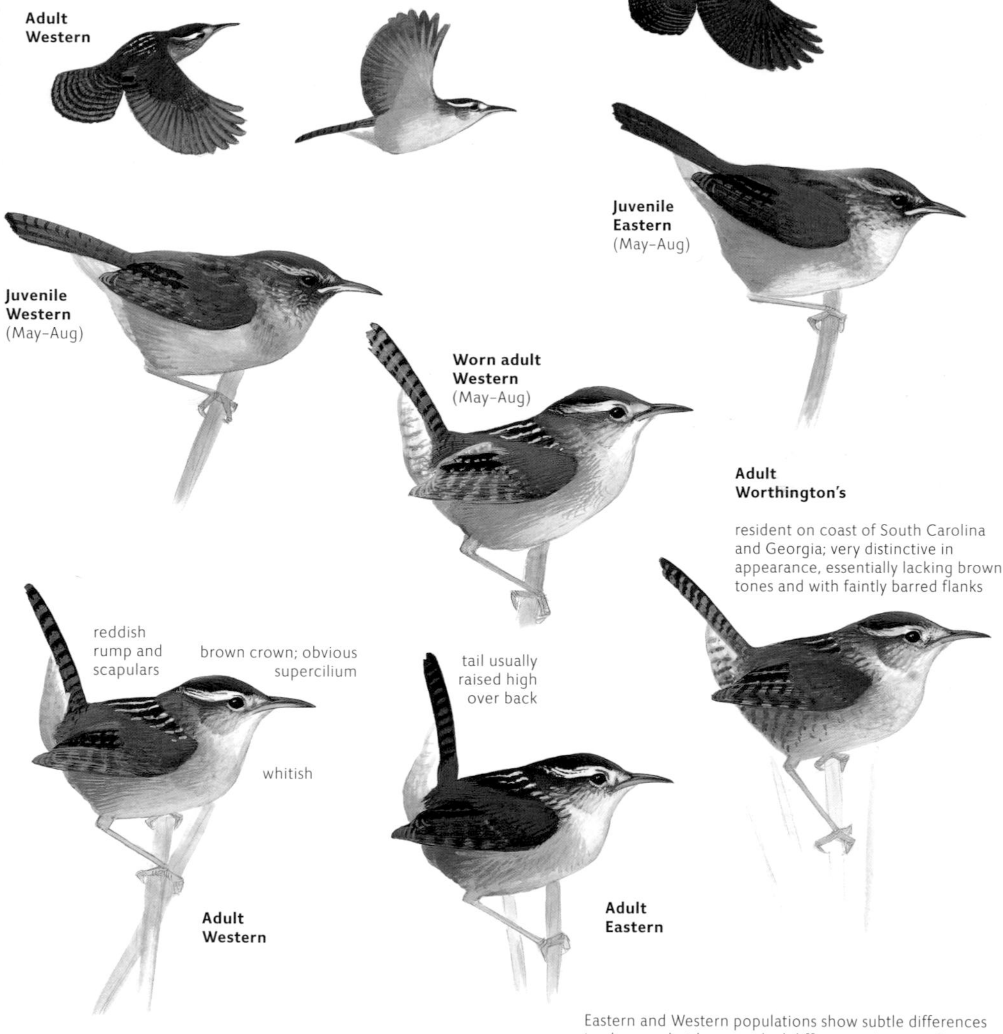

Eastern and Western populations show subtle differences in plumage but have marked differences in song. Comparing mid-continent populations where Eastern and Western meet, Western averages paler and drabber overall. However, individual variation, and the existence of several distinctive local populations along southern coasts, renders sight identification very difficult.

OICE: Song a few chattering or whining notes followed by a omplex gurgling, bubbling, or rattling trill on one pitch *tik k jijijijijiji-jrr* or *jeer jr-gli-gli-gli-gli-gli-gli-jrr*. Call a low, dry *tek* milar to some hard, quiet flight calls of blackbirds. Scold a harsh, escending *shrrrr*; also a low, rolling rattle *chrddd* similar to ouse Wren.

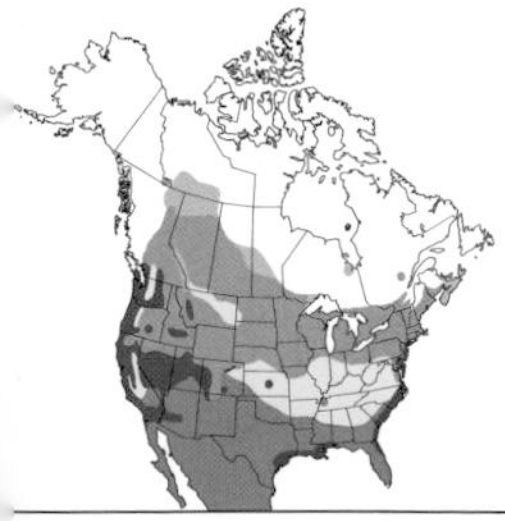

Common locally in marshes of tall cattails, tules, or reeds with standing water below. Secretive and difficult to see as it moves through reeds; much more easily heard. Migrants are occasionally found in weedy or grassy fields.

Two subspecies groups best distinguished by song. Western song a gurgling, rattling trill with distinctive musical and mechanical quality; usually introduced by a few *tek* notes *tik k jijijijijijiji-jrr* or *tuk t jet-t-t-t-t-t-t-trr*. Eastern song similar but more musical, less rattling; often introduced by nasal *gran* note. Individual Western males sing more than 100 song types; Eastern males sing only a few song types.

Cactus Wren

Campylorhynchus brunneicapillus

L 8.5" WS 11" WT 1.4 oz (39 g)

Our largest wren; overall size and habits more similar to thrashers than to other wrens. Long white eyebrow, intricately patterned plumage, and cluster of black spots on breast distinctive. Also note pale spots at tip of tail.

flight similar to thrashers but more flowing

Adult Interior

barred tail with white tips on all feathers (compare Sage Thrasher)

Adult Interior

dense, dark spots

never cocks tail like other wrens

Juvenile Interior (Apr–Aug)

often sits up on conspicuous perch

Adult Coastal California

VOICE: Song low, grating, chugging, unmusical *krrr krrr krrr krrr krrr krrr krrr krrr* slightly quieter at beginning, with little variation in pitch or tempo. Common call a low, hollow knocking *kot* or *kut* repeated in long series. Also a low, coarse, dry *trrk trrk*... or dry, clicking *krrrr*; deep *cheg* notes; series of higher, fairly harsh notes *deeu deeu deeu*... or *raap raap raap*... like a quacking duck.

Common in low cactus and brush in desert or coastal sage-scrub, with open gravely ground. Builds oval nest of sticks in fork of cactus. Usually solitary or in pairs. Hops about on the ground and climbs through low vegetation in search of insects and fruit. Occasionally visits bird feeders.

Most birds along the coastal plain of extreme southern California (Orange County and south) differ slightly from Interior populations (tending towards distinctive populations farther south in Baja) in having more uniform breast pattern with more widely spaced black spots, paler buffy flanks with larger, rounder black spots, and more white barring on tail. Voice is similar.

Canyon Wren

Catherpes mexicanus

L 5.75" WS 7.5" WT 0.37 oz (10.5 g)

A medium-sized wren with extremely long bill. Overall dark rufous color with bright white throat and breast unique; brightest rufous on tail, which also shows a few dark bars.

VOICE: Song a cascading series of clear whistles, falling and slowing down, ending with nasal, hissing notes *twi twi twi towi towi towi toowi toowi jeev jeev.* Call a high, ringing buzz *jiink* or *jeeeet* given with a quick bob; higher and simpler than Rock Wren call.

Uncommon on sheer cliffs and other prominent near-vertical rock features, often near water. Climbs around on rock face and uses long bill to probe crevices for insects and spiders.

Rock Wren

Salpinctes obsoletus

L 6" WS 9" WT 0.58 oz (16.5 g)

Larger than most wrens; plain grayish with very long bill, long legs, and short, broad tail. Almost always found in or near jumbled rocks. Note plain grayish head, faint dark eyeline, buffy flanks, and buffy tips on tail feathers.

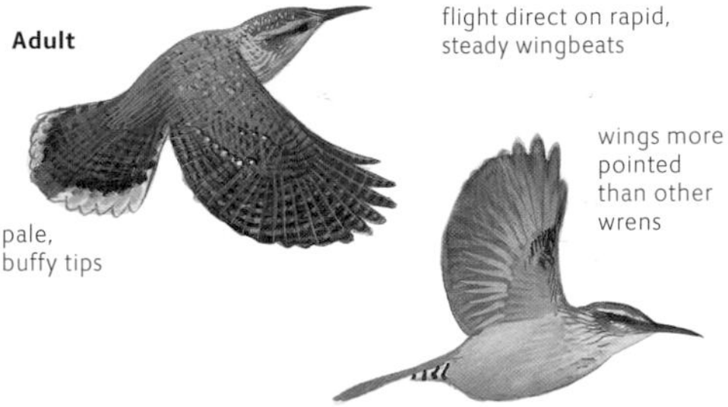

VOICE: Song of buzzy, trilled, ringing phrases, each repeated three to six times: *cheer cheer cheer cheer, krjee krjee krjee krjee, preeyerr preeyerr…*; well-spaced in regular rhythm. Call a ringing, buzzy, trill *pdzeeee* audible at great distance; also a fine, buzzy *deee-dee* and *deee der-dr-dr-dr-dr-dr.*

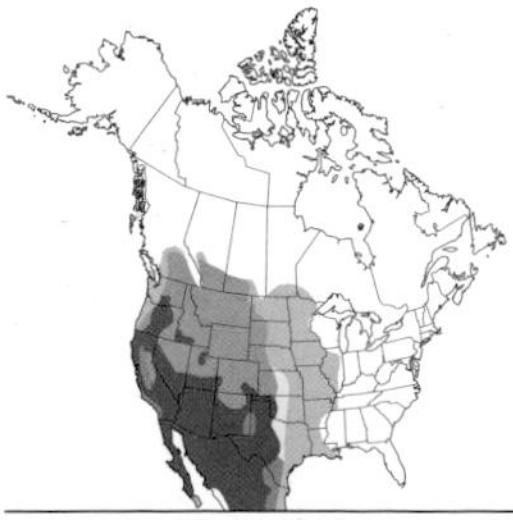

Uncommon on talus slopes and other expanses of jumbled rocks. Usually detected by call. Often stands on top of a rock and bounces (as if doing deep knee-bends) while giving ringing call.

Gnatcatchers, Kinglets, and Others

FAMILY: POLIOPTILIDAE, CINCLIDAE, PYCNONOTIDAE, REGULIDAE, PHYLLOSCOPIDAE, SYLVIIDAE, MEGALURIDAE, MUSCICAPIDAE

24 species in 15 genera in 8 families (11 rare species not shown here). This diverse grouping includes primarily Old World families, represented here by one or two species or as rare visitors, except gnatcatchers, which are found exclusively in the Americas. Some of these families (such as dipper and gnatcatchers) are distinctive in appearance and behavior. Kinglets and the three Eurasian warbler families all resemble wood-warblers (family Parulidae). Some Old World flycatchers closely resemble tyrant flycatchers (family Tyrannidae) while others resemble thrushes. Adult females are shown.

Genus *Polioptila*

Blue-gray Gnatcatcher, page 421

California Gnatcatcher, page 420

Black-tailed Gnatcatcher, page 420

Black-capped Gnatcatcher, page 419

Genus *Cinclus*

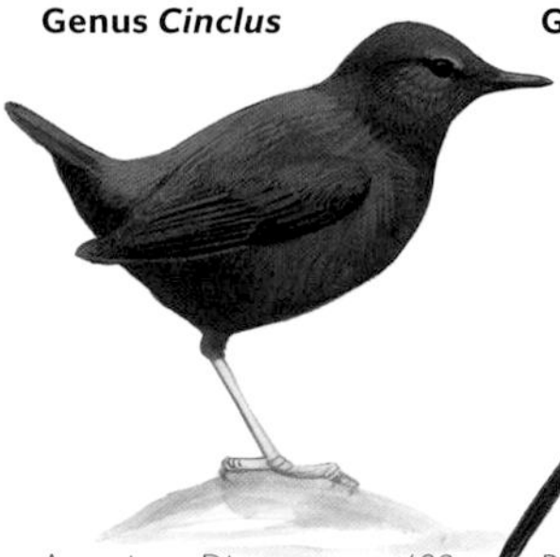

American Dipper, page 422

Genus *Pycnonotus*

Red-whiskered Bulbul, page 424

Genus *Regulus*

Golden-crowned Kinglet, page 423

Ruby-crowned Kinglet, page 423

Genus *Phylloscopus*

Dusky Warbler, page 426

Arctic Warbler, page 425

Genus *Chamaea*

Wrentit, page 422

Genus *Oenanthe*

Northern Wheatear, page 429

Genus *Luscinia*

Bluethroat, page 429

ʒlack-capped Gnatcatcher

olioptila nigriceps

4.25"   WS 6"   WT 0.2 oz (5.6 g)

reeding male distinctive combining black cap with extensive ʼhite in tail. Other plumages very similar to Blue-gray Gnatatcher, best identified by voice; also note slightly longer bill, ıore rounded tail, shorter wings.

Adult ♀

Adult ♀

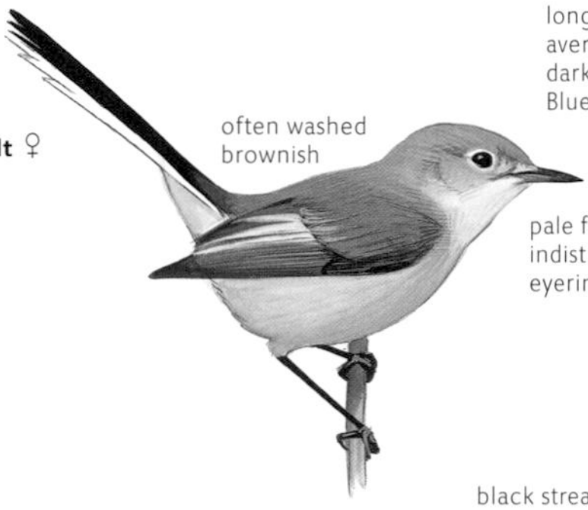

Adult ♂ nonbreeding (Aug–Feb)

Adult ♂ breeding (Feb–Aug)

ɔICE: Song often heard: slow, low-pitched, and varied; includes ırd *che che che* or *trk trk*, husky *jip* like Wilson's Warbler, and ցh, sharp *tip*. Common call rather harsh with rising then falling flection like California *je-eew*; a variety of other notes given, cluding a soft *dear-dear-dear*... series like California.

Rare visitor from Mexico to low hackberry thickets along streams in arid foothill canyons; has nested in region.

Identification of Gnatcatchers

Distinguishing a gnatcatcher from any other bird is normally a simple task, but separating gnatcatcher species from one another can be difficult.

Assessing tail pattern is critical. Keep in mind that when the tail is folded closed, the outer tail feathers are visible only from below.

Gnatcatchers molt their tail feathers just once a year (July–August). During this time it is not unusual to see a gnatcatcher with missing tail feathers. A Blue-gray Gnatcatcher with the mostly white outermost tail feathers missing will show the mostly black next-to-outer feathers and easily could be mistaken for a Black-tailed Gnatcatcher. Also note that in the summer months before molt begins, the outer tail feathers can be very worn or broken, again creating the appearance of a mostly black tail.

Black-capped Gnatcatcher has a more graduated tail than Blue-gray; the distance from the tip of the outer (shortest) tail feather to the tip of the central (longest) tail feather averages 11 mm in Black-capped (7 mm in Blue-gray). Under exceptional conditions this may be visible in the field.

California Gnatcatcher

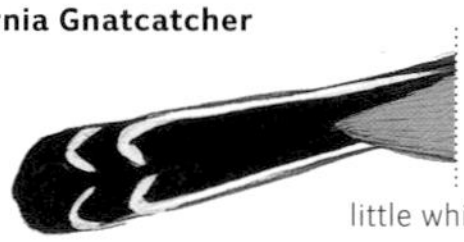

Western Black-tailed Gnatcatcher

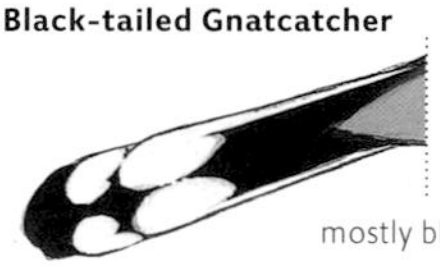

Eastern Black-tailed Gnatcatcher

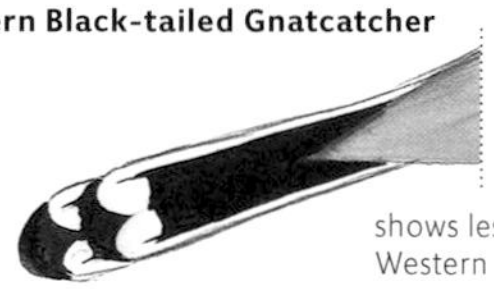

Western Blue-gray Gnatcatcher

Eastern Blue-gray Gnatcatcher

Black-capped Gnatcatcher

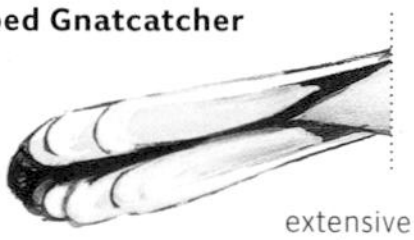

California Gnatcatcher

Polioptila californica

L 4.5" WS 5.5" WT 0.18 oz (5 g)

The darkest and drabbest gnatcatcher, even darker than Black-tailed Gnatcatcher, with dusky gray underparts and almost no white in tail. Much darker than Blue-gray. Range barely overlaps with Black-tailed, best distinguished by call.

VOICE: Common call a nasal, mewing *mee-eeew* or *vwee-eeew*; kittenlike when relaxed, harsher when agitated; always with distinctive rising then falling pattern. Other calls include a soft *dear dear dear...*; a harsh *tssshh*; and *jew jew jew jew* similar to Black-tailed.

Uncommon and local in coastal sagebrush scrub. Only about 3000 pairs estimated in US. Usually seen in pairs or family groups year-round. Habits like Black-tailed.

Black-tailed Gnatcatcher

Polioptila melanura

L 4.5" WS 5.5" WT 0.18 oz (5 g)

Smaller than Blue-gray Gnatcatcher and drabber gray-brow overall with very little white on tail and different voice. Breedin male has black cap. Paler than California, best distinguished b range and voice.

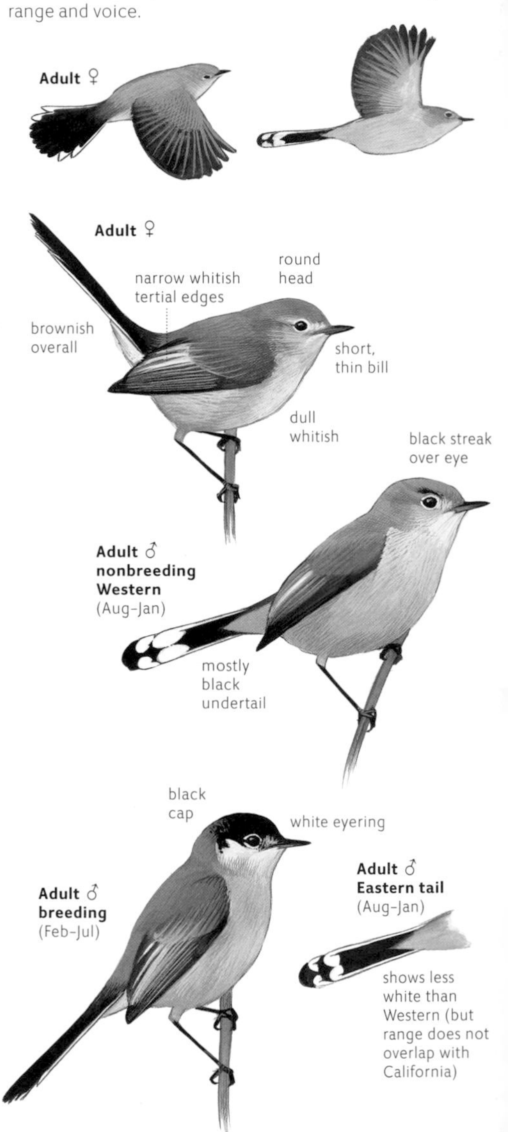

VOICE: Calls include dry, hissing *psssssh* like harsher scold House Wren and *jeew jif jif* like Blue-gray but harsher. Male giv a very harsh series *tssh tssh tssh tssh...* and much more rap *ch-ch-ch-ch....* Other calls include Verdin-like chips.

Uncommon and local in dry desert-scrub habitat. Almost always in pairs year-round, moving through low bushes. Feeds on small insects. Where range overlaps with Blue-gray, Black-tailed is found in low elevation arid desert scrub, not riparian thickets or foothills.

lue-gray Gnatcatcher

olioptila caerulea

4.5" WS 6" WT 0.21 oz (6 g)

natcatchers are tiny and active, with long black-and-white tail
onstantly fanned and waved, and obvious white eyering. This
ecies has more extensive white in tail, cleaner blue-gray color,
d more obvious white secondary edges than other species.

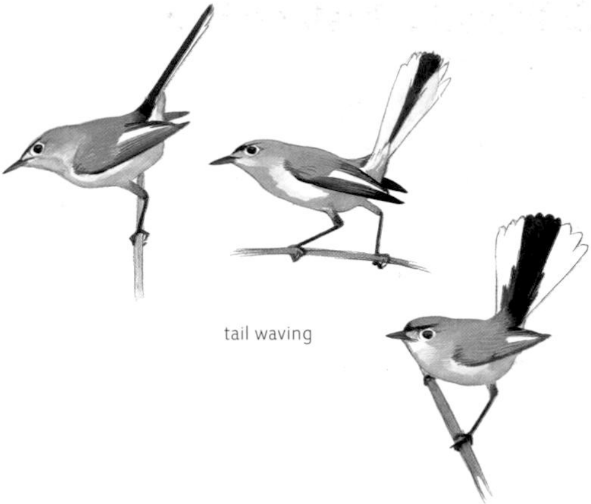

Western

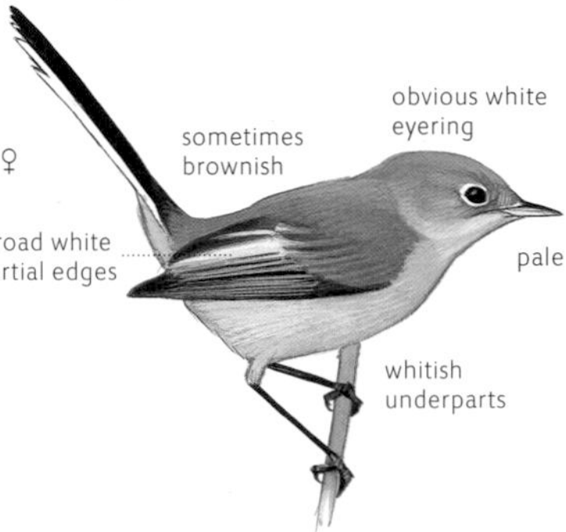

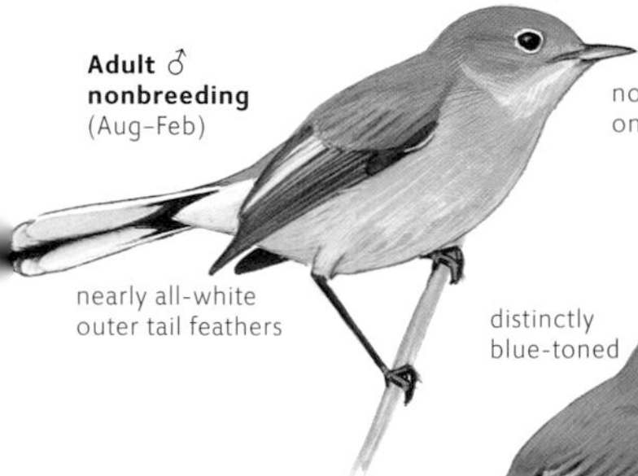

Eastern

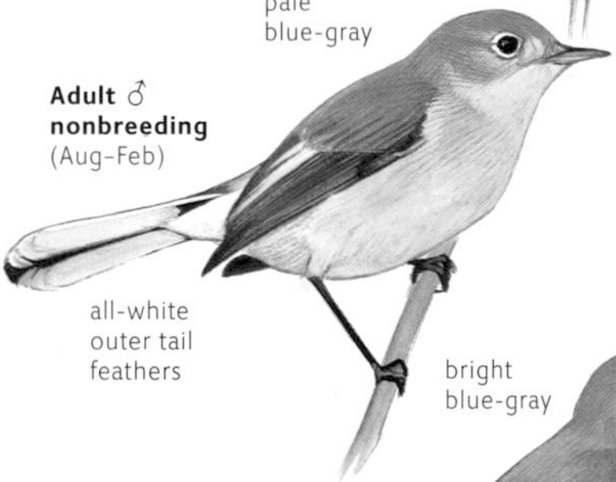

ICE: Song mainly thin, wheezy notes in steady series, inter-
ersed with short bunches of high chips and slurs *zeee zeet zeet*
l zill zwee zwee.... Western song significantly different: lower,
rsher, less varied *jeew jeew bidi bilf....* Common call a very
in, nasal, variable buzz *speee, szeeewv,* or *zeeewv zeef zeef.*
estern birds average lower and harsher calls than Eastern;
orter and more strongly descending *jeewf* (vs. *szeeee* of East-
n); more similar to House Wren and all other gnatcatchers.

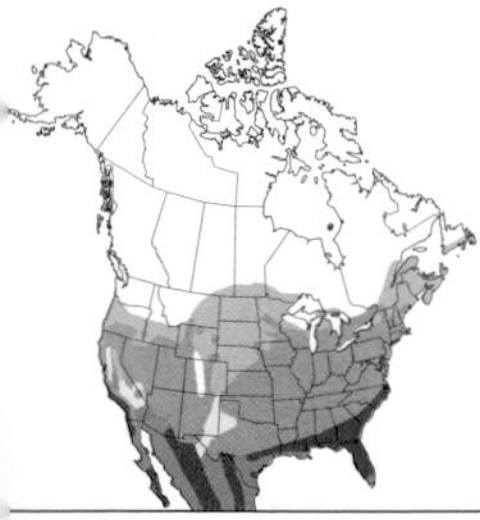

Common and widespread in rich deciduous woods and thickets in East, often near water; in West found in brushy woods or thickets, including chapparal and arid oak-juniper woods. Often solitary, but also joins loose flocks of warblers and kinglets. Moves briskly and erratically through middle- to upper-level twigs in search of small insects.

Western populations average less white on tail than Eastern, with black usually visible on the base of the outer tail feathers and only small white tips on the next-to-outermost feathers; drabber overall, males have gray back with dark bluish crown and broader, shorter black forehead mark; females may have darker grayish lores and cheeks, with less obvious eyering than Eastern; voice is harsher overall. Habitat choice of breeding birds is also quite different: Eastern birds nest in swampy woods; Western nest in arid, dense brush, pinyon-juniper, or open woods.

American Dipper

Cinclus mexicanus

L 7.5" WS 11" WT 2 oz (58 g) ♂>♀

Dark gray and plump overall, with long legs and short tail. Note striking white eyelid. The only songbird that regularly swims. Look for it perched close to water on rocks or logs, or flying very low with buzzing wingbeats, following the curves of a stream.

VOICE: Song of high, whistled or trilled phrases repeated two to four times in thrasherlike pattern *k-tee k-tee wij-ij-ij-ij treeoo treeoo tsebrr tsebrr tsebrr tsebrr...*; has steady rhythm and is much higher and clearer than any thrasher. Call a high, buzzy, metallic *dzeet* often doubled or tripled in rapid series *dzik-dzik*.

Uncommon along clear, fast-flowing mountain streams with lots of exposed rocks and logs alongside. Solitary. Perches on rocks along streams and dives underwater for aquatic insect larvae. Birds make regular forays up and down streams. Look for droppings on exposed rocks to find favored foraging areas, then watch and listen.

Wrentit

Chamaea fasciata

L 6.5" WS 7" WT 0.49 oz (14 g)

A distinctive species, combining characteristics of wrens ar titmice but actually related to Old World warblers; note lar head, long tail often raised, short bill, pale iris, and plain brow ish color.

Variation in color is clinal: darker and redder coastally and north in wetter climates; paler and grayer inland and south in drier climates.

VOICE: Song of clear, popping whistles *pwip pwip pwip pw pwip* or *pwid pwid pwidwidrdrdrdrdr*. Accelerating series giv by male; slow, regular, short series by female. Call a dry, ratche ing *trrrk*.

Common but difficult to see in chaparral and other dense brushy habitats. Usually solitar or in pairs. Feeds on insects an seeds. Not attracted to bird feeders.

olden-crowned Kinglet

gulus satrapa

" WS 7" WT 0.21 oz (6 g)

y; even smaller than Ruby-crowned, with boldly striped face, re gray-green (less buff) overall color, grayish flanks, bolder ng pattern, and slightly shorter tail. Voice is also very different.

ICE: Song two-part: a rising series of very high, thin notes llowed by a lower, tumbling chickadee-like chatter *see see see si si tititichichi-chichi.* Typical call a very high, thin, slightly izzing *zree* or *zee-zee-zee*; also very high, weak *tip* notes. Juvele begs with sharp, high chips.

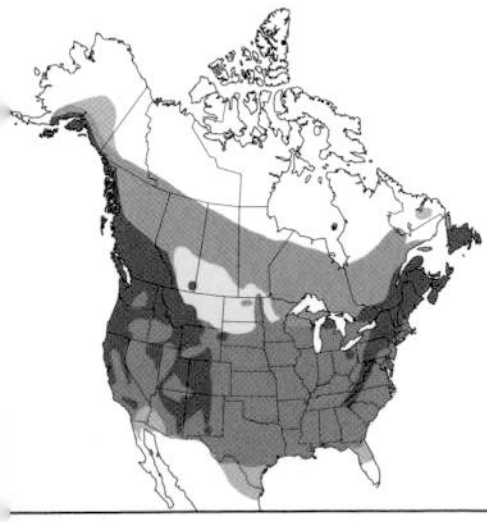

Common in mature trees, usually high in spruces and other conifers. Almost always in small groups of three to eight; often mixes with chickadees, creepers, and others. Gleans tiny insects from branches; often hangs upside-down while foraging or hovers at tips of branches.

Ruby-crowned Kinglet

Regulus calendula

L 4.25" WS 7.5" WT 0.23 oz (6.5 g)

Tiny and active, with constant wing-flicking. Distinguished from Golden-crowned by pale eyering. Similar to drab warblers and flycatchers, note very thin dark bill and legs, horizontal posture, dark bar across base of secondaries, and call. See Hutton's Vireo.

VOICE: Song lively, varied, and loud: begins with high, clear notes and ends with low, whistled chant *sii si sisisi berr berr berr pudi pudi pudi see.* Call a low, husky, dry *jidit*; often a single *jit* or in long series when agitated. Adult in summer gives low, laughing *gido.* Juvenile gives very high, trilled *sreeet.*

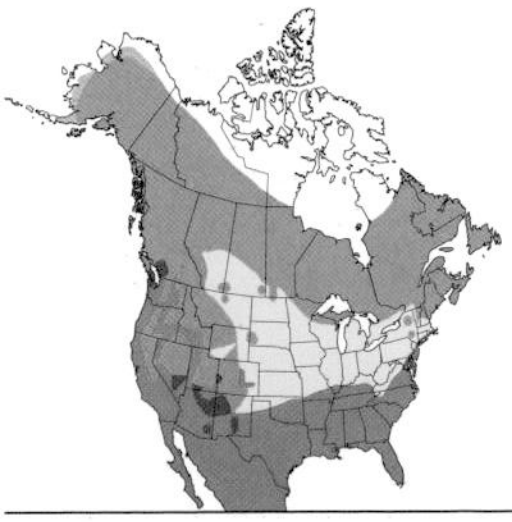

Common in wooded areas. Nests in tall conifers in spruce woods; at other times found in low brush or deciduous woods. Generally solitary, but often joins foraging flocks of warblers, chickadees, and others. Feeds on tiny insects; foraging style less acrobatic than Golden-crowned Kinglet.

Red-whiskered Bulbul

Pycnonotus jocosus

L 7" WS 11" WT 1 oz (29 g)

Smaller than Mockingbird. Note contrasting dark brown upperside and mostly white underside, pointed black crest, and black spur on side of breast.

VOICE: Song a chattering and musical scolding. Call a staccato *kink-a-jou.*

Introduced from Asia. Uncommon and very local in wooded suburban habitat in southern Florida and southern California. In small groups; usually inconspicuous, but sometimes perches on wires or twigs in open areas. Feeds on berries and insects.

Red-vented Bulbul

Pycnonotus cafer

L 8.5" WS 12.5" WT 1.5 oz (43 g)

Blackish overall, with short bushy crest. White rump and tail ti red undertail coverts.

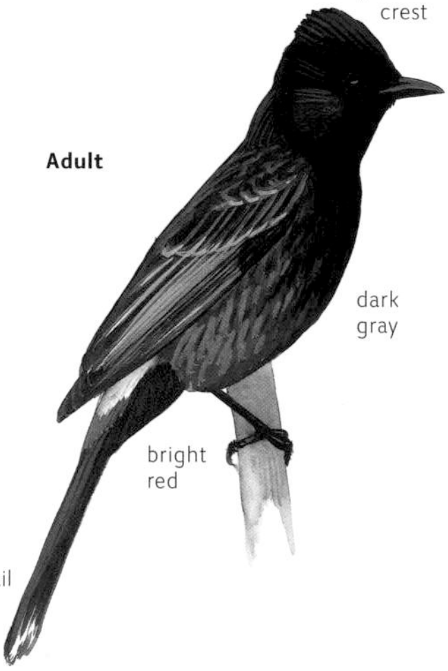

VOICE: Calls musical and chattering. A sharp note followed by a husky warble *PEEK-turreoo,* often repeated.

Introduced from Asia. Very sma numbers are escaped and nesting in suburban neighborhoods with mature broadleaf trees around Houston, Texas.

ɿrctic Warbler

hylloscopus borealis

5" WS 8" WT 0.32 oz (9 g)

ɿperficially similar to wood-warblers such as Tennessee War-er, but drabber, with olive-gray upperside, dusky flanks, pale gs, and long, narrow, pale supercilium. Similar to rarer Willow 'arbler, but larger with longer wings.

Population breeding in Kamchatka is often considered a separate species, Kamchatka Leaf Warbler, recorded rarely in western Alaska. On average brighter yellowish than Alaska breeders, slightly larger and larger-billed, with call a sharp, crackling rattle.

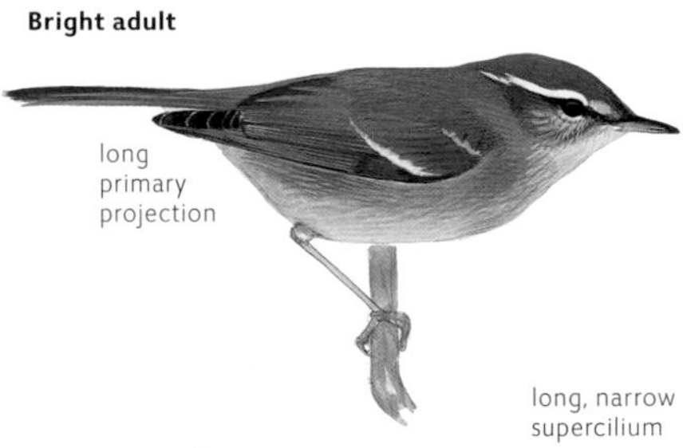

ɔICE: Song fast, mechanical, and rhythmic: a hammering trill *ɿrchrchrchrchrchrchr-chrchr* or *chingingingingingingi̇ng* with no ɿange in pitch or rhythm; reminiscent of redpoll trill. Call a ɿort, penetrating, buzzy *jeet* or *dzrk* with little musical quality; miniscent of American Dipper.

Uncommon breeder in dense streamside willow thickets. Migrates across Bering Strait to winter in Asia. Solitary and rather secretive, skulking in low dense vegetation. Feeds on small insects.

Yellow-browed Warbler

Phylloscopus inornatus

L 4" WS 6.7" WT 0.25 oz (7 g)

Kinglet-sized, with boldly-patterned head and wings. Told from Arctic Warbler by smaller size, broad wingbars and white edges on tertials, mostly whitish underparts, and broad pale supercilium.

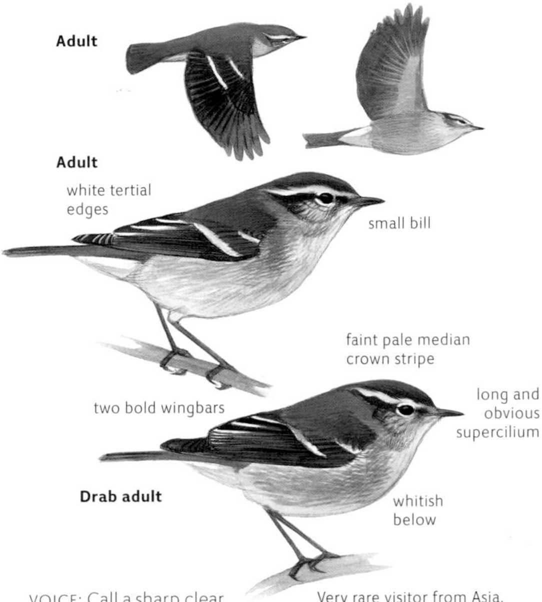

VOICE: Call a sharp clear whistled *tuEES*. Song unlikely to be heard in North America a series of very thin, high, slurred two-syllabled notes with long pauses between each.

Very rare visitor from Asia, recorded a few times in western Alaska in fall, also once in Baja California, Mexico. Found in thickets, gleans tiny insects from leaves and twigs.

Willow Warbler

Phylloscopus trochilus

l 4.3" ws 7.5" wt 0.35 oz (10 g)

Very similar to Arctic Warbler, but smaller, with smaller bill and plain wings, pale flanks, shorter pale eyebrow crosses forehead. Told from Dusky Warbler by paler color overall, usually yellowish throat, pale flanks and undertail coverts.

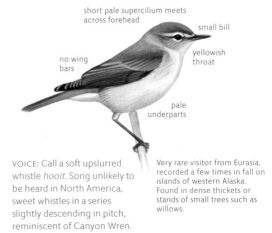

VOICE: Call a soft upslurred whistle *hooit*. Song unlikely to be heard in North America, sweet whistles in a series slightly descending in pitch, reminiscent of Canyon Wren.

Very rare visitor from Eurasia, recorded a few times in fall on islands of western Alaska. Found in dense thickets or stands of small trees such as willows.

Dusky Warbler

Phylloscopus fuscatus

L 5.3" WS 7.5" WT 0.31 oz (8.8 g)

Similar to Arctic Warbler, but slightly darker and browner overall, with slightly longer tail, prominent lower eye-arc, plain dark wings lacking wingbar, shorter primary projection.

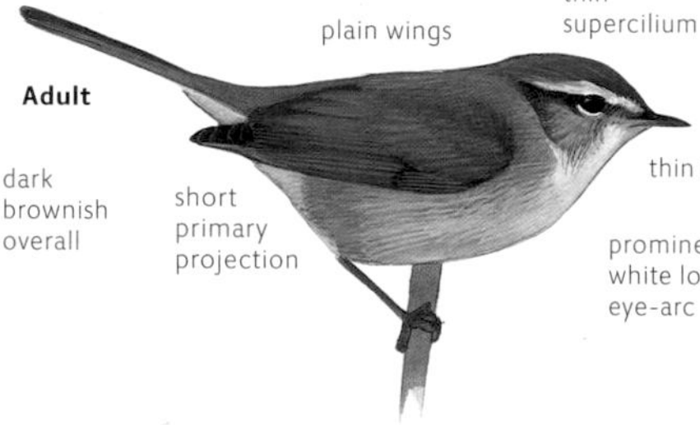

VOICE: Song a varied, short series of high, slow trills. Call a sharp, hard *stak* like Lincoln's Sparrow but drier; repeated in rapid rattle when agitated.

Very rare visitor from Siberia; only a few records (mostly in fall) from Alaska and California. Found in rank weedy vegetation. Solitary. Feeds on small insects.

Lanceolated Warbler

Locustella lanceolata

L 4.7" WS 5.9" WT 0.35 oz (10 g)

Small, stocky, and short-tailed; plain brownish and finely streake overall. Told from sparrows by thin bill, plain face, short rounde tail. From wrens by streaked breast and flanks, longer wings, la of barring.

VOICE: Call a sharp *chik.* Song a sustained pulsing grasshopper-like buzz or rattle, up to a minute long.

Very rare visitor from Asia, recorded several times in far western Alaska, once in California. Preferred habitat is i swampy thickets, or in thick weedy and grassy marshes with scattered bushes, but so secreti it is rarely seen unless singing. Walks or runs through thick vegetation, nervously flicking wings.

Middendorff's Grasshopper-Warbler

Locustella ochotensis

L 5.9" WS 19" WT 0.6 oz (17 g)

Larger, heavier, and darker than Lanceolated Warbler. Note re dish-brown upperparts with indistinct markings, drab bu underparts with very faint streaks, rounded dark tail with pa tips.

VOICE: Call a hard *chik.* Song of hard rattling sounds with piercing clear *seewee-see-wee*... series interspersed.

Very rare vistor from Asia, recorded several times from far western Alaska, mostly in fall. Habits and habitat like Lanceolated Warbler, extremely skulking; rarely seen except when singing.

)ark-sided Flycatcher

luscicapa sibirica

5" WS 7.7" WT 0.4 oz (11 g)

uperficially similar to New World flycatchers such as *Empi-onax*, but with faint wing bars, relatively short tail, suggestion f streaks on sides, and stout, mostly dark bill. Very similar to ray-streaked Flycatcher, best distinguished by unstreaked flanks nd white band on neck.

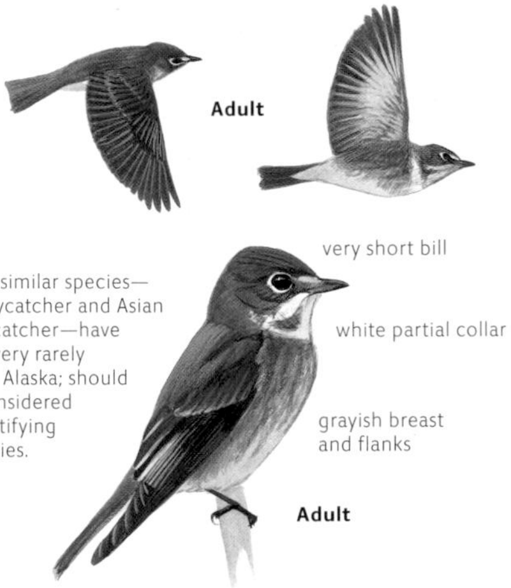

vo other similar species—potted Flycatcher and Asian rown Flycatcher—have ccurred very rarely western Alaska; should so be considered hen identifying nese species.

OICE: Vagrants are generally lent.

Very rare visitor from Siberia to islands of western Alaska, with about 20 records, mostly in spring. Habits like Gray-streaked.

Jray-streaked Flycatcher

luscicapa griseisticta

5.3" WS 8" WT 0.5 oz (14 g)

ery similar to Dark-sided Flycatcher, but always distinctly streaked n flanks, lateral throat stripe merges with streaked breast, and nmatures in fall have more prominent white wingbars.

OICE: Vagrants are generally ilent.

Rare visitor from Siberia to islands of western Alaska, recorded spring and fall. Habits and habitat most like wood-pewees, perching on open weed tops or branches and flying out to capture insects.

Taiga Flycatcher

Ficedula albicilla

L 4.7" WS 7.9" WT 0.35 oz (10 g)

Small and dapper. Breeding male distinctive with gray face and orange throat. On other plumages note very plain gray-brown color, pale eyering, and contrasting black tail with white at base. Tail is often flicked sharply up.

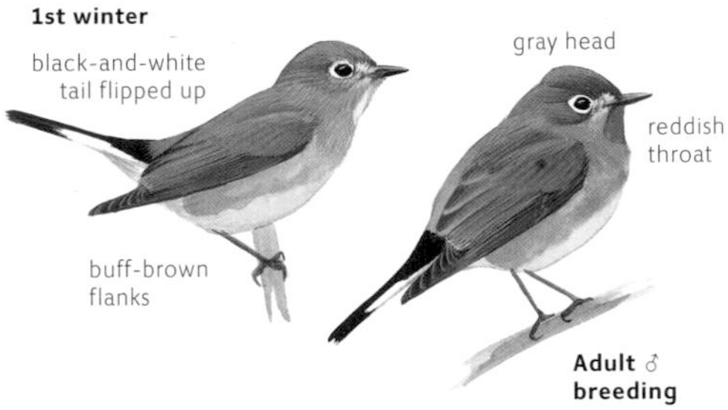

VOICE: Call a slightly descending buzzy rattle *jerrt*.

Very rare visitor from Asia, recorded occasionally in far western Alaska (mainly in spring), once in California (fall). Found in low vegetation such as open understory of woods, hedgerows, etc.

Siberian Rubythroat

Luscinia calliope

L 5.5" WS 9.4" WT 0.8 oz (23 g)

Slightly smaller than Hermit Thrush, with long legs and short tail. Very plain gray-brown overall with no pattern, except distinct white eyebrow and malar stripe, black lores. Brilliant ruby throat of male is obvious only at certain angles.

VOICE: Calls include a two-note whistled *ti-too* and an abrupt dry *chak*. Song a relaxed chattering and creaking warble, often includes mimicry of other species.

Very rare visitor from Asia to far western Alaska, in spring and fall. One record in Ontario (winter). Found in dense brushy or weedy undergrowth, usually stays hidden.

Stonechat

Saxicola torquatus

L 4.8" WS 8" WT 0.5 oz (14 g)

Very small, relatively large-headed and short-tailed. Male is striking with black head and back, pale orange breast. On other plumages note unstreaked pale buff underparts, dark streaked back with pale spot at base of wing, and large pale rump contrasting with all-dark tail

VOICE: Calls include a thin whistled fees and a dry crackling *tak tak*, often combined in series. Song unlikely to be heard in North America, a simple high squeaky twittering, may recall Horned Lark.

Very rare visitor from Asia to Alaska (several records in spring and fall), California (fall), and New Brunswick (fall). Usually seen perched conspicuously on fences, bushes, or tall weeds in open pastures or marshes.

Red-flanked Bluetail

Tarsiger cyanurus

L 5.5" WS 9" WT 0.5 oz (14 g)

Chickadee-sized, stocky, short-tailed, and small-billed. Male ha drab dusty blue upperparts, female just faintly gray-blue on ta On all note contrasting small white throat, pale orange flank habit of flicking tail sharply downwards.

VOICE: Calls include a sharp whistled *peef* and a harsh crackling *track*. Song a short rapid warble.

Very rare visitor from Asia, recorded rarely in far western Alaska, mainly in spring. Also recorded in southern British Columbia (winter) and Californi (fall).

Bluethroat

uscinia svecica

5.75" WS 9" WT 0.74 oz (21 g)

small, short-tailed, and secretive thrush; difficult to detect xcept when singing. Male's blue and rufous throat distinctive. n all plumages note bold white eyebrow and dark tail with ufous at base.

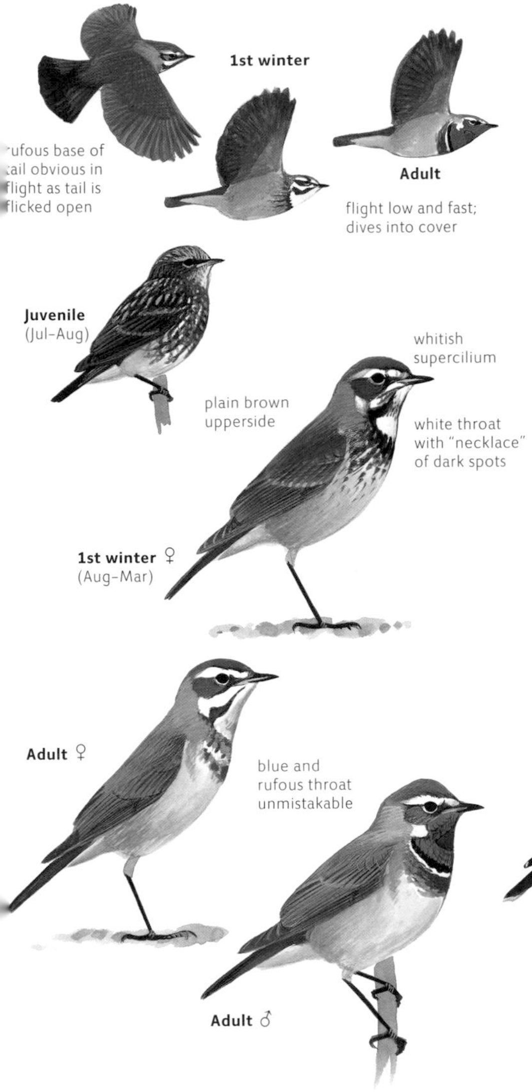

OICE: Song begins with a slowly accelerating series of high, hort whistles, then a series of trilled or buzzy notes; includes mitations of other species and distinctive, bell-like notes. Call a harp, dry *chak* and high, whistled *heet* like Northern Wheatear.

Uncommon and local breeder in patches of shrubby willows at tree line. Migrates across Bering Strait to winter in Asia. Solitary. Feeds on small insects.

Northern Wheatear

Oenanthe oenanthe

L 5.75" WS 12" WT 0.81 oz (23 g)

Slightly smaller than bluebirds; long-legged and short-tailed. White rump and tail with black T pattern unique and conspicuous in flight. Also note overall pale buff and gray color with black wings and dark mask.

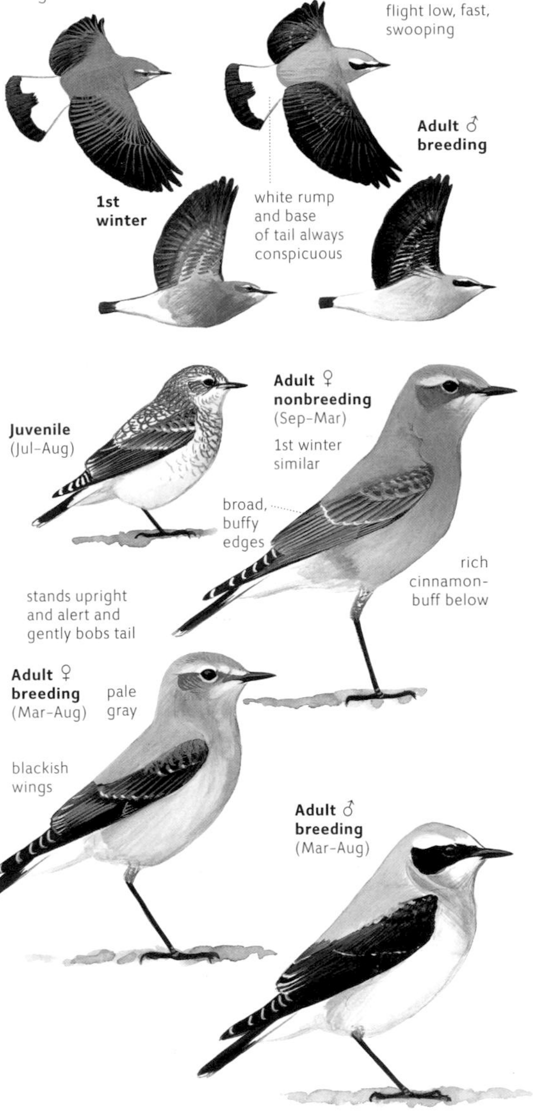

VOICE: Song often given in flight, an unpatterned, level, rapid warbling; overall longspur-like; combines husky, sliding whistles with dry, crackling, toneless phrases. Call a weak, high whistle *heet* and dry, clicking *tek*.

Uncommon and local breeder in rocky outcrops on tundra; very rare visitor farther south (mainly in fall) where it may be seen on sand dunes, fields, and other open ground. Found on the ground or perching on posts or buildings; shuns bushes. Solitary. Feeds mainly on insects. Nearly all winter in Africa.

Thrushes

FAMILY: TURDIDAE

22 species in 5 genera (8 rare species not shown here). Most are long-legged, with relatively blunt-tipped bills, forage mainly on the ground for insects, worms, snails, as well as fallen fruit, or in trees and shrubs for fruit. Three main groups: Bluebirds and solitaires are found in open areas with scattered trees, often in small flocks; spotted thrushes are generally solitary in understory of forest and have intricate fluting songs; robins are relative large and colorful, often found on open ground and form flock in nonbreeding season. Bluebirds nest in cavities or birdhouse all other species build cup nest, often mixing mud and grass, i low shrub or tree. Adults are shown.

Genus *Sialia*

Eastern Bluebird, page 433

Western Bluebird, page 432

Mountain Bluebird, page 432

Genus *Myadestes*

Townsend's Solitaire, page 43

Genus *Catharus*

Veery, page 434

Gray-cheeked Thrush, page 436

Bicknell's Thrush, page 436

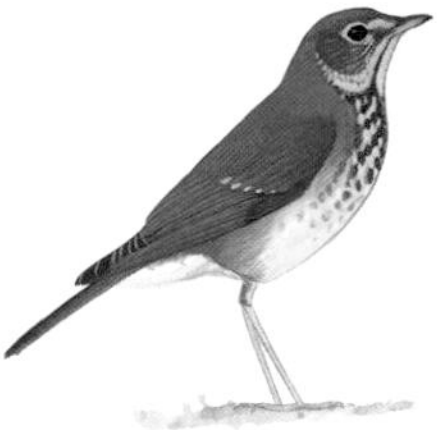

Swainson's Thrush, page 437

Hermit Thrush, page 435

Genus *Hylocichla*

Wood Thrush, page 434

Clay-colored Thrush, page 440

Genus *Ixoreus*

Varied Thrush, page 443

Rufous-backed Robin, page 440

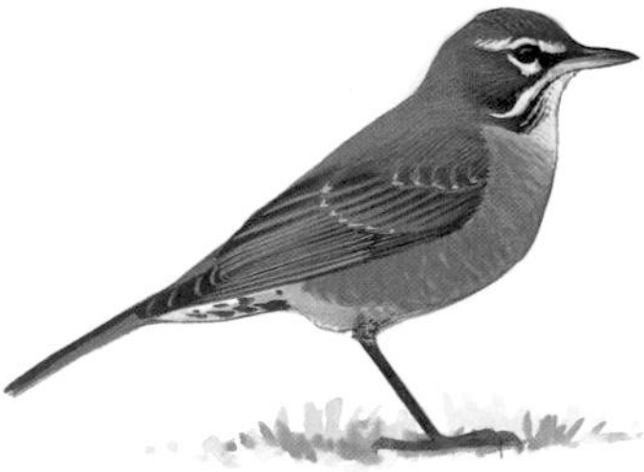

American Robin, page 442

Townsend's Solitaire

Myadestes townsendi

L 8.5" WS 14.5" WT 1.2 oz (34 g)

Plain gray color and long tail may recall Northern Mockingbird; Note white eyering, uniform gray body, and bold buffy wingstripe. Told from female Mountain Bluebird by longer tail with white sides, bold wingstripe, lack of blue.

VOICE: Song a continuous, disjointed, finchlike warble; clear, whistled notes with low, husky notes interspersed but no distinct pattern. Often sings in flight from high in the air. Call a clear, soft whistle *heeh*, reminiscent of a single toot of Northern Pygmy-Owl.

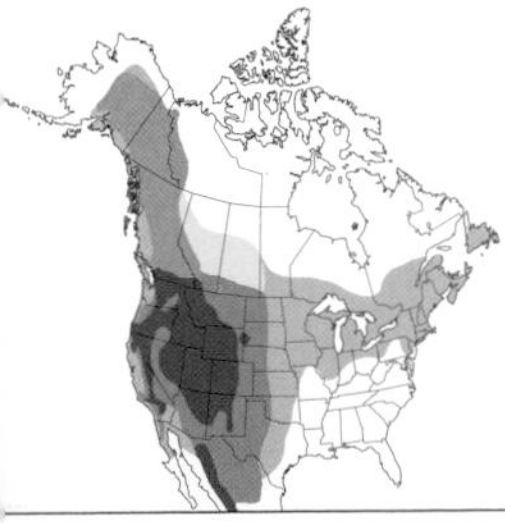

Uncommon. Nests in open coniferous forest (nest on the ground on rocky slope such as road embankment). Winters at lower elevations in pinyon-juniper and other habitats with scattered trees and abundant fruit such as juniper, mistetoe, or russian-olive. Usually solitary, perching inconspicuously in trees. Feeds mainly on insects and berries.

Brown-backed Solitaire

Myadestes occidentalis

L 8.4" WS 14.5" WT 1.2 oz (34 g)

Size and shape like Townsend's Solitaire, but with rich brownish back and wings, grayer head and breast, white spectacles, pale malar, different voice.

VOICE: Song begins with widely spaced squeaky metallic notes, then accelerates into rapid jangling warble, the whole performance slightly descending in pitch. Call a metallic squeaking whistle *hweeh*.

Very rare visitor from Mexico. Several records in pine-oak forest of southeastern Arizona mountains, some could represent escaped cage birds.

Mountain Bluebird

Sialia currucoides

L 7.25" WS 14" WT 1 oz (29 g)

Male pale blue overall. Female usually grayish overall; note gray flanks and belly. Many show a trace of rufous on breast, then distinguished by long wings and tail, thin bill with little or no yellow at base, and details of plumage.

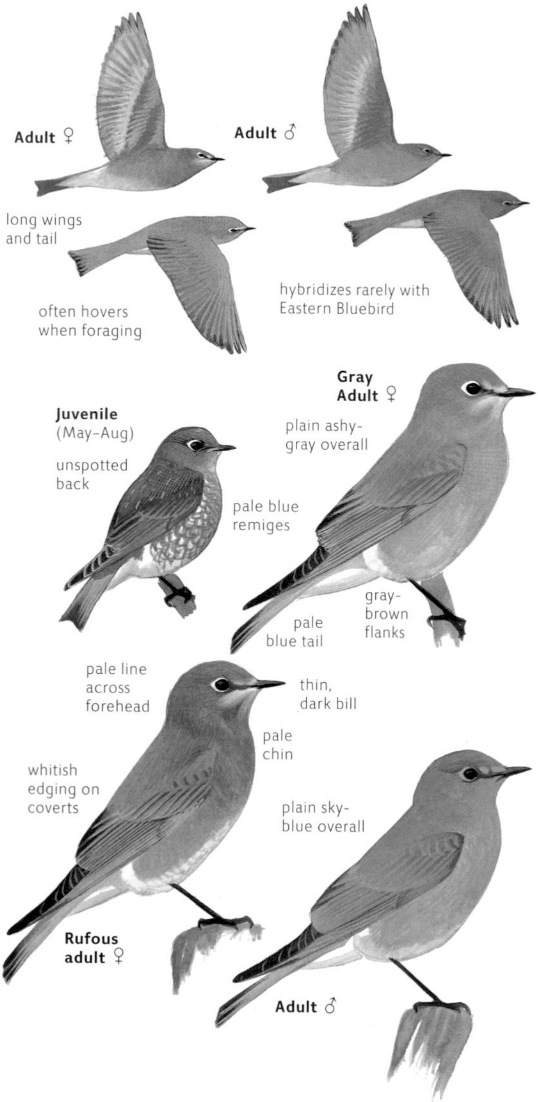

VOICE: Song a series of low, burry whistles like call *jerrf jerrf jewr jipo jerrf.* Call a soft whistle similar to other bluebirds but thinner and clearer: *feeer* or a mellow, muffled *perf,* always descending; also a short, harsh *chik* or *chak.*

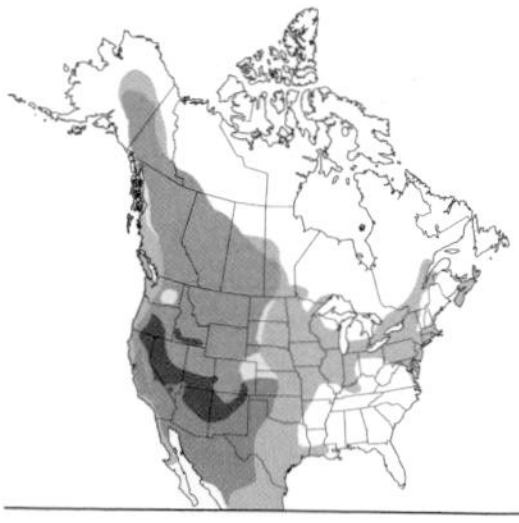

Nests in cavities (and bluebird boxes) in mountain grasslands or sagebrush with widely scattered trees or shrubs. Winters most abundantly in pinyon-juniper woodlands, in large loose flocks up to hundreds feeding on fruit, but also in any open habitat from shrubland or grassland to expansive agricultural fields, where it often hovers to survey open ground below.

Western Bluebird

Sialia mexicana

L 7" WS 13.5" WT 1 oz (29 g)

Rather small and stocky, with bright blue wings and tail an rufous breast. Similar to Eastern Bluebird, but range barely over laps; note blue or gray throat, sides of neck never tinged orange and drab belly not contrasting white as on Eastern.

VOICE: Song heard infrequently, mainly at dawn; simply a series of call notes. Call a fairly hard, low whistle *jewf* or *pew pew pew* shorter and lower than Eastern; also a short, dry chatter.

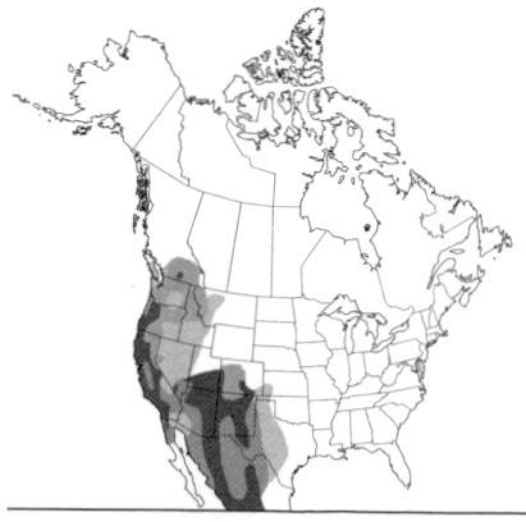

Common in small groups in any open wooded habitat with areas of open or grassy ground, perching on low branches or fenceposts to watch for prey; especially parks, golf courses, mature and well-spaced cottonwoods, Ponderosa Pine forest, oak savannah, etc. In winter also found in pinyon-juniper, especially where fruit is abundant.

Eastern Bluebird

Sialia sialis

L 7" WS 13" WT 1.1 oz (31 g)

A rather small and stocky thrush, always bright blue at least on wings and tail. No other eastern bird has bright blue wings and tail with rufous breast. Distinguished from Western Bluebird by orange throat and sides of neck, brighter white belly.

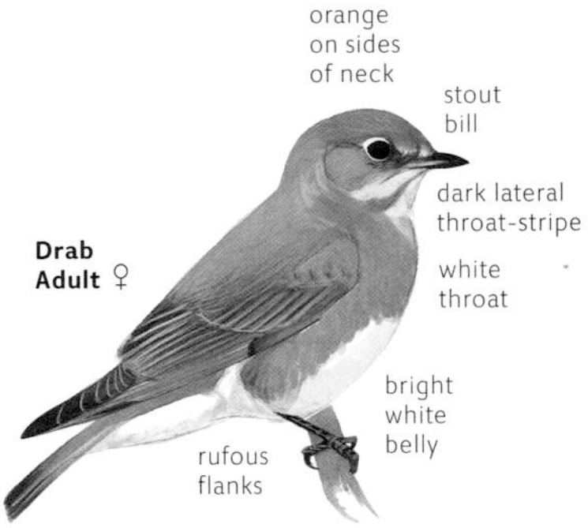

VOICE: Song a pleasing soft phrase of mellow whistles *chiti WEEW wewidoo* and variations. Call of similar pleasant musical quality: a soft, husky whistle *jeew* or *jeew wiwi*; also a short, dry chatter.

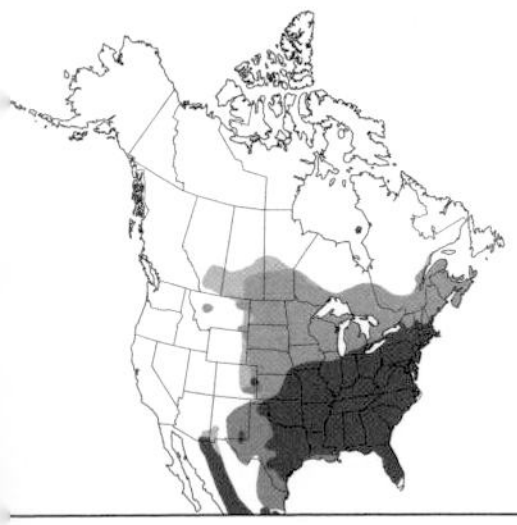

Uncommon, usually seen low on fences or twigs where open grassy areas mix with scattered trees, such as orchards, golf courses, and parks. Nests in tree cavities or manmade boxes. Usually in small groups of five to ten that roam a wide area in search of food. Feeds on insects and fruit gleaned from the ground or vegetation.

Identification of Bluebirds

Distinguishing female Mountain Bluebirds from drab female Eastern and Western can be difficult. Mountain is slightly longer and slimmer overall, but checking details of wing proportions allows a more objective assessment. Mountain Bluebird has a longer primary projection, distinctly longer than the length of the tertials or the length of exposed tail beyond the primary tips. On Eastern and Western Bluebirds these measurements are more nearly equal.

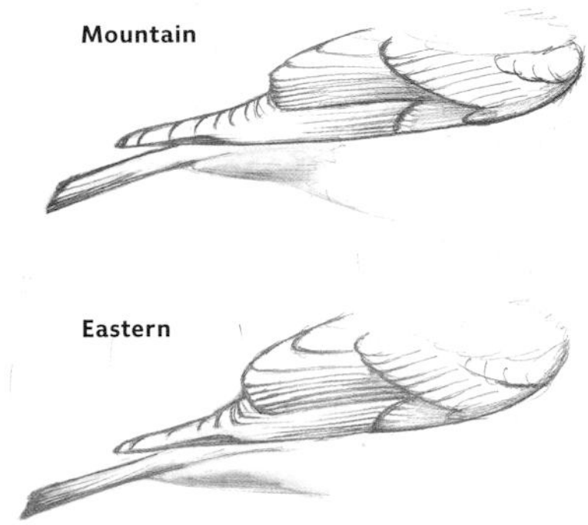

Arizona population rare and local in foothill riparian and pine-oak forest males; readily distinguished from widespread Eastern birds by paler and more orange underparts, orange throat and chin (vs. speckled blue and white malar and chin), paler blue upperparts (vs. darker violet-blue), belly not as bright white. Female also averages paler and drabber below, but may not be distinguishable from Eastern females.

Wood Thrush

Hylocichla mustelina

L 7.75" WS 13" WT 1.6 oz (47 g)

Larger than other spotted thrushes, with bolder black spots on clean white underparts, reddish-brown upperside brightest on nape, bold white eyering. Compare Brown Thrasher, which is streaked below, and has longer tail.

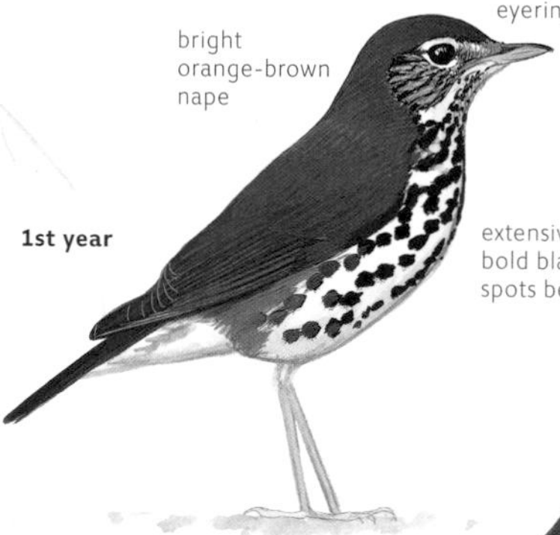

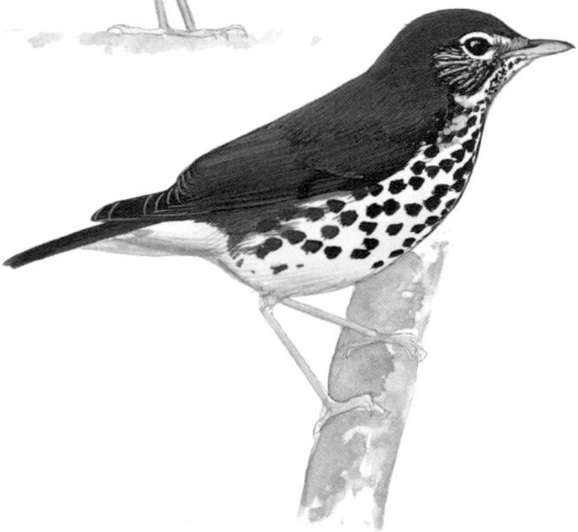

VOICE: Song rich, fluting, and varied: begins with low, soft *po po po* notes; climbs through short, gurgling phrase; ends with rich, buzzy or trilled whistle. Call a gentle, rolling *popopopo* and explosive, staccato *pit pit pit*. Flight call like *Catharus* thrushes but buzzier, more vibrant and nasal: a sharp, nasal *jeeen*.

Common in shaded understory with damp ground in mature deciduous forests. Forages mainly on the ground.

Veery

Catharus fuscescens

L 7" WS 12" WT 1.1 oz (31 g)

Uniformly reddish above and faintly spotted below, with pale face lacking distinct pattern around eye. Very similar to Russet-backed Swainson's Thrush, but with cleaner white belly and gray flanks.

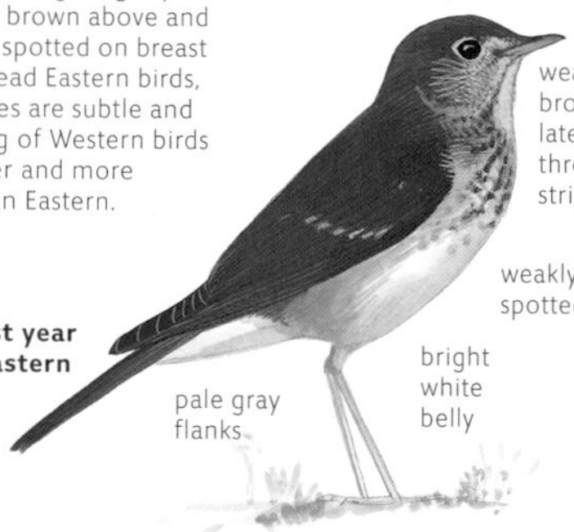

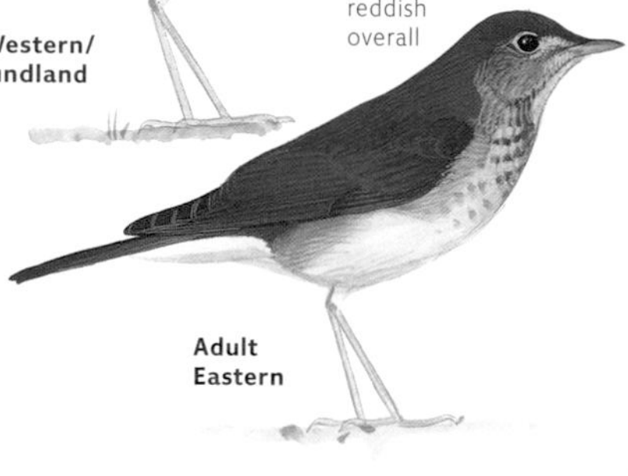

VOICE: Song descending in two stages, a smooth, rolling, somewhat nasal, fluting *vrdi vrreed vreed vreer vreer*. Call a nasal, rough, braying *jerrr* and calls resembling flight call; also a very rapid, harsh chuckle *ho-ch-ch-ch-ch*. Flight call *veer, veerre* or *veeyer*; relatively gentle and low-pitched; more variable from the ground.

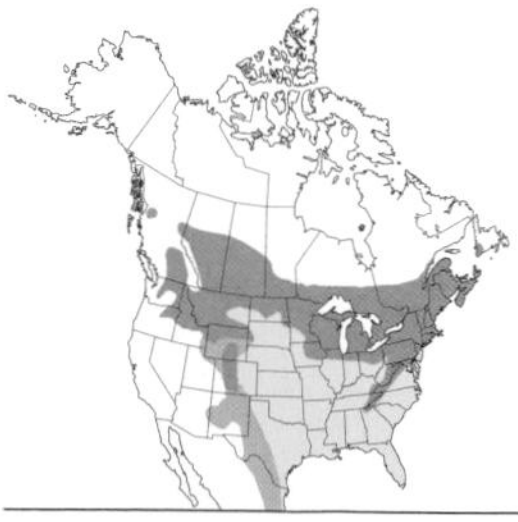

Common in willow thickets or other dense shrubby understory in wet woods. Habits similar to Swainson's Thrush.

Hermit Thrush

Catharus guttatus

6.75" WS 11.5" WT 1.1 oz (31 g)

)istinguished from other spotted thrushes by contrasting reddish ail and narrow but complete white eyering. Averages slightly maller and stockier than other thrushes (less streamlined), and he only species likely to be seen in winter.

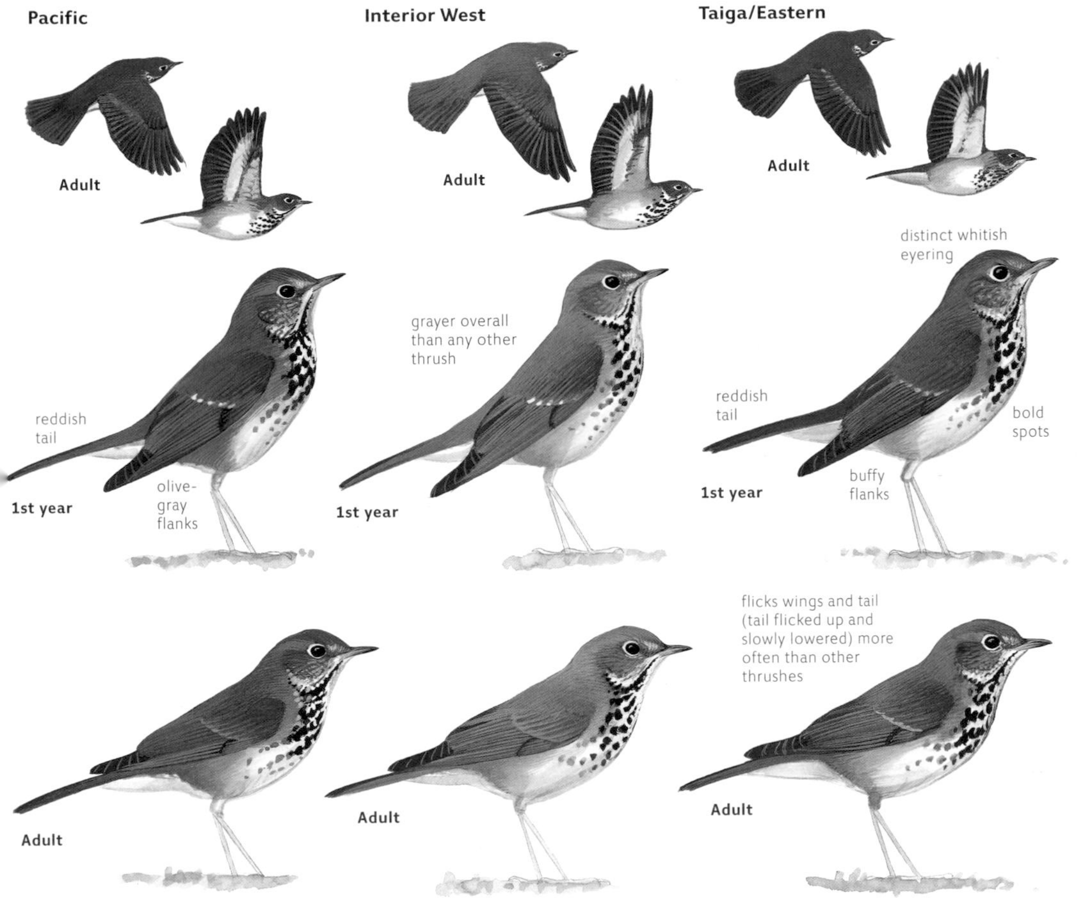

VOICE: Song ethereal, fluting, without clear rising or falling trend; begins with a long whistle followed by two or three higher twirling phrases fading at end; successive songs differ. Taiga/Eastern and Interior West birds sing liquid pure tones *seeeeeee freediila fridla-fridla*. Pacific song a little higher, harsher, more mechanical-sounding *zreeeeew cheedila chli-chli-chli*; introductory note often downslurred and buzzy. Call a low, soft, dry *chup* reminiscent of muffled blackbird call; slightly higher and sharper by Pacific and Interior West birds; also a whining, rising *zhweeee*. Flight call a clear, plaintive whistle *pee* without husky or buzzy quality of other thrushes.

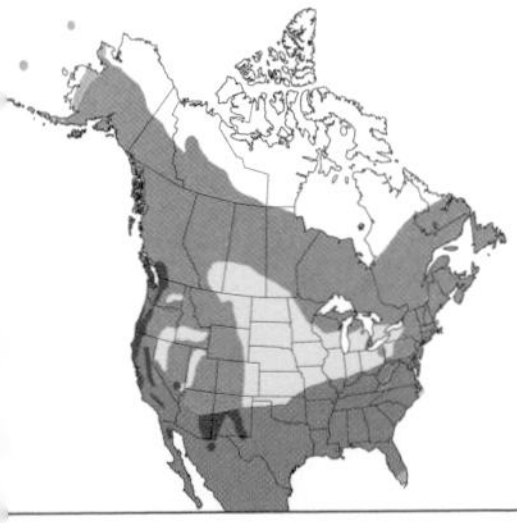

Common in brushy understory of forests, especially pine-oak woods, and often in drier and brushier habitat than other thrushes, foraging mainly on the ground; the only thrush normally seen in North America in winter. Habits similar to Swainson's Thrush.

Populations sort into three main groups, but there is extensive intergradation. Pacific birds are small, slender, and thin-billed; they are dingy brown with olive-gray flanks, white undertail coverts, and more spotted throat than Taiga/Eastern. Interior West birds are larger, longer-winged, and thin-billed; they are pale and grayish overall with gray flanks and limited reddish wash on flight feathers. Taiga/Eastern birds are medium-size, stocky, and thick-billed; they are clean and brightly colored overall with buffy flanks and undertail coverts and have pale buffy tips on the greater coverts at all ages.

Gray-cheeked Thrush

Catharus minimus

L 7.25" WS 13" WT 1.1 oz (32 g)

Similar to other spotted thrushes. Identified by grayish face, partial pale eyering, heavily spotted breast, drab gray-brown upperside and extensively dusky flanks. Nearly identical to Bicknell's Thrush; best distinguished by song.

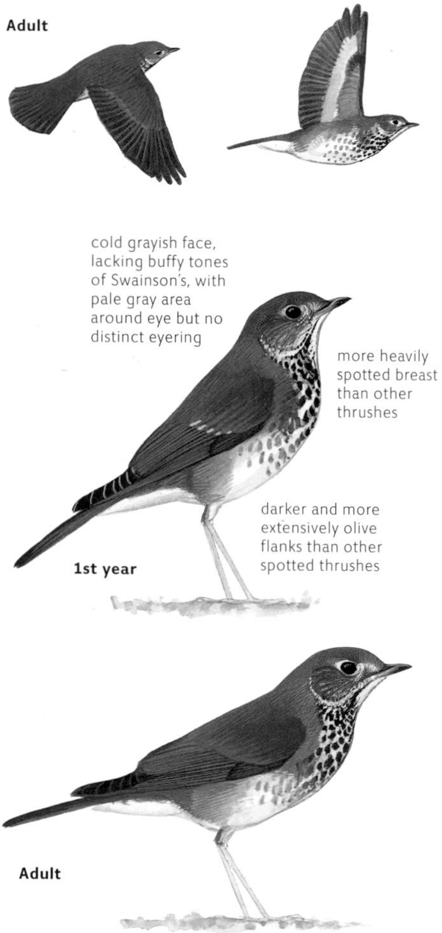

VOICE: Song a descending spiral like Veery but higher, thinner, and nasal with stuttering pauses *ch-ch zreeew zi-zi-zreeee zi-zreeew*; middle phrase rising. Flight call a high, penetrating, nasal *jee-er* or *queer*; many other variations given by perched birds; generally higher and more nasal than Veery.

Common in breeding range in spruce woods, but can be difficult to see. Migrants are uncommon to rare in most regions and can be in any wooded habitat. Habits similar to Swainson's Thrush.

Bicknell's Thrush

Catharus bicknelli

L 6.75" WS 11.5" WT 1.1 oz (30 g)

Nearly identical to Gray-cheeked Thrush and reliably distin guished only by voice; on average slightly smaller size and usuall more reddish plumage, but this is difficult to judge with som overlap.

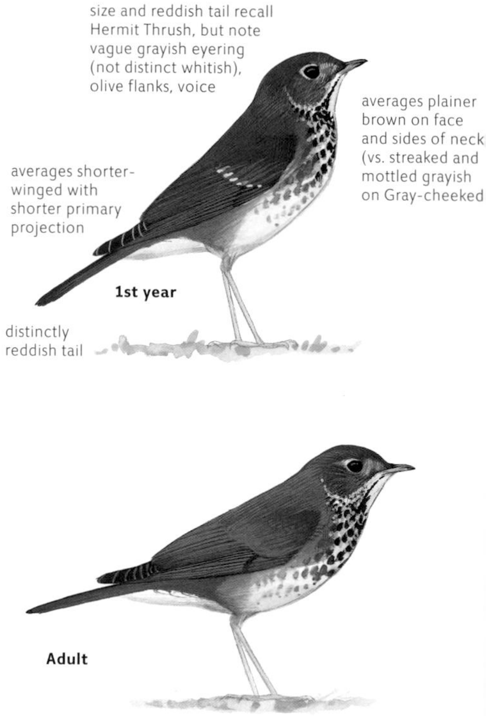

VOICE: Song similar to Gray-cheeked but higher-pitched, even more nasal and wiry; middle phrase descending and last phrase often rising *ch-ch zreee p-zreeew p-p-zreee*. Flight call a sharp, buzzy, descending *peeez* similar to Gray-cheeked but higher and less slurred; many other variations given by perched birds.

Common locally in stunted spruce or fir forests near treeline, also in second-growth forest in clear cuts. Rarely seen on migration. Habits like other thrushes.

Swainson's Thrush

Catharus ustulatus

L 7" WS 12" WT 1.1 oz (31 g)

Olive-backed form distinguished from other spotted thrushes by bold buffy spectacles, uniform gray-brown back, and heavily spotted breast. Russet-backed very similar to Veery but with extensive brownish wash on flanks.

VOICE: Song fluting, smooth, and rolling like Veery but rising. Russet-backed song may be slightly more nasal, longer, and more complex than Olive-backed birds: *po po tu tu tu tureel tureel tiree tree tree*. Olive-backed usually gives one introductory note (sometimes none) followed by single-syllable slurred phrases *po rer reer reeer re-e-e-e-e-e*. Call a low, liquid *pwip* or *quip*; also rough, nasal, braying or chatter usually introduced by call note *qui-brrrrr*. Flight call a mostly clear, level, emphatic *heep* or *queev* reminiscent of Spring Peeper (treefrog) call.

Russet-backed birds occur in Pacific region, wintering in Mexico. Olive-backed birds occupy remainder of range and winter in South America. Russet-backed is warmer reddish-brown overall, with less bold spectacles and less distinct spotting on breast, all tending toward Veery.

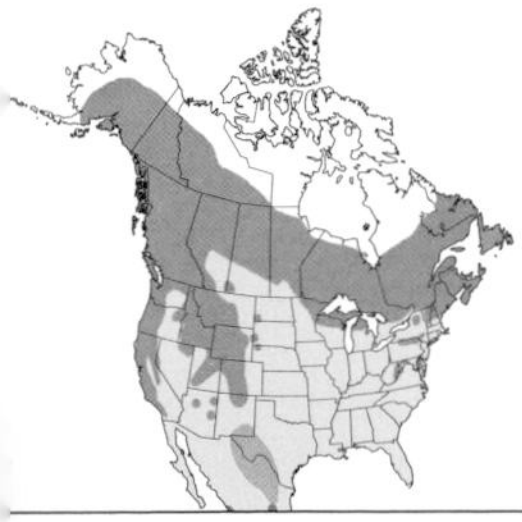

Common breeder in streamside willow and alder thickets, or dense understory near wet areas in mixed or coniferous forest, especially with tangles of blackberry, ferns, etc. Migrants prefer similar brushy understory in wet areas, foraging mainly on the ground; often attracted to fruiting trees and shrubs.

Dusky Thrush

Turdus naumanni

L 9"   WS 15"   WT 2.9 oz (82 g)

Robin-size, very distinctive with bold white eyebrow, black and white breastbands, spotted flanks, and rufous wings.

Several records in western Aleutians referable to the Naumann's subspecies, which has dark markings on underparts replaced by rufous, and less reddish wings. Intergrades between the subspecies are common in Asia.

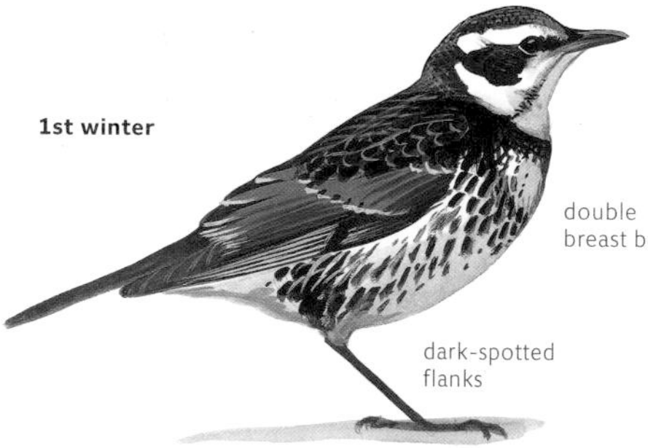

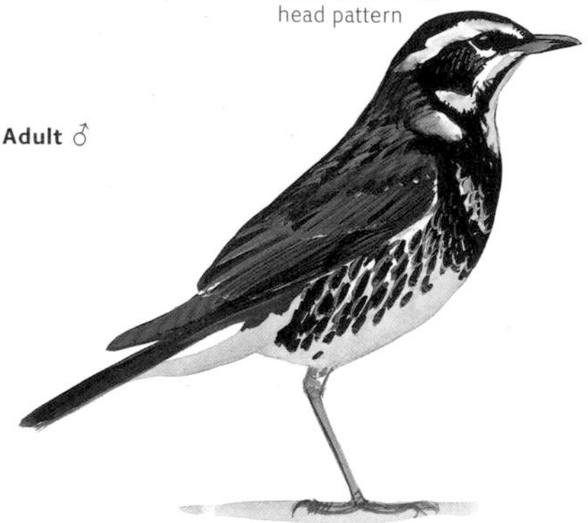

VOICE: Calls include a nasal, complaining *dwee-dwee*; a lower robin-like *pup-pep-pep*; and a thin rolling *shreel*.

Very rare visitor from Asia to islands of western Alaska, recorded a few times on Alaskan mainland, also south to British Columbia and Washington.

Eyebrowed Thrush

Turdus obscurus

L 9"   WS 14.5"   WT 2.6 oz (74 g)

Similar to pale female American Robin, some of which show distinct white eyebrow. Male is distinctive with gray hood and throat and small white chin. Female told from American Robin by grayish upper breast, more extensive white belly, uniform brownish upperparts.

VOICE: Song similar to American Robin but phrases higher-pitched, sweeter sounding, with longer pauses between phrases. Calls include a low *bup bup bup* and sharp dry *chik,* and flight call a high thin *sreee,* similar to American Robin.

Very rare visitor from Asia to western Alaska, mainly in spring in western Aleutians. One record in California (May).

Fieldfare

Turdus pilaris

L 10" WS 18" WT 3.5 oz (99 g)

Slightly larger than American Robin and easily distinguished by dark brown back contrasting with gray head and rump, black tail, densely spotted buffy breast, and bright white underwing coverts.

VOICE: Song an unmusical, squeaky chatter delivered without pauses. Call a loud, harsh *chr-chr-chr-chr* in descending series like scold of Blue-headed Vireo; often given in flight. Also a dry rattle reminiscent of Gray Catbird. Flight call a thin, nasal, rising *zreep*.

Very rare visitor from Eurasia; most records are of birds traveling with flocks of American Robin from Nov–Apr in eastern Canada. Habitat and habits similar to American Robin, and often found in flocks with robins.

Redwing

Turdus iliacus

L 8.3" WS 13.4" WT 2.3 oz (65 g)

Smaller than American Robin, with bold white eyebrow, white malar, streaked underparts, rufous flanks. In flight note uniform dark brown upperside and reddish underwing coverts.

VOICE: Song a jumbled, run-on series of scratchy or chattering notes, most notes repeated several times before switching to the next, finch-like. Calls include a hard *tuk*, robin-like. Flight call a long drawn-out, buzzy *zreeeee*, shrill and slightly descending.

Very rare visitor from Eurasia, most records in winter from Newfoundland, a few records south to Pennsylvania, and also single records in Alaska and Washington. Habits like American Robin, and vagrants are usually found with flocks of American Robins.

Rufous-backed Robin

Turdus rufopalliatus

L 9.25" WS 16" WT 2.7 oz (77 g)

Very similar in size and shape to American Robin, differs especially in rufous back and wing coverts contrasting with gray nape and rump, and all-dark face without white markings around eye.

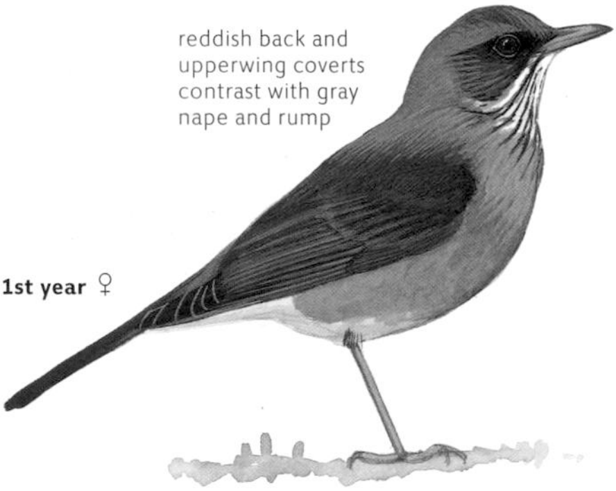

VOICE: Song low-pitched, slow, warbling *weedele loo loo freerlii*... with simple, repeated pattern. Call a short, descending *cherrp* and a rapid series *che che che che* analogous to American. Alarm a long, mellow, descending whistle. Flight call a high, short, rising *zeep*, thinner than American.

Very rare visitor from Mexico to lowland riparian habitat; recorded annually, mainly from Dec–Mar. Look for it in brushy wooded areas near water, usually foraging on the ground under trees or shrubs, or feeding on fruit. Sometimes joins flocks of American Robins but more often solitary.

Clay-colored Thrush

Turdus grayi

L 9" WS 15.5" WT 2.6 oz (74 g)

Size and shape like American Robin, but uniform drab clay-colored overall, just slightly paler below, and relatively dark around eye without any pale markings. Compare White-throated Thrush.

VOICE: Song a slow, low-pitched warbling with steady, monotonous tempo and repeated phrases, such as mellow, slurred *tooowiip tooowip*. Call a long, husky, sliding whistle with nasal mewing quality *teeweeooip*; similar to call given frequently by Long-billed Thrasher. Flight call weaker and buzzier than American Robin.

Rare but regular visitor from Mexico to extreme southern Texas, most numerous in winter, a few pairs nest in brushy woods. More secretive than American Robin; stays low in dense brush within woods, usually near water. Solitary or in pairs. Feeds on insects, worms, snails, and berries.

White-throated Thrush

Turdus assimilis

L 9" WS 15.5" WT 2.6 oz (74 g)

Similar to Clay-colored Thrush, but darker and colder brown overall, with obvious white band across lower throat, bold black streaks on throat, white belly and undertail coverts, and darker brownish underwing coverts.

VOICE: Song a series of low mellow whistled phrases, lower-pitched and slower than American Robin, many phrases with slow gradual slurred pitch change like *oweeeooo*. Call a rising nasal whistle *teweeep*, similar to Clay-colored but simpler. Also a low soft *pep*, and other calls.

Very rare visitor from Mexico to extreme southern Texas, fewer than ten records, but may be increasing in frequency. Habitat and habits similar to Clay-colored Thrush.

Orange-billed Nightingale-Thrush

Catharus aurantiirostris

L 9" WS 15" WT 2.9 oz (82 g)

A relatively small and plain thrush with short tail and long legs; reddish brown above and grayish below, with plain face and no spots. Bill, legs, and orbital ring bright orange on adult male, drabber on others.

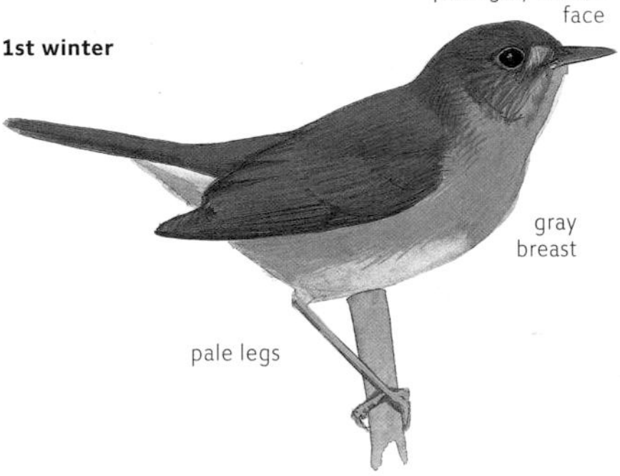

VOICE: Song a short jumbled, halting series of liquid notes and thin scratchy or buzzy whistles *sklideyo-tidee-klidee-treeedoo*; recalls Warbling Vireo (but higher-pitched) rather than other thrushes. Call a thin, nasal, rasping *zweeeee*, usually strongly rising.

Very rare visitor from Mexico, recorded twice in Texas and once in South Dakota. Found in dense shady undergrowth in forest, fairly secretive and difficult to see. Forages mainly on the ground.

American Robin

Turdus migratorius

L 10" WS 17" WT 2.7 oz (77 g)

One of our most familiar birds; relatively large and conspicuous. Uniform dark gray upperside with darker head, white markings around eye, rusty-orange breast and flanks, and dark tail with white corners distinctive.

wingbeats smooth, flicking; short glides with wings held close to body

Adult ♀

Adult ♂

typical feeding motions; shared by all thrushes

Juvenile (May–Sep)

Geographic variation is limited and clinal. Most western populations average paler and drabber than eastern and nearly lack white corners on tail. Some in East are richly colored with extensive black on nape and mantle.

Pale adult ♀

Adult ♂ typical

nape and upper back blackish like crown

limite white

much individual variation

Adult ♂ dark

VOICE: Song a series of low, whistled phrases with liquid quality typical of thrushes; each phrase delivered rather quickly but with long pauses between phrases; often two or three phrases alternately repeated over and over *plurrri, kliwi, plurrri, kliwi....* Call varies from a low, mellow *pup* or a sharp, clucking, often doubled *piik* to a sharper, rapid, urgent series *kli quiquiquiqui koo*; also a lower, softer *puk puk puk* and a harsh, high, descending *shheerr.* Flight call a very high, trilled, descending *srreel*; often combined with other calls such as *srreel puk puk puk.* Alarm like other thrushes: a very high, thin *tseeew* or shorter *seew.*

American Robin sings in a pattern of several short, warbled phrases, followed by a pause and then another set of phrases, and so on. This song pattern is shared in varying degrees by many other species of birds, and references to "robin-like songs" are frequent. Songs most often compared to robins include those of the Red-eyed Vireo group, tanagers, Pheucticus grosbeaks, and some orioles.

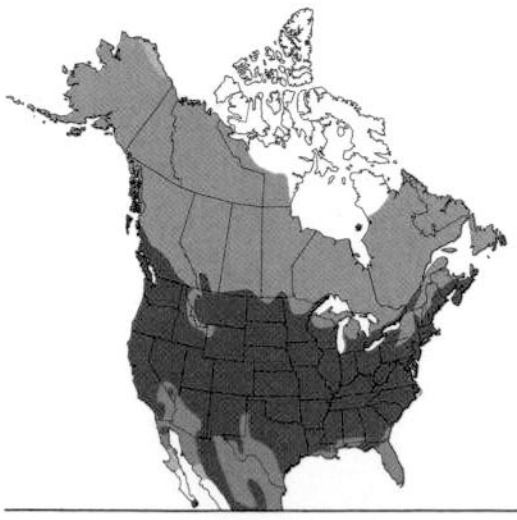

Common and widespread. Nests in any open woodland habitat from coniferous or deciduous forests to suburban neighborhoods. Most often seen on grassy lawns and fields searching for earthworms. Gathers in large foraging flocks and communal roosts in winter that may number in the hundreds or even thousands; winter diet mainly berries.

Varied Thrush

xoreus naevius

9.5" WS 16" WT 2.7 oz (78 g)

uperficially robin-like, but more reclusive and relatively short-ailed. Differs mainly in orange eyebrow, dark breastband, calico ving pattern; lacks white markings around eye. In flight pale vingstripe is usually obvious.

VOICE: Song a single, long whistle on one pitch: one and a half econds long and repeated about every ten seconds; each whis-e on a different pitch; some notes trilled or buzzy. Call a short, ow, dry *chup* very similar to Hermit Thrush but harder; also a ard, high *gipf* and a soft, short *tiup*. Flight call a short, humming histle.

Common but inconspicuous. In summer found in dense mature coniferous forest. In winter often in loose flocks in any dense wooded habitat, foraging on the ground while staying hidden in dense understory of forest.

Aztec Thrush

Ridgwayia pinicola

L 9.25" WS 16" WT 2.7 oz (78 g)

Robin-like in size, but shorter-tailed, with pied dark and white plumage. Main confusion is with juvenile Spotted Towhee; note wing and tail pattern, with white tips on all tail and secondary feathers, thin bill.

VOICE: Generally silent. Song a quiet series of hurried Robin-like phrases without pauses and with a complex metallic burry quality reminiscent of Townsend's Solitaire; *preeep* calls are often incorporated into the song, and buzzy *wheeer* sometimes given between songs. Calls include an explosive, grating *preeep* or *preeer*; a harsh, buzzy, whining *wheeeer* may be the flight call; also a soft, upslurred *seeep*.

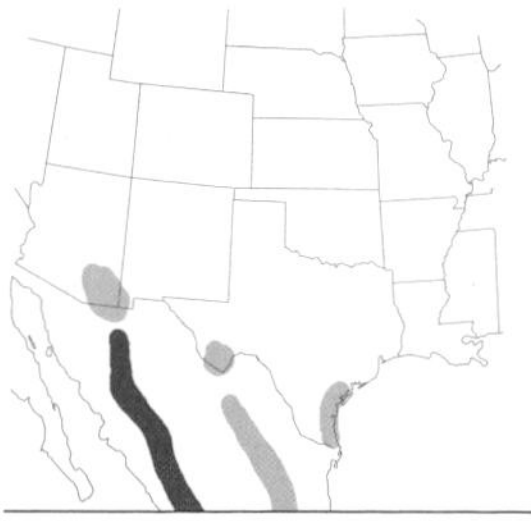

Very rare visitor from Mexico to pine-oak woods in canyons near water and fruiting shrubs, mainly from Jul–Aug; not recorded annually. Secretive and usually solitary, but multiple birds may gather at good food sources.

Mockingbirds, Thrashers, Starlings, Accentors, Wagtails, and Pipits

FAMILY: MIMIDAE, STURNIDAE, PRUNELLIDAE, MOTACILLIDAE

24 species in 11 genera in 4 families (5 rare species not included here). Mockingbirds and thrashers (family Mimidae) are mostly long-tailed, solitary, and secretive, and forage mainly on the ground, using long, sturdy bills to rake through leaf litter. Nest is a bulky cup in a tree or shrub. Starlings and mynas (family Sturnidae) are gregarious, short-tailed, often perch in trees or on buildings, and nest in cavities. Wagtails and pipits (family Moticil lidae) are small, thin-billed, and forage almost exclusively on th ground on open or grassy ground. Accentors (family Prunellidae are Eurasian species with habits somewhat like juncos, foragin mainly on the ground. Adults are shown.

Genus *Dumetella*

Gray Catbird, page 447

Genus *Mimus*

Northern Mockingbird, page 446

Bahama Mockingbird, page 446

Genus *Toxostoma*

Brown Thrasher, page 448

Long-billed Thrasher, page 448

Bendire's Thrasher, page 450

Curve-billed Thrasher, page 451

California Thrasher, page 449

Crissal Thrasher, page 449

Le Conte's Thrasher, page 450

Genus *Oreoscoptes*

Sage Thrasher, page 447

enus *Acridotheres*

Common Myna, page 452

Genus *Sturnus*

European Starling, page 453

enus *Prunella*

iberian Accentor, page 458

Genus *Motacilla*

Eastern Yellow Wagtail, page 454

White Wagtail, page 455

enus *Anthus*

ed-throated Pipit, page 456

American Pipit, page 457

Sprague's Pipit, page 456

Northern Mockingbird

Mimus polyglottos

L 10" WS 14" WT 1.7 oz (49 g)

Pale gray overall, more slender and longer-tailed than American Robin, with flashing white wing patches and white sides on tail. Superficially similar to shrikes, but note smaller head without obvious dark mask, longer tail, larger white wing patch, and slower flight.

VOICE: Song of varied phrases in regimented series: each phrase repeated two to six times, then an obvious pause followed by a different series *krrDEE-krrDEE-krrDEE, jeurrrdi jeurrrdi jeurrrdi...*; most phrases musical; many imitations of other species. Call a harsh, dry *chak*; harsher and longer than blackbirds; aggressive call a high, wheezy *skeeeh*.

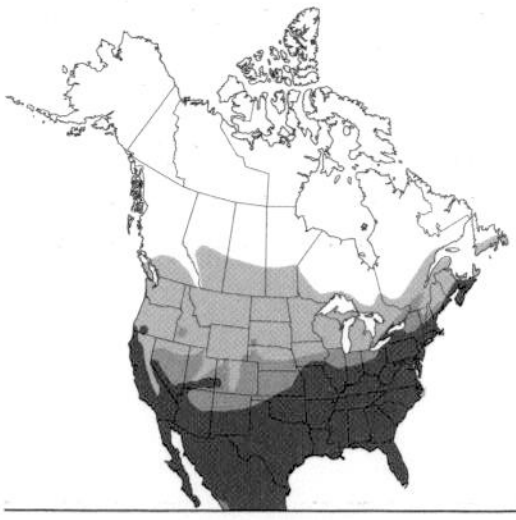

Common and conspicuous in suburban neighborhoods and brushy fields; foraging on the ground or in low brush for insects and fruit. Sings incessantly day and night from high exposed perches. Highly territorial and nearly always seen singly, or in family groups in summer; often defends fruiting trees and shrubs from other birds.

Bahama Mockingbird

Mimus gundlachii

L 11" WS 14.5" wt 2.3 oz (67 g)

Slightly larger and bulkier than Northern Mockingbird, and over all darker and browner, more thrasher-like. Confirm by lack o white in wings, tail with white only on tip, and finely streake pattern on head, back, and flanks.

VOICE: Song similar to Northern but less regimented in structur and with less varied phrases; overall more thrasher-like. Phrase nearly all two-syllabled; average lower and harsher than North ern with shorter pauses and little mimicry. Call a loud *czaa* slightly longer and rougher than Northern.

Very rare visitor from Bahamas t coastal areas of extreme southeastern Florida. Found in dense brush with clearings, such as well-vegetated suburban neighborhoods. More secretive then Northern Mockingbird.

Gray Catbird

umetella carolinensis

8.5" WS 11" WT 1.3 oz (37 g)

maller than American Robin, and relatively long-tailed, nearly niform dark gray, with distinctive black cap and rufous undertail overts. Seen briefly look for all gray body and wings, long, xpressive, all-black tail.

OICE: Song a rambling, halting warble with slow tempo; distincve mewing quality of low, hoarse notes with high, sharp chips nd squeaks interspersed; little repetition and little mimicry. Call hoarse, catlike mewing *mwee* or *meeurr*; also a loud crackling *edekekek* when startled and a low *whurf* deeper and softer than ermit Thrush.

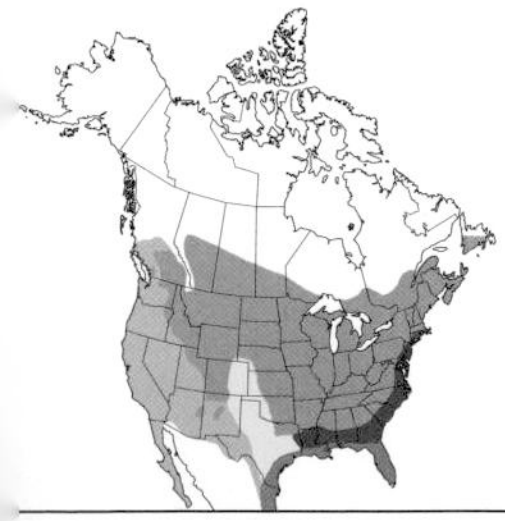

Common in brushy understory of woods, often in damp shaded areas. Generally solitary. Stays within brushy vegetation (except males sing from exposed perch above understory), foraging on the ground or in low shrubs for insects, other invertebrates, and berries.

Sage Thrasher

Oreoscoptes montanus

L 8.5" WS 12" WT 1.5 oz (43 g)

Our smallest thrasher, with relatively short bill and tail. Note overall gray-brown color with distinct blackish streaks above and below, and dark tail with clear white tips. Extremely worn birds can be similar to Bendire's Thrasher.

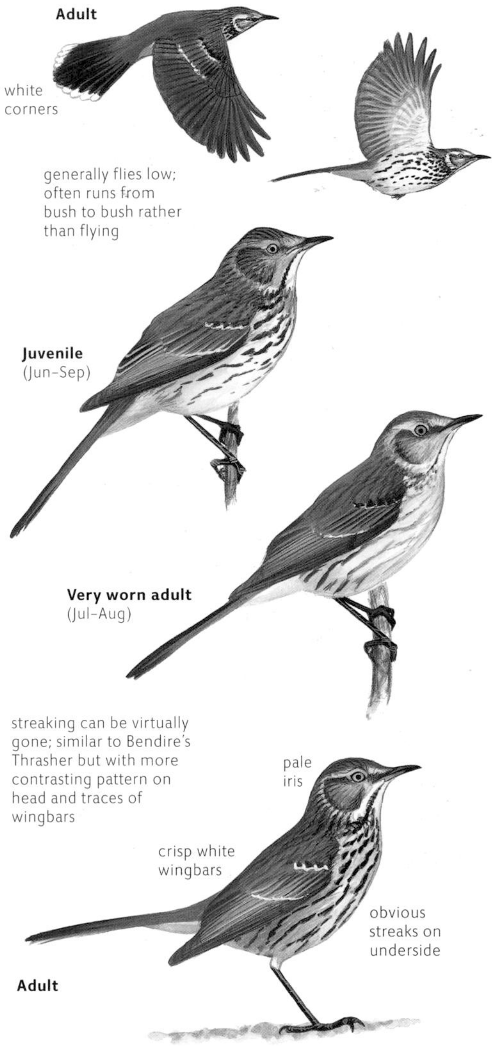

VOICE: Song a run-on warble of mellow, rolling or churring whistles with changeable tempo but very little pitch change; one accented phrase may be interspersed repeatedly; distinguished from Bendire's by clearer, less husky quality. Often gives a sweet, high *wheeurr*. Call a low *chup* like Hermit Thrush but harder.

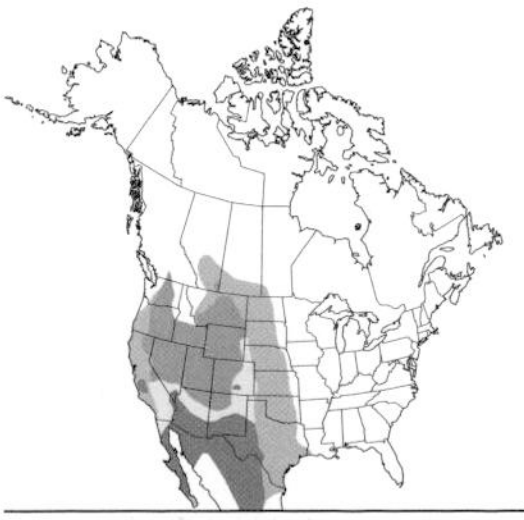

Common in sagebrush plains and similar expansive sparse brushlands with patches of bare ground. Usually solitary, but may gather in loose groups on wintering grounds. Feeds on insects and other invertebrates.

Brown Thrasher

Toxostoma rufum

L 11.5"   WS 13"   WT 2.4 oz (69 g)

Bright rufous upperparts, streaked underparts, and long rufous tail distinctive. Superficially similar to Wood Thrush, but with much longer and brighter tail, streaks (not spots) below, longer bill, different habitat. Most similar to Long-billed Thrasher.

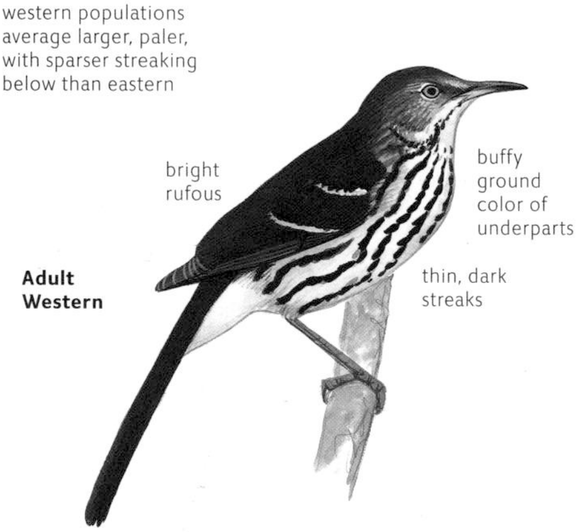

VOICE: Song of rich, musical phrases, each repeated two or three times with pause between each set; no other species has such clearly paired rhythm; often gives partial phrase such as *whichoo-which*. Calls include a loud, sharp *chak* like Fox Sparrow; a low, toneless growl *chhhr*; a sharp *tsssuk*; a rich, low whistle *peeooori* or *breeeew*.

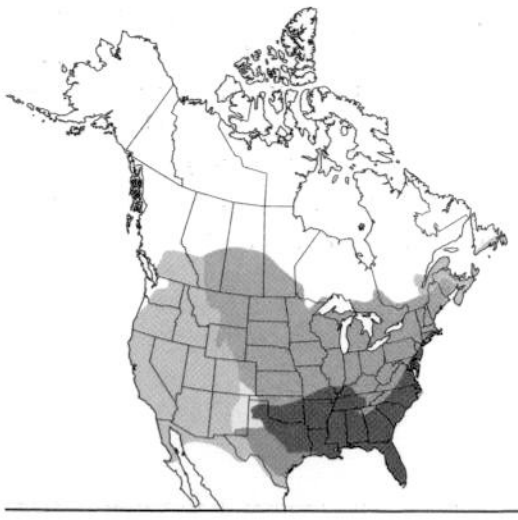

Uncommon, solitary and inconspicuous; usually hidden in dense brush, where it forages on the ground, tossing leaves and debris with bill to expose insects and other invertebrates.

Long-billed Thrasher

Toxostoma longirostre

L 11.5"   WS 12"   WT 2.5 oz (70 g)

Very similar to Brown Thrasher in all respects. Drabber brow above, especially on tail. Best distinguished by upperparts colo darker grayish cheeks, and white underparts with all blackis streaks (vs. off-white underparts with rufous and black streaks)

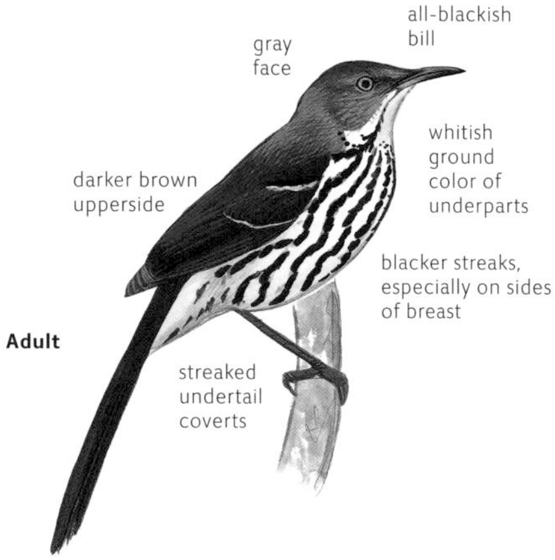

VOICE: Song of rather rich, musical phrases; harsher and mor rambling than Brown with less clearly paired phrases; distin guished from Curve-billed by slower tempo, more musica overall. Calls similar to Brown: a loud, sharp *chak*; a mellov whistled *tweeooip* or *oooeh*; also a very rapid, sharp rattle *chttt*

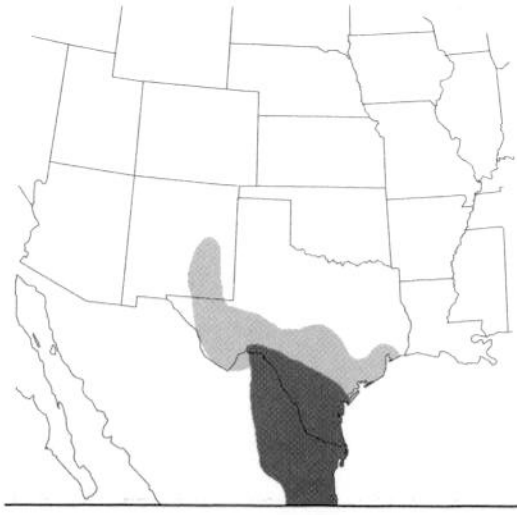

Common but inconspicuous in dense brushy woods. Habits and appearance very similar to Brow Thrasher, but range barely overlaps.

;alifornia Thrasher

'oxostoma redivivum

12" WS 12.5" WT 2.9 oz (84 g)

arge, long-billed, and long-tailed. Within most of its range only ;alifornia Towhee is similar in shape and color. Very similar to ;rissal Thrasher, but almost no range overlap; distinguished by ark iris, dark eyeline, buffy belly and undertail coverts.

OICE: Song of rather low, harsh notes repeated two to three mes but not strongly patterned; overall quality husky, slightly ;cratchy; often incorporates very high, thin notes like Cedar Vaxwing. Call a musical *dlulit* sometimes doubled; also a hard, omewhat musical *djik* or *djik-djik.*

Common but difficult to see in chaparral and other dense brush. Habits similar to other thrashers.

Crissal Thrasher

Toxostoma crissale

L 11.5" WS 12.5" WT 2.2 oz (62 g)

Range and habitat overlaps with Curve-billed Thrasher, note longer tail without pale tips, dark stripes on throat, smooth gray color, and rufous undertail coverts. Most similar to California Thrasher, but almost no range overlap.

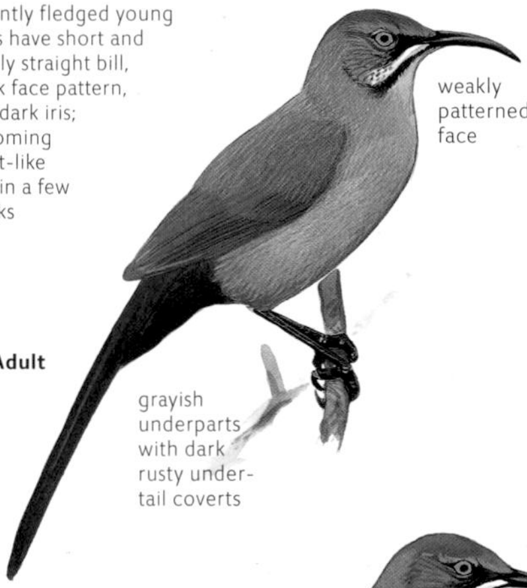

VOICE: Song relatively soft and musical with a few *quit* notes interspersed; not as harsh as California or Curve-billed. Call a rather soft, rolling, rising *pjurrre-durrre.*

Uncommon and difficult to see in dense mesquite and other tall dense brush along desert washes. Range and habitat overlaps with Curve-billed Thrasher, but more secretive, found in larger patches of dense brush, not in open desert scrub.

Le Conte's Thrasher

Toxostoma lecontei

L 11"   WS 12"   WT 2.2 oz (62 g)

Range and habitat usually does not overlap with other thrashers. Paler than all other thrashers, with dark tail and pale buffy undertail coverts, very pale head with dark lores. Bill strongly curved, more slender than other thrashers.

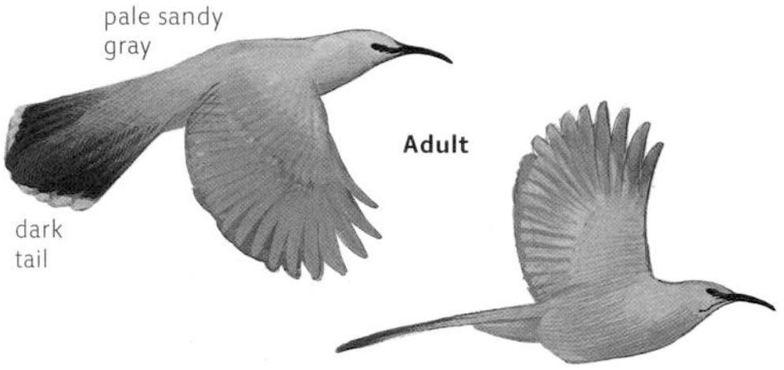

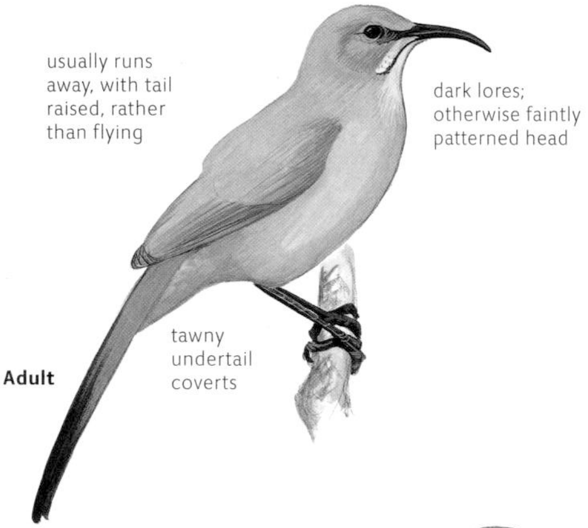

VOICE: Song similar to other thrashers but relatively soft and husky; fairly high-pitched, long, slurred notes distinctive; distinguished from Crissal by more relaxed, smooth rhythm and husky, slurred notes. Call a simple, questioning *weeip* or a quick, husky *kooi-dwid,* like Crissal but simpler.

Uncommon, local, and secretive on extremely arid and sparsely vegetated plains with saltbush and creosote bush and lots of bare sandy ground. Solitary; most often seen running across the ground from one bush to another.

Bendire's Thrasher

Toxostoma bendirei

L 9.75"   WS 13"   WT 2.2 oz (62 g)

Very similar to Curve-billed Thrasher; best distinguished by voic smaller triangular spots on breast and shorter straighter bill (ca be difficult to judge) with pale base of lower mandible. Habita also differs, overlapping only locally in open desert scrub.

VOICE: Song a rapid, husky warble building from a few short, hig whistles to a rapid, jumbled warble with many husky, whinin slurred notes; most phrases repeated but difficult to discern ind vidual phrases; husky quality and slurred delivery with no pause distinctive. Call a low, husky *chuk.*

Uncommon and local in open desert scrub, arid grasslands wit scattered brush or yuccas, and hedgerows along agricultural fields (Curve-billed overlaps onl in desert scrub). Usually solitary and rather secretive, foraging o the ground for insects and seed

Curve-billed Thrasher

Toxostoma curvirostre

L 11"   WS 13.5"   WT 2.8 oz (79 g)

Large, bulky, and pale gray-brown overall. Distinguished from other large thrashers (except Bendire's) by weak face pattern, spotted breast, orange iris, relatively short tail. Loud whistled call is frequent and very distinctive.

Two populations fairly distinctive and identifiable, but intergrade around Arizona-New Mexico border. Sonoran (Western) birds, found in most of Arizona, have overall plainer and less contrasting plumage than Chihuahuan (Eastern) birds found in New Mexico and Texas. Sonoran has spots on underparts less distinct and more evenly distributed (vs. more distinct spots concentrated on the breast on Chihuahuan); pale throat blends into gray breast (vs. more contrasting white throat on Chihuahuan); weak wingbars and poorly defined grayish tail spots (vs. distinct white wingbars and tail spots on Chihuahuan). Call may differ slightly: *wit-WEET* in Sonoran, and a less emphatic *WIT-WIT* in Chihuahuan.

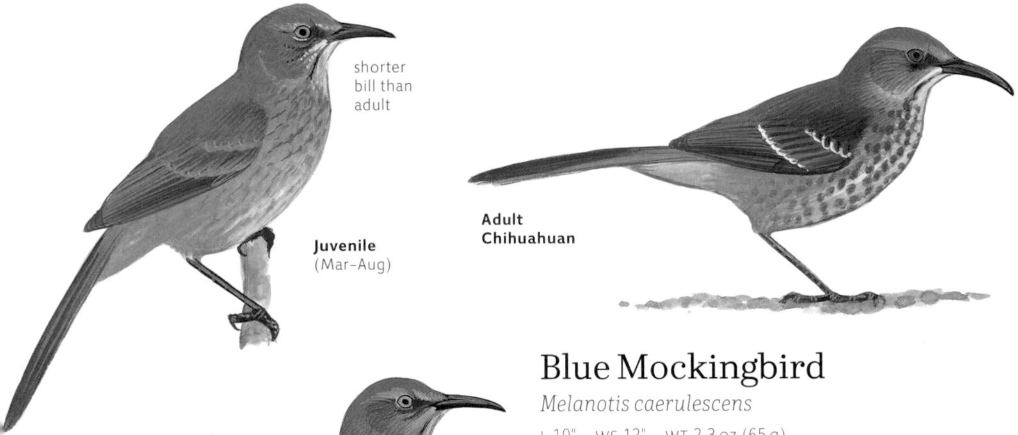

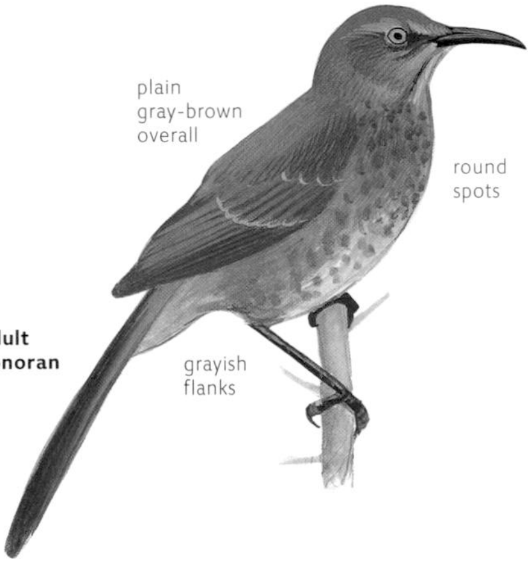

VOICE: Song rather harsh, crisp, and hurried with many short, sharp notes such as *quit-quit* and *weet*; more rattling or trilled phrases than other thrashers (e.g., *kitkitkitkitkit*). Call a very distinctive sharp, liquid whistle *wit-WEET-wit*; also a sharp, dry *pitpitpitpit* and a low, harsh chuck.

Common in variety of open habitats, especially suburban yards and parks (where Bendire's is not found), desert scrub with varied plants and open patches. Within its range this is the most frequently seen thrasher, often foraging in the open. Generally solitary. Forages on the ground, tossing leaves and debris with bill to expose insects and other invertebrates.

Blue Mockingbird

Melanotis caerulescens

L 10"   WS 12"   WT 2.3 oz (65 g)

Bulkier than Northern Mockingbird, dark bluish overall, with broad black mask, very broad and all-dark tail, long and heavy bill.

VOICE: Song thrasher-like, of rich whistles and other phrases, similar to Curve-billed Thrasher but with very long pauses between phrases; much mimicry. Calls extremely varied, include rich whistled *pweeeoo* or *too-lip*, nasal rising *rreeeee*, low *chuk*, sharp *ki-pik*, and more.

Very rare visitor from Mexico to Arizona, Texas, New Mexico, and California. Some records may represent escaped cage birds. Found in dense tall brush such as mesquite and hackberry thickets along streams. Generally quite secretive and difficult to see, thrasher-like.

Common Myna

Acridotheres tristis

L 9.75" WS 18" WT 3.7 oz (106 g)

Starling-size, but distinctively patterned. Note short tail and broad wings with bold black and white pattern. Yellow bill and legs, blackish head, and brown body distinctive.

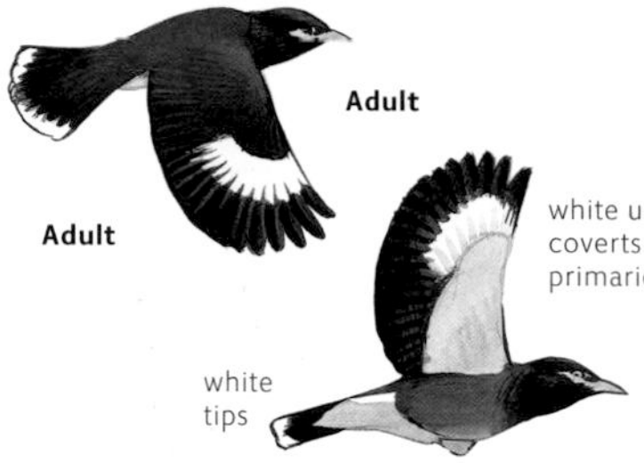

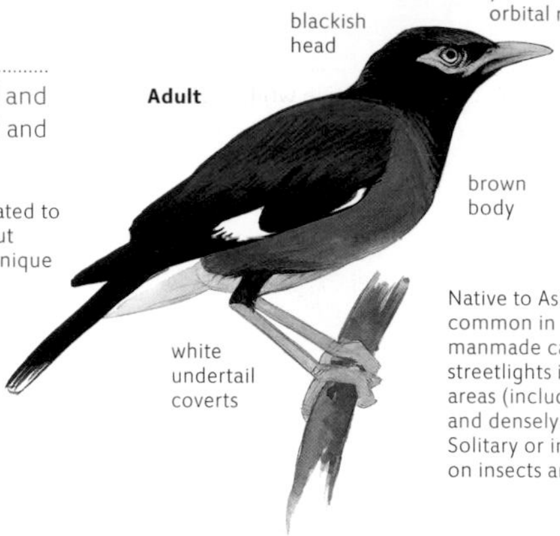

Native to Asia. Introduced and common in Florida. Nests in manmade cavities, such as streetlights in relatively urban areas (including shopping malls and densely settled places). Solitary or in small groups. Feeds on insects and fruit.

VOICE: Song includes drawn-out low grating sound, and loud ringing series such as *keew-keew-keew-keew klee-klee-klee-klee* similar to Northern Mockingbird (but gives a few phrases at a time, then long pause). Some sounds are imitations of other species, such as Boat-tailed Grackle, Purple Martin, and others.

Crested Myna

Acridotheres cristatellus

L 9.75" WS 18" WT 4 oz (113 g)

All-black plumage with white tail tip and white base of primaries distinctive; also note orange bill and yellow iris and legs. Short bushy crest conspicuous on adult.

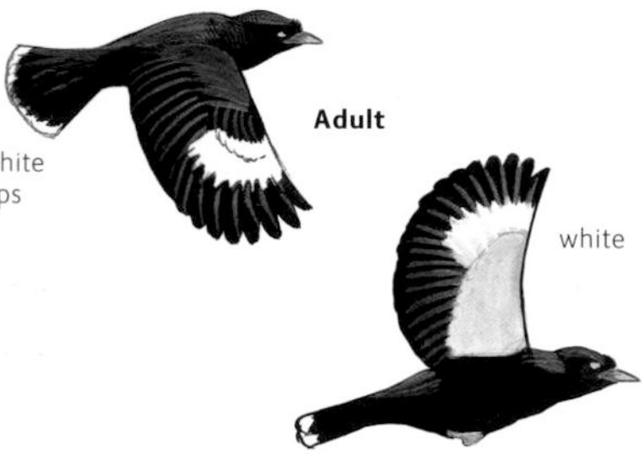

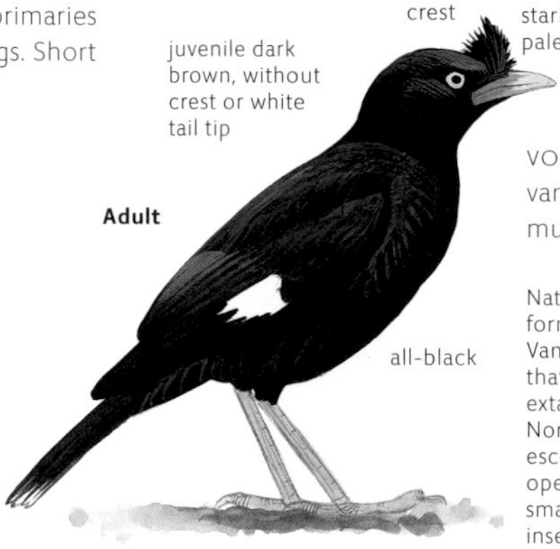

VOICE: Similar to Common; varied harsh chattering and musical phrases.

Native to Asia. Introduced and formerly established around Vancouver, British Columbia, but that population is no longer extant and the species is seen in North America only as occasional escapes from captivity. Found in open areas near trees, often in small groups. Feeds on fruit, insects, and invertebrates.

Hill Myna

Gracula religiosa

L 10.5" WS 20" WT 7 oz (205 g)

A large, short-tailed black bird, intermediate in size between starling and crow. Large white wing patches and yellow face wattles distinctive.

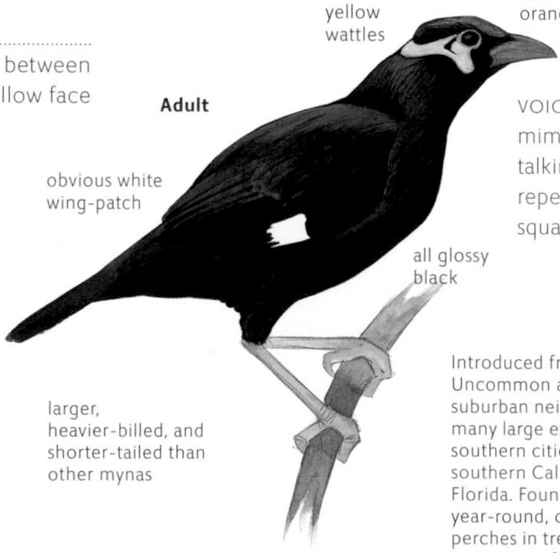

VOICE: An expert mimic, the familiar talking myna; varied repertoire of whistles, squawks, and chirps.

Introduced from Asia. Uncommon and local in suburban neighborhoods with many large exotic trees in southern cities, especially in southern California and Florida. Found in small groups year-round, often on exposed perches in treetops. Feeds on variety of insects and fruit.

:uropean Starling

turnus vulgaris

8.5" WS 16" WT 2.9 oz (82 g) ♂>♀

ote overall blackish color and distinctive white dots of non-reeding plumage. Distinguished from other black birds by short juare tail, long tapered bill, pointed wings and translucent flight :athers, yellowish bill and pale-edged wing coverts.

When disturbed by a raptor, flocks form a very tight, cohesive ball, visible at a great distance and a good clue to the presence of a hawk.

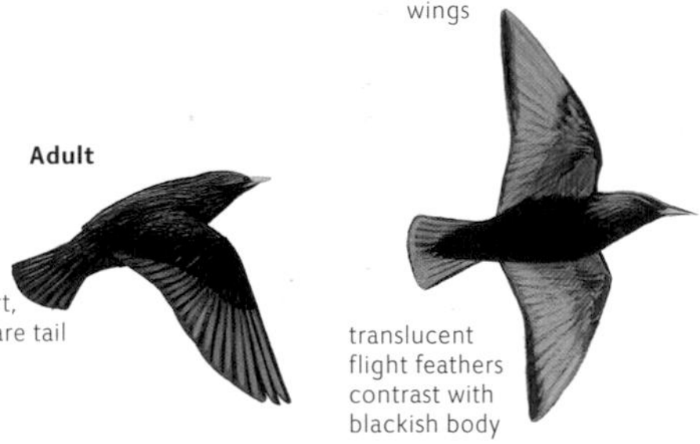

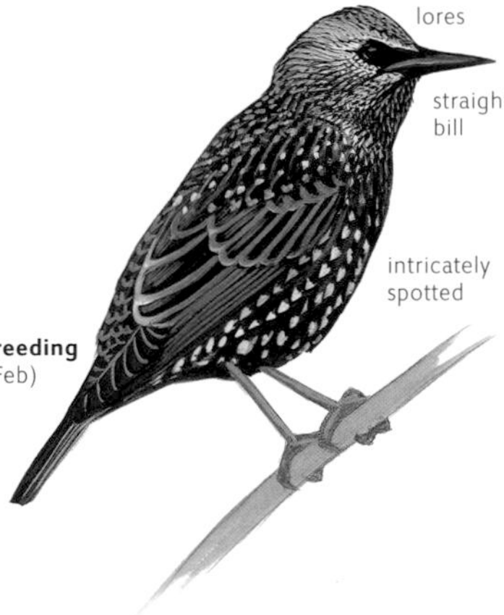

breeding appearance acquired by wear, as pale tips of fresh (nonbreeding) feathers wear off during winter to expose glossy black

OICE: Song mainly quiet, harsh rattling with some high, thin, urred whistles; overall a mushy, gurgling, hissing chatter with igh, sliding whistles. Often includes imitations of other birds' alls. Common call a harsh chatter *che-che-che-che*. Flight call muffled, dry, buzzing *wrrsh*.

Common and widespread. Introduced from Europe to New York in late 1800s; now found throughout North America and one of the most common birds wherever human settlement occurs. Nests in birdhouses, crevices in buildings, and tree cavities. Found in large flocks almost year-round. Forages on the ground for grubs, worms, insects, seeds, and the like, or in trees for fruit.

Eastern Yellow Wagtail

Motacilla tschutschensis

L 6.5" WS 9.5" WT 0.56 oz (16 g)

More slender, longer-tailed than pipits, with relatively long and narrow wings. Adult distinctive bright yellow below, with dark mask. Immature can lack yellow, dark gray above, with contrasting head pattern.

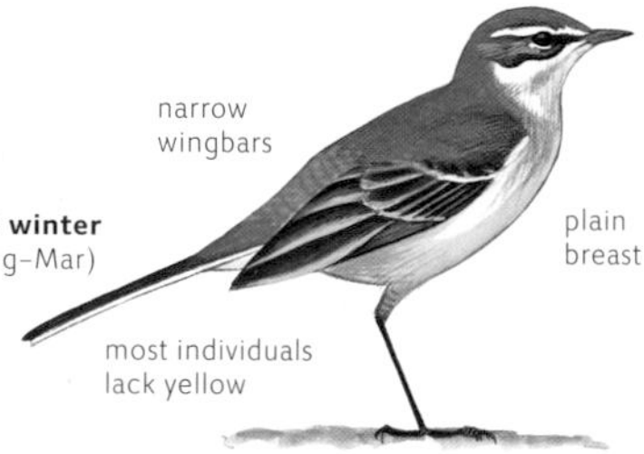

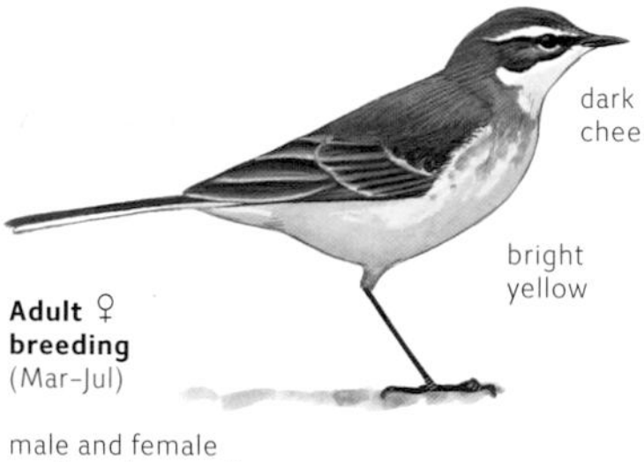

Individuals from Asia, seen rarely in far western Alaska, are brighter yellow below, with cleaner breast and bolder face pattern, including broader white supercilium.

VOICE: Song repetitious phrases resembling calls *tzeeu tzeeu tzeek*.... Flight call an explosive, vibrant, buzzy *tzeer*; occasionally a higher, clear *tsewee*; also a clear, ringing *tzeen tzeen tzeen*.

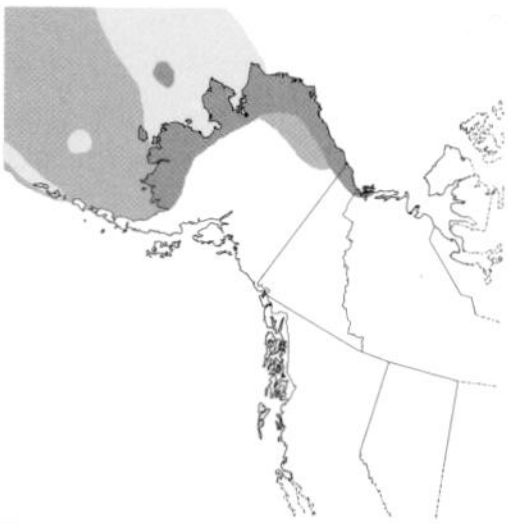

Common locally on coastal tundra in western and northern Alaska. Very rare along Pacific coast to California (mainly Sep). Generally solitary, but small flocks form on migration and winter in normal range. Always in the open, forages for insects on open ground in fields, yards, beaches, or shorelines of lagoons or ponds.

Vhite Wagtail

lotacilla alba

7.25" WS 10.5" WT 0.63 oz (18 g)

ery long-tailed with obvious white outer tail feathers. Note lean black and white plumage with black breastband, gray back, nd clean white underparts, unlike other wagtails. In flight note ery long and narrow tail and wings, sharp calls.

iberian

OICE: Song a regular series of high, thin, finch-like phrases of wo or three syllables, each phrase short and clear with little pitch hange. Flight call a staccato, harsh *jijik*. Call a musical, finch-like hrase *didleer*.

Uncommon and local in western Alaska, where it nests among large boulders. Very rare farther south along Pacific coast with a few records east to Atlantic coast. Generally solitary, but forms small flocks in normal along edges of ponds and lagoons.

Kamchatka (Black-backed)

Nesting in Kamchatka. Adult males are distinctive with black back and mostly white wings. Other plumages closer to Siberian, but always with more white in wings than corresponding plumage of Siberian. Rare visitor to far western Alaska, very rare farther south to California and a few records east to Atlantic coast. Formerly considered a separate species, Black-backed Wagtail.

European

Subspecies nesting in Greenland, Iceland, and Europe has white face lacking dark eyeline, recorded several times along East coast from Québec to Florida.

Sprague's Pipit

Anthus spragueii

L 6.5" WS 10" WT 0.88 oz (25 g)

Stocky and short-tailed compared to other pipits, with more white in tail, more secretive habits. Most often confused with juvenile Horned Lark but note longer pale legs, thinner bill, little or no primary projection, and no white edge on tertials.

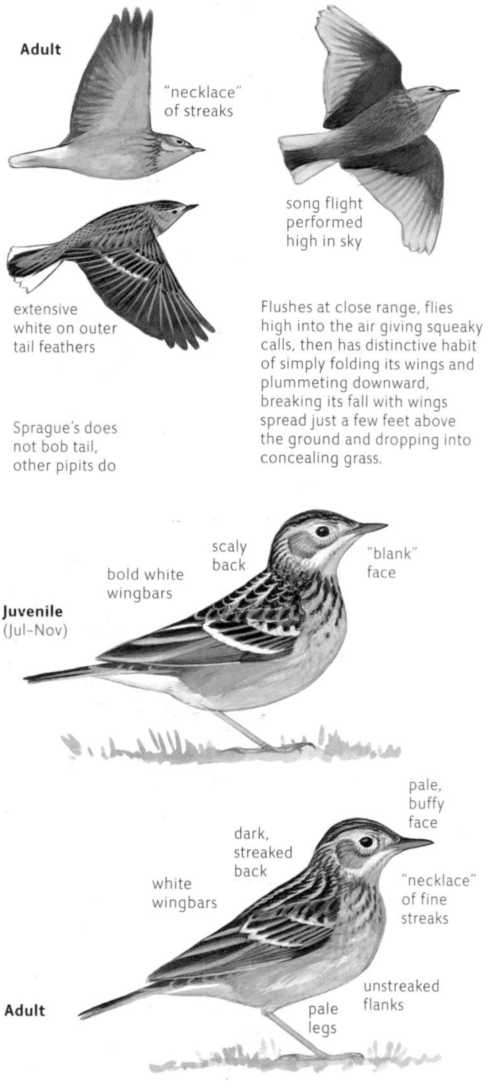

VOICE: Song given in flight from high in the air a descending, jingling cascade of high, dry whistles *shirl shirl shirrl*.... Flight call a high, sharp, slightly nasal *squeet* often repeated; reminiscent of sharper calls of Barn Swallow.

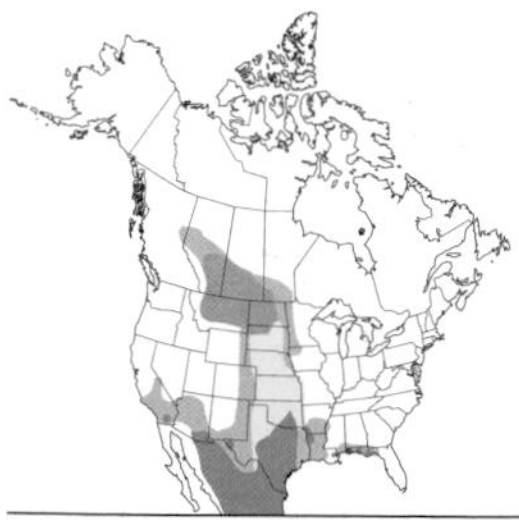

Uncommon and local. Nests and winters on expansive shortgrass prairies or fields with very short and sparse grass. Solitary and secretive, never flocks with other pipits; very difficult to see as it crouches in grass foraging for insects and seeds.

Red-throated Pipit

Anthus cervinus

L 6.25" WS 10.5" WT 0.74 oz (21 g)

Similar to American Pipit, very slightly stockier. Note bold streaked back with pale "braces". Underparts also more bold streaked than American, and legs very pale. Adult has distinctiv reddish face. Often detected by flight call.

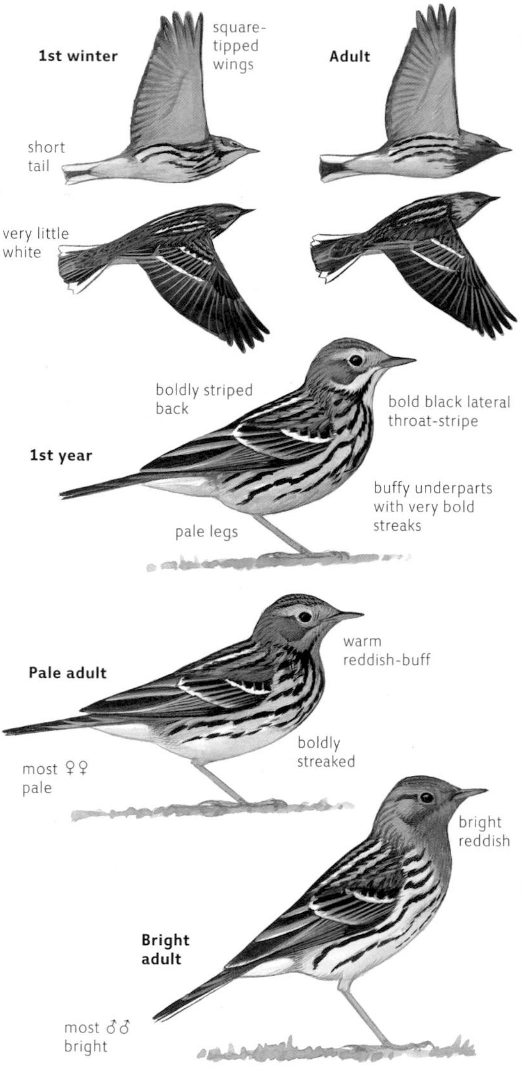

VOICE: Song a long series of jingling phrases, all about the sam pitch and rhythm, each one repeated about five times *jrr jrr jr jrr jrr jrr tree tree tree tree tree tseew tseew tseew*. Flight call very high, thin, drawn-out *psssss* reminiscent of Yellow Wagta but thinner, not buzzy.

Rare visitor from Siberia, very small numbers nest in far wester Alaska, and a few occur each fall along Pacific coast south to California, mainly in Oct. Found on open ground, grassy tundra, and plowed or shortgrass fields. Often flocks with American Pipit but tends to be slightly more secretive. Feeds on insects and seeds.

American Pipit

nthus rubescens

6.5" WS 10.5" WT 0.74 oz (21 g)

mall and superficially sparrow-like, but slender with long legs nd thin bill. Walks delicately with head high. Note gray-brown pperside, pale lores, and faintly or boldly streaked underside. Vhite outer tail feathers show in flight.

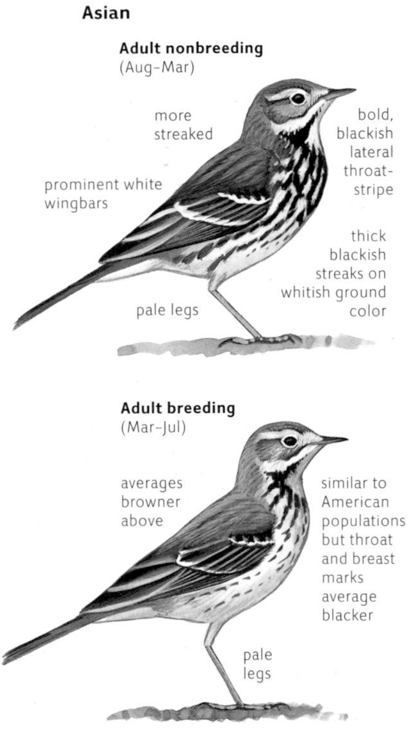

OICE: Song a slow series of high, clear or jingling phrases *seewl-tseewl-tseewl*... or *pleetrr-pleetrr-pleetrr* and other ariations; given in flight for up to 15 seconds. All populations imilar. Flight call a high, squeaky chirp *slip* or *slip-ip*; when ushed a higher *tseep* or *tsitiip*. Alarm call near nest a much ower, rising *pwisp*.

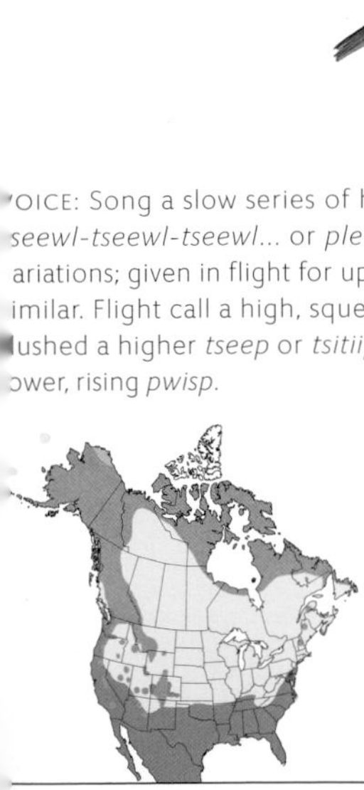

Common. Nests on tundra; winters on expansive open ground, fields, and beaches. Almost always in small flocks, walking briskly in all directions across the ground; flushes readily and frequently. Feeds on insects and seeds.

Paler birds are typical of populations nesting in the Rocky Mountain region; most Arctic breeders are darker and more heavily streaked. Asian birds (seen regularly on Bering Sea islands, with a few records father south on Pacific coast) have pale legs and plumage differences as noted above; they are most distinctive in nonbreeding plumage.

Olive-backed Pipit

Anthus hodgsoni

L 5.7" WS 10.2" WT 0.74 oz (21 g)

Forages on open ground or short grass, often near or under scattered trees, and often flies up into trees when disturbed (unlike other pipits).

VOICE: Song a rapid series of high trills and warbles, reminiscent of Winter Wren but shorter and with some lower phrases. Flight call a high, buzzy *tzzeew*, similar to Red-throated Pipit but lower, shorter, and buzzy. Weaker than Yellow Wagtail.

Very rare visitor from Asia to far western Alaska in spring and fall, with single records in California (fall) and Nevada (spring).

Pechora Pipit

Anthus gustavi

L 5.6" WS 10.2" WT 0.74 oz (21 g)

A secretive and solitary pipit with very bold streaks above and below, dark patch on side of throat, obvious white wingbars, and pale legs.

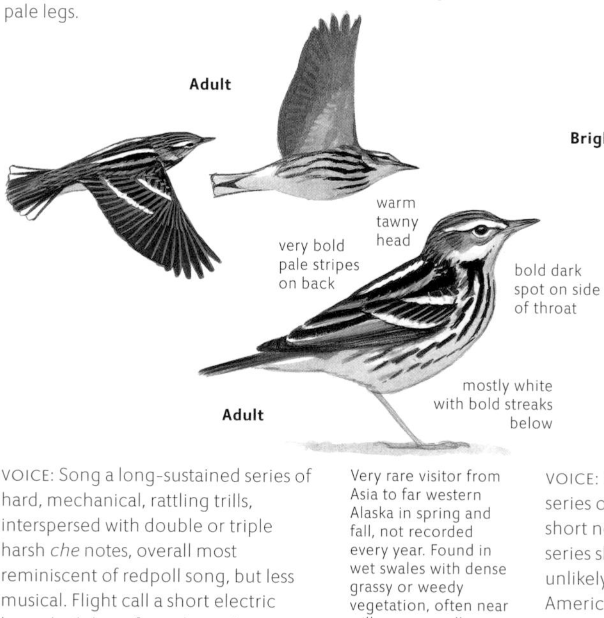

VOICE: Song a long-sustained series of hard, mechanical, rattling trills, interspersed with double or triple harsh *che* notes, overall most reminiscent of redpoll song, but less musical. Flight call a short electric buzz *dzzik*, but often silent when flushed (unlike other pipits).

Very rare visitor from Asia to far western Alaska in spring and fall, not recorded every year. Found in wet swales with dense grassy or weedy vegetation, often near willows or small trees; secretive and difficult to see.

Siberian Accentor

Prunella montanella

L 5.5" WS 8.5" WT 0.6 oz (17 g)

Superficially sparrow-like, but with thinner pointed bill, ric tawny underparts, dark mask and crown with broad and clea tawny eyebrow stripe, fairly short tail.

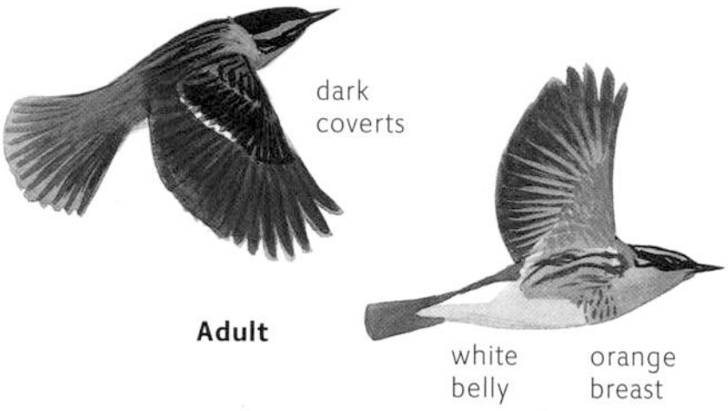

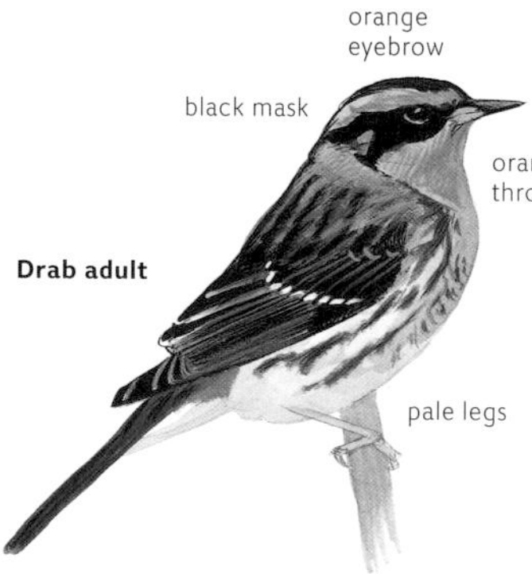

VOICE: Call a rapid insistent series of about five very high short notes *ti-ti-ti-ti-ti*, the series slightly descending. Song unlikely to be heard in North America, a high-pitched monotone chattering or twittering series.

Very rare visitor from Asia to western Alaska, mostly in late fall and winter. Several records in winter south to British Columbia, Washington, Idaho, and Montana. Found in patches of small trees or brush, vagrants in North America sometimes visit bird feeders.

Waxwings, Silky-flycatchers, Olive Warbler, and Longspurs

AMILY: BOMBYCILLIDAE, PTILIOGONATIDAE, PEUCEDRAMIDAE, CALCARIIDAE

1 species in 7 genera in 4 families (1 rare species not included ere). Waxwings (family Bombycillidae) and Phainopepla (in the lky-flycatcher family Ptiliogonatidae) are fairly closely related, nd share silky plumage and similar head shape. Both families orage mainly on berries and are often seen in flocks perched in treetops. Longspurs (family Calcariidae) are sparrow-like, and until recently were included in the family Emberizidae, but have habits more like larks (family Alaudidae). Olive Warbler (family Peucedramidae) is superficially similar to wood-warblers, but differs in bill shape, tail shape, voice, and more. Adult females are shown.

enus *Bombycilla*

ohemian Waxwing, page 460

Cedar Waxwing, page 460

Genus *Phainopepla*

Phainopepla, page 461

Genus *Peucedramus*

Olive Warbler, page 465

enus *Rhynchophanes*

AcCown's Longspur, page 462

Genus *Calcarius*

Lapland Longspur, page 463

Smith's Longspur, page 463

Chestnut-collared Longspur, page 462

enus *Plectrophenax*

now Bunting, page 464

McKay's Bunting, page 464

Bohemian Waxwing

Bombycilla garrulus

L 8.25" WS 14.5" WT 2 oz (56 g)

Larger than Cedar, darker and more grayish overall; distinguished by gray belly, rufous undertail coverts, and white markings on wings. Short tail and triangular wings reminiscent of European Starling in flight; but smaller and with constant flight calls.

VOICE: Similar to Cedar but calls lower-pitched and more clearly trilled; trill slower, more like a rattle.

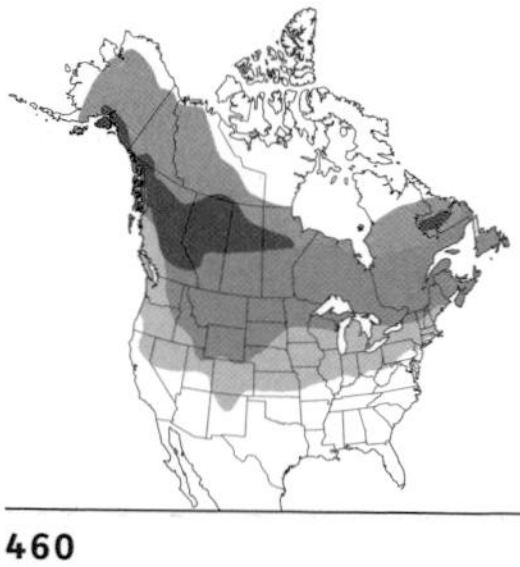

Common but irregular. Nests in spruce forests; winters in flocks that wander widely in search of fruit.

Cedar Waxwing

Bombycilla cedrorum

L 7.25" WS 12" WT 1.1 oz (32 g)

Silky brown color and crest, yellow-tipped tail, and dark mas distinctive. Smaller than Bohemian and browner overall, wit pale yellowish belly, white undertail coverts. Flocks fly in tigh groups like European Starling.

VOICE: Song simply a series of high *sreee* notes in irregula rhythm. Call a very high, thin, clear or slightly trilled *sreee*. Aerial predator alarm a piercing *seeeew* similar to thrushes.

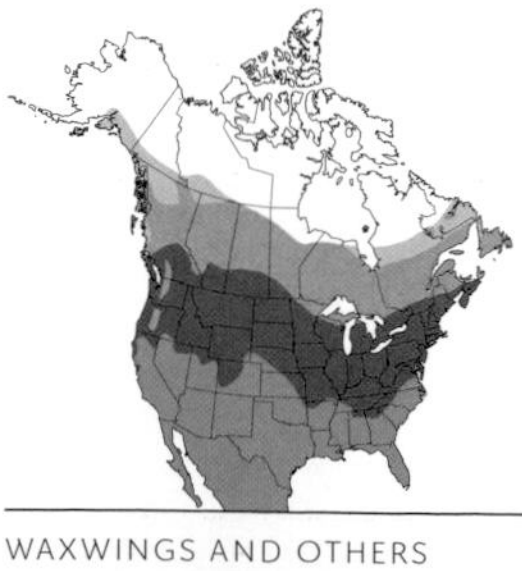

Common but irregular in any open wooded or brushy habitat where fruit or other food (such as tree buds, flowers, and insects is found. Nests in brushy areas such as old fields and stream edges. Winters in brushy hedgerows, open woods, or suburbs where berries are plentiful. Almost always in small or large flocks, except when nesting.

Gray Silky-flycatcher

Ptiliogonys cinereus

L 7.9" WS 11" WT 0.84 oz (24 g)

Size and shape similar to Phainopepla, but with short bushy crest, pale gray body, yellow undertail coverts, and black-and-white tail.

VOICE: Common call a two-syllabled finch-like *chu-pik*, with second syllable higher. Other short chattering, gurgling or warbling phrases also heard from groups.

Very rare visitor from Mexico to Texas and California. Some records may represent escaped cage birds. Found in open pine-oak forest, often perching on treetops. Vagrants could occur in any open woods or suburbs with scattered large trees.

Phainopepla

Phainopepla nitens

L 7.75" WS 11" WT 0.84 oz (24 g)

Sleek, long-tailed, and round-winged, with ragged crest. Plumage dark and unpatterned, with entirely dark underparts and tail; note pale edges of wing feathers on female. Male's broad white wing-patch distinctive in flight.

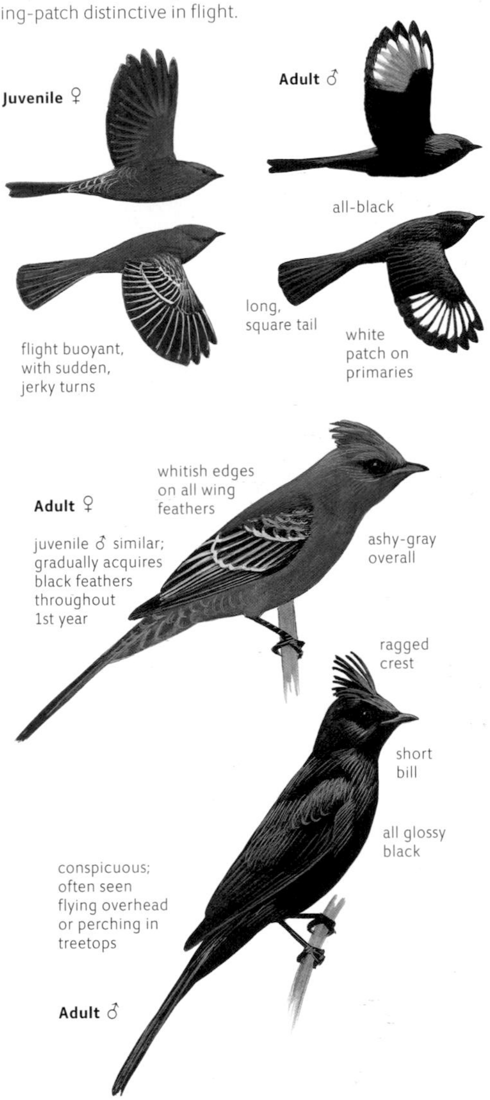

VOICE: Song a series of short phrases with long pauses between; phrases rather low, liquid whistles mixed with crackling, grating sounds (e.g., *krrtiiilwa*); also mimics other species; often accompanied by somersaulting flight display. Call a distinctive, soft, rising whistle *hoi*, low and questioning.

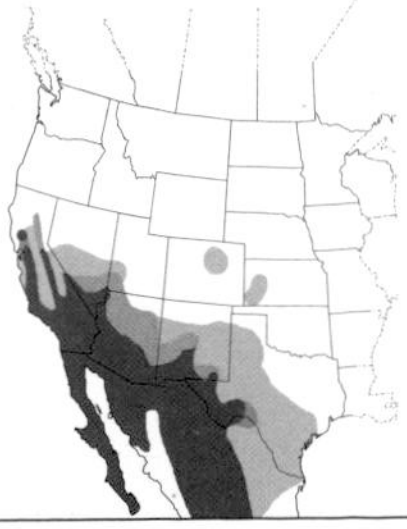

Common in open oak woods in foothills, or mesquite lowlands, wherever mistletoe is common. Usually in small flocks foraging for berries in bushes and trees.

McCown's Longspur

Rhynchophanes mccownii

L 6" WS 11" WT 0.81 oz (23 g)

Short-tailed and large-billed. Note heavy bill, plain face, and unstreaked underparts. Head pattern of drab individuals recalls female House Sparrow. In flight large-headed, and shows mostly white tail, more white than other longspurs.

VOICE: Song a formless, soft, liquid warble of short, rapid phrases with halting pauses *flideli fledeli fleedlili freew*; lower than other longspurs and with different rhythm. Call a liquid, musical rattle like Lapland but softer and shorter, often shortened to *kittip*; also a single or double *poik*, a metallic *pink*, and a sharp, whistled *teep*.

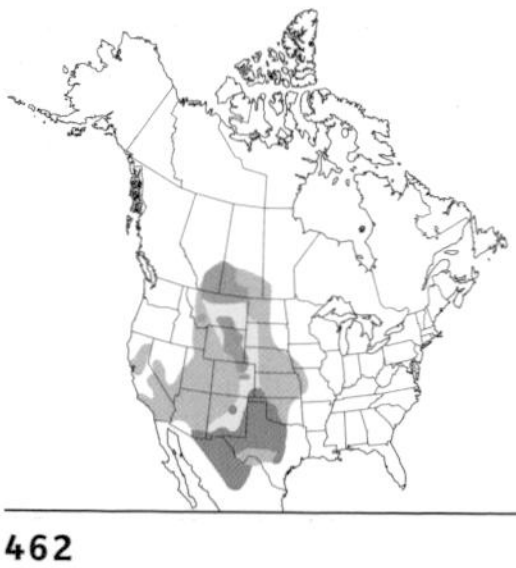

Uncommon and local. Nests in very dry, shortgrass prairies. Winters on barren ground, such as dry lakebeds and bare dirt fields, the same habitat favored by Horned Lark. Usually in small flocks that do not mix freely with Horned Larks or other longspurs.

Chestnut-collared Longspur

Calcarius ornatus

L 6" WS 10.5" WT 0.67 oz (19 g)

Our smallest longspur, with relatively small bill; in winter ver drab gray-brown with mostly white tail. Told from McCown's b smaller, grayish bill, stronger face pattern, streaked breast.

VOICE: Song a sweet warble *seet sidee tidee zeek zeerdi* beginning high and clear, ending lower and buzzy; pattern of gurgling pitch changes and falling trend reminiscent of Western Meadowlark but much higher-pitched. Flight call a soft, husky, two- or three-note *kiddle* or *kidedel*; also a buzz and a soft rattle.

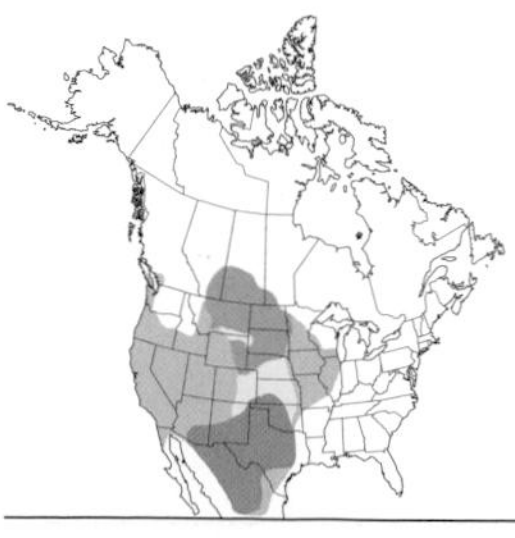

Common locally to uncommon. Nests in dry prairies, but favors slightly wetter and more vegetated sites than McCown's Longspur. Winters in short grass, where it is often difficult to see on the ground, and avoids the more barren places favored by McCown's and Lapland. Usually in flocks; seldom mixes with other longspurs.

Smith's Longspur

Calcarius pictus

L 6.25" WS 11.25" WT 0.91 oz (26 g)

Size and shape like Lapland Longspur, but shows more white on sides of tail, uniform buff color on underside, finely streaked breast, drab greater covert edges, and different head pattern.

VOICE: Song a high, sweet warble *seet seet seeo see zeet zidi zeet zeet zeeo*; similar to American Tree Sparrow but with rising trend. Call on breeding grounds a nasal, buzzy *goeet*. Flight call a staccato rattle similar to Lapland but sharp clicks more widely spaced and falling slightly at end; whole pattern reminiscent of rattle of cowbirds.

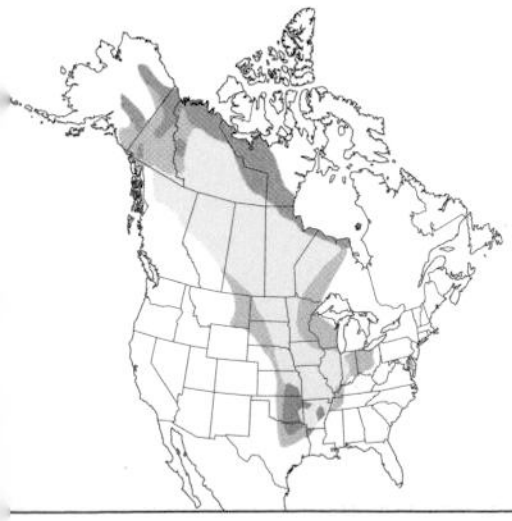

Uncommon and local. Nests on grassy tundra at edges of tree line. Winters on dry hilltops with particular types of short grass; very difficult to see on the ground. Usually in small flocks; does not mix with other longspurs.

Lapland Longspur

Calcarius lapponicus

L 6.25" WS 11.5" WT 0.95 oz (27 g)

Sparrow-like, but with shorter legs, white outer tail feathers, and different habits. Distinguished from other longspurs by rufous-edged greater coverts and strongly patterned face.

VOICE: Song a gentle, jingling warble *freew didi freer di fridi fideew* with rich, husky quality. Common call a husky, whistled *tleew* similar to Snow Bunting; common calls in summer include sharp, dry, whistled *chich* and *chi-kewoo*. Flight call a dry, mechanical rattle similar to Smith's and McCown's and to Snow Bunting.

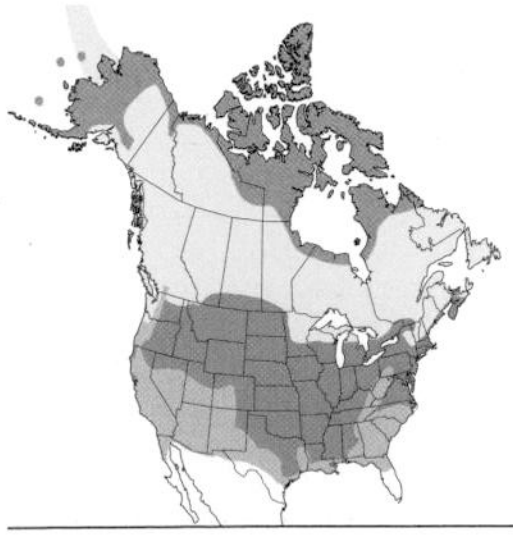

Common. Nests on open tundra. Winters on open ground, such as plowed fields. Usually found in flocks; often shares habitat with Horned Lark and Snow Bunting.

McKay's Bunting

Plectrophenax hyperboreus

L 6.75" WS 14" WT 1.9 oz (54 g)

Very similar to Snow Bunting (with less black) and sometimes considered the same species. Adult male nearly all-white, female and immature harder to distinguish from Snow Bunting, check details of wing and tail pattern.

VOICE: All vocalizations apparently identical to Snow Bunting.

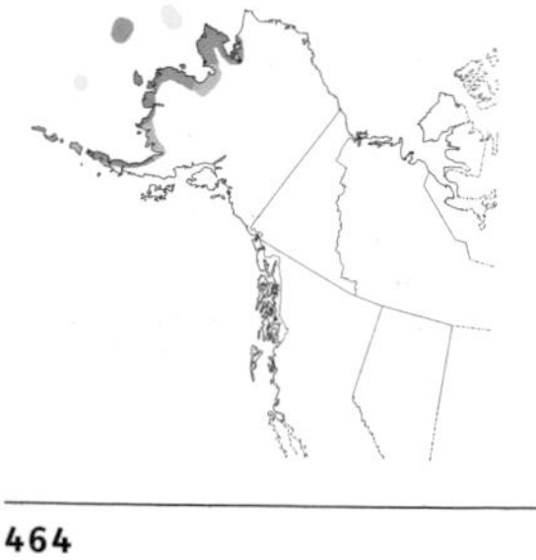

Uncommon to rare. Nests on islands off western Alaska, mainly on St. Matthew Island, with very small numbers recorded nesting most years on St. Lawrence and St. Paul Islands, where apparent hybrids also occur. Winters along beaches in western Alaska. Habits identical to Snow Bunting.

Snow Bunting

Plectrophenax nivalis

L 6.75" WS 14" WT 1.5 oz (42 g)

Larger than most sparrows, with shuffling gait, pale cinnanmor and white in winter with large white wing patches. No othe songbird (except McKay's Bunting) is normally so extensivel white, but partial albinos of any species can occur; check detail of color and shape.

VOICE: Song a rather low, husky warbling with repeated patterr *hudidi feet feet feew hudidi feet feet feew hudidi*; similar tc Lapland Longspur but a bit lower-pitched and less flowing; repetition distinctive. Calls include a soft, husky rattle *didididi* softe than Lapland Longspur, and a clear, descending whistle *cheew* clearer and sweeter than Lapland Longspur; when in flock alsc gives a short, nasal buzz *zrrt*.

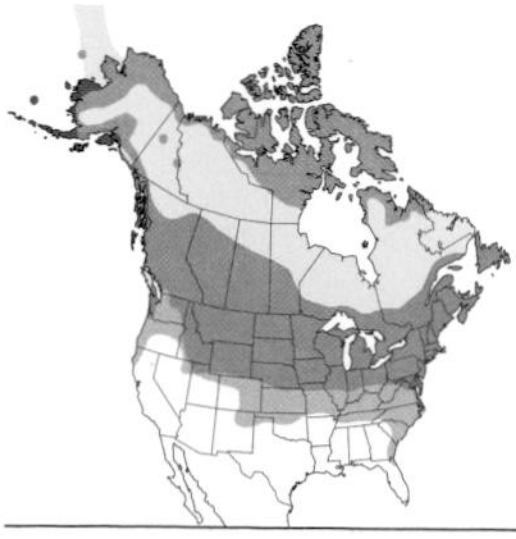

Common. Nests on high-Arctic tundra in boulder fields, among driftwood along rivers or beaches, or around towns or old buildings where debris provides nest sites. Winters on open ground such as beaches and fields. Usually seen in flocks that sweep along low to the ground like larks; often mixes with Horned Lark and Lapland Longspur, but larger than both species.

[dentification of Snow and McKay's Buntings

AcKay's Bunting differs from Snow only in having less black in the plumage. Males with an intermediate streaked pattern on the ack are seen regularly on the Bering Sea Islands, and are presum-.bly hybrids. In both species females show more dark color than nales, and immatures more than adults. Thus, the extent of dark n an immature female McKay's Bunting resembles an adult male now Bunting. Also note the similarity of adult female McKay's to he adult male hybrid. In such cases determining the sex of the ird is critical to identification. Males show clean white head, and harply contrasting black-and-white wing and tail pattern. emales show at least some dusky speckling on the forehead, and ess clean-cut dark and white pattern on wings and tail.

uvenal plumage of both species is gray on head and body. McK-ay's Bunting is generally pale gray, Snow Bunting varies from pale o dark, and birds in this plumage are probably not safely identi-iable.

Olive Warbler

Peucedramus taeniatus

L 5.25" WS 9.25" WT 0.39 oz (11 g)

Superficially similar to wood-warblers; differs by long slender bill, flared and notched tail, white at base of primaries, dark mask, and unique call, and now placed in its own family. Drab individuals most similar to Hermit Warbler.

VOICE: Song of simple, whistled, rapidly repeated phrases *hirrJl hirrJl hirrJl* or *plida plida plida plida*; often a two-part *plida plida plida chir chir*; clear, rich whistle like titmice. Call a soft, clear whistle *teew* or *tewp*; also a hard *pit*.

Common in fir, pine, and pine-oak forests at high elevations, usually staying in upper branches of tall trees. Usually in small groups, often joins mixed flocks with chickadees and other small birds. Gleans insects from leaves and twigs, often in clusters of pine needles in crowns of tall trees.

Wood-Warblers

FAMILY: PARULIDAE

56 species in 15 genera (5 rare species not shown). All are small and active birds with sharp, pointed bills. Plumage is varied; most are brightly colored, but many (especially ground-dwelling species) are relatively drab olive or brownish. Flight of all is strong and slightly undulating, and high sharp flight calls are frequent. Nest of most species is a cup, some species nest on the ground others in shrubs or up to treetops; Lucy's and Prothonotary warblers nest in cavities. Compare vireos, orioles, kinglets, and tanagers. First winter females are shown.

Genus *Seiurus*

Ovenbird, page 469

Genus *Helmitheros*

Worm-eating Warbler, page 469

Genus *Parkesia*

Louisiana Waterthrush, page 470

Northern Waterthrush, page 470

Genus *Mniotilta*

Black-and-white Warbler, page 472

Genus *Vermivora*

Golden-winged Warbler, page 473

Blue-winged Warbler, page 473

Genus *Protonotaria*

Prothonotary Warbler, page 471

Genus *Limnothlypis*

Swainson's Warbler, page 471

Genus *Basileuterus*

Rufous-capped Warbler, page 499

Genus *Oreothlypis*

Tennessee Warbler, page 475

Orange-crowned Warbler, page 474

Colima Warbler, page 476

Nashville Warbler, page 477

Virginia's Warbler, page 477

Lucy's Warbler, page 476

Genus *Geothlypis*

MacGillivray's Warbler, page 479

Mourning Warbler, page 479

Kentucky Warbler, page 478

Common Yellowthroat, page 480

Genus *Oporornis*

Connecticut Warbler, page 478

Genus *Setophaga*

Hooded Warbler, page 481

American Redstart, page 482

Kirtland's Warbler, page 483

Cape May Warbler, page 482

Cerulean Warbler, page 483

Northern Parula, page 484

Tropical Parula, page 484

Magnolia Warbler, page 485

Bay-breasted Warbler, page 488

Blackburnian Warbler, page 485

Yellow Warbler, page 486

Chestnut-sided Warbler, page 487

Blackpoll Warbler, page 488

Black-throated Blue Warbler, page 487

Palm Warbler, page 490

Pine Warbler, page 489

Yellow-rumped Warbler, page 492

Yellow-throated Warbler, page 491

Prairie Warbler, page 489

Grace's Warbler, page 491

Black-throated Gray Warbler, page 493

Townsend's Warbler, page 494

Hermit Warbler, page 494

Golden-cheeked Warbler, page 495

Black-throated Green Warbler, page 495

Genus *Cardellina*

Canada Warbler, page 497

Wilson's Warbler, page 496

Red-faced Warbler, page 497

Genus *Myioborus*

Painted Redstart, page 498

Genus *Icteria*

Yellow-breasted Chat, page 501

Warbler-like Birds

Before you can identify a warbler, you must determine if it is actually a warbler you are observing. The following families include common species that are often mistaken for warblers.

Kinglets

Ruby-crowned Kinglet, page 423

plain olive,
bold wing pattern,
short tail,
tiny dark bill

Vireos

White-eyed Vireo, page 367

stockier,
with thicker bill and legs,
often less active,
scolding calls

Flycatchers

Yellow-bellied Flycatcher, page 343

drab, no tail spots,
perch upright
and stationary

Orioles

Orchard Oriole, page 560

larger (but size
is easily misjudged),
very pointed bill,
greenish tail,
low-pitched calls

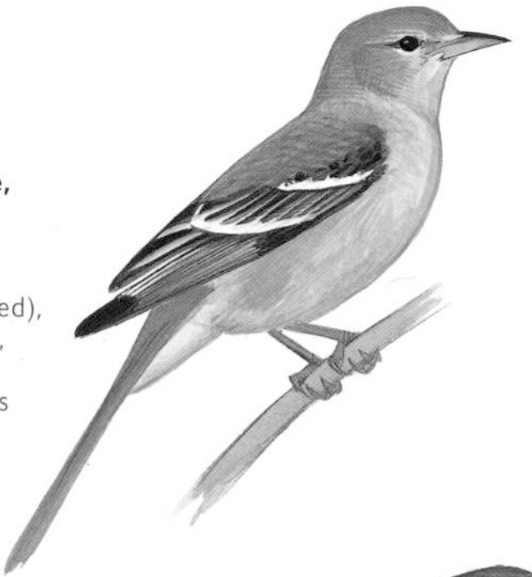

Tanagers

Western Tanager, page 541

larger, with thick bill,
shorter legs,
low-pitched
calls

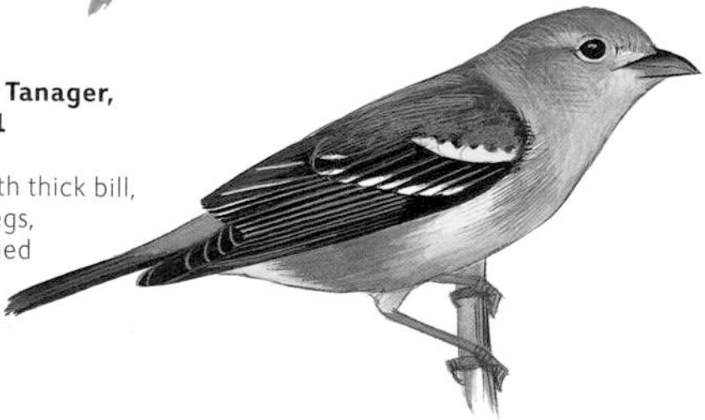

Goldfinches

American Goldfinch, page 579

triangular bill,
short legs,
stationary on perch,
often in flocks
and at bird feeders

Warbler Identification

Identifying warblers can be daunting simply because of the number and variety of species. Paying attention to genera (as shown on the preceding pages) is a helpful way to sort the species into smaller groups with shared characteristics. Look for wingbars, tail spots, streaks, leg color, bill shape, etc., all of which are fairly consistent within each genus.

When studying an unknown warbler in the field, the best feature to focus on is the face pattern. Every individual can be identified to species by details of color patterns around the face. Look for an eye ring (and whether it is complete or broken), a contrasting dark eyeline or pale eyebrow stripe, contrast between the cheeks and throat, or between the cheeks and sides of the neck. The bold and bright colors of adult males may seem to have little resemblance to immatures of the same species, but in most cases a careful study will reveal that the fundamental color pattern is similar even on the drabbest immatures.

Cape May Warbler

1st winter ♀

Adult ♂ breeding

Don't ignore leg color, which is pale flesh-colored on many species, dark on many others. Leg color provides a quick and easy way to narrow the possibilities. It can also resolve some perennial pitfalls, for example distinguishing Nashville and Orange-crowned Warblers (dark legs) from Mourning, MacGillivray's, and Connecticut Warblers (pale legs).

Tail pattern is a very useful clue for identifying wood-warblers. A good view of the spread tail in flight or the underside of the tail on a perched bird can greatly narrow the choices. On Pine Warbler, Prairie Warbler, and others, white outer tail feathers create white sides on the spread tail seen from above, and a mostly white surface on the closed tail from below. The smaller white tips on the outer tail feathers of Blackpoll, Bay-breasted, and other warblers create white corners on the spread tail, and dark color visible on the base of the closed tail from below.

Pine warbler adult ♂

white sides

Blackpoll Warbler 1st winter ♀

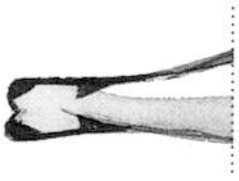

white corners

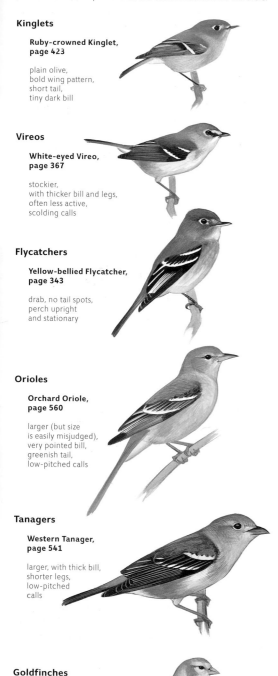

Ovenbird

Seiurus aurocapilla

L 6" WS 9.5" WT 0.68 oz (19.5 g)

Larger than most warblers, and walks on the forest floor. Olive above; white below with distinct black streaks; thrush-like but note smaller size, white breast with black streaks, greenish back, bold white eyering, and dark crown-stripes.

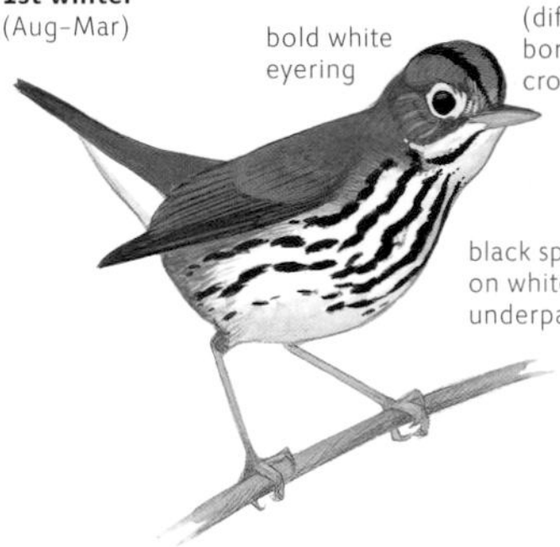

VOICE: Song of explosive two-syllable phrases increasing in volume *chertee chertee cherTEE CHERTEE CHERTEE CHERTEE*; sometimes simpler *chreet chreet CHREET CHREET CHREET*. Call a rather hard and unmusical chip: varies from high, hard *chap* to low, flat *chup* or *dik* when agitated. Complex flight song often heard at night over wooded areas, a jumbled series of chirps and whistles with a short series of strong *teecher* phrases inserted.

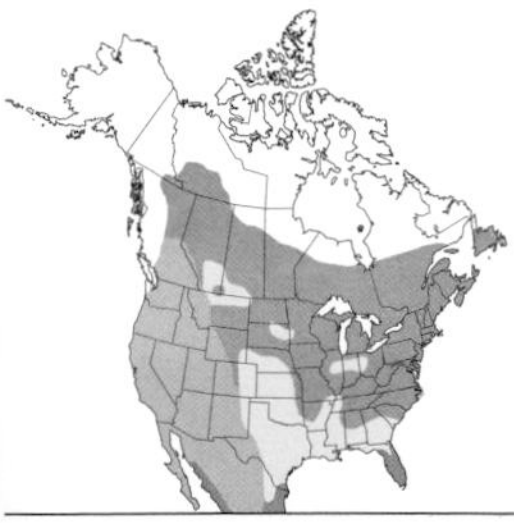

Uncommon in mature deciduous or mixed forests with sparse shaded undergrowth. Seen mainly on the ground, walking with high-stepping gait and head-bobbing, often with tail raised. Perches on larger mid-level branches when singing or when alarmed.

Worm-eating Warbler

Helmitheros vermivorum

L 5.25" WS 8.5" WT 0.46 oz (13 g)

A large warbler, large-billed and short-tailed. Overall a distinctive rich caramel and olive color; note dark stripes on head, unpatterned wings and tail, plain underparts.

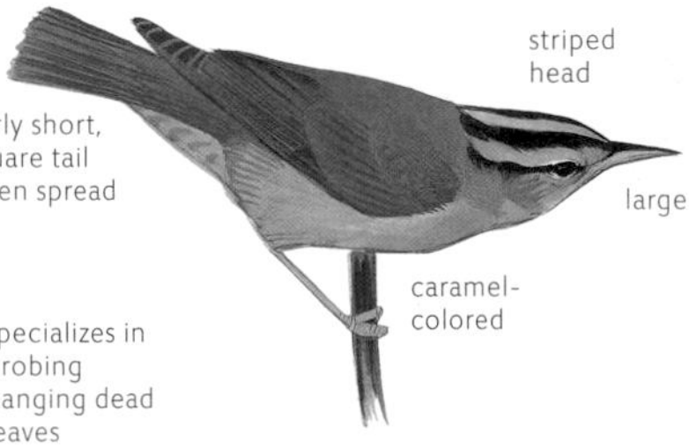

VOICE: Song a rapid, flat buzz (like Chipping Sparrow but usually faster and more insect-like); usually begins softly and builds up, often introduced by chips. Call a loud, clear chip. Flight call a very short, sharp, high buzz *dzt*; repeated rapidly in series of two or three.

Uncommon. Nests in mature deciduous forest, often in oak woods on relatively dry rocky slopes with dense understory. Migrants found mainly in wooded understory close to ground, moving acrobatically to probe vine tangles and clusters of dead leaves for insects and spiders.

Louisiana Waterthrush

Parkesia motacilla

L 6" WS 10" WT 0.72 oz (20.5 g)

Habits and appearance very similar to Northern Waterthrush; best distinguished by broader whiter eyebrow, buffy flanks, and sparser brownish streaking on white underparts.

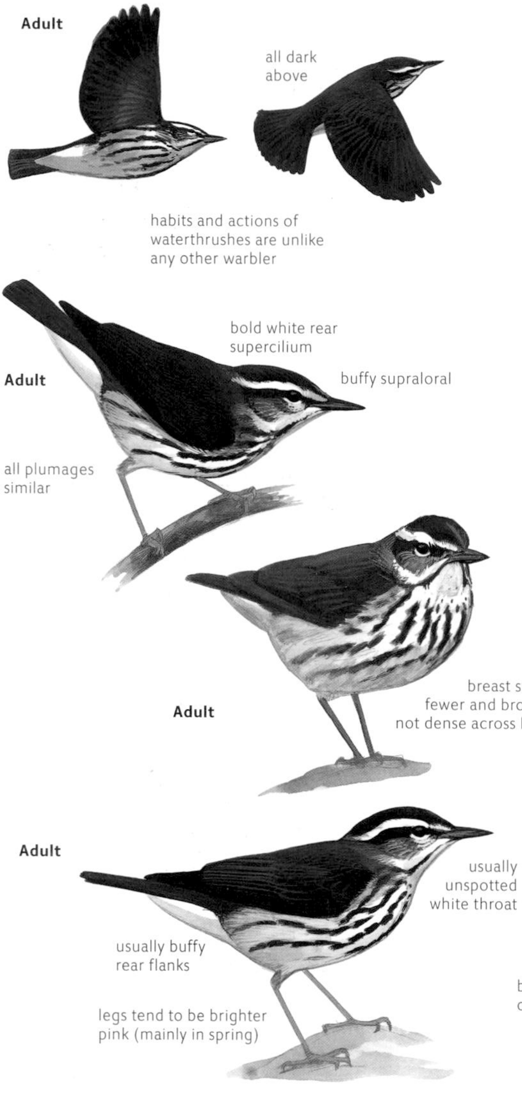

VOICE: Song musical, clear, and sweet: beginning with three or four high, clear, slurred whistles, then a series of jumbled, descending chips and chirps. Alternate song similar but much longer and rambling. Call a loud, strong *spich* not as hard as Northern. Flight call like Northern.

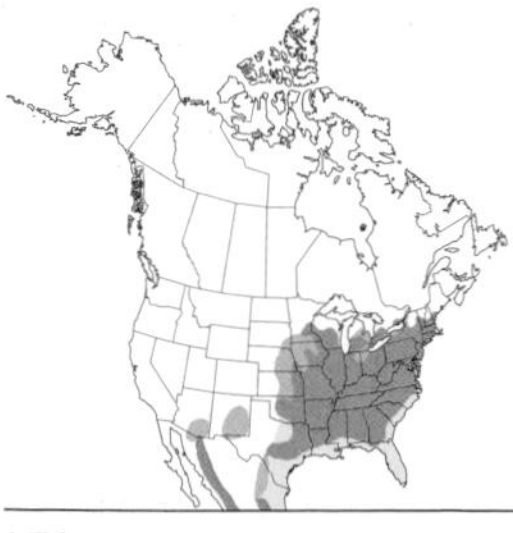

Uncommon. In northern parts of its range nests mainly along clear flowing streams in shaded ravines, in southern areas found in undergrowth around standing water in wooded swamps of bald-cypress, tupelo, etc. Migrants can be found in any dense brushy habitat adjacent to water.

Northern Waterthrush

Parkesia noveboracensis

L 6" WS 9.5" WT 0.63 oz (18 g)

A large brownish warbler, usually seen on the ground pumping its tail and rear body up and down. Note dark brownish upperside, densely-streaked underside, and narrow pale eyebrow.

VOICE: Song of loud, emphatic, clear chirping notes generally falling in pitch and accelerating; loosely paired or tripled, with little variation. Call a loud, hard *spwik* rising with strong *k* sound. Flight call a buzzy, high, slightly rising *zzip*.

Common in dense shrubs and small trees near slow-moving or standing water. Forages for insects on the ground along water's edge.

Prothonotary Warbler

Protonotaria citrea

L 5.5" WS 8.75" WT 0.56 oz (16 g)

Male has brilliant orange-yellow head and blue-gray wings. Even the drabbest birds are bright yellow on head, and all have large bill, plain gray wings, extensive white in tail and mostly white undertail.

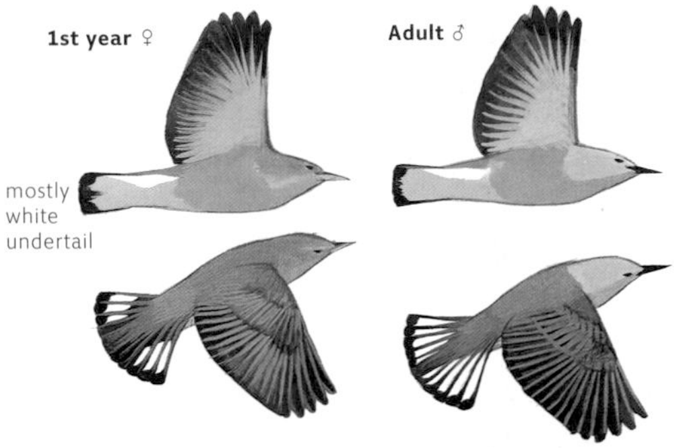

VOICE: Song high, clear, and metallic with emphatic rising notes *tsweet tsweet tsweet tsweet tsweet*; little variation. Call a clear, metallic squeak *tsiip*. Flight call a loud, clear, high *swiit*; rising like American Redstart, but less squeaky.

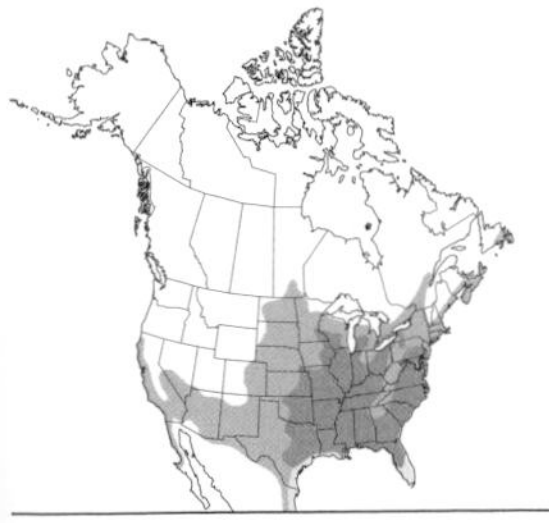

Common in wooded swamps or lowland deciduous forests with ponds or other standing water; usually stays low in understory. Migrants can be found in any brushy understory, usually near water.

Swainson's Warbler

Limnothlypis swainsonii

L 5.5" WS 9" WT 0.67 oz (19 g)

Plain brown color with reddish crown, dark eyeline, and very large pale bill unique. Told from waterthrushes by unstreaked underparts and never bobs tail. From Carolina Wren by drabber color, olive flanks, straight bill, and behavior.

VOICE: Song of strong, clear, slurred notes *seew seew seee SIS-Terville*; downslurred notes at beginning with emphatic ending (compare Louisiana Waterthrush, Hooded Warbler). Call varies from squeaky *teep* like Hooded to strong, descending chip like Louisiana Waterthrush. Flight call a very high, thin *sees* often repeated.

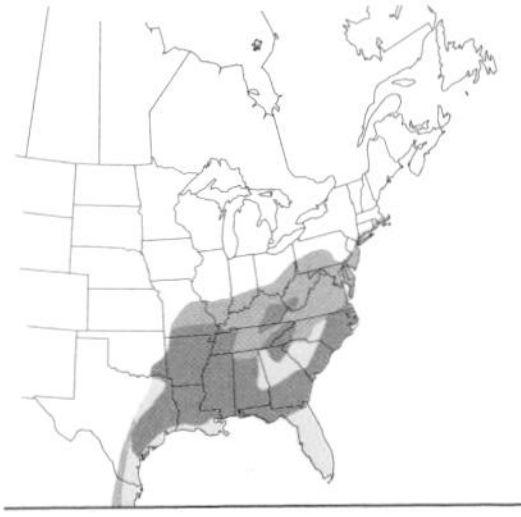

Uncommon and local in patches of brush within mature deciduous forests in lowland areas; not always near water. Forages on the ground within dense undergrowth, turning leaves with bill in search of invertebrates.

Black-and-white Warbler

Mniotilta varia

L 5.25" WS 8.25" WT 0.37 oz (10.7 g)

Distinctively black and white overall, some have buffy flanks. Forages nuthatch-like, gleaning insects from bark along trunks and large limbs. Note striped head, short broad tail, and long bill.

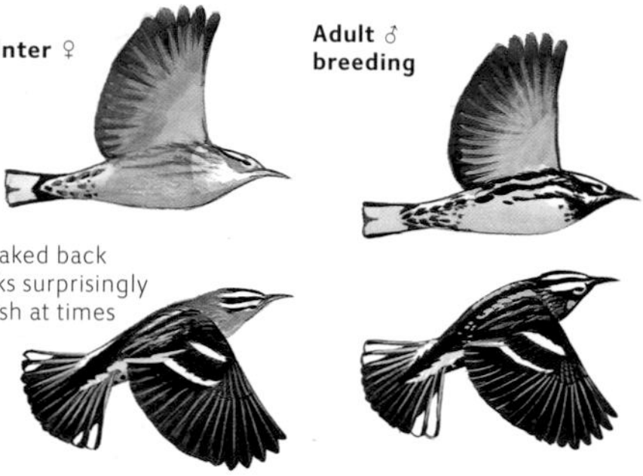

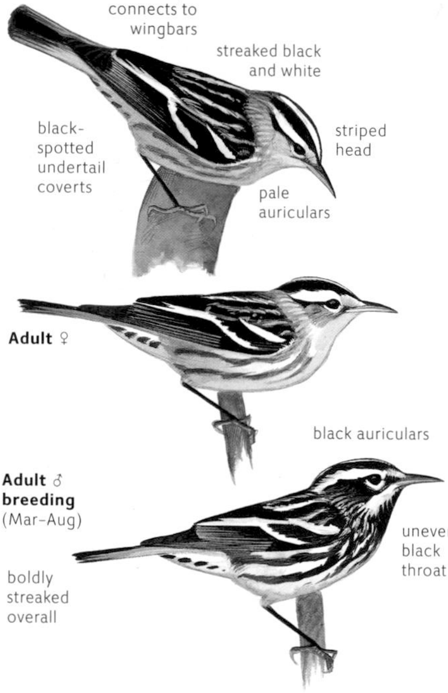

VOICE: Song a high, thin, simple series of two-syllable phrases with five to ten repetitions *weesa weesa weesa weesa weesa weesa weesa*; may be two- or three-part *weesa weesa weetee weetee weetee weetee weet weet weet.* Call a sharp, rattling *stick*. Flight call a high, hissing, rising *fsss.*

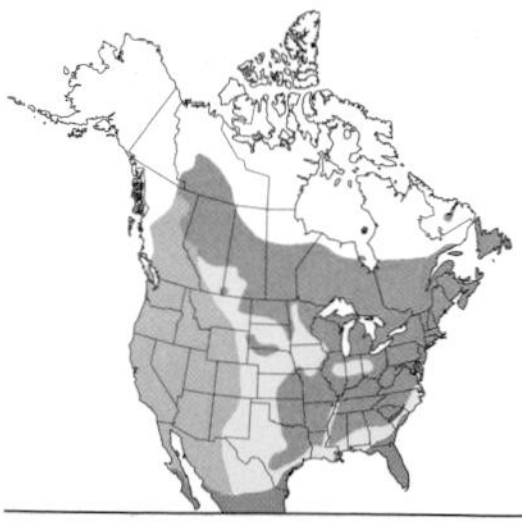

Common. Nests in mature deciduous or mixed forests with large tree trunks for foraging. Migrants also found mainly in woods.

Blue-winged × Golden-winged Warbler Hybrids

These two species hybridize regularly where they meet, and the hybrid zone has been steadily shifting north, with Blue-winged Warbler overtaking the range of Golden-winged. All first-generation hybrids show the dominant traits of black eyeline, whitish throat, and yellowish wingbars (known as "Brewster's" Warbler). Hybrids are fertile and when paired with either species produce backcrosses showing a variety of features. Birds resembling Blue-winged, but with the recessive black throat and mask of Golden-winged are rare, and are known as "Lawrence's" Warbler. Song can be like either parent species, or (more often) an abnormal song combining characteristics of both species.

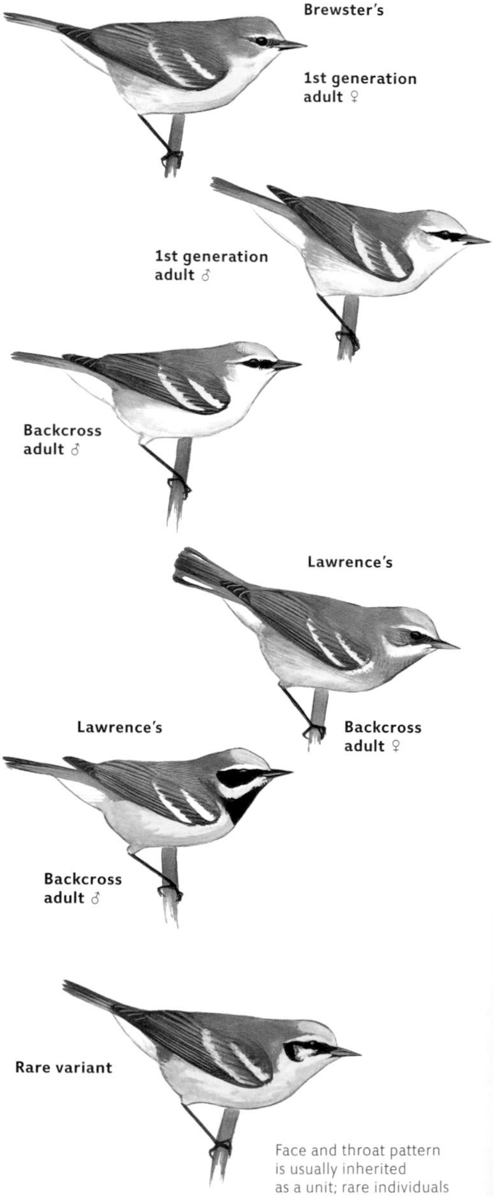

Face and throat pattern is usually inherited as a unit; rare individuals show only part of the black mask and/ or throat.

Golden-winged Warbler

Vermivora chrysoptera

L 4.75" WS 7.5" WT 0.31 oz (8.8 g)

Small and grayish overall, with bold head pattern reminiscent of chickadee, but with dark mask and bright yellow on crown. Also note pale gray flanks and bright yellow wing-patch. Hybridizes commonly with Blue-winged Warbler where range overlaps.

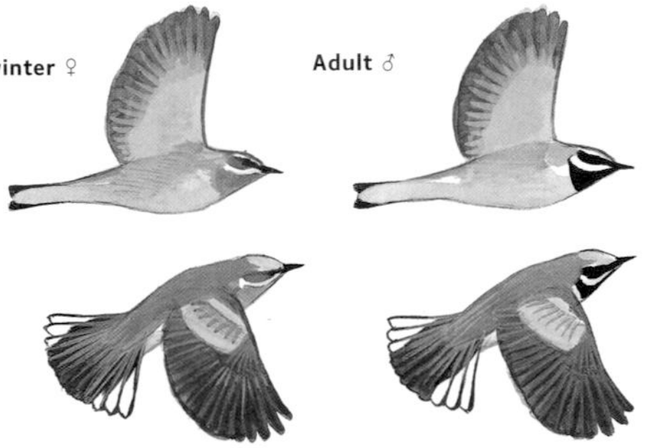

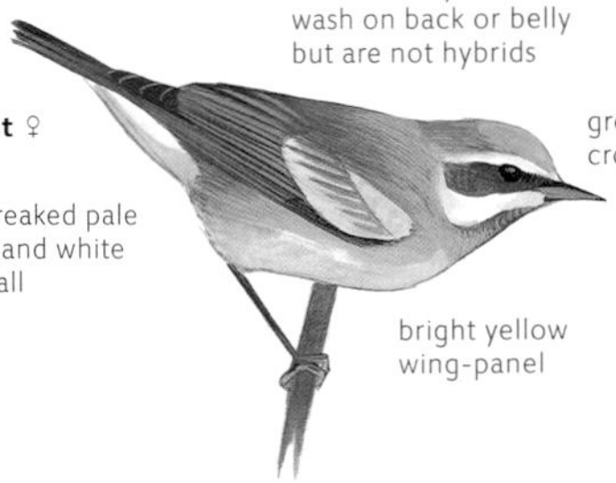

VOICE: Song a very fine, high buzz *zeee zaa-zaa-zaa* with first note higher than following; all higher and finer than Blue-winged. Also a longer, varied song like Blue-winged. Calls like Blue-winged.

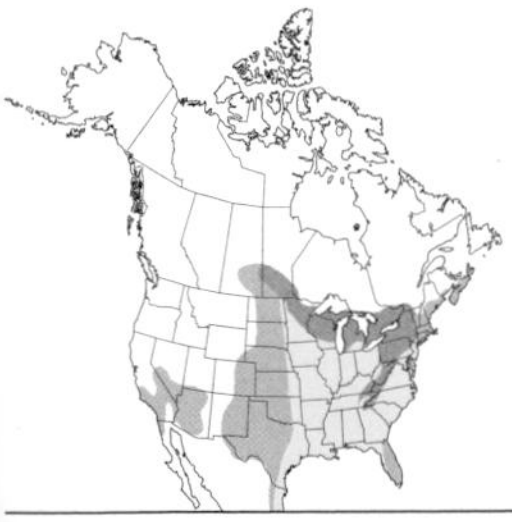

Uncommon and declining. Found in brushy forest edges or habitats with weedy areas, patches of dense brush, and a few taller trees for singing perches.

Blue-winged Warbler

Vermivora cyanoptera

L 4.75" WS 7.5" WT 0.3 oz (8.5 g)

Small and bright yellow, with thin dark eyeline, two white wingbars on grayish wings, and white undertail coverts. Note grayish tail with large white spots, often fanned or flicked open while foraging.

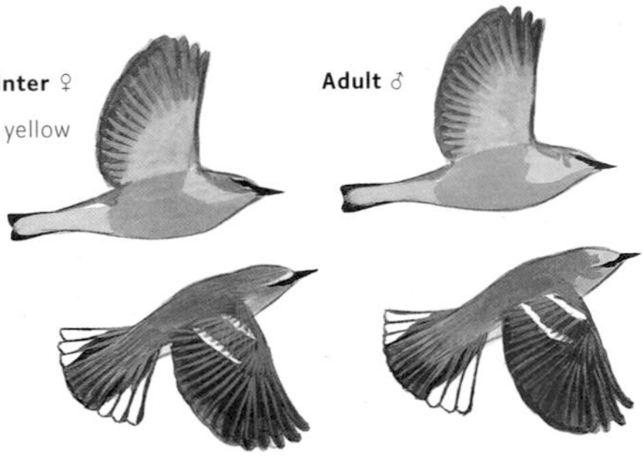

VOICE: Song a rather harsh, buzzy *beeee-BZZZZZ* like a deep sigh; first part high and thin, second part low and rough. Also a long, high buzz with stuttering notes at beginning and end *tsi tsi tsi tsi tsi zweeeeeee zt zt zt zt*. Call a sharp, dry *snik* or *chik*. Flight call a short, high, slightly buzzy *dzit* or *zzip*.

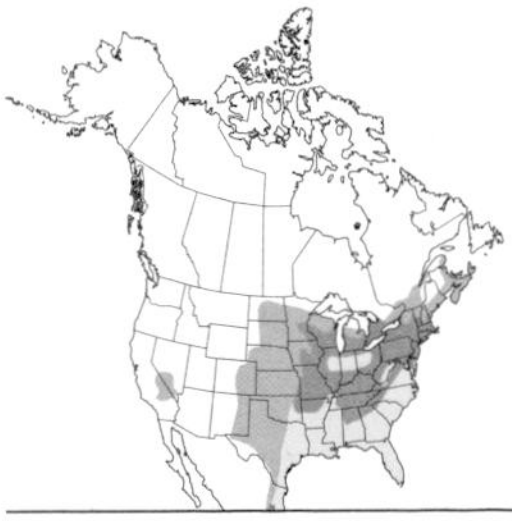

Uncommon. Found in open, second growth woodlands or in clearings with dense but varied undergrowth of weeds and shrubs.

Orange-crowned Warbler

Oreothlypis celata

L 5" WS 7.25" WT 0.32 oz (9 g)

A small, sharp-billed, drab-colored warbler. Note broken eyering, blurry streaks on breast, and yellow undertail coverts. Told from Tennessee Warbler by yellow undertail coverts, drabber colors, faint streaks below; from Yellow by short dark eyeline, pointed bill, longer tail without yellow spots.

Drabber gray-headed individuals are easily confused with Mourning, MacGillivray's, and Connecticut Warblers, but smaller, with smaller bill, less sturdy appearance, drab yellow underparts with faint streaks and dark legs.

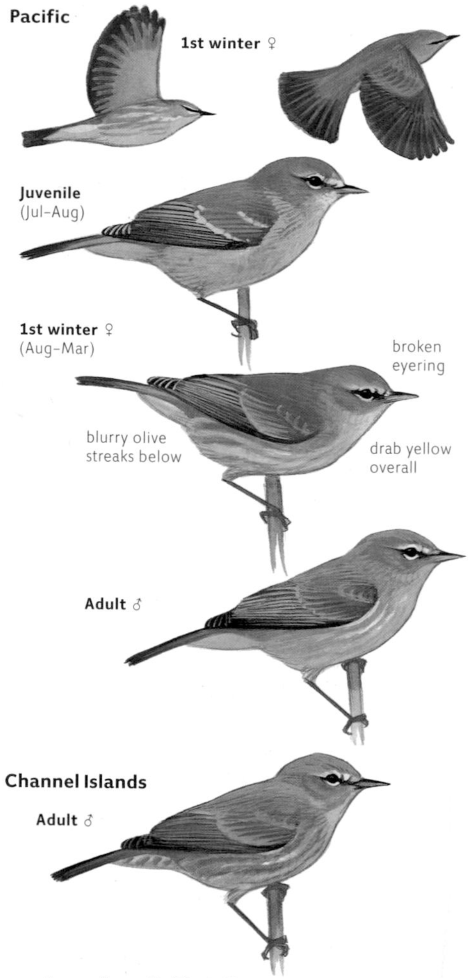

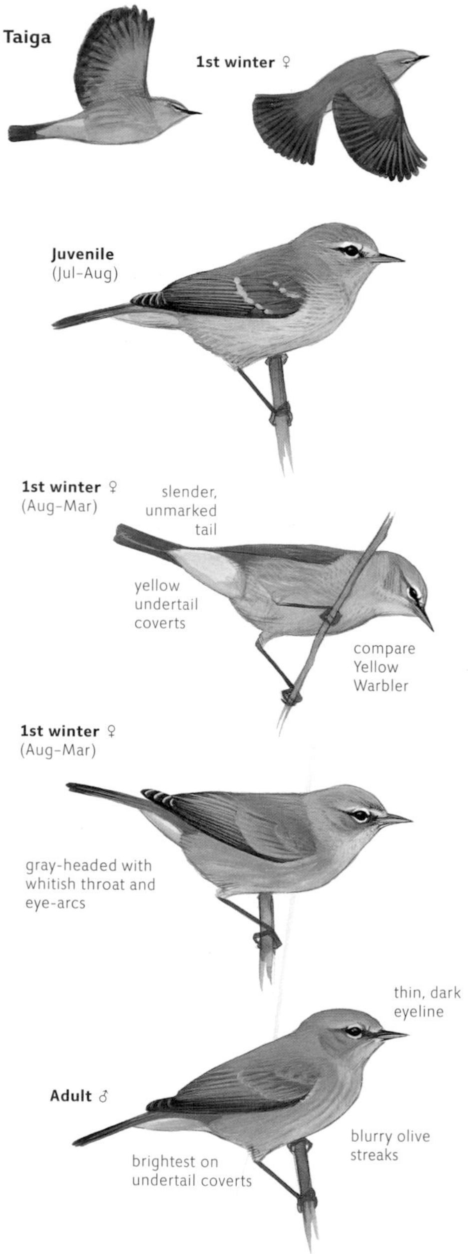

VOICE: Song a fast trill of fairly flat notes; individual notes not very distinct, last few notes on lower pitch *titititititititititututu*; notes generally downslurred but so sharp this is hard to hear. Taiga song trill soft and slow (averages 11 notes/sec); Pacific song trill fast and hard (averages 17 notes/sec); Channel Islands song trill slower and lower-pitched than Pacific. Call distinctive: a simple, clear, high, sharp chip very similar to Field Sparrow but a little richer. Flight call a short, high, clear or slightly husky, rising *seet*.

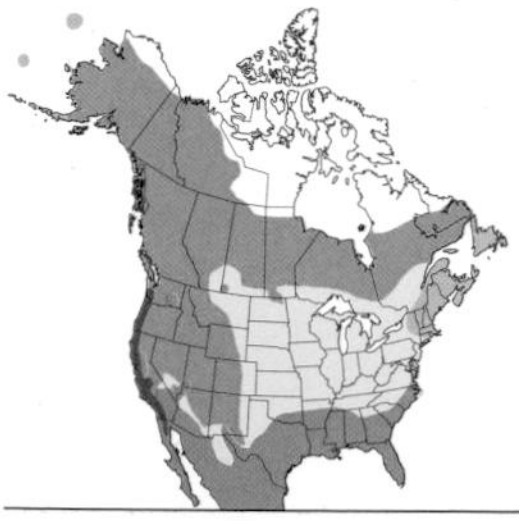

Common and widespread, especially in West, generally uncommon in East. Breeds in dense deciduous brush such as willow and alder thickets. Winters in many weedy and brushy habitats, including hedgerows and gardens, open woods. Often joins mixed foraging flocks with other warblers, vireos, etc.

Four populations differ slightly in plumage and song. Pacific is brightest overall and yellow in all plumages. Channel Islands is bright but larger and darker and more heavily streaked olive below. Taiga is drabbest. Interior West is intermediate between Pacific and Taiga, few can be safely identified in the field. Song differs slightly.

Crescent-chested Warbler

Oreothlypis superciliosa

L 4.5" WS 6.5" WT 0.3 oz (8.5 g)

Small and short-tailed, with blue-gray head, bright yellow throat and breast, and rusty breastband, like parulas, but lacks wingbars and tail spots and shows broad and distinctive white eyebrow.

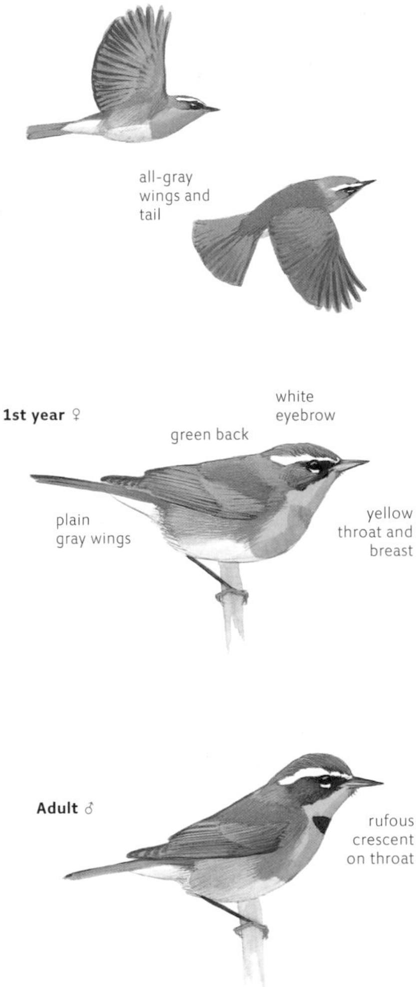

VOICE: Song a short, mechanical rattle or buzzy trill, all on one pitch and sustained less than one second, fading slightly at end. Call a sharp *tsik*.

Very rare visitor from Mexico to montane pine-oak forests of southeastern Arizona, where it has attempted nesting, and two winter records in lowland riparian habitat. Often joins mixed foraging flocks of other warblers, kinglets, and others.

Tennessee Warbler

Oreothlypis peregrina

L 4.75" WS 7.75" WT 0.35 oz (10 g)

A small, sharp-billed, short-tailed warbler with relatively plain plumage. Told from Orange-crowned by stronger eyeline, brighter colors, whitish undertail coverts; from vireos by pointed bill, more active behavior.

VOICE: Song a trill of sharp, spitting, high chips, usually in three parts each with slightly different pitch and tempo *tip tip tip tip teepit teepit teepit teepit ti ti ti ti ti ti ti*, the final series generally faster. Call a sharp, high, smacking *stik*. Flight call a slightly husky *tseet*.

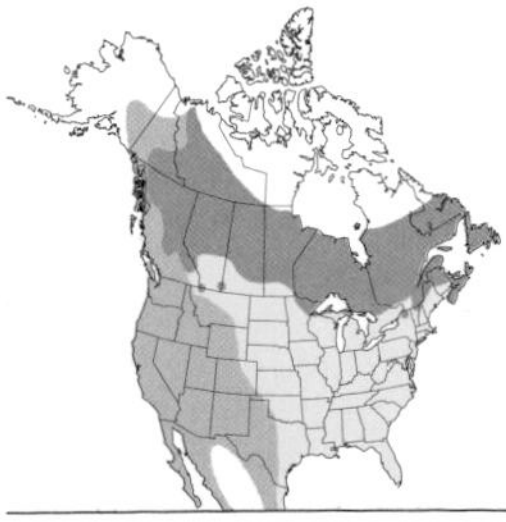

Common. Breeds in open or young spruce-fir woodlands with mossy or weedy groundcover. Migrants found in variety of habitats, including brush, and often concentrated around flowering trees.

Lucy's Warbler

Oreothlypis luciae

L 4.25" WS 7" WT 0.23 oz (6.6 g)

One of our smallest warblers, with sharply pointed bill. Overall pale gray color, cream-colored breast, and rusty rump unique. Could be confused with gnatcatchers (but with shorter, all-gray tail) or Verdin (but smaller and grayer without yellow head).

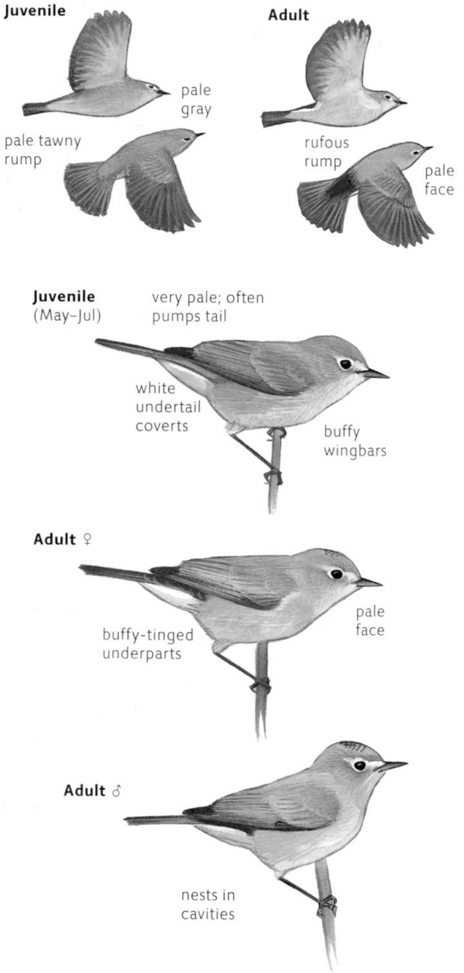

VOICE: Song a simple trill with several pitch changes; higher-pitched than Virginia's with simpler, sharper notes. More complex songs of clear, high whistles *sweeo sweeo sweeo seet seet seet seet sit it it it it,* like unaccented songs of Yellow but usually simpler, more rapid. Call a slightly husky, rattling *vink.* Flight call a high, clear, weak *tsiit.*

Uncommon and local in riparian habitat; in dense mesquite, cottonwoods and willows along streams, where it forages actively among twigs and leaves.

Colima Warbler

Oreothlypis crissalis

L 5.5" WS 7.75" WT 0.3 oz (8.5 g)

Similar to Nashville and Virginia's Warblers, with grayish head and complete white eyering; note distinctly larger size, brownish back and flanks, and orange-yellow upper- and undertail coverts.

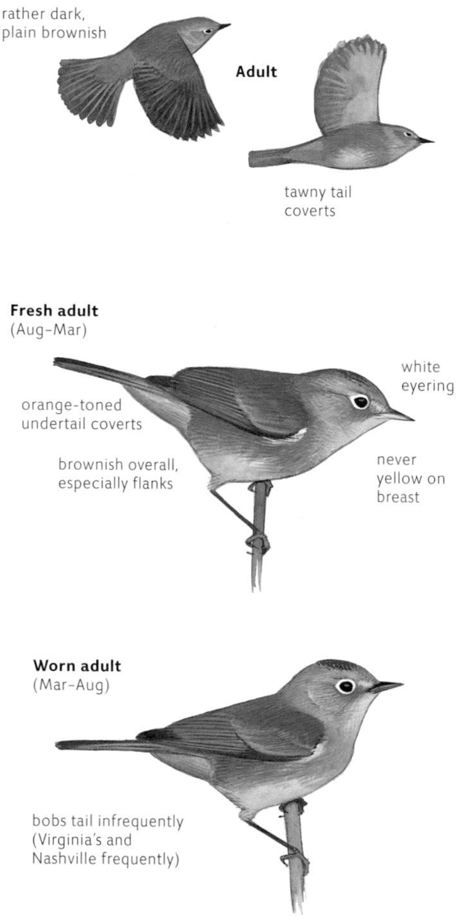

VOICE: Song a hard, chattering, wavering trill, sometimes with high, clear *tew* note at end; notes clearly upslurred, unlike Orange-crowned and most other similar species. Song usually rises slightly near end then drops *tetetetetetetetetititi-tew.* Call similar to close relatives (Virginia's, Nashville, Lucy's) but may be huskier. Flight call a high, rising *tseet.*

Rare and very local within very limited range in North America with about 300 nesting pairs. Found only in oak woods in higher elevations of Chisos Mountains of western Texas.

Virginia's Warbler

Oreothlypis virginiae

L 4.75" WS 7.5" WT 0.27 oz (7.8 g)

Very similar to Nashville Warbler, but overall gray with limited yellow. Told from very drab Nashville by grayish wing edges (even the drabbest, most grayish Nashville always shows some greenish on the wings).

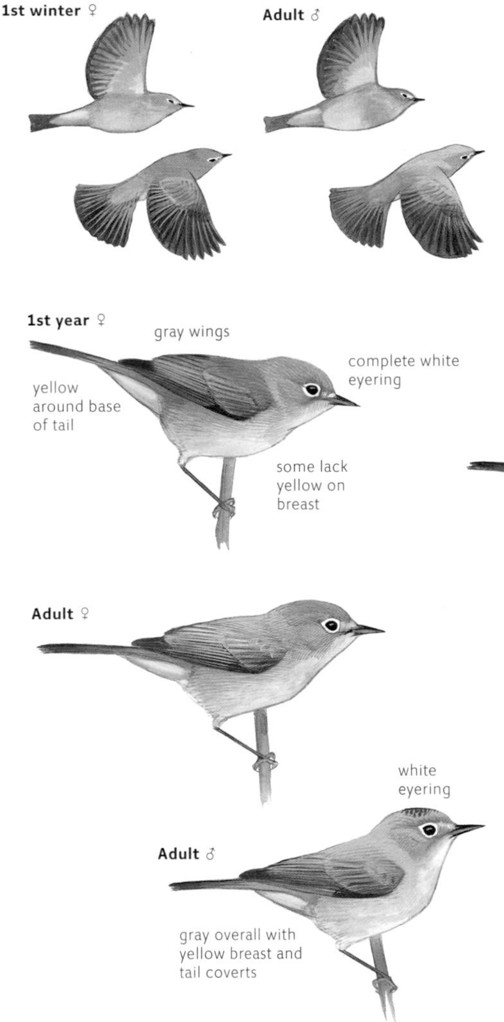

VOICE: Song a fairly weak, clear warble; unaccented, not crisply delivered, usually two-part; resembles some songs of Western Nashville but lower and less structured; some also resemble some songs of Yellow. Often a three-part *seedi seedi seedi seedi silp silp suwi suwi*. Calls like Nashville.

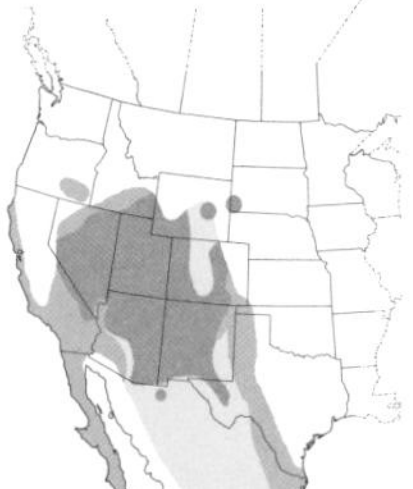

Uncommon. Breeds in dense brushy undergrowth with scattered trees or open woods on arid slopes. Migrants and wintering birds found in wooded or brushy habitats.

Nashville Warbler

Oreothlypis ruficapilla

L 4.75" WS 7.5" WT 0.3 oz (8.7 g)

A very small, sharp-billed, short-tailed warbler, with relatively plain plumage and complete white eyering. Note gray hood, yellow throat, and unmarked greenish wings and tail. Often confused with larger and sturdier Mourning, MacGillivray's and Connecticut Warblers.

VOICE: Song of Eastern a fairly slow, simple, two-part, musical trill *seeta seeta seeta seeta pli pli pli pli* softer than Tennessee. Alternate song a series of single or double notes irregularly descending in pitch *tee tee tee tee tay tay tay tay tati toti toti to*. Call a sharp, rattling, metallic *spink* like a small Northern Waterthrush. Flight call a high, clear *swit*.

Uncommon. Breeds in open coniferous woodlands with patches of brush. Winters in shrubby woods or gardens, often near flowering trees.

Connecticut Warbler

Oporornis agilis

L 5.75" WS 9" WT 0.53 oz (15 g)

A large, sturdy, and thrush-like warbler, with short tail and long wings. Note plain face with complete white eyering, pale yellow underparts, smooth gray or brownish hood, and uniform olive upperparts.

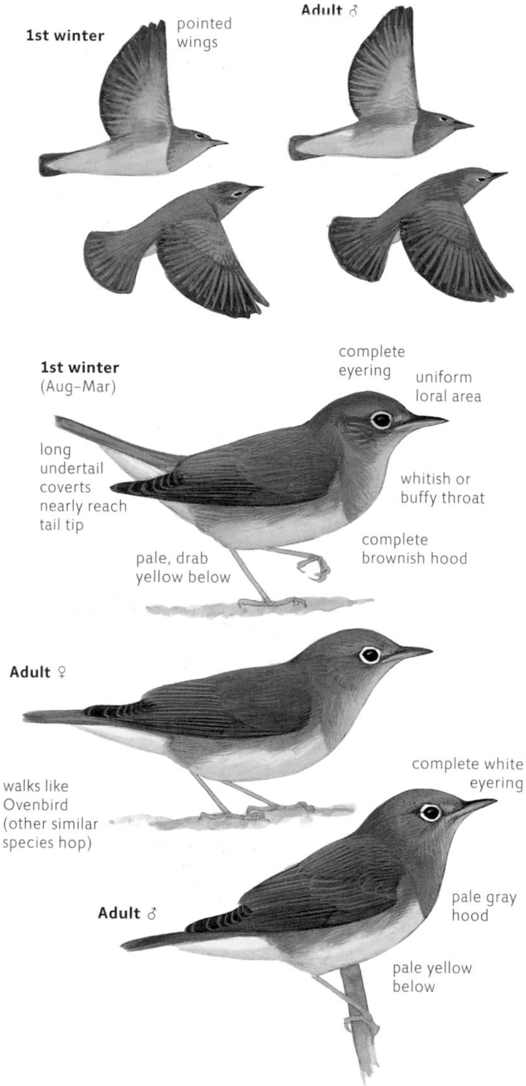

VOICE: Song a series of four-syllable phrases with strong, clear, chirping quality like Northern Waterthrush *tup-a-teepo tup-a-teepo tupateepo-tupateepo*; accelerating tempo unlike all similar species. Call rarely heard; a rather soft *pwik*. Flight call a rough buzz like many *Setophaga* warblers.

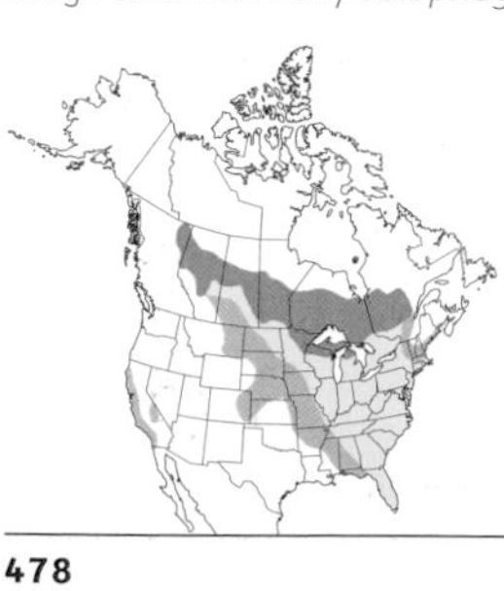

Uncommon; secretive and difficult to spot. Nests in damp mixed coniferous forest such as around spruce bogs, with relatively open woodlands and patches of dense understory. Migrants choose rank weedy edges in wet swales and forage mainly on the ground. Migrates very late in spring.

Kentucky Warbler

Geothlypis formosa

L 5.25" WS 8.5" WT 0.49 oz (14 g)

Relatively large and short-tailed. In all plumages unmarked olive above and bright yellow below, brightest on throat, with dark mask curving onto breast and angular yellow spectacles.

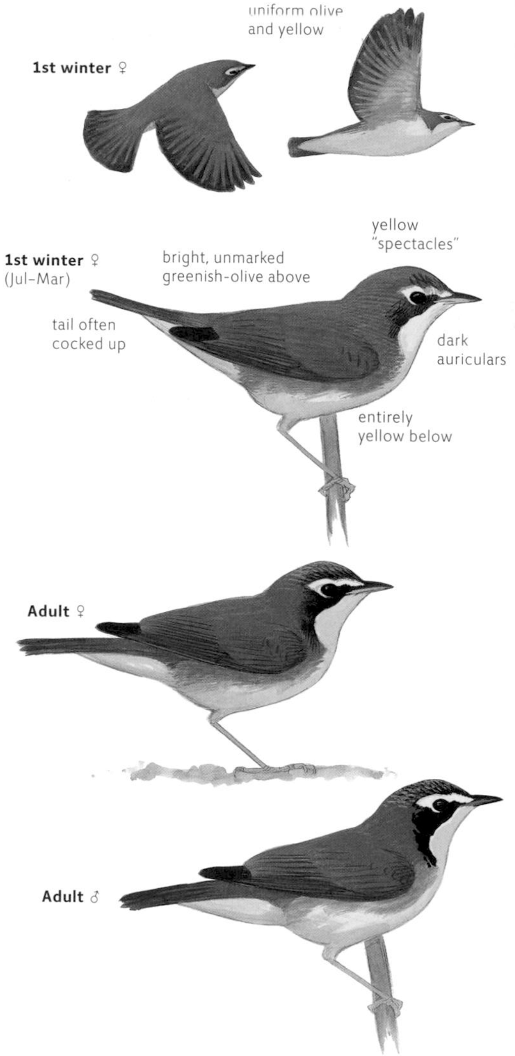

VOICE: Song a rolling series of vaguely two-syllable phrases *prr-reet prr-reet prr-reet prr-reet prr-reet prr-reet* similar to Carolina Wren or Ovenbird; distinguished by rich quality, steady rolling tempo, and lack of clearly defined syllables. Call a low, hollow *chok* to higher, sharper *chuk* when agitated. Flight call a short, rough buzz *drrt*.

Uncommon in dense shaded understory of mature deciduous forests. Forages mainly on the ground, difficult to see in dense cover.

MacGillivray's Warbler

Geothlypis tolmiei

L 5.25" WS 7.5" WT 0.37 oz (10.5 g)

Slightly larger than Common Yellowthroat, with grayish hood and never yellow on throat. Very similar to Mourning Warbler, best distinguished by bold white eye-arcs, whitish throat of female and immature, longer tail and slightly different call.

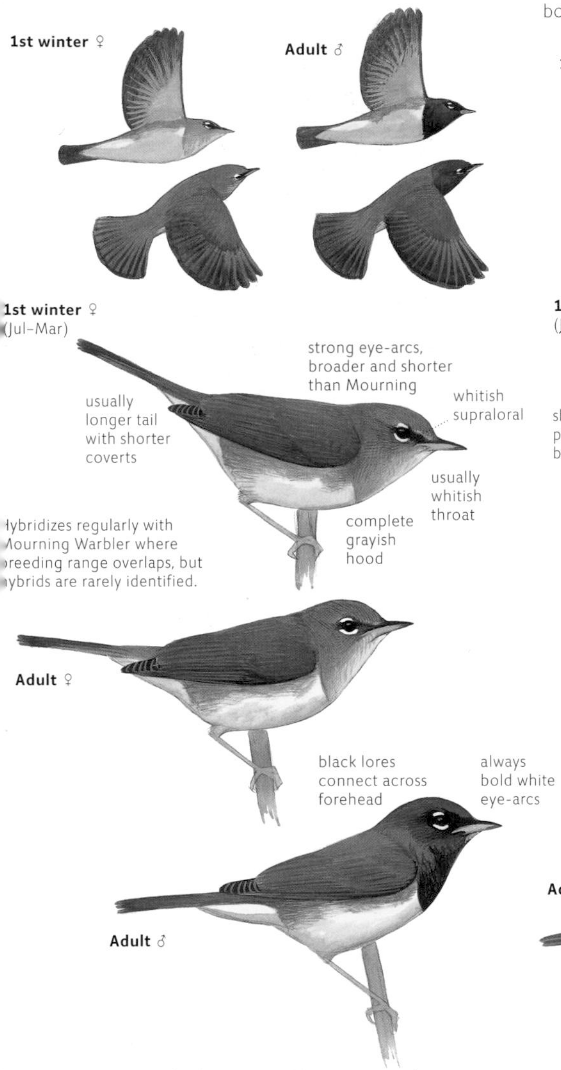

VOICE: Song a short, rhythmic series like Mourning but averages higher-pitched and less rich in quality, more buzzy and less rolling; final phrases usually buzzy and not lower than rest of song. Some songs can be very similar. Call a hard, dry *chik* or *twik* sharper than Mourning; similar to Common Yellowthroat. Flight call flatter and more scratchy than Mourning.

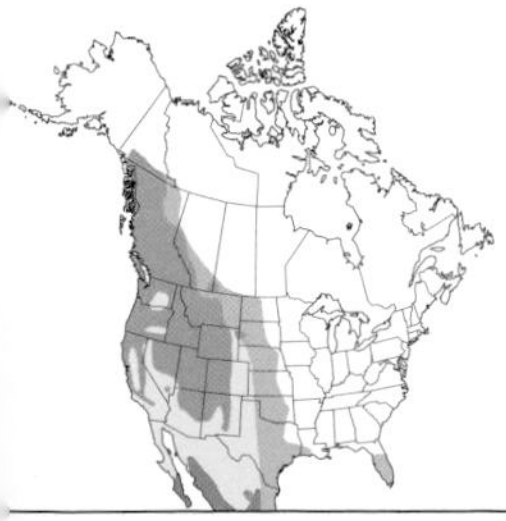

Uncommon. Nests in dense brushy deciduous patches, usually near water.

Mourning Warbler

Geothlypis philadelphia

L 5.25" WS 7.5" WT 0.44 oz (12.5 g)

Intermediate in size between very similar Connecticut and MacGillivray's Warblers; distinguished by having thin or no pale eye-arcs and by call. Drabbest individuals show broken breastband and yellowish throat, while Connecticut and MacGillivray's both show whitish throat.

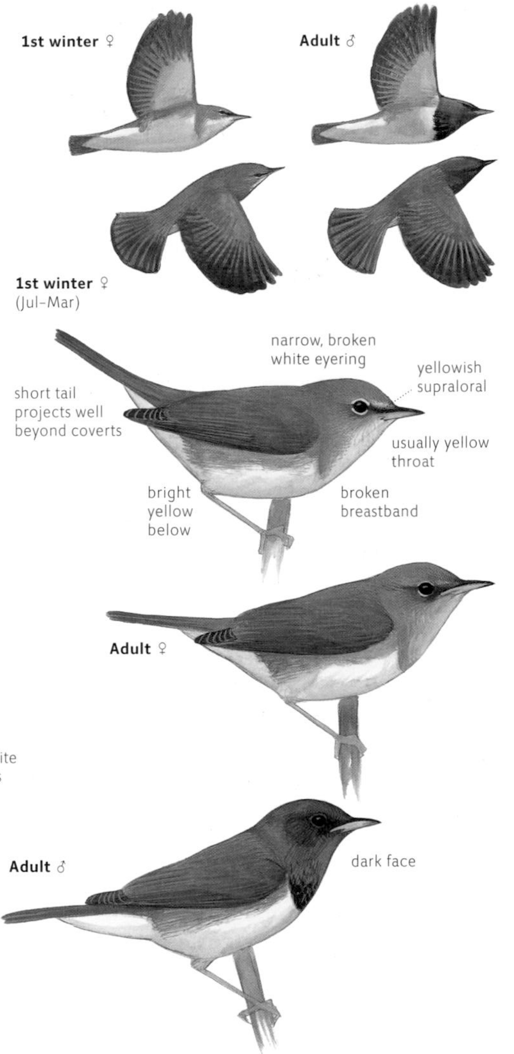

VOICE: Song short and rhythmic with rich, churring quality *churree churree churree turi turi*; last phrases lower, weaker, shorter. Song less variable than MacGillivray's: usually two-part in eastern birds and one-part in western. Call a dry, flat, husky *pwich*. Flight call a clear, high *svit*.

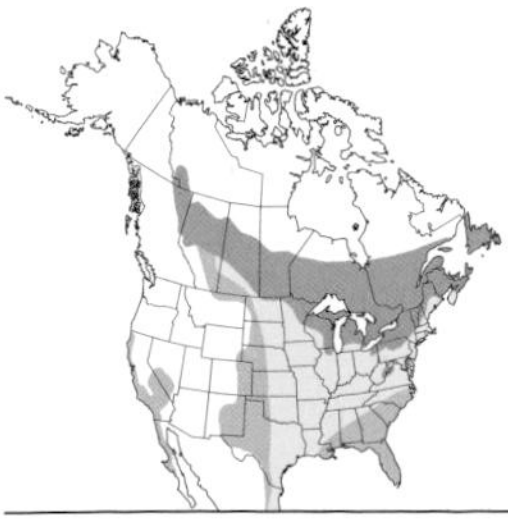

Uncommon. Nests in dense undergrowth, such as alders and brambles within forest clearings, usually in wet areas. Hops on the ground.

Common Yellowthroat

Geothlypis trichas

L 5" WS 6.75" WT 0.35 oz (10 g)

Small and stocky, with short neck, crouching posture and tail often raised. Black mask of male distinctive. Female has yellowish throat contrasting sharply with dark cheek. In most areas this is by far the most frequently seen small green-backed bird.

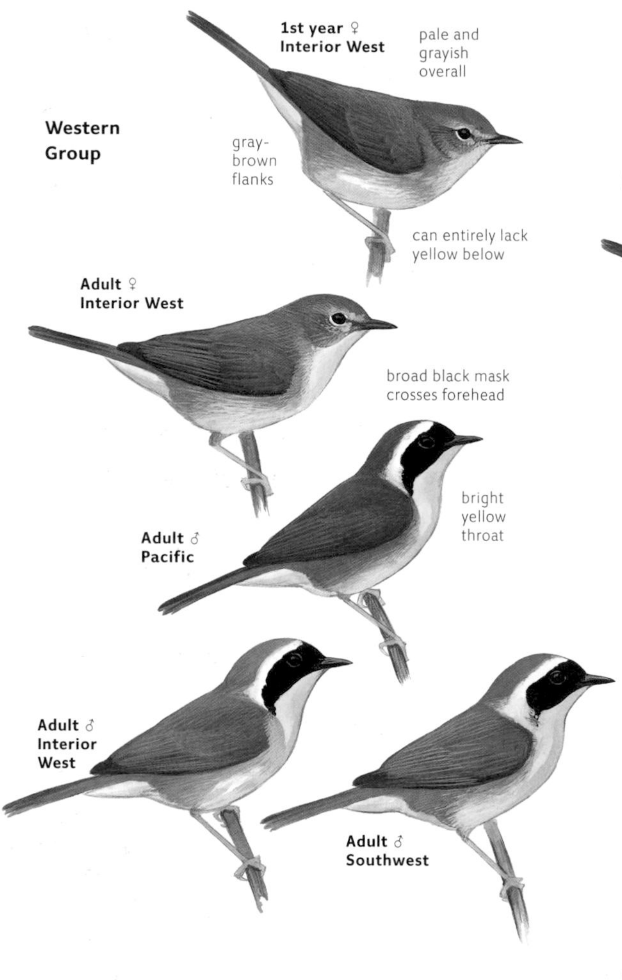

VOICE: Song a gentle, musical whistle in repeated phrases of three to five syllables *wichety wichety wichety*; many variations in pattern but pitch, quality, and rhythmic pattern distinctive. Eastern population has softer, less forceful song and averages more syllables per phrase than all birds to the west (4–5 vs. 2–3 in West). Staccato rattle on one pitch often heard in breeding season (compare Sedge Wren). Call a dry *chedp*; variable from sharp *pik* to softer, longer, descending *jierrk*. Flight call a short, nasal, electric buzz *dzik*.

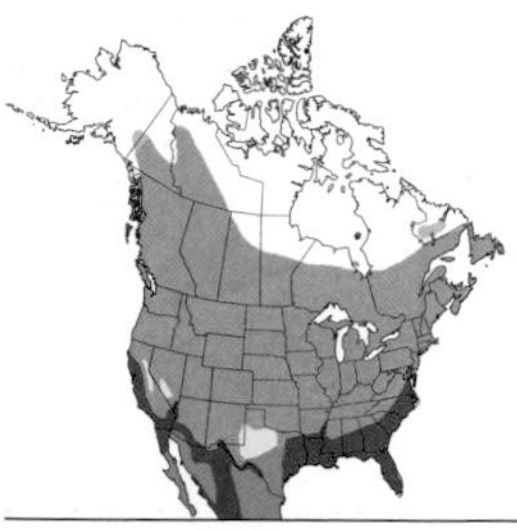

Common in wide variety of weedy, brushy, and marshy habitats; nearly always in low wet areas. Secretive and can be difficult to see, but singing males perch on exposed twigs. When flushed from low grassy or weedy areas; flies low into nearby brush.

The four regional variations of adult males differ mainly in extent of yellow on underparts and in color of the pale forehead band. Eastern; Fairly dark brownish with medium extent of yellow, grayish frontal band. Interior West: Paler and grayer with limited yellow on throat; whitish frontal band. Pacific: Small and dark brown with whitish frontal band and extensive yellow below. Southwest: Relatively large and bright olive; can be entirely yellow below; black mask may be reduced behind auriculars with yellow throat wrapping around rear auriculars; frontal band white with yellow tinge.

Gray-crowned Yellowthroat

Geothlypis poliocephala

L 5.5" WS 8" WT 0.51 oz (14.6 g)

Larger, longer-tailed, and much thicker-billed than Common Yellowthroat. Note pale lower mandible, black lores, gray or brownish forehead, broken eyering, and yellow throat (including malar feathers).

Several recent reports of this species in Texas involve apparent hybrids with Common Yellowthroat. Check for thinner and darker bill, more black on cheeks below eye, and black band continuing above bill across forehead (pure Gray-crowned has no black on forehead).

VOICE: Song bunting-like: a bright, varied warble but somewhat halting and undefined. Calls include a nasal, grating *cher-dlee*; also a series of descending, plaintive whistles *teeu teeu teeu*.

Very rare visitor from Mexico with only a few confirmed records since 1950; formerly resident in southern Texas in dense grassy patches, stands of cane, and overgrown fields.

Hooded Warbler

Setophaga citrina

L 5.25" WS 7" WT 0.37 oz (10.5 g)

Overall bright olive and yellow plumage with plain yellow face and mostly white tail distinctive. Males and some adult females have distinctive black hood and throat; even the drabbest females show suggestion of this unique hooded pattern.

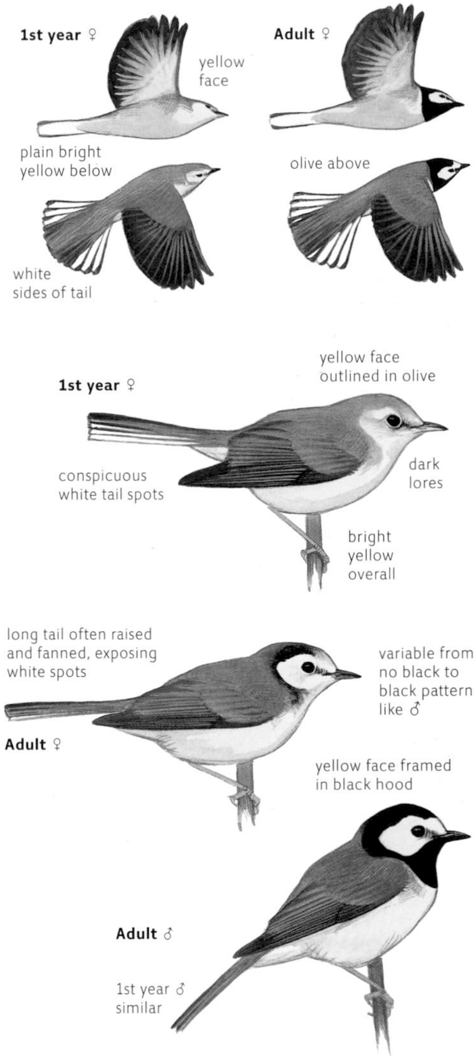

VOICE: Song loud, clear, and musical, without sharp notes: a short series of slurred notes with emphatic end *tawee tawee tawee-teeoo* or *tew tew tew teew teo twee tweee teew*. Song most like Chestnut-sided, Magnolia, and Swainson's. Call a flat, squeaky *tiip*. Flight call a clear, rising *tsiip* like American Redstart.

Common in shady undergrowth within mature deciduous forests. Usually seen flitting through low vegetation with tail raised and fanned, gleaning small insects from leaves and twigs as well as pursuing them in the air in short flights.

American Redstart

Setophaga ruticilla

L 5.25" WS 7.75" WT 0.29 oz (8.3 g)

A small, long-tailed warbler. Broadly fanned and raised tail is distinctive, whether orange and black (adult male) or yellow and gray. Drabbest females still easily identified by all-gray head, yellow sides, and yellow patches in tail.

VOICE: Song of high and rather sharp notes: one distinctive pattern with emphatic buzzy, down-slurred ending *tsee tsee tsee tsee tzirr*; also a softer, lower *tseeta tseeta tseeta tseet* and many variations. Often alternates between two different songs during bouts of singing. Call a clear, high, squeaky chip. Flight call a high, squeaky, rising *tsweet.*

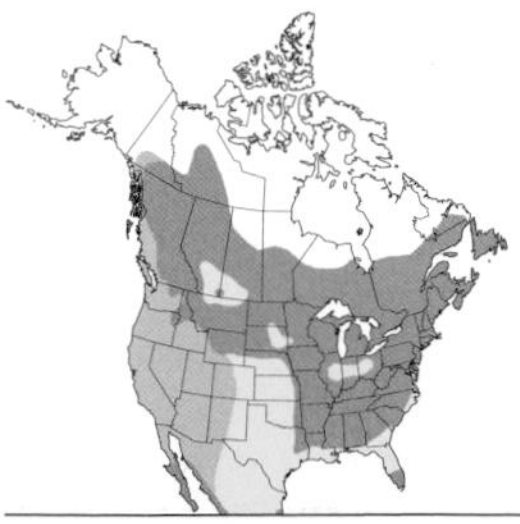

Common in deciduous forests with understory of small trees, particularly in wet areas. Forages mainly among open twigs and leaves at mid-levels of trees or at forest edges. Often fans and raises tail to flush insects, which it chases in short flights through understory.

Cape May Warbler

Setophaga tigrina

L 5" WS 8.25" WT 0.39 oz (11 g)

A small, warbler with very pointed, slightly decurved bill. Immature females are extremely drab, identified by blurry gray streak on breast, hint of yellow on pale sides of neck and on rump, and pale-edged greater coverts.

VOICE: Song very high and thin; may sound slightly buzzy; four to seven upslurred notes (5/sec) *seet seet seet seet seet* or slightly lower-pitched, faster, more complex *seeo seeo seeo seeo seeo* or *witse witse witse wit*. Call a very high, hard, short *ti*. Flight call a very high, slightly buzzy *tzew* or *tzee* slightly descending.

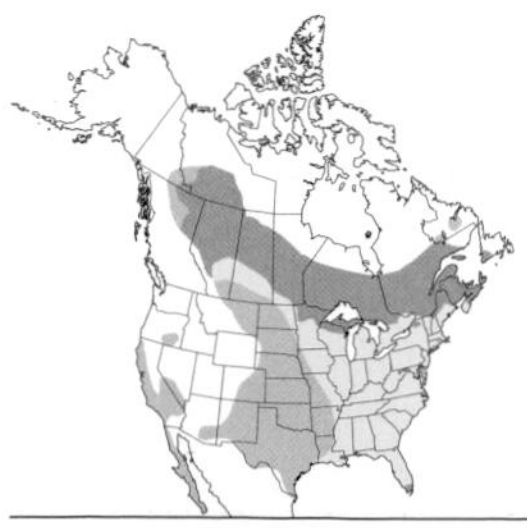

Uncommon and somewhat irregular in mature coniferous forests; most numerous where spruce budworms are abundant. In winter and spring visits flowering trees to feed on nectar; often very territorial, defending flowers against other birds.

Kirtland's Warbler

etophaga kirtlandii

5.75" WS 8.75" WT 0.48 oz (13.8 g)

large and stocky warbler; constantly pumps tail. Dark gray or rown above, pale yellow below, and band of small dark spots cross breast distinctive. Similar to Magnolia Warbler but larger, vith darker face, gray rump, different tail pattern.

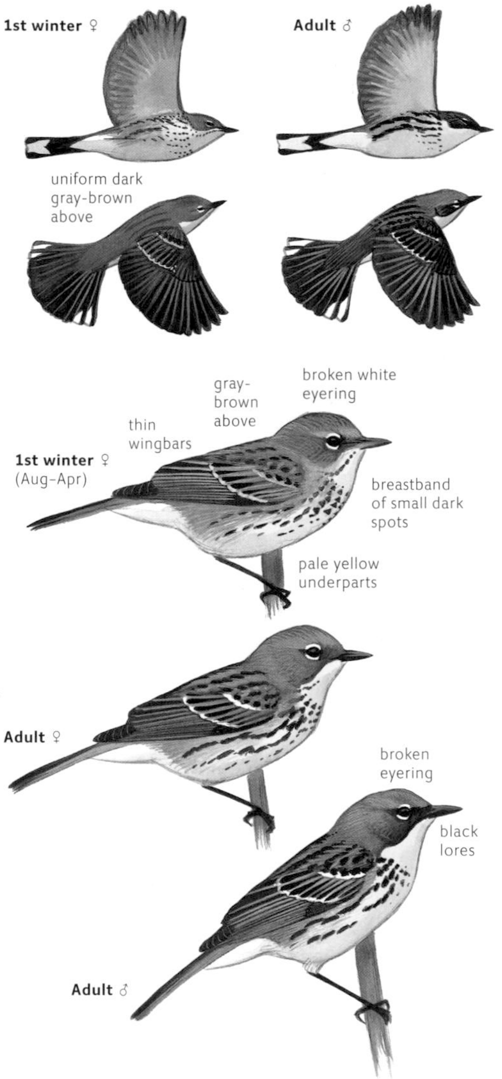

OICE: Song of low, rich notes; loud and emphatic, rising in pitch nd intensity *flip lip lip-lip-tip-tip-CHIDIP.* Alternate song a short hatter reminiscent of House Wren. Call a strong, clear, descending chip like Ovenbird. Flight call a short, high buzz.

Uncommon and very local with total population about 4000. Nesting habitat very specialized: extensive stands of young jack pines, 3–15 feet (1–4.5 m) tall, with small openings and patches of dense groundcover.

Cerulean Warbler

Setophaga cerulea

L 4.75" WS 7.75" WT 0.33 oz (9.3 g)

Small, long-winged, and short-tailed. Adult male unique, with white underparts and blue upperparts. Drab individuals less obvious, but note unique blue-green color above and broad pale eyebrow.

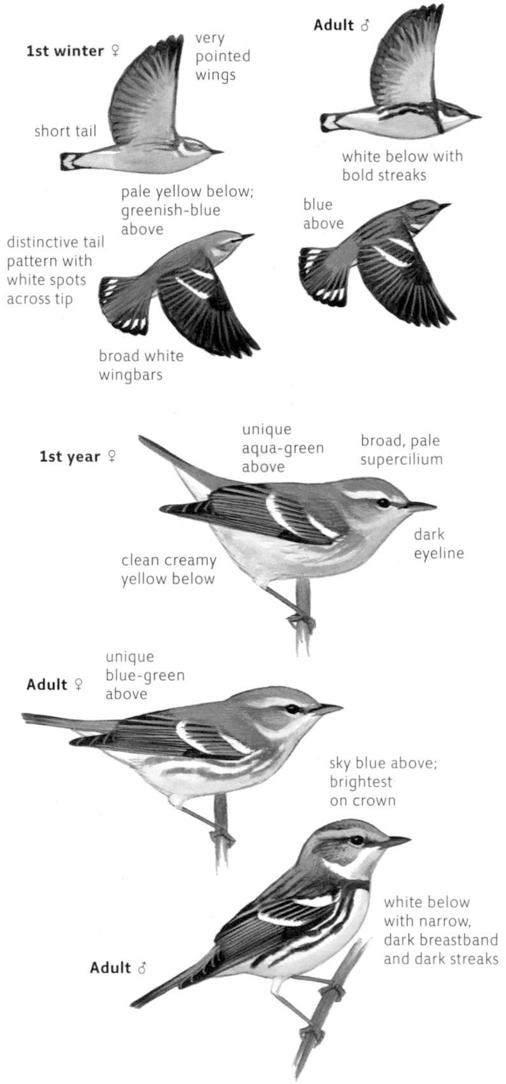

VOICE: Song a high, musical buzz *tzeedl tzeedl tzeedl ti ti ti tzeeeeee*; generally three-part, each part higher than the preceding one; more musical than parulas with no slurred phrases; pattern very similar to Blackburnian but with buzzy quality. Call a clear chip. Flight call a short buzz *dzzt.*

Uncommon, local and declining. Nesting birds prefer mature deciduous forests in both wet lowland settings and mountain slopes. Migrants also prefer mature forests.

Northern Parula

Setophaga americana

L 4.5" WS 7" WT 0.3 oz (8.6 g)

Very small with sharp bill and short tail that is often raised. Blue-gray above, with green mantle, two short white wingbars, limited bright yellow on throat, orange lower mandible, and dark necklace.

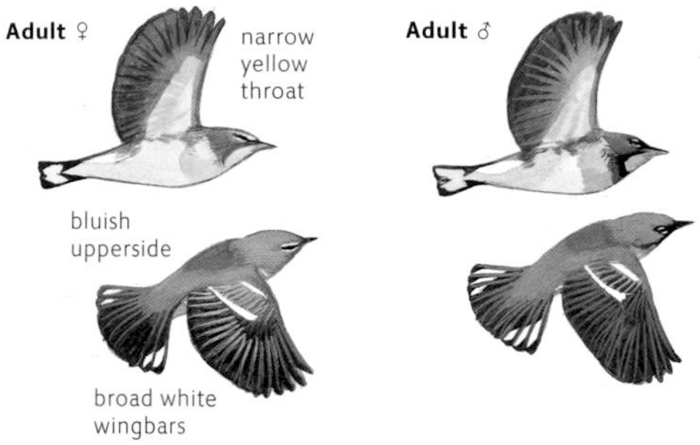

VOICE: Song a rather unmusical, rising buzz usually with distinctive, sharp final note *zeeeeeeeeeeee-tsup* or *zid zid zid zeeeeee tsup* (final note lower and buzzy in western birds). Call a surprisingly strong, clear chip. Flight call a high, clear, descending, frequently repeated *tsif* or *tsiip*.

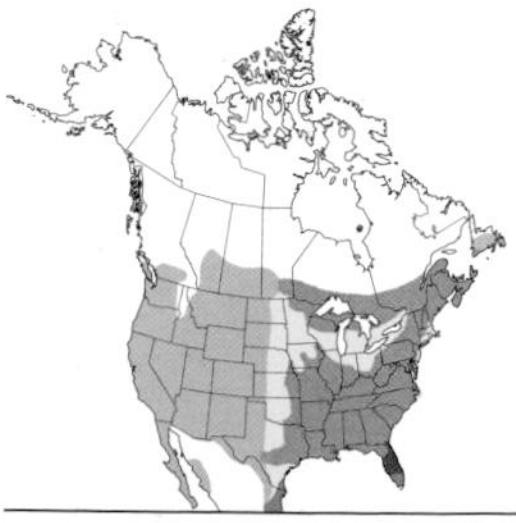

Common in mature woods near water. Nests in hanging moss. Migrants are found in any wooded area, usually high in trees.

Tropical Parula

Setophaga pitiayumi

L 4.5" WS 6.25" WT 0.25 oz (7 g)

Similar to Northern Parula, but bright yellow on throat and breas more extensive; also lacks dark necklace and has all-dark fac without pale eye-arcs.

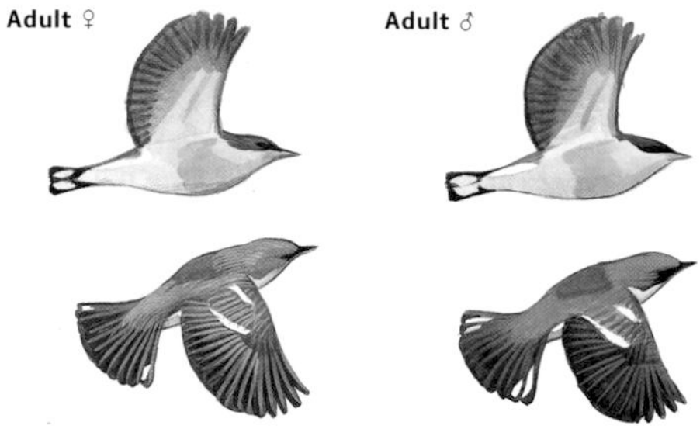

VOICE: Song essentially identical to Northern, but final not usually buzzy (like western populations of Northern); may be slightly higher, more insectlike overall. Call and flight call simila to Northern.

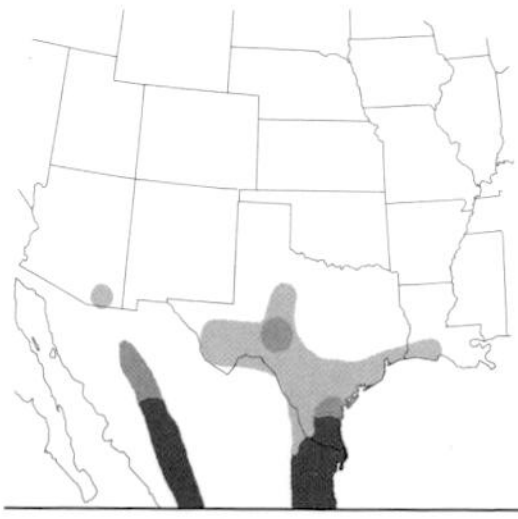

Rare; only a few pairs nest in southern Texas, in mature live oaks with abundant Spanish moss.

Magnolia Warbler

etophaga magnolia

5" WS 7.5" WT 0.3 oz (8.7 g)

mall and relatively long-tailed; tail often raised and fanned, owing off unique pattern. Drab individuals can be similar to anada or Prairie Warblers; note thin wingbars, yellow rump, hite undertail coverts, and bright yellow breast.

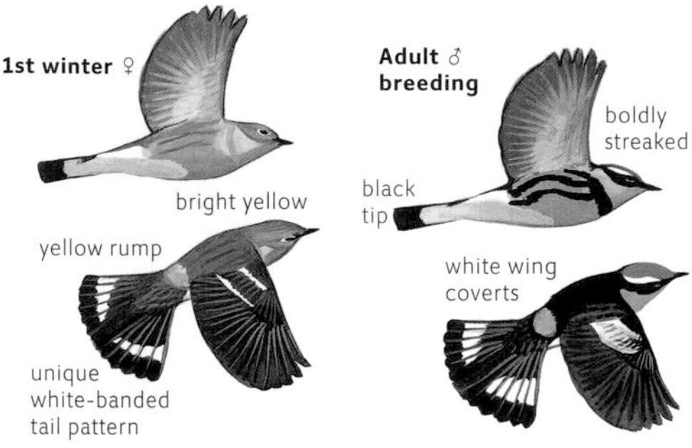

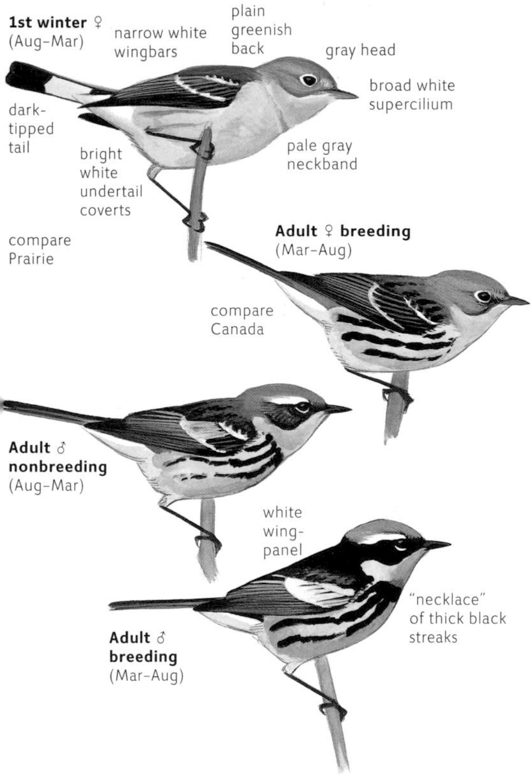

OICE: Song short, musical but rather weak and simple *sweeter veeter SWEETEST*; variable; most similar to Hooded but much ss emphatic. Call a unique, tinny, hoarse *vint* or *chuif*. Flight all a very high, weak, lightly trilled, rather soft *zzip*.

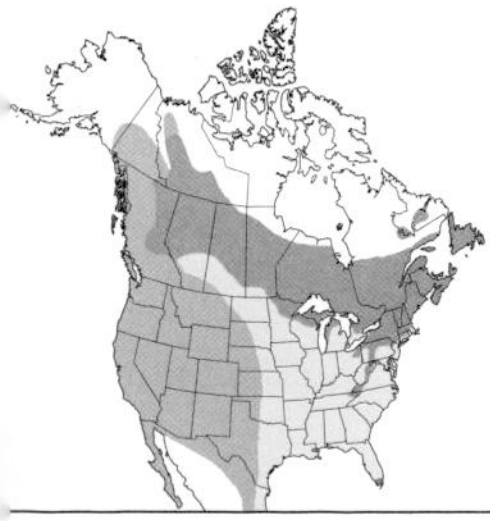

Common in coniferous forests, especially dense, second growth stands. Migrants found in any wooded or brushy area.

Blackburnian Warbler

Setophaga fusca

L 5" WS 8.5" WT 0.34 oz (9.8 g)

A relatively slender, streamlined warbler. Male unmistakable, with flaming-orange throat. Drabbest females distinguished from other warblers by dark cheek surrounded by yellow, broad yellow partial eyering, and pale stripes on back.

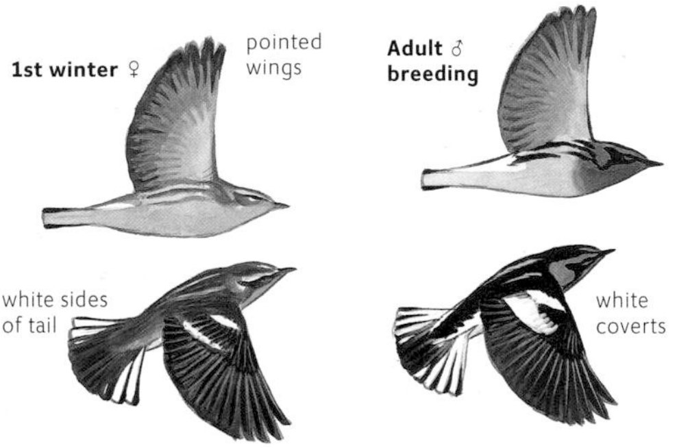

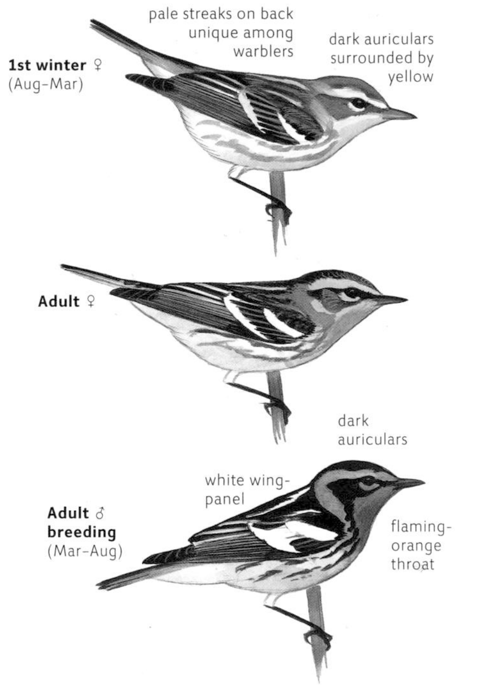

VOICE: Song has sharp, dry quality and usually includes some incredibly high notes *tsi tsi tsi tsi tsi ti ti ti ti ti seeeeee* ending extremely high and thin. Alternate song very high, slow, rising series with rattling quality *tseekut tseekut tseekut tsee*; compare Golden-crowned Kinglet. Call a sharp *tsick*. Flight call a high, thin buzz.

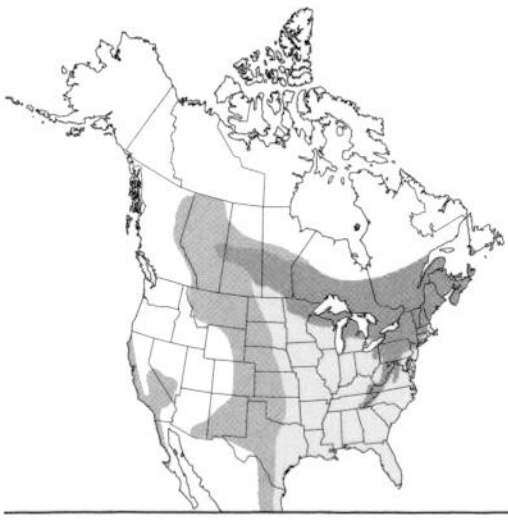

Common in mature coniferous or mixed woodlands; tends to stay in treetops. Migrants also prefer mature woodlands.

Yellow Warbler

Setophaga petechia

L 5" WS 8" WT 0.33 oz (9.5 g)

Most are bright yellow overall, distinctive. Drabber individuals can be confused with Orange-crowned, Nashville, and other warblers. Note stout bill, plain pale face, pale legs, pale-edged wing feathers, and some yellow in tail.

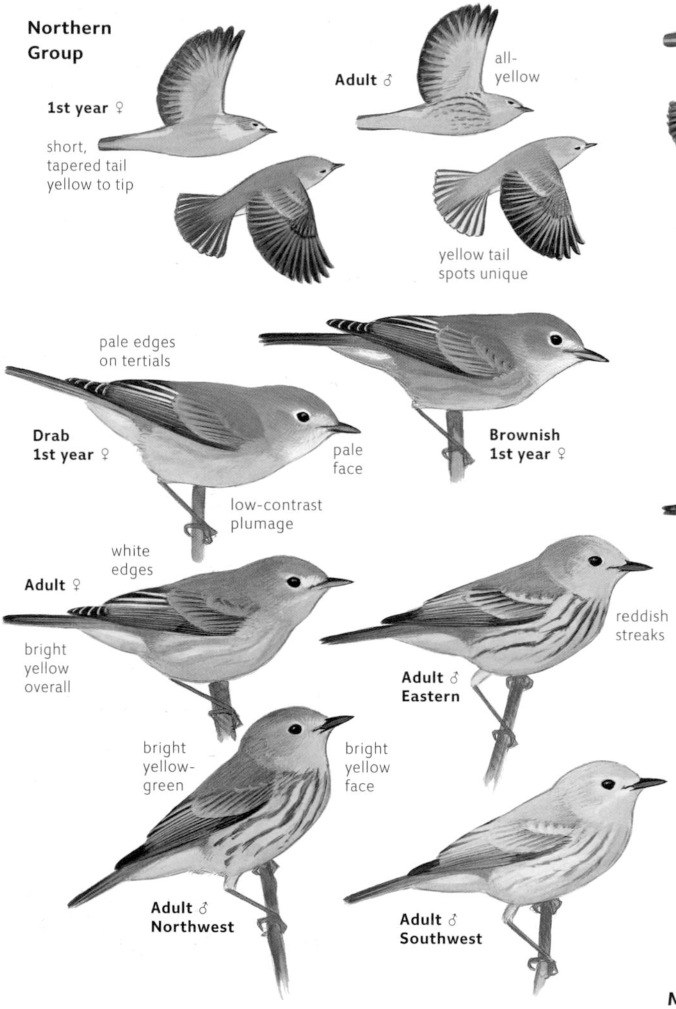

VOICE: Song of sweet, high, clear notes: sharp upslurs followed by sharp downslurs, then emphatic ending *sweet sweet sweet ti ti ti to soo* or *swee swee swee ti ti ti swee*; variable in details of phrasing. Alternate song longer and less structured without emphatic ending *seedl seedl seedl seedl sitew sitew sitew*; variable, resembles Chestnut-sided. Call a clear, loud chip. Flight call a high, clear trill *tzip*.

Common and widespread in any wet brushy habitat, such as willow thickets, field edges. Migrates very early in fall, in Jul–Aug.

Geographic variation within the Northern Group is fairly weak and clinal, with drabber plumage to the north and west, much paler in the desert Southwest. The disjunct Golden population (resident in the Florida Keys) has rounder wings and shorter primary projection; bill averages thinner with slight decurve. Male is relatively rich golden yellow with broader red streaks on underparts but darker olive crown; some 1st winter birds are nearly plain gray, lacking yellow, and 1st summer males can have distinctive pattern of gray head and yellow body. A few pairs of the Mangrove population nest on the southernmost coast of Texas (also recorded in Arizona and southern California). Males have rufous head, females and immatures resemble Golden but often show some rufous on head.

Chestnut-sided Warbler

Setophaga pensylvanica

L 5" WS 7.75" WT 0.34 oz (9.6 g)

A rather stout-billed warbler; nearly always holds tail raised at an angle. Striking seasonal plumage changes, but always distinctive. In fall and winter gray face contrasts with bold white eyrering and green crown and back.

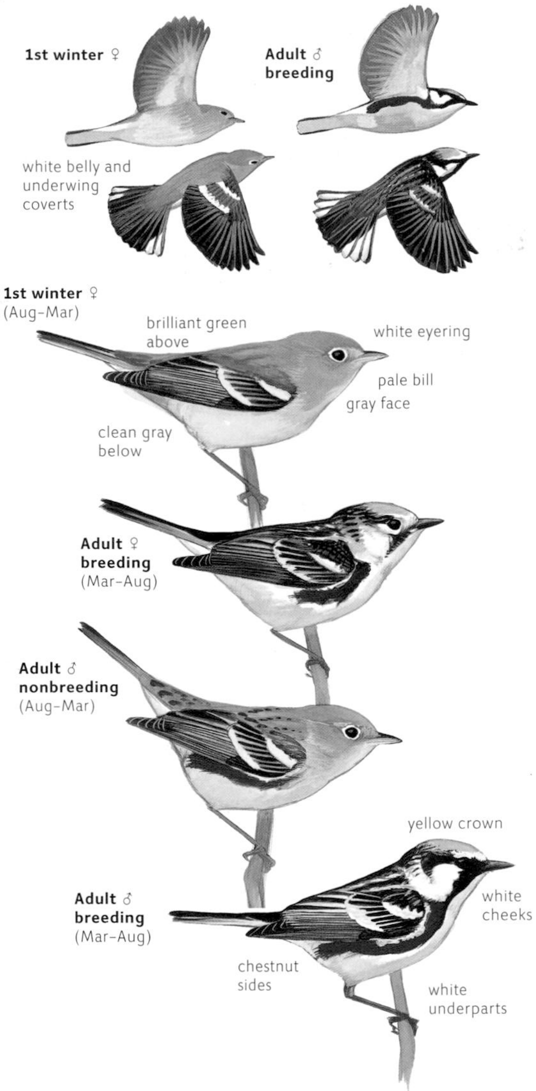

VOICE: Song clear, musical with emphatic ending *witew witew witew WEECHEW* (compare Hooded). Alternate song longer, rambling like Yellow. Both song types average lower-pitched than Yellow; can be matched by some Yellow. Call a low, flat *chidp*. Flight call a rather low, buzzy, nasal *jrrt*.

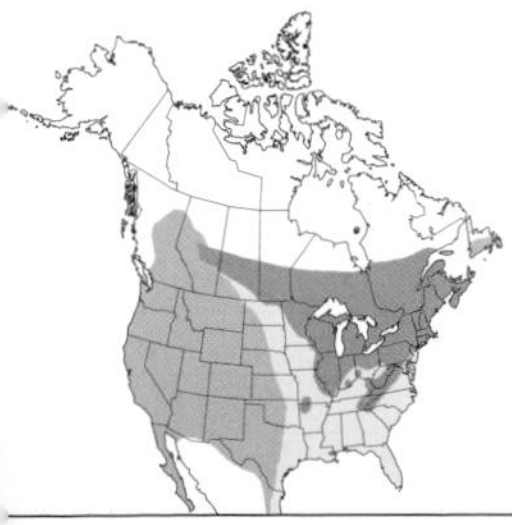

Common locally in second-growth brushlands, in orchards, and along roadsides; often in drier settings than Yellow Warbler.

Black-throated Blue Warbler

Setophaga caerulescens

L 5.25" WS 7.75" WT 0.36 oz (10.2 g)

Male unmistakable. On female note plain drab olive color with whitish undertail coverts, dark cheek, narrow pale eyebrow, and pale lower eye-arc. Most females have small white patch at base of primaries, distinctive when present.

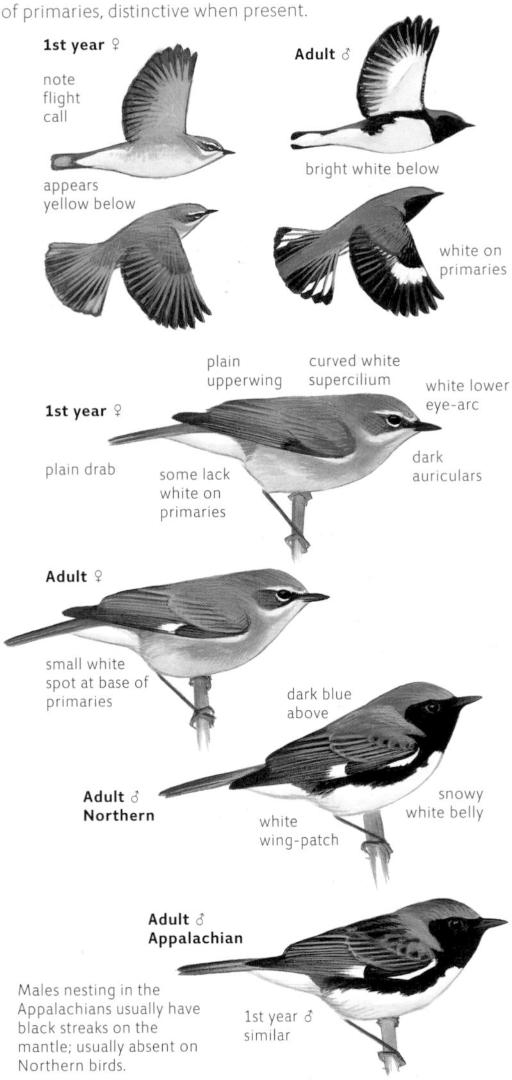

VOICE: Song a husky but musical buzz, lazy and drawling; several introductory notes followed by rather harsh, slow, rising buzz *zheew zheew zheeeeeee* or *zo zo zo zo zo zo zo zeeeeeee*. Call a very high, sharp smack *stip* like juncos. Flight call a sharp, dry *twik* reminiscent of soft, sharp notes of cardinals; given frequently.

Common. Nearly always found in shady understory of woodlands. Nests in mixed coniferous and deciduous woodlands.

Blackpoll Warbler

Setophaga striata

L 5.5" WS 9" WT 0.46 oz (13 g)

A large, robust, streamlined warbler; often dips tail gently. Fall plumages very similar to Bay-breasted; note yellow feet (or all-yellow legs), whitish undertail coverts, faintly streaked breast. Told from Tennessee by wingbars, larger size, faint streaks.

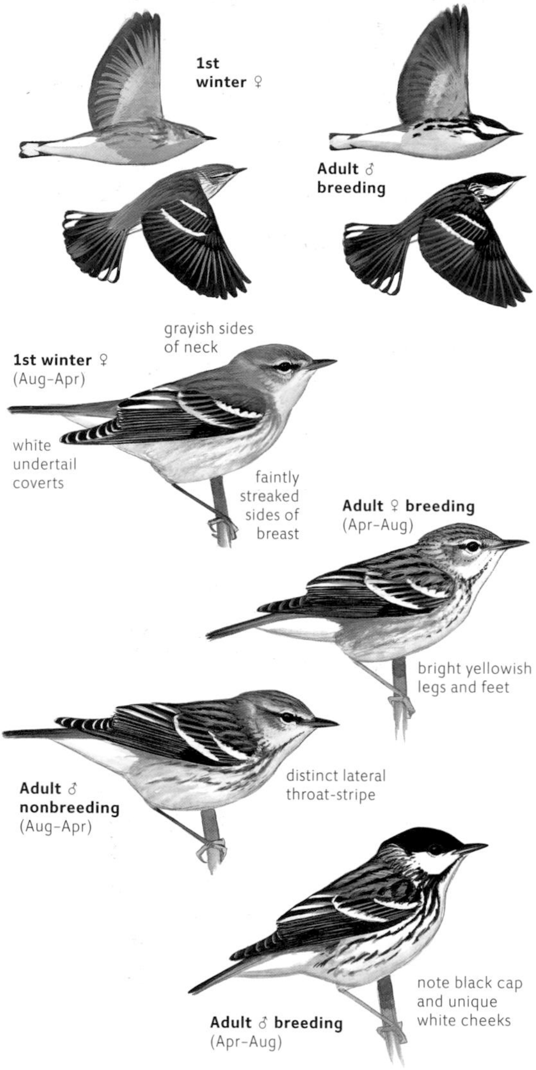

VOICE: Song a very high, rapid series of short notes all on one pitch, strongest in middle *sisisisisiSISISISISIsisisisisis*. Some give a very rapid, insectlike trill *tttttTTTTTTTtttttt*. Call a sharp, clear chip. Flight call a high, buzz similar to other *Setophaga* warblers.

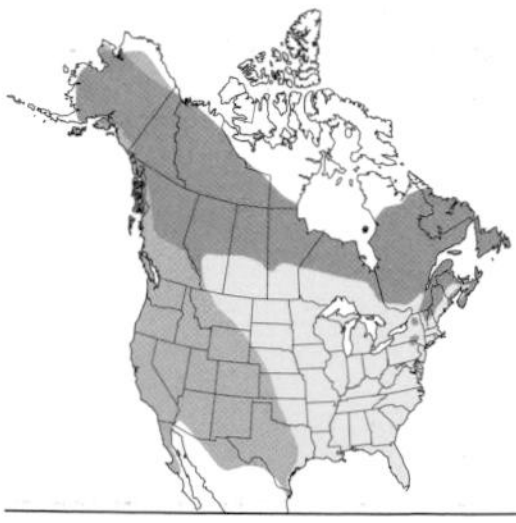

Common. Breeds in stunted coniferous forests on mountaintops or near tree line. Migrants found in any wooded or edge habitat.

Bay-breasted Warbler

Setophaga castanea

L 5.5" WS 9" WT 0.44 oz (12.5 g)

A large and stocky warbler. Breeding male distinctive. Fall plum-ages can be confusingly similar to Blackpoll Warbler, but with grayish legs and feet; also note buffy undertail coverts and flanks, unstreaked breast, vague spectacles.

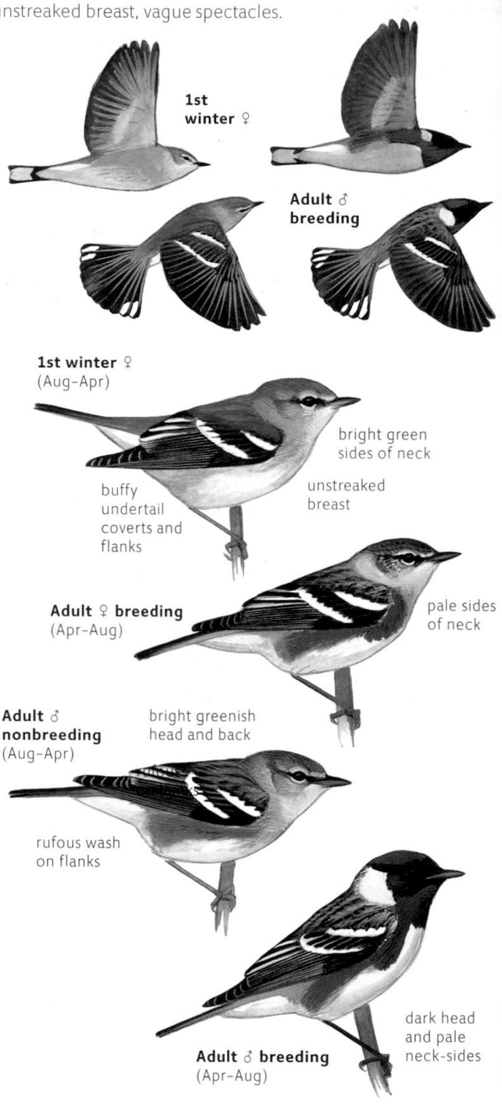

VOICE: Song high and thin but musical like Black-and-white usually short with poorly defined syllables all on one pitch or slightly rising overall *se-seew se-seew se-seew* or *teete teete teete tee tee tee*. Call a clear chip. Flight call a short buzz like other *Setophaga* warblers.

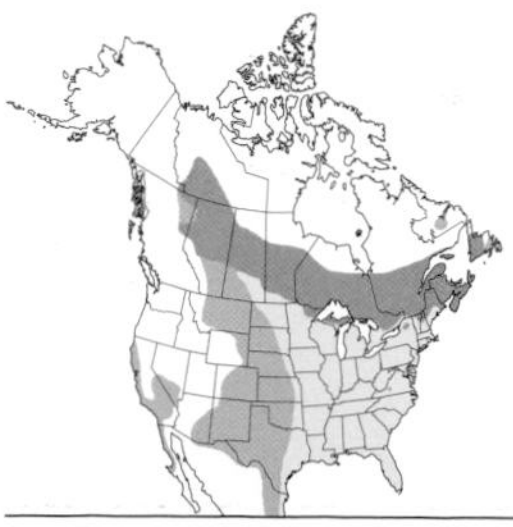

Uncommon in dense coniferous forests with small openings. Migrants found in any woodland habitat.

Pine Warbler

Setophaga pinus

L 5.5"   WS 8.75"   WT 0.42 oz (12 g)

A relatively large, long-tailed warbler. Adult male is bright yellow on head and breast, with dark olive cheeks. On very drab birds note weak wingbars, pale patch on sides of neck, long narrow tail with poorly-defined white spots.

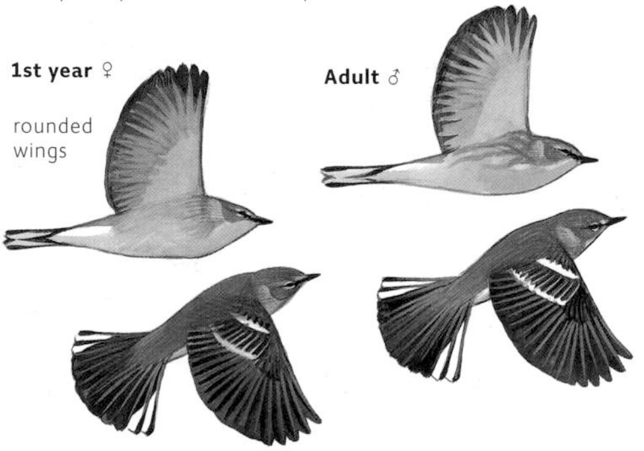

VOICE: Song a rapid trill of simple upslurred notes (similar to but more musical than Chipping Sparrow). Also a slower trill of two-syllable, more musical whistles; whole trill slightly varying in pitch. Call a high, flat chip. Flight call a high, clear, descending *seet*.

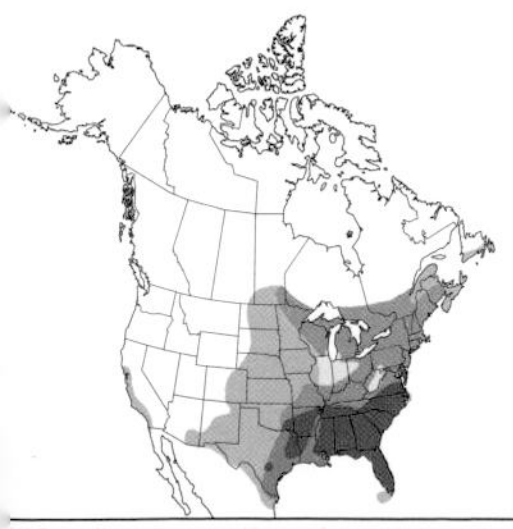

Common and almost always closely associated with pine trees. Nests in pine forests. In winter may be found in other habitats or visit bird feeders; sometimes joins bluebirds and Chipping Sparrows to forage on the ground in suburban settings.

Prairie Warbler

Setophaga discolor

L 4.75"   WS 7"   WT 0.27 oz (7.7 g)

A small, rather long-tailed warbler; often pumps or flicks tail. Always bright yellow below, with dark semicircle below eye and dark spot on sides of neck. Note adult's arched yellow eyebrow.

VOICE: Song a series of high, fine, musical buzzes nearly always steadily rising in pitch; may be fast or slow; also series with first few notes level and later notes rising *zooo zoo zo zozozozoZEEEET.* Call a musical chip with a hint of the huskiness of Palm. Flight call a high, slightly husky, level *tss.*

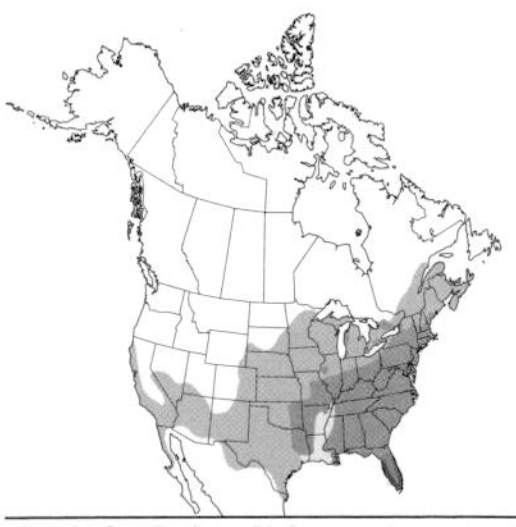

Common in open, sunny, second-growth habitats, such as in old fields or among dune vegetation.

Palm Warbler

Setophaga palmarum

L 5.5" WS 8" WT 0.36 oz (10.3 g)

Relatively long-tailed; constantly pumps tail up and down. Ground-foraging habits and drab streaked plumage may recall sparrows or pipits; note dark eyeline and long pale eyebrow. Always has yellow around base of tail, white corners on dark tail.

Brown (Western)

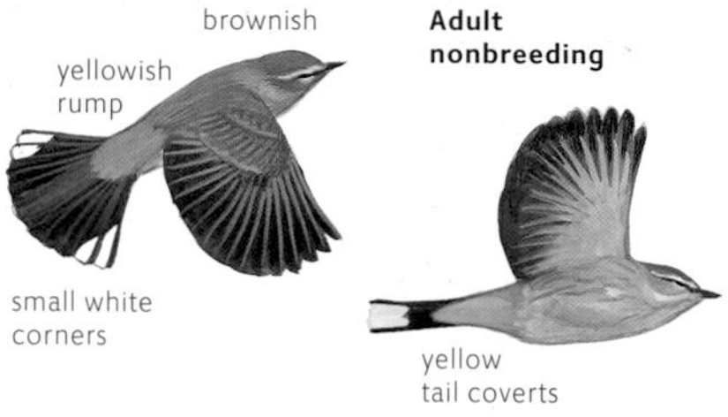

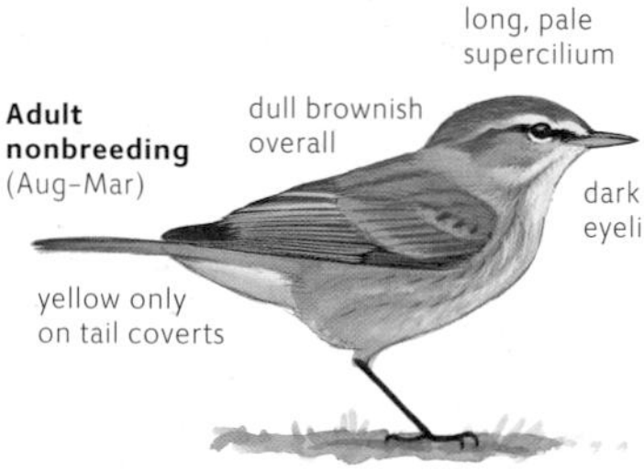

Yellow (Eastern)

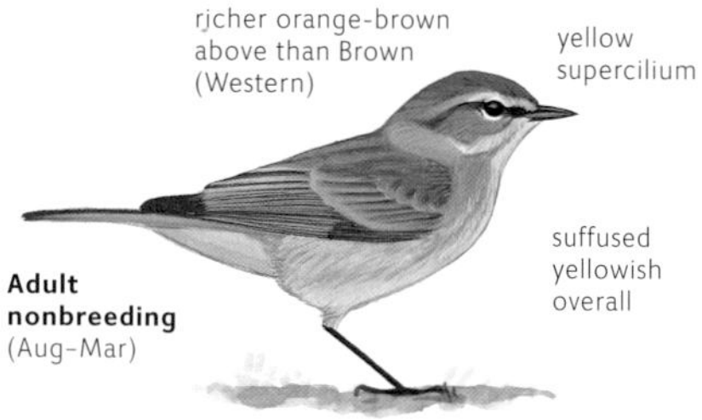

VOICE: Song a rather dull, uneven, buzzy trill *zzizzizzizzizzizzi* slightly changing in pitch and volume but overall steady; rather weak and infrequently heard. Call a sharp, husky, distinctive *chik*; closest to Prairie but stronger, deeper. Flight call a husky *sink*.

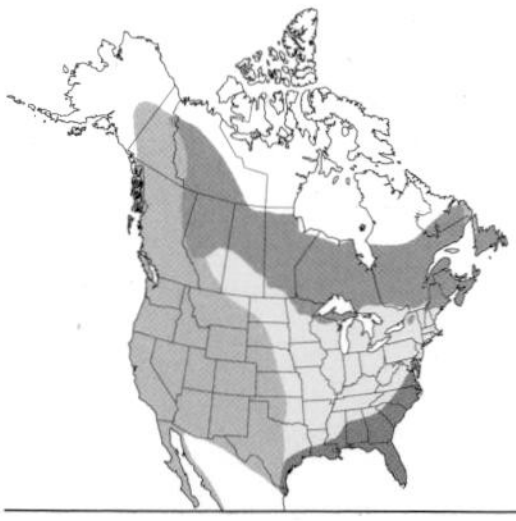

Common. Nests in open bogs with border of spruce or other trees; at other times found in weedy fields, brushy hedges, dunes, and other open habitat. Often in small flocks that forage in open grassy areas and perch on fences or low shrubs.

Population nesting east of Ontario (wintering mainly Georgia to Louisiana) are washed yellow overall, with bright rufous crown and rufous breast streaks. Western populations (wintering from Virginia to Texas, as well as Caribbean islands and Mexico) drabber brownish overall, with yellow restricted to throat and tail coverts. Nearly all are readily identifiable, but the two populations apparently intergrade in western Québec.

Grace's Warbler

Setophaga graciae

L 5" WS 8" WT 0.28 oz (8.1 g)

A small and slender warbler, with short bill, round head, and long tail. Plain gray upperside and bright yellow throat like Yellow-throated (very rarely overlaps in range) but with broad yellow eyebrow and entirely gray sides of neck.

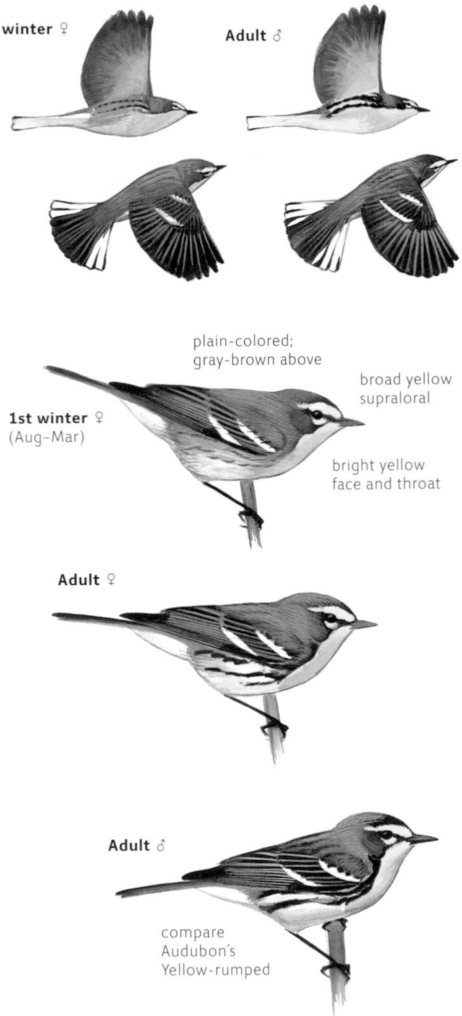

VOICE: Song a slow, choppy trill: usually higher and faster at end, two- or three-part, downslurred *tew tew tew tew tew tee tee tee tee*; resembles some variations of Audubon's Yellow-rumped and Virginia's Warblers, as well as Dark-eyed Junco. Call a soft chip. Flight call a very high, thin, short *fss*.

Common in tall pines, foraging for small insects among needles at branch tips.

Yellow-throated Warbler

Setophaga dominica

L 5.5" WS 8" WT 0.33 oz (9.4 g)

Strikingly patterned, with entirely gray upperside, bright yellow throat; bold white eyebrow and white patch on sides of neck sets off triangular black mask, which extends down sides of breast to frame yellow throat.

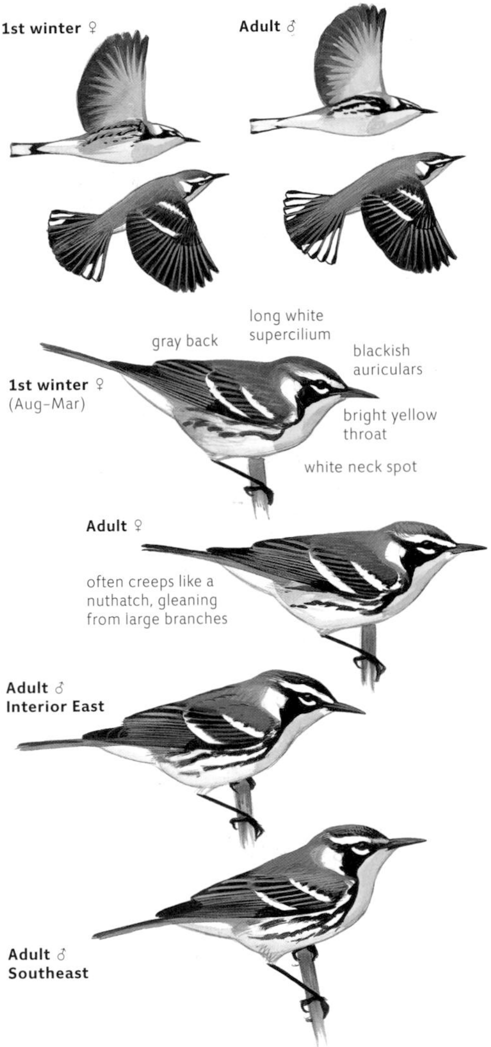

VOICE: Song a simple and rather gentle series of sweet, clear whistles *teedl teedl teedl teedl teedl teedl teedl tew tew twee* steadily descending with a couple of rising notes at end; may suggest Louisiana Waterthrush or Yellow Warbler. Call a soft, clear, descending chip reminiscent of Eastern Phoebe. Flight call a high, clear *seet*.

Common locally in large trees of mature lowland forests; in some areas prefers pines, in others sycamore. Winters in palm trees. Often forages for insects upside-down (nuthatch-like), peering under large limbs, and frequently searches under the eaves of buildings for insects and spiders.

Yellow-rumped Warbler

Setophaga coronata

L 5.5" WS 9.25" WT 0.43 oz (12.3 g)

A rather large, long-tailed warbler, with stout dark bill. Bright yellow rump distinctive and often obvious. Also note pale throat, streaked breast, dark cheeks, and on winter birds overall brownish color.

VOICE: Song a clear warble, soft and rather flat, weaker at beginning and end; usually two-part with last few phrases lower or higher in pitch *sidl sidl sidl sidl sidl seedl seedl seedl seedl*. Song of Myrtle averages slightly higher-pitched with shorter phrases. Call of Myrtle a low, flat *chep* or hard *tep* without rising inflection. Call of Audubon's a dry, husky *chwit* with slightly rising inflection. Flight call a clear rising *svit* or *ssit*.

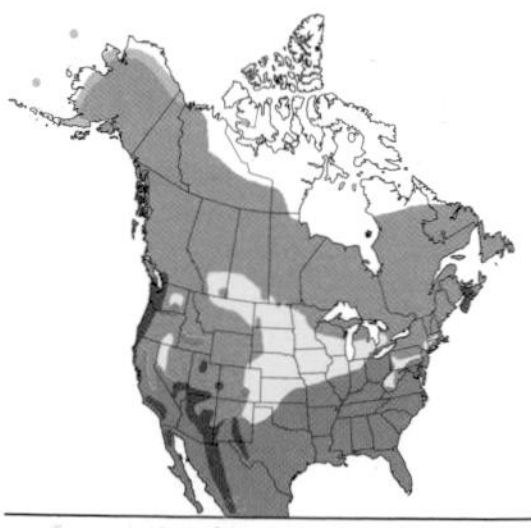

Common and conspicuous; in many areas the only warbler likely to be seen in winter. Nests in relatively open coniferous forests and edges. Winters in open brushy habitats, such as dunes, willow thickets, and field edges, especially among bayberry and juniper. Often in small loose flocks. Perches upright on relatively exposed perches, often flying up to catch passing insects.

Taiga-breeding population known as "Myrtle" warbler, can be distinguished reliably from western "Audubon's" warbler by plumage and call. Myrtle has more contrasting white throat that wraps around onto sides of neck; more contrasting dark cheek, pale eyebrow, and less black at tips of tail feathers. See inergrade Myrtle × Audubon's Yellow-rumped, page 493.

Black-throated Gray Warbler

Setophaga nigrescens

L 5" WS 7.75" WT 0.29 oz (8.4 g)

Plain black, gray, and white color unlike most other warblers. Note dark cheek, broad white eyebrow, and plain gray back. Black throat reminiscent of chickadees and Bridled Titmouse, but note bold dark streaks on flanks, white outer tail feathers.

VOICE: Song of mostly musical, buzzy notes with fast tempo and emphatic ending *zeea zeea zeea ZEEE zaa zoo* or simple, repeated, buzzy phrases *zidza zidza zidza zidza*. Call a low, dull *tip* or *tep* similar to Myrtle Yellow-rumped; lower than other black-throated warblers. Flight call like Black-throated Green.

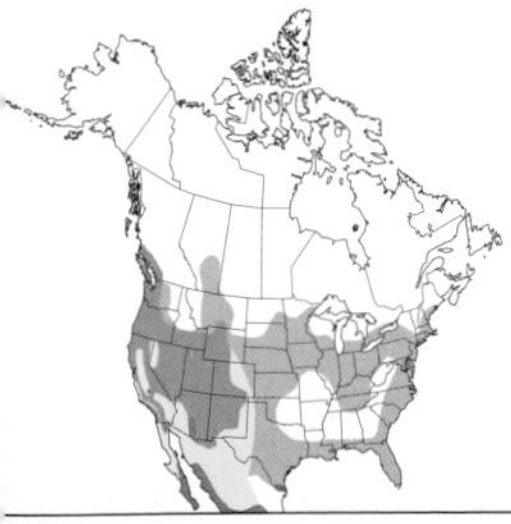

Common in dry oak or juniper woodlands. Migrants often in riparian deciduous trees.

Hybrid Warblers

The two populations of Yellow-rumped Warbler, sometimes considered separate species, interbreed commonly where their range overlaps in the Canadian Rockies, but intergrades are infrequently identified away from the breeding grounds. Intergrades are best identified by an intermediate or mixed face pattern. A "mismatch" can also reveal a likely intergrade, e.g., if a bird shows a Myrtle-like face pattern, but Audubon's-like tail pattern or call.

Hermit and Townsend's Warblers hybridize commonly where their range overlaps in Washington and Oregon, but hybrids are rarely identified elsewhere. Typical hybrids are shown here, but the features of the two species can be combined in many ways. Hybrids generally have face and head pattern closer to Hermit, breast and flank pattern closer to Townsend's. Female and first-winter male hybrids can closely resemble Black-throated Green Warbler, but watch for white vent, pale forehead, yellow breast, and streaked back.

Townsend's Warbler also hybridizes regularly with Black-throated Green Warbler in the Canadian Rockies, but these hybrids are difficult to identify and almost never reported elsewhere.

Hermit Warbler

Setophaga occidentalis

L 5" WS 8" WT 0.32 oz (9.2 g)

Note yellow restricted to face, plain face lacking dark eyeline, drab pale underside without streaks. Drab individuals told from Townsend's and Black-throated Green Warblers by drab gray-brown upperside, weak face pattern, unstreaked flanks.

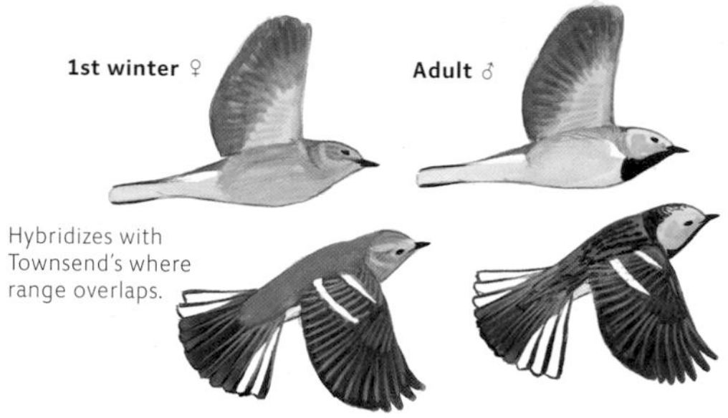

VOICE: Song variable: a rapid series of high, buzzy phrases, typically accelerating and ending with abruptly higher or lower notes *ze ze ze ze ze ze zee sitew*; some songs similar to Townsend's but generally softer and clearer; other songs long and repetitive *ze ze ze zeea zeea zeea zeea ZEEA ZEEA tleep*. Calls similar to Townsend's and Black-throated Green.

Common locally in relatively dense and shady forest of tall pines and firs, foraging mainly at mid-levels of trees. In migration and winter can be found in any wooded habitat, often joining foraging groups of chickadees and other species in oaks and conifers.

Townsend's Warbler

Setophaga townsendi

L 5" WS 8" WT 0.31 oz (8.8 g)

Note bright yellow from face to breast, and strongly contrasting dark olive cheek patch. Drab individuals told from Hermit by streaked flanks, well-defined dark cheeks; from Black-throated Green by more extensive yellow below, contrasting dark cheeks, no yellow on vent.

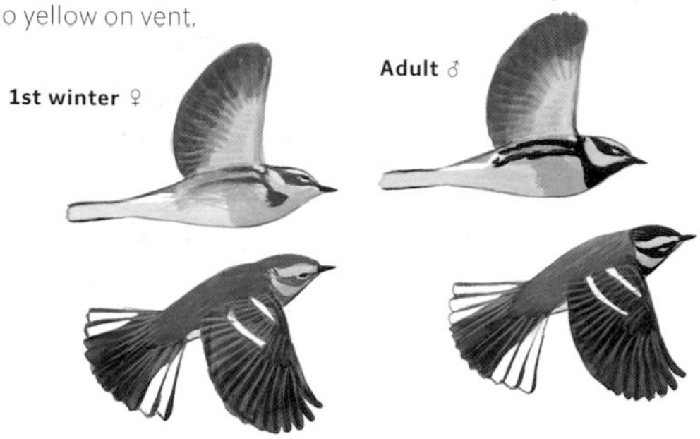

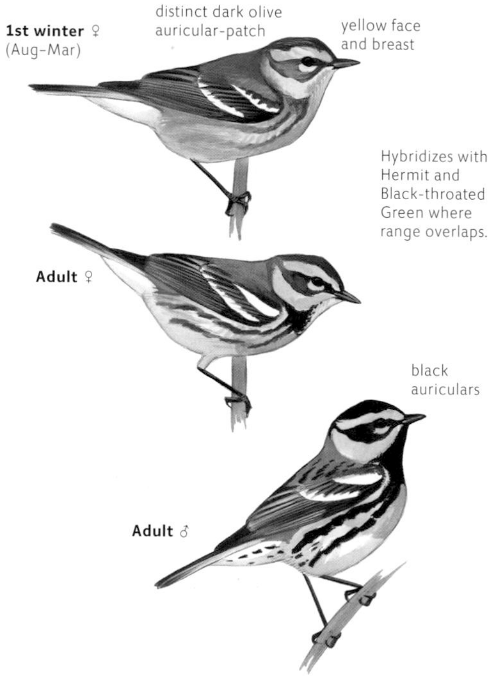

VOICE: Song a rapid series of mainly buzzy notes; variable in pattern but often rising; higher and thinner than Black-throated Gray: *zoo zoo zoo zeeee skeea skeea* or *zi zi zi zi zi zeedl zeedl* or *weezy weezy weezy dzeee*. Some songs very similar to Black-throated Green. Call averages slightly sharper than other black-throated warblers. Flight call like Hermit.

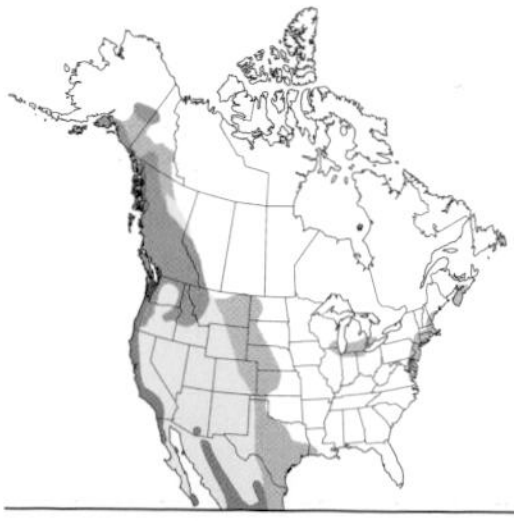

Common in mature coniferous forests, nesting and foraging high in trees. In migration and winter found in any wooded habitat, but prefers relatively dense foliage of mature oaks and conifers.

Golden-cheeked Warbler

Setophaga chrysoparia

L 5" WS 7.75" WT 0.34 oz (9.8 g)

Very similar to Black-throated Green Warbler, but note distinct dark eyeline with yellow cheeks, darker crown and back, and lack of yellow across vent.

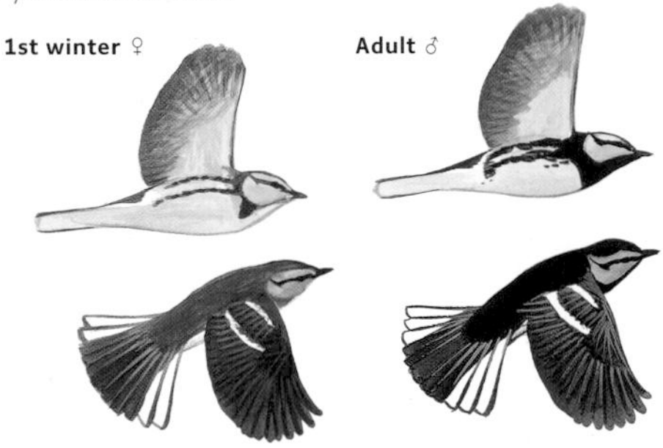

These four species on these two pages are very similar and closely-related. Small, with relatively long tail; tail mostly white when seen from below.

VOICE: Song buzzy like other black-throated warblers; slightly harsher than Black-throated Green; relatively low-pitched, slow, lazy *zrr zooo zeedl zeeee twip* or *brrr zweee seezle zeeeee titip* or *zeedl zeedl zeedl zeedl zweeee tsip*. Calls like Black-throated Green and its other close relatives.

Uncommon, local, and declining. Breeds only in dry open oak woodlands with some mature ashe junipers. Very rarely seen away from known nesting sites.

Black-throated Green Warbler

Setophaga virens

L 5" WS 7.75" WT 0.31 oz (8.8 g)

Note bright yellow face, whitish underparts with distinctly streaked flanks, plain greenish back. Drab individuals told from Townsend's Warbler by whitish breast, yellow across vent, smudgy olive on cheeks.

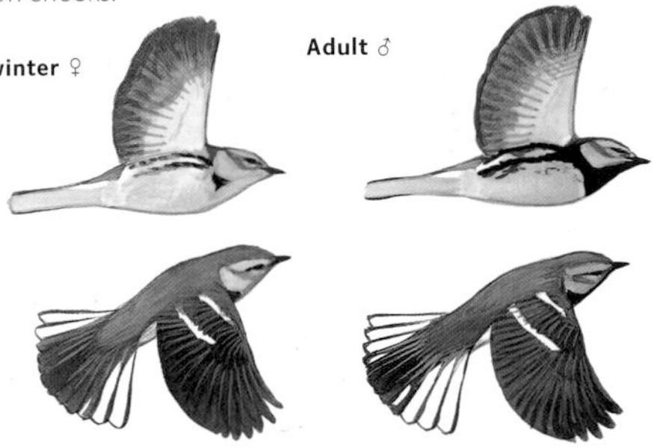

VOICE: Song a series of short, level buzzes; two commonly heard patterns: a rather fast *zee zee zee zee zo zeet* and a more relaxed *zoooo zeee zo zo zeet*; some Townsend's songs similar. Call a sharp *tsik* or *tek* like Hermit; sharper and higher than Myrtle Yellow-rumped. Flight call a clear, high, rising *swit*; shorter than Yellow-rumped.

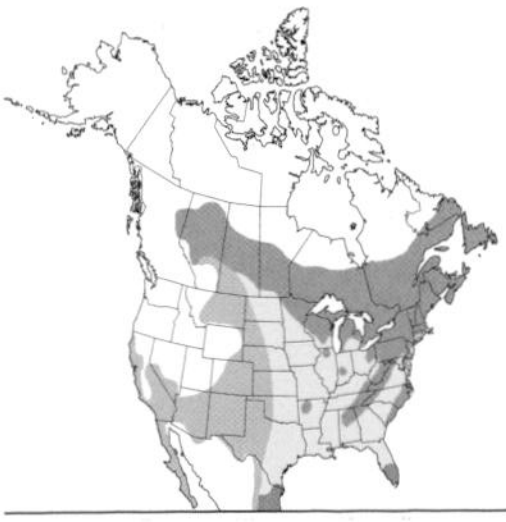

Common. Nests in a variety of mature coniferous and mixed woodlands, especially among white pines and hemlocks. Migrants and wintering birds found in any woodland habitat.

Wilson's Warbler

Cardellina pusilla

L 4.75"   WS 7"   WT 0.27 oz (7.7 g)

A very small, long-tailed warbler; often raises tail and flips it side to side. Unmarked bright yellow below from chin to undertail coverts with all-dark tail; note pale yellow face, dark cap, and large eye; plain yellow-olive above.

Pacific

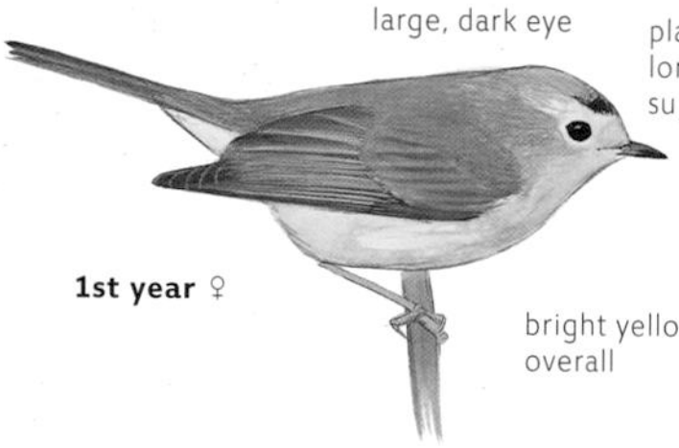

Taiga

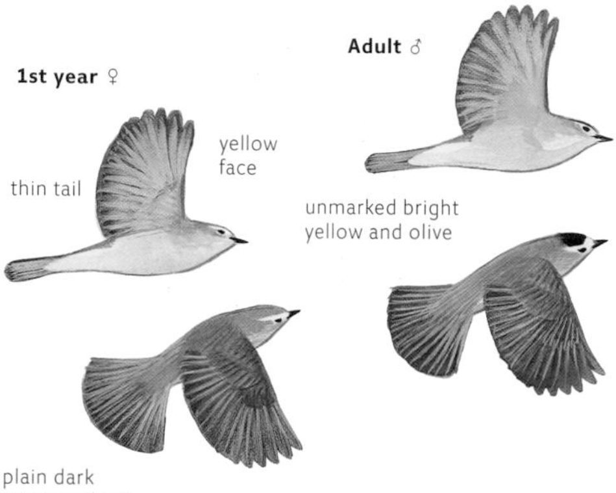

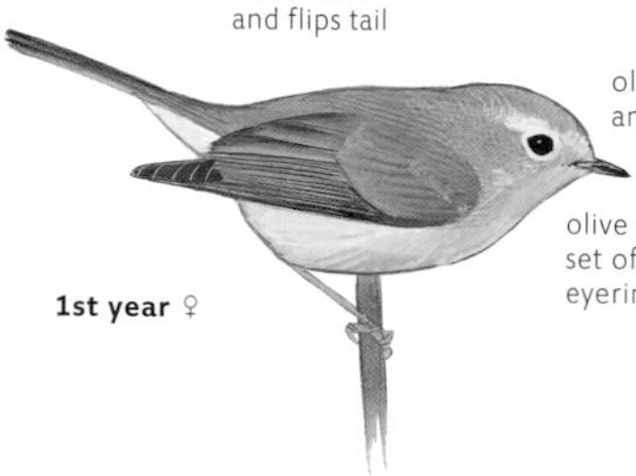

VOICE: Song a rapid series of 10–15 short, whistled notes *chchchchchchchch* with sharp, chattery quality; last few notes usually lower and faster, generally downslurred, sometimes two-syllable. Trill slower than Orange-crowned, notes sharp and clear. Call a husky, sharp *jimp* or *jip* like Pacific Wren. Flight call a clear, abrupt *tilk*.

Common in West, uncommon in East. Nests in wet, sunny shrub thickets such as willow or alder; migrants choose similar settings, but can be found in any wooded or brushy habitat.

Two populations shown are distinctive, but an Interior West population is intermediate in plumage, and therefore few are identifiable in the field. Pacific has brighter golden-yellow on its face and breast and always bright yellow on its forehead and auriculars. Taiga is paler lemon-yellow overall with darker olive auriculars and forehead; 1st year female Taiga can be very drab-faced. Song of Taiga population may average a little sweeter, with slightly longer slurred notes (Pacific birds with sharper, more abrupt slurs) and call of Taiga population may be slightly shorter and sharper than Pacific and Interior West, but all populations are variable, and any vocal differences are very subtle.

Red-faced Warbler

Cardellina rubrifrons

L 5.5" WS 8.5" WT 0.34 oz (9.8 g)

Fairly long and slender, with long tail often raised and flipped up and sideways. Plumage unmistakable: gray back, white rump, black crown and cheek, and red face. Also note relatively short, thick bill.

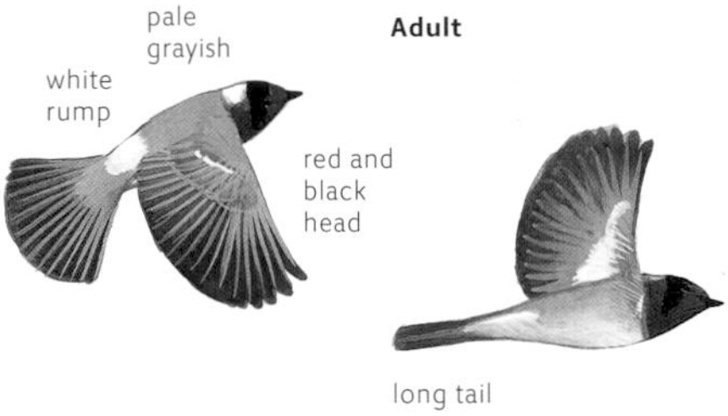

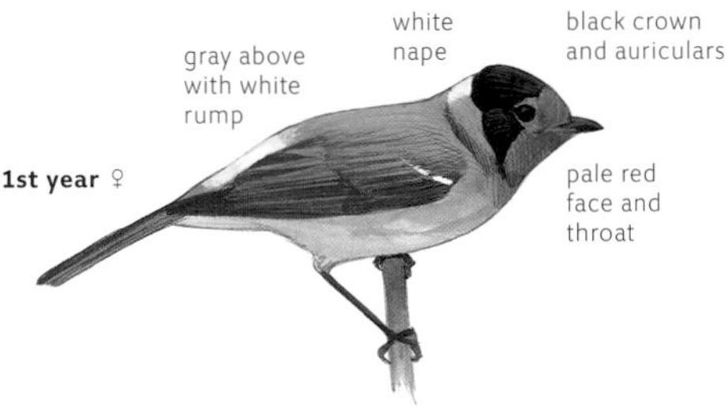

VOICE: Song high, thin, sweet *towee towee towee tsew tsew wetoo weeeeew* with overall descending trend. Call a sharp *tuk* like Canada. Flight call high, sharp, slightly rattling.

Uncommon in shaded canyons along streams within montane pine-oak and fir forests, often where maples grow in canyon bottom.

Canada Warbler

Cardellina canadensis

L 5.25" WS 8" WT 0.36 oz (10.3 g)

A relatively small warbler with long thin tail. Note plain gray upperside, complete white eyring, yellow supraloral, and necklace of black or gray streaks on breast. Similar to drab Magnolia Warbler, but note all-gray tail, bold eyering, plain gray wings.

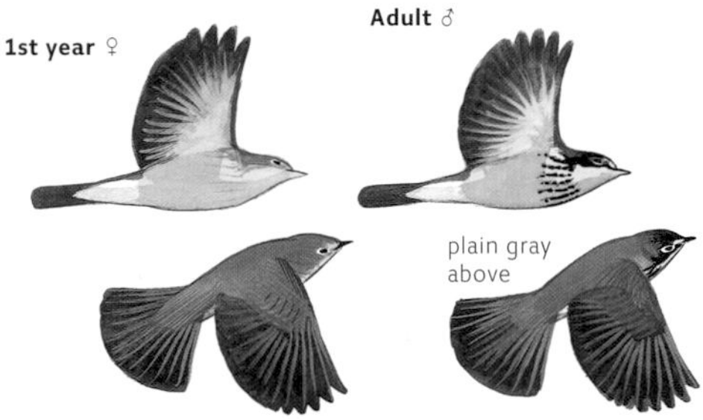

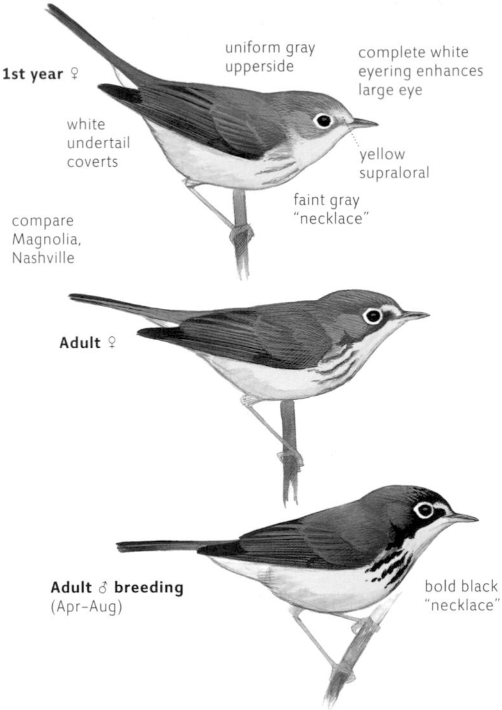

VOICE: Song of high, clear, liquid notes; varied: sputtery, descending, and ending loudly; all notes different; tempo sometimes suggests Common Yellowthroat, but erratic with scattered, sharp chips inserted. Call a sharp, dry, slightly squeaky *tyup*. Flight call a sharp, smacking, upslurred *chwit*.

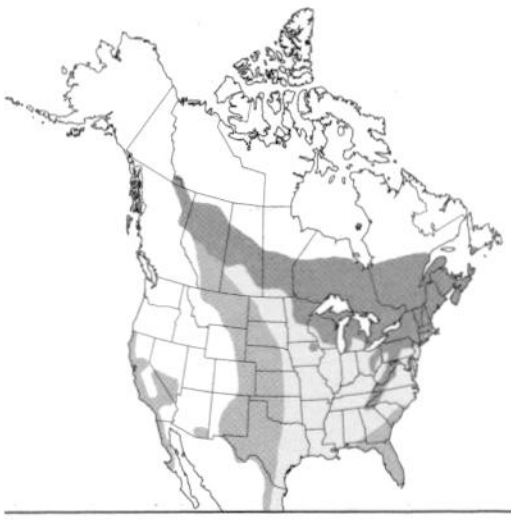

Uncommon in shaded deciduous undergrowth within mature forests, often along streams or in low areas.

Painted Redstart

Myioborus pictus

L 5.2" WS 8.75" WT 0.28 oz (8 g)

Unmistakable, very active and acrobatic, constantly fanning and flicking wings and tail to show off large white patches. Mostly black and white, with small but obvious white mark below eye, and bright red belly.

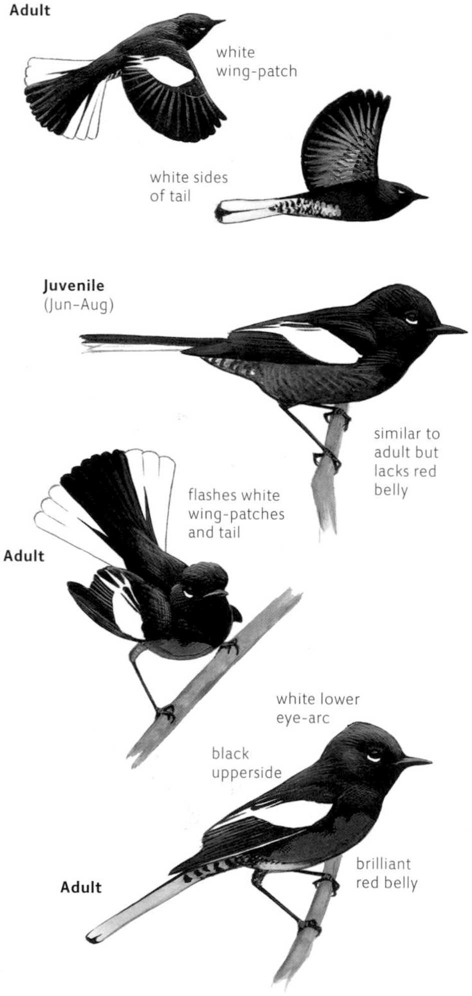

VOICE: Song a relatively low, soft, musical warble *che-wee che-wee weeta weeta witi wi*; phrases given in pairs but in run-on tempo and with little difference between phrases. Call a loud, relatively low *chidi-ew* or *bdeeyu* reminiscent of Pine Siskin; very dissimilar to other warblers. Juvenile gives very high, thin, weak, slightly descending *tsee*.

Common in oak and pine-oak woods along mountain canyons.

Slate-throated Redstart

Myioborus miniatus

L 5.4" WS 8.75" WT 0.4 oz (11 g)

Similar to Painted Redstart but wings all-dark, white on corners of tail (not sides), overall dark gray (not black), and less intense red on belly.

VOICE: Song a rapid series of sharp clear whistles, reminiscent of Yellow Warbler song but faster and weaker, much lower and faster than the song of Painted Redstart. Call a very high weak *tsit*, very different from Painted Redstart call.

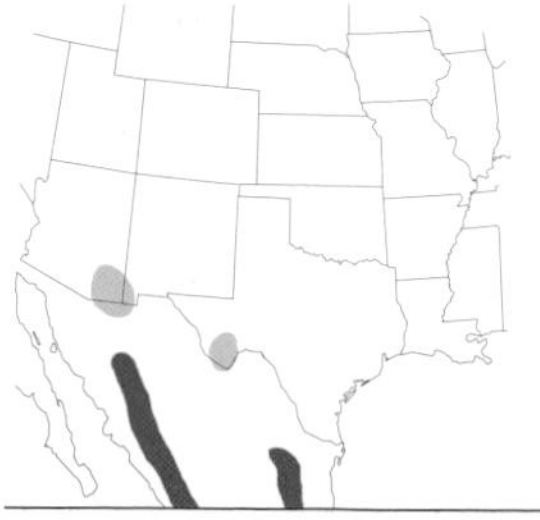

Very rare visitor from Mexico to montane pine-oak forest in Arizona, New Mexico, and Texas. Recorded mainly in spring.

Rufous-capped Warbler

Basileuterus rufifrons

L 5.25" WS 7" WT 0.39 oz (11 g)

A small warbler with large head, stout dark bill, long and oddly slender tail often held up nearly vertical. Bright yellow throat and breast, dark eyeline, and rufous on crown and cheek distinctive.

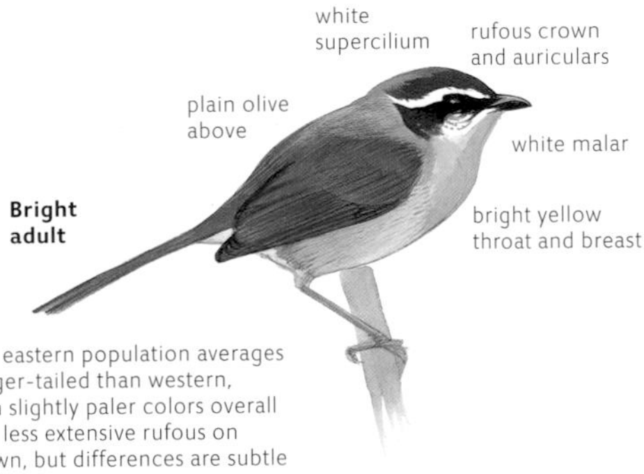

The eastern population averages longer-tailed than western, with slightly paler colors overall and less extensive rufous on crown, but differences are subtle and overlapping.

VOICE: Song a rapid series of chips with jumbled tempo, often accelerating; reminiscent of Rufous-crowned Sparrow. Call a hard *chik* often in series; also a high *tsik*.

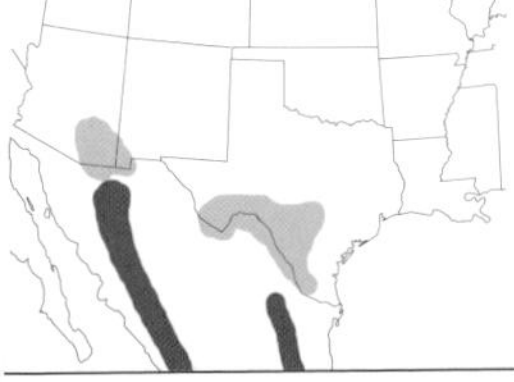

Very rare visitor from Mexico to Arizona, and to southern and western Texas; has nested. Found in sunny brushy habitat with oaks in foothill canyons, usually near water. Stays low in dense vegetation with long tail raised and flicked sideways, wren-like.

Golden-crowned Warbler

Basileuterus culicivorus

L 5" WS 7.5" WT 0.37 oz (10.5 g)

Plain drab plumage, grayish upperside with dull yellow-olive underside recalls Orange-crowned Warbler, but note stout bill, dark crown-stripes, different call.

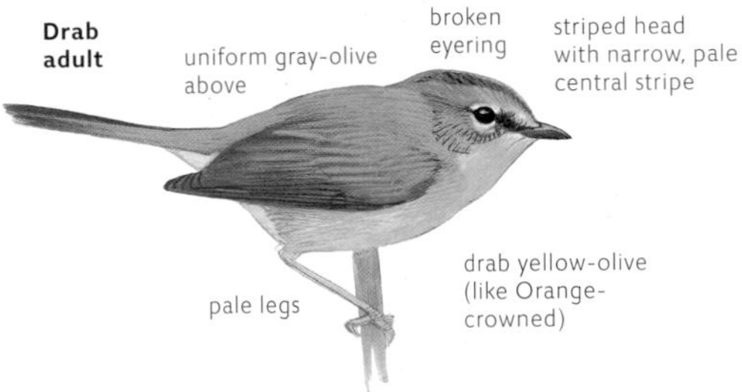

VOICE: Song of clear, slurred whistles reminiscent of Hooded but not quite as emphatic. Call a hard, dry *tek*, much like Ruby-crowned Kinglet but more musical; also a long, loose rattle.

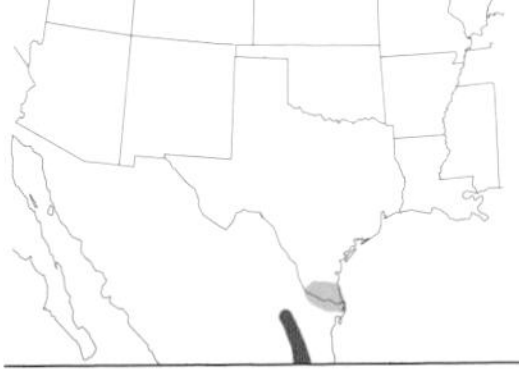

Very rare visitor from Mexico; only a few records in extreme southern Texas, one in New Mexico. Found in low brushy vegetation in woods, often in brush adjacent to wet areas with dense grass. Joins mixed foraging flocks of other small songbirds.

Fan-tailed Warbler

Basileuterus lachrymosus

L 5.8" WS 8.5" WT 0.4 oz (11 g)

Fairly large, with very long and broad tail, which is expressively fanned and raised, flipped sideways, and pumped down during foraging. Note uniform gray upperside and wings, tawny yellow underside, white spots around eye, and white tips on nearly all tail feathers.

Adult

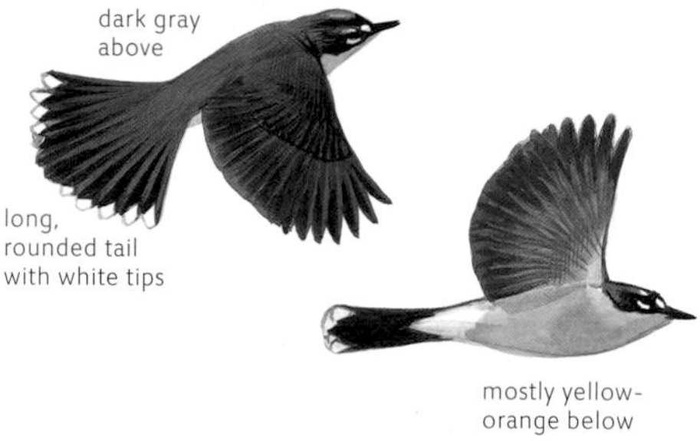

Drab adult

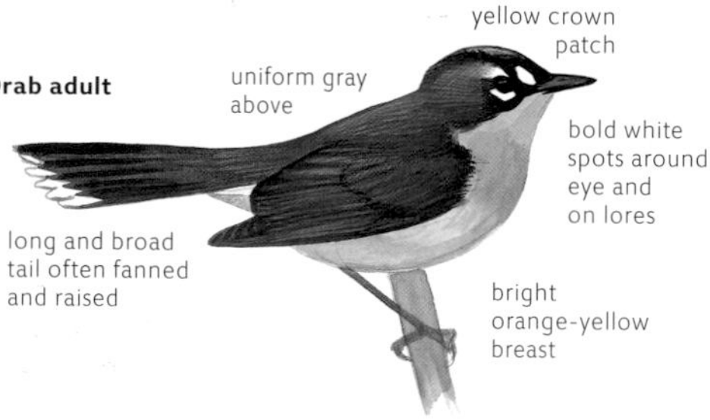

Bright adult

VOICE: Song a simple series of two-syllabled whistled phrases, sweet and gentle sounding *tu hEEder hEEder hEEder heder hedeer*, the last note lower and downslurred. Call a long, thin, high, descending *tseeer.*

Very rare visitor in spring and summer from Mexico to Arizona, New Mexico, and Texas. Found in oak woods along foothill streams, often with open and rocky ground interspersed with patches of brush. Forages mainly on the ground, chasing insects.

Taxonomic Outliers

A number of species with anomalous features have been vexing taxonomists for centuries. Recent DNA studies have solved some of those questions and revealed other misconceptions in the taxonomy, but there are still mysteries to be solved.

Yellow-breasted Chat has been considered a wood-warbler (family Parulidae), but always recognized as an outlier—it is much larger and thicker-billed than other warblers, and with a unique voice. DNA studies show it to be an early branch off of the group that gave rise to wood-warblers, sparrows, and icterids (blackbirds and orioles). Whether it is an offshoot of one of those families, or diverged sometime before those families split, is still unknown (and it is still "on hold" in the wood-warbler family). In any case it is a unique species distantly related to all three of those families.

Olive Warbler was, understandably, first classified as a wood-warbler, but decades ago, taxonomists recognized that it was distinctive and many placed it instead in the unrelated family of Old World warblers. DNA evidence shows that it is related to neither. It is the only species in another family (Peucedramidae), with closest relatives being the accentors of Eurasia (one species is a vagrant to North America).

Sparrow-like birds have been difficult for taxonomists to sort out. The features of a conical seed-cracking bill and streaked plumage are shared by many unrelated species. It is only recently, with biochemical studies, that the divergence between longspurs (family Calcariidae), finches (family Fringillidae), sparrows (family Emberizidae), and cardinals and grosbeaks (family Cardinalidae) has become clear.

Fringillidae

Pine Siskin

Cardinalidae

Indigo Bunting

Emberizidae

Chipping Sparrow

Calcariidae

Smith's Longspur

Longspurs and Snow Bunting, very sparrow-like in appearance and historically considered part of the family Emberizidae, have now been placed in a new family Calcariidae. This family apparently diverged before the whole group of wood-warblers, sparrows, and blackbirds, and therefore now appears before all of those species in the sequence.

Changes like this might be an unwelcome surprise to some readers of bird guides, but it is important to understand that they are based on the best interpretation of careful research. More importantly, each of these discoveries reveals interesting connections and dissimilarities in the species involved. For example, now that DNA has clarified the distinction between longspurs and sparrows, it is easier to appreciate just how different they are.

Yellow-breasted Chat

Icteria virens

L 7.5" WS 9.75" WT 0.88 oz (25 g)

Traditionally classified with warblers, even though much larger, with stout bill and different voice. Note long tail, heavy gray bill, plain gray-olive upperparts, white spectacles, and bright yellow throat and breast.

VOICE: Song extremely varied: mostly simple notes repeated in slow, decelerating series or given singly but with long pause between each utterance; distinctive low, liquid whistled or harsh, rasping quality *toop-toop-toop-toop toop toop toop*; *chook*; *terp*; *jedek*; *chrrr chrrr chrrr chrrr chrrr...*; slow series and long pauses distinctive. Often includes mimicry of other species and even mechanical sounds such as woodpecker drumming. Song of Western higher-pitched with more rapid rattle than Eastern (most rapid trill 20 notes/sec vs. 10 notes/sec for Eastern). Call a harsh, nasal *cheewb*; also a low, soft, unmusical *tuk* or *ka*.

Western populations differ slightly in appearance and song, but many birds are intermediate. Western averages slightly longer-tailed than Eastern and slightly grayer above, with mostly white malar (vs. mostly yellow). Western birds also average deeper yellow-orange on the throat and breast, but some individuals of all populations acquire partly or completely orange throat (presumably diet-related as in Cedar Waxwing variant with orange-tipped tail).

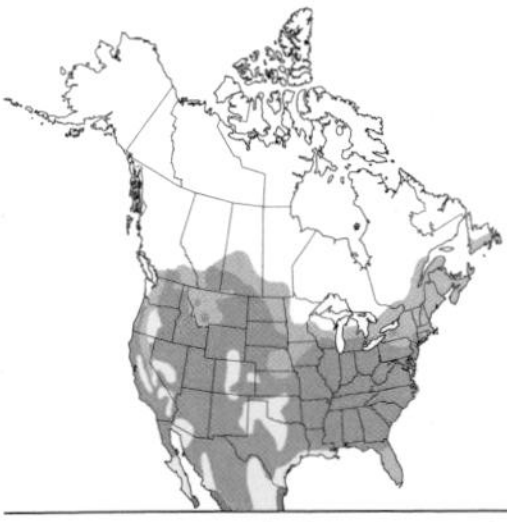

Uncommon and secretive in dense tangled brushy patches, hedgerows and woods edges, in open sunny areas. Usually stays hidden in vegetation, but singing males perch on higher branches and occasionally perform flight display.

True Tanagers

FAMILY: THRAUPIDAE

3 species in 3 genera. This is a large and diverse neotropical family represented here by two rare visitors and one species rarely escaped from captivity. Western Spindalis and Bananaquit are Caribbean species with uncertain affinities, only distantly related to other species in the family. They are apparently early branches off of the group, each perhaps distinct enough to be placed in separate families. North American tanagers are now placed in the cardinal family. First-winter females are shown.

Genus *Spindalis*

Western Spindalis, page 503

Genus *Coereba*

Bananaquit, page 502

Red-crested Cardinal

Paroaria coronata

L 7.1" WS 11" WT 0.7 oz (20 g)

Actually related to true tanagers in the family Thraupidae, not to the familiar Northern Cardinal. Pale gray and white body, bright red head and crest, and pale bill distinctive.

VOICE: Song of rich whistled phrases, somewhat robin-like, repeated alternately with fairly long pauses between.

Native to South America. Occasionally seen as escapes from captivity in southern cities. Found in shrubby habitat, forages mainly on the ground.

Bananaquit

Coereba flaveola

L 4.5" WS 7.75" WT 0.33 oz (9.5 g)

Warbler-sized, with short broad tail and strongly decurved bill. Yellow rump, mostly white underparts and broad white eyebrow distinctive. Juvenile is much duller but still easily identified by shape, pale eyebrow, yellowish rump.

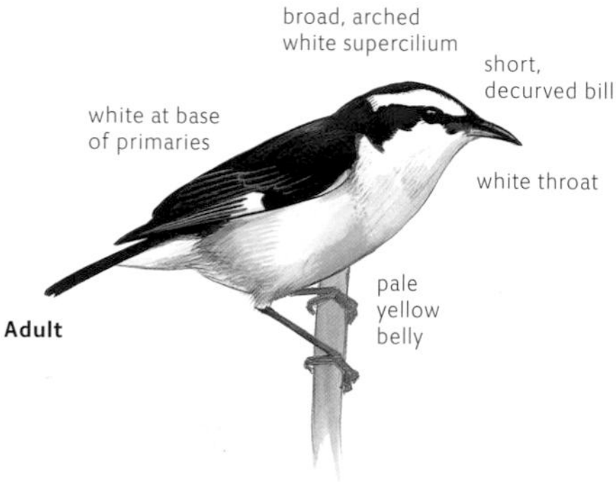

VOICE: Song of high, hissing squeaks and buzzes that sound as if they are squeezed out with great effort *ezeereezee eyteer eyteer sizit zet*; ends with dry, insectlike crackling. Call quite warbler-like: a slightly rising, metallic *ssint*.

Very rare visitor from Bahamas to extreme southeast Florida; recorded less than annually. Usually found in brushy coastal woods near flowers where it feeds, drinking nectar and eating insects or fruit. Usually solitary.

Western Spindalis

Spindalis zena

L 6.75" WS 9.5" WT 0.74 oz (21 g)

Smaller than other tanagers, relatively long-tailed and small-billed. Male distinctive; female drab olive overall, note stout grayish bill, pale-edged wing feathers, suggestion of striped head pattern, and size (larger than warblers).

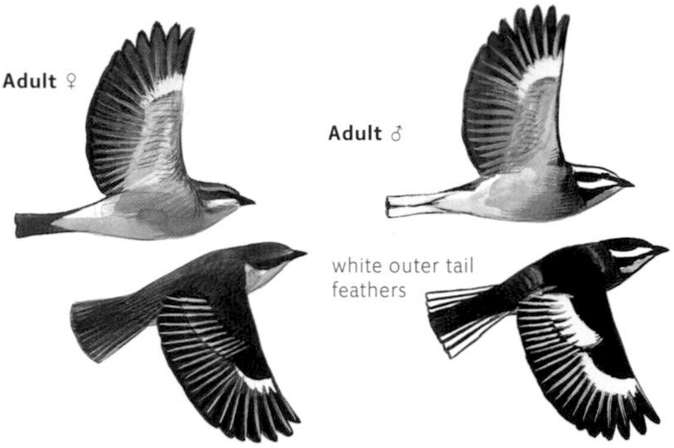

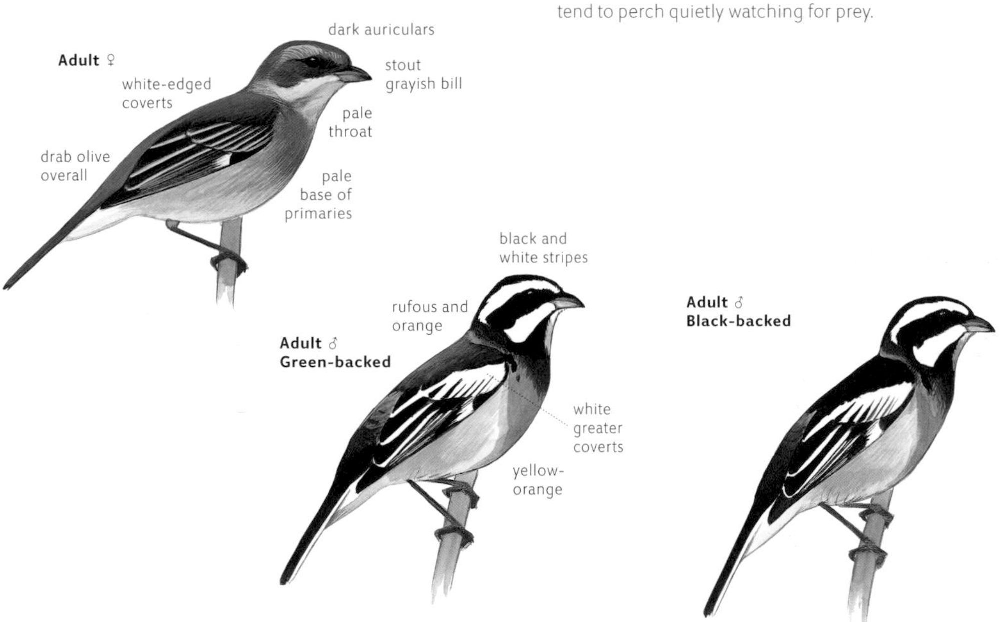

VOICE: Song a series of high, thin notes beginning like Black-and-white Warbler but changing to buzzier phrases. Calls varied but all very high and thin: a series of descending notes *see see see see see*; a rather strong *seeee*; rapid, high twittering; very high, sharp *tit*.

Very rare visitor from the Bahamas; recorded annually in brushy wooded areas in southeast Florida where it has nested. Found in small groups in normal range, but usually solitary in eastern North America. Feeds on insects, larvae, and fruit.

Bill Shapes of Songbirds

The two species shown here both have distinctive bill shapes: Western Spindalis feeds mainly on small fruit, and Bananaquit mainly on nectar and insects. Bill shape is highly consistent within a species, and therefore very useful for identification of songbirds. It quickly places a species in a broad group, and is directly related to the typical diet and foraging habits of each species. Below are some very broad generalizations; with experience you will be able to distinguish very subtle differences in the relative thickness, taper, and pointedness of bills.

- Stout, conical bills indicate a bird that eats primarily seeds. These species are relatively sedate and often gather in flocks.
- Thin, pointed bills indicate a bird that eats primarily small insects and larvae. These species are usually very active, flitting quickly from perch to perch, and do not form flocks.
- Intermediate bill shapes, often blunt-tipped, indicate a bird that eats fruit and slow-moving invertebrates, such as worms or insect larvae. These species are not very active, and form flocks mainly when eating fruit.
- Broad, flattened bills indicate a bird that eats insects captured in flight. These species are always solitary, and tend to perch quietly watching for prey.

Most males in North America have been of the black-backed form from the central and southern Bahamas. Green-backed males predominate in the northern Bahamas and have been recorded rarely in Florida. Some records of Green-backed males (e.g. at Key West) may represent the Cuban subspecies.

Emberizine Sparrows

FAMILY: EMBERIZIDAE

48 species in 20 genera (3 rare species not shown here). Emberizidae are mainly ground-dwelling, often secretive, brownish and streaked with short, conical bills. All feed on seeds in winter and mainly insects in summer. Some species use both feet simultaneously (a hop-scratch) to kick leaves back and uncover food. Nest is a cup in a low bush or on the ground. Some species form large mixed flocks in grassy or weedy areas in winter, other species have very specific habitat preferences and do not flock. Longspurs and Snow Bunting are now placed in a separate family Calcariidae. First-winter females are shown.

Genus *Pipilo*

Green-tailed Towhee, page 507

Spotted Towhee, page 508

Eastern Towhee, page 509

Genus *Melozone*

Canyon Towhee, page 510

California Towhee, page 510

Abert's Towhee, page 511

Genus *Sporophila*

White-collared Seedeater, page 506

Genus *Arremonops*

Olive Sparrow, page 507

Genus *Aimophila*

Rufous-crowned Sparrow, page 511

Genus *Pooecetes*

Vesper Sparrow, page 521

Genus *Peucaea*

Rufous-winged Sparrow, page 512

Cassin's Sparrow, page 512

Bachman's Sparrow, page 513

Botteri's Sparrow, page 513

Genus *Spizella*

American Tree Sparrow, page 515

Chipping Sparrow, page 517

Clay-colored Sparrow, page 517

Brewer's Sparrow, page 516

Field Sparrow, page 515

Black-chinned Sparrow, page 514

Genus *Chondestes*

Lark Sparrow, page 536

Genus *Amphispiza*

Five-striped Sparrow, page 519

Black-throated Sparrow, page 519

Genus *Artemisiospiza*

Sagebrush Sparrow, page 518

Bell's Sparrow, page 518

Genus *Calamospiza*

Lark Bunting, page 536

Genus *Passerculus*

Savannah Sparrow, page 520

Genus *Melospiza*

Song Sparrow, page 528

Lincoln's Sparrow, page 529

Swamp Sparrow, page 529

Genus *Ammodramus*

Grasshopper Sparrow, page 522

Baird's Sparrow, page 522

Henslow's Sparrow, page 523

Le Conte's Sparrow, page 523

Nelson's Sparrow, page 524

Saltmarsh Sparrow, page 524

Seaside Sparrow, page 525

Genus *Passerella*

Fox Sparrow, page 526

Genus *Junco*

Dark-eyed Junco, page 530

Yellow-eyed Junco, page 532

Genus *Zonotrichia*

White-throated Sparrow, page 535

Harris's Sparrow, page 533

White-crowned Sparrow, page 534

Golden-crowned Sparrow, page 535

Genus *Emberiza*

Rustic Bunting, page 537

White-collared Seedeater

Sporophila torqueola

L 4.5" WS 6.25" WT 0.32 oz (9 g)

Very small, with relatively large and stout bill. Drab females most similar to Lazuli Bunting (also goldfinches) distinguished by small size, large bill with curved culmen, shorter, rounded tail, overall plain buffy-olive color.

VOICE: Song of high, clear, sweet whistles in series *sweet sweet tew tew tew tew sit sit*; reminiscent of Yellow Warbler but simpler, thinner, with jingling quality. Common call a husky, rising *quit* or *quitl* or hard, rising *dwink* like Bewick's Wren; also a high, clear, descending whistle *cheew.*

Rare and local resident of extreme southern Texas, mostly associated with riverside patches of giant cane. Usually in small groups of a few birds, forages mainly on grass seeds. A few records of the brighter west Mexican subspecies in California, Arizona, and southern Texas are presumed to represent escaped cage birds.

Yellow-faced Grassquit

Tiaris olivaceus

L 4" WS 6" WS 0.3 oz (8.5 g)

Tiny, with stout bill and short, rounded, dark tail. All plumages dark gray and olive overall, with yellow chin and eyebrow. Male has black breast. Small size, plain dark wings and tail, and yellow patches on face distinctive.

VOICE: Song a short, very high-pitched trill on steady pitch, sometimes weak and insect-like, sometimes reminiscent of Chipping Sparrow song, but shorter and higher. Call a high sharp *tsik.*

Very rare visitor from Mexico to southern Texas and from the West Indies to southern Florida. Found in open weedy fields, hedgerows, brushy edges. Some records could refer to escaped cage birds. The similar Cuban Grassquit has also been recorded in Florida, presumed escaped from captivity.

Black-faced Grassquit

Tiaris bicolor

L 4" WS 6" WS 0.3 oz (8.5 g)

Tiny and plain. Male distinctive with black face and underparts, green back. Female resembles Indigo Bunting, but smaller, with more olive-drab plumage, no pattern on wing coverts, relatively large and dark bill.

VOICE: Song several low introductory notes followed by a pulsing high scratchy buzz *pik, pik, pik-tzizzizzizzizzizz*. Call a very sharp dry *tsik.*

Very rare visitor from the West Indies to extreme southern Florida. Found in dense brushy habitat, hedgerows, weedy edges, can be secretive and hard to see.

Olive Sparrow

Arremonops rufivirgatus

L 6.25" WS 7.75" WT 0.84 oz (24 g)

A small but stocky sparrow, with very plain coloring: drab olive above and unstreaked grayish and cream-colored below with dull brown crown-stripes. Smaller than Green-tailed Towhee, without reddish cap or contrasting white throat.

VOICE: Song an accelerating series of sharp, hard chips ending with fading trill *tsip tsip tsiptsiptsiptiptiptptptptp* (see Botteri's Sparrow). Call a very high, sharp, clicking *stip*. Flight call a high, thin buzz *seere*; towheelike but plainer and thinner.

Common in patches of dense grass and brush within open woods; forages mainly on the ground. Usually in pairs. Secretive and difficult to see, usually stays within forest and is detected by sound.

Green-tailed Towhee

Pipilo chlorurus

L 7.25" WS 9.75" WT 1 oz (29 g)

Slightly smaller than other towhees. Note dark gray head and breast with strongly contrasting white throat, rufous cap, and fairly bright greenish wings and tail.

VOICE: Song typically several short introductory notes followed by two or more trills *tip seeo see tweeeee chchchch* or *tlip tseet-seetsee tlitlitli chrrrr.* Call a mewing, nasal *meewe*. Flight call a long, thin buzz *zeereesh*; rougher and more level than other towhees.

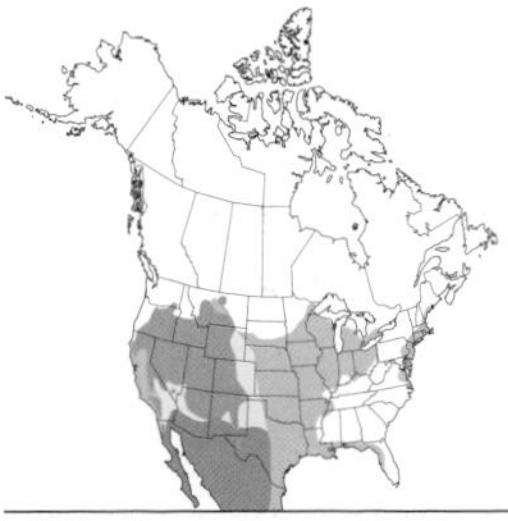

Common in arid brushy habitats. Nests in sagebrush and associated dense shrubs. Solitary. Forages mainly for seeds on the ground.

Spotted Towhee

Pipilo maculatus

L 8.5" WS 10.5" WT 1.4 oz (40 g)

Larger and stockier than sparrows, with long tail. Note rufous flanks, all-dark head, dark tail with bright white corners. Habits and appearance similar to Eastern Towhee, and until recently considered the same species (Rufous-sided Towhee).

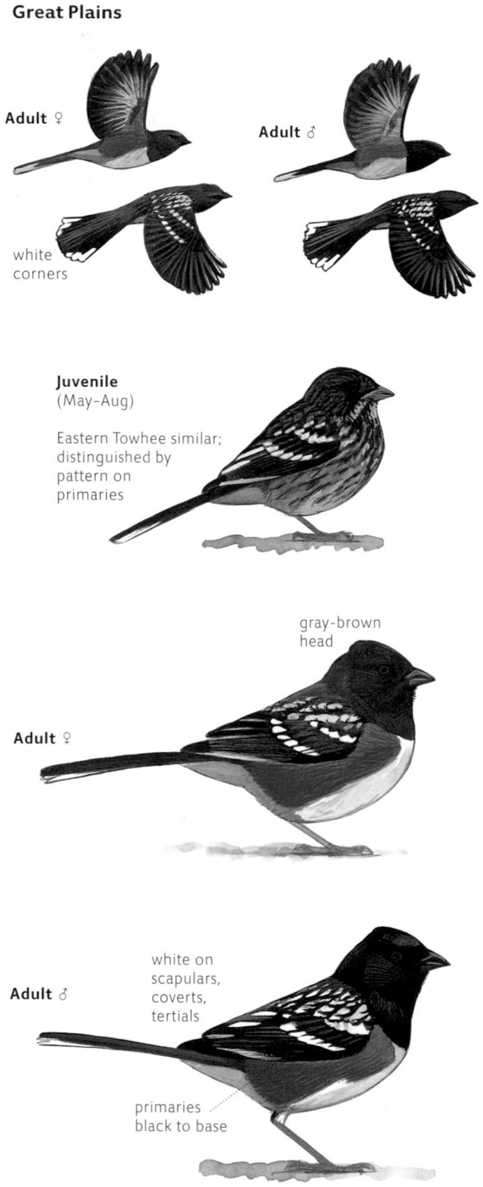

VOICE: Song varies geographically; zero to eight identical introductory notes followed by a harsh or buzzy, rapid trill; buzzier and less varied than Eastern; Pacific Northwest gives a simple buzzy trill *che zheeeee* or *chzchzchzchzchz*; Great Plains one to eight quick notes followed by a buzzy trill *che che che che zheeee*; Southwest similar to Great Plains but averages fewer introductory notes followed by slower trill. Calls of two distinct types: Pacific Northwest and Great Plains give a harsh, rising, growling *zhreeee*; Southwest a harsher, descending *grreeer*. Flight call a high, thin buzz *zeeeeweee* like Eastern.

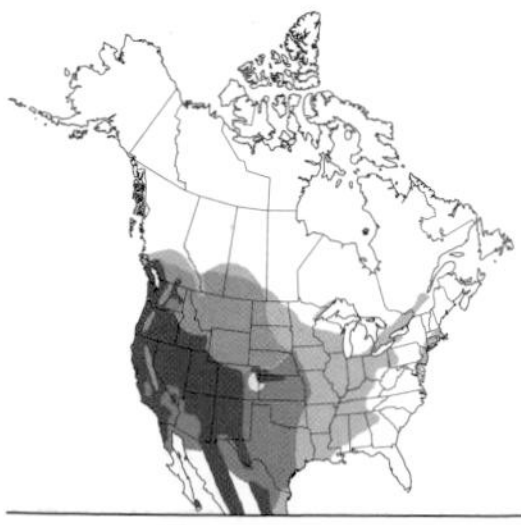

Common in brushy habitats, sunny clearings, and shrublands or brushy undergrowth within open forests, but fairly secretive and usually stays in cover. Solitary. Forages on the ground by scratching vigorously in leaf litter for seeds and insects.

Geographic variation in voice and plumage is complex and poorly understood. There are two distinct call types but less distinct differences in song and plumage. More study is needed. Great Plains is palest, with most extensive white markings; Pacific Northwest is darkest. Great Plains females are gray-brown; other females are very dark, similar to males.

Eastern Towhee

Pipilo erythrophthalmus

L 8.5" WS 10.5" WT 1.4 oz (40 g)

Habits and appearance similar to Spotted Towhee. Note all-black scapulars and wing coverts on male, more brownish upperside of female, and white patch at base of primaries on both sexes; also note differences in voice.

White-eyed is found in the Southeast, with a clinal transition to Red-eyed in the North. It averages less white in the tail and gives a simpler upslurred *zwink* call and more variable song than Red-eyed. Also note that some individuals on the outer banks of North Carolina give a hoarse *merrre* call like Southwest Spotted Towhee.

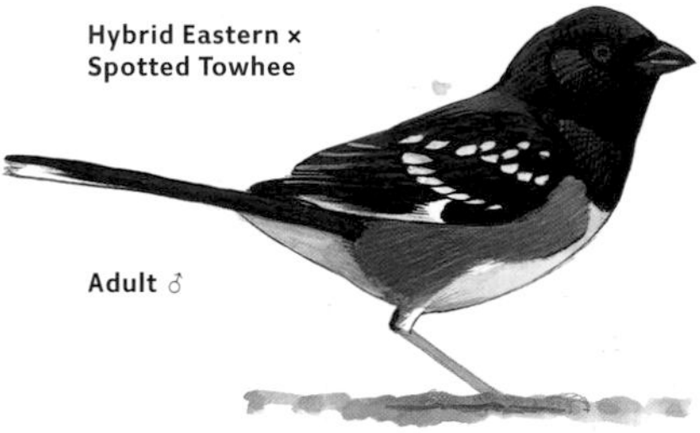

VOICE: Song typically one to three short, husky introductory notes followed by slow, musical trill *jink denk te-e-e-e-e-e-e* ("*drink your teeeea*"); much variation in details. Call typically a strongly rising *chewink* or *zhwink* with husky, nasal quality. Flight call a long, thin buzz *zeeeeweee.*

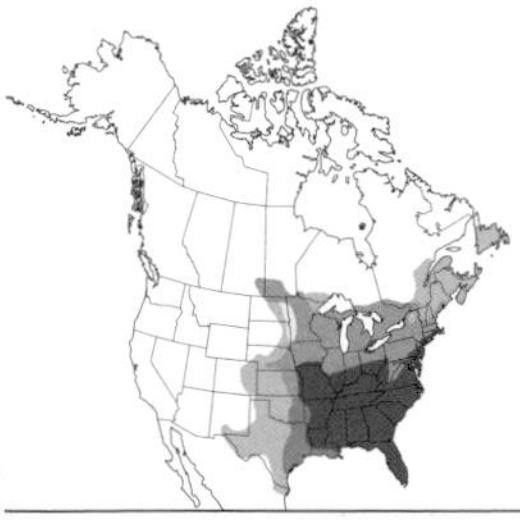

Common in brushy habitats, sunny clearings, and shrublands or brushy undergrowth within open forests, secretive and usually stays hidden in vegetation. Solitary. Forages on the ground by scratching vigorously in leaf litter for seeds and insects.

These two species hybridize regularly where range overlaps in northern Great Plains. Hybrids show pale spotting on the back and wing coverts (like Spotted Towhee but reduced), with a white patch at the base of the primaries (like Eastern but reduced). Call can be like either parent species, or blended. Beware that juvenile Eastern Towhee has pale tips on the wing coverts, forming wingbars, and these feathers can be retained rarely into early winter, suggesting a hybrid.

California Towhee

Melozone crissalis

L 9" WS 11.5" WT 1.5 oz (44 g)

Habits and appearance very similar to Canyon Towhee, but no range overlap. Note overall dark brown color and cinnamon markings on face, but only faint streaking across breast and throat, much higher call.

VOICE: Song an accelerating series of high, flat *teek* notes, sometimes with lower notes near end: *teek teek teek eek eekeekeekeek t-t-t-t-teek*. Call a high, hard, flat *teek*. Flight call a high, buzzy *zeeeee*.

Common. Found in dense brush including chapparal, brushy patches in open woods, parks, or suburbs. Usually in pairs. Forages mainly on the ground, often seen in the open on roadsides, fenceposts, lawns.

Canyon Towhee

Melozone fusca

L 9" WS 11.5" WT 1.9 oz (53 g)

Similar to Abert's Towhee, but usually segregated by habitat, and with very different call. Range does not overlap with California Towhee. Note reddish crown and necklace of dark streaks, dark gray bill, distinctive nasal or tinny call.

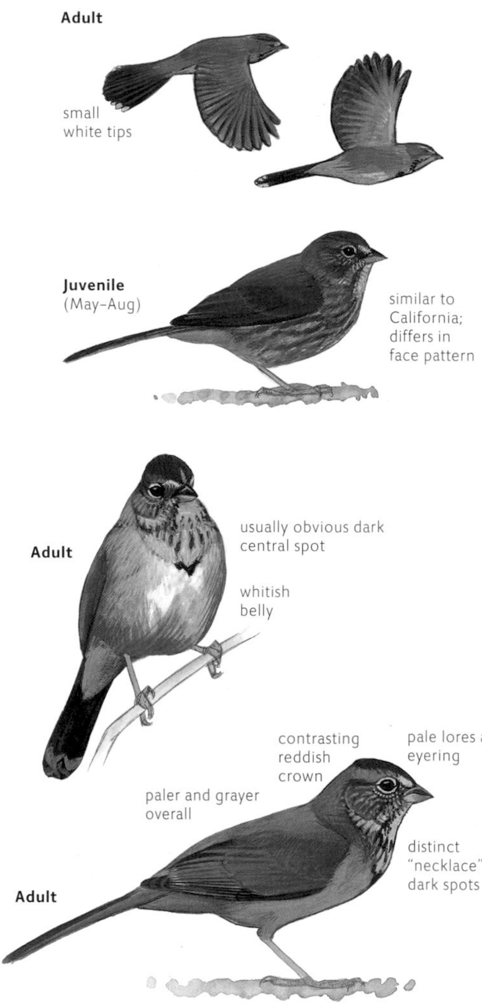

VOICE: Song a simple, methodical, slow trill of short, whistled notes introduced by call note *kild ti ti ti ti ti ti ti kil* or *kild tiwi tiwi tiwi tiwi tiwi*; lower than California and Abert's. Call a nasal, dry *kidl* or loud, tinny *kilt*; also a dry, clicking *ch-ch-ch-ch*. Flight call a high, buzzy *zeeeee*.

Common in relatively open, arid areas with patches of brush, including suburban yards in desert scrub. Solitary or in pairs. Feeds on seeds, fruit, and insects.

Abert's Towhee

Melozone aberti

L 9.5" WS 11" WT 1.6 oz (46 g)t

Similar to Canyon Towhee, but little range overlap. Note uniform pinkish-brown color and pale bill contrasting sharply with blackish face. Call is also very different from Canyon Towhee.

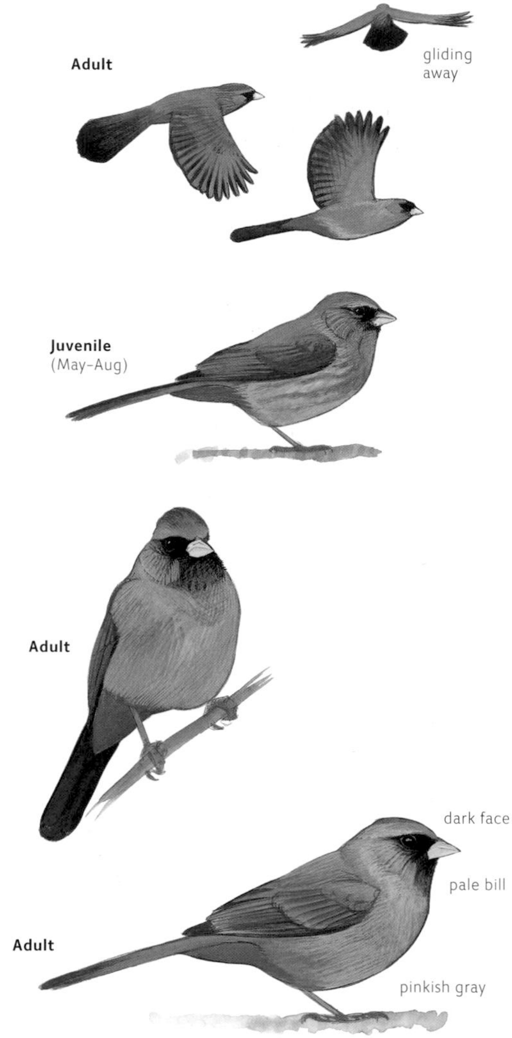

VOICE: Song of high, flat notes followed by an accelerating jumble of stuttering, harsh notes *tuk teek teek teek chi-chi-chi-chi-kll*. Call a high, flat *teek* lower than California; also a very high, clear *seeeep*. Flight call a high buzz *zeeoeeet*.

Common in dense riparian brush, mostly at lower elevations than Canyon Towhee, and tends to be more secretive. Solitary or in pairs. Forages mainly for seeds on the ground in dense brush.

Rufous-crowned Sparrow

Aimophila ruficeps

L 6" WS 7.75" WT 0.65 oz (18.5 g)

A rather large stocky sparrow. Note unpatterned gray breast, rufous crown and eyeline, and pale malar. Stocky shape, ground-dwelling habits, and plain breast may recall towhees, but note smaller size, throat pattern, and streaked back.

VOICE: Song a slightly husky, mumbled, descending chatter; reminiscent of House Wren in pattern and tempo but huskier, less gurgling. Call a nasal, laughing *deer deer deer deer*; also a long chatter and a sharp, high *zeeet*.

Uncommon on arid rocky hillsides with patches of shrubs and grass. Usually solitary or in pairs. Somewhat secretive, does not flock.

Rufous-winged Sparrow

Peucaea carpalis

L 5.75" WS 7.5" WT 0.53 oz (15 g)

Superficially similar to Chipping Sparrow, but heavier, with stout bill and rounded tail. Note pale rufous crown-stripes, clear grayish breast, and rather pale and neatly patterned overall appearance.

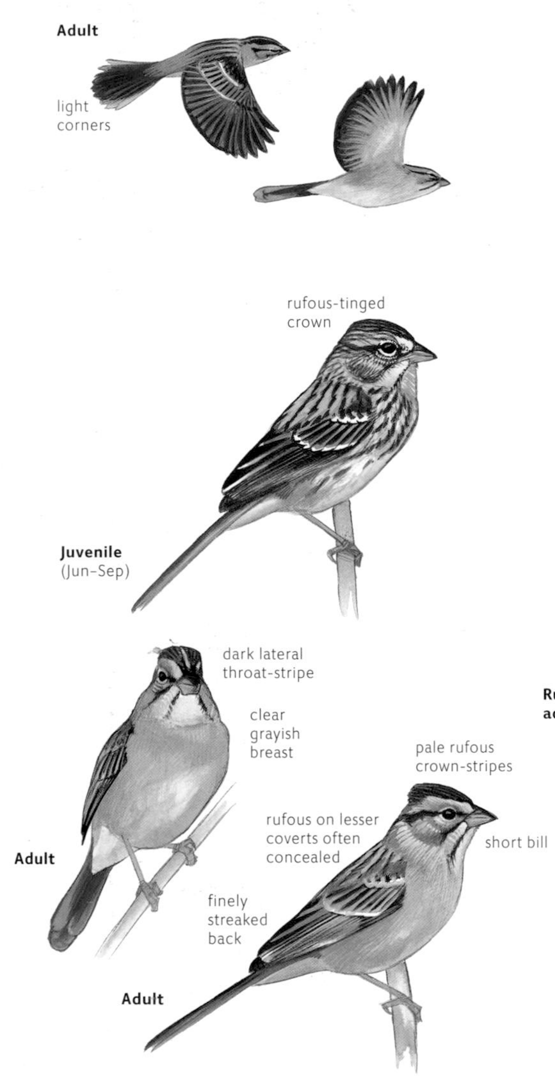

VOICE: Song variable in pattern but always sweet, clear notes; two common patterns: an accelerating series of chips *tip tip-tiptptptptp* and a simple *tip tip tee trrrrrrrrr*. Alarm call a very high, piercing *tsiddp*.

Uncommon and local in flat, low-elevation mesquite grasslands with sparse vegetation and some cactus. Usually in small groups, never large flocks, and usually does not join flocks of other sparrows.

Cassin's Sparrow

Peucaea cassinii

L 6" WS 7.75" WT 0.67 oz (19 g)

Size and shape similar to Song Sparrow. Usually paler and grayer overall and mostly unstreaked below, with unique speckled pattern on upperside; note whitish eyering. Compare Botteri's.

VOICE: Song given mainly in flight: four plaintive, trilled whistles with long, level second note *tsisi seeeeeeeee ssoot ssiit*. Call a very high, abrupt *teep*; also very high, sharp chips; sometimes a series of squeaks when agitated.

Common locally in arid grasslands with scattered shrubs, including mesquite, yucca, or cactus. Solitary and quite secretive, difficult to see except when males sing in flight display. Does not mix with flocks of other sparrows.

Botteri's Sparrow

Peucaea botterii

L 6" WS 7.75" WT 0.7 oz (20 g)

Size and shape similar to Cassin's Sparrow, but longer-billed and with flatter crown. Note entirely unmarked buffy-gray underside, strongly streaked upperside, plainer wings, and lack of eyering.

VOICE: Song a varied, slow series of sharp, whistled notes ending with accelerating trill *tik tik swidi trrr trik tidik tew tew twit-wititititttttttt*; each note in trill sharp and rising (vs. lower and descending notes of Olive Sparrow). Call a high, sharp chip or rapid chatter.

Uncommon and local in extensive dense grass with scattered mesquite, yuccas, or other low shrubs. Population in grassland of Texas coastal prairie is declining, now rare and very local. Solitary and very secretive, never flocks.

Bachman's Sparrow

Peucaea aestivalis

L 6" WS 7.25" WT 0.68 oz (19.5 g)

Note weakly patterned cheek and throat and unstreaked buffy color on breast and flanks contrasting with whitish belly. Similar to Botteri's and Cassin's Sparrows, but usually no range overlap.

VOICE: Song a simple, clear whistle with a musical trill on a different pitch; successive songs differ slightly in pitch and trill: *feeeee-trrrr, sooo-treee*.... Call of high *tsip* notes. When flushed may give piercing, sharp *tsees*; when agitated an extremely high-pitched *tsisisisisi*.

Uncommon and local in open pinewoods with patchy understory of grass, brush, and palmetto. Solitary and secretive; difficult to see except when singing. Does not form flocks, or join with flocks of other sparrows.

Black-chinned Sparrow

Spizella atrogularis

L 5.75" WS 7.75" WT 0.42 oz (12 g)

A small and slender sparrow, with long tail and small bill. Plain gray head and underparts, pink bill, and streaked brown back distinctive. Superficially similar to juncos, but smaller, with long all-dark tail, pale gray belly, streaked back.

VOICE: Song of high, sharp, slurred notes accelerating to rapid trill; higher and more mechanical than Field Sparrow; final trill usually rising. Call a high, weak *stip*. Flight call a soft *ssip* similar to Chipping Sparrow.

Uncommon and local on arid hillsides with dense patches of brushy vegetation such as chaparral. Quick to occupy new brushlands after fires. Usually in small flocks, not mixing with other sparrows.

Sparrow-like Birds

Many more or less unrelated species of small, brownish, streaked birds are frequently mistaken for Emberizine sparrows; some are shown below. In particular, note habits: many of these species fly high and strongly, perch in treetops, and frequently give calls in flight, all behaviors that are not typical of Emberizine sparrows.

Adult nonbreeding (Aug–Mar)

American Pipit (and other pipits)
Thin bill; walks on ground; strong flight with squeaky flight calls.

Adult ♀ breeding (Apr–Sep)

Indigo Bunting (and other *Passerina* buntings)
Rounded tail flicked and wagged; strong flight with buzzy calls.

Adult nonbreeding (Aug–Mar)

Bobolink
Pointed wings; strong flight with soft calls.

Adult ♀

Red-winged Blackbird
Larger than most sparrows; walks on ground; strong flight with sharp calls.

Adult ♀ Pacific

Purple Finch (and other finches)
Notched tail; strong flight with sharp calls.

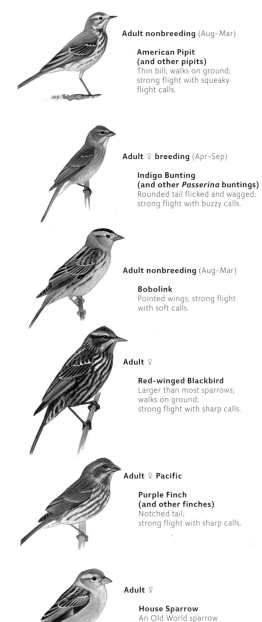

Adult ♀

House Sparrow
An Old World sparrow with stocky shape; chirping calls.

American Tree Sparrow

Spizella arborea

L 6.25" WS 9.5" WT 0.7 oz (20 g)

Larger than Chipping Sparrow, with larger bicolored bill, bold white wingbars, clear rufous crown, and dark spot on center of breast. Recent research suggest that this species is related to the genus *Zonotrichia*, not *Spizella*.

Western populations average slightly larger, paler, and grayer than eastern, but the color variations illustrated here can be found within each population. Note also that worn and faded birds are paler and grayer than fresh birds.

VOICE: Song a very sweet, clear, high warble with descending trend *swee swee ti sidi see zidi zidi zew*; slightly buzzy at end; occasionally heard in winter. Common call a unique soft, jingling *teedleoo*, forming a delicate chorus from winter flocks. Flight call a high, sharp *tsiiw*; similar to Field Sparrow but only slightly descending.

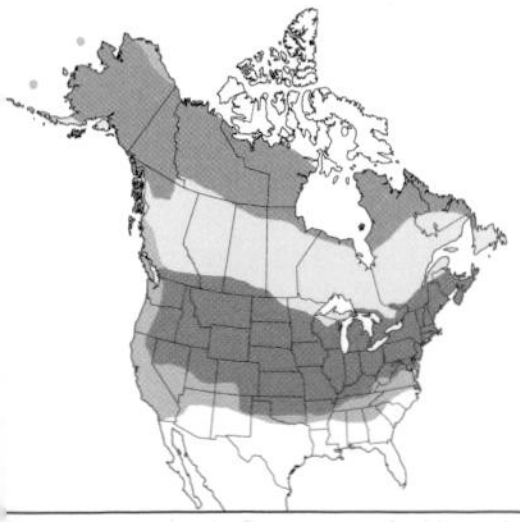

Common. Nests in open shrubby vegetation on tundra. Winters in any brushy or weedy habitat, often near trees. Forms large flocks in winter that wander freely through suitable habitat, sometimes mixing with other sparrows.

Field Sparrow

Spizella pusilla

L 5.75" WS 8" WT 0.44 oz (12.5 g)

Slightly larger and longer-tailed than Chipping Sparrow. Overall buff, rufous and gray plumage with complete white eyering and entirely pink bill distinctive.

Western populations average slightly paler, and grayer than eastern, but the color variations can be found within each population. Note also that worn and faded birds are paler and grayer than fresh birds.

VOICE: Song an accelerating series of soft, sweet whistles *teew teew tew tew tewtewtetetetetitititititi*. Call clear and rather strong like Orange-crowned Warbler; stronger than Chipping Sparrow. Flight call a distinctive, descending, clear *tseeew*.

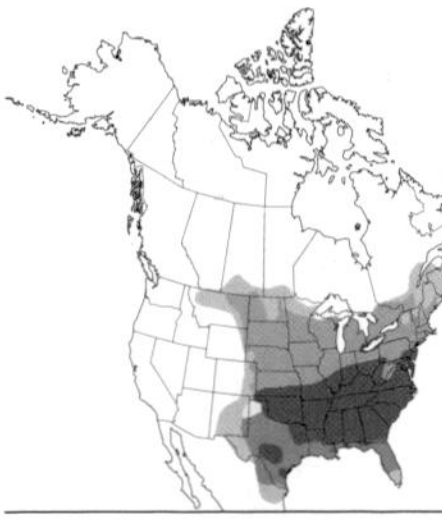

Common but declining in weedy fields with scattered bushes and trees, especially hedgerows. Usually forms small flocks separate from other sparrows. Individuals will join flocks of other sparrows, but do not mix freely.

Brewer's Sparrow

Spizella breweri

L 5.5" WS 7.5" WT 0.37 oz (10.5 g)

Similar to Clay-colored Sparrow; typically plainer and drabber, more uniform gray-brown on breast, head, and nape, with very fine and distinct dark streaks, and more conspicuous white eyering (some birds are intermediate).

Identification of *Spizella* Sparrows

Brewer's, Clay-colored, and Chipping Sparrows are distinctive in breeding plumage (April to August), but the patterns and colors of fall and winter birds (particularly immatures) can be very similar. Identification requires careful study of head pattern. Chipping is most distinctive, with a dark loral-stripe, but this may be faint; more useful is the fact that the face pattern is dominated by the dark eyeline (eyeline of other species is about as dark as other facial markings). Clay-colored is usually more colorful and more contrasting than Brewer's, but some drab Clay-coloreds can overlap with bright Brewer's, and rare individuals may be unidentifiable. Concentrate on the cleaner and more colorful face pattern of Clay-colored, with clean buffy supercilium, contrasting light malar, and buffy breast. Brewer's is less contrasting overall, patterned weakly in grayish-brown.

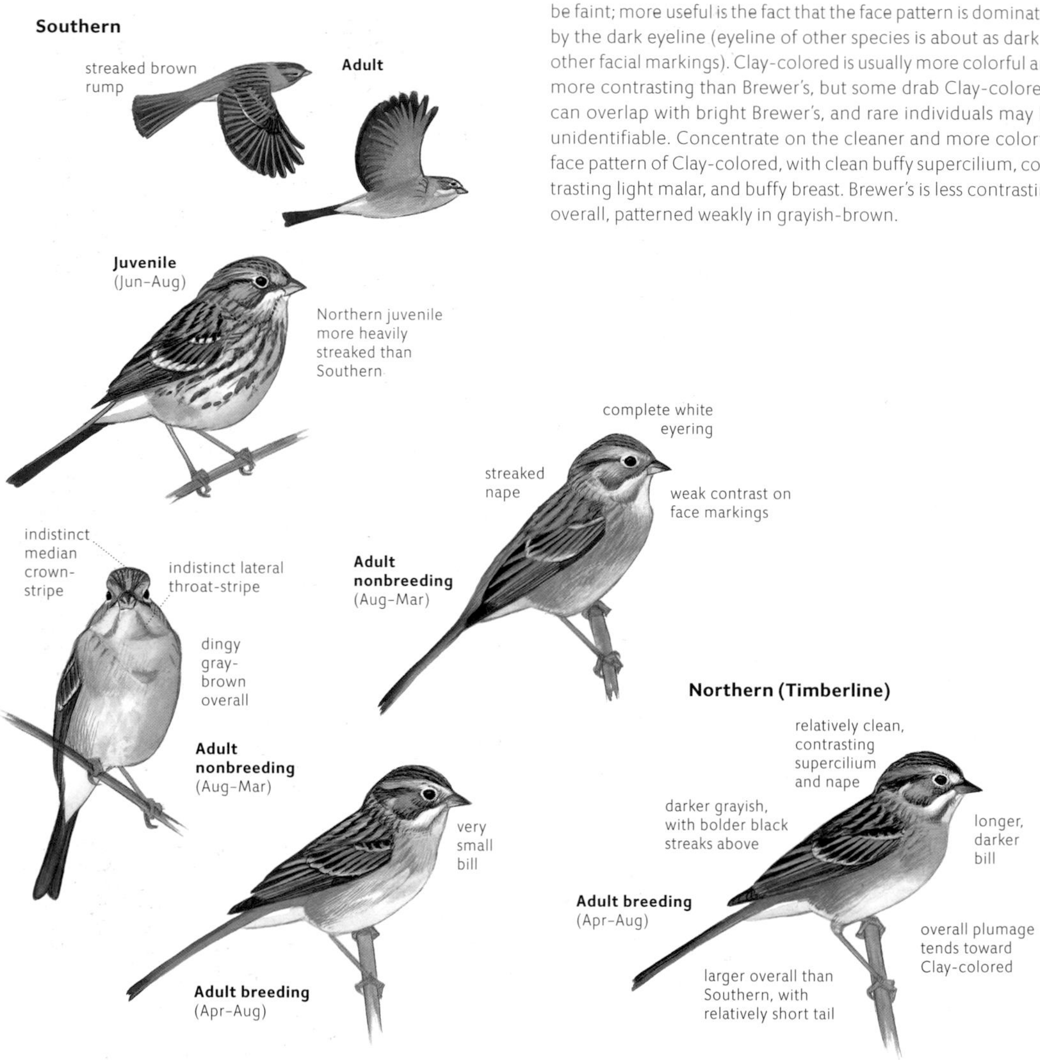

VOICE: Song a long, varied series of trills and buzzes, overall descending: *zerr-zerr-zerr tir-tir-tir-tir cheeeeeeee deee-deee-deee zrr-zrr-zrr-zrr zreeeee...*; some notes are high, clear, and musical, and others are low and rasping like Clay-colored. Call a high, sharp *tsip* like other *Spizella* sparrows. Flight call rising, short, and weak, with abrupt ending: *swit*.

Common. Nests in flat dry expanses with well-spaced shrubs and little grass, such as sagebrush habitats. Winters in loose flocks in similar arid brushy habitats with patches of grass. Often joins flocks of other sparrows, but forms large pure flocks where common.

Two populations with separate breeding ranges differ in subtleties of plumage and song but may not be reliably identified in the field. More study is needed. Northern population (breeds in brushy habitat at treeline in mountains of Canada and Alaska and possibly farther south; winter range overlaps with Southern) averages slightly larger and darker overall with broader black streaks on the back and crown and darker gray breast contrasting slightly with the paler belly; bill may average darker and longer; head pattern is more contrasting, with possibly cleaner gray nape, and may suggest Clay-colored Sparrow. Northern song averages lower, clearer, and less buzzy with slower, more musical trills but lacks slower series of high, clear notes.

Clay-colored Sparrow

Spizella pallida

. 5.5" WS 7.5" WT 0.42 oz (12 g)

Similar to drab Chipping Sparrow, but paler and more buffy overall, with contrasting gray nape, strong dark mustache, pale ores, and broad pale eyering. Also similar to Brewer's but with nore contrasting face pattern, brighter buff colors.

OICE: Song a series of two to five rasping buzzes on one pitch *heee zheee zheee.* Call a high, sharp *tsip* like other *Spizella* parrows. Flight call a short, rising *swit* similar to Brewer's.

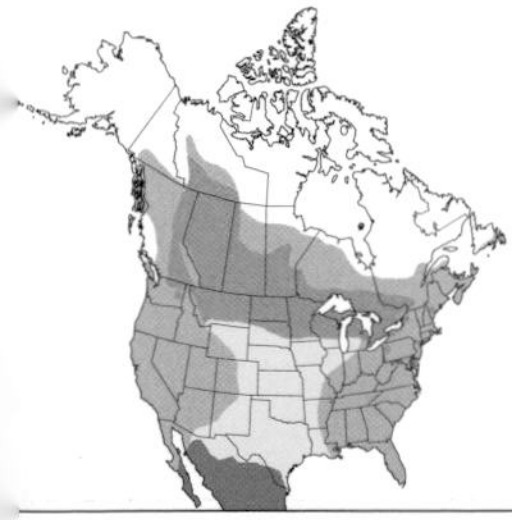

Common. Nests in open areas with scattered bushes. Winters in open brushy areas mixed with grass. Forms loose flocks in winter that often mix with Chipping or Brewer's Sparrows, and individuals may join flocks of other sparrows such as White-crowned.

Chipping Sparrow

Spizella passerina

L 5.5" WS 8.5" WT 0.42 oz (12 g)

A small sparrow with long thin tail, small pinkish bill, and unstreaked breast, like others in the genus. Drabbest individuals told from Clay-colored and Brewer's by distinct dark eyeline extending to bill, pale eye arcs, and usually grayish rump.

VOICE: Song a simple, rather long and mechanical trill; usually longer and more rattling than similar songs of other species. Call a sharp chip like other *Spizella* sparrows. Flight call a high, thin, slightly rising *tsiis*.

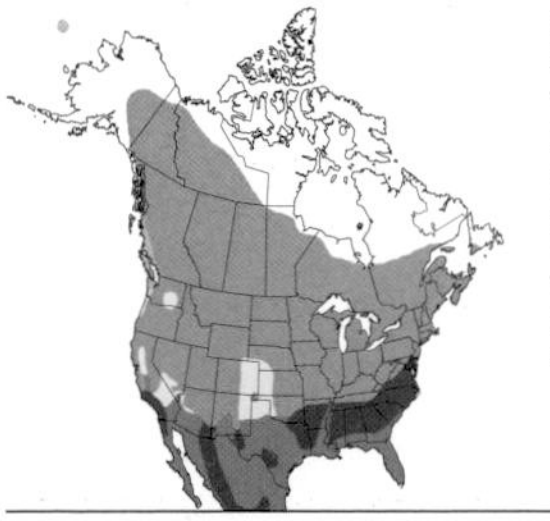

Common and widespread. Summers in open woodlands or woodland edges with grassy understory, such as suburban parks and neighborhoods, open pine forest, oak savannah, etc. Winters in open shortgrass areas with scattered trees or brush. Usually in small flocks, individuals often mix with other sparrows.

Bell's Sparrow

Artemisiospiza belli

L 5.5" WS 7.75" WT 0.5 oz (14.2 g)

Until recently considered the same species as Sagebrush Sparrow, both have relatively long tail, gray-brown upperside, mostly gray head with complete white eyeing and bold white malar, and dark central spot on mostly white breast.

VOICE: Song a bright, finch-like warble, rising and falling rapidly with notes run together in tempo reminiscent of Blue Grosbeak, no sharp notes or accented phrases; higher-pitched and more rapid than Sagebrush. Calls include very high, thin *tip* or *tink* notes singly or in series.

Interior subspecies locally common, breeding in flat expanses of arid sagebrush or saltbush plains and migrating upslope in early summer. Pacific subspecies an uncommon and local resident in dense coastal and foothill chaparral. Habits like Sagebrush, secretive and usually solitary, but interior subspecies may be seen in loose groups in winter.

Sagebrush Sparrow

Artemisiospiza nevadensis

L 6" WS 8.25" WT 0.58 oz (16.5 g)

Very similar to interior subspecies of Bell's Sparrow (much pale than coastal subspecies of Bell's). Best distinguished by distinc streaking on back and weaker dark lateral throat stripes, but some individuals may not be safely identifiable.

VOICE: Song finch-like, a short, mumbled, sing-song warble o relatively low-pitched trills, rising and falling in pitch with nc strong pattern; similar to Bell's but with slower, jerky and halting tempo. Calls like Bell's.

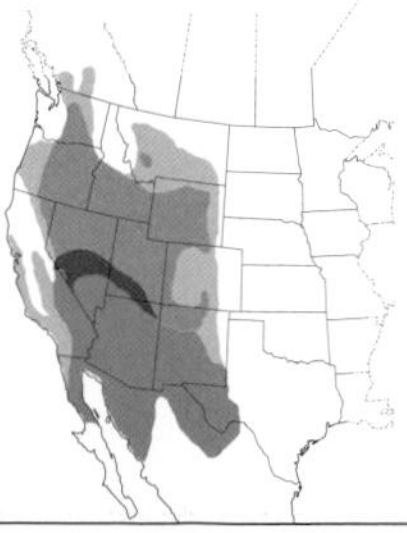

Locally common in flat expanses of arid sagebrush or saltbush plains. Secretive and hard to see, often running between bushes with tail raised high. Usually solitary but may be seen in loose groups in winter.

Black-throated Sparrow

Amphispiza bilineata

L 5.5" WS 7.75" WT 0.47 oz (13.5 g)

A relatively small sparrow of arid southwestern desert scrub. Striking black throat with white stripes on face and whitish belly is unmistakable. Juvenile lacks black throat but still shows bold white eyebrow, dark tail with white margins.

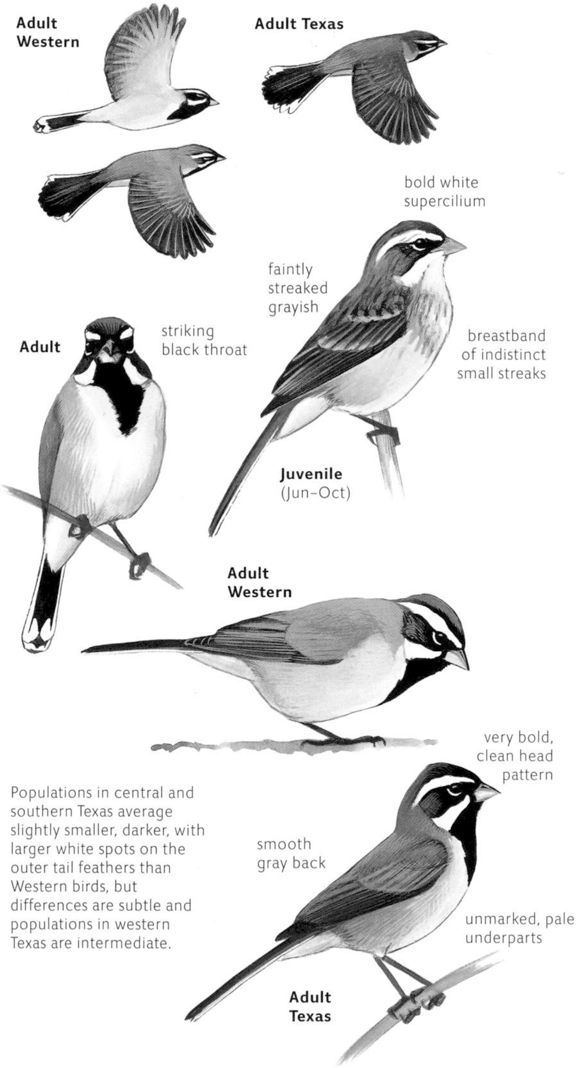

Populations in central and southern Texas average slightly smaller, darker, with larger white spots on the outer tail feathers than Western birds, but differences are subtle and populations in western Texas are intermediate.

VOICE: Song a short, simple, mechanical tinkling *swik swik sweeee te-errrr* or *tip tik* to *tik tik trr tredrrrrr*; also rapid repetitions of tinkling phrases on different pitches *teeteetee, tototo, tletletle*.... Call of high, weak, tinkling notes: a hard, bell-like *tip*; also series of high *tee* notes.

Common in arid desert scrub and sparse shrubby vegetation with patches of open ground. Usually in pairs or small groups. Quite bold, often foraging calmly in the open on the ground, or sitting up on exposed twigs.

Five-striped Sparrow

Amphispiza quinquestriata

L 6" WS 8" WT 0.7 oz (20 g)

A large stocky sparrow with relatively long bill. Dark gray and brown overall, with neat white lines across face, white throat and belly, and dark spot on breast. Note extremely limited range and habitat.

VOICE: Song of short, high, liquid or tinkling phrases, each repeated two or more times, then a pause and another series *tlik, kleesh kleesh; tlees tlees; chik sedlik sedlik sedlik; kwij kwij kwij*.... Call a husky *terp*; occasionally a higher, dry *chik* or very high *tip*.

Rare and local; found at only a few sites in Arizona where dense brushy vegetation covers steep slopes above permanent streams. Solitary or in pairs year-round. Secretive and difficult to see in dense brush except when male is singing.

Savannah Sparrow

Passerculus sandwichensis

L 5.5" WS 8.75" WT 0.7 oz (20 g)

Similar to some subspecies of Song Sparrow, but yellow wash on lores distinctive and usually present. Also note shorter tail, subtle crest, smaller and pinkish bill, paler eyebrow, and usually more crisply streaked overall.

VOICE: Song of high, fine buzzes, each one lower-pitched than the preceding *ti ti ti tseeeeeee tisoooo*; harsher and lower than Grasshopper, with final low buzz; often a simpler *t t t tzeeeeeeee tzz*. Call a very high, sharp *stip*. Flight call a high, thin, weak *tsiw* similar to *Spizella* sparrows but descending and fading. Limited geographic variation in song.

Common and widespread in open grassy or weedy habitats, including marshes, fields, and dunes; less numerous in brushy habitat. Often forms loose flocks in winter, staying mostly separate from other sparrows. In most areas the most numerous streaked sparrow, and the one most likely to be seen in open fields, either on the ground or perched on top of weeds or fences.

Overall color varies regionally from paler grayish to darker reddish-brown. Most geographic variation is subtle, involving size and average color, with clinal transitions from reddish to grayish birds found throughout the range. Most distinctive is Large-billed (next page) with weaker streaking and large bill. Belding's (resident in southern California saltmarshes) is dark and heavily streaked; averages 5 percent smaller overall but with 10 percent longer bill than typical birds. Ipswich (nests on Sable Island, Nova Scotia, and winters in coastal dunes south to Florida) is much paler overall; averages 10 percent larger than typical birds. Aleutian breeders (not shown, winter south to California) are as large as Ipswich but colored like typical birds.

Savannah Sparrow (continued)

Large-billed

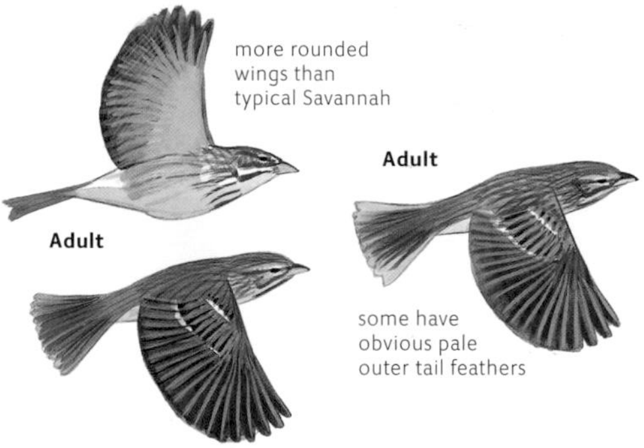

LARGE-BILLED VOICE: Song three rich buzzes *zaaaaaa zooooooo zeeee* bearing little resemblance to other populations of Savannah Sparrow; other songs closer to typical Savannah in pattern but fuller-sounding. Flight call a little longer and lower-pitched.

Uncommon and local visitor from coastal Mexico to dry mudflats, dunes, and other sparsely-vegetated shorelines in southern California, mainly Aug–Feb.

Vesper Sparrow

Pooecetes gramineus

L 6.25" WS 10" WT 0.91 oz (26 g)

Gray-brown and streaked overall, similar to Savannah Sparrow, but larger with longer tail, white outer tail feathers, and complete white eyering. Also note off-white belly, unlike Savannah's clean white belly.

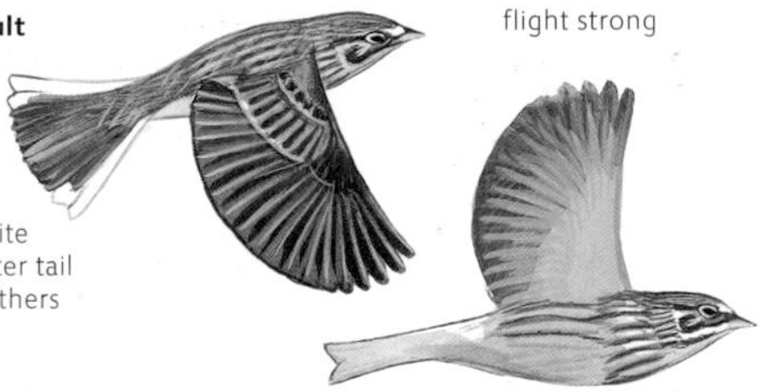

VOICE: Song begins with paired low whistles then slow, musical trills accelerating and descending *too too tee tee chididididi swiswi-swiswiteew*; similar to Song Sparrow. Call a sharp chip, lower and harder than Savannah. Flight call a slightly buzzy, rising, sharp *ssit* or *seeet.*

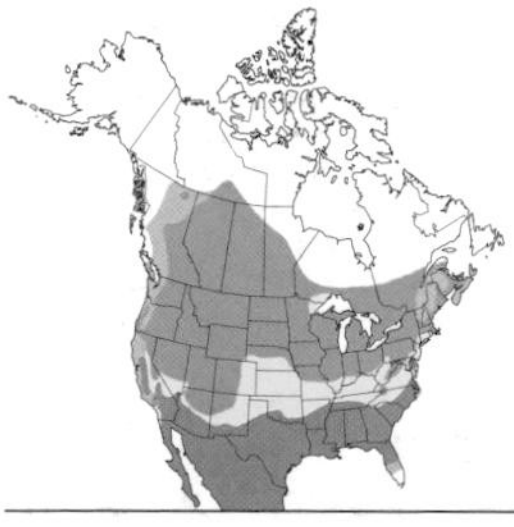

Common in the West, uncommon in East. Nests in extensive grasslands including short-grass prairie. In winter found in loose flocks in relatively dry and sparsely vegetated pastures or agricultural fields with nearby trees. Often perches in the open on wires or twigs. Generally doesn't mix with other sparrows.

Grasshopper Sparrow

Ammodramus savannarum

L 5" WS 7.75" WT 0.6 oz (17 g)

A small buffy sparrow with relatively large head and bill and short tail. Note plain face, complete white eyering, and intricately patterned upperside. Told from LeConte's Sparrow by larger bill, unstreaked breast, pale eyering.

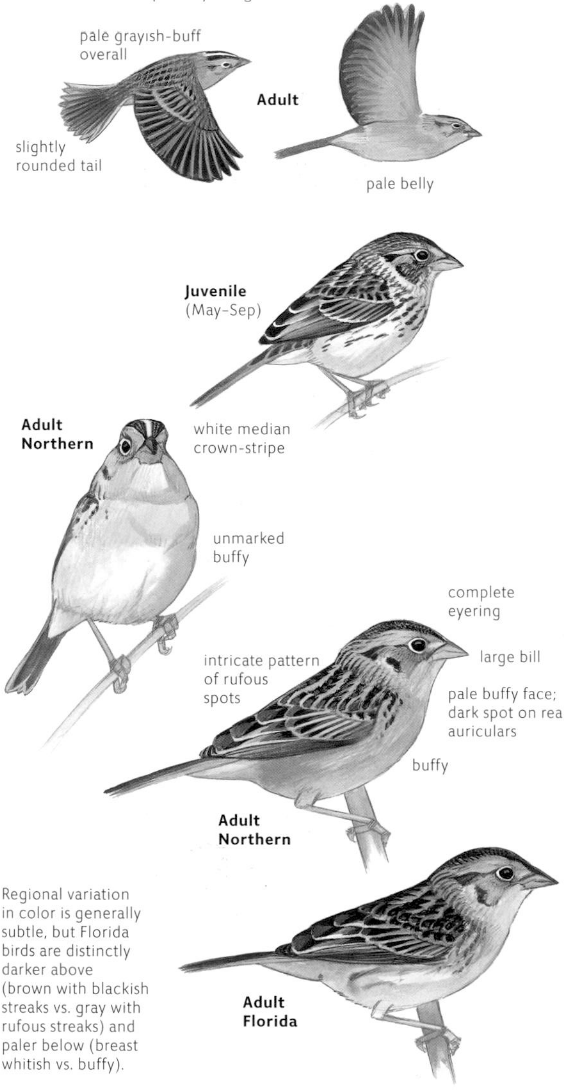

VOICE: Song a very high, hissing, insectlike buzz preceded by weak *tik* notes *tik tuk tikeeeeeeeeeeez*; also a rolling jumble of high, buzzy, slurred phrases. Call a very high, thin, sharp *tip*; usually rapidly doubled or tripled *titip*. Flight call a sharp, high, rising *tswees*.

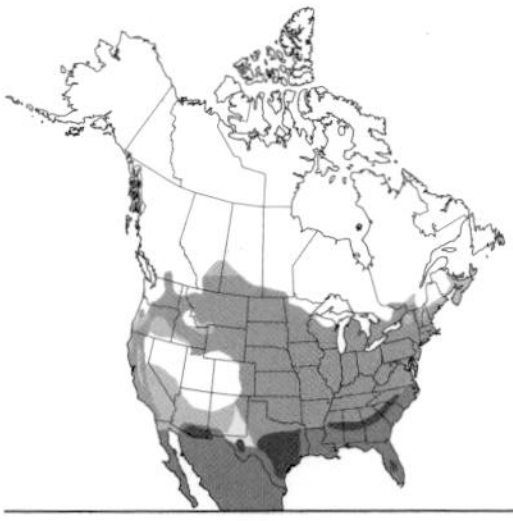

Uncommon and local in large, expanses of dense tall grass with scattered shrubs or weeds on dry ground, avoids damp areas favored by LeConte's. Solitary and secretive; difficult to see except when singing. Never forms flocks, but individuals may be associated with concentrations of other sparrows.

Baird's Sparrow

Ammodramus bairdii

L 5.5" WS 8.75" WT 0.61 oz (17.5 g)

Streaked plumage similar to Savannah Sparrow, but more stocky with orange-buff ground color of head, including median crown stripe, very weak dark eyeline, and narrow breastband of dark streaks.

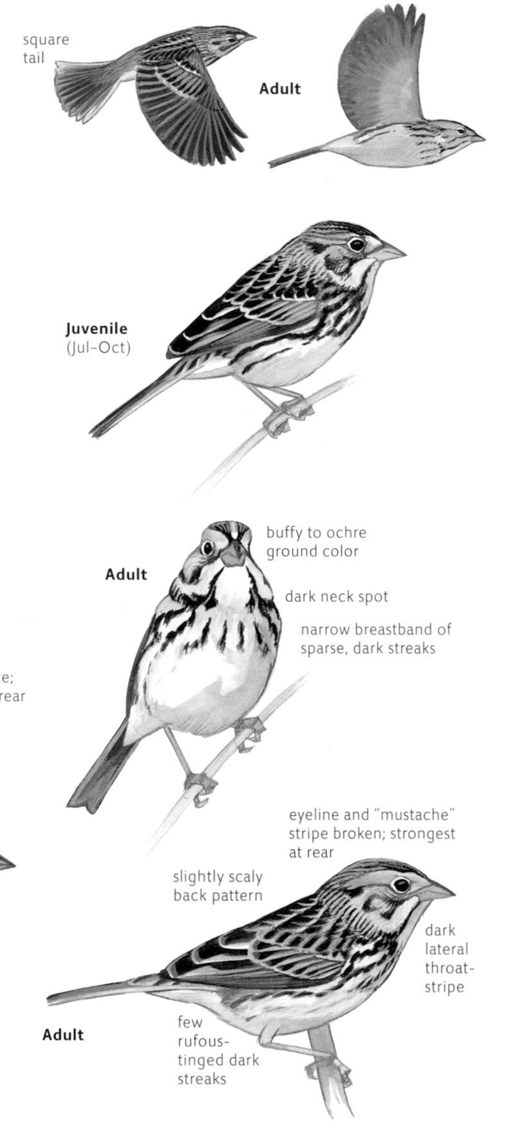

VOICE: Song high, clear jingling; several high, clear *tink* notes followed by clear, musical trill *tik a tl tleeeeee*. Call a very high, weak *teep*. Flight call a high, thin *tsee*; higher than Grasshopper.

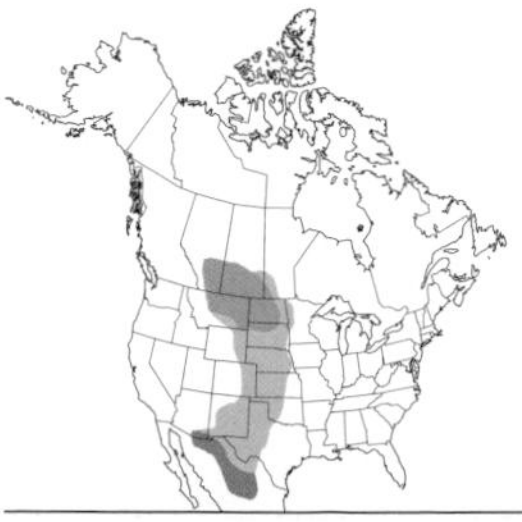

Uncommon and local. Nests in large expanses of relatively lush dense grass in prairie. Winters in similar unbroken expanses of dense grass. Solitary and secretive; difficult to see except when singing.

Henslow's Sparrow

Ammodramus henslowii

L 5" WS 6.5" WT 0.46 oz (13 g)

A very small, short-tailed, large-headed sparrow with relatively large bill. Dark overall; note olive-tinged head, narrow breast-band of fine streaks on buffy ground color, and rufous wings and back.

VOICE: Song a dry, insect-like, feeble hiccup *tsezlik* or *tsillik*. Call a high, sharp *tsik*. Flight call a high, almost waxwinglike trill *sreee*.

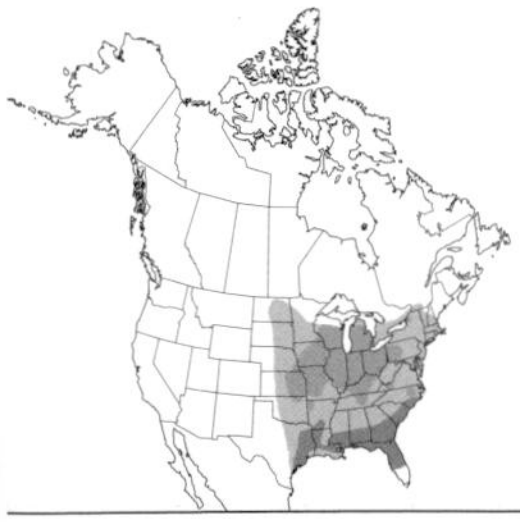

Uncommon, local, and declining. Restricted to tall-grass prairie and damp grassy meadows with old matted vegetation and a variety of weeds and other groundcover. Solitary and very secretive; difficult to see except when singing.

Le Conte's Sparrow

Ammodramus leconteii

L 5" WS 6.5" WT 0.46 oz (13 g)

A very small, short-tailed sparrow; often found with Sedge Wren, and the two can be confused in flight. Told from Grasshopper and other sparrows by clean yellow-buff face and breast, pale lores, crisp dark streaks on flanks, and small bill.

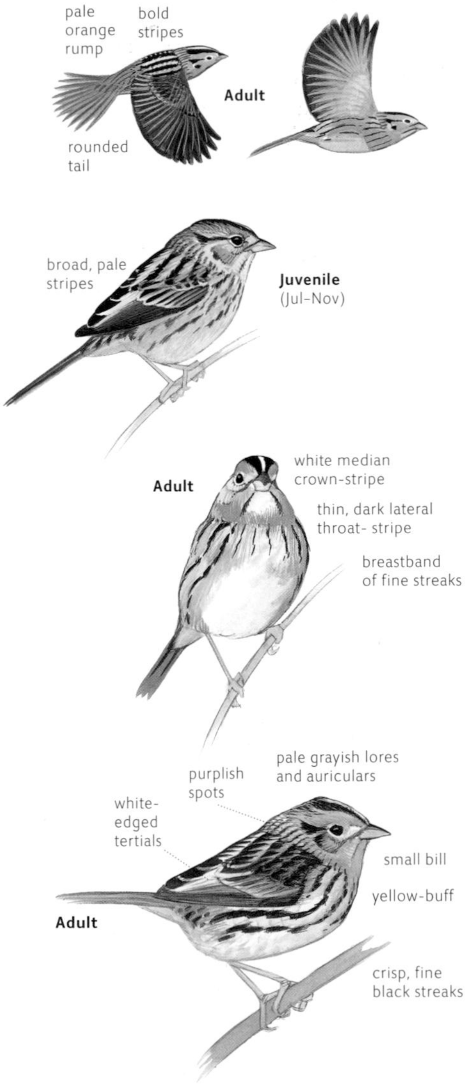

VOICE: Song a hissing, unmusical buzz *tik a t-sshhhhhhhh-t*; softer, higher, more hissing than Grasshopper Sparrow; sharp introductory notes unlike Nelson's. Call a high, thin, descending *tseeez*.

Uncommon and local in dense wet grasslands or sedge marshes, chooses wetter habitat than Grasshopper Sparrow, drier than Nelson's. Solitary and secretive, sometimes perches in open on shrubs or weed tops, but generally difficult to see except when singing.

Nelson's Sparrow

Ammodramus nelsoni

L 5" WS 7" WT 0.6 oz (17 g)

Similar to Le Conte's Sparrow, with short tail and orange triangle on face; distinguished by slightly longer bill, darker overall color. Similar to Saltmarsh Sparrow and until recently considered the same species (Sharp-tailed Sparrow).

VOICE: Song a weak, soft, airy, fading hiss with slightly lower notes at beginning and end *pl-teshhhhhh-ush* (compare Le Conte's). Call a hard *tek* higher and harder than Seaside. Flight call a high, lisping *ssis*. Flight song a series of sharp chips followed by typical song during higher flight.

Common in coastal saltmarsh habitat; uncommon and local inland in grassy freshwater marshes. Mostly solitary and secretive, but numbers may cluster in small patches of suitable habitat. Generally chooses wetter habitat than LeConte's Sparrow.

Saltmarsh Sparrow

Ammodramus caudacutus

L 5.25" WS 7" WT 0.67 oz (19 g)

A rather large, long-billed sparrow; shares saltmarsh habitat with Nelson's and Seaside Sparrows. Note bright orange triangle on face, usually distinct dark streaks on breast, relatively dark gray-brown back with white stripes, and gray crown.

VOICE: Song much softer than Nelson's, less frequently heard; includes sweet gurgling notes and lacks final lower note; usually repeated in rapid sequence. Calls presumably like Nelson's. Flight song a series of weak songs, all different; delivered rapidly during low flight.

Common within narrow strip of coastal saltmarsh habitat. Generally found in slightly drier and grassier habitat than Seaside Sparrow, often seen foraging at edge of grass, on mud exposed by low tide. Solitary and secretive, most easily seen when singing.

Seaside Sparrow

Ammodramus maritimus

L 6" WS 7.5" WT 0.81 oz (23 g)

A large, stocky, very long-billed sparrow. Told from Saltmarsh Sparrow by larger size, plainer and darker gray overall color, yellow supraloral, contrasting white throat, and blurry gray streaks below covering most of belly.

VOICE: Song a rather muffled *tup teetle-zhrrrr*; more complex and fuller-sounding than Sharp-tailed Sparrows; reminiscent of song of distant Red-winged Blackbird. Call a low, husky *tup*. Flight call a long, thin, towheelike buzz *zeeeooee*.

Common in coastal saltmarsh habitat; favors wetter areas and taller vegetation than Saltmarsh Sparrow. Solitary and secretive, but congregates in small patches of suitable habitat.

Three populations differ in range and plumage, but some Atlantic birds are similar to some Gulf Coast in appearance. Song varies geographically, with complex local dialects; more study is needed to determine whether voice differs consistently among the three main populations. Gulf Coast song may be more complex: three- to four-part, with descending trend. Song of Cape Sable population in southern Florida a simple, long, nasal buzz *tli-zheeeeee*. Flight song may also differ among populations: Atlantic gives a series of high, thin, wispy trills and rattles sometimes followed by normal song; Gulf Coast gives a series of chips usually followed by normal song.

Fox Sparrow

Passerella iliaca

L 7" WS 10.5" WT 1.1 oz (32 g)

Four subspecies groups in North America, sometimes considered four separate species. All forms are relatively large and stocky, with rounded head, relatively plain and grayish face, plain dark rufous to gray-brown upperside, and heavily spotted underside.

Sooty (Pacific)

Shorter-tailed than Thick-billed and Slate-colored; plumage uniform brownish with densely spotted breast.

Thick-Billed (California)

This population and Slate-colored both relatively long-tailed with plain gray back; note massive bill.

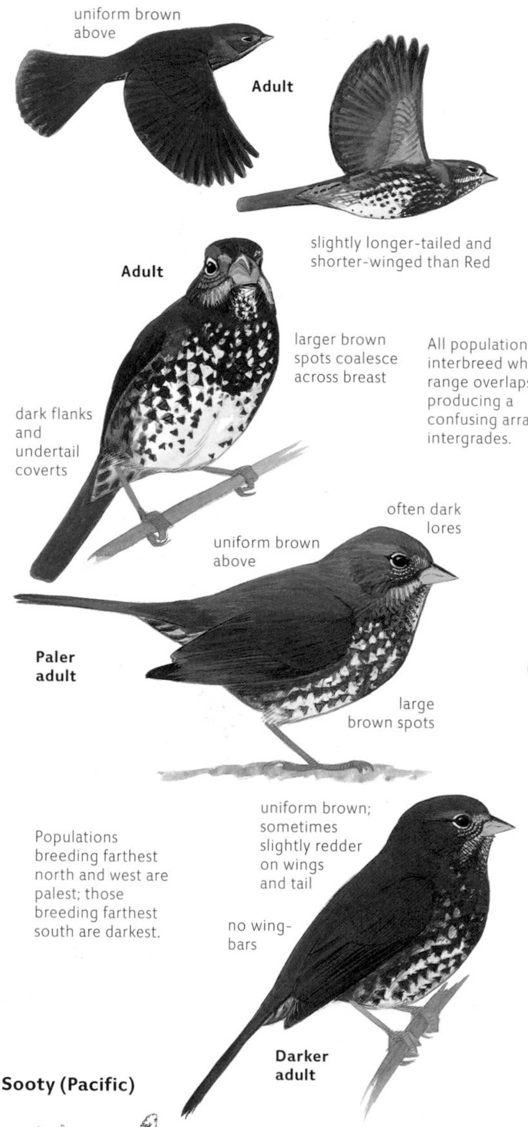

VOICE: Song a long, irregular series with widely separated notes at beginning *wit; tip; swit, wit swit-swit teer zeep-zet-zet-zweeer*; buzzier, thinner, and more staccato than Red; individuals sing same song repeatedly. Calls like Red.

Sooty (Pacific)

VOICE: Song variable but basically similar to other forms: one to three syllables with bubbling warble and one to four ending notes; individuals sing multiple songs, usually alternating among different songs; quality may suggest House Finch or Blue Grosbeak. Call a high, flat squeak *teep* like California Towhee.

All forms occupy similar habitats: nest in dense brushy patches; winter singly or in small groups in brushy patches and thickets usually within woodlands. Never in large flocks, but individuals often associate with flocks of *Zonotrichia* sparrows or groups of Song Sparrows.

Thick-Billed (California)

Slate-Colored (Interior West)

Plumage like Thick-billed but heavier spotting on underparts, smaller bill.

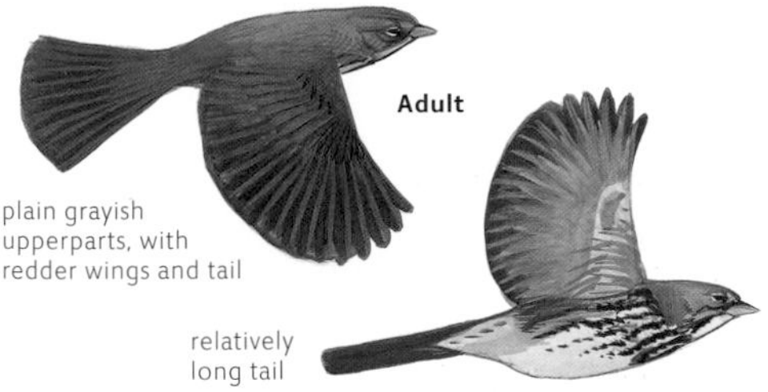

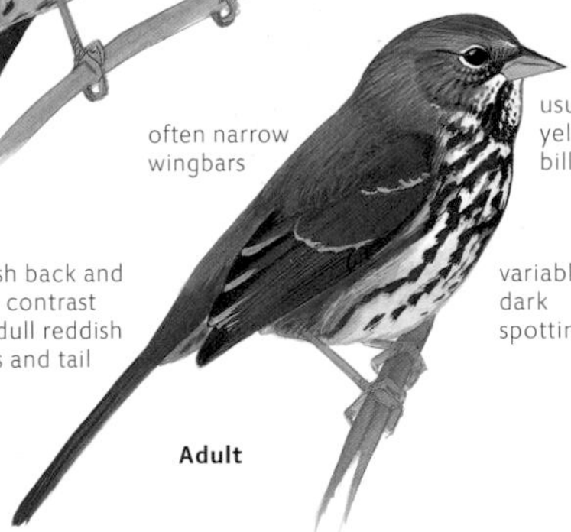

Frequent where breeding range overlaps in southern British Columbia, occasionally identified southwards in winter. Red and Sooty also interbreed from British Columbia to Alaska.

VOICE: Song clear and ringing; every other note emphasized, some buzzy or trilled; often similar to Green-tailed Towhee; individuals sing two to five different songs, never the same song twice in a row. Call a sharp smack like Sooty and Red populations.

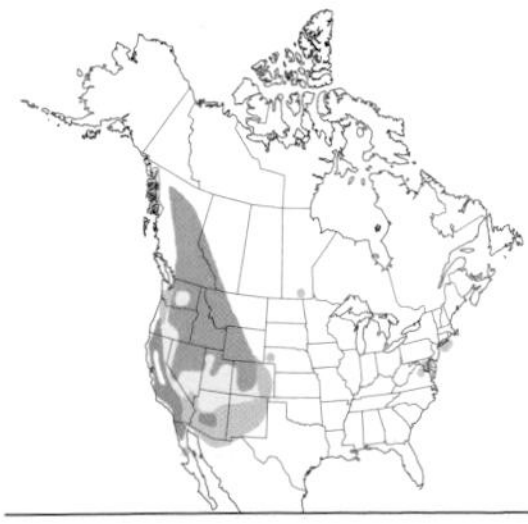

Red (Taiga)

Slightly shorter tailed and longer-winged than Sooty; the most brightly marked Fox Sparrow.

Populations breeding in far West are slightly darker and grayer than those in the East.

VOICE: Song a somewhat halting, relatively low, rich warble; richest and most melodious of all sparrows; mainly clear whistles, lacking trills or rapidly repeated notes *weet weeto teeoo teeo tzee tzer zezer reep*; individuals sing only one song. Call a very hard, sharp smack like Brown Thrasher. Flight call a high, sharp, rising *seeeep*.

Song Sparrow

Melospiza melodia

L 6.25" WS 8.25" WT 0.7 oz (20 g)

The common sparrow of gardens and hedgerows in most of North America; rather stocky and long-tailed. Note bold coarse streaks above and below, converging in large spot on center of breast, grayish face, and rufous-tinged wings and tail.

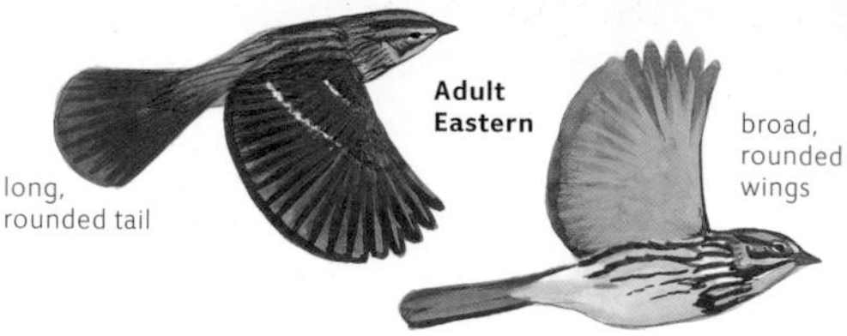

VOICE: Song a variable series of trills and clear notes with slightly husky quality and pleasant gentle rhythm; begins with several short, sharp notes, usually one long trill in middle of song *seet seet seet te zeeeeeee tipo zeet zeet*. Call a very distinctive, husky *jimp*. Alarm a very high, hard *tik*. Flight call a high, thin, level *seeet*. Chase call a rapid series of rising then falling sharp call notes.

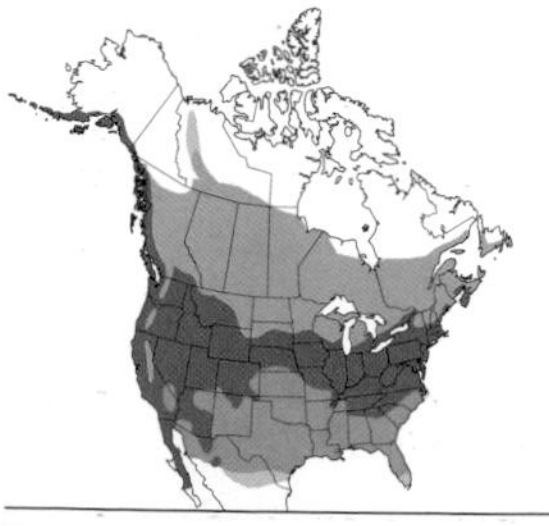

Common and widespread in brushy areas near water; in most areas the most frequently seen streaked sparrow. Found in open brushy areas and edges, such as gardens and hedgerows in suburbs, where it hops around on grass at edges of lawns and fields or sings from top of bush. Does not really form flocks, but often in loose groups where numerous.

Typical regional variations are shown, but all populations are connected by an unbroken cline of intergrades; many birds seen will be intermediate. Variation is most pronounced in overall color as well as color and thickness of streaking on underparts. Aleutian, Pacific Northwest, and California Coast populations average longer- and thinner-billed than Eastern; Aleutian breeders are 25 percent larger than the general average, while California Coast breeders are 10 percent smaller. There is essentially no variation in voice throughout the range.

Lincoln's Sparrow

Melospiza lincolnii

L 5.75" WS 7.5" WT 0.6 oz (17 g)

Similar to Song Sparrow but smaller with smaller bill and shorter tail; also note more grayish color with very crisp dark streaking above, fine streaks on buffy breast, yellowish at base of bill. Also similar to Swamp Sparrow.

VOICE: Song a continuous jumble of husky, chirping trills with several pitch changes *jew-jew-jew-jew-je-eeeeeeeee-do-je-e-e-e-to*; bubbling quality and pattern reminiscent of House Wren. Call a sharp, light chip. Flight call a high, buzzy *zeeet* like Swamp but finer and rising.

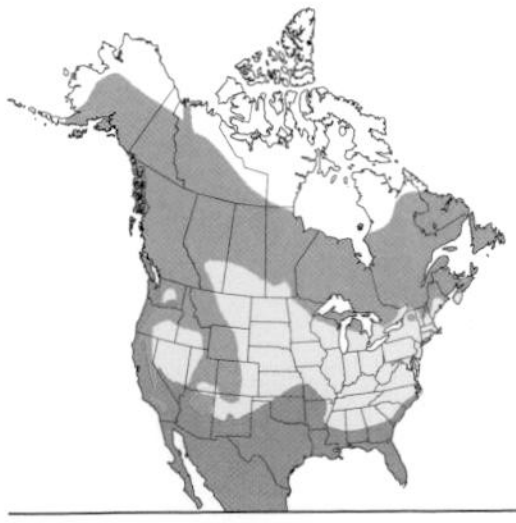

Common but in West, uncommon in East somewhat secretive and inconspicuous. Nests in damp, dense brushy areas in sunny clearings, such as willow and alder thickets, margins of bogs. Winters in grassy patches around brush and trees, often near water. Usually solitary, but may form loose groups or mix with flocks of other sparrows.

Swamp Sparrow

Melospiza georgiana

L 5.75" WS 7.25" WT 0.6 oz (17 g)

Small and dark, with small bill. Note solid rufous wing coverts, blurry streaks (or no streaks) on grayish breast, buffy-rufous flanks, and gray to olive head and breast. Told from White-throated Sparrow by smaller size and lack of wingbars .

VOICE: Song a simple, musical trill with slow tempo *chinga chinga chinga*... fading at end. Call a loud, hard chip; not as metallic as White-throated. Flight call a high, buzzy *zeeet* like Lincoln's but coarser and level.

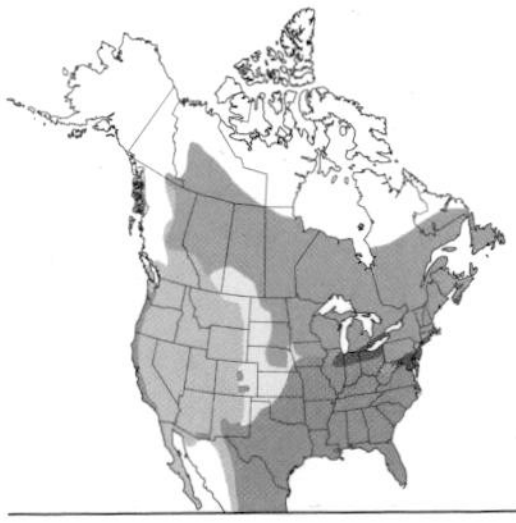

Common in wet marshes or at pond edges in dense vegetation such as cattails, grass, or shrubs. During migration and winter also inhabits other weedy or grassy habitats, such as old fields, usually but not always in wet areas. Usually stays low in dense cover. Often in small loose groups, and often mixes with other sparrows in same habitat, especially Song Sparrow.

Dark-eyed Junco

Junco hyemalis

L 6.25" WS 9.25" WT 0.67 oz (19 g)

One of the most familiar winter feeder visitors across the continent. Six regional populations all share distinctively simple color pattern, with unmarked grayish to blackish head, small pink bill, and flashing white outer tail feathers.

Oregon

Marginally the smallest junco. Dark, dull-gray hood contrasts sharply with brown back and flanks.

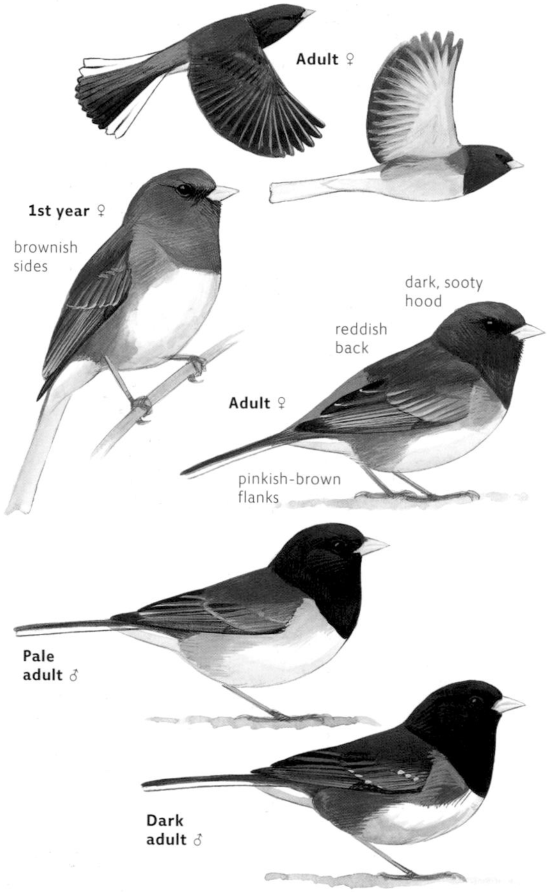

VOICE: Songs of most populations indistinguishable: a short trill averaging slower and more musical than Chipping Sparrow; Slate-colored may average longer song with more rapid tempo and smaller repertoire than Oregon; all populations sing quiet, varied warbling phrases in early spring. Call a very high, hard, smacking *stip*. Flight call a sharp, buzzy *tzeet*; also high, tinkling chips when flushed *tsititit tit*. Chase call a series of high, clear *keew* notes.

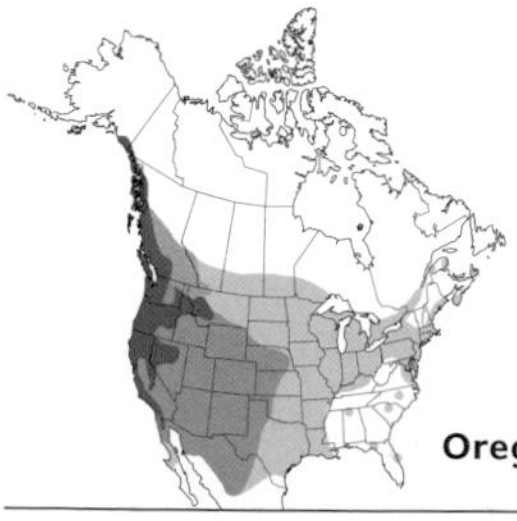

Common and widespread. Nests in relatively open mature coniferous or mixed woods with patches of open ground and brush; winters in flocks in patchy wooded areas such as suburban neighborhoods, roaming widely in search of food; less secretive than most sparrows, foraging on open ground, flying into brush or trees when alarmed.

Pink-Sided

Averages 5 percent larger than Oregon. Clean blue-gray hood (palest on throat) contrasts with blackish lores and extensive pinkish flanks.

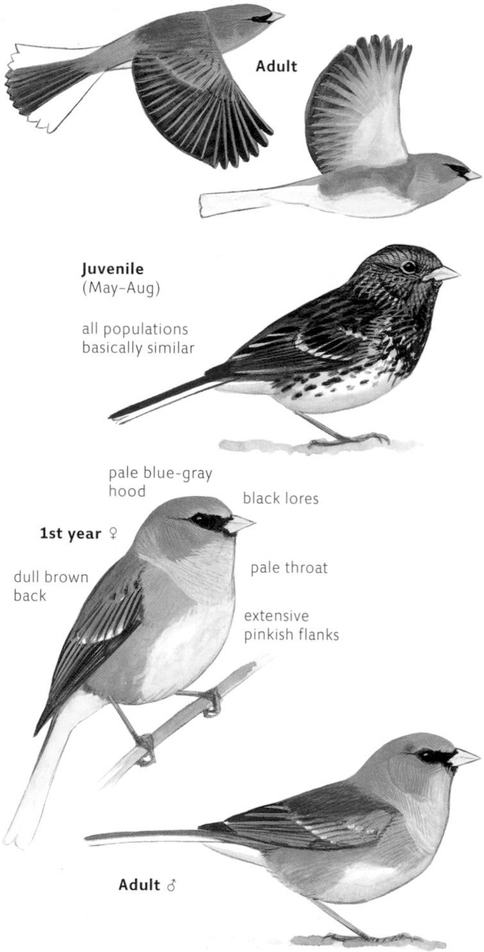

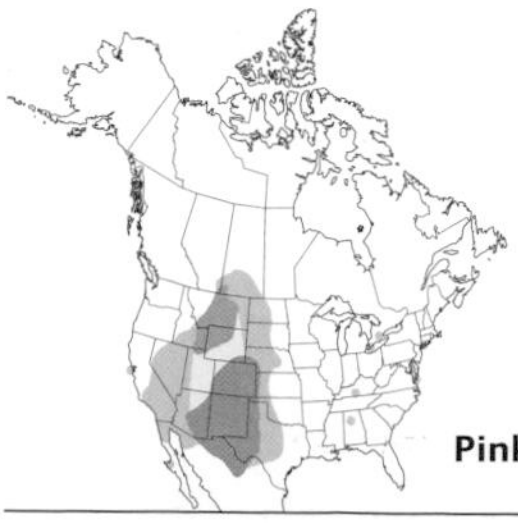

Six subspecies groups in North America all share basic similarities: plain gray head and breast contrasting with pale pinkish-white bill; unstreaked gray, brown, and white body and contrasting white belly; and white outer tail feathers that flash conspicuously in flight.

White-Winged

The largest junco; averages 12 percent larger than Slate-colored, with relatively larger bill. Rather pale gray overall (palest on throat), with weak wingbars and extensive white on tail.

Slate-Colored

Size and shape like Oregon. Overall color varies from pale brown to dark gray; little or no contrast between head and body.

Some Slate-colored individuals (1 in 200) have white wingbars as prominent as White-winged; identification of White-winged must be based on all characteristics combined.

White-Winged

Slate-Colored

Dark-eyed Junco (continued)

Junco hyemalis

Both of these populations average 5 percent larger than Oregon. Note low-contrast gray plumage with dark rufous mantle.

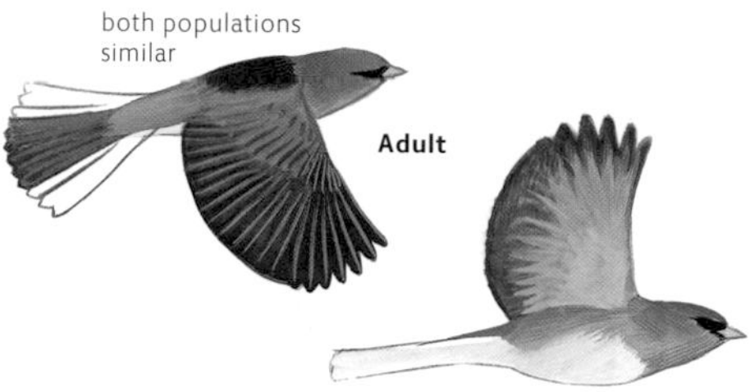

Gray-Headed

Red-Backed

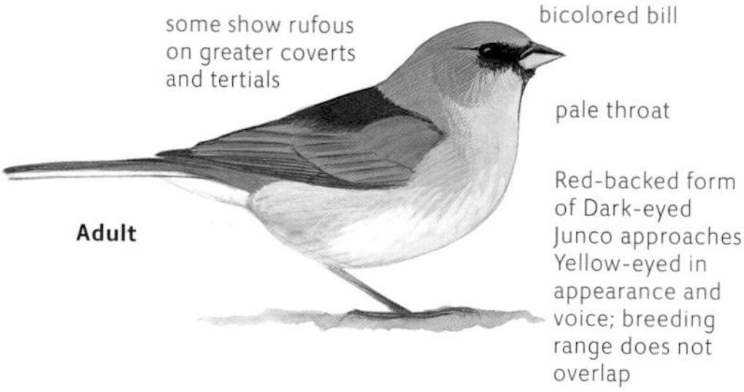

Gray-Headed

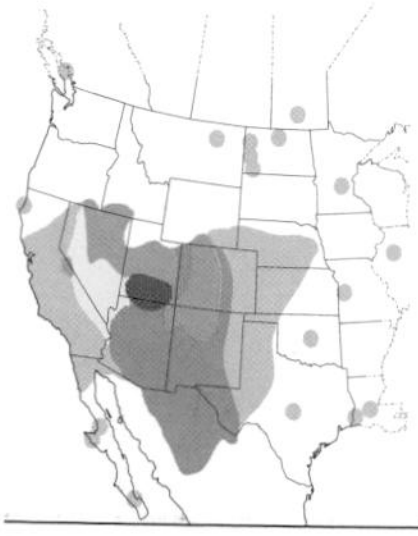

Red-Backed

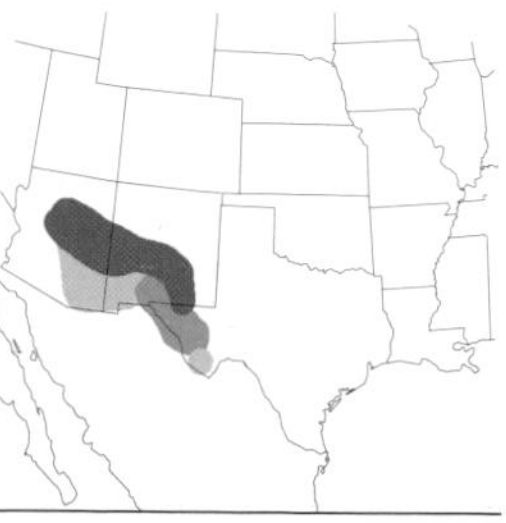

Yellow-eyed Junco

Junco phaeonotus

L 6.25" WS 10" WT 0.74 oz (21 g)

Habits and appearance very similar to Dark-eyed Junco, but walks on the ground while foraging (Dark-eyed hops). Note pale yellow iris, rufous on wing coverts and tertials, off-white belly, and bicolored bill. Some of these features matched by southernmost subspecies of Dark-eyed.

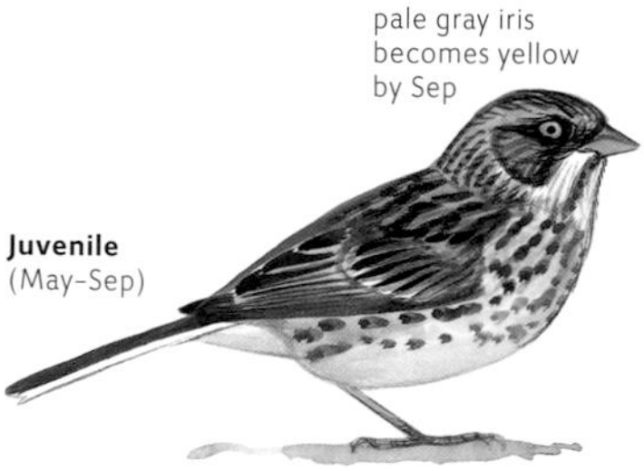

VOICE: Song a high, whistled *tzew tzew titititi tsidip* or *shidle shidle shidle shidle titititi*; first part slurred and bouncy, trill short and rattling (but may be left out). Song unlike Dark-eyed (except Red-backed form), but compare Bewick's Wren and Spotted Towhee. Call slightly lower and fuller than Dark-eyed.

Common in open pine forests in mountains. Often seen in small groups foraging mainly on the ground.

Harris's Sparrow

Zonotrichia querula

L 7.5" WS 10.5" WT 1.3 oz (36 g)

Our largest sparrow; similar in shape, habits, and some plumage to White-crowned Sparrow. Always shows pink bill, extensive bright white on belly, and pale gray to brown cheeks. Adult has striking black face and throat.

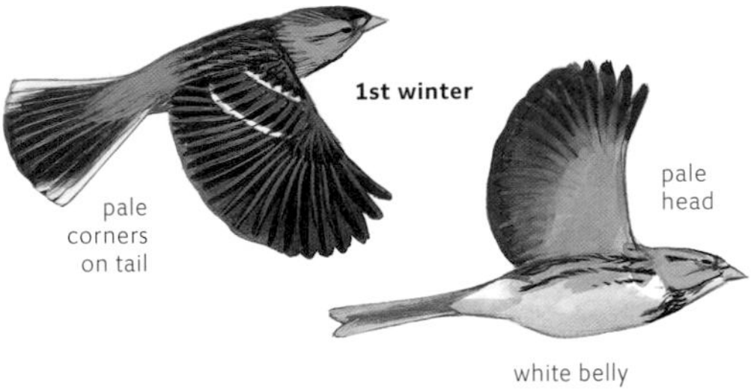

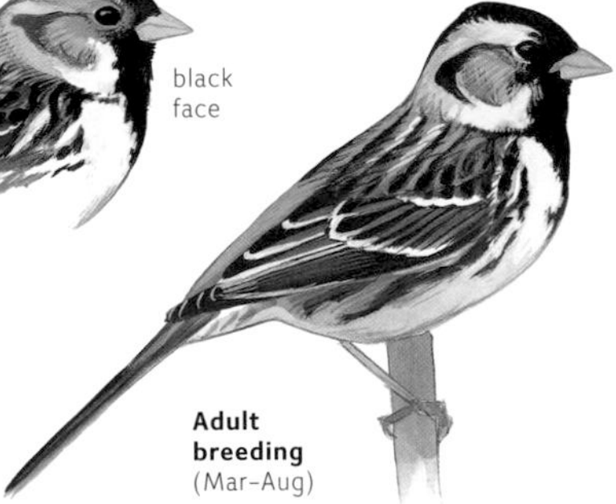

The crown feathers of all sparrows can be raised and lowered at will. Some species have relatively long crown feathers that create a crested appearance when raised, others have uniformly short feathers and look round-headed at all times. The white and black crown stripes of White-crowned Sparrow are very obvious when feathers are raised.

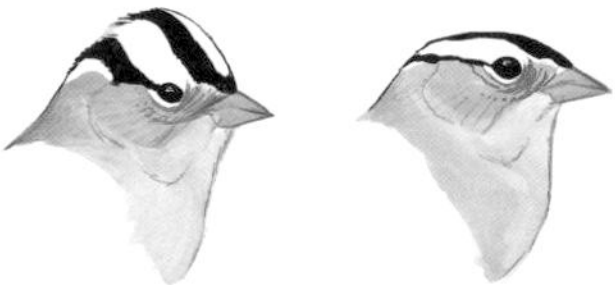

White-throated Sparrow × Dark-eyed Junco hybrid

This combination is rare but regular in the East, and quite variable. Other hybrid sparrows that are seen rarely include White-crowned Sparrow × Golden-crowned Sparrow.

VOICE: Song of high, clear whistles like White-throated, but typically two or three notes with no pitch change *seeeeeeee seeee seeee*; some White-throated sing identical songs. Call a rather harsh *cheek* like White-throated but less sharp.

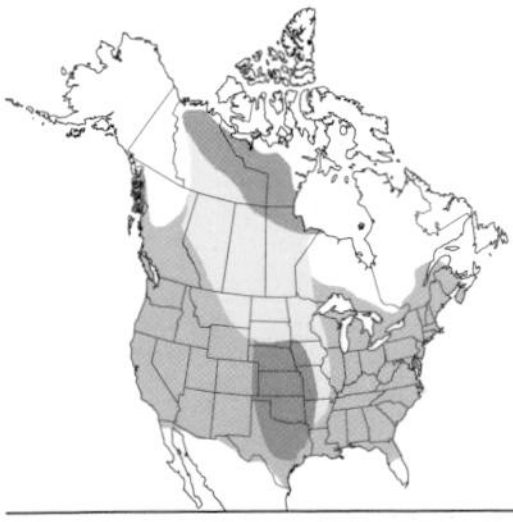

Uncommon and local. Nests in patches of brushy willows and other shrubs among stunted, open spruce woods near treeline. Winters in flocks in hedgerows and brushy areas, often at woodland edges or riparian corridors along streams. Often mixes with related species such as White-crowned and White-throated sparrows.

White-crowned Sparrow

Zonotrichia leucophrys

L 7" WS 9.5" WT 1 oz (29 g)

A relatively large, long-tailed, long-necked sparrow. Adult distinctive, with extensive white on crown, gray cheeks and neck. Immature relatively plain gray and brown overall, with pale grayish cheeks and nape, pinkish-orange bill, and brown crown-stripes.

Taiga/Interior West Group

VOICE: Song begins with clear whistles like White-throated, then a series of several buzzes or trills on different pitches; varies regionally. Call a sharp *pink* lower and less metallic than White-throated Sparrow; flatter in Pacific, lower and drier in Interior West, higher and sharper in northern breeders. Flight call a high, thin, rising *seeep*.

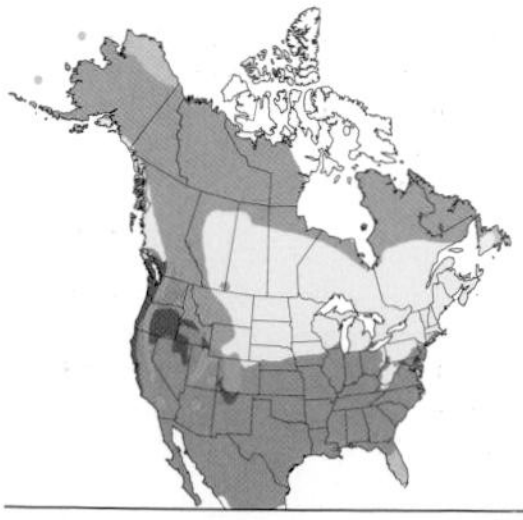

Common. Nests in brushy patches surrounded by open habitat such as tundra, alpine meadows, or floodplain or in chaparral. Winters in flocks in weedy or brushy areas, moving out to forage on the adjacent ground or lawn, and when alarmed all fly at once back into weedy cover. Mixes freely with Golden-crowned Sparrow in winter.

Pacific population is most distinctive, with yellowish bill, drabber head pattern, brownish breast-sides with short streaks, faint lateral throat stripes, less distinct streaking on the back, and the bend of the wing yellow (vs whitish). Other populations are very similar to each other. Western Taiga breeders have pale lores and slightly orange bill color, but intergrade over a broad area with both Eastern and Interior West populations. Interior West might be distinguishable from Eastern Taiga by slightly darker gray breast, more obvious wingbars, and by voice.

Song varies regionally. Taiga breeders from Alaska to Ontario sing a lazy *feee odi-odi zeeeee zaaaa zooo* with little variation; birds in eastern Canada sing slightly different songs. Pacific coast breeders sing clearer and more rapid phases with shorter buzzes, such as *seeee sitli-sitli ti-ti-ti-ti-ti-zrrrr* or *seeee zreee chidli-chidli chi-chi-chi teew*, phrases often repeated, and with many local dialects. Interior West birds sing varied songs more like Pacific, and also have many local dialects.

Golden-crowned Sparrow

Zonotrichia atricapilla

L 7.25" WS 9.5" WT 1 oz (29 g)

One of our largest sparrows. Adult distinctive, with extensive black on head and yellow crown. Immature similar to White-crowned, but more uniform gray-brown overall, especially on face and underparts, with darker gray bill.

VOICE: Song of high, clear whistles: typically only three notes with one or more notes slurred *seeeea seeeeeew soooo* ("oh deear mee"); birds in Canadian Rockies sing more complex song *seeeoo tooo teeee tetetetetete*. Call a loud, hard, clear, sharp *bink*. Flight call a high, level, rather short *seeep*.

Common. Nests in dense brush or stunted trees. Winters in any weedy or brushy area. Forms flocks, often mixed with White-crowned Sparrow but often more secretive and in denser thickets, such as unbroken chapparal or brushy understory within woodland.

White-throated Sparrow

Zonotrichia albicollis

L 6.75" WS 9" WT 0.91 oz (26 g)

A rather stocky, dark reddish-brown sparrow. Gray breast is unstreaked or with coarse mottled streaks, and always shows strong contrast with white throat. Also note dark lateral crown stripes, yellow lores, white wingbars.

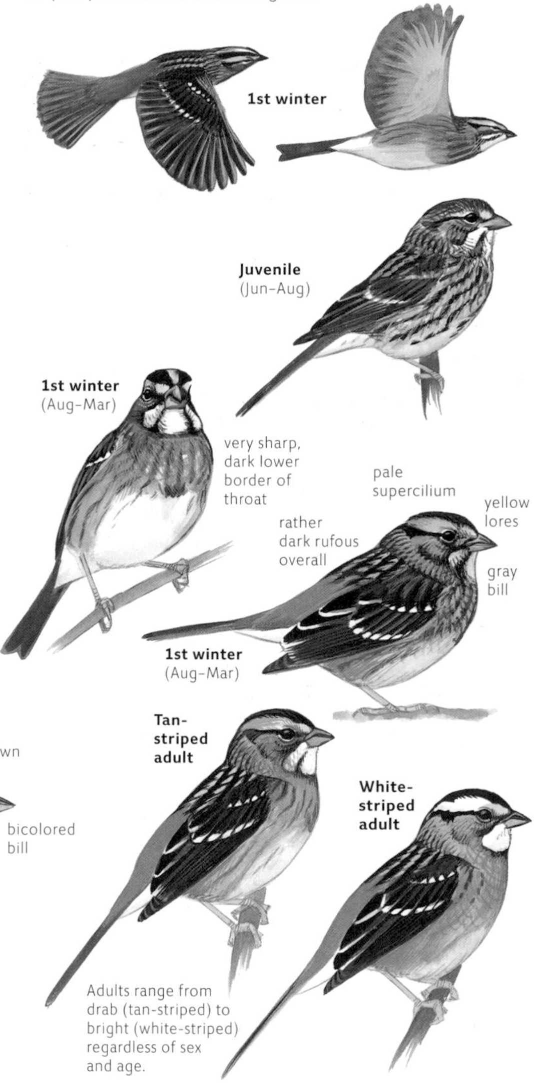

VOICE: Song a high, pure whistle *sooo seeeeeee dididi dididi dididi* ("Old Sam Peabody Peabody Peabody") with little or no pitch change. Call a loud, sharp, metallic *chink*. Flight call a high, level, long *seeeet*, often with slight trill. Flock call a relatively low, laughing *kll kll kll kll*....

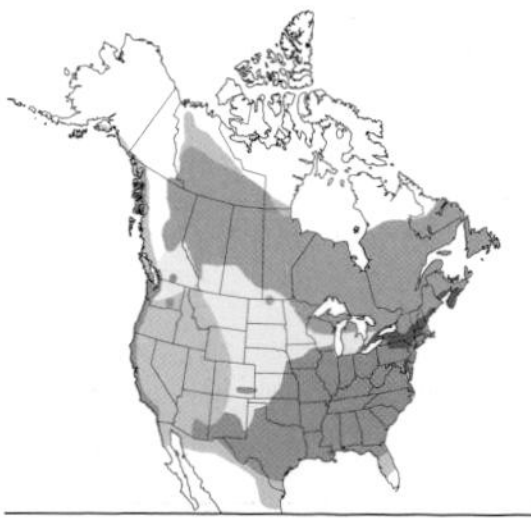

Common in brushy patches in or near openings in mixed woods. Winters in hedgerows, woodland edges, and brushy understory of open woods. Usually in flocks, often mixed with Song and other sparrows. The most frequent sparrow at woodland bird feeders in the East. Rare in the West, joins winter flocks of White-crowned and Golden-crowned Sparrows.

Lark Bunting

Calamospiza melanocorys

L 7" WS 10.5" WT 1.3 oz (38 g)

Stocky, large-headed, large-billed, and short-tailed. Male distinctive. Female differs from sparrows by heavy blue-gray bill, thick dark lateral throat-stripe, white-edged wing coverts, and white-tipped tail feathers.

VOICE: Song of repeated low, liquid, whistled notes *pwid pwid pwid pwid too too too too kree kree kree kree pwido pwido...*; interspersed and overlaid with high, silvery rattles *tt tt tt tt*; entire song rich, complex, and repetitious, with relatively slow tempo. Call a low, soft, whistled *heew* or *howik*.

Common in arid open shortgrass prairies with few or no bushes, in large numbers where habitat is good. Males sing in flight. In winter forms large flocks in arid brushlands mixed with grasslands, and tight flocks can be seen flying long distances low over this habitat, especially dawn and dusk. Generally does not mix with other sparrows, vagrants usually solitary.

Lark Sparrow

Chondestes grammacus

L 6.5" WS 11" WT 1 oz (29 g)

Large, long-necked and long-tailed. Striking head pattern, clean whitish underparts with dark spot on breast, and towhee-like white tail corners distinctive.

VOICE: Song rather slow-paced and varied with choppy rhythm; phrases generally high, mechanical rattling with some long trills *zeer puk treeeeeee chido chido kreet-kreet-kreet-kreet trrrrrrrrrr....* Flight call a high, sharp, metallic *tink*. Alarm a high, piercing *tsewp*.

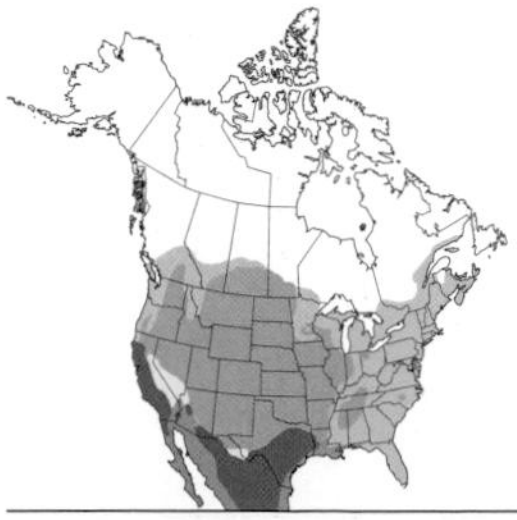

Common locally in open grassy areas with scattered trees or short grass adjacent to hedgerows and trees, often seen foraging on the ground in open short grass such as lawns. Often forms flocks in winter; usually not mixed with other sparrows but individuals sometimes join flocks of other species.

Rustic Bunting

Emberiza rustica

L 6" WS 9.5" WT 0.67 oz (19 g)

Relatively long-tailed, with small pinkish bill and short ragged crest. Note rufous streaks on flanks, white outer tail feathers, and dark V on breast.

VOICE: Song a fairly short, mellow warble reminiscent of Lapland Longspur. Call a short, hard *tick*.

Very rare visitor from Eurasia; recorded annually (mainly in spring) in far western Alaska; a few fall and winter records farther south to California. Found in brushy or weedy edges, especialy in wet areas. Usually solitary, but forms flocks in normal range, and may join feeding flocks of sparrows. Forages on the ground or in low weeds for seeds.

Little Bunting

Emberiza pusilla

L 5.5" WS 8.7" WT 0.6 oz (17 g)

Small and very sparrow-like, with small, pointed, dark gray bill. Note rufous wash on face and crown (strongest in breeding plumage), distinct pale eyering, gray nape, finely streaked underparts, and narrow white sides of tail.

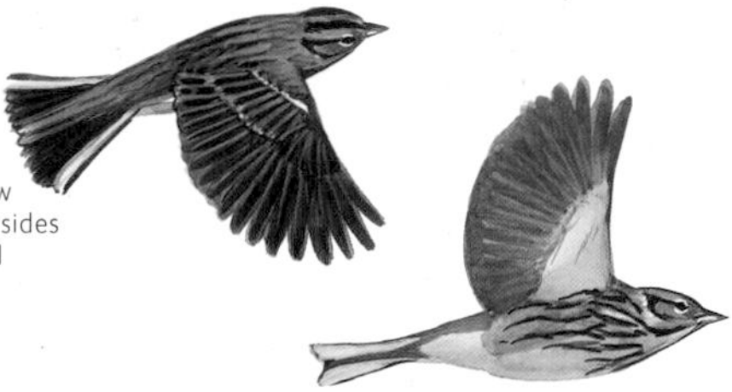

VOICE: Call a sharp clicking *tzik*.

Very rare visitor from Eurasia to far western Alaska, mainly in fall. Also recorded several times in California, once in Oregon. Found in brushy and weedy areas, often near trees. Forages mainly in the open on the ground. Solitary.

Cardinals

FAMILY: CARDINALIDAE

18 species in 7 genera (1 rare species not included here). Most species have very brightly colored male plumage and stout to very stout bill. Most are found in woodlands, hedgerows or brushy areas. All species feed primarily on insects and larvae in summer, gleaned from trees, shrubs, or grass. Most eat seeds and/or fruit in winter. All species build cup-shaped nests in shrubs or trees. First winter females are shown.

Genus *Piranga*

Hepatic Tanager, page 539

Summer Tanager, page 540

Scarlet Tanager, page 541

Western Tanager, page 541

Flame-colored Tanager, page 539

Genus *Pheucticus*

Rose-breasted Grosbeak, page 544

Black-headed Grosbeak, page 544

Genus *Rhodothraupis*

Crimson-collared Grosbeak, page 543

Genus *Cardinalis*

Northern Cardinal, page 542

Pyrrhuloxia, page 542

Genus *Cyanocompsa*

Blue Bunting, page 548

Genus *Passerina*

Blue Grosbeak, page 548

Lazuli Bunting, page 547

Indigo Bunting, page 547

Varied Bunting, page 546

Painted Bunting, page 546

Genus *Spiza*

Dickcissel, page 545

Hepatic Tanager

Piranga flava

L 8" WS 12.5" WT 1.3 oz (38 g)

Larger and stockier than other tanagers, with relatively short tail and stout gray bill. In all plumages note grayish flanks, dark eyeline, and dusky cheek; brightest color always on forehead and throat (orange-red on male, yellow-orange on female).

VOICE: Song clearer than other tanagers; more like Black-headed Grosbeak; delivered slowly with distinct pauses; softer and less metallic than Black-headed Grosbeak, with a few hoarse notes. Call a low, dry *chup* like Hermit Thrush. Flight call a husky, rising *weet*.

Uncommon in montane pine-oak and other coniferous forest. Habits similar to Western Tanager. Very little overlap in breeding habitat/elevation with Summer Tanager. Generally solitary or in pairs.

Flame-colored Tanager

Piranga bidentata

L 7.75" WS 12" WT 1.2 oz (35 g)

Distinguished from Western Tanager by streaked back and rump, orange-yellow plumage, dark border on rear cheek, dark bill, and large white spots on tertials. Hybrids occur, but resemble one-year-old male Flame-colored.

VOICE: Song similar to Western but slower, rougher, with longer pauses between phrases and little pitch change *zheer, zheeree, zhrree, zherri*. All calls apparently very similar to Western.

Very rare visitor from Mexico to pine-oak woods along mountain canyons in Arizona and western Texas; one or two individuals recorded annually and has nested in region. Habits similar to Western Tanager, and the two species have hybridized.

Summer Tanager

Piranga rubra

L 7.75" WS 12" WT 1 oz (29 g)

Larger than Scarlet or Western Tanagers, with larger bill; often raises tail. Female color extremely variable but often more orange-yellow than Scarlet or Western, with paler wings that contrast little with back.

Overall plumage color of female tanagers is variable, complicating identification. Summer is most variable, from pale grayish to orange-red, and identification must focus less on overall color and more on body proportions, calls, and color contrasts such as between wings and body.

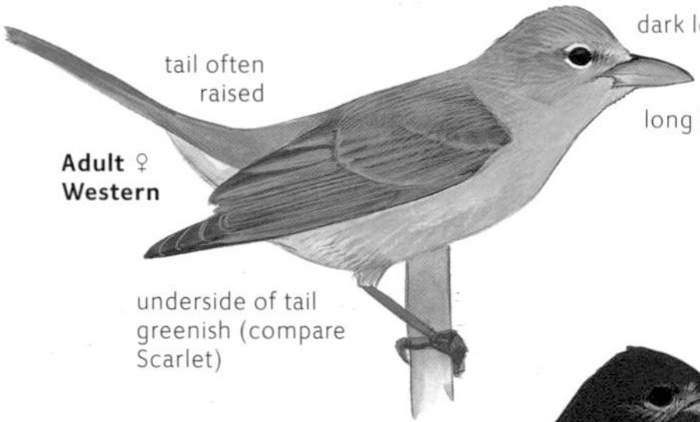

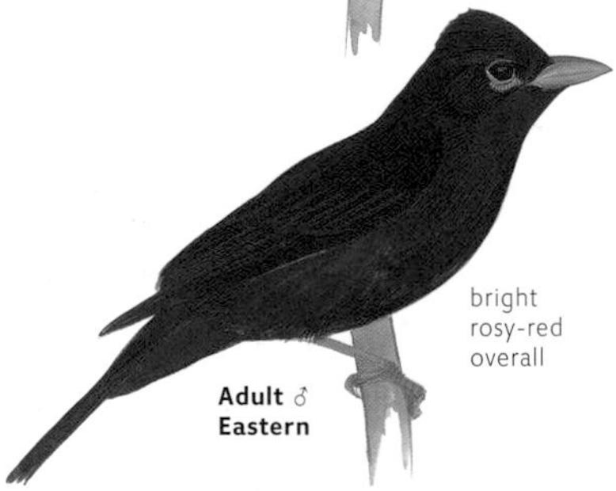

VOICE: Song of five to ten robinlike, musical, three-syllable phrases; some hoarse with brief but distinct pauses between phrases. Call a descending series of hard, unmusical notes *pituk* to *pikitukituk*; also a more rapid, descending rattle *kdddd- rrrddi*. Flight call a soft, wheezy *veedrr* or *verree*.

Common. In West found in mature riparian cottonwood forests; in East prefers relatively open mixed lowland forests of pines and hardwoods. Usually solitary, in upper levels of trees. Bees and wasps are favorite foods.

Western populations (west of central Texas) are paler overall, especially on the sides of the neck, and have slightly longer bill than Eastern. Overall plumage color is extremely variable in both populations, but contrast between neck sides and back is always stronger in western birds. Song of Western birds may be slightly slower tempo and lower-pitched.

Western Tanager

Piranga ludoviciana

L 7.25" WS 11.5" WT 0.98 oz (28 g)

Male distinctive yellow and black with red head. Female similar to Scarlet Tanager, but note dusky back contrasting with yellowish head and rump, and obvious pale wingbars (rarely shown by Scarlet).

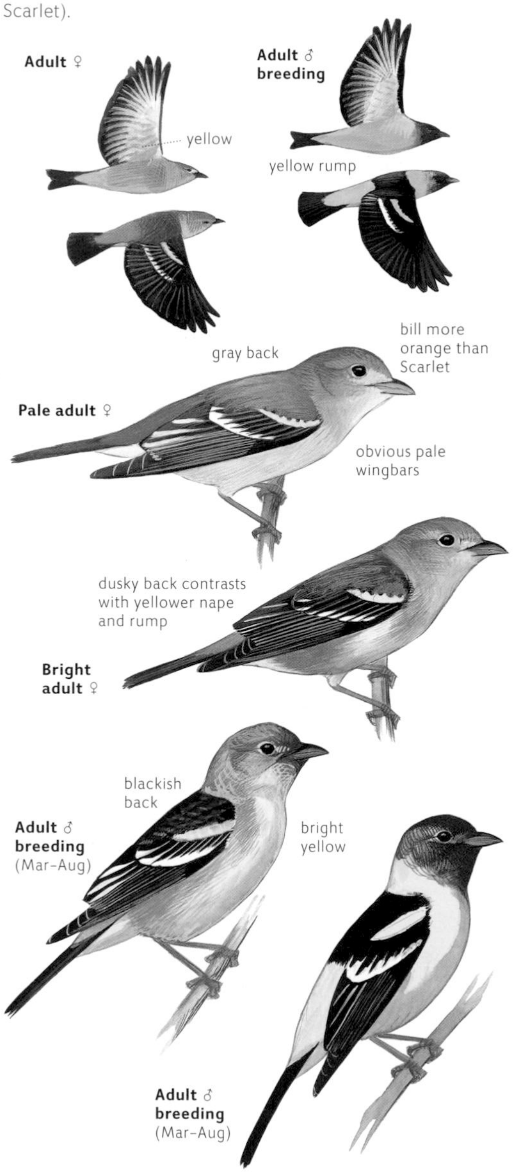

VOICE: Song similar to Scarlet. Call a quick, soft, rising rattle *pridídit*. Flight call a soft whistle *howee* or *weet*.

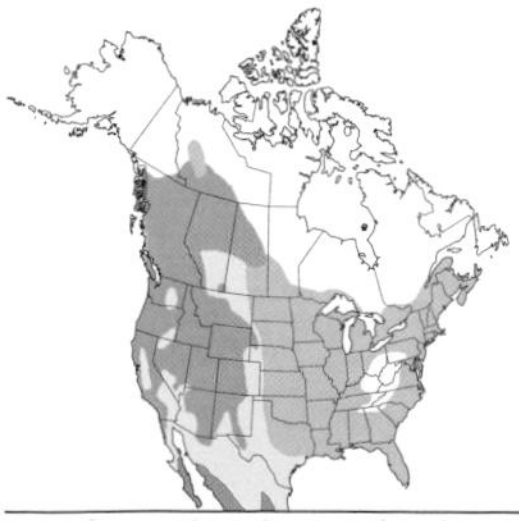

Common in both coniferous (such as Ponderosa pine) and mixed woods. Usually solitary, in upper levels of trees. Feeds on insects and larvae gleaned from leaves.

Scarlet Tanager

Piranga olivacea

L 7" WS 11.5" WT 0.98 oz (28 g)

Breeding male unmistakable. In other plumages very similar to Western Tanager but note uniform greenish upperside, and wings usually without wingbars. Drab females can be similar to Summer Tanager; note dark wings and tail and smaller bill.

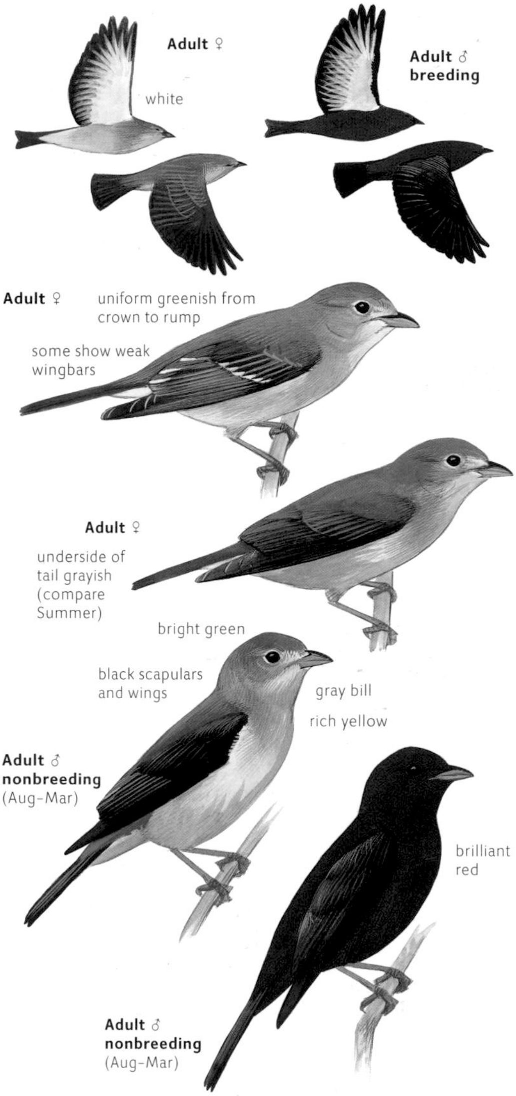

VOICE: Song of about five phrases in fairly rapid, continuous series; pattern reminiscent of American Robin but phrases hoarse, notes more slurred. Call a hard *chik-brrr*; may give single or double *chik* note without *brr*. Flight call a clear whistle *puwi*.

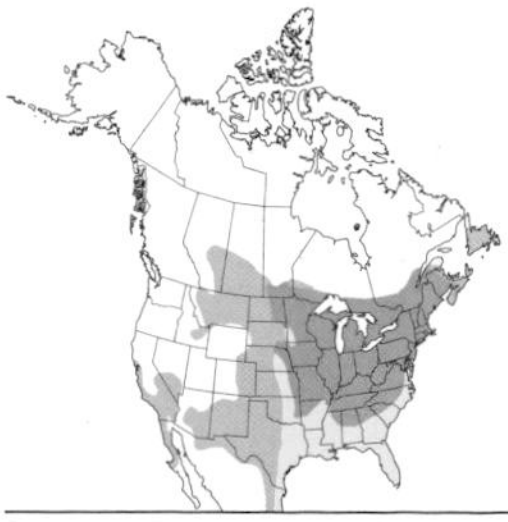

Common in mature deciduous forests. Usually solitary, in upper levels of trees. Feeds on insects and larvae gleaned from leaves.

Northern Cardinal

Cardinalis cardinalis

L 8.75" WS 12" WT 1.6 oz (45 g)

Male brilliant red with black face and red bill; longer-tailed and with very different bill and habits than tanagers. Female also has large triangular red-orange bill, with reddish wings and tail, dark face, brownish overall.

Southwest populations (mainly Arizona) have a larger bill with slightly decurved culmen, a longer and bushier crest, paler overall with less black on the face, particularly across the forehead.

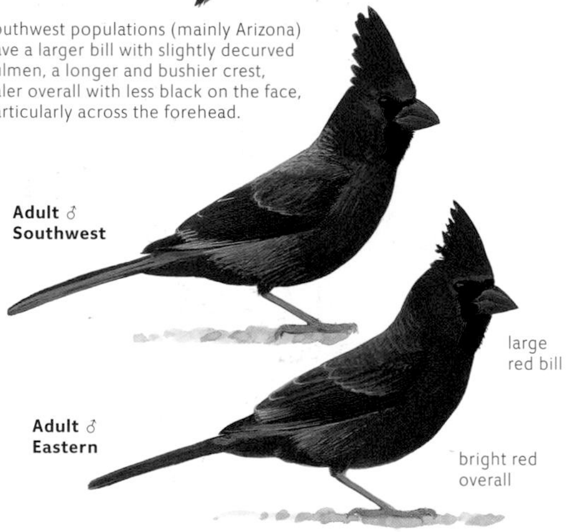

VOICE: Song a series of high, clear, sharp, mostly slurred whistles *woit woit woit chew chew chew chew chew* or *pichew pichew tiw tiw tiw tiw tiw tiw*; many variations. Call a high, hard *tik*; also a softer, rising *twik*.

Common in brushy understory or forest edges, hedgerows, dense thickets. In pairs or small groups year-round. Feeds on seeds, fruit, and insect larvae.

Pyrrhuloxia

Cardinalis sinuatus

L 8.75" WS 12" WT 1.3 oz (36 g)

Similar to Northern Cardinal, most easily distinguished by shape and color of bill: stubby, curved, and yellowish to dusky. Also has longer, pointed crest; less red in wings, grayer overall color with rosy-red highlights, and no dark on face.

VOICE: Song similar to Northern Cardinal, but averages slightly sharper and higher. Call notes slightly longer and squeakier than Northern Cardinal.

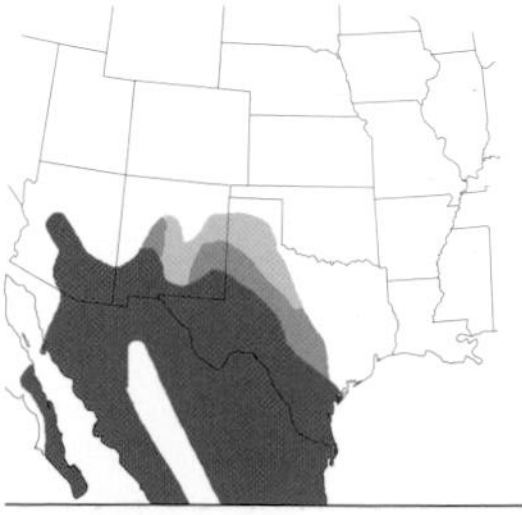

Common in brushy desert habitat. Habits similar to Northern Cardinal, including moving in small groups through brushy vegetation. Where range overlaps in Southwest, this species is found in arid desert scrub, Northern Cardinal in riparian woods.

Crimson-collared Grosbeak

Rhodothraupis celaeno

L 8.75" WS 13" WT 1.9 oz (54 g)

Cardinal-size. Note stout dark bill with curved culmen, long tail, generally dark and uniform coloration on body, wings, and tail. All have black head, more sharply defined on adults; male has wine-red collar, female and immature pale yellow-olive.

VOICE: Song a varied, low, husky warble with quality like Black-headed but more varied with accelerating tempo; ending with bouncy phrases, dry rattle, and upslurred *weeee*. Call a strong, clear, piercing *pweees* or *seeeuw*.

Very rare visitor from Mexico; usually single birds, but sometimes in small groups. Found in dense woods, where it usually stays close to the ground in brushy cover. Generally quiet and sedate, most often detected by call. Feeds mainly on leaves and fruit.

Yellow Grosbeak

Pheucticus chrysopeplus

L 9.25" WS 14.5" WT 2.2 oz (62 g)

Much larger than other grosbeaks, with large head and massive, dark gray bill. Pale yellow color with black-and-white wings and tail most similar to Western Tanager and Scott's Oriole, which are both smaller with very different shape and habits.

VOICE: Song of rich, clear whistles resembling Black-headed but lower, slower, simpler, with halting rhythm: oriole-like *toodi todi toweeoo*. Call a sharp, metallic *piik* intermediate between other *Pheucticus* grosbeaks. Flight call a soft, whistled *hoee*.

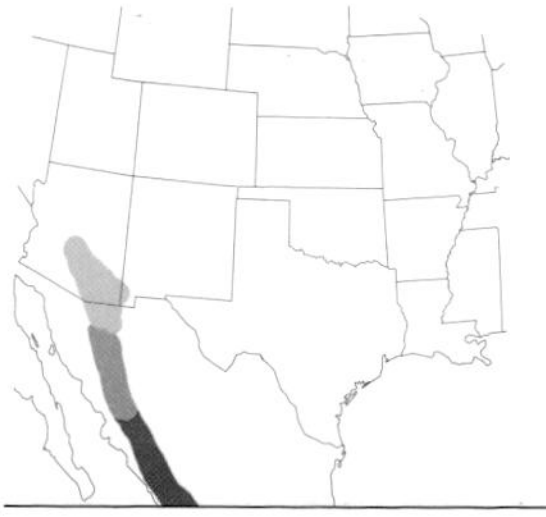

Very rare visitor from Mexico to Arizona, mostly in lower elevations of oak woods along streams in canyons in Jun–Aug. Habits similar to Black-headed Grosbeak.

Black-headed Grosbeak

Pheucticus melanocephalus

L 8.25" WS 12.5" WT 1.6 oz (45 g)

Male distinctive. Female very similar to Rose-breasted Grosbeak; note sparser, finer streaks across buffy breast, and darker bicolored bill. Immature male Rose-breasted in fall especially similar to Black-headed, but with pink underwing coverts.

VOICE: Song a whistled warble, faster, higher, and choppier than Rose-breasted. Call a high, sharp *pik*; more wooden and less squeaky than Rose-breasted; recalls Downy Woodpecker. Flight call like Rose-breasted. Juvenile begs with plaintive, low whistle *weeoo* with wide pitch change.

Common in mature deciduous woods; also found in other wooded or brushy habitats, especially during migration. Usually solitary, staying mostly in upper levels of trees. Feeds on insects, fruit, and seeds.

Rose-breasted Grosbeak

Pheucticus ludovicianus

L 8" WS 12.5" WT 1.6 oz (45 g)

Black, white, and red adult male unmistakable. Female's boldly striped head pattern recalls smaller Purple Finch, but note white wing markings and very large bill. Female very similar to Black-headed Grosbeak but with coarser streaks on whiter breast.

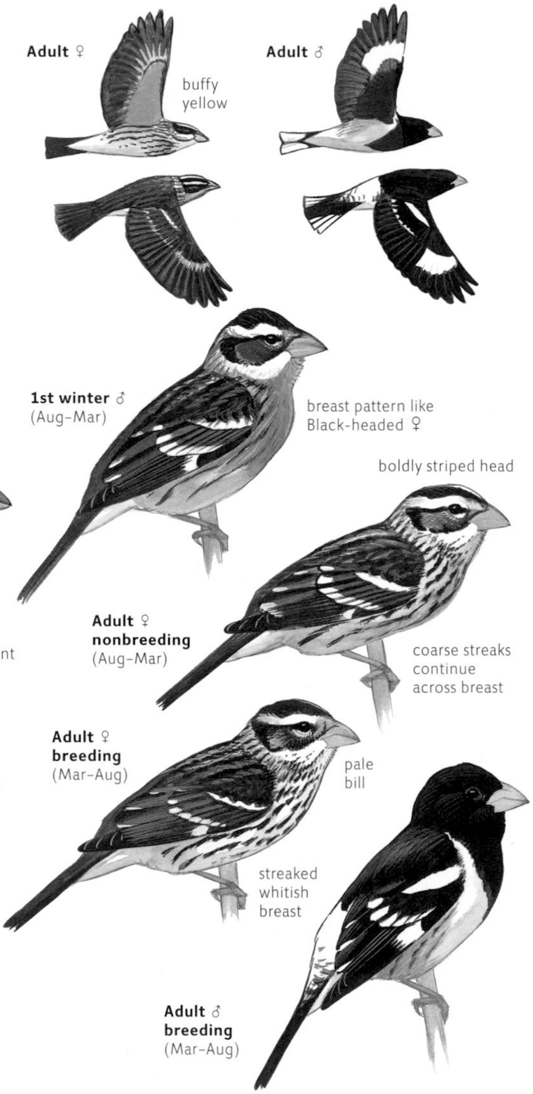

VOICE: Song a slow, whistled warble, robin-like but slightly husky in quality, without gurgling notes; pace steady, slow. Call a sharp, squeaky *iik* like sneakers on a gym floor. Flight call a soft, wheezy *wheek*; thrushlike, with airy quality (unlike husky trumpet sound of Baltimore Oriole).

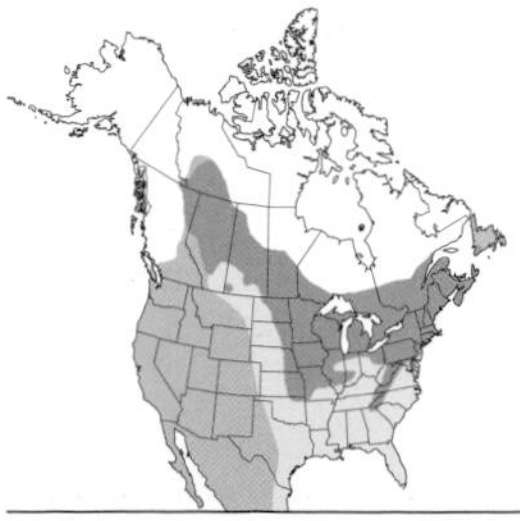

Common in mature deciduous forests. Solitary or in small groups, staying mostly in upper levels of trees. Feeds on insects, fruit and seeds. Migrants found in variety of wooded or brushy habitats. Habits similar to Black-headed Grosbeak.

Identification of Grosbeaks

Rose-breasted Grosbeak and Black-headed Grosbeak are very similar in female and immature plumages, best distinguished (with care) by breast pattern and color. A common source of confusion is first-winter male Rose-breasted Grosbeak, which has unstreaked orange breast identical to female Black-headed Grosbeak. Check the color of underwing coverts—rose-red on male Rose-breasted Grosbeak, lemon yellow on all Black-headed Grosbeak.

Black-headed × Rose-breasted Grosbeak hybrid Adult male breeding (Mar–Aug)

Another pitfall involves hybrids, which are uncommon in the limited area where breeding range overlaps, rarely identified elsewhere. Note that first summer male of both species can show some of the same features. Check age and assess all features before identifying a hybrid. Female hybrids are nearly impossible to identify.

Identification of *Passerina* Buntings

Distinguishing drab female and immature plumages of the four species of buntings requires careful attention to overall plumage color, bill shape, and wing pattern. The most difficult species to distinguish are Indigo and Lazuli Buntings, which are very closely related and differ only in plumage, and all features of female and immature plumage overlap to some extent. Most can be identified with reasonable confidence by breast pattern. Indigo Bunting usually shows at least faint streaking on the breast, Lazuli Bunting (except in juvenal plumage) lacks streaking.

Indigo × Lazuli Bunting hybrid Adult ♂ breeding (Apr–Sep)

twitching sideways tail-flick characteristic of all *Passerina* buntings, including Blue Grosbeak

These two species hybridize frequently in a broad area of range overlap, producing a variety of intermediate plumages. Male hybrids are fairly easy to recognize by extensive blue color with white belly and thin white wingbars. Female hybrids are nearly impossible to identify.

Dickcissel

Spiza americana

L 6.25" WS 9.75" WT 0.95 oz (27 g) ♂>♀

Size and shape similar to House Sparrow, but sleeker with slightly longer bill. Note plain gray-brown cheeks, black breastband, yellow breast. On drab individuals look for large pale bill, plain cheek, and hint of yellow on breast.

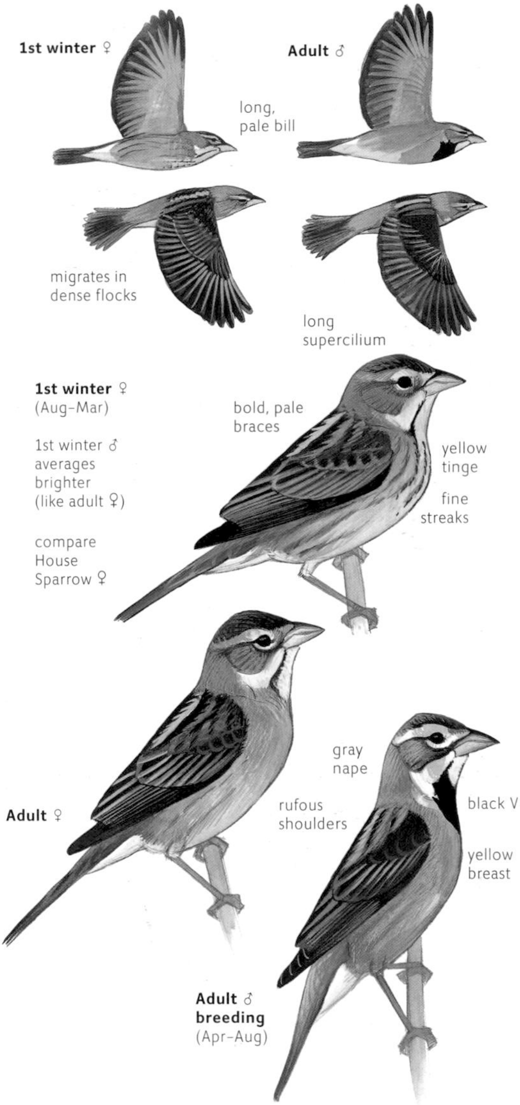

VOICE: Song a series of short notes with dry, insectlike quality *skee-dlees chis chis chis* or *dick dick ciss ciss ciss*; quality reminiscent of song of Henslow's Sparrow. Call a dry, husky *chek* or *pwik*. Flight call a very distinctive, low, electric buzz *fpppt*.

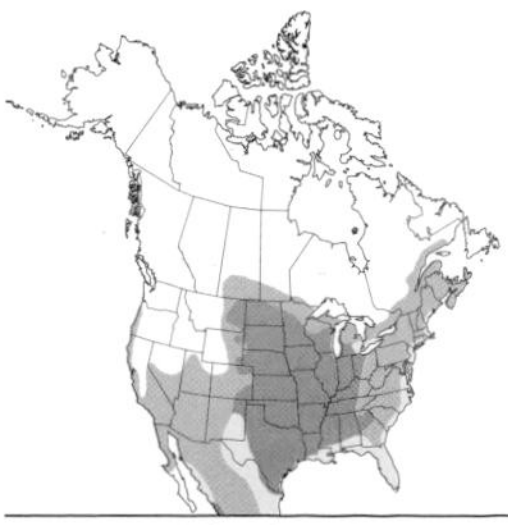

Common in grassy or weedy fallow fields and tallgrass prairies with scattered shrubs, trees, or hedgerows. May form large flocks in brushy or weedy habitats when not breeding; vagrants often flock with House Sparrows. Feeds on insects and seeds.

Varied Bunting

Passerina versicolor

L 5.5" WS 7.75" WT 0.42 oz (12 g)

Male dark overall, often appearing black. Female similar to other buntings but very plain and drab overall; note curved culmen, unpatterned wings, and unstreaked, pale brownish underparts.

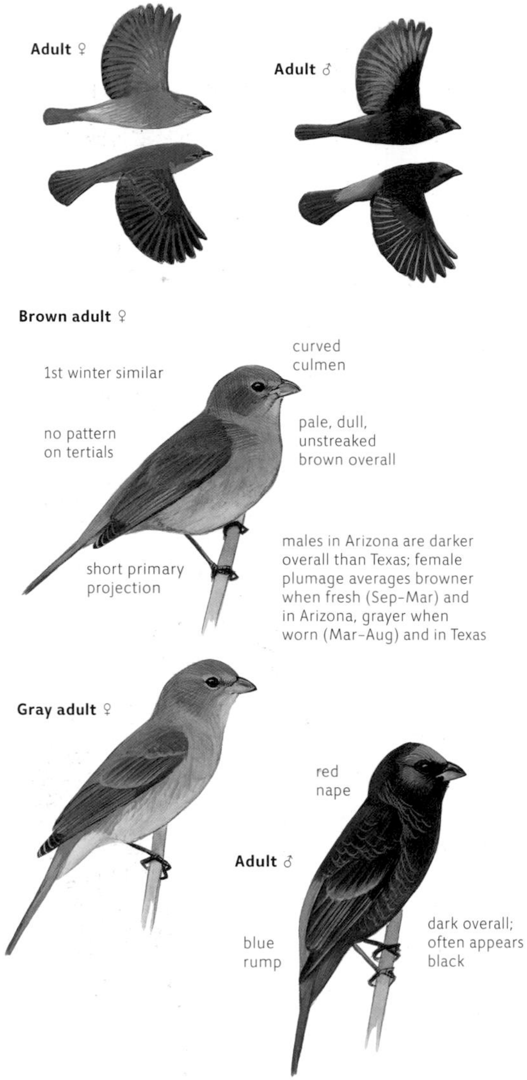

VOICE: Song lower-pitched and slightly harsher than other buntings, without repeated phrases; often includes hoarse, descending *veer*; rhythm intermediate between choppy (like Lazuli) and smooth (like Painted). Call a dry *spik* and flight call a long buzz, both like other buntings.

Uncommon and local in dense mesquite brush along desert washes. Habits similar to other buntings.

Painted Bunting

Passerina ciris

L 5.5" WS 8.75" WT 0.54 oz (15.5 g)

Adult male unmistakable, with red rump and underparts. Other plumages plain greenish; our only green finch-like bird. Immature females are drab buffy olive, immature males and some adult females brilliant green above and yellow-olive below.

VOICE: Song a sweet, continuous warble; quality like Indigo but with unbroken singsong rhythm like Blue Grosbeak. Call *pwich* averages lower and softer than Indigo, with rising inflection. Flight call a buzzing, slightly rising *vvvit*; not as strong or musical as other buntings.

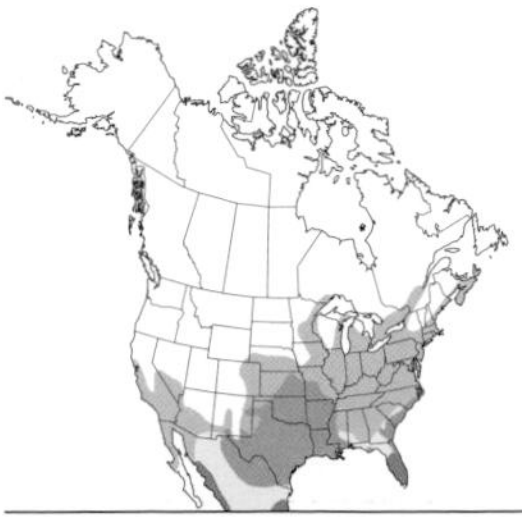

Uncommon and local. Found in brushy lowlands at forest edges or with scattered tall trees. Habits similar to Indigo Bunting, but slightly more secretive.

Lazuli Bunting

Passerina amoena

L 5.5" WS 8.75" WT 0.54 oz (15.5 g)

Male distinctive (but beware confusing hybrids with Indigo). Female difficult to distinguish from Indigo; note distinct wingbars and grayish throat contrasting with buffy breast that has no trace of streaking.

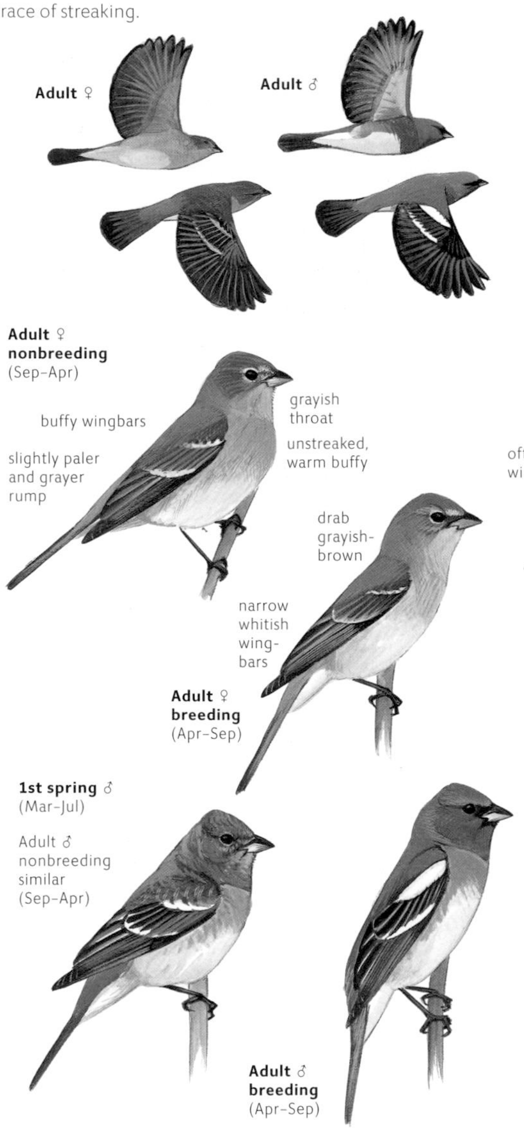

VOICE: Song a high, sharp warble; averages slightly longer, higher, faster, and perhaps less repetitive than Indigo. Call a dry *pik*; may average slightly higher and harder than Indigo. Flight call averages higher and clearer than Indigo.

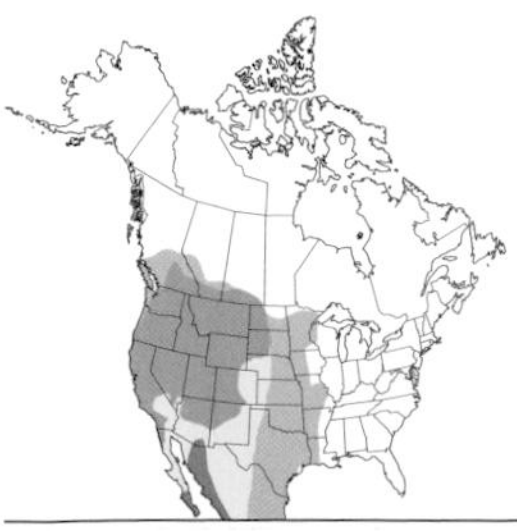

Common in any brushy or weedy habitat, especially along streams in arid regions. Habits similar to Indigo Bunting.

Indigo Bunting

Passerina cyanea

L 5.5" WS 8" WT 0.51 oz (14.5 g)

Bright blue male distinctive. Female easily confused with sparrows; note warm brownish color, plain face, bicolored bill, rounded tail with hints of blue, habits, and calls. Very similar to Lazuli and hybridizes extensively where range overlaps.

VOICE: Song a high, sharp warble with most phrases repeated; quality musical and metallic *ti ti whee whee zerre zerre* ("fire fire where where here here"); similar to other buntings and American Goldfinch. Call a dry, sharp *spik* or *pwik*. Flight call a relatively long, shrill buzz.

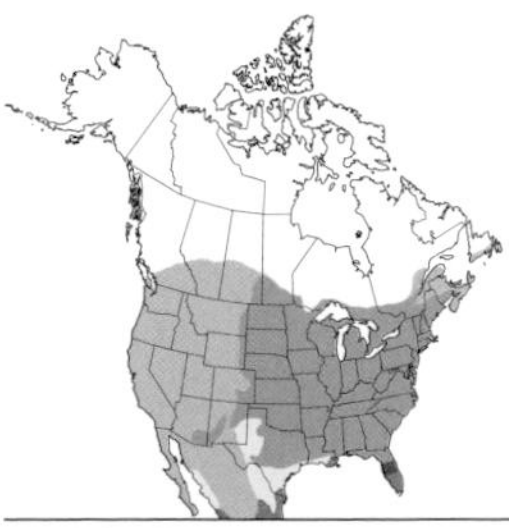

Common in open brushy areas, including weedy fields and hedgerows, with trees nearby. Found mainly in low vegetation, but sings from high in trees. Usually solitary or in small groups, but may form flocks of dozens at favored migration stopovers. Feeds on insects and seeds.

Blue Bunting

Cyanocompsa parellina

L 5.5" WS 8.5" WT 0.53 oz (15 g)

More secretive than Indigo Bunting, stockier with thicker bill and curved culmen. Male dark overall with bright blue highlights. Female dark reddish-brown overall, with plain wings, unstreaked breast, and stout dark bill.

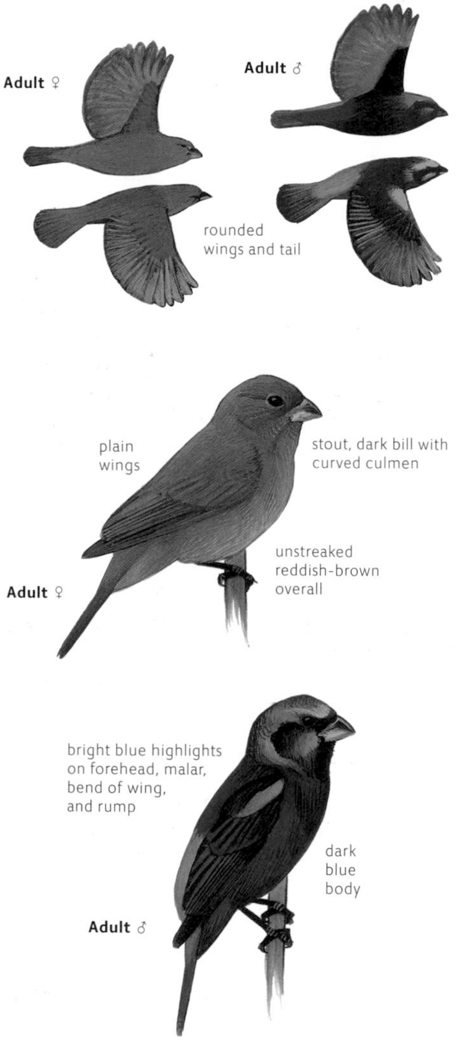

VOICE: Song a high, sweet, tinkling warble of clear phrases: beginning with a couple of separate notes; rhythm jumbled without pauses; fading at end. Call a clear, simple, metallic *chip* reminiscent of Eastern Phoebe or Hooded Warbler, unlike other buntings.

Very rare visitor from Mexico to extreme southern Texas; recorded annually in dense brushy woods, with most records of single birds at feeders from Dec–Mar. Feeds on seeds and insects.

Blue Grosbeak

Passerina caerulea

L 6.75" WS 11" WT 0.98 oz (28 g)

In all plumages similar to Indigo Bunting, but larger with relatively large head and bill, longer, rounded tail, and rufous wingbars. Male darker blue than Indigo, female and immature never streaked on breast.

VOICE: Song a rich, husky warble with distinctive unbroken tempo and mumbled quality; steady tempo recalls Warbling Vireo and Painted Bunting. Call a very metallic, hard *tink* or *chink*. Flight call a harsh buzz like buntings but stronger, rougher, lower.

Uncommon in open weedy fields with brushy patches and hedgerows; sings from tall weeds or bushes. Often in small groups of two to five. Forages for seeds and insects low in weeds.

Orioles and Blackbirds

FAMILY: ICTERIDAE

26 species in 8 genera (3 rare species not shown here). All icterids have rather slender, pointed bills; and most have relatively long tails (meadowlarks a notable exception). All can be found in groups when not breeding, and some species (especially blackbirds, cowbirds, and grackles) can form large flocks and gather in immense nighttime roosts in trees or reeds. Most species are found in open areas such as farmland, pastures, and lawns, feeding on seeds and insects; orioles are found in woodlands and feed mainly on insects, fruit, and nectar. Many species build cup-shaped nest in trees or shrubs; meadowlarks and Bobolink on the ground; orioles build pendulous hanging nest high in trees; cowbirds are brood parasites and do not build nests. Adult females are shown.

Genus *Agelaius*

Red-winged Blackbird, page 554

Tricolored Blackbird, page 555

Genus *Sturnella*

Eastern Meadowlark, page 559

Western Meadowlark, page 558

Genus *Dolichonyx*

Bobolink, page 558

Genus *Quiscalus*

Common Grackle, page 551

Boat-tailed Grackle, page 553

Great-tailed Grackle, page 552

Genus *Xanthocephalus*

Yellow-headed Blackbird, page 555

Genus *Molothrus*

Shiny Cowbird, page 557

Bronzed Cowbird, page 556

Brown-headed Cowbird, page 557

Genus *Icterus*

Orchard Oriole, page 560

Hooded Oriole, page 560

Streak-backed Oriole, page 564

Bullock's Oriole, page 565

Spot-breasted Oriole, page 561

Altamira Oriole, page 561

Audubon's Oriole, page 562

Baltimore Oriole, page 565

Scott's Oriole, page 562

Genus *Euphagus*

Rusty Blackbird, page 550

Brewer's Blackbird, page 550

Brewer's Blackbird

Euphagus cyanocephalus

L 9" WS 15.5" WT 2.2 oz (63 g) ♂>♀

More slender than Red-winged Blackbird, with longer tail and thinner bill; distinguished from Rusty Blackbird by thicker bill, dark eye of female, and habits. Distinguished from grackles by smaller size, thinner bill, and square tail.

VOICE: Song a short, high, crackling *t-kzzzz* or *t-zherr*; usually buzzy (never so in Rusty). Flight call a dry *ket*; averages higher, harder, and more nasal than other blackbirds.

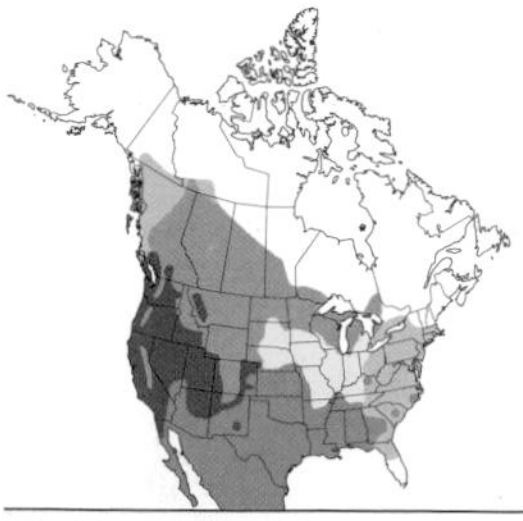

Common and widespread; often in parks and parking lots as well as agricultural fields, livestock pastures, lawns, and other open areas. Little habitat overlap with Rusty Blackbird. Usually found in small flocks, not mixing freely with other blackbirds. Forages for seeds and insects on open ground.

Rusty Blackbird

Euphagus carolinus

L 9" WS 14" WT 2.1 oz (60 g) ♂>♀

Slender and long-tailed; pale iris always obvious. Distinguished from grackles by smaller size, rounded tail, thinner bill, and (in fall and winter) rusty feather edges. Very similar to Brewer's Blackbird, note thinner bill, more rusty plumage.

VOICE: Song a soft gurgle followed by high, thin whistle *ktlr-teee*; often alternated with a soft, rustling or gurgling, descending *chrtldltlr*; similar to Brewer's but softer, more gurgling. Flight call a low *tyuk*; lower, longer, more descending than Red-winged; not as sharp as Brewer's.

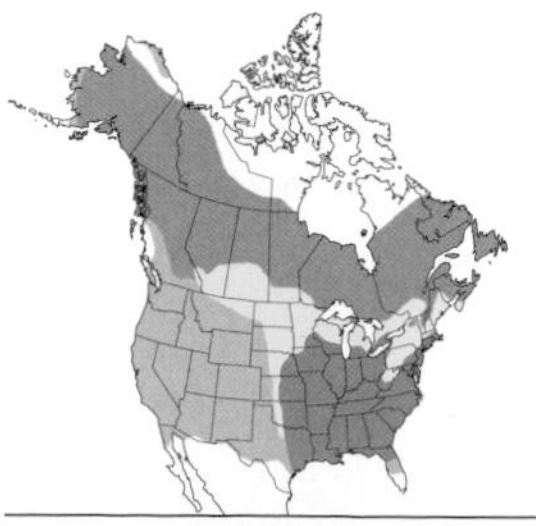

Uncommon. Nests in spruce trees near bogs within boreal forests. In winter found in wooded swamps such as in river floodplains, seldom in open fields. Found in small flocks, usually not mixed with other blackbirds. Forages for insects and seeds on the ground.

Common Grackle

Quiscalus quiscula

L 12.5" WS 17" WT 4 oz (115 g) ♂>♀

Larger and heavier than blackbirds, with longer and thicker bill, much larger tail with distinctive keel shape, pale iris, and varied blue, green, and bronze iridescence. Females are smaller and drabber than males, juveniles drab dark brown with dark eye.

VOICE: Song an unmusical, harsh, metallic hiss *kh-sheee* or *khr-reezzh*. Call a very harsh, toneless, dry *karrz* or *kerrrr* or *karrrg*. Male call a thin, wheezy, toneless *zweeesh* and nasal, hoarse *krrrjk*. Flight call a low, dry *kek*; deeper than all blackbirds, with distinctive harsh quality.

Common and widespread. Nests in trees in small woodlots or edges, often in suburban neighborhoods, especially in dense evergreens such as spruce or juniper. Forages for seeds and invertebrates on the ground in agricultural fields, lawns, or within open woods, often in very large flocks, sometimes mixed with blackbirds.

Interior or Bronzed form found in most of range, with Coastal or Purple form from New York to Louisiana. Differ mainly in color of iridescence of male plumage, and intergrade commonly where range overlaps. Florida population similar to Coastal but slightly smaller with relatively large bill.

Great-tailed Grackle

Quiscalus mexicanus

♂ L 18"   WS 23"   WT 7 oz (190 g)
♀ L 15"   WS 19"   WT 3.7 oz (105 g)

The largest and longest-tailed grackle (but some western females quite small). Told from Boat-tailed Grackle by pale iris, longer tail, voice, and habitat. Female averages grayer than Boat-tailed, but some overlap.

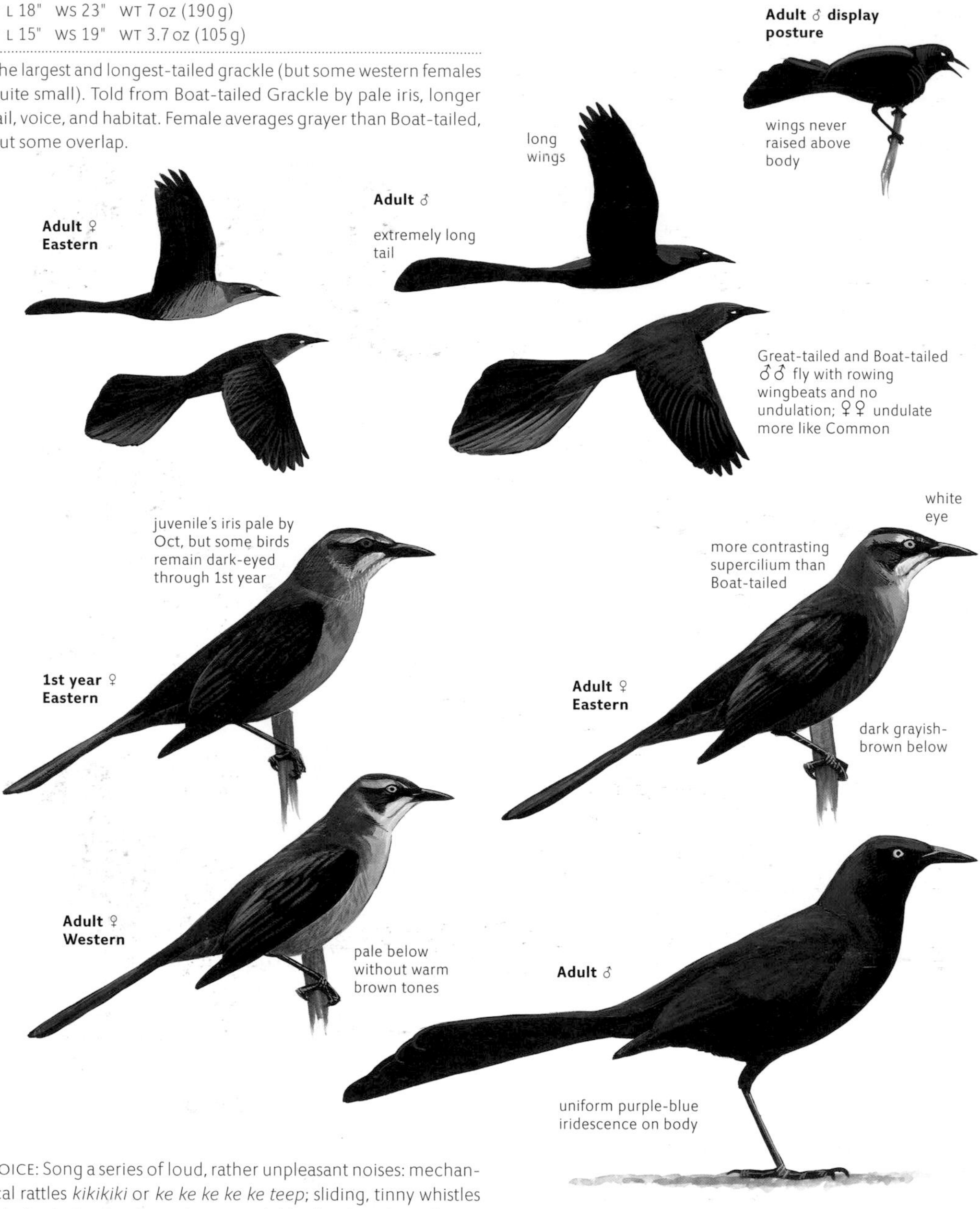

VOICE: Song a series of loud, rather unpleasant noises: mechanical rattles *kikikiki* or *ke ke ke ke ke teep*; sliding, tinny whistles *whoit whoit...*; harsh, rustling sounds like thrashing branches or flushing toilet; loud, hard *keek keek...* or *kidi kidi....* Common call of male a low, hard *chuk* or *kuk*; female call a softer, husky *whidik* or *whid*.

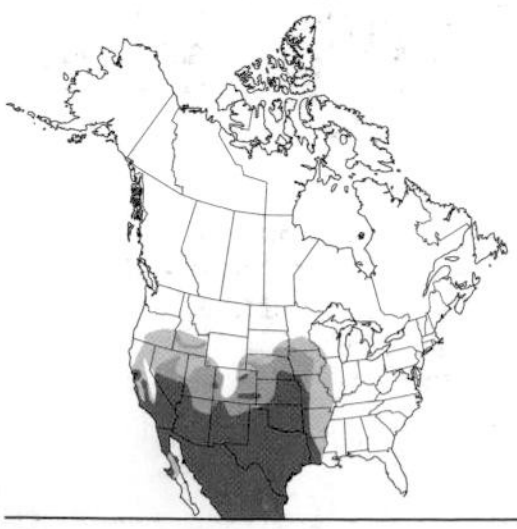

Common and increasing. Nests and roosts in trees and bushes, often in very noisy nocturnal gatherings in ornamental trees along city streets. Forages for seeds and invertebrates mainly on the ground in open areas such as agricultural fields, parking lots, lawns; often in large flocks.

Eastern and Western populations differ in size and female plumage color. Western birds average 15 percent smaller, with relatively short tail and long bill; some females are very pale grayish. From central Arizona eastward all are larger, averaging grayer in west, browner in Texas. Song differs geographically; more study is needed.

Boat-tailed Grackle

Quiscalus major

♂ L 16.5" WS 23" WT 8 oz (215 g)
♀ L 14.5" WS 17.5" WT 4.2 oz (120 g)

Larger, more slender, and longer-tailed than Common Grackle; male has more uniform blue-green iridescence, female rich brown. Very similar to Great-tailed, but little range overlap; note dark iris of Gulf Coast birds, saltmarsh habitat, and voice.

VOICE: Song a varied series of mostly high, ringing notes *kent kent*... or *teer teer*... or *shreet shreet shreet shreet KEET*; most distinctive a ringing *kreen kreen*.... Also a lower, harsher series mixed with dry, rustling sounds; a very loud, clear whistle *teewp*; a harsh, trilling *kjaaaaar*; and a throaty, rattling *klukluklk*. Call of male a deep *chuk*; female call a rather soft, low *chenk* or *chuup*, may average softer than Great-tailed female.

Common within limited range. Nests and roosts in bushes around marshes; restricted to coastal saltmarsh except in Florida. Forages for seeds and invertebrates on the ground close to water, usually in small flocks.

Iris color varies regionally. Pale along Atlantic coast south to northern Florida, dark in most of Florida and west to Texas, except a small area around Mobile, Alabama where many are pale-eyed. In Louisiana and Texas, where range overlaps with Great-tailed Grackle, all are dark-eyed.

Red-winged Blackbird

Agelaius phoeniceus

L 8.75" WS 13" WT 1.8 oz (52 g) ♂>♀

Male distinctive, black with red shoulder. Female streaked brownish, often mistaken for sparrows; note larger size, often pinkish-orange throat, thin bill, very dense streaks below, blackish tail, and habits.

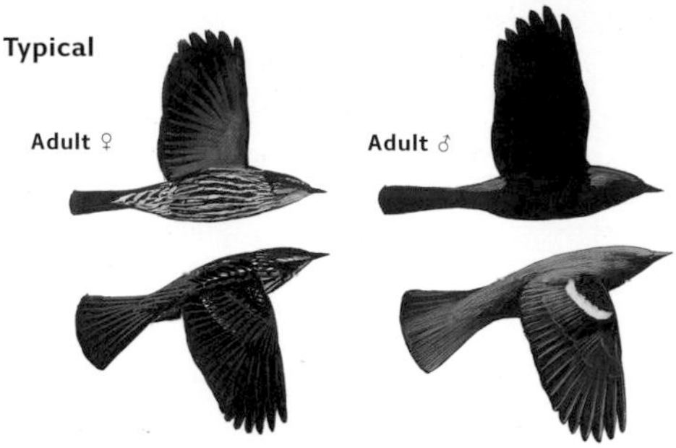

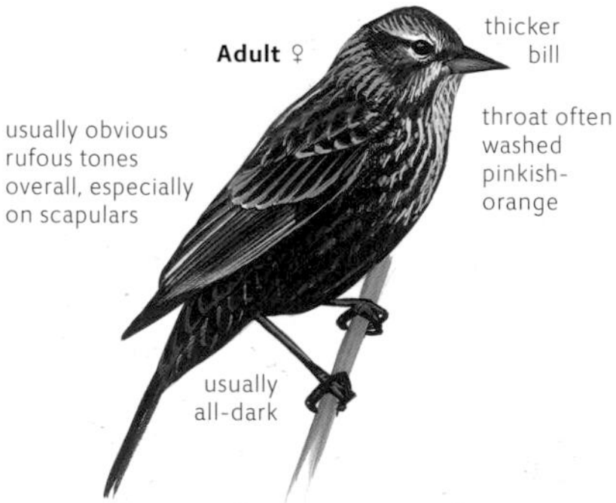

VOICE: Song of several liquid introductory notes followed by variable, harsh, gurgling trill *kon-ka-reeeee*; many western birds (including Bicolored population) sing a less musical *ooPRE-EEEEom*; female song an explosive, harsh rattle (rarely heard from Bicolored). Flight call a low, dry *kek* or *chek*. Alarm call a high, clear, descending *teeeew* or buzzy *zeeer* given by male. Variety of other calls heard mainly from male in nesting season.

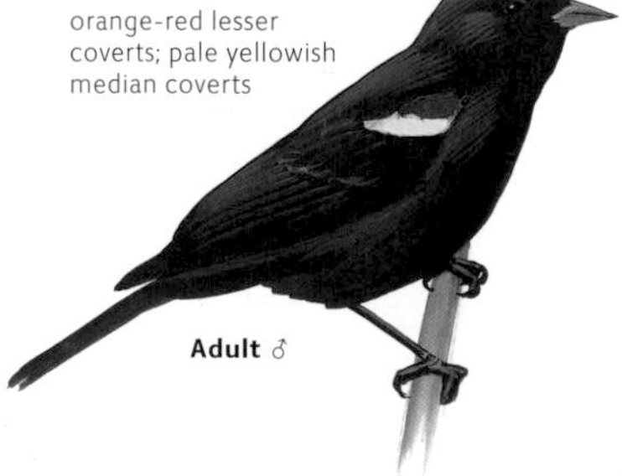

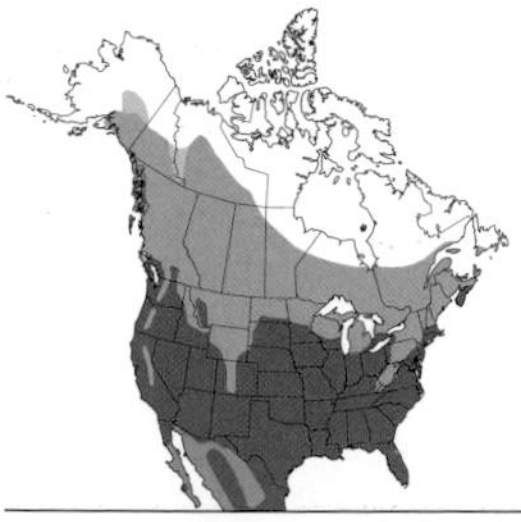

Common; our most widespread blackbird. Nests and roosts in wet, marshy or brushy habitats; almost any small weedy ditch or wet hayfield harbors a breeding pair, and many pairs nest in close proximity in reed beds and similar habitats. Forages for seeds and invertebrates in open fields. Forms very large flocks when not nesting, often segregated by sex; sometimes mixed with grackles, cowbirds, or other blackbirds.

Bicolored population resident in California, Typical forms found throughout range. Male and female Bicolored are distinctive in plumage, and voice differs slightly, but intergrade at edges of range.

Tricolored Blackbird

Agelaius tricolor

L 8.75"   WS 14"   WT 2.1 oz (59 g)   ♂>♀

Habits and appearance very similar to Red-winged Blackbird. Best distinguished by male's dark red shoulder with broad white median coverts, harsh song, and female's lack of rufous tones. Also note more pointed wings, slightly thinner bill.

VOICE: Song a harsh, nasal, descending *oo-grreee drdodrp*; lower-pitched, more nasal, and less musical than Red-winged; flock in song sounds like cats fighting. All calls lower-pitched and some more nasal than analogous calls of Red-winged. Descending, whistled alarm call absent.

Common but very local. Nests in colonies in extensive reedy marshes and rice fields. Winters and forages in large flocks in marshes and on farmland, often with Red-winged Blackbirds.

Yellow-headed Blackbird

Xanthocephalus xanthocephalus

L 9.5"   WS 15"   WT 2.3 oz (65 g)   ♂>♀

Slightly larger than Red-winged Blackbird. Male unique with black body, white primary coverts, and yellow head. Female more subtle, note uniform dark brown body, bright yellow on face and breast.

VOICE: Song extremely harsh, unmusical; a few hard, clacking notes on different pitches followed by wavering raucous wail like chainsaw. Varied calls in colony include a slow, raucous *rad rad rad rad,* a rasping rattle, and a descending, whistled trill. Flight call a low, dry *kuduk* or *kek*.

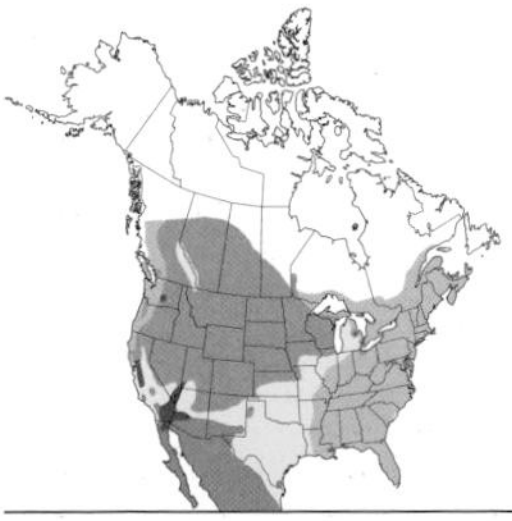

Common locally. Nests and roosts in dense reedy marshes. Forages for seeds and invertebrates in open habitats such as fields and pastures, often in flocks, sometimes mixed with other blackbirds.

Bronzed Cowbird

Molothrus aeneus

L 8.75" WS 14" WT 2.2 oz (62 g) ♂>♀

Stockier than Brown-headed Cowbird, with very thick neck and short tail creating a front-heavy appearance; also note relatively long, heavy bill, glossy wings contrasting with darker head and neck, and red eye.

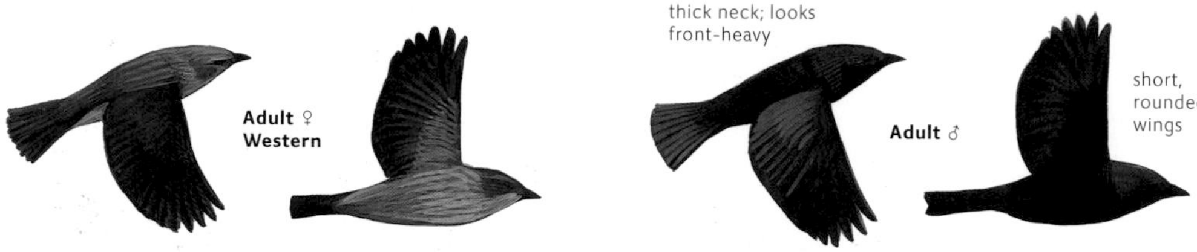

VOICE: Song a soft, tinny, rising whistle with quiet gurgling or rustling noise at beginning and end *kweee-lk*; often repeated on slightly different pitches. Whistle call of male 6–10 seconds long and infrequently heard, a series of tinny or wheezy whistles and grating trills; several regional dialects. Rattle call of female similar to Brown-headed but slower and sharper. Chorus of roosting flock may suggest European Starling.

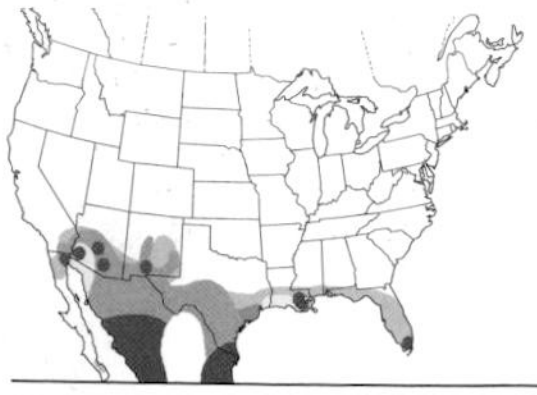

Common locally. Found in open woodlands and brushy forest edges in breeding season; otherwise in flocks that forage on open ground in agricultural fields, lawns, and roadsides. Usually in flocks, sometimes very large, and often associated with other blackbirds. Roosts in trees. Forages for seeds and invertebrates on open ground.

Two distinctive populations differ primarily in female and juvenal plumage; ranges meet in western Texas. Eastern female and juvenile are dark blackish-brown overall, similar to male; Western female and juvenile are pale gray-brown, similar to female Brown-headed. At least three regional dialects in male whistle call are most easily distinguished by the introductory note. Eastern birds give a clear, level whistle about one second long *pseeeeeee*. Western birds in west Texas give a slow, slightly rising trill up to one second long *brrrrreet*. Birds from there west through Arizona give a short upslurred whistle *wink*, and a fourth dialect may exist in western Arizona.

Shiny Cowbird

Molothrus bonariensis

L 7.5" WS 11.5" WT 1.3 oz (36 g) ♂>♀

Slightly longer-tailed than Brown-headed Cowbird, with more rounded wings and thinner bill. Male uniform glossy black. Female averages darker brown than Brown-headed; best distinguished by longer and thinner bill.

VOICE: Song a high, squeaky, clear, liquid series of notes; more varied, not gurgling like Brown-headed; slightly accelerating and descending; ends with one to three liquid *quit* notes. Call a rolling rattle like Brown-headed but slower and more metallic. No other flight call.

Rare visitor from South America, mainly in spring; small numbers now resident in southern Florida. Habits similar to Brown-headed Cowbird and often found with that species.

Brown-headed Cowbird

Molothrus ater

L 7.5" WS 12" WT 1.5 oz (44 g) ♂>♀

Smaller and shorter-tailed than blackbirds, with stout bill and more pointed wings. Male distinctive, with glossy black body and brown head. Female paler gray-brown than other blackbirds, with faint streaks on breast, pale throat.

VOICE: Song low, gurgling notes followed by thin, slurred whistles. Flight whistle of Eastern male high, thin whistles *tseeeee-teeea* or *seeeeeetiti*; flight whistle of Pacific/Mexican more varied, with many local dialects. Call a flat, hard, rising rattle *kkkkkk*. No other flight calls.

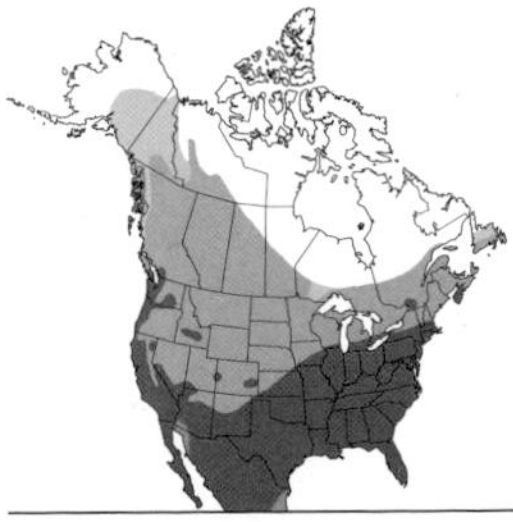

Common in open or patchy woodlands in breeding season, when several males will follow a female, perching in treetops or on poles. At other seasons usually in small flocks on open fields and lawns, often with blackbirds but cowbirds often form tight groups within larger mixed flock. Forages for seeds and invertebrates on open ground.

Bobolink

Dolichonyx oryzivorus

L 7" WS 11.5" WT 1.5 oz (43 g) ♂>♀

Black underside and white on upperside of breeding male unique. Female and nonbreeding male distinguished from sparrows by rich yellowish color, plain nape, pale lores, pointed wings, and habits.

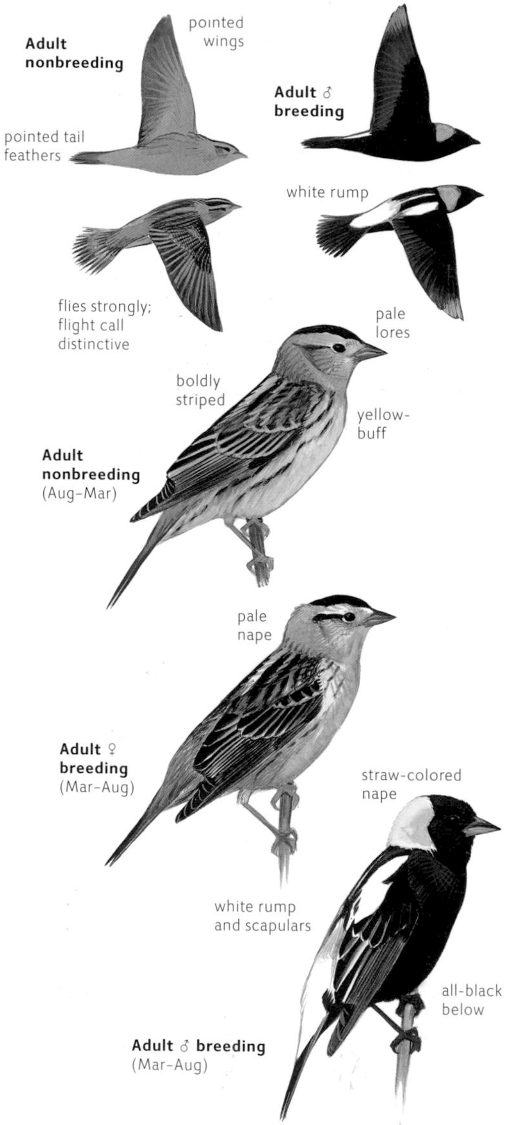

VOICE: Song given in flight a cheerful, bubbling, jangling warble with short notes on widely different pitches; ending faster, fuller, higher. Call a soft, low *chuk* similar to blackbirds. Flight call a soft, musical *bink* or *bwink* similar to some short notes of House Finch.

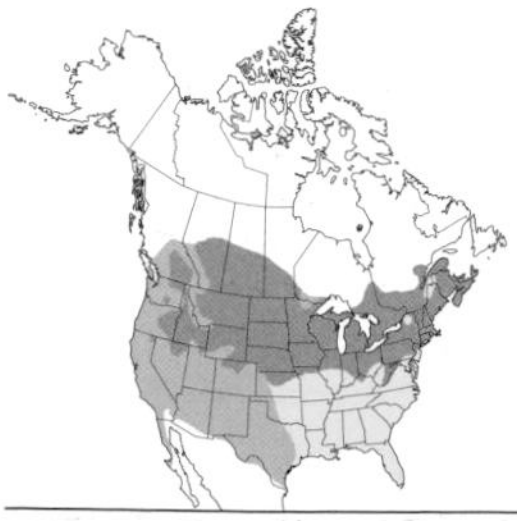

Common locally; nests in open fallow fields, tallgrass prairies, and damp meadows; migrants forage in any weedy field habitat, roost in dense reed beds. Winters and migrates in large flocks, which do not mix with other species. Feeds on seeds and insects.

Western Meadowlark

Sturnella neglecta

L 9.5" WS 14.5" WT 3.4 oz (97 g) ♂>♀

Very similar to Eastern Meadowlark, averages paler gray-brown overall, with less contrasting face pattern, more yellow on malar, and less white in tail, dark bars on feathers of upperside thinner and discrete. Best distinguished by voice.

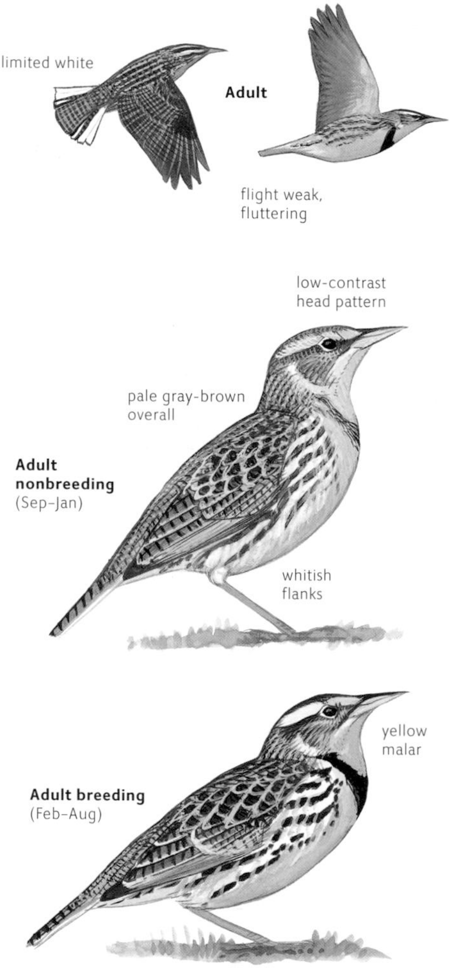

VOICE: Song a rich, low, descending warble *sleep loo lidi lidijuvi*; begins with well-spaced, clear, short whistles and ends with rapid gurgle. Common call a low, bell-like *pluk*; blackbird-like but more musical; also a slow, dull rattle *vidididididididi*. Flight call slightly lower than Eastern.

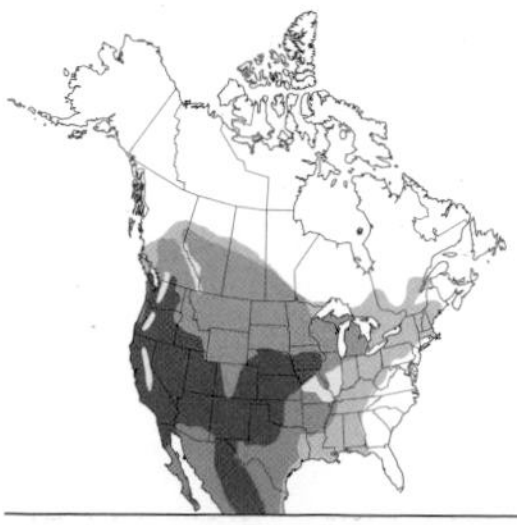

Common in grasslands, agricultural fields, prairies. Winters in loose flocks, often mixed with Eastern Meadowlark where range overlaps. Feeds on seeds, insects, and worms.

Eastern Meadowlark

Sturnella magna

L 9.5" WS 14" WT 3.2 oz (90 g) ♂>♀

Stocky and somewhat awkward-looking with short tail, long bill and legs. Note bright yellow underparts with black breastband, boldly striped head, white outer tail feathers. Appearance very similar to Western; reliably distinguished only by voice.

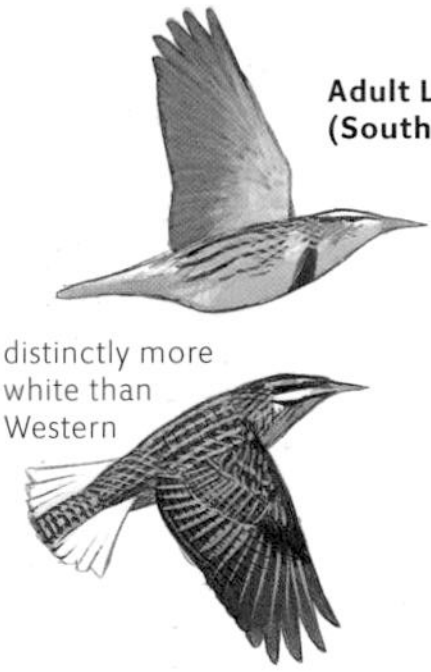

Population nesting in arid grasslands from western Texas to Arizona may be a separate species—Lilian's Meadowlark—with paler overall plumage, more white in tail, and slightly lower-pitched song than Eastern birds. In breeding season, Lilian's is found in dry native grassland, Western Meadowlarks in the same region frequent wetter irrigated land such as alfalfa fields.

VOICE: Song of simple, clear, slurred whistles *seeeooaaa seeee-adoo* with many variations; higher and clearer than Western with no complex gurgling phrases. Call a sharp, electric *dziit* or *jerZIK* and hard, mechanical rattle *ztttttttttt*. Flight call a thin, rising *veeet* or *rrink*.

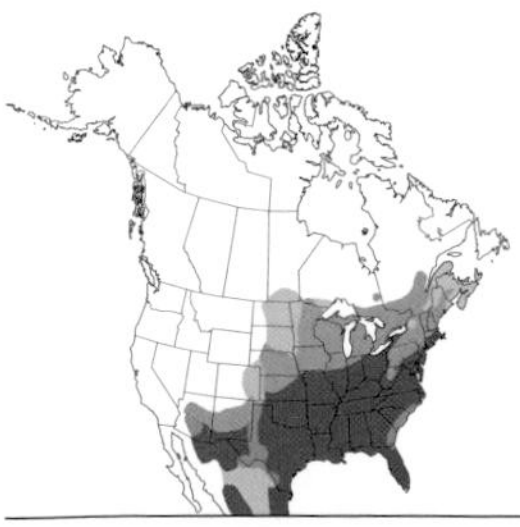

Common locally but declining due to loss of open grassy habitat. Found in expansive grassy fields and prairie, with scattered bushes or hedgerows. Habitat and habits similar to Western Meadowlark. Winters in loose flocks, often mixed with Western where range overlaps.

Voice may be the most reliable clue for identifying meadowlarks, but individual birds can learn the "wrong" song. Listen for details that are less likely to be copied, such as pitch (higher in Eastern) and repertoire size (larger in Eastern: 50 to 100 songs given by each male, but fewer than 10 songs by each male Western). A bird usually uses the same song several times before switching, so extended listening may be required to assess repertoire size. Imitated songs are generally intermediate in pitch, not perfect imitations, and "wrong" songs may comprise only a small part of an individual's repertoire. Although much less variable than songs, "wrong" or intermediate calls are sometimes heard.

Hooded Oriole

Icterus cucullatus

L 8" WS 10.5" WT 0.84 oz (24 g)

A small, slender, long-tailed oriole with curved bill. Male shows orange crown and extensive black face. Female similar to Orchard; note longer tail, drabber overall color, longer bill, more extensive pale edging on wings, and different calls.

VOICE: Song rapid, varied, choppy; combining short, slurred whistles, call notes, and imitations of other species. Call a high, hard *chet* and high, rapid chatter; also a very hard, descending *chairr.* Flight call a sharp, rising, metallic *veek* reminiscent of meadowlarks.

Uncommon in open wooded or brushy habitats, often near fan palms. Habits similar to Orchard Oriole.

Orchard Oriole

Icterus spurius

L 7.25" WS 9.5" WT 0.67 oz (19 g)

Smaller than other orioles and sometimes mistaken for a warbler. Male a distinctive dark chestnut. Female yellowish overall (never orange), very similar to female Hooded Oriole; note smaller size, shorter bill and tail, and narrow wingbars.

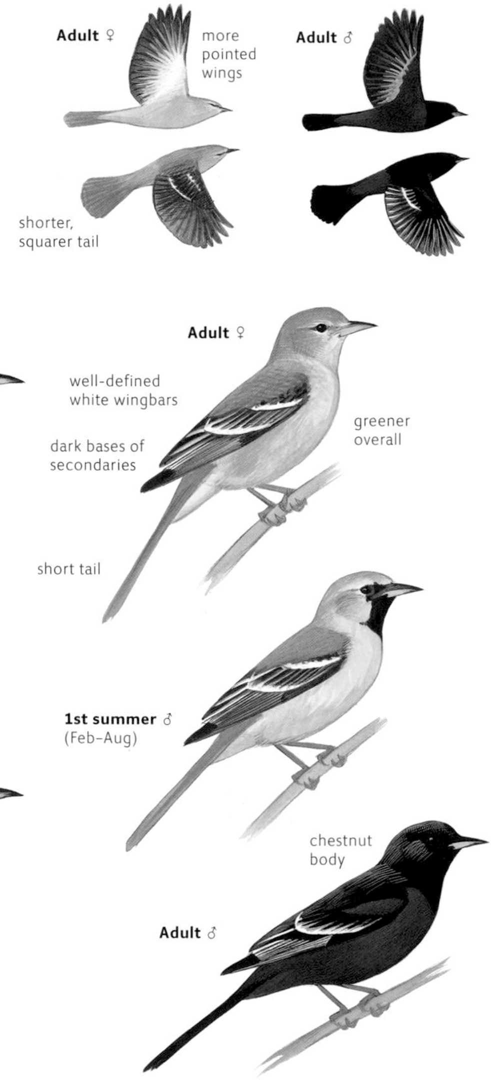

VOICE: Song a rich, lively warbling with wide pitch range; higher than other orioles; distinctive ending a ringing *pli titi zheeeer.* Common call of male a clear, whistled *tweeo.* Call a low, soft *chut* sometimes in slow chatter; also a rasping scold *jarrsh.* Flight call a relatively low, soft, level *yeeep.*

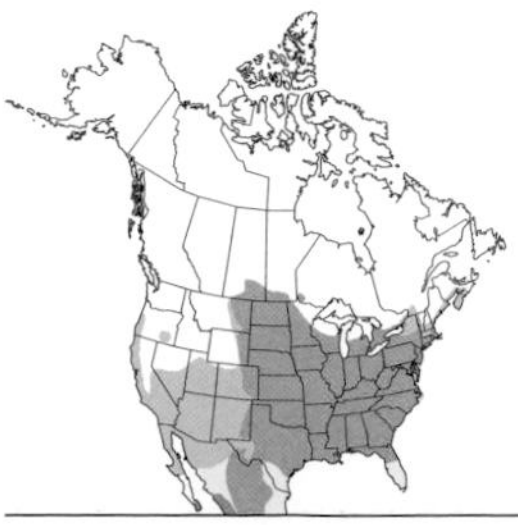

Uncommon in scrubby woods and hedgegrows with isolated tall trees; or neighborhoods with scattered large shade trees, especially near water. Solitary or in small groups. Feeds on insects, larvae, fruit, and nectar in mid to upper levels of trees.

Altamira Oriole

Icterus gularis

L 10" WS 14" WT 2 oz (58 g)

Our largest and heaviest oriole, with very stout and mostly black bill. Plumage similar to Hooded Oriole, but note much heavier bill, shorter tail, less black on face, orange lesser coverts, white patch at base of primaries, and different voice.

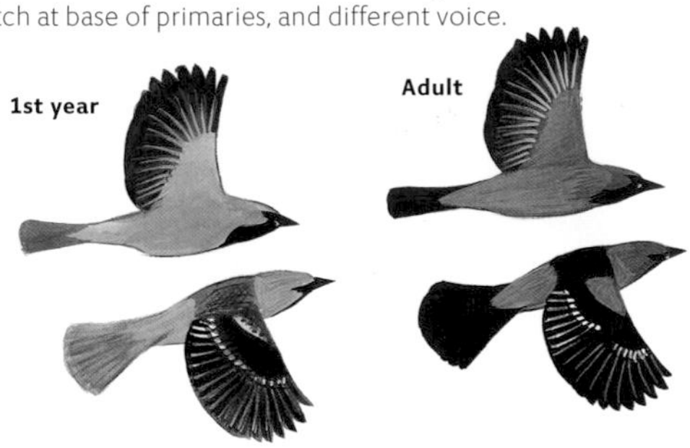

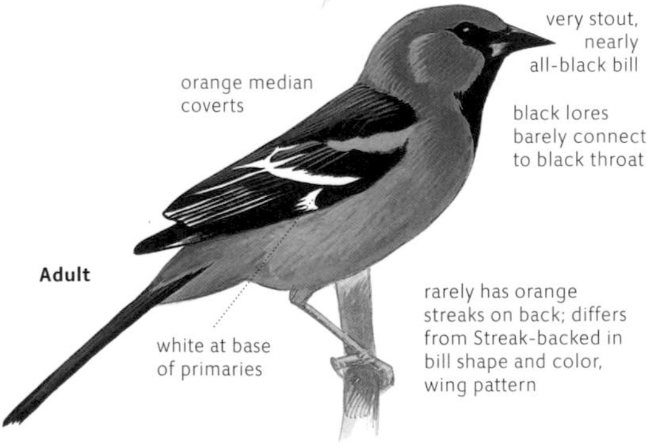

VOICE: Song of low, clear, short, well-spaced one-syllable whistled notes; slow and deliberate with very simple pattern *tooo tooo tooo teeeo tow tow.* Common call a short, whistled note like a single syllable from song *TIHoo* or *teeu* and variations. Flight call a hoarse, nasal, rising *griink* repeated.

Rare within very limited range. Found in relatively open, arid woodlands with shrubby understory along Rio Grande. Solitary or in pairs. Feeds on insects, larvae, fruit, and nectar.

Spot-breasted Oriole

Icterus pectoralis

L 9.5" WS 13" WT 1.5 oz (42 g)

Larger and with heavier bill than Baltimore Oriole, with mostly orange head and all-black tail. Black spots on sides of breast and broad white edges on tertials always distinctive. Like many tropical orioles, male and female are similar in appearance.

VOICE: Song a rich, melodious whistle; rather long, relaxed, and repetitive; doubling of some phrases conspicuous and distinctive. Calls include a nasal *jaaa,* a sharp *whip,* and a chattering *ptcheck.*

Uncommon and very local in lush shade trees and flower gardens of suburban neighborhoods. Introduced from Central America to southeastern Florida. Usually solitary. Feeds on insects, larvae, fruit, and nectar.

Audubon's Oriole

Icterus graduacauda

L 9.5" WS 12" WT 1.5 oz (42 g)

Relatively large and long-tailed. Extensive black hood, green back, and yellow collar distinctive. No other oriole combines all-black hood with green back

VOICE: Song of very low, slow, melancholy, slurred whistles *hooooo, heeeowee, heeew, heweee*; like a person just learning to whistle. Call an unenthusiastic whistle *tooo* or *oooeh*; also a wrenlike series of harsh, husky, rising notes *jeeek jeeek*....

Uncommon and local within limited range. Found in dense wooded areas with brushy understory, often near patches of giant cane. Solitary or in pairs, foraging low in brush and generally more secretive than other orioles. Feeds on insects, larvae, fruit, and nectar.

Scott's Oriole

Icterus parisorum

L 9" WS 12.5" WT 1.3 oz (37 g)

Relatively large, with very fine-tipped, curved bill. Male distinctive, black and yellow with extensive dark hood. Female and immature drab dusky yellow, especially dark and mottled on back, with brightest yellow on belly.

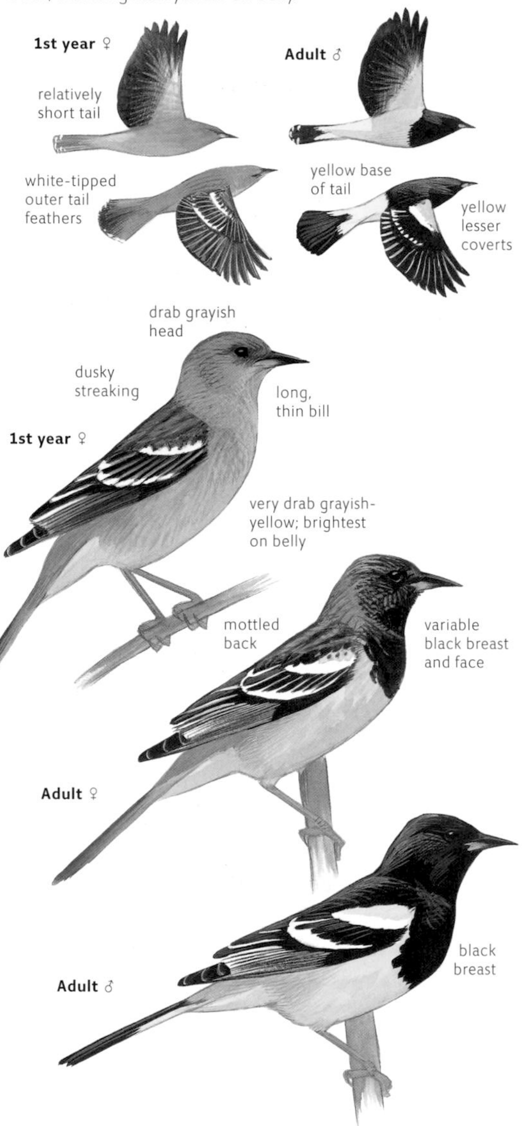

VOICE: Song of low, clear whistles with slightly gurgling quality; reminiscent of Western Meadowlark but pitch level or slightly rising overall. Call a harsh, relatively low-pitched *cherk* or *jug*. Flight call a husky, low *zhet*.

Uncommon on open arid hillsides where agaves and yuccas mix with oak or pine woodlands. Solitary or in pairs. Feeds on insects, larvae, fruit, and nectar.

Black-vented Oriole

Icterus wagleri

L 8.8" WS 11.5" WT 1.3 oz (37 g)

A very slender oriole, with long tail held tightly closed and tapering to a point. Adult has extensive black hood, no white on wings, black undertail coverts and all-black tail. Immature identified by shape and size, dark wings.

VOICE: Song unlikely to be heard in North America, a strong gurgling warble, with some wheezy and nasal notes. Call a short, harsh, nasal *jerrf,* slightly descending.

Very rare visitor from Mexico to Texas and Arizona, recorded several times at various seasons. Found in open woods, brushy edges, suburban neighborhoods; attracted to flowering shrubs or hummingbird feeders.

Old World and New World Names

When the first European explorers and colonists arrived in North America they named many of the local birds based on a resemblance to familiar species from the other side of the Atlantic. In the case of orioles, an Old World family by that name includes many brilliant yellow species. The name oriole has its origin in the Latin word for golden, *aureole*, referring to the birds' color. When early explorers saw the brilliant orange North American species they called them "orioles." Ornithologists would later determine that the two groups are not related, Old World orioles are close to vireos, and the New World orioles are in the Blackbird family, but the names are established.

Similarly, the name robin—originally used for the familiar small, orange-breasted garden bird of Britain and Europe—was applied by British colonists to a wide range of unrelated species worldwide, usually species with reddish breast. The American Robin is much larger than the European Robin and in a different family, but similar in general color and habits. Early settlers in North America also used the name "robin" for Eastern Bluebird. Eastern Towhee was called "Ground Robin," and the name "Robin Snipe" was used for Red Knot and dowitchers, both with red breasts.

Many other familiar bird names in North America also have namesakes in different families in Europe—buntings, sparrows, warblers, flycatchers, and more.

Plumage Variation in Orioles

The American orioles (genus *Icterus*) are primarily a tropical group, with only a few migratory species found widely in North America. Those migratory species molt twice a year and are the most variable in plumage color, with distinct male and female, winter and summer, immature and adult plumages. In contrast, resident tropical species, such as Altamira and Black-vented Orioles, molt just once a year and have one adult plumage, which is essentially the same year-round in males and females. A similar pattern is seen in many other groups as diverse as hummingbirds and wood-warblers.

Streak-backed Oriole

Icterus pustulatus

L 8.25" WS 12.5" WT 1.3 oz (37 g)

A fairly large, stocky oriole; note streaks on back, bright orange-red head, and extensive white in wings. Similar to winter male Hooded Oriole, but larger, with heavier bill, different wing pattern, more distinctly streaked back.

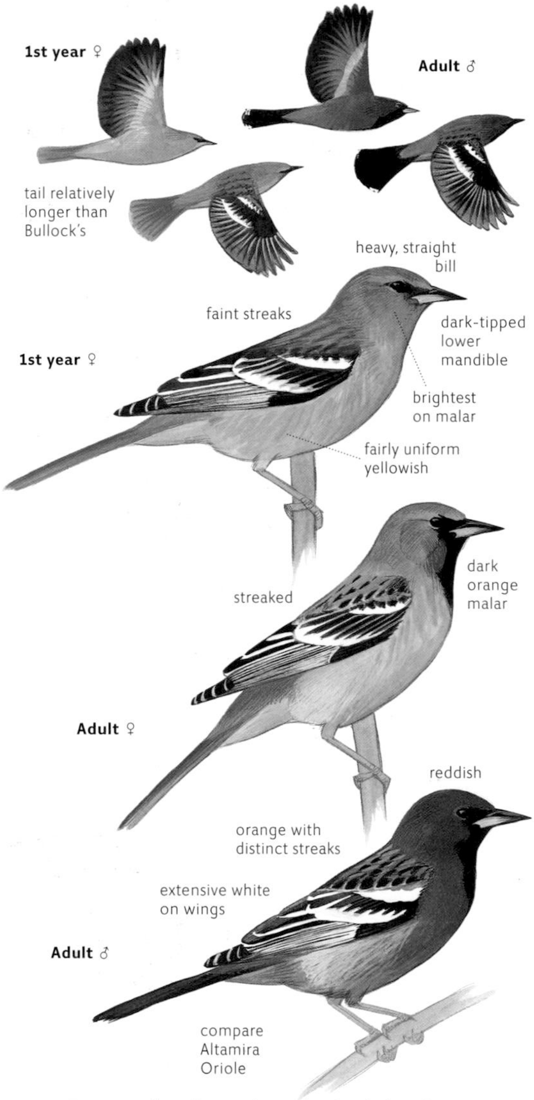

VOICE: Song a rather thin and simple whistled *to do, teweeep, yoo teewi*; quality like Bullock's but with short pauses and halting rhythm. Common call an upslurred whistle *toeep*, also a dry rattle. Flight call a soft *wheet*; lower and hoarser than Hooded and not strongly rising.

Very rare visitor from Mexico; recorded annually, mostly from Dec–Mar, and has nested in region. Most records from suburban yards where flowering plants and hummingbird feeders provide food. Habits similar to Bullock's Oriole.

Identification of Orioles

Male Bullock's and Baltimore orioles are easily distinguished by color (except hybrids, see below). Females and immature males are less distinctive.

First year male Bullock's develops a black throat by October and is subsequently easily identified.

Female Bullock's varies little from the pattern illustrated here of gray body with yellow ends, but female Baltimore is extremely variable. The drabbest female Baltimore is pale gray and yellowish, very similar to typical female Bullock's.

Careful study of head pattern reveals that on Baltimore the brightest color is on the breast (usually tinged orange, unlike the paler yellow of Bullock's), while on Bullock's the brightest color is on the malar.

Baltimore has dusky brownish cheeks washed with orange-yellow and about the same color as the crown, while on Bullock's the cheeks and eyebrow are clearer and brighter yellow, contrasting with the dusky crown. Bullock's tends to have broader whitish edging on the greater coverts, suggesting the white wing panel of adult males, and pointed dark centers on the median coverts forming a jagged border. Many female Bullock's Orioles have grayish undertail coverts, unlike the yellow-orange shown by all Baltimores.

Note also that molt timing differs: Bullock's molts later, during or after fall migration (Sep–Nov); Baltimore molts on the breeding grounds before migration (Jul–Aug).

Bullock's × Baltimore Oriole hybrid adult ♂

intermediate pattern on head, wing coverts, and tail; backcrosses produce complete range of variation between parent species

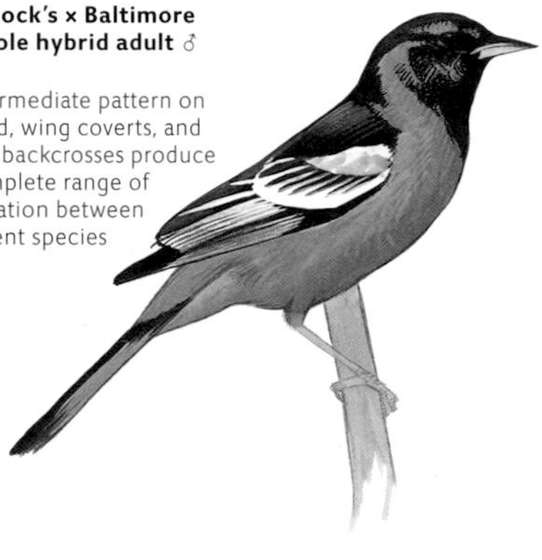

Hybrids are fairly common in limited area of range overlap, seen occasionally elsewhere. Hybrids show intermediate color pattern on head, wings, and tail, and numerous backcrosses produce a full range of appearance from Baltimore to Bullock's. Female hybrids are generally not identifiable.

Bullock's Oriole

Icterus bullockii

L 9" WS 12" WT 1.3 oz (36 g)

Very similar to Baltimore Oriole, female Bullock's generally has grayer body with yellower head and tail, brightest yellow on cheeks. This species and Baltimore have more extensive blue-gray on bill than other orioles, and males have partly orange tails.

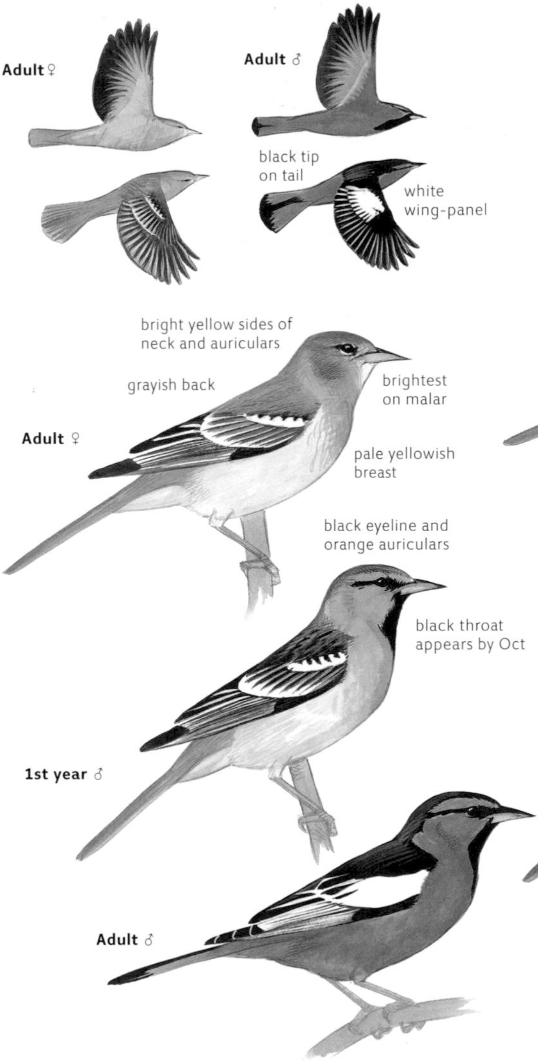

VOICE: Song whistled like Baltimore but less rich, more nasal and barking, with shorter notes and sharper pitch changes *goo gidoo goo peeka peeka*. Simple whistled calls less frequent than Baltimore; a dry, husky chatter or slow, muffled *cheg cheg*... slower than Baltimore. Flight call like Baltimore.

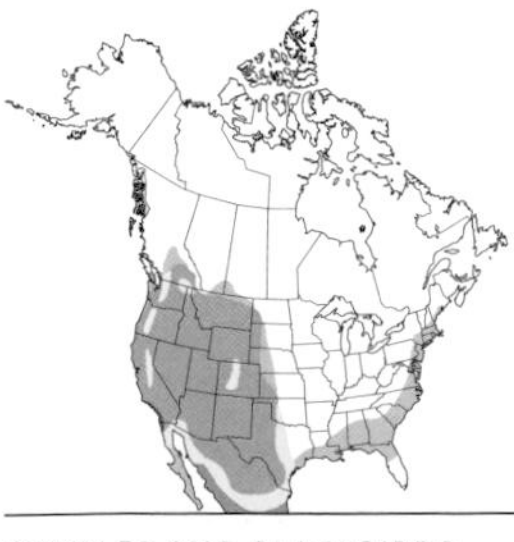

Common in deciduous trees in or near open areas, characteristically in large cottonwood trees near water. Solitary or in small groups. Forages widely for caterpillars, fruit, and nectar in low brush and in trees.

Baltimore Oriole

Icterus galbula

L 8.75" WS 11.5" WT 1.2 oz (33 g)

The only bright orange bird in most of its range. Very similar to Bullock's Oriole. Male differs in head, wing, and tail pattern. Female usually more extensively orange than Bullock's with brightest orange on breast.

VOICE: Song a short series of rich, clear, whistled notes *pidoo tewdi tewdi yewdi tew tidew*; variable in pattern, with slight pauses between each phrase; often gives simple two-note whistle *hulee* and variations. Call a dry, harsh, uneven rattle. Flight call a husky, tinny, trumpeting *veeet*.

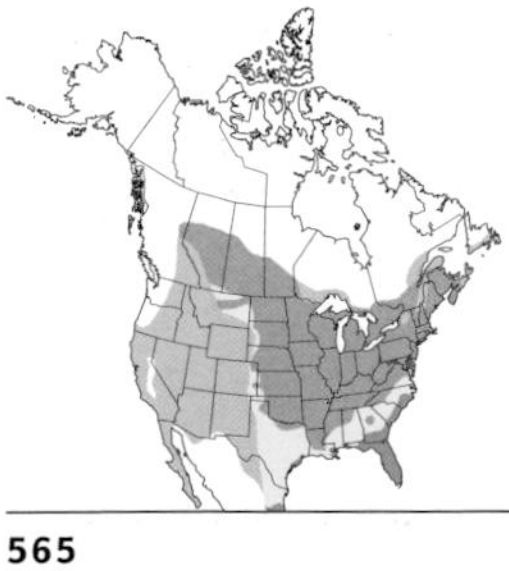

Common in open deciduous woodlands or among scattered tall trees; suburban neighborhoods with large shade trees provide ideal habitat. Solitary or in small groups. Forages mainly in mid to upper levels of trees, gleaning insects and larvae from foliage. In fall and winter also eats fruit.

Finches and Allies

FAMILY: FRINGILLIDAE, PASSERIDAE, PLOCEIDAE, ESTRILDIDAE

30 species in 17 genera in 4 families (9 rare species not shown here). All have stout triangular bills like sparrows and cardinals. Most are in family Fringillidae (Finches), with introduced species in three other families. Finches are small to medium size, distinguished from sparrows by relatively pointed wings, notched tails, strongly undulating flight, and distinctive flight calls. Males of all species are brightly colored with red or yellow. Nest is a cup in a tree or shrub. House Sparrow and Eurasian Tree Sparrow are in the Old World sparrow family (Passeridae), usually found in small flocks near buildings, nest in cavities, often hover before landing, and give hoarse chirping calls. Orange Bishop is in the weaver family (Ploceidae), Mannikins, Munias, and Java Sparrow in the family Estrildidae. First year females are shown.

Genus *Fringilla*

Brambling, page 567

Genus *Leucosticte*

Gray-crowned Rosy-Finch, page 568

Black Rosy-Finch, page 569

Brown-capped Rosy-Finch, page 569

Genus *Pinicola*

Pine Grosbeak, page 570

Genus *Haemorhous*

Purple Finch, page 572

Cassin's Finch, page 573

House Finch, page 573

Genus *Loxia*

Red Crossbill, page 574

White-winged Crossbill, page 575

Genus *Acanthis*

Common Redpoll, page 576

Hoary Redpoll, page 577

Genus *Spinus*

Pine Siskin, page 578

Lesser Goldfinch, page 579

Lawrence's Goldfinch, page 578

American Goldfinch, page 579

Genus *Coccothraustes*

Evening Grosbeak, page 571

Genus *Passer*

House Sparrow, page 582

Eurasian Tree Sparrow, page 582

Genus *Euplectes*

Orange Bishop, page 583

Genus *Lonchura*

Nutmeg Mannikin, page 583

Brambling

Fringilla montifringilla

L 6.25" WS 11" WT 0.74 oz (21 g)

Slightly larger than House Finch, with bright plumage and deeply-notched tail. Plumage distinctive, with unstreaked orange breast, orange shoulder, and white rump. In winter note gray head and yellow bill.

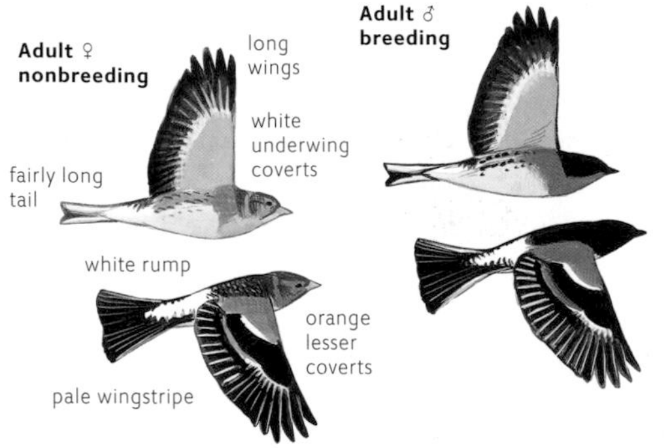

VOICE: Song (unlikely to be heard in North America) a simple, wheezy or rattling, nasal trill *dzhreeeeee*. Call a distinctive, harsh, rising *jaaaek* reminiscent of Gray Catbird. Flight call a short, low, hard, often doubled *tup*.

Very rare visitor from Eurasia. Occurs in flocks in normal range and occasionally in spring in western Alaska, very rare in winter elsewhere in North America south to California and east to New Jersey; usually found at feeders, sometimes associated with flocks of juncos. Feeds on seeds and some insects.

Common Chaffinch

Fringilla coelebs

L 5.5" WS 10.2" WT 0.85 oz (24 g)

Slightly larger than House Finch, pinkish-brown and gray without streaks. Note very broad and obvious white wingbars, easily visible in flight and at rest, white sides on tail.

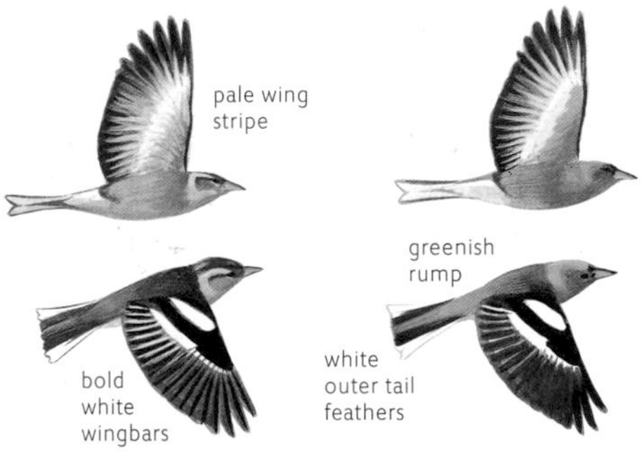

VOICE: Song (unlikely to be heard in North America) a series of sharp rattling phrases. Call a loud sharp *wink*, and in flight a softer *pep*.

Very rare visitor from Europe to the Northeast, with at least one record in Newfoundland likely a natural vagrant. Many other records presumed to be escapes from captivity.

Gray-crowned Rosy-Finch

Leucosticte tephrocotis

L 6.25" WS 13" WT 0.91 oz (26 g)

Slightly larger and more slender than Purple Finch, with long tail and wings. Dark brownish overall, with gray hind-crown and pale translucent flight feathers, pink on wings and belly. Very similar to other rosy-finches.

VOICE: Song heard infrequently; a slow, descending series of husky, whistled notes *jeew jeew jeew...*; a more varied Purple Finch-like song has also been described. Flight call a rather soft, husky chirp *jeewf* or *cheew* or a buzzy *jeerf*; may recall House Sparrow or Evening Grosbeak but softer. Many variations on this call create a chorus from winter flocks.

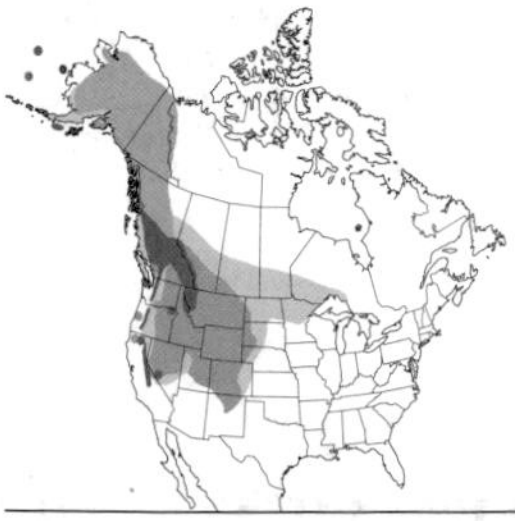

Common locally on open rocky hillsides, often near cliffs, above treeline, often closely associated with lingering snowfields in summer. Nests on alpine or Arctic tundra. Almost always occurs in flocks, and the three species flock together indiscriminately. Feeds on seeds and insects. In winter often moves to slightly lower elevations.

Three recognizable populations of this species, distinguished by head pattern, size, and body color; Gray-crowned nests in the Rocky Mountains and interior ranges from Alaska to Montana, Gray-cheeked in coastal mountain ranges from Alaska to California, and Bering Sea population is resident on the Pribilof and Aleutian Islands.

Black Rosy-Finch

Leucosticte atrata

L 6.25" WS 13" WT 0.91 oz (26 g)

Very similar to other rosy-finches, distinguished by overall darker blackish-brown plumage, with sharply contrasting gray hindcrown. Drab winter birds have cold grayish feather edges and average less pink than other species.

both species identical to Gray-crowned in flight; distinguished only by head pattern

Juvenile (Jun–Aug)

buffy cinnamon wingbars

uniform gray-brown

all rosy-finch juveniles similar; differ only in body color

1st winter ♀ (Aug–Mar)

grayish overall

Adult ♂ breeding (Mar–Aug)

blackish

averages less rosy on belly than other rosy-finches

Hybridizes with Gray-crowned in a few locations along Montana/Idaho border.

VOICE: Similar to Gray-crowned Rosy-Finch.

Uncommon and local on alpine tundra near talus slopes and cliff faces, above treeline, lower in winter. Habits like other rosy-finches.

Brown-capped Rosy-Finch

Leucosticte australis

L 6.25" WS 13" WT 0.91 oz (26 g)

Very similar to other rosy-finches, distinguished by little or no gray on crown, with dark forehead blending gradually to dark brown nape, richer brown color on back, and brighter and more extensive pink on wings and belly than other species.

1st winter ♀ (Aug–Mar)

pale gray head with dark-centered crown feathers

pale mouse-gray overall

Adult ♂ breeding (Mar–Aug)

dark-centered crown feathers become paler to rear

Variant adult ♂ breeding (Mar–Aug)

gray supercilium

some have conspicuous gray supercilium, closely resembling Gray-crowned, but with less contrast between black forecrown and gray supercilium; some may be indistinguishable

VOICE: Similar to Gray-crowned Rosy-Finch.

Uncommon and local within its very limited range on alpine tundra, lower in winter. Habits like other rosy-finches.

Pine Grosbeak

Pinicola enucleator

L 9" WS 14.5" WT 2 oz (56 g)

Large and long-tailed; size and shape can recall American Robin. Subtly colored gray and pink or greenish. Note very stubby bill, long notched tail, white wingbars.

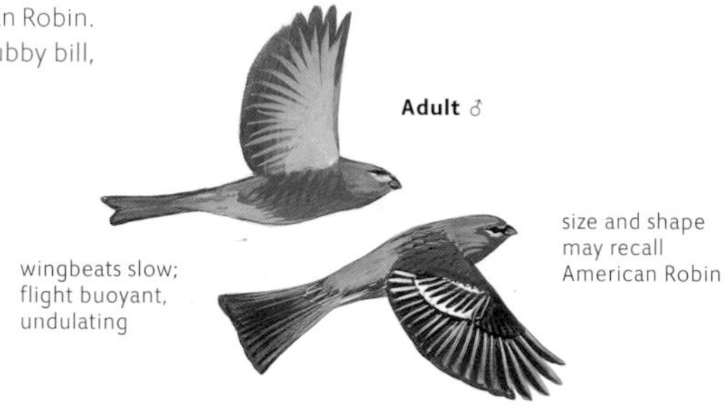

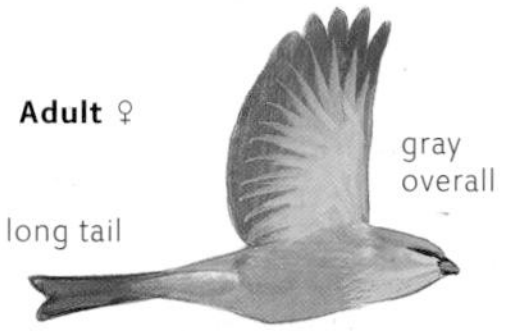

VOICE: Song a relatively low, lazy, unaccented warble of soft, whistled notes *fillip illy dilly didalidoo* usually with descending trend; metallic quality reminiscent of *Pheucticus* grosbeaks (Rose-breasted and Black-headed). Quieter whisper song often includes imitations of other species. Flight call generally several soft husky whistles, pattern and quality varies regionally.

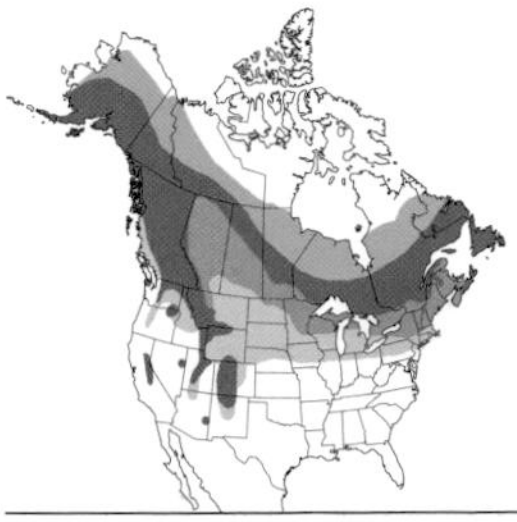

Uncommon and somewhat irregular; nests in spruce-fir forests, and often seen at edges of clearings in deciduous shrubs; in winter attracted to fruit trees. Usually in small flocks. Feeds on seeds and fruit.

Subtle regional variation in size and male plumage is partly overshadowed by variation. Regional variation in flight calls and song combined with plumage may allow identification of populations, but more study is needed. Pacific male has dark and extensive red on underparts; Interior West male has duller red, with much gray mottling on the breast and entirely gray flanks, as well as little or no dark spotting on the back; Taiga male has pinkish-red of intermediate extent.

Flight call varies geographically. Taiga gives rather lethargic, soft, whistled notes *peew* or *po peew peew*; vaguely reminiscent of Greater Yellowlegs. Pacific and Interior West give more complex, harder, husky notes *quid quid quid* or *quidip quidip* generally rising in pitch; reminiscent of Western Tanager. All populations also give quiet, low calls such as *pidididid* or *ip ip*.

Evening Grosbeak

Coccothraustes vespertinus

L 8"   WS 14"   WT 2.1 oz (60 g)

Large and short-tailed, with massive head and bill. Male unmistakable. On female note large pale bill, complex white wing markings, overall gray-brown color with greenish nape, bright yellow underwing coverts.

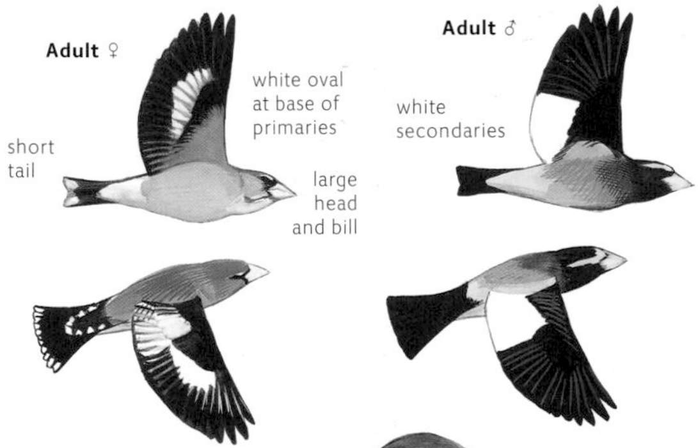

VOICE: Song apparently a regular repetition of call notes. Call varies regionally, a high, sharp, ringing trill *kleerr* reminiscent of House Sparrow or clear descending *teew*; in flocks a low, dry rattle or buzz *thirrr*.

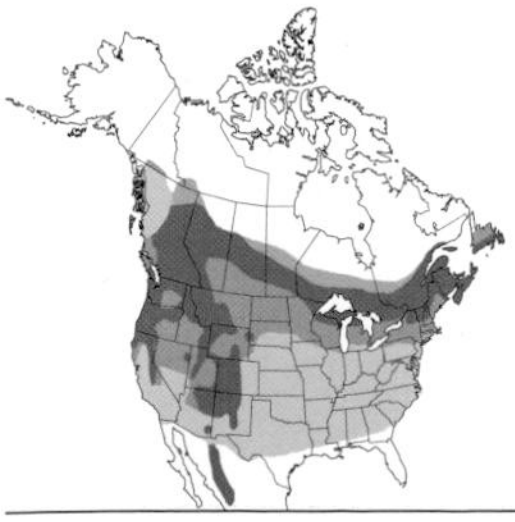

Common but irregular in mixed forests. Almost always seen flying overhead, perched in treetops, or at bird feeders. Usually in tight flocks, does not mix with other species. Feeds mainly on insects and on seeds of trees.

Five regional variations of the high descending *kleew* call are distinguishable, with very small visual differences between subspecies.

Eastern type found east of the Rocky Mountains has relatively low-pitched and distinctly trilled *kleerrr* call, bill is short and pale yellowish, yellow band on forehead of male is broad.

All four western types have clearer call, longer and darker greenish bill, and narrower yellow forehead band. Birds from British Columbia to Oregon and Wyoming give a simple clear *teew*, while very similar subspecies from California and from the southern Rocky Mountains give slightly longer and higher-sounding call.

Mexican population found in southeastern Arizona has a more distinctive call, longer and slightly trilled (higher-pitched and less trilled than Eastern birds). Currently the best way to identify these subspecies is by analysis of sound recordings.

Hawfinch

Coccothraustes coccothraustes

L 7"   WS 12"   WT 2 oz (57 g)

Size and shape similar to Evening Grosbeak, but with plainer brown and gray body, dark blue-black secondaries and whitish coverts, massive dark gray bill.

VOICE: Flight call very high, a thin, dry trill or abrupt mechanical clicking reminiscent of waxwings but drier, often shorter.

Very rare visitor from Eurasia to islands of far western Alaska.

Purple Finch

Haemorhous purpureus

L 5.7"   WS 10"   WT 0.88 oz (25 g)

Stockier than House Finch, with shorter and deeply notched tail, triangular bill with straighter culmen, and longer and more pointed wings. On female, note bold head pattern and whitish underparts with distinct, short dark streaks. Adult male has reddish wash over back, wings, and flanks.

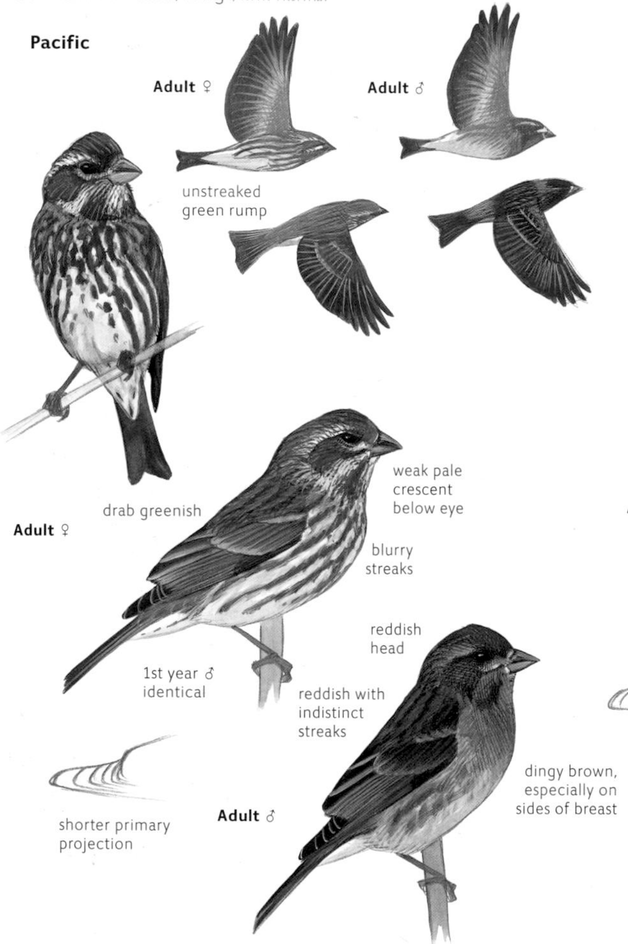

Pacific
Adult ♀
Adult ♂
unstreaked green rump
drab greenish
weak pale crescent below eye
blurry streaks
Adult ♀
1st year ♂ identical
reddish head
reddish with indistinct streaks
dingy brown, especially on sides of breast
shorter primary projection
Adult ♂

Eastern
Adult ♀
distinctly notched tail
bold brown and white head pattern
slight crest
brownish with distinct streaks
Adult ♀
short, dark streaks
1st year ♂ identical
long primary projection
extensive red color on head and back
Adult ♂

VOICE: Pacific: Song a mumbled, low, unstructured *fridi ferdi frididifri fridi frrr*; more rapid than House Finch. Call a husky, muffled whistle *wheeoo* or *fwidowip*; quality like song, unlike the clearer, vireolike calls of Eastern. Flight call a *bik* slightly lower than Eastern with hard overtones like Brewer's Blackbird. Eastern: Song a slightly hoarse, warbled *plidi tididi preete plidi tititi preeer*; bright, lively, and clearly structured with accented ending; generally ends with strongly descending trill *cheeeer*; overall trend rising. Call a short, whistled phrase like vireo song *tweeyoo*. Flight call a light, hard *pik* with musical overtones.

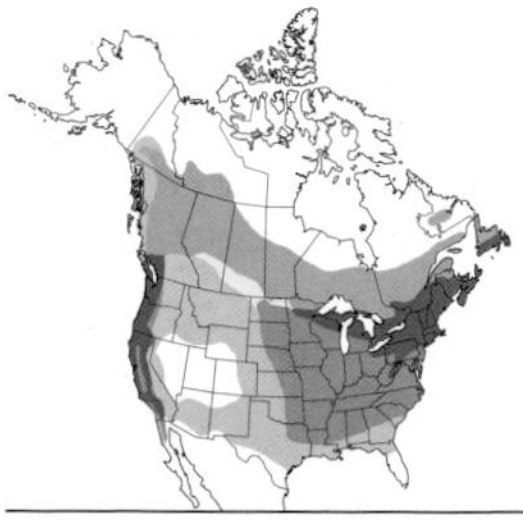

Uncommon; found in a variety of wooded areas, especially in low shrubs on the margins of clearings, roads, or ponds, where it forages quietly for fruit, seeds, and insects. Usually in small flocks and commonly visits feeders like House Finch, but less frequent in suburban habitats, and not attracted to buildings.

Pacific and Eastern populations are moderately distinctive. Pacific birds average slightly rounder-winged with shorter primary projection and are longer-tailed with more curved culmen than Eastern (tending toward House Finch in shape). Pacific females are greenish above, have indistinct streaks, and are washed yellowish below with longer, paler, more blurry streaks (Eastern are brownish above and white below with shorter, darker streaks). Underparts of Pacific males are washed dull brown and rump is dark red (Eastern males have cleaner, brighter colors overall).

Cassin's Finch

Haemorhous cassinii

L 6" WS 11.5" WT 0.91 oz (26 g)

Similar to Purple Finch, slightly longer-tailed with more pointed bill. Female has very crisp streaks overall, paler and less boldly-patterned than Purple Finch. Male has bright red crown contrasting with distinctly streaked, gray-brown back.

VOICE: Song a brighter and higher warble than Pacific Purple Finch; higher and more rapid than House Finch with descending trend slightly rising at end. Call similar to Purple but drier *tedeyo* or *widee-ooli*. Flight call a distinctive, high, dry warble *krdlii* or *chidilip*.

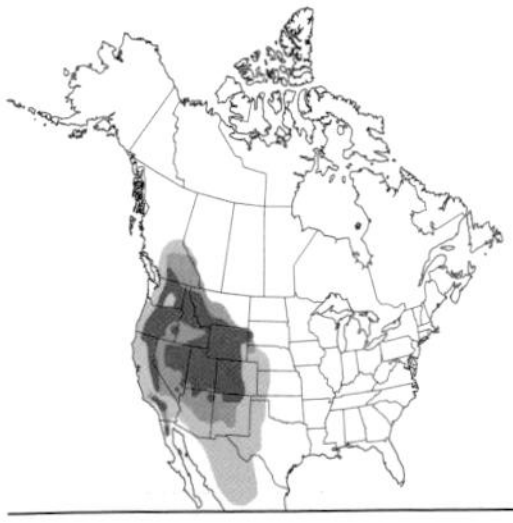

Uncommon in montane coniferous forest, especially open, dry pine forests. Usually in small flocks. Feeds on seeds, fruit, and some insects, often foraging on the ground.

House Finch

Haemorhous mexicanus

L 5.7" WS 10" WT 0.74 oz (21 g)

Longer-tailed than Purple Finch, with round head and short bill with curved culmen. Female drab overall; weakly patterned gray-brown on head and pale grayish below with blurry gray-brown streaks. Male shows brightest red on eyebrow and breast.

VOICE: Song a varied warble; generally begins with husky, whistled notes and ends with slightly lower, burry notes; often a long *veeerrr* note; steady tempo with distinct downward trend. Call a soft, mellow *fillp* or *fiidlp*. Flight call a soft, husky *vweet* like softer notes of House Sparrow, but rising.

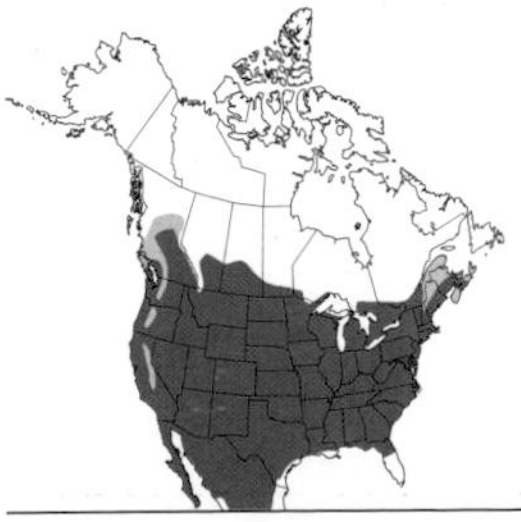

Common and widespread in suburbs and patchy, brushy wooded areas, often foraging in weedy fields where Purple Finch would not be expected. Nearly always in small flocks, flying above treetops and swooping down to land in tops of bushes or weeds. A regular visitor to bird feeders and often nests on or near buildings. Feeds on seeds, fruit, and some insects.

Red Crossbill

Loxia curvirostra

L 6.25" WS 11" WT 1.3 oz (36 g)

Larger than House Finch; relatively large-headed and short-tailed. Overall size and bill size varies with population. Rather dark reddish or greenish overall, told from White-winged Crossbill by call, usually plain dark wings, more uniform head color.

VOICE: Song a series of short, hard or clicking phrases, some buzzy, with many call notes interspersed: *tikuti ti chupity chupity chupity tokit kyip kyip kyip jree-jree-jree*.... Flight calls are fundamentally similar in all types: very hard, sharp *gyp* or *kip* notes usually repeated in short series *gyp-gyp-gyp*.

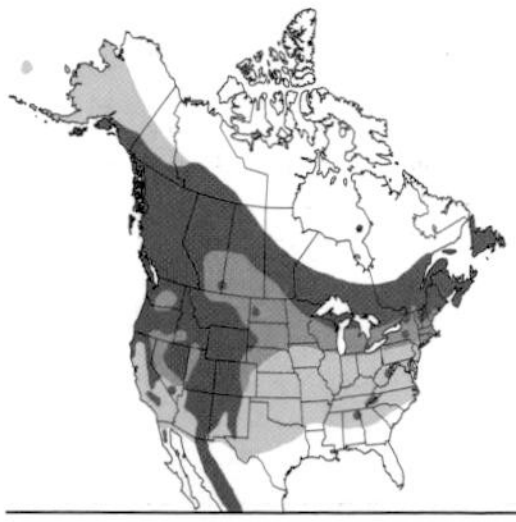

Uncommon and very irregular, sometimes locally common. Always in flocks, wandering widely in search of pines or other conifers. Foraging flocks are generally tame and quiet, climbing around cones and working to extract seeds, sometimes on ground for fallen seeds or cones. Strongly attracted to salt, along salted roads in winter or at salt blocks for livestock.

Red Crossbill Types

Recent research has identified ten call types among North American Red Crossbills, which can be interpreted as ten species. Subtle variations in bill size and overall size correspond to the call types, and (as bill size is optimized for efficiently extracting seeds from certain cones) birds of each call type tend to flock together and show a preference for a few species of trees. Each call type has a "core" range where it occurs consistently, but coniferous trees do not produce cones every year, and crossbills wander nomadically when food is scarce in their core range.

In general the larger- and stouter-billed types forage most efficiently on larger and harder cones (pines) while the smaller-billed types forage most efficiently on smaller and softer cones (spruce, douglas-fir, and hemlock). These preferences may provide clues to identification, but there is extensive overlap, especially with wandering flocks where the choice of cones may be very limited.

Experienced observers are able to identify some call types with reasonable confidence in the field, but positive identification generally requires analysis of recorded sounds (a cell phone or the video setting on a digital camera usually produces a recording of sufficient quality). Further complicating the situation, these calls are learned and variable, and atypical or intermediate calls are sometimes encountered.

Type 3
(smallest)

Type 6
(largest)

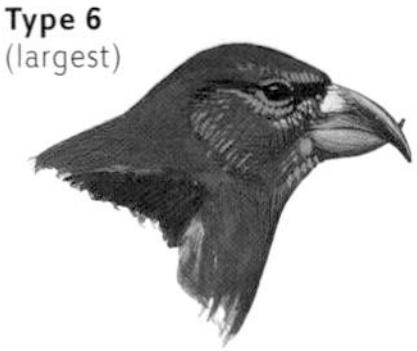

Type 1
Core range in Appalachians, wanders throughout East
Bill size medium
Call *kyep* descending, similar to type 2

Type 2
Core range in West, wanders widely
Bill size second largest
Call *kewp* descending, similar to Type 1 but deeper, huskier, longer

Type 3
Core range in Pacific Northwest, wanders widely
Bill size smallest
Call *chik* squeaky or scratchy, relatively high, without strong inflection

Type 4
Core range in Pacific Northwest, wanders some in West
Bill size medium
Call *kwit* bouncy, rising, similar to Type 10 but lower and less squeaky

Type 5
Core range in Interior West, wanders some in West
Bill size medium-large
Call *clip* springy or twangy, similar to type 1 but higher, weaker

Type 6
Core range in Mexico, wanders to southwestern US
Bill size largest
Call *teew* clear and almost whistled

Type 7
Core range in Pacific Northwest, wanders little
Bill size medium
Call *chek* without strong inflection, similar to Type 3 but lower and flatter

Type 8
Newfoundland only
Bill size medium-large
Call *chilp* ringing and complex

Type 9
South Hills only (southern Idaho)
Bill size medium-large
Call *pik* low, dry, very hard

Type 10
Core range on coast of Oregon and northern California, wanders widely
Bill size medium-small
Call *pweet*, strongly rising, squeaky, and relatively high-pitched

White-winged Crossbill

Loxia leucoptera

L 6.5" WS 10.5" WT 0.91 oz (26 g)

Slightly smaller than most Red Crossbills, more slender with smaller bill and longer tail. Always shows broad white wingbars (Red Crossbill rarely shows narrow wingbars), dark lores, and dark mark on rear cheek contrasting with pale neck sides.

VOICE: Song a series of nervous, rattling, mechanical trills on different pitches *jrrr jrrr jrrr treeeeeee kerrrrrr treeeeeee krrr.* Call *tyik-tyik* weaker and thinner than Red; a rising *veeeht* singly or in series unlike Red; also a redpoll-like *chut-chut.* Flock produces a dry, rattling chorus.

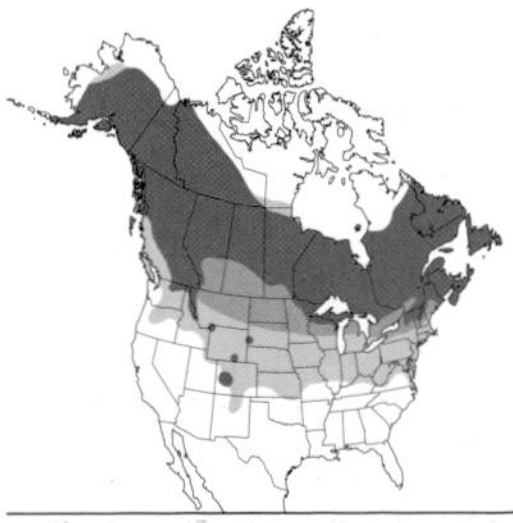

Uncommon and very irregular; sometimes common locally. Always in flocks like Red Crossbill. Feeds on conifer seeds, especially trees with smaller and softer cones such as hemlocks and spruces, but will use pines when necessary. Almost never seen away from coniferous trees.

Common Redpoll

Acanthis flammea

L 5.25" WS 9" WT 0.46 oz (13 g)

A tiny, streaked finch, similar to Pine Siskin but note dark face, red crown, stubby yellow bill, usually cleaner white breast with distinct streaks, different voice; adult male has pink breast. Very similar to Hoary Redpoll.

Southern

Greenland (Greater)

VOICE: Song a series of short, repeated notes; mainly call notes and short trills chit *chit chit twirrrrrr toweeoweeowee chrrr chit chit chit tiree tiree...*; in flight a long, rattling buzz *chrrrrrr.* Call a husky, wiry, rising *tweweee* and a soft, rising *vweeii.* Feeding flock gives constant soft *tip* notes; harder than American Goldfinch. Flight call a hard, rapid, bouncy *chid chid chid* or *tjip tjip* varying from soft to hard and dry (higher, more ringing versions on breeding grounds) or a rapid rattle *jijijiji*; also a strong, nasal, husky *tew* or *kewp.*

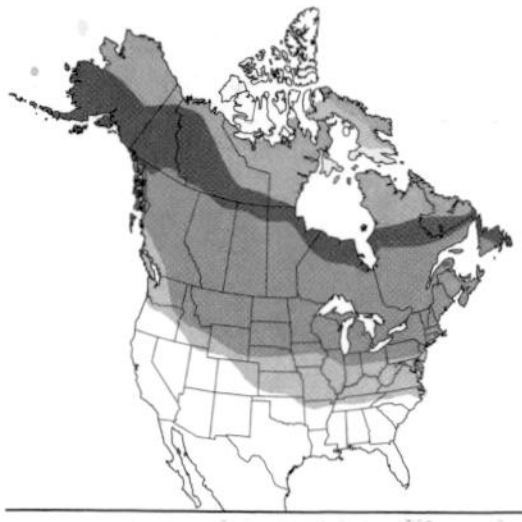

Common but nomadic; numbers vary from year to year. Nests in boreal foest in clearings, edges, and stream corridors with willow and birch trees; in winter feeds primarily on birch catkins and weed seeds. Nearly always in flocks.

Greenland population is scarce southward in winter; averages 10 percent larger than Southern and is heavier-billed, darker with heavier streaking overall, more extensive black on throat, and darker buffy-brown wash on head; adult male has limited red on breast.

Hoary Redpoll

Acanthis hornemanni

L 5.5" WS 9" WT 0.46 oz (13 g)

Averages paler, fluffier, and smaller-billed than Common Redpoll, but not always identifiable. Palest birds are distinctive—overall whitish with faint streaks on flanks. On darker individuals note paler rear scapulars; reduced dark streaking on underparts.

Southern

Adult ♀

Adult ♂

usually unstreaked rump

pale rear scapulars

1st year ♀

short-billed appearance emphasized by fluffy nasal feathering

pale "frosty" overall

averages more white on secondaries and coverts than Common

Adult ♀

stubby bill with straight culmen

fine streaking

Adult ♂

limited pale pink on breast

faintly streaked

Greenland (Hornemann's)

Adult ♀

Adult ♂

1st year ♀

Adult ♀

Greenland population scarce southward in winter; averages 12 percent larger and overall paler than Southern

Adult ♂

whitish overall

very little pale pink

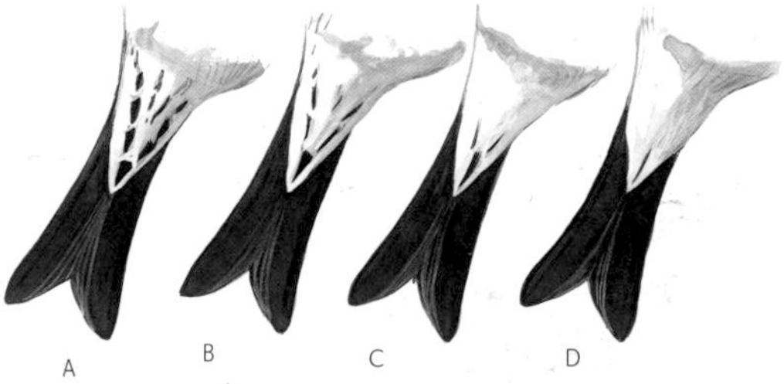

VOICE: Essentially identical to Common. Call a wiry *juwee*; may be lower and simpler than Common. Flight call a *chif chif chif*; each note slightly descending; may average lower and softer than Common.

Uncommon and irregular. Nests on Arctic tundra in stunted willows. Most winter in north-central Canada; farther south generally occurs only in very small numbers among flocks of Common Redpoll.

Pattern of streaks on undertail coverts are often useful for identifying redpolls. Common averages more streaks (from A to C), Hoary fewer (from C and D to entirely unstreaked). In both species males average fewer streaks than females.

A   B   C   D

Pine Siskin

Spinus pinus

L 5" WS 9" WT 0.53 oz (15 g)

Size and habits similar to goldfinches, note slender bill and fairly uniform brown streaking. Yellow wing and tail markings conspicuous on some individuals. Smaller than *Haemorhous* finches, with strongly notched tail and slender bill.

VOICE: Song a rapid, run-on jumble of fairly low, husky notes; generally a string of call notes like goldfinches but husky and harsh, often includes the characteristic rising buzzy call and imitations of other species' calls. Common call a rough, rising buzz *zhreeeeee*. Flight call a high, sharp *kdeew* and a dull *bid bid* or *ji ji ji*.

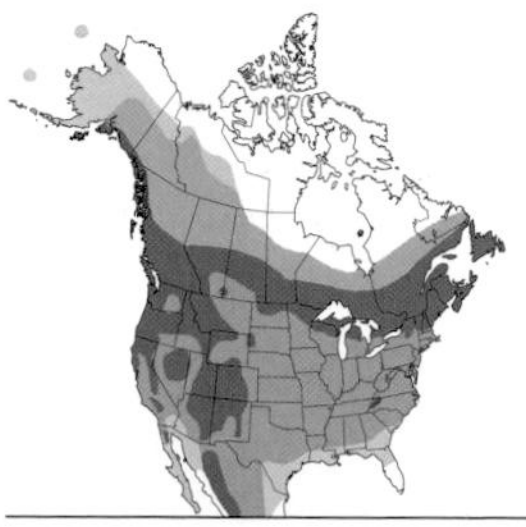

Common but nomadic; numbers vary from year to year at any given location. Found in open coniferous or mixed forests where it feeds on buds and seeds of alders, birches, pines, hemlocks, and other trees; also takes small insects. Almost always in flocks, sometimes with goldfinches.

Lawrence's Goldfinch

Spinus lawrencei

L 4.75" WS 8.25" WT 0.4 oz (11.5 g)

Slightly smaller than American Goldfinch, with relatively small bill and long tail. Note plain gray-brown head and back with broad yellowish or grayish-yellow wingbars. Male has distinctive black face contrasting with pink bill.

VOICE: Song a high, extended, tinkling warble; rapid and varied without repeated phrases; composed almost entirely of imitations of call notes of other species. Call a nasal *too-err*; also a sharp, high *Plti* and *Itititi*. Flight call a high, clear *ti-too*.

Uncommon and local in open oak grasslands or riparian corridors of dry foothills and valleys. Nearly always seen in small flocks feeding on weed seeds, sometimes mixed with other goldfinches. In some winters large numbers move east as far as southern Arizona.

Lesser Goldfinch

Spinus psaltria

L 4.5" WS 8" WT 0.33 oz (9.5 g)

Our smallest goldfinch, with relatively short tail and large bill, always shows white patch at base of primaries, extensive on male. Female told from American by shape, call, more uniform olive head and neck, and weak wingbars.

VOICE: Song slower, hoarser, and more disjointed than American; little repetition of notes; includes many imitations of call notes of other species. Call a distinctive, very high, clear, wiry *tleeee, teeeeyEE,* and *tseee-eeeew* and variations. Flight call a hoarse, grating *chig chig chig.*

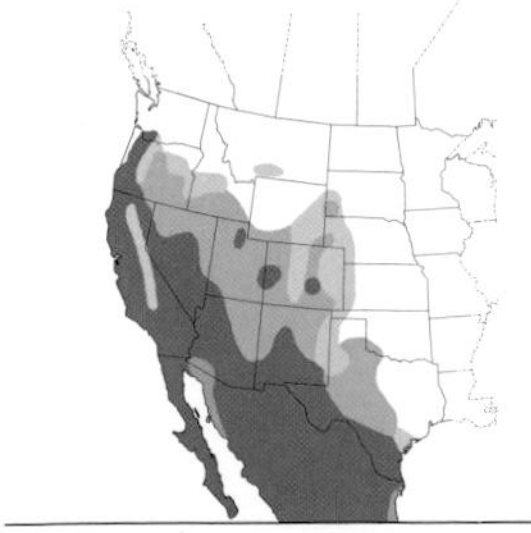

Common in patchy open habitat such as brushy fields, woods edges, riparian thickets, and gardens. Habits similar to American Goldfinch. Nearly always in flocks, forages mainly on weed and grass seeds.

American Goldfinch

Spinus tristis

L 5" WS 9" WT 0.46 oz (13 g)

Breeding male brilliant yellow with black wings and cap. Female distinguished from other finches by small size, unstreaked gray-brown to yellow plumage, dark wings with obvious pale wingbars.

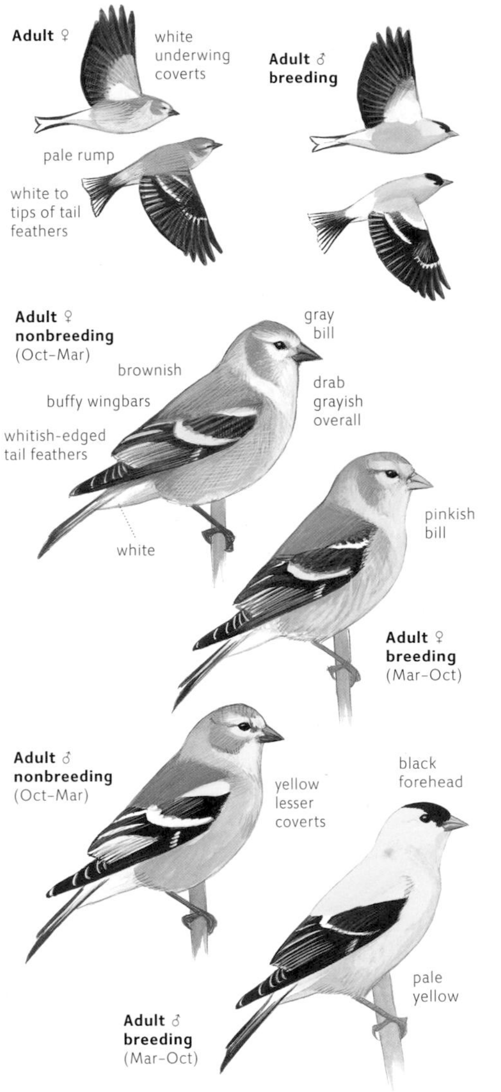

VOICE: Song high, musical, rapidly repeated phrases *toWEE toWEE toWEEto tweer tweer tweer ti ti ti ti*; may suggest buntings but less stereotyped; fading at end. Call a thin, wiry *toweeeowee* or *tweeee*; also a soft *tihoo* and variations. Flight call a soft, whistled, descending series *ti di di di.*

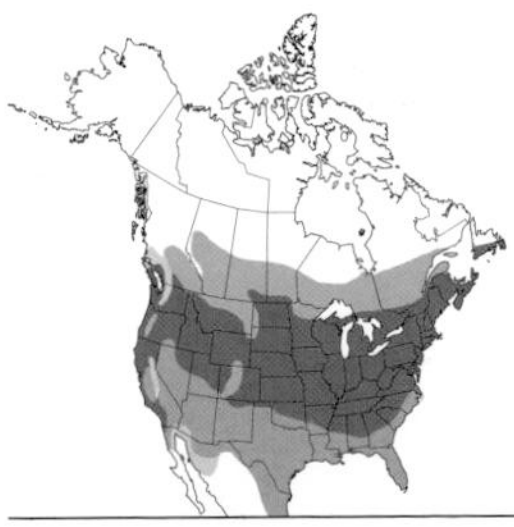

Common and widespread in orchards, hedgerows, overgrown fields, and suburban gardens. Almost always seen in flocks, feeding on tree buds, grass and weed seeds (especially thistles), and some insects.

Common Rosefinch

Carpodacus erythrinus

L 5.5" WS 10.25" WT 0.8 oz (23 g)

Very similar to House Finch. Male has extensive red on head and breast, without obvious pattern. Female and immature have plain face like House Finch, but more olive-tinged above, more distinct fine streaks on paler ground color below.

VOICE: Song a clear whistled warble, quality similar to House Finch but much shorter and simpler, without husky or trilled phrases. Call a husky, rising *hueee* very similar to call of House Finch.

Very rare visitor from Eurasia to far western Alaska, mainly in spring. One fall record in California.

Oriental Greenfinch

Chloris sinica

L 5.9" WS 10.25" WT 1 oz (28 g)

Much stockier and shorter-tailed than House Finch, reminiscent of a miniature Evening Grosbeak. Note large pale bill, bold yellow wing stripe, and subtle pattern of gray, brownish, and olive on body.

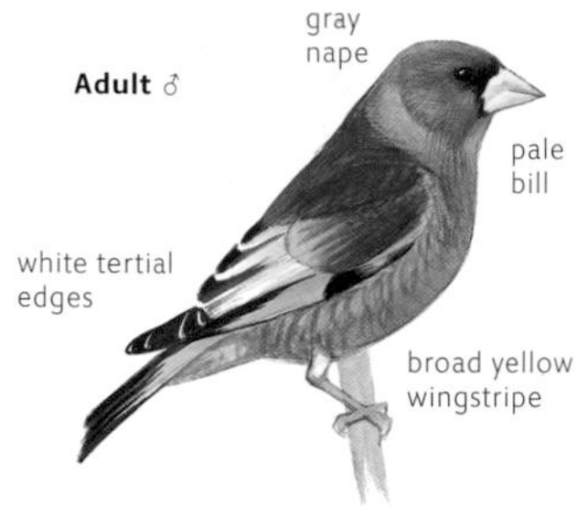

VOICE: Song a simple rasping, whining phrase, slurred up or down *zherrrrree* or *zhrrrreew*, may recall Western Wood-Pewee. Call a very rapid series of high sharp notes *tititititi*, almost a rattle, slightly descending in pitch.

Very rare visitor from Asia to far western Alaska, recorded a few times, mainly in spring. One record in northern California may have been an escape from captivity.

Eurasian Siskin

Spinus spinus

L 4.7" WS 8.7" WT 0.5 oz (14 g)

Very similar to Pine Siskin. Male distinctive, with bright yellow breast, black crown and chin. On female note yellow on face and neck, finer and more distinct streaks below. Some bright male Pine Siskins can be similar.

VOICE: Song a jumble of stuttering, twangy notes, similar to Pine Siskin. Call in flight a very distinctive high whistled *tee-oo* with clear bell-like quality, two-syllabled with second lower. Other calls include a husky rising *zhweeee*, reminiscent of House Finch but longer.

Very rare visitor from Eurasia. Two records in far western Alaska are accepted as natural vagrants, and several records in the Northeast may also be wild, but others are likely escapes from captivity. Habits similar to Pine Siskin, and vagrants usually flock together.

Eurasian Bullfinch

Pyrrhula pyrrhula

L 6.3" WS 10.25" WT 0.75 oz (21 g)

Slightly larger than House Finch, with relatively long tail. Note black crown curving smoothly into stubby black bill, unstreaked grayish back,white rump, black tail and wings with broad pale wingbar. Male rosy pink below, female pinkish-brown.

VOICE: Song a very odd, slow series of mechanical buzzy whistles, level or sliding up or down in pitch, interspersed with clearer *heew* call notes. Call a very soft, airy, husky whistle, level or slightly descending *hweeh*.

Very rare visitor from Eurasia, most records from far western Alaska, especially in late fall and winter. Also recorded several times in winter in mainland Alaska as far southeast as Petersburg, Florida.

European Goldfinch

Carduelis carduelis

L 5.25" WS 10" WT 0.6 oz (17 g)

Slightly larger and more elongated than American Goldfinch. Pale body, dark red face, and brilliant yellow wingstripe are distinctive.

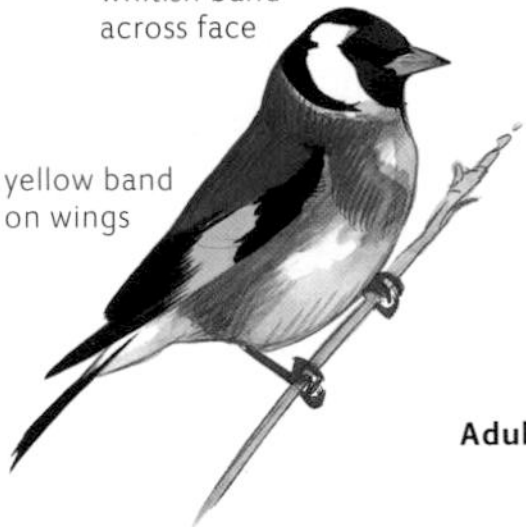

VOICE: Song similar to American Goldfinch but more rapid, higher-pitched, with light tinkling quality. Call a sharp *pit*, often repeated in rapid series rising and falling in pitch *po-pi-pi-po*; also a high, thin *tsii*.

Native to Eurasia. Occasionally escaped from captivity in North America. Habits like American Goldfinch. Escaped birds often join flocks of native goldfinches and visit bird feeders.

Yellow-fronted Canary

Serinus mozambicus

L 4.5" WS 8.5" WT 0.4 oz (11 g)

Closely related to the Atlantic Canary (*Serinus canaria*) and various domesticated varieties of both could escape. Slightly stockier and larger-billed than goldfinches, with bold face pattern and weak wingbars.

VOICE: Song bright and clear, quality like American Goldfinch, but tempo much slower, many notes repeated. Call a short twittering rattle, reminiscent of Snow Bunting.

Native to Africa; occasionally escapes from captivity in North America. Escaped birds are variable in appearance, may be partly or mostly white or pale yellow.

Tricolored Munia

Lonchura malacca

L 4.5" WS 7" WT 0.4 oz (11 g)

Tiny, strikingly colored with chestnut, black and white pattern. Black head contrasts sharply with very large and pale, bluish bill.

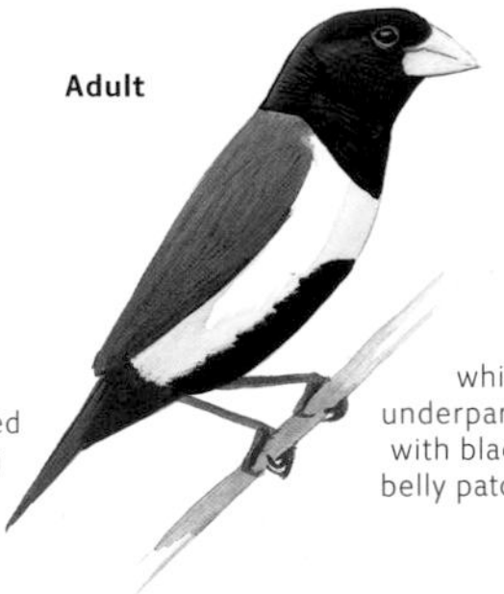

VOICE: Calls include tinny, nasal *beep*.

Native to Asia. Occasionally escaped from captivity in North America, and small numbers may persist in southern cities.

Java Sparrow

Padda oryzivora

L 6" WS 8.5" WT 0.9 oz (26 g)

A small gray bird, unmistakeable with massive pink bill, oval white cheek patch.

VOICE: Calls include a hard *pit*, repeated in rapid chorus by flock. Also a high squealing whistle *skwee*.

Native to Asia. Occasionally escaped from captivity in California, Texas and Florida.

House Sparrow

Passer domesticus

L 6.25" WS 9.5" WT 0.98 oz (28 g)

Stocky, short-tailed, and large-headed, with blunt-tipped bill. Male strikingly patterned: black throat and lores, mostly rufous upperparts, and one broad white wingbar. Female drab gray-brown overall, with plain face.

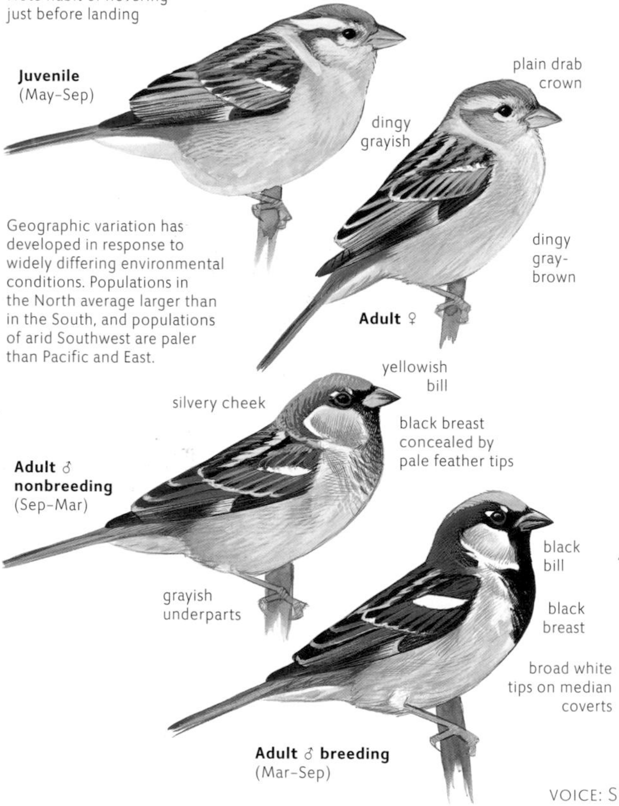

VOICE: Song a monotonous series of nearly identical chirps. Call a husky *fillip*, similar to House Finch but not rising; also a low, rattling series in excitement; and constant chatter from flocks. Flight call a soft, husky *pido* and high *pirv.*

Common and widespread; the ubiquitous sparrow of cities, towns, parking lots, and farms. Introduced from Europe in the mid-1800s. Nests in any sheltered cavity from birdhouses to streetlights to crevices in buildings. Almost always in small flocks. Feeds on seeds and insects; avidly seeks handouts such as bread crumbs or french fries at parks and parking lots.

Eurasian Tree Sparrow

Passer montanus

L 6" WS 8.75" WT 0.77 oz (22 g)

Similar to House Sparrow; slightly smaller, with smaller body and longer tail. Note paler underside with small black chin spot, dark bill, maroon crown, white collar, and dark spot on cheek.

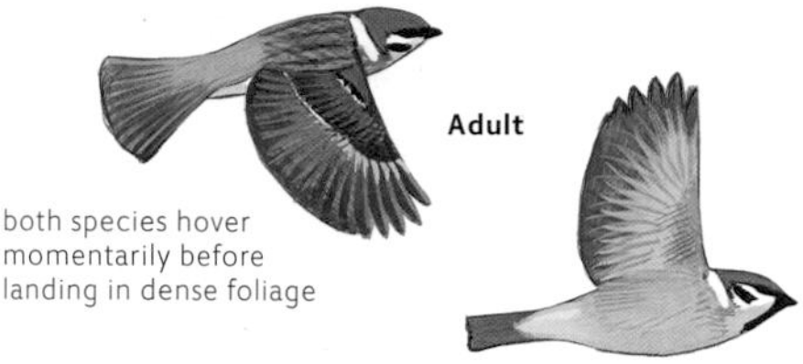

VOICE: Similar to House Sparrow, but all calls sharper, higher-pitched, more musical; chorus from flock more varied, higher, and lilting than House Sparrows. One common call a sharp *pli-pli* usually doubled or tripled, with sharp rattling husky quality. Flight call a hard *pik, pik, pik.*

Uncommon and very local in and around St. Louis, Missouri. Introduced from Europe in 1870. Usually found in more natural brushy and woods-edge habitats than House Sparrow, in flocks, not mixed with House Sparrows.

Orange Bishop

Euplectes franciscanus

L 4.5" WS 6.5" WT 0.8 oz (23 g)

Adult males in breeding plumage are distinctive with brilliant orange-red body, but females and immatures are easily confused with sparrows (especially Grasshopper Sparrow). Note pale buff color, relatively large, pinkish bill, and very short tail, which is often flicked open.

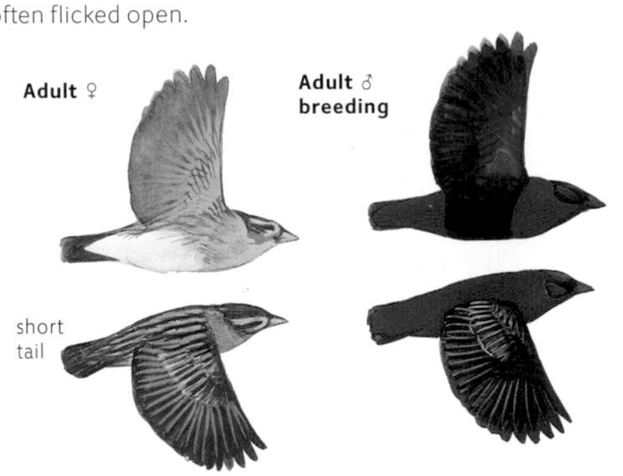

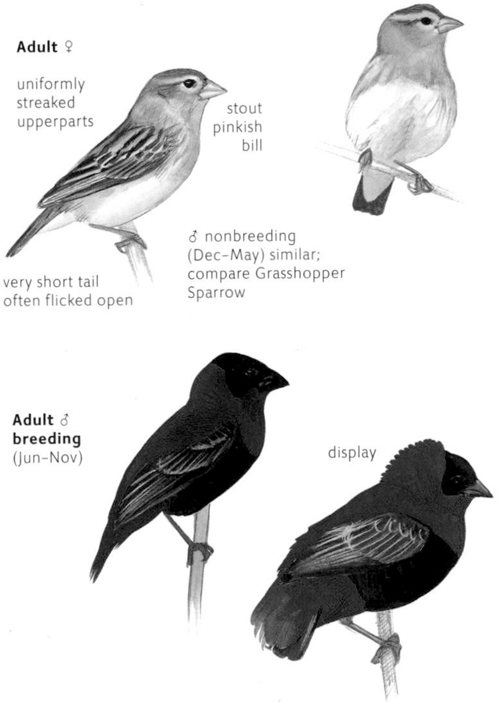

VOICE: Song a rhythmic, insect-like series of very high and scratchy two-syllabled phrases *tsee kisee kisee kisee kisee kisee.* Calls include sharp and mechanical *tsik* notes.

Native to Africa. Occasionally escaped in North America, with a substantial feral population established and breeding in southern California and around Phoenix, Arizona, also regularly seen in Texas. Found in any weedy habitat, especially river floodplains. Also known as Northern Red Bishop.

Nutmeg Mannikin

Lonchura punctulata

L 4" WS 7" WT 0.5 oz (14 g)

Tiny, with very large dark bill and thin pointed tail. Adult dark chestnut-brown with scaly belly. Juvenile plain brownish, told from buntings and others by small size, large bill, pointed tail, and usually associated with distinctive adults.

VOICE: Song of high *tik* notes and thin scratchy whistles, is very inconspicuous. Call a sharp high *pidee* repeated in chorus by flock and reminiscent of Least Tern. Also quiet, thin, scratchy or nasal notes.

Native to Asia. Small populations established and spreading in southern California since 1980s, also increasing around Houston, Texas. Found in weedy areas such as vacant lots and stream floodplains. Also known as Spice Finch or Scaly-breasted Munia.

The Sibley Guide to Birds Checklist

The following checklist includes all species included in the second edition of *The Sibley Guide to Birds*. Arranged by family in taxonomic order, each entry includes common name and corresponding page number in the guide. Of the total 930 species in 87 families, 53 are considered exotic in our region (not accepted as an established part of the avifauna) and are appended with an asterisk (*); 7 species appended with a dagger (†) are considered extinct.

Family Anatidae

☐ Black-bellied Whistling-Duck, 14
☐ Fulvous Whistling-Duck, 14
☐ Taiga Bean-Goose, 5
☐ Tundra Bean-Goose, 5
☐ Pink-footed Goose, 5
☐ Greater White-fronted Goose, 4
☐ Bar-headed Goose*, 17
☐ Graylag Goose, 5
☐ Swan Goose*, 16
☐ Emperor Goose, 9
☐ Snow Goose, 10
☐ Ross's Goose, 11
☐ Brant, 8
☐ Barnacle Goose, 9
☐ Cackling Goose, 6
☐ Canada Goose, 7
☐ Mute Swan, 12
☐ Trumpeter Swan, 12
☐ Tundra Swan, 13
☐ Whooper Swan, 13
☐ Black Swan*, 17
☐ Egyptian Goose*, 16
☐ Common Shelduck*, 16
☐ Ruddy Shelduck*, 16
☐ Muscovy Duck, 15
☐ Wood Duck, 15
☐ Mandarin Duck*, 17
☐ Ringed Teal*, 17
☐ Gadwall, 20
☐ Falcated Duck, 25
☐ Eurasian Wigeon, 21
☐ American Wigeon, 21
☐ American Black Duck, 19
☐ Mallard, 18
☐ Mottled Duck, 19
☐ Blue-winged Teal, 22
☐ Cinnamon Teal, 22
☐ Northern Shoveler, 23
☐ White-cheeked Pintail, 17
☐ Northern Pintail, 20
☐ Garganey, 23
☐ Baikal Teal, 25
☐ Green-winged Teal, 24
☐ Red-crested Pochard*, 25
☐ Rosy-billed Pochard*, 25
☐ Canvasback, 26
☐ Redhead, 26
☐ Common Pochard, 25
☐ Ring-necked Duck, 27
☐ Tufted Duck, 27
☐ Greater Scaup, 28
☐ Lesser Scaup, 28
☐ Steller's Eider, 32
☐ Spectacled Eider, 31
☐ King Eider, 31
☐ Common Eider, 30
☐ Harlequin Duck, 32
☐ Labrador Duck†, xv
☐ Surf Scoter, 34
☐ White-winged Scoter, 35
☐ Black Scoter, 34
☐ Long-tailed Duck, 33
☐ Bufflehead, 37
☐ Common Goldeneye, 36
☐ Barrow's Goldeneye, 36
☐ Smew, 35
☐ Hooded Merganser, 37
☐ Common Merganser, 38
☐ Red-breasted Merganser, 38
☐ Masked Duck, 39
☐ Ruddy Duck, 39

Family Cracidae

☐ Plain Chachalaca, 42

Family Numididae

☐ Helmeted Guineafowl*, 41

Family Odontophoridae

☐ Mountain Quail, 45
☐ Scaled Quail, 45
☐ California Quail, 44
☐ Gambel's Quail, 44
☐ Northern Bobwhite, 43
☐ Montezuma Quail, 42

Family Phasianidae

☐ Chukar, 46
☐ Himalayan Snowcock, 41
☐ Gray Partridge, 46
☐ Red Junglefowl*, 41
☐ Ring-necked Pheasant, 47
☐ Golden Pheasant*, 41
☐ Indian Peafowl, 41
☐ Ruffed Grouse, 52
☐ Greater Sage-Grouse, 54
☐ Gunnison Sage-Grouse, 54
☐ Spruce Grouse, 48
☐ Willow Ptarmigan, 51
☐ Rock Ptarmigan, 50
☐ White-tailed Ptarmigan, 51
☐ Dusky Grouse, 49
☐ Sooty Grouse, 49
☐ Sharp-tailed Grouse, 52
☐ Greater Prairie-Chicken, 53
☐ Lesser Prairie-Chicken, 53
☐ Wild Turkey, 55

Family Gaviidae

☐ Red-throated Loon, 59
☐ Arctic Loon, 58
☐ Pacific Loon, 58
☐ Common Loon, 57
☐ Yellow-billed Loon, 57

Family Podicipedidae

☐ Least Grebe, 63
☐ Pied-billed Grebe, 63
☐ Horned Grebe, 60
☐ Red-necked Grebe, 61
☐ Eared Grebe, 60
☐ Western Grebe, 62
☐ Clark's Grebe, 62

Family Phoenicopteridae

☐ American Flamingo, 119
☐ Greater Flamingo*, 119
☐ Chilean Flamingo*, 119
☐ Lesser Flamingo*, 119

Family Diomedeidae

☐ Yellow-nosed Albatross, 68
☐ Shy Albatross, 68
☐ Black-browed Albatross, 68
☐ Laysan Albatross, 66
☐ Black-footed Albatross, 66
☐ Short-tailed Albatross, 67

Family Procellariidae

☐ Northern Fulmar, 69
☐ Great-winged Petrel, 75
☐ Herald Petrel, 74
☐ Murphy's Petrel, 71
☐ Mottled Petrel, 70
☐ Bermuda Petrel, 73
☐ Black-capped Petrel, 72
☐ Hawaiian Petrel, 71
☐ Fea's Petrel, 73
☐ Cook's Petrel, 70
☐ Bulwer's Petrel, 75
☐ White-chinned Petrel, 75
☐ Streaked Shearwater, 82
☐ Cory's Shearwater, 76
☐ Pink-footed Shearwater, 77
☐ Flesh-footed Shearwater, 78
☐ Great Shearwater, 76
☐ Wedge-tailed Shearwater, 82
☐ Buller's Shearwater, 77
☐ Sooty Shearwater, 79
☐ Short-tailed Shearwater, 79
☐ Manx Shearwater, 80
☐ Black-vented Shearwater, 80
☐ Audubon's Shearwater, 81
☐ Barolo Shearwater, 81

Family Hydrobatidae

- ☐ Wilson's Storm-Petrel, 84
- ☐ White-faced Storm-Petrel, 89
- ☐ European Storm-Petrel, 88
- ☐ Black-bellied Storm-Petrel, 89
- ☐ Fork-tailed Storm-Petrel, 87
- ☐ Swinhoe's Storm-Petrel, 88
- ☐ Leach's Storm-Petrel, 85
- ☐ Ashy Storm-Petrel, 86
- ☐ Band-rumped Storm-Petrel, 84
- ☐ Wedge-rumped Storm-Petrel, 88
- ☐ Black Storm-Petrel, 86
- ☐ Least Storm-Petrel, 87

Family Phaethontidae

- ☐ White-tailed Tropicbird, 103
- ☐ Red-billed Tropicbird, 103
- ☐ Red-tailed Tropicbird, 102

Family Ciconiidae

- ☐ Jabiru, 106
- ☐ Wood Stork, 106

Family Fregatidae

- ☐ Magnificent Frigatebird, 96
- ☐ Great Frigatebird, 97
- ☐ Lesser Frigatebird, 97

Family Sulidae

- ☐ Masked Booby, 93
- ☐ Blue-footed Booby, 93
- ☐ Brown Booby, 92
- ☐ Red-footed Booby, 94
- ☐ Northern Gannet, 95

Family Phalacrocoracidae

- ☐ Brandt's Cormorant, 98
- ☐ Neotropic Cormorant, 100
- ☐ Double-crested Cormorant, 101
- ☐ Great Cormorant, 100
- ☐ Red-faced Cormorant, 99
- ☐ Pelagic Cormorant, 99

Family Anhingidae

- ☐ Anhinga, 980

Family Pelecanidae

- ☐ American White Pelican, 90
- ☐ Brown Pelican, 91

Family Ardeidae

- ☐ American Bittern, 107
- ☐ Least Bittern, 107
- ☐ Great Blue Heron, 108
- ☐ Gray Heron, 113
- ☐ Great Egret, 109
- ☐ Little Egret, 110
- ☐ Western Reef-Heron, 110
- ☐ Snowy Egret, 109
- ☐ Little Blue Heron, 111
- ☐ Tricolored Heron, 111
- ☐ Reddish Egret, 112
- ☐ Cattle Egret, 114
- ☐ Green Heron, 114
- ☐ Black-crowned Night-Heron, 115
- ☐ Yellow-crowned Night-Heron, 115

Family Threskiornithidae

- ☐ White Ibis, 116
- ☐ Scarlet Ibis, 116
- ☐ Glossy Ibis, 117
- ☐ White-faced Ibis, 117
- ☐ Sacred Ibis*, 118
- ☐ Roseate Spoonbill, 118

Family Cathartidae

- ☐ Black Vulture, 123
- ☐ Turkey Vulture, 123
- ☐ California Condor, 122

Family Pandionidae

- ☐ Osprey, 125

Family Accipitridae

- ☐ Hook-billed Kite, 125
- ☐ Swallow-tailed Kite, 127
- ☐ White-tailed Kite, 126
- ☐ Snail Kite, 126
- ☐ Mississippi Kite, 127
- ☐ Bald Eagle, 143
- ☐ Northern Harrier, 124
- ☐ Sharp-shinned Hawk, 128
- ☐ Cooper's Hawk, 128
- ☐ Northern Goshawk, 129
- ☐ Common Black-Hawk, 130
- ☐ Harris's Hawk, 130
- ☐ Roadside Hawk, 132
- ☐ Red-shouldered Hawk, 133
- ☐ Broad-winged Hawk, 135
- ☐ Gray Hawk, 132
- ☐ Short-tailed Hawk, 134
- ☐ Swainson's Hawk, 136
- ☐ White-tailed Hawk, 137
- ☐ Zone-tailed Hawk, 131
- ☐ Red-tailed Hawk, 138
- ☐ Ferruginous Hawk, 140
- ☐ Rough-legged Hawk, 141
- ☐ Golden Eagle, 142

Family Rallidae

- ☐ Yellow Rail, 153
- ☐ Black Rail, 153
- ☐ Clapper Rail, 150
- ☐ King Rail, 151
- ☐ Virginia Rail, 152
- ☐ Sora, 152
- ☐ Purple Swamphen, 146
- ☐ Purple Gallinule, 146
- ☐ Common Gallinule, 147
- ☐ American Coot, 147

Family Aramidae

- ☐ Limpkin, 145

Family Gruidae

- ☐ Sandhill Crane, 149
- ☐ Common Crane, 148
- ☐ Whooping Crane, 148

Family Recurvirostridae

- ☐ Black-necked Stilt, 165
- ☐ American Avocet, 165

Family Haematopodidae

- ☐ Eurasian Oystercatcher, 163
- ☐ American Oystercatcher, 164
- ☐ Black Oystercatcher, 164

Family Charadriidae

- ☐ Northern Lapwing, 161
- ☐ Black-bellied Plover, 156
- ☐ European Golden-Plover, 156
- ☐ American Golden-Plover, 157
- ☐ Pacific Golden-Plover, 157
- ☐ Lesser Sand-Plover, 162
- ☐ Greater Sand-Plover, 163
- ☐ Snowy Plover, 158
- ☐ Wilson's Plover, 158
- ☐ Common Ringed Plover, 162
- ☐ Semipalmated Plover, 159
- ☐ Piping Plover, 159
- ☐ Killdeer, 160
- ☐ Mountain Plover, 160
- ☐ Eurasian Dotterel, 163

Family Jacanidae

- ☐ Northern Jacana, 161

Family Scolopacidae

- ☐ Terek Sandpiper, 170
- ☐ Common Sandpiper, 170
- ☐ Spotted Sandpiper, 170
- ☐ Solitary Sandpiper, 167
- ☐ Gray-tailed Tattler, 177
- ☐ Wandering Tattler, 177
- ☐ Spotted Redshank, 169
- ☐ Greater Yellowlegs, 166
- ☐ Common Greenshank, 167
- ☐ Willet, 168
- ☐ Lesser Yellowlegs, 166
- ☐ Wood Sandpiper, 169
- ☐ Upland Sandpiper, 171
- ☐ Little Curlew, 171
- ☐ Eskimo Curlew†, xv
- ☐ Whimbrel, 172
- ☐ Bristle-thighed Curlew, 173
- ☐ Long-billed Curlew, 173
- ☐ Black-tailed Godwit, 174
- ☐ Hudsonian Godwit, 175
- ☐ Bar-tailed Godwit, 174
- ☐ Marbled Godwit, 175
- ☐ Ruddy Turnstone, 176
- ☐ Black Turnstone, 176
- ☐ Great Knot, 177
- ☐ Red Knot, 186
- ☐ Surfbird, 179
- ☐ Ruff, 189
- ☐ Sharp-tailed Sandpiper, 184
- ☐ Stilt Sandpiper, 188
- ☐ Curlew Sandpiper, 186
- ☐ Temminck's Stint, 181
- ☐ Long-toed Stint, 181
- ☐ Red-necked Stint, 180
- ☐ Sanderling, 183
- ☐ Dunlin, 187
- ☐ Rock Sandpiper, 178
- ☐ Purple Sandpiper, 179
- ☐ Baird's Sandpiper, 185
- ☐ Little Stint, 180

- ☐ Least Sandpiper, 183
- ☐ White-rumped Sandpiper, 185
- ☐ Buff-breasted Sandpiper, 188
- ☐ Pectoral Sandpiper, 184
- ☐ Semipalmated Sandpiper, 182
- ☐ Western Sandpiper, 182
- ☐ Short-billed Dowitcher, 191
- ☐ Long-billed Dowitcher, 190
- ☐ Jack Snipe, 193
- ☐ Wilson's Snipe, 192
- ☐ Common Snipe, 192
- ☐ American Woodcock, 193
- ☐ Wilson's Phalarope, 194
- ☐ Red-necked Phalarope, 195
- ☐ Red Phalarope, 195

Family Stercorariidae

- ☐ Great Skua, 241
- ☐ South Polar Skua, 240
- ☐ Pomarine Jaeger, 239
- ☐ Parasitic Jaeger, 238
- ☐ Long-tailed Jaeger, 237

Family Alcidae

- ☐ Dovekie, 245
- ☐ Common Murre, 244
- ☐ Thick-billed Murre, 244
- ☐ Razorbill, 245
- ☐ Great Auk†, xv
- ☐ Black Guillemot, 243
- ☐ Pigeon Guillemot, 243
- ☐ Long-billed Murrelet, 249
- ☐ Marbled Murrelet, 248
- ☐ Kittlitz's Murrelet, 248
- ☐ Scripps's Murrelet, 246
- ☐ Guadalupe Murrelet, 246
- ☐ Craveri's Murrelet, 247
- ☐ Ancient Murrelet, 247
- ☐ Cassin's Auklet, 249
- ☐ Parakeet Auklet, 250
- ☐ Least Auklet, 250
- ☐ Whiskered Auklet, 251
- ☐ Crested Auklet, 251
- ☐ Rhinoceros Auklet, 252
- ☐ Atlantic Puffin, 253
- ☐ Horned Puffin, 253
- ☐ Tufted Puffin, 252

Family Laridae

- ☐ Black-legged Kittiwake, 199
- ☐ Red-legged Kittiwake, 199
- ☐ Ivory Gull, 220
- ☐ Sabine's Gull, 198
- ☐ Bonaparte's Gull, 201
- ☐ Black-headed Gull, 201
- ☐ Little Gull, 200
- ☐ Ross's Gull, 198
- ☐ Laughing Gull, 203
- ☐ Franklin's Gull, 202
- ☐ Black-tailed Gull, 220
- ☐ Heermann's Gull, 219
- ☐ Mew Gull, 204
- ☐ Ring-billed Gull, 206
- ☐ Western Gull, 215
- ☐ Yellow-footed Gull, 216
- ☐ California Gull, 207
- ☐ Herring Gull, 208
- ☐ Yellow-legged Gull, 221
- ☐ Thayer's Gull, 210
- ☐ Iceland Gull, 211
- ☐ Lesser Black-backed Gull, 217
- ☐ Slaty-backed Gull, 214
- ☐ Glaucous-winged Gull, 213
- ☐ Glaucous Gull, 212
- ☐ Great Black-backed Gull, 218
- ☐ Kelp Gull, 221
- ☐ Brown Noddy, 235
- ☐ Black Noddy, 235
- ☐ Sooty Tern, 233
- ☐ Bridled Tern, 233
- ☐ Aleutian Tern, 232
- ☐ Least Tern, 231
- ☐ Gull-billed Tern, 232
- ☐ Caspian Tern, 225
- ☐ Black Tern, 234
- ☐ White-winged Tern, 234
- ☐ Roseate Tern, 231
- ☐ Common Tern, 228
- ☐ Arctic Tern, 229
- ☐ Forster's Tern, 230
- ☐ Royal Tern, 226
- ☐ Sandwich Tern, 227
- ☐ Elegant Tern, 226
- ☐ Black Skimmer, 236

Family Columbidae

- ☐ Rock Pigeon, 255
- ☐ White-crowned Pigeon, 257
- ☐ Red-billed Pigeon, 257
- ☐ Band-tailed Pigeon, 256
- ☐ Oriental Turtle-Dove, 259
- ☐ African Collared-Dove*, 260
- ☐ European Turtle-Dove, 259
- ☐ Eurasian Collared-Dove, 260
- ☐ Spotted Dove, 261
- ☐ White-winged Dove, 258
- ☐ Zenaida Dove, 259
- ☐ Mourning Dove, 258
- ☐ Passenger Pigeon†, xv
- ☐ Inca Dove, 262
- ☐ Common Ground-Dove, 263
- ☐ Ruddy Ground-Dove, 263
- ☐ White-tipped Dove, 261
- ☐ Key West Quail-Dove, 262

Family Cuculidae

- ☐ Common Cuckoo, 266
- ☐ Yellow-billed Cuckoo, 265
- ☐ Mangrove Cuckoo, 264
- ☐ Black-billed Cuckoo, 265
- ☐ Greater Roadrunner, 266
- ☐ Smooth-billed Ani, 267
- ☐ Groove-billed Ani, 267

Family Tytonidae

- ☐ Barn Owl, 270

Family Strigidae

- ☐ Flammulated Owl, 277
- ☐ Western Screech-Owl, 278
- ☐ Eastern Screech-Owl, 279
- ☐ Whiskered Screech-Owl, 278
- ☐ Great Horned Owl, 272
- ☐ Snowy Owl, 273
- ☐ Northern Hawk Owl, 281
- ☐ Northern Pygmy-Owl, 280
- ☐ Ferruginous Pygmy-Owl, 281
- ☐ Elf Owl, 277
- ☐ Burrowing Owl, 276
- ☐ Spotted Owl, 274
- ☐ Barred Owl, 274
- ☐ Great Gray Owl, 273
- ☐ Long-eared Owl, 270
- ☐ Short-eared Owl, 271
- ☐ Boreal Owl, 275
- ☐ Northern Saw-whet Owl, 275

Family Caprimulgidae

- ☐ Lesser Nighthawk, 286
- ☐ Common Nighthawk, 287
- ☐ Antillean Nighthawk, 286
- ☐ Common Pauraque, 283
- ☐ Common Poorwill, 283
- ☐ Chuck-will's-widow, 284
- ☐ Buff-collared Nightjar, 284
- ☐ Eastern Whip-poor-will, 285
- ☐ Mexican Whip-poor-will, 285

Family Apodidae

- ☐ Black Swift, 290
- ☐ White-collared Swift, 288
- ☐ Chimney Swift, 291
- ☐ Vaux's Swift, 291
- ☐ Common Swift, 289
- ☐ Fork-tailed Swift, 289
- ☐ White-throated Swift, 290

Family Trochilidae

- ☐ Green Violetear, 303
- ☐ Green-breasted Mango, 303
- ☐ Magnificent Hummingbird, 302
- ☐ Plain-capped Starthroat, 300
- ☐ Blue-throated Hummingbird, 302
- ☐ Lucifer Hummingbird, 298
- ☐ Ruby-throated Hummingbird, 295
- ☐ Black-chinned Hummingbird, 295
- ☐ Anna's Hummingbird, 294
- ☐ Costa's Hummingbird, 294
- ☐ Broad-tailed Hummingbird, 296
- ☐ Rufous Hummingbird, 297
- ☐ Allen's Hummingbird, 297
- ☐ Calliope Hummingbird, 296
- ☐ Broad-billed Hummingbird, 299
- ☐ Berylline Hummingbird, 301
- ☐ Buff-bellied Hummingbird, 301
- ☐ Violet-crowned Hummingbird, 300
- ☐ White-eared Hummingbird, 299

Family Trogonidae

- ☐ Elegant Trogon, 305
- ☐ Eared Quetzal, 305

Family Alcedinidae

- ☐ Ringed Kingfisher, 307
- ☐ Belted Kingfisher, 306
- ☐ Green Kingfisher, 307

Family Picidae

- ☐ Lewis's Woodpecker, 309
- ☐ Red-headed Woodpecker, 310
- ☐ Acorn Woodpecker, 309
- ☐ Gila Woodpecker, 310
- ☐ Golden-fronted Woodpecker, 311
- ☐ Red-bellied Woodpecker, 311
- ☐ Williamson's Sapsucker, 312
- ☐ Yellow-bellied Sapsucker, 313
- ☐ Red-naped Sapsucker, 313
- ☐ Red-breasted Sapsucker, 312
- ☐ Ladder-backed Woodpecker, 314
- ☐ Nuttall's Woodpecker, 314
- ☐ Downy Woodpecker, 316
- ☐ Hairy Woodpecker, 317
- ☐ Arizona Woodpecker, 315
- ☐ Red-cockaded Woodpecker, 319
- ☐ White-headed Woodpecker, 315
- ☐ American Three-toed Woodpecker, 318
- ☐ Black-backed Woodpecker, 319
- ☐ Northern Flicker, 320
- ☐ Gilded Flicker, 321
- ☐ Pileated Woodpecker, 322
- ☐ Ivory-billed Woodpecker†, xv

Family Falconidae

- ☐ Crested Caracara, 329
- ☐ Eurasian Kestrel, 328
- ☐ American Kestrel, 326
- ☐ Merlin, 327
- ☐ Eurasian Hobby, 328
- ☐ Aplomado Falcon, 329
- ☐ Gyrfalcon, 325
- ☐ Peregrine Falcon, 324
- ☐ Prairie Falcon, 326

Family Cacatuidae

- ☐ Sulphur-crested Cockatoo*, 330
- ☐ Cockatiel*, 330

Family Psittacidae

- ☐ Budgerigar, 330
- ☐ Rosy-faced Lovebird, 330
- ☐ Rose-ringed Parakeet*, 331
- ☐ Monk Parakeet, 331
- ☐ Carolina Parakeet†, xv
- ☐ Blue-crowned Parakeet*, 332
- ☐ Green Parakeet, 332
- ☐ Red-masked Parakeet*, 332
- ☐ White-eyed Parakeet*, 333
- ☐ Mitred Parakeet*, 332
- ☐ Dusky-headed Parakeet*, 333
- ☐ Black-hooded Parakeet*, 331
- ☐ Chestnut-fronted Macaw*, 335
- ☐ Blue-and-Yellow Macaw*, 335
- ☐ Thick-billed Parrot, 333
- ☐ White-winged Parakeet, 333
- ☐ Yellow-chevroned Parakeet*, 333
- ☐ White-fronted Parrot*, 334
- ☐ Red-crowned Parrot, 334
- ☐ Lilac-crowned Parrot*, 334
- ☐ Red-lored Parrot*, 335
- ☐ Orange-winged Parrot*, 335
- ☐ Blue-fronted Parrot*, 335
- ☐ Mealy Parrot*, 335
- ☐ Yellow-headed Parrot*, 334
- ☐ Yellow-naped Parrot*, 335
- ☐ Yellow-crowned Parrot*, 335

Family Tyrannidae

- ☐ Northern Beardless-Tyrannulet, 360
- ☐ White-crested Elaenia, 360
- ☐ Tufted Flycatcher, 340
- ☐ Olive-sided Flycatcher, 338
- ☐ Greater Pewee, 338
- ☐ Western Wood-Pewee, 339
- ☐ Eastern Wood-Pewee, 339
- ☐ Cuban Pewee, 340
- ☐ Yellow-bellied Flycatcher, 343
- ☐ Acadian Flycatcher, 343
- ☐ Alder Flycatcher, 345
- ☐ Willow Flycatcher, 344
- ☐ Least Flycatcher, 345
- ☐ Hammond's Flycatcher, 346
- ☐ Gray Flycatcher, 347
- ☐ Dusky Flycatcher, 346
- ☐ Pacific-slope Flycatcher, 342
- ☐ Cordilleran Flycatcher, 342
- ☐ Buff-breasted Flycatcher, 347
- ☐ Black Phoebe, 348
- ☐ Eastern Phoebe, 348
- ☐ Say's Phoebe, 349
- ☐ Vermilion Flycatcher, 349
- ☐ Dusky-capped Flycatcher, 350
- ☐ Ash-throated Flycatcher, 350
- ☐ Nutting's Flycatcher, 352
- ☐ Great Crested Flycatcher, 351
- ☐ Brown-crested Flycatcher, 351
- ☐ La Sagra's Flycatcher, 352
- ☐ Great Kiskadee, 359
- ☐ Sulphur-bellied Flycatcher, 359
- ☐ Piratic Flycatcher, 358
- ☐ Variegated Flycatcher, 358
- ☐ Tropical Kingbird, 356
- ☐ Couch's Kingbird, 356
- ☐ Cassin's Kingbird, 357
- ☐ Thick-billed Kingbird, 354
- ☐ Western Kingbird, 357
- ☐ Eastern Kingbird, 355
- ☐ Gray Kingbird, 355
- ☐ Loggerhead Kingbird, 354
- ☐ Scissor-tailed Flycatcher, 353
- ☐ Fork-tailed Flycatcher, 353

Family Tityridae

- ☐ Rose-throated Becard, 361

Family Laniidae

- ☐ Brown Shrike, 362
- ☐ Loggerhead Shrike, 363
- ☐ Northern Shrike, 363

Family Vireonidae

- ☐ White-eyed Vireo, 367
- ☐ Thick-billed Vireo, 367
- ☐ Bell's Vireo, 365
- ☐ Black-capped Vireo, 366
- ☐ Gray Vireo, 366
- ☐ Yellow-throated Vireo, 369
- ☐ Plumbeous Vireo, 368
- ☐ Cassin's Vireo, 368
- ☐ Blue-headed Vireo, 369
- ☐ Hutton's Vireo, 372
- ☐ Warbling Vireo, 370
- ☐ Philadelphia Vireo, 370
- ☐ Red-eyed Vireo, 371
- ☐ Yellow-green Vireo, 372
- ☐ Black-whiskered Vireo, 371

Family Corvidae

- ☐ Gray Jay, 381
- ☐ Black-throated Magpie-Jay*, 375
- ☐ Brown Jay, 379
- ☐ Green Jay, 379
- ☐ Pinyon Jay, 380
- ☐ Steller's Jay, 374
- ☐ Blue Jay, 375
- ☐ Florida Scrub-Jay, 377
- ☐ Island Scrub-Jay, 377
- ☐ Western Scrub-Jay, 376
- ☐ Mexican Jay, 378
- ☐ Clark's Nutcracker, 380
- ☐ Black-billed Magpie, 382
- ☐ Yellow-billed Magpie, 382
- ☐ Eurasian Jackdaw, 385
- ☐ American Crow, 384
- ☐ Northwestern Crow, 384
- ☐ Tamaulipas Crow, 385
- ☐ Fish Crow, 385
- ☐ Hooded Crow*, 384
- ☐ House Crow*, 384
- ☐ Chihuahuan Raven, 383
- ☐ Common Raven, 383

Family Alaudidae

- ☐ Sky Lark, 386
- ☐ Horned Lark, 387

Family Hirundinidae

- ☐ Purple Martin, 389
- ☐ Brown-chested Martin, 394
- ☐ Tree Swallow, 390
- ☐ Mangrove Swallow, 395
- ☐ Violet-green Swallow, 390
- ☐ Bahama Swallow, 395
- ☐ Northern Rough-winged Swallow, 391
- ☐ Bank Swallow, 391
- ☐ Cliff Swallow, 392
- ☐ Cave Swallow, 392
- ☐ Barn Swallow, 393
- ☐ Common House-Martin, 394

Family Paridae

- ☐ Carolina Chickadee, 398
- ☐ Black-capped Chickadee, 399
- ☐ Mountain Chickadee, 398
- ☐ Mexican Chickadee, 400
- ☐ Chestnut-backed Chickadee, 400
- ☐ Boreal Chickadee, 401
- ☐ Gray-headed Chickadee, 401
- ☐ Eurasian Blue Tit*, 397
- ☐ Great Tit*, 397
- ☐ Bridled Titmouse, 404
- ☐ Oak Titmouse, 402
- ☐ Juniper Titmouse, 402
- ☐ Tufted Titmouse, 403
- ☐ Black-crested Titmouse, 403

Family Remizidae

- ☐ Verdin, 404

Family Aegithalidae

- ☐ Bushtit, 405

Family Sittidae

- ☐ Red-breasted Nuthatch, 406
- ☐ White-breasted Nuthatch, 407
- ☐ Pygmy Nuthatch, 408
- ☐ Brown-headed Nuthatch, 408

Family Certhiidae

- ☐ Brown Creeper, 409

Family Troglodytidae

- ☐ Rock Wren, 417
- ☐ Canyon Wren, 417
- ☐ House Wren, 412
- ☐ Pacific Wren, 413
- ☐ Winter Wren, 413
- ☐ Sedge Wren, 414
- ☐ Marsh Wren, 415
- ☐ Carolina Wren, 410
- ☐ Bewick's Wren, 411
- ☐ Cactus Wren, 416
- ☐ Sinaloa Wren, 411

Family Polioptilidae

- ☐ Blue-gray Gnatcatcher, 421
- ☐ California Gnatcatcher, 420
- ☐ Black-tailed Gnatcatcher, 420
- ☐ Black-capped Gnatcatcher, 419

Family Cinclidae

- ☐ American Dipper, 422

Family Pycnonotidae

- ☐ Red-vented Bulbul*, 424
- ☐ Red-whiskered Bulbul, 424

Family Regulidae

- ☐ Golden-crowned Kinglet, 423
- ☐ Ruby-crowned Kinglet, 423

Family Phylloscopidae

- ☐ Willow Warbler, 425
- ☐ Dusky Warbler, 426
- ☐ Yellow-browed Warbler, 425
- ☐ Arctic Warbler, 425

Family Sylviidae

- ☐ Wrentit, 422

Family Megaluridae

- ☐ Middendorff's Grasshopper-Warbler, 426
- ☐ Lanceolated Warbler, 426

Family Muscicapidae

- ☐ Gray-streaked Flycatcher, 427
- ☐ Dark-sided Flycatcher, 427
- ☐ Siberian Rubythroat, 427
- ☐ Bluethroat, 429
- ☐ Red-flanked Bluetail, 428
- ☐ Taiga Flycatcher, 427
- ☐ Northern Wheatear, 429
- ☐ Stonechat, 428

Family Turdidae

- ☐ Eastern Bluebird, 433
- ☐ Western Bluebird, 432
- ☐ Mountain Bluebird, 432
- ☐ Townsend's Solitaire, 431
- ☐ Brown-backed Solitaire, 431
- ☐ Orange-billed Nightingale-Thrush, 441
- ☐ Veery, 434
- ☐ Gray-cheeked Thrush, 436
- ☐ Bicknell's Thrush, 436
- ☐ Swainson's Thrush, 437
- ☐ Hermit Thrush, 435
- ☐ Wood Thrush, 434
- ☐ Eyebrowed Thrush, 438
- ☐ Dusky Thrush, 438
- ☐ Fieldfare, 439
- ☐ Redwing, 439
- ☐ Clay-colored Thrush, 440
- ☐ White-throated Thrush, 441
- ☐ Rufous-backed Robin, 440
- ☐ American Robin, 442
- ☐ Varied Thrush, 443
- ☐ Aztec Thrush, 443

Family Mimidae

- ☐ Blue Mockingbird, 451
- ☐ Gray Catbird, 447
- ☐ Curve-billed Thrasher, 451
- ☐ Brown Thrasher, 448
- ☐ Long-billed Thrasher, 448
- ☐ Bendire's Thrasher, 450
- ☐ California Thrasher, 449
- ☐ Le Conte's Thrasher, 450
- ☐ Crissal Thrasher, 449
- ☐ Sage Thrasher, 447
- ☐ Bahama Mockingbird, 446
- ☐ Northern Mockingbird, 446

Family Sturnidae

- ☐ European Starling, 453
- ☐ Common Myna, 452
- ☐ Crested Myna*, 452
- ☐ Hill Myna, 452

Family Prunellidae

- ☐ Siberian Accentor, 458

Family Motacillidae

- ☐ Eastern Yellow Wagtail, 454
- ☐ White Wagtail, 455
- ☐ Olive-backed Pipit, 458
- ☐ Pechora Pipit, 458
- ☐ Red-throated Pipit, 456
- ☐ American Pipit, 457
- ☐ Sprague's Pipit, 456

Family Bombycillidae

- ☐ Bohemian Waxwing, 460
- ☐ Cedar Waxwing, 460

Family Ptiliogonatidae

- ☐ Gray Silky-Flycatcher, 461
- ☐ Phainopepla, 461

Family Peucedramidae

- ☐ Olive Warbler, 465

Family Calcariidae

- ☐ Lapland Longspur, 463
- ☐ Chestnut-collared Longspur, 462
- ☐ Smith's Longspur, 463
- ☐ McCown's Longspur, 462
- ☐ Snow Bunting, 464
- ☐ McKay's Bunting, 464

Family Parulidae

- ☐ Ovenbird, 469
- ☐ Worm-eating Warbler, 469
- ☐ Louisiana Waterthrush, 470
- ☐ Northern Waterthrush, 470
- ☐ Bachman's Warbler†, xv
- ☐ Golden-winged Warbler, 473
- ☐ Blue-winged Warbler, 473
- ☐ Black-and-white Warbler, 472
- ☐ Prothonotary Warbler, 471
- ☐ Swainson's Warbler, 471
- ☐ Crescent-chested Warbler, 475
- ☐ Tennessee Warbler, 475
- ☐ Orange-crowned Warbler, 474
- ☐ Colima Warbler, 476
- ☐ Lucy's Warbler, 476
- ☐ Nashville Warbler, 477
- ☐ Virginia's Warbler, 477
- ☐ Connecticut Warbler, 478
- ☐ Gray-crowned Yellowthroat, 481
- ☐ MacGillivray's Warbler, 479
- ☐ Mourning Warbler, 479
- ☐ Kentucky Warbler, 478
- ☐ Common Yellowthroat, 480
- ☐ Hooded Warbler, 481
- ☐ American Redstart, 482
- ☐ Kirtland's Warbler, 483
- ☐ Cape May Warbler, 482
- ☐ Cerulean Warbler, 483
- ☐ Northern Parula, 484
- ☐ Tropical Parula, 484
- ☐ Magnolia Warbler, 485
- ☐ Bay-breasted Warbler, 488
- ☐ Blackburnian Warbler, 485
- ☐ Yellow Warbler, 486
- ☐ Chestnut-sided Warbler, 487
- ☐ Blackpoll Warbler, 488
- ☐ Black-throated Blue Warbler, 487

☐ Palm Warbler, 490
☐ Pine Warbler, 489
☐ Yellow-rumped Warbler, 492
☐ Yellow-throated Warbler, 491
☐ Prairie Warbler, 489
☐ Grace's Warbler, 491
☐ Black-throated Gray Warbler, 493
☐ Townsend's Warbler, 494
☐ Hermit Warbler, 494
☐ Golden-cheeked Warbler, 495
☐ Black-throated Green Warbler, 495
☐ Fan-tailed Warbler, 500
☐ Rufous-capped Warbler, 499
☐ Golden-crowned Warbler, 499
☐ Canada Warbler, 497
☐ Wilson's Warbler, 496
☐ Red-faced Warbler, 497
☐ Painted Redstart, 498
☐ Slate-throated Redstart, 498
☐ Yellow-breasted Chat, 501

Family Undetermined

☐ Bananaquit, 502

Family Thraupidae

☐ Red-crested Cardinal*, 502
☐ Western Spindalis, 503

Family Emberizidae

☐ White-collared Seedeater, 506
☐ Yellow-faced Grassquit, 506
☐ Black-faced Grassquit, 506
☐ Olive Sparrow, 507
☐ Green-tailed Towhee, 507
☐ Spotted Towhee, 508
☐ Eastern Towhee, 509
☐ Rufous-crowned Sparrow, 511
☐ Canyon Towhee, 510
☐ California Towhee, 510
☐ Abert's Towhee, 511
☐ Rufous-winged Sparrow, 512
☐ Botteri's Sparrow, 513
☐ Cassin's Sparrow, 512
☐ Bachman's Sparrow, 513
☐ American Tree Sparrow, 515
☐ Chipping Sparrow, 517
☐ Clay-colored Sparrow, 517
☐ Brewer's Sparrow, 516
☐ Field Sparrow, 515
☐ Black-chinned Sparrow, 514
☐ Vesper Sparrow, 521
☐ Lark Sparrow, 536
☐ Five-striped Sparrow, 519
☐ Black-throated Sparrow, 519
☐ Sagebrush Sparrow, 518
☐ Bell's Sparrow, 518
☐ Lark Bunting, 536
☐ Savannah Sparrow, 520
☐ Grasshopper Sparrow, 522
☐ Baird's Sparrow, 522
☐ Henslow's Sparrow, 523
☐ Le Conte's Sparrow, 523
☐ Nelson's Sparrow, 524
☐ Saltmarsh Sparrow, 524
☐ Seaside Sparrow, 525
☐ Fox Sparrow, 526
☐ Song Sparrow, 528
☐ Lincoln's Sparrow, 529
☐ Swamp Sparrow, 529
☐ White-throated Sparrow, 535
☐ Harris's Sparrow, 533
☐ White-crowned Sparrow, 534
☐ Golden-crowned Sparrow, 535
☐ Dark-eyed Junco, 530
☐ Yellow-eyed Junco, 532
☐ Little Bunting, 537
☐ Rustic Bunting, 537

Family Cardinalidae

☐ Hepatic Tanager, 539
☐ Summer Tanager, 540
☐ Scarlet Tanager, 541
☐ Western Tanager, 541
☐ Flame-colored Tanager, 539
☐ Crimson-collared Grosbeak, 543
☐ Northern Cardinal, 542
☐ Pyrrhuloxia, 542
☐ Yellow Grosbeak, 543
☐ Rose-breasted Grosbeak, 544
☐ Black-headed Grosbeak, 544
☐ Blue Bunting, 548
☐ Blue Grosbeak, 548
☐ Lazuli Bunting, 547
☐ Indigo Bunting, 547
☐ Varied Bunting, 546
☐ Painted Bunting, 546
☐ Dickcissel, 545

Family Icteridae

☐ Bobolink, 558
☐ Red-winged Blackbird, 554
☐ Tricolored Blackbird, 555
☐ Eastern Meadowlark, 559
☐ Western Meadowlark, 558
☐ Yellow-headed Blackbird, 555
☐ Rusty Blackbird, 550
☐ Brewer's Blackbird, 550
☐ Common Grackle, 551
☐ Boat-tailed Grackle, 553
☐ Great-tailed Grackle, 552
☐ Shiny Cowbird, 557
☐ Bronzed Cowbird, 556
☐ Brown-headed Cowbird, 557
☐ Black-vented Oriole, 563
☐ Orchard Oriole, 560
☐ Hooded Oriole, 560
☐ Streak-backed Oriole, 564
☐ Bullock's Oriole, 565
☐ Spot-breasted Oriole, 561
☐ Altamira Oriole, 561
☐ Audubon's Oriole, 562
☐ Baltimore Oriole, 565
☐ Scott's Oriole, 562

Family Fringillidae

☐ Common Chaffinch, 567
☐ Brambling, 567
☐ Gray-crowned Rosy-Finch, 568
☐ Black Rosy-Finch, 569
☐ Brown-capped Rosy-Finch, 569
☐ Pine Grosbeak, 570
☐ Eurasian Bullfinch, 580
☐ Common Rosefinch, 580
☐ Purple Finch, 572
☐ Cassin's Finch, 573
☐ House Finch, 573
☐ Red Crossbill, 574
☐ White-winged Crossbill, 575
☐ Common Redpoll, 576
☐ Hoary Redpoll, 577
☐ Eurasian Siskin, 580
☐ Pine Siskin, 578
☐ Lesser Goldfinch, 579
☐ Lawrence's Goldfinch, 578
☐ American Goldfinch, 579
☐ European Goldfinch*, 581
☐ Oriental Greenfinch, 580
☐ Yellow-fronted Canary*, 581
☐ Evening Grosbeak, 571
☐ Hawfinch, 571

Family Passeridae

☐ House Sparrow, 582
☐ Eurasian Tree Sparrow, 582

Family Ploceidae

☐ Orange Bishop*, 583

Family Estrildidae

☐ Nutmeg Mannikin, 583
☐ Tricolored Munia*, 581
☐ Java Sparrow*, 581

Index

Q

R

S

Quick Index

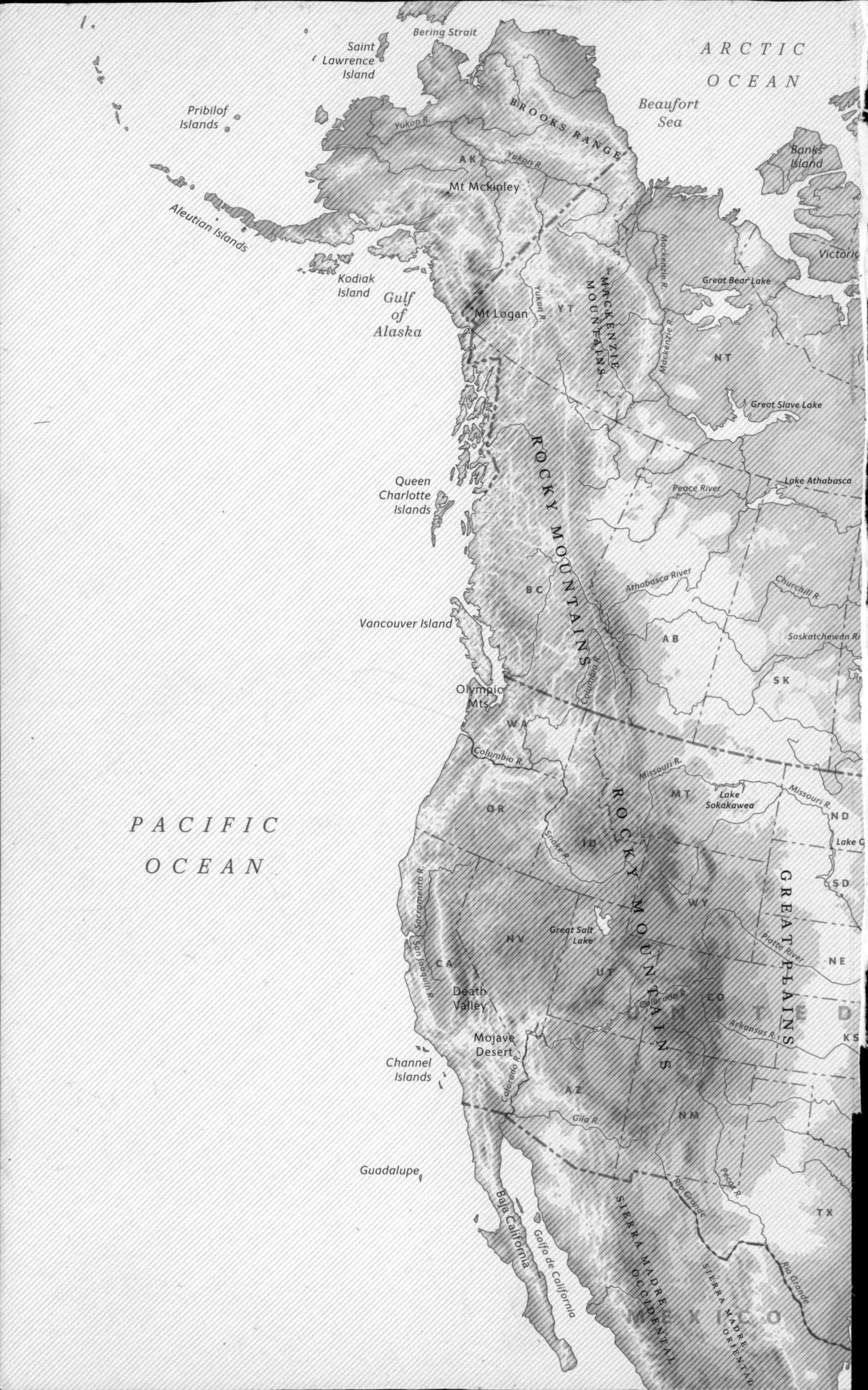

Bering Strait
ARCTIC OCEAN
Saint Lawrence Island
Pribilof Islands
Beaufort Sea
BROOKS RANGE
Banks Island
Yukon R.
AK
Mt Mckinley
Aleutian Islands
Victoria
Kodiak Island
Gulf of Alaska
Mt Logan
YT
MACKENZIE MOUNTAINS
Mackenzie R.
Great Bear Lake
NT
Great Slave Lake
ROCKY MOUNTAINS
Queen Charlotte Islands
Peace River
Lake Athabasca
Athabasca River
Churchill R.
BC
Vancouver Island
AB
Saskatchewan R.
SK
Olympic Mts.
Columbia R.
WA
Missouri R.
MT
Lake Sakakawea
ND
OR
ID
Snake R.
PACIFIC OCEAN
SD
WY
GREAT PLAINS
Sacramento R.
Great Salt Lake
NV
San Joaquin R.
CA
UT
Platte River
NE
Death Valley
Colorado R.
CO
UNITED
Arkansas R.
KS
Mojave Desert
Channel Islands
AZ
Gila R.
NM
Guadalupe
Rio Grande
Pecos R.
TX
Baja California
Golfo de California
SIERRA MADRE OCCIDENTAL
SIERRA MADRE ORIENTAL
MEXICO